INFLUENCE OF RADIATION ON MATERIAL PROPERTIES: 13th INTERNATIONAL SYMPOSIUM (PART II)

A symposium sponsored by ASTM
Committee E-10 on Nuclear Technology
and Applications
Seattle, WA, 23–25 June 1986

ASTM SPECIAL TECHNICAL PUBLICATION 956

F. A. Garner, Westinghouse Hanford Company,
C. H. Henager, Jr., Battelle Pacific Northwest Laboratory, and N. Igata, University of Tokyo, editors

ASTM Publication Code Number (PCN)
04-956000-35

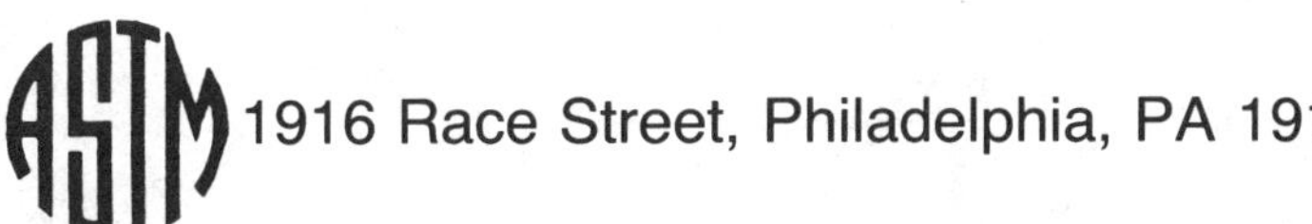

1916 Race Street, Philadelphia, PA 19103

Library of Congress Cataloging-in-Publication Data

Influence of radiation on material properties.

(ASTM special technical publication; 956)
"ASTM publication code number (PCN) 04-956000-35."
Companion to: Radiation induced changes in microstructure (part I).
Includes bibliographies and index.
1. Materials—Effect of radiation on—Congresses.
I. Garner, F. A. II. Henager, C. H. III. Igata, N.
(Naohiro), 1926– . IV. ASTM Committee
E-10 on Nuclear Technology and Applications.
V. Radiation induces changes in microstructure.
VI. Series.
TA418.6.J54 1987 620.1′1228 87-26992
ISBN 0-8031-0963-6

Library of Congress Catalog Card Number: 87-26992

NOTE

The Society is not responsible, as a body,
for the statements and opinions
advanced in this publication.

Printed in Ann Arbor, MI
December 1987

Foreword

Effects of Radiation on Materials: 13th International Symposium was presented at Seattle, WA, 23–25 June 1986. The symposium was sponsored by ASTM Committee E-10 on Nuclear Technology and Applications. F. A. Garner, Westinghouse Hanford Company, served as chairman of the symposium; and N. H. Packan, Oak Ridge National Laboratory, C. H. Henager, Jr., Battelle Pacific Northwest Laboratory, N. Igata, University of Tokyo, and A. S. Kumar, University of Missouri-Rolla, served as vice-chairmen. There are two resulting special technical publications (STPs) from the symposium: *Radiation Induced Changes in Microstructure: 13th International Symposium (Part I), STP 955*, and *Influence of Radiation on Material Properties: 13th International Symposium (Part II), STP 956.*

Related ASTM Publications

Radiation Induced Changes in Microstructure: 13th International Symposium (Part I), STP 955 (1987), 04-955000-35

Effects of Radiation on Materials: Twelfth International Symposium (Volumes I and II), STP 870 (1985), 04-8700000-35

Effects of Radiation on Materials—11th International Symposium, STP 782, 04-7820000-35

Effects of Radiation on Materials—10th International Symposium, STP 725, 04-7250000-35

Effects of Radiation on Structural Materials (9th Conference), STP 683, 04-683000-35

A Note of Appreciation to Reviewers

The quality of the papers that appear in this publication reflects not only the obvious efforts of the authors but also the unheralded, though essential, work of the reviewers. On behalf of ASTM we acknowledge with appreciation their dedication to high professional standards and their sacrifice of time and effort.

ASTM Committee on Publications

ASTM Editorial Staff

Contents

DOSIMETRY OF RADIATION ENVIRONMENTS

Overview

The Thirteenth International Symposium on Effects of Radiation on Materials was held in Seattle, WA on 23–25 June 1986. This biennial symposium series was begun in 1960 and has served as a major international forum for the exchange and discussion of both the fundamental and technological aspects of behavior change in materials exposed to radiation environments. The thirteenth symposium reached a record level of participation, necessitating its publication in two separate Special Technical Publications (STPs).

This volume contains the majority of the papers presented in those sessions devoted to creep, creep rupture, and changes in mechanical properties of metals and alloys. Also included are papers on radiation damage in nonmetals, papers that describe irradiation facilities, and papers on the dosimetry of radiation environments. Most of the other papers presented at this symposium are published separately in ASTM STP 955, *Radiation-Induced Changes in Microstructure,* which addresses the radiation-induced production of point defects, their subsequent diffusion and their consequences on microstructural evolution, phase changes, and dimensional stability.

The first section on *Irradiation Creep and Creep Rupture* focuses on the in-reactor creep and creep rupture behavior of both simple metals and engineering alloys. It is argued in one paper that the use of unirradiated and post irradiation data for predicting in-reactor behavior is invalid. Another paper discusses low-dose irradiation creep mechanisms in model ferritic alloys and demonstrates that intrinsic bias differences between face centered cubic (fcc) and body centered cubic (bcc) metals do not affect climb-glide creep rates. One new conclusion of another study was that irradiation creep appears to disappear at swelling levels in the range of 5 to 10%.

The second section on *Changes in Mechanical Properties of Alloys* explores changes in strength, fracture behavior, fracture toughness, fatigue crack growth, and fatigue crack initiation occurring in a wide variety of important engineering alloys caused by irradiation. Because of their unique problems, pressure vessel steels are not included in this section.

In *Pressure Vessel Steels,* both fundamental data on irradiation embrittlement and statistical studies of data bases are presented and discussed in a number of papers. Positron annihilation and small angle neutron scattering techniques are explored, which provide information on submicroscopic clustering in these alloys. Mechanical property data are also presented, which address the possibility of thermal recovery of damage.

The section on *Radiation Damage in Nonmetals* is significantly larger than that of previous symposia and contains discussions on polymeric, glass, lithium-containing ceramic, and elastomeric materials in radiation environments. Papers were presented on cable connectors, glass fibers, polyimide film, ophthalmic preservatives, reactor seals, and radioactive waste forms.

The Irradiation Facilities section contains discussions on major irradiation facilities available to qualified users. Facilities at Idaho National Engineering Laboratory are described, as well as the Los Alamos Spallation Radiation Effects Facility, and the National Low-Temperature Neutron Irradiation Facility.

The section on *Dosimetry of Radiation Environments* explores topics in neutron dosimetry and damage analysis, radiation damage prediction, and neutronics of fusion environments.

Frank A. Garner
Westinghouse Hanford Company, P.O. Box 1970,W/A-58, Richland, WA 99352; symposium chairman and editor.

C. H. Henager, Jr.
Battelle Pacific Northwest Laboratory, P.O. Box 999, 306/300, Richland, WA 99352; symposium vice-chairman and editor.

N. Igata
University of Tokyo, 7-3-1 Hongo, Bunkyo-ku, Tokyo 133, Japan; symposium vice-chairman and editor.

Irradiation Creep and Creep Rupture

Shavkat Sh. Ibragimov,[1] Emil S. Aitkhozhin,[1] and Yuri S. Pyatiletov[1]

Radiation-Induced Creep of Aluminum and Copper

REFERENCE: Ibragimov, Sh. Sh., Aitkhozhin, E. S., and Pyatiletov, Yu. S., "**Radiation-Induced Creep of Aluminum and Copper,**" *Influence of Radiation on Material Properties: 13th International Symposium (Part II), ASTM STP 956,* F. A Garner, C. H. Henager, Jr., and N. Igata, Eds., American Society for Testing and Materials, Philadelphia, 1987, pp. 5–10.

ABSTRACT: Results of experimental and theoretical investigations of in-reactor creep of metals are presented in this paper. Experiments have been made in the WWR-K reactor at a neutron flux density of 1.4×10^{16} n/m$^2 \cdot$ s (2.5×10^{15} n/m$^2 \cdot$ s, $E > 0.1$ MeV). Polycrystalline aluminum (99.99%) and copper (99.99%) have been investigated. Aluminum specimens have been examined at temperature levels of 333 to 473 K and stresses of 9.8 to 24.5 MPa, and copper specimens at 423 to 773 K and 20 to 69 MPa, respectively.

Experiments indicate the steady state creep rate depends on stress as $\dot{\epsilon} \sim \sigma^n$ ($n \approx 3$ for aluminum and $n \approx 5$ for copper) and increases with increasing test temperature. Three regions with different slopes are observed on the temperature versus creep rate $\dot{\epsilon}(1/T)$ curve.

Theoretical calculations of the creep rate of irradiated metal as a function of the temperature have been carried out within the framework of the model based on the dislocation climb and glide. Calculation of the temperature dependence of the creep rate established that three temperature regions were recognized as in the case of experiments on copper.

KEY WORDS: creep, irradiation, reactor, dislocation, climb, glide, irradiation creep, copper, aluminum

The phenomenon of the creep enhancement during neutron irradiation is practically observed in all reactor materials. This phenomenon has been intensively investigated during the last ten years [*1–4*]. At the same time the greatest part of experimental data concerning this problem are obtained on alloys with inhomogeneous structure. It complicates the interpretation of results. Therefore, the radiation creep investigation of pure metals is desirable. Results of experimental investigations of polycrystalline copper and aluminum radiation creep are presented. The discussion of these results is carried out in terms of a dislocation climb-glide model.

The observation of aluminum (99.99%) and copper (99.99%) creep has been carried out on planar specimens 20 by 4 by 0.5 mm. They have been annealed at 1023 and 623 K for copper and aluminum, respectively. After the heat treatment, the average grain size was 50 μm and 40 to 50 μm for copper and aluminum, respectively. Experiments have been held in channels of the WWR-K reactor (water-cooled and water-moderated reactor) at neutron flux density 1.4×10^{16} m^{-2}s^{-1} (2.5×10^{15} Nm^{-2}s^{-1}, $E > 0.1$ MeV). Temperature and stress were 423 to 773 K and 20 to 65 MPa for copper and 333 to 473 K and 9.8 to 24.5 MPa for aluminum. The reported temperature was determined with an accuracy of 2°; deformation was measured with an accuracy of 1%. Test duration was 100 to 400 h. As in-reactor creep was measured, out-of-reactors (or thermal) creep tests were performed under analogous conditions.

[1] Professor and director, doctor of sciences, and doctor of sciences, respectively, Institute of Nuclear Physics, Academy of Sciences of KazSSR, Alma-Ata, 480082, Union of Soviet Socialist Republics.

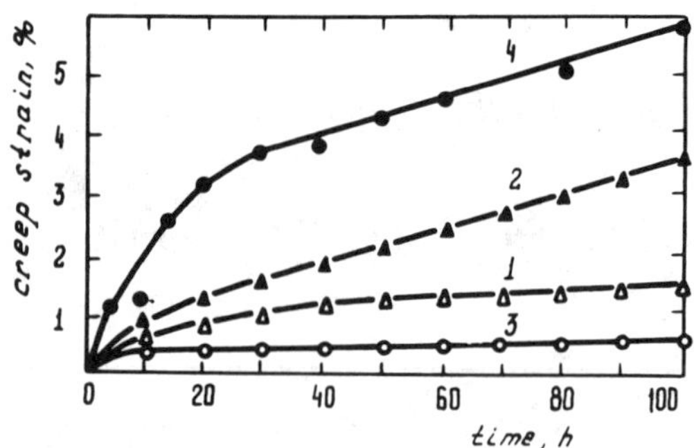

FIG. 1—*Copper* (1,2) *and aluminum* (3,4) *curves at stresses 57 and 14.7 MPa and temperature 523 and 393 K, respectively.* (2,4) *in reactor; neutron flux density is* 1.4×10^{16} $Nm^{-2}s^{-1}$.

Typical curves of aluminum and copper creep are given in Fig. 1. As is seen, the steady state in-reactor creep rate is 4 and 21 times as high as the thermal one for copper (Curves 1 and 2) and aluminum (Curves 3 and 4), respectively. Moreover, the in-reactor irradiation brings out much more clearly the creep transition stage. The deformation for equal time periods during transition and steady state stages are much higher in the case of in-reactor irradiation compared to out-of-reactor tests. In-reactor irradiation also influences τ, the transition stage duration. τ is recognized as inversely proportional to neutron flux density in the case of aluminum.

Creep rates are determined during the steady state stage from experimental curves at different test conditions. The temperature dependence of the in-reactor $\dot{\epsilon}$ and the thermal $\dot{\epsilon}_T$ creep rate of aluminum and copper is given in Figs. 2 and 3. Above all for both materials $\dot{\epsilon}$ is higher than $\dot{\epsilon}_T$

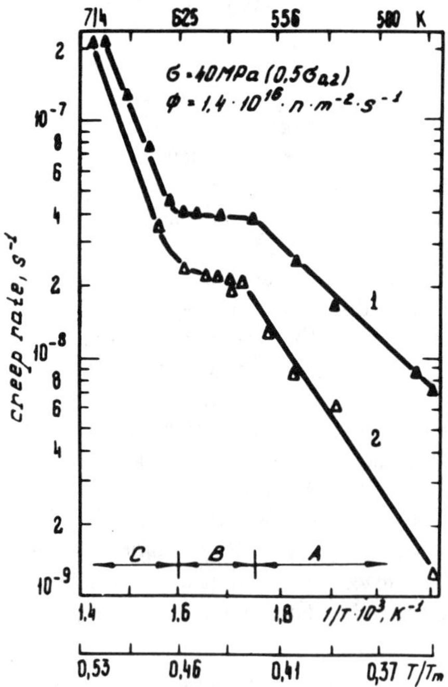

FIG. 2—*Test temperature dependence of the reactor* (1) *and the thermal* (2) *copper creep rate.*

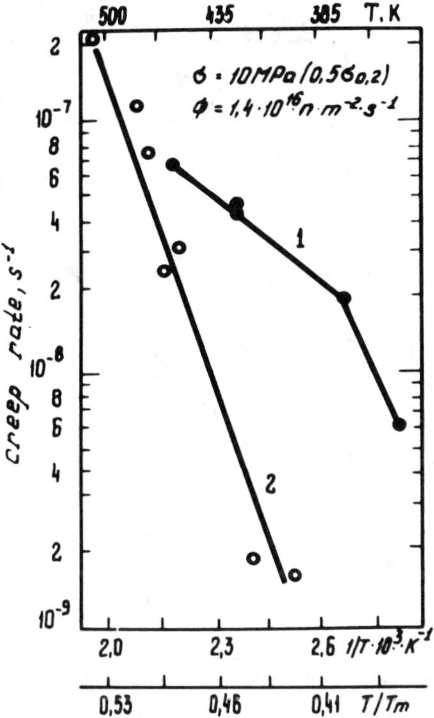

FIG. 3—*Test temperature dependence of the reactor* (1) *and the thermal* (2) *aluminum creep rate.*

in the studied temperature interval, but the magnitude of the irradiation effect depends on the test temperature. The studied temperature interval can be conventionally divided into three regions. The first region A is a low-temperature one (lower than 0.4 T_m) when the rate of material deformation under stress is mainly determined by the radiation creep. For example, the in-reactor creep rate is 6.4 times as high as the thermal one for copper at 473 K (Fig. 2). The second one is an intermediate region B (0.4 $T_m \lesssim T \lesssim 0.5$ T_m) where the in-reactor and the thermal creep rates of copper are comparable and do not significantly change with the temperature increase. The radiation effect is negligible in the third high-temperature region C (higher than 0.5 T_m).

Figures 4 and 5 present the stress dependence of the in-reactor and the thermal creep rate for copper and aluminum at 0.38 T_m. As is seen, $\dot{\epsilon}$ and $\dot{\epsilon}_T$ values depend on stress as $\dot{\epsilon} \sim \sigma^n$ ($n \approx 3$ for aluminum and $n \approx 5$ for copper) in the observed interval and $\dot{\epsilon} > \dot{\epsilon}_T$.

Because of the data obtained for copper when the shape of curves of the in-reactor and the thermal creep are similar (Fig. 2), a supposition about the similarity of deformation mechanisms in initial and irradiated samples can be made. At the same time, it is known that the thermal creep is due to dislocation glide in a studied temperature region. Consequently, it is likely that dislocation glide contributes during radiation creep. Thus, the experimental data will be discussed according to the dislocation climb-glide mechanism of radiation creep [*1*].

For the sake of clarity, we consider rectilinear boundary dislocations in a given crystal forming a rectangular network and being oriented along the coordinate axes. Under the applied stress, dislocations pass the distance L in the glide plane until they stop at barriers such as clusters of point defects of radiation origin (vacancy and interstitial loops, vacancy voids), dislocation

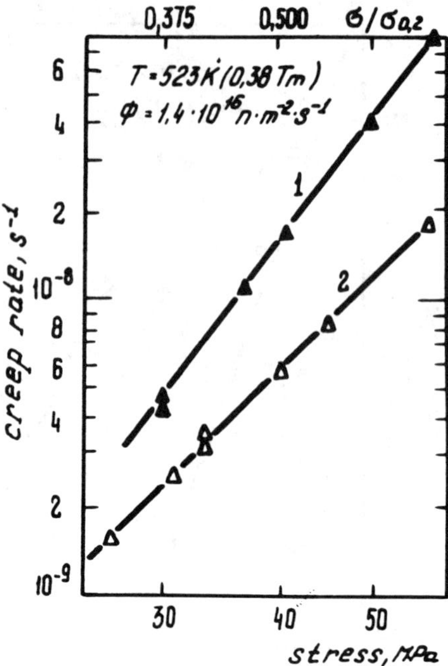

FIG. 4—*Stress dependence of the reactor* (1) *and the thermal* (2) *copper creep rate.*

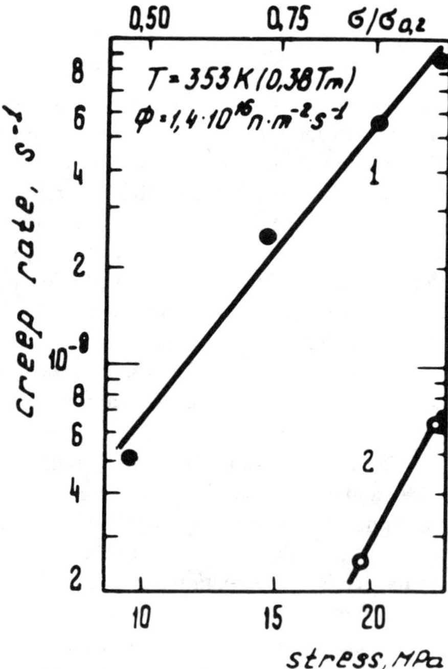

FIG. 5—*Stress dependence of the reactor* (1) *and the thermal* (2) *aluminum creep rate.*

"forest," and other lattice defects. Since interstitials and vacancies are continuously produced during irradiation, dislocations climb over barriers to a new glide plane. Let λ denote the dislocation climb distance. In the new plane the dislocation glides until it stops at a new barrier. Then the climb-glide process is repeated.

From the above described scheme we deduce a formula for the creep rate $\dot{\epsilon}$ suggesting that the stress is operating along X_1 axis. It is based on the general expression

$$\dot{\epsilon} = b\rho_n^{(1)}V \tag{1}$$

where b is the Burgers vector modulus and $\rho_n^{(1)}$ is the dislocation density, where the Burgers vectors are oriented along the X_1 axis, V is the dislocation motion rate in the X_1 direction

$$V = \frac{L}{\tau_l + \tau_n} \tag{2}$$

where τ_l is the time of dislocation glide through the distance L; τ_n is the time of dislocation climb over barriers. Usually, $\tau_l << \tau_n$, so the approximate equality $V \approx L/\tau_n$ is true. Time τ_n is calculated by formula

$$\tau_n = \lambda/\upsilon_n \tag{3}$$

but the dislocation climb rate is determined by the difference between fluxes of interstitials and vacancies on the dislocation

$$\upsilon_n = \frac{1}{b}[Z_I^{(1)}D_IC_I^0 - Z_V^{(1)}D_V(C_V^0 - C_V^{(1)})] \tag{4}$$

where $Z_I^{(1)}$ and $Z_V^{(1)}$ are the parameters characterizing the absorption efficiency of interstitials and vacancies by the dislocation with the Burgers vector aligned along X_1; D_I and D_V are the diffusion coefficients of interstitials and vacancies: C_I^0 and C_V^0 are their concentrations midway between sinks; $C_V^{(1)}$ is the thermal equilibrium concentration of vacancies near the dislocation of type 1.

Using Formulae 1 through 4, we obtain the final equation for the creep rate

$$\dot{\epsilon} = A\rho_n^{(1)}(D_IC_I^0Z_I^{(1)} - D_VC_V^0Z_V^{(1)} + D_VC_V^{(1)}Z_V^{(1)}) \tag{5}$$

where $A = L/\lambda$. The L value calculation can be made by a computer simulation method by the investigation of dislocation motion through a network of randomly distributed obstacles [5,6]. Quantitative data for glide barriers and glide dislocations (dislocation density, sizes and density of point defect clusters, and strength of all types of obstacles on the path of dislocation motion) are necessary. Parameters to put in the brackets of the right-hand side of Eq 5 are found in Ref 7. Hence they will not be reproduced here.

Unfortunately, at this time we do not have the necessary information about the structure of radiation-induced defects in copper and aluminum to make a quantitative evaluation of the creep rate using Formula 5 and to make a comparison of calculated data with experimental ones. Therefore, a calculation using known parameters for iron-based alloys [7] is carried out. Conditions are similar to the ones in the case of the copper experiment (temperature region is $0.25\ T_m - 0.55\ T_m$, stress is 98 MPa, about half of the yield stress). Calculation results are given in Fig. 6. Note the calculated temperature dependencies are of the same type as for the experimental data (Fig. 2, Line 1). As in the case of the experiment, there are three regions on the calculated curve: low ($T/T_m \lesssim 0.4$), intermediate ($0.4 \lesssim T/T_m \lesssim 0.5$), and high homologous temperatures that coincide practically with the experimental data for copper (Fig. 2).

Conclusions

Thus, temperature and stress dependencies of the in-reactor creep rate of pure metals (copper and aluminum) are obtained. The temperature dependence of radiation creep rate has the same

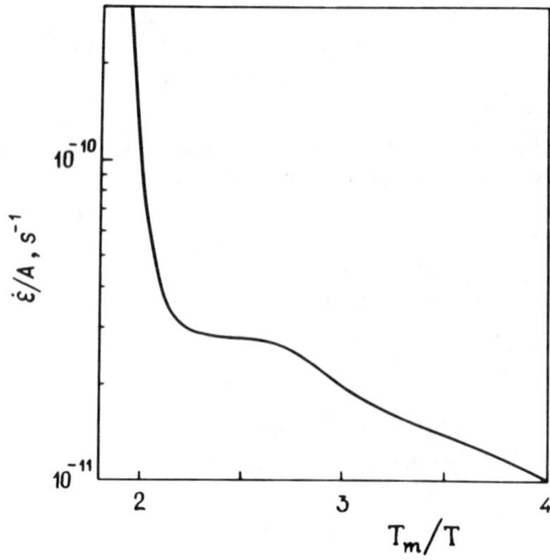

FIG. 6—*Temperature dependence of creep rate (calculation by formula* (5) *using the parameters for iron-based alloys).*

form as that of the climb-glide mechanism. The comparison of the calculated and experimental results shows that the $\dot{\epsilon}(T)$ curves are similar. This allows the conclusion that the climb-glide mechanism is one of the more possible mechanisms of copper and aluminum radiation creep.

References

[*1*] Ibragimov Sh. Sh., Kirsanov V. V., and Pyatiletov, Yu. S., "Radiation Damage of Metals and Alloys," *M., Energoatomizdat,* 1985, p. 240.

[*2*] Ibragimov Sh. Sh., "Radiation Strengthening and Creep of Metals," *Voprosi Atomnoi Nauki i Techniki. Seriya: Fizika Radiatsion-nikh Povrezhdeniji i Radiatsionnoe Materialovedenie,* Vol. 19, No. 5, 1981, pp. 25–36.

[*3*] Harries, D. R., "Irradiation Creep in Non-Fissile Metals and Alloys," *Journal of Nuclear Materials,* Vol. 65, 1977, pp. 157–173.

[*4*] Ehrlich, K., "Irradiation Creep and Interrelation with Swelling in Austenitic Stainless Steels," *Journal of Nuclear Materials,* Vol. 100, 1981, pp. 149–166.

[*5*] Foreman, J. E. and Makin, M. J., "Dislocation Movement through Random Array of Obstacles," *Philosophical Magazine,* 1966, Vol. 14, pp. 911–924.

[*6*] Kirsanov, V. V., Pyatiletov, Yu. S., and Tyupkina, O. G., "Irradiation Creep Due to Dislocation Glide," *Physics Status Solidi (a),* 1981, Vol. 64, pp. 735–740.

[*7*] Pyatiletov, Yu. S., "Temperature Dependence of Radiation Creep Rate," *Fizika Metallov i Metallovedenie,* Vol. 54, 1982, pp. 1080–1087.

Douglas L. Porter[1] *and Frank A. Garner*[2]

Cessation of Irradiation Creep in AISI 316 Concurrent with High Levels of Swelling

REFERENCE: Porter, D. L. and Garner, F. A., **"Cessation of Irradiation Creep in AISI 316 Concurrent with High Levels of Swelling,"** *Influence of Radiation on Material Properties: 13th International Symposium (Part II), ASTM STP 956,* F. A. Garner, C. H. Henager, Jr., and N. Igata, Eds., American Society for Testing and Materials, Philadelphia, 1987, pp. 11–21.

ABSTRACT: At high neutron exposures where the swelling of AISI 316 exceeds ~5% the irradiation creep rate at 550°C begins to decline and eventually disappear. Swelling continues to proceed, however, leading to a maximum deformation rate dictated only by that of the steady-state swelling rate or ~0.33%/displacements per atom (dpa). It is not clear at this point whether large levels of voidage are directly responsible for the decrease in creep rate. There is some evidence to support the possible role of intermetallic precipitate formation, although the mechanism of creep suppression by such precipitates is not known.

KEY WORDS: 316 stainless steel, irradiation creep, thermal creep, swelling, voids, intermetallic precipitates

In a series of recently published studies it was shown that the relationship of creep to concurrent void swelling is more complex than previously envisioned [*1–3*]. Whereas it has been commonly accepted that the irradiation-induced creep rate accelerates with the onset of swelling and would continue at a rate proportional to that of the swelling, it now appears that the creep rate in AISI 316 stainless steel subsequently declines as swelling approaches levels in excess of ~5%. Creep also appears to be coupled to swelling in a manner such that the total rate of diametral deformation in pressurized tubes does not exceed ~0.33%/displacements per atom (dpa), regardless of how the swelling and creep strains are partitioned. Since 0.33%/dpa represents the anticipated steady-state swelling rate [*4*], this implies that the creep rate must eventually vanish.

These somewhat perplexing developments first were brought to light when the measured creep strains of fuel pin cladding at high neutron fluence often fell substantially below that predicted by creep correlations developed at lower fluence but higher stress levels [*1,3*]. The tendency toward underprediction appears to increase as the fluence to burn-up ratio is increased. (Burn-up refers to the percentage of fissile atoms which have undergone fission). This ratio is controlled primarily by the fissile enrichment of the fuel and governs the rate of increase of fission gas pressure. Since fuel cladding usually has very low initial stress levels which then increase primarily due to fission gas pressure as burn-up proceeds, the fluence to burn-up ratio determines the relative amounts of creep and swelling strain. Although the onset of swelling is somewhat sensitive to the stress level, swelling does not require stress to proceed in AISI 316 [*5,6*].

At this point it is relevant to note that there is a fundamental difference in the conditions under which stress-driven creep data are collected and the manner in which stresses arise in most reactor components. On the whole, creep data are very expensive to obtain compared to swelling data.

[1] Metallurgist, Argonne National Laboratory, EBR-II Project, Idaho Falls, ID.
[2] Fellow scientist, Hanford Engineering Development Laboratory, Richland, WA.

This cost and the space constraints in available reactors necessitates several compromises in the experimental derivation of a comprehensive creep data base. First, creep experiments usually proceed on small creep capsules in order to cover the range of fluence and temperature desired. While the use of short tubes has been shown to yield results representative of longer tubes with minimum end effects, the gas pressures are chosen to maximize the creep strain, using constant stress levels that usually exceed the time-dependent levels encountered in fuel pin cladding and other reactor components [7]. This approach guarantees that creep strains develop in pressurized tubes prior to the onset of swelling. The opposite situation, where swelling develops before much creep occurs, would never happen in a typical highly-pressurized tube. However, swelling can start prior to substantial creep in fuel pins that have a high fluence to burn-up ratio.

The second compromise involves the method of calculating the creep strain. The usual practice is to calculate the creep strain by subtracting from the diametral change of the stressed tube the diametral change of a companion tube without stress. After measurement of both tubes, they are returned to the reactor for further irradiation and subsequent measurements. The alternative approach would be much more expensive and requires the destructive examination of a stressed tube at every fluence and stress level. One consequence of the current approach is the inclusion of some non-creep components of deformation in the calculated creep rate. These non-creep strains are the stress-affected component of swelling and the texture and stress-dependent anisotropies associated with the formation of various precipitate phases [8,9]. As long as these non-creep components are small relative to that of creep, and providing that these non-creep strains saturate with continued exposure, the errors associated with the trade-off between cost and necessity are small and can be compensated partially by using an appropriately coupled set of creep, swelling, and densification equations. However, this approach also requires that the creep rate continue to develop in a manner which is proportional to the swelling rate. Otherwise, the potential exists for a misprediction of creep strains subject either to large levels of void swelling or phase-related volume changes. The latter consideration has been demonstrated to have a large effect on creep strains and creep predictions in AISI 316 [9]. It may also be involved in the decrease of creep rates at high fluence, as discussed later.

In this paper we examine the relationship between strains arising from precipitation, irradiation creep, and swelling in AISI 316 using rather long creep tubes irradiated to higher fluence levels than currently probed by most short creep tube experiments. One can also make use of the length of the tube to study at one stress level the influence of both fluence and displacement rate on the various components of strain, using the neutron flux and energy gradients along the tube axis.

Experimental Details

Another way to assess the interrelationships between swelling and creep strains is to employ different thermal-mechanical starting conditions that affect the duration of the transient regime of swelling. The steel employed was the V87210 reference heat of AISI 316 used in the U.S Breeder Reactor Program. Its composition in percent by weight was 13.57 Ni, 16.36 Cr, 2.88 Mo, 1.42 Mn, 0.47 Si, 0.07 C, 0.02 P, 0.01 S with <0.005 B and the balance Fe. This steel was irradiated in five metallurgical conditions: solution annealed, 10% cold-worked, 20% cold-worked, 20% cold-worked followed by aging for 24 h at 482°C (designated Heat Treat C), and Heat Treat C followed by another aging for 216 h at 704°C to cause extensive carbide precipitation. This latter heat treatment is hereafter designated as Heat Treat D but is often called the Garafalo treatment. All conditions were prepared using vendor-produced 20% cold-worked steel as the starting condition. Thus the 10% cold-worked condition was prepared by working the annealed condition prepared in the laboratory from the 20% cold-worked condition.

The creep capsules have been described in detail elsewhere [10,11]. In brief, the capsules are 1.02 m long and have an outer diameter of 0.584 cm and a wall thickness of 0.038 cm, but only the top 0.28 m length of the capsules is pressurized with helium to yield hoop stresses varying from 0 to 200 MPa (0–30 ksi). The pressurized portion is welded to the lower portion of the capsule which contains a tantalum rod to heat the sodium to ~550°C ± 10°C as it flows upward. The irradiations proceeded in row 7 of the Experimental Breeder Reactor (EBR-II) in Idaho Falls, Idaho. Approximately 5 dpa (± 10%) are produced in this reactor for each 1.0×10^{22} n · cm^{-2} ($E > 0.1$ MeV). The dpa levels cited for each datum are derived from the neutron spectrum and flux for the reactor position examined.

Diameter measurements were made with a contact profilometer on a spiral trace along the entire length of the capsule after each irradiation period. Some tubes had obviously failed during a given irradiation period and were withdrawn. Those that did not fail were returned for reirradiation until a decision was made to terminate the irradiation of that particular pin. At that point the tube was punctured and the gas volume measured to confirm that the design stress level had been maintained throughout the irradiation. Some pins, particularly those that were in the annealed condition, were found not to have gas pressure and thus to have failed by some not quite so obvious mode such as a pin-hole failure. The stress history of these pins was therefore unknown and they were not used in this analysis.

Some pins were sectioned into 25.4 mm (1 in.) axial increments and immersion density measurements performed on them in order to determine the contribution of the various strain components.

Results

Figure 1 shows typical strain profiles for both an unstressed and a stressed pin, as well as a photograph of a typical pin in the unirradiated condition. The discontinuity in each profile shows the location of the weld at the bottom of the pressurized section. The oscillations in the spiral trace reflect a slight ovality that often develops in the pin during irradiation. The swelling of the five metallurgical conditions of this steel in general has been found to be Heat Treat D >annealed> 20% cold-worked> Heat Treat C >10% cold-worked [*10,11*]. The lesser swelling of the 10% cold worked condition probably reflects the different annealing conditions it received in the laboratory compared to that of the mill-prepared 20% cold-worked condition. Normally, the swelling decreases as the cold-worked level increases, providing that the intermediate annealing conditions between cold-worked passes does not change. The onset of swelling for AISI 316 is known to be sensitive to many production variables including the rate of feed of the tube through the annealing furnace [*6*]. Laboratory anneals are in general conducted for times that exceed the residence time of mill-prepared steels.

Since the creep rate is thought to increase as swelling accumulates, one would expect that the total deformation would exhibit the same dependence on starting condition as seen in the swelling behavior. Figure 2 shows that the total deformation behavior of a typical series of pins follows the trend observed in the swelling behavior of earlier studies.

Figure 3*a* shows the total diametral strains measured at the position of maximum displacement rate for three high fluence capsules in the Heat Treat D condition. Note that while the transient regime of diametral deformation decreases in duration with increasing stress, the post-transient deformation rate does not appear to change. Even more significantly, the total deformation rate does not exceed the 0.33%/dpa expected from steady-state swelling alone. Instead it appears that the well-known linear stress dependence of creep rate has disappeared somewhere between hoop stress levels of 0 and 100 MPa (15 ksi). Based on the current conception of irradiation creep where the largest component of the creep rate is proportional to the swelling rate [*4*] one would expect at 200 MPa a creep rate at high fluence that approaches the swelling-induced deformation rate. The swelling of these three pins was not measured so we cannot at this time separate the swelling and creep strains.

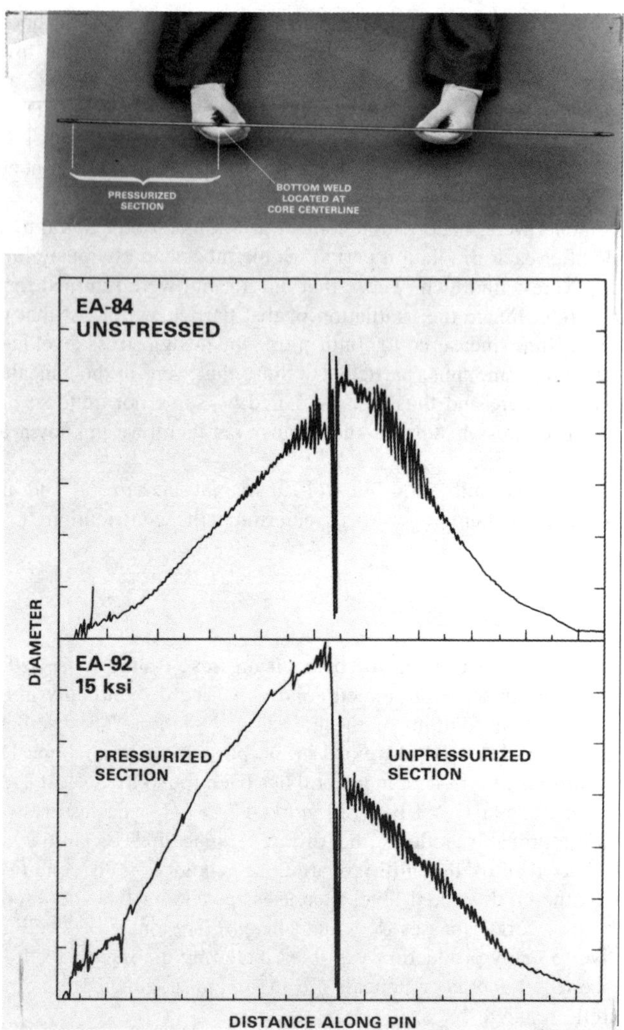

FIG. 1—*Picture of creep capsule and typical profilometer traces of stressed and unstressed capsules.*

Figure 3*b* shows three pins which were removed from the reactor at a lower fluence level when the 200 MPa (30 ksi) pin (designated EA-223) failed. The creep rates for the two pressurized pins were calculated two ways as shown in Fig. 4. First, the swelling-induced change in diameter of the *stress-free* tube was subtracted from the total diametral change of the pressurized tubes. Note, however, that the total deformation data was taken as a function of time at one position while the swelling-induced deformation data was taken as a function of position, all at one time in-reactor, representing the time of removal of the capsule. This approach assumes that there is no dependence of swelling on displacement rate. In the second calculation, the swelling-induced deformation of the *pressurized* tube was used, once again assuming that there is no effect of displacement rate. The swelling-induced strain of the pressurized tube is obviously a better choice than that of the unpressurized tube, but as explained previously, such data are very expensive to obtain and are usually not available.

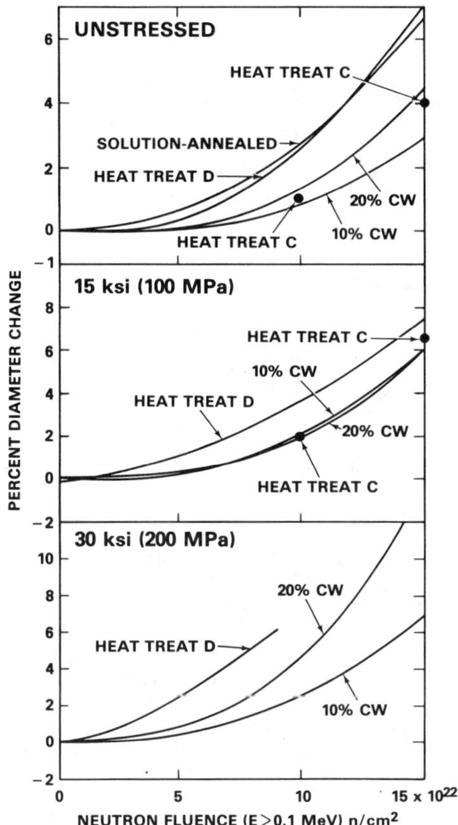

FIG. 2—*Typical deformation behavior at core centerline of creep capsules with different stress levels and starting metallurgical conditions.*

Figure 4 shows that the inclusion of the stress-affected portion of swelling in the creep rate camouflages a trend toward saturation in the creep rate. While the high fluence portion of these data obviously have no influence of flux differences, one can question whether the low fluence (and therefore low flux) results reflect some interference from flux effects. Since capsule EA-223 failed it was not used to answer this question but we can address it using a pin that did not fail.

Figure 5 shows the total deformation measured for another Heat Treat D pin. One trace is that recorded for the core center position as a function of time, the other is that determined as a function of position along the pin upon its removal. There obviously is a small but discernible influence of displacement rate on the total deformation of the Heat Treat D pin at lower flux levels. It is important to note, however, that the trend shown in Fig. 5 would tend to further accentuate the tendency toward saturation when applied to Fig. 4 in that it should steepen the slope of the 100 MPa (15 ksi) curve at lower fluence while not affecting the higher fluence portion of the curve.

We can see the trend toward saturation of the creep rate even more clearly in the data derived from the 20% cold-worked condition. As shown in Fig. 6 there is once again a strain rate that approaches but does not exceed 0.33%/dpa. Capsule EA-38 at 200 MPa (30 ksi) also failed after the last examination shown in Fig. 6 but once again we calculate for the 200 MPa (30 ksi) capsule

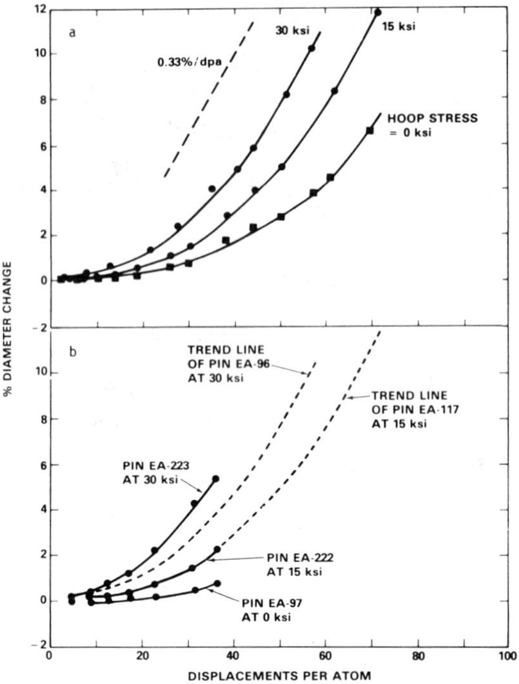

FIG. 3—(a) *Diametral strains observed at core centerline position of EBR-II for three creep capsules irradiated at 550°C in the Heat Treat D condition,* (b) *Comparison of the strains above with those of another set of Heat Treat D capsules removed at lower fluence.*

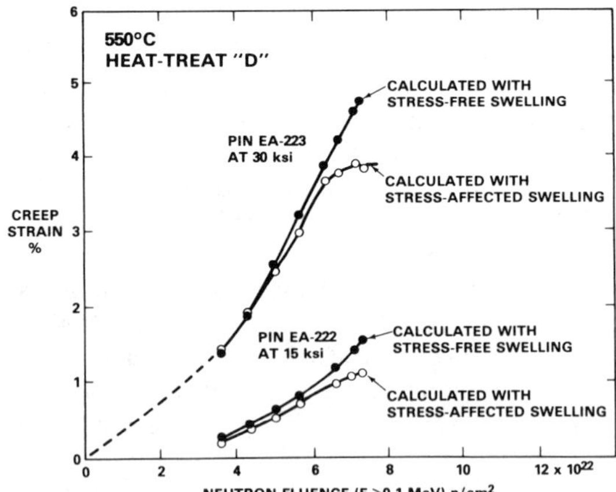

FIG. 4—*Calculated creep strains for the two pressurized pins shown in Fig.* 3b, *showing how the use of stress-free swelling camouflages the onset of the saturation stage of creep.*

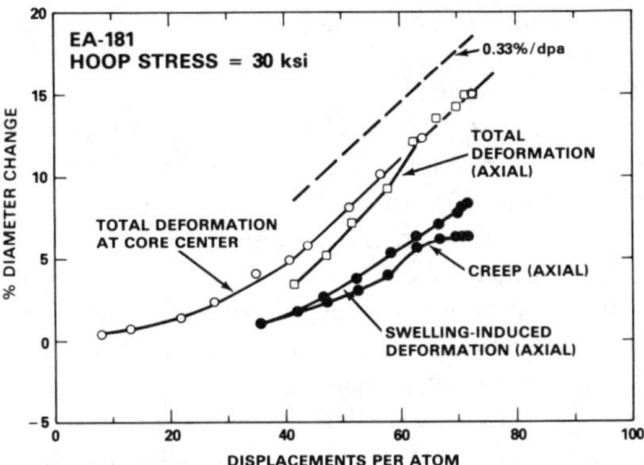

FIG. 5—*Analysis of strains of another capsule in the Heat Treat D condition, showing influence of flux effects on analysis, but only at the lower displacement levels.*

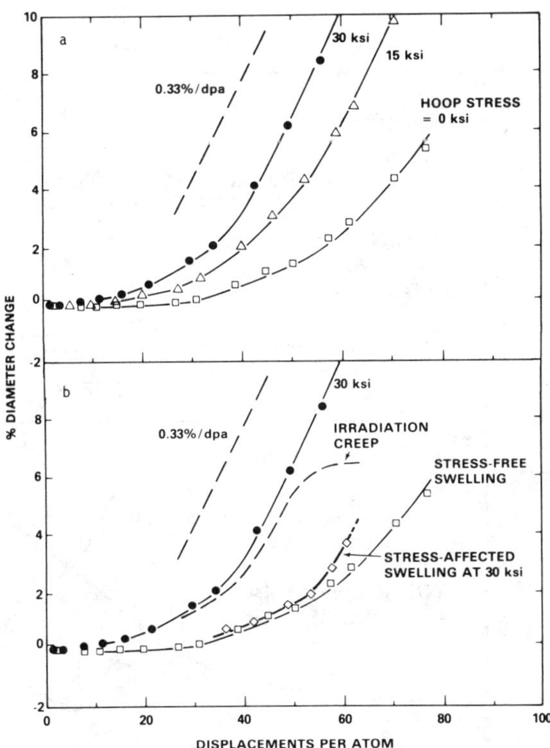

FIG. 6—(a) *Deformation observed in pressurized tubes of 20% cold-worked AISI 316 irradiated in EBR-II at 550°C. Note that the total strain rate does not exceed 0.33%/dpa, even at very high stress levels.* (b) *Density measurements on the 30 ksi (200 MPa) pressurized tube show that stress accelerates the rate of swelling and its approach to 1%/dpa, but also causes the creep rate to approach zero at high swelling levels.*

an irradiation creep behavior that exhibits saturation. In this case, however, the actual stress-affected swelling of EA-38 was measured below the region where failure occurred and was included in the calculation of creep rate.

Similar behavior is seen in the Heat Treat C condition at 100 MPa (15 ksi) as shown in Fig. *7b*. In the Heat Treat C case the trend toward saturation is not obscured by flux effects. Note that the axial and core center determinations of total deformation are nearly identical, indicating that flux effects do not operate on the Heat Treat C condition.

Figure *7a* shows an apparent flux effect on the stress-free swelling portion of the deformation, however. While the swelling and swelling-induced deformation agree at core center, they do not agree as the dpa level decreases. Note, however, that the difference quickly becomes constant at ~1%. It has earlier been shown that intermetallic phase formation at relatively high temperatures can lead to increases in lattice parameter and an apparent swelling that is on the order of several percent [*9*]. Most importantly, this volume increase is distributed anisotropically, with the diameter absorbing most of the strain [*9*]. The difference seen in Fig. *7a* probably represents not only the anisotropy of intermetallic-induced volume changes but also the fact that intermetallic formation requires thousands of hours to occur. At a given low displacement level the axial data represent much longer times than that of the core center data.

The total deformation data for annealed and 10% cold-worked material have been presented in a previous paper [*3*] and will not be reproduced here. The annealed tubes had all failed in a

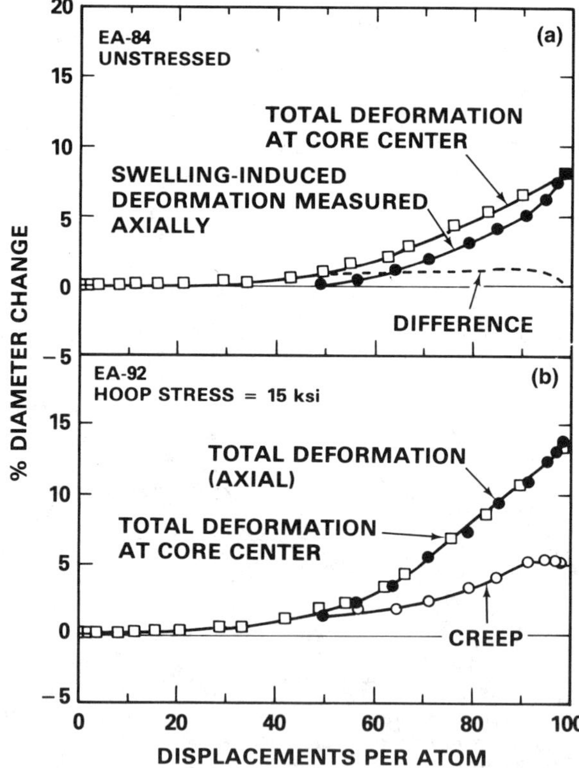

FIG. 7—*Analysis of strains developed in capsules constructed of AISI 316 irradiated in the Heat Treat C condition.*

manner which rendered their stress history to be indeterminate. The 10% cold-worked tubes were not subjected to sufficient tests to separate the creep and swelling components of strain.

Discussion

There are two ways in which to characterize the major findings of this study. One is to assume that the emergence of voids at relatively large volumes is directly responsible for the decline in creep rate. This would be a natural extension of the current creep model which states that the creep rate has a small athermal component unrelated to swelling and a much larger second component which is proportional to swelling [*4*]. This second component is often interpreted to imply a direct coupling of swelling and creep.

Another interpretation would be that creep and swelling both respond to the same developments in microstructure and microchemistry and therefore are not directly coupled to each other [*4*]. In this viewpoint creep might respond to late-term changes in microchemistry or microstructure that do not affect swelling. Post-transient swelling has recently been shown to be a phenomenon which is dictated only by the crystal structure. This occurs after the completion of a transient regime in which only void nucleation has been found to be sensitive to all swelling-relevant variables [*4*]. Post-transient swelling of fcc metals does not seem to be affected by the softness or hardness of the alloy [*2–4*] while one would expect the irradiation creep rate to be related to the yield strength of the alloy [*12*]. The increase in hardness due to swelling at higher levels has been determined [*13*], however, and is insufficient in itself to account for the near cessation of creep.

There are other direct-coupled types of mechanisms which might be invoked to partially explain the observed results. First of all, at these swelling levels the overwhelming majority of dislocations terminate at void surfaces, a situation in which the climb rate of dislocations can be strongly affected by pipe diffusion of defects into the voids. This might tend to change the bias of the dislocation network. Second, the glide distance of dislocations will be decreased as the voids become the dominant obstacle. This will make dislocation climb the dominant creep mechanism and reduce the amount of creep that can occur for a given level of irradiation. Third, if the creep process is now restricted primarily to irradiation-induced climb of dislocations, one can envision a situation in which the stress-induced preferential absorption (SIPA) creep mechanism [*14*] may no longer operate effectively. This mechanism requires that the climb of dislocations lying on some planes be enhanced by the applied stress state while other less favorably oriented dislocations are inhibited in their climb rate. In short, this mechanism requires that dislocations lie on different planes and have different Burgers vectors. Gelles and coworkers have recently shown that large applied stresses lead to a strong anisotropy of dislocations such that unfavorably oriented dislocations exist at substantially reduced densities while favorably oriented dislocations increase in density [*15,16*]. Similar stress-induced anisotropies have been observed in the Frank loop population [*17,18*]. Hence, the SIPA creep mechanism may not function very effectively after the stress state has substantially altered the dislocation and loop microstructure.

There is one indication in the data presented in this paper that suggests that the near-cessation of creep may be only coincident with large swelling levels and not directly related to void growth. This is the suggestion that intermetallic formation may be occurring and affecting the creep rate, while not affecting the swelling rate. In an earlier paper it was shown that intermetallic formation during thermal annealing of AISI 316 [*9*] leads to an abrupt decrease in the thermal creep rate at 649, 705, and 760°C as shown in Fig. 8. While the thermal creep experiment did not proceed for a sufficient time at 594°C to cause a convincing down-turn in creep rate it is reasonable to assume that longer times might cause a similar effect to occur at 594°C and that the additional influence of irradiation-enhanced diffusion might cause precipitation to occur at the 550°C temperature of this experiment.

Whatever the cause or causes of creep cessation it appears to be a reproducible phenomenon.

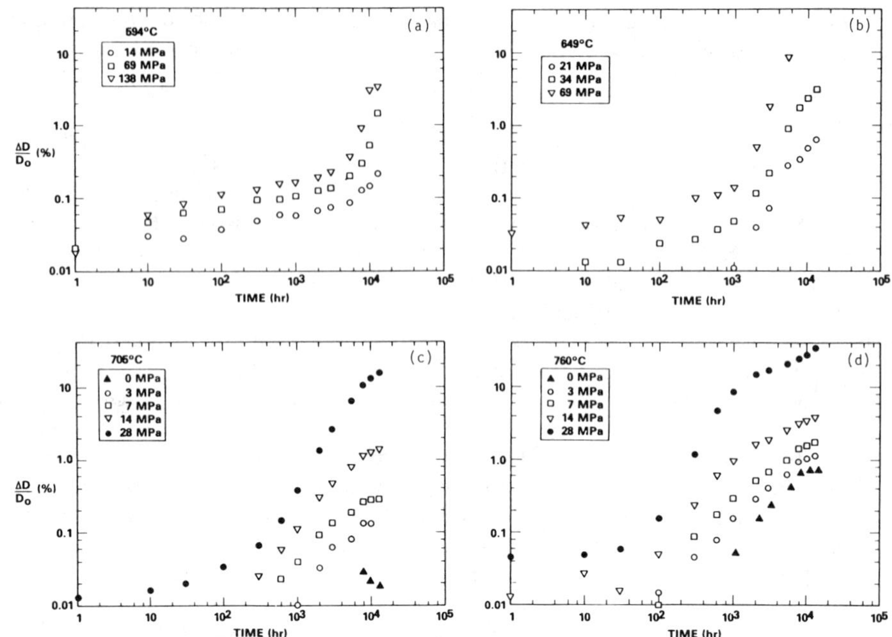

FIG. 8—*Thermal creep strains observed in the diameter of gas-pressurized tubes constructed from AISI 316 heat K81581* [9]. *The stresses shown are midwall hoop stress levels.*

Attempts to confirm this observation using current short tube experiments were unsuccessful, however. These experiments have diametral deformation rates that have currently reached only ~0.2%/dpa.

Since the diametral creep rate does saturate at approximately one-third of 1%/dpa, the latter being the steady-state swelling rate, this implies that stress-affected swelling itself is isotropic. Increases in volume obviously require dislocation motion but the apparent disappearance of creep implies a greatly reduced mobility of dislocations, or at least a reduction in their ability to sense and react to the stress state. This apparent paradox awaits the acquisition of more data before it can be resolved.

It is difficult to imagine, however, how the applied and swelling-generated stresses are being relieved on a microscopic level when the macroscopic creep rate seems to have disappeared.

Conclusions

At displacement levels of ≲50 dpa, the creep rate at 550°C of AISI 316 in a variety of metallurgical starting conditions begins to decline and eventually disappear. One consequence of this is that the total diametral deformation rate of pressurized tubes cannot exceed that dictated by steady-state swelling (0.33%/dpa). It is not yet clear whether the cessation of creep arises as a direct consequence of large swelling levels or whether other late-term microstructural developments are responsible.

References

[1] Makenas, B. J., in *Effects of Radiation on Materials: Twelfth International Symposium, ASTM STP 870*, F. A. Garner and J. S. Perrin, Eds., American Society for Testing and Materials, Philadelphia, 1985, pp. 202–211.

[2] Garner, F. A., Brager, H. R., Hamilton, M. L., Dodd, R. A., and Porter, D. L., *Radiation Effects*, Vol. 101, 1987, pp. 37–53.
[3] Garner, F. A., Porter, D. L., and Makenas, B. J., *Journal of Nuclear Materials*, Vols. 122 and 123, 1984, pp. 459–471.
[4] Garner, F. A., *Journal of Nuclear Materials*, Vols. 122 and 123, 1984, pp. 459–471.
[5] Garner, F. A., Gilbert, E. R., and Porter, D. L., in *Effects of Radiation on Materials: Tenth Conference, ASTM STP 725*, D. Kramer, H. R. Brager, and J. S. Perrin, Eds., American Society for Testing and Materials, Philadelphia, 1981, pp. 680–697.
[6] Garner, F. A., in *Optimizing Materials for Nuclear Applications*, F. A. Garner, D. S. Gelles, and F. W. Wiffen, Eds.; TMS-AIME, Warrendale, PA, 1985, pp. 111–139.
[7] Straalsund, J. L., in *Radiation Effects in Breeder Reactor Structural Materials*, M. L. Bleiberg and J. W. Bennett, Eds., The Metallurgical Society of AIME, New York, 1977, pp. 191–207.
[8] Garner, F. A., Cummings, W. V., Bates, J. F., and Gilbert, E. R., "Densification-Induced Strains in 20% Cold Worked 316 Stainless Steel During Neutron Irradiation," HEDL-TME 78-9, Hanford Engineering Development Laboratory, Richland, WA, June 1978.
[9] Puigh, R. J., Lovell, A. J., and Garner, F. A., *Journal of Nuclear Materials*, Vols. 122 and 123, 1984, pp. 242–245.
[10] Walters, L. C., McVay, G. L., and Hudman, G. D., in *Radiation Effects in Breeder Reactor Structural Materials*, M. L. Bleiberg and J. W. Bennett, Eds., The Metallurgical Society of AIME, New York, 1977, pp. 277–294.
[11] McVay, G. L., Einziger, R. E., Hofman, G. L., and Walters, L. C., *Journal of Nuclear Materials*, Vol. 78, 1978, pp. 201–209.
[12] Jung, P., and Ansari, M. I., *Journal of Nuclear Materials*, Vol. 138, 1986, pp. 40–45.
[13] Garner, F. A., Hamilton, M. L., Panayotou, N. F., and Johnson, G. D., *Journal of Nuclear Materials*, Vols. 103 and 104, 1981, pp. 803–808.
[14] Wolfer, W. G., *Journal of Nuclear Materials*, Vol. 90, 1980, p. 175.
[15] Gelles, D. S., in *Effects of Radiation on Materials: Twelfth International Symposium, ASTM STP 870*, American Society for Testing and Materials, Philadelphia, pp. 98–112.
[16] Gelles, D. S., Garner, F. A., and Adams, B. L., in *Damage Analysis and Fundamental Studies Quarterly Progress Report*, DOE/ER-0046/11, August 1982, pp. 81–100.
[17] Brager, H. R., Garner, F. A., and Guthrie, G. L., *Journal of Nuclear Materials*, Vol. 66, 1977, pp. 301–321.
[18] Gelles, D. S., Garner, F. A., and Brager, H. R., in *Effects of Radiation on Materials: Tenth Conference, STP 725*, American Society for Testing and Materials, Philadelphia, 1981, pp. 735–753.

Raymond J. Puigh[1] *and Margaret L. Hamilton*[1]

In-Reactor Creep Rupture Behavior of the D19 and 316 Alloys

REFERENCE: Puigh, R. J. and Hamilton, M. L., "**In-Reactor Creep Rupture Behavior of the D19 and 316 Alloys,**" *Influence of Radiation on Material Properties: 13th International Symposium (Part II), ASTM STP 956,* F. A. Garner, C. H. Henager, Jr., and N. Igata, Eds., American Society for Testing and Materials, Philadelphia, 1987, pp. 22–29.

ABSTRACT: In-reactor creep rupture data on one heat and two cold work levels of the D9-type alloy and on two heats of cold worked 316 stainless steel have been obtained from the Materials Open Test Assembly (MOTA) experiment in the Fast Flux Test Facility (FFTF). The on-line detection of creep ruptures using a tag gas technique has provided in-reactor creep rupture data for irradiation temperatures of 575, 605, 670, and 750°C to a peak fluence of 9×10^{22} n/cm^2 ($E > 0.1$ MeV). Similar thermal creep rupture data have been accumulated for exposure temperatures of 650, 704, and 760°C to exposure times of 10 000 h. In general, the in-reactor rupture times are comparable to or less than those exhibited by the thermal testing. The 10% cold worked condition of D9 exhibits slightly better creep rupture resistance when compared to the 20% cold worked condition. The in-reactor failure strains are less than those of the thermal controls for a given temperature.

KEY WORDS: neutron irradiation, stainless steels, 316 stainless steels, D9 stainless steels, in-reactor creep, creep rupture, stress rupture

The in-reactor stress rupture behavior of structural materials has long been an important issue in optimizing the design of both fission and fusion reactor designs. Early work in EBR-II on the austenitic steels [*1,2*] has shown that in-situ stress rupture tests must be conducted in order to obtain valid in-reactor stress rupture data. Subsequent instrumented neutron irradiation experiments on cold worked 316 stainless steel (SS) showed that the in-reactor stress rupture behavior at 650°C was better than the thermal behavior for neutron fluences up to 5×10^{22} n/cm^2 ($E > 0.1$ MeV), which corresponds to an exposure time of ~8000 h. More recent data on this material to fluences of 8×10^{22} n/cm^2 ($E > 0.1$ MeV), or exposure times of ~11 000 hours, suggested that for these conditions neutron irradiation does not further change rupture times [*3*]. The major problems associated with these results, however, are the uncertainties in the actual irradiation temperatures and the assignment of rupture times. The uncertainty in irradiation temperature for the EBR-II experiment was on the order of ±25°C. The rupture times were assigned by extrapolating the strains measured as a function of time to the failure strain. Caution is needed when applying this technique, however, since this material has been observed to exhibit significant volume changes at higher irradiation temperatures, which has been attributed to phase changes in the material [*4*].

The uncertainties in the stress rupture data obtained in-reactor have been significantly reduced with the use of the Materials Open Test Assembly (MOTA) for testing of materials in the Fast Flux Test Facility (FFTF) [*5*]. The temperature uncertainty associated with irradiation in this vehicle is ±5°C. Moreover, through the use of tag gases and an on-line cover gas monitoring system, on-line detection of specimen ruptures is possible during irradiation, thereby significantly reducing the uncertainty associated with the rupture times.

[1] Managers, Westinghouse Hanford Company, Richland, WA 99352.

Titanium additions, increases in nickel content, and decreases in chromium content, which were made to improve the swelling response of 316 SS, resulted in an alloy class referred to as "D9." In-reactor stress rupture data from the MOTA experiment have been reported [*5*] on two conditions of the D9-type alloys for neutron fluences to 3×10^{22} n/cm^2 ($E > 0.1$ MeV) (exposure times to 2400 h) at irradiation temperatures of 575, 605, 670 and 750°C. For these conditions the in-reactor rupture times were similar to those observed in thermal control tests.

This report will describe both the in-reactor stress rupture behavior and the thermal control data for 20% cold worked (CW) 316 SS and for 10 and 20% CW D9 over a similar temperature range for in-reactor exposure times up to 13 170 h and peak fluences of 17×10^{22} n/cm^2 ($E > 0.1$ MeV). The inclusion of two cold work levels of D9 was based on prior experience with 316 SS, which showed that a slight advantage was gained at higher temperatures with the lower cold-work level [*6*].

Experimental Procedure

Thermal Control Tests

Thermal control tests were performed on 76.2-mm (3-in.) long sections of tubing that were either 5.84 mm outside diameter (OD) by 5.08 mm inside diameter (ID) (0.230 in. OD by 0.200 in. ID) or 4.57 mm OD by 4.17 mm ID (0.180 in. OD by 0.164 in. ID). The compositions of the D9 and 316 SS heats are given in Table 1. Two heats of 20% CW 316 were included, made to identical 316 SS specifications. Two heats of 20% CW D9 were also studied, again made to identical specifications. Only one of the two D9 heats was available in the 10% CW condition. Specimen loading was accomplished by gas pressurization.

Reference *1* documents the control tests performed on 316 tubing. D9 tubing was tested using identical procedures. Machined end caps of 316 SS were tungsten-inert-gas (TIG) welded to one end of each D9 or 316 specimen, and a gas inlet tube was similarly welded to the other end. Rupture tests were conducted at constant pressure in static argon at temperatures ranging from 538 to 760°C and stresses ranging from 20 to 400 MPa. Pressure was checked daily and adjusted if necessary. Failure was detected by a sudden rise in retort pressure, which actuated a solenoid valve, shutting off an electric timer. Strains were measured only after specimen failure.

In-Reactor Tests

Pressurized tubes were also used to provide in-reactor creep and creep rupture data in the MOTA experiment. Specimens were fabricated from cladding segments having the following dimensions: either 19.8 mm (0.78 in.) long and 4.57 mm OD by 4.17 mm ID (0.180 in. OD by 0.164 in. ID), or 25.7 mm (1.01 in.) long and 5.84 mm OD by 5.08 ID (0.230 in. OD by 0.200 in. ID). The same heats of steel were used for both thermal control and in-reactor testing. End caps made of a D9-type alloy were electron beam welded to the cladding segments. Unique isotopic mixtures of krypton and xenon gases were used with helium to fill the specimens to the pressures required to produce the desired stresses in the cladding at the intended irradiation temperatures, which were 575, 605, 670, and 750°C. A different isotopic mixture was used for each combination of alloy and irradiation temperature. Empirical correlations were developed from experimental measurements to relate the room temperature fill pressure to the pressure at temperature. After gas filling, the specimens were sealed with a laser weld of the fill hole located in the specimen top end cap, and the specimens were leak checked with helium. Finally, the specimen preirradiation diameters were measured using a noncontacting laser system before being loaded into the MOTA irradiation vehicle. Additional details of specimen fabrication may be found in Ref *5*.

TABLE 1—*Creep rupture specimen chemical compositions, wt %.*

Alloy (Heat)	Fe	C	Mn	P	S	Si	Ni	Cr	Mo	V	Ti	Co	Cu	Al	B	N
1st core 316 (81600)	balance	0.047	1.57	0.003	0.005	0.52	13.76	17.56	2.34	0.02	. . .	0.016	0.01	<0.005	<0.0005	0.004
4th core 316 (93591)	balance	0.047	1.65	0.002	0.006	0.58	13.30	17.29	2.61	0.01	. . .	0.03	0.01	<0.01	<0.0005	0.003
D9 (83508)	balance	0.039	2.03	<0.005	0.004	0.80	15.77	13.70	1.65	0.01	.34	<0.01	<0.01	<0.01	<0.0005	0.004
D9 (83510)	balance	0.036	2.06	<0.005	0.004	0.82	15.61	13.53	1.68	<0.01	.34	<0.01	<0.01	<0.01	<0.0005	0.003

Irradiation of the creep rupture specimens began in Jan. 1982 in the FFTF. The specimens were examined three times before obtaining a peak neutron exposure of 17 $\times$ 10^{22} n/cm^2 ($E > 0.1$ MeV). During the course of irradiation, the specimen temperatures were maintained within 5°C of the design temperatures except for brief periods when the canister temperatures were reduced by more than 75°C to identify gas releases from ruptured experimental fuel pins. Typically these periods totaled less than 72 h out of approximately 2600 h associated with a given irradiation cycle. During the course of irradiation, a number of gas releases were detected by the Cover Gas Monitoring System described in Ref *5*. Spectrographic analysis of cover gas samples taken during these events confirmed the releases of tag gases from the creep rupture specimens and uniquely identified the alloy and irradiation temperature. The test matrix contained five to nine specimens at different stresses for a given combination of alloy and irradiation temperature. For a given gas release, it was initially assumed that the most highly stressed specimen had failed for the alloy and irradiation temperature which had been identified. This assumption was later verified during the interim examinations performed during reconstitutions of the MOTA experiment.

Results

Comparison Between Thermal Behaviors

Figure 1 summarizes the overall behavior of the thermal control data on a Larson Miller Parameter (LMP) plot. Individual data points will be shown in later figures. The data are plotted as a function of LMP to facilitate comparisons between the thermal and the in-reactor stress rupture data, which are at slightly different but overlapping exposure temperatures. The specific form for the LMP that was used in this report is given by

$$\text{LMP} = T[13.5 + \log(t_R)] \tag{1}$$

where T is the temperature in K, and t_R is the time to rupture in hours. The lines shown in Fig. 1

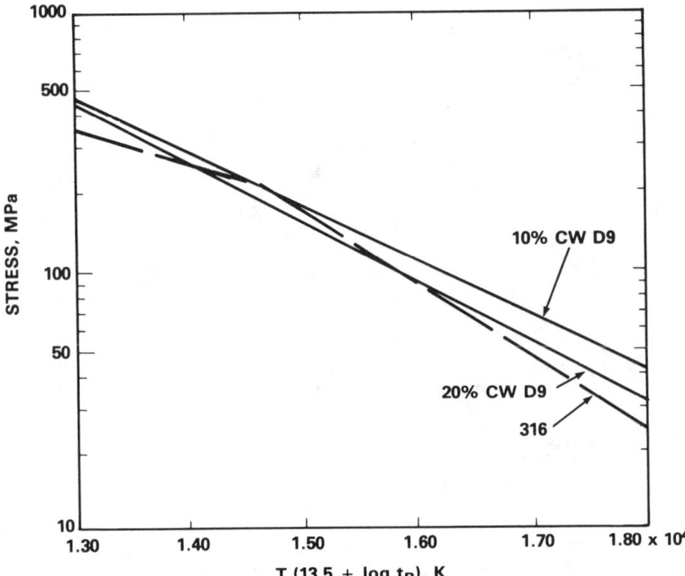

FIG. 1—*Summary of the thermal stress rupture behavior of 20% CW 316 SS and 10 and 20% CW D9 cladding. The rupture time* t_R *is in hours and the temperature* T *is in Kelvin in the Larson Miller parameter.*

represent correlations developed to describe the data using a multiple linear regression analysis. The value of the constant, 13.5, was chosen on the basis of the 316 SS data, for which the largest data base existed [*1–3*]. The constants that best collapsed the data on the 10 and 20% CW D9 were only slightly different; they were 13.6 and 16.2, respectively. For the purposes of this paper, equations were therefore developed for the D9 data using a constant of 13.5.

Identical behavior of the two heats of 316 SS was observed in the thermal control tests. Identical behavior was also observed in the thermal control tests on the two heats of 20% CW D9. Figure 1 indicates that the thermal behavior of 20% CW D9 and 20% CW 316 are similar, while that of 10% CW D9 is slightly better, particularly at longer times or higher temperatures. The similarity between the 20% CW D9 and 316 sets of data indicate that the compositional modifications that were made to reduce swelling in D9 had virtually no effect on the stress rupture behavior of the alloy. The difference in slope between the two cold-work levels of D9 suggests that the 20% CW steel is recovering more quickly than the 10% CW steel, a result consistent with previous observations in 316 SS [*6*].

The similarity in the rupture strengths of the D9 and 316 types of alloys has not been observed elsewhere. It has been shown by other researchers that similar compositional modifications to 316 SS can indeed produce significant improvements in creep rupture strength [*2*]. The improvement is attributed to finely dispersed precipitates that retard the recovery of the dislocation structures during creep at elevated temperatures. The effectiveness of the compositional modifications at improving the rupture strength is probably very sensitive to the preirradiation microstructure.

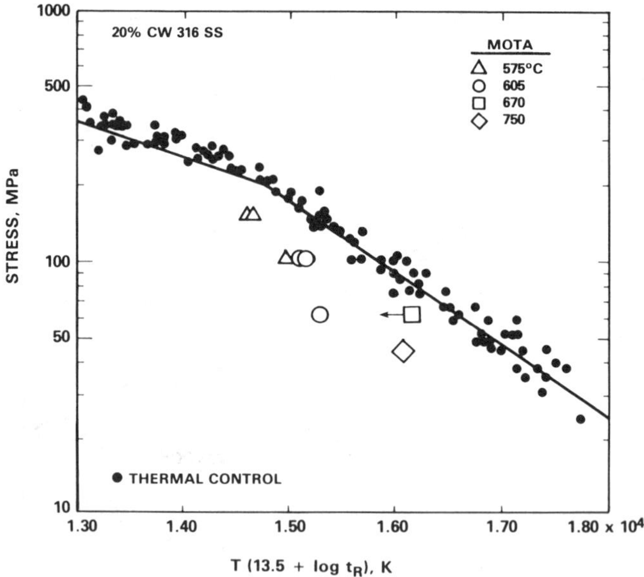

FIG. 2—*Comparison between the thermal control (solid circles) and in-reactor (open symbols) stress rupture data on 20% CW 316 SS. The arrow on the in-reactor data at 670°C indicates that the rupture life shown is an estimated upper bound. The rupture time* t_R *is in hours, and the temperature* T *is in Kelvin in the Larson Miller parameter.*

Comparison Between Thermal and In-Reactor Behaviors

A comparison between the thermal control and in-reactor creep rupture data for 20% CW 316 is shown in Fig. 2. The range of in-reactor rupture times extends to greater than 4800 h. The in-reactor stress rupture lifetimes are significantly shorter than the comparable thermal control data. A comparison between the thermal control and in-reactor creep rupture data for the 10% CW D9 cladding is shown in Fig. 3. For the 10% CW D9, the in-reactor and thermal control stress rupture behaviors are similar for short exposure times, that is, less than 2000 h. For longer tests, however, the in-reactor data exhibit shorter rupture lives than the thermal control data. A comparison between the thermal control and the in-reactor stress rupture data is shown in Fig. 4 for the 20% CW D9. In this alloy condition, the in-reactor data are also similar to the thermal control data for exposure times up to 2000 h. The data suggest, however, that the rupture life decreases with increasing exposure times in-reactor.

The creep rupture data in this report suggest that neutron irradiation does affect the stress rupture behavior of these alloys, contrary to earlier indications [*3,5*]. For exposure times of less than approximately 2000 h the in-reactor and thermal control rupture times are similar. However, as the length of neutron exposure is increased, the in-reactor rupture times become significantly less than those observed in the thermal control data. Moreover, the data suggest that the difference in rupture times between the in-reactor and thermal control data may actually increase with increasing exposure times.

The in-reactor stress rupture behavior of D9 and 316 are compared in Fig. 5 for irradiation temperatures of 575, 605, and 670°C. At 575°C the 10% CW D9 exhibits better in-reactor stress rupture resistance than the 20% CW condition. At 605°C the stress rupture data at stresses above

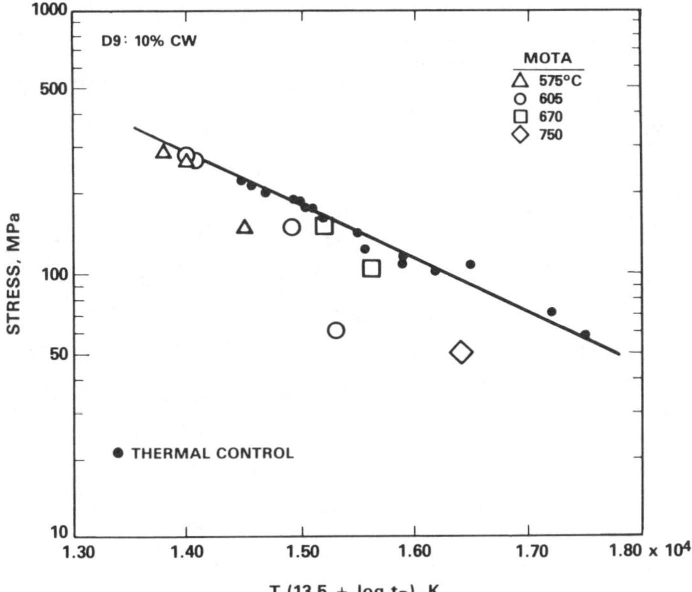

FIG. 3—*Comparison between the thermal control (solid circles) and in-reactor (open symbols) stress rupture data on 10% CW D9. The rupture time* t_R *is in hours, and the temperature* T *is in Kelvin in the Larson Miller parameter.*

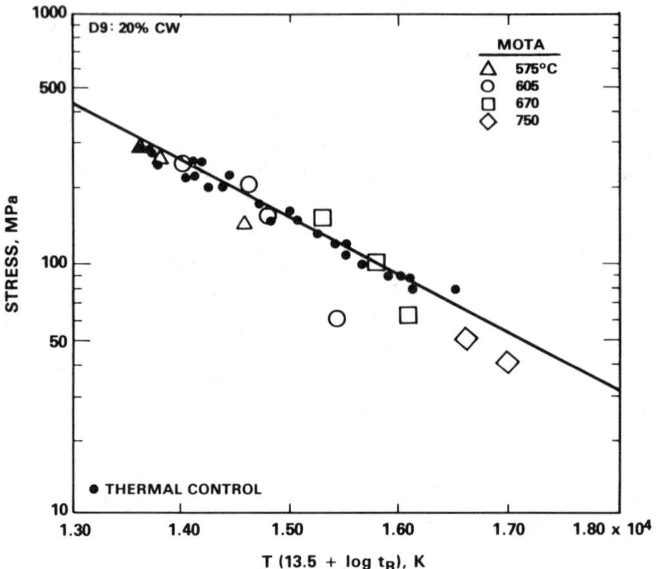

FIG. 4—*Comparison between the thermal control (solid circles) and in-reactor (open symbols) stress rupture data on 20% CW D9. The rupture time* t_R *is in hours, and the temperature* T *is in Kelvin in the Larson Miller parameter.*

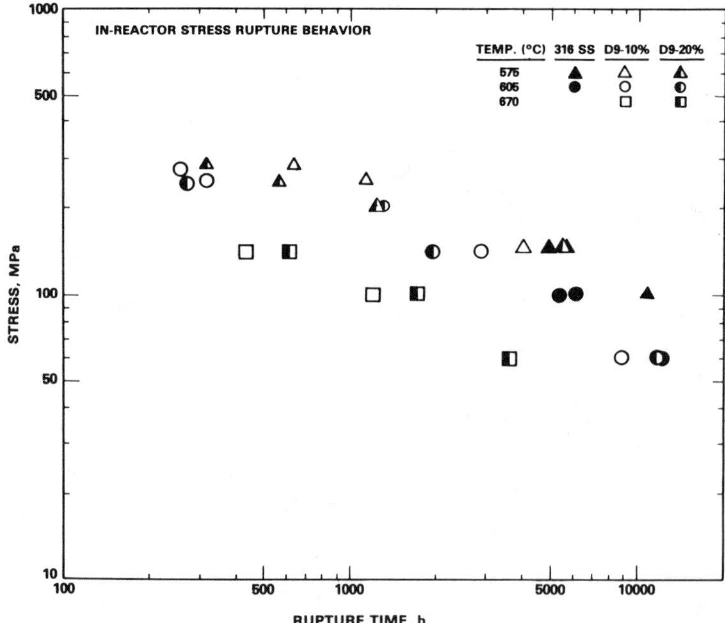

FIG. 5—*Comparison of the in-reactor stress rupture behavior of 20% CW 316 and 10 and 20% CW D9 cladding. The rupture time* t_R *is in hours, and the temperature* T *is in Kelvin in the Larson Miller parameter.*

100 MPa also indicate that the 10% CW material has a longer rupture life than 20% CW D9. At the lower stresses, however, the limited data suggest that the 20% CW condition may exhibit in-reactor stress rupture times superior to those of the 10% CW condition. Finally, at 670°C the 20% CW condition exhibits longer rupture lifetimes than the 10% CW condition.

Conclusions

Both thermal and in-reactor creep rupture data have been obtained for 20% CW 316 SS, and 10 and 20% CW D9. The thermal data suggest the 10% CW condition of D9 exhibits the best thermal stress rupture behavior and that the 20% CW conditions of 316 SS and D9 have similar thermal stress rupture behavior. The in-reactor stress rupture data for each alloy/condition indicate that neutron irradiation eventually leads to a decrease in the rupture life when compared to the thermal control data. The in-reactor stress rupture data also suggest that for irradiation temperatures greater than 605°C, the 20% CW condition of D9 exhibits slightly better stress rupture behavior than the 10% CW condition.

References

[1] Fish, R. L., "Creep Rupture Properties of 20% Cold Worked Type 316 Stainless Steel After High Fluence Neutron Irradiation," *Nuclear Technology*, Vol. 35, Aug. 1977, pp. 9–11.

[2] Lovell, A. J., Chin, B. A., and Gilbert, E. R., "In-Reactor Creep Rupture of 20% CW AISI 316 Stainless Steel," *Journal of Materials Science*, Vol. 16, 1981, pp. 870–876.

[3] Gilbert, E. R., Chin, B. A., and Duncan, D. R., "Effect of Irradiation on Crack Propagation During Creep," *Metallurgical Transactions*, to be published.

[4] Puigh, R. J., Lovell, A. J., and Garner, F. A., "Thermal Creep and Stress-Affected Precipitation of 20% Cold Worked 316 Stainless Steel," *Journal of Nuclear Materials*, Vols. 122 and 123, 1984, pp. 242–245.

[5] Puigh, R. J. and Schenter, R. E., "In-Reactor Creep Rupture Experiment in the Materials Open Test Assembly (MOTA)," *Effects of Radiation on Materials: Twelfth International Symposium, STP 870*, F. A. Garner and J. S. Perrin, Eds., American Society for Testing and Materials, Philadelphia, 1985, pp. 795–802.

[6] Moen, R. A. and Duncan, D. R., "Cold Work Effects: A Compilation of Data for Types 304 and 316 Stainless Steel," Hanford Enginering Development Laboratory Report, HEDL-TI-76005, March 1976.

Christo Wassilew,[1] *Karl Ehrlich,*[1] *and Hans-Jürgen Bergmann*[2]

Analysis of the In-Reactor Creep and Rupture Life Behavior of Stabilized Austenitic Stainless Steels and the Nickel-Base Alloy Hastelloy-X

REFERENCE: Wassilew, C., Ehrlich, K., and Bergmann, H. J., "**Analysis of the In-Reactor Creep and Rupture Life Behavior of Stabilized Austenitic Stainless Steels and the Nickel-Base Alloy Hastelloy-X,**" *Influence of Radiation on Material Properties: 13th International Symposium (Part II), ASTM STP 956,* F. A. Garner, C. H. Henager, Jr., and N. Igata, Eds., American Society for Testing and Materials, Philadelphia, 1987, pp. 30–53.

ABSTRACT: An overall evaluation of several in-reactor creep and creep-rupture experiments in BR2 and FFTF on pressurized tubes of the stabilized austenitic stainless steels 1.4970, 1.4981, 1.4988 and the Ni-base alloy Hastelloy-X is given in this report. Even at temperatures around and above 0.5 T_m the analysis of the data implies that for low stress levels the stress-induced preferred absorption (SIPA) creep process is the predominant deformation mechanism. The results show that the in-reactor rupture lives of the austenitic steels as well as the Ni-base alloy Hastelloy-X are reduced compared with the rupture lives of the corresponding material in the unirradiated and post-irradiation tested conditions, so that an additional damage mechanism via helium has to be assumed to account for the reduction in rupture life.

Analytical models have been developed, which accounts for the critical variables stress, temperature, and dpa rate and which describes in a more general way the in-reactor creep and creep-rupture life behavior of the materials mentioned above. The analysis of the results indicates clearly that the use of the unirradiated and post-irradiation data for predicting in-reactor creep and failures is invalid for these materials.

KEY WORDS: in-pile creep, elastodiffusion, in-pile creep rupture, SIPA mechanisms, temperature dependencies, activation energies, dpa rate dependency, stress dependency, dynamic resolution, helium-bubble growth

Since the early observations of the irradiation-induced helium embrittlement phenomena [*1,2*], numerous experiments have been initiated to investigate their influence on the mechanical properties of alloys selected for use in fast fission reactors.

The reason for the observed degradation of the mechanical properties, in particular the embrittlement, has been considered to be the formation of helium bubbles on the grain boundaries perpendicular to the applied stress axis. It has been reported that in austenitic stainless steels, He-embrittlement is observed at temperatures around and above 0.45 of the absolute melting temperature,

[1] Research scientist and division leader respectively, Kernforschungszentrum, Karlsruhe, D-7500 Karlsruhle, Federal Republic of Germany.

[2] Research engineer, Fa. INTERATOM, D-5060 Bergisch Gladbach 1, Federal Republic of Germany.

[3] Wassilew, C., Dr-Thesis Fakultät für Maschinenbau der Universität Karlsruhe (TH), Karlsruhe, Federal Republic of Germany.

T_m [3,4]. It intensifies with increasing temperature. The extremely low solubility of the helium contributes to its very strong tendency to precipitate as clusters or bubbles in the matrix and on the grain boundaries. This is ultimately the reason for its harmful effect on the mechanical properties.

The creep-rupture properties of helium-containing specimens have been studied under various experimental conditions. Most of these studies have concentrated on obtaining rupture life data by post-irradiation experiments [5–16]. Only limited data from in-reactor creep-rupture experiments have been reported [17–23]. In our results a completely different behaviour has been found for the two type of experiments. In particular below a given stress level, the stress dependence of the in-pile creep rate and the stress dependence of the in-pile creep-rupture strength were found to be characterized by totally different stress exponents as compared with those of the unirradiated and the pre irradiated material, respectively [19,20].

This paper covers investigations concerning the creep, creep-rupture life, and ductility behavior of unirradiated, pre irradiated and in-reactor tested samples of 1.4970, 1.4981, and 1.4988 steels, and the nickel-base alloy Hastelloy-X. The purpose of the present paper is to contribute to a qualitative and a quantitative understanding of the basic mechanisms controlling the in-reactor creep and creep-rupture behavior.

Experimental Procedures

The chemical composition of a variety of heats of the steels 1.4970, 1.4988, and 1.4981, and the nickel-base alloy Hastelloy-X used in our investigations are given in Table 1. The irradiation conditions are summarized in Table 2. The specimens for the Mol-2 and MONIX in-pile experiments were taken from fast breeder cladding tubes. Because of its strong sensitivity to changes in microstructure, the 1.4970 steel was chosen for more thorough investigations concerning the influence of the micro-structure and neutron spectrum on the post-irradiation and in-pile creep-

TABLE 1—*Chemical composition of the investigated steels and the Ni-based alloy Hastelloy-X, wt%.*

	DIN							
	1.4981	1.4981	1.4981	1.4988	1.4970	1.4970	1.4970	Hastelloy-X
	Heat Number							
Component	HV139	70015	51857	25116	29052	22075	21557	X4-4493
C	0.06	0.05	0.06	0.05	0.10	0.095	0.11	0.10
Si	0.58	0.37	0.44	0.63	0.40	0.31	0.42	0.86
Mn	0.97	1.11	1.37	1.35	1.75	1.81	1.81	0.56
P	0.02	0.01	0.01	...	0.009	...	...	...
S	0.007	0.01	0.01	...	0.008	...	...	...
Cr	17.0	16.5	15.7	16.2	14.9	15.1	14.6	21.92
Ni	16.6	16.5	16.0	13.7	15.2	15.0	15.2	47.00
Mo	1.64	1.76	1.78	1.44	1.24	1.29	1.22	9.00
V	...	...	...	0.75	...	...	...	...
Nb	0.70	0.81	0.72	0.65	...	...	...	...
Ti	...	...	...	...	0.48	0.3	0.3	...
B	0.0001	0.0018	0.0010	0.0002	0.0090	0.0050	0.0052	...
N	0.02	0.01	0.007	0.09	0.006	0.01	0.016	...
W	...	...	...	...	...	...	...	0.63
Co	...	...	...	...	...	...	...	1.60
Fe	balance	balance	balance	balance	balance	balance	balance	18.29

TABLE 2—*Irradiation conditions.*

Experiment	Reactor	Shape of Specimens	Fluence $E > 0.1$ MeV, $n \cdot m^{-2}$	He-Cont (calculation), appm	Irradiation Temperature, °C
Mol-2	BR-2	tubes	maximum 0.70×10^{26}	maximum 140	615, 650, 670 700, 720
MOTA	FFTF	tubes	maximum 1.01×10^{27}	maximum 20	605, 670

rupture behavior. Therefore, specimens were irradiated in different metallurgical conditions. The thermal-mechanical pretreatments of the specimens chosen are summarized in Table 3.

Some features of the various experiments will be briefly described below. Mol-2 and MONIX are in-pile creep-rupture experiments. Tube sections within the inert gas filled Mol-2 rig were loaded with controlled, constant, internal gas pressures at controlled test temperatures. Heating was achieved by γ-absorption and by resistance heater coils, which were positioned within the tubular specimens. Chromel-alumel thermocouples provided on-line temperature control for each of the specimens with a temperature uncertainty of ±2 K.

The MONIX experiment was performed in the Material Open Test Assembly, MOTA, an irradiation vehicle used in the Fast Flux Test Facility (Hanford) for in-reactor material experiments [*22*]. The MOTA irradiation vehicle provides the following experimental capabilities:

- on-line temperature control for each of its canisters within ±5 K, and
- the cover gas sampling capabilities of FFTF permit the determination of the exact time-to-rupture for a given creep-rupture specimen.

The equivalent stress σ_{eq} in the middle of the tube wall was calculated from the internal pressure p, the wall thickness s_o, and the radius of the wall centerline r_m according to the following formula

$$\sigma_{eq} = 3^{1/2} p r_m / 2 s_o \tag{1}$$

The equivalent strain ϵ_{eq} in the middle of the tube wall was calculated according to the following formula

$$\epsilon_{eq} = 2/3^{1/2}[1 + s_o/(d_o - s_o)]^2 \Delta d/d_o \tag{2}$$

where d_o is the initial outer diameter, Δd is the diameter increase, and s_o is the initial tube wall thickness.

Experimental Results and Discussion of the In-Pile Creep Process

The Stress Dependence of the In-Pile Creep Rate

In Fig. 1 the logarithm of the creep rate of steel 1.4970 is plotted as a function of the equivalent stress σ_{eq} for temperatures between 593 and 993 K. The data were obtained for specimens in the cold worked (cw) and the cold-worked and aged (cw + a) metallurgical conditions. Details are listed in Table 3.

TABLE 3—*Heat treatments of the investigated materials.*

Material	Treatment
1.4981	sa; sa + cw
1.4988	sa; sa + 750°C, 3h; sa + 800°C, 1 h
1.4970	sa + cw; sa + cw + 800°C, 2 h; sa + cw + 800°C, 4 h; sa + cw + 800°C, 23 h
Hastelloy-X	sa + a

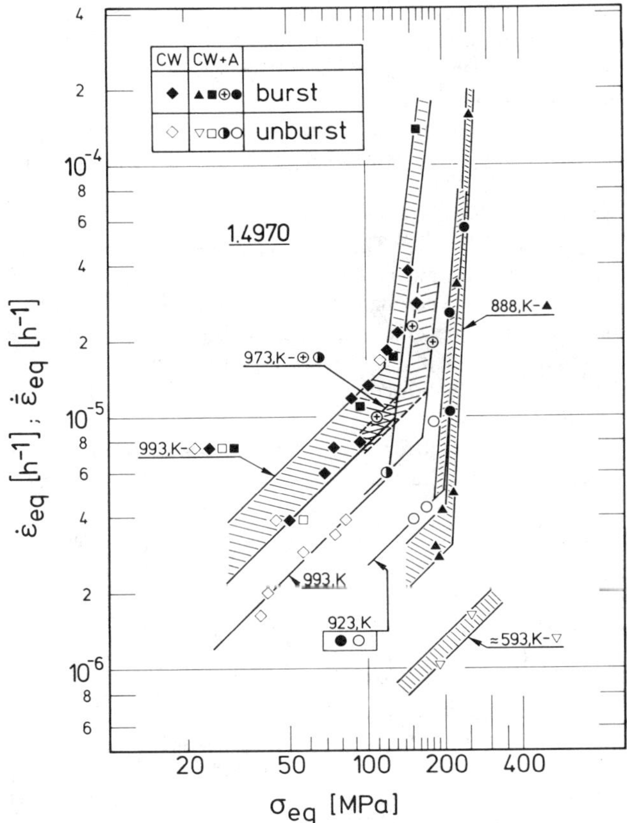

FIG. 1—*Stress dependence of the creep rate for steel 1.4970 during neutron irradiation.*

Since intermediate strain measurements were not possible for the specimens assembled in the Mol-2 device, one may distinguish between measured strains for burst and for unburst specimens. The following cases can occur

- The measured strains of the unburst or burst specimens include possible strain elements of different creep processes so that the creep rate corresponds to an averaged creep rate $\dot{\bar{\epsilon}}$.
- The measured strains of the unburst specimens possibly reflect only a single deformation process with a strain rate $\dot{\epsilon}_p$ of a steady state creep process.
- The measured strains of the unburst specimens may also be due to a single creep process with a decelerating or an accelerating creep rate so that the creep rate corresponds once more to an averaged creep rate $\dot{\bar{\epsilon}}$.

This is a disadvantage of the device, which can be compensated by a careful interpretative analysis of the experimental results obtained.

Another important fact is that, depending on the specimens' position with respect to the core centerline, different dpa rates are effective. This spreads the scatterbands of the determined creep rates, unless the influence of the dpa rates is normalized.

Provided that the characteristic stress range is well covered in the experiment, the data plotted in Fig. 1 show that for all test temperatures at a given stress level (called transition stress σ_{tr}), an abrupt change in the stress dependence of the creep rate occurs. Furthermore, the data clearly indicate that the transition stress depends on the test temperature. This figure also shows that below the transition stress, the stress dependence of the creep rate is linear with stress whereas above the transition stress it is nonlinear.

Below the transition stress the average creep rates as determined from the measured strains of the burst specimens are approximately twice as high as those determined by the measured strains of the unburst specimens.

In Fig. 2 the creep rate of steel 1.4988 is plotted as a function of the equivalent stress for temperatures between 888 and 993 K. The data were obtained on specimens in the solution annealed (sa) and solution annealed and aged (sa + a) metallurgical conditions. If the characteristic stress-range is well covered in the experiment, the same conclusions can be drawn for this material concerning transition stress, stress dependence, and the existence of an additional creep mechanism with a fast accelerating creep rate within a short period of time before burst.

A creep behavior equivalent to that discussed above for steels 1.4970 and 1.4988 has also been observed for steel 1.4981.

It can, therefore, be concluded that for steels 1.4970, 1.4981, and 1.4988 an additional creep process with a fast accelerating creep rate is effective below the transition stress and within a short period of time before burst.

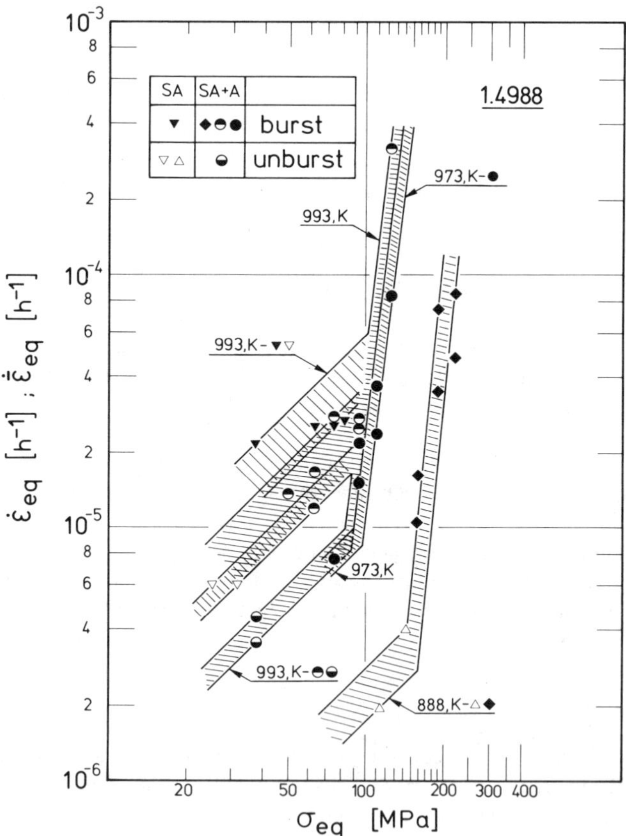

FIG. 2—*Stress dependence of the creep rate for steel 1.4988 during neutron irradiation.*

Dependence of the Operating Irradiation-Induced Creep Mechanism on the Displacement Damage Rate

Below the transition stress the creep data obtained between 320 and 750°C and the data available in the literature at temperatures far below $T_m/2$ can be used to analyze the dpa rate dependence of the observed irradiation-induced creep process.

Some of the Mol-2 experimental results for steel 1.4970 tested at 720°C as well as Ripcex-I experimental results for steel 1.4981 tested at approximately 420°C [*24*] are plotted in Fig. 3. To present the data of steel 1.4981 tested at 420°C in the same figure as the data of steel 1.4970 tested at 720°C, the former have been multiplied by a factor of 30, since only the slope is of importance. The figure shows the observed creep rates, normalized with stress, versus the displacement damage rate $\dot{k}$ for the steels 1.4970 and 1.4981 in the cw metallurgical conditions and for steel 1.4970 in two different (cw + a) metallurgical conditions. The data plotted in this figure show that the slope of all curves representing two different steels (steel 1.4970 under three different metallurgical conditions) is roughly equal to ½. This indicates a square-root dependence of the creep rate on the dpa rate. These observations are comparable and in complete agreement with those reported by Mosedall et al. [*25*] on the cold-worked steel, En58B, which are reproduced in Fig. 4.

It can clearly be concluded from the graphs in Figs. 3 and 4 that in the covered temperature range between, approximately, 300 and 750°C the creep rate for the steady state irradiation-induced creep mechanism is proportional to the square-root of the dpa rate.

If mutual recombination is the rate-controlling process for the loss-of-point defects formed by displacement events, then according to Bullough and Hayns [*26*], a square-root dependence of the SIPA creep rate on the dpa rate has to be considered. If mutual recombination is neglected, a linear dependence of the SIPA creep rate on the dpa rate will exist.

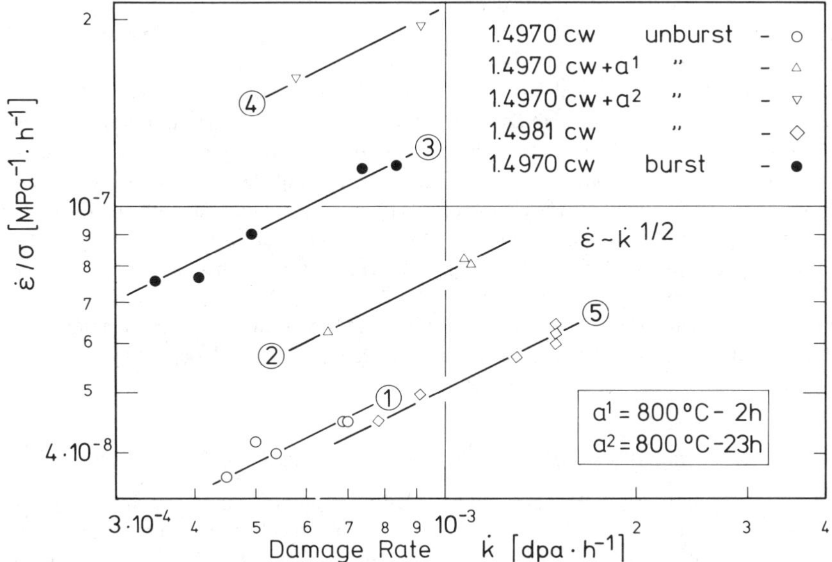

FIG. 3—*Displacement damage-rate dependence of the creep rate for steel 1.4970 in the cold-worked and the cold-worked and aged condition. To present the data of steel 1.4981 cw tested at 420°C in this diagram, the measured values have been multiplied by a factor of 30.*

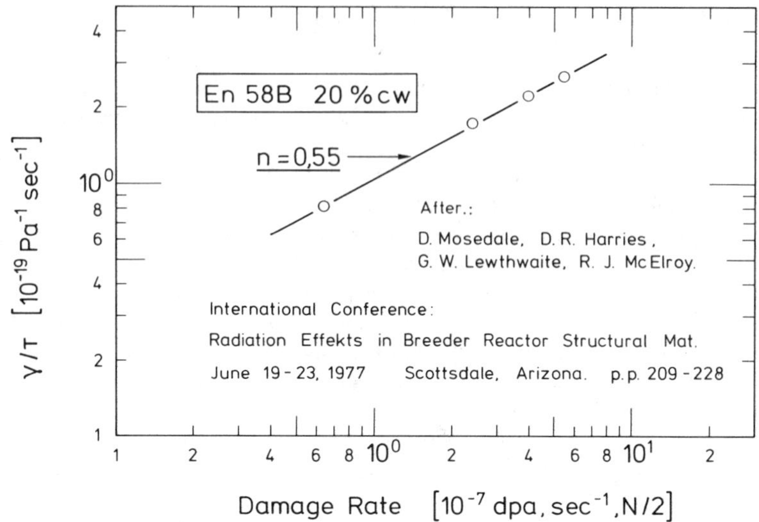

FIG. 4—*Displacement damage-rate dependence of the creep rate for the cold worked En 58B steel.*

The Dependence of the In-Pile Creep Rate on the Dislocation Density and Dislocation Configuration

Comparing the experimental results obtained on steel 1.4970 in the three different metallurgical conditions comprised in the groups labeled 1, 2, and 4 in Fig. 3, it can be clearly seen that the creep rate of the operating irradiation-induced creep mechanism is a function of the metallurgical condition. The steel in the cold-worked condition shows the lowest creep rate. The creep rate increases with decreasing dislocation density. The decrease in dislocation density results from the aging heat-treatment performed on the initially cold-worked material.

The following explains the similarity of the creep behavior observed between the cold-worked and cold-worked and aged materials. The individual dislocations in the cold-worked materials are members of dense tangles. Their heterogeneous distribution and high-dislocation density is less capable of climbing by absorption of irradiation induced point defects. The dislocation structure of a cold-worked and aged material is characterized by a lower dislocation density and an alignment (polygonization) of the dislocations. The dislocations in such a more orderly structure are considerably capable of climbing. The creep rate dependence of the external variables such as stress, dpa rate, and temperature remains unchanged.

In this context it should be pointed out that for the steels in the solution annealed, and the solution-annealed and aged metallurgical conditions, a square-root dependency of the creep rate on the dpa rate cannot be established experimentally.

Temperature Dependence of the Operating Irradiation-Induced Creep Mechanism

Since the formation of point defects during irradiation is a constrained process, only the velocity of the migrating point defects will be the rate-controlling mechanism for the climb velocity of dislocations participating in the irradiation-induced creep. Thus, their mobility will be ultimately the rate controlling mechanism. The temperature dependence of such diffusion-controlled processes is usually given by the Arrhenius expression.

The plot in Fig. 5 shows the temperature dependence of the creep rate on the macroscale θ_c

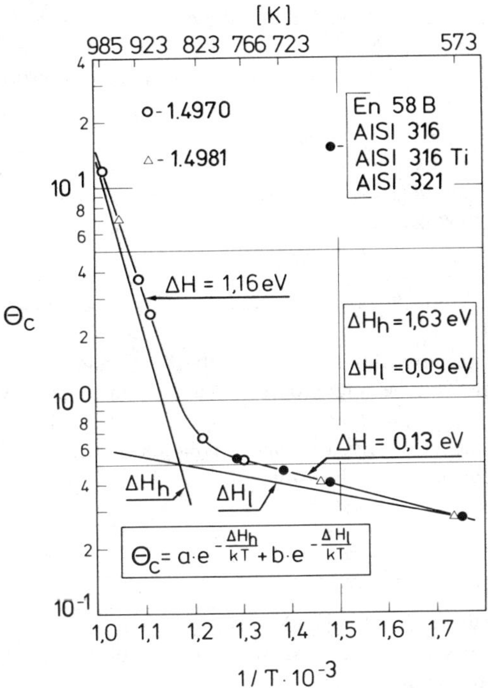

FIG. 5—*Temperature dependence of the irradiation induced creep mechanism.*

versus the reciprocal of temperature for various austenitic steels. A word should be said about how θ_c has been obtained from the creep rate measured on specimen tested at the various experimental conditions. Since the steady state creep rate of the irradiation-induced creep process is linear with stress and is proportional to the square-root of the dpa rate, thus the θ_c values are equal to the normalized creep rate.

$$\theta_c = \epsilon C_o/(k^{1/2}\sigma) \quad (3)$$

The constant C_o has been calculated by fixing θ_c equal to 1 for $T_{test} \simeq 0.5T_m$. Later on we will recognize that C_o is the product of the initial entropy of the body and the efficiency factor. It should be emphasized that only the data of cold-worked steels have been used in this graph, since only in the cw conditions a nearly steady state creep regime is effective in the beginning. The θ_c corresponds to the temperature dependence of the point defect velocity on the microscale. It is evident from this figure that the resulting response function θ_c, the creep rate, has a strong temperature dependence at high temperatures whereas, at low temperatures this dependence is weak. For the austenitic steels investigated the temperature dependence of the in-pile creep rate seems to be independent of the type of steel. The experimentally obtained temperature dependence for the irradiation-induced creep mechanism has been decomposed to determine the true activation energy values of the participating individual rate-controlling processes. The value of the true activation energy at low temperatures is calculated to be less than 0.1 eV, whereas at high temperatures it is calculated to be 1.63 eV. These perceptions allow the interpretation of the obtained true activation energies as follows:

- The rate-controlling process at temperatures below about 850 K is the migration of self interstitials.

- The rate-controlling process at temperatures above about 850 K is the migration of vacancies.
- At a given temperature, which is approximately 850 K for austenitic steels, the vacancy contribution to the creep deformation is equal to that of the self interstitials.

Thus, the experimentally obtained apparent activation energy seems to be temperature dependent on contrary to the true activation energies for migration of vacancies and interstitials, respectively.

The sum of the two individual rate-controlling processes dominating at different temperature levels is given by the following equation

$$\theta_c = a_{\exp}(-\Delta H_h/kT) + b_{\exp}(-\Delta H_1/kT) \quad (4)$$

which represents the creep rate θ_c on the macroscale. θ_c is equivalent to the apparent diffusion coefficient on the microscale.

In this equation, T is the temperature in K, k is the Bolzmann constant, ΔH_h and ΔH_1 are the activation energies for the volume migration of vacancies and for the volume migration of self interstitials, respectively, and a and b are the corresponding frequency factors.

The temperature dependence obtained for the irradiation-induced creep process differs considerably from the theoretical prediction by Bullough and Hayns [*27*] according to which the SIPA creep is a temperature independent mechanism.

The Elements of the Irradiation-Induced Creep Process

The basic relations obtained for the dependence of the creep rate of the irradiation-induced creep mechanism on the external variables such as stress, temperature, and dpa rate have been used to analyze the results in greater detail in order to understand the individual rate-controlling mechanisms. The most important perceptions will be briefly discussed below.

Figure 6 shows the synthetic creep curves at 720°C for the steel 1.4970 in the cold-worked (cw) condition, obtained by normalizing the influence of the external variables σ and k. It can be recognized from this figure that the unburst specimens seem to be members of a steady state creep

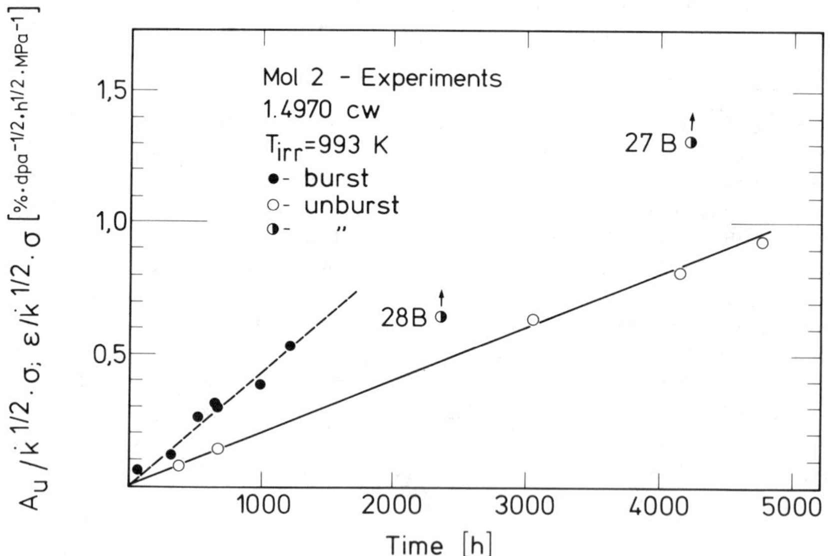

FIG. 6—*Synthetic in-pile creep curve for steel 1.4970 cw.*

process. As has recently been discussed by Ehrlich [*28*], two important observations have been made: Cloß and Herschbach [*29*] have shown that the actual creep rate for steel 1.4981 at 700°C is steady state over a reasonable period of time, as determined by continuous measurement of the deformation during irradiation. The in pile creep rate is considerably higher than the creep rate of the material tested in the unirradiated condition. These observations have been interpreted by assuming microstructural changes, for example, precipitate coarsening, which happens under irradiation.

Further it can be seen from Fig. 6 that the unburst specimens labeled 27B and 28B are outside the steady state creep stage. The burst specimens show an additional increase in creep strain. Thus, it is apparent that a dpa rate dependent "tertiary" creep strain element contributes to the total creep strain. These observations allow the following conclusions:

1. A primary creep stage does not exist.
2. The irradiation-induced creep process of austenitic steels in the cold-worked metallurgical condition, that is, with a high dislocation density, starts with a nearly steady state creep rate.
3. A creep process with a monotonically accelerating strain rate follows the steady state creep stage.
4. Beyond the creep process with a monotonically accelerating strain rate and until rupture occurs, a "tertiary" creep stage with a rapidly increasing strain rate exists.

The existence of a steady state creep stage is observed only for cold worked steels. The steels in the solution annealed (sa) and solution annealed and aged (sa + a) metallurgical conditions, however, do not show of such a steady state creep stage that exists. These important facts are confirmed in Figs. 7 and 8. Figure 7 shows the accumulated creep strain normalized with stress for steel 1.4981 at 720°C in the (sa) condition as a function of the dpa. Figure 8 represents the

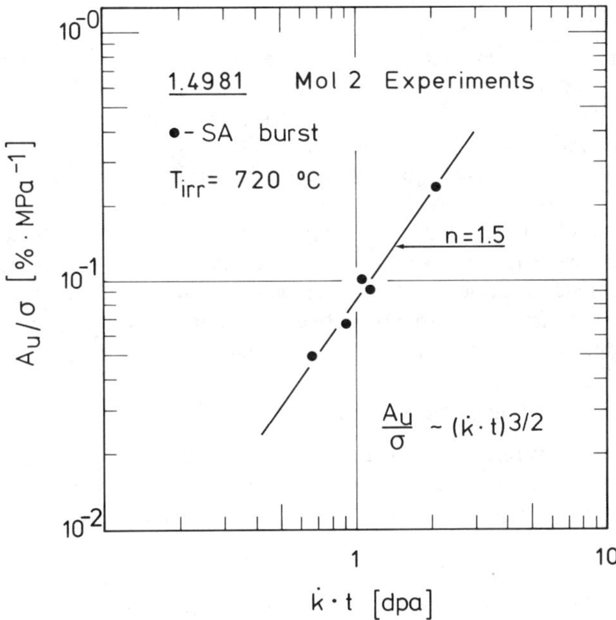

FIG. 7—*Dose-rate dependence of the accumulated in-pile creep-strain for the solution annealed steel 1.4981.*

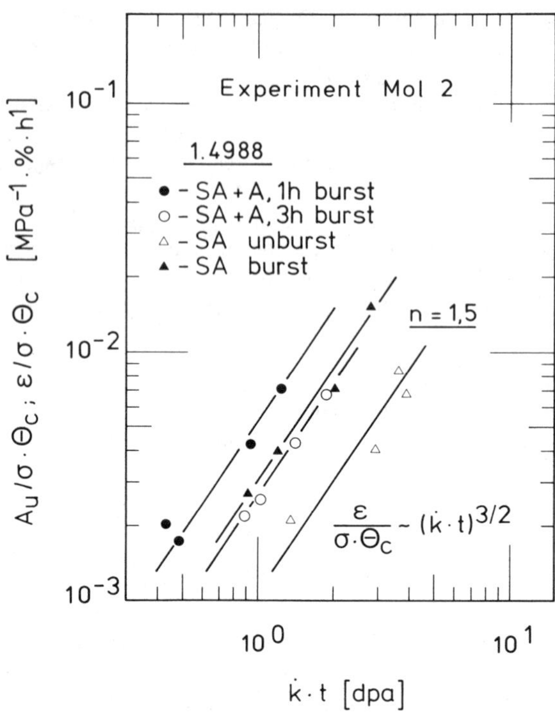

FIG. 8—*Dose-rate dependence of the accumulated in-pile creep-strain for the solution annealed and solution annealed and aged steel 1.4988.*

accumulated creep strain normalized with stress and θ_c (temperature dependence of the creep process) for steel 1.4988 in the (sa) and (sa + a) condition as a function of the dpa. The accumulated strain normalized with stress and temperature-stress, respectively, for the burst and unburst $3/2$ power of the dpa-rate [*30*]. This is an important observation which leads to the conclusion that $3/2$ power of the dpa-rate [*30*]. This is an important observation which leads to the conclusion that divacancies and di-interstitials may control the climb process of steels in the sa and sa + a metallurgical conditions. It can also be established furthermore that a $t^{3/2}$ dependence of the irradiation-induced creep strain exists. Taking into account the dislocation density for the solution annealed and the solution annealed and aged materials, which is extremely low, it is obvious that the corresponding mean free path of the migrating point defects will be longer. This promotes the probability of the formation of divacancies and di-interstitials [*30*].

The Dependence of the In-Pile Creep Rate on the Grain Size

The grain size of the various steels investigated varied approximately between 5 and 70 μm. However, the grain size seems to have little effect on the creep rate of the irradiation-induced creep mechanism. Only the strains accumulated in the irradiation-induced tertiary stage of creep are found to be inversely proportional to the square root of the grain diameter.

The Nature of the Observed Irradiation-Induced Creep Process

From the basic relations obtained for the dependence of the creep rate on the external variables such as stress, temperature, and dpa rate, and on the internal variables like density and configuration of the dislocations, structure and structural changes, it can be concluded that the observed deformation mechanism belongs to the category of the stress-induced preferred absorption (SIPA) [*31,32*] and the stress-induced preferred nucleation (SIPN) [*33*] mechanisms. Nevertheless, the basic relations discussed above are in more accordance with the elastodiffusion-SIPA creep mechanism]34–37] and show some discrepancies with respect to the usual SIPA mechanisms. It was shown recently by Woo [*37*] that elastodiffusion induced by an external stress gives rise to an anisotropy in an otherwise isotropic diffusional field, thereby causing a SIPA effect. Elastodiffusion is independent of the point-defect/dislocation interaction and arises from the difference in the elastodiffusivities of the vacancies and the interstitials. Elastodiffusion is capable of generating irradiation creep similar to, but much larger than, that resulting from the usual SIPA mechanism. The elastodiffusion effect is, according to Woo [*37*] a first-order effect, in contrast to the second-order effects for the usual SIPA mechanism proposed by Wolfer and Ashkin [*31*] and Heald and Speight [*32*]. Therefore, it has been assumed that elastodiffusion governs the irradiation induced deformation mechanism. To distinguish elastodiffusion SIPA from the usual SIPA originating from the inhomogeneity interaction, we will call it SIPA-AD, according to Tome et al. [*36*].

The Irradiation-Induced Creep Mechanisms Operating Above the Transition Stress

Figures 9*a* and 9*b* exhibit the creep rates normalized with the dpa rate and the creep rates normalized with stress, respectively, for steel 1.4970 cw at 720°C and at stress levels above the transition stress. It can be inferred from Fig. 9*a* that above the transition stress, the dpa rate normalized creep rate shows a quadratic stress dependence. Such a stress dependence of the creep rate has been theoretically predicted by Mansur [*38*] for irradiation-induced creep deformation by climb-enabled glide of dislocations.

Figure 9*b* shows the dependence of the stress normalized creep rate on the dpa rate. The creep rate which seems to be proportional to the 3/2 power of the dpa rate, as indicated already before for the SIPA-AD creep of austenitic steels with a low dislocation density, that is, in the sa and sa + a conditions. It can be inferred that the irradiation-induced climb-enabled glide of dislocations may likewise be controlled by the absorption of divacancies and di-interstitials.

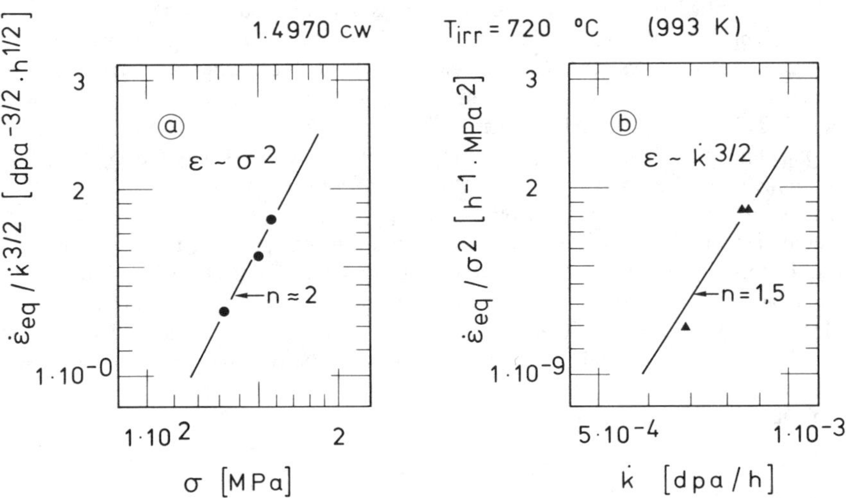

FIG. 9—*Stress and dpa-rate dependence of the creep rate above the transition stress for steel 1.4970 cw.*

Since the climb-controlled SIPA-AD mechanism and the glide-climb controlled mechanisms are mutually exclusive, only the faster one can control the creep rate. Consequently, it is anticipated that at high stress values the climb mechanisms will revert to glide-climb controlled mechanisms. A convincing proof of such a transition is provided by the experimental data presented in Fig. *9a* and Fig. *9b* which have been selected from Fig. 1 containing the data which belong to the SIPA-AD mechanism and the climb-enabled glide mechanism proposed by Mansur [*38*].

In concluding it can be noted that climb-enabled glide of dislocations is indeed an important mechanism for irradiation creep above the transition stress.

Constitutive Equation of the SIPA-AD Creep

The observations and the derived dependencies of the creep rate of the irradiation-induced creep mechanism on the external variables could be considered as essential input to the modelling and formulation of a more realistic constitutive equation for describing the irradiation-induced creep mechanisms. The indispensible requirement of the model is to predict the creep deformation by considering continuing structure variations in the material that develop during the creep process. The knowledge obtained allows the derivation of the constitutive equation for the SIPA-AD creep mechanisms operating exclusively during irradiation by modelling the structural development in the material. Since there are different dependencies of the SIPA-AD creep rate on the dpa rate or on the time, which are preferably dependent on the dislocation structure and configuration, the total irradiation induced strain rate will be the sum of two individual processes. Depending on the initial dislocation structure of the material, the steady state or the accelerating SIPA-AD creep process will be dominant in the beginning. The steady state creep rate of materials in the cold-worked metallurgical condition, which is dominant in the beginning, will be gradually dominated by the time-dependent irradiation-induced creep process. Besides, this process is proportional to $\dot{k}^{3/2}$.

The constitutive equation for the in-pile creep process has been developed by application of the thermodynamics, the mechanical statistics, and the kinetics. In this way, it was possible to derive the constitutive equation for the total creep rate of the SIPA-AD creep consisting of the two fundamental rate-controlling processes discussed above

$$\dot{\epsilon} = \dot{k}^{1/2}\sigma\theta_c\Omega_1/S_o + \dot{k}^{3/2}\sigma\theta_c\Omega_2C_1t^{1/2}\exp\left(-\ C_2S_o/\dot{k}\sigma^{1/2}\theta_ct^{1/2}\right) \tag{5}$$

where σ is the applied stress, $\dot{k}$ is the dpa rate, θ_c as given by Eq 4, is the temperature function for the rate controlling-process which are effective exclusively during irradiation, Ω_1 is the efficiency factor for single vacancies and interstitials, and Ω_2 is the efficiency factor for divacancies and di-interstitials, S_o is the initial entropy of the material, which corresponds to the metallurgical condition, C_1 is a material dependent constant, and C_2 is a proportionality factor.

The exponential function in this equation describes the time-dependent entropy evolution of the body as a function of the external variables stress, temperature, and dpa rate [*30*]. Since the creep rate depends on the structure, that is, on the entropy state of the body, the exponential function consequently describes the time-dependent alteration of the creep rate as a function of the external variables. In this context it should be pointed out that the initial entropy for the steels in the solution-annealed and solution-annealed and aged conditions is lower than the equilibrium entropy and will increase with time. The initial entropy for the cold-worked and the cold-worked and aged material is higher than the equilibrium entropy and will decrease with time to reach the equilibrium entropy state.

The advantage of the formalism of Eq 5 is that it can be applied to austenitic steels

- at stress levels below the corresponding transition stress only,
- in different metallurgical conditions, and
- at temperatures and dpa rates prevailing in the operating fusion and fission facilities.

This equation describes rather well the creep deformation behavior of the experimental data on pressurized tubes and real fuel pins [30].

Comparison of the Observed Irradiation-Induced Creep Mechanism with the Diffusion-Controlled Nabarro-Herring and Coble Creep Mechanisms

Since at temperatures around and above $T_m/2$ diffusion becomes reasonably rapid, dislocations acquire a new degree of freedom in their motion. A number of different diffusion-controlled mechanisms are imaginable and will be briefly discussed in the following, as possible rate-controlling creep mechanisms during irradiation. This will be done in such a way as to provide a unified quantitative picture of the creep and creep-rupture behavior of austenitic steels when irradiation induced diffusion controlled creep mechanisms are operative. In order to provide a comparison of the irradiation induced creep with other similar creep mechanisms, two well known mechanisms, Nabarro-Herring creep [39] and Coble creep [40], will be presented subsequently.

The Nabarro-Herring creep results from the diffusion of vacancies from regions of high chemical potential at grain boundaries subject to normal tensile stresses to regions of lower chemical potential where the average tensile stresses across the grain boundaries are zero. Atoms migrating in the opposite direction account for the creep strain. When volume diffusion controls the tensile creep rate, $\dot{\epsilon}_s$ is given by

$$\dot{\epsilon}_s = A_n(D\mathbf{b}^3\sigma/d^2kT) \tag{6}$$

where d is the average grain diameter, $\mathbf{b}$ is the Burgers-vector, k is the Boltzmann constant, T is the absolute temperature, σ is the stress, and D is the diffusivity. The dimensionless constant A_n depends insensitively on the geometry of grains, but is generally estimated to have a value of 5 to 8.

The major characteristics of this type of diffusion creep are as follows:

1. The Nabarro-Herring creep does not involve the motion of dislocations.
2. It predominates over high-temperature dislocation-dependent mechanisms only at low stress levels, and then only for fine grained materials.
3. Since it is unaffected by substructural changes, primary creep does not occur.
4. It is characterized by creep rates that increase linearly with the stress.
5. The creep rate is inversely proportional to the square of the grain diameter.
6. The temperature dependency is characterized by an activation energy that is nearly equal to that for volume self diffusion.

Coble [40] has shown that stress-directed diffusion of vacancies can take place by grain boundary diffusion as well as by Nabarro volume diffusion. The creep rate $\dot{\epsilon}_s$ is given by

$$\dot{\epsilon}_s kT/D_b G_b = 48(\mathbf{b}/d)^3\,(\sigma/G) \tag{7}$$

where d is the average grain diameter, $\mathbf{b}$ is the Burgers-vector, k is the Boltzmann constant, T is the absolute temperature, σ is the stress, G is the shear modulus, G_b is the grain boundary shear modulus, and D_b is the grain boundary diffusivity. The thickness of the grain boundary is assumed to be about $2b$. The creep rate caused by stress directed diffusion of vacancies along the grain boundary, that is, the Coble creep, is characterized by

(1) a temperature dependence corresponding to the activation energy for grain boundary diffusion, which is assumed to be equal to one-half of that for lattice diffusion,
(2) a linear stress dependence, and
(3) a grain-size dependence reciprocal to the cube of the grain diameter.

The main characteristics of the SIPA-AD creep mechanisms based on Eq 5 are compared with the characteristic elements of the Nabarro-Herring creep mechanism given by Eq 6 and the Coble creep mechanism given by Eq 7 in order to provide evidence that even at temperatures around and above 0.5 T_m the SIPA-AD creep is an independent process and by no means linked to the Nabarro-Herring or the Coble creep mechanisms:

1. The SIPA-AD creep mechanism involves dislocation climb processes to produce deformation whereas the Nabarro-Herring and the Coble creep processes do not involve dislocation climb and motion.
2. The SIPA-AD creep mechanism as well as the Nabaro-Herring and the Coble creep mechanisms show a creep rate linear with stress.
3. The creep rate of the SIPA-AD mechanism is independent of the grain-size, whereas the creep rate of the Nabarro-Herring mechanism is inversely proportional to the square of the grain diameter, and the creep rate of the Coble creep mechanism is inversely proportional to the cube of the grain diameter.
4. The temperature dependence of the SIPA-AD creep rate is characterized by the migration of interstitials at low temperatures and the migration of vacancies at high temperatures (Fig. 5). On the other hand, the rate-controlling mechanisms for the Nabarro-Herring and the Coble creep processes are only the formation and migration of vacancies.
5. The SIPA-AD creep rate depends on the dpa rate.
6. The SIPA-AD creep rate is usually nonlinear with respect to time, whereas the characteristic of the Nabarro-Herring and the Coble creep mechanisms is a steady state creep rate.
7. Only materials with a high dislocation density start with a steady state SIPA-AD creep rate, which gradually increases to be ultimately transformed into a $t^{1/2}$ time-dependent creep rate.
8. Swelling seems to have little effect on SIPA-AD creep.

This list of the preceeding items shows that, with the exception of a linear stress dependence and steady state creep rate occurring in the beginning, there is no further similarity between the SIPA-AD creep and the Nabarro-Herring or the Coble creep mechanisms. Thus, there is also no evidence for the mechanism proposed by Jones [*41*], which explains the enhanced in-pile creep rates if a combination of SIPA and Coble creep with a d^{-3} grain-size dependence is assumed.

The conclusion of Boutard et al. [*23*] that for test temperatures higher than 600°C the cold-worked steels 316 and 321 have a dose rate independent thermally activated characteristic of creep is likewise in contradiction to our experimental observations and basic relations derived.

Experimental Results and Discussion of the In-Pile Creep-Rupture Mechanisms

In-Pile Creep-Rupture Properties of the Steels 1.4970, 1.4981 and 1.4988 and the Nickel-Base Alloy Hastelloy-X

The analysis of the data has shown that for stress levels below the limiting transition stress the SIPA-AD mechanism is the predominant in-pile deformation process, even at temperatures above 0.57 T_m. However, a limitation of the rupture life and ductility is being observed so that an additional, helium-induced damage mechanism has to be assumed to account for the reduction in rupture life.

Because of the obvious importance of the in-pile rupture life properties, a more physically based analysis is required with regard to the rate-controlling mechanisms of nucleation and growth behavior of helium bubbles on the grain boundaries perpendicular to the stress axis when simultaneously subjected to uniaxial tensile stress, displacement events, and a continuous helium gas supply.

Bullough et al. [42] considered the effect of stress on the growth of gas bubbles in stainless steels during irradiation. They argued that the growth of gas bubbles at the grain boundaries perpendicular to the applied stress is driven by the helium produced during irradiation (n,α) reactions and that the residual interstitials form suitably oriented interstitial platelets which extend the grains by the "jacking mechanism" according to Harris [43]. This eventually gives rise to irradiation creep and to the rupture of the material along the grain boundaries. Using the ideal gas law to simplify their analysis, they deduced a time-to-rupture inversely proportional to the square of the stress.

Wood and Kear [44] have extended the work of Bullough et al. [42] with the aim to predict the density of bubbles nucleated on the grain boundaries in order to derive an expression for the time-to-rupture. The expression of Wood and Kear shows the same dependence on stress as the model of Bullough et al. [41]; however, it is improved because it includes the temperature dependence. The function for the temperature dependence is similar to the function that describes the nucleation of fission-gas bubbles within fuel grains [45,46].

For a detailed understanding of the various rate-controlling rupture life mechanisms that are possible, our experimental results for steel 1.4970 have been chosen for discussion. Results of our own on the steels 1.4981, 1.4988, and the nickel-base alloy Hastelloy-X as well as results available from the literature will be included for further support of our view.

In Fig. 10, the rupture life properties of the steel 1.4970 tested in absence of irradiation and after irradiation in BR-2 at 700°C are compared with the in-pile rupture life properties obtained from the Mol-2 in pile creep rupture experiments with pressurized tubes at 720°C in Br-2. The comparison of the results in this figure shows that the post-irradiation rupture lives are reduced by

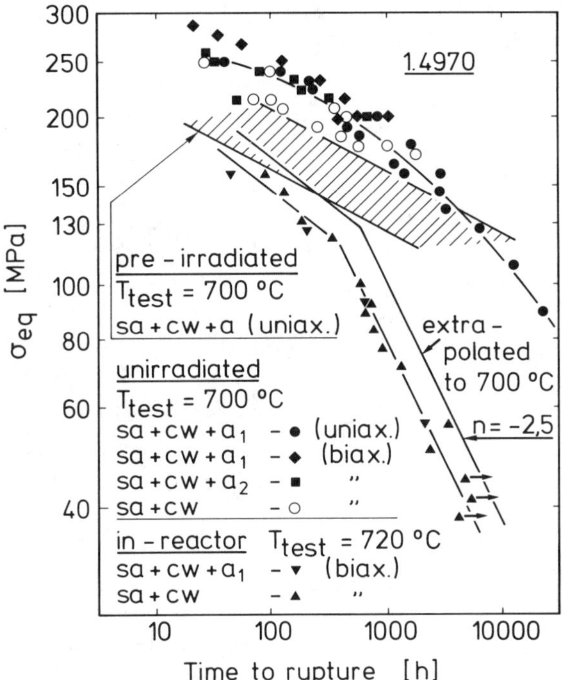

FIG. 10—*Creep-rupture strength before, during, and after neutron irradiation for steel 1.4970.*

a factor of approximately 3 to 5 at higher stress levels. With decreasing stresses the effect of irradiation decreases. The in-reactor creep-rupture strength shows clearly an additional, remarkable reduction compared with the creep rupture strength of the unirradiated material and the material tested after irradiation. The data plotted in Fig. 10 show that an abrupt change in the stress dependence of the creep-rupture strength occurs for the in-pile rupture life at the same stress level as already indicated for the in-pile creep rate, and was called transition stress σ_{tr}. The transition stress corresponding to the damage accumulating processes is consequently equivalent to that one for the in-pile creep processes. Below the transition stress the stress exponent of the rupture life is found to be equal to -2.5 as has been already found in 1979 [*20*], whereas above the transition stress it is twice as high. At stresses above the transition stress the irradiation-induced rupture life behavior is believed to be controlled by the climb-enabled glide creep mechanism. This paper only considers and discusses the in-pile rupture life behavior below the transition stress. The transition stress depends on the temperature as well as on the type of steel, and on the metallurgical condition. In order to analyze, as far as possible, the influence of the chemical composition, the neutron flux and spectrum, the helium generation rate, the helium to dpa ratio, and the temperature on the in-pile rupture life behavior, the creep rupture results on the 1.4981, 1.4988 and 1.4970 steels, obtained from experiments in the BR-2 and the FFTF reactors as well as Lovell's creep-rupture results from experiments in EBR-2 on American Iron and Steel Institute Type (AISI) 316 steel [*21*] have been evaluated. These data have been chosen because of the variations in helium content, chemical composition, metallurgical conditions, and the different neutron spectra and neutron fluxes of the BR-2, FFTF, and EBR-2 reactors.

In Figure 11 are compiled the in-pile creep-rupture data of the analyzed BR-2 results on the 1.4981, 1.4988, and 1.4970 steels and the nickel-base alloy, Hastelloy-X, at temperatures of 615, 650, 670, and 720°C. This diagram contains only the data below the appropriate transition stress to facilitate the discussion of the in-pile creep-rupture behavior. This figure includes the creep-rupture data on 1.4970 steel tested at 670°C in the FFTF reactor as well as Lovell's data on AISI 316 cw steel tested at 650°C in EBR-2 [*21*]. The austenitic steels mentioned above have been

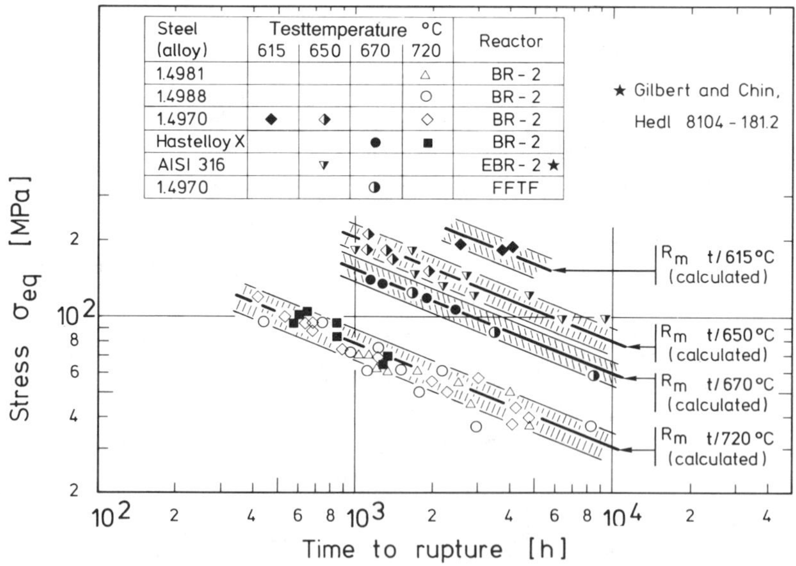

FIG. 11—*In-pile rupture-life of austenitic steels and Hastelloy-X tested in different reactors at various temperatures.*

tested in various metallurgical conditions, which are listed in Table 2. These allow to evaluate a possible influence of the metallurgical conditions on the rupture life. Since the austenitic steels contain boron quantities between about 1 and 90 appm, as shown in Table 1, a possible influence of the helium to dpa ratio on the rupture life behavior could be detected. A possible effect of the neutron flux and neutron spectrum on the rupture life behavior can be recognized by comparison of data obtained at the various reactors.

The results reveal that below the transition stress for the investigated austenitic stainless steels and the nickel-base alloy Hastelloy-X the in-pile rupture life is

- independent of the type of alloy,
- independent of the metallurgical conditions,
- not affected by the neutron flux and neutron spectrum,
- not affected by the helium gas production rate, and
- not affected by the helium to dpa ratio,
- characterized by a stress dependence inversely proportional to the 5⁄2 power, and
- characterized by a temperature independent stress exponent.

Temperature Dependence of the Operating Irradiation-Induced Rupture Life Mechanism

The presented and numerous individual experimental results allow the determination of the temperature dependence of the in-pile creep-rupture mechanism for the austenitic steels analyzed.

In Fig. 12, the reciprocal temperature dependence of the rupture life on the macroscale, θ_{tm}, which is equivalent to the diffusivity on the microscale, is plotted versus the reciprocal temperature. This figure indicates that the rupture life at high temperatures has a relatively strong temperature dependence whereas at low temperatures this dependence is weak. The temperature dependence of the rupture life seems to be independent of the steel type for the austenitic steels investigated.

As shown in Fig. 12, the experimentally obtained temperature dependence of the operating creep-rupture mechanisms has been decomposed, in the same way as done for the creep rate of the SIPA-AD deformation process, in order to derive the true activation energies for the individual rate-controlling processes. The true activation energy at low temperatures is calculated to be 0.127 eV whereas at high temperatures it is calculated to be 2.85 eV. The observations allow the rate-controlling process at low temperatures to be interpreted as the migration of helium atoms in the interstitial mode and at high temperatures the migration of substitutional helium atoms, that is, by vacancy-helium complexes. Thus, the apparent activation energy seems to be temperature dependent. The sum of the two individual rate-controlling processes which are effective at different temperature levels is given by the following equation

$$\theta_{tm} = a_\alpha \exp(-\Delta H_{h,\alpha}/kT) + b_\alpha \exp(-\Delta H_{l,\alpha}/kT) \qquad (7)$$

which represents the apparent diffusion coefficient θ_{tm} on the microscale. In this equation $\Delta H_{h,\alpha}$ and $\Delta \mathrm{H}_{l,\alpha}$ are the activation energies for the volume migration of substitutional helium atoms and for the migration of interstitial helium atoms, respectively, k is the Boltzmann constant, T is the temperature in K, and a_α and b_α are the corresponding proportionality (frequency) factors.

The results discussed above differ from the observations by Lovell et al. [*21*], Puigh et al. [*22*], and Boutard et al. [*23*]. Analyzing the creep-rupture results for steel AISI 316 they came to the conclusion that in-pile time-to-rupture does not change compared with that for the unirradiated material. The authors inferred further that the damage mechanism effective during irradiation is identical to the thermally activated mechanism. Lovell et al. [*21*] analyzed the temperature dependencies of the in-pile damage mechanism and of the mechanisms effective in the absence of irradiation, respectively, for AISI 316 steel applying the Sherby-Dorn parameter [*47*]. They found that the apparent activation energy, equal to 2.85 eV, is identical for the thermally activated creep-

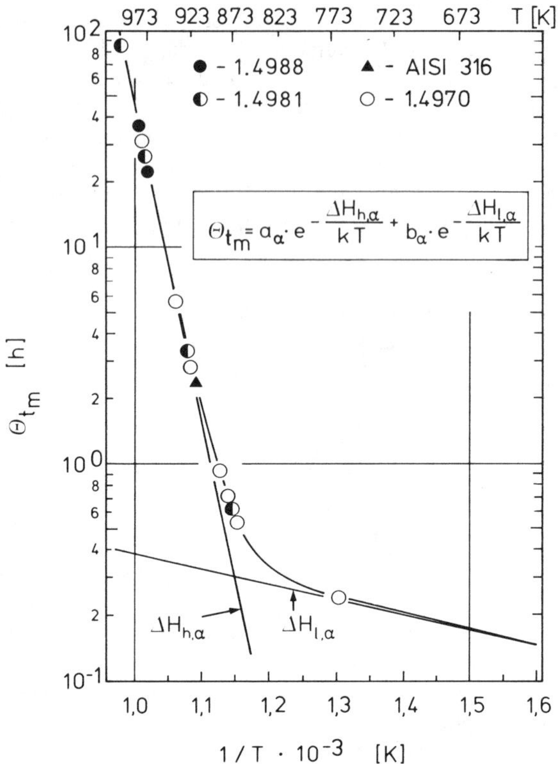

FIG. 12—*Temperature dependence for the damage accumulating mechanism.*

rupture mechanism and for the in-pile damage mechanism. Ehrlich [*28*], also analyzed the temperature dependence of the creep-rupture mechanisms which becomes effective in the absence of irradiation and in-pile, respectively, for steel 1.4970, likewise by applying the Sherby-Dorn parameter [*47*]. Using an activation energy equal to 2.8 eV, Ehrlich [*28*] came to the same conclusion, namely that the rate controlling in-pile damage mechanism does not change compared with the thermally activated mechanism. Nevertheless, according to Wassilew [*30*], the apparent activation energy for the thermally activated damage accumulating mechanism is equal to 4.1 eV within the considered field of the external and internal variables, i.e. below the transition stress. This value is considerably higher as compared with the "variable" apparent activation energy within the limits of 2.85 and 0.127 eV for the in-pile damage accumulating mechanism presented above. The conclusion by Lovell et al. [*21*], Boutard et al. [*23*] and Ehrlich [*28*] that the effective in-pile creep-rupture mechanism has a thermally activated damage characteristic is consequently in contradiction to our experimental observations and to the basic relations derived.

Correlations with the Sherby-Dorn parameter [*47*] are extremely useful if used over a range including the same, well defined, damage mechanism. However, their application is very dubious when parameters change such that different damage accumulating mechanisms occur. If creep-rupture correlations should be applied, this has to be done with some caution when external or internal variables change. Nevertheless, the results and the analyses presented above demonstrate that creep deformation during irradiation has a large effect on the rupture life properties.

In this connection, the following important differences between high-temperature in-pile creep-rupture experiments and post-irradiation creep-rupture experiments should be emphasized:

1. For in-pile below a given transition stress, depending on material and temperature, SIPA-AD creep with a linear stress dependence is the rate-controlling mechanism. It is the direct consequence of the interaction between dislocations and loops, and the interstitials and vacancies continuously produced by the displacement events.
2. During irradiation helium-bubbles are supplied with continuously generated helium atoms. As a consequence of the displacement events, dynamic resolution is efficient and additional free helium atoms are formed. In contrast, after irradiation gas bubbles can only be supplied with helium atoms from other bubbles, since no free helium atoms can remain in the lattice.

If we accept that during irradiation even, at temperatures above $T_m/2$, irradiation induced creep occurs as discussed above and already stated by Wassilew et al. [*20,48*], then we should accept that an interaction between the irradiation-induced creep mechanism and the helium-bubble nucleation and growth at grain boundaries perpendicular to the applied stress axis may influence the rate-controlling process for the in-pile rupture life behavior. Consequently, it follows that neither the in-pile creep behavior nor the in-pile creep-rupture properties can be related to the creep-rupture properties of the material tested in the unirradiated and irradiated conditions since they are controlled by thermally activated mechanisms. The observed stress dependencies are also different. According to Wassilew et al. [*20*], the stress exponent for the in-reactor rupture life is equal to -2.5, whereas for the material tested in the unirradiated and preirradiated conditions the stress exponent is approximately twice as high.

Nevertheless, Gilbert and Lovell [*18*] have suggested that creep deformation during irradiation delays the onset of tertiary creep resulting in a stress rupture life equal to or greater than the rupture life of the unirradiated material. This retardation of tertiary creep has been assumed for temperatures above 620°C. Our experimental results show that not even indications for such an in-pile creep-rupture behavior as quoted by Gilbert and Lovell [*18*], Lovell et al. [*21*], and Puigh and Schenter [*22*] are perceptible. The exact cause of the observed differences between the two data sets is not yet clear.

Dynamic Resolution Effects and Damage Mechanisms

The activation energies obtained for the temperature dependence of the in-pile creep-rupture mechanism allow to state that the considered damage mechanism is a diffusion-controlled process. Since the observed damage behavior, as shown in Fig. 11, is evidently independent of the external helium production by (n, α) reactions, it can be concluded that a monotonic helium-atom flux into the grain boundaries, caused by dynamic resolution of helium in clusters and bubbles as a consequence of the displacement events, will be the rate-controlling damage accumulating mechanism.

Stable Helium-Bubble Growth

If helium atoms react with vacancies, vacancy-helium complexes are formed and a net flux of substitutional helium atoms will end up at grain boundaries and give rise to nucleation and growth of helium bubbles by the absorption of such substitutional helium atoms. This can be compared with nucleation and growth of precipitates.

Provided that an interstitial helium atom is absorbed in a bubble, its appertaining vacancy has to be produced to prevent excessive bubble pressure. The growth behavior in this case can be described as constrained growth mode. This mechanism involves vacancies in ample quantities. Since vacancies are readily available at grain boundaries and dislocations, nucleation and growth of helium-bubbles at grain boundaries or at dislocations by interstitial helium atoms is possible.

The observed low activation energies that have been correlated to the migration energy of interstitial helium atoms, suggest that a net flux of interstitial helium atoms will end up at grain boundaries and result in bubble nucleation and growth. This mechanism was suggested by Trinkaus [49] but not implemented in his theoretical treatment. An implementation of this idea has been carried out recently by Al-Hajji [50].

It can be concluded from our results, the obtained activation energies, and the proposed mechanisms that the rate of nucleation and growth of helium bubbles at grain boundaries is influenced mainly by the effective migration energy of the substitutional or the interstitial helium atoms. The stable bubble growth at grain boundaries is evidently the rate controlling damage accumulating process that determines the in-pile rupture life behavior.

Such a damage accumulating process is evidently operative only during irradiation, and it cannot be compared with any other damage accumulating processes that are efficient in the absence of the rate-controlling dynamic resolution process. Thus, the dynamic resolution is the main rate-controlling in-pile damage accumulating process, which exclusively controls the in-reactor creep-rupture behavior. Since the binding energy of helium with a dislocation is high, the dislocations cutting a helium-bubble will absorb in core helium atoms. The climbing dislocations will sweep likewise helium atoms to the grain boundaries.

Rupture will occur after a short period of time provided that the critical bubble radius r_c for the corresponding applied stress is attained, and unstable bubble growth and coalescence give rise to a rapid crack propagation.

In conclusion it can be said that the in-pile creep-rupture behavior of the analyzed austenitic steels and the nickel-base alloy Hastelloy-X is determined by the intrinsic properties of the helium-atoms, the grain boundaries, and the helium dynamic resolution mechanism, leading to a stable helium bubble growth at the grain boundaries perpendicular to the applied stress axis. The observed differences in the stress and temperature dependencies of the creep-rupture strength of the materials tested after irradiation and during irradiation, respectively, are attributed to the formation of free helium atoms during irradiation by dynamic resolution in consequence of the displacement events.

Constitutive Equation for In-Pile Creep-Rupture

Analysis of the experiments allows us to recognize and understand the differences between the various rate controlling mechanisms. The knowledge obtained allows derivation of a simple constitutive equation for the in-reactor time-to-rupture in terms of the external variables: stress and temperature

$$t_m = \frac{t_{m,o}}{\theta_{tm}\,\sigma^{2.5}} \tag{8}$$

where σ is the applied stress, θ_{tm} is the temperature function for the damage accumulating process effective exclusively during irradiation, which is given by Eq 7, and $t_{m,o}$ is the rupture life for θ_{tm} and σ equal to 1. This equation describes rather well the experimental rupture life data on pressurized tubes as shown in Fig. 11.

Comparison of the Post-Irradiation and In-Pile Creep Ductilities

Figure 13 represents the dependence of the post-irradiation and in-reactor accumulated creep-rupture strains, respectively, as a function of the creep rate. The dependence of the in-reactor creep-rupture strain on the creep rate inverts at a given transition creep rate value. Above the

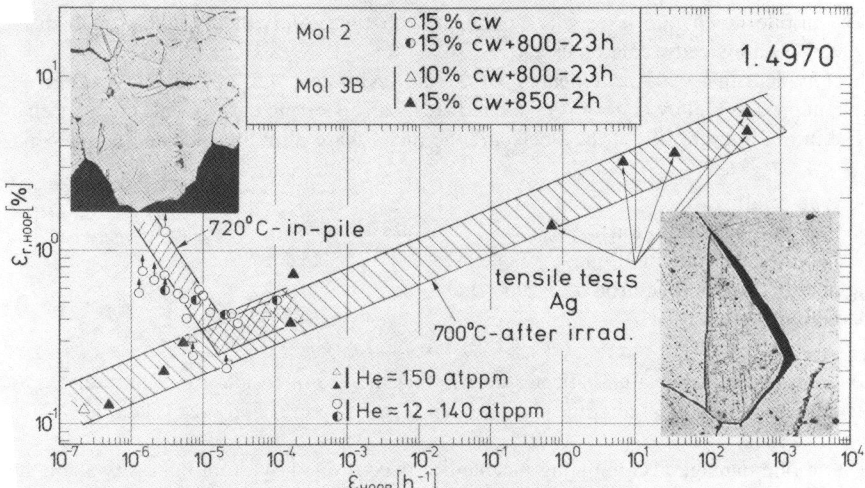

FIG. 13—*Comparison of the in-pile creep-rupture ductility with the creep-rupture ductility of preirradiated specimens.*

transition creep rate, the dependence of the in-reactor and post-irradiation creep-rupture strains is similar. Below the transition creep rate the in-pile creep-rupture strain increases with decreasing creep rate, whereas the creep-rupture strain of the material tested after irradiation still decreases with decreasing creep rate. The transition creep rate corresponds to the transition stress found for the stress dependence of the in-pile creep rate and for the stress dependence of the in-pile creep-rupture strength as stated by Wassilew et al. [*20*].

The fracture surface morphology of the in-pile tested specimens was metallographically examined. The fracture can be classified into two classical types of failure cracks: w cracks (triple-point wedge cracks) and r cracks, spherical cavities at the high angle, grain-boundaries perpendicular to the applied stress axis. The different failure crack types are shown in the photographs in Fig. 13.

In accordance with Garofalo [*51*], the w cracks were found predominantly at high strain rates while the r cracks were found at low strain rates.

It is now evident that in general the rupture lives and ductilities obtained from in-pile creep-rupture tests cannot be reproduced by post-irradiation tests. The in-reactor experiments discussed above show that in-situ tests yield lower absolute values and a weaker temperature and stress dependence of the creep-rupture strength than observed for experiments on unirradiated and pre irradiated specimens.

Conclusions

The following general conclusions can be drawn concerning the creep and creep-rupture mechanisms which are effective during irradiation:

1. Based on the basic relations obtained for the dependence of the creep rate on the external variables stress, temperature, and dpa rate, and on the internal variables density and configuration of the dislocations, structure as well as structural changes, it can be concluded that the observed deformation mechanism is in good accordance with the elastodiffusion-SIPA creep mechanism.
2. The temperature dependence of the irradiation-induced creep rate is characterized by the migration of interstitials at low temperatures and the migration of vacancies at high temperatures.

3. The in-pile creep rate is usually nonlinear with time. Only steels with a high dislocation density start with a steady state creep rate.

4. It is evident that even at temperatures around and above 0.5 T_m the SIPA-AD creep is an independent process and by no means linked to the Nabarro-Herring or the Coble creep mechanisms.

5. The in-pile rupture life of the steels and the nickel-base alloy investigated is independent of

- the type of alloy,
- the metallurgical conditions,
- the helium gas production rate,
- the neutron flux and neutron spectrum, and
- the helium to dpa ratio.

6. The analysis of the results indicates clearly that the use of results from unirradiated and pre irradiated experiments for predicting in-pile creep, creep-rupture, and creep ductility is not valid for these materials.

7. The in-pile damage accumulating mechanism is controlled by grain boundary stable helium-bubble growth supplied with helium by dynamic resolution.

References

[1] Collins, C. G., Hammons, G. L., Robertshaw, F. C., and Kouts, W. H., *APEX 676,* NMPO, General Electric Co., Dec. 1961.

[2] Robertshaw, F. C. and Collins, C. G., *Transactions of the American Society of Metals,* Vol. 55, 1962.

[3] Barnes, R. S., *Nature,* Vol. 206, 1965, p. 1307.

[4] Harries, D. R., *Journal of the British Nuclear Energy Society,* Vol. 5, 1966, p. 1509.

[5] Martin, W. R. and Weir, J. R., *Effect of Radiation on Structural Materials, STP 426,* American Society for Testing and Materials, Philadelphia, 1967.

[6] Williams, J. A. and Carter, J. W., *Irradiation Effects in Alloys for Thermal and Fast Reactors, STP 457,* American Society for Testing and Materials, Philadelphia, 1969.

[7] Lovell, A. J., *Transactions of the American Nuclear Society,* Vol. 12, 1969.

[8] Ehrlich, K., Böhm, H., and Wassilew, C., *Irradiation Effects on Structural Alloys for Nuclear Reactor Applications, STP 484,* American Society for Testing and Materials, Philadelphia, 1971, pp. 495–508.

[9] Bloom, E. E. and Stiegler, J. O., *Irradiation Effects on Structural Alloys for Nuclear Reactor Applications, STP 484,* 1971, pp. 451–467.

[10] Bagley, K. Q., Barnaby, J. W., and Fraser, A. S., Irradiation Embrittlement and Creep in Fuel Cladding and Core Components, British Nuclear Energy Society, London, 1972.

[11] Lovell, A. J., *Nuclear Technology,* Vol. 16, 1972, pp. 323–331.

[12] Böhm, H. and Wassilew, C., *Effects of Radiation on Substructure and Mechanical Properties of Metals and Alloys, STP 529,* American Society for Testing and Materials, Philadelphia, 1973, pp. 437–450.

[13] Bloom, E. E. and Stiegler, I. O., *Effects of Radiation on Substructure and Mechanical Properties of Metals and Alloys, STP 529,* American Society for Testing and Materials, Philadelphia, 1973, pp. 360–380.

[14] Lovell, A. J., *Nuclear Technology,* Vol. 26, No. 3, 1975, pp. 297–306.

[15] Anderko, K., Schäfer, L., Wassilew, C., Ehrlich, K., and Bergmann, H. J., *Radiation Effects in Breeder Reactor Structural Materials,* American Institute of Mechanical Engineers, New York, 1977.

[16] Maziasz, P. J., *Scripta Metallurgica,* Vol. 14, 1980, pp. 65–82.

[17] Closs, K. D. and Schäfer, L., *Effects of Radiation on Substructure and Mechanical Properties of Metals and Alloys, STP 529,* American Society for Testing and Materials, Philadelphia, 1973, pp. 460–472.

[18] Gilbert, E. R. and Lovell, A. J., *Radiation Effects in Breeder Reactor Structural Materials,* American Institute of Mechanical Engineers, New York, 1977, pp. 269–276.

[19] Wassilew, C., Schäfer, L., and Anderko, K., *Reaktortagung,* Hannover, Deutsches Atomforum e.V., 1978, p. 609.

[20] Wassilew, C., Anderko, K., and Schäfer, L., *Irradiation Behaviour of Metallic Materials for Fast Reactor Core Components,* Proceedings of the International Conference, Ajaccio, France, 1979, pp. 419–425.

[21] Lovell, A. J., Chin, B. A., and Gilbert, E. R., *Journal of Material Science,* Vol. 16, 1981.

[22] Puigh, R. J. and Schenter, R. E., *Effects of Radiation on Materials: Twelfth International Symposium, STP 870,* American Society for Testing and Materials, Philadelphia, 1985, pp. 795–802.
[23] Boutard, J. L., Maillard, A., Carteret, Y., Levy, V., and Boyer, J. M., *Effects of Radiation on Materials: Twelfth International Symposium, STP 870,* American Society for Testing and Materials, Philadelphia, 1985, p. 61.
[24] Bergmann, H. J., Haas, D., and Herschbach, K., *Radiation Effects in Breeder Reactor Structural Materials,* American Institute of Mechanical Engineers, New York, 1977, pp. 241–251.
[25] Mosedale, D., Harries, D. R., Hudson, J. A., Lewthwaite, G. W., and Mcelroy, R. J., *Radiation Effects in Breeder Reactor Structural Materials,* American Institute of Mechanical Engineers, New York, 1977, pp. 209–228.
[26] Bullough, R. and Hayns, M. R., *Journal of Nuclear Materials,* Vol. 57, 1975, p. 348.
[27] Bullough, R. and Hayns, M. R., *Journal of Nuclear Materials,* Vol. 65, 1977, pp. 184–191.
[28] Ehrlich, K., *Journal of Nuclear Materials,* Vols. 133 and 134, 1985, pp. 119–126.
[29] Cloß, K. D. and Herschbach, K., *Reaktortagung,* Düsseldorf, 1976.
[30] Wassilew, C., Dr-Thesis, Fakultät für Maschinenbau an der Universität Karlsruhe (TH), Karlsruhe.
[31] Wolfer, W. G. and Ashkin, M., *Journal Applied Physics,* Vol. 47, 1976, p. 791.
[32] Heald, P. T. and Speight, M. V., *Philosophy Magazine,* Vol. 29, 1974, p. 1075.
[33] Hesketh, R. V., *Philosophy Magazine,* Vol. 7, 1962, p. 1417.
[34] Savino, E. J., *Philosophy Magazine,* Vol. 36, 1977, p. 323.
[35] Dederichs, P. H. and Schroeder, K., *Physics Review,* Vol. B 17, 1978, p. 2524.
[36] Tome, C. N., Ceccatto, H. A., and Savino, E. J., PMTM/1–2, 1981, CNEA Report, *Physics Review,* Vol. B 25, 1982, p. 7428.
[37] Woo, C. H., *Journal of Nuclear Materials,* Vol. 120, 1984, pp. 55–64.
[38] Mansur, L. K., *Philosophy Magazine A,* Vol. 39, No. 4, 1979, pp. 497–506.
[39] Nabarro, F. R., Reports of the *Conference on the Strength of Solids,* The Physical Society, London, 1948, pp. 75–90.
[40] Coble, R. L., *Journal Applied Physics,* Vol. 34, 1963, p. 1679.
[41] Jones, R. B., *Proceedings of the Conference on Dimension Stability and Mechanical Behaviour of Irradiated Metals and Alloys,* British Nuclear Energy Society, London, 1984.
[42] Bullough, R., Harries, D. R., and Haynes, M. R., *Journal of Nuclear Materials,* Vol. 88, 1980, pp. 312–314.
[43] Harris, J. E., *Proceedings of the Conference on Vacancy '76,* Metal Society, Bristol, 1976, p. 170.
[44] Wood, M. H. and Kear, K. L., *Journal of Nuclear Materials,* Vol. 118, 1983, pp. 320–324.
[45] Greenwood, G. W., Foreman, A. J. E., and Rimmer, D. E., *Journal of Nuclear Materials,* Vol. 4, 1959, p. 305.
[46] Whapham, A. D., *Philosophy Magazine,* Vol. 23, 1971, p. 987.
[47] Orr, L. R., Sherby, O. D., Dorn, J. E., *Transactions of the American Society of Metals,* Vol. 46, 1954, pp. 113–128.
[48] Wassilew, C., Schneider, W., and Ehrlich, K., Radiation Effects, Vol. 101, pp. 201–219.
[49] Trinkaus, H., *Journal of Nuclear Materials,* Vol. 118, 1983, p. 39.
[50] Al-Hajji, J. N., "Theoretical Modeling of the High Temperature Helium Embrittlement in Structural Alloys," PPG-894, UCLA-ENG-8533, 1985.
[51] Garofalo, F., *Fundamentals of Creep and Creep Rupture in Metals,* MacMillan, New York, 1965.

Allan R. Causey, [1] *Vilius Fidleris,*[1] *Stuart R. MacEwen,*[1] *and Chris W. Schulte*[2]

In-Reactor Deformation of Zr-2.5 wt% Nb Pressure Tubes

REFERENCE: Causey, A. R., Fidleris, V., MacEwen, S. R., and Schulte, C. W., **"In-Reactor Deformation of Zr-2.5 wt% Nb Pressure Tubes,"** *Influence of Radiation on Material Properties: 13th International Symposium (Part II), ASTM STP 956,* F. A. Garner, C. H. Henager, Jr., and N. Igata, Eds., American Society for Testing and Materials, Philadelphia, 1987, pp. 54–68.

ABSTRACT: Changes in shape of internally pressurized tubes caused by operating temperatures and pressures are enhanced by fast neutron irradiation. Lengths and diameters of Zr-2.5 wt% Nb pressure tubes in CANada Deuterium Uranium-Pressurized Heavy Water (CANDU-PHW) power reactors and test reactors have been monitored periodically over the past 15 years. Axial and transverse strain rates have been evaluated in terms of the operating variables and the crystallographic texture and anisotropic microstructure of the extruded and cold-drawn tubes. The anisotropic deformation can be described by models for creep and irradiation growth in which the anisotropy factors are calculated from texture. It is assumed that prismatic slip is the dominant creep mode and that growth occurs by net fluxes of interstitials to a nonrandom distribution of ⟨**a**⟩ type edge dislocations and vacancies to ⟨**a**⟩ type screw dislocations, ⟨**c**⟩ type edge dislocations, and grain boundaries. The equations based on data from the Pickering Generating Station and WR1 test reactors give good agreement with measurements on internally pressurized tubes in Bruce Generating Station and the National Reactor Universal (NRU) test reactor and uniaxially stressed specimen in NRU.

KEY WORDS: zirconium alloys, nuclear industry, pressure tubes, deformation, Zr-2.5 wt% Nb, in-reactor deformation, empirical equation

Internally pressurized zirconium alloy tubes in service at operating temperatures and pressures undergo enhanced changes in shape when irradiated by fast neutrons. Several equations have been proposed to account for the changes in length and diameter of cold-worked Zr-2.5 wt% Nb pressure tubes in CANada Deuterium Uranium-Pressurized Heavy Water (CANDU-PHW) reactors in terms of the operating environment and the microstructures produced during tube fabrication [*1–3*]. These equations have generally described the anisotropic deformation using the concepts of additive, separable components [*4*] of in-reactor thermal creep, irradiation-induced creep, and irradiation growth (shape change in the absence of applied stress, at constant volume) with the strain models related to the crystallogaphic texture. The equation to be presented here differs from previous equations in the models proposed to account for the anisotropy of the creep and growth strains [*5,6*], and in the calculation of polycrystalline behavior by accounting for the effects of interactions between grains of different orientations [*7*].

Results of monitoring the dimensional changes in pressure tubes in the CANDU-PHW power reactors at the Pickering and Bruce Nuclear Generating Stations and the test reactors, Whiteshell Reactor 1 (WR1) at Whiteshell Nuclear Research Establishment (WNRE), and National Reactor

[1] Atomic Energy of Canada, Ltd., Chalk River Nuclear Laboratories, Chalk River, Ontario KOJ 1J0, Canada.

[2] Ontario Hydro, 700 University Ave., Toronto, Ontario M5G 1X6, Canada.

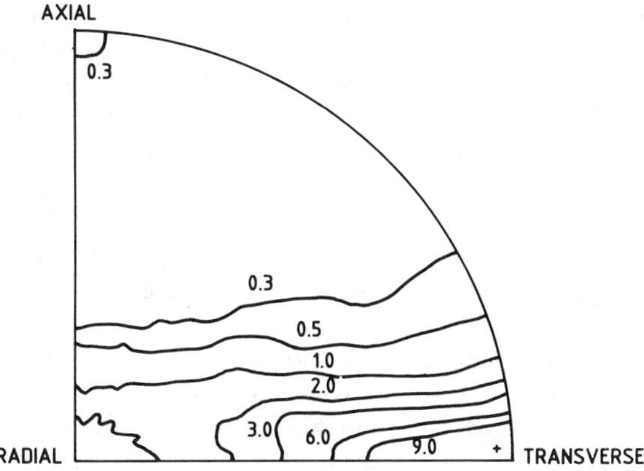

FIG. 1—*Typical basal plane pole figure for cold-worked Zr-2.5 wt% Nb pressure tube: numbers on contours are times random, + is maximum, typically 12–18 R.*

Universal (NRU) at Chalk River Nuclear Laboratories (CRNL), will be presented. A constitutive equation based on functional relationships between temperature, stress, flux, and microstructural parameters has been developed to interpret the measurements and to predict behaviors of tubes fabricated by slightly different procedures and operated at different conditions.

Pressure Tube Materials

Fabrication procedures and the resulting microstructures for the extruded and cold-worked Zr-2.5 wt% Nb tubes have been described in detail elsewhere [*8*]. The major parameters that affect the deformation behavior of the tubes in-reactor are the crystallographic texture and the density and Burger's vectors of the dislocations. The texture can be described by basal-plane pole figures determined by X-ray diffraction measurements, such as shown in Fig. 1. The majority of the basal poles are oriented in the transverse (circumferential) direction in the radial-transverse plane. The resolved fractions of basal poles f_d in the radial, transverse, and axial directions for the tubes studied in this report are given in Table 1. The dislocation densities, as measured by X-ray diffraction line-broadening [*9*], are included in Table 1.

Deformation Measurements

Elongation of the 6-m-long pressure tubes at Pickering and Bruce reactors is measured by a tool attached to the remotely operated on-power fuelling machines which measure the displacement of

TABLE 1—*Texture and dislocation densities of cold-worked Zr-2.5 wt% Nb pressure tubes.*

	Resolved Fraction of Basal Poles			
Reactor	f_r	f_t	f_a	Dislocation Density, m^{-2}
Pickering	0.35	0.57	0.08	4.6×10^{14}
Bruce	0.33	0.61	0.06	6.5
WR1	0.30	0.63	0.07	4.6
NRU	0.32	0.59	0.09	5.3

the end-fittings joined to each end of the pressure tube [*10*]. Each measurement is accurate to ±1 mm and thus the corresponding axial strain has an error bar of $\pm 2.5 \times 10^{-4}$. A typical plot of elongation against integrated channel power of all 390 tubes in Pickering Unit 3 measured four times during its twelve years of operation, is shown in Fig. 2. The average axial strain rate for tubes in central, high flux locations derived from four measurements of each of the tubes in Pickering Units 3 and 4 is 8.5×10^{-8} h^{-1} ± 15% (±2 standard deviations from regression statistics). The scatter in Fig. 2 is five times that of the measurement of a single tube because of variations in the initial lengths. The average strain rate for central core tubes in the Bruce reactors is 9.4×10^{-8} h^{-1} ± 15%. Length measurements in the test reactors are made using the diameter gauging tool.

Diameters are measured with the reactor shutdown by passing a gauging tool similar to that described by Ross-Ross and Hunt [*11*], but containing two rotating linear variable differential transformer (LVDT) probes, through the empty tube. Nine readings taken at 20° intervals around the tube are averaged to give one diameter every 12.7 mm of axial travel accurate to ±2.5 μm. Typical diameter profiles for a pre-service and in-service measurement after nine years are shown in Fig. 3 for a pressure tube irradiated in a Bruce reactor. The pre-service inside diameter and wall thickness of power reactor tubes are about 103.8 and 4.2 mm, respectively. The dips every 0.5 m correspond to the location of the ends of the fuel bundles. Plots of strain at the mid points of the bundles versus time, similar to the data in Fig. 4, show that the strain increases nearly linearly with time and that the intercepts at time zero range from 0.02 to 0.06%. Typical transverse

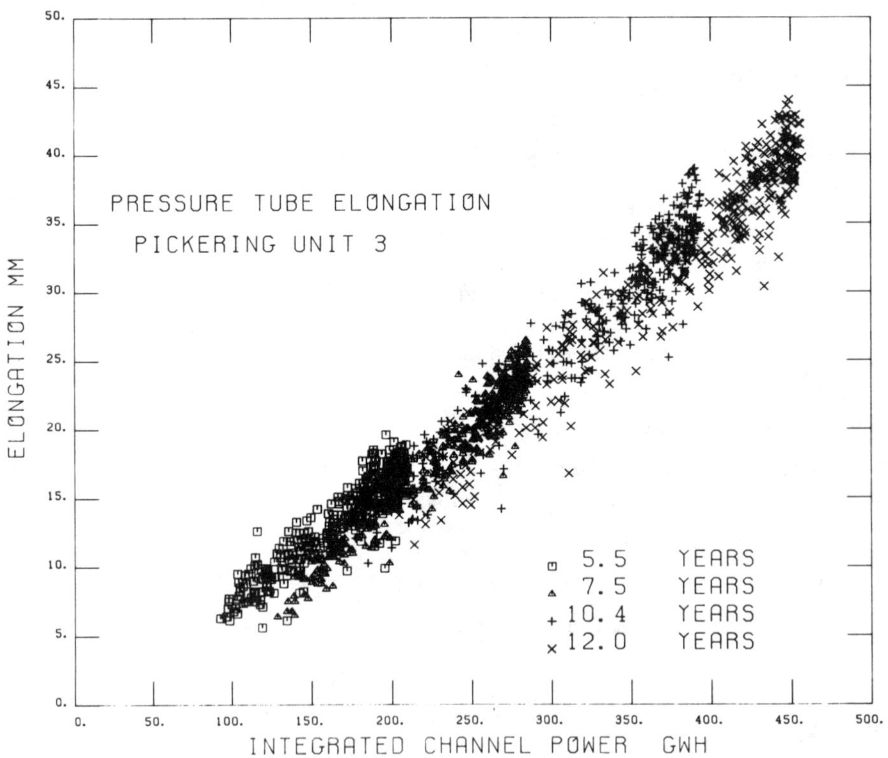

FIG. 2—*Elongation of Zr-2.5 wt% Nb pressure tubes in a Pickering reactor, showing linear dependence of elongation on integrated channel power.*

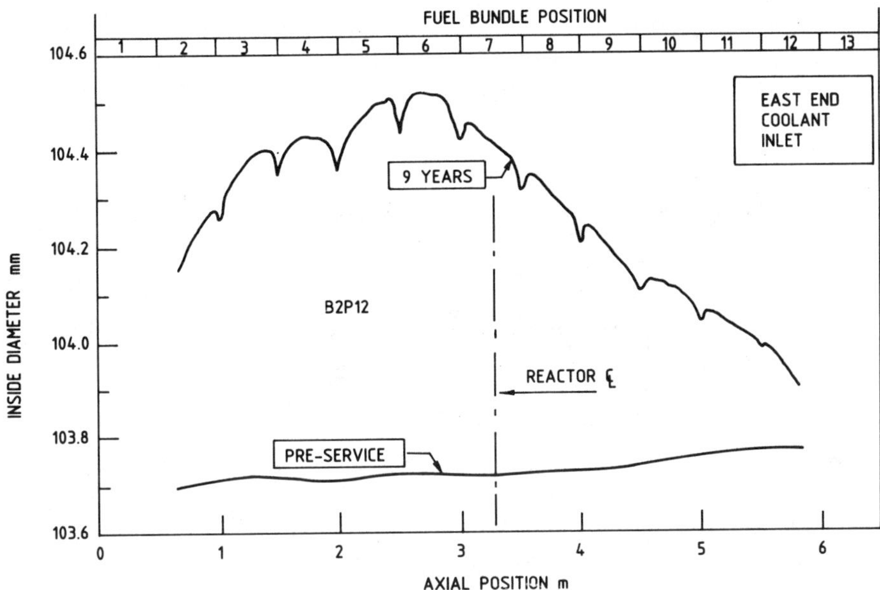

FIG. 3—*Diameter profiles of Zr-2.5 wt% Nb pressure tube from a Bruce reactor before and after nine years service: inlet temperature 524 K, outlet 569 K, and peak fast neutron flux* $3.1 \times 10^{17}\ n \cdot m^{-2} \cdot s^{-1}$ E > 1 *MeV.*

strain rates derived from diameter measurements after 1 or 2, 7 and 10 years operation for two tubes from a Pickering reactor are shown in Fig. 5, and the average values for four tubes at each temperature are plotted against fast neutron flux in Fig. 6. The strain rates increase along the tube from inlet (at 523 K) to outlet (at about 567 K) and are higher in the high flux region of the

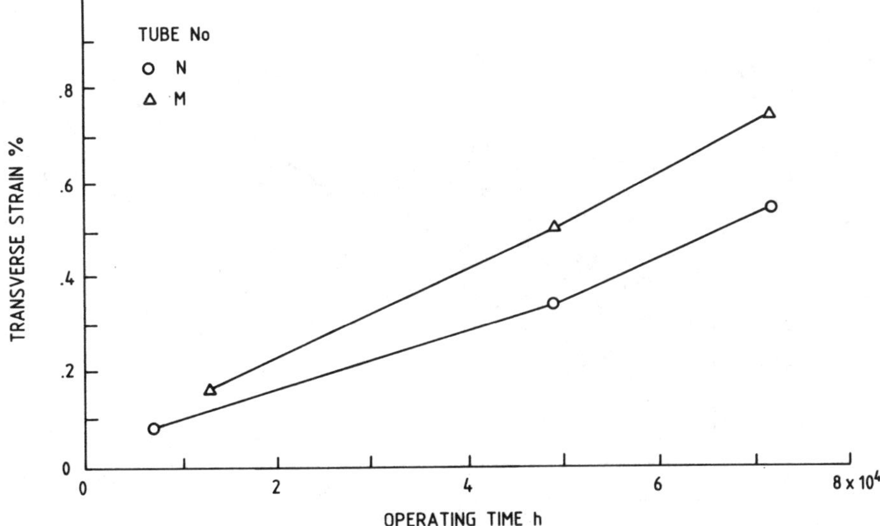

FIG. 4—*Transverse strain in pressure tubes in a Pickering reactor at position along tube where maximum strain occurs, as function of operating time.*

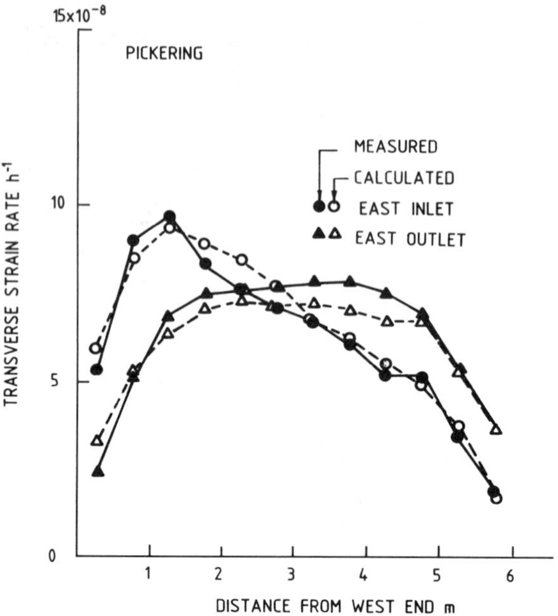

FIG. 5—*Measured transverse strain rates for pressure tubes in a Pickering reactor showing two typical types of strain rate behaviors, and a comparison with rates calculated using Eq 1.*

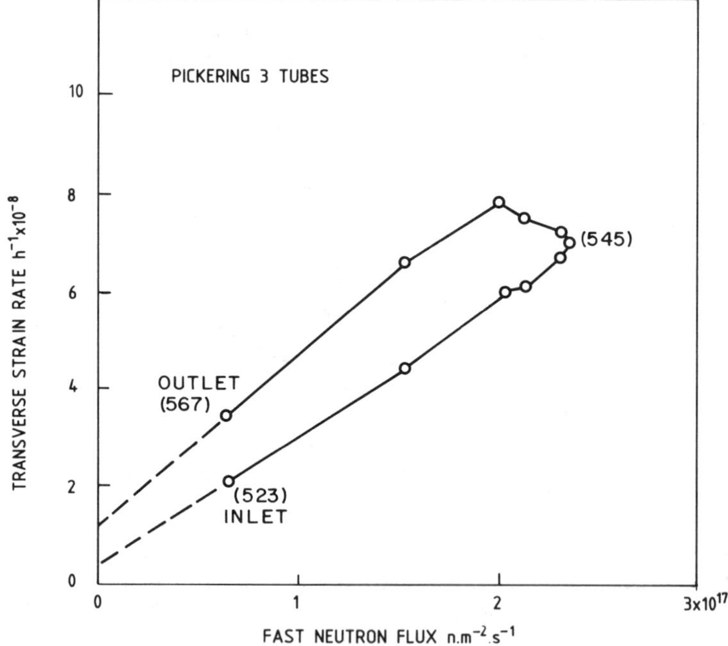

FIG. 6—*Average transverse strain rates of four pressure tubes in a Pickering reactor as a function of fast neutron flux* E > *1 MeV (Numbers in brackets are temperatures in degrees Kelvin).*

flattened cosine shape flux distribution, which has a peak flux of about $2.4 \times 10^{17}\ n \cdot m^{-2} \cdot s^{-1}$ ($E > 1$ MeV).

The profiles in Fig. 5 illustrate two types of strain patterns; tubes with their coolant outlets at the west end of the reactor have pronounced peaks in strain near the outlet end, while tubes with their flow outlets at the east end have more uniform strain profiles. This pattern is related to the fact that all tubes were installed with the back end, that is, the end extruded last, at the west end of the reactor. Hot extrusion results in thinner α-phase grains and a lower dislocation density at the end exiting the extrusion press first (front end) than at the back end.

A typical series of diameter profiles for a tube in WR1 is shown in Fig. 7. Nominal initial diameter and wall thickness of WR1 tubes are 82.6 and 3.1 mm, respectively, with some zones of reduced wall thickness to give different stress zones. The axial stress in an internally pressurized, closed end tube remains constant along the tube at one-half that of the hoop stress at the inlet. The hoop stresses, however, decrease along the tube as a result of the pressure drop, which is less than 5% in water cooled tubes, but is significantly more than this in the organic cooled loops in the WR1 reactor. Thus the ratio of transverse (hoop) to axial stress decreases from 2.0 at the inlet to 1.9 in water cooled tubes and to about 1.2 in organic cooled tubes at the outlet.

Typical strain rates from pressure tubes operated in WR1 and NRU and from uniaxial tests on sections from transverse and axial sections in NRU are listed in Tables 2 and 3 with the appropriate stresses. Test temperatures ranged from 564 to 666 K in WR1 and from 539 to 648 K in NRU, with fast neutron fluxes up to $3.3 \times 10^{17}\ n \cdot m^{-2} \cdot s^{-1}$ $E > 1$ MeV.

Analysis

The complex interaction between the effects of temperature and fast neutron flux on the deformation of zirconium alloys [3] has led to the development of analyses that assume that long-term steady state deformation consists of separable, additive components from thermal creep, irradiation creep, and irradiation growth [4]. In-reactor thermal creep rates in Zr-2.5 wt% Nb do not appear to diminish with time like those in laboratory tests [12]. All components are anisotropic

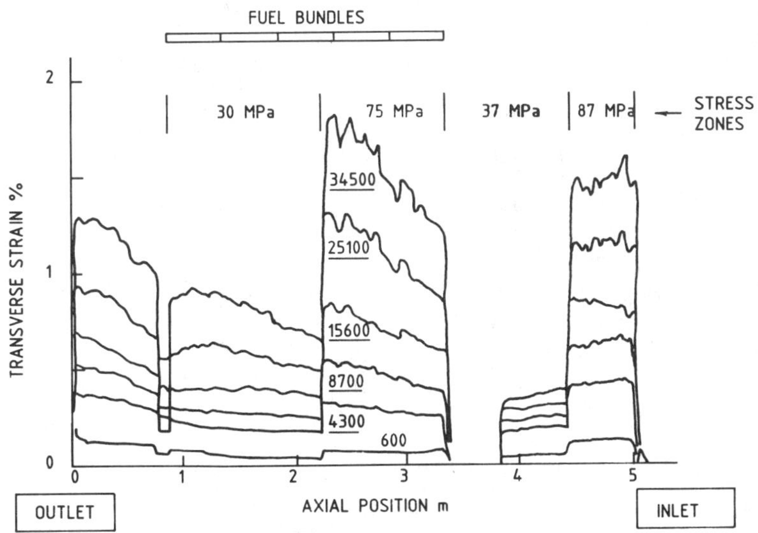

FIG. 7—*Transverse strain profiles of a pressure tube in WR1.*

TABLE 2—*Typical transverse strain rate data and stresses for cold-worked Zr-2.5 wt% Nb pressure tubes in WR1 reactor.*

Tube	Hoop Stress, MPa	Stress Ratio[a]	Strain Rates $h^{-1} \times 10^{-8}$ Measured	Calculated
13	108	2.0	2.3	2.0
	91	1.7	4.8	5.1
	73	1.6	2.7	2.6
7	70	1.5	0.9	0.9
	47	1.3	−0.2	−1.8
	43	1.2	−0.6	−2.2
16	30	1.8	−5.8	−4.1
	47	1.7	−3.8	−1.2
	42	1.6	21.0	19.0
22	84	2.0	35.0	32.0
	72	1.7	45.0	48.0
	34	1.6	17.0	18.0
2	50	[b]	3.0	2.9
	38	[b]	3.4	4.2
	36	1.3	0.6	2.9
11	79	1.8	6.0	5.4
	75	1.7	6.0	5.5
	35	1.6	−0.5	−0.8

[a] Stress ratio is ratio of transverse to axial stresses.
[b] Tube was under axial compressive load to reduce axial stress to zero.

TABLE 3—*Typical strain rate data and stresses for uniaxial and biaxial tests on cold-worked Zr-2.5 wt% Nb pressure tubes in NRU reactor.*

Tube Number-Loading and Measuring Direction[a]	Hoop Stress, MPa	Strain Rates, $h^{-1} \times 10^{-7}$ Measured	Calculated
1-BT	128	1.16	1.17
BT	127	1.20	1.08
BT	123	0.19	0.20
BA	128	0.60	0.72
1-UT	138	2.0	1.6
UT	138	6.4	5.4
UT	138	17.5	15.9
UA	138	4.6	5.4
UA	138	5.0	7.2
UA	138	39.0	36.4
3-BT	154	1.9	1.6
BT	236	3.8	3.0
BT	122	1.4	0.9
4-BT	113	0.82	0.60
BT	138	1.01	1.00
BT	112	0.91	0.76
BT	111	0.79	0.61
BA	112	0.78	0.81

[a] Loading and measuring direction: BT = biaxial/transverse, BA = biaxial/axial, UT = uniaxil/transverse, and UA = uniaxial/axial. Stress ratios are approximately 2.0 in biaxial tests.

and contribute to length as well as diameter changes. The equation is based on functional relationships between deformation rates, operating variables, and microstructural parameters. Some of these relationships have been well documented, such as the linear dependence of strain rate on flux and stress at temperatures below 570 K [3], while some will be described here.

It is proposed that the deformation of pressure tubes can be described by an equation of the form

$$\begin{aligned}\dot{\epsilon}_d = {} & [A_1 C_d \sigma + A_2 C_d' \sigma'^2]\exp(-Q_1/T) \\ & + A_3(x) C_d \sigma \exp(-Q_2/T) \\ & + A_4 C_d \sigma \phi \exp(-Q_3/T) \\ & + A_5 G_d \phi \exp(-Q_4/T)\end{aligned} \tag{1}$$

The first two terms represent in-reactor thermal creep at temperatures above and below 570 K, respectively, while the last two terms describe flux dependent creep and irradiation growth, respectively. The parameters are defined as follows:

$\dot{\epsilon}_d$ = the strain rate in direction d (radial, transverse, or axial), h^{-1},
A_1 and A_2 = constants for high-temperature in-reactor thermal creep,
$A_3(x)$ = function describing the variation of low-temperature thermal creep along the tube,
A_4 and A_5 = constants for irradiation creep and growth,
Q_1, Q_2, Q_3 and Q_4 = activation temperatures,
C_d and C'_d = anisotropy factors for creep for stress exponents of 1 and 2, respectively, for direction d,
σ and σ' = effective stresses for stress exponents of 1 and 2, MPa,
G_d = the anisotropy factors for growth in direction d,
T = temperature, K, and
ϕ = fast neutron flux, $n \cdot m^{-2} \cdot s^{-1}$ ($E > 1$ MeV).

The effective stresses and creep anisotropy factors are defined in terms of Hill's [13] anisotropy constants F, G, H, and F', G', H', for the stress exponents of 1 and 2, respectively, and the principal stresses, σ_r, σ_t, and σ_a, as follows

$$\sigma = [F(\sigma_a - \sigma_t)^2 + G(\sigma_t - \sigma_r)^2 + H(\sigma_r - \sigma_a)^2]^{1/2} \tag{2}$$

and

$$\begin{aligned}C_r &= [H(\sigma_r - \sigma_a) - G(\sigma_t - \sigma_r)]/\sigma \\ C_t &= [G(\sigma_t - \sigma_r) - F(\sigma_a - \sigma_t)]/\sigma \\ C_a &= [F(\sigma_a - \sigma_t) - H(\sigma_r - \sigma_a)]/\sigma\end{aligned} \tag{3}$$

The creep and growth anisotropy constants have been calculated from texture, as expressed by the basal pole figures, by considering the grains in each orientation to deform on specific slip systems in response to the applied stress, temperature, and flux. For creep, the applied stresses in the principal fabrication directions of the specimen are resolved into shear stresses on the slip planes of interest, and a linear or power law relationship between stress and strain rate is assumed to calculate the shape change for each orientation of grain. For growth, the shape change is independent of stress, and is calculated from the annihilation of point defects at sinks. The total shape change in the grain is tensor-rotated to the specimen axes and the shape change of the component is calculated from the volume-fraction-weighted sum of the strains for all orientations in the pole figure. The effects of interactions between the grains on polycrystalline behavior have been accounted for similar to the manner suggested by Adams et al. [14], but including anisotropic elastic and thermal as well as plastic strain, as described in Ref 7.

Hill's anisotropy constants are calculated using a model, which assumes that the dominant creep mechanism is slip of ⟨**a**⟩ type dislocations on prism planes with secondary slip of ⟨**c**+**a**⟩ type dislocations on pyramidal planes. The anisotropy of experimental in-reactor creep data at 540 to 570 K [*6*] can be described by this model when the critical resolved shear stresses for slip on prism and pyramidal systems are taken as 50 and 300 MPa, respectively.

The growth anisotropy factors have been calculated using a model that attributes the strain in each grain to annihilation of the interstitials and vacancies, produced in equal numbers during neutron irradiation, at different sinks. The model assumes a net flux of interstitials to a nonrandom distribution of ⟨**a**⟩ type edge dislocations with a corresponding net flux of vacancies to ⟨**a**⟩ type screw dislocations, ⟨**c**⟩ type edge dislocations, and grain boundaries [*15*]. If grain interactions are neglected the growth anisotropy is related to texture by the equation

$$G_d = (1 - 3f_d) \tag{4}$$

With the resolved fraction of basal poles given in Table 1, Eq 4 predicts a ratio of transverse to axial growth of about −0.9. Recent measurements by Parker et al. [*16*] and Fidleris [*17*] have found a ratio of transverse to axial growth rates of about −0.6 in pressure tube material. The difference between the measured ratio and that calculated with Eq 4 is attributed to grain interaction, which generates internal stresses, and to the nonrandom distribution of dislocation Burger's vectors produced during the tube fabrication processes. Both of these factors are accounted for in the calculations used to evaluate the growth anisotropy factors of Eq 1. Calculated anisotropy factors for creep and growth for tubes in this study are given in Table 4.

The constants and temperature dependencies for each term in Eq 1 have been obtained from the dimensional mesurements of pressure tubes in Pickering and WR1 reactors. The stress exponent for in-reactor thermal creep, which differs from laboratory creep caused by the effects of irradiation hardening, for stresses up to about 150 MPa is equal to 1 up to 570 K and then increases to about 2 with increasing temperature [*3*]. The high-temperature thermal creep term reflects the high temperature dependence of the creep rate exhibited by the WR1 tubes, Q_1 = 25 000 K.

The constants for the irradiation creep and growth terms were derived from the Pickering pressure tube data. The transverse strain rates for the four Pickering tubes (Fig. 5) were averaged at each temperature and flux to eliminate the end to end effect noted earlier (Fig. 6). Average transverse strain rates, the total axial strain rate, and the average anisotropy factors for the four tubes, were used to obtain A_4 and A_5. The temperature dependence for growth was taken from previous studies [*17*] to be low (Q_4 = 600 K), while that for creep was fitted to the axial variation of the transverse strain rates, with the thermal creep rate removed, and found to give Q_3 = 2200 K.

With the constants defined above, the transverse strain rate at the middle of the tubes was calculated to comprise a positive creep rate of 13.1×10^{-8} h^{-1} and a negative growth rate of -6.0×10^{-8} h^{-1}. The growth rate compares well with that deduced from the WR1 and NRU

TABLE 4—*Creep and growth anisotropy factors for Zr-2.5 wt% Nb pressure tubes.*

	Creep Factors						Growth Factors		
Reactor	F	G	H	F'	G'	H'	G_r	G_t	G_a
Pickering	0.85	0.24	0.91	1.01	0.10	1.13	−0.42	−0.58	1.00
Bruce	0.89	0.21	0.96	1.05	0.06	1.18	−0.48	−0.63	1.11
WR1	0.90	0.22	1.01	1.09	0.06	1.28	−0.44	−0.68	1.12
NRU	0.87	0.26	0.93	1.05	0.11	1.19	−0.43	−0.60	1.03
Route 1 [*16*]	0.99	0.12	1.00	1.10	0.02	1.14	−0.62	−0.66	1.28
Heat-treated [22]	0.61	0.47	0.58	0.58	0.40	0.57	−0.16	−0.20	0.36

data shown in Fig. 8, where the extrapolated strain rates at zero stress are of the order of -6×10^{-8} h^{-1} at a flux of about 2×10^{17} n · m^{-2} · s^{-1}. The axial strain rate in the Pickering tubes, which is obtained by intergratng the axial rates at each fuel bundle position over the length of the tube, comprises about 90% growth and 10% creep.

In the plot of the average transverse strain rates versus flux (Fig. 6), extrapolation of the last two data points on each part of the loop back to zero flux gives an estimate of the thermal creep rates at 523 and 567 K, from which an activation temperature Q_2 = 5500 K is deduced for the low temperature thermal creep term. The function $A_3(x)$ has been approximated by the sigmoidal expression

$$A_3(x) = B_1 - B_2L/[1 + L \exp(-B_3(x - B_4))] \tag{5}$$

where x is the distance from the back end of the extrusion, L is the tube length, and B_1, B_2, B_3, and B_4 are the average values of the constants determined for each of the four Pickering tubes by assuming average flux-induced creep and growth parameters and fitting the transverse strain rate data.

The creep and growth rate constants were assumed to vary with dislocation density according to the relation $A \; \alpha \; \rho^n$, where n is 0.4 and 1.0 for creep and growth, respectively [*18*].

The good fit of Eq 1 to the Pickering and WR1 pressure tube data from which it was derived is illustrated in Fig. 9 where calculated transverse and axial strain rates are compared with the measured values. Transverse strain rates in some of the WR1 tubes at the lower stresses are calculated to be negative, as given in Table 3. The equation gives good agreement with the axial variation in transverse strain rates on the Pickering tubes, as shown in Fig. 5, where the calculated transverse strain rates from inlet to outlet ends of two Pickering tubes are compared with the

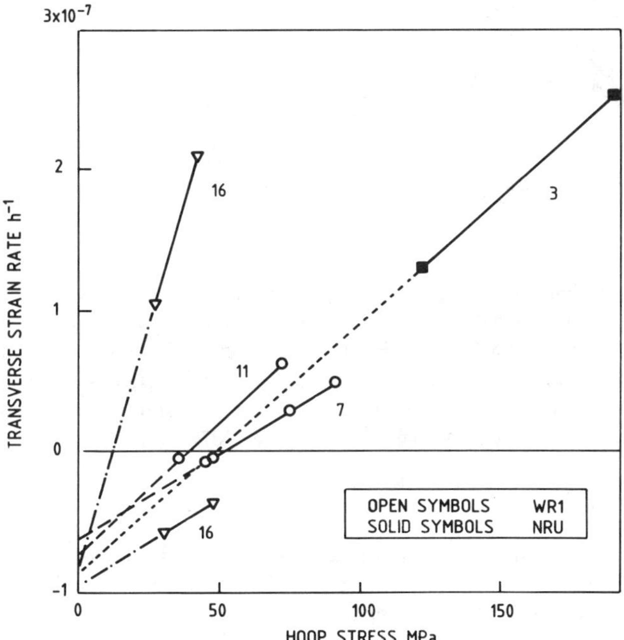

FIG. 8—*Transverse strain rates of pressure tubes in WR1 and NRU as a function of hoop stress. Numbers identify tubes.*

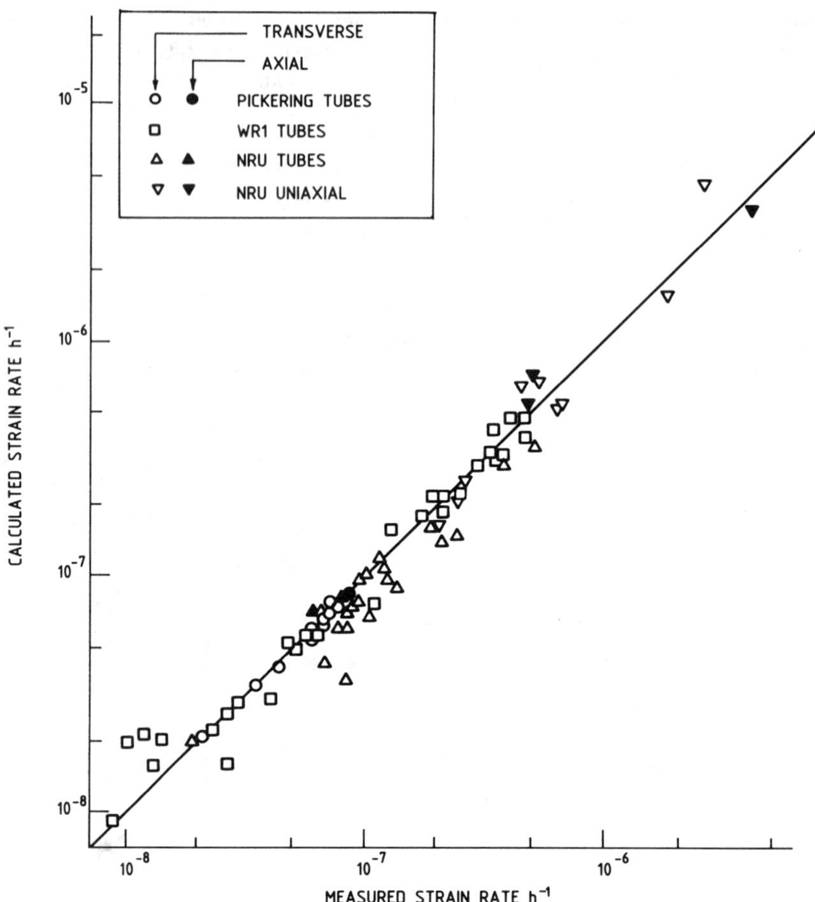

FIG. 9—*Comparison of measured strain rates with rates calculated using Eq 1, straight line is 1:1 agreement between rates.*

measurements. A similar plot for two tubes irradiated in the Bruce reactors (Fig. 10) shows that the equation can be extrapolated from the Pickering operating conditions.

A test of the multiaxial flexibility of Eq 1 is shown by the good agreement in Fig. 9 and in Table 4 of the uniaxial and biaxial data from experiments in NRU, which were not used to derive the equation. The equation is applicable over a range of temperatures from 520 to 670 K, fluxes up to about $4.0 \times 10^{17}\ \text{n} \cdot \text{m}^{-2} \cdot \text{s}^{-1}$, stresses up to 140 MPa and stress ratios from 0 to 2.0.

Discussion

Equation 1 represents a significant departure from previous depictions of the creep and growth behavior in that while the strain producing steps are still assumed to be slip, primarily on the prism planes with some degree of secondary slip on pyramidal systems, the use of Holt and Causey's [7] technique to account for the interactions between grains results in altered creep anisotropy parameters. There is, in effect, a hardening or softening relative to the behavior expected in a material with random texture. When Hill's anisotropy factors are calculated ignoring the grain

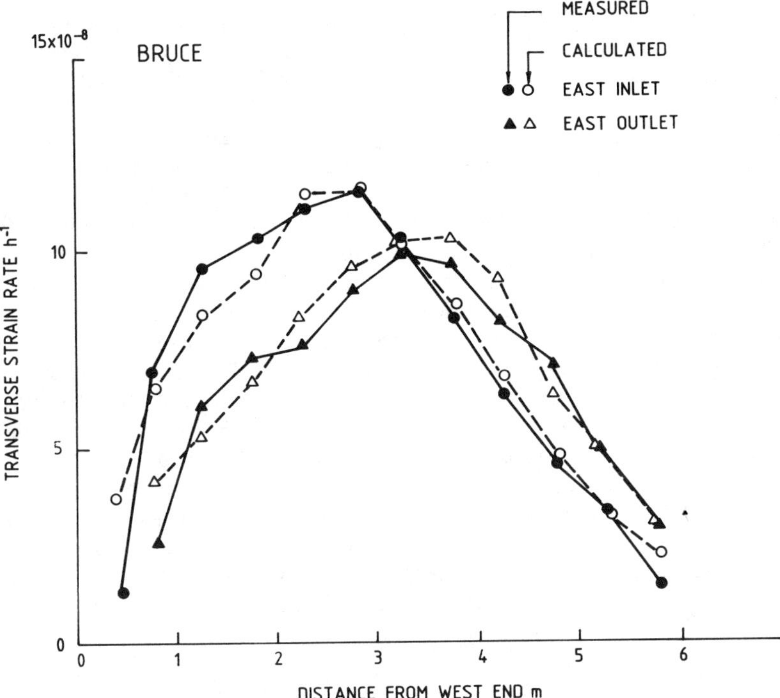

FIG. 10—*Comparison of measured transverse strain rates for pressure tubes in a Bruce reactor with rates calculated from Eq 1.*

interactions, as done originally by Holt and Ibrahim [*1*], and normalized to the random case, where $F=G=H=0.5$, the sum of F, G, and H equals 1.5, for all textures. The sum of F, G, and H (Table 4) does not equal 1.5 when they are calculated while accounting for the grain interactions, because the stress tensors in each grain are modified by the elastic strains necessary to produce the same total elastic plus plastic strain in each grain required for compatibility. When grain interactions are considered, specimens with randomly oriented grains have the minimum average strain rates. The nonrandom texture imparts strengthening in one or two directions and weakening in the other directions.

The model for the growth anisotropy in Eq 1 also represents a major change from that used in previous analyses for cold-worked Zr-2.5 wt% Nb tubes. Holt and Ibrahim [*1*] proposed a growth model based on net fluxes of interstitials to $\langle \mathbf{a} \rangle$ type edge dislocations and vacancies to grain boundaries via $\langle \mathbf{a} \rangle$ type screw dislocations, with the anisotropy described by the equation

$$G_d = (1 - f_d - 2A_d) \tag{6}$$

where A_d is the grain boundary anisotropy. The model was based on growth results for Zircaloy and stainless steel, both of which are almost equiaxed and have easily defineable grain sizes and shapes. The difficulty in defining an effective grain shape within the α-phase grains in Zr-2.5 wt% Nb probably accounts for the lack of agreement between Eq 6 and the recent growth data quoted above. At the time Eq 6 was proposed, electron microscopy observations had found only $\langle \mathbf{a} \rangle$ type dislocations, with no $\langle \mathbf{c} \rangle$ component dislocations, in neutron irradiated zirconium alloys [*19*]. However, recent studies show that dislocations with $\langle \mathbf{c} \rangle$ and $\langle \mathbf{c}+\mathbf{a} \rangle$ component Burger's vectors are present in the cold-worked materials before irradiation [*16,20*] and are generated during irradiation in recrystallized materials [*21*]. Thus, contraction along the **c** axis and expansion along the **a** axes, which requires the presence of ⟨**c**⟩ component dislocations and is approximated in a textured material by Eq 4, is now supported by electron microscopy, as well as by the growth results noted above.

The term with $A_3(x)$ ascribes the axial variation in transverse strain rates shown in Figs. 5 and 10 to thermal creep; however, the microstructural feature(s) that are responsible have not yet been identified. The increase in dislocation density, as measured by X-ray diffraction, from the front to the back end is usually less than 20%, and the change in calculated anisotropy factors is less than 5%, thus these differences do not account for the observed behavior.

There are important implications arising from the WR1 creep data and the creep and growth anisotropies of this new equation for the behavior of Zr-2.5 wt% Nb pressure tubes in CANDU-PHW reactors. The large negative growth component in the transverse direction belies a significant effect of the hoop stress on the transverse strain, as shown in Fig. 11 where the transverse strain rate is plotted as a function of inverse temperature for hoop stresses of 60 and 120 MPa. Under low stress conditions the strain rate is negative. With some slight increase in wall thickness and thus lower operating stress the problems associated with increasing tube diameter can be reduced. The transverse and axial strain rates, which would result in Zr-2.5 wt% Nb tubes with different textures and dislocation densities than those in the current Pickering reactors, are shown in Table 5. The rates for a cold-worked Zr-2.5 wt% Nb tube produced by a modified fabrication process [*16*] and for a beta heat-treated tube with very little cold work [*22*], illustrate the improvements possible from "texture-tailoring" and reducing the dislocation density.

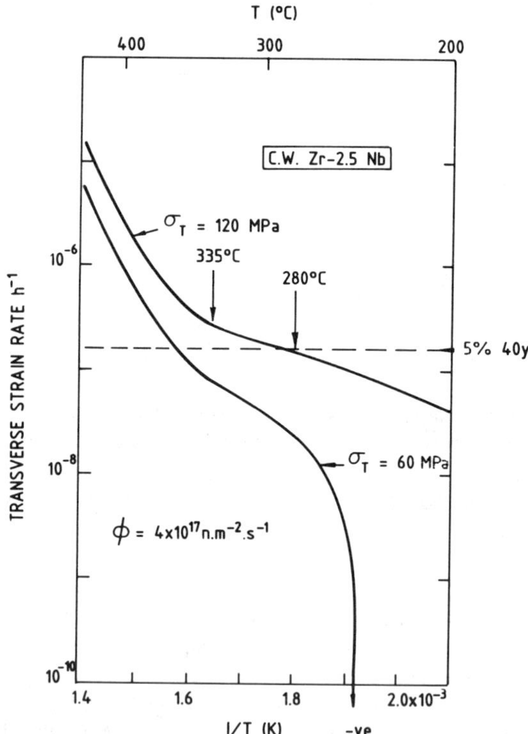

FIG. 11—*Temperature dependence of the transverse strain rate of Zr-2.5 wt% Nb pressure tubes with different hoop stresses calculated with Eq 1.*

TABLE 5—*Calculated transverse and axial strain rates for Zr-2.5 wt% Nb pressure tubes operating at 120-MPa hoop stress, peak flux of 3.6 × 10^{17} $n \cdot m^{-2} \cdot s^{-1}$, inlet temperature of 536 K and outlet temperature of 588 K.*

Pressure Tube Material	Strain Rates, h^{-1}	
	Maximum Transverse	Axial
Current cold-worked tube	1.5×10^{-7}	1.4×10^{-7}
Modified tube [*16*]	1.2×10^{-7}	0.8×10^{-7}
Heat-treated tube [*22*]	1.4×10^{-7}	0.1×10^{-7}

Summary

1. Changes in length and diameter of cold-worked Zr-2.5 wt% Nb pressure tubes during service have been measured in power and test reactors and have been correlated to the operating conditions and tube microstructure.

2. Expressions to describe anisotropic creep and irradiation growth strain have been derived using models for creep based on the slip of ⟨**a**⟩ type dislocations on prism planes and ⟨**c**+**a**⟩ type dislocations on pyramidal planes, and growth based on the $(1 - 3f_d)$ relation and the dislocation structure.

3. Grain interactions and the resulting residual stresses have been accounted for and result in creep anisotropy factors, which indicate higher creep rates in some directions in strongly textured materials, such as the pressure tubes, relative to a material with a random texture. The grain interactions also contribute to a reduction in the ratio of transverse to axial growth anisotropy compared with the ratio predicted by the $(1 - 3f_d)$ relation.

4. The multiaxial strain rate equations describe behavior in uniaxial tests as well as in the internally pressurized tubes from which they were derived.

References

[*1*] Holt, R. A. and Ibrahim, E. F., *Acta Metallurgica,* Vol. 27, 1979, p. 1319.
[*2*] Ibrahim, E. F. and Holt, R. A., *Journal of Nuclear Materials,* Vol. 91, 1980, p. 311.
[*3*] Holt, R. A., Causey, A. R., and Fidleris, V., *Proceedings of the British Nuclear Society,* London, 1983, p. 175.
[*4*] Nichols, F. A., *Journal of Nuclear Materials,* Vol. 30, 1969, p. 249.
[*5*] MacEwen, S. R. and Carpenter, G. J. C., *Journal of Nuclear Materials,* Vol. 90, 1980, p. 108.
[*6*] Causey, A. R., Holt, R. A., and MacEwen, S. R., *Zirconium in the Nuclear Industry: Sixth International Symposium, STP 824,* American Society for Testing and Materials, Philadelphia, 1984, pp. 269–288.
[*7*] Holt, R. A. and Causey, A. R., "Effects of Intergranular Constraints On Irradiation Growth of Zircaloy-2 at 320 K," to be published in *Journal of Nuclear Materials.*
[*8*] Cheadle, B. A., Coleman, C. E., and Licht, H., *Nuclear Technology,* Vol. 57, 1982, p. 413.
[*9*] Holt, R. A., *Journal of Nuclear Materials,* Vol. 59, 1976, p. 234.
[*10*] Causey, A. R., MacEwen, S. R., Jamieson, H. C., and Mitchell, A. B., *Nuclear Engineering Design,* Vol. 58, 1980, p. 367.
[*11*] Ross-Ross, P. A. and Hunt, C. E. L., *Journal of Nuclear Materials,* Vol. 26, 1968, p. 2.
[*12*] Fidleris, V., *Atomic Energy Review,* Vol. 13, 1975, p. 51.
[*13*] Hill, R., *The Mathematical Theory of Plasticity,* Oxford University Press, London, 1954.
[*14*] Adams, B. L., Clevinger, G. S., and Hirth, J. P., *Journal of Nuclear Materials,* Vol. 90, 1980, p. 75.
[*15*] Woo, C. H., and Gosele, U., *Journal of Nuclear Materials,* Vol. 119, 1983, p. 219.
[*16*] Parker, J. D., Perovic, V., Leger, M., and Fleck, R. G., *Zirconium in the Nuclear Industry: Sixth International Symposium, STP 939,* American Society for Testing and Materials, Philadelphia, 1987, pp. 86–100.
[*17*] Fidleris, V., to be published.

[*18*] Fidleris, V., Tucker, R. P., and Adamson, R. B., *Zirconium in the Nuclear Industry: Sixth International Symposium, STP 939,* American Society for Testing and Materials, Philadelphia, 1987, pp. 49–85.
[*19*] Northwood, D. O., *Atomic Energy Review,* Vol. 15, 1977, p. 547.
[*20*] Woo, O. T., Carpenter, G. J. C., and MacEwen, S. R., *Journal of Nuclear Materials,* Vol. 87, 1979, p. 70.
[*21*] Holt, R. A. and Gilbert, R. W., *Journal of Nuclear Materials,* Vol. 137, 1986, p. 185.
[*22.*] Fidleris, V., *Journal of Nuclear Materials,* Vol. 26, 1968, p. 51.

Charles H. Henager, Jr. [1] *and Edward P. Simonen*[1]

Light-Ion Irradiation Creep and Hardening of Model Ferritic Alloys

REFERENCE: Henager, C. H., Jr. and Simonen, E. P., "**Light-Ion Irradiation Creep and Hardening of Model Ferritic Alloys,**" *Influence of Radiation on Material Properties: 13th International Symposium (Part II), ASTM STP 956,* F. A. Garner, C. H. Henager, Jr., and N. Igata, Eds., American Society for Testing and Materials, Philadelphia, 1987, pp. 69–80.

ABSTRACT: Light-ion irradiation creep testing was combined with microhardness testing to study changes in creep strength of two model ferritic alloys. Creep enhancement followed by radiation hardening was observed for both alloys. Post-irradiation creep rates and post-irradiation microhardness data confirm that significant radiation hardening occurred due to the 673 K deuteron bombardment of 0.05 displacements per atom (dpa). Measured rapid creep rates were consistent with a climb-glide creep model developed for pure Ni after dislocation bias factors were reduced by a factor of five to account for the body centered cubic (bcc) structure. This was attributed to the relative insensitivity of climb glide creep to the interstitial bias, and to the fact that the small dislocation loops were unfaulted in the ferritic alloy but were initially faulted loops in the Ni. Electron microscopy revealed that a high density of small dislocation loops were formed during the irradiation. Microhardness testing results were consistent with irradiation hardening due to dislocation loops, as were the creep results.

KEY WORDS: irradiation creep, dislocation loops, ferritic alloys, microhardness, climb-glide creep, radiation, deuterons, dislocation bias, dislocation glide, dislocation climb, hardening

Ferritic alloys are widely studied for advanced reactor applications due to their apparent intrinsic radiation damage resistance compared to austenitic alloys [*1–3*]. Fundamentally, this has been attributed to a difference in the properties of self-interstitial atoms (SIAs) in the body centered cubic (bcc) lattice relative to that of the face centered cubic (fcc) lattice. A lesser bias for SIA capture by extended defect sinks occurs for bcc alloys compared to fcc alloys [*4*]. Void swelling and in-reactor creep rates for ferritic alloys at high doses are reduced compared to 316 stainless steel [*4*]. These results are consistent with the lesser SIA bias for bcc alloys. Yet, other deformation processes, such as climb-glide creep, may exist that are insensitive to the SIA bias and the ferritic alloys may deform as readily as fcc alloys during these processes.

A climb-glide creep model was previously developed to account for light-ion irradiation creep measured in pure nickel for low doses (< 0.1 displacements per atom (dpa)) at 473 K [*5,6*]. The model expressed the creep rate dependence on the ratio of the network dislocation climb rate to the interstitial dislocation loop growth rate. Because both the climb rate and the loop growth rate depend linearly on the SIA bias, the model predicted a general insensitivity to the SIA bias. This is in contrast to the stress-induced preferred absorption (SIPA) creep model, which predicts that creep rates are linearly proportional to the SIA bias [*6,7,8*].

Light-ion irradiation creep testing of model ferritic alloys was performed to study the low-dose creep mechanism of these alloys in comparison to pure nickel and other fcc alloys. Measured

[1] Senior research scientists, Pacific Northwest Laboratory, P.O. Box 999, Richland, WA 99352.

creep rates were consistent with the climb-glide model developed earlier, and the measured creep rates were as rapid for the ferritic alloys as measured for nickel, suggesting that the deformation mechanism is insensitive to the dislocation bias at low doses. Transmission electron microscopy (TEM) results show that a high density of small dislocation loops is produced by the irradiation and the creep hardening and microhardness tests are consistent with loop hardening.

Experimental Techniques and Results

Techniques

The two ferritic alloy compositions are given in Table 1. These alloys were chosen to study the effects of copper and nickel additions on irradiation hardening kinetics [9]. As discussed in the previous work [9], no copper precipitation during irradiation was observed, and the two alloys behaved similarly during irradiation creep. The alloys were vacuum arc-melted at Metal Samples Company and were received as hot-rolled sheets 0.1-cm thick. Strips approximately 2 cm by 10 cm were sheared from the plates, surface cleaned by grit blasting, and cold-rolled to 175 μm thick. Rectangular sections, 5 cm by 2 cm, were cut from the rolled strips and polished on 2/0 grit SiC paper. These sections were then machined into 5-cm-long creep specimens having gage dimensions of 6.35 mm by 3.18 mm with a thickness of 150 μm.

The machined creep specimens were heat treated in vacuum for 2 h at 973 K at better than 10^{-4} Pa followed by a fast furnace cool sufficient to retain the solute in solution. The heat treatment below the austenite phase field recrystallized the heavily cold-worked alloys. The heat-treated creep specimens were electropolished in a 5% perchloric acid, 50%:50% methanol-butanol solution held at 218 K using 40 to 70 V. TEM disks (0.3 cm diameter by 50 μm or 175 μm thick) were simultaneously heat treated and reserved for TEM examination and microhardness testing. Optical metallography was performed to determine grain size.

Irradiation creep testing was performed using the Pacific Northwest Laboratory light-ion irradiation creep apparatus, with the basic technique unchanged from previous experiments [5]. Briefly, the creep apparatus consists of a vacuum chamber in which the creep specimen can be stressed in tension, heated by Joule heating with feedback control, and irradiated with high-energy particles from a Van de Graaf accelerator. The specimen strain is measured simultaneously with the deuteron irradiation using a noncontacting laser extensometer. Irradiations were performed at 673 K using 15 MeV deuterons (14 MeV at the specimen front surface) and 35–45 MPa tensile stresses. Prior to irradiation, the specimens were thermally crept for about 12 h. The specimens were irradiated for 24 h at a displacement rate of 6×10^{-7} dpa/s, followed by 12 h of post-irradiation thermal creep. Thermal creep rates were measured to establish the thermal creep contribution to the measured irradiation creep.

Tensile testing prior to irradiation was performed on an Instron 1125 at room temperature and at 673 K in air utilizing the same specimen geometry as the creep tests. A crosshead speed of 0.01 cm/min was used for the tensile tests. Microhardness tests were performed on a Page-Wilson 300 hardness tester using a Vickers indenter and 100-g or 50-g loads. Specimen surfaces for microhardness tests were prepared by metallographic polishing with 1-μm diamond, followed by a slight etch in 3% nital and a careful repolish with 1-μm diamond. Irradiated microhardness

TABLE 1—*Composition of ferritic alloys (% by weight)*

Alloy	Cu	Ni	C	O	N	P	S	Si	Mn	Cr	Al
F5	0.29	1.08	0.032	0.033	0.001	0.005	0.002	0.02	<0.01	<0.01	0.03
F7	0.01	1.08	0.054	0.013	0.001	0.006	0.002	0.03	<0.01	<0.01	0.01

specimens were prepared by punching two TEM disks from each gage section and prepared as above. Unirradiated microhardness specimens were punched from the grip ends of the creep samples, as well as obtained from the stock of heat-treated disks mentioned previously. Thicknesses of all the microhardness tested TEM disks were enough to ensure that the Vickers indent depth was less than one-tenth of the specimen thickness. Hardness standards were obtained for the range of Vickers hardness observed in testing the original heat-treated disks.

Pre-Irradiation Characterization

Mechanical property and microstructural information was obtained for each alloy in the heat-treated condition (Table 2) for comparison with post-irradiation properties. The Vickers hardness data agrees well with the room-temperature tensile data. TEM and optical microscopy reveal a low dislocation density, fine-grained material with a dispersion of fine epsilon carbide (20-nm radius). The fine grain size ensured 8 to 15 grains through the specimen thickness, sufficient to approximate bulk mechanical behavior. Comparison of room-temperature and elevated-temperature yield strengths reveals differences attributable to the amount of carbon in solution and to the epsilon carbide density (Table 2). Observed differences in grain size, carbide density, and carbon content can be expected to influence measured strength.

Irradiation Testing

Creep enhancement and radiation hardening were observed in each of the irradiated alloys (Table 3). Five creep specimens, two F5 and three F7 alloy specimens, were irradiated from 35 to 45 MPa at 673 K using 15-MeV deuterons. An average ion current of 9.3 $\mu A/cm^2$ was achieved to give a damage rate of 5.8×10^{-7} dpa/s, resulting in 0.05 dpa for a 24-h irradiation. This displacement rate was calculated using the EDEP code [*10*] with a displacement threshold energy of 40 eV.

An increased creep rate at the onset of irradiation was typically observed, followed by a steadily decreasing creep rate (Fig. 1). A primary creep rate, $\dot{\varepsilon}_{pr}$, was measured during the first 2 to 3 h of the test, while the irradiation creep rate, $\dot{\varepsilon}_{irr}$, was determined by a least squares fit to the creep curve during the last 6 to 10 h of the test (Fig. 2). Calculated uncertainties in this data suggest that the error on the reported creep rates is $\pm 35\%$ [*5*]. Irradiation creep enhancement is apparent when comparing the measured final irradiation creep rate to the post-irradiation thermal creep rate, $\dot{\varepsilon}_{p\text{-}th}$.

Irradiation hardening was observed in the strain rate data by comparing initial irradiation creep rates to final irradiation creep rates, $\dot{\varepsilon}_{pr}$ versus $\dot{\varepsilon}_{irr}$. Hardening is also evident when comparing the thermal creep rates before ($\dot{\varepsilon}_{th}$) and after ($\dot{\varepsilon}_{p\text{-}th}$) irradiation. These comparisons reveal significant hardening in each of the alloys, with Specimen F7-5 showing an unusually large creep rate reduction.

TABLE 2—*Pre-irradiation characterization of ferritic alloys.*

Property	Value: F5	F7
Composition	Fe-1Ni-0.3Cu	Fe-1Ni
Grain size, μm	20	10
RT σ_y, MPa	204	244
673K σ_y, MPa	106	97
RT HV, HV	90	105
ρ_d, 10^{12} m^{-2}	1	1
ρ_{ppt}, 10^{19} m^{-3}	4	2
Carbon content, % by weight	0.032	0.054

TABLE 3—*Irradiation creep data summary.*

Specimen	σ, MPa	K, dpa/s	$\dot{\varepsilon}_{th}$ s^{-1}[a]	$\dot{\varepsilon}_{pr}$, s^{-1}[b]	$\dot{\varepsilon}_{irr}$, s^{-1}[c]	$\dot{\varepsilon}_{p\text{-}th}$, s^{-1}[d]	$\dot{\varepsilon}_{irr}/\dot{\varepsilon}_{pr}$	$\dot{\varepsilon}_{p\text{-}th}/\dot{\varepsilon}_{th}$
F5-5	45	5.8×10^{-7}	2.1×10^{-8}	5.8×10^{-8}	2.8×10^{-8}	6.8×10^{-9}	0.48	0.32
F5-6	35	5.8×10^{-7}	2.3×10^{-8}	3.5×10^{-8}	1.3×10^{-8}	5.3×10^{-9}	0.37	0.23
F7-1	35	5.8×10^{-7}	2.1×10^{-7}	1.2×10^{-7}	6.4×10^{-8}	4.4×10^{-8}	0.53	0.21
F7-5	35	5.8×10^{-7}	1.0×10^{-7}	1.8×10^{-7}	1.5×10^{-8}	5.7×10^{-9}	0.08	0.06
F7-11	40	5.2×10^{-7}	7.2×10^{-8}	2.6×10^{-8}	1.9×10^{-8}	1.4×10^{-8}	0.73	0.19

[a] ε_{th}—Thermal creep rate.
[b] ε_{pr}—Primary irradiation creep rate.
[c] ε_{irr}—Final irradiation creep rate.
[d] $\varepsilon_{p\text{-}th}$—Post-irradiation thermal creep rate.

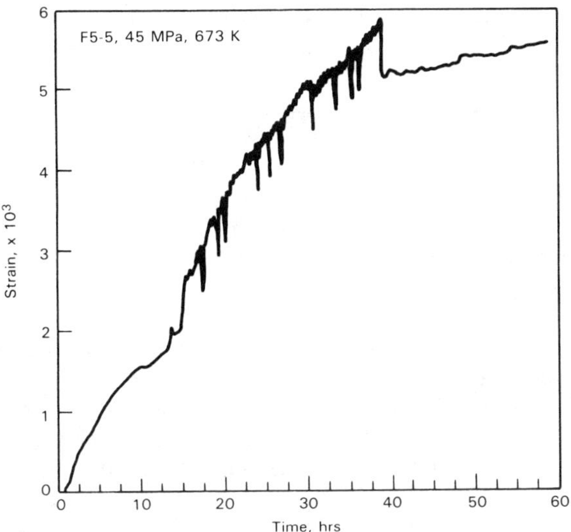

FIG. 1—*Light-ion irradiation creep curve for specimen F5-5 for an applied stress of 45 MPa at 673 K.*

Post-Irradiation Testing and Examinations

Post-irradiation microhardness testing and TEM examinations were performed on selected specimens. Qualitative agreement was observed between the measured microhardness increase, ΔHV, and the hardening inferred from the creep testing (Table 4). The microhardness increase was used to calculate a yield strength increase due to irradiation (see discussion). TEM specimens were punched from the irradiated gage sections and examined in a Philips EM400T at 120 kV.

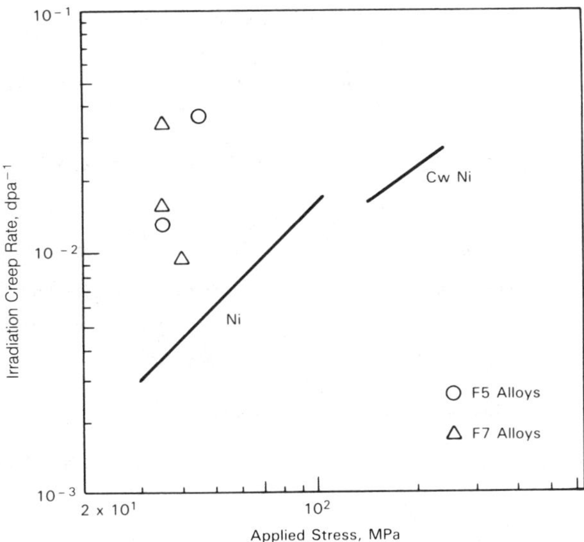

FIG. 2—*Irradiation creep rate, $\dot{\varepsilon}_{irr}$, as a function of the applied stress. Data for annealed and cold-worked nickel is included for comparison.*

TABLE 4—*Irradiation hardening comparison.*

Specimen	ΔHV	$\Delta\sigma_y$,[a] MPa	$\varepsilon_{p\text{-}th}/\varepsilon_{th}$ (Creep Ratio)
F5-5	21	57	0.32
F5-6	19	51	0.23
F7-1	. . .	. . .	0.21
F7-5	65	175	0.06
F7-11	9	24	0.19

[a] $\Delta\sigma_y = 2.7\ \Delta HV$

Dislocation densities, loop densities, and loop radii were measured for the two alloys (Table 5, Figs. 3–4). Microstructural differences were observed between the F5 and F7 alloys (Table 5), and specimen F7-5 contained a fine carbide dispersion not observed in other irradiated specimens. The TEM micrographs reveal the small dislocation loops that formed during the irradiation, network segments bowed between loops (Fig. 3), and carbides decorating network dislocations for Specimen F7-5 (Fig. 4).

Analysis and Discussion

Calculated loop growth rates are consistent with observed loop growth rates when a low SIA dislocation bias appropriate for bcc alloys was assumed. The loop growth calculations and low bias were used in a cimb-glide creep model that had been developed for irradiation creep of nickel. Application of that model was to establish whether irradiation creep of ferritics also occurs by a climb-glide mechanism. Calculated yield strength increases based on dislocation loop hardening compared well with measured hardness increases.

Theory

A system of rate equations was solved to calculate the growth of dislocation loops in ferritic alloys. Loop growth was calculated from numerical solutions of the following three equations:

$$\frac{dC_v}{dt} = K_v - \alpha\, C_iC_v - D_vC_v[Z_v\rho_d + 4\pi r_{c,v}\rho_l] + D_v[C^e_{v,d}Z_v\rho_d + C^e_{v,l}4\pi r_{c,v}\rho_l] \tag{1}$$

$$\frac{dC_i}{dt} = K_i - \alpha C_iC_v - D_iC_i[Z^N_i\rho^N_d + Z^A_i\rho^A_d + 4\pi r_{c,i}\rho_l] \tag{2}$$

$$\frac{dr_l}{dt} = \frac{2}{r_lb}[D_iC_ir_{c,i} - D_v(C_v - C^e_{v,l})r_{c,v}] \tag{3}$$

The subscripts refer to interstitial (i), vacancy (v), dislocation (d), and dislocation loop (l) parameters. The $r_{c,v}$ and $r_{c,i}$ refer to the capture radii of interstitial loops for vacancies and

TABLE 5—*TEM data summary.*

Specimen	ρ_d, m^{-2}	ρ_l, m^{-3}	r_l, nm
F5 Alloys	4×10^{13}	7.5×10^{20}	10.2
F7 Alloys	2×10^{13}	2.0×10^{20}	19.0
F7-5		$\rho_{ppt} = 1 \times 10^{20}$ m^{-3}, $r_{ppt} = 30$ nm carbides decorating dislocations, not observed in other F7 alloys.	

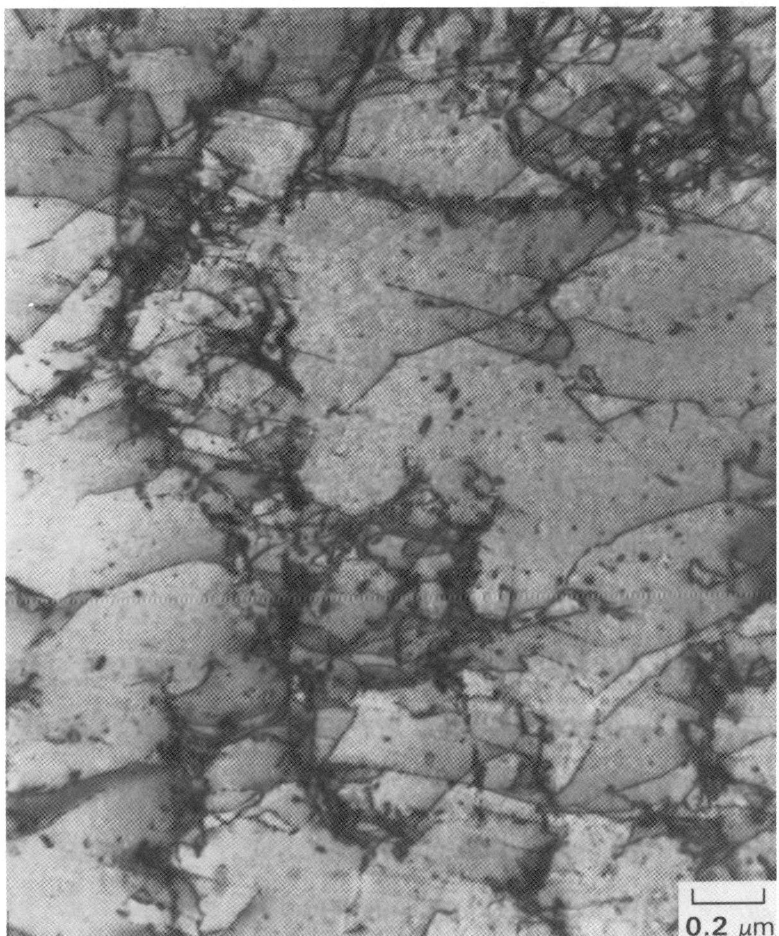

FIG. 3—*Electron micrograph showing microstructure of specimen F5-5 following irradiation to 0.05 dpa at 673 K. Note dislocation segments bowing between small dislocation loops.*

interstitials, respectively. The term, C_v^e, refers to the concentration of vacancies in equilibrium with dislocations ($C_{v,d}^e$) or loops ($C_{v,l}^e$). The defect production rates, K, biases, Z, and diffusivities, D, are defined in Table 6.

A parametric study of the SIA bias was performed for this work, and the stress-free interstitial bias was varied from 0.15 to 0.03 (from 15% to 3%). The dislocation loop capture radii were computed based on the loop radius and were scaled to approach the stress-free interstitial bias at large loop radii [*6*]. Loop growth was calculated for $K = 6 \times 10^{-7}$ dpa/s out to 0.05 dpa using Eqs 1–3 (Fig. 5). Good results were obtained for $\Delta Z_i^o = 0.03$ for the ferritic alloys, compared to $\Delta Z_i^o = 0.15$, which gave good results for loop growth in pure Ni [*5*].

Sniegowski and Wolfer [*4*] have discussed the bias difference between fcc and bcc metals based on point defect relaxation volumes. The net dislocation bias for SIA capture in bcc metals is generally less than for fcc metals. Calculations based on current values of relaxation volumes for Ni and α-Fe show that the SIA bias of α-Fe is about one-fifth that of pure Ni [*4*]. The parametric

FIG. 4—*Electron micrograph showing microstructure of specimen F7-5 following irradiation to 0.05 dpa at 673 K. Note the extensive carbide decoration of dislocation segments.*

TABLE 6—*Parameter values for ferritic alloys.*

Parameter	Symbol	Value
Shear modulus	G	8.6×10^{10} Pa
Burgers vector	**b**	2.48×10^{-10} m
Recombination constant	α	$1.2 \times 10^{21} (D_i + D_v)$ s^{-1}
Equilibrium vacancy concentration	C_v^e	$\exp(1.5)\exp(-1.7/kT)$
Vacancy diffusivity	D_v	$5 \times 10^{-5} \exp(-1.3/kT)$ m^2/s
Interstitial diffusivity	D_i	$1 \times 10^{-4} \exp(-0.3/k/T)$ m^2/s
Dislocation bias factors:		
Vacancy	Z_v	$2\pi/ln(\lambda_d/2b)$, $\lambda_d = (\pi\rho_d)^{-1/2}$
Interstitial	Z_i	$Z_v(1 + \Delta Z_i)$
Stress-free interstitial bias	ΔZ_i^0	0.03
Defect production rates	$K_i = K_v$	6×10^{-7} dpa/s

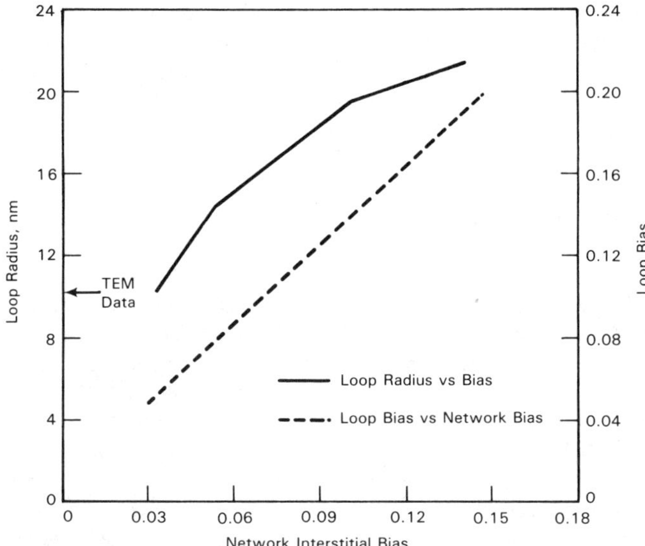

FIG. 5—*Parametric plot of predicted dislocation loop radius as a function of the SIA dislocation bias. Also shown is the SIA loop bias as a function of the network bias and the measured average loop radius. The measured radius is indicated by the arrow.*

study presented here demonstrates that this bias reduction of one-fifth (3% versus 15%) is necessary to achieve good agreement between loop growth modeling and experiment for these ferritic alloys compared to pure Ni.

Creep Modeling

Creep modeling was performed by calculating loop growth rates and dislocation climb velocities in the context of a climb-glide model developed previously [*6*]. The network dislocation climb velocity was calculated from

$$V_c = \frac{1}{b}\left(\frac{1}{3}\left| D_i C_i Z_i^A - D_v(C_v - C_{v,d}^{e,A})Z_v \right| + \frac{2}{3}\left| D_i C_i Z_i^N - D_v(C_v - C_v^e)Z_v \right|\right) \tag{4}$$

where $C_{v,d}^{e,A}$ is given by $C_v^e \cdot \exp(\sigma\Omega/kT)$. The first term within the absolute value brackets of Eq 4 represents the climb rate of network dislocations aligned with the applied stress, while the second term represents the climb rate of nonaligned network dislocations.

The climb-glide creep rate was calculated from the average glide distance between glide barriers and the average time the glide dislocation spends at the barrier before gliding to the next barrier. The waiting time at the barrier depends on the dislocation climb velocity and the characteristics of the barrier [*6*]. The climb-glide creep rate was calculated from:

$$\dot{\varepsilon}_{CG} = \beta b \rho_d^m (\lambda/\theta) \tag{5}$$

where β is the Taylor factor ($\beta = 1/3$), b is the Burgers vector, ρ_d^m is the mobile dislocation density ($\rho_d^m = \rho_d$), λ is the dislocation glide distance, and θ is the loop reaction wait time. The glide distance, λ, is calculated based on the average spacing of mono-sized spherical particles, while the loop reaction wait time is given by

$$\theta = r_l/(V_c + \beta \frac{dr_l}{dt}) \tag{6}$$

Interstitial loop growth and creep strain as functions of fluence were modeled using a simple algorithm. The system of rate equations was solved by assuming steady-state conditions for point defect concentrations, $dC_i/dt = dC_v/dt = 0$. Sink densities were assumed to be independent of fluence, and an initial interstitial loop radius equal to twice the Burgers vector was chosen. Loop growth rates were calculated from Eq 3, and a new loop radius was calculated from:

$$r_l(t_i) = r_l(t_{i-1}) + \dot{r}_l(t_i)\Delta t \tag{7}$$

Creep strains were calculated from

$$\varepsilon(t_i) = \varepsilon(t_{i-1}) + \dot{\varepsilon}(t_i)\Delta t \tag{8}$$

The new loop radius was used to recalculate the loop sink strength, and the system of equations was solved again. One thousand iterations were taken to simulate the irradiation to 0.05 dpa.

The predicted creep strain and loop radii are shown in Fig. 6 and compared to the strain-time data of Specimen F5-6. The model underpredicts the creep strain, and the calculated strain rate of 1.5×10^{-3} dpa^{-1} is a factor of ten less than the measured value of 2.2×10^{-2} dpa^{-1}. The loop radius prediction of 11 nm agrees well with the observed value of 10.2 nm as discussed previously.

The model strain predictions are consistent with the calculated dislocation loop growth since the creep model used the loop growth calculations as input and the loops are the dominant glide obstacles in the irradiated alloys. As the loops continue to grow, the material becomes harder, and the creep strength is predicted to increase with increasing loop radius. The observed strain-time behavior of Specimen F5-6 reflects initial hardening consistent with the model, but the observed rate of hardening is less than that predicted using the calculated loop growth rate. The creep model assumes an average, geometric glide distance corresponding to a random distribution of dislocation loops. The observed creep behavior is consistent with a gradual increase in the average dislocation glide distance λ as deformation proceeds. Various mechanisms have been discussed [*11*], including loop annihilation and loop sweeping, to account for barrier removal from

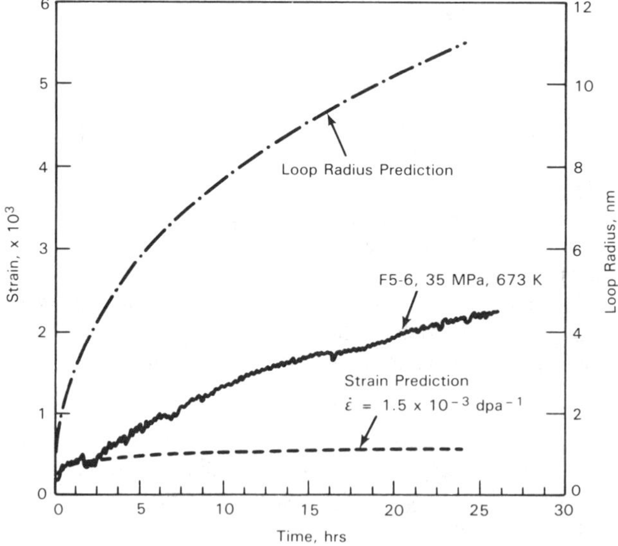

FIG. 6—*Model prediction of irradiation creep strain and dislocation loop radius as a function of irradiation time. Also shown is the irradiation creep data from specimen F5-6 for an applied stress of 35 MPa at 673 K.*

active slip planes that would lead to an increase in λ. The unfaulted loops in the bcc alloys make the sweeping and clearing processes easier compared to the faulted loops in the fcc alloys, and the bcc alloys would be expected to have less creep strength than fcc alloys. These effects are not easily quantified for the present case. The glide distance would have to increase by a factor of 10 to account for the discrepancy between measurement and theory.

Creep rate predictions from the climb-glide model are essentially independent of the SIA bias. The model predicts creep rates only 11% faster using a 15% bias as compared to the 3% bias. For comparison, predicted loop radii are twice as large for the 15% bias. However, network dislocation climb velocities are about three times faster for the larger bias value. Since the climb-glide creep rate is proportional to the ratio of the climb velocity to the loop radius, the bias has relatively little effect on the creep rate. This is in contrast to SIPA creep, which is linearly proportional to the SIA bias [*6,7,8*].

Hardening Calculations

Dislocation loop hardening was calculated using an Orowan hardening model to compare with measured yield strength increases [*9*]. Increases in yield stress, $\Delta\sigma_y$, were calculated using:

$$\Delta\sigma_y = 0.5 \text{ Gb } (2r_l\rho_l)^{1/2} \tag{9}$$

Changes in Vickers microhardness were related to changes in yield strength using:

$$\Delta\sigma_y = 2.7\ \Delta HV \tag{10}$$

where the proportionality constant was determined based on the measured Vickers hardness and tensile data of the two alloys prior to irradiation (Table 2).

The change in Vickers microhardness due to irradiation creep exposure is listed in Table 4. Yield strength increases were calculated using Eq 10 and compared to the creep rate ratio of $\dot{\varepsilon}_{p\text{-}th}$ to ε_{th}. Moderate hardening is observed for all alloys except F7-5, which exhibits a large strength increase based on the microhardness measurements. These measured strength increases are qualitatively consistent with the hardening observed during irradiation creep as evidenced by the creep rate ratio [*9*]. Specimen F7-5, which exhibited the largest increase in microhardness, also exhibited the largest amount of hardening during irradiation creep.

Yield strength increases due to dislocation loops calculated using Eq 9 are listed in Table 7 and are consistent with observations. The additional hardening due to the difference in network dislocation density between F5 alloys and F7 alloys was calculated using a forest hardening law. The computed total hardening qualitatively agrees with the measured values with the exception of Specimen F7-5. The Orowan hardening due to the small carbides is not sufficient to explain the observed hardness values. This suggests that dislocation pinning has occurred and that large stresses are necessary to cause dislocations to be mobile in this possibly anomalous specimen.

TABLE 7—*Calculated irradiation hardening.*

	$\Delta\sigma_y$, MPa			
Specimen	Loops	Network Difference	Carbides	Total
F5 Alloys	42	11	. . .	53
F7 Alloys	30	0	. . .	30
F7-5, Carbides	30	0	25	55

Summary and Conclusions

Model ferritic alloys were irradiated to 0.05 dpa with 15 MeV deuterons at 673 K in an irradiation creep apparatus. Irradiation creep rates were slightly faster than rates measured for pure nickel under similar conditions. Creep rates were consistent with a climb-glide creep model formulated for fcc metals. Dislocation loop growth was consistent with a reduced SIA dislocation bias for bcc alloys, one-fifth of that for fcc alloys. The similarity between the measured creep rates for the ferritic alloys and pure nickel requires that the creep mechanism be insensitive to the SIA bias, which is true of the proposed climb-glide model.

Yield strength increases due to irradiation were calculated from measured microhardness increases and were consistent with dislocation loop hardening. Hardening observed during irradiation creep testing was also consistent with dislocation loop hardening provided that the glide distance between loop obstacles was about ten times larger than the measured loop spacing. This is consistent with loop sweeping or clearing mechanisms operating in the ferritic alloys as was suggested for fcc metals.

Acknowledgments

The authors would like to thank L. A. Charlot for performing the TEM. A special thanks is given to the staff of the Nuclear Physics Laboratory, University of Washington, for assisting with the irradiation experiments. This work was supported by the Office of Basic Energy Sciences, Division of Materials Sciences, U.S. Department of Energy under Contract DE-AC06-76RLO 1830.

References

[*1*] Harries, D. R., in *Proceedings,* Topical Conference on Ferritic Alloys for Use in Nuclear Energy Technologies, J. W. Davis and D. J. Michel, Eds., TMS-AIME, 1984, pp. 141–155.

[*2*] Gelles, D. S. and Puigh, R. J., in *Effects of Radiation on Materials: Twelfth International Symposium, ASTM STP 870,* F. A. Garner and J. S. Perrin, Eds., American Society for Testing and Materials, Philadelphia, 1985, pp. 19–37.

[*3*] Klueh, R. L. and Vitek, J. M., *Journal of Nuclear Materials,* Vol. 126, 1984, pp. 9–17.

[*4*] Sniegowski, J. J. and Wolfer, W. G., in *Proceedings,* Topical Conference on Ferritic Alloys for Use in Nuclear Energy Technologies, J. W. Davis and D. J. Michel, Eds., TMS-AIME, 1984, pp. 579–586.

[*5*] Henager, C. H., Jr., PhD Thesis, University of Washington, Seattle, WA, 1983.

[*6*] Henager, C. H., Jr. and Simonen, E. P., in *Effects of Radiation on Materials: Twelfth International Symposium, ASTM STP 870,* F. A. Garner and J. S. Perrin, Eds., American Society for Testing and Materials, Philadelphia, 1985, pp. 75–97.

[*7*] Bullough, R. and Haynes, M. R., *Journal of Nuclear Materials,* Vol. 57, 1975, pp. 348–352.

[*8*] Wolfer, W. G., *Journal of Nuclear Materials,* Vol. 90, 1980, pp. 175–192.

[*9*] Henager, C. H., Jr. and Simonen, E. P., in *Proceedings,* Second International Symposium on Environmental Degradation of Materials in Nuclear Power Systems—Water Reactors, ANS, 1986, pp. 377–384.

[*10*] Manning, I. and Mueller, G. P., *Computer Physics Communication,* Vol. 7, 1974, p. 85.

[*11*] Henager, C. H., Jr. and Simonen, E. P., *Journal of Nuclear Materials,* Vols. 122 and 123, 1984, pp. 413–417.

Changes in Mechanical Properties of Alloys

Wan-liang Hu[1] *and David S. Gelles*[1]

The Ductile-to-Brittle Transition Behavior of Martensitic Steels Neutron Irradiated to 26 dpa

REFERENCE: Hu, W.-L. and Gelles, D. S., "**The Ductile-to-Brittle Transition Behavior of Martensitic Steels Neutron Irradiated to 26 dpa,**" *Influence of Radiation on Material Properties: 13th International Symposium (Part II), ASTM STP 956,* F. A. Garner, C. H. Henager, Jr., and N. Igata, Eds., American Society for Testing and Materials, Philadelphia, 1987, pp. 83–97.

ABSTRACT: Charpy impact tests were conducted on specimens made of HT-9 and 9Cr-1Mo in various heat treatment conditions that were irradiated in EBR-II to 26 dpa at 390 to 500°C. The results are compared with previous results on specimens irradiated to 13 dpa. HT-9 base metal irradiated at low temperatures showed a small additional increase in ductile brittle transition temperature and a decrease in upper shelf energy from 13 to 26 dpa. No fluence effect was observed in 9Cr 1Mo base metal. The 9Cr-1Mo weldment showed degraded ductile-to-brittle transition temperature (DBTT) but improved upper shelf energy (USE) response compared to base metal, contrary to previous findings on HT-9. Significant differences were observed in HT-9 base metal between mill annealed material and normalized and tempered material. The highest DBTT for HT-9 alloys was 50°C higher than for the worst case in 9Cr-1Mo alloys. Fractography and hardness measurements were also obtained. Significant differences in fracture appearance were observed in different product forms, although no dependence on fluence was observed. Failure was controlled by the preirradiation microstructure.

KEY WORDS: steels, impact tests, alloys, fractography, irradiation

Martensitic steels have been considered as a prime candidate for the first wall of a fusion reactor. However, these steels are susceptible to brittle failure at low service temperatures. It is well known that the temperature at which steels undergo a transition in fracture mode from ductile to brittle shifts toward higher temperatures with increasing neutron exposure. Accordingly, the change in ductile-to-brittle transition temperature (DBTT) as a result of irradiation must be evaluated before the selection and application of this material. Furthermore, the transition behavior of fracture for weldments should also be studied to assure structural integrity. Previous work evaluated the irradiation induced shift in DBTT and the change in the upper shelf energy (USE) of ferritic alloys and weldments at an exposure of 13 dpa. This paper considers the fracture behavior of ferritic alloys and weldments with additional neutron exposure to a fluence of 26 dpa. Fractographic examinations on selected broken specimens were conducted to determine the microstructural effects of further irradiation.

Specimen Preparation

Miniature Charpy impact specimens were prepared to study the effect of irradiation on the transition of fracture mode in steels. Specimen geometry was 5 by 5 by 23.6 mm with a notch

[1] Senior engineer and principal engineer, respectively, Westinghouse Hanford Co., P.O. Box 1970, Richland, WA 99352.

depth of 0.76 mm. Specimens were precracked to an a/W approximately equal to 0.5, where a is the notch depth plus the precrack length and W = 5 mm, the specimen height.

Material stocks involved in this study include the following:

1. HT-9 base metal from heat 91354 in a mill annealed condition (TT series). The bar stock was hot worked after soaking at 1149°C for a minimum of 1 h, slow cooled but allowed to transform to martensite, then tempered at 750°C for 1 h and air cooled.
2. HT-9 base metal from heat 91354 (KT series), with a heat treatment of 1038°C/10 min/AC + 760°C/30 min/AC.
3. 9Cr-1Mo base metal from heat 30182 (TV series). The heat treatment was 1038°C/1 h/AC + 760°C/1 h/AC.
4. 9Cr-1Mo base metal from heat 30176 (AF series). The heat treatment is almost identical to that of the TV series, that is, 1038°C/30 min/AC + 760°C/30 min/AC.
5. 9Cr-1Mo weldment from heat 30182 (NL series), with a post-weld treatment of 1 h at 780°C followed by an air cool.

Specimens of the TT series were fabricated from bar stock in the C-R orientation. Specimens of the KT, TV, and AF series were fabricated from plate stock with the crack plane oriented in the T-L direction. In the case of the 9Cr-1Mo weldment (NL series), the orientation was assumed to be irrelevant.

Specimens of both the TT and the TV series were irradiated in EBR-II to a peak fluence of approximately 6×10^{22} n/cm^2 (~26 dpa). The rest of the specimens, the KT, AF and NL series, were inserted in EBR-II in the unirradiated condition as part of the reconstitution test matrix and hence experienced a peak fluence of only 3×10^{22} n/cm^2 (~13 dpa). The irradiation temperatures ranged from 390 to 550°C. Details of the alloy compositions, specimen preparation, heat treatments, test matrix, and capsule loading were reported previously [*1*,*2*].

Charpy impact test results on the specimens in the TT and TV series were documented previously at a fluence of 3×10^{22} n/cm^2 (~13 dpa) [*3*,*4*]. Those test results are incorporated in this study for comparison.

Experimental Procedures

The impact tests were conducted using a drop tower installed in a hot cell. The specimen heating or cooling or both, transport, and testing were automated to handle the irradiated specimen remotely. Load traces were digitized and stored on both magnetic tape and disk. A computer program was developed to integrate the load data and calculate the fracture energy. The details of the test system were described in Ref *4*.

The specimen temperature, as indicated by a spring loaded contacting thermocouple, was calibrated before the test period versus a thermocouple spot welded on a dummy specimen. The dynamic response of the load cell and the impact velocity were checked daily during the test period. The load cell calibation was conducted dynamically by comparing the maximum load obtained during an impact test on a calibrated half size Charpy specimen with the predetermined maximum load from a three-point bending test. The calibration specimens were made of 6061 aluminum in the T651 heat treated condition, which are nominally insensitive to loading rate. The impact velocity was calculated from the time interval required for a flag on the crosshead with a known gap width to pass through a stationary infrared sensor during free fall.

The total energy absorbed in the impact test was calculated following Newton's second law from the area under the load-time record and the initial impact velocity

$$E = E_a (1 - E_a/4E_o)$$

where $E_a = V_{o/}\int P\,dt$ is the apparent energy absorbed by the specimen during impact, V_o is the initial impact velocity, P is the load, and t is time. $E_0 = \frac{1}{2}mV_o^2$ is the total available energy at impact, where m is the mass of the crosshead.

Several specimens were selected for fractographic examination to determine the effect of irradiation on microstructure. In general, specimens were selected that had been tested at approximately the transition temperature so that both ductile and brittle fracture surface features could be studied. One specimen was also selected for examination from both lots of HT-9 base metal and from 9Cr-1Mo at a higher test temperature to determine upper shelf behavior. Hardness measurements on irradiated specimens were made on a calibrated Wilson Rockwell hardness tester with a diamond braille indenter and 150-kg load, which had been installed in a hot cell and was operated remotely.

Results

Charpy Impact Tests

The calculated impact energies for each material are plotted as a function of temperature in Figs. 1 to 5. Fractographic examinations were performed on those specimens represented by solid symbols for which a specimen identification is given in these figures. The impact energy was normalized by multiplying by a factor of $L/B(W - a)^2$, where $B = 5$ mm is the specimen width, $W = 5$ mm is the specimen height, a is the notch depth plus the precrack length, and $L = 4W$ is the span. Although still preliminary, it is speculated that such normalization could extend the data base to cover other specimen sizes. Test results are tabulated in Ref *5*.

A Gaussian integral curve was fitted through each series of test results, according to a technique described in detail in a previous report [*4*]. The DBTT, which is defined as the inflection point

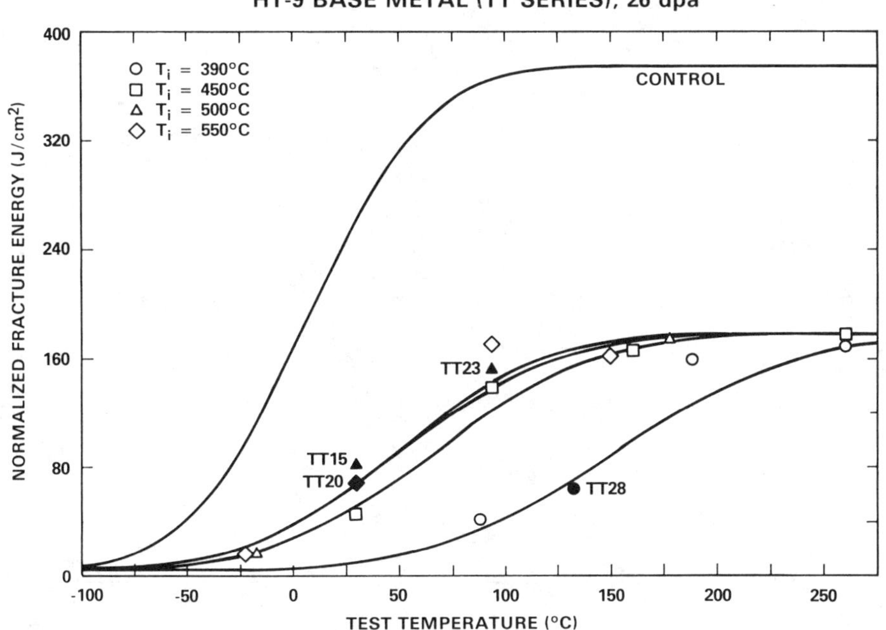

FIG. 1—*Charpy impact test results on HT-9 base metal irradiated to 26 dpa (TT series).*

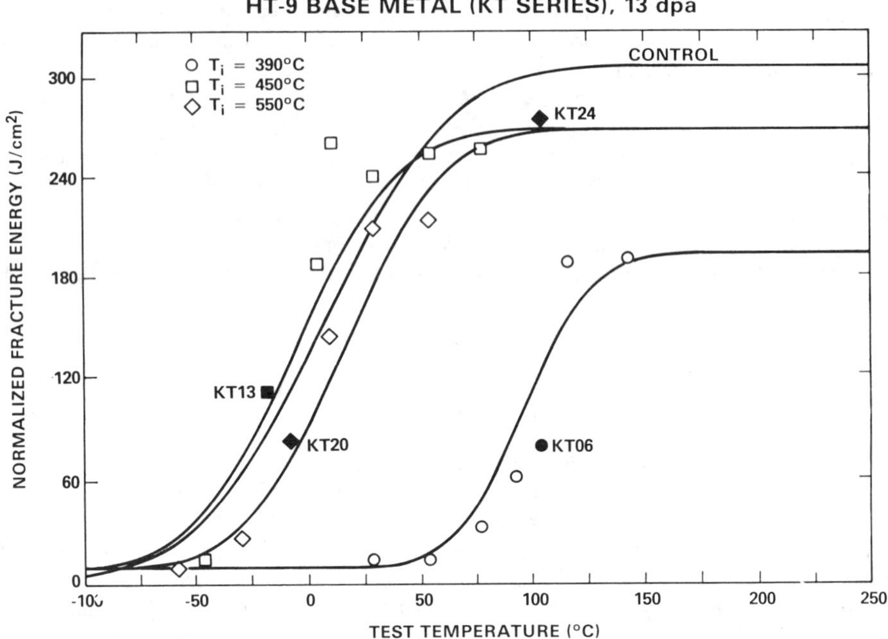

FIG. 2—*Charpy impact test results on HT-9 base metal irradiated to 13 dpa (KT series).*

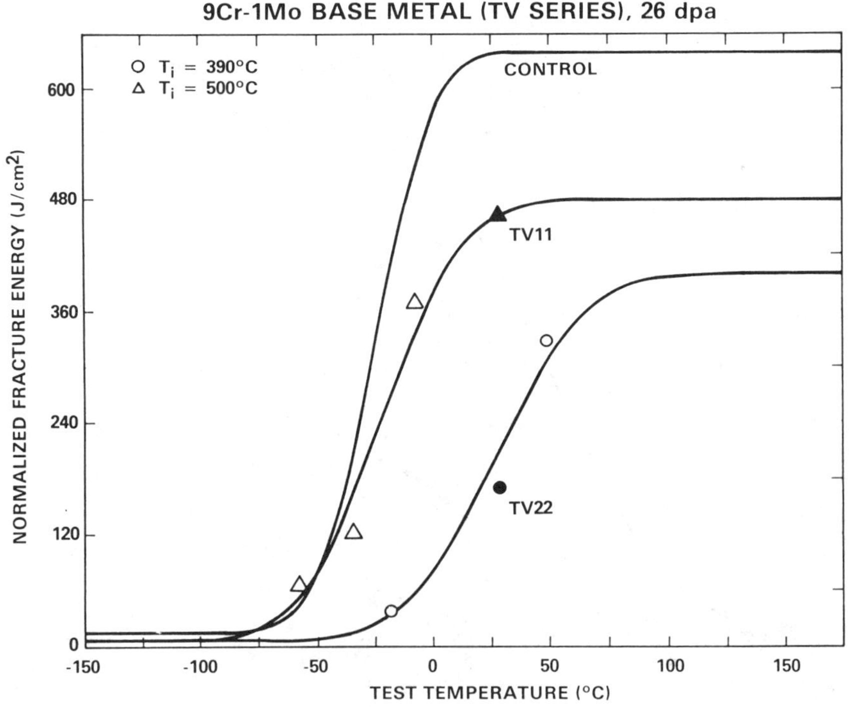

FIG. 3—*Charpy impact test results on 9Cr-1Mo base metal irradiated to 26 dpa (TV series).*

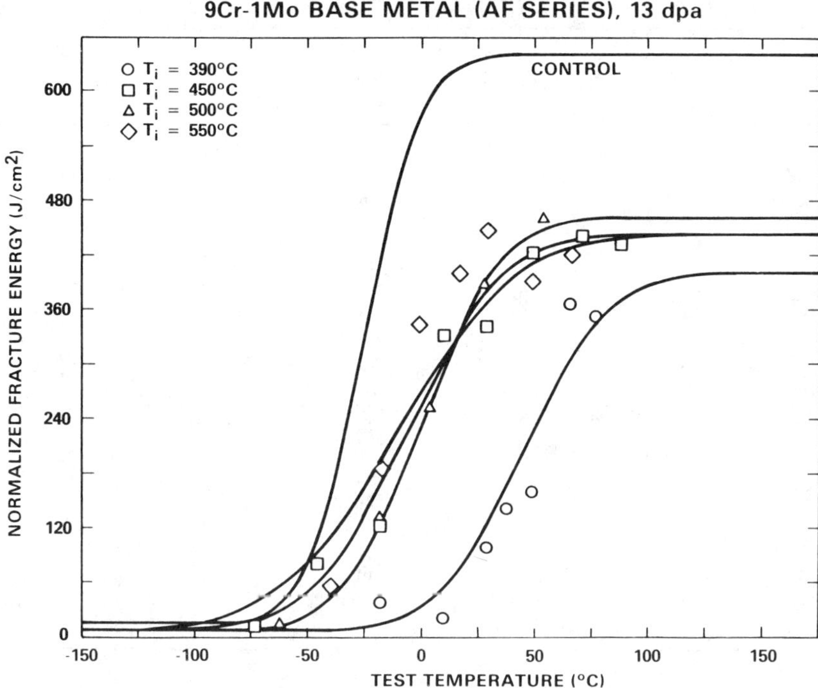

FIG. 4—*Charpy impact test results on 9Cr-1Mo base metal irradiated to 13 dpa (AF series).*

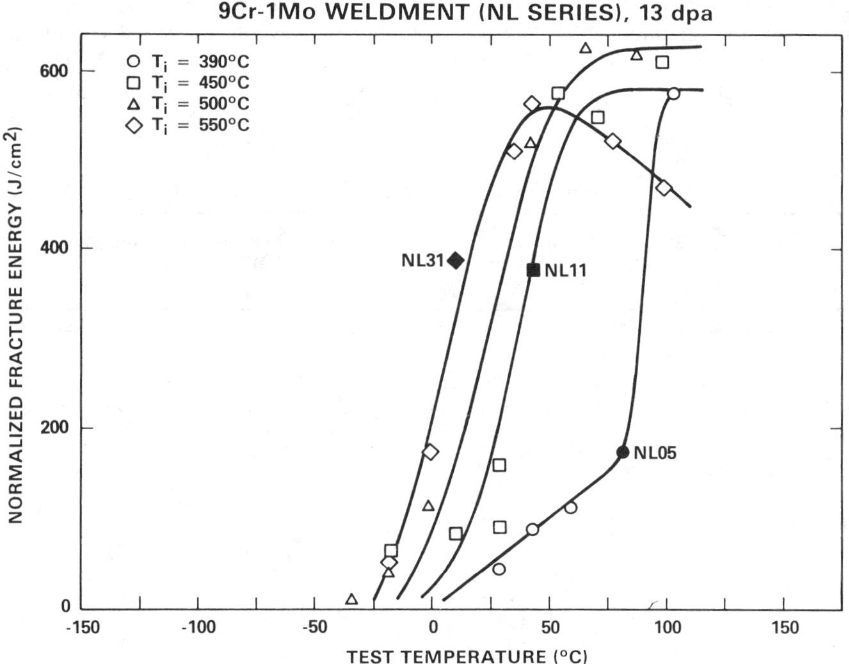

FIG. 5—*Charpy impact test results on 9Cr-1Mo weldment irradiated to 13 dpa (NL series).*

TABLE 1—*USE, DBTT, and shift in DBTT for ferritic alloys in the reconstitution of the AD-2 experiment.*

Material	Fluence $\times 10^{22}$, n/cm^2	Irradiation Temperature, °C	DBTT, °C	Shift in DBTT, °C	Estimated USE, J/cm°	Reduction in USE, %
HT-9	6	390	149	144	89	53
base metal	6	450	63	59	89	53
TT series	6	500	48	43	89	53
	6	550	46	41	89	53
HT-9	3	390	94	89	120	36
base metal	3	450	−8	−13	165	12
KT series	3	550	14	9	165	12
9Cr-1Mo	6	390	27	52	200	37
TV series	6	500	−22	3	240	25
9Cr-1Mo	3	390	45	70	200	37
base metal	3	450	−1	24	220	31
AF series	3	500	0	25	230	28
	3	550	−11	14	220	31
9Cr-1Mo	3	390	91	...	360	...
weldment	3	450	39	...	360	...
NL series	3	500	29	...	390	...
	3	550	13	...	350	...

on the curve, is tabulated in Table 1 together with the shift in the DBTT as compared with the DBTT of unirradiated control specimens. Because of the limited number of specimens available for each irradiation condition, the USE is not well defined. Nevertheless, a reduction in USE can be discerned in most cases. The estimated USE is listed in each case in Table 1. The USE, the DBTT, and the shift in DBTT for other relevant ferritic alloys are tabulated in Table 2 for comparison.

TABLE 2—*Previous impact test results on ferritic alloys* [4].

Material	Fluence $\times 10^{22}$, n/cm^2	Irradiation Temperature, °C	DBTT, °C	Shift in DBTT, °C	Estimated USE, J/cm°	Reduction in USE, %
HT-9 control TT series	0	...	5	0	188	0
HT-9	3	390	129	124	96	49
base metal	3	450	31	27	122	35
TT series	3	500	37	33	120	36
	3	550	61	57	108	43
9Cr-1 Mo control AF series	0	...	−25	54	320	0
9Cr-1Mo	3	390	29	54	188	41
base metal	3	450	−23	2	240	25
TV series	3	500	−27	−2	212	34
	3	550	−33	−8	208	35

TABLE 3—*Results of Rockwell C hardness measurements on irradiated base metal specimens.*

Alloy	Specimen	Fluence, dpa	Irradiation Temperature, °C	Test Temperature, °C	R_c
HT-9	TT28	26	390	132	28.7
HT-9	TT15	26	500	29	22.5
HT-9	TT20	26	550	29	21.4
HT-9	KT06	13	390	104	33.4
HT-9	KT13	13	450	−18	23.5
HT-9	KT24	13	550	104	22.3
9Cr-1Mo	TV22	26	390	29	21.4
9Cr-1Mo	TV11	26	500	29	12.1

Hardness Measurements

Results of Rockwell C hardness measurements on irradiated specimens are given in Table 3. Specimen identification is shown in Figs. 1, 2, 3, and 5. These values are plotted in Fig. 6 and compared with previous results. Examination of Fig. 6 reveals several trends. Significant hardening occurred in both HT-9 and 9Cr-1Mo following irradiation at 390°C, whereas following irradiation at temperatures of 450°C and above, hardness remained constant or decreased slightly. Hardness decreased with increasing fluence for HT-9 irradiated at 390°C and for 9Cr-1Mo irradiated at both 390 and 500°C. The reduction in hardness with increased fluence suggests a saturation in the degradation in DBTT caused by neutron exposure and indicates a softening effect caused by

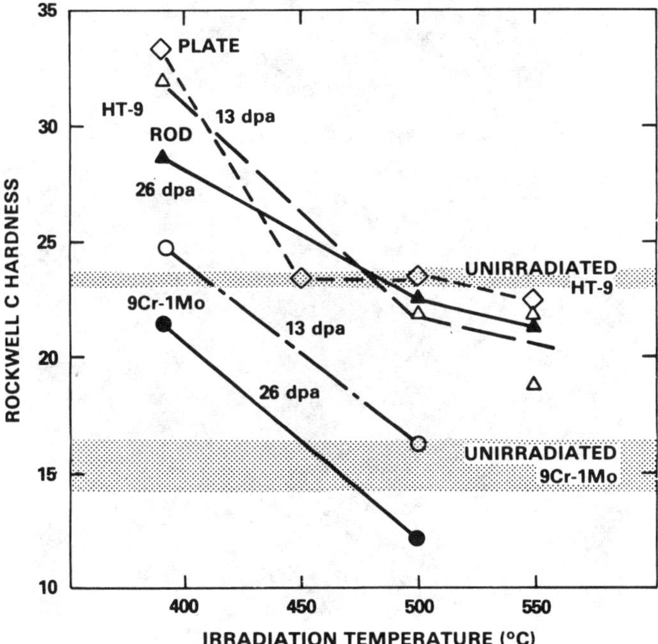

FIG. 6—*Hardness measurement as a function of irradiation temperature.*

prolonged aging at the irradiation temperature. Hardness values for HT-9 were stable at 500 and 550°C. Finally, HT-9 plate stock was slightly harder than bar stock at all temperatures where direct comparison was possible.

Fractography

Specimens selected for fractographic examination were indicated by a specimen identification number in Figures 1, 2, 3, and 5. Results of fractographic examinations are presented as stereo pairs in Figs. 7 through 10. In each case, a low magnification photo of the specimen is given in Part a, showing a small portion of the fatigue surface on the left but primarily showing the fracture surface created following irradiation. Part b displays a region in the center of the fracture surface immediately adjacent to the fatigue surface with the fatigue surface again shown on the left. The reader is encouraged to view these fractographs with a small stereo viewer in order to appreciate the fracture surface morphology. The reader is also encouraged to compare the present fractographs with those given in Ref *6*.

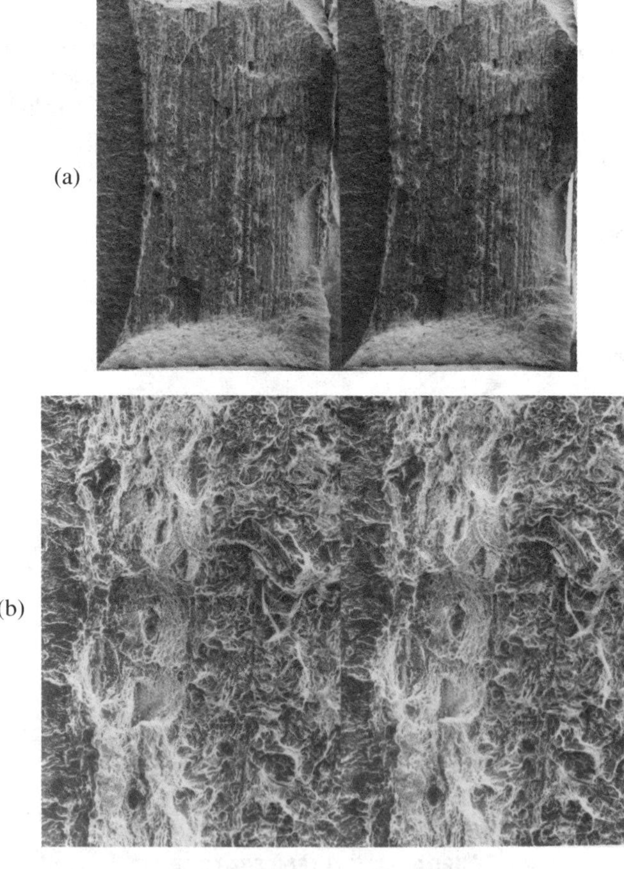

FIG. 7—*Stereo pair fractographs of HT-9 Specimen TT28 irradiated at 390°C to 26 dpa and tested at 132°C.*

(a)

(b)

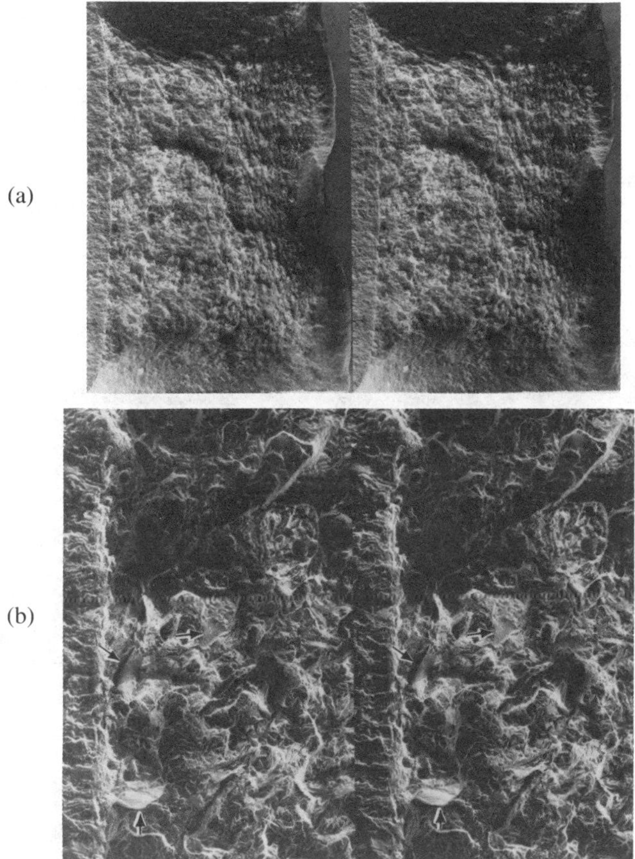

FIG. 8—*Stereo pair fractographs of HT-9 specimen KT06 irradiated at 390°C to 13 dpa and tested at 104°C. Arrows point to features discussed in the text.*

HT-9 Base Metal, TT Series—The specimens of series TT that are examined were very similar to those examined previously. All specimens showed evidence of delta ferrite stringers oriented parallel to the initial crack front. The stringers appeared in both ductile and brittle regions. For example, the photos in Fig. 7 were taken of specimen TT28, which was irradiated at 390°C to 26 dpa and tested at 132°C, in the transition temperature range.

The fracture surface created during the Charpy test can be divided into three parts: (1) a ductile region located immediately adjacent to the fatigue surface, which forms a narrow band, most easily recognized at the lower left, (2) a brittle region adjacent to this extending approximately halfway across the fracture surface and consisting of three distinct levels that are relatively flat, and (3) the remainder, which is a ductile region consisting mainly of a ''corrugated'' structure. These three regions can be identified in Fig. 7*a;* in Fig. 7*b,* the transition from Region 1 to Region 2 is shown at higher magnification. In each region evidence of the elongated structure caused by the delta ferrite stringers is visible.

Comparison of the fracture surfaces of Specimens TT15 (in the transition region) and TT2 (close to the upper shelf) revealed that both contain a similar corrugated structure of ductile failure in

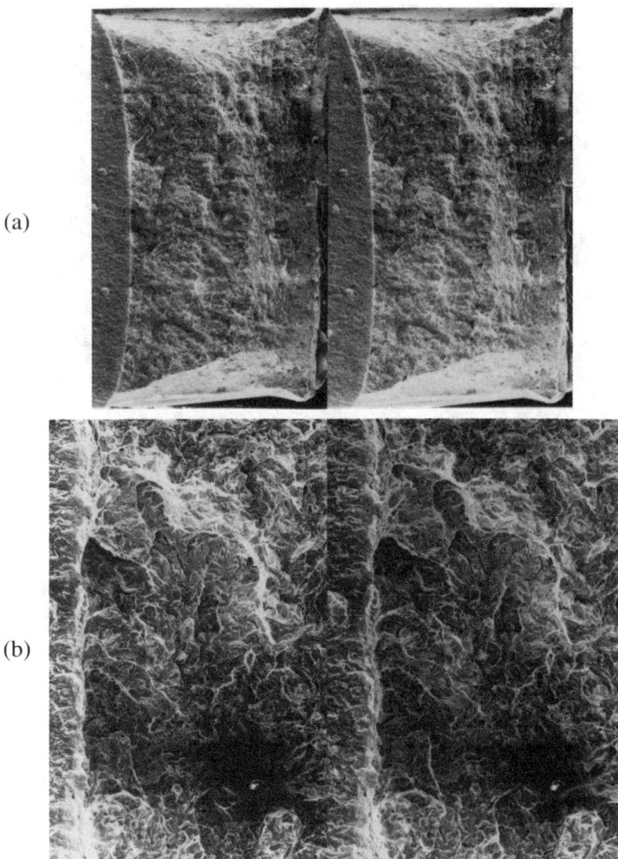

FIG. 9—*Stereo pair fractographs of modified 9Cr-1Mo base metal Specimen TV22 irradiated at 390°C to 26 dpa and tested at 29°C.*

association with delta ferrite stringers, but only specimen TT15 contained a region typical of brittle failure, which extended approximately halfway across the fracture surface. Therefore, as expected, the difference in fracture energy between these test conditions is due to brittle failure over almost half the fracture surface area. The fracture surfaces of Specimens TT28, TT15, and TT20 were similar, illustrating fracture behavior at the transition temperature for a range of irradiation temperatures. The major difference between the three specimens was that the initial ductile region adjacent to the fatigue surface was wider in Specimen TT20, that is, for the highest irradiation temperature.

Comparison of the present results with previous results [*7*] leads to two significant conclusions: (1) Upper shelf failure following irradiation is very similar to upper shelf response in unirradiated specimens, and therefore, the reduction in upper shelf energy caused by irradiation is not a consequence of a change in fracture mechanism. (2) The consequences of irradiation at 390°C, as described in Ref *6*, are not corroborated in the present study. The specimen described in Ref *6* contained few examples of linear structure caused by delta ferrite stringers, and shear lips at specimen surfaces were found to be wider but not as steep as the unirradiated case. Also, the fracture surface appeared to be flatter. These differences were attributed to increased hardness

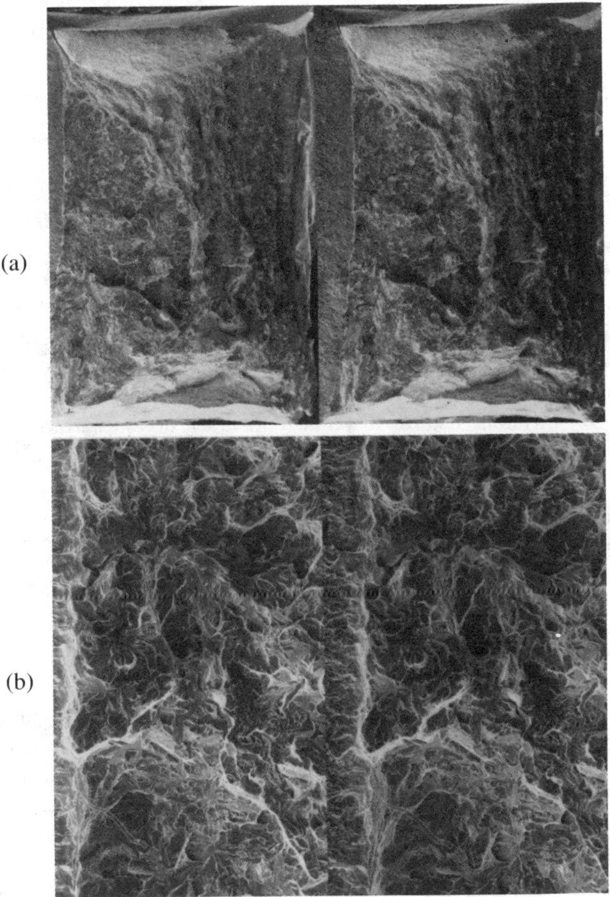

FIG. 10—*Stereo pair fractographs of modified 9Cr-1Mo weld metal Specimen NL05 irradiated at 390°C to 13 dpa and tested at 82°C.*

because of irradiation, which resulted in reduced likelihood of cavitation at delta ferrite stringers. Specimen TT28 in the present study is quite different, however. The shear lips adjacent to specimen surfaces are of normal width and steepness, that is, comparable to those of the unirradiated case, delta ferrite stringer effects are clearly defined, and the fracture surface is corrugated on the same scale as specimens irradiated at higher temperature. A definitive explanation for these differences is not obvious. It may be due in part to reduced yield strength with increasing fluence as indicated by reduced hardness found in Specimen TT28, or it may be due to greater precipitation expected at delta ferrite stringers with increasing fluence, which would compensate for the hardness increases found following irradiation at 390°C.

HT-9 Base Metal, KT Series—Specimens in the KT series exhibited fracture surfaces that were very different from those in the TT series. Three significant differences were identified: (1) The effect of delta ferrite stringers on the fracture surface was not evident. Only specimen KT24, on the upper shelf, showed an indication of delta ferrite stringer formation with the stringers normal to the crack front and in the plane of the fracture surface as expected [*1*]. (2) In regions where brittle failure occurred, the plateau regions in specimens from the KT series were much smaller than in those from the TT series. This difference is probably a consequence of delta ferrite effects on crack propagation. (3) An unusual banded structure was observed on two of the four specimens of condition KT that were examined. This banded structure may be present on the other two specimens but obscured by other features. Such a structure is typical of specimens with nonuniform internal stresses.

Figure 8 shows the fracture surface of Specimen KT06, which was irradiated at 390°C to 13 dpa and tested at 104°C, in the vicinity of the transition temperature. The fracture surface was typical of Charpy behavior at the transition temperature. A narrow ductile region was adjacent to the fatigue surface. Adjacent to it is a brittle region that extended approximately halfway across the specimen, and the remainder consisted of ductile failure or shear lips at specimen surfaces. The differences between this material and the specimens in the TT series consisted of (1) the apparent disappearance of the effect of delta ferrite stringers and (2) the smaller scale of the individual brittle plateau regions. Also of note were three regions in Fig. 8*b* (center to lower left), which appear to have a different texture than the remainder of the surface. The regions are uniformly gray in color and have been marked by arrows. These features are believed to be typical of failure at prior austenite grain boundaries.

The structure on specimen KT24, on the upper shelf, was very similar to that observed in the ductile region of specimen KT06. It can be concluded, therefore, that the reduction in USE following irradiation at 390°C was not a result of a change in fracture mechanism. The reduction in USE was much less severe following irradiation at 550°C.

9Cr-1Mo Base Metal—The Charpy test results for the 9Cr-1Mo base metal specimens indicated that negligible change in fracture behavior occurred as a result of a fluence increase from 13 to 26 dpa. Fractographic examinations in general confirmed that response, but significant differences were found regarding the distribution of large dimples in regions of ductile fracture.

Figure 9 shows the fracture surface of 9Cr-1Mo base metal Specimen TV22, which was irradiated at 390°C to 26 dpa and tested at 29°C in the vicinity of the transition temperature. The fracture surface was typical for such a condition with an initial narrow ductile region, a brittle region covering just over half the fracture ligament and with the remaining surface ductile in appearance. The brittle fracture surface was similar in appearance to that for condition KT, consisting of small plateau regions linked by steeply inclined surfaces. An example of brittle fracture far from the vicinity of the fatigue crack can be seen in Fig. 9*b*. A steep ledge separates the fatigue crack from the bottom of a brittle plateau region. Such topography has been observed in other ferritic specimens.

9Cr-1Mo Weld Metal—The fracture surfaces of 9Cr-1Mo weld metal specimens were more contorted (that is, nonplanar) than either base metal specimens of HT-9 and 9Cr-1Mo or weld metal specimens of HT-9 [*6*]. Also, brittle failure was less extensive than in other conditions, often occurring in localized regions.

Figure 10 shows the fracture surface of modified 9Cr-1MO weld metal specimen NL05, which was irradiated at 390°C to 13 dpa and tested at 82°C, slightly below the transition temperature. Figure 10 reveals that the usual ductile regions are present immediately adjacent to the fatigue surface and across approximately half of the remaining ligament. However, a large brittle region does cover about a third of the specimen in a thumb nail shaped area slightly below the center of the fractograph. Careful inspection showed that brittle failure also occurred in isolated regions beyond the "thumb nail" in the lower part of the fractograph. A deep step is found in the middle of the thumb nail region, and a tall ridge appears to divide the fracture surface into two parts. Also, the shear lip at the lower part of the micrograph is unusually distorted.

Discussion

Irradiation Fluence

The shift in DBTT caused by neutron exposure for both HT-9 and 9Cr-1Mo base metal was plotted in Fig. 11 as a function of irradiation temperature and fluence. The additional shift in DBTT from 13 to 26 dpa for HT-9 is 20, 32, 10, and −16°C for specimens irradiated at 390, 450, 500, and 550°C, respectively. The reduction of USE at 26 dpa is ~53% for all irradiation temperatures. The small increases in DBTT suggest that the effect of fluence is approaching saturation. The 9Cr-1Mo base metal, on the other hand, does not show any additional change in either DBTT or USE as the result of increased neutron exposure.

Comparisons can now be made of fracture surface topography as a function of irradiation fluence for both HT-9 and 9Cr-1Mo base metal. Although irradiation does cause a significant shift in DBTT, a reduction in USE and an increase in hardness at 390°C for HT-9 and 9Cr-1Mo, the fracture appearance remains relatively unchanged. Both ductile and brittle fracture regions on irradiated specimens are similar to comparable surfaces on unirradiated specimens. It can therefore be concluded that other failure mechanisms are not significantly enhanced by irradiation, at least to doses on the order of 26 dpa. Changes in fracture energy must instead be attributed to changes in the microstructure, inducing failure at lower plastic strain but at the same sites that operate in unirradiated material. For example, reductions in USE caused by irradiation are likely a result of enhanced carbide precipitation and intermetallic formation in the vicinity of carbide particles formed during heat treatment, which leads to cavity nucleation and growth at lower plastic strain. The shift in DBTT is probably caused by matrix hardening as a result of G-phase, α', and carbide precipitation and dislocation loop formation. Voids are not present in HT-9 following irradiation in EBR-II to 26 dpa. It can be concluded, therefore, based on fractographic examination, that the

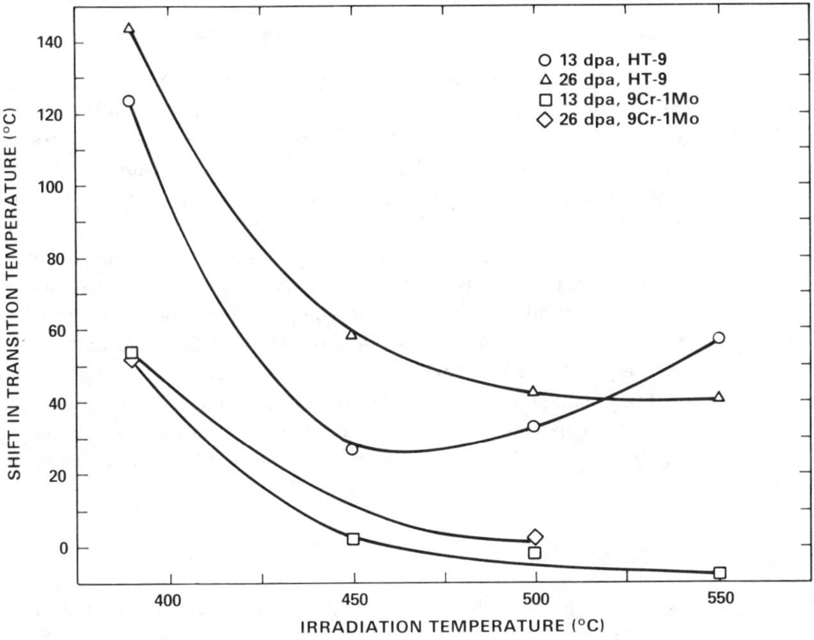

FIG. 11—*Irradiation temperature and fluence effect on the shift of DBTT for ferritic alloys.*

effect of irradiation to 26 dpa on fracture energy is to facilitate cavity nucleation and to reduce the plastic strain before cavity nucleation and coalescence. No new mechanisms, such as temper embrittlement, become dominant.

Weld Metal Response

Weld metal specimens consistently have a slightly higher USE than do base metal specimens. The present series of tests demonstrates this effect for 9Cr-1Mo, and a similar conclusion can be drawn from results for HT-9 at the lower fluence. Test results on 9Cr-1Mo weldment, as shown in Fig. 5, show a high USE, in the range of 400 J/cm². The USE for unirradiated 9Cr-1Mo base metal is only 320 J/cm². This is attributed to the high energy associated with failure propagation. Fractographic examination shows that the surfaces are highly contorted; that is, a larger region is plastically deformed ahead of the propagating rack, and more energy is absorbed. As a result, USE is higher. This leads to a distorted Charpy impact energy curve, often asymmetric about the transition temperature. A good example of this is provided in Fig. 5 for 9Cr-1Mo specimens irradiated at 390°C. The distortion could probably be reduced by using a fracture appearance transition temperature (FATT) rather than an absorbed energy criterion.

The transition temperature of the 9Cr-1Mo weldment is approximately 60°C higher than the base metal that was exposed to the same irradiation conditions. The DBTT of 9Cr-1Mo weldment irradiated in HFIR at 55°C was also found to be 25°C higher than the similarly irradiated base metal [*7*]. This is in contrast with the results of an earlier study of the transition behavior of HT-9, where the DBTT of the weldment as well as the heat-affected zone (HAZ) were lower than that of the base metal [*4*]. The 9Cr-1Mo weldment also shows a less well defined upper and lower shelf. In particular, specimens irradiated at 550°C show a USE that decreases as test temperature increases.

Processing Effects for HT-9

Two conditions of HT-9 can be compared from the tested Charpy specimen matrix. Series TT represents a mill annealed condition, which differs from the standard condition in that it was austenitized at 1149°C for 1 h and then hot worked before cooling to room temperature. It did undergo a full martensite transformation, and the delta ferrite content remained in the acceptable range of 1%. In comparison, the KT series was given a lower annealing treatment at 1038°C for 4 min, but the tempering conditions were similar (760°C for 30 min versus 750°C for 1 h). Both series came from the same alloy heat. The difference in Charpy fracture behavior can be ascribed, based on the present effort, to the effects on fracture of delta ferrite stringers (and indirectly perhaps on rolling direction). The delta ferrite stringers appear to promote the growth of brittle cracks over large areas, crossing prior austenite grain boundaries. If the delta ferrite stringers are broken up by cross rolling, brittle cracks extend only to prior austenite boundaries; therefore, energy must be absorbed in breaking the ligaments between brittle cracks. Similar arguments apply to ductile failure (upper shelf response); therefore, it can be concluded that fabrication procedures can have a measurable effect on toughness and that cross rolling is likely to be beneficial if delta ferrite is present.

Conclusions

Charpy impact tests were conducted on ferritic specimens irradiated in EBR-II to a fluence of 26 dpa. It was concluded that

1. The effect of neutron irradiation on the DBTT appears to be approaching saturation for HT-

9 base metal specimens irradiaed to 26 dpa at 390°C. The DBTT increased only 20°C as the fluence was raised from 13 to 26 dpa.

2. No effect of the additional neutron exposure was detected on DBTT or USE for 9Cr-1Mo base metal.

3. The 9Cr-1Mo weldment irradiated to 13 dpa exhibited a DBTT, which was a maximum of 62°C higher than that of the base metal irradiated under the same conditions.

4. Significant improvements in both the DBTT and the USE of HT-9 base metal were observed because of the different crack plane orientations.

5. Significant hardening occurs in both HT-9 and 9Cr-1Mo following irradiation at 390°C, whereas following irradiation at temperatures of 450°C and above, hardness remains constant or decreases slightly. Hardness decreases with increasing fluence in HT-9 irradiated at 390°C and in modified 9Cr-1Mo irradiated at 390 and 500°C.

6. Fracture appearance is not altered significantly by irradiation to higher fluence. New fracture mechanisms are therefore not promoted by irradiation. Failure is controlled by the microstructure generated before irradiation, but plastic strain is decreased or the nucleation and coalescence of cavities is accelerated by irradiation.

References

[*1*] Puigh, R. J. and Panayotou, N. F., "Specimen Preparation and Loading for the AD-2 Ferritic Experiment," ADIP Quarterly Progress Report, DOE/ER 0045/3, 30 June 1981, pp. 261 293.

[*2*] Ermi, A. M., "Reconstitution of the AD-2 Ferritic Experiment," ADIP Semiannual Progress Report, DOE/ER-0045/8, 31 March 1982, pp. 431–444.

[*3*] Hu, W.-L. and Panayotou, N. F., "Miniature Charpy Specimen Test Device for the Ferritic Alloy HT-9," ADIP Semiannual Progress Report, DOE/ER-0045/7, 30 Sept. 1981, pp. 235–251.

[*4*] Hu, W.-L., "Miniature Charpy Impact Test Results for Irradiated Ferritic Alloys," ADIP Semiannual Progress Report, DOE/ER-0045/9, 30 Sept. 1981, pp. 255–272.

[*5*] Hu, W.-L., "Charpy Impact Test Reslts of Ferritic Alloys at a Fluence of 6×10^{22} n/cm^2," ADIP Semiannual Progress Report, DOE/ER-0045/13, 30 Sept. 1984, pp. 106–119.

[*6*] Hu, W.-L. and Gelles, D. S., "Miniature Charpy Impact Test Results for the Irradiated Ferritic Alloys HT-9 and Modified 9Cr-1Mo," *Proceedings of Topical Conference on Ferritic Alloys for Use in Nuclear Energy Technologies,* Snowbird, UT, 19–23 June 1984, pp. 631–646.

[*7*] Gelles, D. S., Hu, W. L., Huang, F. H., and Johnson, G. D., "Effects of HFIR Irradiation at 55°C on Microstructure and Toughness of HT-9," DOE/ER-0045-11, 30 Sept. 1983, pp. 115–127.

Masahide Suzuki,[1] Kiyoshi Fukaya,[1] Tsuneo Kodaira,[1] and Tatsuo Oku[1]

Effects of Microstructure and Thermal Aging on Neutron Irradiation Embrittlement of 2¼Cr-1Mo Steel for a Nuclear Pressure Vessel

REFERENCE: Suzuki, M., Fukaya, K., Kodaira, T., and Oku, T., **"Effects of Microstructure and Thermal Aging on Neutron Irradiation Embrittlement of 2¼Cr-1Mo Steel for a Nuclear Pressure Vessel,"** *Influence of Radiation on Material Properties: 13th International Symposium (Part II), ASTM STP 956,* F. A. Garner, C. H. Henager, Jr., and N. Igata, Eds., American Society for Testing and Materials, Philadelphia, 1987, pp. 98–110.

ABSTRACT: Plate, weld, and heat-affected zone (HAZ) materials of 2¼Cr-1Mo steels, designated as ASTM A387 Grade 22 Cl.1, and Cl.2 were irradiated at around 400°C to a fluence of about 2×10^{18} n/cm^2 ($E > 1$ MeV). A387 Grade 22 Cl.2 with bainite has a lower susceptibility to neutron irradiation embrittlement than A387 Grade 22 Cl.1 with ferrite plus bainite. The anomalous ductile to brittle transition behavior of A387 Grade 22 Cl.2 was explained by taking into account the parameter k_y, which is a constant in the Cottrell-Petch equation. It is concluded that there does not exist any strong interaction between thermal aging embrittlement and neutron irradiation embrittlement.

KEY WORDS: 2¼Cr-1Mo steel, neutron irradiation embrittlement, radiation anneal hardening, A387 Grade 22 Cl.1, A387 Grade 22 Cl.2, temper embrittlement, phosphorus, intergranular fracture

2¼Cr-1Mo steel is expected to be used for the pressure vessel of the experimental Very High Temperature Gas Cooled Reactor (VHTR), on which the Japan Atomic Energy Research Institute (JAERI) has carried out research and development work for the past decade. Though this material has not been used as a nuclear reactor pressure vessel, good mechanical properties at elevated temperature imply the possibility of its usage in an advanced nuclear system. Since the design temperature and neutron exposure in the VHTR pressure vessel are 440°C and 1×10^{18} n/cm^2 ($E > 1$ MeV), respectively, a combined effect of thermal aging and neutron irradiation on the embrittlement is of principal interest for assuring structural integrity. Experimental data concerning the toughness degradation for 2¼Cr-1Mo steel are limited, and existing data were mostly obtained in the temperature regime for the light water reactor vessels [*1–4*]. A typical example was that of Hawthorne et al., which gave the following experimental formula for Charpy impact properties of quenched and tempered 2¼Cr-1Mo steel irradiated at 288°C to a fluence of 2.8×10^{19} n/cm^2 ($E > 1$ MeV) [*1*]

$$``A" = -118 + 14800\,(\text{wt\%P}) + 990\,(\text{wt\%Cu}) \quad (1)$$

$$\Delta E = -2 - 100\,(\text{wt\%Cu}) \quad (2)$$

Here, "*A*" is the increment of the ductile to brittle transition temperature (DBTT, °F) and ΔE is the change in the upper shelf energy (ft · lbf).

[1] Research engineer, chief, principal engineer, and head, respectively, Japan Atomic Energy Research Institute, Department of High Temperature Engineering, Tokai-mura, Naka-gun, Ibaraki-ken 319-11, Japan.

TABLE 1—*Chemical composition and the materials used, wt%.*

Designation	C	Si	Mn	P	S	Ni	Cr	Mo	Cu	As	Sn	Sb
AN	0.13	0.07	0.52	0.009	0.009	0.12	2.32	1.03	0.09	0.007	0.007	0.0027
NT-1	0.14	0.08	0.54	0.009	0.009	0.12	2.36	1.04	0.09	0.008	0.008	0.0027
NT-1 weld metal	0.11	0.22	0.75	0.015	0.014	0.04	2.37	1.07	0.03	0.005	0.005	0.0044
NT-2	0.15	0.05	0.55	0.008	0.01	0.11	2.33	0.9	0.07	0.006	0.008	0.002

In the higher temperature regime, thermal aging affects the toughness of the 2¼Cr-1Mo steel by impurity segregation to grain boundaries or by microstructural changes [*5*]. Little is known about the toughness degradation under irradiation, although some information is available on the damage structure and tensile properties [*6,7*].

The objective of the present study is to examine the characteristics of irradiation embrittlement in 2¼Cr-1Mo steels and to clarify the interaction between the embrittlement caused by thermal aging and neutron irradiation in the low fluence regime.

Materials and Experimental Procedure

Materials

Materials used in the present experiment are 2¼Cr-1Mo steels, annealed (AN), and normalized and tempered (NT), which correspond to A387 Grade 22 Cl.1 and Cl.2, respectively, in the ASTM specification. These are industrial plates, 160 mm in thickness, including the welds. Tables 1 and 2 show the chemical compositions at the ¼*T* (*T* is thickness) position, the heat treatment conditions, and the post weld heat treatment (PWHT) conditions for these plates with their designations. The microstructures are ferrite plus pearlite in the AN and bainite in the NT material. Both exhibit a fine grain size of about 7.5. Welding was conducted by the submerged arc welding method.

Experimental Procedure

Neutron irradiations were carried out using the Japan Material Testing Reactor (JMTR). Charpy impact and tension specimens with dimensions shown in Fig. 1 were mounted in an irradiation capsule filled with helium gas. Neutron fluences were about 1 to 6 $\times$ 10^{18} n/cm^2 ($E > 1$ MeV), and irradiation temperatures ranged from 330 to 410°C. Control of the irradiation temperature was conducted by adjusting the gas pressure between the inner and outer tubes of the irradiation

TABLE 2—*Heat treatment of materials used.*

Designation	Type of Steel	Treatment Type	Details[a]
AN	A387Grade 22Cl.1	annealed	austenized at 900/925°C for 5 h, cooled at a rate of 60°C/h, PWHT
NT-1	A387Grade 22Cl.2	normalized and tempered	austenized at 900/925°C for 5 hr, quenching tempered at 640/660°C for 6 h, air cooled PWHT
NT-2	A387Grade 22Cl.2	normalized and tempered	normalized at 900/930°C for 6.5 h, water cooling, tempered at 677/690°C for 7 h, air cooled, PWHT(1)

[a] PWHT: 677/690°C for 22 h and PWHT(1): 685/690°C for 20 h.

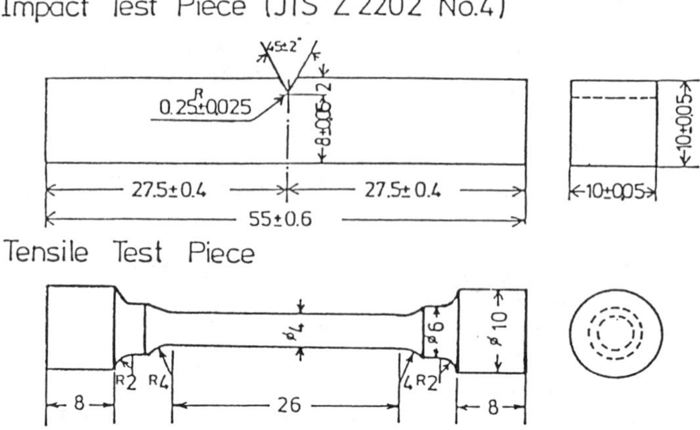

FIG. 1—*Configuration of the Charpy impact and tension specimens used.*

capsule. The post irradiation tests were conducted at the hot laboratory of the Oarai Research Establishment, JAERI. Besides Charpy impact and tension tests, hardness measurements and fractographic observation by a scanning electron microscope were also performed.

Results and Discussion

Microstructure Dependence of Irradiation Embrittlement

Effects of neutron irradiation on the mechanical properties of steels are known to be very sensitive to the microstructure. The microstructure dependence of neutron irradiation embrittlement in 2¼Cr-1Mo steels having different heat treatments and different thermal histories with welding were observed. Figure 2 shows the absorbed energy transition curves of four kinds of specimens (AN, NT-1, NT-W, and NT-H) in the as-received, thermal controlled, and irradiated conditions, and these curves are drawn using a curve-fitting routine. The following facts can be seen. When comparison is made between the base metals, it is clear that the AN with ferrite plus pearlite structure has a higher susceptibility to neutron irradiation embrittlement than the NT with a bainite structure. Both shifts in the upper shelf energy and the DBTT were higher in the AN material. Numerical values are presented in Table 3. The NT material even exhibited a slight decrease in DBTT. In general, the shift of the DBTT caused by neutron irradiation can be explained clearly by the Ludwig-Davidenkov relationship illustrated in Fig. 3. That is, an increase in yield stress or radiation hardening results in a DBTT shift. In the present experiment with the NT material, however, the increase in yield stress did not lead to a DBTT increase. The results of the tension tests are shown in Fig. 4. This anomaly was more obvious in the weld heat-affected zone (NT-H). The DBTT decreased more than 30°C after neutron irradiation. In the weld metal (NT-W), on the other hand, a DBTT increase of over 10°C was observed. This higher susceptibility to neutron irradiation embrittlement in the weld metal is probably caused by the higher phosphorus (P) content, judging from Eq (1).

To summarize, the annealed material has a higher susceptibility than the normalized and tempered one to neutron irradiation embrittlement. Moreover, irradiation actually lowers DBTT for plate and heat-affected zone (HAZ) in the NT material. Detailed discussion will be presented concerning this property later.

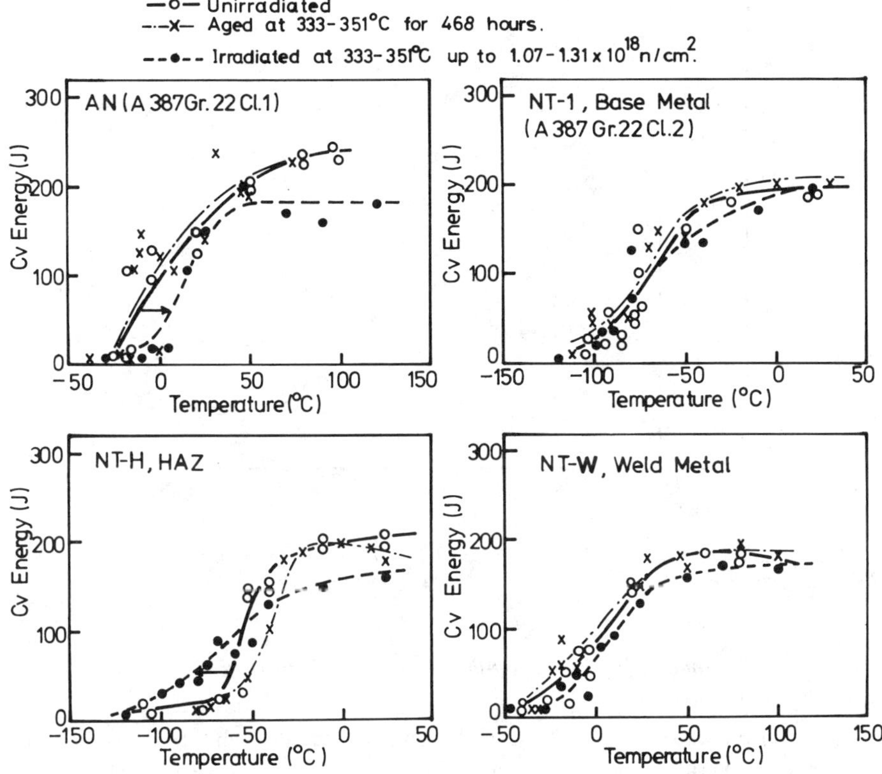

FIG. 2—*Absorbed energy transition curves before and after neutron irradiation of four kinds of materials: A387 Grade 22 Cl.1 base metal, A387 Grade 22 Cl.2 base metal, weld heat-affected zone, and weld metal.*

TABLE 3—*Ductile to brittle transition temperature and upper shelf energy before and after neutron irradiation. Neutron fluence: 1 × 10^{18} n/cm² (E > 1 MeV), irradiation temperature: 330 ~ 350°C.*

Specimen	vTr30, °C	vTr50, °C	vE, J
As received	−16	−9	227
AN thermal controlled[a]	−25	−15	223
Irradiated	+4	+12	182
As received	−89	−81	191
NT-1 thermal controlled[a]	−94	−81	206
Irradiated	−94	−81	185
As received	−63	−59	195
NT-H thermal controlled[a]	−53	−47	196
Irradiated	−89	−70	167
As received	−18	−6	188
NT-W thermal controlled[a]	−25	−14	195
Irradiated	−9	+3	165

NOTE: vTr30: DBTT at 30 ft · lb level, vTr50: DBTT at 50 ft · lb level, and vE: upper shelf energy.
[a] Thermal controlled at 350°C for 468 h.

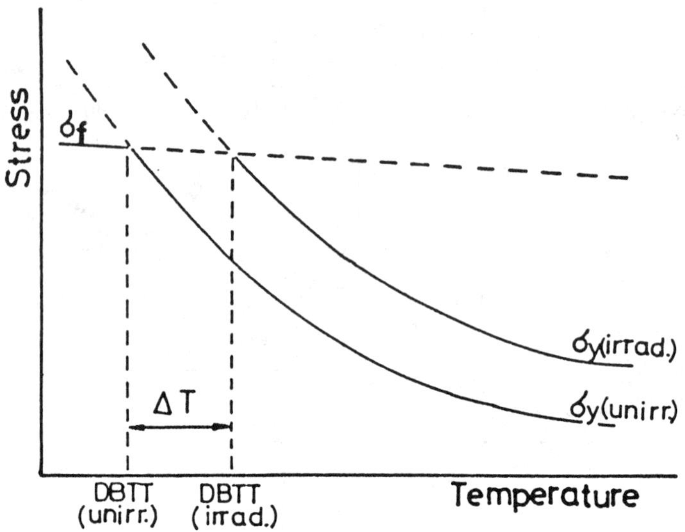

FIG. 3—*Schematic illustration of the Ludwig-Davidenkov approach to the relationship between strength and DBTT.* σ_y *is the yield stress, and* σ_f *is the cleavage fracture stress.*

Characteristics of Thermal Aging Effects on the Mechanical Properties of Normalized and Tempered 2¼Cr-1Mo Steel

1. Nonirradiated Material—The NT material becomes brittle through thermal aging by what is known as temper embrittlement. This phenomenon is caused by holding or slowly cooling the material in the temperature region from about 375 to 550°C. Figure 5 shows the time-temperature diagram of the DBTT increase for the NT-1. This diagram shows the increase in the DBTT as a function of aging time and temperature. It is clearly seen from this figure that thermal control at 350°C for 468 h does not affect the DBTT. Changes in the tensile properties caused by the thermal aging are very small. The relative change in the yield stress (0.2% offset stress) at room temperature is shown in Fig. 6, indicating that below 500°C the change is less than 5% even after aging for 10 000 h.

Therefore, the ductile to brittle transition behavior is more important than the aging stability. Figure 7 shows the scanning electron micrographs of the fracture surface for the broken Charpy specimens tested at −100°C. The appearance of the fracture surface was altered by thermal aging from the entirely transgranular to mixed transgranular plus intergranular fracture. Chemical analysis by auger electron microscopy revealed that phosphorus segregated in the grain boundary region. A typical auger spectrum is shown in Fig. 8, indicating only phosphorus as an impurity element. We can conclude that the effect of thermal aging on the mechanical properties of the normalized and tempered material is most probably caused by the intergranular segregation of phosphorus.

2. Irradiated Material (Radiation Anneal Hardening)—The effect of thermal aging on the mechanical properties of the irradiated material was examined by hardness measurements. Figure 9 shows the hardness change of the irradiated NT material with annealing. Each annealing time is 1 h. Hardness was measured by a Vickers hardness testing machine with a load of 10 kg. Hardening occurred by annealing at around 400°C. Isothermal annealing after irradiation was conducted at temperatures from 350 to 450°C using almost the same material irradiated to the fluence of 1×10^{19} n/cm^2 ($E > 1$ MeV). Figure 10 shows the time to reach the maximum

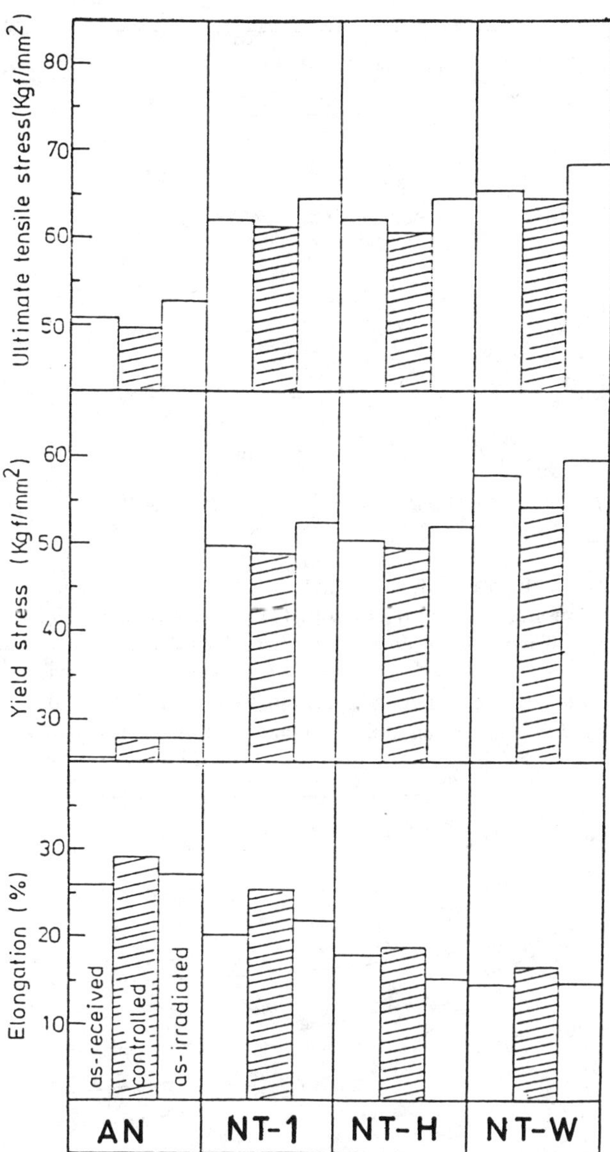

FIG. 4—*Tensile properties of as-received, thermal controlled, and neutron irradiated specimens for the AN and NT materials. Irradiation temperature: ~361°C and neutron fluence: 1 × 10^{18} n/cm^2* (E > *1 MeV).*

hardening in each annealing as a function of the reciprocal of the annealing temperature. From the slope of the line an apparent activation energy of about 8.4 × 10^4 J/mole (20 kcal/mole) was obtained. Since this value is close to the migration energy of carbon or nitrogen in α-iron, migration of the interstitial atoms is considered to occur during post-irradiation annealing. It might be also pointed out that existence of radiation induced precipitate (M_3P) and complexes between phosphorus and copper are almost unlikely, judging from the facts such as the value of activation

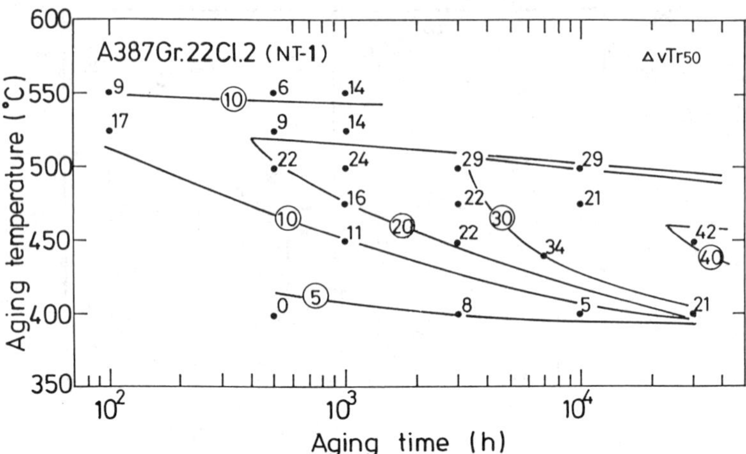

FIG. 5—*Time-temperature curve for the constant increase in the transition temperature for NT-1. Numbers denote the shift of the 50ft · lbf transition temperature, °C.*

energy and almost no shift in DBTT at as irradiated state. Microstructural changes could not be revealed by the transmission electron microscopy (TEM), despite these changes in the mechanical properties. The detailed mechanism of radiation anneal hardening is not certain, but we would like to point out an important role of a high density of preexisting dislocations, which are pinned by a complex atmosphere. According to Baird and Jamieson [8], solid solution hardening by alloying elements (molybdenum or chromium) is enhanced by the addition of interstitial impurities

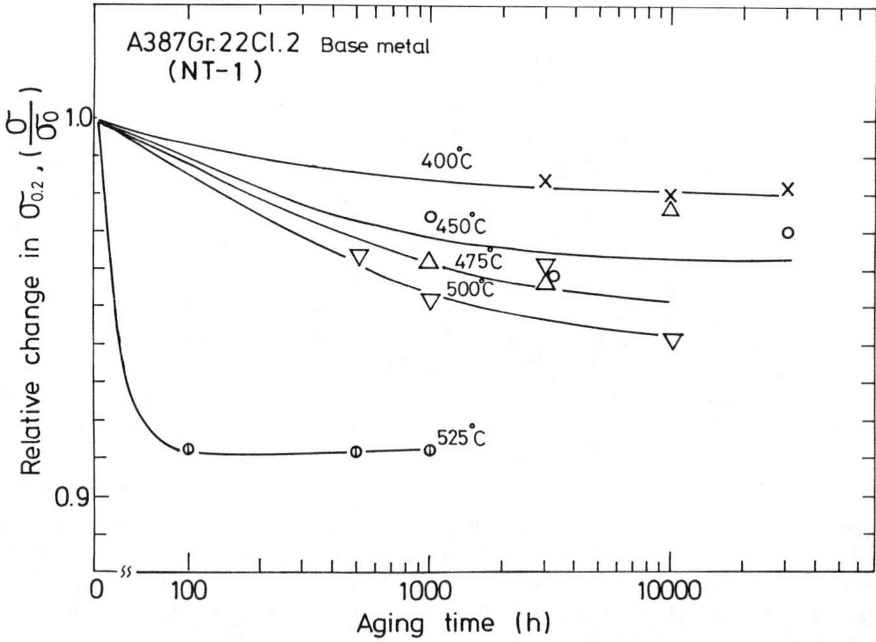

FIG. 6—*Thermal aging curves of the 0.2% offset stress for the NT-1 material.*

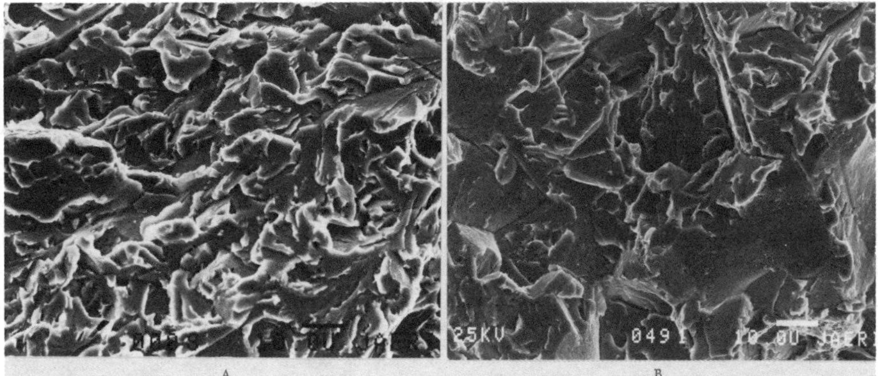

FIG. 7—*Scanning electron micrographs of fracture surfaces in Charpy specimens for the NT-1:* (a) *as received and* (b) *isothermal aged (450°C, 30 000 h), both tested in the lower shelf region.*

(carbon or nitrogen). They considered that the interaction between the dislocation and a complex atmosphere comprising the alloying elements and the interstitial impurities was responsible for the strengthening, and this was called "interaction solid solution hardening." The dislocation density of the normalized and tempered 2¼Cr-1Mo steel is very high even in the as received condition (around $10^{11}/cm^2$), and the strength is strongly governed by these dislocation [*9*], which are considered to be pinned by the complex atmosphere. During irradiation, decomposition of the

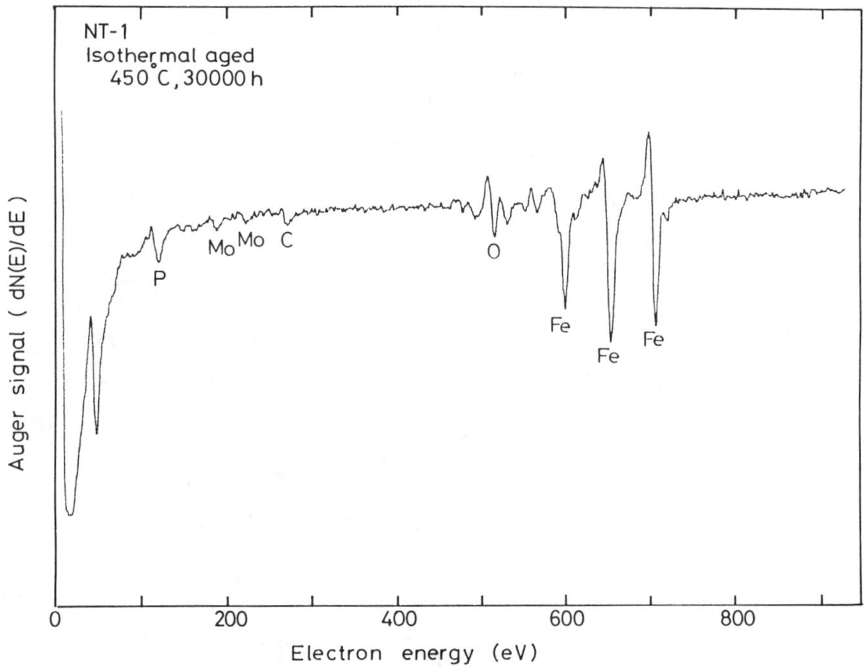

FIG. 8—*Auger spectrum of a grain boundary facet of the NT-1 material, isothermally aged for 30 000 h at 450°C.*

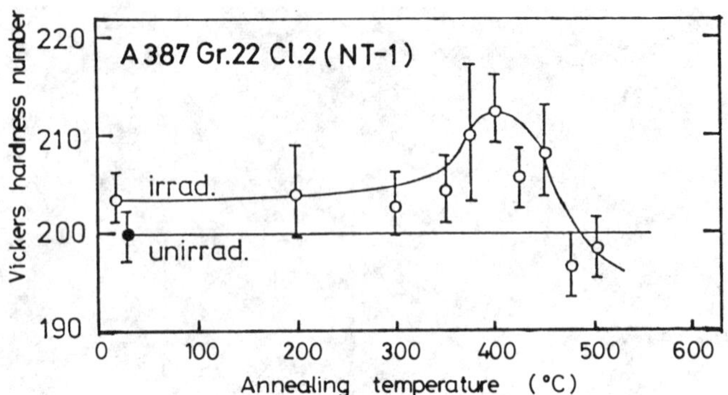

FIG. 9—*Change in the Vickers hardness caused by annealing after neutron irradiation for the NT-1 material. Neutron fluence: 1 × 10*^{18} *n/cm*^{2} (E > *1 MeV) and irradiation temperature: 320 ~ 380°C.*

complex atmosphere could take place. The process may be indicated as follows

$$\begin{aligned} \text{M-I-dislocation} &\rightarrow \text{M} + \text{I} + \text{dislocation} \\ &\rightarrow \text{I} + \text{M-dislocation} \\ \text{M-I} &\rightarrow \text{M} + \text{I} \end{aligned} \tag{3}$$

where M indicates alloying element (chromium or molybdenum) and I indicates interstitial impurities. Next, during post-irradiation annealing, reformation of the complex could take place by

$$\begin{aligned} \text{M} + \text{I} &\rightarrow \text{M-I} \\ \text{I} + \text{M-dislocation} &\rightarrow \text{M-I-dislocation} \end{aligned} \tag{4}$$

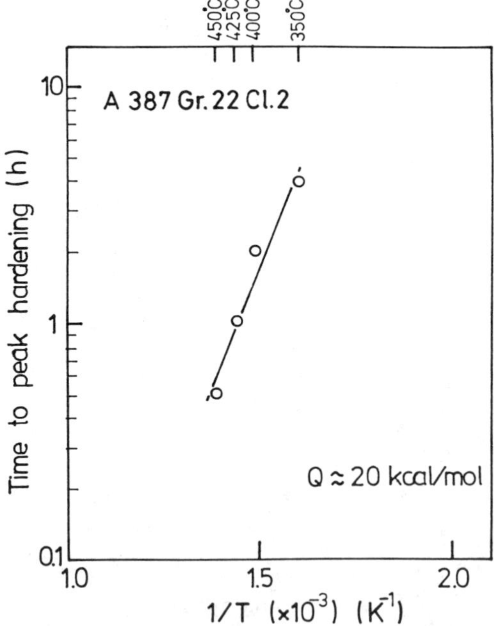

FIG. 10—*Time to reach the peak hardening versus the reciprocal of the annealing temperature,* K^{-1}. *Neutron fluence: 1 × 10*^{19} *n/cm*^{2} (E > *1 MeV) and irradiation temperature: ~360°C.*

If such is the case, Process 3 would correspond to a decrease of the DBTT, and Process 4 would correspond to radiation anneal hardening.

Combined Effects of Neutron Irradiation Embrittlement and Thermal Aging Embrittlement

The 2¼Cr-1Mo steel in the normalized and tempered condition has an excellent radiation embrittlement resistance when considering the DBTT. However, if the material receives a certain thermal aging treatment, this advantage can be decreased. We will consider briefly the theoretical aspects of the ductile to brittle transition behavior [*10*].

The DBTT (Tc) is defined by the relationship

$$\sigma_f = m\sigma_y \tag{5}$$

for the testing temperature T (Tc), where σ_f is the cleavage fracture stress, σ_y is the yield stress, and m is the constraint factor (~2.2), and they are given by the following expressions according to Cottrell or Cottrell and Petch [*10*]

$$\sigma_f = \frac{2G\gamma}{k_y} d^{-\frac{1}{2}} \tag{6}$$

$$\sigma_y = \sigma_i + k_y d^{-\frac{1}{2}} \tag{7}$$

where G is the shear stress, γ is the effective surface energy, k_y is a material constant, d is the grain size, and σ_i is the friction stress. The parameters, G and d, are considered to be constant during irradiation. γ is also thought to be almost constant in spite of the fear that γ may decrease because of phosphorus segregation. Because the fracture appearance of all irradiated Charpy specimens shows entirely transgranular mode, and this fact tells us the degree of phosphorus segregation would be inappreciable. Then $\sigma_y k_y$ yields a constant from Eq 5. Thus

$$\sigma_y dk_y + k_y d\sigma_y = 0 \tag{8}$$

$$dk_y = \left(\frac{\partial k_y}{\partial T}\right) dT + \left(\frac{\partial k_y}{\partial \sigma_i}\right) d\sigma_i \tag{9}$$

$$d\sigma_y = \left(\frac{\partial \sigma_y}{\partial T}\right) dT + \left(\frac{\partial \sigma_y}{\partial \sigma_i}\right) d\sigma_i \tag{10}$$

Using the Eqs 8 through 10, the increase of the DBTT, ΔT can be estimated by the following equation

$$\Delta T = -\int \frac{1}{\left(\frac{\partial \sigma_y}{\partial T}\right)} \left\{ \frac{\sigma_y}{k_y}\left(\frac{\partial k_y}{\partial \sigma_i}\right) + \left(\frac{\partial \sigma_y}{\partial \sigma_i}\right) \right\} d\sigma_i \tag{11}$$

The decrease in DBTT in the present experiment on the NT material can be explained by assuming a change in k_y caused by neutron irradiation. Since k_y is a parameter related to the stress to operate dislocation sources, the decomposition of the complex atmosphere could contribute to a k_y decrease. To make the ΔT be negative, it is necessary that the relative change in k_y caused by neutron irradiation exceeds the change in σ_y, that is

$$-\frac{\Delta k_y}{k_y} > \frac{\Delta \sigma_y}{\sigma_y} \tag{12}$$

As $\Delta\sigma_y/\sigma_y$ is about 10% in the present case, a change in k_y greater than 10% is estimated to have occurred.

If the DBTT change was caused by the mechanism discussed above, we can predict that this favorable property will disappear under some thermal agings. According to the above-mentioned consideration, a decrease in k_y may be easily recovered by subsequent annealing at around 400°C. Furthermore, if the material becomes brittle by impurity segregation to the grain boundary through thermal aging, cleavage fracture is determined by the grain boundary strength. Then a σ_y increase will naturally lead to the DBTT increase.

Some other factors might exist to give an interaction between the thermal aging embrittlement and neutron irradiation embrittlement, such as radiation induced impurity segregation. The next section describes an experiment that was focused on this problem.

Neutron Irradiation Embrittlement of Prethermal Aged Material

Two different sets of specimens were irradiated at around 400°C with a fluence of about 1×10^{18} n/cm^2 ($E > 1$ MeV): one set was in the as-received condition, and the other was treated by pre-irradiation aging at 475°C for 1000 h. The material used was the normalized and tempered 2¼Cr-1Mo steel presented in Tables 1 and 2 with the designation NT-2.

Figures 11 and 12 represent the tensile and Charpy impact properties before and after irradiation including those of thermally controlled specimens. The DBTT of the material without prethermal aging decreased slightly after neutron irradiation. But the thermally treated material showed an increase in the DBTT, although upper shelf energy and tensile properties did not show any difference between the two specimens.

As might be expected from the above arguments, the DBTT in the prethermal aged material increased. The result of the fractographic examination revealed that the nonprethermal aged material exhibited all transgranular type fracture, and the neutron irradiation did not alter the fracture appearance. On the other hand, prethermal aged material showed mixed transgranular and intergranular fracture. Figure 13 shows how the neutron irradiation affected the fracture appearance in terms of the fraction of intergranular fracture (FIF). The FIF of the aged material was found to increase after neutron irradiation. This change can be attributed to an increase in the transgranular

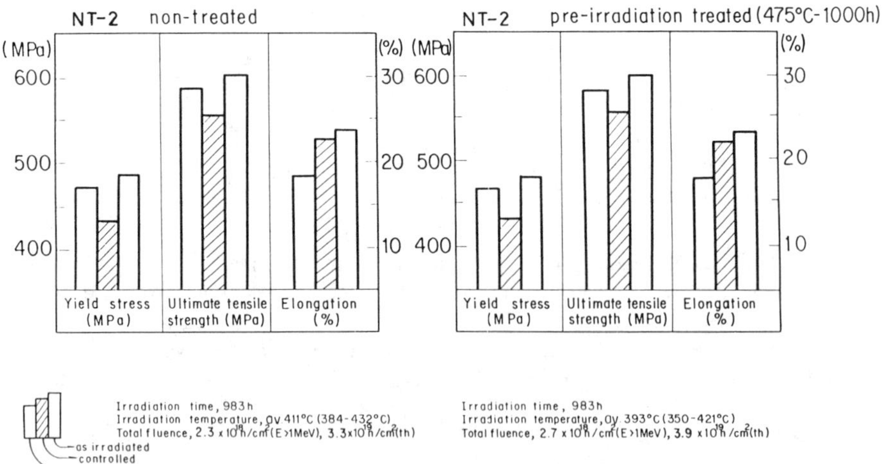

FIG. 11—*Tensile properties of the NT-2 material with and without prethermal aging and before and after neutron irradiation.*

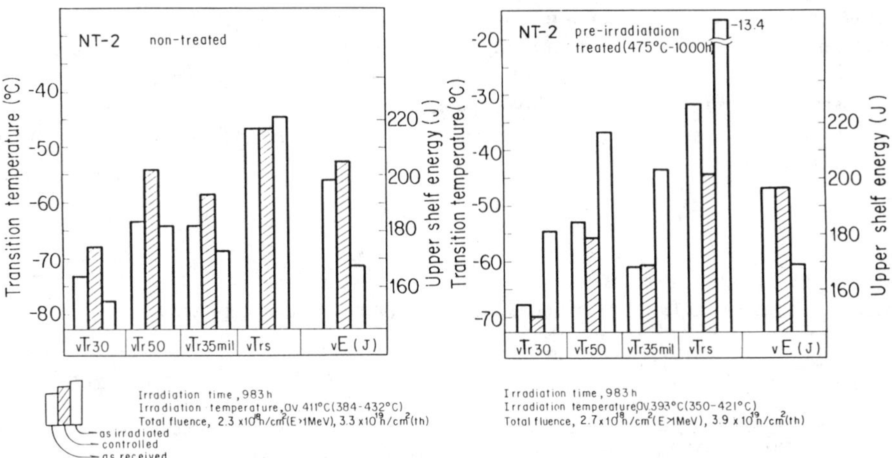

FIG. 12—*Charpy impact properties of the NT-2 material before and after neutron irradiation, showing the effect of prethermal aging on the susceptibility to neutron irradiation embrittlement.*

cleavage fracture stress caused by a decrease of k_y during irradiation and not considered to be due to the decrease of the grain boundary strength induced by neutron irradiation. We conclude that a strong interaction does not exist between the thermal aging embrittlement and neutron irradiation embrittlement within our results.

Conclusions

1. The annealed 2¼Cr-1Mo steel with ferrite plus pearlite has a higher susceptibility to neutron irradiation embrittlement than the normalized and tempered steel with bainite.

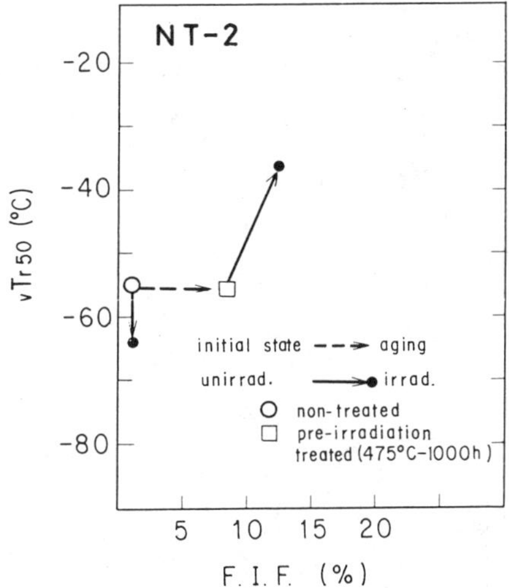

FIG. 13—*Variation of vTr50 and the fraction of intergranular fracture caused by prethermal aging and neutron irradiation.*

2. The normalized and tempered material exhibits a decrease in the ductile to brittle transition temperature after exposure to neutron irradiation.

3. The behavior described in above (2) disappears by subjecting the material to thermal aging. The possible mechanism for this was investigated with the aid of hardness measurements.

4. Strong interaction between the thermal aging embrittlement and neutron irradiation embrittlement was not apparent.

Acknowledgments

The authors wish to express their thanks to the staff of Irradiation Division I and the Hot Laboratory at the Oarai Research Establishment for the support on this work.

References

[1] Hawthorne, J. R., Fortner, E., and Grant, S. P., "Radiation Resistant Experimental Weld Metals for Advanced Reactor Vessel Steels," *Welding Journal,* Vol. 453-S, Oct. 1970.

[2] Kawasaki, M., Fujimura, T., Suzuki, K., Namatame, H., and Kawasaki, M., "Evaluation of the Embrittlement of Pressure Vessel Steels Irradiated in JPDR," *Irradiation Effects on Structural Alloys for Nuclear Reactor Applications, STP 484,* American Society for Testing and Materials, Philadelphia, 1971, pp. 74–95.

[3] Hobsons, R. R. and Wotton, B. L., "The Effect of Fast Neutron Irradiation on the Mechanical Properties of Some Quenched and Tempered Steels," *Irradiation Effects on Structural Alloys for Nuclear Reactor Applications, STP 484,* American Society for Testing and Materials, Philadelphia, 1971, pp. 42–73.

[4] Steel, L. E., Hawthorne, J. R., and Gray, Jr., R. A., "Neutron Irradiation Embrittlement of Several High Strength Steels," NRL-6419, Naval Research Laboratory, Sept. 1965.

[5] Erwin, W. E. and Kerr, J. G., "The Use of Quenched and Tempered 2¼Cr-1Mo Steel for Thick Wall Reactor Vessels in Petroleum Refinery Processes: An Interpretive Review of 25 Years of Research and Application," *WRC Bulletin 275,* Feb. 1982.

[6] Klueh, R. L. and Vitek, J. M., "Elevated-Temperature Tensile Properties of Irradiated 2¼Cr-1Mo Steel," *Journal of Nuclear Materials,* Vol. 126, 1984, p. 9.

[7] Gelles, D. S., "Evaluation of Ferritic Alloy Fe-2¼Cr-1Mo After Neutron Irradiation—Microstructural Development," HEDL-7447, May 1984.

[8] Baird, J. D. and Jamieson, A., "High-Temperature Tensile Properties of Some Synthesized Iron Alloys Containing Molybdenum and Chromium," *Journal of Iron and Steel Institute,* Nov. 1972, p. 841.

[9] Hilton, S. O. and Wilis, F. V., "Effect of Heat Treatment on the Microstructure and Properties of 2¼Cr-1Mo Steel at 1050°F (566°C)," MPC-6, 1980, p. 235.

[10] Olander, D. R., "Fundamental Aspects of Nuclear Fuel Elements," TID-ERDA, 1976, pp. 456–458.

Akira Kohyama,[1] Kyowichi Asano,[1] and Naohiro Igata[1]

Effects of 14-MeV Neutron Irradiations on Mechanical Properties of Ferritic Steels

REFERENCE: Kohyama, A., Asano, K., and Igata, N., "**Effects of 14-MeV Neutron Irradiations on Mechanical Properties of Ferritic Steels,**" *Influence of Radiation on Material Properties: 13th International Symposium (Part II), ASTM STP 956,* F. A. Garner, C. H. Henager, Jr., and N. Igata, Eds., American Society for Testing and Materials, Philadelphia, 1987, pp. 111–122.

ABSTRACT: Seven kinds of 9Cr-2Mo type ferritic steels were irradiated at 300, 473, and 673 K with 14-MeV neutrons from Rotating Target Neutron Source II (RTNS-II) at Lawrence Livermore National Laboratory (LLNL). The maximum neutron fluence was 4×10^{22} n/m^2. Post-irradiation mechanical property tests, microVickers hardness tests, microtension tests, and micro-bulge tests were performed. Microstructural evolutions under irradiation were examined by transmission electron microscopy.

Irradiation hardening and softening were observed at 300 K and at elevated temperatures, respectively. Radiation induced defect cluster formation was observed. The clusters were identified as mostly vacancy type complex defects from their fluence dependence. The irradiation hardening can be reasonably explained by the dispersion hardening from these complex defects. The correlation of mechanical properties among miniaturized sample tests, which are initially derived for unirradiated materials, have proved to be valid for the neutron irradiated specimens with fluences up to 4×10^{22} n/m^2. Specimen size effects on tension test results were investigated.

KEY WORDS: radiation, ferritic stainless steels, dual phase steel, 14-MeV neutron, miniaturized sample testing, size effect, radiation hardening, defect cluster, Vickers hardness, microbulge tests

The research and development (R&D) program of ferritic stainless steels in Japan is important because it is anticipated that these steels are strongly swelling resistant and at the same time an enormous engineering data base exists for such steels. Based on many previous studies [*1,2*], we have selected seven kinds of ferritic stainless steels for irradiation with 14-MeV neutrons and fission neutrons to low fluences. The objectives of this work are to establish the evaluation method of the mechanical property change by means of miniaturized sample testing methods and to evaluate the influence of the cascade damage produced by the high energy neutrons on microstructural evolution and to understand mechanical property changes in terms of these microstructures using industrial materials [*3,4*]. Mechanical properties of welded joints are also a concern, and they are also investigated.

Experimental Procedure

Seven ferritic steels have been selected as shown in Table 1. They were heat treated to provide optimum mechanical properties for the first wall of fusion reactors [*4*]. Metal inert gas (MIG) welded joints of the steels were made, and mechanical properties of their weld metals were also investigated. Neutron irradiations were carried out using 14-MeV neutrons from the Rotating

[1] Associate professor, graduate student, and professor, respectively, Department of Materials Science, The University of Tokyo, Hongo, Bunkyo-ku, Tokyo 113, Japan.

TABLE 1—*Chemical composition of specimens used, wt.%.*

Mark	C	Si	Mn	P	S	Ni	Cr	Mo	V	Nb	Fe
JFMS	0.050	0.67	0.58	0.009	0.006	0.94	9.85	2.31	0.12	0.06	balance
J-1	0.068	0.017	0.50	0.002	0.002	. . .	9.24	1.76	0.16	0.05	balance
45	0.05	0.16	0.55	0.015	0.21	. . .	9.27	1.82	0.11	0.055	balance
29	0.054	0.17	0.63	0.010	0.005	. . .	9.17	0.96	0.15	0.05	balance
C-8	0.045	0.41	0.51	0.007	0.002	. . .	9.10	1.08	0.20	0.25	balance
C-9	0.02	0.35	0.70	0.005	0.007	0.56	11.40	1.01	0.30	. . .	balance
C-12	0.06	0.34	0.51	0.022	0.004	0.14	9.48	2.1	. . .	. . .	balance

NOTE: Mark 29: martensite; others: martensite + ferrite.

TYPE	Specimen Dimensions (mm)			Gage (mm)	
	Width	Length	Thickness	Width	Length
T-1	2.3	12.5	0.15	1.2	5.0
T-2	4.6	25.0	0.30	2.4	10.0
T-3	6.9	37.5	0.45	3.6	15.0

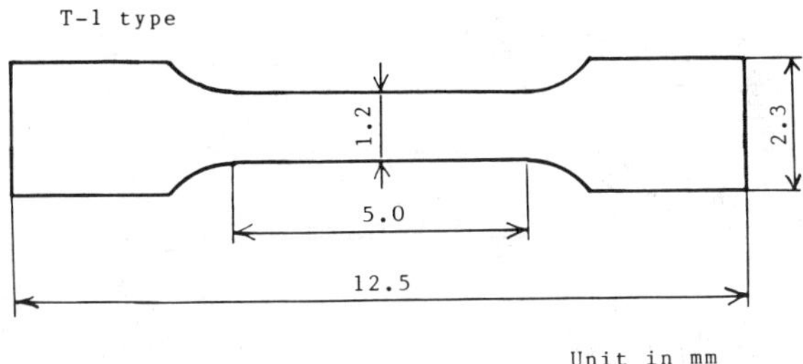

FIG. 1—*Dimensions of tension specimens used.*

Target Neutron Source II (RTNS-II) at Lawrence Livermore National Laboratory (LLNL) and by fission neutrons from Kyoto University Research Reactor (KUR). The 14-MeV neutron irradiation at 300 K was done on the 3-mm-diameter transmission electron microscopes (TEM) disks and three types of tension specimens as shown in Fig. 1. Those at elevated temperatures (473 and 673 K) were done on 3-mm-diameter TEM disks, which were 0.1 mm in thickness. The fission neutron irradiation was performed at about 600 K. Microtension specimens with the same dimension (T1 in Fig. 1) were also irradiated in KUR. Post irradiation tests were carried out at The Research Institute of Iron, Steel and Other Metals, Tohoku University, Ooarai Branch. High resolution TEM observations of 14-MeV neutron irradiated steels were performed using a JEM-200CS at Ooarai and a JEM 200 CX at LLNL. Mechanical properties were measured by microtension tests, microbulge tests, and microVickers hardness tests on the RTNS-II irradiated specimens and by microtension tests and microVickers tests on the KUR irradiated specimens [*3*].

Experimental Results and Discussion

Size Effects in Microtension Tests for Irradiated Specimens

In the microtension tests, a strong thickness dependence on test results was observed when the specimen thickness became less than about 80 μm. Above 80 μm, test data showed only a slight thickness dependence and values were very close to bulk data. This dependence for the C-8 material is shown in Fig. 2. The strong specimen thickness dependence can be understood mainly by interaction among grains and specimen surface effects [*4*]. Another important factor for miniaturized specimens is the occurrence of anomalous large grains, which can often happen for industrial materials with complex microstructures. In case of the dual phase steels studied in this work, anomalous large ferritic grains influenced the ultimate strength, ultimate strain, and total elongation but did not influence 0.2% proof stress [*5*].

The three kinds of tension specimens had their dimensions proportional to each other as indicated in Fig. 1. Figure 3 shows the fluence dependence of the 0.2% proof stress for T1, T2, and T3 specimens. The figure indicates that the 0.2% proof stress is independent of specimen size among those specimens. Figures 4 and 5 show the fluence dependence of ultimate strength and that of ultimate strain for T1, T2, and T3 specimens. Although in case of the T1 specimens, increase of

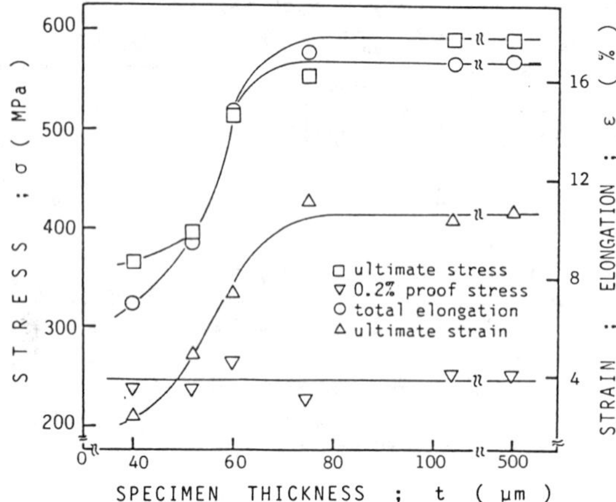

FIG. 2—*Specimen thickness dependence of tensile properties for the C-8 material obtained from miniaturized tension specimens.*

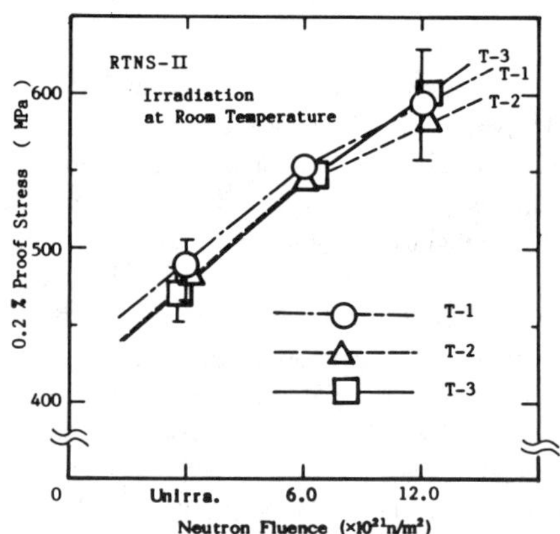

FIG. 3—*Neutron fluence dependence of 0.2% proof stress indicating specimen size effect.*

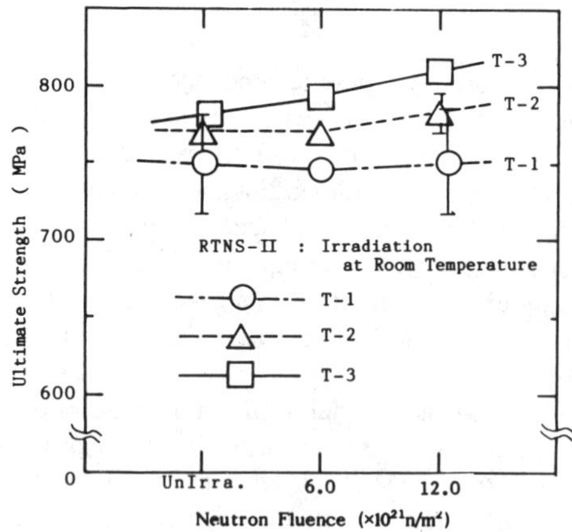

FIG. 4—*Neutron fluence dependence of ultimate stress indicating specimen size effect.*

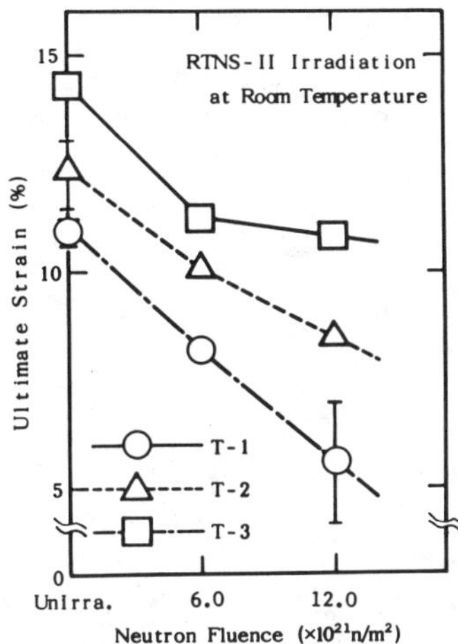

FIG. 5—*Neutron fluence dependence of ultimate strain indicating specimen size effect.*

ultimate strength by neutron irradiation was small and decrease of ultimate strain was large, in case of the T3 specimen, the fluence dependences were reversed. Differences in the ultimate strength and ultimate strain of T1, T2, and T3 specimens showed tendencies to increase with increasing neutron fluence. This means that when heavily irradiated, size effects tended to be enhanced.

The origins of the size effect can be divided into three groups: (1) effects of surface finishing and effect of plastic deformation during specimen preparation, (2) effect of surface on damage microstructure development, and (3) effect of surface on plastic deformation during tension testing. The effects of (1) can be mostly eliminated by annealing and surface finishing after specimen sampling, but these could not be applied to ferritics because these treatments destroy the microstructure. All tension specimens used in this work were obtained from 5 to 10-mm thickness plates by mechanical thinning and final electro-polishing followed by specimen cut-out by shear punch. Therefore, surface condition and microstructures should be identical for all specimens representing the properties of bulk materials. The differences in TEM observed microstructure development in pre-thinned specimens and bulk specimens [*1,2*] suggest that the effect of (2) should be taken into account for high purity materials and for elevated temperature irradiation. But in case of ferritics irradiated at 300 K, this may not be the case. In case of ferritics, the effect of (3) would be the dominant factor. Scanning electron microscope (SEM) observation revealed that there were no significant differences in fracture mode (ductile dimple mode) among unirradiated T1, T2, and T3 specimens. However, in case of neutron irradiated specimens, coarse slip band formation was observed and lath boundary or grain boundary dimple fractures were partly observed, and the fraction of this fracture mode decreased with increasing specimen size. At present we have no indication of saturation of this kind of size effect, and thus further irradiations to examine the fluence dependence at higher fluence should be carried out.

Mechanical Property Change Caused By Cascade Damage

For the materials used in this experiment, mechanical property changes by 14-MeV neutron irradiation were detectable above the fluence of 2×10^{21} n/m^2.

In case of 300 K irradiation, irradiation hardening was clearly observed in every steel as shown in Fig. 6. The increase in yield stress was more significant than that in ultimate stress, reflecting the formation of very fine defect clusters as hardening obstacles. A decrease in uniform elongation

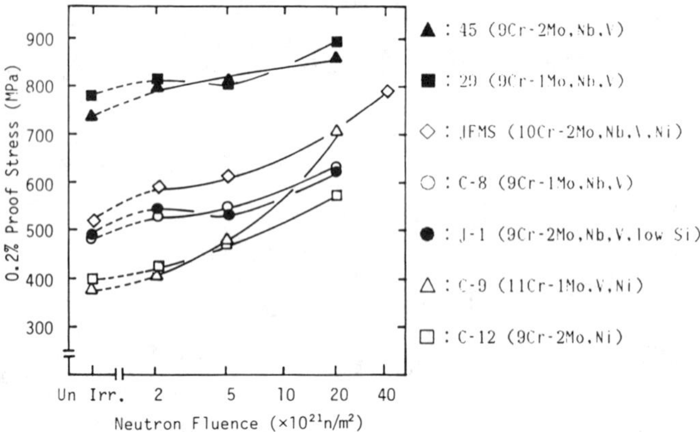

FIG. 6—*Neutron fluence dependence of ultimate strength for steels investigated.*

was also observed in most of the steels. The irradiation hardening and embrittlement was suppressed by additions of niobium and vanadium. The same tendencies were observed using microVickers hardness measurements and also microbulge tests.

The elevated temperature irradiated TEM disks were microVickers hardness tested and were successively microbulge tested. The maximum load/thickness value obtained from the microbulge test is proposed for evaluating the ultimate tensile stress from the tension test in Ref *4* and *6*. The microbulge test results can be seen in Ref *6*, where good correlation among ultimate tensile strength, Vickers hardness value, and the maximum load/thickness value obtained from the microbulge test are demonstrated. The microVickers hardness test and microbulge test results show a trend of irradiation softening with increasing neutron fluence for the elevated temperature irradiated steels and a trend of irradiation hardening for the 300 K irradiated specimens as shown in Fig. 7.

The irradiation induced increases of 0.2% proof stress are plotted as a function of neutron fluence in the Fig. 8. The gradient of the lines in the figure was named as irradiation hardening index n. The n index is related to the following equation.

$$\text{(increment of 0.2\% proof stress)} = \text{(constant)} \times \text{(neutron fluence)}^n \quad (1)$$

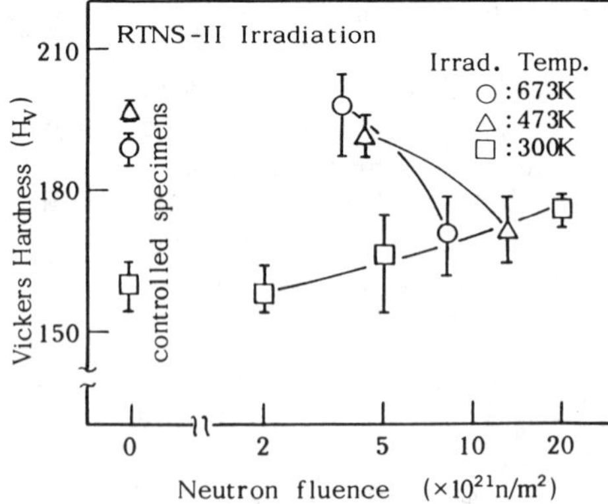

FIG. 7—*Neutron fluence dependence of Vickers hardness for 9Cr-2Mo steel (indentation load: 200 g).*

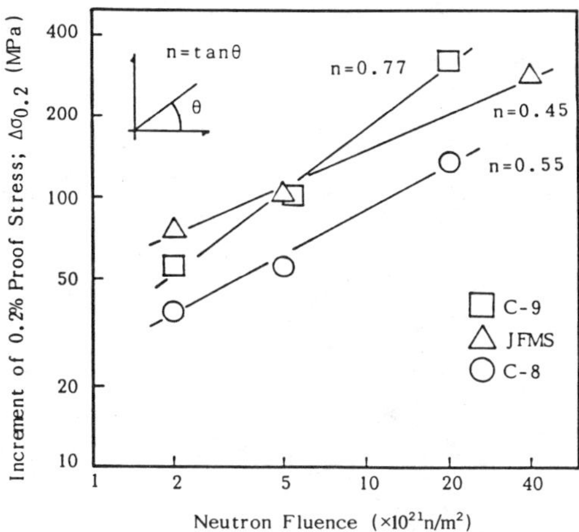

FIG. 8—*Neutron fluence dependence of 0.2% proof stress increase by 14 MeV neutron irradiation at 300 K.*

The observed n values, shown in Fig. 8, ranged mostly from 0.45 to 0.55 with an exception of 0.77 for Material C-9. The index value of 0.5 suggests that the irradiation induced yield stress increases are due to dispersion hardening from the irradiation induced defect clusters where the extent of the increase depends on the average defect cluster spacing. Thus, the linear increase of yield stress as the function of neutron fluence corresponds to the lack of cascade overlapping at low neutron fluence. This is also proved by the linear increase of number density of the defect clusters with increasing neutron fluence, observed. The slight shift of n value from 0.5 would be reflected in the resolution of fine carbide precipitates by cascade damage and local dislocation unpinning by climb motion. Wavy and uniform slip bands observed in unirradiated steels were changed to coarse crossing slip bands with increasing neutron fluence. This change in the deformation characteristic was interpreted to be introduced by dislocation channelings. The microVickers test to the irradiated samples provided informations about the individual phases. A typical example is shown in Fig. 9. In this figure, irradiation induced hardening was observed in base metal and weld metal, indicating that the JFMS material has the same hardening behavior in weld metal and base metal. Wih 200-g indentation load, the measured hardness value correlated well with the tension test results. At the lower indentation loads, less than 100 g, significant irradiation hardening was observed both in ferrite grains and martensite grains. However, the C-9 and C-12 steels, which did not include niobium or vanadium, showed irradiation induced hardening only in ferrite grains and not in martensite grains. The wider data scatter band in the irradiated martensitic phase as opposed to that in the unirradiated martensitic phase may reflect some of the characteristics of the irradiation effects, dislocation channeling, for example, but at present we do not have enough information to understand them.

Many investigations demonstrate correlation rules for hardness data, uniaxial tensile data, bend test data, and bulge test data for simple materials, but the applicability to complex materials, such as dual phase steels, and for irradiated materials are yet known. We found fairly good correlations between tension tests and Vickers hardness tests for the unirradiated steels. These are

$$H_v = 0.31\sigma_u \text{ and } H_v = 0.35\sigma_y \tag{2}$$

Figure 10 shows the correlation between tension test results (0.2% proof stress) and Vickers hardness test results in the 14-MeV neutron irradiated steels. The results obtained in this work indicate that the correlation of the Vickers hardness and the tensile strength for unirradiated ferritic

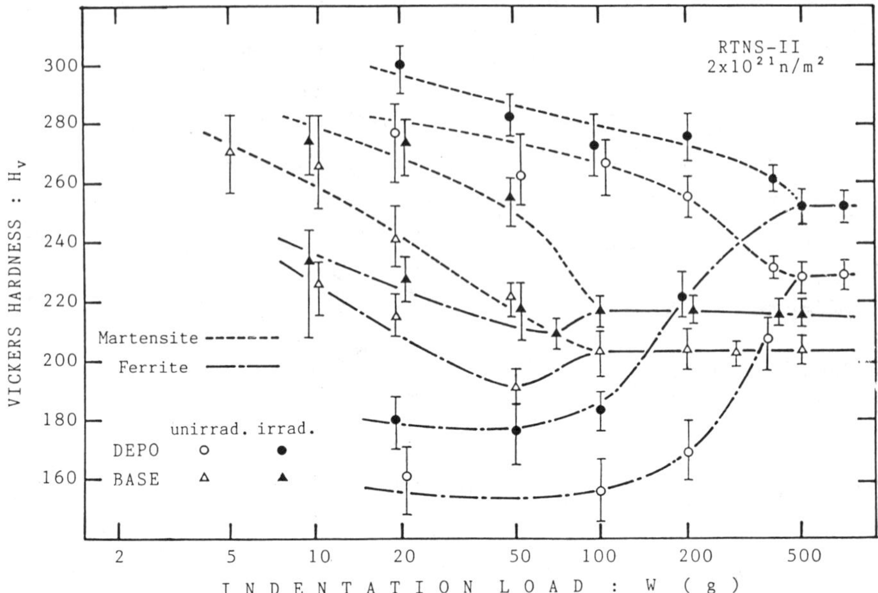

FIG. 9—*Indentation load dependence of Vickers hardness in MIG welded joint of JFMS.*

stainless steel changes for neutron irradiated specimens for doses up to 4 × 10^{22} n/m². The slight deviation of the data from the correlation lines at small hardness values reflect post yield deformation and nonuniform deformation in the Vickers hardness test.

Microstructural Evolution in Ferritics

1. Microstructure Change from Cascade Damage—The significant feature of defect cluster formation in the specimens irradiated by 14 MeV neutrons at 300 K is the small vacancy type defect cluster formation with a size very close to the resolution limit of TEM [*3*]. The density of

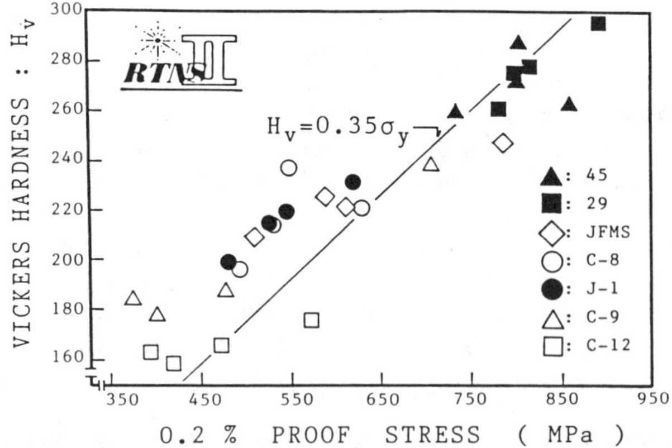

FIG. 10—*Relation between Vickers hardness and 0.2% proof stress for irradiated and unirradiated specimens.*

defect clusters observed in the 4.7×10^{21} n/m^2 irradiated specimens is of order from 10^{24} to 10^{25} m^{-3}. A typical microstructure observed is shown in Fig. 11. These defect clusters were rather uniformly distributed in both phases, and no significant effect of the dislocations or precipitates on cluster formation could be seen. This feature is very similar to that observed in pre-thinned nickel specimens irradiated at room temperature [7].

Two characteristic features of the microstructural evolution in the specimens irradiated at elevated temperatures can be extracted from the results obtained. One is the rather uniform distribution of the vacancy type defect clusters in the matrix with large defect clusters in the vicinity of dislocations mainly observed in steels without niobium and vanadium additions. The other characteristic feature is the local breakaway of the dislocation lines from carbide pinning points and the enhancement of carbide formation. The microstructures of the ferritic stainless steels with niobium and vanadium before neutron irradiation are characterized by fine carbide formation along dislocations. After the 14-MeV neutron irradiation to 3.6×10^{21} n/m^2, some of the carbide decorated dislocations had locally broken away, and at the same time enhancement of carbide formation and growth was observed along dislocation lines before dislocation break away. As shown in Fig. 12, the defect clusters formed by 14-MeV neutrons are remarkably dependent on the irradiation temperature. In case of the 300 K irradiated specimens, the defect cluster density is considerably higher than the others. Many studies of the recovery stage of point defects in pure iron suggest that the free vacancies migrate below 300 K and detrapping takes place between 300 and 600 K. Large recovery stages at 370 and 500 K have been found [8] in pure iron, and these stages can be correlated with the temperature dependence of the defect cluster formation in the ferritic stainless steels investigated here. Although dynamical point defect reactions during the collision process and the higher point defect mobility in the point defect condensed region should be taken into account, the dominant factor that determines the defect cluster microstructure may be a flow of vacancies back to sub cascades. This explains why defect clusters larger in size and lower in number density at 673 K than those at 473 K were produced.

2. *Fluence Dependence of Microstructural Evolution*—Neutron fluence dependences of microstructure development were investigated and were quantitatively analyzed at two fluences. At the

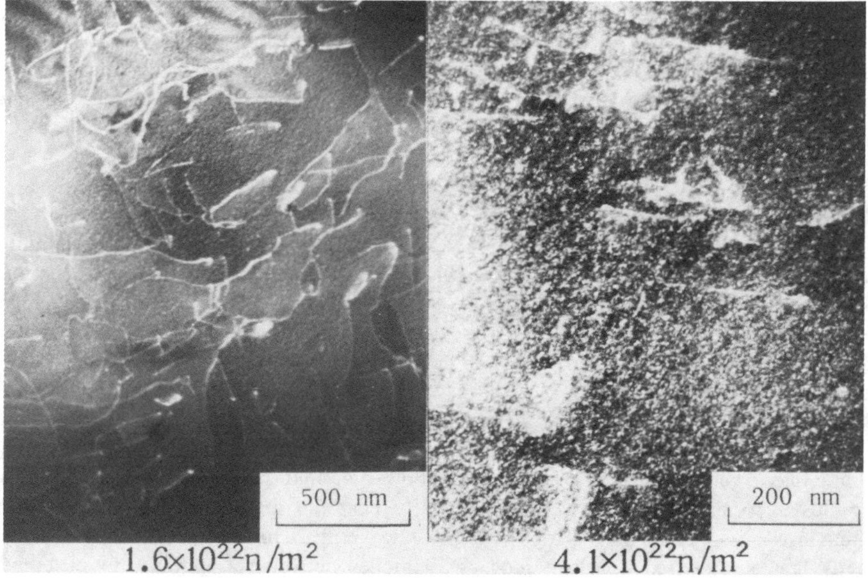

FIG. 11—*Neutron damaged defect structure observed in C-12.*

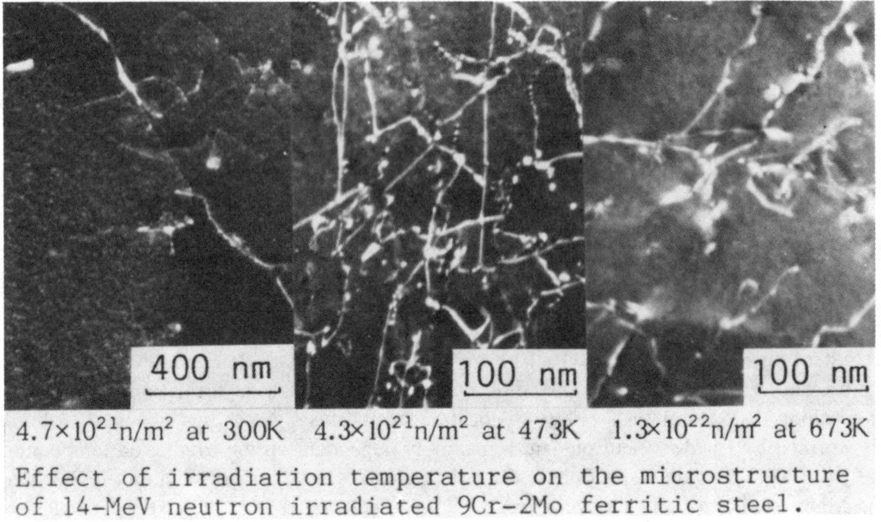

FIG. 12—*Irradiation temperature dependences of defect cluster formation.*

lower fluence, size distribution of defect clusters had an average cluster size of 4 nm with a long tail out to 10 nm. At higher fluence the average size of clusters is reduced to 3 nm with an upper tail out to 7 nm. The density of clusters becomes 2.5 times larger than at the lower fluence, as shown in Fig. 13. Comparing the defect cluster density to the neutron fluence ratio, 1.3×10^{22} n/m^2, is not a sufficiently high fluence for cascade overlapping to affect the elementary collision processes. The change of cluster density and size distribution with the neutron fluence variation can be understood in terms of the destruction of clusters by interstitials released from other cascades. The other characteristic feature is the local dislocation breakaway from carbide pinning where dislocation climb is thought to be the dominant mechanism. Another possible mechanism, such as stress assisted dislocation breakaway, was considered not to prevail here.

Effect of Alloying Elements on the Radiation Responses

In the ferritic stainless steels used in this work, the addition of niobium and vanadium and a reduction in silicon content enhanced the carbide formation, mainly along the dislocation lines and loops. In this case, the microstructural evolution arising from cascade damage, characterized by the local breakaway of dislocations from carbides rather than defect cluster formation and irradiation hardening was observed by fission neutron irradiation at about 600 K in the KUR to 5×10^{22} n/m^2 [*3*]. In these materials most of the vanadium exists as solute or small clusters, which tends to precipitate as V_4C_3, VC, or their agglomerates. Thus at the beginning of this process, scavenging of carbon and vanadium may result in softening, and then progressive precipitation hardening may take place. In the absence of addition of niobium and vanadium, radiation induced defect cluster formation is observed preferentially near the dislocations, and only irradiation softening was observed up to a fluence of 5×10^{22} n/m^2. This feature is thought to have an indirect effect on the irradiation softening as mentioned above. But as to the room temperature irradiation, we could not find significant differences in microstructural evolution by the neutron

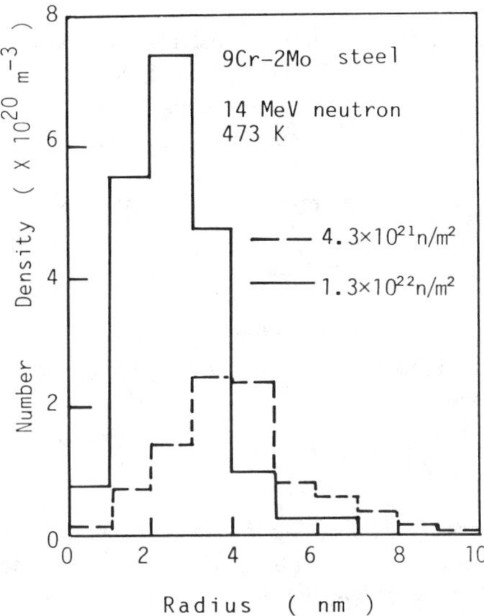

FIG. 13—*Size distribution of defect clusters observed in 14-MeV neutron irradiated ferritic steel.*

irradiation. Only a slight difference in mechanical property changes, as mentioned above, were found among the seven kinds of steels. The detailed inspection of the irradiated microstructures and additional experiments would lead us to better understanding.

Correlation Between 14-MeV Neutron and Fission Neutron

The microstructures observed in the specimens, 14-MeV neutron irradiated at 673 K, gave basically the same radiation induced microstructure as those observed in fast neutron irradiated specimens at the same temperature in JOYO [*9*] and as those observed in 9Cr ferritic steels irradiated with fast neutrons in FFTF [*10*]. The microstructure consisted of a very fine defect cluster formation uniformly in matrix. Mechanical property change results from JOYO and FFTF experiments showed a trend of irradiation softening at low fluences followed by irradiation hardening after critical neutron doses. This is the same trend as was observed in 14-MeV neutron irradiated materials. Fission neutron irradiation in KUR was compared with fusion neutron irradiation [*3*], and the data from KUR coincided well with the extrapolated values of RTNS-II data. At present, within a few sets of data, fusion neutron and fission neutron experiments show no qualitative discrepancy as to irradiation effects.

Conclusions

1. Radiation induced defect cluster formation is observed, and clusters are interpreted to be mostly vacancy type. They are distributed uniformly at 300 K irradiation and are preferentially produced near dislocations at elevated temperatures.

2. The temperature dependence of defect cluster formation suggests that effective vacancy migration takes place between 473 and 673 K.

3. The fluence dependence of mechanical property shows a trend for irradiation softening for 473 and 673 K irradiations and irradiation hardening for 300 K irradiation.

4. The addition of niobium and vanadium and a reduction in silicon content enhance carbide formation, where the microstructural evolution by cascade damage is characterized by the local

breakaway of dislocations, rather than defect cluster formation; also irradiation hardening is observed after fission neutron irradiation in the KUR to 5 $\times$ 10^{22} n/m^2.

5. The correlations of mechanical properties among miniaturized specimen tests, which are initially derived for unirradiated materials, have proved to be valid for 14-MeV neutron fluences up to 4 $\times$ 10^{22} n/m^2.

6. Specimen size effects on tensile properties have been investigated and 0.2% proof strength was found to be insensitive to specimen size, whereas ultimate strength and ultimate stress were strongly dependent on specimen size, and the dependences were accelerated at higher neutron fluences.

7. As to fission-fusion correlation, within a few sets of data, fusion neutron and fission neutron experiments show no qualitative discrepancy as to irradiation effects, both for microstructural and mechanical property changes.

Acknowledgments

The authors wish to express their appreciation to Dr. Y. Kohno, Mr. K. Komamura, and Dr. K. Asakura for their kind assistance throughout this work. They are grateful to Professors K. Kawamura and K. Sumita for organizing the program and to Drs. C. Logan, D. Heikkinen, and D. Short of LLNL for their great help in the D-T neutron irradiation procedure. They wish to express their appreciation to Professors M. Kiritani, N. Yoshida, K. Abe, and H. Matsui for help with the irradiations and the arrangements for post irradiation experiments. They are also grateful to The Research Institute of Iron, Steel and Other Metals, Tohoku University, Ooarai Branch for facilitating the post irradiation inspection.

Japan-US Fusion Cooperation Program, Collaboration on the RTNS-II Utilization was supported by The Ministry of Education, Science and Culture, Japan.

References

[1] Fujita, T., Asakura, K., Sawada, T., Takamatsu, T., and Otoguro, Y., *Metallurgical Transactions,* Vol. 12A, 1981, p. 1071.

[2] Igata, N., *Journal of Nuclear Materials,* Vols. 133/134, 1985.

[3] Kohyama, A., Asakura, K., Komamura, K., Igata, N., Kiritani, M., and Fujita, T., *Journal of Nuclear Materials,* Vols. 133/134, 1985, p. 628.

[4] Kohyama, A., Komamura, K., and Igata, N., *Proceedings of the University of Tokyo—Harbin Institute of Technology Symposium on Materials Sciences,* N. Igata and T. C. Lei, Eds., University of Tokyo, 1985, pp. 119–128.

[5] Igata, N., Miyahara, K., Uda, T., and Asada, S., *The Use of Small-Scale Specimens for Testing Irradiated Materials, STP 888,* American Society for Testing and Materials, Philadelphia, 1986, pp. 161–170.

[6] Kohyama, A., Asakura, K., Asano, K., and Igata N., *Journal of Nuclear Materials,* Vols. 141–143, 1986, p. 921.

[7] "RTNS-II 1983 Annual Report," UCID-1983783, Lawrence Livermore National Laboratory, Livermore, CA, 1984.

[8] Kiritani, M., Takata, H., Moriyama, K., and Fujita, F. E., *Philosophy Magazine,* Vol. A40, 1979, p. 779.

[9] Kohyama, A., Asakura, K., Igata, N., and Tujita, T., Abstract for Annual Meeting of The Japan Institute of Metals, Tokyo, Japan, April, 1986, p. 111.

[10] Gelles, D. S. and Hamilton, M. L., ADIP Semiannual Progress Report, DOE/ER-0045/13, 1985, pp. 128–140.

Shanghsien H. Rou,[1] *Kay J. Harrelson,*[1] *and Roy C. Wilcox*[1]

Impurity Element Effects on the Toughness of 12 Cr-1 Mo Steel

REFERENCE: Rou, S. H., Harrelson, K. J., and Wilcox, R. C., "**Impurity Element Effects of the Toughness of 12 Cr-1 Mo Steel,**" *Influence of Radiation on Material Properties: 13th International Symposium (Part II), ASTM STP 956,* F. A. Garner, C. H. Henagar, Jr., and N. Igata, Eds., American Society for Testing and Materials, Philadelphia, 1987, pp. 123–130.

ABSTRACT: The effects of the trace impurity elements, phosphorus, sulfur, and silicon, on the impact toughness of a 12 Cr-1 Mo steel were investigated. The results of Charpy impact tests showed that the additions of all three elements produced an increase in the ductile-brittle transition temperature and the addition of phosphorus or sulfur caused a decrease in the upper shelf energy. Impurity segregation is used to explain the effect of phosphorus. The higher silicon concentrations produced increased amounts of delta ferrite, which is detrimental to toughness. Sulfur promotes nonmetallic inclusions and delta ferrite, both of which can result in embrittlement.

KEY WORDS: ferritic steels, impact toughness, phosphorus, sulfur, silicon, mechanical properties

The resistance to irradiation induced swelling [*1–4*] and high temperature creep resistance [*5–8*] make ferritic steels excellent candidate materials for applications in nuclear power plants. However, marginal fracture toughness at low temperatures and the significant transition from ductile to brittle behavior near 0°C limit the applications of these alloys [*9*]. This work investigates the effects of doped impurity elements, sulfur, phosphorus, and silicon on the impact behavior of a ferritic steel known as Sandvik alloy HT-9 (12 Cr-1 Mo-V-W). The present study investigated which impurity element has the greatest effect on the impact properties and to what degree the impact properties might be affected as a function of impurity concentration.

Experimental Procedure

Two bars of 12 Cr-1 Mo steel manufactured by Carpenter Technology Corporation (Heat 84425) were used as the basis of this investigation. The compositions of these bars are given in Table 1. Phosphorus and sulfur additions were made to remelted ingots of Bar 1 while silicon was added to Bar 2. Raw materials were vacuum melted. Phosphorus and sulfur were varied by adding ferrous phosphide (Fe_2P) and ferrous sulfide (FeS), respectively, to the base steel. Additions of silicon were made as the pure metal. Ingots of 300 g were obtained. The variation in P-content was 0.014, 0.02, 0.027, 0.036, and 0.044%. The S-content varied as 0.003, 0.01, 0.027, and 0.034% while silicon varied as 0.27, 0.38, 0.47, 0.54, and 0.82%. The smallest amount in each case corresponds to the original or control bar composition. An independent source determined the composition of each ingot by the use of a vacuum spectrometer.

[1] Graduate students and professor, respectively, Materials Engineering, Department of Mechanical Engineering, Auburn University, Auburn, AL 36849.

TABLE 1—*Composition of 12 Cr-1 Mo control material, wt%.*

Element	Bar 1	Bar 2
Cr	12.1	11.84
Mo	1.03	0.99
W	0.52	0.59
V	0.33	0.32
C	0.21	0.19
Ni	0.75	0.51
Mn	0.50	0.58
P	0.012	0.010
S	0.003	0.004
Cu	0.04	0.01
Si	0.21	0.26
Fe	balance	balance

Each ingot was preheated to 760°C and hot rolled to a thickness of 12 mm. Standard size V-notched Charpy specimens were machined parallel to the rolling direction. Specimens were then placed with tantalum and zirconium foils in sealed fused quartz capsules at 4 psia of argon. The heat treatment for each specimen consisted of 1100°C/30 min/AC + 700°C/2 h/AC [*10*]. Impact tests were conducted at temperatures between −196 to 500°C to investigate the ductile-brittle

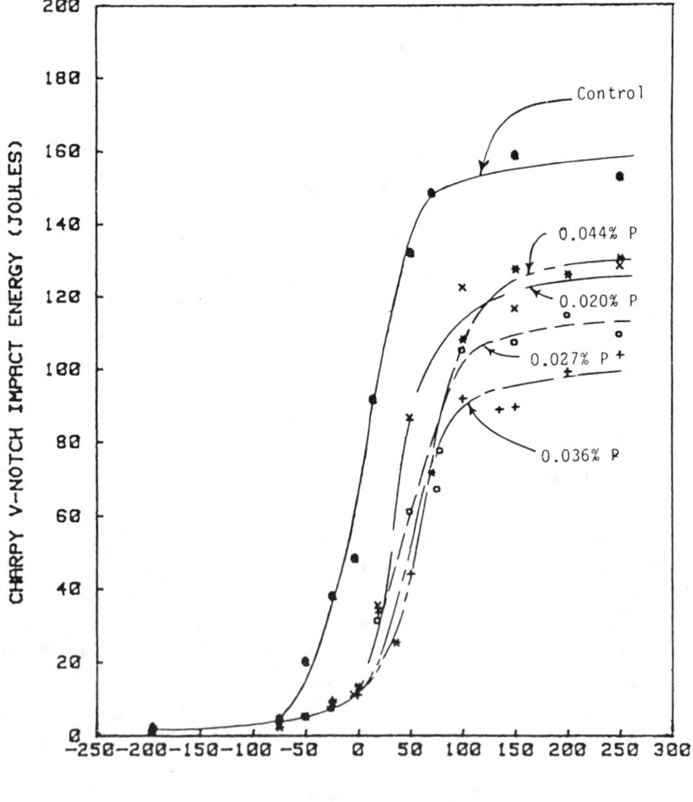

FIG. 1—*Effect of phosphorus on impact energy.*

transition behavior. The ductile-brittle transition temperature (DBTT) was considered to be the temperature corresponding to an impact energy of 50 J.

Results and Discussion

The typical heat treated microstructure of both the original and the remelted undoped material consists of a matrix of tempered martensite with regions of elongated delta ferrite. Two types of carbides generally are present. Small primary carbides resulting from tempering are dispersed throughout the matrix while larger precipitated secondary carbides are present mainly at the delta ferrite-matrix interface.

Impact values plotted as a function of temperature yielded sigmoidal curves. Such curves clearly depict the upper shelf region, the lower shelf region, and the transition region. The average upper shelf energies were calculated by averaging data points of the upper shelf. Average lower shelf energies were calculated in a similar manner. The 50-J ductile-brittle transition temperatures were determined by interpolating the corresponding temperature from the sigmoidal curves at this impact energy.

Phosphorus

Figure 1 shows the impact curves obtained with varying phosphorus contents. This figure shows that increasing phosphorus concentrations shift the impact curves to the right. In general, the average upper shelf energies decrease as the DBTT increases. The average upper shelf energy (Fig. 2) decreased almost 26 J per 0.01 w/o phosphorus increment up to 0.036 w/o. However, at

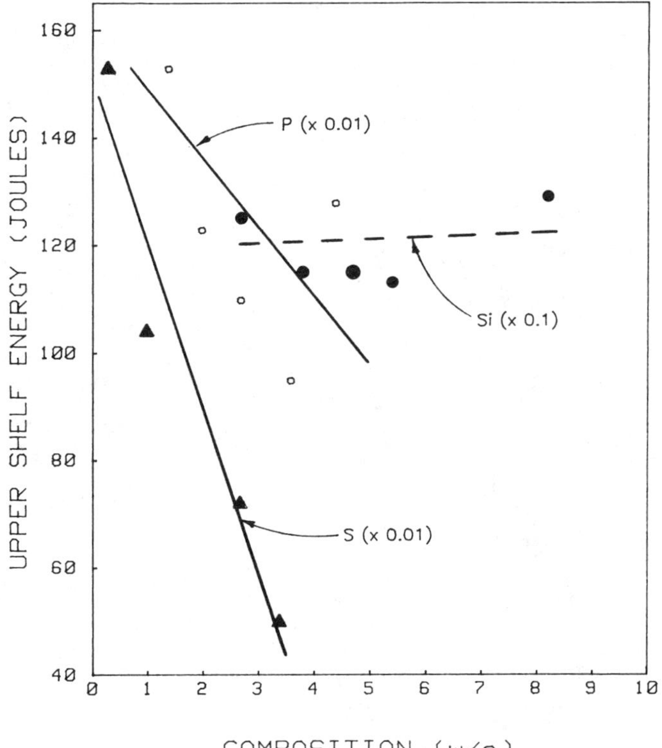

FIG. 2—*Effect of impurity content on the upper shelf energies.*

the 0.044 w/o phosphorus impurity level, the average upper shelf energy increased to 128 J. The unchanged lower shelf energies depict that phosphorus additions did not affect the already poor intrinsic impact toughness at low temperatures.

The DBTT at 50-J impact energy showed an approximate 28°C increase in each 0.01% phosphorus increment (Fig. 3). Nevertheless, there was an increasing tendency of the DBTT to level off at the higher phosphorus contents (0.037 and 0.044%). Basically, increasing phosphorus concentrations exhibited an embrittling effect on the HT-9 alloy and thereby led to a decrease in the average upper shelf energies and an increase in the DBTT.

The basic microstructure, tempered martensite plus delta ferrite [*10*], was not changed with the addition of phosphorus up to 0.044 w/o. Precipitated secondary carbides were still evident at delta ferrite grain boundaries with primary carbides within the tempered martensitic matrix. The tendency of the DBTT to level off at high phosphorus concentrations has been reported [*11*] and possibly can be explained by the equilibrium segregation model [*12*]. Phosphorus prefers to segregate to regions of high lattice misfit such as secondary phase interfaces and grain boundaries. These imperfections can only provide a fixed number of segregation sites. As more and more segregation sites are occupied, the movement of phosphorus to these sites is progressively reduced. With additional concentrations of phosphorus, there is less of an influence on phosphorus segregation, and the rate of the detrimental effects of phosphorus on toughness properties should gradually be reduced. As the number of phosphorus atoms exceeds the available segregation sites, uniform distribution of phosphorus is more likely to occur.

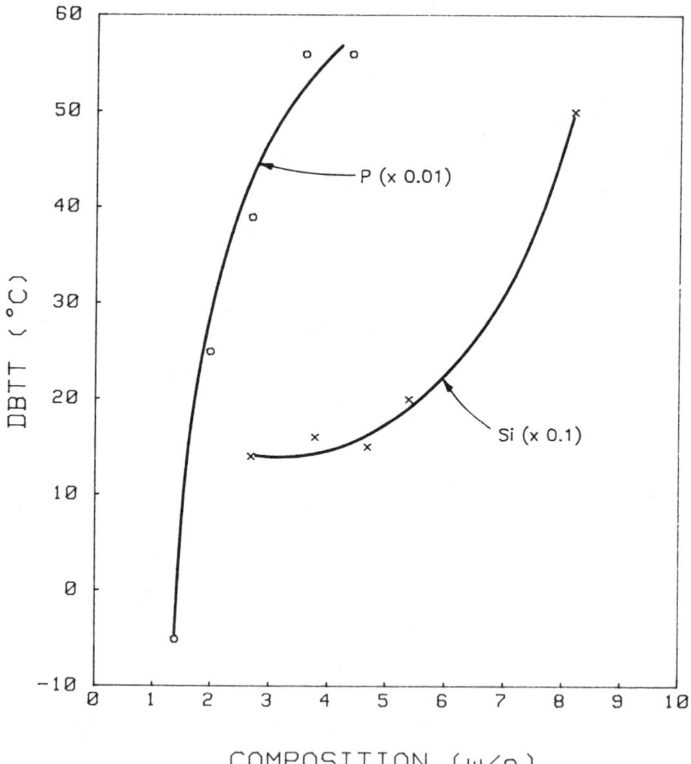

FIG. 3—*Effect of impurity content on the DBTT.*

Eventually, equilibrium segregation results in reduced local plastic incompatibilities, which then makes nucleation of microcracks easier. Since as phosphorus increases, the influence on segregation is progressively reduced, the rate of increase in the DBTT also is expected to diminish gradually. This hypothesis is consistent with the results of this investigation.

Phosphorus segregates at various interfaces [*12*]. Such segregation would reduce the allowable plastic flow of the alloy. Therefore, the upper shelf energies should degrade with increasing phosphorus concentrations and is consistent with the results of this investigation.

Sulfur

Impact curves for the different sulfur contents are shown in Fig. 4. As sulfur content was increased the upper shelf energy decreased drastically (Fig. 2). The addition of 0.01 w/o sulfur to the control composition decreased the upper shelf energy from 153 to 104 J. The 0.027 and 0.034% sulfur samples yielded upper shelf energies of 72 and 50 J, respectively. The sulfur content of 0.034 represents a ten-fold increase over that in the control material of 0.003%. An increase in the DBTT over that of the control material is evident when sulfur is added (Fig. 4). Sulfur has little or no effect on the lower shelf energy.

The addition of sulfur produced an increase in the amount of nonmetallic inclusions along with an increase in delta ferrite. These changes in microstructure are considered to be the possible cause for the observed embrittlement of the 12 Cr-1 Mo steel used in this investigation.

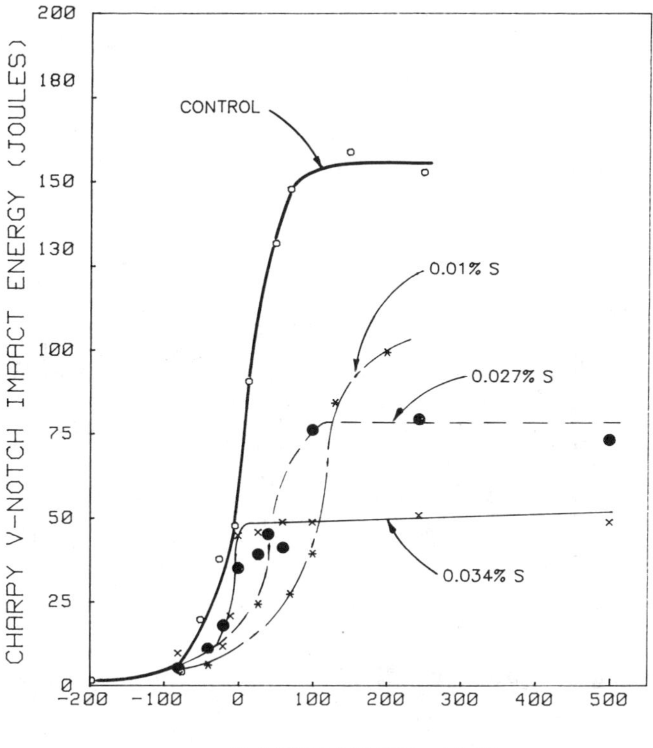

FIG. 4—*Effect of sulfur on impact energy.*

Silicon

The impact energies of the five 12 Cr-1 Mo steels with 0.27, 0.38, 0.47, 0.54, and 0.82% silicon concentrations are illustrated in Fig. 5. Generally, the element silicon has a mild effect on the impact properties of 12 Cr-1 Mo steels. The impact energies fluctuate slightly, and the DBTTs are gradually shifted upward with increasing silicon contents. The 50-J DBTTs (Fig. 3) show a large increase between 0.54 and 0.82% silicon. However, the fluctuation of the 50-J DBTTs in the lower silicon content alloys indicate that there is little effect below 0.5% silicon on the DBTT.

No noticeable change in the average upper shelf energies was observed in the silicon-doped alloys (Fig. 2). The addition of 0.82% silicon only resulted in shifting the transition range to a higher temperature. Although silicon does not have much of an effect on the upper shelf energy, the addition of this element gradually shifts the impact curves to higher temperatures. The average lower shelf energies were not affected by the addition of silicon.

The addition of silicon caused an increase in the amount of delta ferrite present in the microstructure. The typical microstructure consists of tempered martensite and elongated delta ferrite. Precipitated secondary carbide particles were present along the delta ferrite grain boundaries with primary carbides in the tempered martensitic matrix.

Silicon is a ferrite stabilizer [*13*] and serves to balance the volume fraction of ferrite and austenite/martensite in steels. The addition of silicon causes the shrinkage of the austenite region,

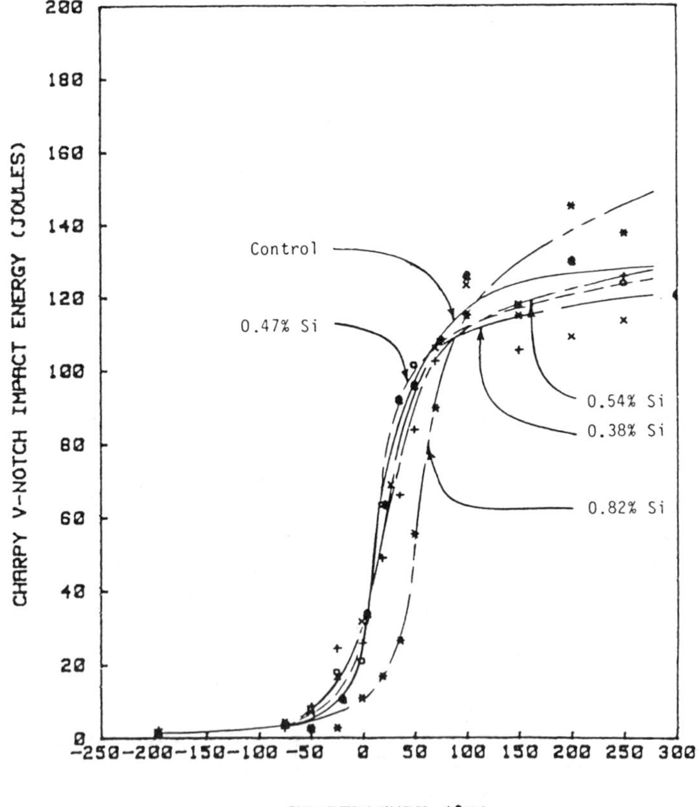

FIG. 5—*Effect of silicon on impact energy.*

which in effect, changes the fraction of ferrite and austenite/martensite during the cooling of the casting, during annealing both before and after hot-rolling, and during the final heat treatment. Kinetically, at high silicon contents the delta ferrite is more stable, thus reducing the tendency to transform to austenite in the equilibrium austenite region. The increase in the DBTT is directly attributed to this increase in delta ferrite. A previous investigation found that delta ferrite is detrimental to the impact properties [*14*].

A model is postulated to explain the shift of the DBTT of the higher silicon concentration alloys. As delta ferrite increases with increasing silicon, more grain boundary regions are generated and thereby provide extra places for carbides to precipitate. The total quantity of all precipitated carbides, however, remain unchanged in the silicon-doped alloys. Thus, fewer carbides can precipitate in the matrix. Usually, these secondary carbides forming along the delta ferrite grain boundaries are larger than those in the tempered martensitic matrix. These larger secondary carbides provide favorable places for critical microcracks to nucleate, and hence the DBTT is degraded.

Two contradictory effects seem to govern the behavior of the upper shelf energies. First, high silicon concentrations have been reported to result in tougher ferrite [*15*]. Second, the secondary carbides along the delta ferrite boundaries should make the generation of microcracks easier and be detrimental to the upper shelf energy. These two contradictory effects likely compete with each other and result in the observed fluctuations of the upper shelf energies.

Conclusions

The following conclusions resulted from the investigation of trace impurity elements on the impact properties of 12 Cr-1 Mo steels:

1. Phosphorus additions increase the DBTT and decrease the upper shelf energy thus causing embrittlement. Impurity segregation of phosphorus at positions of lattice misfits is considered to be responsible for the degradation of the impact properties.
2. Sulfur additions result in drastic embrittlement of 12 Cr-1 Mo steels. Sulfur promotes the formation of nonmetallic inclusions and an increase in delta ferrite both of which can reduce toughness.
3. Silicon causes embrittlement with an increase in the DBTT. A model involving the increase in delta ferrite is postulated to explain the upward shift in the DBTT by silicon.

References

[*1*] Powell, R. W., Peterson, D. F., Zimmershield, M. K., and Bates, J. F., "Swelling of Several Commercial Alloys Following High Fluence Neutron Irradiation," *Journal of Nuclear Materials,* Vols. 103 and 104, 1981, pp. 969–974.

[*2*] Erler, J., Maillard, A., Bru, G., Lehman, J., and Dupouy, J. M., "The Behavior of Ferritic Steels under Irradiation with Fast Neutrons," *Irradiation Behavior of Metallic Materials for Fast Reactor Core Components,* proceedings of the conference held 4–8 June 1979 in Ajaccio, France, p. 11.

[*3*] Gelles, D. S., "Microstructural Examination of Several Commercial Ferritic Alloys Irradiated to High Fluence," *Journal of Nuclear Materials,* Vols. 103 and 104, 1981, pp. 975–980.

[*4*] Little, E. A. and Stow, D. A., "Void Swelling in Irons and Ferritic Steels II. An Experimental Survey of Materials Irradiated in a Fast Reactor," *Journal of Nuclear Materials,* Vol. 87, 1979, p. 25.

[*5*] Paxton, M. M., Chin, B. A., and Gilbert E. R., "The In-Reactor Creep of Selected Ferritic, Solid Solution Strengthened, and Precipitation Hardened Alloys," *Journal of Nuclear Materials,* Vol. 95, 1980, p. 185.

[*6*] Paxton, M. M., Chin, B. A., Gilbert, E. R., and Nygren, R. E., "Comparison of the In-Reactor Creep of Selected Ferritic, Solid Solution Strengthened and Precipitation Hardened Commercial Alloys," *Journal of Nuclear Materials,* Vol. 80, 1979, p. 144.

[*7*] Harries, D. R., Standring, J., Barnes, W. D., and Lloyd, G. L., "The U. K. Fast Rcactor Materials Programme," *Effects of Radiation on Materials: Eleventh Conference, STP 782,* H. R. Brager and J. S. Perrin, Eds., American Society for Testing and Materials, Philadelphia, 1982, pp. 1197–1217.

[8] Hasiguti, R. R., "Japanese Program of Materials Research for Fusion Reactors," *Journal of Nuclear Materials,* Vols. 103 and 104, 1981, pp. 51–56.

[9] Smidt, F. A., Jr., Hawthorne, J. R., and Provenzano, V., "Fracture Resistance of HT-9 After Irradiation at Elevated Temperature," *Effects of Radiation on Materials: Tenth Conference, STP 725,* D. Kramer, H. R. Brager, and J. S. Perrin, Eds., American Society for Testing and Materials, Phialdelphia, p. 269.

[10] Wilcox, R. C. and Chin, B. A., "Austenitizing and Microstructure of a HT-9 Steel," *Metallography,* Vol. 17, 1984, pp. 285–298.

[11] McMahan, C. J., Jr., "Solute Segregation and Intergranular Fracture in Steels: A Status Report," *Materials Science and Engineering,* Vol. 42, 1980, p. 215.

[12] McLean, D., *Grain Boundaries in Metals,* Clarenda Press, Oxford, United Kingdom, 1957.

[13] Fidler, R. S. and Gooch, D. J., "The Hot Tensile Properties of Simulated Heat Affected Zone Structures in 9CrMo and 12 CrMoV Steels," *Ferritic Steels for Fast Reactor Steam Generators,* Proceedings of the British Nuclear Energy Society (BNES) International Conference, London, 1978, pp. 128–135.

[14] Wilcox R. C. and Chin, B. A., "Reduction of the Ductile Brittle Transition Temperature of the Ferritic-Martensitic Alloy HT-9," *Journal of Nuclear Materials,* Aug. 1982, p. 52.

[15] Davies, G. R., "Influence of Silicon and Phosphorus on the Mechanical Properties of Both Ferrite Dual Phase Steels," *Metallurgical Transactions A,* Vol. 10A, 1979, p. 113.

Robert D. Brown,[1] Monroe S. Wechsler,[2] and Christoph Tschalär[3]

Tensile Properties of Several 800 MeV Proton-Irradiated bcc Metals and Alloys

REFERENCE: Brown, R. D., Wechsler, M. S., and Tschalär, C., "**Tensile Properties of Several 800 MeV Proton-Irradiated bcc Metals and Alloys**" *Effects of Radiation on Material Properties: 13th International Symposium (Part II), ASTM STP 956,* F. A. Garner, C. H. Henager, Jr., and N. Igata, Eds., American Society for Testing and Materials, Philadelphia, 1987, pp. 131–140.

ABSTRACT: A spallation neutron source for the 600-MeV proton accelerator facility at the Swiss Institute for Nuclear Research (SIN) consists of a vertical cylinder filled with molten Pb-Bi. The proton beam enters the cylinder, passing upward through a window in contact with the Pb-Bi eutectic liquid that must retain reasonable strength and ductility upon irradiation at about 673 K to fluence of about 1×10^{25} protons/m^2.

Investigations are underway at the 800-MeV proton accelerator at the Los Alamos Meson Physics Facility (LAMPF) to test the performance of candidate SIN window materials under appropriate conditions of temperature, irradiation, and environment. Based on considerations of chemical compatibility with molten Pb-Bi, as well as interest in identifying fundamental radiation damage mechanisms, Fe, Ta, Fe-2.25Cr-1Mo, and Fe-12Cr-1Mo (HT-9) were chosen as candidate materials.

Sheet tensile samples, 0.5-mm thick, of the four materials were fabricated and heat treated. The samples were sealed inside capsules containing Pb-Bi and were proton-irradiated at LAMPF to two fluences, 4.8 and 54×10^{23} p/m^2. The beam current was approximately equal to the 1 mA anticipated for the upgraded SIN accelerator. The power deposited by the proton beam in the capsules was sufficient to maintain sample temperatures of about 673 K. Post-irradiation tensile tests were conducted at room temperature at a strain rate of 9×10^{-4}s^{-1}. The yield and ultimate strengths increased upon irradiation in all materials, while the ductility decreased, as indicated by the uniform strain. The pure metals, Ta and Fe, exhibited the greatest radiation hardening and embrittlement. The HT-9 alloy showed the smallest changes in strength and ductility.

The increase in strength following irradiation is discussed in terms of a dispersed-barrier hardening model, for which the barrier sizes and formation cross sections are calculated.

KEY WORDS: accelerator, ferritic steels, iron, proton irradiation, tantalum, tension test

The Swiss Institute for Nuclear Research (SIN) is studying the possibility of constructing a neutron source/beam stop that consists of a vertical cylinder of liquid Pb-Bi. The Pb-Bi will be irradiated by a 600 MeV proton beam which will pass through a window in the base of the cylinder. At the design current, the beam current density will be about 0.2 μA/mm^2, essentially that presently available (at 800 MeV) at the Los Alamos Meson Physics Facility (LAMPF) A-6 irradiation effects area. The window in the cylinder base must be made of an alloy chemically compatible with the molten Pb-Bi and able to maintain reasonable ductility and strength following proton irradiation to fluences of about 10^{25} protons/m^2 at a temperature of about 673 K. Initial examination of materials for chemical compatibility with the Pb-Bi indicated that iron-base alloys and tantalum could be candidates. From the standpoint of alloys showing reasonable strength, it

[1] Staff member, Los Alamos National Laboratory, P.O. Box 1663, MSH840, Los Alamos, NM 87545.
[2] Professor of materials science, Iowa State University, Ames, IA 50011.
[3] Swiss Institute for Nuclear Research.

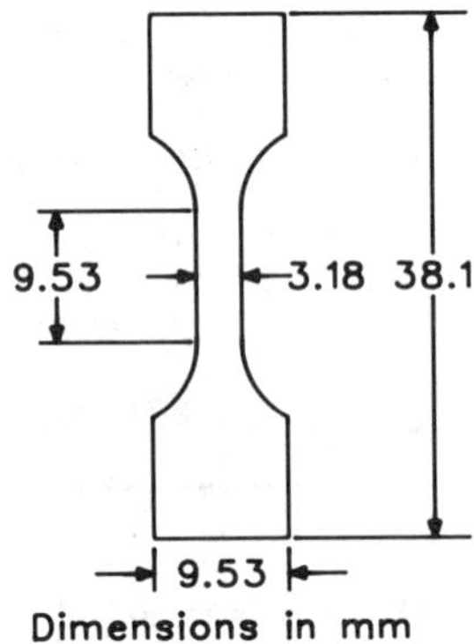

FIG. 1—*The sheet tensile sample, 0.5-mm thick.*

was decided to study Fe-2.25Cr-1Mo. Corrosion studies of liquid Pb-Bi in contact with several ferritic alloys have shown that higher chromium content may result in lower corrosion resistance [*1,2*], so we decided to include samples of pure iron and a higher chromium alloy, HT-9 (Fe-12Cr-1Mo). Samples of tantalum were also included to determine whether their corrosion resistance might be superior to that of the iron-base alloys. The performance of these candidate window materials for SIN was evaluated under appropriate conditions of temperature, irradiation, and environment at LAMPF.

Experimental Procedure

Tensile samples were prepared from 0.5-mm thick sheet (Fig. 1) and were heat treated as described in Table 1, which also includes brief comments on the resulting microstructures. Heat treatments were done in a high-vacuum furnace from which it was not possible to quench the samples. Of the two iron-base alloys, the HT-9 developed a fully martensitic structure, while the Fe-2.25 Cr-1Mo alloy developed a complex microstructure that is being investigated further.

At the LAMPF beam energy of 800 MeV, the proton range is about 3400 kg/m^2 in copper, corresponding to 0.38 m. Therefore, thick samples may be penetrated with little energy loss. Calculations of 800-MeV proton radiation damage parameters, such as displacement per atom concentration (dpa) and transmutation atom concentration, have been reported previously for

TABLE 1—*Sample heat treatments.*

Material	Temperature, K	Time, s	Comments
Fe	1193	300	about 0.5-mm diameter grains
Ta	1523	3600	not recrystallized
Fe-2.25Cr-1Mo	1213	420	10–20 μm grain size; not fully hardened
	873	1800	
HT-9	1323	420	martensitic, < 5% ferrite
	973	3600	

copper [3] and are also available for iron and HT-9.[4] The beam full width at half maximum is about 20 mm at the irradiation location. To bring the samples to temperature and maintain them in contact with the molten Pb-Bi simultaneously, the samples were placed inside short thin-walled cylindrical capsules with exterior dimensions of 50-mm diameter and 8-mm length. The wall thickness was about 0.5-mm. The capsules were machined from the Fe-2.25Cr-1Mo alloy. After the samples were mounted in a sheet metal holder inside each capsule, the capsule was filled with molten Pb-Bi, which was allowed to solidify, with the excess being machined away. The capsule lid was then welded in place, and the capsule was tested for leaks. For each of the two irradiations, two capsules were placed inside the isotope production target holder so that the outside of the capsules was directly cooled by circulating water.

Although the A-6 area has now been rebuilt to allow instrumentation of proton irradiation samples, the tests reported herein were conducted prior to this renovation. The approximate maximum sample temperatures were deduced from a preliminary irradiation, in which the tensile samples were replaced with melt wires. It was found that the melt wires reached temperatures of 658 to 673 K at proton beam currents typical of those used during the later tensile sample irradiations. This indicated that the combination of beam heating in the samples and Pb-Bi together with heat transfer through the Pb-Bi to the cooling water would provide temperatures in the range desired.

The proton fluence was determined from the integrated beam current and profile area (full width at half maximum). The fluences were 4.8×10^{23} p/m^2 for the low-fluence samples and 5.4×10^{24} p/m^2 for the high-fluence samples. These proton fluences correspond, respectively, to about 0.14 dpa and 1.6 dpa in iron samples, assuming a damage energy cross section of 300 barn · keV[4] and a displacement energy of 40 eV, the latter recommended in the ASTM Standard Recommended Practice for Neutron Radiation Damage Simulation by Charged-Particle Irradiation (E 521-77). Based on similar calculations for tungsten with a damage energy cross section of 1480 barn · keV and a displacement energy of 90 eV, the displacement levels for tantalum were inferred to be about 0.26 and 2.9 dpa at the low and high fluence, respectively. Irradiation with 800 MeV protons results in a high rate of helium production via the spallation process [4]. Irradiation of iron samples to our high fluence is calculated to have produced about 290 ppm helium. Similar information is not available for tantalum.

Post-Irradiation Examination and Testing

Examination of the capsules following irradiation showed some corrosion and minor pitting of the surfaces exposed to the cooling water. The capsules were cut open in an alpha-containment hot cell to contain the ^{210}Po formed by spallation from the Bi during irradiation. The opened capsules were placed on a hot plate where the Pb-Bi was melted and the sample holder removed. After some minor cleaning of the sample holder, it was possible to remove the individual samples. At this point, a radiation survey of the samples revealed high levels of alpha, beta, and gamma activity. Repeated scubbing with organic solvents proved insufficient to reduce the contamination to levels that could be tolerated outside an alpha-containment box, and it was decided to clean the tensile samples in acid followed by rinses in water and alcohol to remove the contaminated surface layers. This approach succeeded in adequately reducing the surface contamination so that the samples could be removed from the alpha-containment cell for tensile testing. Following acid cleaning, the cross-sectional area of each sample was remeasured in the hot cell using a micrometer. This reduced area was used to calculate the tensile stresses. For the Fe and Fe-2.25Cr-1Mo samples, the decrease in area was about 25%, while for the HT-9 and Ta, the decrease was about 4%. In addition, more careful checks of a few samples showed that the cross-sectional area was

[4] Information provided by D. R. Davidson, Group MP-3. Los Alamos National Laboratory, June 1985.

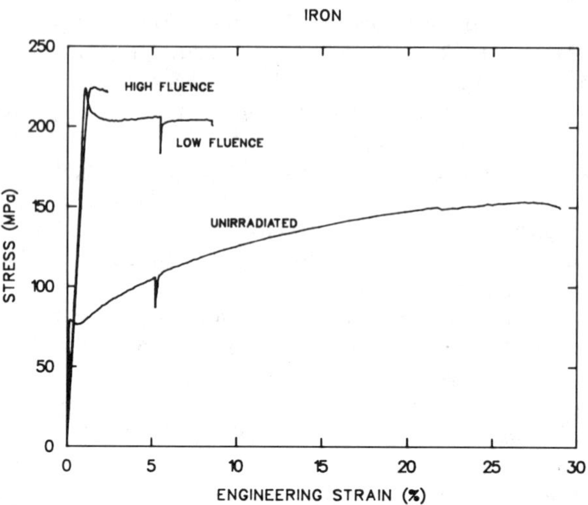

FIG. 2—*Stress-strain curves for representative iron samples.*

no longer uniform, but varied locally by about 5 to 10% along the sample. Since it was not possible to measure this variation routinely or to find the smallest cross-sectional area on the sample, the tensile strengths for the irradiated samples may be understated. Mercury is expected to be produced due to transmutations in the Pb-Bi eutectic during irradiation. However, in accord with a previous report, the evaporation of mercury during the opening of the capsules was not found to be a problem.

Tensile tests were conducted at a nominal crosshead speed of 8.5×10^{-3} mm/s, which for the gage length of 9.5 mm resulted in a strain rate of 8.9×10^{-4} s^{-1}. Prior to pulling each sample, marks were made near the ends of the gage section and the distance between them measured. Following the test, which typically did not result in breaking the sample, this distance was remeasured and used to check the total plastic strain determined from the load-extension curve. A clip-on extensometer was used to monitor strain during the first portion of the test. At about 5% strain it was necessary to remove the extensometer, and the test was continued with crosshead displacement as the measure of strain.

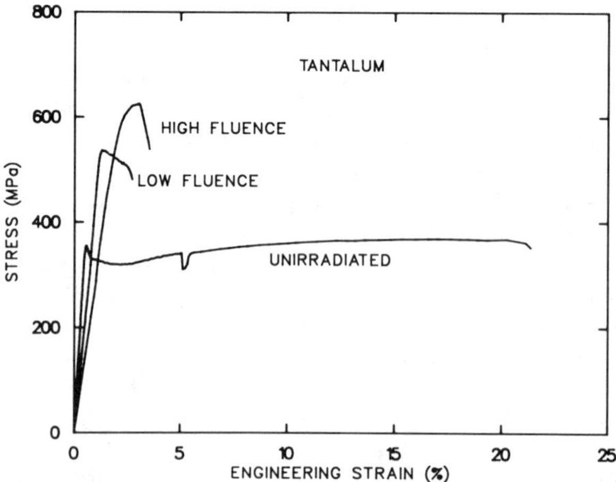

FIG. 3—*Stress-strain curves for representative tantalum samples.*

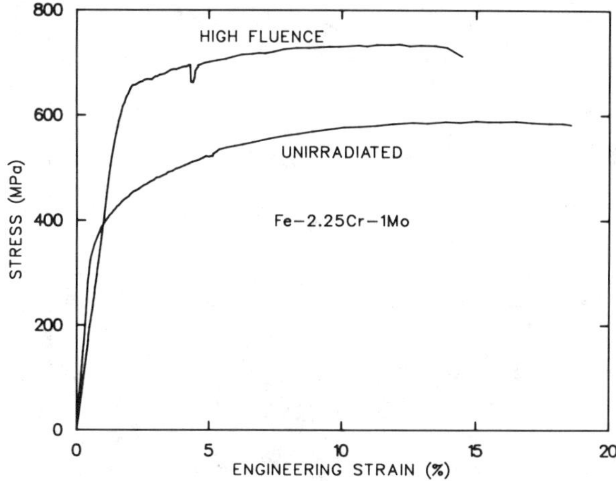

FIG. 4—*Stress-strain curves for representative Fe-2.25Cr-1Mo samples.*

Results

Following irradiation to 5.4 × 10^{24} p/m^2, no corrosion was observed on any of the samples. When the Pb-Bi was melted to remove the samples in the hot cell, only a small amount of Pb-Bi was observed to adhere to the samples. Typically the samples did not appear to have been wetted by the Pb-Bi.

Figures 2 and 3 show typical stress-strain curves for Fe and Ta, respectively. The pre-irradiation stress-strain curves typically exhibited a yield drop, and the yield stress was taken to be the lower yield stress. The sharp discontinuity in the flow curve at about 5% strain is an artifact associated with the change in the basis of strain measurement from the extensometer to crosshead displacement. There is a marked decrease in the rate of work hardening upon irradiation, so much so that for the higher fluence the Fe and Ta samples exhibited plastic instability immediately upon plastic deformation and the yield and ultimate stresses were coincident. The two iron-base alloys, Fe-2.25Cr-1Mo and Fe-12Cr-1Mo (HT-9) show much smaller fractional changes in yield stress and ultimate tensile stress than do the pure materials (see Figs. 4 and 5). These alloys undergo

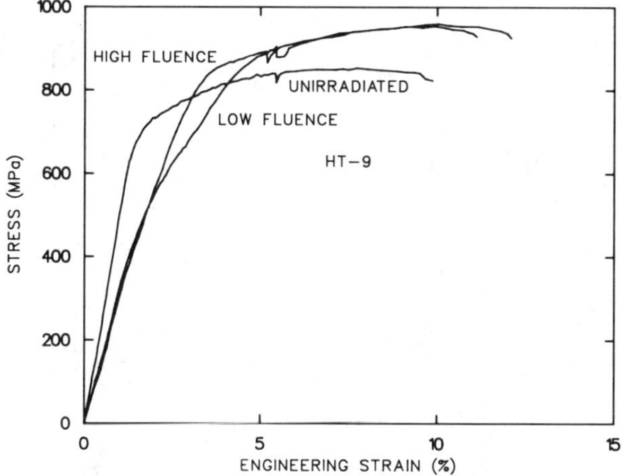

FIG. 5—*Stress-strain curves for representative HT-9 samples.*

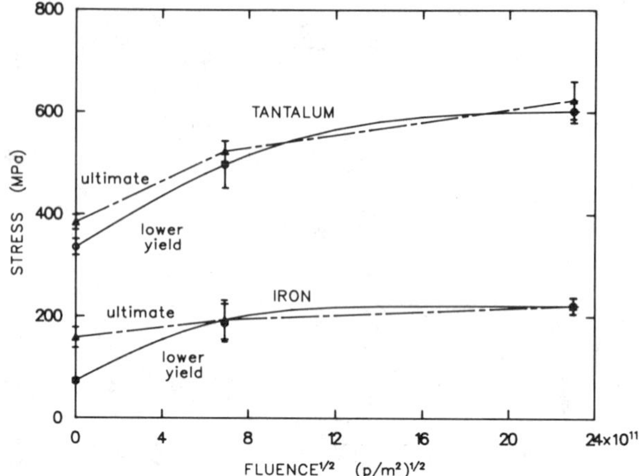

FIG. 6—*The yield stress and ultimate tensile stress versus the square root of the proton fluence for Fe and Ta samples. The solid curve for yield stress represents the fit to Eq 1.*

some work hardening following even the higher proton fluence, and the ductilities are not greatly reduced compared to the pre-irradiation values. The data for the yield stress, ultimate tensile stress, and uniform strain are presented in Figs. 6 through 9, where each point represents the average of the data from all the samples tested and the standard deviation for each point is shown. For Fe and Ta, the increases in strength are shown in Fig. 6, while the decreases in ductility are displayed in Fig. 7. Both figures present the data as a function of the square root of the fluence. It can be seen that both strength and ductility show large changes at the lower fluence and a tendency to saturate at the higher fluence. Figures 8 and 9 present the iron-base alloy data. The fractional increases in strength and decreases in ductility with fluence are much smaller than for the Fe and Ta. Data for the Fe and Ta, as well as for the alloys, show the decrease in work hardening as the irradiation fluence increases. In the case of the HT-9, the changes in ductility with fluence are small compared to the scatter in the results at the two fluences. Non-uniform

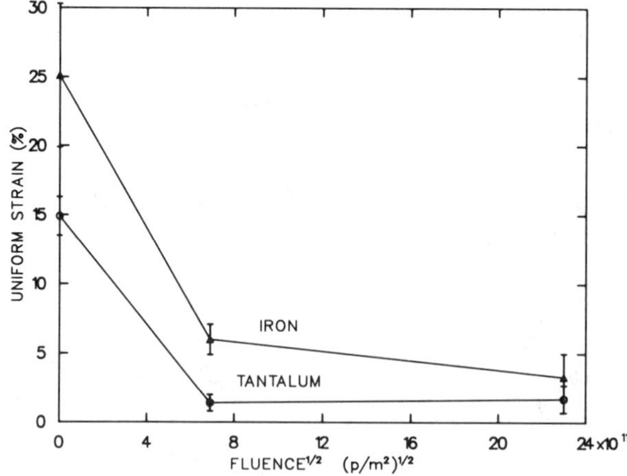

FIG. 7—*Uniform strain versus the square root of proton fluence for iron and tantalum samples.*

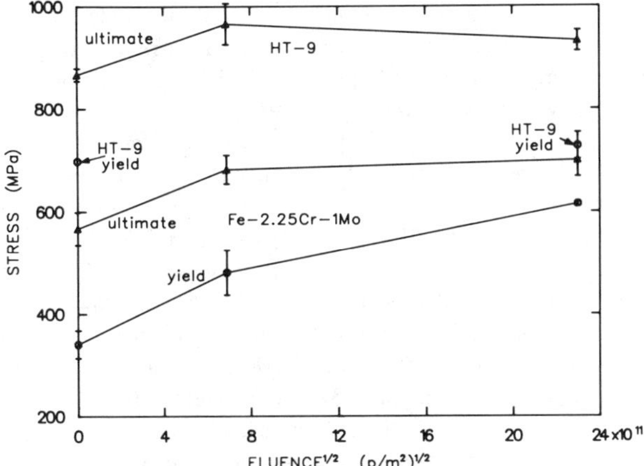

FIG. 8—*The yield stress and ultimate tensile stress versus the square root of proton fluence for Fe-2.25Cr-1Mo and HT-9 samples.*

reduction of the sample area during the acid cleaning may have led to loading difficulties and premature sample failure in certain cases. No data have been presented for the low-fluence strength of HT-9 samples due to loading difficulties, nor is the uniform strain shown for the Fe-2.25Cr-1Mo samples at the low fluence, because the values obtained were believed to be anomalously low.

Discussion

The mechanism of radiation hardening is usually analyzed in terms of a dispersed barrier model. Radiation-produced microstructural irregularities serve as barriers to the motion of slip dislocations, and this causes an increase in yield stress. Of special interest here is the dependence of the increase in yield stress on radiation fluence and the tendency toward saturation of the radiation hardening at higher fluence, as seen in Fig. 6. Tucker and Wechsler [*6*] discussed the neutron fluence

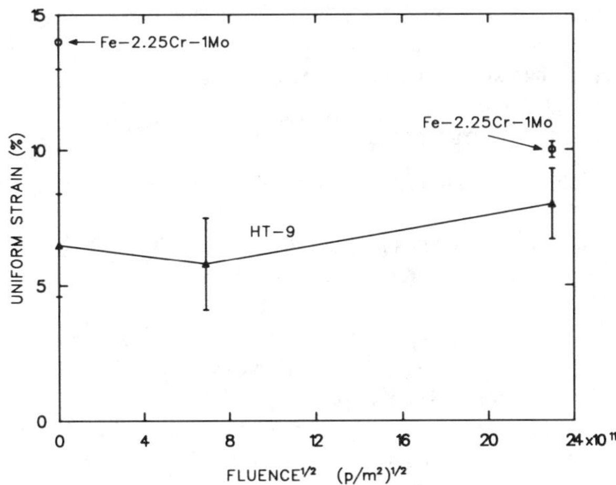

FIG. 9—*Uniform strain versus the square root of proton fluence for Fe-2.25Cr-1Mo and HT-9 samples.*

dependence and saturation of hardening in unalloyed metals irradiated at moderate temperatures and fluences. Under these conditions, the key microstructural irregularities are radiation-produced defect clusters. More recently, Simons and Hulbert [7] discussed radiation hardening in neutron-irradiated stainless steel, where the hardening agents are precipitates or voids. Also, radiation hardening was treated by Lucas et al. [8] for neutron-irradiated pressure-vessel steels, where the hardening was considered to be due to radiation-produced copper precipitation.

In one of the first studies of the effect of spallation radiation damage on tensile properties, Brown and Cost [9] irradiated Ta, Mo, 304 stainless steel, and Alloy 718 sheet tensile samples in the LAMPF 800 MeV proton beam. Large increases in yield stress and decreases in ductility were observed for a proton fluence of about 2×10^{23} p/m^2. The present study is a preliminary attempt to study the effect of 800-MeV proton fluence on tensile properties. Displacement levels, as mentioned earlier, did not exceed 1.6 dpa for Fe or 2.9 dpa for Ta, while the homologous temperatures during irradiation were about 0.37 for Fe and 0.20 for Ta, corresponding to our irradiation temperature of about 673 K. At the moderate displacement concentrations and irradiation temperatures employed, the hardening is therefore assumed to be due chiefly to radiation-produced defect clusters. Consideration of the possible influence of other radiation-produced hardening agents, such as helium and voids, is deferred until the further electron microscope work on the irradiated transmission electron microscope (TEM) disks is completed. Void-induced swelling in Ta has been observed at 773 K [10], but not at 698 K [11].

An equation has been presented by Makin and Minter [12] which describes the saturation in yield strength by assuming that associated with each defect cluster producing the hardening is an exclusion volume within which no other defect cluster may form. The yield strength, σ_Y, as a function of fluence, Φ, may be written as

$$\sigma_Y = \sigma_u + A(1 - \exp(-B\Phi))^{\frac{1}{2}} \quad (1)$$

$$A = \left(\frac{24}{\pi}\right)^{\frac{1}{2}} \frac{F}{bD} \quad (2)$$

$$B = \frac{\pi}{6} n_o \sigma_B D^3 \quad (3)$$

where

σ_u = unirradiated yield stress,
F = critical force to break through or bow around the barrier,
b = Burgers vector,
D = diameter of the exclusion volume (barrier size),
n_o = atomic density, and
σ_B = cross section for barrier production.

Applying Eq 1 to the yield-stress data for Fe and Ta in Fig. 6 gives the curves shown by the solid lines in that figure.

In earlier work on neutron-irradiated niobium [6], the density and size distribution of defect clusters were determined by transmission electron microscopy as a function of fluence, and the increase in yield stress was also measured as a function of fluence. Based on those measurements, the critical force F was found to be approximately 0.5–0.8 Gb^2, where G is the shear modulus. This corresponds to a strong barrier model. Calculations of the force for planar dislocation motion through infinitely strong barriers [13,14] also resulted in values of about 0.8 Gb^2. In keeping with these earlier results, we have used a strong-barrier model with $F = 0.8\ Gb^2$ to calculate the barrier size, D, and the barrier-production cross section, σ_B. Table 2 gives our results for the parameters A and B, together with the barrier sizes and production cross sections.

TABLE 2—*Barrier-model parameters.*

	Ta	Fe
b, nm	0.285	0.248
n_o, m^{-3}	5.59×10^{28}	8.50×10^{28}
G, GPa [15]	68.6	83.1
$F = 0.8\ Gb^2$, nN	4.47	4.09
A, MPa	267	148
B, m^2	9.59×10^{-25}	22.5×10^{-25}
D, nm	162	309
σ_B, μbarn	77	17

We see that the barrier sizes, D, are of the order of several hundred nanometers, which is roughly consistent with the expected size of defect clusters. Also, the barrier cross sections, σ_B, are many orders of magnitude lower than the displacement cross sections, as expected. These conclusions must be regarded as preliminary. For one thing, there is implicit in the above analysis the assumption that the exclusion diameter is the same as the barrier size, which is not necessarily the case. Also, our determinations of A and B in Eq 1 are based on the use of only two fluence levels, which does not permit an estimate of how well the radiation hardening fits Eq 1. Thus, no independent judgement can be made, based on the preliminary experimental information, of the validity of the radiation-hardening model employed in the present work.

Conclusion

Room-temperature tensile tests on Fe, Ta, and two iron-base alloys indicate that the strength increases and the ductility decreases upon 800-MeV proton irradiation at about 673 K. These results are consistent with the well-known effects of fission-neutron irradiations. The application of a strong-barrier hardening model suggests barrier sizes of several hundred nanometers. Work is now underway to determine the microstructure in both the unirradiated and irradiated samples using transmission electron microscopy, following which a more precise determination of the strengthening will be possible.

On the basis of the measured strength and ductility, it appears that either Fe-2.25Cr-1Mo or Fe-12Cr-1Mo would be a suitable alloy for a proton beam window in contact with Pb-Bi. As long as window temperatures do not exceed our maximum irradiation temperature of about 673 K, corrosion appears unlikely to be a serious problem for these alloys.

Acknowledgment

We thank A. G. Nicol for his help in decontaminating and tensile-testing the samples.

References

[1] Romano, A. J., Klamut, C. J., and Gurinsky, D. H., "The Investigation of Container Materials for Bi and Pb Alloys; Part I, Thermal Convection Loops," Report BNL 811 (T313), Brookhaven National Laboratory, Upton, New York, July 1963.

[2] Weeks, J. R., *Nuclear Engineering and Design*, Vol. 15, 1971, pp. 363–372.

[3] Coulter, C. A., Parkin, D. M., and Green, W. V., *Journal of Nuclear Materials*, Vol. 67, 1977, pp. 140–154.

[4] Wechsler, M. S. and Sommer, W. F., *Journal of Nuclear Materials*, Vols. 122 and 123, 1978, pp. 1078–1083.

[5] Brown, R. D., "Examination of Proton-Irradiated Melt Wires," Group MP-7 Technical Note TN-19, Los Alamos National Laboratory, Los Alamos, New Mexico, Nov. 1983.
[6] Tucker, R. P. and Wechsler, M. S., *Radiation Effects,* Vol. 3, 1970, pp. 73–87.
[7] Simons, R. L. and Hulbert, L. A., in *Effects of Radiation on Materials: Twelfth International Symposium, STP 870,* F. A. Garner and J. S. Perrin, Eds., American Society for Testing and Materials, Philadelphia, 1985, pp. 820–839.
[8] Lucas, G. E., Odette, G. R., Lombrozo, P. M., and Sheckherd, J. W., in *Effects of Radiation on Materials: Twelfth International Symposium, STP 870,* F. A. Garner and J. S. Perrin, Eds., American Society for Testing and Materials, Philadelphia, 1985, pp. 900–930.
[9] Brown, R. D. and Cost, J. R., in *Effects of Radiation on Materials: Eleventh Conference, STP 782,* H. R. Brager and J. S. Perrin, Eds., American Society for Testing and Materials, Philadelphia, 1982, pp. 917–926.
[10] Murgatroyd, R. A., Bell, I. P., and Bland, J. T., in *Properties of Reactor Structural Alloys After Neutron or Particle Irradiation, STP 570,* American Society for Testing and Materials, Philadelphia, 1975, pp. 421–432.
[11] Wiffen, F. W., *Journal of Nuclear Materials,* Vol. 67, 1977, pp. 119–130.
[12] Makin, M. J. and Minter, F. J., *Acta Metallurgica,* Vol. 8, 1960, pp. 691–699.
[13] Foreman, A. J. E. and Makin, M. J., *Philosophical Magazine,* Vol. 14, 1966, pp. 911–923.
[14] Kocks, U. F., *Philosophical Magazine,* Vol. 13, 1966, pp. 541–566.
[15] Tegart, W. J., *Elements of Mechanical Metallurgy,* Macmillan, New York, 1966, p. 91.

Fan-Hsiung Huang[1]

Post-Irradiation Fracture Properties of Precipitation-Strengthened Alloy D21

REFERENCE: Huang, F. H., "**Post-Irradiation Fracture Properties of Precipitation-Strengthened Alloy D21,**" *Influence of Radiation on Material Properties: 13th International Symposium (Part II), ASTM STP 956,* F. A. Garner, C. H. Henager, Jr., and N. Igata, Eds., American Society for Testing and Materials, Philadelphia, 1987, pp. 141–150.

ABSTRACT: The precipitation strengthened alloys have the potential for use in fuel cladding and duct applications for liquid metal reactors due to their high strength and low swelling rate. Unfortunately, these high strength alloys tend to exhibit poor fracture toughness, and the effects of neutron irradiation on the fracture properties of the material are of concern. Compact tension specimens of alloy D21 were irradiated in the Experimental Breeder Reactor II to a fluence of 2.7 $\times$ 10^{22} neutrons (n)/cm^2 ($E > 0.1$ MeV) at 425, 500, 550, and 600°C. Fracture toughness tests on these specimens were performed using electric potential techniques at temperatures ranging from 205 to 425°C. The material exhibited low post-irradiation fracture toughness which increased with either increasing test or irradiation temperature. The tearing modulus, however, increased with increasing irradiation temperature but decreased with increasing test temperature. Results were analyzed using the *J*-integral approach. The fracture toughness of irradiated D21 was evaluated essentially following the procedure recommended in ASTM Test Method E813. It was found that the data elimination limits illustrated in E813 were too large for the specimens tested, although the thickness criterion was satisfied. The precautions needed to determine J_{Ic} based on a reduced data qualification range were discussed.

KEY WORDS: fracture toughness, precipitation strengthened alloy, irradiation specimen size effects, data validity, data exclusion zone

Materials exposed to irradiation environments are susceptible to irradiation-induced embrittlement and swelling. It has been shown that austenitic steels used as cladding and duct materials for liquid metal reactors suffer severe swelling. In search of low swelling metals, the high-nickel, precipitation strengthened alloys and the ferritic steels are attractive. The precipitation strengthened alloys have high strength, however, they exhibit low ductility in certain temperature ranges and poor fracture toughness. In an effort to improve the mechanical properties of the material for potential core component applications, a series of commercial-like developmental alloys were designed. Among the alloys, D21 has proved to be superior in mechanical properties to the others. However, irradiation embrittlement is still of concern for this alloy. To study the effects of neutron fluence on the fracture properties of the material, fracture toughness specimens were prepared and irradiated in the Experimental Breeder Reactor II (EBR-II).

Due to the limited irradiation space available, small compact tension specimens were designed. In addition, single specimen electric potential techniques were used to reduce the number of specimens needed to obtain sufficient fracture toughness data.

Since the specimens were small, failure by ductile fracture with some plastic deformation had to be analyzed by elastic-plastic fracture mechanics. The *J*-integral approach used in ASTM Standard E813 is suitable for data analysis in small specimen testing if some precautions are taken. The objectives of the present investigation were to evaluate the effect of irradiation on the fracture

[1] Senior scientist, Westinghouse Hanford Company, P.O. Box 1970, Richland, WA 99352.

TABLE 1—*Nominal chemical composition (percent by weight) of alloy D21.*

Fe	Ni	Cr	Mo	Ti	Al	Mn	Si	C	B
58.6	35	8.5	1.0	3.3	1.7	1.0	1.0	0.05	0.005

properties of a precipitation strengthed alloy; to demonstrate the capability of using single specimen, electric potential techniques to obtain post-irradiation fracture toughness data; and to discuss the precautions required in using ASTM Standard E813 for small specimen testing.

Experimental Procedure

Disk-shaped compact tension specimens which were 2.54 mm thick were fabricated from D21 in two different thermo-mechanical treatments (TMT). One was given a final heat treatment of 775°C/8 h/AC + 700°C/2 h/AC (AC = Air Cooled) while the other was in a condition consisting of 1025°C/15 min/AC + 30% cold work. Specimens in the latter TMT were irradiated in EBR-II to a fluence of 2.7×10^{22} neutrons (n)/cm² ($E > 0.1$ MeV). Table 1 shows the chemical composition of D21. The specimen configuration and test procedure were essentially the same as reported earlier [*1*]. Prior to the test, specimens were fatigue precracked using a servo-hydraulic system and the electric potential technique to monitor the crack length. A constant d-c current of 13 amp was applied to the specimens during both the precracking and the test. The test specimen assembly was electrically insulated by a ceramic spacer contained in the couplings of the pull rods. After each test was completed, the cracked specimens were heat tinted at 500°C for 1 h to reveal the amount of crack extension, which provided the data point required for calibration of the potential versus crack extension curve.

Since the specimen dimensions are relatively small, the fracture properties are best studied in terms of the *J*-integral. The value of J_{Ic} was determined as the value of the intersection of a *J* versus crack extension (Δa) curve and a blunting line. Such a plot can be obtained by either a multi-specimen method or a single specimen method. The crack extension Δa was calculated from an electric potential calibration curve, determined as V/V_o versus a/a_0. The curve was normalized to offset the variations due to the test system and specimen. The value of *J* was calculated by

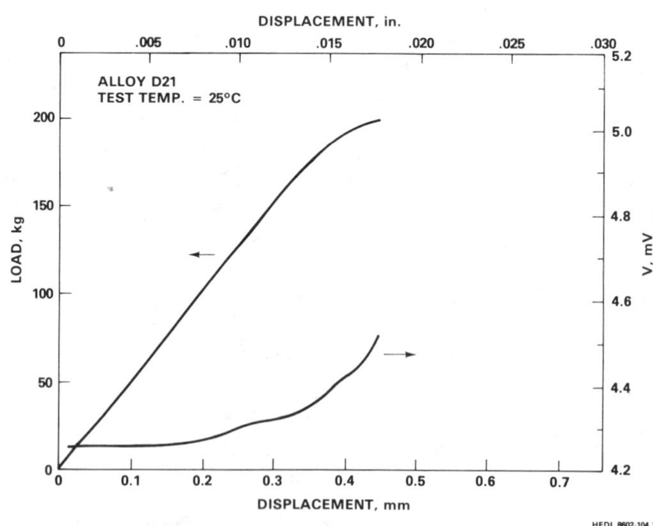

FIG. 1—*Potential output and load versus load-line displacement for unirradiated aged D21 tested at 25°C.*

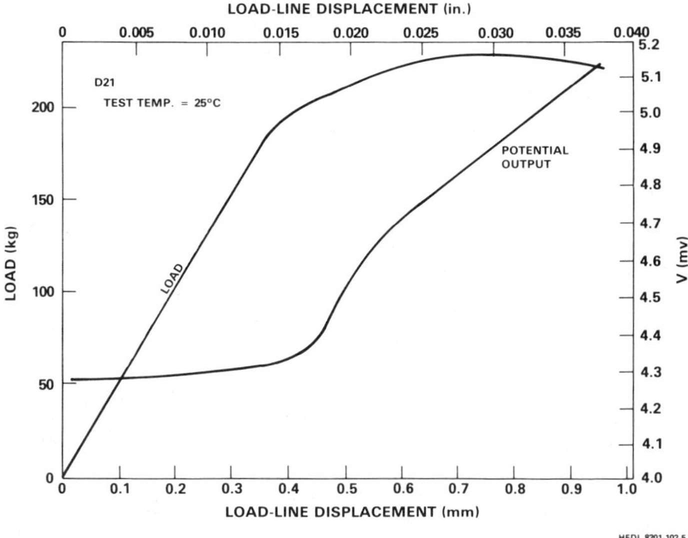

FIG. 2—*Potential output and load versus load-line displacement for unirradiated cold worked D21 tested at 25°C.*

using the area under the load-displacement record up to the corresponding Δa following the procedure recommended in ASTM Standard E813, except that the maximum Δa was smaller than 1.5 mm. The tests were stopped at smaller crack extensions in the region of flat fracture [2].

Results

Figures 1, 2, and 3 show the potential output and the load versus load-line displacement curves of unirradiated aged D21, unirradiated cold worked, and irradiated D21, respectively. As shown

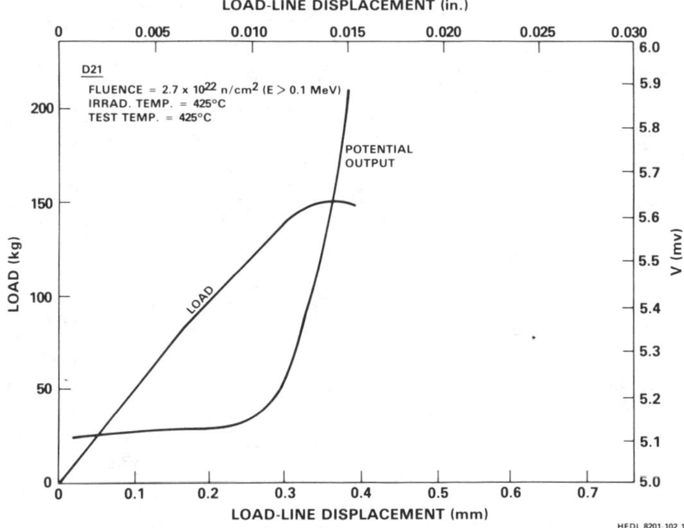

FIG. 3—*Potential output and load versus load-line displacement for irradiated D21 tested at 425°C.*

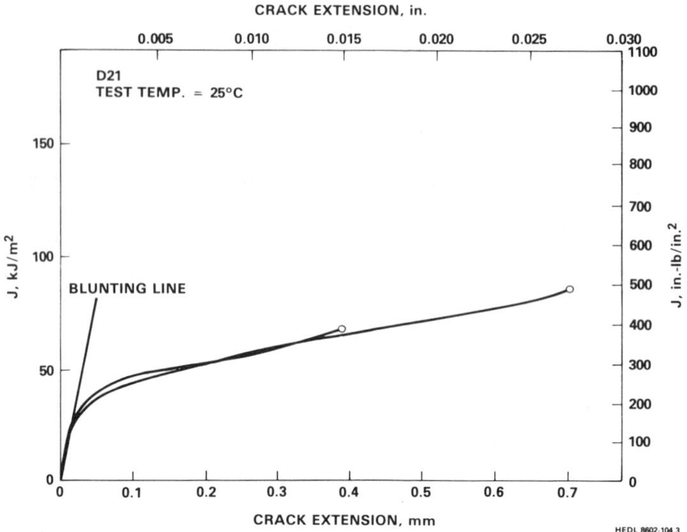

FIG. 4—J *versus* Δa *curves obtained via an electric calibration curve for unirradiated aged D21.*

in Fig. 1, the slope of the electric potential curve changes gradually, indicating that there was no distinct point of crack initiation. Figure 2 or 3 shows that the point of initiation is confined to a narrow range of displacement in these two cases. Two specimens irradiated at the same condition were tested at the same temperature to establish the electric potential curve.

Continuous J versus Δa curves can be obtained from each specimen using a semi-empirical expression [*1*] which relates electric potential to crack extension. Such curves are shown in Figs. 4 and 5 for unirradiated D21 specimens. It can be seen in Fig. 5 that J values increase rapidly

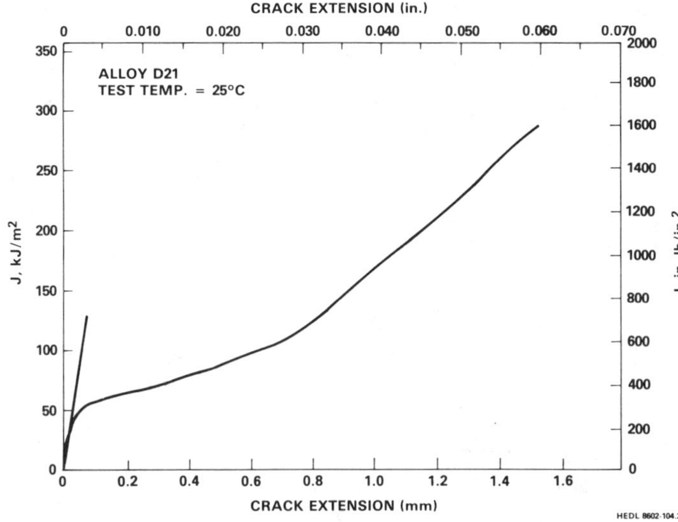

FIG. 5—*Continuous* J *versus* Δa *curve for unirradiated cold worked D21 tested at 21°C.*

with increasing Δa for crack extensions larger than 0.6 mm (in the slant fracture region). Evidently, the data exclusion zone of 1.5 mm recommended in the ASTM Standard E 813 is too large for subsize specimens. This will be discussed later. The ligament of a subsize specimen is torn under a combination of plane strain and plane stress conditions when the crack extension becomes large; the J values so obtained would have been larger than plane strain J values. Figure 6 presents a scanning electron microscope fractograph for unirradiated D21 tests at 25°C. It reveals a shallow dimpled fracture indicating that the material exhibits ductile fracture.

Figures 7 and 8 show J versus Δa curves obtained from each irradiated specimen of D21 with the same irradiation and test temperatures. Also plotted in the figures are dashed lines fitted to the experimental data points. The dashed line provides the value of J_{Ic} at its intersection with the blunting line. The values of J_{Ic} so determined and those obtained from single specimen methods are in good agreement. Figure 9 shows the J versus Δa curves for other irradiation and test conditions. Test results are listed in Table 2 along with the irradiation and test conditions. Also listed in the table are the values of the tearing modulus, T, which is defined as $(E/\sigma_f^2)(dJ/da)$, where dJ/da is the slope of the J versus crack-extension curve, E is Young's modulus, and σ_f is the flow stress of the material.

The effects of test temperature on the fracture toughness and tearing modulus are shown in Figs. 10 and 11. As shown in Fig. 10, the fracture toughness of D21 irradiated at 500°C increases with increasing test temperature. For D21 irradiated at temperatures ranging from 400 to 600°C, the tearing modulus decreases with increasing test temperature.

The irradiation temperature dependence of J_{Ic} and T are plotted in Figs. 12 and 13. As shown in these figures, the fracture toughness and tearing modulus of D21 increase with irradiation

FIG. 6—*SEM fractography for unirradiated D21 tested at 25°C (400 × magnification).*

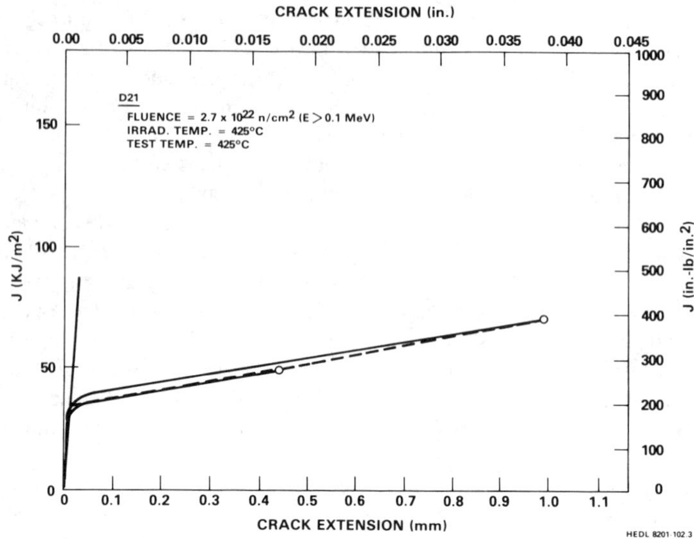

FIG. 7—*Continuous* J *versus* Δa *curves for D21 irradiated and tested at 425°C.*

temperature, but the rate of change of J_{Ic} and T with irradiation temperature is steeper for specimens tested at a lower temperature. It is interesting to note from Fig. 12 that, while the fracture toughness of D21 irradiated at a temperature ≤500°C increases with increasing test temperature, the trend is reversed if the material is irradiated at a temperature ≥500°C.

It is known that precipitation strengthened alloys like D21 suffer a significant loss of ductility after neutron irradiation [*3*]. The present results also show that the alloy exhibits poor fracture toughness. The ductility loss induced by irradiation is believed to be a result of the precipitation

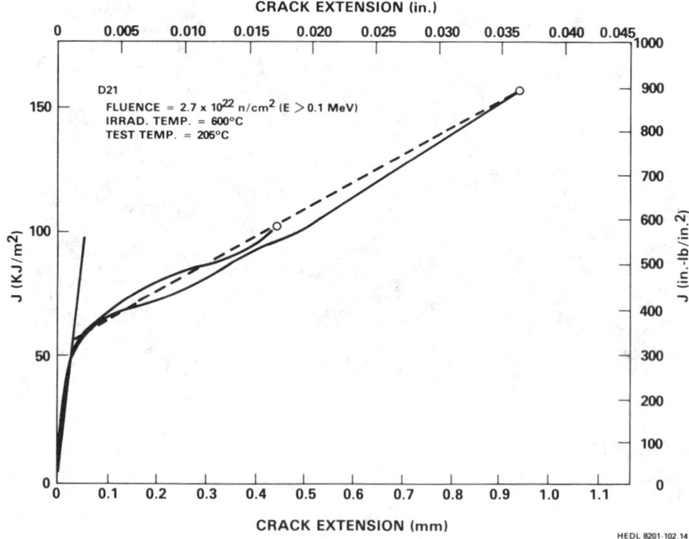

FIG. 8—*Continous* J *versus* Δa *curves for D21 irradiated at 600°C and tested at 205°C.*

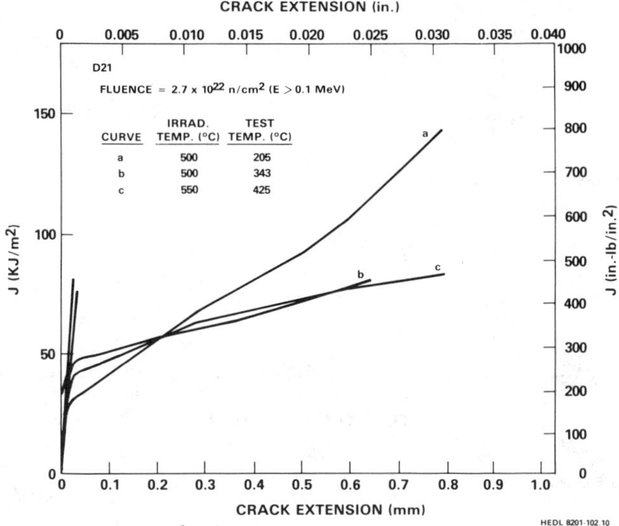

FIG. 9—*Continuous* J *versus* Δa *curves for D21 irradiated at various temperatures.*

TABLE 2—*Fracture toughness test results of alloy D21.*

Test Temperature, °C	Irradiation, Temperature, °C	J_{Ic} kJ/m^2	T	Fluence, 10^{22} n/cm^2
25[a]	. . .	45.5	24.5	. . .
25[b]	. . .	53.9	20.7	. . .
205[b]	500	29.8	25.3	2.7
205[b]	600	59.5	34.3	2.7
343[b]	500	43.8	9.1	2.7
425[b]	425	34.7	3.2	2.7
425[b]	550	46.4	5.0	2.7

[a] TMT: 775°C/8 h/AC + 700°C/2 h/AC.
[b] TMT: 1025°C/15 min/AC + 30% cold work.

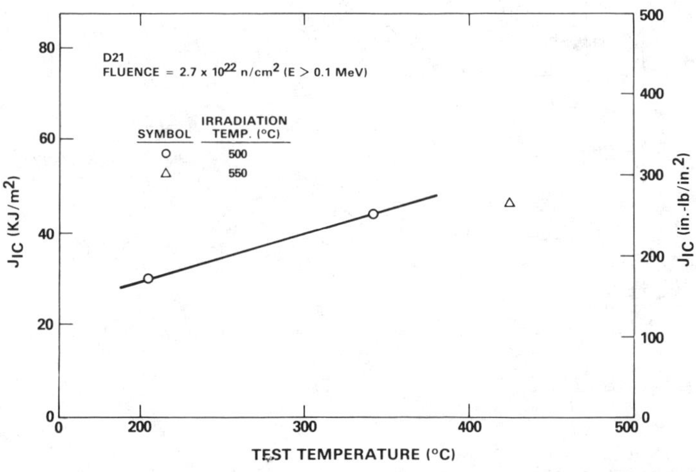

FIG. 10—*Test temperature dependence of fracture toughness for D21 irradiated at 500°C and 550°C.*

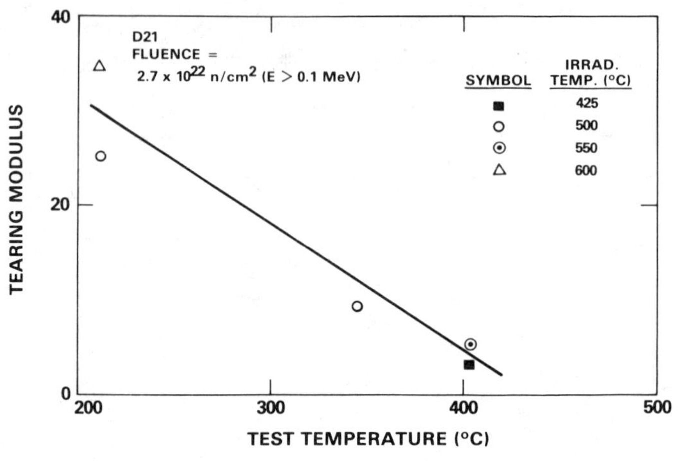

FIG. 11—*Test temperature dependence of tearing modulus for D21 irradiated at various temperatures.*

of γ' along the grain boundary in these alloys [*4*]. The fracture toughness of irradiated D21 is about 50% lower than that of irradiated HT-9. According to the fracture toughness guideline proposed in Ref *5*, with such low post-irradiation fracture toughness, a thin (a few millimeters) structure of D21 could not tolerate a 1 mm crack without failure.

Discussion

The advantage of using the electric potential method is that the fracture toughness at the crack initiation is readily determined if the potential output increases drastically. In this case the determination of J_{Ic} is defined as the deviation from linearity of the potential output. However, as shown in Fig. 1 the potential output from unirradiated aged specimens increases smoothly with crack extension. Such specimens require a calibration curve relating potential output and crack

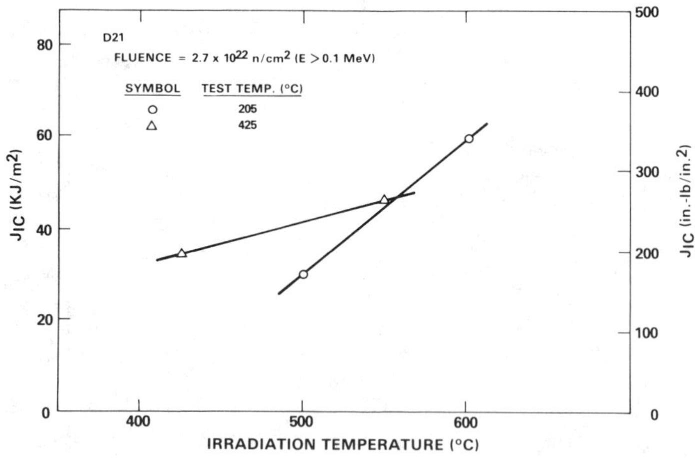

FIG. 12—*Irradiation temperature dependence of fracture toughness for D21 tested at 205°C and 425°C.*

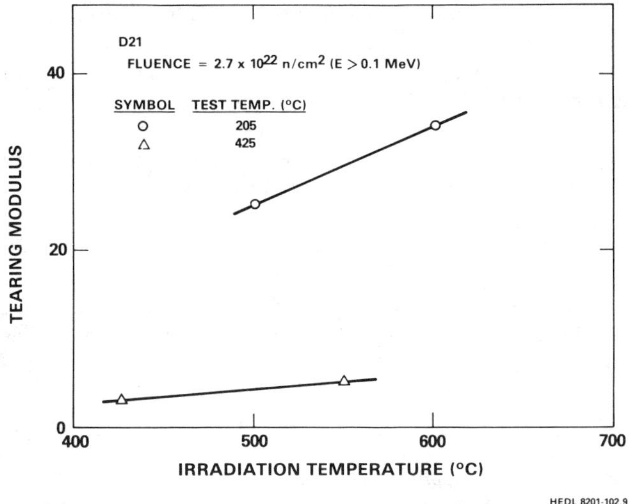

FIG. 13—*Irradiation temperature dependence of tearing modulus for D21 tested at 205°C and 425°C.*

extension to determine J_{Ic}. In comparison, unirradiated, cold worked D21 (Fig. 2) and irradiated D21 (Fig. 3) displayed a rapid increase in potential output in the region of crack initiation, although the point of initiation is not apparent. All data in this report were therefore analyzed for J_{Ic} values through the use of electric potential calibration curves. The semi-empirical expression used to calculate continuous crack extensions was formulated on the assumption that the electric potential output increases linearly with blunting prior to crack initiation.

The value of J_{Ic} is a material property, like K_{IC} and is obtained under conditions of "plane strain." Small specimens tend to fail by ductile fracture and there are, therefore, limitations on the minimum thickness of the specimen which is acceptable for valid J_{Ic} measurements. The minimum thickness recommended by ASTM Standard E813 is 25 J_{Ic}/σ_y. According to this thickness criterion, valid J_{Ic} tests require the thickness of D21 specimens to be greater than 1.8 mm for the unirradiated condition and greater than 0.8 mm for the irradiated condition. The 2.54 mm thick specimens used in this work therefore satisfied the thickness requirement. Moreover, an additional validity requirement is needed for a growing crack to maintain *J*-controlled conditions [*6*]. For specimens subject to bending, this requirement limits crack extension to about 6% of the ligament length. Since the E813 standard recommended 1.5 mm as the size of the data exclusion zone, the standard should be used only for specimens 25.4 mm (1 in.) or larger in thickness. Otherwise, if the standard is followed literally, specimens having thicknesses less than 25 mm but larger than 25 J_{Ic}/σ_y might cause J_{Ic} values to be overestimated. In order to obtain valid J_{Ic} values for such a small specimen, the size of the data exclusion zone must be reduced accordingly so that *J*-controlled conditions can be satisfied.

Conclusions

The following conclusions are drawn from the fracture toughness test results of alloy D21:

1. Alloy D21 exhibited high strength but low fracture toughness as expected.
2. Increasing the test temperature resulted in an increase in fracture toughness but a decrease in crack propagation resistance.

3. Both fracture toughness and tearing modulus increased with increasing irradiation temperature.

4. Although a significant increase in electric potential occurred during crack growth in the irradiated specimen of D21, the point of crack initiation is not readily apparent; an electric potential calibration curve is needed for calculating continuous crack extension and determining J_{IC}.

5. The thickness criterion for a valid J_{Ic} test was satisfied by the small specimens used. Results showed that valid J_{Ic} values could be obtained based on the J versus Δa curve with Δa smaller than the 1.5 mm recommended in ASTM Standard E813.

Acknowledgments

This paper is based on work performed under U. S. Department of Energy Contract DE-AC06-76FF02170 with Westinghouse Hanford Company, a subsidiary of Westinghouse Electric Corporation.

References

[1] Huang, F. H., *Journal of Testing and Evaluation,* Vol. 13, 1985, pp. 257–264.
[2] Huang, F. H., in *The Use of Small-Scale Specimens for Testing Irradiated Material, STP 888,* American Society for Testing and Materials, Philadelphia, 1985, pp. 290–304.
[3] Bajaj, R., Shogan, R. P., DeFlitch, C., Fish, R. L., Paxton, M. M., and Bleiberg, M. L., in *Effects of Radiation on Materials: Tenth Conference, STP 725,* American Society for Testing and Materials, Philadelphia, 1981, pp. 326–381.
[4] Yang, W. J. S., *Journal of Nuclear Materials,* Vols. 108 and 109, 1982, pp. 339–346.
[5] Huang, F. H., *Nuclear Engineering and Design,* Vol. 90, 1985, pp. 13–23.
[6] Hutchinson, J. W. and Parison, P. C., in *Elastic-Plastic Fracture, STP 668,* American Society for Testing and Materials, Philadelphia, 1979, pp. 37–64.

Carlo Albertini,[1] A. Del Grande,[1] M. Montagnani,[1] and A. Pachera[1]

Mechanical Properties at Low and High Strain Rates of PE 16 Alloys Irradiated to 9.2 dpa

REFERENCE: Albertini, C., Del Grande, A., Montagnani, M., and Pachera, A., **"Mechanical Properties at Low and High Strain Rates of PE 16 Alloys Irradiated to 9.2 dpa,"** *Influences of Radiation on Material Properties: 13th International Symposium (Part II), ASTM STP 956,* F. A. Garner, C. H. Henager, Jr., and N. Igata, Eds., American Society for Testing and Materials, Philadelphia, 1987, pp. 151–161.

ABSTRACT: The mechanical properties of nimonic alloy PE 16 are reported, for the strain-rate range 10^{-3} to 10^{3} s^{-1}, at ambient temperature, 400 and 500°C, on as-received, thermally aged (9864 h in sodium at 500°C), and irradiated to 9.2 dpa materials. Regarding the irradiated material one observes: (1) the effect of irradiation consisted of a marked increase of flow stress ranging between 100 and 30% and a reduction of ductility of about 20% points with respect to the as-received material; (2) decrease of flow stress with increasing strain-rate of small value at ambient temperature while it reaches the value of about 15% at high temperature; (3) dynamic strain aging with serrations along the flow curves at high temperature and low strain-rate.

The decrease of the flow stress with increasing strain-rate of the irradiated material seems to be caused by (1) the dynamic strain aging that is present at low strain rate and not at high strain rate; (2) the thermal softening caused by the temperature rise during the adiabatic dynamic testing; and (3) the accentuation of the effects of the two preceeding phenomena caused by the damage introduced by irradiation.

KEY WORDS: nimonic alloys, mechanical properties, strain rate, irradiation, Hopkinson's bar, yield stress, ultimate tensile strength, uniform deformation, fracture strain

The design of reactor structures for containment under extreme dynamic loading conditions requires knowledge of the constitutive equations of the materials, that is, the relationship between stress and strain, as a function not only of the temperature but also of the strain-rate, which in the case of thermal excursions of the core of a fast reactor can reach values of up to $\varepsilon = 10^{3}$ s^{-1}, and in the case of hard missile impact, values higher than $\dot{\varepsilon} = 10^{5}$ s^{-1}.

The constitutive equations must be determined for the materials damaged up to end-of-life conditions by thermal aging, creep, low cycle fatigue, and irradiation. In order to determine the stress-strain/strain-rate relationships, which will be used for the calibration of constitutive equations, we have developed some experimental devices based on the split Hopkinson bar technique. These devices are installed in the hot cells and allow the performance of uniaxial tension tests on irradiated material at strain-rates of up to 10^{3} s^{-1} and temperatures of up to 650°C.

In recent years we have determined the mechanical properties of American Iron and Steel Institute (AISI) 304L and 316L (Unified Numbering System [UNS] 530400 and 531603) irradiated steels up to a dose of 9.2 dpa, tested at strain rates of up to $\sim 10^{3}$ s^{-1} at 400 and 550°C [*1–3*], using the technique mentioned above.

[1] Engineers, Commission of the European Communities, Joint Research Centre, Ispra Establishment, 21020 Ispra (Va), Italy.

Another structural material proposed for fast breeder reactor core structure is the nimonic alloy, PE 16, whose mechanical properties have been measured at strain rates ranging over six orders of magnitude (10^{-3} to 10^3 s^{-1}), at ambient and high temperatures (400, 500°C), after having been irradiated to a dose of 9.2 dpa. The results of this research are the subject of this paper.

Materials and Experimental Procedures

Equipment

In order to determine the stress-strain/strain-rate relationship we used the following devices to perform uniaxial tension tests at constant strain rate:

(1) an Instron machine and an Hounsfield machine for the lowest strain rate of 3.5×10^{-3} s^{-1};

(2) a hydropneumatic machine [*1,3*] giving strain rates ranging from 10^{-1} and 10^2 s^{-1}, installed in a hot cell of the Medium Activity Laboratory of JRC-Ispra; and

(3) a modified Hopkinson bar with a prestressed loading device [*1,3*] that allows testing at strain rates ranging from 10^2 and 10^3 s^{-1}, installed in a hot cell of the High Activity Laboratory of JRC-Ispra.

Specimen Preparation

A short cylindrical specimen of 3 mm diameter and 5 mm gage length was developed [*4*], as required by the Hopkinson's bar principle, which permits the homogeneous stress distribution in the specimen by successive reflections of the elastoplastic waves travelling through the specimen itself.

The grain size of the as-received material was of 70 to 90 μm. The comparison of grain size to specimen size insures that the specimen can be considered a representative volume [*5*] of the bulk material. The chemical composition of the PE 16 nimonic alloy is shown in Table 1.

The solution of annealed specimens of as-received material were subjected to a fast fluence of 7×10^{21} n/cm² ($E > 0.1$ MeV), instantaneous flux 2.05×10^{14} n/cm² s, reaching a dose of 9.2 displacements per atom (dpa), in a sodium environment, at a temperature of 500°C. The irradiation took place in the High Flux Reactor (HFR) of the Petten Establishment of the Joint Research Centre. An equal number of control specimens were subjected out-of-pile to the same heating cycle (9864 h) in a sodium environment at 500°C as were the irradiated specimens. These specimens will be called thermally aged.

Procedure

As received, thermally aged, irradiated specimens of PE 16 were tested at strain-rates ranging from 10^{-3} and 10^3 s^{-1}, at temperatures of 20, 400, and 500°C, which were reached in about 15 min and maintained for 5 min before testing.

TABLE 1—*Chemical composition of the PE nimonic alloy, weight percent.*

C	Mn	S	Si	Ni	Cr	Mo	Ti	B	Co	Cu	Al	Fe
0.05	0.05	0.003	0.21	43.7	16.75	3.73	1.23	0.002	0.09	0.06	1.30	balance

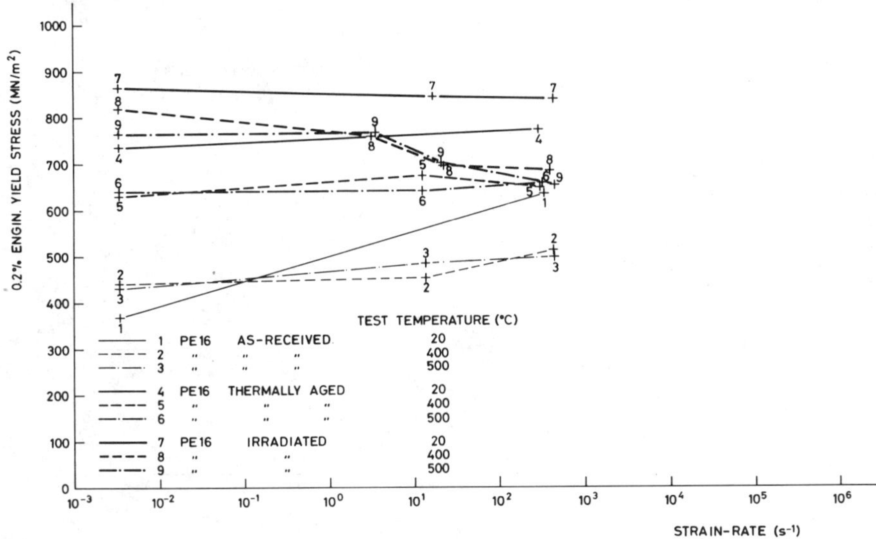

FIG. 1—*Effect of strain-rate on the 0.2% yield stress of PE 16 nimonic alloy irradiated to 9.2 dpa in sodium at 500°C.*

Test Results

The main engineering mechanical properties of as-received, thermally aged, and irradiated PE 16 (0.2% yield stress, ultimate tensile strength, uniform elongation, and fracture elongation) are reported as a function of strain rate in Figs. 1 through 4. Each curve on these figures has been traced by connecting three or four experimental points, given in Table 2, each determined at low

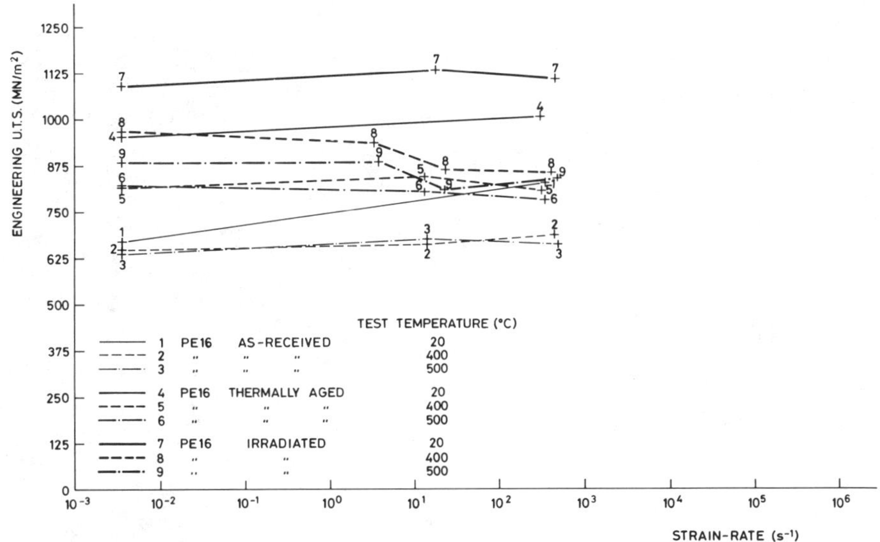

FIG. 2—*Effect of strain-rate on the ultimate tensile stress of PE 16 nimonic alloy irradiated to 9.2 dpa in sodium at 500°C.*

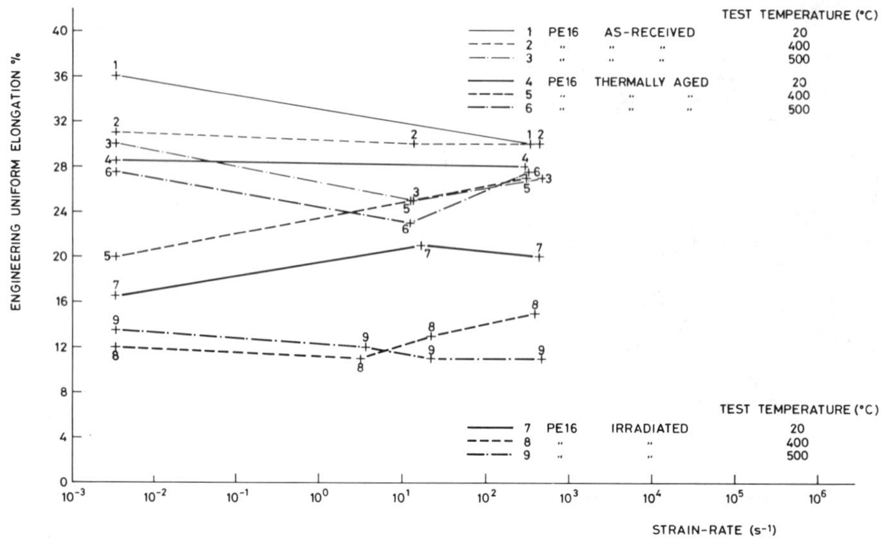

FIG. 3—*Effect of strain rate on the uniform elongation of PE 16 nimonic alloy irradiated to 9.2 dpa in sodium at 500°C.*

strain rate (~0.35 × 10^{-2} s^{-1}), medium strain rate (~10 s^{-1}), and at high strain rate (~600 s^{-1}), respectively. Each of these experimental points is the average of two or three specimens tested. The engineering flow curves shown in Figs. 5 through 7 are the average of the flow curves of the single tests on the irradiated specimens. These diagrams of the irradiated material show a strain-rate softening where the flow stress at a given strain decreases with increasing strain rate. This phenomenon is more enhanced at the high temperatures of 400 and 500°C (Figs. 6 and 7).

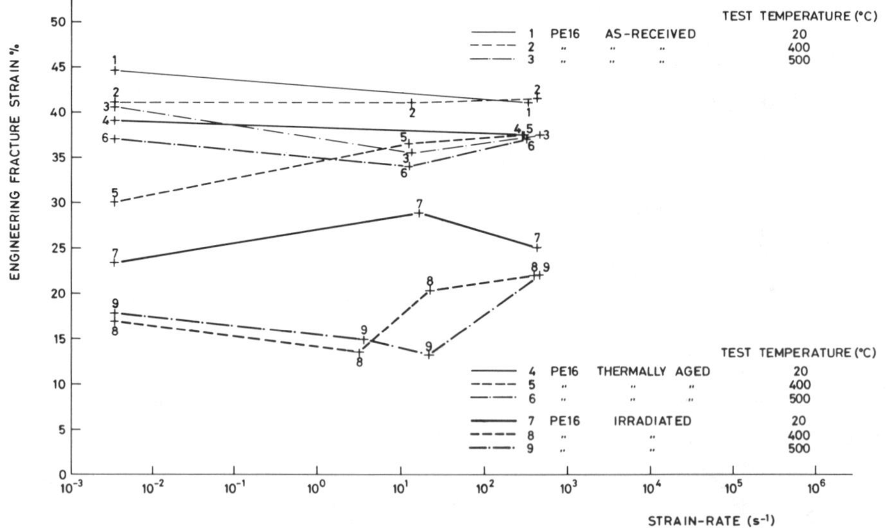

FIG. 4—*Effect of strain-rate on the engineering fracture strain of PE 16 nimonic alloy irradiated to 9.2 dpa in sodium at 500°C.*

TABLE 2—*Test results.*

Test Temperature, °C	Strain Rate, s^{-1}	0.2% Yield Stress, MN/m^2	UTS, MN/m^2	Uniform Strain, %	Fracture Strain, %
		PE-16 AS-RECEIVED			
20	0.35×10^{-2}	369	672	36	44.5
20	380	634	832	30	41
400	0.35×10^{-2}	441	650	31	41
400	15	452	665	30	41
400	490	511	690	30	41.5
500	0.35×10^{-2}	431	640	30	40.5
500	15	484	680	25	35.5
500	530	497	665	27	37.5
		PE-16 THERMALLY AGED IN SODIUM AT 500°C (9864 H)			
20	0.35×10^{-2}	735	957	28.5	39
20	330	771	1012	28	37.5
400	0.35×10^{-2}	630	819.5	20	30
400	14	673	849.2	25	36.5
400	340	647	811	27	37.5
500	0.35×10^{-2}	640	825	27.5	37
500	14	640.5	808.5	23	34
500	370	657	786.5	27.5	37
		PE-16 IRRADIATED TO 9.2 DPA IN SODIUM AT 500°C			
20	0.35×10^{-2}	864	1094	16.5	23.3
20	19	843	1137.5	21	28.8
20	500	837	1112.5	20	25
400	0.35×10^{-2}	818	972	12	16.9
400	3.5	758	940	11	13.5
400	25	694	868	13	20.3
400	450	683	860	15	22
500	0.35×10^{-2}	763	888	13.5	17.8
500	4	765	890	12	14.9
500	24	700	814	11	13.2
500	520	651	845	11	22

A slight tendency to strain-rate softening is also shown by the thermally aged PE 16 (Fig. 8) while the as-received material shows at all temperatures a strain-rate hardening (Fig. 9) where the flow stress at a given strain increases with increasing strain rate.

Discussion of Results

The comparison of the main dynamic mechanical properties of the as-received, thermally aged, and irradiated materials can be obtained from Figs. 1 through 4. We have summarized the comparison as follows:

0.2% Yield Stress

At all test temperatures and at all strain rates the irradiation provokes an increase of the 0.2% yield stress with respect to the as-received material. This increase at room temperature varies between ~100% at low strain rate and ~30% at high strain rate because of the strain-rate hardening shown by the as-received material. At both high test temperatures (400 and 500°C) the increase

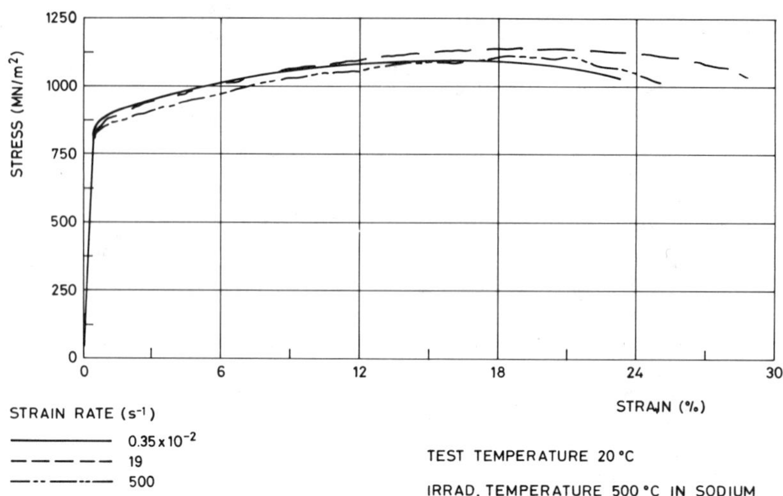

FIG. 5—*Engineering stress-strain curves for PE 16 nimonic alloy irradiated to 9.2 dpa. Test temperature 20°C.*

varies between ~100% at low strain-rate and ~30% at high strain rate caused by the marked strain-rate softening shown by the irradiated material. The 0.2% yield stress values of the thermally aged material lies in between the values of as-received and irradiated materials at low strain rate, while they approximate those of the irradiated material at high strain rate.

Ultimate Tensile Strength (UTS)

At all test temperatures and at all strain rates the ultimate tensile strength of the irradiated material is higher than that of the as-received material. At room temperature this increase varies

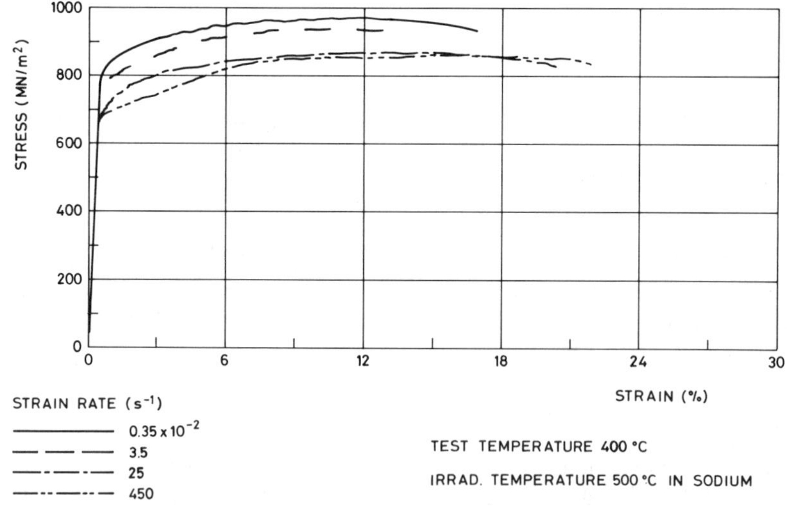

FIG. 6—*Engineering stress-strain curves for PE 16 nimonic alloy irradiated to 9.2 dpa. Test temperature 400°C.*

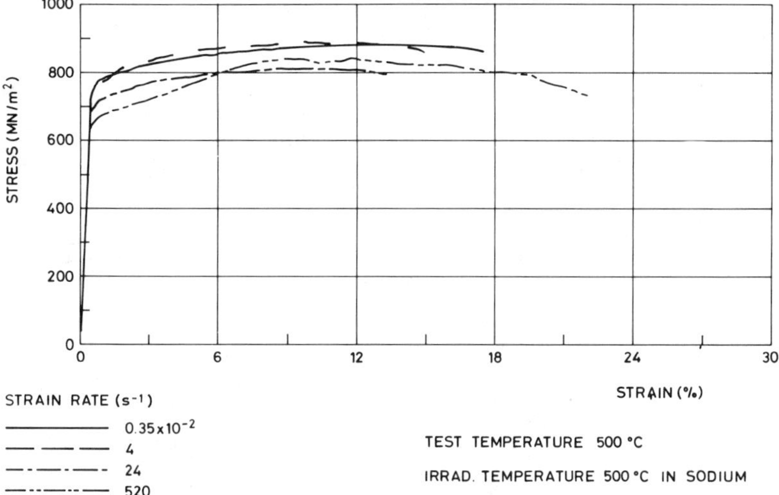

FIG. 7—*Engineering stress-strain curves for PE 16 nimonic alloy irradiated to 9.2 dpa. Test temperature 500°C.*

between ~50% at low strain-rate and ~30% at high strain rate because of the marked strain-rate hardening shown by the as-received material.

At the temperatures of 400 and 500°C the increase of UTS varies between ~50% at low strain-rate and ~25% at high strain-rate because of the strain-rate softening shown by the irradiated material. The UTS values of the thermally aged material lie in between the values of the as-received and irradiated materials at low strain rate, while they approximate those of the irradiated material at high strain rate. The UTS values of thermally aged materials slightly decrease with increasing strain rate. From optical observation it has been noted that thermal aging strongly

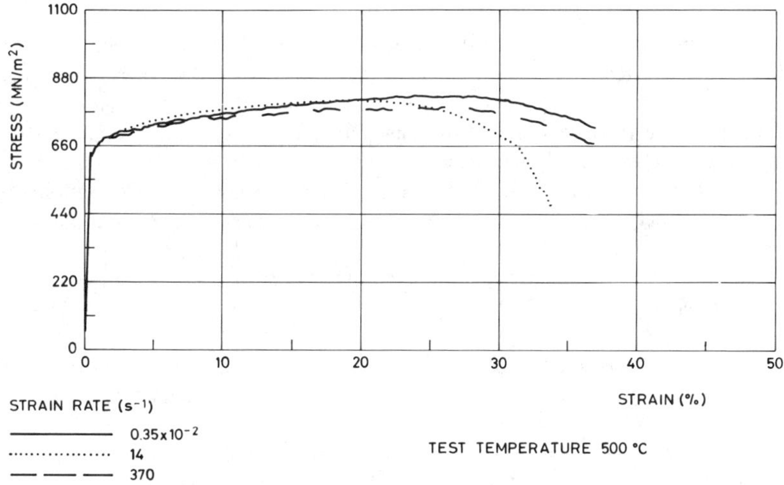

FIG. 8—*Engineering stress-strain curves for PE 16 nimonic alloy thermally aged for 9864 h in sodium at 500°C.*

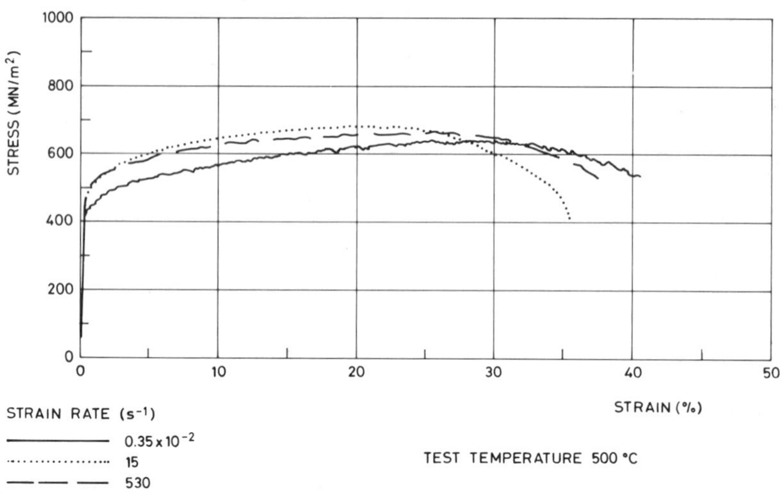

FIG. 9—*Engineering stress-strain curves for PE 16 nimonic alloy as-received.*

reduces the grain size, from 70 ÷ 90 μm to 42 ÷ 60 μm. Together with the reduction of grain size, the formation of twinning is observed. (Fig. 10*a* and *b*).

Uniform Strain

The irradiation provokes a decrease of uniform strain of irradiated material of about 15% points with respect to the as-received material at all test temperatures and at all strain rates. The thermally aged material presents uniform strain values that are just several percent points below the as-received values. The strain-rate sensitivity of uniform strain is not significant.

Fracture Strain

The fracture strain of irradiated material is reduced by about 20% points with respect to the as-received material at all test temperatures and at all strain rates. The thermally aged material presents fracture strain values a few percent points lower than those of the as-received material. At high temperature the fracture strain of irradiated material increases a few percent points with increasing strain rate. Qualitatively the present effects of irradiation on the mechanical properties of PE 16 at low strain rates are confirmed in Ref *6*, where an increase of strength and reduction of ductility have also been found at the temperatures of 400 and 500°C and at a fluence of 1.8×10^{22} n/cm².

The strain-rate softening, shown by the irradiated material at high temperature, seems to be due to the following reasons:

1. The presence at low strain-rate of dynamic strain aging (aging during straining) shown by the serrations along the flow curves (Figs. 6 and 7), which give an increase of flow stress at low rates.

2. The averaging technique (on two or three flow curves) masks often the serrations as in the case of the low strain-rate curve of Fig. 7. These serrations are nevertheless clearly observable along the single flow curves used to build up the average curve of Fig. 7. Negative rate sensitivities are reported by Campbell [*7*] and Manjoine [*8*] for mild steel at high temperatures, for aluminum

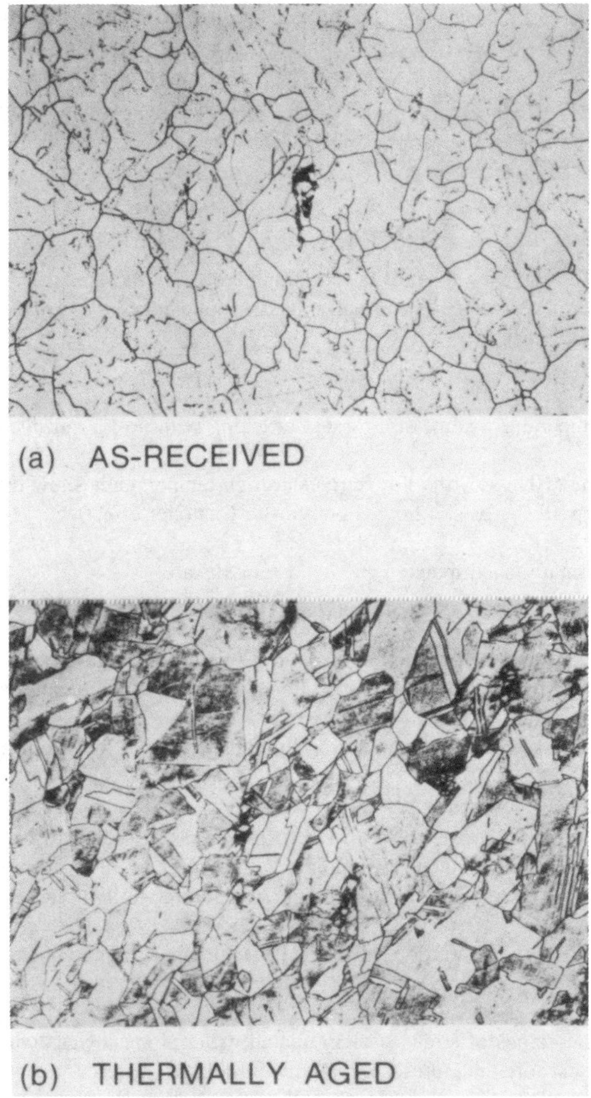

FIG. 10—*Grain size of PE 16, ×200.*

alloys in certain states of particle precipitation [*9*], and for AISI 316L irradiated to 9.2 dpa [*3*], where the dynamic strain aging phenomenon was also present at low strain rate.

3. Stress softening at high strain rate caused by the temperature increase caused by the adiabatic deformation conditions, phenomenon considered also by other authors [*10*]. We have measured on the same nonirradiated specimens a temperature increase near to 100°C when tested at 500°C and at strain rates of 500 s^{-1}. Stress softening at low strain rate has been shown in Ref *6* for the irradiated PE 16 at temperatures higher than 500°C.

4. The adiabatic temperature increase seems to have an accentuated effect on the materials damaged by irradiation and thermal aging. In fact the as-received material (Fig. 9) shows, instead of strain-rate softening, a slight strain-rate hardening. Testing techniques at strain-rates of about

10^3 s^{-1}, based on the Hopkinson's bar [*4*], require the use of short specimens in order to avoid stress gradients along the specimen gage length caused by the wave propagation phenomena. In order to simplify the comparison of mechanical properties the same short specimen has been used in the tests at low and high strain rate. We are now checking if there is a size effect on the mechanical properties by testing ten times larger specimens at low and medium (~10 s^{-1}) strain rate using a large dynamic testing facility [*11*].

Conclusions

The results of the tension tests at strain-rates ranging between 10^{-3} and 10^3 s^{-1}, test temperatures 20, 400, and 500°C, on as-received, thermally aged, irradiated to 9.2-dpa PE 16 nimonic alloy, can be summarized as follows:

Regarding the as-received material:

- At all testing temperatures some strain-rate hardening with no loss of ductility at high strain-rate.
- At low strain rate (10^{-3} s^{-1}) the flow curves at high temperatures show dynamic strain aging with serrations during all the plastic flow (Portevin-Le Chatelier effect).

Regarding the thermally aged material the main remarks are

- The effect of thermal aging consists in a marked increase of the flow stress at a given strain (strain rate 10^{-3} s^{-1}) and a small reduction of ductility with respect to the as-received material.
- No strain-rate sensitivity at ambient temperature while at high temperature and at large strains a slight strain-rate softening appears.
- Dynamic strain aging with serrations during the plastic flow at low strain-rate.

Regarding the irradiated material one notices

- The effect of irradiation consisted in a marked increase of the flow stress, at a given strain, and the reduction of ductility of about 20% points with respect to the as-received material.
- Strain-rate softening of small value at ambient temperature while it reaches the value of about 15% at high temperatures.
- Dynamic strain aging with serrations along the flow curves at high temperatures and low strain rate.

The high strain-rate values of strength of irradiated material are lower than the low strain rate values. The strain-rate softening presented by the irradiated material seems to be due to the following two phenomena:

- The dynamic strain aging presented by the flow curves at low strain rate and not by the flow curves at high strain rate.
- The thermal softening caused by the temperature rise in the specimen during the adiabatic deformation at high strain rate.
- The accentuation of the two preceding phenomena caused by the damage introduced by irradiation.

Acknowledgments

The authors wish to thank Mr. F. Genet and the team of the High Flux Reactor—Petten, for successfully performing the irradiation. Thanks are also due to Messrs. A. Drago, D. Bazzoni and

E. Ghezzi of the Medium Activity Laboratory; and F. Daniele and M. Murarotto, of the High Activity Laboratory for their generous help in preparing and operating the hot cells in the Ispra Establishment. Messrs. K. H. Schrader and A. Pachera performed the "static" tension tests and the automatic data evaluation, respectively. Mr. H. A. Weir performed the microstructure observations. The authors wish to thank the United Kingdom Atomic Energy Authority (UKAEA) for their collaboration in this research.

References

[*1*] Albertini, C., Del Grande, A., and Montagnani, M., Effect of Irradiation on the Mechanical Properties of Austenitic Stainless Steels under Dynamic Loading," *Effects of Radiation on Structural Materials, STP 683,* J. A. Sprague and D. Kramer, Eds., American Society for Testing and Materials, Philadelphia, 1979, pp. 546–556.

[*2*] Albertini, C., Del Grande, A., and Montagnani, M., "Dynamic Mechanical Properties of AISI 304L Irradiated to 2.2 dpa," *Effects of Radiation on Material: 10th Conference, STP 725,* D. Kramer, H. R. Brager, and J. S. Perrin, Eds., American Society for Testing and Materials, Philadelphia, 1981, pp. 431–442.

[*3*] Albertini, C., Del Grande, A., Montagnani, M., and Pachera, A., "Mechanical Properties at High Strain-Rate of AISI Type 316L Stainless Steel Irradiated to 9.2 dpa," *Effects of Radiation on Structural Materials, STP 870,* American Society for Testing and Materials, Philadelphia, 1984, pp. 783–794.

[*4*] Albertini, C. and Montagnani, M., "Dynamic Material Properties of Several Steels for Fast Breeder Safety Analysis," EUR 5787 EN, JRC-Ispra (Va), Italy.

[*5*] Lemaitre, J. and Chaboche, J. L., *Mécanique des Materiaux Solides,* Dunod, Paris, 1985, pp. 72–73.

[*6*] Bajaj, R., Shogan, R. P., De Flitch, C., Fish, R. L., Paxton, M. M., and Bleiberg, M. L., "Tensile Properties of Neutron Irradiated Nimonic PE 16," *Effects of Radiation Material: 10th Conference, STP 725,* American Society for Testing and Materials, Philadelphia, pp. 326–351.

[*7*] Campbell, J. D., *Material Science and Engineering,* Vol. 12, 1973, pp. 3–21.

[*8*] Manjoine, M. J., *Journal of Applied Mechanics,* Vol. 11, 1944, p. A–211.

[*9*] Morris, J. G. and Howard, H. K., *Journal of Applied Physics,* Vol. 42, 1971, p. 3252.

[*10*] Zener, C. and Hollomon, J. H., *Journal of Applied Physics,* Vol. 12, No. 22 (1944).

[*11*] Albertini, C. and Montagnani, M., *NED 68,* 1981, pp. 115–128.

Martin I. de Vries[1]

J-Integral Toughness of Low Fluence Neutron-Irradiated Stainless Steel DIN 1.4948

REFERENCE: de Vries, M. I., "***J*-Integral Toughness of Low Fluence Neutron-Irradiated Stainless Steel DIN 1.4948,**" *Influence of Radiation on Material Properties: 13th International Symposium (Part II), ASTM STP 956,* F. A. Garner, C. H. Henager, Jr., and N. Igata, Eds., American Society for Testing and Materials, Philadelphia, 1987, pp. 162–173.

ABSTRACT: Small compact-tension specimens of stainless steel German Industrial Standard (DIN) 1.4948, which is similar to American Iron and Steel Institute (AISI) Type 304, have been irradiated in the high flux reactor (HFR) at Petten up to a fluence level of 2×10^{24} n · m^{-2} ($E > 0.1$ MeV) at 823 K. Post-irradiation *J*-integral toughness tests have been performed with constant displacement rates ranging over four decades from 10^{-8} to 10^{-4} m · s^{-1} at the irradiation temperature.

The test method is based on a combination of the multiple-specimen interrupted loading method with the direct current potential drop technique (DCPD) for continuous monitoring of the crack length. The *J,R* curves, obtained by this method, show good reproducibility, and consequently the scatter of the *J*-toughness parameters is small.

The irradiation caused modest reduction, by about 30%, of the characteristic *J*-toughness parameters $J_{0.3}$ and dJ/da at the displacement rate of 2×10^{-6} m · s^{-1}. Low toughness values were measured at the lowest applied displacement rate. Results from additional tests showed that the effect of (slow) prior-fatigue loading at the low frequency of 10^{-2} Hz is minor compared to the effects of the lower displacement rates.

KEY WORDS: neutron irradiation, stainless steels, elevated temperature, elastic-plastic fracture mechanics, fracture tests, *J* toughness, crack-growth resistance (*J,R* curve)

Post-irradition mechanical properties of stainless steel German Industrial Standard (DIN) 1.4948, which is similar to American Iron and Steel Institute (AISI) Type 304, have been extensively reported in the period 1975 through 1985 [*1–7*]. The mechanical testing encompassed the standard classical tension and creep testing as well as advanced test techniques to measure the low cycle fatigue and fatigue crack growth properties. Fracture toughness testing was excluded because the standard test method to measure the conventional toughness parameter K_{Ic} is not applicable for austenitic stainless steels at elevated temperatures.

This paper reports on recent tests to measure the monotonic-loading crack growth resistance of irradiated stainless steel at elevated temperature in terms of the elastic-plastic fracture mechanics parameter *J* integral (*J,R* curve). Highly reproducible resistance curves and toughness parameters were measured with small compact-tension specimens at 823 K. The *J*-toughness tests were performed at constant displacement rates ranging from 10^{-8} to 10^{-4} m · s^{-1}. The results complete the prior reported information on the effects of low-fluence irradiations on the mechanical properties of DIN 1.4948. Unfortunately the corresponding control (reference) tests with unirradiated parallel heat-treated specimens are not completed yet.

[1] Research scientist, Netherlands Energy Research Foundation, ECN, P.O. Box 1, 1755 ZG Petten, The Netherlands.

TABLE 1—*Chemical composition of the material.*

C	Cr	Ni	Mn	Mo	Ti	Si	B
0.055	18.0	10.8	1.43	0.002	0.002	0.39	0.0004

Experimental Procedure

Material

The steel, with German designation DIN X6 CrNi 1811 (Werkstoff Nr. 1.4948), is similar to AISI Type 304 stainless steel, as shown by the chemical composition in Table 1. The material has a fully austenitic microstructure (Fig. 1) with an average grain diameter of about 150 μm and a (low) dislocation density, which is typical for a mill-annealed and stretched plate. The main characteristics of the material are listed in Table 2. The information in this table, particularly the homogeneous distribution of the small inclusions and the high RA-value, indicate excellent fracture toughness properties for this material. Experimental J-toughness measurements have been reported recently by de Vries and Schaap [*8*]. They report values of 300 and 750 kJ $\cdot$ m^{-2} for "first-dimple-fracture" (J-initiation, J_i) and "onset-of-stable-crack-growth" J_Q, respectively.

Specimens

The specimen is shown in Fig. 2. The specimen-size was restricted by the need to accommodate a relatively large number of 20 specimens within the dimensional limitations of the irradiation capsules. The dimensions of the specimen, 30.0 by 28.8 by 12.0 mm, are somewhat smaller than for the proportional compact-tension specimen (½TCT) with standard configuration. On the front face there are four little holes for threaded connections of the electrical wires for continuous crack length monitoring with the direct current potential drop (DCPD) technique.

The specimens were taken from either sides of the mid-plane of the 40-mm-thick plate. The orientation of the notch-plane was parallel to the rolling direction (T-L). The specimens were not precracked or heat-treated or both before the irradiations (mill-annealed condition).

Irradiations

Forty fracture-toughness specimens were irradiated at 823 K in two extensively instrumented capsules, placed in an outer position of the core of the HFR at Petten, The Netherlands. The irradiations formed part of a series of similar experiments in the frame of a large program to measure a broad range of post-irradiation mechanical properties of the stainless steel DIN 1.4948. The specimens were irradiated in liquid sodium, using standard facilities and routines for detailed monitoring, analyses, and documentation of the irradiation conditions [*9*]. The temperature was carefully controlled in order to keep the temperature differences in the specimen-stacks within small margins of 20 K from the target irradiation temperature during the full irradiation period.

TABLE 2—*Main characteristics of the material.*

Product Form	Condition	Grain Size, μm	Inclusions (ASTM E 45 Method C)[a]	Hardness HVN, MPa	0.2-Yield Stress MPa[a]	UTS,[a] MPa	Elongation,[a] %	RA, %
Plate 40 mm	Mill-annealed	150	O-2, S-3	1800	132	405	33	65

[a] ASTM Practice for Determining the Inclusion Content of Steel (E 45).

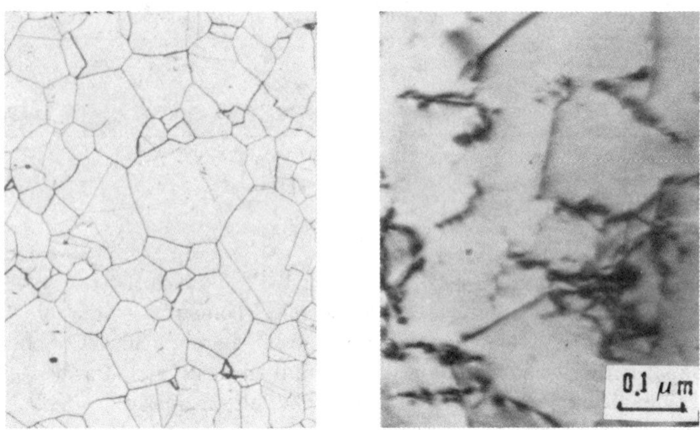

FIG. 1—*Microstructures of the fully austenitic stainless steel DIN 1.4948 (AISI Type 304).*

The specimens were irradiated for 1300 h up to a fluence level of 2×10^{24} n · m^{-2} ($E > 0.1$ MeV) at a fast to thermal flux density ratio of about 0.9. The unirradiated control specimens for the reference tests were heat-treated at 823 K for 1300 h in air.

Method

The experimental test method is based on the combination of the direct current potential drop technique with the multiple-specimen interrupted-loading method.

Series of six specimens are loaded at constant displacement rate under crack-extension Δa control up to selected Δa values ranging between 0.3 and 1.8 mm with constant intervals of 0.3 mm.

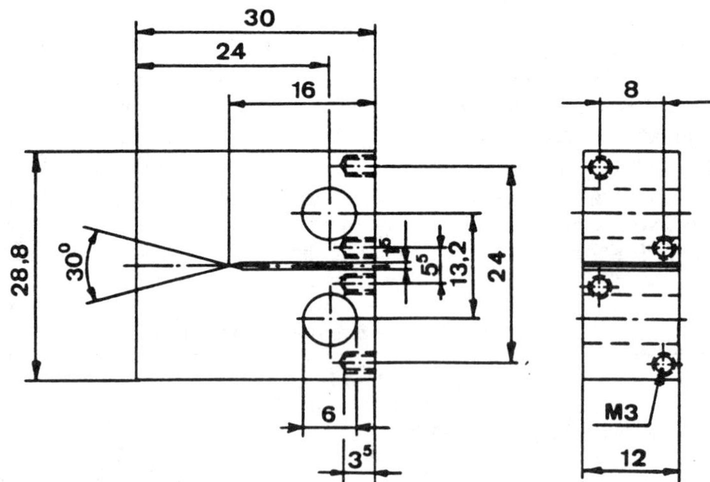

FIG. 2—*Dimensions of the small compact-tension specimen.*

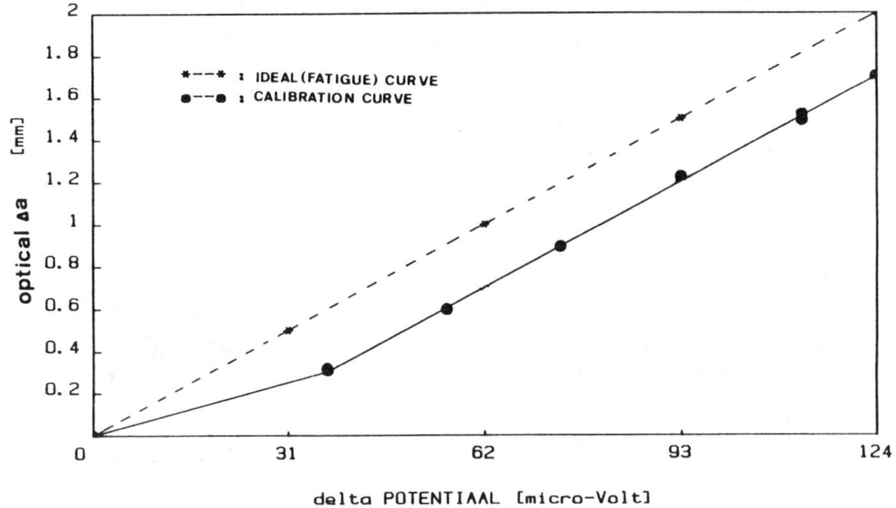

FIG. 3—*Calibration curves for DCPD-Δa conversion.*

From the direct measurements of Δa on the fracture surfaces of the six broken specimens, a calibration curve is derived for conversion of the DCPD-signal into Δa data. Figure 3 shows a curve for DCPD-Δa (stable crack growth) conversion at 823 K together with a conversion curve for fatigue crack growth measurements under linear elastic fracture mechanics (LEFM) conditions. The difference between the two curves is due to the excessive plasticity during the initial tensile deformation before the onset of real stable crack growth. The DCPD contribution from this plasticity is shown to amount to about 18 μV.

The method has the advantage that a series of five single-specimen J,R curves is obtained as well as six regularly distributed data points for a multiple-specimen J-Δa curve based on the final crack length values of the individual specimens.

Equipment and Test Procedure

The tests were performed on a servo-hydraulic testing machine, adapted for remote handling in a lead-shielded facility. The electronic system of the testing machine and the DCPD electronics were installed outside the lead cell. A microcomputer, connected with these electronic systems, enabled automatic machine control and data mesurement but no real-time data processing.

The specimens were precracked, tensile loaded, and finally post-cracked in one loading sequence, as schematically shown in Fig. 4.

The precracking stage consists of a succession of four Δa increments of about 1 mm each. Precracking was performed at 823 K under fast (10 Hz) cyclic loading with maximum load values of 4.0, 3.2, 2.6, and 1.8 kN successively. Additionally a limited number of specimens have been fatigue precracked at the low frequency of 10^{-2} Hz.

The specimens were tensile-loaded under displacement control at 823 K. The specimen-deflection was measured indirectly by means of measurements of the relative displacement of the clevises. The measurements were corrected for additional displacements caused by extraneous sources. This correction was validated by comparison with data from system-stiffness measurements.

Post-cracking was done at room temperature by means of fatigue-loading in order to separate the specimen halves without distortion of the final crack front. The details of the equipment, the

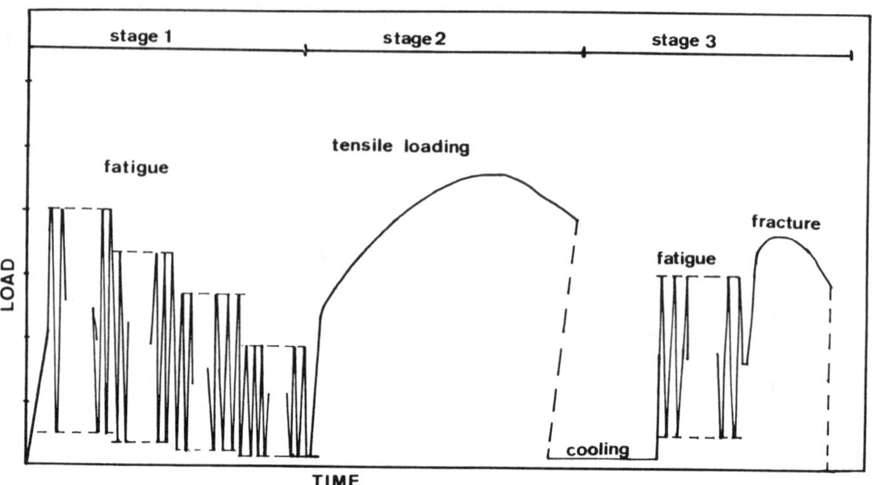

FIG. 4—*Schematic representation of the loading sequence.*

measuring methods, and the software package have been extensively reported by Tjoa et al. [*10,11*].

Irradiated specimens have been tested at four displacement rates ranging from 2×10^{-8} to 2×10^{-5} m · s^{-1}. Unfortunately, the reference (unirradiated control) tests have not been completed yet. However, some reference tests have been performed, but at the displacement rate of 2×10^{-6} m · s^{-1} only.

All the fracture surfaces were visually inspected and recorded by means of a video color-camera. Initial fatigue crack size and final crack extension were optically measured according to the weighted nine-points averaging method. These measurements were made on photographic pictures at $\times 10$ magnification.

Selected specimens were inspected by means of scanning electron microscopy (SEM) for detailed observations of the fracture surfaces. Thin samples for transmission electron microscopy (TEM) were taken from an area located near the lower-edge of the backface of the specimens.

Data Processing and Analysis

Three test-parameters, load P, displacement V, and potential drop (DCPD) are measured and stored for post-test data-processing by the microcomputer. The data reduction is based on sampling increments of 0.005 mm for the displacement and 3 μV for the potential drop. The latter corresponds to a crack-length increment of 0.05 mm.

The displacement data are automatically converted to load-line displacement data after completion of each individual test. The correction for extraneous displacement is calculated from the difference between the slope of the initial linear portion of the original P,V curve and the theoretical compliance of the specimen, taking into account the actual measured fatigue precrack length. This correction is applied on displacement values for which the area A under the corrected P,V curve is less than 95% of the area under the original curve.

The DCPD data are converted to Δa data after completion of the testing (including the optical measurements of the final crack length) of all the specimens from a test series. The calibration curve for this conversion is based on a linear regression fit for the optical measurements versus the corresponding DCPD-data. This linear calibration curve is valid within the range between the minimum (0.3 mm) and the maximum (1.8 mm) of the applied crack extensions in a test series.

For each tested specimen a series of $(J,\Delta a)$ data-pairs, with Δa intervals of 0.05 mm, are calculated according to the incremental J_{i+1} formula

$$J_{i+1} = \left[J_i + \left(\frac{f\,a/W}{b}\right)_i \frac{A_{i,i+1}}{B}\right] \times \left\{1 - \left(\frac{1 + 0.76\, b_i/W}{b}\right)_i [(a_p)_{i+1} - (a_p)_i]\right\} \quad (1)$$

where A is the area under the corrected P,V curve, B is thickness, W is width, A is crack length, and b is $(W - a)$. The data pairs with Δa less than 0.3 mm, being data outside the validity range of the Δa calibration, are excluded for further analyses.

It is interesting to note that this 0.3 mm "exclusion limit" corresponds closely with the lower exclusion limit $\Delta a_{p(min)}$, resulting from the procedure with the 0.15 mm parallel to the blunting line. The blunting line being based on the empirical relationship from Ref *8*

$$J = 4\sigma_y \times \Delta a \quad (2)$$

with σ_y being the effective yield stress.

The $(J,\Delta a)$ data-pairs from each test are further analyzed in terms of a linear regression fit

$$J = dJ/da \times \Delta a + C \quad (3)$$

with C being a numerical constant without any physical meaning and dJ/da being the crack growth resistance for Δa between 0.3 and 1.8 mm.

Two J-toughness parameters are derived from the regression analyses: the engineering J-toughness $J_{0.3}$ (the level of J at 0.3 mm crack extension) and the $(J_Q,\Delta a_Q)$ data pair (after validation: J_{Ic}) from the intercept of the regression line and the blunting line.

Results and Discussion

Load-displacement curves from a test series with irradiated specimens, tested at a constant displacement rate of 2×10^{-6} m · s^{-1}, are shown in Fig. 5. The corresponding crack extension

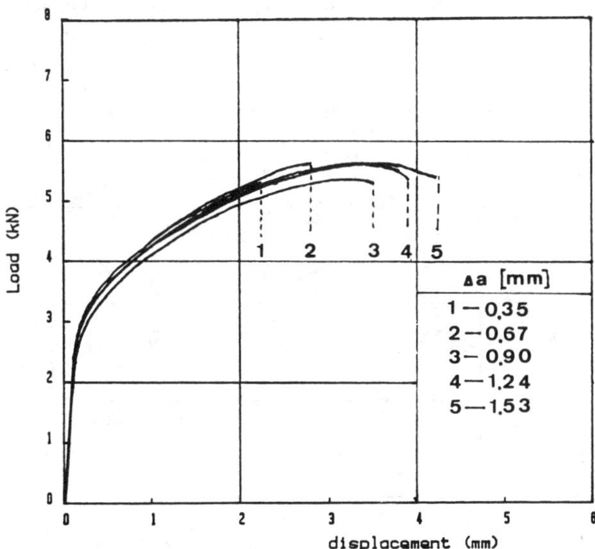

FIG. 5—*Series of load-deflection curves for irradiated specimens tested up to various final crack extension values ranging from 0.3 to 1.5 mm at 823 K.*

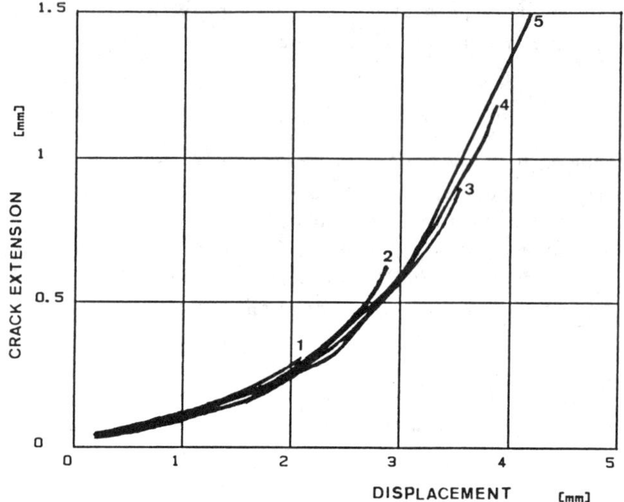

FIG. 6—*Curves of crack-extension versus displacement for the test series in Fig. 5.*

curves (Δa versus displacement) are shown in Fig. 6. The scatter of the tensile curves is small, and good reproducibility of the crack extension (DCPD) measurements is demonstrated too. This was observed for all the test series, consequently the *J,R* curves exhibited no large scatter.

The initial part of the tensile curves was not affected by the irradiation, indicating there was no irradiation hardening. TEM of small samples, taken from undisturbed areas of the CT specimens revealed no displacement damage. This is in accordance with previous observations after similar low-fluence irradiations of the same material at temperatures from 723 to 923 K showing no displacement damage and no hardening [*1,4*].

Figure 7 shows the load-deflection curves at various displacement rates for irradiated specimens tested up to equal amounts of about 1.5-mm final crack extension. The figure shows that tensile

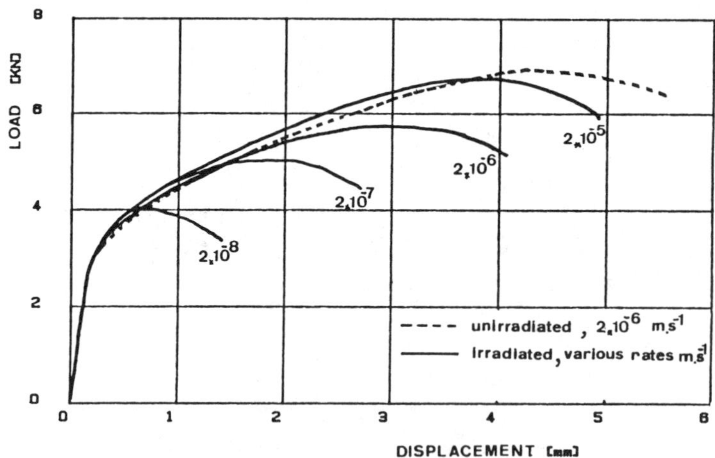

FIG. 7—*Load-deflection curves of irradiated specimens tested up to constant crack-extension of 1.5 mm at various displacement rates ranging from 2×10^{-8} to 2×10^{-5} m $\cdot$ s^{-1} at 823 K.*

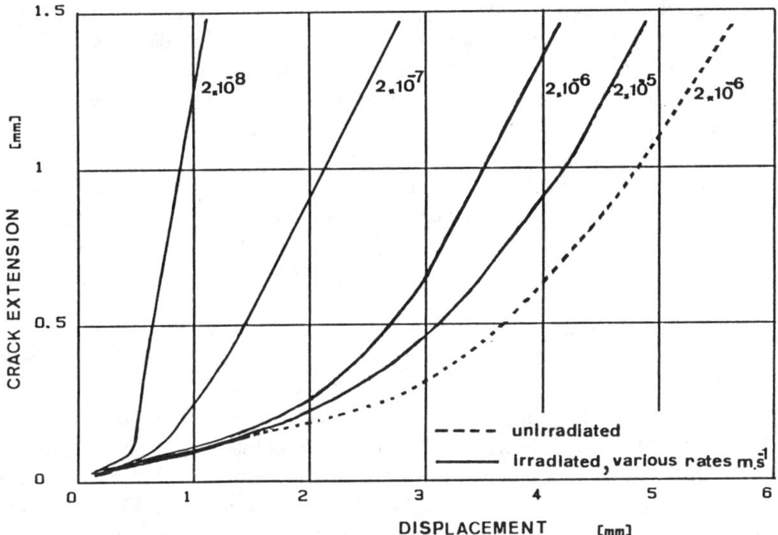

FIG. 8—*The Δa-versus-deflection curves for the tests in Fig. 7.*

curves reduce systematically at decreasing displacement rates ranging from 2×10^{-5} to 2×10^{-8} $m \cdot s^{-1}$. The corresponding crack growth curves are plotted in Fig. 8. Crack extension is shown to occur at lower deflections if the displacement rate is reduced. Obviously the systematic rate effect on the crack growth behavior is reflected in the *J,R* curves too.

The results of the analysis of the data from three test series, two with irradiated specimens and one with unirradiated (reference) ones, are listed in Table 3. The table gives the constant *C*, dJ/da, and the coefficient of determination r^2 for the linear regression analysis of the three test series. The high r^2 values justify the linear fit. The table also gives the two *J*-toughness parameters and the T-modulus, calculated according to the formula

$$T = \frac{4E}{(\sigma_{0.2} + \mathrm{UTS})^2} \times [dJ/da] \tag{4}$$

The parameters $J_{0.3}$ and dJ/da have been chosen to quantify the irradiation effect on the resistance curve. This choice was made for reasons of consistency as both parameters have unambiguous

TABLE 3—*J-toughness parameters for irradiated and unirradiated Type 304 plate at 823 K. Irradiation:* 2×10^{24} $n \cdot m^{-2}$ (E > *0.1 MeV) at 823 K.*

Parameters	Unirradiated Rate, 2×10^{-6} $m \cdot s^{-1}$	Irradiated Rate, 2×10^{-6} $m \cdot s^{-1}$	Irradiated Rate, 2×10^{-7} $m \cdot s^{-1}$
C, kJ/m²	213	131	79
dJ/da, MN/m²	214	146	72
r^2 [a]	0.91	0.96	0.97
$J_{0.3}$, kJ/m²	277	175	101
T [b]	530	360	210
J_Q (J_{Ic}), kJ/m²	267	150	86
Δa_Q, mm	0.26	0.14	0.09

[a] Coefficient of determination for linear regression analyses.
[b] T = tearing modulus.

well-defined physical meaning. In this respect J_{Ic} and T are considered to be less suitable for comparisons of the resistance curves from the various test series in this work. J_Q (J_{Ic}) represents an undefined (not constant) amount of crack extension, which in Table 3 is shown to vary already by a factor of 2 for unirradiated and irradiated material. T is an even more complex parameter as it includes dJ/da, $\sigma_{0.2}$, and UTS, the latter has been shown in Ref *1* to be affected by irradiation for tensile strain rates less than 10^{-3} s^{-1}.

For tests at a constant displacement rate of 2×10^{-6} $m \cdot s^{-1}$, the irradiation caused modest reduction by about 30% of $J_{0.3}$ and dJ/da. The reduction factors are respectively

$$D_{0.3} = \frac{J_{0.3}, \text{irr.}}{J_{0.3}, \text{unirr.}} = 0.63 \tag{5}$$

$$D_R = \frac{dJ/da, \text{irr.}}{dJ/da, \text{unirr.}} = 0.68 \tag{6}$$

The $J_{0.3}$ and dJ/da values from the four test series with irradiated specimens are plotted versus the applied displacement rate in Fig. 9. Both curves show the same trend of large increase at increasing rate. Similar trends have been measured for tensile ductility, low cycle fatigue life, and fatigue crack growth rate after identical irradiations of the same material. This has been attributed to enhanced intergranular crack formation caused by helium from the thermal neutron reaction with the boron in the steel [*1,4,5*].

Details of the fracture surfaces of the toughness specimens are shown in Fig. 10. The irradiated material shows a complete transition from fully ductile dimple-mode fracture at the highest applied displacement rate of 2×10^{-5} $m \cdot s^{-1}$ to fully intergranular fracture at the lowest rate of 2×10^{-8} $m \cdot s^{-1}$. This is consistent with earlier observations from tensile tested specimens in Ref *1*, showing a transition from 0 to 100 percentage points intergranular fracture contribution over the strain rate range covering three decades from 10^{-3} to 10^{-6} s^{-1}. A similar rate dependent transition to intergranular fracture has been reported for low cycle fatigue and fatigue crack propagation after identical irradiations of Type 304 stainless steel [*4,5*].

Low frequency loading is considered to be a realistic loading condition in practice for nuclear

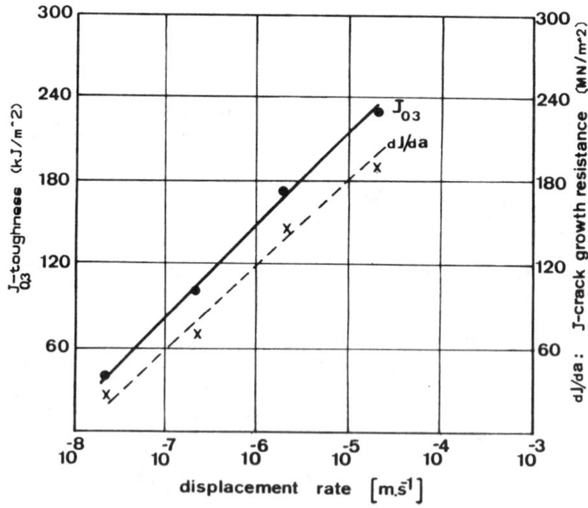

FIG. 9—*The characteristic resistance-curve parameters* $J_{0.3}$ *and* dJ/da *versus applied displacement rate for irradiated Type 304 at 823 K.*

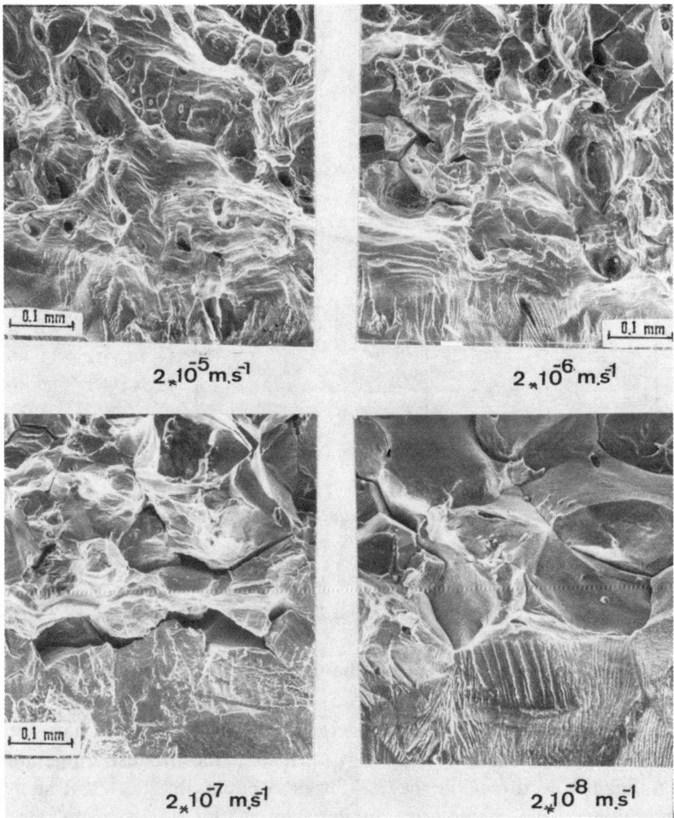

FIG. 10—*Details of the fracture surfaces of irradiated specimens tested at various displacement rates.*

components of advanced reactors. Consequently, results from J,R-curve testing after low-frequency precracking are of practical interest for fracture mechanics analysis. Therefore, some irradiated specimens have been tested after precracking at 10^{-2} Hz. The results are shown in Fig. 11. The lower curves in this figure show no effect from low frequency precracking with K_{max} of 11 MPa $\cdot$ m$^{1/2}$. This K_{max} value is in accordance with the standard conditions for fatigue precracking.

For ΔK of about 24 MPa $\cdot$ m$^{1/2}$ and above, it has been reported in Ref *6* that the fatigue crack growth rate of irradiated stainless steel increased because of enhanced inergranular fracture contribution. In order to look for effects from intergranular fatigue precracking we have also performed two additional tests after high- and low-frequency precracking under the nonstandard high K_{max} value of 24 MPa $\cdot$ m$^{1/2}$. The small difference between the upper and mid curve in Fig. 11, is attributed to the limited intergranular fracture contribution after 0.01 Hz precracking at 823 K. This effect is minor compared to the rate effect on the J,R-curve parameters shown in Fig. 9.

Summary and Conclusions

The combination of the multiple-specimen interrupted loading method with the single specimen DCPD-technique provides a reliable test method for fracture toughness measurements of irradiated stainless steel at elevated temperature.

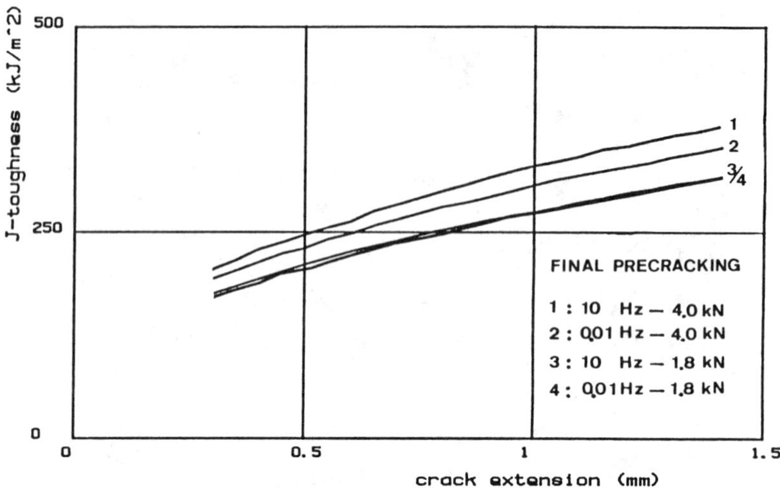

FIG. 11—*Crack growth resistance curves for irradiated Type 304 tested at the displacement rate of* 2×10^{-6} $m \cdot s^{-1}$ *after high and low frequency precracking with two final* K_{max}*-loading values: 11 and 24 MPa* $\cdot$ $m^{\frac{1}{2}}$.

Highly reproducible *J,R* curves are measured at 823 K with small (½TCT, 30.0-28.8-12.0 mm) compact-tension specimens.

The irradiation effect on the *J,R*-curve behavior can be unambiguously quantified by the "engineering toughness" $J_{0.3}$ and the resistance parameter dJ/da.

The low fluence irradiation caused modest reduction by about 30% of the characteristic parameters $J_{0.3}$ and dJ/da at the displacement rate of 2×10^{-6} m $\cdot$ s^{-1} and the test temperature of 823 K.

There is a significant rate effect on the *J,R*-curve behavior of irradiated stainless steel Type 304, causing low values of the characteristic parameters at low displacement rates.

The rate effect is quite similar to the effect of slow loading on tensile ductility and fatigue properties, attributed to enhanced intergranular fracture caused by helium from the thermal neutron reaction $^{10}B(n,\alpha)^{7}$Li.

The effect of low frequency precracking (10^{-2} Hz) is minor compared to the rate effect on the resistance curves of irradiated stainless steel Type 304 at 823 K.

Acknowledgments

I wish to express my appreciation for the skillful technical assistance from Mr. F. P. van den Broek and Mr. B. A. J. Schaap, in particular for developing and programming the computerized system for fully autmatic test control and data handling.

References

[1] Vries, M. I. de, Schaaf, B. van der, Staal, H. U., and Elen, J. D., *Effects of Radiation on Structural Materials: Ninth International Symposium, STP 683,* J. A. Sprague and D. Kramer, Eds., American Society for Testing and Materials, Philadelphia, 1979, pp. 477–489.

[2] Vries, M. I. de and Schaaf, B. van der, *Effects of Radiation on Materials: Tenth International Symposium, STP 725,* D. Kramer, H. R. Brager, and J. S. Perrin, Eds., American Society for Testing and Materials, Philadelphia, 1981, pp. 303–325.

[3] Schaaf, B. van der, in *Effects of Radiation on Materials: Eleventh International Symposium, STP 782,* H. R. Brager and J. S. Perrin, Eds., American Society for Testing and Materials, Philadelphia, 1982, pp. 597–618.

[4] Vries, M. I. de, *Effects of Radiation on Materials: Eleventh International Symposium, STP 782,* H. R. Brager and J. S. Perrin, Eds., American Society for Testing and Materials, Philadelphia, 1982, pp. 665–689.

[5] Vries, M. I. de, *Effects of Radiation on Materials: Eleventh International Symposium, STP 782,* H. R. Brager and J. S. Perrin, Eds., American Society for Testing and Materials, Philadelphia, 1982, pp. 720–735.

[6] Schaaf, B. van der, *Effects of Radiation on Materials: Twelfth International Symposium, STP 870,* F. A. Garner and J. S. Perrin, Eds., American Society for Testing and Materials, Philadelphia, 1985, pp. 703–719.

[7] Vries, M. I. de and Michel, D. J., *Effects of Radiation on Materials: Twelfth International Symposium, STP 870,* F. A. Garner and J. S. Perrin, Eds., American Society for Testing and Materials, Philadelphia, 1985, pp. 803–819.

[8] Vries, M. I. de and Schaap, B. A. J., *Elastic-Plastic Fracture Test Methods: The User's Experience, STP 856,* E. T. Wessel and F. J. Loss, Eds., American Society for Testing and Materials, Philadelphia, 1985, pp. 183–195.

[9] Röttger, H., Tas, A., Hardt, P. von der, and Voorbraak, W. B., High Flux Materials Testing HFR, Petten, *EUR 5700,* 1981 edition.

[10] Tjoa, G. L., Broek, F. P. van den, and Hoepen, J. van, "Fatigue Testing of Irradiated Steels at High Temperatures," ECN-Report *ECN-183,* Petten, The Netherlands, 1983.

[11] Tjoa, G. L., Broek, F. P. van den, and Schaap, B. A. J., *Automated Test Methods for Fracture and Fatigue Crack Growth, STP 877,* W. H. Cullen, R. W. Landgraf, L. R. Kraisund, and J. H. Underwood, Eds., American Society for Testing and Materials, Philadelphia, 1985, pp. 197–212.

Martin I. de Vries[1]

Fatigue Crack Growth and Fracture Toughness Properties of Low Fluence Neutron-Irradiated Type 316 and Type 304 Stainless Steels

REFERENCE: de Vries, M. I., "**Fatigue Crack Growth and Fracture Toughness Properties of Low Fluence Neutron-Irradiated Type 316 and Type 304 Stainless Steels,**" *Influence of Radiation on Material Properties: 13th International Symposium (Part II), ASTM STP 956,* F. A. Garner, C. H. Henager, Jr., and N. Igata, Eds., American Society for Testing and Materials, Philadelphia, 1987, pp. 174–190.

ABSTRACT: Small compact-tension specimens of Type 316 plate and Type 304 forging have been irradiated in the High Flux Reactor (HFR) at Petten, The Netherlands, up to a fluence level of 2×10^{24} neutrons (n) $\cdot$ m^{-2} ($E > 0.1$ MeV) at 573 K. Post-irradiation fatigue crack propagation tests and J-integral fracture toughness tests have been performed at the irradiation temperature. Additional tests were made at the higher temperatures of 723 and 823 K.

The two materials exhibited almost identical crack growth curves after irradiation, despite the chemical and microstructural differences between the materials in the as-fabricated condition.

The irradiation caused a slight increase of the resistance to crack growth under cyclic and monotonic loading at 573 K. This is attributed to irradiation hardening from displacement damage, the latter being observed in the form of the well-known black dots. The additional tests showed modest degradation of the resistance at 723 and 823 K, but the fatigue crack growth resistance decreased significantly under low frequency loading at 823 K because of helium-enhanced intergranular crack formation.

The results are compared with fatigue and toughness curves for Type 304 and Type 316 plate and welds after similar irradiations in the HFR and in thermal material test reactors in the United Kingdom (UKMTR). The closely corresponding curves show the same type of results after the various irradiations.

It is concluded that low fluence (less than 1 dpa, displacements per atom) irradiations at temperatures below 700 K, have a slight beneficial effect on the crack growth resistance of austenitic stainless steels. Nevertheless, the fatigue crack growth rates of Type 316 plate and Type 304 forging are still higher than for weld metal, although the toughness properties are better than for weld metal.

KEY WORDS: neutron irradiation, radiation effects, stainless steels, fatigue (materials), crack propagation, elastic-plastic fracture toughness, J-integral test, elevated temperatures

The contribution from fracture mechanics in the design and safety analyses of nuclear components for elevated temperature application in advanced reactors, such as liquid metal fast breeder reactors (LMFBR) and systems for magnetic fusion energy (MFE), is rapidly growing [*1,2*]. For future design of core-peripheral permanent components of compact LMFBRs and for successful design of MFE first wall structures, it is even considered that design against crack growth under monotonic and cyclic loading is of major importance. Consequently there is a need to establish a data base on the fracture mechanics properties of the construction materials with particular emphasis on the

[1] Research scientist, Netherlands Energy Research Foundation, P.O. Box 1, 1755 ZG Petten, Netherlands.

TABLE 1—*Chemical composition of Type 316 plate and Type 304 forging stainless steels in weight percentages.*

Steel	C	Cr	Ni	Mo	Si	Mn	P	S	N	B
Type 316	0.07	16.6	11.4	2.35	0.19	1.78	0.02	0.013	0.027	0.0035
Type 304	0.05	17.4	11.2	0.04	0.44	1.57	0.01	0.018	0.035	0.0006

effects of neutron irradiation. In this respect there is also a need for information on the comparability and transferability of data from various irradiation and testing programs.

This paper gives results from crack growth tests with low fluence HFR-irradiated compact-tension specimens of Type 304 forging and Type 316 plate stainless steels. The results from the corresponding control (reference) tests with unirradiated parallel heat-treated specimens are not available yet. Consequently, comparisons can only be made with results from similar irradiations of Type 316 material in UKMTRs and with results from HFR irradiations of Type 304 plate material, in conjunction with the results from the respective control tests. The small differences in experimental conditions are considered to be acceptable for these comparisons.

Experimental Procedure

The chemical composition of the materials is given in Table 1. The composition of the forging is in accordance with the German specification for German Industrial Standard (DIN) 1.4948 stainless steel.

The materials have been used in the as-fabricated condition: mill-annealed after hot-rolling to 28 mm thickness for the Type 316 and forging to 500 mm thickness for the Type 304 material. The different microstructures, resulting from the fabrication procedures, are clearly seen in Fig. 1. The Type 316 material shows a fine austenite grain size with precipitate-free grain boundaries

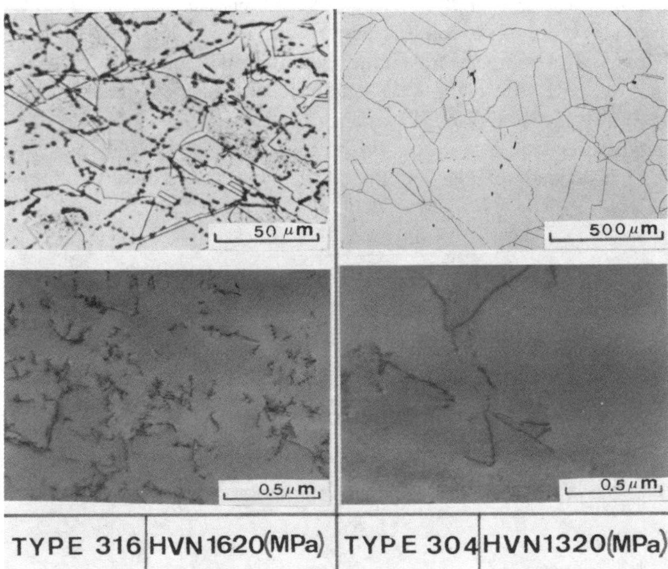

FIG. 1—*Microstructures of Type 316 plate and Type 304 forging as fabricated.*

TABLE 2—*Characteristics of the Type 316 and Type 304 materials.*

Product Form	Thickness, mm	Grain Diameter, μm	Hardness HVN, MPa	0.2-Proof Stress (MPa) at 700 K	UTS (MPa) 700 K	Elongation, %, 700 K
316 plate	26	30	1620	160[a]	500[a]	45[a]
304 forging	500	300	1320	90[b]	380[b]	40[b]

[a] Interpolation between 650 and 850 K.
[b] Backwards extrapolation from data between 725 and 925 K.

but with heavily decorated original austenite boundaries elongated in the rolling direction. The dislocation density is typical for mill-annealed and stretched plate but higher than is normally associated with fully annealed material. The Type 304 material shows a very large grain size and a low dislocation density (fully annealed). Clusters of manganese sulfides, locally distributed, could be frequently observed in this material. The main characteristics of the materials are summarized in Table 2.

The specimen, shown in Fig. 2, is a proportional compact-tension type (CT) with small dimensions of 30.0 by 28.8 by 12.0 mm. On the front face of the specimen are four little holes for threaded connections of the electrical wires for continuous crack length monitoring with the direct current potential drop (PD) technique. The length of the mechanical notch is 10.0 mm. The specimens have been precracked for 0.5- and 4.0-mm crack extension before fatigue crack growth testing and *J*-toughness testing, respectively.

The 316-specimens were located with the notch-plane perpendicular to the rolling direction (L-T orientation) of the plate, whereas the 304-specimens were orientated with the notch-plane perpendicular to the thickness direction (TS-L) of the forging. Two additional specimens for *J*-toughness tests were orientated parallel to the thickness direction (L-TS).

The irradiation experiment is one from a large series of well-documented similar experiments in the HFR, using standard facilities and routines for detailed monitoring, analyses and control of the irradiation conditions [*3*]. A total number of 20 specimens, stacked in the center of the capsule, was irradiated in liquid sodium at a nominal temperature of 573 K. The extensively instrumented capsule allowed for careful temperature control within small margins of 20 K from the target irradiation temperature. The specimens were irradiated up to about 0.2 dpa (fast fluence level of 2.2×10^{24} n · m^{-2}) at a fast to thermal flux density ratio of 0.9.

The post-irradiation tests were performed with an Instron servohydraulic testing machine, adapted for remote-handling in a lead-shielded facility. The experimental details have been extensively

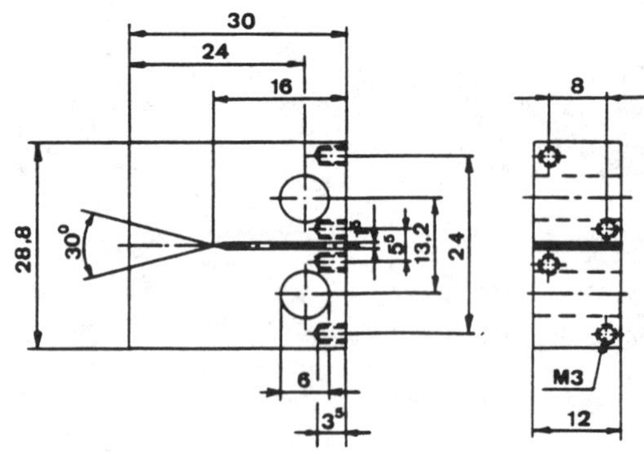

FIG. 2—*Dimensions of the small compact-tension specimen.*

reported by Tjoa et al. [*4,5*], de Vries and Schaap [*6*], and de Vries et al. [*7*]. The applied crack measuring method, by means of the direct current potential drop technique (DCPD), has the advantage that the same test equipment could be used for the fatigue testing and for the *J*-toughness testing.

The specimens were tested in air at the irradiation temperature (573 K) and at two higher temperatures: 723 and 823 K.

The fatigue crack growth tests were performed under constant-amplitude cyclic loading (triangular waveform) with an *R* ratio of 0.1. The cyclic frequency was 10 Hz; additionally a low frequency of 0.01 Hz was applied at 823 K.

About 100 data pairs of crack-length *a* and corresponding number of cycles *N*, based on crack-length increments of 0.05 mm, were collected during a fatigue test. These data were stored for post-test data processing and data analyses.

From the series of *N*,*a*-data pairs the fatigue crack growth rate (*da*/*dN*) and the corresponding stress intensity factor range ΔK were calculated, using the equation for the *K* solution for CT specimens given in the ASTM Test Method For Constant-Load-Amplitude Fatigue Crack Growth Rates Above 10^{-8} m/Cycle (E 647)

$$\Delta K = \frac{\Delta P}{B\sqrt{W}} \times \frac{(2 + \alpha)}{(1 - \alpha)^{3/2}} [0.866 + 4.64\alpha - 13.32\alpha^2 + 14.72\alpha^3 - 5.6\alpha^4] \tag{1}$$

where ΔP is the load range ($P_{max} - P_{min}$), *B* is thickness, $\alpha = a/W$ with *a* being crack length, and *W* is width.

The results were further analyzed in terms of the empirical ''Paris''-relationship

$$da/dN = C(\Delta K)^n \tag{2}$$

where *C* and *n* are constants depending on material and testing conditions.

The fracture toughness (*J*-) tests were performed under displacement control with a constant displacement rate of 2×10^{-6} m · s^{-1}, after fast (20 Hz) precracking at the test temperature.

Up to 1200 data pairs of load *P* and displacement (*V*, increments of 0.005 mm), together with a series of 40 corresponding crack extension data (Δa, increments of 0.05 mm), were collected during a *J* test. These data were stored for post-test data processing and calculations of the *J*-data for the crack growth resistance ($J,\Delta a$) curve. The *J* data were calculated according to the incremental J_{i+1} formula

$$J_{i+1} = \left[J_i + \left(\frac{f(a/W)}{b}\right)_i \frac{A_{i,i+1}}{B}\right] \times \left\{1 - \left(\frac{1 + 0.76(b_i/W)}{b}\right)_i [(a_p)_{i+1} - (a_p)_i]\right\} \tag{3}$$

where *A* is area under the load-loadline displacement (P,V) curve, *B* is thickness, *W* is width, *a* is crack length, and *b* is ($W - a$).

The $J,\Delta a$-data pairs with Δa less than 0.3 mm were excluded for further analyses, because of the high relative inaccuracy of Δa based on the estimated absolute inaccuracy of 0.1 mm for the crack length measurements. It is interesting to note that this 0.3-mm exclusion limit nearly coincides with the limit $a_{p_{min}}$ from the application of the ''0.15 mm parallel blunting line exclusion procedure.''

The data were further analyzed in terms of a linear regression fit for the relationship

$$J = m \times \Delta a + C_o \tag{4}$$

where C_o is a numerical constant without physical meaning and *m* is a constant (the crack growth resistance parameter *dJ*/*da*, used in crack instability analyses) for the validity range of the linear regression.

Two *J* values were calculated for the characterization of the toughness near the onset of crack extension: $J_{0.3}$ and J_Q. The ''engineering initiation toughness'' $J_{0.3}$ is the calculated *J* value at 0.3-mm total crack extension. The ''approximate initiation toughness'' J_Q (J_{Ic} after validation) and the corresponding Δa_Q are the $J,\Delta a$-data pair at the intercept of the regression line with the blunting line, the latter being based on the empirical relationship from Ref *6*

$$J = 4\sigma_y \times \Delta a \tag{5}$$

where σ_y is the effective yield stress

$$\sigma_y = \frac{\sigma_{0.2} + \text{UTS}}{2} \tag{6}$$

The post-test examinations included optical metallography, microVickers hardness measurements, and scanning electron fractography of complete fracture surfaces. Thin specimens for transmission electron microscopy (TEM) were taken from an area located near the lower-edge of the back-face of the CT specimens.

Results

Fatigue Tests

The results of the fatigue tests are shown in the form of double-logarithmic plots of *da/dN* versus ΔK in Figs. 3*a* and *b* for Type 316 plate and Type 304 forging, respectively. The figures show higher crack growth rates at higher temperatures and an additional reduction of the fatigue crack growth resistance under low frequency (0.01 Hz) loading.

The results of the Paris analyses are listed in Table 3. The data for the coefficient of determination r^2, with very high mean value of 0.99 for Type 316, indicate good fit for the Paris relationship. The variability of *da/dN* is shown to be limited to factors of 1.2 and 1.8 for the Type 316 and Type 304 materials, respectively. Both values are less than the factor of 2, which is stated in ASTM E 647. The table shows the "Paris coefficient" n to range between 2.3 and 4.1 with a mean value of 3.3 This is in good accordance with the usually observed values of the Paris coefficient, ranging between 2 and 4.

Unfortunately not all results from the control tests are available yet. However, comparisons can be made with results for Type 304 plate after identical HFR irradiations [*8*]. Further, there are results for the same Type 316 material as used in the actual experiment, but after UKMTR

TABLE 3—*Constants and statistical data of the Paris analyses.*

Parameters	Type 316 (L-T)				Type 304 (TS-L)			
	573 K 10 Hz	723 K 10 Hz	823 K 10 Hz	823 K 10^{-2} Hz	573 K 10 Hz	723 K 10 Hz	823 K 10 Hz	823 K 10^{-2} Hz
Number of data pairs	87	89	90	89	92	91	85	91
Domain ΔK_{min}, MPa · $m^{1/2}$	15.6	15.4	15.3	15.4	15.2	15.2	15.2	15.2
Domain ΔK_{max}, MPa · $m^{1/2}$	28.3	28.2	28.2	28.1	28.4	28.5	27.3	28.2
Center ΔK, MPa · $m^{1/2}$	21.0	20.8	20.8	20.8	20.8	20.8	20.4	20.7
Center *da/dN*, 10^{-4} mm/cycle	0.47	0.81	1.4	5.8	0.54	0.97	1.6	6.1
Constant[a] $C \times 10^{-9}$	0.47	1.97	144	41	0.63	2.99	10.7	2.20
Exponent[a] n	3.78	2.74	2.26	3.15	3.74	3.42	3.19	4.14
Coefficient of determination	0.99	0.99	0.98	0.99	0.99	0.98	0.87	0.97
Estimated standard error	0.03	0.03	0.03	0.03	0.04	0.05	0.13	0.07
Vertical half-width of 95% confidence	0.04	0.04	0.04	0.04	0.06	0.07	0.18	0.11

[a] $da/dN = C(\Delta K)^n$ with *da/dN* in mm/cycle and ΔK in MPa · $m^{1/2}$.

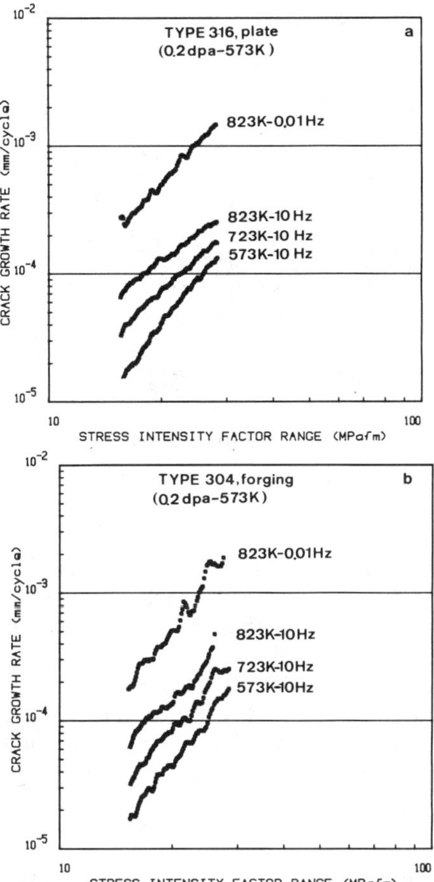

FIG. 3—(a) *Fatigue crack growth of irradiated Type 316 plate and* (b) *fatigue crack growth of irradiated Type 304 forging.*

irradiations [*9*].[2] There are results from the respective control tests too. The slightly different experimental conditions are considered to be acceptable for comparisons with the actual results. These comparisons are shown in the Figs. 4 through 7.

Trends for HFR irradiated (0.2 dpa, 570 K) and UKMTR irradiated (2 dpa, 660 K) materials are shown in Fig. 4. The arrows in this figure indicate reduction of the fatigue crack growth rate after low fluence irradiations at temperatures up to 660 K. A trend-band for UKMTR-irradiated and reference Type 316 weld metal is shown too. Even after irradiation, the fatigue crack growth rates of plate and forging stainless steels are still higher than for weld metal.

Paris curves at 723 K are shown in Fig. 5. The 573 K irradiated materials show slightly lower crack growth rates than the 723 K irradiated Type 304 plate, which shows no irradiation effect at all.

Paris curves for HFR- and UKMTR-irradiated materials under high frequency loading (10 Hz) at 823 K are plotted in Fig. 6, together with a trend-band for HFR-irradiated Type 304 weld and

[2] Sanderson, S. J., Walls, J. D., and Douglas, J., United Kingdom Atomic Energy Authority, 1985, unpublished work.

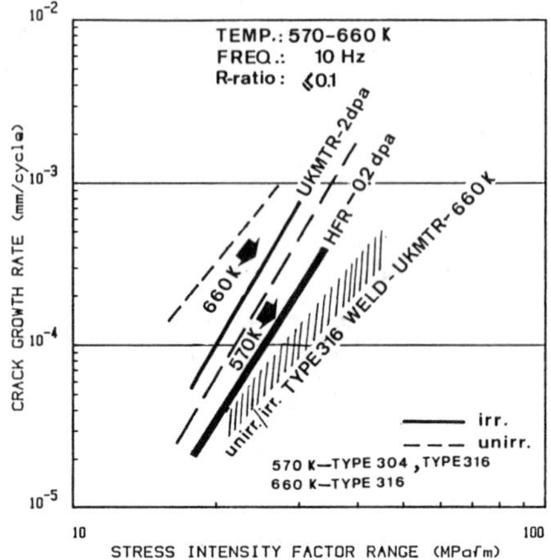

FIG. 4—*Trend-curves for the fatigue crack growth of irradiated and unirradiated stainless steels at elevated temperatures (570 to 660 K).*

UKMTR-irradiated Type 316 weld metal. The figure shows no irradiation effect, neither for the 573 K irradiation nor for the 823 K irradiations. Further it is shown that the fatigue crack growth rates of forging and plate materials are higher than for Type 304 and Type 316 irradiated weld metal.

Results from low frequency tests (<0.05 Hz) at 823 K are compared in Fig. 7. The Paris curves for the 573 K (HFR) irradiated materials nearly coincide with the curves for 823 K (HFR) irradiated Type 304 plate. Especially at higher ΔK values (>20 $MPa \cdot m^{1/2}$) the crack growth rate is significantly increased after irradiation.

Fracture Toughness Tests

Results from the *J*-toughness tests, in the form of load-displacement (*P,V*) curves and corresponding curves for the crack extension ($\Delta a, V$ curves), are shown in Figs. 8*a* through *c* for

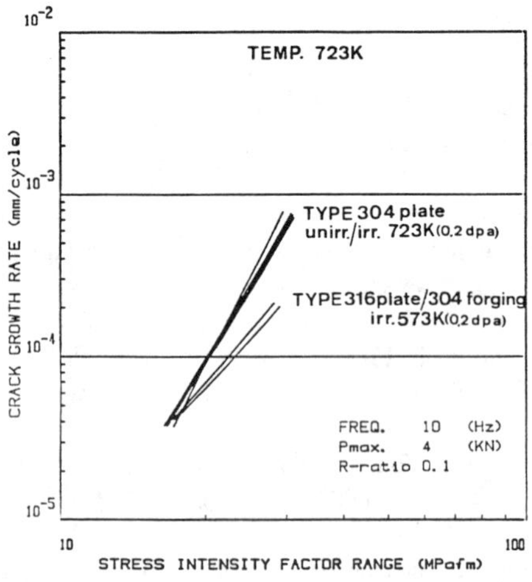

FIG. 5—*Paris curves for irradiated and unirradiated stainless steels, tested at 723 K.*

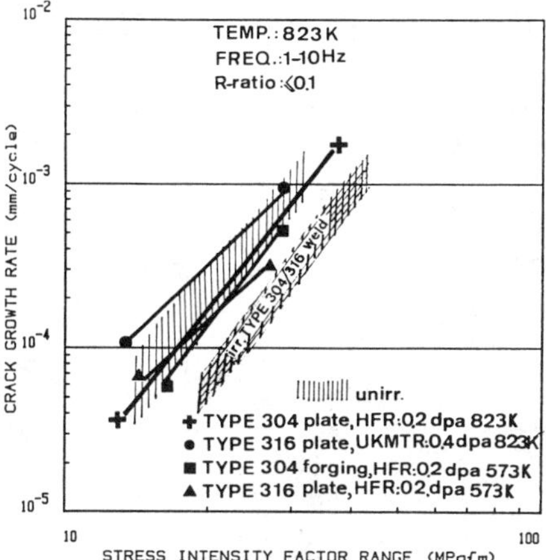

FIG. 6—*Paris curves for low fluence (<1 dpa) irradiated stainless steels, tested at 823 K under high frequency (10 Hz) loading.*

Type 316 plate with L-T orientation and Type 304 forging with TS-L and L-TS orientations, respectively. At higher test temperatures the curves are lower. This trend is even more accentuated if it is noted that the precrack length for Type 316 at 573 K was 0.7 mm longer than for the other tests, considering that the load-bearing capacity decreases with increasing crack length.

The results from the linear regression analyses of the J,R curves are listed in Table 4. The table gives the m (dJ/da) and C_o values, J_Q and Δa_Q and the "engineering toughness" $J_{0.3}$. The values for the tearing modulus T

$$T = \frac{E}{\sigma_y^2}(dJ/da) \tag{7}$$

where E = Young's modulus, were calculated using the estimated effective yield stress σ_y (Table 4). The σ_y values were obtained by applying a correction, based on the post-test hardness-

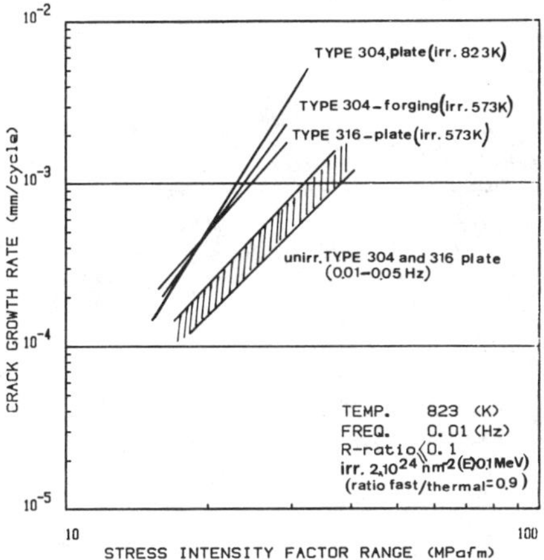

FIG. 7—*Paris curves for low frequency (0.01 Hz) loaded stainless steels at 823 K, after irradiation at 573 and 823 K up to* 2×10^{24} $n \cdot m^{-2}$ (E > *0.1 MeV) at a fast to thermal flux density ratio of 0.9.*

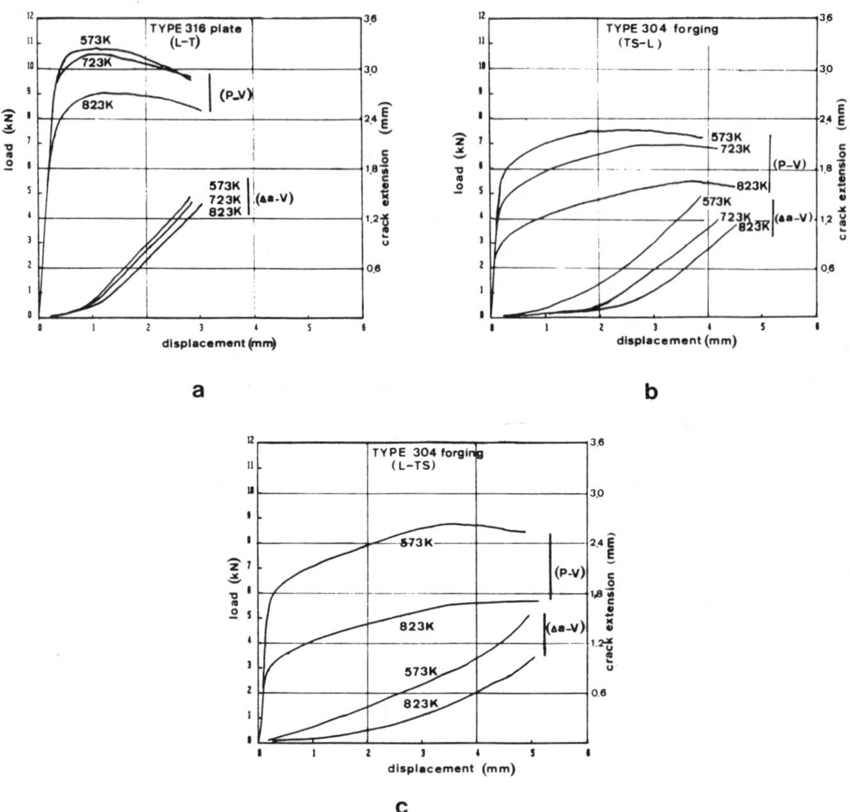

FIG. 8—(a) *Tensile curves and corresponding crack growth curves for* (a) *irradiated Type 316 (L-T orientation),* (b) *irradiated Type 304 (TS-L orientation), and* (c) *irradiated Type 304 (L-TS orientation).*

TABLE 4—*Values of the* J-*toughness parameters for irradiated Type 316 plate and Type 304 forging after irradiation (0.2 dpa) at 573 K.*

Material, Type Orientation Test temperature	316[a] (L-T) 573K	316[a] (L-T) 723 K	316[a] (L-T) 823 K	304 (TS-L) 573 K	304 (TS-L) 723 K	304 (TS-L) 823 K	304 (L-TS) 573 K	304 (L-TS) 823 K
$m(dJ/da)$, N/mm²	189	203	197	189	223	160	375	267
C, kJ/m²	185	153	138	174	180	178	225	151
J_Q, kJ/m²	206	175	162	200	215	204	166	190
Δa_Q, mm	0.12	0.11	0.13	0.14	0.16	0.16	0.11	0.15
$J_{0.3}$, kJ/m²	241	214	196	231	247	226	237	231
T modulus, (-)	172	209	280	243	316	242	482	404
σ_y estimated, MPa	440	400	330	370	340	320	370	320

[a] Average value, from two tests.

measurements, on the tensile properties of the unirradiated materials to account for the irradiation hardening at 573 K and the subsequent annealing at the higher temperatures.

Comparing the $J_{0.3}$ data, which are considered to be the most appropriate data for comparisons because of the fact that they represent a constant amount (0.3 mm) of physical crack extension, there is a minor temperature effect (decrease of $J_{0.3}$ with increasing temperature) for Type 316, whereas there is no temperature effect for Type 304.

In Fig. 9 the $(J, \Delta a)$ trend bands are shown for the three material conditions (316 L-T, 304 TS-L, and 304 L-TS). The bands are formed from the resistance curves at the test temperature range of 573 to 823 K. The band for Type 316 (L-T orientation) nearly coincides with the band for the Type 304 forging with the TS-L orientation. The slopes of the resistance curves in these bands range from dJ/da of 160 to 223 $N \cdot mm^{-2}$ with a mean value of 195 $N \cdot mm^{-2}$. The Type 304 with the notch-plane orientation parallel to the thickness direction of the forging (L-TS) shows a much steeper trend band with a significantly higher dJ/da value of about 330 $N \cdot mm^{-2}$. The notch-plane orientation appears to affect the crack growth resistance more than the test temperatures over the range of 573 to 823 K.

The measured J,R curves for Type 316 at 573 and 723 K are compared in Fig. 10 with trend curves for the same but unirradiated material at 643 K. Also included in this figure are the lower band reference curves and the trend-curves for Type 316 wrought material and weld metal after irradiation in UKMTRs at 643 K from the work of Picker [*10,11*]. The figure indicates a small beneficial effect of low fluence (less than 1 dpa) irradiations at temperatures up to 643 K. Significant reduction is shown for the weld metal after irradiation to higher fluences (4 dpa). Further it is shown that toughness properties of Type 316 plate are superior to the toughness properties of weld metal.

In Fig. 11 the results from the tests at 823 K are compared with the trend curves for 823 K irradiated and reference Type 304 plate.[3] Irradiation at 823 K significantly reduces the crack

[3] de Vries, M. I., in this publication, pp. 162–173.

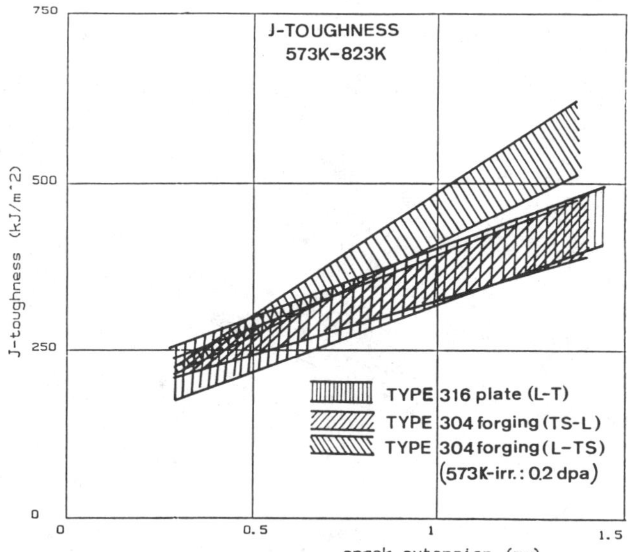

FIG. 9—*Trend-bands for the crack growth resistance (J,R-curves) or irradiated Type 304 forging and Type 316 plate for the temperature range of 573 to 823 K, after irradiation up to 0.2 dpa at 573 K.*

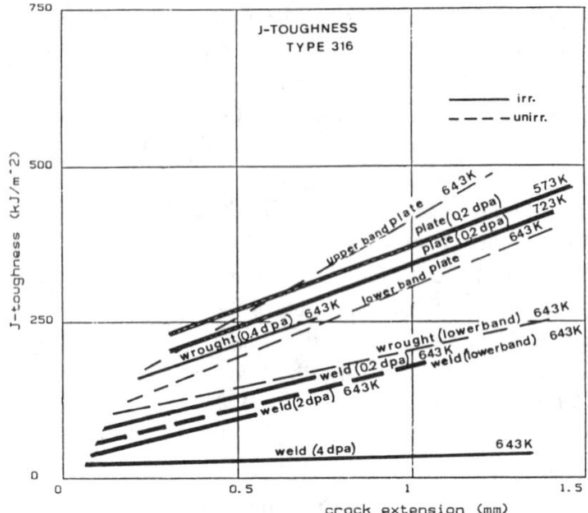

FIG. 10—*Comparison of the* J,R-*curves for irradiated Type 316 plate at 573 and 723 K with trend-curves for unirradiated Type 316 plate and Type 316 wrought and weld metal at 643 K.*

growth resistance of Type 304 plate at 823 K. The resistance curves for the 573 K irradiated materials are between the curves for unirradiated Type 304 and the 823 K irradiated Type 304. The 573 K irradiation seems to affect the crack growth resistance at 823 K less than the 823 K irradiation.

Post-Test Examinations

Transmission electron microscopy revealed severe displacement damage in the form of a high density of black dots in both materials. The microstructures and corresponding hardness values of the toughness specimens after testing at various temperatures are shown in Fig. 12. The irradiation at 573 K caused an increase of the microVickers hardness by about 500 (MPa) HVN points. Partial

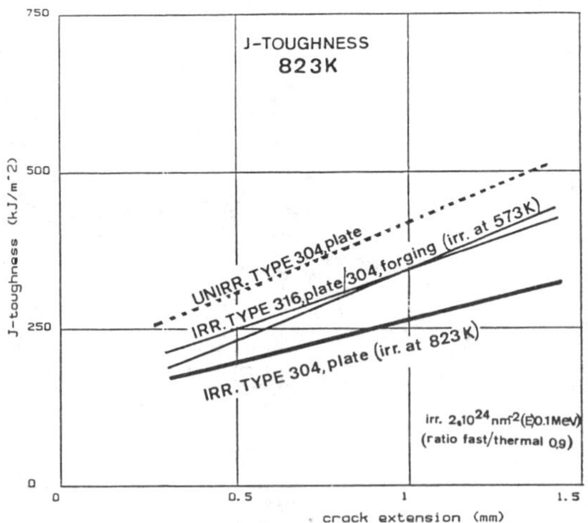

FIG. 11—*Comparison of the* J,R-*curves for unirradiated and 823 K irradiated Type 304 with the trend-curves for 573 K irradiated stainless steels, tested at 823 K.*

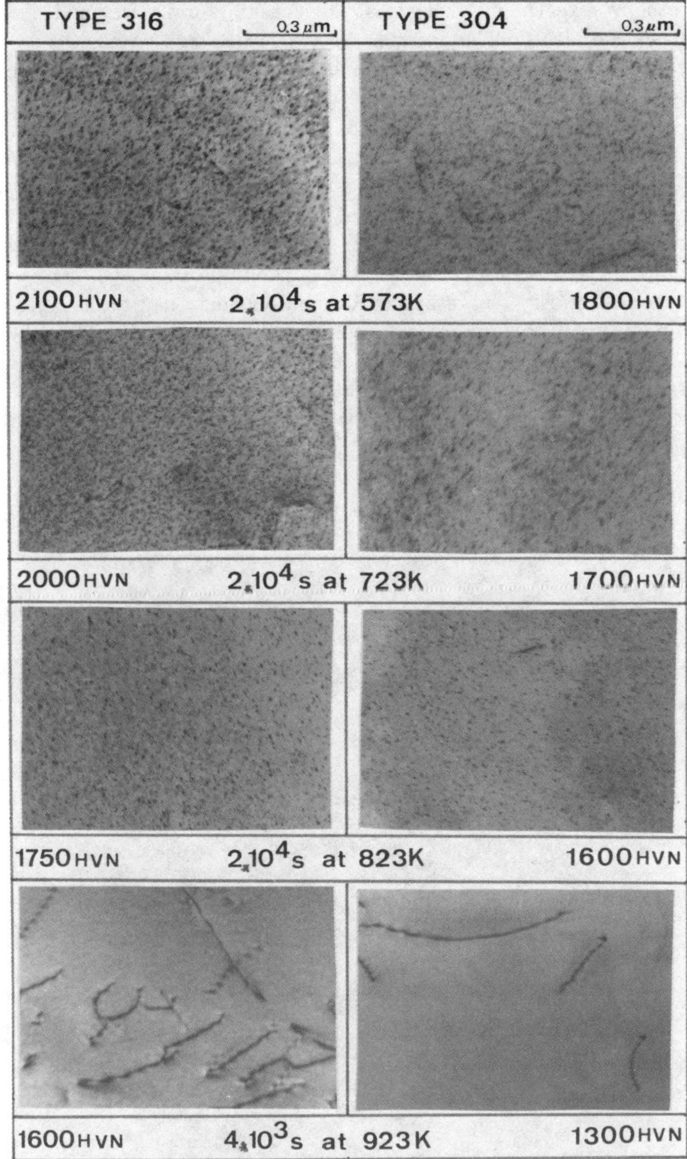

FIG. 12—*Microstructures and corresponding hardness values for the toughness specimens after testing at various temperatures.*

annealing of the displacement damage and recovery of the irradiation hardening is shown after testing at 723 and 823 K. The displacement damage disappeared completely after additional annealing of the TEM specimens from the 573 K tests for 1 h at 923 K.

Partial annealing of the displacement damage was also observed after high frequency fatigue testing at 723 and 823 K. However, complete annealing was observed after low frequency fatigue testing (periods $> 10^6$ s) at 823 K.

Optical microscopy revealed no changes of the density, form, size, and distribution of the original precipitates and inclusions after irradiations at 573 K.

Fractography revealed slight differences between Type 316 and Type 304 specimens. The fatigue cracks of the Type 316 specimens were smooth with straight crack fronts, whereas the Type 304 specimens exhibited rough fatigue cracks with somewhat irregular crack fronts. All fatigue specimens showed characteristic fatigue striations; very small striation spacings ($\leqq 0.1$ μm) could be resolved on the fracture surfaces of the Type 304 specimens in particular. The specimens tested under low-frequency loading at 823 K exhibited additionally intergranular fatigue fracture. The intergranular contribution started at positions on the fracture surfaces for which ΔK values were about 23 MPa · $m^{1/2}$. Type 304 showed intergranular fracture of individual large grains whereas in Type 316 elongated secondary cracks of linked grain-boundary-like facets were observed (Fig. 13). These intergranular substructures are consistent with the large grain size structure of the Type 304 forging and the elongated original boundaries of the Type 316 plate, respectively.

The toughness specimens of Type 316 material (L-T orientation) exhibited severe crack-tunnelling whereas the Type 304 toughness specimens showed crack extension fairly parallel to the fatigue crack front (no tunnelling). The well-known dimple structures, characteristic for ductile tearing, were observed on the fracture surfaces of all toughness specimens. The Type 316 specimens showed large dimples surrounded by a network with a high density of small dimples, reflecting

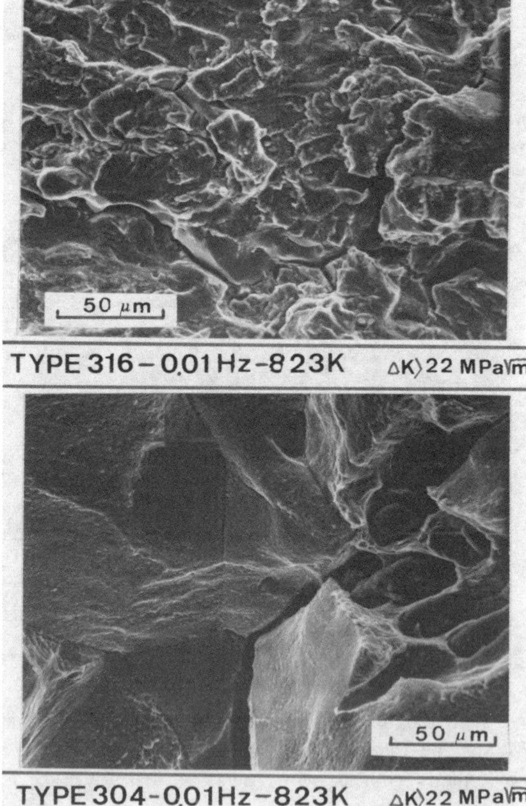

FIG. 13—*Fractographs of irradiated Type 316 and Type 304 after low frequency (0.01 Hz) testing at 823 K.*

the original precipitate-network, whereas the Type 304 specimens exhibited very large dimples with fragmented inclusions (Fig. 14). Careful inspection of the 823 K tested specimens revealed no intergranular crack features.

Discussion

It is well established that the post-irradiation classical mechanical properties, such as tensile, creep, and fatigue life properties, of austenitic stainless steels strongly depend on testing and irradiation conditions [*12–14*]. The irradiation effects are generally attributed to a complex contribution from displacement damage or transmutation reactions, or both, being primarily a function of neutron energy, relevant cross section, neutron fluence, and irradiation temperature.

However, the effects of low fluence (<1 dpa) irradiations can be schematically simplified. At low temperatures (below 0.4 T_m where T_m is the absolute melting temperature) the steel is hardened and concomitantly the (high) ductility is reduced. This is caused by displacement damage in the form of point defect clusters ("black dots"), which act as barriers for the movement of glissile dislocations. In the temperature range between 0.3 T_m and 0.6 T_m the increased atomic mobility causes annealing of the displacement damage. At elevated temperatures (above 0.4 T_m) enhanced intergranular fracture caused by helium embrittlement becomes the dominant phenomenon. This is primarily caused by the helium from the thermal neutron transmutation reaction $^{10}B(n,\alpha)Li^7$. Recent publications have paid attention to the effect of test temperature and strain rate with particular emphasis on the critical dislocation velocity for the onset of intergranular fracture caused by helium contents as low as 0.1 appm (atomic parts per million) or even less [*15,16*].

Our observations of a high density of point defect clusters after irradiation at 573 K, thermal annealing, and recovery in the temperature range of 723 to 923 K, and enhanced intergranular fatigue fracture under low frequency loading at 823 K, strongly support this view.

The irradiation at 573 K reduced the fatigue crack growth rate at this temperature. This is attributed to the reduced crack-tip plasticity (reduction of the plastic zone-size r_p) caused by irradiation hardening by the displacement damage.

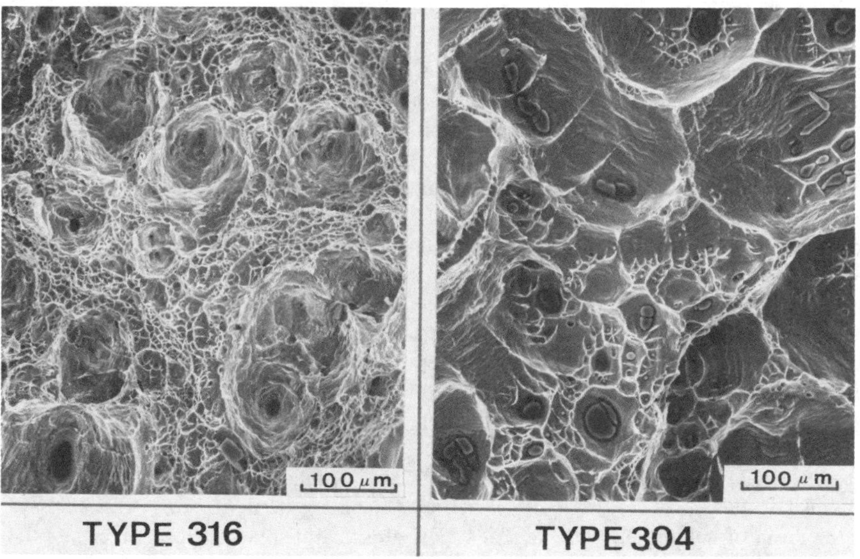

FIG. 14—*Fractographs of irradiated Type 316 and Type 304 after* J*-toughness testing at 823 K.*

Fatigue crack growth rates at 723 K were somewhat lower after 573 K irradiation than after 723 K irradiation. Damage-free microstructures have been reported after the 723 K irradiation [*17*]. However, the testing periods at 723 K appear to be too short for significant annealing of the displacement damage for the 573 K irradiated materials. Consequently the lower fatigue crack growth rates after irradiation at 573 K than after irradiation at 723 K are attributed to the remaining displacement damage.

Irradiations at 573 and 823 K cause equally increased crack growth rates of the low frequency tests at 823 K. The testing periods of about two weeks are long enough for complete annealing of the displacement damage. In this respect it is important to recall the observations of intergranular fatigue fracture at higher ΔK values. Similar observations were previously made for low frequency tested Type 304 and Type 316 stainless steels after HFR-irradiation at 823 K [*18*]. De Vries and Michel attribute this irradiation effect to enhanced intergranular crack formation by helium from the thermal neutron $^{10}B(n,\alpha)$-reaction. The increased crack growth rates after irradiation at 573 K are attributed similarly to the contribution from enhanced intergranular crack formation caused by helium (up to 15 appm for our Type 316 material).

Low fluence irradiations (<1 dpa) have a marginal beneficial effect on the J-toughness properties at temperatures below 700 K. This is attributed to the counterbalancing effects of irradiation hardening and concomitant reduction of ductility by the displacement damage, as J,R behavior depends on plasticity as well as on critical crack-tip strain. However, the effects might be somewhat obscured by the relative high inaccuracy of the crack-extension data, especially at low Δa values. This is due to the variability of the optical crack extension measurements for the calibration of the electrical PD signals, amounting to about 0.1-mm crack extension.

The effect of notch-plane orientation seems to be much stronger than the effects of test temperature for the range of 573 to 823 K and low fluence irradiation at 573 K. Balladon has reported a similar orientation dependency for unirradiated Type 316 plate [*19*]. He reports dJ/da values of 160 and 82 $N \cdot mm^{-2}$ at 823 K for the (T-L) and (S-L) orientations, respectively, whereas a much higher value of 460 $N \cdot mm^{-2}$ is reported for the orientation with the notch-plane perpendicular to the rolling direction (L-T). According to Balladon the crack growth resistance of austenitic stainless steel strongly depends on the presence of transgranular precipitates and on the content and shape of the inclusions in relation to the notch-plane orientation.

In this respect it is important to refer to the fractographs of Fig. 14, reflecting to original structures consisting of local manganese-sulfide conglomerates for Type 304 and a very high density of transgranular precipitates orientated parallel to the rolling direction in the form of linked-up chains of elongated networks bounding the original austenite grains for Type 316. The 573 K irradiation did not affect these original microstructures.

In view of the above argument it can be considered that the dJ/da value of 195 $N \cdot mm^{-2}$, from the trend-bands for Type 316 (L-T) and Type 304 (TS-L) in Fig. 9, represents the most favorable ("nonconservative resistance curve") orientation for the Type 316 plate whereas this value represents the least favorable ("conservative resistance curve") orientation for the Type 304 forging.

The crack growth resistance at 823 K seems to be hardly affected after irradiation at 573 K whereas a significant reduction has been reported for Type 304 plate after irradiation at 823 K.[3] This strong effect has been reported on p. 170[3] to be associated with intergranular crack formation attributed to helium from the $^{10}B(n,\alpha)$-reaction. In this respect it is interesting to recall the observations of partial annealing of the displacement damage after the relatively short testing periods of the J tests at 823 K. Further the 573 K irradiated specimens did not show any intergranular features.

For the onset of helium embrittlement at 823 K a critical displacement rate of about 0.1 $mm \cdot s^{-1}$ is estimated from results on p. 170[3] for Type 304 plate after irradiation at 823 K. Consequently, the actual applied displacement rate of 2×10^{-3} $mm \cdot s^{-1}$ could be low enough for helium enhanced intergranular crack formation. It is supposed that the displacement damage, which only partly anneals during the short testing periods of the J-tests at 823 K, causes an increase of the effective strain rate at the crack-tip and consequently the velocity of the dislocations exceeds the critical value for the onset of intergranular fracture due to helium from the ^{10}B-transmutation reaction. Effectively the critical displacement rate for the onset of helium embrittlement is thus reduced.

Summary and Conclusions

Low fluence irradiations (<1 dpa) at temperatures below 700 K cause decrease of the fatigue crack growth rate, attributed to irradiation hardening by the displacement damage.

The increase of the crack growth rate under slow cyclic loading (0.01 Hz) at 823 K is equal to the increase after similar irradiations at 823 K, due to enhanced intergranular crack formation attributed to helium, primarily from the ^{10}B-transmutation reaction.

The crack growth resistance under monotonic loading is slightly improved by low fluence irradiations at temperatures below 700 K. This is attributed to the limited beneficial effects of irradiation hardening by the displacement damage.

The J-toughness properties strongly depend on the appearance of inclusions and precipitates in relation to the orientation of the notch plane. These material inhomogeneities are not significantly affected by low fluence irradiations at temperatures below 0.4 T_m.

Results from high rate loading tests at temperatures above the irradiation temperature are nonconservative as far as there is incomplete annealing of the displacement damage caused by the relatively short annealing periods at the higher test temperatures.

The J-toughness properties of plate and forging Type 316 and Type 304 stainless steels are better than for weld metal, however, the fatigue crack growth properties of weld metal are better. This includes irradiated and nonirradiated (reference) material.

Acknowledgments

The author wishes to acknowledge the contribution from the Risley Nuclear Laboratories, UKAEA, England, by providing the Type 316 steel and the information on that material. Particularly the helpful discussions with C. J. Sanderson and C. Picker of that organization, during the preparation of the manuscript, are highly appreciated.

References

[1] Dufresne, J., Henry, B., and Larsson, H., *Effects of Radiation on Materials: Proceedings of the Tenth International Symposium, STP 725,* J. A. Sprague and D. Kramer, Eds., American Society for Testing and Materials, Philadelphia, 1979, pp. 511–528.

[2] Reuter, W. G., Rahl, T. E., and Merrill, B. J., *Effects of Radiation on Materials: Twelfth International Symposium, STP 870,* F. A. Garner and J. S. Perrin, Eds., American Society for Testing and Materials, Philadelphia, 1985, pp. 537–547.

[3] Röttger, H., Tas, A., Hardt, P. von der, and Voorbraak, W. P., *High Flux Materials Testing HFR Petten, EUR 5700,* 1981 version.

[4] Tjoa, G. L., Broek, F. P. van den, and Hoepen, J. van, "Fatigue Testing of Irradiated Steels at High Temperatures," ECN-report, ECN-183, Petten, The Netherlands, 1983.

[5] Tjoa, G. L., Broek, F. P. van den, and Schaap, B. A. J., in *Automated Test Methods for Fracture and Fatigue Crack Growth, STP 877,* W. H. Cullen, R. W. Landgraf, L. R. Kaisand, and J. H. Underwood, Eds., American Society for Testing and Materials, Philadelphia, 1985, pp. 197–212.

[6] de Vries, M. I. and Schaap, B. A. J., *Elastic-Plastic Fracture Test Methods: The User's Experience, ASTM STP 856,* E. T. Wessel and F. J. Loss, Eds., American Society for Testing and Materials, Philadelphia, 1985, pp. 183–195.

[7] de Vries, M. I., Schaap, B. A. J., and Tjoa, G. L., *Proceedings of the OECD-CSNI Workshop on Ductile Fracture Test Methods,* to be issued in 1986 as a CSNI-NRC, NUREG-report.

[8] de Vries, M. I., *Effects of Radiation on Materials: Eleventh International Symposium, STP 782,* H R. Brager and J. S. Perrin, Eds., American Society for Testing and Materials, Philadelphia, 1982, pp. 720–734.
[9] Lloyd, G. J., Walls, J. D., and Gravenor, J., *"Low Temperature Fatigue Crack Propagation in Neutron-Irradiated Type 316 Steel and Weld Metal,"* UKAEA-report ND-R-506 (R), Risley, 1982.
[10] Picker, C., "The Fracture Toughness of Type 316 Steel and Weld Metal," in *Proceedings of IAEA Specialists Meeting on Mechanical Properties of Fast Reactor Structural Materials,* Paper IWGFR 49/420-4, Chester, 1983.
[11] Picker, C., Stott, A. L., and Cocks, H., "Effects of Low Dose Fast Neutron Irradiation on the Fracture Toughness of Type 316 Stainless Steel and Weld Metal," in *Proceedings of IAEA Specialists Meeting on Mechanical Properties of Fast Reactor Structural Materials,* Paper IWGFR 49/440-4, Chester, 1983.
[12] Schaaf, B. van der and de Vries, M. I., "Low Dose Irradiation Effects on DIN 1.4948 Mechanical Properties," in *Proceedings of IAEA Specialists Meeting on Mechanical Properties of Fast Reactor Structural Materials,* Paper IWGFR 49/370-3, Chester, 1983.
[13] Simons, R. L. and Hulbert, L. A., in *Effects of Radiation on Materials: Twelfth International Symposium, STP 870,* F. A. Garner and J. S. Perrin, Eds., American Society for Testing and Materials, Philadelphia, 1985, pp. 820–839.
[14] Schaaf, B. van der and de Vries, M. I., "Fatigue and Crack Growth Properties of Type 316 Steel for Fusion Applications," *Proceedings of International Workshop on "The Relation Between Mechanical Properties and Microstructure under Fusion Irradiation Conditions,* Ebeltoft, Denmark, 1985 (to be issued).
[15] de Vries, M. I., Schaaf, B. van der, Staal, H. U., and Elen, J. D., *Effects of Radiation on Structural Materials: Proceedings of the Ninth International Symposium, STP 683,* J. A. Sprague and D. Kramer, Eds., American Society for Testing and Materials, Philadelphia, 1979, pp. 477–489.
[16] Gerke, R. D. and Jesser, W. A., *Effects of Radiation on Materials: Twelfth International Symposium, STP 870,* F. A. Garner and J. S. Perrin, Eds., American Society for Testing and Materials, Philadelphia, 1985, pp. 605–618.
[17] de Vries, M. I., *Effects of Radiation on Materials: Proceedings of the Eleventh International Symposium, STP 782,* H. R. Brager and J. S. Perrin, Eds., American Society for Testing and Materials, Philadelphia,1982, pp. 665–689.
[18] de Vries, M. I. and Michel, D. J., in *Effects of Radiation on Materials: Twelfth International Symposium, STP 870,* F. A. Garner and J. S. Perrin, Eds., American Society for Testing and Materials, Philadelphia, 1985, pp. 803–819.
[19] Balladon, P., Heritier, J., and Rabbe, P., *Fracture Mechanics, Fourteenth Symposium, STP 791,* J. C. Lewis and G. Sines, Eds., American Society for Testing and Materials, Philadelphia, 1983, pp. II-496–513.

J. Russell Hawthorne,[1] *Blaine H. Menke,*[1] *Nabil G. Awadalla,*[2] *and Kevin R. O'Kula*[2]

Experimental Assessments of Notch Ductility and Tensile Strength of Stainless Steel Weldments After 120°C Neutron Irradiation

REFERENCE: Hawthorne, J. R., Menke, B. H., Awadalla, N. G., and O'Kula, K. R., **"Experimental Assessments of Notch Ductility and Tensile Strength of Stainless Steel Weldments After 120°C Neutron Irradiation,"** *Influence of Radiation on Material Properties: 13th International Symposium (Part II), ASTM STP 956,* F. A. Garner, C. H. Henager, Jr., and N. Igata, Eds., American Society for Testing and Materials, Philadelphia, 1987, pp. 191–206.

ABSTRACT: The Charpy-V (C_v) properties of American Iron and Steel Institute (AISI) 300 series stainless steel plate, weld, and weld heat-affected zone (HAZ) materials from commercial production weldments in 406-mm-diameter pipe (12.7-mm wall) were investigated in unirradiated and irradiated conditions. Weld and HAZ tensile properties were also assessed in the two conditions. The plates and weld filler wires represent different steel melts; the welds were produced using the multipass metal inert gas (MIG) process. Weldment properties in two test orientations were evaluated. Specimens were irradiated in a light water cooled and moderated reactor to 1×10^{20} n/cm², $E > 0.1$ MeV, using a controlled temperature assembly. Specimen tests were performed at 25 and 125°C.

The radiation-induced reductions in C_v energy absorption at 25°C were about 42% for the weld and the HAZ materials evaluated. A trend of energy increase with temperature was observed. The concomitant elevation in yield strength was about 53%. The increase in tensile strength in contrast was only 16%. The postirradiation yield strength of the axial test orientation in the pipe was less than that of the circumferential test orientation.

Results for the HAZ indicate that this component may be the weakest link in the weldment from a fracture resistance viewpoint.

KEY WORDS: stainless steels, weld deposits, weld heat affected zone, radiation embrittlement, nuclear irradiation, notch ductility, tensile strength, fracture resistance, 120°C irradiation

This study represents one phase of a large investigation into the fracture resistance of austenitic stainless steel weldments, both as-fabricated and after irradiation at 120°C. The general objective was to explore notch ductility and tensile strength changes produced by a fluence of $\sim 1 \times 10^{20}$ n/cm², $E > 0.1$ MeV, as a precursor to a study involving much higher fluences at the same nominal temperature. The total investigation is intended to develop a broad base of information on fracture toughness properties as well as notch ductility and tensile properties where research variables are test temperature, loading rate, neutron fluence, and weldment component, that is, base metal versus weld metal versus weld heat-affected zone (HAZ).

[1] Research metallurgist and senior engineer, respectively, Materials Engineering Associates, Inc., 9700-B Martin Luther King, Jr. Highway, Lanham, MD 20706-1837.

[2] Research staff engineer and research engineer respectively, E. I. du Pont de Nemours & Co., Savannah River Laboratory, P.O. Box A, Aiken, SC 29808-001.

In the literature, a large volume of data exists for stainless steel materials irradiated at high temperature (>370°C), especially for thin sheet and plate forms. The information was developed primarily in support of advanced reactor concepts, such as breeder and fusion systems that would involve high-temperature service. For lower temperature irradiation conditions, data on mechanical properties and property changes wrought by neutron exposure are sparse. A paucity of data is particularly evident for irradiation temperatures less than 200°C. The present study was undertaken to help fill the void in knowledge on the fracture resistance of American Iron and Steel Institute (AISI) Type 304 weldments for "low-temperature" radiation service applications. Materials of interest were those produced commercially in the early 1950s.

Materials

The test materials were 1950 vintage, rolled and welded pipe sections having an outer diameter (OD) of 406 mm (16 in.) and a nominal wall thickness of 12.7 mm (0.5 in.). A total of eight pipe sections was evaluated. In this report, the individual materials are referenced to the arbitrarily assigned pipe ring number (1 through 8).

Each pipe test section contained a circumferential weld made by the metal inert gas (MIG) process and one or more longitudinal welds. Only the circumferential welds were evaluated here. The weld joint was a single Vee; the joint preparation contained a small land on the inner diameter (ID) side to aid preweld fitup. Figure 1 is an etched cross section of one typical weld. The joint was filled from the OD side using several weld passes; a root pass made from the ID side was evident in most joints as well. The width of the deposit on the OD surface was on the order of 15 mm (0.6 in.) and was 2.5 mm (0.1 in.) on the ID surface. Based on weld bead patterns, the welds were not filled using the same sequence in all cases.

The chemical compositions of the base metals and the weld metals are given in Table 1. Note that the base compositions indicate different steel melts as their source; the weld metal compositions likewise infer that the weld fillers were from different melts. Thus, the materials provide a basis for statistical evaluation of properties of weldments produced in the 1950s timeframe.

Test Specimens

Full size Charpy-V C_v specimens (ASTM Type A) and threaded-end tension specimens having a gage diameter of 5.08 mm and a gage length of 20.3 mm were used for the notch ductility and strength determinations, respectively. Specimen blanks were cut in two orientations. One group of specimens had their long dimension parallel to the original pipe axis and are identified here as "axial orientation specimens." The second group had their long dimension perpendicular to the pipe axis and tangential to the OD; these are termed "circumferential (CMFL) specimens." In all cases, the notch of the C_v specimen was made perpendicular to the pipe ring surfaces (OD and ID).

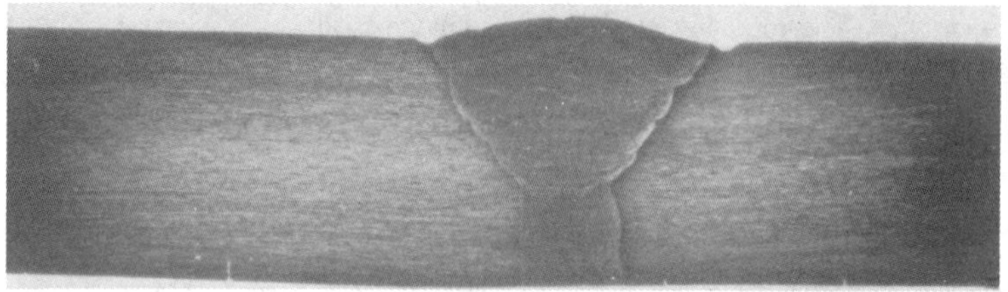

FIG. 1—*Etched cross section of Weld 5 (12.7-mm base plate thickness).*

For best control over the location of weld and HAZ specimens in the stock, the blanks were first rough-machined to approximately 11.5 mm square in cross section and then individually etched to reveal the weld fusion lines. For axial weld specimens, that is, those which spanned the deposit, the specimen midlength was placed on the weld center line. For the CMFL weld specimens, the cross section was centered on the weld nugget. Because of the Vee-shape of the deposit and its limited width on the ID surface, the C_v specimens of this orientation contained a small triangular volume of HAZ material on the front face and on the rear face of the specimen. This nonweld material is not believed to have influenced the test result.

Figure 2 shows schematically, the placement of the HAZ specimen relative to the weld deposit. For the CMFL orientation, specimen locations were indexed to the intersection of the adjacent fusion line with the ID surface side of the rough-machined blank. The axial orientation HAZ C_v specimens were also indexed in this manner. Although arbitrarily chosen, the reference distance from the fusion line was intended to make the specimen "bracket" the heat-affected zone. Typically, the C_v specimens but not the tension HAZ specimens of the CMFL orientation contained a small volume of weld metal in the test section.

It will be noted that the axial orientation "weld" tension specimens, in fact, were composite specimens since the gage length of 20.3-mm spanned not only the weld nugget but also the HAZ on either side. Tests of these specimens would produce failure in the weakest component.

Material Irradiation

The irradiation of C_v and tension specimens was performed in the 2-MW light-water-cooled and moderated UBR test reactor located in the Buffalo Materials Research Center. The 93 specimens were contained in three, independently temperature-controlled capsules designated A, B, and C, which together formed one irradiation assembly. Thermocouples welded to specimen midsections were used for temperature monitoring and control. The target irradiation temperature was 120°C (248°F). At full reactor power, temperature differences were less than ±15°C.

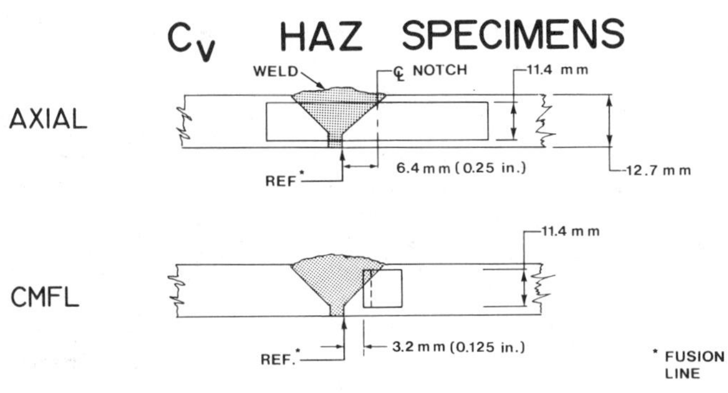

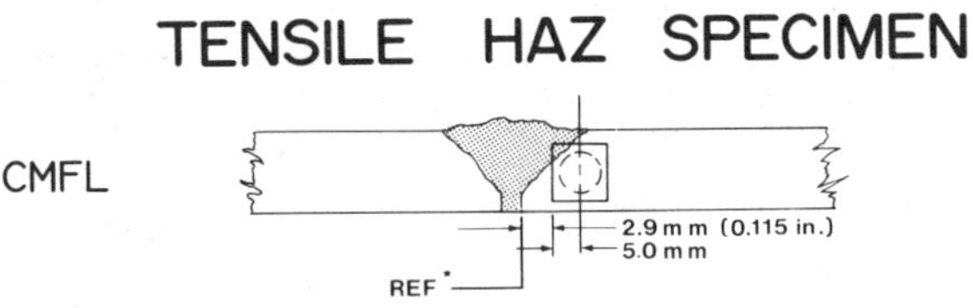

FIG. 2—*Placements of* C_v *and tension specimen blanks for HAZ studies.*

TABLE 1—*Base metal and weld compositions.*

Ring	Base Metal (Side)	Composition, wt%											
		C	Mn	Si	P	S	Ni	Cr	Mo	B	Co	Cu	N[a]
						BASE METAL							
1	A	0.079	1.60	0.79	0.031	0.011	9.63	18.79	0.41	0.001	0.11	0.29	0.047
	B	0.035	1.56	0.58	0.024	0.016	9.19	18.44	0.25	0.002	0.10	0.24	0.036
2	A	0.079	1.50	0.34	0.031	0.024	9.65	18.27	0.45	0.002	0.13	0.42	0.043
	B	0.052	1.41	0.38	0.031	0.025	8.50	19.40	0.39	<0.001	0.15	0.42	0.036
3	A	0.063	1.30	0.31	0.028	0.024	9.38	18.59	0.40	0.001	0.12	0.38	0.044
	B	0.048	1.33	0.39	0.027	0.025	9.13	18.67	0.36	0.002	0.13	0.39	0.034
4	A	0.053	1.81	0.33	0.026	0.017	8.75	18.97	0.35	0.002	0.11	0.28	0.033
	B	0.083	1.75	0.74	0.033	0.017	9.60	18.88	0.46	0.002	0.13	0.32	0.043
5	A	0.041	1.39	0.67	0.026	0.024	9.64	18.88	0.52	0.002	0.12	0.28	0.035
	B	0.080	1.25	0.32	0.026	0.016	10.00	19.05	0.44	0.001	0.13	0.41	0.043
6	A	0.058	1.44	0.49	0.027	0.017	9.65	19.05	0.43	0.001	0.15	0.62	0.044
	B	0.046	1.46	0.66	0.026	0.024	8.48	18.88	0.22	0.001	0.13	0.17	0.034
7	A	0.052	1.30	0.55	0.028	0.016	9.35	18.65	0.38	0.002	0.12	0.26	0.039
	B	0.047	1.33	0.34	0.027	0.019	9.15	18.50	0.21	0.001	0.08	0.20	0.037
8	A	0.055	1.30	0.40	0.030	0.026	8.72	19.05	0.42	0.002	0.16	0.45	0.036
	B	0.078	1.75	0.40	0.033	0.018	8.30	19.66	0.44	0.003	0.54	0.34	0.043

						Weld Metal						
1	2PW216W3	0.038	1.39	0.41	0.023	0.018	9.65	20.15	0.23	0.002	0.11	0.21
											0.11[a]	0.20[a]
2	6PW1816W3	0.052	1.45	0.41	0.022	0.019	10.50	19.20	0.20	0.005	0.10	0.22
3	4PW16W5	0.039	1.25	0.39	0.020	0.017	10.16	19.56	0.21	0.004	0.20	0.21
4	1P1W1316W3	0.047	1.41	0.43	0.022	0.018	10.75	19.29	0.17	0.005	0.094	0.20
5	2PW1716W2	0.048	1.52	0.42	0.023	0.010	10.15	19.96	0.26	0.001	0.16	0.23
											0.17[a]	0.18[a]
6	3PW1516W5	0.050	1.56	0.49	0.024	0.008	10.12	19.87	0.24	<0.001	0.18	0.19
					0.022[a]	0.010[a]						0.19[a]
7	4PW416W4	0.042	1.47	0.43	0.020	0.009	9.88	19.47	0.24	0.003	0.15	0.21
8	2PW216W5	0.045	1.52	0.37	0.022	0.018	9.70	20.15	0.21	0.002	0.22	0.18
												0.16[a]

[a] Total nitrogen, wt%.

[b] Duplicate analysis using separate stock.

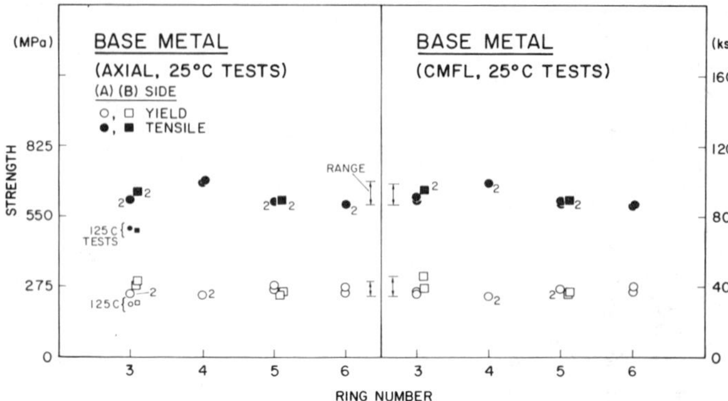

FIG. 3—*Individual yield and tensile strength determinations at 25°C for base metals from four pipe rings (axial and circumferential test orientations). Results for 125°C tests of base metals from Ring 3 are also shown.*

The target fluence was 1 × 10^{20} n/cm², $E > 0.1$ MeV. Actual fluences were determined from iron and nickel dosimeter wires placed in the V-notches of the C_v specimens and from ^{238}U dosimeter capsules nested among the tension specimen gage sections. The average measured specimen fluence was 1.1 × 10^{20} n/cm². Neutron spectrum calculations indicate that the fluence, n/cm², $E > 0.1$ MeV, is 2.91 times the fluence, n/cm², $E > 1$ MeV [*1,2*]. The exposure equivalent, in terms of displacements per atom (dpa), was 0.063 dpa ($E > 1$ MeV).

Results

Unirradiated Condition—Tensile Properties

The test matrix for tension specimens encompassed two specimen orientations, three temperatures, and two loading rates (static and dynamic). Findings from the dynamic tests will be treated in a future report. Figures 3 and 4 are plots summarizing the individual strength determinations for the base metal, the weld, and the weld HAZ materials, respectively.

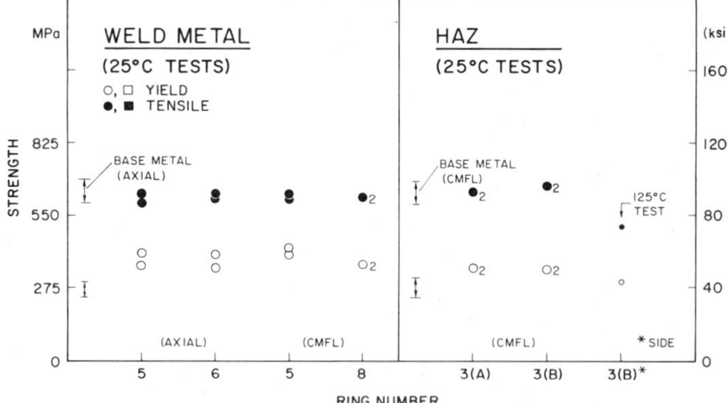

FIG. 4—*Individual yield and tensile strength determinations at 25°C for weld deposits and weld heat-affected zone materials (axial and circumferential test orientations). The range of base metal values from Fig. 3 is indicated for comparison. Results of one 125°C test of the HAZ is also shown.*

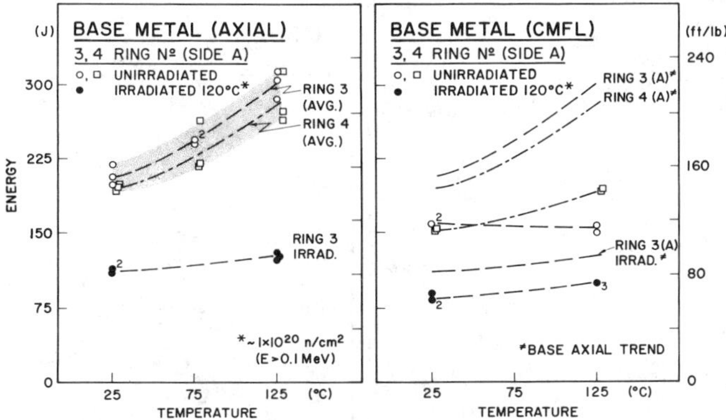

FIG. 5—*Charpy-V notch ductility of base materials from Ring 3 and Ring 4 in unirradiated and 120°C irradiated conditions (axial and circumferential test orientations) at three temperatures. Data trends from the left panel are reproduced in the right panel to simplify data comparisons.*

The data for the base and weld metals do not show a pronounced effect of test orientation. The yield strength levels of the weld and HAZ are about the same but are somewhat higher than the yield strengths of the two base metals evaluated. However, the tensile strengths of the three components were found to be about equal. An increase in test temperature from 25 to 125°C resulted in a lowering of the tensile strengths but did not change appreciably the yield strengths of either the base metals or the HAZ materials. Weld metal tension tests at 125°C were not performed.

Unirradiated Condition—Notch Ductility Properties

The test data for the unirradiated condition C_v specimens are shown graphically in Figs. 5 through 9 (see open symbols). All C_v specimens were tested on the same (hot cell) impact tester.

In general, specimen energy absorption was greater at 125 than at 25°C, suggesting an increase in fracture resistance with temperature (see Figs. 5 to 7). The increase in energy absorption with temperature is not the same for all materials, however. For example, weld 1 showed a much

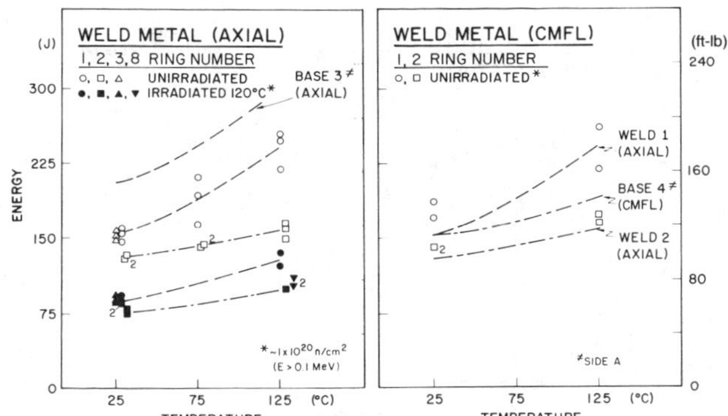

FIG. 6—*Charpy-V notch ductility of weld metals from several rings in unirradiated and 120°C irradiated conditions at three temperatures. Unirradiated condition data trends for base metals 3 and 4 from Fig. 6 are given for reference.*

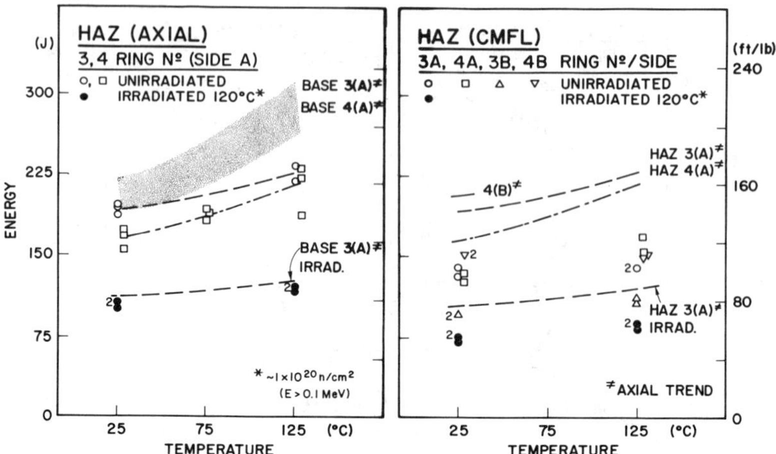

FIG. 7—*Charpy-V notch ductility of HAZ materials from Rings 3 and 4 in unirradiated and 120°C irradiated conditions (axial and circumferential test orientations) at three temperatures. Data trends for the base (parent) metals for both conditions are also shown in the left panel.*

greater temperature effect than weld 2 (Fig. 6). Also for reasons unclear at this time, the increase in energy absorption with temperature was much greater for the axial orientation than the CMFL orientation for some base and HAZ materials. For example, compare the results for HAZ 3(A) and HAZ 4(A) (Fig. 7). Moreover, in some cases, such as HAZ 3(A) and HAZ 4(B), the CMFL orientation data for 125 versus 25°C form a common scatter band.

At 25°C, the base material appears to have a higher energy absorption than its adjoining HAZ or the weld metals in general. In turn, the base metal on balance would appear to be the most fracture-resistant component of the test weldments for a given orientation. Weld 6 was found to have a notch ductility comparable to that of the base metal 4 (Side A) (Fig. 8), but this appears to be the exception rather than the norm. The axial and CMFL test orientations of the base and the HAZ materials describe a pronounced difference in notch ductility at 25°C (high and low, respectively). This was not the case for the two weld metals tested (<26 J or 19 ft · lb). Also,

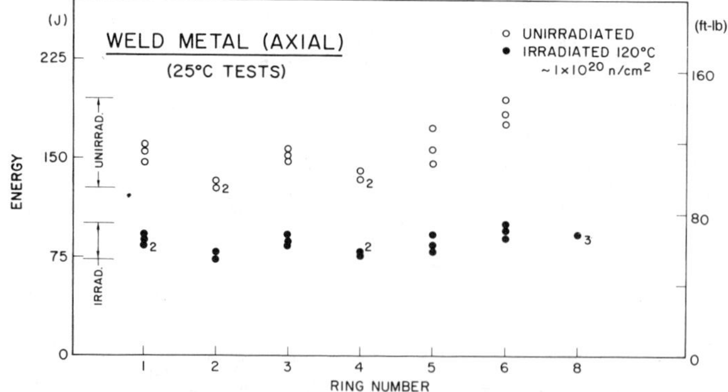

FIG. 8—*Charpy-V notch ductility of the weld deposits from seven pipe rings in unirradiated and 120°C irradiated conditions (axial orientation, 25°C tests). The bands at the left show the range of properties observed.*

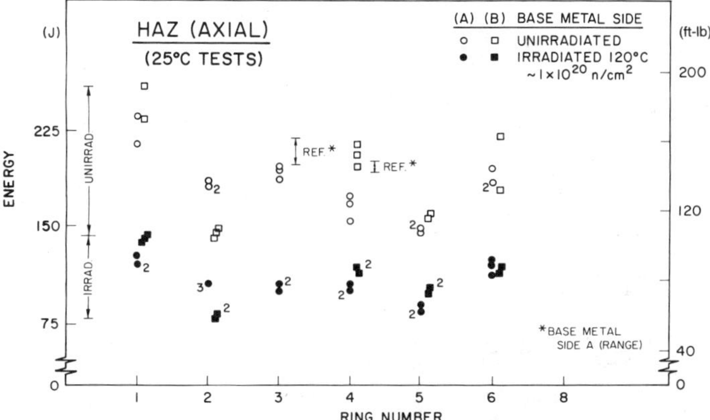

FIG. 9—*Charpy-V notch ductility of weld heat-affected zone materials from seven pipe rings in unirradiated and 120°C irradiated conditions (axial orientation, 25°C tests). The range of properties observed for two base (parent) metals are also shown.*

notice the wide variability in energy absorption for the HAZs in the axial orientation (117 J or 86 ft · lb), with the lowest HAZ value being just slightly higher than the lowest value for the welds. The HAZ can be judged superior to the weld in this orientation; but because of the test orientation sensitivity of the HAZs, this does not appear to hold in each case for the CMFL orientation. Notice that the lowest value of the HAZ 3(B) is much lower than the trend band for the welds, and accordingly, the former could be the weak link of the weldment for the CMFL orientation.

Irradiation Condition—Tensile Properties

Strength determinations for the irradiated condition are summarized in Table 2 and are illustrated in Fig. 10.

Very little data scatter is observed for each of the specimen groups even though the gage sections were composed wholly or partly of weld metal. All specimens exhibited a pronounced elevation

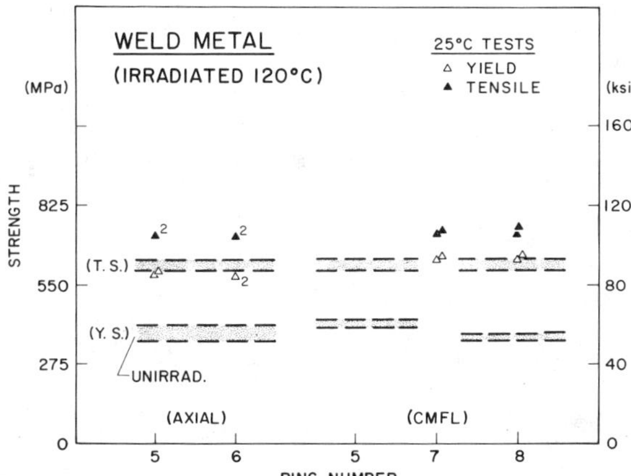

FIG. 10—*Yield and tensile strengths observed with axial and circumferential test orientation weld specimens irradiated at 120°C to ~1×10^{20} n/cm², E > 0.1 MeV. Data trend bands for the unirradiated condition from Fig. 9 are shown for reference.*

TABLE 2—*Postirradiation tensile properties determinations (weld metal specimens).*

				Yield Strength[a]			Tensile Strength		
Ring	Orientation	Test Temperature, °C	Specimen	MPa	ksi	Increment, %[b]	MPa	ksi	Increment, %
5	axial	25	5W23	586	85.0	. . .	722	104.7	
			5W25	598	86.7	. . .	717	104.0	
			(avg)	(592)	(85.8)	53.7	(720)	(104.4)	16.7
		125	5W24[d]	390	56.6	. . .	450	65.2	
6	axial	25	6W50	579	84.0	. . .	718	104.1	
			6W51	576	83.6	. . .	719	104.3	
			(avg)	(578)	(83.8)	52.9	(718)	(104.2)	14.8
		125	6W52[d]	380	55.1	. . .	441	63.9	
7	CMFL	25	7W7	646	93.7	. . .	739	107.2	
			7W8	638	92.6	. . .	731	106.2	
			(avg)	(642)	(93.1)	53.3[c]	(736)	(106.7)	17.9
		125	7W9	401	58.2	. . .	430	62.4	
8	CMFL	25	8W8	641	93.0	. . .	732	106.2	
			8W9	651	95.0	. . .	751	109.0	
			(avg)	(646)	(94.0)	77.3	(742)	(107.6)	19.4
		125	8W7	428	62.1	. . .	468	67.9	

[a] 0.2% Offset, static tests.
[b] Increase over unirradiated condition average.
[c] Referenced to unirradiated ring no. 5 weld properties (CMFL).
[d] Specimen broke outside of extensometer gage length of 13.4 mm.

in yield strength. Specimens from Rings 5, 6, and 7 described a yield strength elevation by irradiation on the order of 205 MPa or 53%. For Ring 8, the elevation was much greater, 283 MPa or 77%. Radiation-induced elevations in tensile strength were much less, being on the order of 90 to 120 MPa or 16%. (For weld metal 7, preirradiation condition properties were not available; properties of weld metal 5 [same orientation] were referenced as an alternative.)

Based on the tensile strength data, the radiation effect on the CMFL orientation was greater than that of the axial orientation. Specifically, tensile strength elevations for Welds 7 and 8 were in the range of 110 to 125 MPa while those for Welds 5 and 6 were in the range of 90 to 105 MPa. Postirradiation yield strengths for the CMFL orientation were much greater than those for the axial orientation in each case (Table 2).

For the composite (axial orientation) specimens, the point of final specimen separation was not consistent for Weld 5 or for Weld 6 specimens. One specimen of each failed at the midpoint of the gage length, that is, at the weld center line; the second specimen failed at a point significantly displaced from the weld center line and inferred failure in the HAZ. This behavior inconsistency was not mirrored in the apparent postirradiation strength level, perhaps because the strengths of the weld and the HAZ in the unirradiated condition were about the same (see Fig. 3). Further evaluation of the relative load carrying capability of individual weldment components was not possible with the particular specimens available.

Irradiation Condition—Notch Ductility Properties

Individual C_v test determinations are included in Figs. 5 through 9 as filled symbols. Table 3 compares average values for the irradiated versus unirradiated conditions.

Referring first to the 25°C data for the axial orientation of the weld deposits, the reductions in energy absorption range from 54 to 67 J for all but two of the welds, Welds 5 and 6. Higher reductions were recorded for the latter in terms of both absolute energy and percentage reduction. Percentage reductions for the bulk of the welds ranged from 41.5 to 43.7%. The greatest reduction in energy, 90 J, was observed for Weld 6, which also had the highest preirradiation energy absorption. The lowest reduction was found for Weld 2, which exhibited the lowest preirradiation test value. Except for Weld 6, the spread in absolute values for the axial orientation is considered quite small: 76 versus 88 J.

At 125°C, postirradiation values tend to be higher than values for the 25°C tests, paralleling the preirradiation energy trend with temperature. Reductions in energy absorption with irradiation appear to be much higher at 125°C; but, percentage-wise, the reductions are about equal to those at the lower test temperature. Notice that the increase in energy absorption with temperature again is greater for Weld 1 than Weld 2. On the other hand, the difference in energy absorption between the two test temperatures is less for the irradiated condition than the unirradiated condition. For example, Weld 1 data trends describe increases of 42 and 87 J (31 and 64 ft · lb) for the irradiated and unirradiated conditions, respectively.

The CMFL orientation was evaluated with Weld 8 only. Because of stock limitations, preirradiation data could not be developed for this weld. However, the CMFL data for this weld do match the axial orientation data for Weld 6 at both test temperatures.

Axial orientation data developed for base metal 3 at 25°C indicate a percentage decrease in energy absorption that is comparable to that described by the weld metals (46% versus 41 to 49%). The energy decrease is much greater at 125°C than at 25°C in terms of percentage reduction or absolute reduction. For the CMFL orientation, however, the decreases in energy absorption are approximately the same at both temperatures. In fact, the data suggest a lesser irradiation effect for the 125°C condition. An orientation dependence is clearly indicated in this case, for reasons unclear at this time.

TABLE 3—*Radiation-induced changes in average C_v energy absorption of materials.*

Ring	Base Metal (Side)	Preirradiation		Postirradiation		Decrease		
		J[a]	ft · lb	J[a]	ft · lb	ΔJ	Δft · lb	%
				WELD METAL, 25°C				
1	. . .	154	113.3	87	63.8	67	49.5	43.7
2	. . .	130	95.7	76	56.0	54	39.7	41.5
3	. . .	152	112.3	88	64.7	64	47.6	42.4
4	. . .	137	100.7	78	57.3	59	43.4	43.1
5	. . .	159	117.3	85	62.7	74	54.6	46.6
6	. . .	185	136.7	95	70.0	90	66.7	48.8
				WELD METAL, 125°C				
1	. . .	241	177.7	129	95.0	112	82.7	46.5
2	. . .	158	116.7	99	73.0	59	43.7	37.5
				BASE METAL, 25°C				
3	A	207	153.0	113	83.3	94	69.7	45.6
3[b]	A	160	118.0	85	63.0	75	55.0	46.6

				BASE METAL, 125°C				
3	A	300	221.3	126	93.0	174	128.3	58.0
3[b]	A	154	113.5	100	73.7	54	39.8	35.1
				HAZ, 25°C				
1	A	225	166.0	123	91.0	102	75.0	45.2
1	B	246	181.5	141	104.0	105	77.5	42.7
2	A	183	135.0	105	77.7	78	57.3	42.4
2	B	145	107.3	81	60.0	64	47.3	44.1
3	A	192	142.0	103	76.3	89	65.7	46.3
3[b]	A	136	100.0	73	54.0	62	46.0	46.0
4	A	165	122.0	102	75.3	63	46.7	38.3
4	B	206	151.7	118	86.7	88	65.0	42.9
5	A	149	109.7	86	63.7	62	46.0	41.9
5	B	159	117.0	101	74.3	58	42.7	36.5
6	A	188	138.7	119	88.0	69	50.7	36.6
6	B	201	148.0	116	85.7	85	62.3	42.1
				HAZ, 125°C				
3	A	226	167.0	119	87.7	107	79.3	47.5
3[b]	A	139	102.5	86	63.3	53	39.2	38.2

[a] Axial orientation unless noted.
[b] CMFL orientation.

Referring next to the HAZ, the percentage decrease in energy caused by irradiation (36.6 to 46.3%) is on the order of that shown by the weld deposits for the axial orientation, but a much greater range in energy reductions is evident: 58 to 105 J. Again the greater reductions appear associated with the higher preirradiation values. Significantly, postirradiation values for the HAZ are equal to or higher than weld deposit values (compare Fig. 9 versus Fig. 8). On the other hand, average HAZ energy levels are consistently lower (by 7 to 14 J) than the base metal in the irradiated condition.

Discussion

The C_v notch ductility values reported here for the preirradiation and postirradiation conditions of the materials are indicative of high fracture resistance, but the radiation-induced reductions themselves were appreciable. Whether or not the materials have strong tendencies toward embrittlement saturation at much higher fluence accumulations is a key question to be addressed by the continuing program. Data suggesting that saturation tendencies may be found for the base metals do exist [*3–5*]. On the other hand, reductions of notch ductility to low levels by irradiation have been observed for some stainless steel welds. Specifically, C_v notch ductility values of ~20 J and dynamic fracture toughness values of ~60 MPa · $m^{1/2}$ have been found for AISI Type 308-16 submerged-arc welds after ~1.5 × 10^{22} n/cm^2, $E > 0.1$ MeV, at 427°C [*6*]. The higher irradiation temperature in this instance would be expected to reduce the fluence effect through a self-annealing of the displacement damage.

The metallurgical cause of the difference in energy absorption versus test temperature trends for the axial versus CMFL orientation in some cases is not known. Where the two orientations exhibit about the same preirradiation energy absorption at 25°C, trend differences would not be expected. Scanning electron microscopy of fracture surfaces is planned to gain insight on the probable cause.

Referring to Table 1, dissimilarities in composition among the base materials (and HAZs) are noted, especially in copper content. Copper is known to have a highly detrimental effect on the radiation resistance of low alloy steels (Ref. *7*); an influence on radiation resistance has been observed for an exposure temperature of 288°C and temperatures approaching that of this study (120°C). For the weld deposits examined here, the range of copper contents is small and would not be a factor in their radiation resistances. For the welds and the base materials, the ranges of phosphorus content also are narrow; although, on balance, the phosphorus content of the base materials is higher than that of the welds. The HAZ data do not provide clear examples of a dependence of measured irradiation effect on material composition. As illustrated in Fig. 11, however, a relationship of the absolute radiation-induced decrease to the preirradiation energy absorption level is apparent. The percentage decreases in energy absorption in contrast are about equal (~42%) for all eleven HAZ materials. These observations, in turn, should simplify the estimating of notch ductility change for other, comparable materials.

Fracture toughness determinatons, using 0.4T-CT specimens and *J*-integral assessment procedures, are being developed for the materials for the unirradiated condition and will be reported in 1987. Further details of the present investigation are given in Ref *8*.

Conclusion

Primary observations and conclusions drawn from the materials and test conditions of this investigation are as follows:

Preirradiation Condition

- Yield strengths of the weld and HAZ materials were about the same but were somewhat higher than the yield strengths of the base metals.

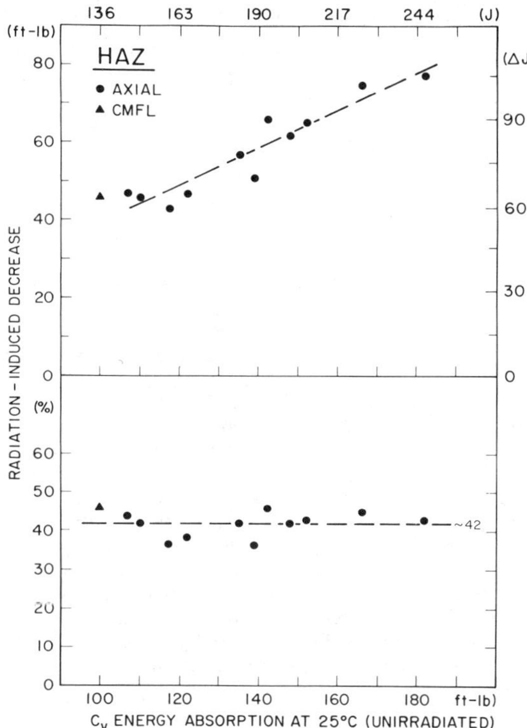

FIG. 11—*Dependence of radiation-induced reduction in* C_v *energy absorption on preirradiation* C_v *energy level.*

- Tensile strengths of the base metal and HAZ materials were lower at 125 than at 25°C; yield strengths were approximately the same (within ~ 55 MPa) at the two temperatures.
- The base metal and weld strengths did not exhibit a pronounced sensitivity to test orientation (axial versus circumferential).
- C_v energy absorption tended to be higher at 125°C than 25°C.
- C_v energy absorption of the base metal (at 25°C) was higher than the C_v energy absorption of the adjoining HAZ or the C_v energy absorption range for the weld metals.
- A pronounced difference in C_v energy absorption was observed between the axial orientation (high) and the CMFL orientation (low) of base metal and HAZ materials but not the weld.
- Wide variability in energy absorption (117 J) was observed for the HAZ axial orientation. HAZ values were superior to weld values in this orientation, but HAZ values could be inferior to those of the weld in the CMFL orientation.

Irradiated Condition

- Yield strength elevations on the order of 53% were found for specimens spanning the weld metal (axial orientation); tensile strength elevations by comparison were on the order of 16%.
- For those tension specimens that spanned the weld joint (axial orientation), the number that broke in the weld metal was about equal to the number that broke in the HAZ.
- Yield strength levels in the CMFL orientation were greater than those for the axial orientation after irradiation.

• C_v energy absorption reductions by irradiation were on the order of 42% for weld and HAZ materials at 25°C. Absolute energy reductions tended to be greater for a test temperature of 125°C.

• The magnitude of radiation-induced decrease in energy absorption appears related to the preirradiation energy level but not to composition.

• Postirradiation C_v and yield strength values are indicative of good fracture resistance for the fluence condition evaluated.

References

[*1*] Lippincott, E. P., *Buffalo Light Water Reactor Calculation,* Hanford Engineering Development Laboratory, Richland, WA, 15 Nov. 1977.

[*2*] Lippincott, E. P., Kellogg, L. S., and McElroy, W. N., "Evaluation of Neutron Exposure Conditions for the Buffalo Reactor," HEDL-SA-3101, Hanford Engineering Development Laboratory, Richland, WA, Aug. 1984.

[*3*] Steele, L. E., Hawthorne, J. R., Serpan, Jr., C. Z., and Gray, Jr., R. A., "Notch Ductility Characteristics of Irradiated AISI 304L and 347 Stainless Steels," *Irradiation Effects on Reactor Structural Materials, Quarterly Progress Report, 1 August–31 October 1966,* NRL Memorandum Report 1731, Naval Research Laboratory, Washington, DC, 15 Nov. 1966, pp. 10–12.

[*4*] Steele, L. E., Hawthorne, J. R., and Serpan, Jr., C. Z., "Notch Ductility Characteristics of Irradiated AISI 304L and 347 Stainless Steels After a High Neutron Exposure," *Irradiation Effects on Reactor Structural Materials, Quarterly Progress Report, 1 November 1965–31 January 1966,* NRL Memorandum Report 1676, Naval Research Laboratory, Washington, DC, 15 Feb. 1966, pp. 23–26.

[*5*] Joseph, J. W., Jr., "Stress Relaxation in Stainless Steel During Irradiation," DP-369, E. I. du Pont de Nemours & Co., Inc., Savannah River Laboratory, Aiken, SC, June 1959.

[*6*] Hawthorne, J. R., "Fatigue and Fracture Resistance of Stainless Steel Weld Deposits After Elevated-Temperature Irradiation," NRL Report 8451, Naval Research Laboratory, Washington, DC, 18 Nov. 1980.

[*7*] Hawthorne, J. R., "Chapter 10, Irradiation Embrittlement," *Treatise on Materials Science and Technology, Vol. 25, Embrittlement of Engineering Alloys,* C. L. Briant and S. K. Banerji, Eds., Academic Press, NY, 1983, pp. 462–518.

[*8*] Hawthorne, J. R., Menke, B. H., Awadalla, N. G., and O'Kula, K. R., "Experimental Assessment of Notch Ductility and Tensile Strength of Stainless Steel Weldments After 120°C Neutron Irradiation," MEA-2133, Materials Engineering Associates, Inc., Lanham, MD, Dec. 1985.

R. L. Klueh[1] *and P. J. Maziasz*[1]

Effect of Cold Work on Tensile Behavior of Irradiated Type 316 Stainless Steel

REFERENCE: Klueh, R. L. and Maziasz, P. J., "**Effect of Cold Work on Tensile Behavior of Irradiated Type 316 Stainless Steel,**" *Influence of Radiation on Material Properties: 13th International Symposium (Part II), ASTM STP 956,* F. A. Garner, C. H. Henagar, Jr., and N. Igata, Eds., American Society for Testing and Materials, Philadelphia, 1987, pp. 207–222.

ABSTRACT: The effect of various levels of cold work on the tensile behavior of Type 316 stainless steel was investigated. Tensile specimens were irradiated in the Oak Ridge Research Reactor (ORR) at 250, 290, 450, and 500°C to produce a displacement damage of ~5 displacements per atom (dpa) and 40 atomic ppm He. Irradiation at 250 and 290°C caused an increase in yield stress and ultimate tensile strength and a decrease in ductility relative to unaged and thermally aged controls. The changes were greatest for the 20%-cold-worked steel and lowest for the 50%-cold-worked steel. Irradiation at 450°C caused a slight decrease in strength for all cold-worked conditions relative to the aged material. A large decrease was observed at 500°C, with the largest decrease occurring for the 50%-cold-worked specimen. No bubble, void, or precipitate formation was observed for specimens examined by transmission electron microscopy (TEM). The irradiation hardening was correlated with Frank-loop and "black-dot" damage. A strength decrease at 500°C was correlated with dislocation network recovery. Comparison of tensile and TEM results from ORR-irradiated steel with those from steels irradiated in the High Flux Isotope Reactor and the Experimental Breeder Reactor indicated consistent strength and microstructure changes.

KEY WORDS: stainless steel, cold work, neutron irradiation, microstructure, tensile properties, ductility, elevated temperature

Twenty-percent cold-worked Type 316 stainless steel was chosen as the structural material for several core components of first-generation fast breeder reactors. It is also a candidate for the construction of the first-wall and blanket structure of fusion reactors. The choice of the cold-worked structure followed from the theoretical prediction [*1*] and experimental [*2*] observations that high dislocation densities should [*1*] and did [*2*] have an inhibiting effect on the void density, which should and did lead to a decrease in irradiation-produced swelling compared with a solution-annealed structure. Several studies have shown that cold working stainless steel leads to a reduction in void swelling when the steel is irradiated with fast neutrons between 0.3 and 0.5 T_m (T_m is the absolute melting point) [*3–9*]. These investigations generally considered cold-work levels up to about 25 or 30%, although levels up to 50% have been considered [*9*].

The choice of a 20%-cold-work level (as opposed to a higher or lower level) for breeder applications was apparently based on the observations on void swelling. The properties of an unirradiated cold-worked steel may also have been taken into account when this decision was made, although no property studies that considered the effect of radiation on various cold-worked levels have been found in the literature.

Brager [*6*] showed that for Type 316 stainless steel a 10%-cold-work level caused nearly as great a suppression of void formation as did 20% cold work. Above about 475°C, he found little

[1] Research metallurgists, Metals and Ceramics Division, Oak Ridge National Laboratory, Oak Ridge, TN 37831.

difference in the swelling behavior of steels with 10 or 20% cold work; at 420°C (the lowest temperature investigated), there was very little difference between steel cold worked 20 and 30%. Appleby et al. [*7*] and Uematsu et al. [*8*] both found a significant decrease in swelling in going from cold-work levels of 10 to 30% at 550°C; Uematsu et al. found very little further reduction in going from 30 to 50%.

Most irradiation studies on the effect of thermomechanical treatment (cold-work level and solution-anneal temperature) on Type 316 stainless steel have examined the effect of irradiation on swelling. Little information has appeared on the effect of preirradiation treatment on the irradiated tensile properties [*10–12*]. There is, however, considerable information available on the properties of solution-annealed (1 h at 1050°C) and 20%-cold-worked Type 316 stainless steel irradiated in a fast reactor environment [*11–16*]. These irradiations in the Experimental Breeder Reactor (EBR-II) extend to $\sim 13 \times 10^{26}$ neutrons (n)/m^2 (>0.1 MeV). In addition to the irradiation in fast reactors, properties of solution-annealed and 20%-cold-worked Type 316 stainless steel irradiated in the High Flux Isotope Reactor (HFIR), a mixed-spectrum reactor, have been reported [*17–19*].

In a fusion reactor, two types of irradiation effects are expected: displacement damage and a large amount of transmutation helium are produced by the high-energy neutrons. It is, therefore, of interest to determine the effect of both displacement damage and helium on properties. The effects of displacement damage alone can be studied by irradiation in a fast-spectrum fission reactor. The HFIR studies mentioned above were carried out as part of the fusion-reactor materials studies. HFIR has a mixed spectrum that contains both fast and thermal neutrons; thermal neutrons produce helium by a two-step reaction with ^{58}Ni in a nickel-containing alloy. Thus, irradiation of stainless steel in a mixed-spectrum reactor can produce both displacement damage and helium to approximately simulate fusion irradiations.

Helium can have pronounced effects on the mechanisms of void swelling, precipitation, and radiation-induced solute segregation in austenitic stainless steels under either ion or neutron irradiation [*20–22*]. Helium effects for fusion applications have been inferred from a comparison of the results from the EBR-II and HFIR irradiation of the same heat of steel. However, neither of these reactor irradiations produces helium and displacement damage in Type 316 stainless steel at the ratio to match fusion first-wall service (10–15 at. ppm He/dpa); HFIR produces far too much helium (20–70 at. ppm He/dpa) and EBR-II far too little (0.5–1 at. ppm He/dpa). The Oak Ridge Research Reactor (ORR) has a lower flux, but gives a better match to the helium/dpa (~10 at. ppm He/dpa) generation ratio of a fusion first wall. This paper presents tensile and microstructural data obtained for Type 316 stainless steel irradiated in the ORR.

Experimental Procedure

The Type 316 stainless steel used in this study was taken from the Magnetic Fusion Enegy (MFE) reference heat (X-15893); the chemical composition is given in Table 1.

Tensile specimens were machined from 0.76-mm-thick sheet. Specimens were obtained from sheet reduced by 20, 30, and 50% by rolling; prior to the final cold work, the steel was solution annealed 1 h at 1050°C. Material was also irradiated and tested that was aged for 10 h at 800°C after the 1 h at 1050°C anneal and prior to the 20%-cold work.

Sheet tensile specimens in this experiment were of the SS-1 type with a gage section of 20.3-mm long by 1.52-mm wide by 0.76-mm thick. Specimens were irradiated in the E-7 position of the ORR in experiment ORR-MFE-2. These specimens were irradiated in capsules that contained 22 sheet samples that were sealed with a helium environment. The capsules were contained in a water-cooled aluminum block; each holder contained a central hole that contained an electric heater. Temperature was measured and controlled by two thermocouples located at the position of the center of the gage section in two unused sample positions located 180° apart.

TABLE 1—*Chemical composition of MFE reference heat X15893.*

Element	Percent by Weight	Element	Percent by Weight
Cr	17.28	Nb	<0.05
Mn	1.70	Ta	<0.05
Ni	12.44	Ti	<0.05
Mo	2.10	B	0.004
Co	0.3	C	0.061
Cu	0.3	S	0.018
Si	0.67	P	0.037

Irradiation temperatures obtained in this experiment were approximately 250, 290, 450, and 500°C. The temperature variations as measured by the thermocouples were approximately 250 ± 5°C, 290 ±20°C, 450 ± 5°C, and 500 ±40°C. The neutron fluence of ~6.8 × 10^{25} n/m^2 (>0.1 MeV) produced ~5 dpa and the associated thermal neutron fluence of ~4.7 × 10^{25} n/m^2 produced ~40 atomic parts per million (at. ppm) He.

Tensile tests were made on unirradiated and irradiated specimens at the irradiation temperatures. Tests were conducted in a vacuum chamber on a 44-kN capacity Instron universal test machine at a strain rate of 4.2 × 10^{-5}/s.

Transmission electron microscopy (TEM) studies were made on the 20%-cold-worked material. Disks about 3 mm in diameter were electrodischarge machined from the shoulders of the tensile specimens that had been tensile tested. Previous work has established that the tensile testing does not disturb the as-irradiated microstructure in the shoulder region [*23*]. These samples were examined with the TEM techniques described previously [*23,24*]. The relatively high cobalt content of this alloy resulted in high levels of radioactivity after the ORR irradiation and made specimen handling difficult.

Results

The metallographic appearance of the unirradiated cold-worked structure depended on the amount of cold work (Fig. 1); the grain structure was still clearly visible for the material deformed only

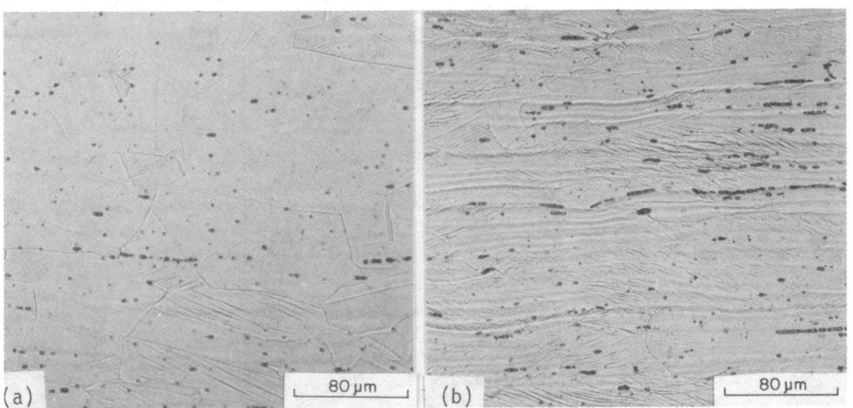

FIG. 1—*Microstructure of Type 316 stainless steel cold worked* (a) *20% and* (b) *50%.*

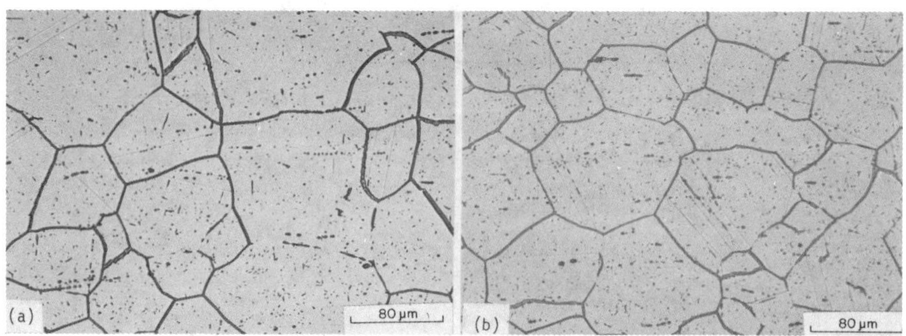

FIG. 2—*Microstructure of Type 316 stainless steel* (a) *solution annealed 1 h at 1050°C and then aged 10 h at 800°C, and* (b) *this steel after rolling 20%.*

20%. When the solution anneal for 1 h at 1050°C was followed by a thermal age of 10 h at 800°C, considerable precipitate formed on grain boundaries and within the matrix (Fig. 2*a*). After this structure was cold worked 20%, the grain structure was still easily seen by optical microscopy (Fig. 2*b*).

For control specimens, tensile specimens were aged 4000 h (the approximate time of the irradiation experiment) at 300, 450, 500, and 650°C to determine the effect of the thermal exposure on the properties (the higher temperature was used because these thermal aging results are to be used as controls for other experiments). The aged specimens were tested at the aging temperature and the properties obtained were compared with the as-cold-worked material (Figs. 3–5).

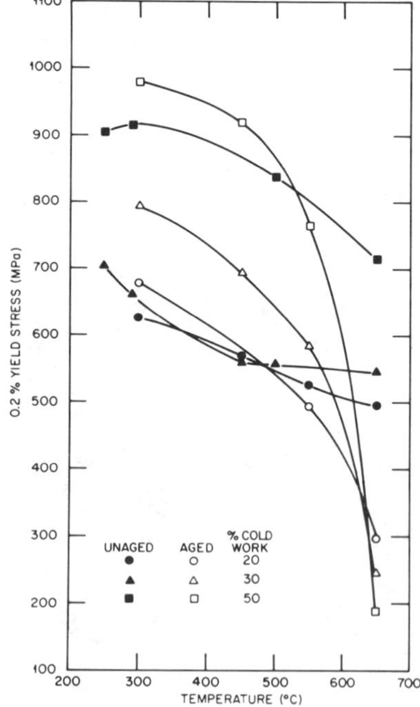

FIG. 3—*The 0.2% yield stress of cold-worked and cold-worked and thermally aged Type 316 stainless steel. Specimens were aged 4000 h at the same temperature used for the tests.*

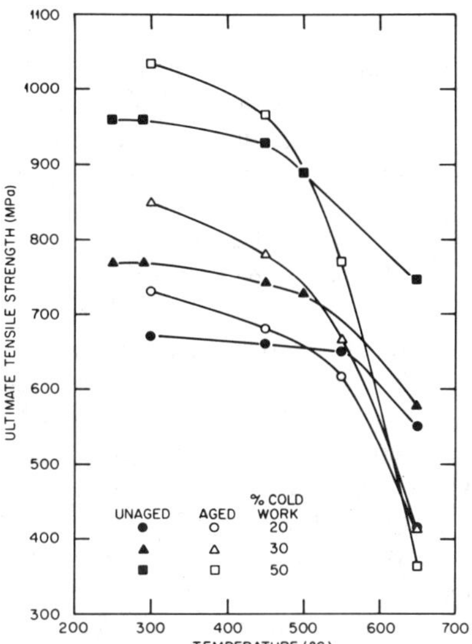

FIG. 4—*The ultimate tensile strength of cold-worked and cold-worked and thermally aged Type 316 stainless steel. Specimens were aged 4000 h at the same temperature used for the tests.*

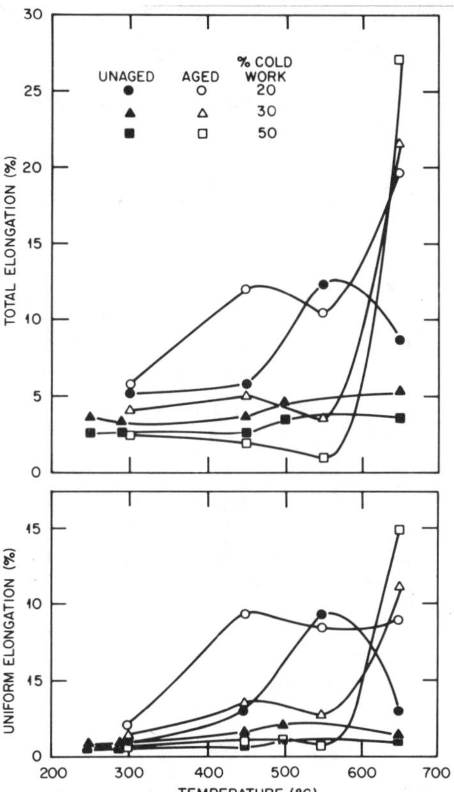

FIG. 5—*The uniform and total elongation of cold-worked and cold-worked and thermally aged Type 316 stainless steel. Specimens were aged 4000 h at the same temperature used for the tests.*

Mechanical Properties

The 0.2% yield stress (*YS*) and ultimate tensile strength (*UTS*) of the unaged specimens showed the normal effects of cold work: The higher the amount of cold work, the higher the strength (Figs. 3 and 4). Thermal aging caused changes in strength, even at temperatures as low as 300°C, where a strength increase was observed over that of the unaged specimens. The strength advantage of the aged material was maintained up to about 500°C. At higher temperatures, however, the aged steel became weaker than the unaged steel, and the strength was significantly less after aging at 650°C for all cold-work levels. In fact, at 650°C, the weakest of the aged materials was the steel with 50% cold work.

The ductility of the unaged material inversely reflected the strength properties, with the strongest material (50% cold work) having the lowest ductility and the weakest having the highest ductility. The ductility of the unaged material appeared to reach a maximum beween 500 and 600°C. Except at 650°C, the ductilities of the thermally aged specimens did not differ significantly from the unaged specimens. Upon aging, a large increase in ductility occurred at 650°C, which coincided with the large strength decrease. Just as the 50%-cold-worked steel showed the largest strength loss on aging at 650°C, it also showed the largest increase in ductility.

In Figs. 6 through 8, the tensile data for the irradiated specimens tested at 250 to 500°C are compared with the data for the thermally aged specimens. The *YS* and *UTS* of all three irradiated, cold-worked levels increased over the corresponding unirradiated values at 250 and 290°C. The relative increase was inversely related to the level of prior cold work. Between 290 and 450°C there was a large decrease in the irradiation-produced strength; at 450°C the strength of the irradiated steel fell below that of the thermally aged steel. When aged and irradiated strengths were compared, the strength of the irradiated steel became less than that of the unirradiated steel at 500°C for the 20- and 30% cold-worked material. The crossover for the 50%-cold-worked steel occurred at 450°C. Although the originally strongest 50%-cold-worked steel remained strongest

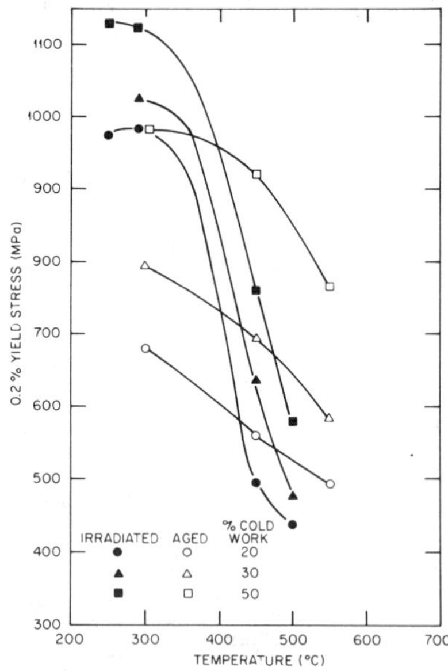

FIG. 6—*The 0.2% yield stress of the thermally aged and the irradiated cold-worked Type 316 stainless steel. Tests were at the aging and irradiation temperatures.*

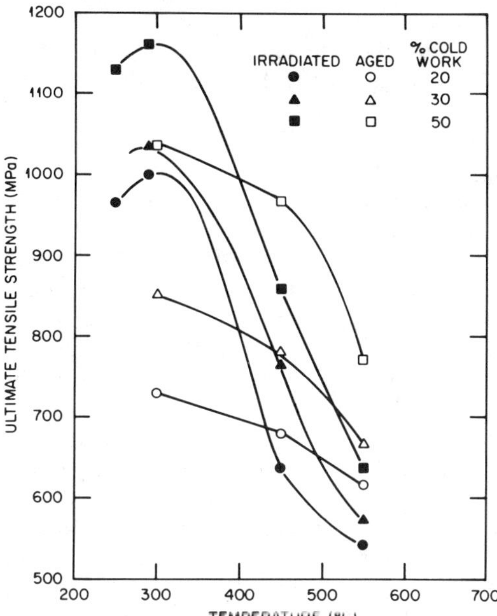

FIG. 7—*The ultimate tensile strength of the thermally aged and the irradiated Type 316 stainless steel. Tests were at the aging and irradiation temperatures.*

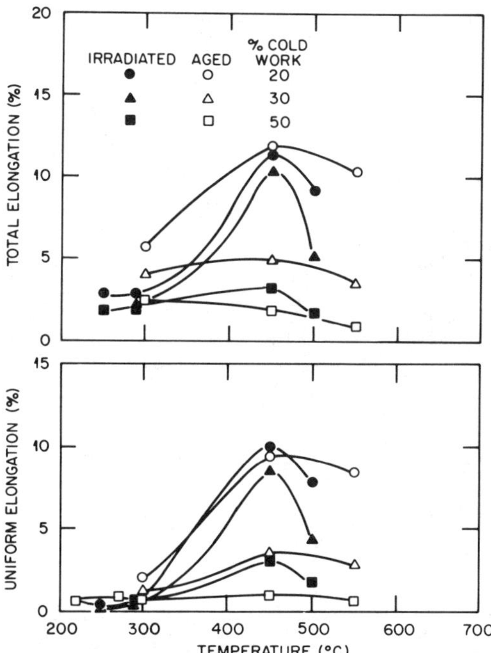

FIG. 8—*The uniform and total elongations of the thermally aged and irradiated Type 316 stainless steel. Tests were at the aging and irradiation temperatures.*

after irradiation at 500°C, the relative loss of strength at the highest temperatures was greatest for the 50%-cold-worked steel, followed by the 30%-cold-worked steel. Thus, the irradiated strengths for the three cold-worked levels appeared to be approaching a common value with increasing irradiation temperature.

The change in ductility of the irradiated cold-worked Type 316 stainless steel does not directly reflect the increase in strength (Fig. 8). Both uniform and total elongations of the irradiated steels increased with temperature between 290 and 450°C, but they decreased between 450 and 500°C, even though the strength decreased. The steel with 20- and 30%-cold work showed a large elongation increase between 290 and 450°C, whereas the 50%-cold-worked steel showed a much smaller increase. After irradiation at 500°C, the total elongation of the 50%-cold-worked steel was actually less than the value at 250°C. At all temperatures, the ductility reduction of the three steels maintained the inverse relationship with cold-work level that was true for the unirradiated material. However, in all cases the ductility of the irradiated steel was similar to that of the thermally aged steel.

For the three cold-work levels discussed above, the cold deformation followed a 1-h solution anneal at 1050°C. In Figs. 9 through 11, the strength and ductility of the Type 316 stainless steel cold-worked 20% after the 1050° solution anneal are compared with the steel cold-worked 20% following a 1050°C anneal plus a 10-h age a 800°C. When this latter steel was thermally aged, the results were similar to those observed for the other cold-worked steels. Near 300°C, aging increased the strength. At 500°C, there was little difference between aged and unaged steel. Both before and after irradiation, there was very little difference in strength for the two 20%-cold-worked steels given the different heat treatments prior to the cold work (Figs. 9 and 10). The uniform and total elongation values of the steel given the 800°C age did not increase between 290

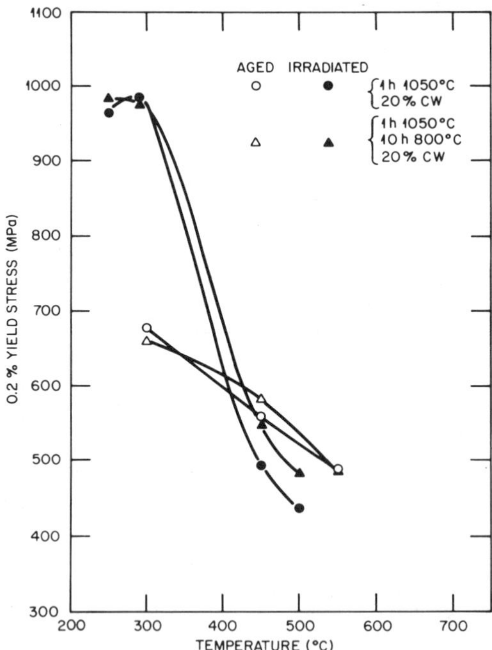

FIG. 9—*The 0.2% yield stress of the thermally aged and the irradiated 20%-cold-worked Type 316 stainless steel; cold work followed two different heat treatments. Tests were at the aging and irradiation temperatures.*

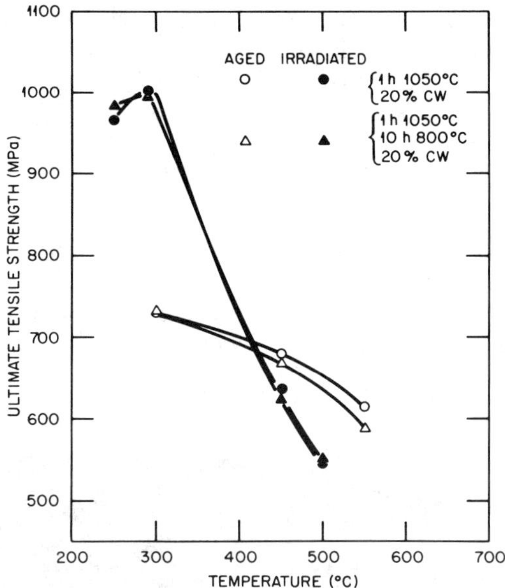

FIG. 10—*The ultimate tensile strength of the thermally aged and the irradiated 20%-cold-worked Type 316 stainless steel; cold work followed two different heat treatments. Tests were at the aging and irradiation temperatures.*

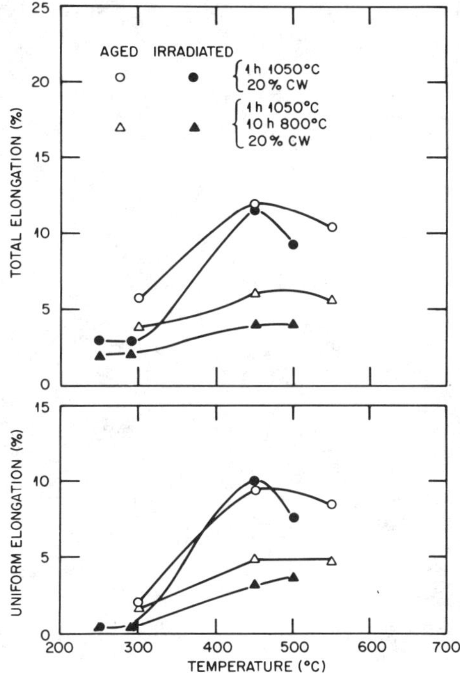

FIG. 11—*The uniform and total elongations of the thermally aged and the irradiated 20%-cold-worked Type 316 stainless steel; cold work followed two different heat treatments. Tests were at the aging and irradiation temperatures.*

and 450°C nearly as much as did the values for the other steel. This was true for both irradiated and unirradiated (aged and unaged) steels. However, the ductility values for both irradiated steels approached those for the unirradiated steels at 450 and 500°C.

Selected fracture surfaces were examined by scanning electron microscopy. All specimens displayed typical ductile shear-type failures. There was no indication of intergranular fracture at any irradiation or aging temperature in any material condition.

Transmission Electron Microscopy

The TEM studies on the irradiated 20%-cold-worked steel did not reveal any bubbles or voids in the microstructure of the steel irradiated in ORR to 5 dpa at temperatures from 250 to 500°C. There was also no evidence of precipitation for any of these irradiation conditions, although there was an effect of the irradiation on the dislocation structure.

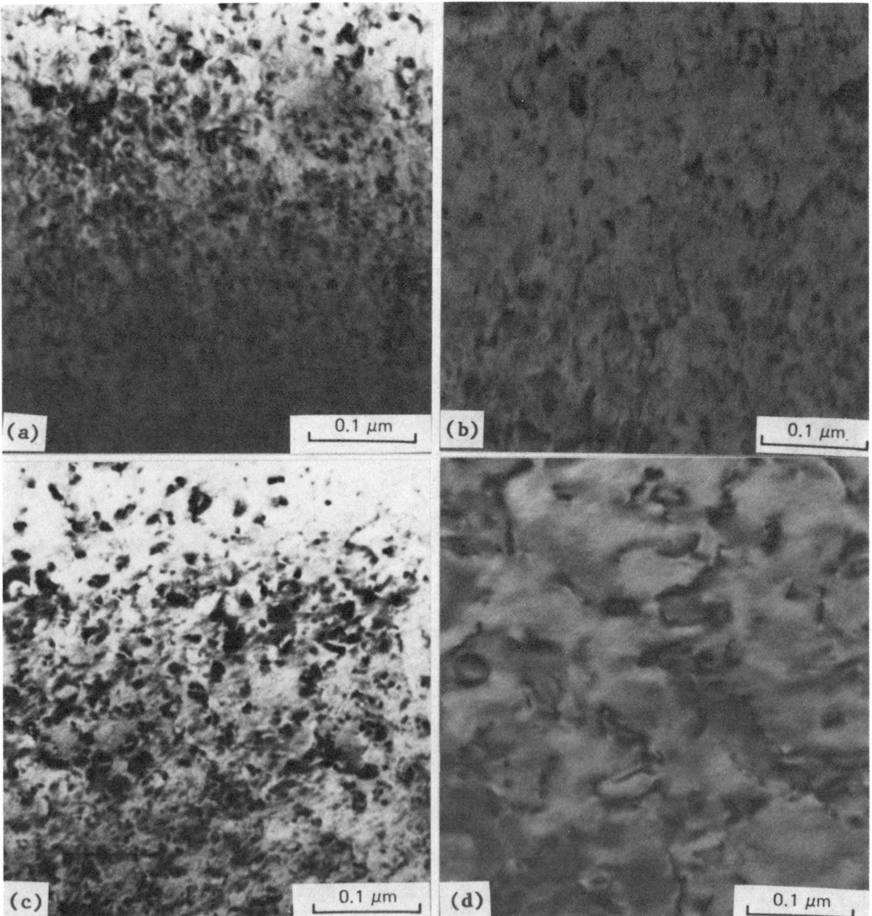

FIG. 12—*Dislocation microstructures of CW 316 (reference heat) irradiated in ORR to 5 dpa (38.7 at. ppm He) at* (a) *250°C,* (b) *290°C,* (c) *450°C, and* (d) *500°C. All are imaged using g_{200} with $s > 0$. Note the more relaxed structure at 500°C* (d).

After irradiation, the overall dislocation concentration was high and relatively independent of temperature from 250 to 450°C, but, by comparison, appeared much less concentrated at 500°C (Fig. 12). Frank loops (assumed to be interstitial) were a significant portion of the overall dislocation microstructure at all temperatures. The loop component at the various temperatures is shown in Fig. 13 for one set of (111) planes [hence one-fourth of a total isotropic population of loops on (111) planes] imaged in dark field via <111> satellite streaks around g_{200} matrix reflections. A high concentration of loops ranging from 8 to 25 nm in diameter was observed from 250 to 450°C. Fewer, much larger (17 to 100 nm diameter) loops were observed at 500°C. The dislocation structures at 250 to 450°C appeared similar to or denser than the network normally found in the as-cold-worked material [*22*], but the dislocation structure at 500°C was somewhat recovered compared to as-cold-worked material. Finally, a high concentration of "black dots" (<5 nm in diameter) can be seen in Fig. 12*a–c*. These features, seen at 250 to 450°C when these samples were imaged in a weak-beam dark-field condition (g_{200}, $+g/3g$), may also be small loops. "Black dots" were not present after irradiation at 500°C.

Selected TEM specimens of the unaged and aged control specimens of the 20- and 50%-cold-worked specimens were examined to determine if the source of hardening during aging at 300 and 450°C could be detected. There was no indication of precipitation during aging nor any indication

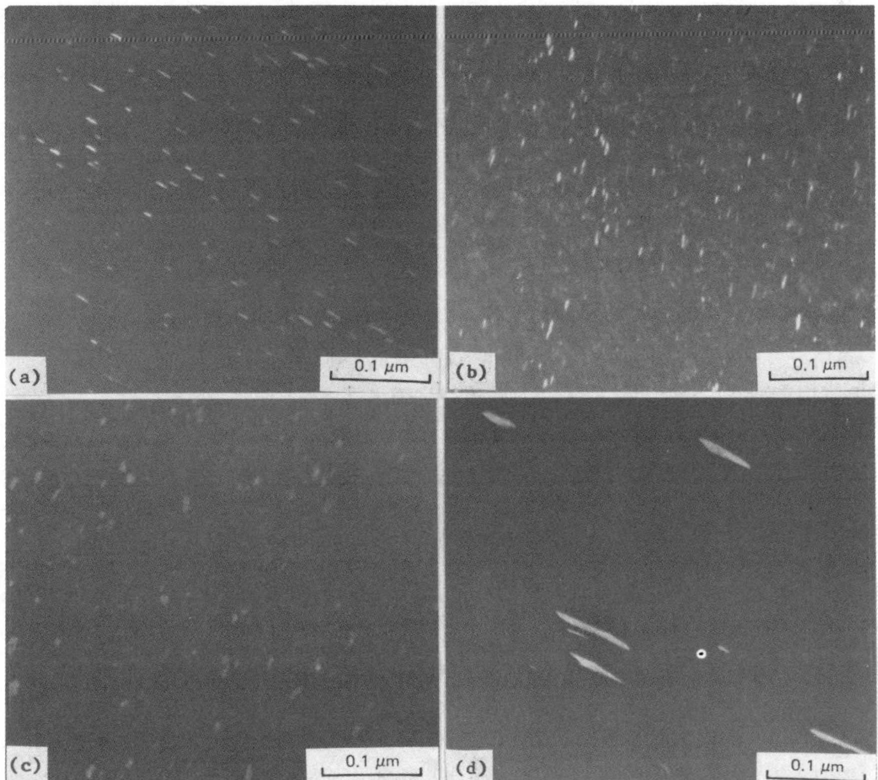

FIG. 13—*Larger Frank loop component of the dislocation microstructure of CW 316 (reference heat) irradiated in ORR to 5 dpa (38.7 at. ppm He) at* (a) *250°C,* (b) *290°C,* (c) *450°C, and* (d) *500°C. All are imaged in dark field using <111> satellite streaks around g_{200} reflections. Note the large increase in size and decrease in density of loops at 500°C, compared with the three lower temperatures.*

of a change in the dislocation structure that could be the cause of the hardening. The hardening must be associated with a solute-dislocation-locking mechanism (probably due to carbon).

Discussion

It is of interest to compare the results of these studies with irradiation effects observed after irradiation in HFIR, where much more helium is generated, and irradiation in EBR-II, where very little helium is generated. The comparison will be restricted to 20%-cold-worked material, as very few data are available for other cold-work levels.

Because most irradiation studies on 20%-cold-worked Type 316 stainless steel have been conducted in conjunction with the fast-breeder program and have generally used EBR-II, few irradiations have been below about 370°C (limited by the EBR-II reactor coolant inlet temperature). Fish et al. [*14,15*] and Hamilton et al. [*16*] tested 20%-cold-worked Type 316 stainless steel irradiated in EBR-II at temperatures between 371 and 816°C. They found an increase in both *YS* and *UTS* for material irradiated and tested below about 483°C, and only a slight amount of hardening was observed at 483°C. The hardening was observed in comparison with unaged material. They reported softening of cold-worked material for irradiation and testing at 538°C and higher temperatures, where comparisons were made with both aged and unaged material [*15*].

These observations on EBR-irradiated 20%-cold-worked steel are similar to observations in the present tests on the 20%-cold-worked steel. When compared to the unaged steel, the irradiated steel was softer than the unirradiated steel at 500°C but not at 450°C. Because of the increase in strength during aging, the 20%-cold-worked steel in the present tests showed a change from hardening to softening at a somewhat lower temperature. However, considering that the heat of steel tested in the present experiment and that tested in EBR-II are different, the agreement is quite good. The 30%-cold-worked steel showed hardening and softening behavior similar to that of the 20%-cold-worked steel, but the steel with 50% cold work softened at a somewhat lower temperature.

Tensile specimens of the 20%-cold-worked reference heat have been irradiated in HFIR, and where the irradiation and test temperatures were similar to those of the present experiment, similar tensile results were observed [*19*]. For tensile tests at 300°C of specimens irradiated at 284°C, the steel hardened to a saturation level. The strengths obtained in that experiment were similar to those obtained in this experiment for the 20%-cold-worked steel irradiated and tested at 290°C. This similarity occurred despite the fact that the HFIR-irradiated steel was irradiated up to 27 dpa and contained 1600 at. ppm He. This observation indicates that saturation of strengthening occurs at relatively low exposure levels.

A cold-worked condition was chosen for breeder reactor applications because the high dislocation density of a cold-worked structure retards void swelling when the dislocations act as sinks for irradiation-produced defects and traps for transmutation-product helium. Although the higher dislocation density of the 50%-cold-worked structure would be expected to provide the highest irradiation resistance, Uematsu et al. [*8*] showed that at 550°C there was only a slight improvement in swelling resistance between 30 and 50% cold work. The present mechanical-property results also provide reasons for the use of a lower cold-work level. Below ~550°C, the ductility of the 50%-cold-worked steel was lower than for the steel rolled 20 and 30% in the as-cold-worked, aged, and irradiated conditions. Also, differences in strength and ductility between the steels with 20 and 30% cold work are relatively small, especially after irradiation; this, despite the fact that these two steels showed relatively more hardening than the 50%-cold-worked steel. It is also evident from the aging results that for temperatures above about 600°C, all of the cold-work levels rapidly lose strength as recovery and recrystallization occur, with the 50%-cold-worked steel losing strength at the highest rate. At 650°C, the 50%-cold-worked steel is the weakest. During irradiation, the 50%-cold-worked steel also softened at a lower temperature than the other two cold-work

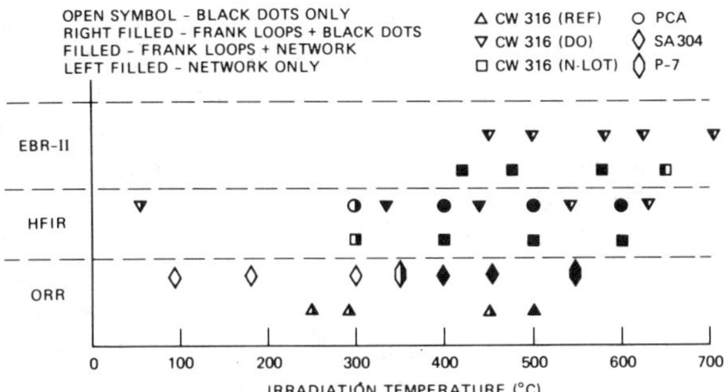

FIG. 14—*Temperature boundaries between the various components of the dislocation microstructure observed in austenitic stainless steels irradiated to fluences producing 12 dpa or lower.*

levels. This indicates that the higher defect concentration for the higher cold-work level can significantly accelerate the recovery and recrystallization processes, thus making the microstructure unstable much sooner than for the steels with a lower cold-work level.

No detailed correlation was attempted between the changes in mechanical properties and changes in microstructure. However, the observations on the 20%-cold-worked steel (Figs. 12 and 13) are qualitatively consistent with the mechanical-property changes with temperature that occurred as a result of irradiation.

There have been extensive microstructural studies performed on 20%-cold-worked Type 316 stainless steel irradiated in HFIR [*22,24–26*] and EBR-II [*6,22,27*], and these can be compared with the TEM studies on the ORR-irradiated material. Key features of the present results and the literature data are summarized in Figs. 14 and 15.

Bloom et al. [*28*] irradiated solution-annealed Type 304 stainless steel in the ORR to fluences producing less than 0.5 dpa and less than about 5 at. ppm He at temperatures of 93 to 454°C. At 93°C, they observed a high density of fine "black-dot" structures that were believed to be loops. With only small changes in size and density, this structure persisted at 177 and 300°C. At 371°C, the black-dot density decreased by several orders of magnitude; it disappeared at 398°C. Larger loops, precipitates, and small helium bubbles began to appear at 454°C. Figures 14 and 15 define the temperature limits of these observations. Brager and Garner [*9*] also observed Frank loops, networks, and "small defect clusters" (≤3 nm in diameter) in solution-annealed and cold-worked high-purity 316 austenitic stainless steel (P7) irradiated in ORR to 3 dpa at 350°C. They found

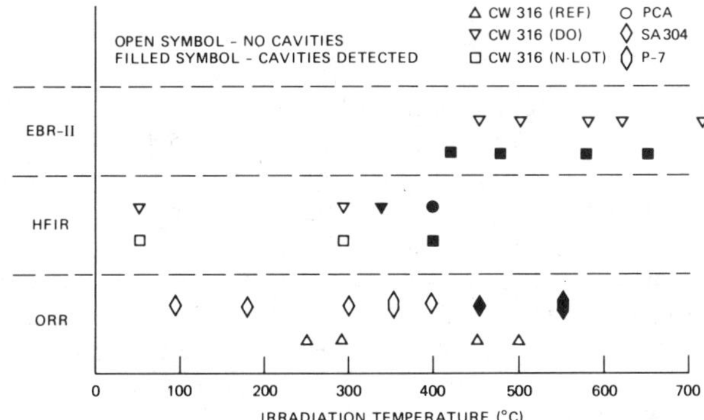

FIG. 15—*Observations at fluences producing 12 dpa or lower that define the lower temperature limit for cavity formation in austenitic stainless steel.*

voids and bubbles in P7 after irradiation to 4 dpa (33 at. ppm He) at 550°C, which is qualitatively consistent with the present results in that a fairly constant, low-temperature microstructure (black dots and dislocation loops) is observed over a range of temperatures before a transition occurs to a microstructure characteristic of higher temperatures (voids and bubbles). The transition (judging from loop and black-dot structures) occurs about 100 to 150°C higher for cold-worked 316 than for the solution-annealed Type 304 of Bloom et al [*28*].

The microstructural results on cold-worked 316 (reference heat) irradiated in ORR are consistent with microstructural data on several heats of steel [the Type 304, the DO and N-lot heats of Type 316, and the Path A Prime Candidate Alloy (PCA) irradiated in HFIR in several microstructural conditions] [*23–26,29,30*] (Figs. 14 and 15). No voids, bubbles, or irradiation-induced or -enhanced precipitates were observed after HFIR irradiation at 55 to 300°C, for fluences producing up to approximately 11 dpa. Regardless of material or pretreatment, the HFIR microstructures consisted of similar-size Frank loops (~10–30 nm in diameter) and a high concentration of the black-dot defects (~3–5 nm in diameter) (Fig. 14). However, dislocation network concentration varied widely and depended strongly on pretreatment. Bubbles were found after HFIR irradiation at 400°C for some of the PCA pretreatments and for cold-worked 316 (N-lot). The very fine black dots were not observed when these materials contained bubbles [*31*]. These black dots were also absent when voids appeared in cold-worked 316 (DO-heat) irradiated in HFIR at 325 to 350°C to 8.4 dpa. Again, this transition from characteristic lower temperature to higher temperature microstructure occurred at about 300 to 350°C in HFIR (Fig. 14), 100 to 150°C lower than for cold-worked 316 (reference heat) irradiated in ORR.

Fast-breeder reactor data do not extend below 300 to 350°C. However, the limited data are also mapped with temperatures in Figs. 14 and 15. Voids were not observed by Bramman et al [*30*] in Type 316 stainless steel irradiated below 350°C to 30 to 40 dpa in the Dounreay Fast Reactor. Voids were also not observed by Bloom and Stiegler [*27*] for cold-worked 316 (DO-heat) irradiated in EBR-II at 450°C to 10 dpa or at 510°C to 6.4 dpa. Maziasz [*22*] found no voids or fine bubbles in the same material irradiated in EBR-II to 8.4 dpa at 500°C, but found both voids and bubbles after 36 dpa at 525°C. Brager [*6*], however, observed voids in another heat of cold-worked 316 after EBR-II irradiation to 12.2 dpa at 420 and 475°C (Fig. 15). The early stages of void formation for Type 316 stainless steel in EBR-II are extremely variable from heat to heat of steel. Within this scatter, however, the ORR results are consistent with EBR-II results (Fig. 15); it is not clear whether or not voids should be expected after only 5 dpa in the ORR at 450 to 500°C. Higher fluence data will be required before confident statements about retardation (or acceleration) of void or bubble formation in ORR can be made. Comparison with the HFIR data shows that bubble swelling at 500°C is retarded in ORR.

Summary and Conclusions

The effect of cold-work level on the tensile behavior of irradiated Type 316 stainless steel was investigated. After irradiation in ORR at 250, 290, 450, and 500°C to ~5 dpa and 40 at. ppm He, tensile specimens with 20-, 30-, and 50%-cold-work levels were tested at the irradiation temperatures. The results indicate that the use of a 20 to 30% cold-worked steel has mechanical-property advantages over a steel cold worked to 50%, even though the latter may have the best swelling resistance. Although the steels with 20 and 30% cold work are hardened by irradiation relatively more at 250 and 290°C than the steel with 50% cold work, the ductility of the 20 and 30% cold-worked steels remains significantly better than for the steel with 50% cold work. For tests above 450 to 500°C where radiation-enhanced softening occurs, the properties for all three cold-work levels appear to approach a common value.

Transmission electron microscopy observations on the 20%-cold-worked steels were compared with results from EBR-II and HFIR. In EBR-II, little helium is produced relative to ORR; in

HFIR, a much higher He/dpa ratio results than in ORR. No noticeable difference in microstructural evolution occurred for the samples irradiated in ORR and in the other two reactors.

Acknowledgments

We wish to thank the following people who helped in the completion of this work: L. T. Gibson and N. H. Rouse did the tensile tests, N. H. Rouse prepared the TEM specimens, F. W. Wiffen and N. H. Packan reviewed the manuscript, and Frances Scarboro prepared the manuscript.

This research was sponsored by the Office of Fusion Energy, U.S. Department of Energy under Contract No. DE-AC05-84OR21400 with Martin Marietta Energy Systems, Inc.

References

[*1*] Bullough, R., Eyre, B. L., and Perrin, J. S., *Nucl. Appl. Technol.*, Vol. 9, 1970, pp. 346–355.
[*2*] Farrell, K. and Houston, J., *Journal of Nuclear Materials,* Vol. 35, 1970, pp. 352–355.
[*3*] Bramman, J. I. et al., "Voids Formed by Irradiation of Reactor Materials," British Nuclear Energy Society, London, 1971, pp. 27–33.
[*4*] Cawthorne, C. et al., "Voids Formed by Irradiation of Reactor Materials," British Nuclear Energy Society, London, 1971, pp. 35–43.
[*5*] Stiegler, J. O. and Blom, E. E., *Journal of Nuclear Materials,* Vol. 41, 1971, pp. 341–344.
[*6*] Brager, H. R., *Journal of Nuclear Materials,* Vol. 57, 1975, pp. 103–118.
[*7*] Appleby, W. K., Bloom, E. E., Flinn, J. E., and Garner, F. A., in *Radiation Effects in Breeder Reactor Structural Materials,* M. L. Bleiberg and J. W. Bennett, Eds., The Metallurgical Society of the American Institute of Mining, Metallurgical, and Petroleum Engineers, 1977, pp. 509–527.
[*8*] Uematsu, K., Kodama, T., Ishida, U., Suzuki, K., and Koyama, M., in *Radiation Effects in Breeder Reactor Structural Materials,* M. L. Bleiberg and J. W. Bennett, Eds., The Metallurgical Society of the American Institute of Mining, Metallurgical, and Petroleum Engineers, 1977, pp. 571–589.
[*9*] Brager, H. R. and Garner, F. A., in *Effects of Irradiation on Materials, ASTM STP 782,* H. R. Brager and J. S. Perrin, Eds., American Society for Testing and Materials, Philadelphia, 1985, pp. 152–165.
[*10*] Kawasaki, S., Fukaya, K., and Nagasaki, R., *Nippon Genshreryoko Gakkaishi,* Vol. 14, 1972, pp. 283–289.
[*11*] Fahr, D., Bloom, E. E., and Stiegler, J. O., "Irradiation Embrittlement and Creep of Fuel Cladding and Core Components," British Nuclear Energy Society, London, 1973, pp. 167–177.
[*12*] Garr, K. R., Pard, A. G., and Kramer, D., in *Properties of Reactor Structural Alloys after Neutron and Particle Irradiation, ASTM STP 570,* Baroch, C. J., Eds., American Society for Testing and Materials, Philadelphia, 1975, pp. 143–155.
[*13*] Garr, K. R. and Pard, A. G., in *Irradiation Effects on the Microstructure and Properties of Metals, ASTM STP 611,* F. R. Shrober, Eds., American Society for Testing and Materials, Philadelphia, 1976, pp. 79–90.
[*14*] Fish, R. L. and Waltrous, J. D., in *Irradiation Effects on the Microstructure and Properties of Metals, ASTM STP 611,* F. R. Shrober, Ed., American Society for Testing and Materials, Philadelphia, 1976, pp. 91–100.
[*15*] Fish, R. L., Cannon, N. S., and Wire, G. L., in *Effects of Radiation on Structural Materials, ASTM STP 683,* J. A. Sprague and K. Kramer, Eds., American Society for Testing and Materials, Philadelphia, 1979, pp. 450–465.
[*16*] Hamilton, M. L., Cannon, N. S., and Johnson, G. D., in *Effects of Radiation on Materials: Eleventh Conference, ASTM STP 782,* H. R. Brager and J. S. Perrin, Eds., American Society for Testing and Materials, 1982, pp. 636–647.
[*17*] Bloom, E. E. and Wiffen, F. W., *Journal of Nuclear Materials,* Vol. 86, 1979, p. 171–184.
[*18*] Grossbeck, M. L. and Maziasz, P. J., *Journal of Nuclear Materials,* Vol. 86, 1979, pp. 883–887.
[*19*] Klueh, R. L. and Grossbeck, M. L., in *Effects of Irradiation on Materials: Twelfth International Symposium, ASTM STP 870,* F. A. Garner and J. S. Perrin, Eds., American Society for Testing and Materials, Philadelphia, 1985, pp. 768–782.
[*20*] Odette, G. R., Maziasz, P. J., and Spitznagel, J. A., *Journal of Nuclear Materials,* Vols. 103 and 104, 1981, pp. 1289–1304.
[*21*] Lee, E. H., Rowcliffe, A. F., and Mansur, L. K., *Journal of Nuclear Materials,* Vols. 103 and 104, 1981, pp. 1475–1480.
[*22*] Maziasz, P. J., *Journal of Nuclear Materials,* Vols. 108 and 109, 1982, pp. 359–384.

[23] Maziasz, P. J., Horak, J. A., and Cox, B. L., "Phase Stability During Irradiation," J. R. Holland, L. K. Mansur, and D. I. Potter, Eds., The Metallurgical Society of AIME, Warrendale, PA, 1981, pp. 271–292.
[24] Maziasz, P. J. and Grossbeck, M. L., *Journal of Nuclear Materials,* Vol. 104, 1982, pp. 987–992.
[25] Wiffen, F. W. and Maziasz, P. J., *Journal of Nuclear Materials,* Vols. 103 and 104, 1981, pp. 821–826.
[26] Maziasz, P. J., *Transactions American Nuclear Society,* Vol. 39, 1981, pp. 433–485.
[27] Bloom, E. E. and Stiegler, J. O., in *Effects of Radiation on Substructure and Mechanical Properties of Metals and Alloys, ASTM STP 529,* J. Moteff, Eds., American Society for Testing and Materials, Philadelphia, PA, 1973, pp. 360–387.
[28] Bloom, E. E., Martin, W. R., Stiegler, J. O., and Weir, J. R., *Journal of Nuclear Materials,* Vol. 22, 1967, pp. 68–76.
[29] Maziasz, P. J. and Braski, D. N., *Journal of Nuclear Materials,* Vols. 122 and 123, 1984 pp. 305–310.
[30] Bramman, J. E. et al, in *Radiation Effects in Breeder Reactor Structural Materials,* M. L. Bleiberg and J. W. Bennett, Eds., The Metallurgical Society of AIME, Warrendale, PA, 1977, pp. 479–508.
[31] Maziasz, P. J. and Braski, D. N., in *Radiation Effects in Breeder Reactor Structural Materials,* M. L. Bleiberg and J. W. Bennett, Eds., The Metallurgical Society of AIME, Warrendale, PA, 1977, pp. 311–316.

Timothy D. Naughton,[1] *Nasr M. Ghoniem,*[2] *and Tung H. Lin*[2]

Radiation Effects on the Micromechanics of Fatigue Crack Initiation

REFERENCE: Naughton, T. D., Ghoniem, N. M., and Lin, T. H., **"Radiation Effects on the Micromechanics of Fatigue Crack Initiation,"** *Influence of Radiation on Material Properties: 13th International Symposium (Part II), ASTM STP 956,* F. A. Garner, C. H. Henager, Jr., and N. Igata, Eds., American Society for Testing and Materials, Philadelphia, 1987, pp. 223–238.

ABSTRACT: A micromechanical model is employed to investigate parameters that affect the rate of local plastic strain in and around persistent slip bands in cyclically loaded metals. The model utilizes a solution of the stress field in a semi-infinite elastic-plastic medium under generalized plane deformation. By following the evolution of the stress field, a natural gating mechanism is observed to produce regions of localized deformation. The regions of local deformation develop into extrusions and intrusions that are responsible for the formation of fatigue cracks.

By coupling this micromechanical model with an empirical correlation developed to predict the increase in critical shear stress as a function of fluence and temperature, an investigation of radiation effects on fatigue crack initiation is carried out. It is determined that longer fusion reactor burn times and lower operating temperatures inhibit the crack initiation rate. Because of irradiation induced-hardening, increases in the fatigue limit are predicted for long reactor burn times ($>10^3$ s) and low operating temperatures ($<300°C$). Irradiation is found to increase the number of cycles required to obtain a failure strain at low temperatures and high-cycle fatigue conditions.

KEY WORDS: micromechanical model, persistent slip bands, dislocation dipoles, extrusions, intrusions, crack initiation, critical shear stress, radiation hardening

An example of a modern machine with fail-safe and safe-life components is a fusion reactor. In a conventional nuclear power plant, the pressure vessel ensures plant safety by containing the core in over 1-ft (0.3-m) thick. Large heat fluxes, important space constraints, and parasitic neutron absorption forbid the use of excessively thick walls inside the magnet's shield in a fusion reactor. Thus, most components inside the magnet's shield will be designed fail safe. To account for this fact, most fusion blanket designs are usually divided into modules, which are scheduled for periodic replacement.

Within each module, there will be a series of fail-safe and safe-life components. One of the most important safe-life components in a fusion reactor is the first wall, which is a pressure retaining boundary that is located closest to the plasma. Its primary function is to intercept the surface energy flux, which may be up to ≃20% of the total fusion power. To do so, it must be actively cooled. The coolant is usually at a high pressure, or very corrosive, or both. If the first wall develops a leak and loses any coolant into the plasma chamber, the entire fusion reaction will cease until the first wall or the entire module is replaced. Thus, it is essential to plant economics that safe-life module parts, such as the first wall, survive their expected life within the module.

[1] Member of technical staff, TRW Inc., Mail Station 134/9816, One Space Park, Redondo Beach, CA 90278.

[2] Professor and professor emeritus, respectively, School of Engineering and Applied Sciences, Mechanical, Aerospace and Nuclear Engineering Department, University of California, Los Angeles, CA 90024.

Many difficulties arise when predicting the fatigue life of a fusion component. Environmental effects unique to fusion reactors must be taken into account, for example, the 14-MeV neutron spectrum, pulsed surface heat fluxes, surface bombardment by energetic particles, magnetic forces, corrosive materials, vacuum, and high and low temperatures. Present uncertainties in the loading sequence in a fusion device stem from uncertainties in plasma design. Regardless of the loading sequence, a reliable fatigue design is sure to play a major role in the safe and successful operation of a fusion device.

Justifiably, a great deal of effort has been devoted to developing methods of predicting the fatigue life of fusion reactor components as reviewed by Watson [*1*]. In 1984, the failure modes for the first wall and blanket were reviewed by members of the fusion community and ranked by the likelihood of occurrence. The top-ranked issue was the effect of first-wall heat flux and cycling on fatigue or crack-growth-related failures [*2*].

Therefore, it is quite apparent that the fatigue-life issues will be a major concern in fusion reactor designs for many generations. Up until now, the majority of the effort invoked in predicting the fatigue life in fusion reactors has gone into predicting the effects of a fusion environment on fatigue-crack growth rates. However, very little effort has been devoted to understanding how the fusion environment will affect fatigue-crack initiation. This work is a study of how a fusion environment affects crack initiation. To investigate the effect, a crack-initiation mechanism proposed by Lin [*3*] will be coupled with an empirical correlation relating the damage produced by neutron irradiation to the increase in critical shear stress.

Fatigue-Crack Initiation

Forsyth [*4*] and Forsyth and Stubbington [*5*] observed thin sheets of metal extruded from slip bands on material surfaces. Hull [*6–8*] noted that slip band intrusions also occur on the surface of cyclically loaded materials. He showed that extrusions and intrusions formed in copper subjected to cyclic loadings at 4.2 K. Forsyth [*4*] has also found that extrusions form but leave no voids below the surface. Thus, thermally activated processes, surface corrosion, gas adsorption, gas diffusion into the metal, or vacancy diffusion to form voids are not necessary for the formation of extrusions or intrusions [*9*].

The formation of extrusions and intrusions are generally accepted as mechanisms that lead to the formation of transcrystalline (high cycle) fatigue cracks. Therefore, several theories have been proposed to explain the formation of slip-band intrusions and extrusions. Some of the most reknowned theories are reviewed and criticized by Kennedy [*10*]. Most fatigue crack-initiation theories attempt to explain the formation of extrusions and intrusions with dislocation mechanisms.

Dislocation motion is governed by the resolved shear stress in certain crystallographic directions on certain planes. The models reviewed by Kennedy present possible motions of dislocations without considering the resolved shear stress necessary to initiate the motion. The first quantitative theory of fatigue-crack initiation based on the microstress field around slip bands was presented by Lin and Ito [*11*]. The theory is based on the assumption that slip occurs when a critical resolved shear stress is reached. The resolved shear stress is taken to be the sum of an initial, an applied, and a residual shear stress. The theory illustrates a natural gating mechanism provided by the residual stress fields created by slip. The gating mechanism leads to the formation of extrusions and intrusions.

A Quantitative Theory of Fatigue-Crack Initiation

Single crystal tests have shown that dislocations slide in certain directions on certain crystallographic planes [*12,13*]. Dislocations slide when a critical resolved shear stress τ_c is reached along the Burger's vector on a sliding plane. Lin's quantitative fatigue-crack initiation model is based on this facet of dislocation theory. Lin calculates the "local" resolved shear stress in the most favorably oriented crystals and follows the evolution of the stress field as local plastic deformation occurs in persistent slip bands (PSBs).

All metals have imperfections that create equivalent internal forces. The equivalent internal forces result in an initial stress field τ^I_{ij}. An initial stress field favorable for the formation of extrusions and intrusions has been calculated from a strain field formed by aligned dislocation dipoles in PSBs [*14*]. Dislocation diplole structures have been experimentally observed in cyclically loaded materials by Katagiri [*15*] and Jin and Winter [*16,17*].

When the material is loaded, the initial stress field is coupled with the applied stress field τ^A_{ij} to create the local resolved stress field. If the local resolved shear stress in any region of the crystal exceeds the critical resolved shear stress along the Burger's vector in a sliding plane, slip occurs. Now if the material is unloaded, the plastic strain resulting from slip remains and introduces a residual stress field τ^R_{ij}. Under subsequent loading, the resolved stress field becomes the sum of the initial, applied, and residual stresses.

To determine whether a slip system in a region of a crystal will slide, the following three components of the stress field must be accounted for: The applied shear stress can be calculated from the loading conditions. The initial stress field is usually assumed or can be calculated from an assumed initial strain field. The residual stress field must be calculated by an analogy of plastic strain and applied force, similar to Duhammel's analogy for thermal stress analysis [*18*].

Calculation of Residual Stress Field

Figure 1*a* illustrates a rectangular grid deformed by a positive plastic shear strain caused by dislocation glide. The forces necessary to restore the grid to its previous shape are also shown. These forces are the forces necessary to keep the body strain free, $e_{ij} = 0$. Since these forces do not actually exist, the elastic body surrounding the grid supplies the restoring forces and prevents gross plastic deformation. The forces are relaxed in the elastic body by applying equal and opposite resultant forces. The stresses caused by the resultant forces, illustrated in Fig. 1*b*, are continuous and can be related to the finite jump in plastic strain across the surface of the grid.

Adopting a notation for representing the strain field in two parts (an elastic part denoted by a single prime and a plastic part denoted by double primes, $e_{ij} = e'_{ij} + e''_{ij}$) the relationship between stress and strain for an isotropic material is

$$\tau_{ij} = \delta_{ij}\lambda\theta' + 2Ge'_{ij} \tag{1}$$

or

$$\tau_{ij} = \delta_{ij}\lambda(\theta - \theta'') + 2G(e_{ij} - e''_{ij}) \tag{2}$$

where λ and G are Lame's constants, θ is the dialation, and δ_{ij} is the Kronecker delta. The conditions of equilibrium within a body of volume V and surface Γ are

$$\tau_{ij,j} + F_i = 0, \quad \text{in } V \tag{3}$$

$$T^\nu_i = \tau_{ij}\nu_j, \text{ on } \Gamma \tag{4}$$

where the comma denotes differentiation. F_i is the body force per unit volume in the i direction. T^ν_i is the i-component of the traction on the boundary Γ and ν_i is the cosine of the angle between the normal ν and the x_j axis. These equilibrium conditions are combined with the stress-strain relationship to yield

$$\delta_{ij}\lambda\theta, j + 2Ge_{ij,j} - (\delta_{ij}\lambda\theta'', j + 2Ge''_{ij,j}) + \mathrm{F}_i = 0 \tag{5}$$

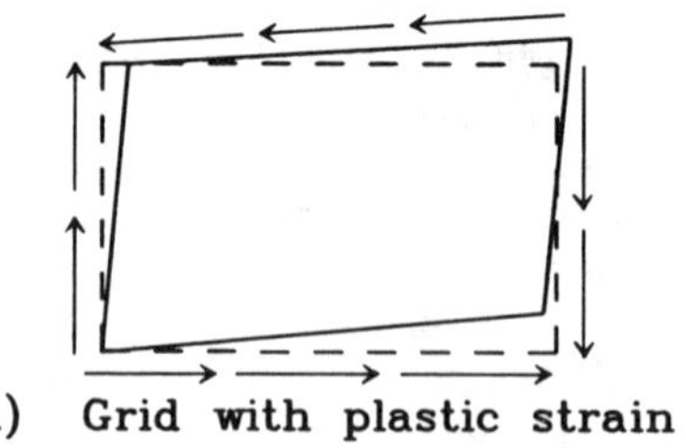

(a) Grid with plastic strain and restoring forces

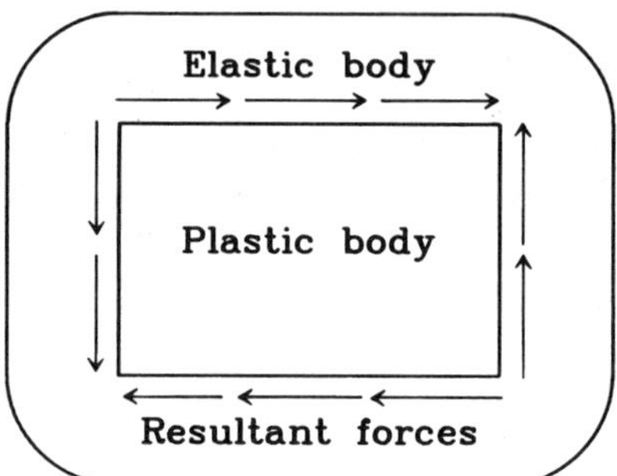

(b) Plastic strain in an elastic medium

FIG. 1—*Restoring forces and resultant forces resulting from slip in a rectangular grid.*

in the interior of the body and

$$T_i^\nu = \nu_j[(\delta_{ij}\lambda\theta + 2Ge_{ij}) - (\delta_{ij}\lambda\theta'' + 2Ge_{ij}'')] \tag{6}$$

on the surface. The plastic strain components can be represented by equivalent body and surface forces F_i and T_i^ν, respectively

$$\delta_{ij}\lambda\theta,j + 2Ge_{ij,j} - F_i + F_i = 0 \tag{7}$$

$$T_i^\nu + T_i^\nu = \nu_j(\delta_{ij}\lambda\theta + 2Ge_{ij}) \tag{8}$$

The residual stress field can be represented by

$$\tau_{ij}^R = \delta_{ij}\lambda(e_{kk} - e_{kk}'') + 2G(e_{ij} - e_{ij}'') \tag{9}$$

where e_{ij} is the strain field caused by slip. Hence, the solution of the stress field of a body with a known plastic strain distribution reduces to the solution of an identical elastic body with an additional set of equivalent body or surface forces or both.

Since most fatigue cracks form at the surface of a material, it is necessary to account for the effect of the free surface when calculating the microstress field initiated by local slip. The plane stress solution of a stress field resulting from a point force applied in a semi-infinite plate was determined by Melan [*19*]. The solution was modified for plane strain by Tung and Lin [*20*].

For numerical calculations, a slip band in the most favorably oriented crystal is divided into thin slices of width $2b$. As shown in Fig. 2, each slice is divided into N thin parallelogram grids. The plane surface bounding each grid is denoted S_r where $r = 1, 2, \ldots, 3N$. The plastic strain

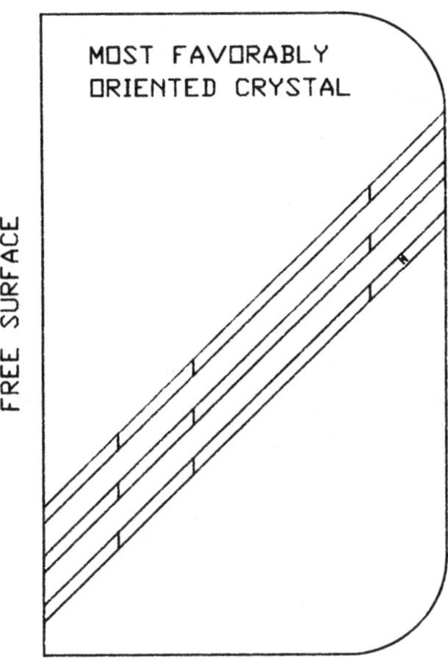

FIG. 2—*Grids in persistent slip bands.*

in each grid is assumed constant. Thus, the equivalent body forces are eliminated. If β is taken to be normal to the direction of slip and α is the direction of slip in the *m*th slip system, then the plastic shear strain caused by slip can be expressed in the *x* coordinates as

$$e''_{ij}(x') = m''_{ij}e(x',m) \tag{10}$$

where

$$m_{ij} = \beta_i(m)\alpha_j(m) + \beta_j(m)\alpha_i(m)$$

where $\alpha_i(m)$ and $\beta_i(m)$ denote the components of the unit vectors along $\alpha(m)$ and $\beta(m)$, respectively. The slip system portrayed in Fig. 2 is oriented at 45° from the free surface. Thus, plastic strain components determined by Eq 10 are

$$e''_{11} = -e''_{\alpha\beta},\ e''_{22} = e''_{\alpha\beta},\ \text{and}\ e''_{12} = 0 \tag{11}$$

From Eqs 4 and 11 the equivalent surface forces on S_n are

$$T_{1n} = 2Ge''_{\alpha\beta r}\nu_1$$

and

$$T_{2n} = 2Ge''_{\alpha\beta r}\nu_2 \tag{12}$$

The residual stress field caused by a uniform plastic strain in an active grid can be obtained by performing the surface integral of the surface tractions coupled with the stress functions for the plane strain solution of a semi-infinite medium. The sum of the integrals of each active grid yields the total residual shear stress at a point resulting from slip in the active grids. Therefore, the total residual stress field can be expressed as

$$\begin{aligned}\tau^R_{\alpha\beta}(x_1,x_2) = & -\sum_{n=1}^{\zeta}\chi\int_{S_n}[K_1(x_1,\ x_2;\ x'_1x'_2)T^{\eta}_{1n} \\ & - K_2(x_1,\ x'_2x'_1,\ x'_2)T^{\eta}_{2n}]dS - 2Ge''_{\alpha\beta}(x_1,\ x_2)\end{aligned} \tag{13}$$

where K_1 and K_2 are given by Lin [*11*].

The integrals are analytically evaluated by Lin [21] for multiple slip systems. Once the stress field at a point x caused by slip in the mth slip system is determined, the resolved shear stress in the nth slip system can be obtained by

$$\tau(x,n) = (\tfrac{1}{2})n_{ij}\tau_{ij}(x) \tag{14}$$

The results of his integration then are reported in the form

$$\tau^R(x,n) = -2Ge''(x',m)\; U(x,n;x',m) \tag{15}$$

where $U(x,n;x',m)$ is called the stress influence coefficient. It is the residual resolved shear stress at x in the nth slip system caused by a constant unit plastic strain distribution in the parallelogram grids S_n with centroid x' caused by slip in the mth slip system.

To determine the plastic strain in the active grids, the following set of simultaneous equations must be solved

$$\Delta\tau_{ex} + 2G\sum_r U(x,n;x',m)\,\Delta e''_r = 0 \tag{16}$$

where

$$\begin{aligned} \Delta\tau_{ex} &= \left|\tau^A + \tau^I + \tau^R\right| - \tau_c, \\ U(x,n;x'm) &= bU'(x,n;x'm), \text{ and} \\ 2b &= \text{grid thickness.} \end{aligned}$$

for r active grids. It should be noted that the relief of resolved shear stress is directly proportional to the plastic strain and the grid thickness $2b$. Once the plastic strain resulting from an applied load is accounted for, the residual stress field caused by the plastic strain in active grids is calculated by means of Eq 15 and used to compute the resolved stress field for subsequent loadings.

Description of a Natural Gating Mechanism

In the current work, the results of Lin's integration are used to calculate the influence coefficients. Thus, the residual stress fields resulting from uniform plastic strain are accurately calculated. The residual stress fields are used to follow the evolution of the microstress field around PSBs as a uniform cyclic load is applied. The continuous local plastic deformation, which occurs as a result of a gating mechanism provided by the residual stress field, leads to the formation of extrusions and intrusions as shown in Fig. 3.

When the positive resolved shear stress in Q reaches the critical shearing stress, positive plastic strain $+e''_{\alpha\beta}$ occurs in Q. After the load is removed, the plastic strain remains. This plastic strain gives rise to equivalent forces as shown by the analogy of plastic strain and applied forces discussed previously. The residual stress field created by the plastic strain will have components of resolved shear stress. Since the resolved shear stress field is continuous, slip in Q relieves the positive resolved shear stress not only in Q but also in P and R. This relief of the positive shear stress in Q increases the magnitude of the negative shear stress in P and R.

This process enhances the slip that occurs in P and R during reversed loading. When the negative resolved shear stress in P and R reaches the critical resolved shear stress, a negative plastic shear strain $-e''_{\alpha\beta}$ is produced. This plastic strain creates a new residual stress field. By relieving the excessive negative shear stress in P and R, the plastic strain also increases the positive shear stress in Q. This process is repeated for every loading cycle. Thus, the residual stress field provides a natural gating mechanism leading to the formation of extrusions and intrusions.

Development of a Fluence- and Temperature-Dependent Critical Shear Stress

Dislocation motion is the principal process contributing to plastic deformation in metals. The total shear stress required to cause dislocation glide is termed the critical shearing stress τ_c. The

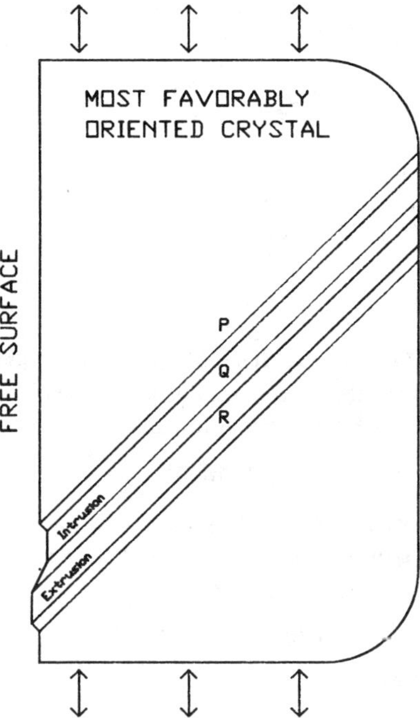

FIG. 3—*Lin's extrusion and intrusion model.*

hardening process can be analytically expressed as an increase in the critical shear stress. The forces responsible for resisting the motion of dislocations can be categorized as long range or short range. The critical shear stress is the stress needed to overcome both forces, $\tau_c = \sigma_{SR} + \sigma_{LR}$.

Long-range forces arise from the interaction of stress fields of gliding dislocations with surrounding impurities. Long-range resistance to dislocation motion may be caused by dislocation pileups, surrounding immobile dislocations, grain boundaries, and general disorder in the crystalline structure. The stress required for a moving dislocation to overcome the long-range forces is

$$\sigma_{LR} = Gb/2\pi l \tag{17}$$

where l is related to the dislocation density ρ_d by

$$l = (3/\rho_d)^{1/2} \tag{18}$$

Radiation has been shown to increase the dislocation density through the formation of networks and loops. Brager and Straalsund [*22*] developed empirical formulas to account for the fluence and temperature dependence of dislocation density. Their empirical formulas for solution annealed 316 stainless steel (SS) are

$$\rho_l + \rho_N = 10^9(\phi t \times 10^{-22})^{F(T)} \exp [G(T)] \tag{19}$$

where

$F(T) = 31.07 - 0.0145T - 13750/T$ and
$G(T) = -47.7 + 0.0193T + 25970/T$.

where the fraction of the total dislocation density, in the form of network dislocations is

$$\rho_N/(\rho_l + \rho_N) = \{1 + \exp [0.11(715\text{-}T)]\}^{-1} \tag{20}$$

The increase in yield strength resulting from radiation-produced long-range forces supplied by dislocation networks is

$$\sigma_{LR} = Gb/2\pi(\rho_N/3)^{1/2} \quad (21)$$

where G is the temperature-dependent shear modulus. This is the dominant hardening mechanism at high temperatures and low fluences.

According to Seeger [*23*], the average spacing between particles is inversely proportional to the square root of the number of particles. Thus, the critical shear stress is proportional to the square root of the number of obstacles

$$\Delta\tau_c \alpha (N)^{1/2} \quad (22)$$

At low fluences the number of obstacles is inversely proportional to the fluence. Since the average obstacle radius does not change drastically with fluence, the increase in critical shearing stress is proportional to the square root of the fluence at low fluences

$$\Delta\tau_c \alpha (\phi t)^{1/2} \quad (23)$$

At higher fluences, the density of obstacles saturates because of cascade overlap. Makin and Minter [*24*] assume the time rate of change in the number of a specific type of obstacle with an average radius r as

$$dN/dt = \kappa\Sigma_s\phi(1 - vN) \quad (24)$$

Solving the differential equation yields

$$N = 1/v[1 - \exp(-\kappa v\Sigma_s\phi t)] \quad (25)$$

Thus the increase in critical shear stress is

$$\Delta\tau_c \alpha [1 - \exp(-\kappa v\Sigma_s\phi t)]^{1/2} \quad (26)$$

At higher temperatures, depleted zones become unstable and anneal out. In the current study the temperature dependence of depleted zones is accounted for by modifying Makin's rate formulation. A temperature-dependent constant $\theta(T)$ is used in Eq 26. The number of depleted zones per collision is assumed to be unity at temperatures below 200°C and decays linearly to zero at 400°C. This phenomenological approach is used to reflect the annealing behavior of the depleted zones as temperature is increased.

Dislocation loops are formed by the condensation of interstitial atoms produced in collision cascades. If the slip plane of a mobile dislocation passes close to or intersects a loop, dislocations gliding on the plane will experience resistance to the motion. The increase in critical shear stress resulting from the formation of loops is given by

$$\Delta\sigma_{SR} = F_{max}/b_e l \quad (27)$$

where F_{max} = maximum force between loop and dislocation line, $\mathbf{b}_e$ = Burgers vector of mobile dislocation, and l = loop spacing. Kroupa and Hirsch [*25*] and Forman [*26*] have developed models to predict the value of F_{max}. Olander [*27*] states that the bulk experimental evidence on loop hardening favors a relationship of the form

$$\Delta\sigma_{SR} = Gb(2R_lN_l)^{1/2}/\omega \quad (28)$$

where N_l is the concentration of loops in the solid, R_l is the average loop radius, and ω is a numerical factor between 2 and 4.

Brager and Straalsund [22] attempted to experimentally determine the temperature and fluence dependence of density of dislocation loops and the average radius of loops in 316 SS. They developed the following empirical formula to predict the faulted loop number density

$$N_l = 10^{15}(\phi t \times 10^{-22})^{0.53} \exp[L(T)] \quad (29)$$

where $L(T) = -203.5 + 0.116T + 85900/T$. Their empirical formula, developed to predict the average radius of faulted loops, is given by

$$R_l = \tfrac{1}{2}(\phi t \times 10^{-22})^{H(T)} \exp[J(T)] \quad (30)$$

where

$H(T) = -6.31 + 0.00262T + 3060/T$ and
$J(T) = 23.89 - 0.0071T - 9040/T$.

Since most irradiated components subject to fatigue failure in fusion reactors are made of high strength alloys, the following calculations will attempt to simulate the high strength alloys. Holmes et al. [*28*] reported the yield strength of American Iron and Steel Institute (AISI) 304 SS to be 12 ksi (82.7 MPa). Since the 304 SS is a polycrystalline material, the critical shear stress (which is determined from single-crystal test) will be considerably less than the yield strength. Taylor [*29*] measures the yield strength of single crystal to be about ⅓ that of a polycrystal. Garber and Mogilnikova [*30*] note that dislocation motion occurs in cyclically loaded materials at stress amplitudes less than ¼ the yield strength. Therefore, the critical shear stress is roughly 1⁄12 the yield strength. Thus, an appropriate critical shear stress corresponding to stainless steel will be assumed to be 1000 psi (6894 kPa). A temperature-dependent shear modulus and Poisson's ratio are also assumed for calculations modeling high-strength steel.

From Holmes' annealing experiments [*28*], it appears reasonable to assume that the increase in yield strength because of depleted zones is 12 ksi (82.7 MPa). This increased value can be thought of as a macroscopic or measured hardening value resulting from irradiation. In reality, the localized plastic strain around slip bands is very large compared to the entire specimen. Therefore, the actual hardening occurring within the local region is much less than that predicted by the measured value. The actual local increase in critical shear stress is assumed to scale similar to the initial measured value of critical shear stress (1⁄12 × yield strength). Thus, a corresponding saturation value for the increase in critical shear stress resulting from depleted zones is assumed to be 1000 psi (6894 kPa).

To simplify the empirical correlation used to determine the increase in yield strength, it is assumed that correlation (Eq 26) can be written in the form

$$\Delta\tau_c = C(T)\,[1 - \exp(-t/t^*)]^{1/2} \quad (31)$$

If it is assumed that the radiation hardening resulting from depleted zones and loops reaches 98% of its saturated value by a displacement damage of ≃10 dpa, then the irradiation time constant t^* can be calculated. For a fusion reactor the displacement damage rate is on the order of 3.4×10^{-7} dpa/s. Thus, t^* is calculated to be 10^7 s. The empirical correlation relating the depleted zone hardening to dose is simplified to a correlation relating the increase in critical shear stress zones to the reactor on-time

$$\Delta\tau_c = C(T)[1 - \exp(-t/10^7)]^{1/2} \quad (32)$$

where t is the reactor on-time.

To calculate the increase in critical shear stress resulting from dislocation loops, the empirical correlation (Eq 28) is modified by assuming a fluence of 10^{22} n/cm^2 is equal to 5 dpa. Then to convert from yield strength to critical shear stress, Eq 28 is divided by 12. This correlation is valid up to ≃500°C where thermal annealing starts to influence loop density. The loop density is assumed to drop off linearly to zero by 600°C. At temperatures above 600°C and low fluences, the dominant hardening mechanism is dislocation networks. The increase in critical shear stress caused by dislocation networks is obtained by dividing Eq 21 by 12.

Irradiation hardening resulting from voids, bubbles, and precipitates occurs over a much longer time scale. Generally, the fluence over which they have measurable action is large enough that by the time the voids and precipitates form, the critical shear stress will have increased to a value close to the saturation value resulting from loops and depleted zones. Since this saturation value in critical shear stress is usually much greater than the excessive shear stress ($|\tau^I + \tau^A + \tau^R| - \tau_c$), the slip necessary to continue the initiation process will not occur unless driven by stress concentrations. The current model has not yet been developed to accurately account for stress concentrations that develop because of large local plastic strains.

Thus, the present investigation will be limited to studying radiation effects at low fluences. The mechanisms contributing to irradiation hardening are depleted zones, dislocation loops, and networks. The increase in critical shear stress because of these mechanisms is dependent on temperature and fluence. The contribution of each mechanism to the total hardening is illustrated in Fig. 4. The fluence dependent increase in critical shear stress is illustrated in Figs. 5 and 6 for various temperatures. The plots show that the increase in critical shear stress versus fluence (in dpa) for various temperatures.

Investigation of Radiation Effects on Initiation

In the current calculation the applied alternating cyclic shear stress is taken as ±995 psi (±6826 kPa). The initial shear stress will be assumed to be −50 psi (−345 kPa) in *P* and 50 psi (345 kPa) in *Q*. Since the critical shear stress is assumed to be 1000 psi (6894 kPa), the excessive shear stress on the first applied loading is 45 psi (310 kPa). The empirical correlation developed to predict the increase in critical shear stress is used to investigate the effect of the local hardening on the local plastic strain in the region of a slip band. Using a reactor on-time of 100 s, the

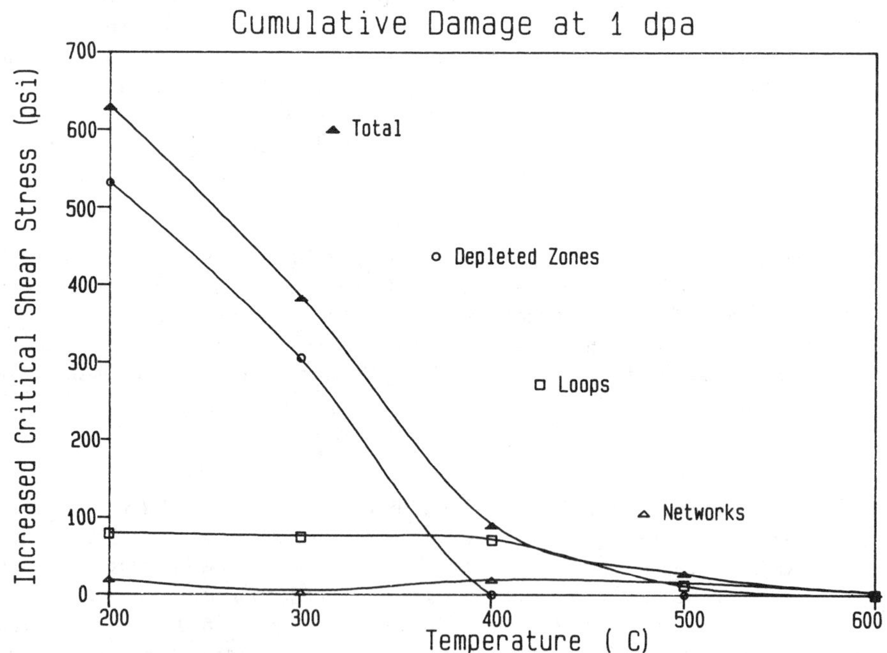

FIG. 4—*Contribution of hardening mechanisms to total hardening.*

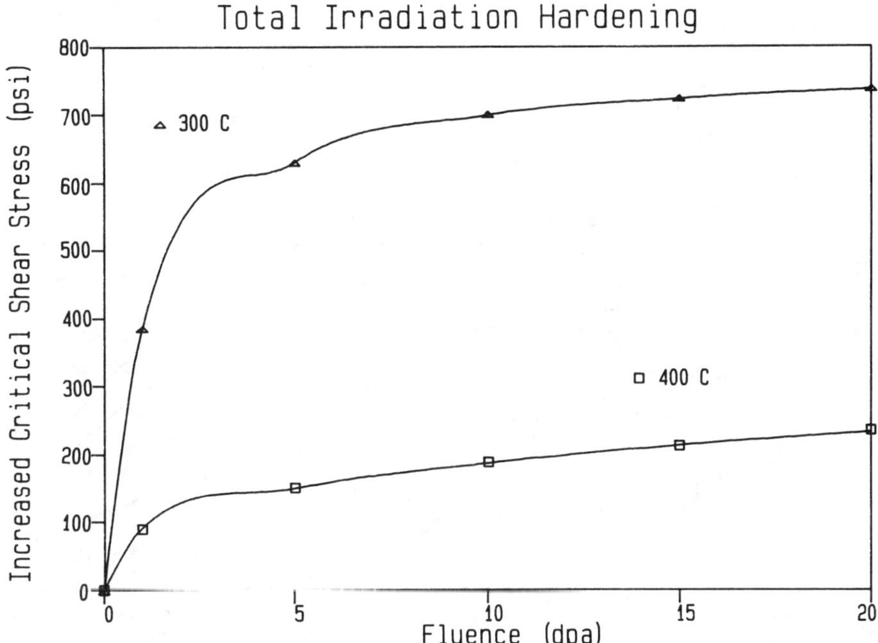

FIG. 5—*Fluence-dependent increase in critical shear stress at low operating temperatures.*

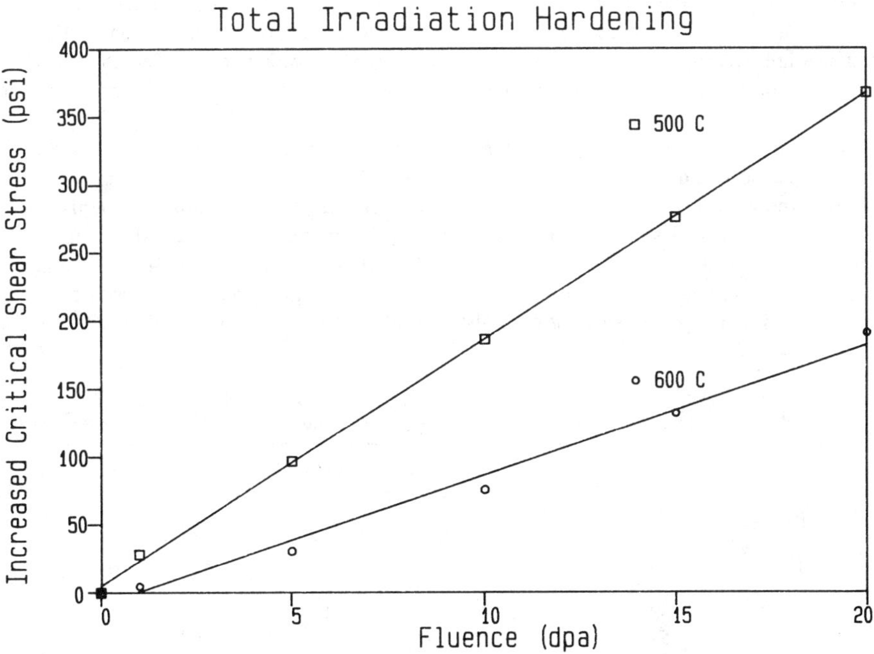

FIG. 6—*Fluence-dependent increase in critical shear stress at high operating temperatures.*

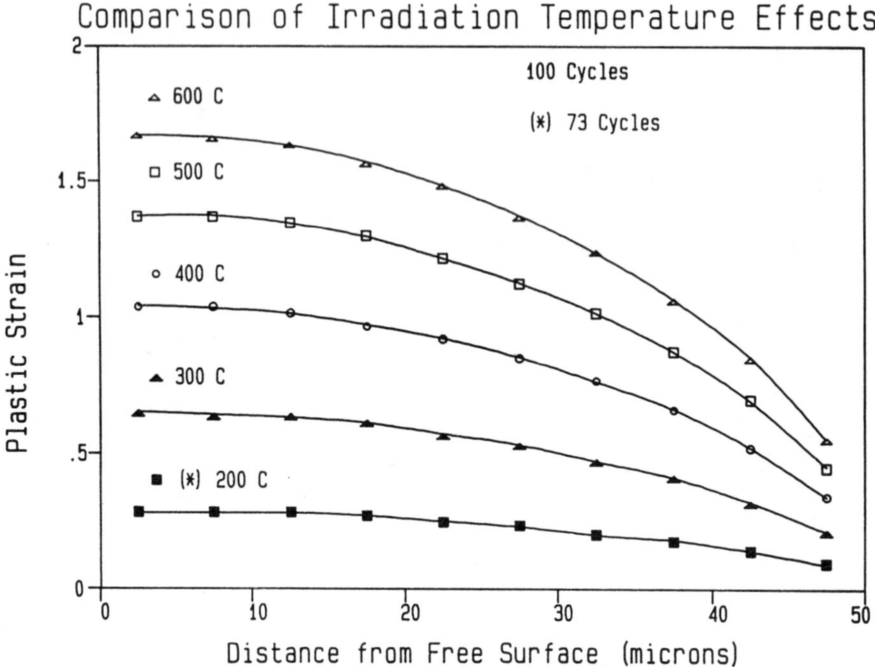

FIG. 7—*Effects of temperature-dependent hardening on rate of plastic strain at the free surface.*

calculations comparing the effect of irradiation at different temperatures were made and the results reported in Fig. 7. (Note: Because only the temperature effects on irradiation-induced hardening mechanisms are investigated, creep is neglected.) Using these calculations, Table 1 was compiled to show how the increase in critical shear stress scales with temperature for a reactor on-time of 100 s.

Figure 7 illustrates that as the temperature is decreased, the plastic strain at the free surface is decreased because of an increase in hardening. Thus, the effect of irradiation can be interpreted as requiring more cycles to initiate a crack. At low temperatures, the gating mechanism is halted by a steep rise in critical shearing stress. For 200°C, the increase in critical shear stress exceeds the excessive shear stress after 73 cycles. This phenomenon can be physically explained by an experimentally observed increase in the fatigue limit. The fatigue limit is increased because after 73 cycles the range in applied shear stress must be greater than +995 psi (+6826 kPa) to continue the initiation process.

TABLE 1—*Increase in critical shear stress for a reactor burn time of 100 s, psi.*[a]

	Temperature, °C				
Cycles	200	300	400	500	600
1	8.6	6.6	4.9	4.1	0.1
5	15.5	11.4	7.6	4.8	0.1
50	38.5	25.9	14.4	6.1	0.1
100	51.4	33.6	17.5	6.7	0.1
200	69.0	44.0	21.3	7.1	0.1

[a] 1 psi = 6.8948 kPa.

The empirical correlation used to express the increase in critical shear stress has been simplified to a function of reactor on-time for a constant temperature. To do so, the critical shear stress is expressed as a function of the number of loadings. The number of cycles is related to the reactor on-time through the cyclic frequency or the reactor burn time. To avoid clouding the issue in the time-dependent analysis, it is necessary to neglect the effects of creep and consider only the irradiation effects.

Early tokamak design-specified burn times were on the order of 100 s. However, the more recent designs are emphasizing the need for a steady state device. Figure 8 compares the effect of burn times on the local plastic strain for an operating temperature of 450°C. The calculations used to obtain Fig. 8 assume an initial stress of ±25 psi (172 kPa) in P and Q. The cyclic-applied shear stress was ±995 psi (±6826 kPa). Thus, the excessive shear stress on the first cycle was +20 psi (138 kPa). For these calculations, Table 2 illustrates how the reactor burn time influences the increase in critical shear stress for given amount of cycles.

Recalling that rate of increase of plastic strain at the free surface is a measure of the crack-initiation rate, the effect of longer burn times is to decrease the initiation rate. Thus, as the burn time is increased the number of cycles to initiate a crack will be increased. For a burn time of 10 000 s, the increase in critical shear stress exceeds the excessive shear stress in 21 cycles. Therefore, it may be expected that fatigue limit is increased for longer burn times. In high-cycle fatigue (where the majority of the fatigue life is spent initiating a crack), the number of cycles to failure can actually be increased even though the crack-propagation rate is increased because of irradiation.

Discussion

The effect of irradiation on fatigue life has been investigated by many authors [*1,2,31–33*]. The typical predicted behavior of an irradiated metal exhibits a decrease in fatigue life in the low-cycle

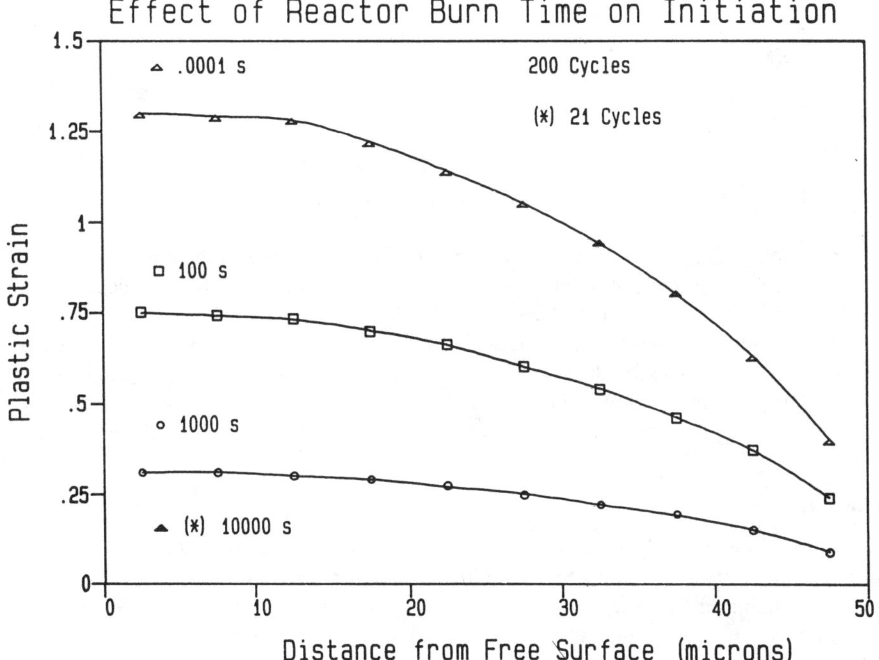

FIG. 8—*Effects of burn time on the rate of plastic strain at the free surface.*

TABLE 2—*Increase in critical shear stress for an operating temperature of 450°C, psi.*[a]

Cycles	Reactor Burn Time, s			
	0.0001	100	1000	10 000
1	0.4	4.0	6.0	9.3
5	0.5	5.3	8.1	13.4
50	0.7	8.1	13.4	27.5
100	0.8	9.3	16.1	37.0
200	0.9	10.8	19.8	52.0

[a] 1 psi = 6.8948 kPa.

regime and an increase in the high-cycle fatigue life. The typical estimated fatigue properties of irradiated 316 SS based on the tensile properties are given by Michel [*31*]. These results compare well with Murty's experimentally determined fatigue properties of 304 SS at room temperature and at 350°C [*32*]. More experimental evidence backing this hypothesis is given by Beeston and Brinkman [*33*].

The typical fatigue life of an irradiated metal can be explained by analyzing the low-cycle and high-cycle fatigue lives separately. In low-cycle fatigue, large plastic strains initiate cracks early in fatigue life, and the majority of the fatigue life is usually consumed in the crack propagation mode (Stage II growth). The investigation of radiation effects on fatigue-crack growth rate indicates that increasing damage rates tend to increase the fatigue crack growth rate by decreasing the fracture toughness or ductility. This has been predicted by using the modified Forman Eq 32 [*34*]

$$da/dN = [C\lambda^m(f\Delta K - \Delta K_0)^n]/[K_{1c} - \lambda f\Delta K] \quad (33)$$

where

a = crack length,
N = number of cycles,
$\lambda = (1 - R)^{-1}$,
$R = K_{min}/K_{max}$,
$f = E(T_0)/E(T)$,
T_o = 25°C.

Thus, the theoretical effect of irradiation on the degradation of the low-cycle fatigue life is predicted.

In high-cycle fatigue, the gross plastic deformation is negligible and the majority of the fatigue life is exhausted during the initiation of a crack. The crack-initiation process occurs through the build up of local plastic strain in PSBs, which leads to the formation of extrusions and intrusions. Until now, no attempt has been made to account for the radiation effects on fatigue-crack initiation. It has been shown in this paper that the theoretical effects of irradiation on fatigue-crack initiation can be investigated by coupling Lin's quantitative crack-initiation model with an empirical correlation developed for estimating the critical shear stress as a function of fluence and temperature.

Conclusions

The results of this investigation on the effect of irradiation on crack initiation indicate that the radiation-produced crystalline defects will retard the initiation process for high-cycle fatigue cracks. When the irradiation-induced increase in critical shear stress is accounted for, a decrease in local plastic strain rate in and around PSBs is observed. This decrease in local plastic strain rate is interpreted as requiring more cycles to develop extrusions and intrusions on metal surfaces. Since extrusions and intrusions are mechanisms leading to the formation of high-cycle fatigue cracks, it is concluded that the fatigue life may be enhanced in irradiated metals through the degradation of the initiation process. Experimental evidence agrees well with these predictions [*32,33*].

If an excessive amount of radiation damage is produced in a metal before a fatigue crack develops, the critical shear stress may become larger than the local resolved shear stress. When these circumstances arise, the excessive shear stress, which drives the initiation processes, is zero and crack initiation will not occur unless the applied shear stress is increased. This behavior occurred in the present theoretical approach for a reactor burn time greater than 1000 s at 450°C. It also occurred for a reactor burn time of 100 s when the temperature was below 300°C. This phenomenon is physically explained by an experimentally observed increase in the fatigue limit of a metal [*32,33*].

At low fluences, a decrease in irradiation temperature causes an increase in the number of crystalline defects frozen in the lattice. The increase in defects is directly related to an increase in the critical shearing stress which leads to a decrease in the local plastic strain rate in PSBs. Thus, decreasing the temperature retards the initiation process through enhanced irradiation effects. If the temperature is fixed and the reactor burn time is increased, the amount of irradiation damage sustained every cycle is increased. Therefore, increases in burn time enhance the fatigue life of a metal through larger induced radiation hardening.

Acknowledgment

This work was suported by the U.S. Department of Energy, Office of Fusion Energy, Grant DE-FG03-84ER52110, with UCLA. This work has been dedicated in memory of Christopher T. Winstead.

References

[*1*] Watson, R. D., "The Impact of Inelastic Deformation, Radiation Effects, and Fatigue Damage on Fusion Reactor First Wall Lifetimes," Ph.D. thesis, University of Wisconsin Report UWFDM-60, 1981.
[*2*] Abdou, M. A et al., "Finesse, Interim Report," UCLA Report UCLA-ENG-84-30, 1984, pp. 2–46.
[*3*] Lin, T. H., *Reviews on the Deformation Behavior of Materials,* Freund Publishing House, 1977, pp. 263–316.
[*4*] Forsyth, P. J. E., *Nature (London),* Vol. 171, 1953, p. 172.
[*5*] Forsyth, P. J. E. and Stubbington, C. A., *Journal of the Institute of Meals,* Vol. 83, 1954–55, p. 395.
[*6*] Hull, D., *Journal of the Institute of Metals,* Vol. 85, 1955–56, p. 527.
[*7*] Cottrell, A. H. and Hull, D., *Proceedings of the Royal Society of London, Series A,* Vol. 242, 1957, p. 211.
[*8*] Hull, D., *Journal of the Institute of Metals,* Vol. 86, 1957–58, p. 425.
[*9*] Grosskreutz, J. C., *Fatigue, An Interdisciplinary Approach,* Syracuse University Press, 1964, p. 27.
[*10*] Kennedy, A. J., *Processes of Creep and Fatigue in Metals,* Wiley, New York, 1963, pp. 331–341.
[*11*] Lin, T. H. and Ito, Y. M., *Journal of the Mechanics and Physics of Solids,* Vol. 17, Dec. 1969, pp. 511–523.
[*12*] Taylor, G. I. and Elam, C. F., *Proceedings of the Royal Society of London, Series A,* Vol. 102, 1923, p. 647.
[*13*] Schmit, E. and Boas, W., *Plasticity of Metals,* Hughes, London, 1950.
[*14*] Naughton, T. D., "Radiation Effects on the Micromechanical Aspects of Fatigue Crack Initiation," Masters thesis, University of California, Los Angeles, CA, Jan. 1986.
[*15*] Katagiri, K., Onura, A., Koyanagi, K., Awatani, J., Shiraishi, T., and Kaneshiro, H. et al., *Metallurgical Transactions,* Vol. 8A, 1977, p. 1769.
[*16*] Jin, N. Y. and Winter, A. T., *Acta Metallurgica,* Vol. 32, No. 7, 1984, pp. 989–995.
[*17*] Jin, N. Y. and Winter, A. T., *Acta Metallurgica,* No. 8, 1984, pp. 1173–1176.
[*18*] Duhamel, J. M., *Cours de Mecanique,* Mallet-Bachelier, Paris, 1862.
[*19*] Melan, E., *Zietschrift fuer Angewandte Mathematik and Mechanik,* Vol. 12, 1932, pp. 343–346. Correction in Vol. 20, Dec. 1940, p. 368.
[*20*] Tung, T. K. and Lin, T. H., *Journal of Applied Mechanics,* Vol. 29, 1966, pp. 363–370.

[21] Lin, S. R., "Effect of Secondary Slip Systems on Early Fatigue Damage," Ph.D. thesis, University of California, Los Angeles, CA, 1971.
[22] Brager, H. R. and Straalsund, J. L., *Journal of Nuclear Materials,* Vol. 46, 1973, p. 134.
[23] Seeger, A., *Proceedings of the 2nd United Nations International Conference on the Peaceful Uses of Nuclear Energy,* Vol. 6, Geneva, Switzerland, 1958, p. 250.
[24] Makin, M. J. and Minter, P. J., *Acta Metallurgica,* Vol. 8, 1960, p. 691.
[25] Kroupa, F. and Hirsch, P. B., *Discussions of the Faraday Society,* Vol. 38, 1964, p. 49.
[26] Forman, A. J., *Philosophy Magazine,* Vol. 17, 1968, p. 353.
[27] Olander, D. R., *Fundamental Aspects of Nuclear Reactor Fuel Elements,* NTIS TID-26711-P1, National Technical Information Service, Springfield, VA, 1976.
[28] Holmes, J. J., Robbins, R. E., Brimhall, J. L., and Mastel, B. et al., *Acta Metallurgica,* Vol. 16, 1968, pp. 955–967.
[29] "Plastic Strain in Metals," *The Scientific Papers of G. I. Tayor,* Mechanics of Solids, Vol. 1, Ed., G. K. Batchelor, Cambridge University Press, 1958, p. 437.
[30] Garber, R. I. and Mogilnikova, T. T., *Mechanics of Internal Friction in Hard Bodies,* Nauka, Moscow, 1976, p. 85.
[31] Michel, D. J., and Korth, G. E., *Proceedings of the Conference on Radiation Effects in Breeder Structural Materials,* Scottsdale, AZ, 1977, p. 117.
[32] Murty, K. L. and Holland, J. R., *Nuclear Technology,* Vol. 58, No. 3, 1982, p. 530.
[33] Beeston, J. M. and Brinkman, C. R., *Irradiation Effects on Structural Alloys for Nuclear Reactor Applications, STP 484,* American Society for Testing and Materials, Philadelphia, 1970, pp. 419–450.
[34] Speidel, M. O., Proceedings of the Symposium on High Temperature Materials in Gas Turbines, Baden, Switzerland, 1974, pp. 207–251.

Alvin J. Jacobs[1]

Hydrogen Buildup in Irradiated Type-304 Stainless Steel

REFERENCE: Jacobs, A. J., "**Hydrogen Buildup in Irradiated Type-304 Stainless Steel,**" *Influence of Radiation on Material Properties: 13th International Symposium (Part II), ASTM STP 956,* F. A. Garner, C. H. Henager, Jr., and N. Igata, Eds., American Society for Testing and Materials, Philadelphia, 1987, pp. 239–244.

ABSTRACT: Type-304 stainless steel (SS), irradiated to a range of fast ($E > 1$ MeV) neutron fluences (1.90×10^{18} to 9×10^{21} n/cm^2) in several different boiling water reactors, was found to contain ≥3 times as much hydrogen as unirradiated material. Hydrogen buildup in irradiated Type-304 SS was independent of fluence and flux in the range of fluences and fluxes investigated. This study was conducted as part of an investigation of irradiation-assisted stress corrosion cracking (IASCC). The accumulation of hydrogen in irradiated Type-304 SS may exacerbate the irradiation-induced loss of ductility, thereby increasing susceptibility to IASCC, or it may contribute to IASCC by other means.

KEY WORDS: austenitic stainless steel, irradiation-assisted stress corrosion cracking, hydrogen solubility, hydrogen traps, hydrogen-assisted cracking

During the course of a transmission electron microscope (TEM) examination of two cracked in-core components of Type-304 stainless steel (SS) [*1*], an epsilon-type martensite phase was observed in the vicinity of some of the fracture surfaces (Fig. 1). It is well known that epsilon-martensite forms by low-temperature (< −100°C) deformation or cooling. Therefore, if this phase was to form between ambient and boiling water reactor (BWR) operating temperatures, as it did in the examined components, then it must have done so in response to some other stimulus. Cathodically produced hydrogen can provide such a stimulus [*2*]. Moreover, hydrogen can help an intergranular crack to initiate and grow without externally applied or residual stresses. Minkovitz and Eliezer [*2*] produced mainly intergranular cracks and some transgranular cracks by hydrogen charging a TEM sample. Epsilon-martensite was present in front of the crack tip and was observed to provide the easiest path for crack propagation.

The present study was motivated by the possibility that hydrogen-induced epsilon-martensite formation is associated with irradiation-assisted stress corrosion cracking (IASCC) found occasionally in certain in-core components. The objectives of this study were as follows: (1) to determine if annealed/irradiated Type-304 SS, which is susceptible to IASCC above a threshold fast neutron fluence, contains significantly more hydrogen than annealed/unirradiated Type-304 SS, and if it does, (2) to determine if there is a fluence and flux dependence of hydrogen content.

Procedure

Type-304 SS specimens were taken from irradiated hardware and analyzed for hydrogen content by vacuum hot extraction using an extraction temperature of 1200°C for 30 min. The irradiated

[1] Principal engineer, General Electric Co., 175 Curtner Ave., San Jose, CA 95125.

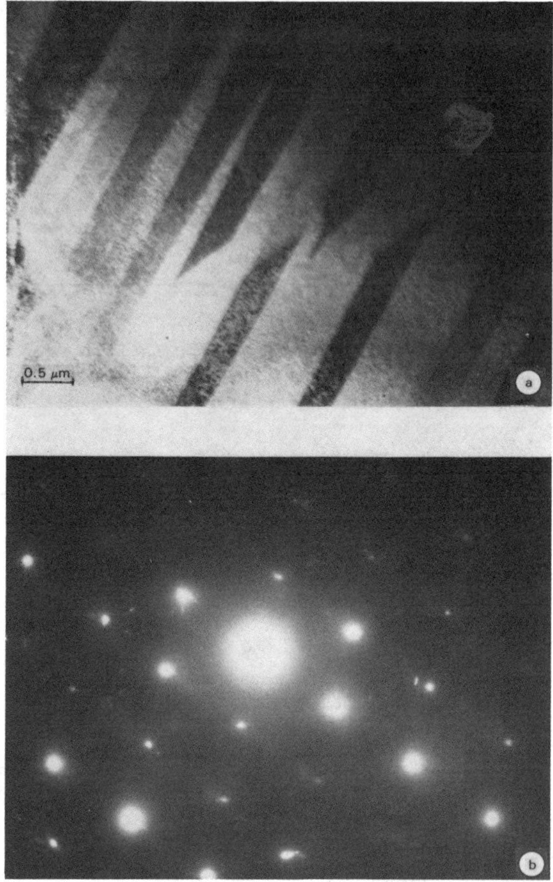

FIG. 1—(a) *Epsilon-type martensite platelets observed near the fracture surface of a cracked in-core component of Type-304 stainless steel.* (b) *Selected area diffraction pattern.*

specimens originated from five different BWRs, hereafter designated by letters "A" through "E." A wide range of fast ($E > 1$ MeV) fluences was covered, *viz.*, from 1.90×10^{18} n/cm^2 to 9×10^{21} n/cm^2. None of the archive material for the irradiated specimens was available, so the unirradiated specimens were taken from other heats.

Results

The hydrogen contents of the various irradiated specimens and unirradiated controls are summarized in Table 1 and plotted in Fig. 2 as a function of fluence. Table 1 shows that the hydrogen contents of the unirradiated specimens vary from 2 to 9 weight ppm, while those of the irradiated specimens range from 4 to 89 weight ppm. The mean hydrogen levels for the unirradiated and irradiated groups of specimens are 5.0 and 19.25 weight ppm, respectively. A two-sample t test performed on the data indicated that the difference between the two mean hydrogen levels was highly significant, *i.e.*, the significance level was less than 0.01% [*3*]. In other words, there is a virtual certainty that the average hydrogen content of irradiated Type-304 SS exceeds that of the unirradiated material.

TABLE 1—*Hydrogen contents of irradiated and unirradiated control specimens of Type-304 SS.*

Material Source	Heat or Specimen	Fluence, n/cm^2	Mass, g[a]	Hydrogen Content, weight ppm
ASEA-Atom	455561	unirradiated(U)	1.8213	4
Earl M. Jorgenson Co.	808228	U	0.5470	6
	808228-A	U	0.0224	9
	808228-B	U	0.0358	7
	808228-C	U	0.0288	7
	808228-D	U	0.0483	5
Coulter Steel Co.	812292	U	0.5945	3
Tube Sales Co.	27388	U	0.7078	2
	78500	U	0.6630	3
	TH6656	U	0.6208	5
Unknown	2P6396	U	0.5651	4
BWR E	B2	1×10^{18}	0.0657	4
BWR A	IV-13-A-F	1.90×10^{18}	0.0717	8
	III-13-A-H	2.33×10^{18}	0.0348	28
	III-13-A-I	2.86×10^{18}	0.0545	30
	III-13-A-J	3.38×10^{18}	0.0929	18
	III-13-A-K	6.64×10^{18}	0.0666	17
	III-13-A-L	9.94×10^{18}	0.0614	11
	IV-13-A-K	1.23×10^{19}	0.0731	6
	III-13-A-M	2.47×10^{19}	0.0759	8
	III-13-B-C	2.36×10^{20}	0.0822	12
	III-13-B-E	3.32×10^{20}	0.0602	34
	IV-13-B-H	5.00×10^{20}	0.2584	9
	III-13-B-G	5.04×10^{20}	0.1098	13
	III-16-B-10	7.03×10^{20}	0.1059	20
	III-13-B-K	9.50×10^{20}	0.1262	15
	III-13-C-B	1.28×10^{21}	0.0343	34
BWR D	12145	2.00×10^{21}	0.3209	14
BWR A	III-13-D-F	2.01×10^{21}	0.0885	89
	III-13-D-G	2.08×10^{21}	0.1531	22
	IV-13-D-H	2.30×10^{21}	0.1134	11
	III-13-D-L	2.77×10^{21}	0.0290	16
BWR E	T1	3×10^{21}	0.0542	23
BWR C	3B-4	4×10^{21}	0.1300	13
BWR B	A2	9×10^{21}	0.1090	7

[a] Uncertainties depend on mass:

Mass, mg	Uncertainty, %
20–30	±25
100	±10
>500	±3

A least squares analysis of the data in Fig. 2 yielded a straight line having the equation, $H = 0.374 \log F + 7.64$, where H is hydrogen content in weight ppm and F is fluence in n/cm^2. (The point $H = 89$ weight ppm at $F = 2.01 \times 10^{21}$ n/cm^2 was omitted from the analysis.) The slope and, hence, the fluence dependence of hydrogen content are nearly zero. To determine if the hydrogen level was dependent on flux, the data were considered for four particular BWR A

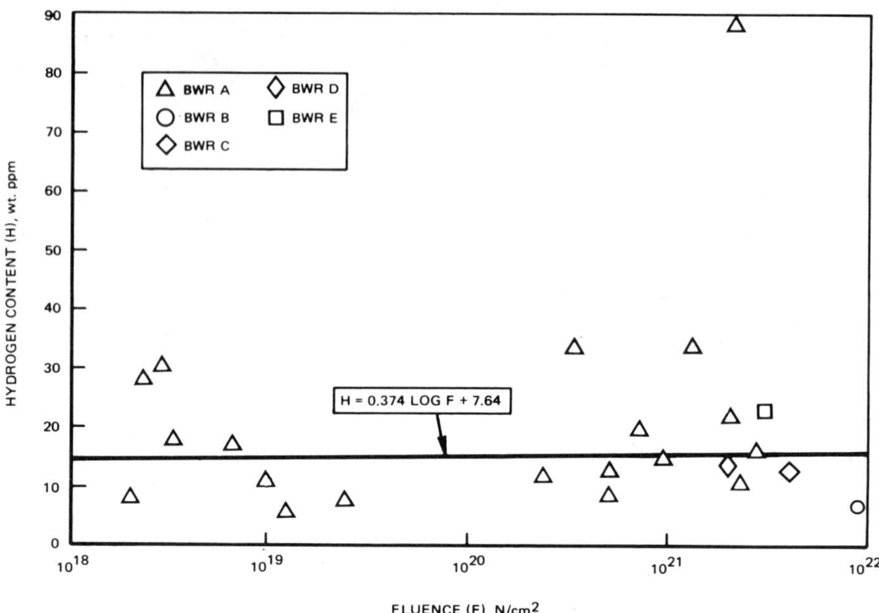

FIG. 2—*Hydrogen content of irradiated Type-304 SS as a function of fast (*E > *1 MeV) neutron fluence.*

components. Hydrogen analyses were available for two or more axial locations (fluxes) in each of these components. Hydrogen content is listed versus flux, *i.e.*, fluence divided by a constant time for a given component, in Table 2. The relationship between hydrogen content and flux is not seen to follow any consistent trend. Thus, hydrogen content is independent of both fast neutron fluence and flux.

Discussion

The most probable sources of hydrogen in an in-core component of Type-304 SS are radiolysis, corrosion processes, and (*n,p*) reactions with the nickel, iron, chromium, and nitrogen present in the stainless steel. These processes and, therefore, hydrogen production depend not only on fast neutron flux, but also on thermal neutron flux and gamma dose. The present investigation has ruled out the dependence of hydrogen buildup on fast neutron fluence and flux. Since thermal neutron flux and gamma dose are not necessarily proportional to fast neutron flux, hydrogen buildup may show a dependence on one or both of these other types of radiation. Unfortunately, insufficient data are available at the present time to correlate hydrogen content with thermal neutron flux and gamma dose.

The solubility of hydrogen in Type-304L SS increases with temperature and pressure according to the equation

$$C_o = 64.7\, p^{1/2} \exp\,(-16/2RT) \tag{1}$$

where C_o is the concentration just inside the alloy surface (mol/m³), 64.7 is the Sievert's law parameter (mol/m³ · $\sqrt{\text{MPa}}$), p is pressure (MPa), 16 is the molar heat of solution of molecular hydrogen (kJ/mol H_2), R is the gas constant (8.315 J/ K/mol), and T is the temperature (K) [*4*]. Using this equation and solubility data reported by Caskey for unirradiated Type-304L exposed to high pressure hydrogen [*4*], a solubility of 19 weight ppm hydrogen is found to correspond to a

TABLE 2—*Hydrogen contents of four irradiated BWR A components at different axial locations (fluxes).*

Component Specimen No.	Flux, n/cm²sec, = Fluence (n/cm²)/ Time, s[a]	Hydrogen Content, weight ppm
(1) III-13-A-H	$2.33 \times 10^{18}/t_1$	28
-I	$2.86 \times 10^{18}/t_1$	30
-J	$3.38 \times 10^{18}/t_1$	18
-K	$6.64 \times 10^{18}/t_1$	17
-L	$9.94 \times 10^{18}/t_1$	11
-M	$1.47 \times 10^{19}/t_1$	8
(2) III-13-B-C	$2.36 \times 10^{20}/t_2$	12
-E	$3.32 \times 10^{20}/t_2$	34
-G	$5.04 \times 10^{20}/t_2$	13
-K	$9.50 \times 10^{20}/t_2$	15
(3) III-13-D-F	$2.01 \times 10^{21}/t_3$	89
-G	$2.08 \times 10^{21}/t_3$	22
-L	$2.77 \times 10^{21}/t_3$	16
(4) IV-13-A-F	$1.9 \times 10^{18}/t_4$	8
-K	$1.23 \times 10^{19}/t_4$	6

[a] Time is constant for a given component: t_1, t_2, t_3, t_4.

pressure of 9.6 MPa (1397 psi) at 562 K (289°C). This "equivalent" pressure must be even higher, since not all of the hydrogen dissolved under BWR operating conditions could have been retained at ambient temperature and pressure.

The retention of hydrogen in irradiated Type-304 SS must be attributable, at least in part, to the presence of irradiation-induced traps in the material. Such traps may include irradiation defects, *e.g.*, defect clusters and dislocation loops; grain boundaries containing segregated solutes that attract hydrogen by virtue of chemical interactions, stress fields, etc.; and interfaces between matrix and precipitates (*e.g.*, carbides) or other transformation products such as martensite resulting from irradiation. Some traps may form in the in-core BWR environment independently of the intense irradiation field. Examples of these are surface oxides, twin boundaries, cracks, and intersections of slip lines with grain boundaries.

Over the past two decades, there has been a profusion of papers dealing with the subject of hydrogen-assisted cracking (HAC) in austenitic stainless steels. A number of reviews have appeared covering the role of hydrogen in the SCC of stainless steels [*5–9*], but nothing specifically relating to IASCC in austenitic stainless steels. Several investigations have yielded results which suggest that a combined irradiation/HAC mechanism may be operating in austenitic [*10*] and martensitic [*11*] stainless steels, and in carbon and low alloy steels [*12–14*]. Schreiber and Engell [*10*] found that at the rest potential, irradiated austenitic stainless steel samples were more susceptible to SCC than the unirradiated ones. However, under anodic polarization (reduced hydrogen production) the irradiated samples were more resistant. Gordon and Blood [*11*] obtained SCC failures in martensitic stainless steels after unexpectedly short in-core exposures. Such steels are known to fail by an HAC mechanism, but in this case, hydrogen and neutron radiation could have been interacting synergistically. Alekseenko and Nikolaev [*12*] found a larger ductility loss in a low alloy steel which had been irradiated in water than in the same steel irradiated in air. Brinkman and Beeston [*13*] were able to show that by irradiating several pressure vessel steels, the loss of notched strength and ductility due to hydrogen could be markedly increased. Takaku and Kayano [*14*] also showed a large ductility loss in an irradiated pressure vessel steel charged with hydrogen. One way in which hydrogen may contribute to IASCC is by enhancing the irradiation-induced loss of ductility. This possibility is discussed elsewhere [*15*]. The elongation of irradiated Type-304 SS has been observed to decrease to less than 3% above fluences of 1×10^{21} n/cm^2 [*16,17*]. Such low ductility can result in enhanced stress concentrations at the tip of a crack, thereby assisting the crack to grow.

Conclusions

1. Annealed Type-304 SS, which has been irradiated to fast ($E > 1$ MeV) neutron fluences in the range from 1.90×10^{18} to 9×10^{21} n/cm^2, contains significantly more ($\geq 3\times$) hydrogen than unirradiated material.

2. Hydrogen buildup in irradiated Type-304 SS is independent of fluence and flux in the range of fluences and fluxes investigated.

3. The accumulation of hydrogen in irradiated Type-304 SS may exacerbate the irradiation-induced loss of ductility, thereby increasing susceptibility to IASCC, or it may contribute to IASCC by other means.

References

[*1*] Bell, W. L., Unpublished results, 1983.

[*2*] Minkovitz, E. and Eliezer, D., *Scripta Metallurgica,* Vol. 16, 1982, p. 981.

[*3*] Wilson, S. A., "Two-Sample t Test of Hydrogen Data," Unpublished calculations, 27 August 1984.

[*4*] Caskey, G. R. and Sisson, R. D., *Scripta Metallurgica,* Vol. 15, 1981, pp. 1187–1190.

[*5*] Latanision, R. M. and Staehle, R. W., in *Fundamental Aspects of Stress Corrosion Cracking,* R. W. Staehle, A. J. Forty, and D. Van Rooyen, Ed., NACE, Houston, 1969, p. 214.

[*6*] Staehle, R. W., in *The Theory of Stress Corrosion Cracking in Alloys,* NATO Scientific Affairs Division, Brussels, 1971, pp. 231–243.

[*7*] Fidelle, J. P., Broudeur, R., Pellissier-Tanon, A., and Roux, C., in *Symposium Corrosion pour Tension et Fragilisation par l'Hydrogene,* Paris, April 1971.

[*8*] Thompson, A. W., in *Environment-Sensitive Fracture of Engineering Materials,* The Metallurgical Society of AIME, Warrendale, PA, 1979, p. 379.

[*9*] Gibala, R. and Hehemann, R. F., Eds., *Hydrogen Embrittlement and Stress Corrosion Cracking,* American Society for Metals, Metals Park, OH, 1984.

[*10*] Schreiber, F. and Engell, H. J., *Werkstoffe und Korrosion,* Vol. 23, No. 4, Apr. 1972, p. 255.

[*11*] Gordon, G. M. and Blood, R. E., in *Symposium on Materials Performance in Operating Nuclear Systems,* Ames, IA, August 1973.

[*12*] Alekseenko, N. N. and Nikolaev, V. A., *Atomnaya Energiya,* Vol. 34, No. 1, 1973, p. 40.

[*13*] Brinkman, C. R. and Beeston, J. M., "The Effects of Hydrogen on the Ductile Properties of Irradiated Pressure Vessel Steels," Idaho Nuclear Corp. Report IN-1359, National Reactor Testing Station, Idaho Falls, ID, 1970.

[*14*] Takaku, H. and Kayano, H., *Journal of Nuclear Materials,* Vol. 110, Nos. 2 and 3, October 1982, pp. 286–295.

[*15*] Jacobs, A. J. and Wozadlo, G. P., in *International Conference on Nuclear Power Plant Aging, Availability Factor and Reliability Analysis,* American Society for Metals, Metals Park, OH, 1985.

[*16*] Irvin, J. E. and Bement, A. L., *Effects of Radiation on Structural Materials, STP 426,* American Society for Testing and Materials, Philadelphia, 1967, p. 278.

[*17*] Holmes, J. J. et al., *Irradiation Effects in Structural Alloys for Thermal and Fast Reactors, STP 457,* American Society for Testing and Materials, Philadelphia, 1970, p. 371.

Margaret L. Hamilton,[1] Fan-Hsiang Huang,[1] Walter J. S. Yang,[2] and Frank A. Garner[1]

Mechanical Properties and Fracture Behavior of 20% Cold-Worked 316 Stainless Steel Irradiated to Very High Neutron Exposures

REFERENCE: Hamilton, M. L., Huang, F.-H., Yang, W. J. S., and Garner, F. A., "**Mechanical Properties and Fracture Behavior of 20% Cold-Worked 316 Stainless Steel Irradiated to Very High Neutron Exposures,**" *Influence of Radiation in Material Properties: 13th International Symposium (Part II), ASTM STP 956,* F. A. Garner, C. H. Henager, Jr., and N. Igata, Eds., American Society for Testing and Materials, Philadelphia, 1987, pp. 245–270.

ABSTRACT: Stainless steels irradiated in EBR-II at temperatures in the range 380 to 500°C tend to exhibit saturation of mechanical properties at relatively low fluence levels. At fluences on the order of 13 to 15 × 10^{22} n/cm^2 ($E > 0.1$ MeV), however, there is a rapid increase in hardness just after the onset of void swelling. The hardness increase is not attributed just to the voids themselves but rather to an indirect effect of voids on nickel depletion in the alloy matrix. The strong dependence of stacking fault energy on nickel and chromium content and particularly on deformation temperature combine to promote extensive stress-induced formation of ε-martensite at room temperature. This leads to a very brittle failure mode characterized as quasi-cleavage. At higher deformation temperatures it leads to a failure mode designated as channel fracture.

KEY WORDS: 316 stainless steels, neutron irradiation, swelling, mechanical properties, fracture toughness, failure modes, ε-martensite, radiation-induced segregation, inverse Kirkendall effect, stacking fault energy

In earlier studies conducted in the Experimental Breeder Reactor EBR-II to moderate levels of neutron fluence it was shown that irradiation-induced changes in the tensile properties of American Iron and Steel Institute (AISI) 316 stainless steel and some other austenitic steels could be related to concurrent changes in the microstructure [*1–5*]. Since the various microstructural components tend to saturate in density with continued exposure, it was expected that the tensile properties would also reach saturation levels that would persist to relatively large neutron fluences. Data on a relatively low swelling heat of 316 stainless steel confirmed the validity of this assumption at neutron fluence levels approaching 13 × 10^{22} n/cm^2 ($E > 0.1$ MeV) [*6*]. Other related studies demonstrated that predictive correlations of post-irradiation fracture toughness could be developed from the more easily obtained tensile properties, presumably because the microstructural origins of radiation-induced toughness degradation should be identical to those responsible for changes in tensile properties [*7,8*].

[1] Manager of Mechanical Properties Section, senior engineer, and fellow scientist, respectively, Hanford Engineering Development Laboratory, Richland, WA 99352.
[2] Principal engineer, General Electric Company, Pleasanton, CA 94566.

TABLE 1—*Alloy composition, wt%.*

Alloy	Ni	Cr	Mo	Si	Mn	Ti	C	P	B	Zr	W	Fe
AISI 316[a]	13.5	17.5	2.5	0.6	1.75	. . .	0.05	0.002	0.0005	. . .	. . .	balance
LS-1[b]	13.0	16.0	1.8	0.9	1.0	0.10	0.05	0.001	. . .	0.05	0.05	balance
LS-2[b]	19.5	8.7	2.3	1.0	1.9	0.27	0.04	. . .	. . .	. . .	. . .	balance
D-18[b]	15.0	12.0	2.5	1.0	2.0	0.25	0.04	. . .	. . .	. . .	. . .	balance

[a] 1050°C/5 min/air cool + 20% cold work.
[b] 1150°C/15 min/air cool + 25% cold work.

When other heats of steel exhibiting an earlier onset of swelling were evaluated, it was obvious that significant changes in fracture mode were occurring at high neutron exposure [*9,10*]. These changes were most pronounced for deformation at low temperatures. At higher test temperatures the fracture mode resembled the channel fracture that was observed earlier both in annealed 304, 308, and 316 stainless steels [*11,12*] and to a limited extent in 20% cold-worked 316, all at fluences less than or equal to 10×10^{22} n/cm^2 ($E > 0.1$ MeV) [*13*]. The channels observed in these studies appeared as narrow bands in which the local deformation was often several 100% or higher, as evidenced by the elongation of voids within the band. This flow localization is thought to be a consequence of dislocation sweeping that significantly reduced the hardness within the channel.

The unanticipated change in behavior at higher fluence was first observed during disassembly and evaluation of a fueled experiment designated the P-53 test which had been irradiated in EBR-II. Both the fuel cladding and duct assembly were constructed from 20% cold-worked American Iron and Steel Institute (AISI) 316 stainless steel. When 1.9-cm-diameter circular specimens were punched for density measurements from the duct that housed the fuel assembly, several cracks developed between adjacent holes. Small pieces of this duct were also easily broken off while using a manipulator in the hot cell. In addition, the fuel pins, which had run at low power to very high fluences ($\sim 18 \times 10^{22}$ n/cm^2, $E > 0.1$ MeV), had survived irradiation with no breaches but sustained numerous breaches during handling of the pins in the hot cell.

It was therefore decided to conduct mechanical property tests, fractography, and microstructural analyses on the duct material over a range of test temperatures. Several 3.0-mm-diameter microscopy specimens of AISI 316 and three titanium-modified 316 stainless steels that were irradiated in other experiments were also broken at room temperature and their fracture behavior examined. Table 1 contains the alloy compositions and initial thermal-mechanical conditions.

Experimental Details

Specimen Preparation

The hexagonal P53 duct was constructed from the Fast Flux Test Facility, Richland, WA (FFTF) First Core steel with the individual flats measuring 3.2 cm in width and 1.02 mm in thickness. Tensile specimens were obtained from rectangular strips having their long dimension parallel to the axis of the duct. The obvious embrittlement of the duct made it imperative to utilize a specimen with the largest possible gage width, gage length, and filet radius. The dimensions previously utilized for flat tensile specimens in the U.S. Breeder Reactor or Fusion Materials programs were judged to be unacceptable. Additional constraints on specimen fabrication arose from the limited amount of material remaining between the hole punched in the duct and the necessity for remote fabrication. Conforming as much as possible to the guidance provided by ASTM Procedure E 8, a new tensile specimen was designed and is shown in Fig. 1.

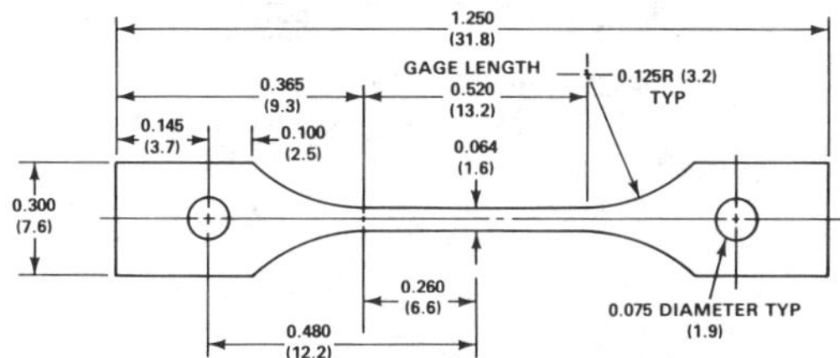

FIG. 1—*Tensile specimen dimensions.*

Most of the tensile blanks were obtained from regions adjacent to the corners of the duct and thus had the potential for being somewhat stronger than the central regions of the duct face. To check for this possibility, hardness measurements were made on sections of unirradiated duct, and it was found that the duct corners were worked only slightly more (3 to 6%) than were the duct flats. Even this small variation in hardness and dislocation density is expected to be relaxed during irradiation [*2,14*], and thus no significant variations in strength are expected to arise because of the location on the duct face from which the specimens were obtained.

Seven tension specimens from the duct and a number of unirradiated control specimens were machined remotely in the hot cell and photographed at a magnification of ~X2.8. The gage dimensions were determined from the photographs. A small amount of variability was observed between specimens but was judged to be insignificant with respect to the test results.

Compact tensile specimens were fabricated from the 1.9-cm diameter disks punched from the center of the duct faces. Density measurements were performed before fabrication of the specimens using an immersion technique. The configuration of the circular compact tension specimen is shown in Fig. 2. Since the strength of the irradiated material was extremely high, it was very difficult to cut a sharp notch in the specimen and to induce a crack by fatigue before testing (precrack). The first few specimens failed during precracking because of the lack of prior knowledge of the material's fracture properties. Successful precracking required careful optical examination of crack extension and careful monitoring of cyclic load reduction.

Fractographic and microstructural analyses were performed on the broken tensile and compact tension specimens as well as on irradiated microscopy disks that were fractured manually. Since the radioactivity of the original tensile and compact tension specimens was substantial, fractographic specimens of smaller volume were prepared by slicing off a thin section containing the fracture surface. The fractographic analysis of microscopy disks was preceded by submerging them in pure ethyl alcohol and breaking them remotely by bending with pliers. The fracture surfaces were examined in a JSM 35C scanning electron microscope. Microscopy foils were prepared from tensile specimens by slicing perpendicularly to the tensile axis. Microstructural studies were performed in either a JEM 100CX or a JEM 1200EX transmission electron microscope.

Test Procedures

Tensile tests were conducted in air at three different temperatures: room temperature, the fuel handling temperature of 205°C, and the irradiation temperature, which varied from 385 to 488°C.

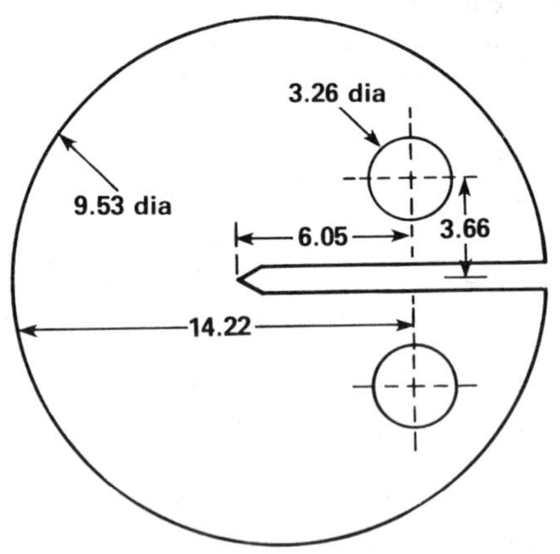

FIG. 2—*Compact tension specimen dimensions.*

All tests were performed at a strain rate of 6.4×10^{-4} s^{-1}. Specimens were tested in a shielded testing facility using the Modular Test Unit (MTU). The MTU is a hydraulically actuated test frame designed to remotely perform tensile, fatigue, and compression type mechanical testing. Tensile specimens were carefully positioned in the load train. A thermocouple in contact with the specimen was used to monitor its temperature. To prevent undesirable specimen straining during heat up to the test temperature, the load was continuously monitored and the position of the actuator was adjusted to maintain loads well below the load required for yielding. Load versus deflection data were obtained using a load cell in series with the load train and a linear variable differential transformer extensometer attached to the load train actuator. Yield strength was determined by the 0.2% offset method.

Details of the compact tension test procedure are described in Ref *13* and *15*. Fracture toughness tests were conducted in air at 32, 205, and at the irradiation temperature (395 to 425°C). The specimen was unloaded after reaching the maximum load, heat tinted at 500°C for about 1 h, and broken at room temperature to allow optical measurement of the corresponding crack extension. The results were analyzed by a *J*-integral approach. A list of the specimens tested is included in Table 2. Note that void swelling in these specimens ranged from 3.4 to 8%.

Test Results

The tensile test results are given in Table 3. Since there were small gradients in neutron flux and temperature along the specimen axis, the irradiation parameters listed were determined for the specimen midpoint. Fluences were based on HIST computer code calculations, while irradiation temperatures were interpolated between the temperature estimated for the bottom of the fuel column (370°C) and the beginning-of-life subchannel outlet temperature adjacent to the relevant duct face. The temperatures of the duct faces at the top of the core were assumed to be the same as the subchannel outlet temperatures, which were calculated with the COBRA computer code. COBRA

TABLE 2—*Density specimens used for compact tension tests.*

Specimen Identity	Fluence 10^{22} n/cm^2 $E > 0.1$ MeV	Irradiation Temperature, °C	Swelling, % $\Delta V/V$
F1C	14.0	395	4.6
F2C	14.2	395	4.3
F3C	14.9	395	4.6
F4C	15.5	395	4.7
F5C	15.3	395	ND
F6C	14.5	395	3.4
F1D	15.3	410	5.6
F2D	15.4	410	5.4
F3D	16.2	410	5.5
F4D	16.9	410	6.0
F5D	16.7	410	4.9
F6D	15.9	410	4.6
F1E	15.9	425	6.7
F2E	16.0	425	7.3
F3E	16.8	425	7.1
F4E	17.5	425	8.0
F5E	17.4	425	6.2
F6E	16.4	425	5.9

NOTE: ND = determined.

TABLE 3—*Tensile data on P53 specimens.*[a]

Specimen	Irradiation Exposure: Temperature, °C	Irradiation Exposure: Fluence, 10^{22} n/cm^2	Test Temperature, °C	Yield Strength, MPa	Ultimate Tensile Strength, MPa	Uniform Elongation, %	Total Elongation, %
4a	460	15.5	20	1293	1432	1.9*	1.9
5	474	14.5	20	1238	1505	6.5*	6.5
7	487	11.3	20	1140	1393	4.6*	4.6
3	442	16.2	205	1350	1492	4.6	6.3
2	385	12.8	205	1627	1703	1.8	5.2
4	460	15.5	460	1127	1223	5.6	7.2
6	488	12.8	488	917	1067	5.8	7.7

[a] $\varepsilon = 6.4 \times 10^{-4}\ s^{-1}$.

[b] Failure occurred at the maximum load.

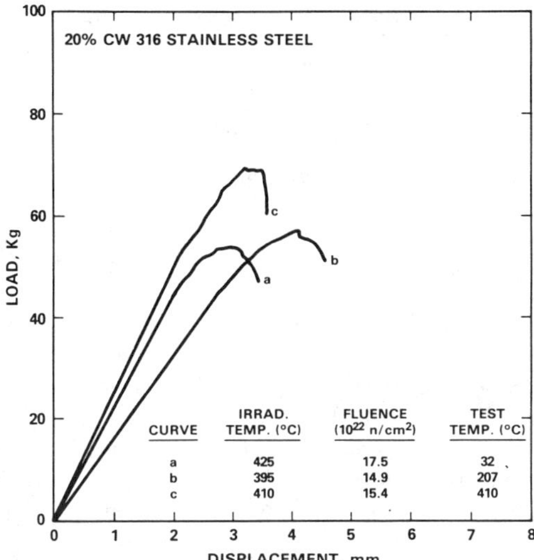

FIG. 3—*Load versus displacement curves for specimens tested at elevated temperatures.*

is a standard thermal-hydraulic code in wide use in the nuclear industry. HIST is a fuel performance predictive code developed by Westinghouse Hanford Company and is a version of the SIEX code.

Typical load-displacement curves of the compact tension specimens are shown in Fig. 3. The material displayed little plasticity before reaching the maximum load, at which point a sudden load-drop was observed for all specimens. Since well-defined crack extension was not observed, the maximum load was taken as the onset point of crack extension, where the critical value of J, J_{Ic}, was calculated.

The compact tension test results are given in Table 4. Only eleven valid tests were performed. Several specimens failed during precracking, and several others were tested before a fatigue crack was produced while precracking methods for these specimens were still being established.

TABLE 4—*Fracture toughness of P53 duct specimens (loading rate = 2.1×10^{-3} mm/s).*

Test Temperature, °C	Irradiation Temperature, °C	Fluence, 10^{22} n/cm^2 $E > 0.1$ MeV	J_{Ic}, kJ/m^2	K_c, MPa · m$^{1/2}$
207	395	14.9	38.6	88.8
395	395	15.5	42.9	90.2
32	410	15.3	35.6	87.0
30	410	16.2	39.7	91.0
205	410	16.9	34.0	83.8[a]
205	410	16.7	46.5	97.5
410	410	15.4	34.0	80.1
32	425	17.5	24.3	71.9
206	425	16.4	37.8	87.8
205	425	16.8	27.4	74.8
425	425	17.4	29.6	74.4

[a] Loading rate = 2.1 mm/s.

NOTE: For the 11 tests: average K_c = 84.3 MPa · m$^{1/2}$; standard deviation = 8.2 MPa · m$^{1/2}$; and range = 71.9 to 91.9 MPa · m$^{1/2}$.

The plane strain fracture toughness K_c can be estimated from the values of J_{Ic} through the following equation

$$K_c = \left[\frac{E J_{Ic}}{1 - \nu^2}\right]^{1/2} \tag{1}$$

where ν is Poisson's ratio and E is Young's modulus. The calculated values of K_c are also listed in Table 4.

Discussion of Test Results

Examination of Figs. 4 through 6 shows that the post-irradiation strength of the P53 First Core duct was significantly higher than was observed for this steel at lower fluences. This was an unexpected result, since the strength and ductility of 20% cold-worked 316 steel were both believed to saturate with fluence at about 5×10^{22} n/cm² ($E > 0.1$ MeV). As shown in Figures 7 and 8 the ductility of this steel depends not only on the fluence and irradiation temperature but also on the test temperature, particularly at lower test temperatures. With one exception (Specimen 4a tested at room temperature) the P53 specimens exhibited ductilities greater than 4.5%. The ductilities found in the P53 tests conducted at room temperature, however, fell substantially under the trend line established at lower fluence.

There was another very pronounced difference in the behavior of First Core steel with respect to the test temperature. As shown in Fig. 9 the fracture surfaces of the specimens tested at room temperature were oriented 90° to the tensile axis, suggesting a brittle mode of failure, while the fracture surfaces of specimens tested at higher temperatures were inclined at angles of 30 to 60° to the tensile axis, a behavior typical of plastic deformation in a ductile material. Faceted regions were also evident on the surface of specimens tested at the higher temperatures.

It was observed after the tensile tests were completed that all seven specimens were strongly attracted to a magnet. Examination of other pieces of duct material showed a variable degree of

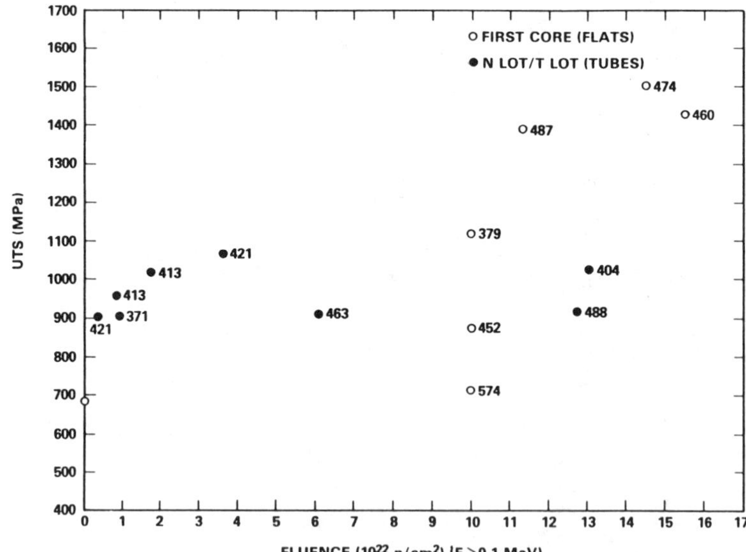

FIG. 4—*Ultimate strength of First Core and N-lot AISI 316 steels tested at room temperature. The irradiation temperatures are shown beside each datum. The N-lot heat does not begin to swell as soon as First Core and does not exhibit an increase in ultimate strength.*

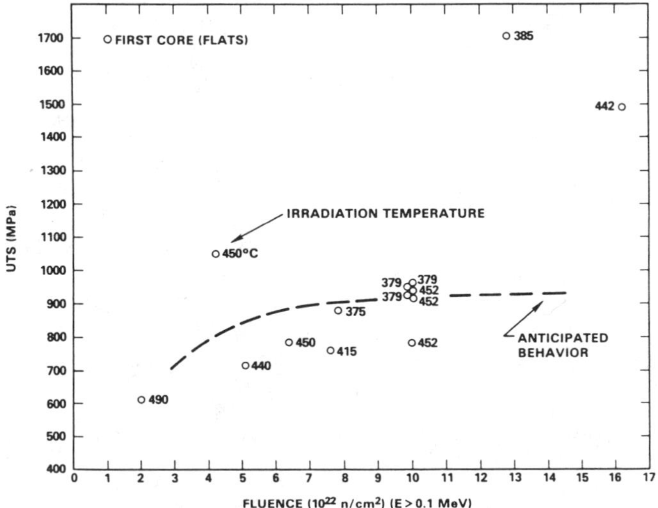

FIG. 5—*Ultimate strength of First Core steel tested at 205 to 232°C. Irradiation temperatures are shown beside each datum.*

magnetism, which was particularly strong in areas deformed during punching. It appears that while the duct specimens may have exhibited some degree of magnetization before fabrication, they were strongly magnetized along their entire length during tensile testing. This suggested that a martensitic transformation had occurred during deformation.

Figures 10 and 11 show the temperature dependence of J_{Ic} and K_c for various irradiation conditions. Previous results [*13*] for another duct designated P14A and irradiated to lower fluence are also plotted for comparison. The dependence of fracture toughness on irradiation temperature

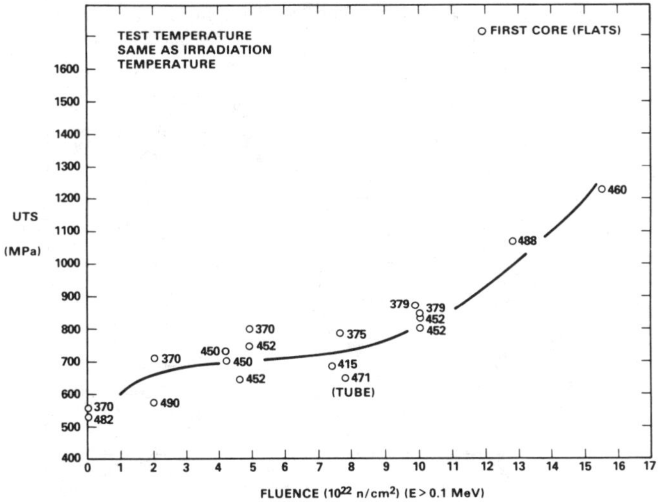

FIG. 6—*Ultimate strength of First Core steel tested at the irradiation temperature. Irradiation temperatures are shown beside each datum.*

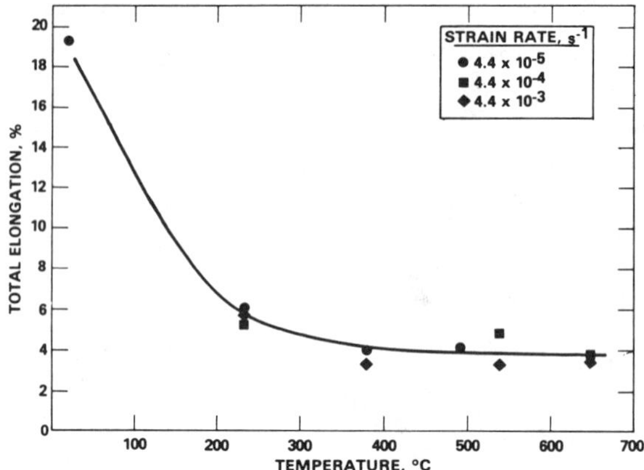

FIG. 7—*Ductility of 20% cold-worked 316 (First Core) irradiated to 1.3 × 10^{23} n/cm² (E > 0.1 MeV) at 380°C.*

is shown in Fig. 12 where the values of J_{Ic} are plotted for specimens with the same test and fluence conditions. Earlier studies have shown that the fracture toughness of 20% CW 316 decreases by a factor of 2 after irradiation to a fluence of 11 × 10^{22} n/cm² ($E > 0.1$ MeV). No further degradation was observed in this study with increasing fluence up to 17.5 × 10^{22} n/cm². Figure 13 shows the absence of any strong effect of fluence on the fracture toughness of the P53 duct material.

The most significant change in fracture behavior following high fluence exposure of 316 is the drastic reduction in resistance to crack propagation. The tearing modulus of the P53 specimens tested at elevated temperatures was estimated to be about 2, compared with a value of 15 observed for the material irradiated to a fluence of 11 × 10^{22} n/cm² [*13*]. For P53 tests conducted at room temperature the tearing modulus approached zero.

As shown in Fig. 10, test temperature did not have a strong effect on the fracture toughness for irradiation temperatures up to 425°C. The toughness decreased somewhat with increasing irradiation temperature in the range of 395 to 425°C, but the trend for higher irradiation temperatures is not known.

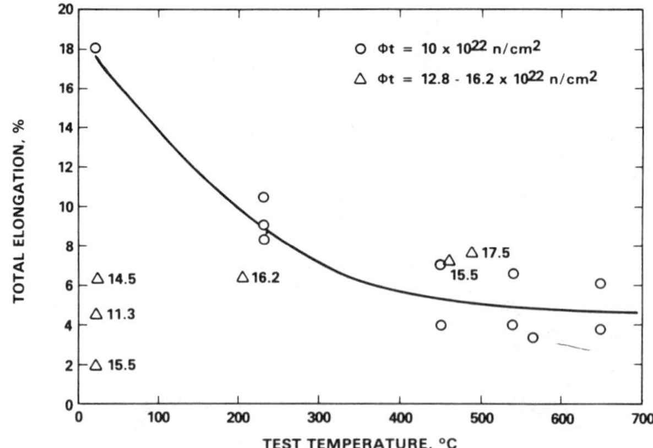

FIG. 8—*Ductility of 20% cold-worked 316 (First Core) irradiated at 450 to 490°C. Neutron fluences are given beside each P53 datum.*

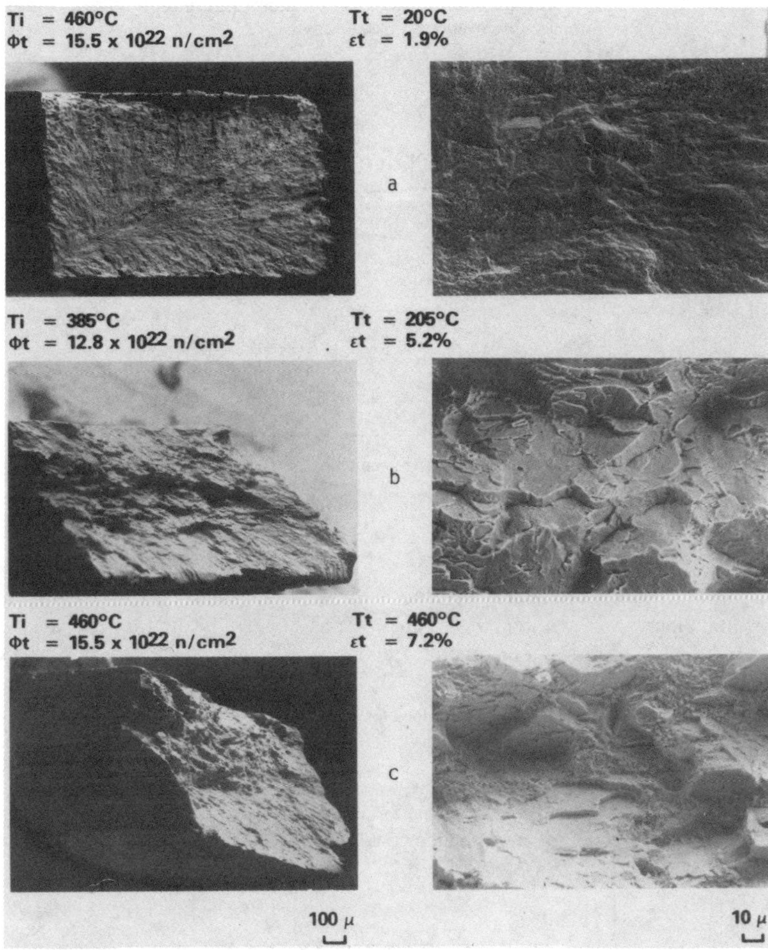

FIG. 9—*Fractography of tensile specimens from P53:* (a) *brittle fracture at room temperature (Specimen 4a);* (b) *channel fracture at 205°C (Specimen 2); and* (c) *channel fracture at 460°C (Specimen 4). The total elongation and irradiation test conditions are shown for each specimen.*

Fractographic and Microstructural Analyses

Tensile Specimens

Specimens 4a and 5 from the P53 duct, irradiated to 16×10^{22} n/cm^2 ($E > 0.1$ MeV), fractured transgranularly at room temperature, exhibiting features on the fracture surface similar to chevron patterns. These specimens had elongations of 1.9 and 6.5%, respectively. As shown in Fig. 14, the fracture was initiated at one corner of Specimen 4a, presumably from a pre-existing flaw, and propagated across the specimen diagonally. Small regions having a typical shear fracture morphology appeared on the edges of the specimen at the end of the crack path. The fracture surface was flat and perpendicular to the tensile axis. Faint river patterns were visible along the diagonal crack path. High magnification pictures of local regions revealed a high density of cavities with a definite crystallographic shape. The size and density of these cavities correspond well with that of the void

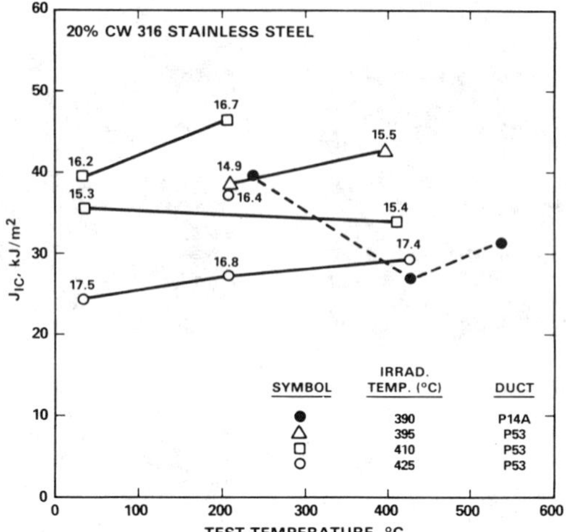

FIG. 10—*Effect of test temperature on* J_{Ic} *of irradiated 20% cold-worked Type 316 stainless steel.*

swelling found in the matrix during examination of a TEM specimen cut from this tensile specimen. The cavities in the fracture surface are therefore believed to be voids. The crystallographic shape of the voids remains intact without any sign of smearing, indicating brittle behavior in the surrounding matrix. The room temperature fracture mode was therefore designated "quasi-cleavage."

The microstructure of the specimen prepared from the section adjacent to the fracture surface exhibited little evidence of plastic deformation. There were no obvious deformation channels in

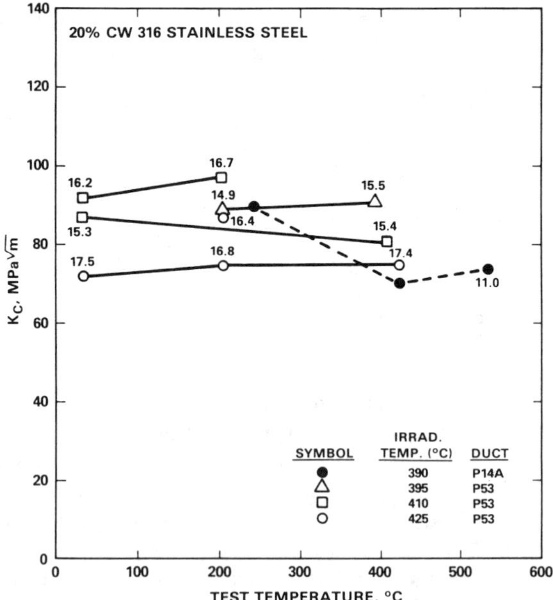

FIG. 11—*Effect of test temperature on* K_c *of irradiated 20% cold-worked Type 316 stainless steel.*

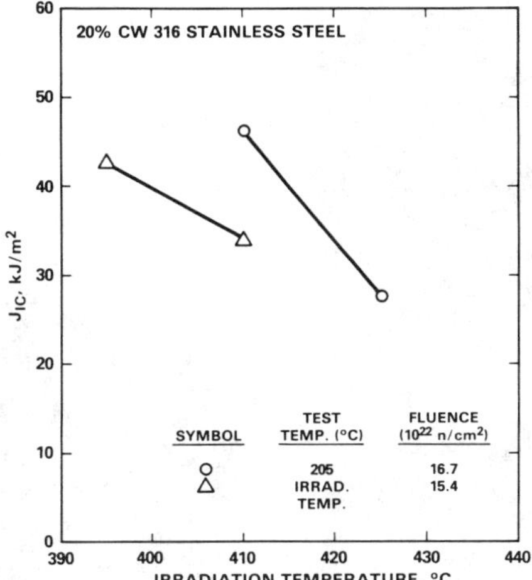

FIG. 12—*Effect of irradiation temperature on the fracture toughness of irradiated 20% cold-worked Type 316 stainless steel.*

this specimen. Occasionally, however, fine slip bands were found, as illustrated in the bright field micrograph shown in Fig. 15*a*. In dark field, however, a very high density of thin stacking fault plates or precipitates was observed throughout the specimen, as can be seen in Fig. 15*b*. Note that only one-fourth of these faults are imaged under the conditions employed to produce this micrograph. The streaks shown in the selected area diffraction pattern (inset, lower left) are without intensity maxima, indicating that these microstructural features are on the order of monolayers in thickness.

Two other P53 tensile specimens, designated Numbers 2 and 4, tested at 205 and 460°C, respectively, fractured transgranularly by a shear mode. Smeared voids resulting from dislocation shearing were revealed at high magnifications of the facet surfaces, providing direct evidence of channel fracture. Figure 16 shows examples of the facets and smeared voids found on the fracture surface and also shows sheared voids found in deformation bands below the fracture surface.

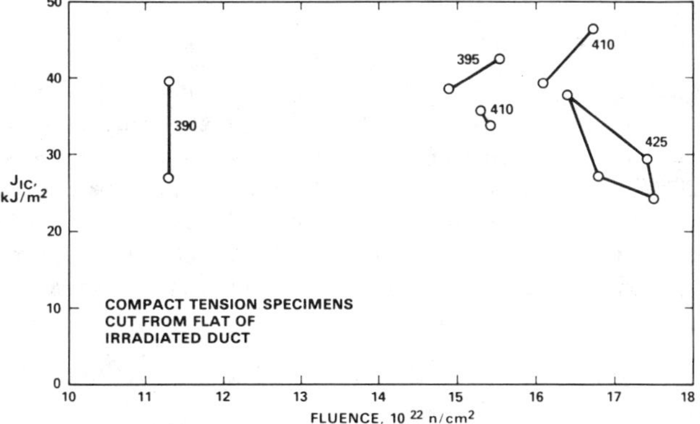

FIG. 13—*Effect of neutron fluence on the fracture toughness of specimens obtained from the P53 duct. Connected data points are obtained from specimens irradiated at the temperature shown (°C) but tested at different temperatures.*

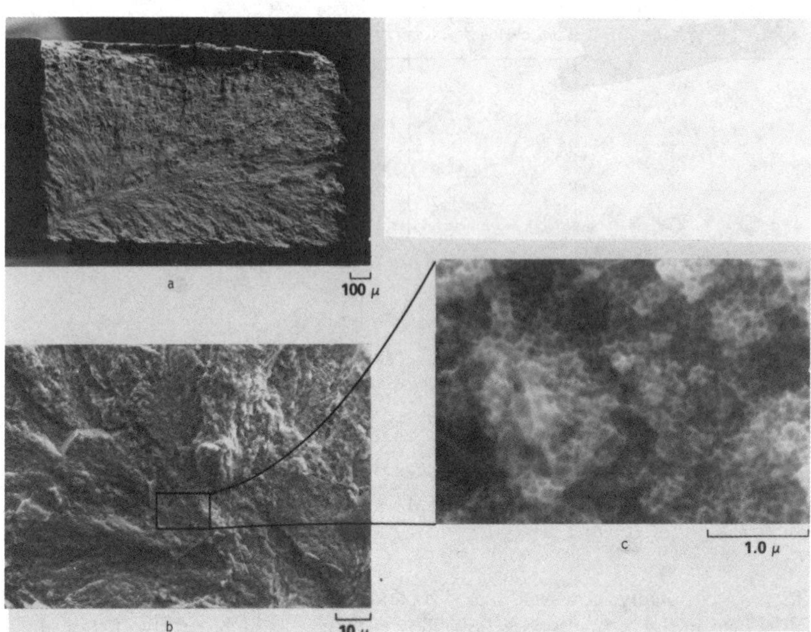

FIG. 14—(a *and* b) *Quasi-cleavage fracture of P53 tensile Specimen 4a and* (c) *voids with their crystallographic shape intact in a high-magnification picture of a local region.*

Burgers vector analyses were performed for the three deformed specimens shown in Fig. 9. No slip systems were detected other than the {111} <110>, the {111} <112> partials, and the {111} <112> twins normally operating in fcc systems. Analysis of the various precipitates (M_6C, $M_{23}C_6$, γ', G-phase, and so forth) were conducted for a variety of specimens from the high fluence P53 test and a lower fluence test designated P14. No dramatic changes were observed in precipitation behavior with increased fluence. Thus the large increase in hardening observed at high fluence cannot be attributed either to irradiation-induced precipitation or changes in dislocation substructure.

Compact Tension Specimens

Examination of the fracture surfaces of the compact tension specimens from the P53 duct revealed that the precrack induced by room temperature fatigue exhibited cleavage-type growth. The fracture surface of this specimen exhibited quasi-cleavage fracture when tested at room temperature and 205°C. The specimen tested at the irradiation temperature, however, showed channel fracture. Fractographs of the precracked region and the tested zone are given in Figs. 17, 18, and 19 for specimens tested at room temperature, 205 and 410°C, respectively. Since the fracture surfaces were oxidized during heat tinting, high magnification examination of the channeling patterns on the fracture surface was not conducted.

Microscopy Disks

Table 5 summarizes the failure modes observed in manually deformed microscopy disks for a variety of titanium-modified stainless steels irradiated to high fluence. Since First Core steel was used as a reference for the evaluation of room temperature embrittlement, a few additional

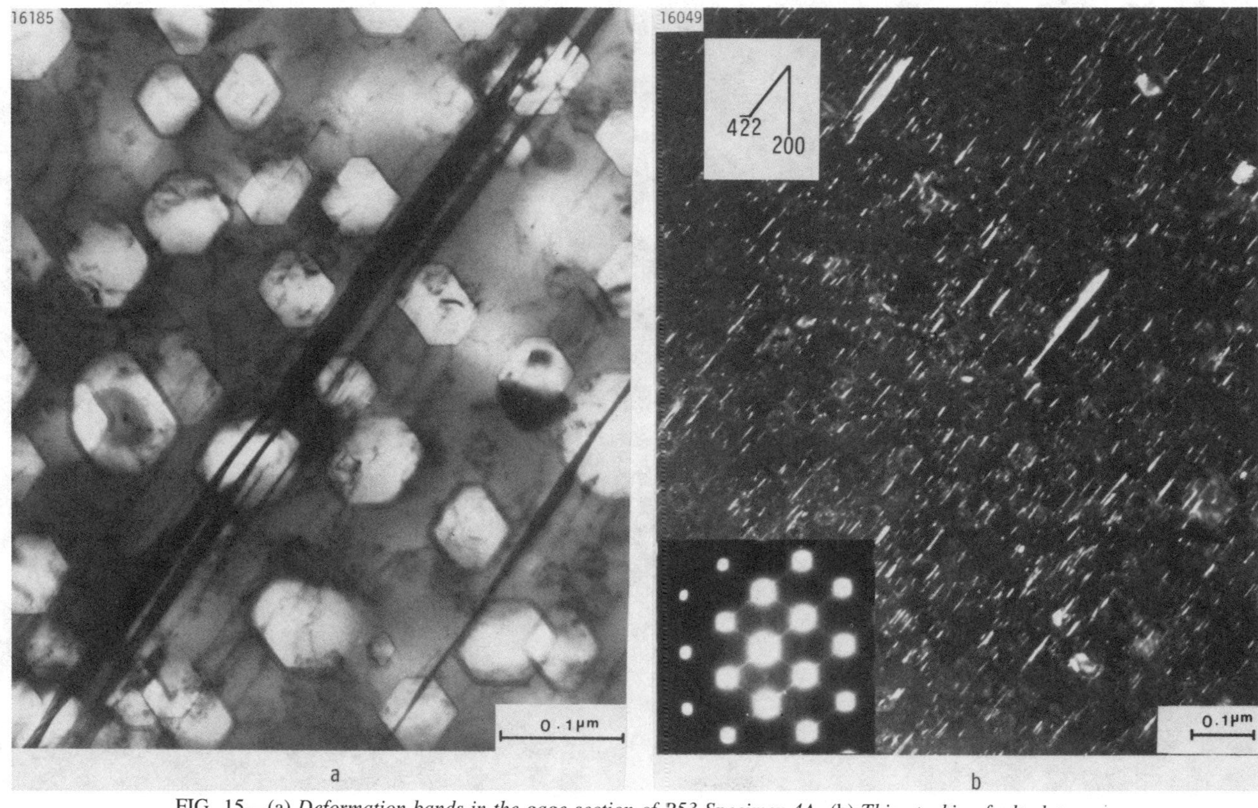

FIG. 15—(a) *Deformation bands in the gage section of P53 Specimen 4A.* (b) *Thin stacking fault plates on one set of [111] planes in P53 Specimen 4A. The inserted limited area diffraction pattern reveals streaks without maxima arising from the thin stacking fault plates.*

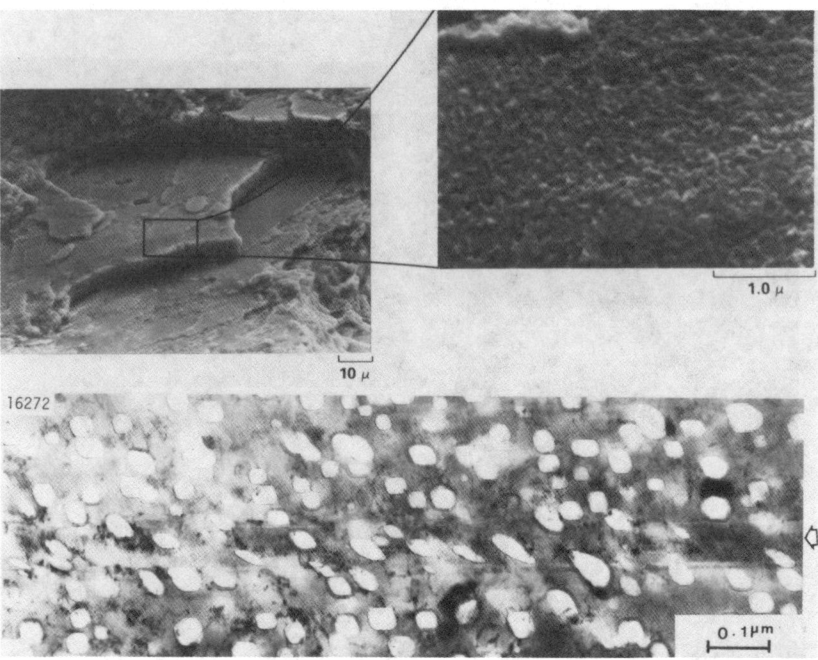

FIG. 16—*High magnification SEM pictures of the channel faceted surface and a TEM picture of deformation bands revealing sheared voids in P53 Specimen 4 tested at 460°C.*

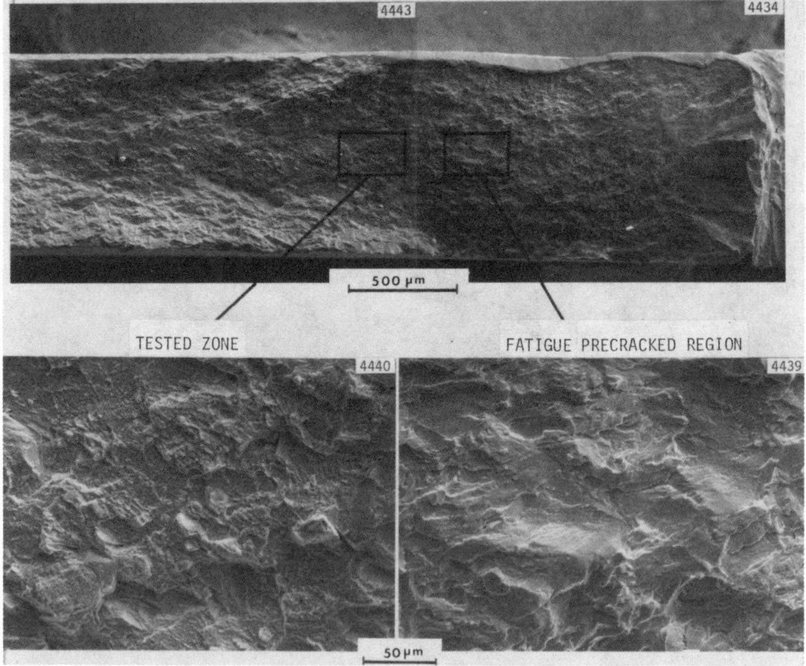

FIG. 17—*Cleavage fracture in fatigue precrack region and quasi-cleavage fracture in tested zone of P53 compact tension specimen tested at room temperature.*

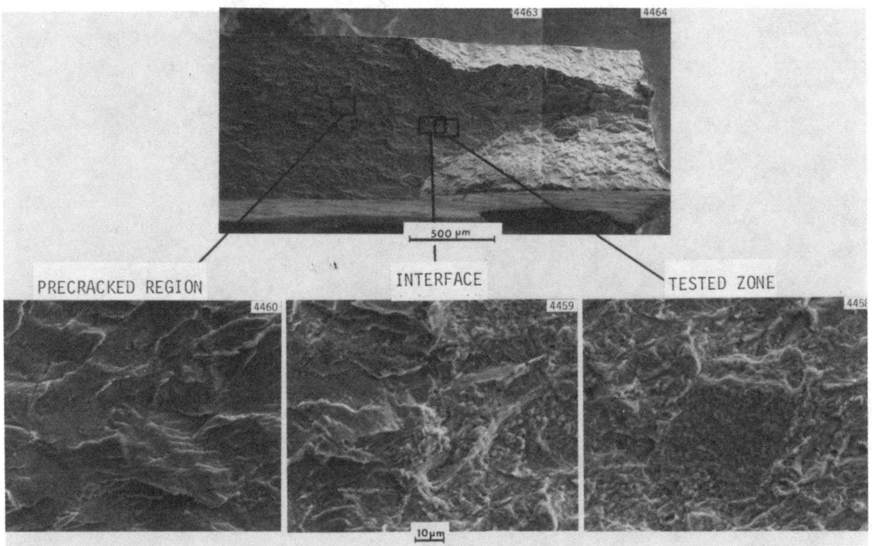

FIG. 18—*Cleavage fracture in fatigue precrack region and quasi-cleavage fracture in tested zone of a P53 compact tension specimen tested at 205°C.*

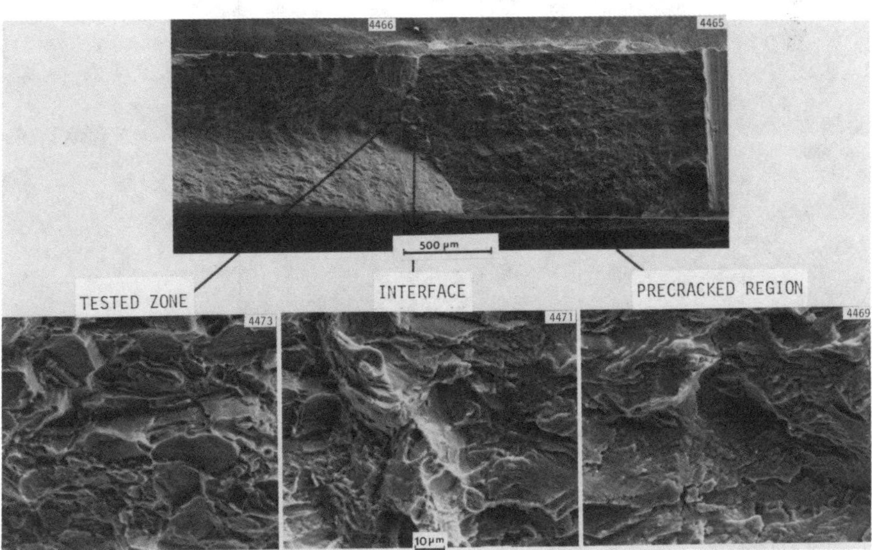

FIG. 19—*Cleavage fracture in fatigue precrack region and channel fracture in tested zone of a P53 compact tension specimen tested at 395°C.*

TABLE 5—*SEM fractography study of TEM disks fractured at room temperature.*

Alloy	Specimen	Irradiation Experiment	Fluence 10^{22} n/cm², $E > 0.1$ MeV	Irradiation Temperature, °C	Density Change, %	Fracture Mode
Core 1	C75	RS-1	10.3	370	0.41	transgranular ductile dimple
	46M7	AA-I	15.6	400	3.09	transgranular channel
Titanium-modified stainless steels	LS-1	AA-VII	15.1	400	1.32	transgranular channel
	LS-2	AA-VII	18.5	427	0.08	transgranular ductile dimple plus small amount of channeling
	D18	AA-VII	18. 5	427	2.08	transgranular fracture with poorly defined dimples and some channeling

specimens of First Core were also evaluated. These were an electropolished disk designated 46M7 from the AA1 experiment and an unpolished fragment of the P53 duct. The two specimens received a similar neutron fluence, 16×10^{22} n/cm^2 ($E > 0.1$ MeV), and both were irradiated at ~400°C. The measured void swelling was 3.5% for the 46M7 specimen and 3.9% for the P53 specimen.

The polished TEM disk, 46M7, was examined by microscopy and then purposely cracked by impact, that is, the fine point of a sharp pair of tweezers was dropped on the foil surface. Cracks opened fro n the perforated foil edge in a brittle zig-zag fashion as shown in Fig. 20. Each sawtooth of the crack edge was aligned with a twin plate in the matrix as demonstrated in Fig. 21. Careful examination of the thin region along the crack path revealed diffraction patterns corresponding to a bcc structure. The lattice parameter of the bcc structure deduced from the diffraction patterns was $a_0 = 0.29$ nm, which is close to the lattice parameter of α-ferrite (0.286 nm). Dark-field pictures using the bcc diffraction spots revealed a thin layer of bcc material along the crack path, 20 to 40 nm in width. Examples are given in Fig. 22. Finding a thin bcc layer along the crack path suggests that a stress-induced martensitic transformation ($\gamma \rightarrow \varepsilon \rightarrow \alpha$) may have occurred along the crack path. EDX analysis indicated that the matrix Ni content varied from 9 to 12 wt%, which is not low enough for the γ matrix to thermally transform into ferrite. Furthermore, no bcc phase was detected in the thin foil before it was damaged.

The same TEM disk was then broken in half by bending, and its fracture surface examined in the scanning electron microscope. This revealed that failure in thin regions of the foil, where the material is essentially a single crystal, was quasi-cleavage in nature, occurring along twin boundaries. The fracture mode changed to channel fracture immediately after the crack met a grain boundary on its way into a thicker region of the foil.

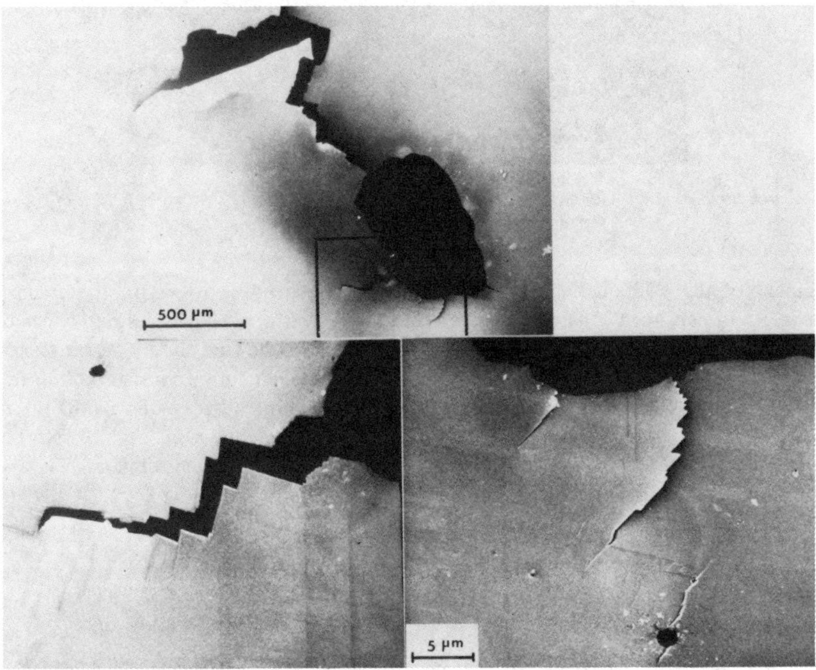

FIG. 20—*Brittle cracks produced at room temperature in a prethinned first core TEM disk designated 46M7.*

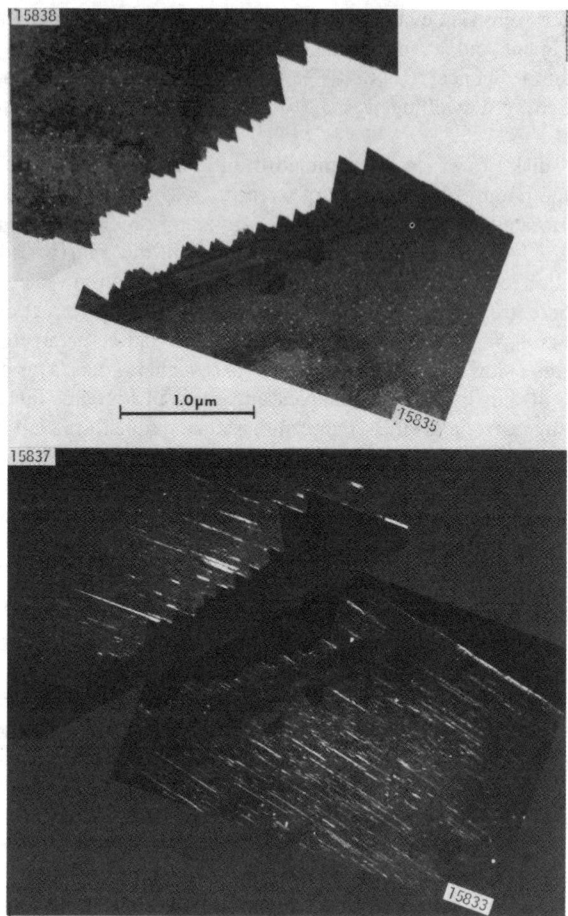

FIG. 21—*Crack teeth aligned with twins in the matrix in Specimen 46M7:* (a) *bright-field and* (b) *dark-field.*

A fragment of the P53 duct specimen was broken by bending to verify the fracture mode observed in Specimen 46M7. The fragment was clamped at one end in a small vice to form a cantilever, and then a small piece of material was broken from the free end by pushing and striking it with the manipulator. The fracture surface showed channel fracture, consistent with the results obained on Specimen 46M7. Microscopy conducted on a thin foil prepared from this fragment revealed a martensitic transformation in some of the grains as shown in Fig. 23. No bcc phase was found in thin foils prepared from the gage section of tensile specimens tested at room temperature. Deformation twins found in some of the grains shown in Fig. 24 had the same appearance as the ε phase earlier reported to form in the absence of irradiation in deformed 304 stainless steel [*16*].

Summary of Failure Modes

Whereas the tensile data seemed to indicate an abrupt change from channel failure to quasi-cleavage failure when the test temperature fell to room temperature, the results of the compact

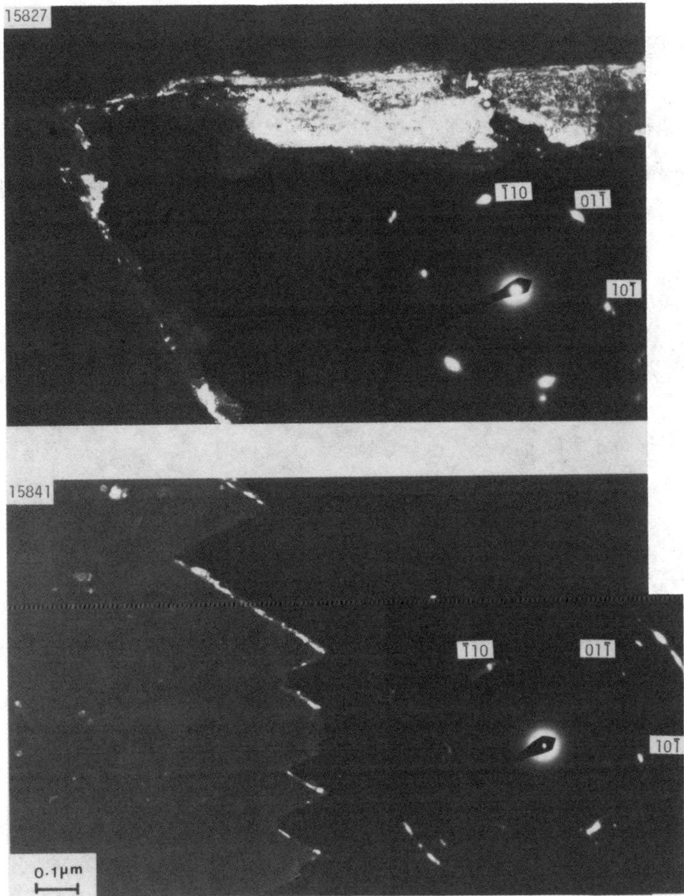

FIG. 22—*α-martensite layer observed along the crack path in 46M7. The inserted diffraction patterns are ⟨111⟩ patterns of a bcc lattice.*

tension and microscopy disk studies show that the situation is somewhat more complex. The First Core steel not only changes its fracture mode as a function of neutron fluence and test temperature, but also as a function of the stress state and the nature of the constraints imposed on the deforming volume.

Consider the various failure modes observed at room temperature. Failure occurred by cleavage at low threshold stress in the precracked zones and by quasi-cleavage in the tested zone of compact tension specimens. Quasi-cleavage was also observed in tensile tests and in the thinned region adjacent to the hole of a microscopy disk. In the latter case the failure mode changed to channel fracture as the crack point encountered different constraints in crossing a grain boundary. Bending of disks, on the other hand, caused only channel fracture to occur.

Channel fracture also occurred in the titanium-modified steels, even though the phase evolution of these steels is somewhat more complex than that of unmodified steels, and swelling is delayed by this difference in evolution.

The amount of channel fracture appears to be related to the amount of swelling. Note in Fig. 25 that comparable levels of faceting on the fracture surface require higher fluences for the LS2

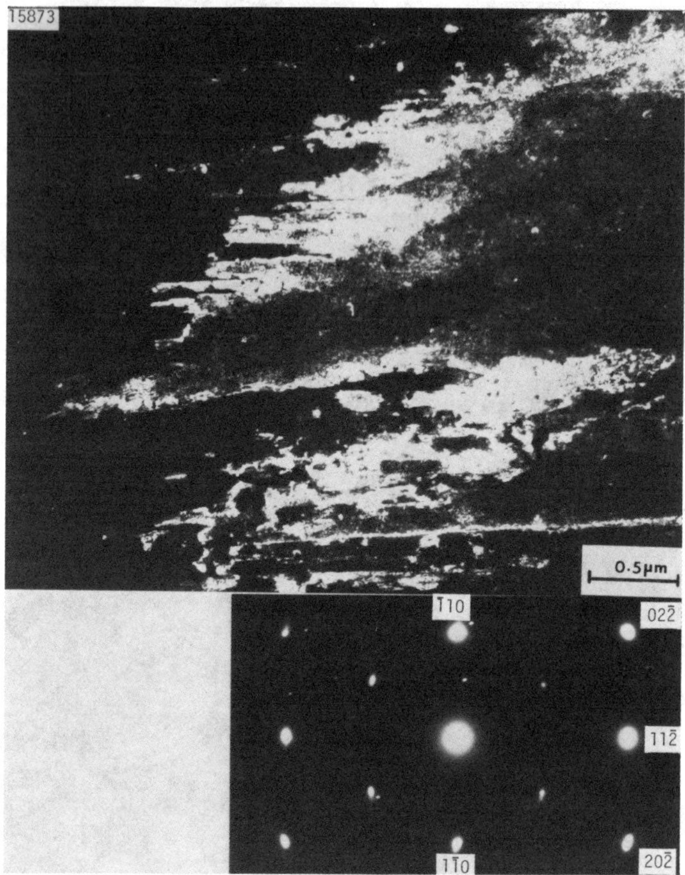

FIG. 23—*Martensitic material and its associated diffraction pattern observed in a P53 duct fragment deformed at room temperature.*

and D18 steels, which swell later than the 316 and LS1 steels. It also appears that the amounts of swelling and faceting each correlate with the nickel level of the steel.

A Proposed Model for the Changes in Fracture Mode

The results of these and other studies [*11–13*] show that the development of both channel fracture and quasi-cleavage is associated with the concurrent development of significant levels of swelling. Both swelling and the changes in failure mode are retarded by cold-work, increases in nickel content (304 versus 316 stainless, D18 versus LS2) and the addition of solutes (316 versus LS1). The current study also indicates that a deformation-induced martensite transformation plays a large role in hardening the alloy and thus contributing to its precipitous decrease in tearing resistance.

It appears that it is possible to link these two factors in a model that will explain the results of this study. Such a model will have its most severe test in providing an explanation for the shift from channel fracture to quasi-cleavage as the temperature is lowered.

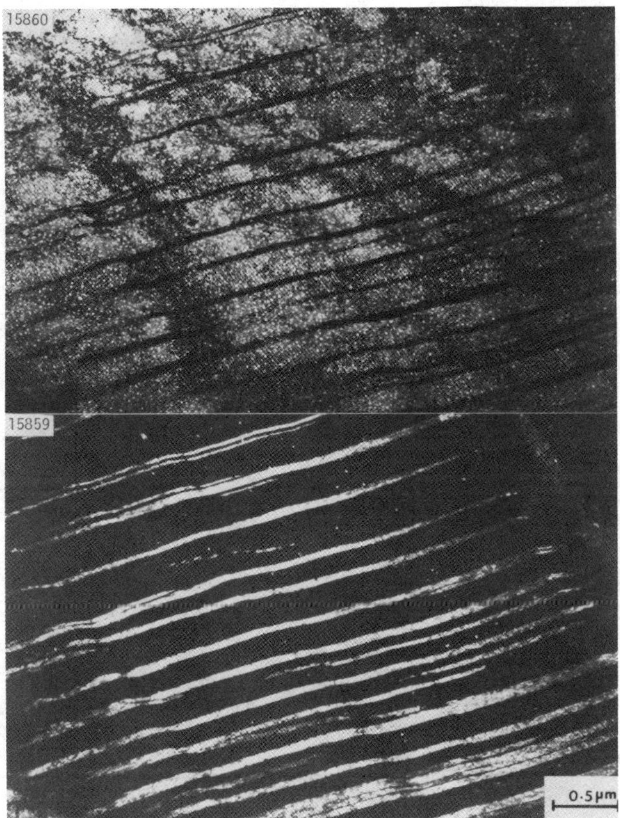

FIG. 24—*Deformation twins observed in the P53 duct that appear to be* ε*-martensite:* (Top) *bright-field and* (Bottom) *dark-field picture of the same area.*

The link between martensite formation and the onset of rapid void swelling is proposed to lie in the radiation-induced changes in composition that occur on a microscopic scale. This microchemical evolution results in the progressive removal of a large fraction of the nickel and most of the silicon from the alloy matrix into a variety of precipitates such as γ′ (Ni_3Si) and G-phase [*17–19*]. Nickel and silicon have a large effect on vacancy diffusivity and their removal from solution increases the supersaturation of vacancies, thus promoting void nucleation [*19–24*]. It has been shown that irradiation of 316 stainless steel in a variety of conditions always leads to a reduction in the matrix nickel content to ~9% [*25*]. Similar behavior has been observed in titanium and phosphorus-modified steels, where the onset of rapid swelling has also been correlated with the drop in nickel content [*26,27*].

Once formed, however, voids strongly segregate nickel at their surface via the inverse Kirkendall effect, in which the slowest diffusing elements segregate by default at the bottom of vacancy gradients [*28,29*]. Thus, as voids become the dominant feature of the microstructure, the spaces between them become increasingly depleted in nickel and enriched in chromium. In several irradiation experiments involving low nickel levels and larger than average silicon levels, the matrix depletion has been so strong as to cause the region between voids to transform to massive ferrite, leaving only the voids coated with austenite [*28–30*].

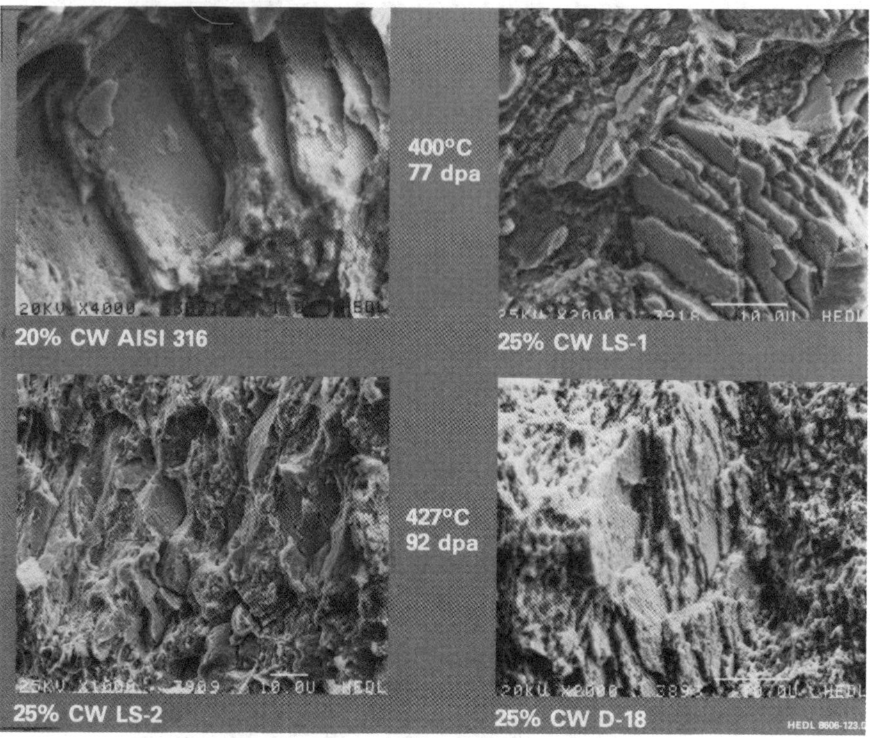

FIG. 25—*Channel fracture observed on microscopy disks of various austenitic alloys irradiated in EBR-II. These disks were broken manually at room temperature.*

The fault structure that occurs on {111} planes in fcc alloys is identical to that of the hcp ε-phase involved in the $\gamma \rightarrow \varepsilon \rightarrow \alpha$ transformation of austenite to ferrite. A drop in stacking fault energy thus predisposes the alloy toward this transformation. While silicon and other solutes influence somewhat the stacking fault energy (SFE) of austenitic alloys [*31,32*], chromium and particularly nickel have a very large influence.

As shown in Table 6, Latanison and Ruff [*33*] have demonstrated that the stacking fault energy at room temperature of Fe-18Cr-XNi alloys drops 31% when X decreases from 15.9 to 10.7%. Lecroisey and Thomas [*34*] noted that in simple ternary alloys the SFE at room temperature fell from 45 ergs/cm^2 in Fe-14.1Ni-17.8Cr to 24 ergs/cm^2 in Fe-12.5Ni-15.9Cr. Bampton et al. [*35*] and Rhodes and Thompson [*36*] confirm the influence of nickel and chromium on SFE, with the latter stating that SFE in mJ/m^2 = 17.0 + 2.29 (Ni) $-$0.9 (Cr) for alloys with relatively low chromium levels.

Table 6 shows one other very important factor that must be taken into consideration. As noted by Latanison and Ruff [*33*] as well as Lecroisey and Thomas [*34*], the stacking fault energy increases with temperature but not at the same rate in each alloy. Table 6 shows that the increase in SFE at 325°C is only 5% when the nickel level increases from 10.7 to 15.9%, while at room temperature a comparable increase in nickel yields an increase of 44%.

This provides a possible explanation of why the formation of ε microplatelets, such as those shown in Fig. 15, is so much more pronounced in the tensile tests conducted at room temperature. Not only is the yield strength larger at the lower temperature but the impact of the reduced matrix nickel levels is much larger at room temperature. At higher temperatures the stacking fault energy

TABLE 6—*Stacking fault energy (SFE) of Fe-Cr-Ni alloys as a function of temperature and composition* [33].

T, °C	Ni, wt%	Cr, wt%	SFE, mJ/m^2	
25	10.7	18.3	16.4	44% difference
25	15.9	18.7	23.6	
325	10.7	18.3	30.4	5% difference
325	15.9	18.7	31.8	

is higher and the impact of nickel segregation is not so pronounced. Thus more localized deformation (channels) is permitted before martensite formation and its associated embrittlement occurs.

If this model is correct, the effect of voids on hardening is more indirect than direct. The incremental hardening attributable to the voids themselves is overshadowed by their influence on matrix composition and the subsequent formation of martensite. Whereas voids, dislocations, and Frank loops represent pre-existing hardening obstacles, the martensite barriers are probably formed within the 0.2% offset used to define the yield strength. As the martensite barriers develop, the local hardening probably shifts the deformation to other areas. When the entire gage length is hardened, failure follows virtually immediately thereafter and the extensive γ/ε boundaries provide low energy paths for crack propagation. This explains how the tearing modulus can decrease so strongly with neutron fluence while the fracture toughness does not decline.

Conclusion

An important conclusion of this study is that the tensile strength, ductility, and fracture toughness of 20% cold-worked AISI 316 at fluences on the order of 15 to 18 $\times$ 10^{22} n/cm^2 ($E > 0.1$ MeV) are sufficient to permit typical in-reactor operations. However, careful handling must be employed in post-irradiation handling operations conducted out of reactor at much lower temperatures.

The severe embrittlement observed at low temperatures is thought to be an indirect consequence of the onset of rapid void swelling. The nickel that segregates to void surfaces via the inverse Kirkendall effect further depletes a matrix already reduced by extensive radiation-induced coprecipitation of nickel and silicon. The loss of nickel and the increase in chromium in the region between voids leads to a substantial decrease in stacking fault energy and causes extensive formation of ε-martensite microplatelets. These harden the matrix so as to increase the strength substantially above the ''saturation levels'' attained at much lower fluence. The γ/ε boundaries also provide a low energy path for crack propagation, reducing the tearing modulus to near zero and producing a very brittle failure mode referred to as quasi-cleavage.

At higher deformation temperatures the influence of nickel segregation on stacking fault energy is much less pronounced and large levels of flow localization are required before failure occurs. In this case failure occurs along the flow localization paths, commonly referred to as channels. These channels survive as facets on the fracture surface.

References

[1] Johnson, G. D., Garner, F. A., Brager, H. R., and Fish, R. L., *Effects of Radiation on Materials: Tenth Conference, STP 725,* D. Kramer and J. S. Perrin, Eds., American Society for Testing and Materials, Philadelphia, 1981, pp. 393–412.

[2] Garner, F. A., Hamilton, M. L., Panayotou, N. F., and Johnson, G. D., *Journal of Nuclear Materials,* Vols. 103 and 104, 1981, pp. 803–807.

[3] Brager, H. R., Blackburn, L. D., and Greenslade, D. L., *Journal of Nuclear Materials,* Vols. 122 and 123, 1984, pp. 332–337.

[4] Brager, H. R., Garner, F. A., and Hamilton, M. L., *Journal of Nuclear Materials,* Vols. 133 and 134, 1985, pp. 594–598.
[5] Simons, R. L. and Hulbert, L. A., *Effects of Radiation on Materials: Twelfth International Symposium, STP 870,* F. A. Garner and J. S. Perrin, Eds., American Society for Testing and Materials, Philadelphia, 1985, pp. 820–839.
[6] Hamilton, M. L., Cannon, N. S., and Johnson, G. D., *Effects of Radiation on Materials: Eleventh Conference, STP 782,* H. R. Brager and J. S. Perrin, Eds., American Society for Testing and Materials, Philadelphia, 1982, pp. 636–647.
[7] Hamilton, M. L., Garner, F. A. and Wolfer, W. G., *Journal of Nuclear Materials,* Vols. 122 and 123, 1984, pp. 106–110.
[8] Hamilton, M. L., Garner, F. A., and Yang, W. J. S., *Fusion Technology,* Vol. 10, 1986, pp. 405–410.
[9] Garner, F. A., *Journal of Nuclear Materials,* Vols. 133 and 134, 1985, pp. 113–118.
[10] Garner, F. A., Brager, H. R., Hamilton, M. L., Dodd, R. A., and Porter, D. L., *Radiation Effects,* Vol. 101, 1986, pp. 37–53.
[11] Fish, R. L., Straalsund, J. L., Hunter, C. W., and Holmes, J. J., *Effects of Radiation on Substructure and Mechanical Properties of Metals and Alloys, STP 529,* American Society for Testing and Materials, Philadelphia, 1973, pp. 149–164.
[12] Mills, W. J., "Fracture Toughness of Irradiated Stainless Steel Alloys," HEDL-SA-3471, Hanford Engineering Laboratory Report, Richland, WA, Jan. 1986.
[13] Huang, F. H., *International Journal of Fracture,* Vol. 25, 1984, p. 181.
[14] Brager, H. R., Garner, F. A., Gilbert, E. R., Flinn, J. E., and Wolfer, W. G., *Radiation Effects in Breeder Reactor Structural Materials,* M. L. Bleiberg and J. W. Bennett, Eds., TMS-AIME, New York, 1977, pp. 479–507.
[15] Huang, F. H. and Wire, G. L., *Journal of Engineering Materials and Technology,* Vol. 101, 1978, p. 403.
[16] Magonon, P. L., Jr. and Thomas, G., *Metallurgical Transactions,* Vol. 1, 1970, p. 1571.
[17] Garner, F. A., *Phase Stability During Radiation,* J. R. Holland, L. K. Mansur and D. I. Potter, Eds., TMS-AIME, Warrendale, PA, 1981, pp. 165–189.
[18] Garner, F. A., *Journal of Nuclear Materials,* Vols. 122 and 123, 1984, pp. 459–471.
[19] Garner, F. A. and Wolfer, W. G., *Journal of Nuclear Materials,* Vols. 122 and 123, 1984, pp. 201–206.
[20] Garner, F. A. and Wolfer, W. G., *Journal of Nuclear Materials,* Vol. 102, 1981, pp. 142–150.
[21] Esmailzadeh, B. and Kumar, A. S., *Effects of Radiation on Materials: Twelfth International Symposium, STP 870,* F. A. Garner and J. S. Perrin, Eds., American Society for Testing and Materials, Philadelphia, 1985, pp. 468–480.
[22] Esmailzadeh, B., Kumar, A. S., and Garner, F. A., *Journal of Nuclear Materials,* Vols. 133 and 134, 1985, pp. 590–593.
[23] Garner, F. A. and Kumar, A. S., *Radiation Induced Changes in Microstructure: 13th International Symposium (Part I), STP 955,* F. A. Garner, N. H. Packan, and A. S. Kumar, Eds., American Society for Testing and Materials, Philadelphia, 1987, pp. 289–314.
[24] Coghlan, W. A. and Garner, F. A., *Radiation Induced Changes in Microstructure: 13th International Symposium (Part I),* F. A. Garner, N. H. Packan, and A. S. Kumar, Eds., Philadelphia, 1987, pp. 315–329.
[25] Brager, H. R. and Garner, F. A., *Journal of Nuclear Materials,* Vol. 117, 1983, pp. 159–176.
[26] Thomas, L. E., *Proceedings, 40th Annual Meeting of the Electron Microscopy Society of America,* G. W. Bailey, Ed., Washington, DC, 1982, p. 597.
[27] Itoh, M., Onose, S., and Yuhara, S., *Phase Stability During Radiation,* J. R. Holland, L. K. Mansur, and D. I. Potter, Eds., TMS-AIME, Warrendale, PA, 1981, pp. 219–235.
[28] Brager, H. R. and Garner, F. A., *Radiation Induced Changes in Microstructure: 13th International Symposium (Part I),* F. A. Garner, N. H. Packan, A. S. Kumar, Eds., Philadelphia, 1987, pp. 195–206.
[29] Porter, D. L. and Wood, E. L., *Journal of Nuclear Materials,* Vol. 83, 1979, pp. 90–97.
[30] Mazey, D. J., Harries, D. R. and Hudson, J. A., *Proceedings, Irradiation Behavior of Metallic Materials for Fast Reactor Core Components,* Ajaccio, Corsica, 1979, pp. 61–67.
[31] Dulieu, D. and Nutting, J., *Proceedings, Metallurgical Developments in High Alloy Steels,* Scarborough, England, June 1964, pp. 140–145.
[32] Thomas, B. and Henry, G., *Memoires Scientifique, Revue Metallurgique,* Vol. LXIV, 1967, pp. 625–636.
[33] Latanison, R. M. and Ruff, A. W., Jr., *Metallurgical Transactions,* Vol. 2, 1971, pp. 505–509.
[34] Lecroisey, F. and Thomas, B., *Physica Status Solidi A,* Vol. 2, 1970, pp. K217–K219.
[35] Bampton, C. C., Jones, I. P., and Loretto, M. H., *Acta Metallurgica,* Vol. 26, 1978, pp. 39–51.
[36] Rhodes, C. G. and Thompson, A. W., *Metallurgical Transactions,* Vol. 8A, 1977, pp. 1901–1906.

David N Braski[1]

The Effect of Neutron Irradiation on the Tensile Properties and Microstructure of Several Vanadium Alloys

REFERENCE: Braski, D. N., "**The Effect of Neutron Irradiation on the Tensile Properties and Microstructure of Several Vanadium Alloys,**" *Influence of Radiation on Material Properties: 13th International Symposium (Part II), ASTM STP 956,* F. A. Garner, C. H. Henager, Jr., and N. Igata, Eds., American Society for Testing and Materials, Philadelphia 1987, pp. 271–290.

ABSTRACT: Specimens of V-15Cr-5Ti, VANSTAR-7, and V-3Ti-1Si were encapsulated in molybdenum alloy tubes containing ^{7}Li to prevent interstitial pickup and irradiated in the Fast Flux Test Facility (FFTF), using Materials Open Test Assembly (MOTA) experiments, to a damage level of 40 displacements per atom (dpa). The irradiation temperatures were 420, 520, and 600°C. For a better simulation of fusion reactor conditions, helium was preimplanted in some specimens using a modified version of the "tritium trick." The V-15Cr-5Ti alloy was most susceptible to irradiation hardening and helium embrittlement, followed by VANSTAR-7, and V-3Ti-1Si. VANSTAR-7 exhibited a relatively high maximum void swelling of ~6% at 520°C while V-15Cr-5Ti and V-3Ti-1Si had values of <0.3% at all three temperatures. The V-3Ti-1Si clearly outperformed the other two vanadium alloys in resisting the effects of neutron irradiation.

KEY WORDS: vanadium alloys, neutron irradiation, irradiation hardening, helium embrittlement, void swelling

Vanadium alloys are currently being evaluated for use in fusion reactor first wall/blanket applications. Their relatively high thermal conductivity and low thermal expansion properties create lower thermal stresses for a given heat flux than the stainless steels, and the vanadium alloys demonstrate good mechanical properties in the design temperature range [*1*]. The vanadium alloys also show good compatibility with lithium (a proposed blanket material [*1*]), excellent resistance to void swelling [2], and very low residual radioactivity after irradiation. In the past, vanadium alloys have been neutron irradiated in several experiments by encapsulating specimens in tubes containing sodium or an inert gas to prevent interstitial pickup that can embrittle the material. Like the other refractory metals (for example, Mo, Ta, and W), vanadium is embrittled by picking up C, N, or O, that fill interstitial positions in its crystal lattice. Unfortunately, previous encapsulation techniques have not been effective in preventing interstitial pickup and to make matters worse the experimental alloys were often contaminated to begin with. In the present experiment, an effort was made to eliminate the factor of interstitial contamination, especially by oxygen. This was accomplished by using alloys with low oxygen contents and encapsulating specimens in molybdenum alloy tubes containing lithium. Unlike sodium, from which dissolved oxygen will be absorbed by a vanadium alloy [*3*], lithium will remove oxygen from vanadium at elevated temperatures [*4*]. In addition, a number of encapsulated specimens were thermally aged in lithium at the same times and temperatures as their counterparts in the reactor. These aged

[1] Research staff member, Metals and Ceramics Division, Oak Ridge National Laboratory, Oak Ridge, TN 37831.

TABLE 1—*Vanadium alloy data.*

Alloy	Heat	Composition, % by weight								Final Heat Treatment	^3H Charging Time, h	^3He Content[a], appm
		Cr	Ti	Fe	Zr	Si	C	O	N			
V-15Cr-5Ti[b]	CAM-834-3	14.5	6.2	. . .	. . .	. . .	0.032	0.031	0.046	1 h @ 1200°C	107	74
VANSTAR-7[b]	CAM-837-7	9.7	. . .	3.4	1.3	. . .	0.064	0.028	0.052	1 h @ 1350°C	206	70
V-3Ti-1Si[c]	11153	. . .	3.4	0.04	. . .	1.28	0.045	0.091	0.026	1 h @ 1050°C	67	82

[a] Analyses performed by Dr. B. M. Oliver, Rockwell International Corporation, Canoga Park, CA.
[b] Source: Westinghouse Electric Corporation.
[c] Source: KFK, Karlsruhe, West Germany (Dr. D. Kaletta).

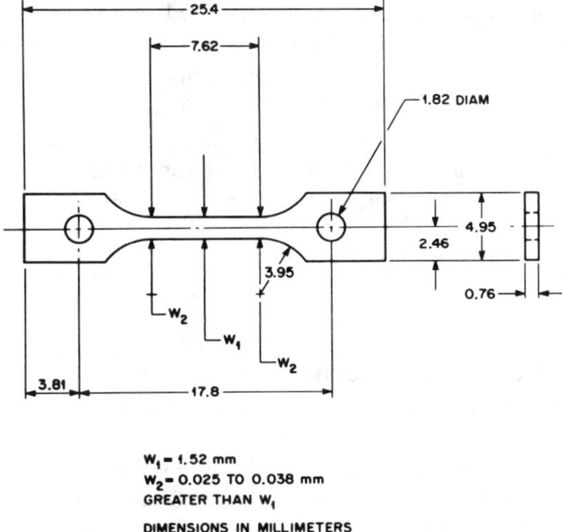

FIG. 1—*SS-3 type sheet tensile specimen.*

controls allow clear differentiation between any effects on tensile properties or microstructures that might be caused by the elevated-temperature exposure (and possible contamination) and the effects of neutron irradiation. The effect of irradiation at 420, 520, and 600°C in the Fast Flux Test Facility (FFTF) on the tensile behavior and microstructure of V-15Cr-5Ti, VANSTAR-7, and V-3Ti-1Si was investigated. The effects of helium, preimplanted to levels of ~80 atomic parts per million (appm) using a modified "tritium trick" technique [*5*], were also evaluated.

Experimental Procedure

The three vanadium alloys used in this experiment were V-15Cr-5Ti, VANSTAR-7, and V-3Ti-1Si. The source, chemistry, and final annealing heat treatment (for 30 μm grain size) for each alloy are listed in Table 1. Miniature tensile specimens (SS-3 type) with the configuration shown in Fig. 1 were machined from 0.76-mm-thick sheet of V-15Cr-5Ti and VANSTAR-7, and from 0.51-mm-thick sheet of V-3Ti-1Si. Disks (3-mm diameter) for examination by transmission electron microscopy (TEM) were punched from 0.25-mm-thick sheet.

Some tensile specimens and disks were implanted with ~80 appm ^3He using a modified version of the tritium trick [*5*]. The tritium charging times and final ^3He contents for each alloy are also given in Table 1. The charging temperature, charging pressure, and pumpout temperature were 400°C, 53 kPa, and 600°C, respectively, for all three alloys.

Selected specimens were encapsulated at Westinghouse Hanford Company (WHC) in 9.525 mm outside diameter × 8.255 mm inside diameter × 38.1 mm long TZM (Mo-.5Ti-.08Zr-.015C) subcapsules containing enriched 99.99% ^7Li [*6*]. The enriched ^7Li was used because the 7.5% ^6Li present in natural lithium would result in unacceptable levels of helium and tritium formation. Each subcapsule was loaded with nine SS-3 tensile specimens and nine disks. The subcapsules were irradiated in the FFTF, Materials Open Test Assembly (MOTA) at 420, 520, and 600°C to a fast fluence of ~7 × 10^{22} neutrons (n)/cm² and a nominal damage level of ~40 displacements per atom (dpa). The results for a few specimens from cycle 4 (10 dpa at 420°C and 17 dpa at 520°C) will also be shown in this report [*7*]. An equal number of control specimens were also encapsulated in capsules with ^7Li and thermally aged for 257 days at the same temperatures used in the irradiation.

Both irradiated and control tensile specimens were tested at the irradiation/aging temperature in a hot cell tensile machine under a reduced pressure of $<10^{-4}$ Pa. The time required to heat a specimen and hold it at the test temperature was approximately 1800 s. The crosshead speed was 8.5 μm/s (strain rate = 1.1×10^{-3} s^{-1}). After testing, the fracture surfaces were examined using a scanning electron microscope (SEM) installed in a radioactive hot cell [*8*]. Microstructural details were studied by TEM using the irradiated disks that were subsequently electropolished in a solution of one part concentrated H_2SO_4 to seven parts methanol, at −30°C.

Results

Tension Tests

The results of tension tests on the aged vanadium alloy control specimens and postirradiation tensile tests on the same alloys are listed in Tables 2 and 3, respectively.

V-15Cr-5Ti—The results of tensile tests conducted on the V-15Cr-5Ti alloy are shown graphically in Fig. 2. Typical stress-strain curves for specimens in various conditions are shown at the three irradiation temperatures. First, note that the thermally aged control specimens with or without the 74 appm of implanted helium exhibited tensile behavior that was virtually the same as the starting

TABLE 2—*Results of tensile tests of vanadium control specimens aged 257 days in* ^7Li.

				Strength, MPa		Elongation, %	
Specimen Number	Alloy	Helium Level, appm	Aging/Test Temperature, °C	Yield	Ultimate Tensile	Uniform	Total
RD40	V-15Cr-5Ti	0	420	399	545	15.0	26.0
RD66	V-15Cr-5Ti	74	420	381	540	14.3	23.7
RA187	V-15Cr-5Ti	74	420	369	539	14.0	21.3
RD38	V-15Cr-5Ti	0	520	387	539	15.3	24.3
RA73	V-15Cr-5Ti	74	520	352	511	12.7	20.7
RA170	V-15Cr-5Ti	74	520	381	578	13.0	20.7
RD03	V-15Cr-5Ti	0	600	390	555	12.3	21.7
RA129	V-15Cr-5Ti	74	600	390	573	11.3	19.0
RA176	V-15Cr-5Ti	74	600	377	574	10.3	17.7
QA71	VANSTAR-7	0	420	256	398	16.0	26.3
QA73	VANSTAR-7	70	420	312	451	14.7	23.7
QA110	VANSTAR-7	70	420	302	448	12.7	22.3
QA08	VANSTAR-7	0	520	282	469	15.7	23.7
QA118	VANSTAR-7	70	520	283	469	13.0	20.7
QA119	VANSTAR-7	70	520	297	480	11.7	20.7
QA69	VANSTAR-7	0	600	298	490	12.3	20.0
QA80	VANSTAR-7	70	600	274	482	13.0	20.3
QA102	VANSTAR-7	70	600	261	473	14.0	22.0
RC54	V-3Ti-1Si	0	420	378	473	14.0	21.0
RC08	V-3Ti-1Si	82	420	444	550	10.7	17.3
RC42	V-3Ti-1Si	82	420	430	542	12.3	19.0
RC30	V-3Ti-1Si	0	520	454	595	11.0	20.0
RC33	V-3Ti-1Si	82	520	452	587	10.0	14.3
RC44	V-3Ti-1Si	82	520	446	570	14.0	20.0
RC55	V-3Ti-1Si	0	600	315	467	8.0	12.3
RC23	V-3Ti-1Si	82	600	300	495	13.3	20.0
RC57	V-3Ti-1Si	82	600	290	498	14.0	20.3

TABLE 3—*Results of tensile tests and void swelling measurements for vanadium alloys irradiated to 40 dpa in FFTF(MOTA).*

Specimen Number	Alloy	Helium Level, appm	Irradiation/Test Temperature, °C	Strength, MPa		Elongation, %		Disk Number	$\Delta V/V$, %
				Yield	Ultimate Tensile	Uniform	Total		
RD14	V-15Cr-5Ti	0	420	825	913	0.8	0.8	RD18	0.003
RD64	V-15Cr-5Ti	74	420	. . .	793	0	0	RD52	0.005
RA34	V-15Cr-5Ti	74	420	692	834	0.8	0.8	. . .	. . .
RD37	V-15Cr-5Ti	0	520	879	989	3.0	9.5	RD65	0.031
RA20	V-15Cr-5Ti	74	520	882	1003	1.8	1.8	RD38	0.046
RA108	V-15Cr-5Ti	74	520	868	956	1.5	1.7	. . .	. . .
RD30	V-15Cr-5Ti	0	600	674	797	4.5	11.9	RD21	0.024
RA68	V-15Cr-5Ti	74	600	broke during loading			. . .	RD16	0.283
RA139	V-15Cr-5Ti	74	600	broke during loading			. . .	. . .	. . .
QA19	VANSTAR-7	0	420	881	896	0.7	6.8	QB55	0.001
QA40	VANSTAR-7	70	420	978	997	0.9	6.0	QB38	0.004
QA71	VANSTAR-7	70	420	987	1002	0.6	5.7	. . .	. . .
QA29	VANSTAR-7	0	520	1058	1138	1.5	4.0	QB60	0.958
QA87	VANSTAR-7	70	520	960	1016	1.3	1.3	QB39	5.940
QA91	VANSTAR-7	70	520	failed at hole in shoulder			. . .	. . .	. . .
QA53	VANSTAR-7	0	600	360	429	2.8	5.5	QB57	0.014
QA83	VANSTAR-7	70	600	651	755	3.2	4.5	QB40	0.065
QA90	VANSTAR-7	70	600	722	790	3.5	5.2	. . .	. . .
RC34	V-3Ti-1Si	0	420	692	780	3.4	4.8	RC40	0.079
RC35	V-3Ti-1Si	82	420	795	884	3.8	6.0	RC18	0.090
RC53	V-3Ti-1Si	82	420	717	880	6.3	11.8	. . .	. . .
RC17	V-3Ti-1Si	0	520	541	682	8.2	13.3	disk lost	
RC11	V-3Ti-1Si	82	520	632	758	6.3	10.3	RC15	0.002
RC29	V-3Ti-1Si	82	520	565	688	6.0	10.9	. . .	. . .
RC27	V-3Ti-1Si	0	600	378	564	8.1	12.1	R12	0
RC41	V-3Ti-1Si	82	600	276	480	7.8	11.3	RC11	0.052
RC43	V-3Ti-1Si	82	600	369	583	8.8	12.3	. . .	. . .

material. This establishes the technique of irradiating vanadium alloys in lithium as a viable one (that is, the exposure to lithium did not affect the tensile properties).

Neutron irradiation hardened the material as shown by the marked increase in yield strengths and reduction in elongation. This was especially true at 420°C, where both implanted and unimplanted specimens failed at very low strains. However, at 520°C, most of the irradiated specimens demonstrated moderate ductility in spite of a doubling of their yield strengths. The irradiated specimen implanted with helium (I40/74) displayed low ductility after 40 dpa. Serrated portions of the stress-strain curves were observed occasionally in tests conducted at 520°C and frequently at 600°C. This phenomenon is not unusual for vanadium and its alloys and is probably due to dynamic strain aging [*9*]. Irradiation hardening was lower at 600°C than at the two lower temperatures as demonstrated by the specimen without helium (I40/0) shown in the upper graph in Fig. 2. Two specimens that contained 74 appm He became very brittle after irradiation at 600°C and broke during subsequent handling.

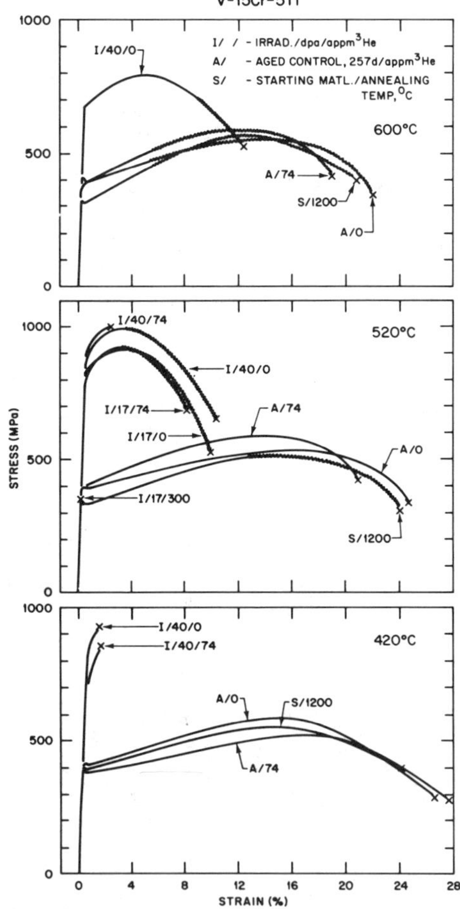

FIG. 2—*Stress-strain curves for V-15Cr-5Ti alloy in the starting condition* (S), *after aging for 257 days in lithium* (A), *and after irradiation in FFTF(MOTA)–(I).*

VANSTAR-7—Typical stress-strain curves for the starting VANSTAR-7 material, aged controls, and irradiated specimens are presented in Fig. 3. As in the case of V-15Cr-5Ti, aging at the three elevated temperatures for 257 days had little effect on the tensile properties of VANSTAR-7. At 420°C, irradiation doubled the yield strength after 10 dpa and tripled it after 40 dpa. In spite of the substantial hardening, the VANSTAR-7 specimen retained excellent ductility. Moreover, the ductility was unaffected by the implanted helium, at least at the levels investigated. At 520°C, the VANSTAR-7 specimens without helium exhibited total elongations of at least 4% along with even greater amounts of irradiation hardening (for example, curves I40/0 and I17/0). However, unlike the results at 420°C, specimens containing helium showed signs of embrittlement (see curves I40/70 and I17/150). At 600°C, irradiation hardening was attenuated and the specimen with 70 appm He (I40/70) had a total elongation comparable to a similar specimen without helium (I40/0). However, the respective yield strengths of these two specimens were quite different. Unfortunately, SEM and TEM investigations of both specimens failed to provide an explanation for the difference. As before, serrated yielding was most evident at the higher temperature, but for some reason was not observed in the irradiated VANSTAR-7 specimens.

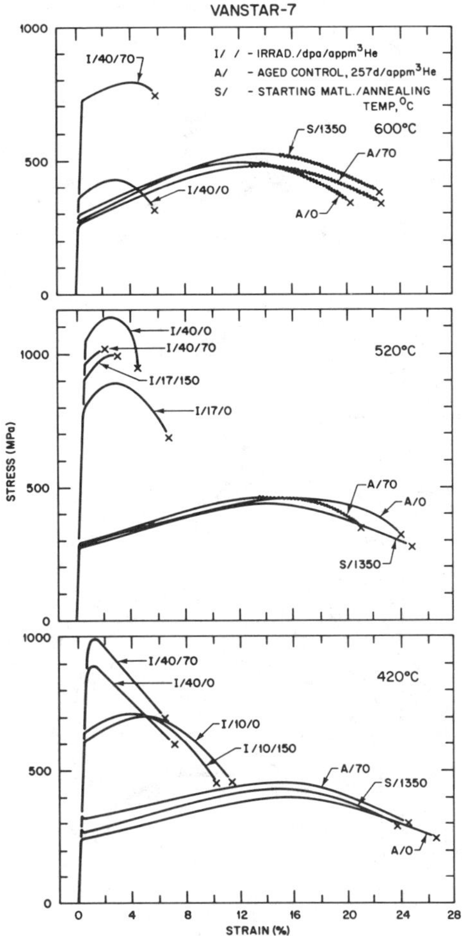

FIG. 3—*Stress-strain curves for VANSTAR-7 alloy in the starting condition* (S), *after aging for 257 days in lithium* (A), *and after irradiation in FFTF(MOTA)–(I).*

before, serrated yielding was most evident at the higher temperature, but for some reason was not observed in the irradiated VANSTAR-7 specimens.

V-3Ti-1Si—Like the first two vanadium alloys, aged V-3Ti-1Si specimens demonstrated the same tensile behavior that was similar to the starting material (Fig. 4). Only one specimen, A/O at 600°C, exhibited a noticeable difference in tensile behavior. The V-3Ti-1Si specimens also showed substantial irradiation hardening at 420°C, but unlike the VANSTAR-7 specimens, this hardening tended to saturate at the relatively low damage level of 10 dpa. Reasonable ductilities were exhibited by all of the irradiated specimens including those containing helium. However, the lower elongation exhibited by the specimen with the higher helium content (I10/135) may be an indication that higher helium levels may not be accommodated very well. At 520 and 600°C, the magnitude of irradiation hardening was markedly lower in the V-3Ti-1Si than the other two alloys and saturated quickly at relatively low damage levels. Along with the reduced hardening, the V-3Ti-1Si specimens exhibited total elongations that were usually above 10% and were quite insensitive to 82-appm helium.

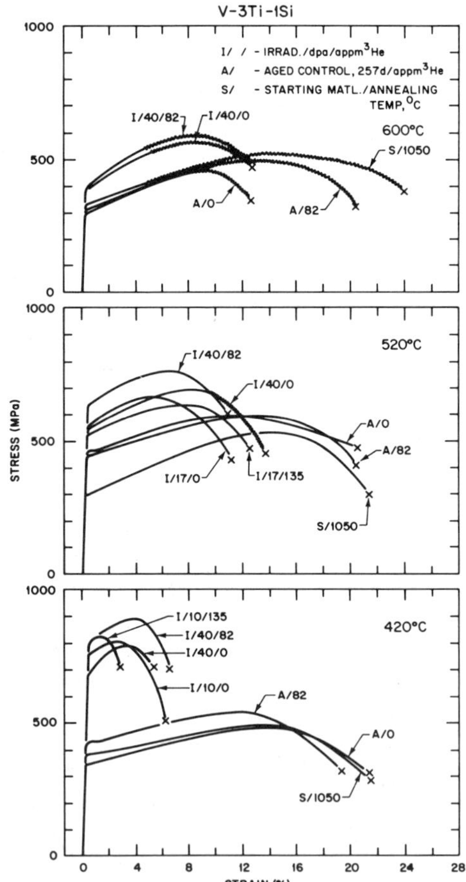

FIG. 4—*Stress-strain curves for V-3Ti-1Si alloy in the starting condition* (S), *after aging for 257 days in lithium* (A), *and after irradiation in FFTF(MOTA)–(I).*

Ductility—The effects of irradiation, irradiation temperature, and helium on the ductility of the three vanadium alloys are shown in Fig. 5, where total elongation is plotted as a function of temperature. Starting at the top of the figure with V-15Cr-5Ti, a small but consistent difference is seen between aged specimens containing helium and those without helium. The difference was probably caused by the slight hardening due to ^{3}He implantation and the introduction of small ^{3}He bubbles in the microstructure. Irradiation to 40 dpa had a drastic effect on elongation at 420°C due to irradiation hardening that abated with increased irradiation temperatures. The irradiated V-15Cr-5Ti specimens were particularly sensitive to the presence of helium in the microstructure as relatively low elongations were measured in all of the specimens containing helium. Thermally aged VANSTAR-7 specimens with helium also had slightly lower elongations at 420 and 520°C than those without helium, but this difference disappeared at 600°C. Irradiation hardening substantially lowered the ductility of VANSTAR-7 across the entire temperature range, but the effect of helium was less than observed for V-15Cr-5Ti. Finally, for V-3Ti-1Si, large scatter in the results for the aged specimens, with and without helium, prohibited any conclusions being made concerning helium and its effects on aged specimens. However, it is seen that both irradiation and helium had less effect on the total elongations of V-3Ti-1Si specimens compared to those of

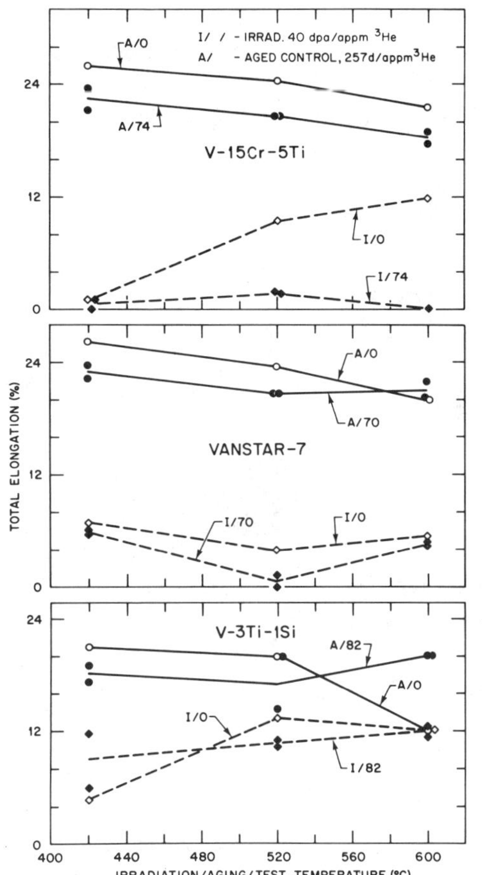

FIG. 5—*Total elongation of vanadium alloy aged control specimens* (A) *and specimens irradiated in FFTF(MOTA) to 40 dpa* (I) *as a function of temperature.*

TABLE 4—*SEM results.*

Alloy	Helium Level, appm	420°C		520°C		600°C
		10 dpa	40 dpa	17 dpa	40 dpa	40 dpa
V-15Cr-5Ti	0	cleavage[a]	6*a*[b]/cleavage	ductile	ductile	7*a*/ductile
V-15Cr-5Ti	14	. . .	. . .	ductile + cleavage + intergranular	. . .	. . .
V-15Cr-5Ti	74	. . .	6*c*/cleavage + intergranular	. . .	6*d*/intergranular + some cleavage	intergranular
V-15Cr-5Ti	300	6*b*/intergranular	. . .	intergranular	. . .	. . .
VANSTAR-7	0	ductile	ductile	ductile	7*b*/ductile + cleavage	ductile
VANSTAR-7	70	. . .	ductile	. . .	7*c*/integranular	ductile
VANSTAR-7	150	ductile	. . .	intergranular + some cleavage	. . .	. . .
V-3Ti-1Si	0	ductile	7*d*/ductile	ductile	ductile	ductile
V-3Ti-1Si	82	ductile	. . .	ductile	. . .	. . .
V-3Ti-1Si	135	ductile	. . .	ductile	. . .	. . .

[a] Broke in grips at ~100°C lower temperature.
[b] Alpha numeric designation is figure number of micrograph.

the other alloys. Of the vanadium alloys tested, the V-3Ti-1Si alloy clearly had the best performance as far as postirradiation ductility is concerned.

Scanning Electron Microscopy

The results of the SEM examinations of irradiated vanadium alloys are given in Table 4. The type of fracture observed by SEM is listed for each alloy as a function of helium level, damage level, and irradiation temperature. These results couple directly with the stress-strain curves presented in Figs. 2 through 5 and help to identify the mechanism whereby each specimen failed. Note that eight representative micrographs have been selected from the table and are referenced in the table to the micrographs in Figs. 6 and 7.

V-15Cr-5Ti—Referring to Table 4, it is seen that V-15Cr-5Ti specimens, without helium and irradiated at 420°C, failed by a cleavage-type of fracture (Fig. 6*a*). This suggests that irradiation hardening has elevated the yield strength of the material above the fracture (cleavage) stress, causing the material to fail transgranularly along cleavage planes. The ''river'' patterns on the cleavage faces (Fig. 6*a*) distinguish this type of fracture from intergranular, in which such patterns are missing from the exposed grain boundary surfaces (see Fig. 6*b*). With 300 appm He, V-15Cr-5Ti specimens failed intergranularly after irradiation to 10 dpa at 420°C (Fig. 6*b*). As will be shown later in TEM micrographs, helium bubbles formed in the grain boundaries and weakened them, thus causing intergranular failure (helium embrittlement) upon testing in tension. With less implanted helium and higher damage level (see 74 appm He in Table 4), the mechanisms of irradiation hardening and helium embrittlement overlap, producing a fracture surface characteristic of both (that is, cleavage and intergranular separation), as shown in Fig. 6*c*. For the V-15Cr-5Ti specimen conditions represented in this figure, the fraction of cleavage-type fracture outweighs that due to intergranular separation. At 520°C, 40 dpa, and 74 appm He, the situation was reversed (refer to Table 4) and the fracture mode was mostly intergranular with a small amount of cleavage as shown in Fig. 6*d*. Three different fracture modes were observed after testing the specimen with 14 appm He, irradiated to 17 dpa at 520°C (see Table 4). Specimens of V-15Cr-5Ti without helium that were irradiated and tested at 520 and 600°C failed in a ductile manner and exhibited the familiar cone-like structure shown in Fig. 7*a*.

VANSTAR-7—All of the VANSTAR-7 specimens irradiated and tested at 420 and 600°C displayed ductile-type fracture surfaces similar to those shown in Fig. 7*a*. At 520°C, irradiation to 40 dpa produced enough hardening in a specimen without helium to cause some cleavage to develop along on a generally ductile fracture surface (Fig. 7*b*). Under these same conditions, a specimen with 70 appm He failed intergranularly as shown in Fig. 7*c*, accounting for the low ductilities shown in Figs. 3 and 5. Intergranular fracture surfaces and some cleavage fracture were found at 17 dpa with 150 appm He.

V-3Ti-1Si—The relatively high ductility of this alloy demonstrated in the postirradiation tensile tests is also reflected in the results of the SEM examinations (Table 4). All of the irradiated V-3Ti-1Si specimens, with and without implanted helium, failed in a ductile manner. As an example, the ductile-type fracture surface of a specimen irradiated at 420°C without helium is shown in Fig. 7*d*.

Transmission Electron Microscopy

The results of postirradiation tensile tests and SEM examinations support two mechanisms by which some of the vanadium alloys were embrittled by neutron irradiation: irradiation hardening

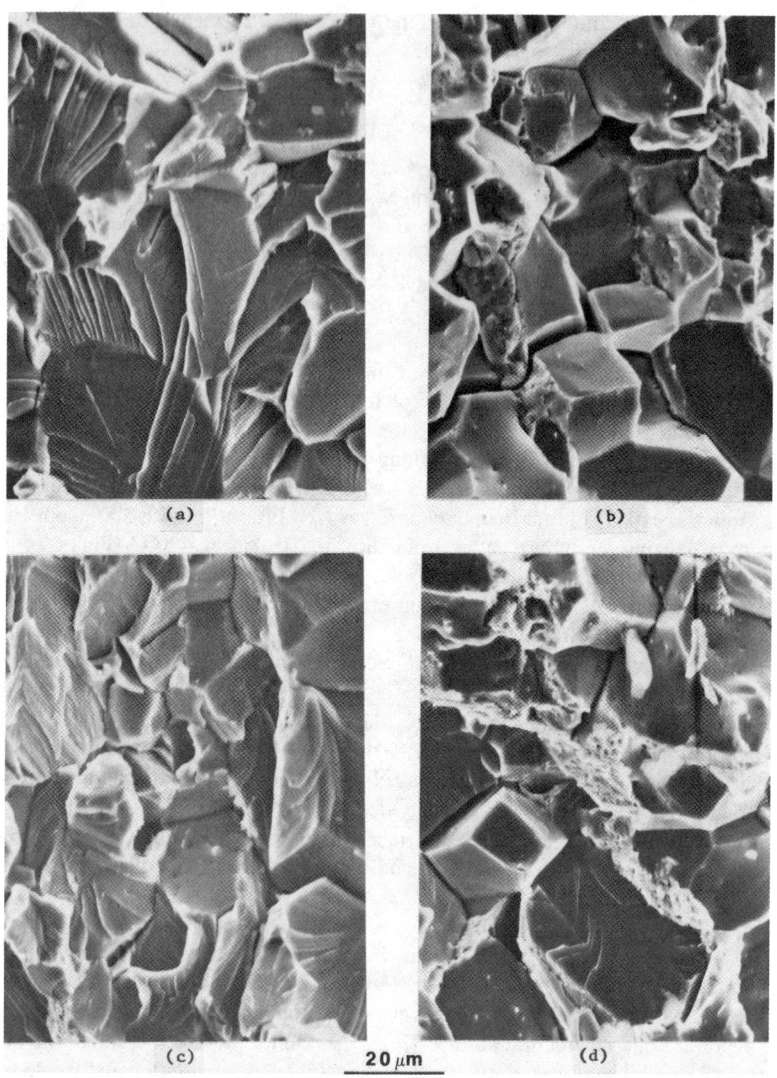

FIG. 6—*Fractographs of V-15Cr-5Ti after irradiation in FFTF(MOTA) and tensile testing at the irradiation temperature (refer to Table 4):* (a) *No helium, irradiated at 420°C to 40 dpa, cleavage.* (b) *300-appm He, irradiated at 420°C to 10 dpa, intergranular.* (c) *74-appm He, irradiated at 420°C to 40 dpa, cleavage plus intergranular.* (d) *74-appm He, irradiated at 520°C to 40 dpa, intergranular plus cleavage.*

and helium embrittlement. Before exploring a third source of material degradation—void swelling—it would be useful to correlate these two forms of embrittlement with the microstructure. The TEM micrographs in Fig. 8 show representative microstructures of the vanadium alloys that produced the three major types of tensile fracture observed: cleavage, intergranular separation, and ductile. Figure 8*a* shows the microstructure of V-15Cr-5Ti, without implanted helium, that was irradiated to 40 dpa at 420°C. Referring back to Figs. 2 and 5 (and Table 4), it is seen that this specimen failed at low strain with a cleavage mode of fracture. The microstructure in Fig. 8*a*

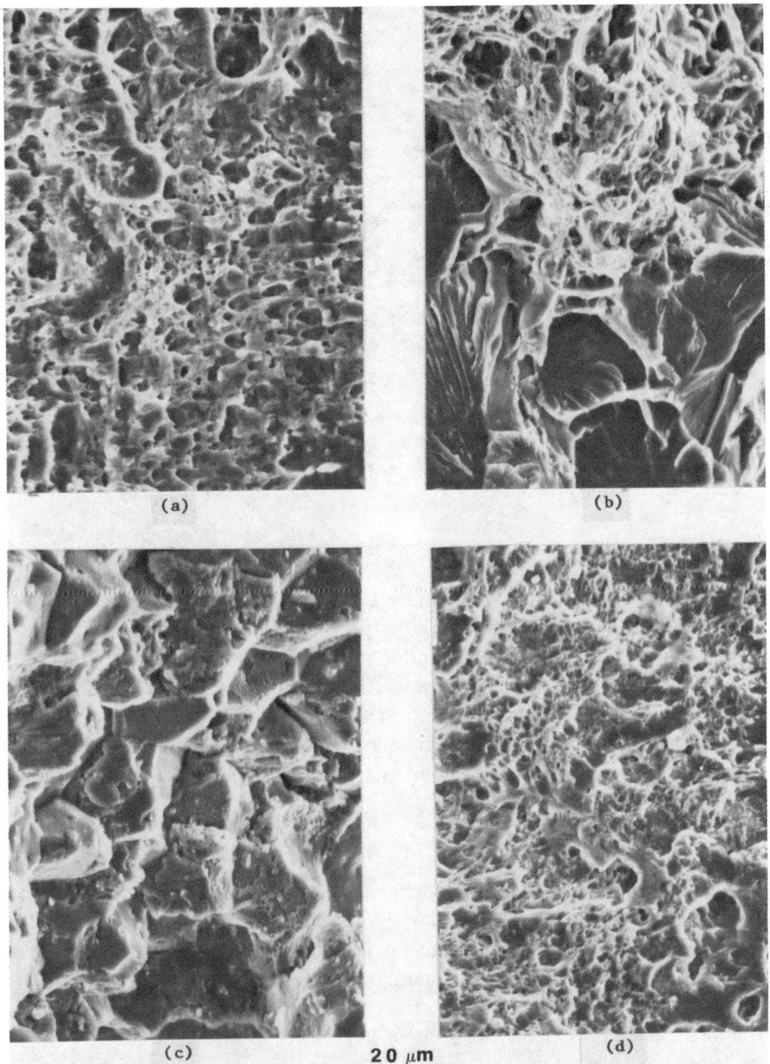

FIG. 7—*Fractographs of vanadium alloys after irradiation in FFTF(MOTA) and tensile testing at the irradition temperature (refer to Table 4):* (a) *V-15Cr-5Ti, no helium, irradiated at 600°C to 40 dpa, ductile.* (b) *VANSTAR-7, no helium, irradiated at 520°C to 40 dpa, ductile plus cleavage.* (c) *VANSTAR-7, 70-appm He, irradiated at 520°C to 40 dpa, intergranular.* (d) *V-3Ti-1Si, no helium, irradiated at 420°C to 40 dpa, ductile.*

is consistent with this mode of failure as it shows a high density of small dislocation loops and segments that are characteristic of irradiation hardening. There were no helium bubbles or voids observed in the structure. Figure 8*b* is a micrograph of the same alloy, with 300 appm He, that was irradiated to 17 dpa at 520°C and failed with only ~2% total elongation (see Fig. 5). It was not surprising that this specimen failed intergranularly along grain because it contained large helium bubbles along the grain boundaries.

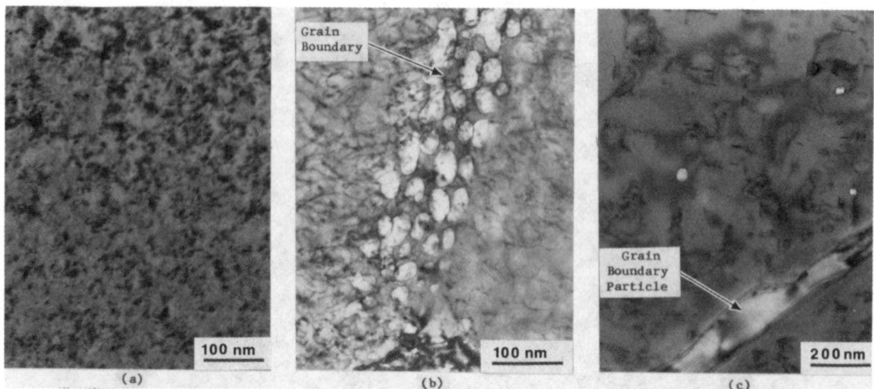

FIG. 8—*Representative vanadium alloy microstructures for different tensile behavior and fracture modes:* (a) *V-15Cr-5Ti, no helium, irradiated at 420°C to 40 dpa, cleavage fracture.* (b) *V-15Cr-5Ti, 300-appm He, irradiated at 520°C to 17 dpa, intergranular fracture.* (c) *V-3Ti-1Si, 82-appm He, irradiated at 600°C to 40 dpa, ductile fracture.*

The shape of many of the bubbles indicates that bubble coalescence had occurred at some stage of the experiment, probably during the tritium trick procedure itself. The ramifications of this result will be discussed later in this paper. Finally, a microstructure that effectively resisted both types of embrittlement and enabled the retention of good tensile ductility is shown in Fig. 8*c*. It shows V-3Ti-1Si with 80 appm He after irradiation to 40 dpa at 600°C. A total elongation of ~12% was measured for this specimen tested at 600°C (Fig. 5). The microstructure is complex: large particles and extremely small helium bubbles in the grain boundaries, dislocation segments in the grain matrices, planar defects or precipitate phase with fringe contrast, an unknown phase with tiny bubbles attached, and a few voids.

Void Swelling—The microstructural response of the three vanadium alloys to irradiation-induced void swelling is summarized in Fig. 9. Representative electron micrographs are shown for each alloy at each of the three irradiation temperatures. In this figure, specimens containing ~80 appm He are compared. The micrographs were taken within grain matrices using absorption contrast imaging conditions and were used subsequently to calculate void swelling ($\Delta V/V$) [*10*]. Void swelling is observed to vary markedly with irradiation temperature. The actual swelling values are listed for irradiated specimens, with and without helium, in Table 3 and are shown graphically as a function of irradiation temperature in Fig. 10. The swelling behavior differs among the alloys. The peak swelling temperature for the highest swelling alloy, VANSTAR-7, was at 520°C while V-15Cr-5Ti had its highest swelling at 600°C. The V-3Ti-1Si specimens containing helium had a swelling minimum at 520°C, and specimens without helium showed no swelling at 600°C. Except for this last instance, the swelling behavior of specimens without helium followed those containing helium in a parallel fashion and in all cases exhibited lower swelling values. Therefore, it was concluded that helium slightly enhanced void swelling. Finally, it should be emphasized that the majority of swelling values measured for the three vanadium alloys are really very low. The swelling data in Fig. 10 was plotted on a logarithmic scale, because it was well-behaved and illustrates some consistent effects of helium. However, this can also be misleading because the lower values are overemphasized. From a technological standpoint it would have been better to use a linear scale for swelling, then all but two or three points would drop to points just slightly above zero swelling. Only the swelling values for VANSTAR-7 at 520°C (~1 and 6%) and perhaps the V-15Cr-5Ti specimens at 600°C (~0.3%) would be considered significant.

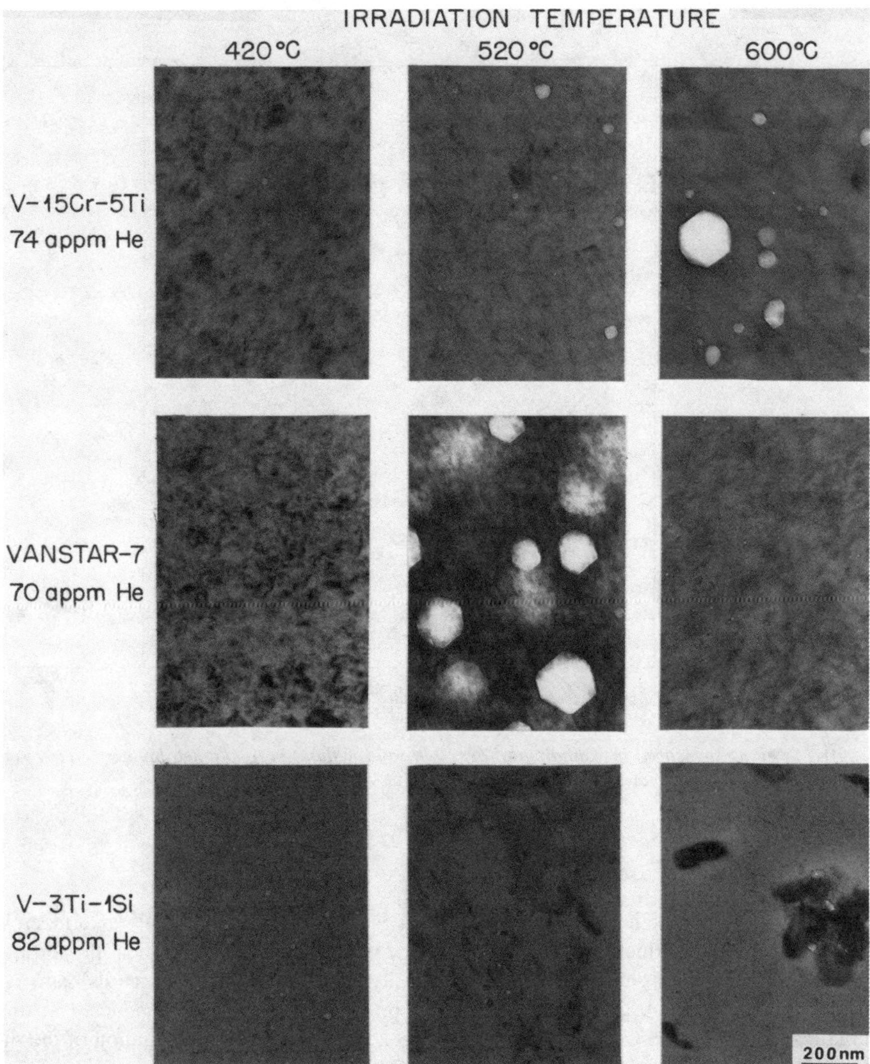

FIG. 9—*Representative micrographs of vanadium alloys after irradiation in FFTF(MOTA) to 40 dpa showing irradiation-produced voids.*

Discussion

Helium Implantation

Helium must be preimplanted in vanadium alloys because nuclear reactors such as FFTF have neutron spectra that will not produce the helium expected in a fusion reactor. A fusion reactor will produce about 5 appm He/dpa [*11*]. Although the tritium trick has been used by a number of investigators to determine the effects of helium on metals [*12,13*], the present experiment may be one of the first where it has been utilized to preimplant helium prior to irradiation. Helium is usually implanted with a cyclotron. There are a number of advantages to using tritium rather than

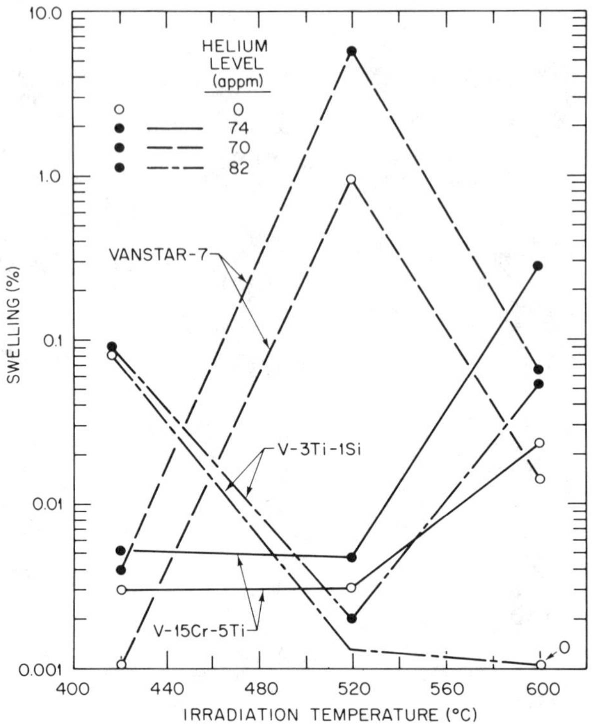

FIG. 10—*Swelling measured in vanadium alloys, with and without preimplanted helium, after irradiation in FFTF(MOTA) to 40 dpa. Note logarithmic scale for void swelling.*

a cyclotron for injecting the helium. The tritium trick will implant ^3He uniformly on a macroscopic scale in large numbers of thick specimens that can be used to provide mechanical properties data. The cyclotron technique is capable of injecting only very thin specimens (foils) and is very expensive. In the tritium trick, the tritium decay process can easily be conducted at elevated temperatures (the subsequent irradiation temperatures), making it a good simulation of the thermal processes occurring in a fusion reactor first wall. At the same time, the reader is reminded that this is probably more severe than actual service because implanted specimens start out with helium bubble distributions that may be representative of several years in the reactor. There are some additional concerns in using the tritium trick. First, tritium constitutes a slight radiation hazard and implanted specimens will usually contain small residual amounts of the isotope. They must be handled and stored in specified ways and tested in dedicated systems. Even more annoying is the fact that some of the ^3He is converted back to ^3H by neutron irradiation. At 40 dpa, only ~60% of the original ^3He is left in the specimen, even accounting for some retransformation of tritium back to ^3He (in a closed system) [*14*]. This last feature may also introduce small amounts of ^3He into unimplanted specimens sharing the same capsule. In spite of these drawbacks, the tritium trick appears to be a very convenient method of implanting helium in bulk specimens of vanadium alloys and lends itself nicely for screening studies and alloy development. Although the topic is not addressed in this paper, the tritium trick can also be used to provide important information on the effects of tritium itself in a vanadium alloy first wall.

Postirradiation Tensile Behavior

The main effects of neutron irradiation on the vanadium alloys studied are irradiation hardening, helium embrittlement, and void swelling. Although all three mechanisms are present through the range of experimental conditions used, one was usually prominent for any set of conditions. Irradiation hardening or irradiation embrittlement has been recognized for some time and is caused by the creation of dislocations, voids, and precipitate particles within the material. These imperfections resist the flow of dislocations through the structure during plastic deformation. Vanadium and other bcc metals already exhibit some unique tensile characteristics, such as yield point behavior, which is believed to be related to the interaction of dislocations with interstitial impurity elements such as C, N, and O. After irradiation, the concentration of defects in the microstructure increases radically, hardening the alloy. This is illustrated clearly in the stress-strain curves for all three alloys at 420 and 520°C as shown in Figs. 2 through 4. In the case of VANSTAR-7 at 420°C (Fig. 3), the amount of hardening was proportional to damage level with little sign of saturation, while at 520°C the hardening appears to have saturated by 40 dpa. The V-3Ti-1Si was most resistant to irradiation hardening (Fig. 4) as increases in yield strength were not as large and total elongations were greater. Furthermore, the hardening quickly saturated with increased damage level. The V-15Cr-5Ti specimens not only were hardened substantially, but at 420°C failed by cleavage at very low strains (Figs. 2 and 6; Table 4). For those specimens the stress needed to cause cleavage fracture (cleavage stress) was exceeded after very little deformation. At 520°C, the cleavage stress was not exceeded, but the specimen containing 74 appm He (curve I40/74) failed prematurely by intergranular separation. In this instance, the hardening of the matrix and weakening of the grain boundaries by helium bubbles acted together, perhaps even synergistically, to reduce specimen ductility. At 600°C, some hardening was observed for V-15Cr-5Ti and VANSTAR-7 (Figs. 2 and 3), but very little hardening and excellent total elongation were exhibited by V-3Ti-1Si specimens with and without helium (Fig. 4). At this highest temperature, VANSTAR-7 specimens began to show some ductility problems (Fig. 3) that may have been related to their high swelling rather than hardening or helium, because the respective fracture surfaces were ductile (Table 4). The helium-implanted V-15Cr-5Ti specimens at 600°C suffered severe helium embrittlement and failed inergranularly during loading into the tensile machine chamber. However, the unimplanted specimen of the same alloy had excellent ductility (Fig. 2—I40/0).

Void Swelling

Both the V-15Cr-5Ti and V-3Ti-1Si were quite resistant to void swelling as shown in the micrographs in Fig. 9, graphically in Fig. 10, and in Table 3. On the other hand, VANSTAR-7 with 70 appm He demonstrated a moderately high swelling of nearly 6% at 520°C. Without helium, the same alloy showed swelling of ~1% under the same conditions. The reason why the VANSTAR-7 produced such high swelling values is not clear. It is unlikely that high levels of nitrogen or oxygen in the alloy alone were responsible, because levels of these gases were considerably lower in VANSTAR-7 than in the V-3Ti-1Si, which exhibited the lowest swelling of the three alloys. However, zirconium may not be as effective in tying up nitrogen and oxygen (which enhance cavity formation) in VANSTAR-7 as titanium is in the other two alloys. Only 1.3% by weight zirconium is present in the alloy to begin with and over one-third of that amount is tied up in the ZrC phase. The other two alloys have lower carbon (less carbides) and substantial titanium contents (see Table 1).

Comparison to Other Irradiation Experiments

The V-15Cr-5Ti alloy has been included in at least two previous neutron irradiation experiments: Magnetic Fusion Energy-2 (MFE-2) in Oak Ridge Research Reactor (ORR) [*15*], and X-287 in

Experimental Breeder Reactor II (EBR-II) [*16*]. In MFE-2, the V-15Cr-5Ti specimens were irradiated in sodium-filled stainless steel capsules at 375, 550, and 600°C to approximately 5 dpa. Tensile tests showed that most of the specimens were severely embrittled. However, it was impossible to determine whether the embrittlement was due to irradiation hardening, interstitial pickup, or perhaps some other segregating element. The latter possibility was studied in some detail by Gold and Bajaj [*15*] and later by Diercks and Smith [*17*] using Auger spectroscopy. Diercks and Smith found that no correlation could be made between sulfur segregation and fracture behavior. To further complicate the analysis of the experiment, the starting material itself exhibited only one-half the total elongation of later heats of V-15Cr-5Ti. The author believes that the present V-15Cr-5Ti specimens performed better than the earlier heat in MFE-2 because of better starting material (lower interstitial content) and the improved encapsulation technique that used lithium instead of sodium.

The V-15Cr-5Ti specimens in the X-287 experiment were sealed in sodium-filled capsules containing a zirconium getter material and were irradiated to 24 to 32 dpa at 400, 525, 625, and 700°C [*16*]. Some specimens were injected with ~80 appm He using a cyclotron. Some of the results for the X-287 experiment agree with the present work (that is, effects of irradiation hardening and helium), but the starting material again showed total elongations that were one-half of the present heat. Therefore, all of the ductilities that could be compared directly were lower in the X-287 experiment. Although the precaution of using a getter material in the capsules was taken, the absence of thermally aged controls makes it difficult to assess the effectiveness of the getter in preventing specimen contamination.

The only results on irradiated VANSTAR-7 that could be found in the literature were by Bentley and Wiffen [*18*], who concentrated on microstructural changes and swelling. In that experiment, VANSTAR-7 was sealed in capsules under inert gas and irradiated to ~17 dpa in EBR-II at 496, 580, 690, and 805°C. The present results for swelling generally agree with those of Bentley and Wiffen, which showed that 496°C was the temperature of highest swelling for VANSTAR-7. The present irradiated microstructures of VANSTAR-7 at 520 and 600°C were also respectively similar to those irradiated in EBR-II at 496 and 580°C, except that the present void sizes and concentrations were larger for the higher damage level of 40 dpa.

Previous results on irradiated V-3Ti-1Si have been reported by Böhm [*19*]. Specimens were irradiated at 640°C in sodium-filled capsules to 1×10^{22} n/cm^2 ($E > 0.1$ MeV) in the BR-2 reactor in Möl/Belgium. Post-irradiation tensile tests showed a ~65% increase in yield strength at 650°C compared to the ~20% measured in this experiment at 600°C (compare results for V-3Ti-1Si in Tables 3 and 4). Böhm attributed the increase to irradiation as well as interstitial pickup in the surface regions of the specimens. The present results that show less total hardening at 600°C are consistent with those of Böhm in the sense that the component of hardening due to interstitial pickup was absent in the present specimens and total hardening would be expected to be less. It is interesting to note that the same planar imperfections or phase with stacking fault fringe contrast observed by Böhm [*19*] in the irradiated V-3Ti-1Si microstructures were also found in this experiment. More work is needed to prove that these imperfections are radiation-induced, as assumed by M. Bocek and J. O. Elen [*20*].

Conclusions

1. The tritium trick provides an excellent technique for preimplanting helium in bulk specimens of vanadium alloys for irradiation performance screening experiments and alloy development.
2. Vanadium alloys encapsulated in TZM tubes filled with ^{7}Li were protected from interstitial (C,N,O) pickup during irradiation.
3. The V-15Cr-5Ti alloy without preimplanted helium was most susceptible to irradiation

hardening and failure by cleavage at 420°C. VANSTAR-7 was less susceptible and V-3Ti-1Si was least affected by this mechanism.

4. The V-15Cr-5Ti alloy was also more sensitive to helium embrittlement, followed by the VANSTAR-7 alloy, with V-3Ti-1Si showing the least sensitivity.

5. The VANSTAR-7 alloy exhibited a relatively high swelling value of nearly 6% at 520°C. V-15Cr-5Ti and V-3Ti-1Si were much more resistant to swelling and exhibited swelling values of <0.3% at all temperatures investigated.

6. After irradiation in FFTF to 40 dpa, V-3Ti-1Si demonstrated better resistance to irradiation hardening, helium embrittlement, and void swelling than either V-15Cr-5Ti or VANSTAR-7.

Acknowledgments

The author wishes to thank D. L. Smith, Argonne National Laboratory, R. E. Gold, Westinghouse Electric Corporation, and F. W. Wiffen and J. H. DeVan, Oak Ridge National Laboratory, for initial planning of the experiment. He also acknowledges D. Kaletta, KFK Karlsruhe, West Germany, for the V-3Ti-1Si material, A. M. Ermi, Westinghouse Hanford Corporation, for the encapsulations, B. M. Oliver, Rockwell International, for helium analyses, D. W. Ramey, for the tritium trick work, E. L. Ryan, N. H. Rouse, L. T. Gibson, and H. Blevins, for specimen preparation and testing, and Frances Scarboro, for manuscript preparation.

Research sponsored by the Office of Fusion Energy, U.S. Department of Energy, under Contract No. DE-AC05-84OR21400 with the Martin Marietta Energy Systems, Inc.

References

[*1*] Smith, D. L., Loomis, B. A., and Diercks, D. R., *Journal of Nuclear Materials,* Vol. 135, 1985, pp. 125–137.

[2] Loomis, B. A., "Comparison of Swelling for Neutron and Ion-Irradiated MFE Structural Materials," ADIP Semiannual Progress Report March 31, 1985, DOE/ER-0045/14, pp. 62–67.

[*3*] Klueh, R. L. and DeVan, J. H., *Journal of Less-Common Metals,* Vol. 30, 1973, pp. 9–24.

[*4*] DeVan, J. H. and Klueh, R. L., Nuclear Technology, Vol. 24, Oct. 1974, pp. 64–72.

[*5*] Braski, D. N. and Ramsey, D. W., "A Modified Tritium Trick Technique for Doping Vanadium Alloys with Helium," in *Effects of Radiation on Materials: 12th International Symposium, STP 870,* F. A. Garner and J. S. Perrin, Eds., American Society for Testing and Materials, Philadelphia, 1985, pp. 1211–1224.

[*6*] Ermi, A. M., "FFTF Fusion Irradiation, FFTF Cycles 4–6," ADIP Semiannual Progress Report Sept. 30, 1984, DOE/ER-0045/13, pp. 21–40.

[*7*] Braski, D. N., "The Tensile Properties of Several Vanadium Alloys after Irradiation in FFTF," ADIP Semiannual Progress Report March 31, 1985, DOE/ER-0045/4, pp. 54–61.

[*8*] Gibson, J. R. and Braski, D. N., "Scanning Electron Microscope Facility for Examination of Radioactive Materials," ORNL/TM-9451, Feb. 1985.

[*9*] Baird, J. D., "The Effects of Strain-Aging Due to Interstitial Solutes on the Mechanical Properties of Metals," *Metals and Materials,* Vol. 5, Review No. 149, Feb. 1971, pp. 14–15.

[*10*] Annual Book of ASTM Standards, American Society for Testing and Materials, Philadelphia, PA, E521-83, 1983, pp. 14–15.

[*11*] McHargue, C. J. and Scott, J. L., *Metallurgical Transactions A,* Vol. 9A, 1978, pp. 151–159.

[*12*] Remark, J. F., Johnson, A. B., Jr., Farrar, H., IV, and Atteridge, D. G., *Nuclear Technology,* Vol. 29, 1976, pp. 369–377.

[*13*] Mattas, R. F., Wiedersich, H., Atteridge, D. G., Johnson, A. B., Jr., and Remark, J. F., "Elevated-Temperature Tensile Properties of V-15Cr-5Ti Containing Helium Introduced by Ion Bombardment and Tritium Decay," *Proceedings,* 2nd Topical Meeting on the Technology of Controlled Nuclear Fusion, Richland, WA, Sept. 21–23, 1976, USERDA CONF-760935-P1, pp. 199–208.

[*14*] Greenwood, L. R., Argonne National Laboratory, private communication, July 19, 1983.

[*15*] Bajaj, R., and Gold, R. E., "Evaluation of V-15Cr-5Ti Specimens After Neutron Irradiation in the MFE-2 Experiment," Phase I Final Report, Westinghouse Electric Corporation, Advanced Energy Systems Division, Aug. 1983.

[16] Grossbeck, M. L. and Horak, J. A., "Tensile Properties of Helium-Injected V-15Cr-5Ti after Irradiation in EBR-II," ADIP Semiannual Progress Report Sept. 30, 1984, DOE/ER-0045/13, pp. 99–103.
[17] Diercks, D. R. and Smith, D. L., "Environmental Effects on the Properties of Vanadium-Base Alloys," ADIP Semiannual Progress Report March 31, 1985, DOE/ER-0045/14, pp. 179–180.
[18] Bentley, J. and Wiffen, F. W., Nuclear Technology, Vol. 30, 1976, pp. 376–384.
[19] Böhm, H., "The Effect of Neutron Irradiation on the High-Temperature Mechanical Properties of Vanadium-Titanium Alloys," *Proceedings,* International Conference on Defects and Defect Clusters in B.C.C. Metals and Their Alloys Gaithersburg, MD, Aug. 1973, R. J. Arsenault, Ed., M. P. Graphics, Inc., Washington, D.C., pp. 163–175.
[20] Bocek, M. and Elen, J. D., *Journal of Nuclear Materials,* Vol. 44, 1972, p. 194.

Martin L. Grossbeck[1] *and James A. Horak*[1]

Tensile and Fracture Properties of EBR-II-Irradiated V-15Cr-5Ti Containing Helium

REFERENCE: Grossbeck, M. L. and Horak, J. A., "**Tensile and Fracture Properties of EBR-II-Irradiated V-15Cr-5Ti Containing Helium,**" *Influence of Radiation on Material Properties: 13th International Symposium (Part II), ASTM STP 956,* F. A. Garner, C. H. Henager, Jr., and N. Igata, Eds., American Society for Testing and Materials, Philadelphia, 1987, pp. 291–309.

ABSTRACT: The alloy V-15Cr-5Ti was cyclotron-implanted with 80 appm He and subsequently irradiated in the Experimental Breeder Reactor (EBR-II) to 30 displacements per atom (dpa). The same alloy was also irradiated in the 10, 20, and 30% cold-worked conditions. Irradiation temperatures ranged from 400 to 700°C. No significant effects of helium on mechanical properties were found in this temperature range although the neutron irradiation shifted the temperature of transition from cleavage to ductile fracture to about 625°C. Ten percent cold work was found to have a beneficial effect in reducing the tendency for cleavage fracture following irradiation, but high levels (20%) were observed to reduce ductility. Still higher levels (30%) improved ductility by inducing recovery during the elevated-temperature irradiation. Swelling was found to be negligible, but precipitates—titanium oxides or carbonitrides—contained substantial cavities.

KEY WORDS: vanadium, helium, ductility, swelling, microstructure, fracture, tensile properties, embrittlement, irradiation

Vanadium alloys have been studied for use as cladding and structural materials for fast reactors [*1*]. Although unalloyed vanadium is not particularly swelling resistant, many of its alloys, and particularly those containing titanium are very swelling resistant [*2,3*]. In addition to high swelling resistance, V-15Cr-5Ti has high creep strength. It is these properties combined with higher melting points than iron- and nickel-based alloys and compatibility with liquid alkali metals that make vanadium alloys, and in particular V-15Cr-5Ti, candidates for first wall and blanket structures in fusion reactors [*4*]. This alloy and several others were investigated for fusion reactor application, but studies of these alloys waned because of high hydrogen diffusivity and solubility, resulting in high tritium inventory in the structures and high tritium leakage. It is now believed that diffusion barriers can be employed to reduce leakage and that tritium can be removed from the first wall and blanket prior to its decommissioning [*5*]. The disadvantage of tritium inventory is believed to be compensated for by the low neutron activation properties of vanadium alloys. Induced radioactivity is sufficiently low to allow direct trench burial upon decommissioning [*6*].

Fission reactors can be used to study the effects of radiation displacement damage on candidate fusion reactor materials, but the production of helium expected in a fusion reactor (~5 appm He/displacements per atom (dpa)) [*7*] must be simulated. Two methods that are employed to deposit helium in vanadium are the tritium trick [*8*] and accelerator implantation. Neither method is an ideal simulation of fusion reactor service, but both methods complement each other and together provide required information on the effects of helium. In the tritium trick, tritium is diffused into the metal where it decays to ^{3}He. This method deposits helium at sites to which tritium diffuses

[1] Metals and Ceramics Division, Oak Ridge National Laboratory, P.O. Box X, Oak Ridge, TN 37831.

and becomes trapped and is limited by the solubility and the decay constant of tritium. However, tritium will be present in vanadium structures in a fusion reactor so that at least some helium will result from just this mechanism. Accelerator implantation limits study to thin specimens and is a difficult and costly procedure, but deposition can be made homogeneous as is the case for helium produced by transmutations in a fusion reactor. Either method can be used to introduce helium, but neither method introduces helium simultaneously with neutron displacement damage. Such an experiment will not be possible until either a high flux 14 MeV neutron accelerator or an actual fusion device becomes available.

In the absence of such a device, the present experiment was designed to evaluate the effects of neutron irradiation and helium on the tensile and fracture properties of V-15Cr-5Ti. Helium was implanted with a cyclotron, then the specimens were irradiated in the Experimental Breeder Reactor (EBR-II). Postirradiation tensile tests were then performed on the specimens. This type of experiment achieves both homogeneously distributed helium and atomic displacement damage thus simulating the two most important components of fusion reactor irradiation.

A second part of this experiment evaluates the effect of cold work on V-15Cr-5Ti. When the effects of irradiation on the swelling and mechanical behavior of nonferrous alloys are evaluated, the nonferrous alloys in the annealed condition are most often compared with Type 316 stainless steel in the 20% cold-worked condition. The cold-work study was done to provide data for a more relevant comparison of vanadium alloys with Type 316 stainless steel. Additionally, the effects of cold work in irradiated V-15Cr-5Ti have not been previously investigated. In this phase of the experiment, 10, 20, and 30% cold-worked material was neutron irradiated, but no helium was implanted.

The postirradiation examination consisted of three phases. Tensile tests at the irradiation temperature were conducted to evaluate strength and ductility. These tests were followed by scanning electron microscopy to determine the mode of fracture. The third phase, transmission electron microscopy, was done to determine the reasons for the fracture behavior, in particular to determine the mechanism of the irradiation embrittlement.

Experimental Procedure

Specimen material was prepared from a heat of V-15Cr-5Ti melted by Westinghouse Electric Corp. (heat HSV-307); the composition is given in Table 1. Rolled sheet 0.25-mm thick was given a recrystallization anneal at 900°C for 0.5 h in a vacuum environment at a pressure of 0.5 to 2×10^{-7} torr ($0.7–3 \times 10^{-5}$ Pa). This heat treatment resulted in a grain size of 30 μm. The cold-worked material was prepared by annealing sheets at 900°C in thicknesses that resulted in 10, 20, and 30% reduction of area upon rolling to provide 0.25-mm-thick sheet.

The sheet was implanted with 52-MeV α-particles in the Oak Ridge Isochronous cyclotron. The sheet formed the front face of the irradiation chamber through which 18 to 20°C water flowed at a rate of 9.5×10^{-4} m^3/s. Since the α-particle penetration range is larger than the specimen

TABLE 1—*Composition of V-15Cr-5Ti Heat HSV-307.*

Element	% by weight	Element	% by weight
Cr	<14.8	Ni	0.004
Cu	0.02	P	0.0003
Fe	0.07	S	<0.001
Mn	0.001	Ti	4.8
Mo	0.03	Zr	0.01
Nb	0.02	V	bal

thickness, a rotating energy degrader was used to produce a uniform helium deposition over the entire thickness of the sheet. The beam was collimated to provide essentially uniform helium deposition over the region which was to become the gage sections of the specimens. In addition, the chamber was oscillated horizontally in front of the beam to irradiate the entire width of the sheet. Chemical analysis later indicated that a helium concentration of 80 appm was achieved.

A tensile specimen was designed to permit uniform α-particle deposition while still providing a valid measure of strength and ductility. The specimen design is shown in Fig. 1. The 0.25-mm thickness also made possible fabrication of tensile specimens by stamping with a precision die. The general reliability of the tensile specimen has been assessed and found satisfactory, although not as good as larger specimens, as reported by Klueh [*9*].

Following helium implantation the specimens were encapsulated in sodium and irradiated in subassembly X-287 in row 7 of EBR-II at 400, 525, 625, and 700°C. The capsules irradiated at the three highest temperatures contained zirconium foil as a getter. Each capsule was designed to operate isothermally, using a helium gas gap at 525°C and heat pipes at 625 and 700°C for temperature control. The 400°C capsules were in direct thermal contact with the reactor coolant. Differential thermal expansion monitors indicated that maximum temperatures of 425 and 525°C were attained in the 400 and 525°C capsules. The monitors in the 625 and 700°C capsules were damaged. A fluence of 4.8 to 5.5×10^{26} neutrons (n)/m^2 ($E > 0.1$ MeV) was obtained resulting in a damage level for the specimens of 28 to 32 dpa. The helium level, 80 appm, corresponds to 16 dpa in a fusion reactor.

Following irradiation, the sodium was removed from the capsules using liquid anhydrous ammonia. Ammonia dissolves sodium without producing hydrogen and thus avoids the danger of introducing hydrogen into the specimens [*10*]. When the specimens appear free of sodium, they are washed with ethyl alcohol followed by water to remove residual sodium compounds.

All specimens were tested at the nominal irradiation temperatures on an Instron Model 1125 testing machine equipped with a vac-ion-pumped vacuum chamber. Pressures below 7×10^{-7} torr (9×10^{-5} Pa) were maintained during the test. The duration of exposure at elevated temperatures was minimized to prevent significant contamination by oxygen during the test. Preliminary tests were conducted to confirm that the vacuum was sufficient to prevent oxygen contamination during testing.

Scanning electron microscope (SEM) fractography was done on an Amray 1200 B SEM extensively modified for remote operation in a radiation hot cell [*11*]. Transmission electron microscopy (TEM) was performed on a JEOL 100CX using an accelerating voltage of 120 kV.

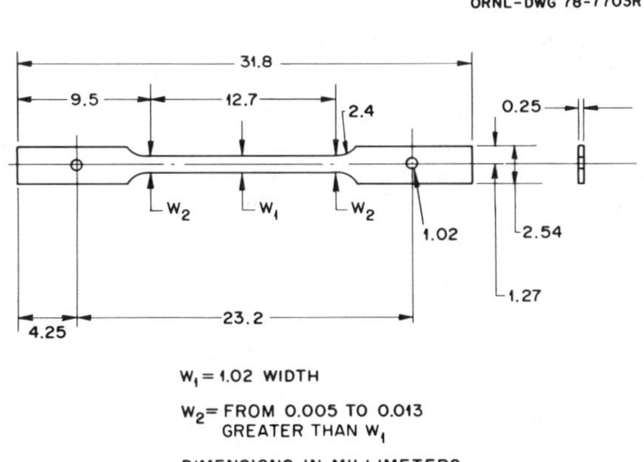

FIG. 1—*Sheet tensile specimen.*

Experimental Results

Tensile Properties

Results of the tensile tests appear in Table 2. Strength levels for annealed material are plotted in Fig. 2 for four conditions: helium-injected, neutron-irradiated, helium-injected plus neutron-irradiated, and control material without either irradition or implantation. Helium implantation does not significantly alter the strength. However, neutron irradiation increases strength by about a

TABLE 2—*Tensile properties.*

Specimen	Condition	Test Temperature, °C	Strength, MPa		Elongation, %	
			Yield	Ultimate Tensile	Uniform	Total
			CONTROLS			
100C	Annealed	400	491	610	7.4	8.9
104C	Annealed	525	422	574	8.0	10.0
106C	Annealed	525	423	586	8.3	9.8
101C	Annealed	625	417	621	8.9	10.3
103C	Annealed	625	405	621	9.3	11.1
102C	Annealed	700	403	574	9.2	15.2
3C	10% CW	400	672	769	1.1	2.6
15C	10% CW	525	655	774	0.86	1.9
14C	10% CW	700	517	683	2.4	6.9
4C	20% CW	400	724	827	1.1	2.1
13C	20% CW	525	603	796	1.0	1.8
2C	20% CW	700	543	707	2.1	5.6
14C	20% CW	700	517	698	2.3	5.3
2C	30% CW	400	724	827	1.1	2.7
8C	30% CW	525	603	814	1.1	2.3
6C	30% CW	700	483	698	2.3	6.1
			HELIUM IMPLANTED			
3-64	Annealed	400	503	574	4.4	5.9
1-8	Annealed	400	512	590	2.1	3.5
1-55	Annealed	525	479	582	7.0	9.2
1-68	Annealed	625	396	581	9.2	11.2
3-57	Annealed	700	336	531	10.0	12.3
1-58	Annealed	700	365	548	12.1	15.3
			NEUTRON IRRADIATED			
7	Annealed	625	776	853	0.38	0.50
8	Annealed	700	738	943	0.94	0.94
9	10% CW	400	1140	1260	1.1	1.2
6	10% CW	525	819	972	4.2	4.9
16	10% CW	700	608	883	1.9	6.6
11	20% CW	400	1130	1220	1.3	1.4
12	20% CW	400	1100	1200	0.94	1.0
7	20% CW	525	710	853	1.0	1.0
19	20% CW	700	612	1000	1.8	1.8
15	30% CW	400	1130	1210	0.92	1.2
11	30% CW	525	664	910	3.3	3.6
17	30% CW	700	434	533	2.3	4.4
			HELIUM IMPLANTED PLUS NEUTRON IRRADIATED			
1-80	Annealed	400	991	1170	0.76	0.76
1-72	Annealed	700	741	822	2.0	3.0

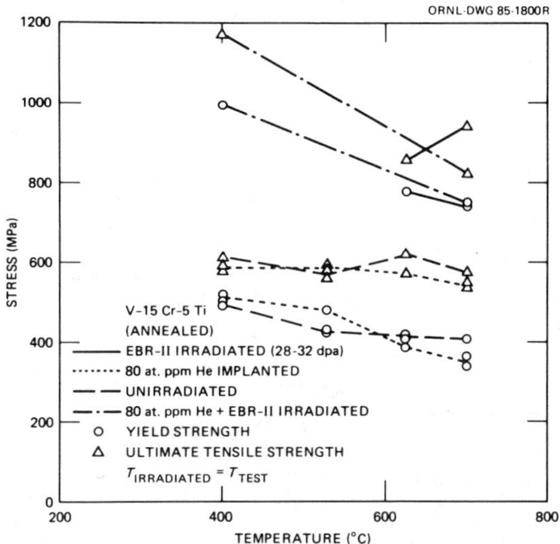

FIG. 2—*Tensile and 0.2% yield stress as a function of test temperature for annealed V-15Cr-5Ti.*

factor of 2 at 400°C and about 50% at 700°C. The neutron-irradiated, helium-injected material has similar strength to the neutron-irradiated material.

Helium has a larger effect on ductility as shown by Fig. 3. Helium alone has an effect only at 400°C where it reduces ductility by about a factor of two; elongations are approximately the same as unimplanted material at higher temperatures. Neutron irradiation reduces tensile elongations to below 1%, and neutron irradiation of helium-implanted material reduces elongation to about 2%. The difference is probably not significant.

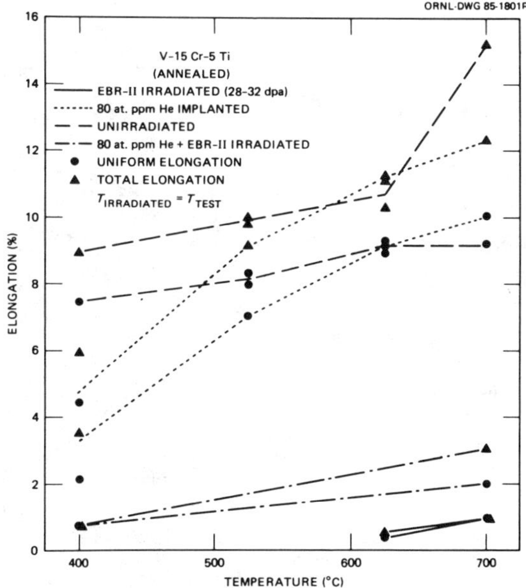

FIG. 3—*Uniform and total elongation as a function of test temperature for annealed V-15Cr-5Ti.*

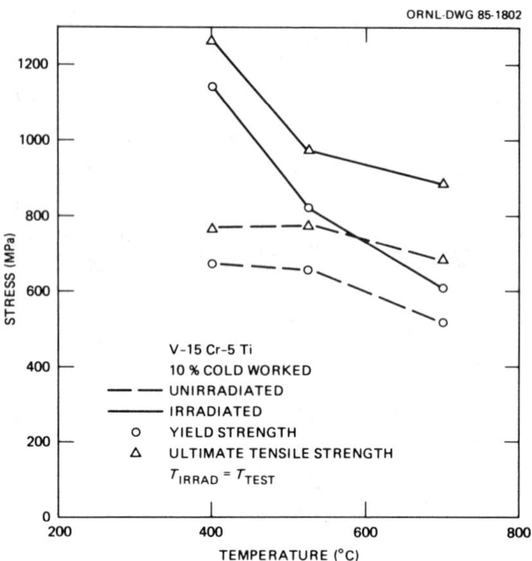

FIG. 4—*Tensile and 0.2% yield stress as a function of test temperature for 10% cold-worked V-15Cr-5Ti.*

The cold-worked material was neutron-irradiated, but none was helium-implanted. By comparison of Figs. 4 through 6, the effect of cold work on strengthening is observed to saturate by a cold-work level of 20%. Since the controls were not aged at the irradiation temperatures, the differences between the curves for irradiated material and the curves for the control material represent a combination of thermal exposure and irradiation. The curves show the effect of radiation strengthening at 400°C and, in some cases, at 525°C. In general, irradiation increased the yield and ultimate strengths for cold-worked material to about the same levels as those for annealed

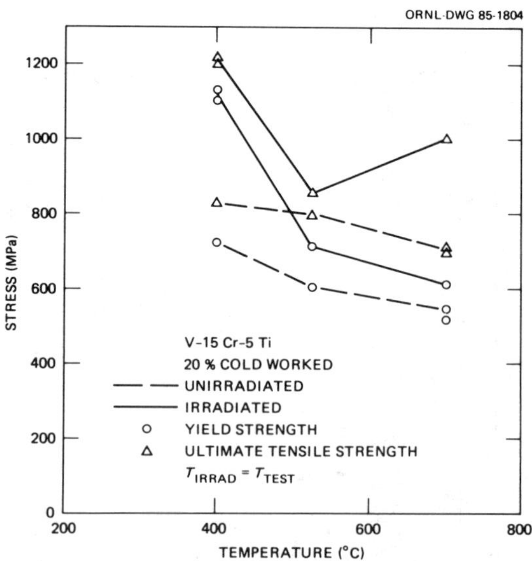

FIG. 5—*Tensile and 0.2% yield stress as a function of test temperature for 20% cold-worked V-15Cr-5Ti.*

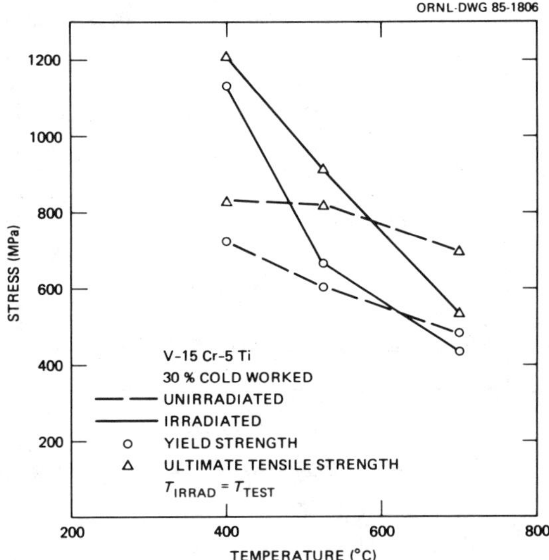

FIG. 6—*Tensile and 0.2% yield stress as a function of test temperature for 30% cold-worked V-15Cr-5Ti.*

material. However, strength decreases rapidly with increasing temperature, and in the case of 30%-cold-worked material, the additional strain energy appears to have been sufficient to induce a degree of recovery resulting in strength levels characteristic of the annealed material. The effect of irradition on ductility appears to be slight, with the exception that uniform and total elongation become nearly equal, especially at the lower temperatures.

Ductility as measured by uniform and total elongation is reduced by about a factor of four by cold working and appears to saturate at a cold-work level of 10% or less (Figs. 7 through 9). What is more significant is that, following neutron irradiation, the cold-worked material is more ductile

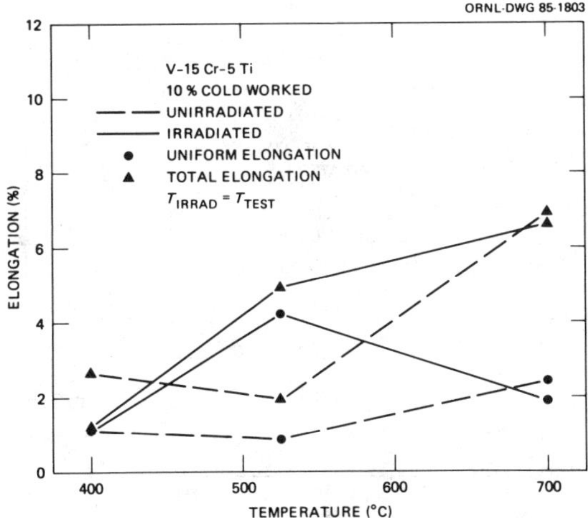

FIG. 7—*Uniform and total elongation as a function of test temperature for 10% cold-worked V-15Cr-5Ti.*

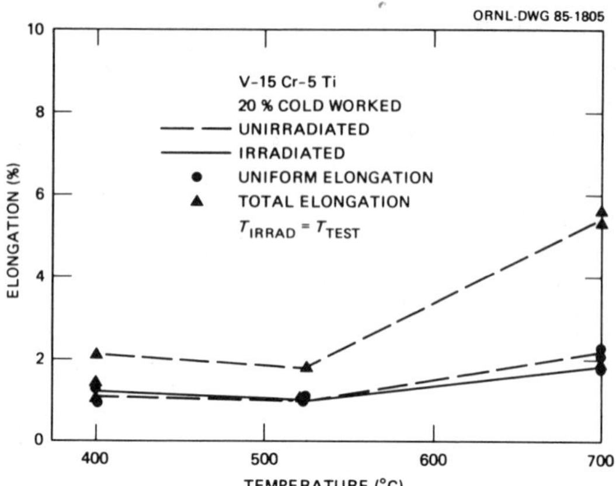

FIG. 8—*Uniform and total elongation as a function of test temperature for 20% cold-worked V-15Cr-5Ti.*

than the annealed material at all temperatures investigated. This is especially true at 700°C where total elongations for cold-worked material are about six times those of annealed material.

The effects of cold work can be seen better by plotting tensile properties as a function of cold-work level. Two properties, 0.2% yield strength and total elongation, have been selected and plotted as a function of degree of cold work. Figure 10 shows yield strength as a function of cold work for unirradiated V-15Cr-5Ti. The curves indicate the expected strengthening and slight recovery during testing at 525 and 700°C at the 30%-cold-work level. The yield strength of the irradiated material (Fig. 11) is also well behaved showing strengthening at 400°C and recovery during irradiation at 525 and 700°C. The recovery becomes more complete as the level of cold work increases.

Ductility of the unirradiated material corresponds to the yield strength in that stronger material is less ductile. Figure 12 illustrates a large reduction in ductility in going from annealed material to 10% cold-worked material. At 10% cold work, total elongation saturates then exhibits slight recovery at a cold-work level of 30%. Following irradiation, the alloy displays rather unusual behavior as shown in Fig. 13.

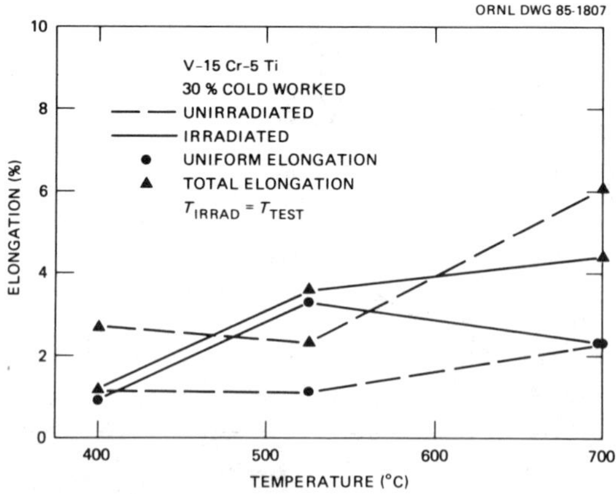

FIG. 9—*Uniform and total elongation as a function of test temperature for 30% cold-worked V-15Cr-5Ti.*

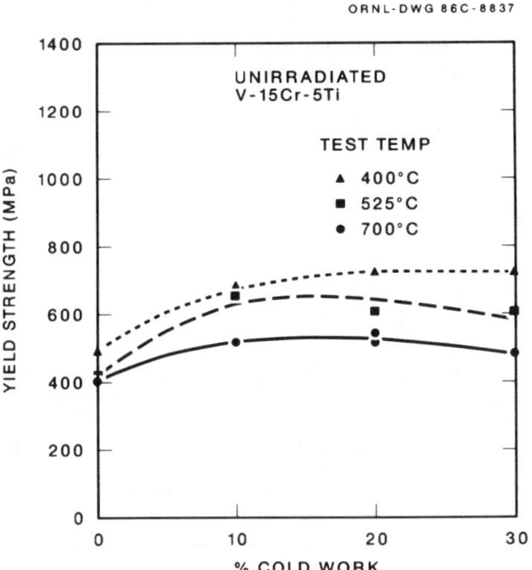

FIG. 10—*Yield strength as a function of cold-work level (reduction of area) for unirradiated V-15Cr-5Ti.*

At 400°C ductility is nearly constant. At 525 and 700°C, adding 10%-cold work results in a very large increase in ductility. This ductility is reduced to that of the annealed material if the alloy is reduced 20% by rolling but returns if the alloy is further reduced by 30%-cold work. This phenomenon is not yet understood but will be discussed later.

Fracture Surface Morphology

The general fracture surface morphology is tabulated in Table 3 and correlates well with tensile ductility. All of the unirradiated specimens failed by ductile rupture, even if implanted with

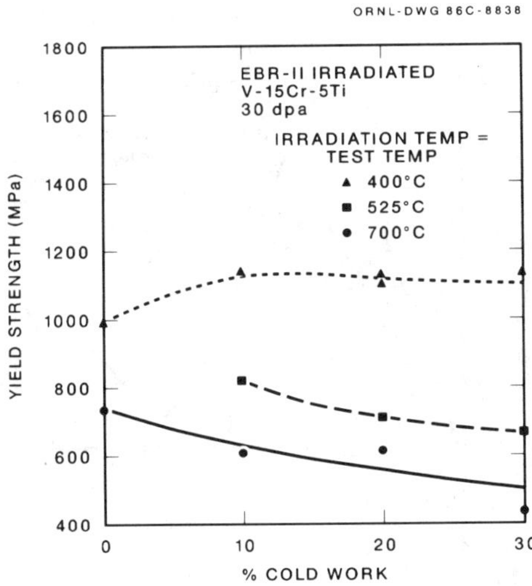

FIG. 11—*Yield strength as a function of cold-work level (reduction of area) for V-15Cr-5Ti irradiated in EBR-II to 28 to 32 dpa.*

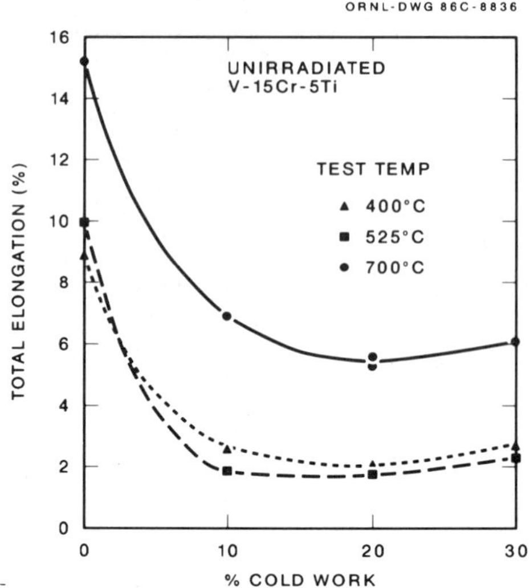

FIG. 12—*Total tensile elongation as a function of cold-work level (reduction of area) for unirradiated V-15Cr-5Ti.*

helium. The fracture surfaces of the neutron-irradiated specimens showed mechanisms ranging from ductile rupture (Fig. 14) to total cleavage in the case of specimens fractured in an uncontrolled fashion at room temperature (Fig. 15). Helium implantation, which was found to have little effect on strength and ductility, did not alter the mode of fracture although ductility appeared to be more limited than for unimplanted, unirradiated specimens. Neutron irradiation, however, resulted in cleavage fracture at 400°C in annealed material, quasi-cleavage at 625°C, and only at 700°C did

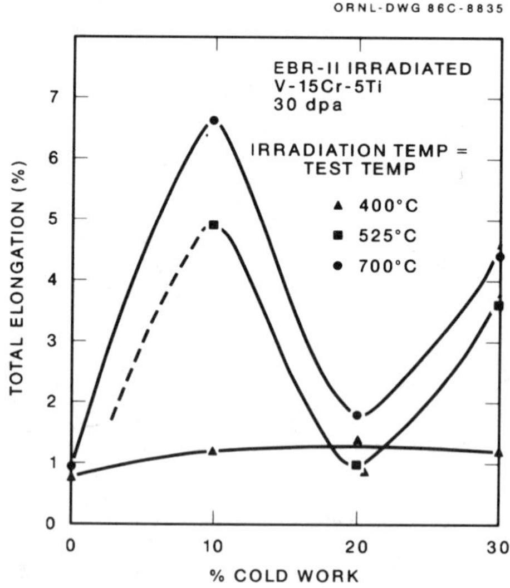

FIG. 13—*Total tensile elongation as a function of cold-work level (reduction of area) for V-15Cr-5Ti irradiated in EBR-II to 28 to 32 dpa.*

TABLE 3—*Fracture surface morphology.*

	400°C	525°C	625°C	700°C
		ANNEALED		
Unirradiated	Ductile	. . .	Ductile	Ductile
Helium implanted	Ductile	. . .	. . .	Ductile
Neutron irradiated	. . .	. . .	Quasi-cleavage	Ductile
Helium implanted plus neutron irradiated	Cleavage	. . .	. . .	Ductile
		10%-COLD WORK		
Unirradiated	Ductile	Ductile	. . .	Ductile
Neutron irradiated	Ductile/cleavage	Ductile	. . .	Ductile
		20%-COLD WORK		
Unirradiated	Ductile	Ductile	. . .	Ductile
Neutron irradiated	Quasi-cleavage/ cleavage	Ductile/cleavage	. . .	Ductile/cleavage
		30%-COLD WORK		
Unirradiated	Ductile	Ductile	. . .	Ductile
Neutron irradiated	Ductile/cleavage	Ductile	. . .	Ductile

ductile rupture become evident (Fig. 16). Note that in discussing this progression the helium-injected, neutron-irradiated material was treated the same as neutron-irradiated material since helium injection appeared to have no significant effect. The presence of cleavage fracture at the lowest temperture and complete disappearance at 700°C indicates a transition in fracture mechanism. The intermediate-type morphology, known as quasi-cleavage, at 625°C is evidence that the transition temperature is near this value.

The fracture morphology of the cold-worked material also reflected mechanical behavior, but the effects were not as apparent. In 10%-cold-worked material at 400°C, the irradiated and unirradiated material had similar ductilities, but the irradiated material had essentially no necking. The fracture surface was completely ductile in the case of unirradiated material but exhibited partial cleavage fracture in the case of irradiated material, consistent with the degradation of plastic deformation. Higher temperatures, however, resulted in failure entirely by ductile rupture. The case of 20%-cold-worked material was the most interesting in that the irradiated material had

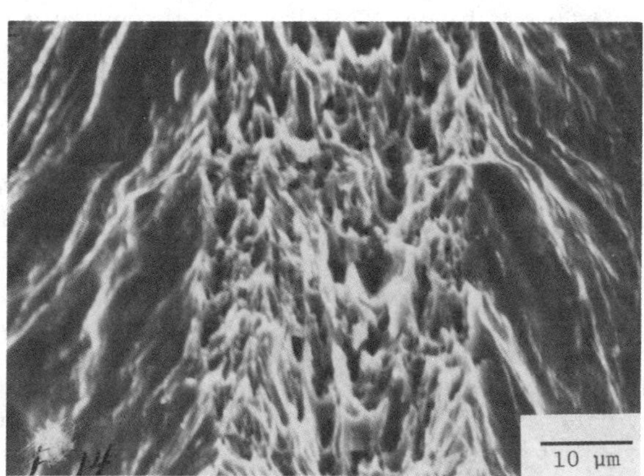

FIG. 14—*Fracture surface of unirradiated 10% cold-worked V-15Cr-5Ti specimen No. 3c fractured at 400°C.*

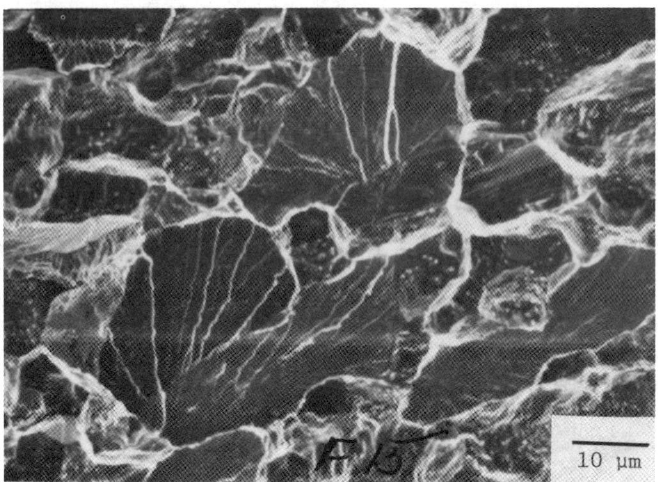

FIG. 15—*Fracture surface of annealed V-15Cr-5Ti irradiated at 625°C and fractured at room temperature.*

unexpectedly low ductilities at 525 and 700°C (Fig. 13). In both of these cases, the fracture mode was primarily ductile rupture, but some cleavage fracture was present (Fig. 17). This behavior was not repeated in the 30%-cold-worked material where complete ductile rupture was observed at 525 and 700°C. The prevalence of ductile rupture in the cold-worked material is evidence that the temperature of transition to ductile rupture is lowered by cold work.

Microstructural Observations

Transmission electron microscopy was performed on the specimens to aid in interpretation of the mechanical and fracture properties. Although no attempt was made to analyze swelling behavior or precipitation phenomena, general observations of these features will be discussed.

Helium-Injected Material—Helium-injected material was examined prior to testing, following testing, and following neutron irradiation. No helium bubbles could be found despite a careful search of grain boundaries and matrix-precipitate interfaces.

Neutron-irradiated material exhibits a dense dislocation structure which coarsens with increasing temperature. This effect may be seen by comparing Figs. 18*a* and *c* where microstructures of specimens irradiated at 400 and 700°C, respectively, are shown. Figure 18 also compares the specimen shoulder and gage sections; recall that helium was injected only into the gage sections. The microstructure appears nearly the same in the injected and noninjected regions, consistent with the similarity in strength. What appears to be a slightly coarser structure in Fig. 18*b* could have resulted from a higher temperature in the gage section during testing.

Cold-Worked Material—Figure 19 shows the structure of the 10, 20, and 30% cold worked material prior to irradiation. Consistent with the saturation in strength at the 10% cold work level, the dislocation density does not appear to vary significantly as cold work increases from 10 to 30%. The study of the effects of irradiation on the cold-worked material is really a study of the combination of cold work and recovery coupled with irradiation rather than a study of the effect of cold work alone. These effects are illustrated by the array of transmission electron micrographs

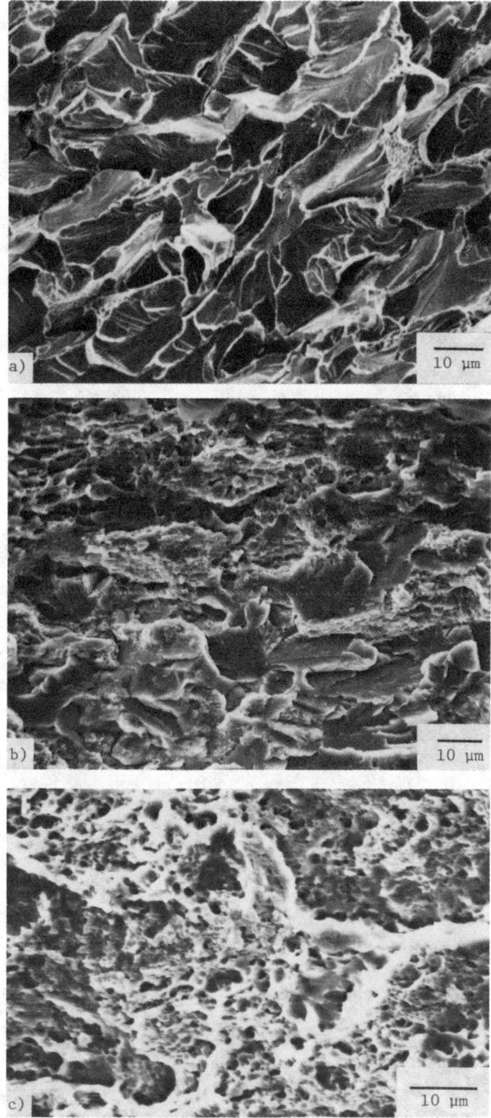

FIG. 16—*Fracture surfaces of annealed V-15Cr-5Ti irradiated to 30 dpa showing range of surface morphologies.* (a) *Specimen No. 1-80 containing 80 appm He irradiated and tested at 400°C;* (b) *specimen No. 7 irradiated and tested at 625°C;* (c) *specimen No. 8 irradiated and tested at 700°C.*

shown in Fig. 20. The micrographs were selected to show representative dislocation structures in diffraction contrast for the three cold-work levels and three temperatures. Quantitative dislocation densities were not determined, but visual comparisons will be discussed. Two general trends in microstructural development are apparent in Fig. 20. As temperature increases, microstructure coarsens. As cold work level increases, the microstructure coarsens. This is true at all three temperatures; defect density decreases moving up the figure. Both trends combine in the diagonal

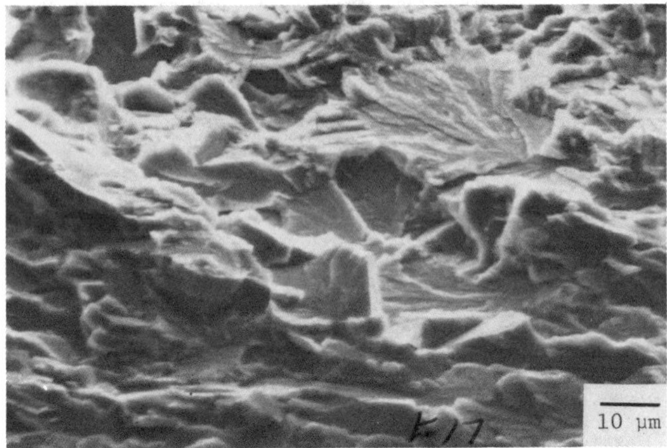

FIG. 17—*Portion of the fracture surface showing cleavage in 20% cold-worked V-15Cr-5Ti irradiated to 30 dpa at 700°C and tensile tested at 700°C.*

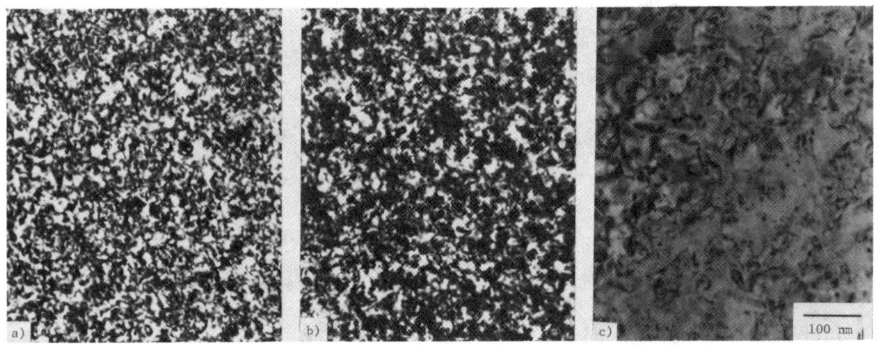

FIG. 18—*Microstructure of V-15Cr-5Ti irradiated to 30 dpa.* (a) *Shoulder of specimen No. 1-80 containing no helium, irradiated and tested at 400°C;* (b) *gage section of specimen No. 1-80 containing 80 appm He;* (c) *specimen No. 8 irradiated and tested at 700°C.*

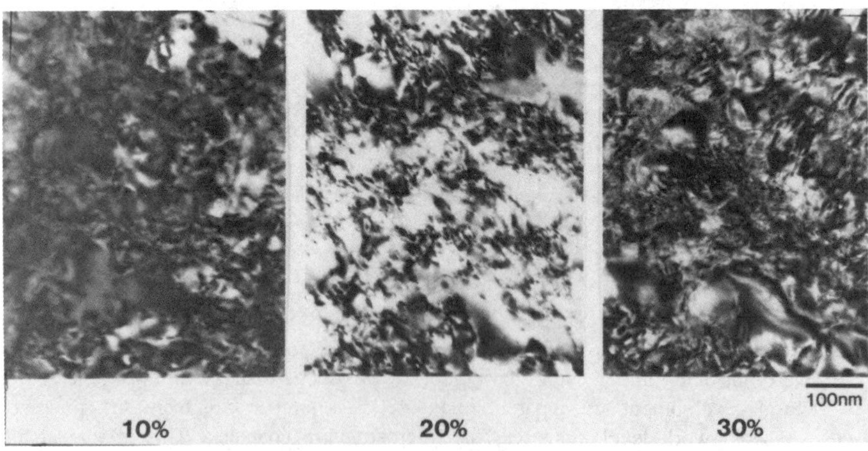

FIG. 19—*Microstructure of unirradiated 10, 20, and 30% cold-worked V-15Cr-5Ti.*

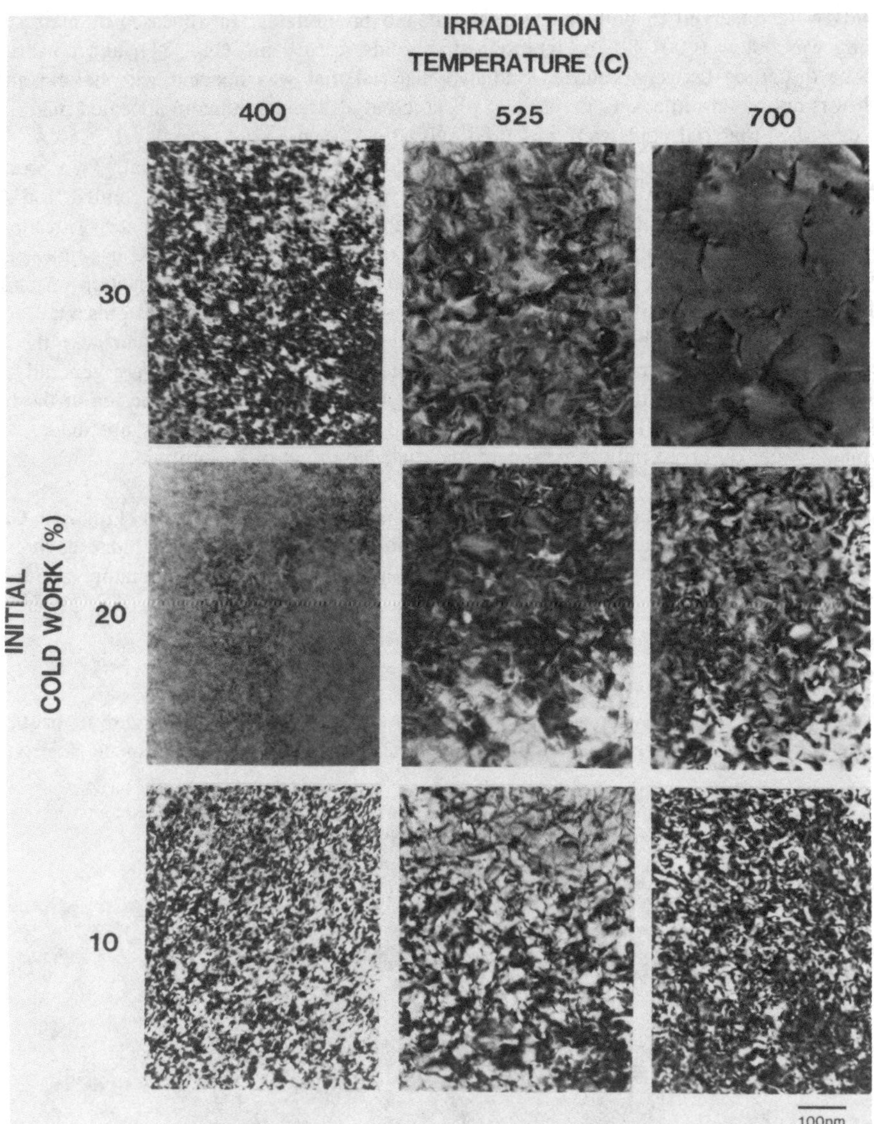

FIG. 20—*Microstructure of cold-worked V-15Cr-5Ti following irradiation to 30 dpa. No helium has been injected into these specimens. The effects of cold work and recovery may be seen.*

direction so that the micrograph in the lower left shows a much higher defect density than that in the upper right.

Swelling and Precipitation—No bubbles could be resolved in any material following helium injection. This is consistent with previous research on cyclotron implanted helium in vanadium and vanadium alloys where the implanation was below 200°C and not followed by annealing [*12*,*13*,*14*].

Voids were observed in both matrix material and precipitates. In all cases of matrix voids, swelling was below 0.001%. The largest matrix voids were 9 nm (Fig. 21), and no difference could be discerned between neutron-irradiated material that was injected with helium and that which was not helium injected. Voids were observed at all temperatures in annealed material, but in cold-worked material voids were observed only at 400°C at a cold-work level of 30%.

Precipitates were observed in all specimens including the starting material. Two precipitate morphologies were observed: blocky and larger rod-shaped precipitates. Similar rod-shaped precipitates were observed by Tanaka [*12*], and both blocky and rod precipitates were observed by Sprague et al [*13*]. Although previously thought to be TiO_2 [*12,13*], Braski has more recently analyzed precipitates using electron energy loss spectrometry and x-ray diffraction on extracted precipitates and identified titanium carbonitride [*15*]. Both precipitates may be present.

In the present study the blocky precipitates ranged in size from 50 to 1500 nm with the largest appearing at the highest irradiation temperature. The rod-like precipitates were generally larger ranging from 200 to 1500 nm as observed by Sprague et al, the largest appearing at the highest irradiation temperature [*13*]. Although a detailed study of precipitation was not made, the rod precipitates were observed only in irradiated material. No effect of helium injection was observed on the precipitates.

Cavities were frequently observed in both blocky and rod-like precipitates (Fig. 22). Cavities were observed in precipitates at all temperatures and cold-work levels. Precipitate cavities were usually comparable in size to matrix voids but exhibited a stronger temperature dependence. Cavities were found as large as 70 nm and were sometimes elongated in shape (Fig. 22) at 700°C.

Discussion

Perhaps one of the most striking features of this study is the absence of helium embrittlement, intergranular fracture, and helium bubbles. The absence of helium bubbles in the matrix or on

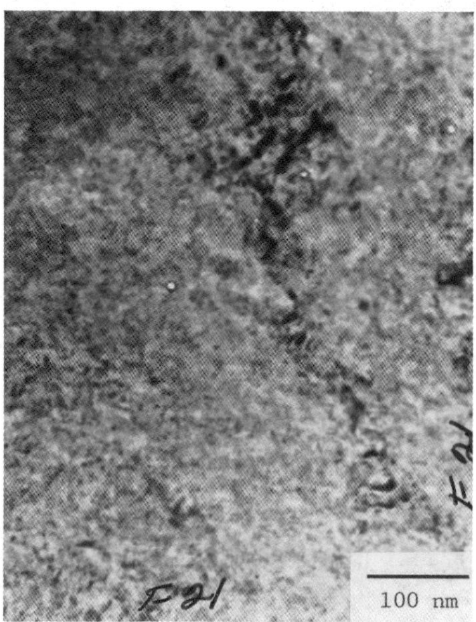

FIG. 21—*Matrix voids in 30%-cold-worked V-15Cr-5Ti (specimen No. 15) irradiated to 30 dpa at 400°C.*

FIG. 22—*Voids in blocky precipitate in 10% cold-worked V-15Cr-5Ti (specimen No. 16) irradiated to 30 dpa at 700°C.*

grain boundaries appears to be a result of the low implantation temperature. Tanaka also did not observe helium bubbles in V-20Ti implanted using the same method as that for V-15Cr-5Ti in the present study [*12*]. Implantation at higher temperature apparently allows the helium to diffuse sufficiently to form bubbles as observed by van Witzenburg et al who implanted helium at 420 K [*16*]. The tritium trick method of doping, which must be done at an elevated temperature, results in numerous helium bubbles in the matrix and at grain boundaries as observed by Braski [*17*]. The helium might have been expected to agglomerate into bubbles on grain boundaries during elevated temperature testing or during the two years of neutron irradiation. However, despite a careful search, no bubbles were found on grain boundaries in any specimens. Although this does not ensure that intergranular fracture resulting from helium will not occur, it is consistent with the observation that no specimens failed intergranularly. The absence of helium embrittlement in V-15Cr-5Ti is consistent with Santhanam et al who observed helium embrittlement only at 750°C and above following cyclotron injection of 25 at. ppm He [*2*]. In V-Nb alloys and V-3Ti-1Si, Ehrlich and Böhn observed intergranular fracture only above 750°C [*18*]. However, Tanaka et al found embrittlement resulting from intergranular fracture at 300°C in V-20Ti [*12*]. It appears that helium embrittlement resulting from cyclotron-implanted helium does not occur in V-15Cr-5Ti at 700°C and below. Nonetheless, since tritium trick doping results in embrittlement and since fusion reactor service has aspects of both helium introduction methods, caution should be used in interpretation of these results for fusion reactor application.

Despite the absence of what could be identified as helium embrittlement, there was clearly irradiation embrittlement. In all cases the low ductilities were accompanied by cleavage fracture. This might result from interstitial impurities or from radiation hardening. Since large amounts of precipitate were not found and since cold-worked specimens in the same irradiation vehicles were not embrittled, contamination by interstitial solutes is not considered most likely. The getter was probably effective. However, the tangles of dislocation loops and segments observed by TEM are consisent with the observed matrix hardening, and the defect density is highest where cleavage fracture is most apparent.

The cleavage fracture observed at 400°C in annealed material becomes quasi-cleavage at 625°C and ductile rupture at 700°C. This is evidence that the cleavage stress was achieved prior to plastic flow until a temperature of about 625°C was reached, or the temperature of transition from brittle to ductile fracture occurred at approximately 625°C.

The mechanism of radiation hardening is also evidenced by the embrittlement observed in the cold worked material. Cold work appears to have the effect of retarding cleavage fracture. Perhaps the deformation and accompanying dislocations result in a higher energy for cleavage crack propagation since additional work is required to move dislocations during fracture, and surface energy is increased by cracking on multiple planes. However, a point is reached where the lattice hardening effect of dislocation tangles raises the yield stress above the cleavage stress and thus promotes cleavage fracture. It is possible that this mechanism is responsible for the abrupt decrease in elongation observed at 625 and 700°C at a cold-worked level of 20%. When the cold-work level is increased still further, to 30%, the additional deformation energy promotes recovery. This was evident in Fig. 20 where it is shown that the dislocation density is lower in 30% cold-worked material than in 10 or 20% cold-worked material. This recovery resulted in sufficient softening to cause a return of ductile rupture at the 30% cold-work level. Further investigation of this phenomenon is required since, at 700°C, there appears from the microstructure to be recovery at 20% cold work. The strength level is higher at the 20% level than at 10 and 30%, but experimental uncertainties do not permit conclusions to be drawn.

Although it is not within the scope of this paper to discuss swelling, it is clear that V-15Cr-5Ti is among the swelling resistant vanadium alloys. This observation agrees with that of Carlander et al who observed no voids at 425°C at a fast fluence of 2×10^{26} n/m^2 [*1*] and Braski who observed less than 0.2% swelling at 40 dpa in helium doped material [*17*].

Tanaka observed minimal swelling in V-20Ti irradiated in the same vehicles as used in the present study [*12*]; Bentley and Wiffen did not find any voids in V-20Ti irradiated in EBR-II in the range 470 to 780°C at fluences in the range of 1.3 to 6.1×10^{26} n/m^2 ($E > 0.1$ MeV) [*3*].

The absence of helium bubbles is more difficult to explain since Tanaka observed them in V-20Ti [*12*]. Perhaps the presence of chromium results in a high degree of helium trapping, but clearly more work is needed to answer this question.

Conclusions

1. Cyclotron-injected helium at a level of 80 appm has no significant effects on tensile, fracture, or microstructural properties of V-15Cr-5Ti between 400 and 700°C.
2. Ten percent cold work has a beneficial effect in reducing the tendency for cleavage fracture in irradiated V-15Cr-5Ti but higher levels (certainly 20%) have a detrimental effect. Still higher levels are beneficial in that they induce recovery.
3. Tensile properties and microstructure indicate a high degree of recovery in 30% cold-worked material following irradiation at 700°C.
4. The temperature for transition from cleavage to ductile failure in neutron-irradiated (30 dpa) V-15Cr-5Ti is increased to approximately 625°C by irradiation hardening.
5. Swelling is negligible in V-15Cr-5Ti in the temperature range of 400 to 700°C at a fluence of 5×10^{26} n/m^2 ($E > 0.1$ MeV) (30 dpa).
6. Swelling was observed in precipitate particles which may be titanium carbonitrides or titanium oxides.

Acknowledgments

The authors wish to thank D. N. Braski for many helpful discussions and assistance with electron microscopy. J. H. DeVan provided much useful information on liquid alkali metals. The authors also wish to thank D. N Braski and R. L. Klueh for reviewing the manuscript. The assistance of B. L. Cox, C. K. Thomas, N. H Rouse, L. T. Gibson, and E. L. Ryan with experimental work is most appreciated. J. W. Woods and A. F. Zulliger led the fabrication and design of the irradiation vehicle. The work of J. Young and F. A. Scarboro in typing the manuscript is also gratefully acknowledged.

This research was sponsored by the Office of Fusion Enery, U.S. Department of Energy, under Contract No. DE-AC05-84OR21400 with the Martin Marietta Energy Systems, Inc.

References

[1] Carlander, R., Harkness, S. D., and Santhanam, A. T., in *Effects of Radiation on Substructure and Mechanical Properties of Metals and Alloys, ASTM STP 529,* American Society for Testing and Materials, Philadelphia, 1973 p. 399.

[2] Santhanam, A. T., Taylor, A., and Harkness, S. D., "Defects and Defect Clusters in *bcc* Metals and Their Alloys," Nucl. Metal., Vol. 18, 1973, 302, *Proceedings,* International Conference, Gaithersburg, Md.

[3] Bentley, J. and Wiffen, F. W., *Nuclear Technology,* Vol. 30, 1976, p. 376.

[4] Smith, D. L., Loomis, B. A., and Diercks, D. R., "Vanadium-Base Alloys for Fusion Reactor Applications—A Review," *Journal Nuclear Materials,* Vol. 135, 1985, pp. 125–139.

[5] Baker, C. C., Brooks, J. N., Ehst, D. A., Smith, D. L., and Sze, D. K., "Tomamak Power System Studies, FY 1985," ANL/FPP-85-2, Argonne National Laboratory, Dec. 1985, pp. 4–85.

[6] Bloom, E. E., Conn, R. W., Davis, J. W., Gold, R. E., Little, R., Schultz, K. R., Smith, D. L., and Wiffen, F. W., "Low Activation Materials for Fusion Applications," *Journal Nuclear Materials,* Vol. 122 and 123, 1984, pp. 17–26.

[7] Gabriel, T. A., Bishop, B. L., and Wiffen, F. W., "Calculated Irradiation Response of Materials Using a Fusion-Reactor First-Wall Neutron Spectrum," ORNL/TM-5956, Oak Ridge National Laboratory, June 1977, p. 5.

[8] Remark, J. F., Johnson, A. B., Jr., Farrar, Harry, IV, and Atteridge, D. G., "Helium Charging of Metals by Tritium Decay," *Nuclear Technology,* Vol. 29, 1976, p. 369.

[9] Klueh, R. L., "Miniature Tensile Test Specimens for Fusion Reactor Irradiation Studies," Nucl. Eng. & Design/Fusion 2, 1985, pp. 407–416.

[10] Johnson, H. E., "Chemistry," in *Sodium-NaK Engineering Handbook,* Vol. I, Gordon and Breach, 1972, pp. 253–254.

[11] Gibson, J. R., and Braski, D. N., "Scanning Electron Microscope Facility for Examination of Radioactive Materials," ORNL/TM-9451, Oak Ridge National Laboratory, Feb. 1985.

[12] Tanaka, M. P., Bloom, E. E., and Horak, J. A., "Tensile Properties and Microstructure of Helium Injected and Reactor Irradiated V-20Ti," *Journal Nuclear Materials,* Vol. 103 and 104, 1981, pp. 895–900.

[13] Sprague, J. A., Smidt, F. A., Jr., and Reed, J. R., *Journal Nuclear Materials,* Vol. 85 and 86, 1979, p. 739.

[14] Smidt, F. A., and Pieper, A. G., "Studies of the Mobility of Helium in Vanadium," *Journal of Nuclear Materials,* Vol. 51, 1974, pp. 361–365.

[15] Braski, D. N., "The Effect of Neutron Irradiation on Vanadium Alloys," Second International Conference on Fusion Reactor Materials, April 13–17, 1986, Chicago.

[16] van Witzenburg, W., Mastenbroek, A., and Elen, J. D., "The Influence of Preimplanted Helium on the Microstructure of Neutron Irradiated Vanadium," *Journal of Nuclear Materials,* Vol. 103 and 104, 1981, pp. 1187–1192.

[17] Braski, D. N., "The Effect of Neutron Irradiation on the Tensile Properties and Microstructure of Several Vanadium Alloys, this conference.

[18] Ehrlich, K., and Böhm, H., *Proceedings,* International Atomic Energy Agency, Conference on Radiation Damage in Reactor Materials 2, Vienna, Austria, 1969, p. 349.

Hua T. Lin,[1] *Roy C. Wilcox,*[1] *and Bryan A. Chin*[1]

The Effects of Temperature and Strain Rate on the Properties of $(Fe,Ni)_3V$ LRO Alloys

REFERENCE: Lin, H. T., Wilcox, R. C., and Chin, B. A., **"The Effects of Temperature and Strain Rate on the Properties of $(Fe,Ni)_3V$ LRO Alloys,"** *Influence of Radiation on Material Properties: 13th International Symposium (Part II), ASTM STP 956,* F. A. Garner, C. H. Henager, Jr., and N. Igata, Eds., American Society for Testing and Materials, Philadelphia, 1987, pp. 310–318.

ABSTRACT: Tension tests with two distinctly different strain rates, 2.8×10^{-5} and 2.8×10^{0} s^{-1}, have been performed on the $(Fe,Ni)_3V$ long-range ordered (LRO) alloys in aged and unaged condition between 20 and 750°C. Tensile results as a function of test temperature indicate that the LRO alloys have good high-temperature strength. The yield strength of the ordered alloys increases with increasing test temperature, rather than decreasing as in conventional alloys. Long-term aging at 600°C has little or no effect on the elevated temperature tensile properties, and tends to reduce the sensitivity of aged materials to strain rate. The deformation mechanisms are changing as a function of test temperature, while there is a minor change in fracture characteristics over test temperature range. Oxidation features show the iron-base LRO alloys have limited oxidation resistance above 600°C.

KEY WORDS: long-range ordered alloys, tension test, strain rate effect, long-term thermal age, mechanical properties, scanning electron microscopy

Long-range-ordered (LRO) alloys offer potential advantages over conventional alloys at elevated temperature [*1–4*]. Atomic ordering produces a pronounced increase in the rate of work hardening [*4–8*] and improves the fatigue resistance [*9*]. As a result of stronger binding and closer packing of atoms, most kinetic processes, such as creep and grain growth, are slower in the ordered structure [*10*]. Also, the unique dislocation dynamics associated with the ordered lattices give the LRO alloys excellent high-temperature strength and fatigue resistance [*11*]. In spite of the above advantages, the LRO alloys have seen only limited application because of a lack of ductility associated with the ordered state. The newly developed $(Fe,Ni)_3V$ LRO alloys show tensile elongations greater than 30% [*12–13*]. This has been achieved by controlling the ordered lattice structure through use of the *e/a* ratio or average electron density per atom outside the inert gas shell.

The excellent mechanical properties exhibited by these ordered alloys have led them to be considered as structural materials in fusion reactors and steam turbine systems. Before these alloys can be used widely for high-temperature applications, the major challenge is to improve and maintain the ductility of LRO alloys in the ordered state. In an effort to clarify the deformation and failure mechanisms, and to optimize the ductility of the iron-base LRO alloys, this investigation examined the effects of strain rate, temperature, and long-term thermal aging on the tensile properties and fracture mechanism of the $(Fe,Ni)_3V$ LRO alloys.

[1] Graduate student, associate professor, and alumni associate professor, Department of Mechanical Engineering, Auburn University, Auburn, AL 36849.

Experimental Procedure

The tensile specimens, LRO-37, of iron-base LRO alloys with gage section 12.7 × 2.8 × 0.76 mm were received from Oak Ridge National Laboratory. The chemical composition analysis of ingots is given in Ref *14*. The specimens were solution-treated for 20 min at 1100°C, followed by an ordering treatment involving step cooling from 600°C to 500°C. Half of the ordered specimens were aged at 600°C for three months. Upon completion of the long-term age, all specimens were electropolished in a solution of 40% water, 40% nitric acid, and 20% hydrofluoric acid. The specimens were polished between 15 and 30 s, with voltage and current settings at 0.7 V and 0.3 A respectively.

Tensile tests were performed with an MTS hydraulic testing machine at two distinctly different strain rates, 2.8×10^{-5} and 2.8×10^{0} s^{-1}. The tensile measurements were digitized by a Nicolet-3091 digital oscilloscope with bubble memory, and an HP-3497A data acquisition unit combined with an HP-9836 minicomputer. To perform tensile tests at elevated temperatures, a special oven consisting of a quartz tube, nichrome wire, and fiberglass was fabricated. To investigate deformation and fracture mechanisms, all failed specimens were examined in detail with an ISI-SS 40 scanning electron microscope (SEM) operated at 20 KV.

Experimental Results

Tension Tests

Figure 1 shows the ultimate tensile strength of the LRO alloys as a function of test temperatures and strain rates. Generally, the tensile strength of the LRO alloys decreases steadily with increasing test temperature and drops dramatically as temperature above critical ordering temperature (T_c = 710°C). For aged and unaged specimens the tensile strength at high strain rates is generally higher than those at lower strain rates, but the behavior of the strength at 650°C is seen to be opposite to the above observation. From 20 to about 600°C, the aged specimens tested at high strain rates have a higher strength than the unaged specimens. There is a little effect of long-term age on elevated temperature tensile strength.

In contrast to the ultimate tensile strength, the yield strength of the LRO alloys increases with test temperature, rather than decreasing, as in conventional alloys, as shown in Fig. 2. The yield

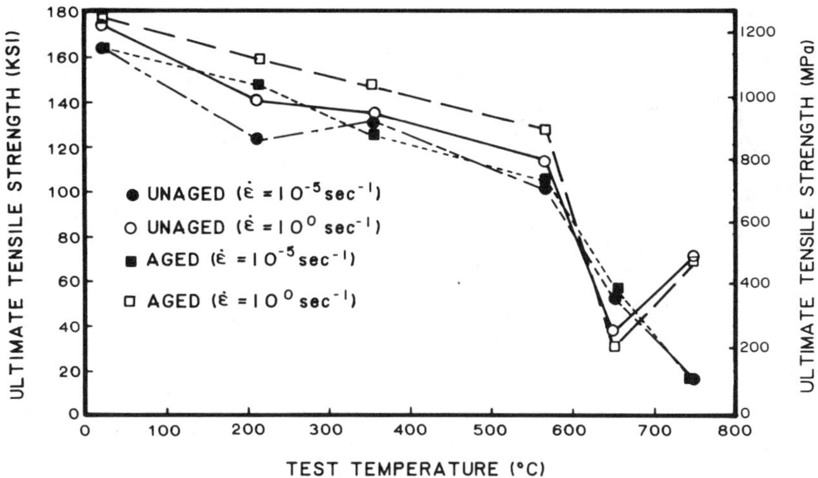

FIG. 1—*Temperature dependence of ultimate tensile strength.*

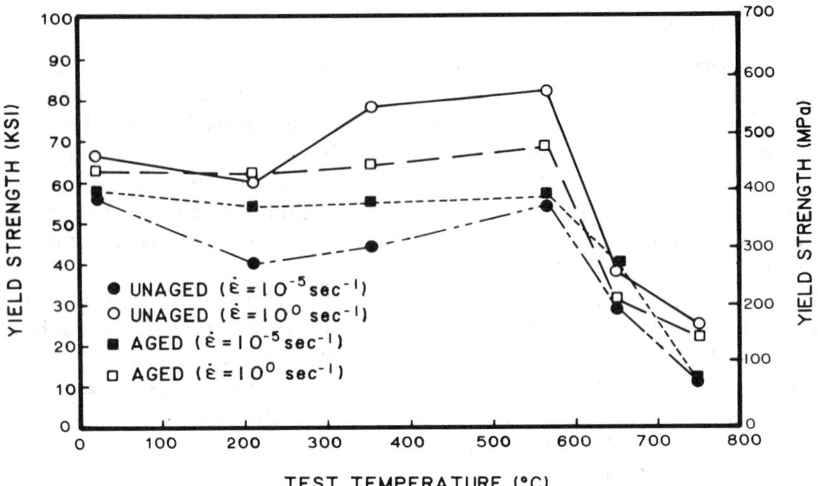

FIG. 2—*Temperature dependence of yield strength.*

strength of unaged specimens increases substantially with temperature above 213°C and reaches a maximum around 568°C, then decreases dramatically above that. The decrease in yield strength is due to the occurrence of order-disorder transformation. For aged specimens, the yield strength is insensitive to test temperature up to 568°C. Both the aged and unaged specimens tested at high strain rate have a higher yield strength than specimens tested at lower strain rate. Additionally, the deviation in yield strength between high and low strain rate of the unaged group is greater than that of the aged group. Aging has a small effect on the yield strength of specimens tested at elevated temperatures.

The effect of temperature and strain rate on the elongation is shown in Fig. 3. For aged and unaged specimens, elongation shows a similar trend with test temperature at either strain rate. The

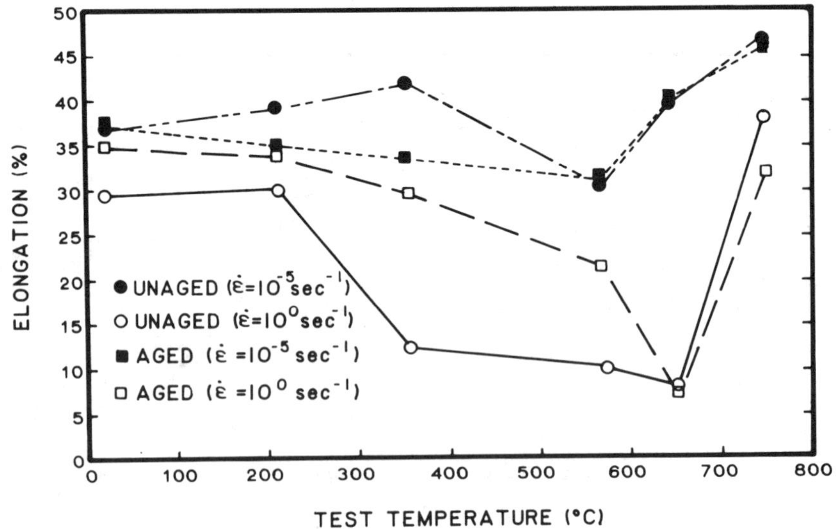

FIG. 3—*Temperature dependence of elongation.*

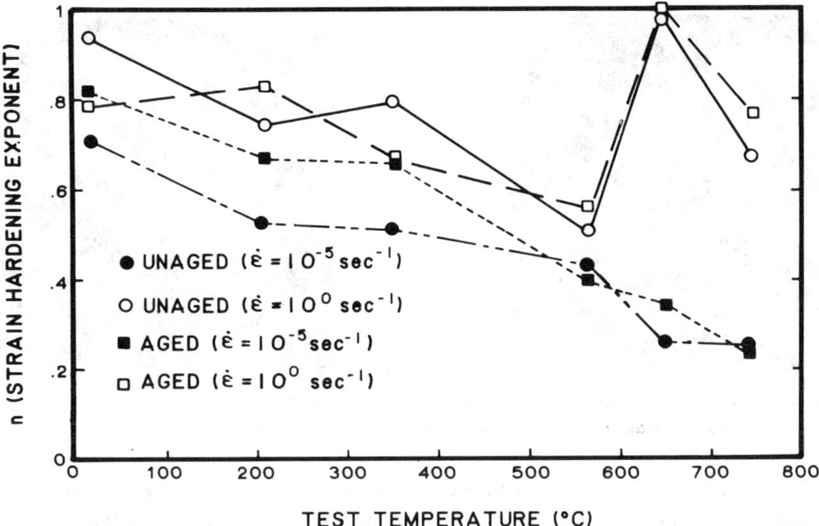

FIG. 4—*Temperature dependence of strain hardening exponent.*

elongation at lower strain rate is higher than that at high strain rate. Also, the minimum in the high strain rate tests occurs at a higher temperature than the minimum in lower strain rate tests. At temperatures below 650°C, unaged specimens tested at high strain rates have less ductility than aged specimens similarly tested. There is little effect of the aging on the elongation.

The strain-hardening exponent as a function of test temperature and strain rate is shown in Fig. 4. The figures show that the exponent decreases with test temperatures up to 750°C, but specimens tested at high strain rate display a minimum at 568°C. Also, the high strain rate exponent is higher than that observed in lower strain rate test. Long-term aging has a little effect on the strain-hardening exponent.

Figure 5 shows the effect of test temperature and aged condition on the strain rate sensitivity.

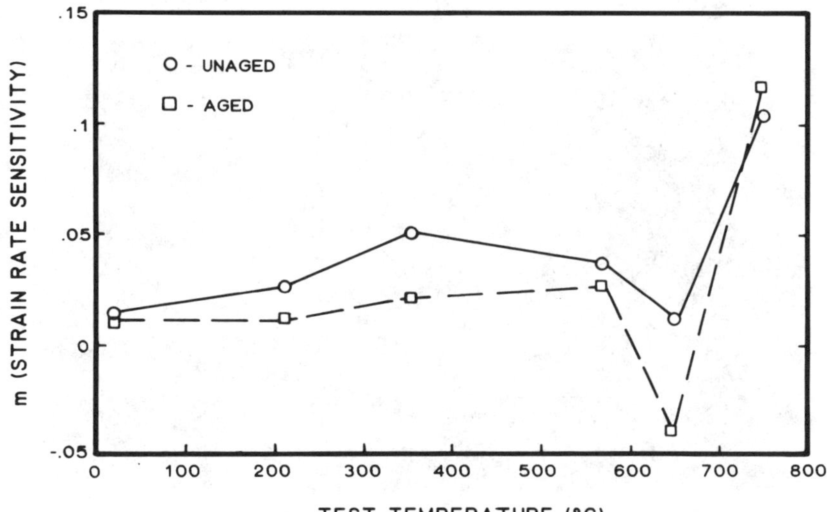

FIG. 5—*Temperature dependence of strain rate sensitivity.*

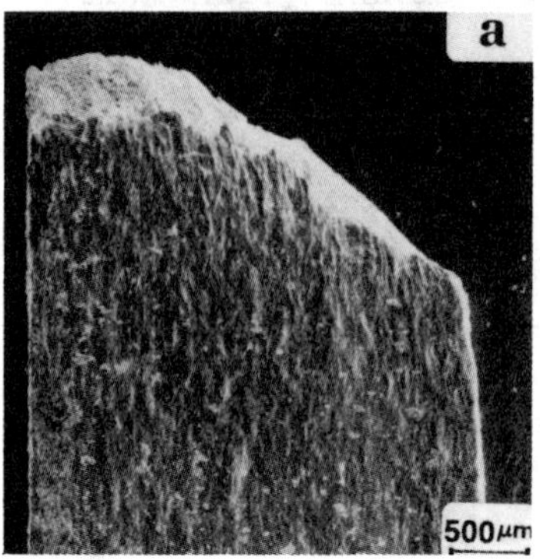

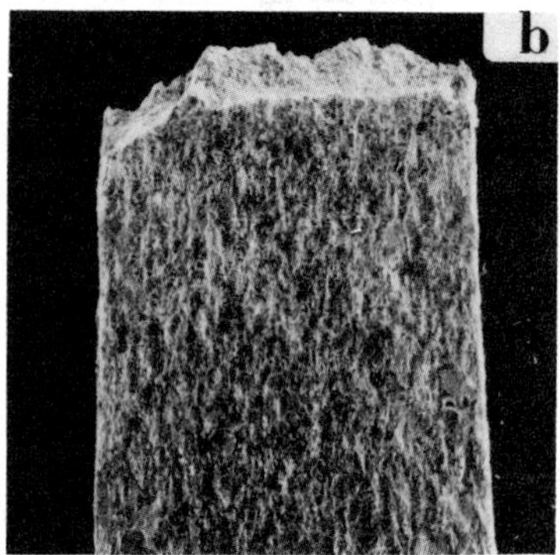

FIG. 6—*Macroscopic fracture features of LRO alloys tested at 650°C.* (a) *High strain rate tests failed at an angle of 90°.* (b) *Low strain rate tests failed at an angle of approximately 45°.*

The results indicate that strain rate sensitivity of both the aged and control group show only minor changes until approaching the critical ordering temperature of 710°C. The strain rate sensitivity coefficient was found to be higher in the unaged group than in the aged group. This suggests that aging the material reduces its sensitivity to strain rate.

Scanning Electron Microscopy Analysis

The morphology of deformation and fracture characteristics as a function of test temperature, strain rate, and aged condition were identified through scanning electron microscopy (SEM) analysis for all tested specimens. Figure 6 shows the macroscopic feature of two specimens tested at 650°C at strain rates of 2.8×10^{0} and 2.8×10^{-5} s^{-1}. In general, the macroscopic fracture features of specimens tested at high strain rates failed at an angle approximately 45° to the tensile

FIG. 7—*Photomicrographs show temperature dependence of slip band density for control specimens. Note intergranular cracking in low strain rate specimens perpendicular to stress axis.*

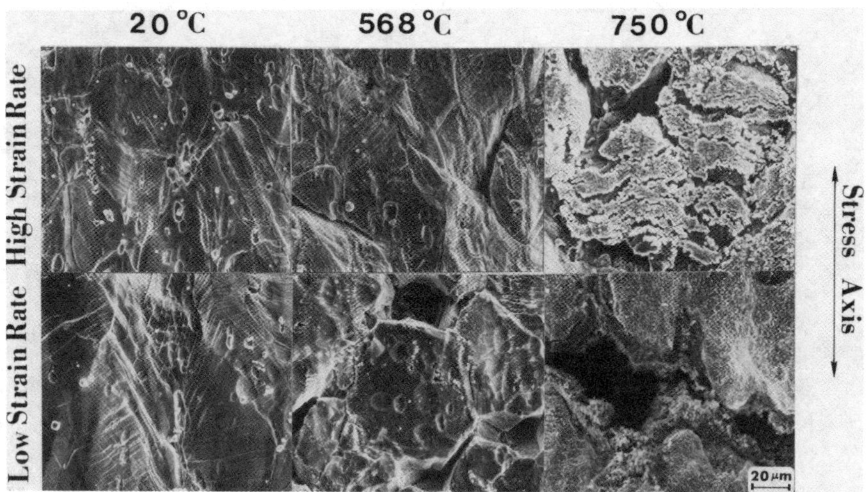

FIG. 8—*Photomicrographs show temperature dependence of slip band density for aged specimens. Note intergranular cracking in low strain rate specimens perpendicular to stress axis.*

axis. In specimens tested at a lower strain rate the fracture occurred at 90° with respect to the tensile axis below the critical ordering temperature and 45° above T_c.

The photomicrographs in Figs. 7 and 8 compare the elevated temperature deformation mechanism of the LRO alloys in the different aged and strain rate conditions. All photomicrographs shown in Figs. 7 and 8 are at the same magnification. The deformation features in the gage section surfaces of aged and unaged specimens at lower strain rates show the number of intergranular microcracks increasing with increasing test temperature. The orientation of these microcracks is perpendicular to the tensile axis. At high strain rates, there is a greater tendency for intergranular fracture with increasing test temperature. Also, the features reveal that there is a greater density of surface microcracks at lower strain rates than that at high strain rates. The slip band density of both the aged and unaged specimens increases with temperature up to 355°C, then decreases above that. In addition, the gage section features indicate that oxidation and surface diffusion mechanisms become important as test temperatures increase above 568°C.

Correspondingly, Figs. 9 and 10 show the fracture surfaces of specimens as a function of aging condition, strain rate, and test temperature. The fracture features of unaged specimens show a mixed mode for test temperatures below 568°C, then become more dimple rupture dominated above that. For aged specimens, the figures show that dimple size is smaller than that of unaged specimens and aged specimens have more dimple rupture than unaged specimens below 568°C. A larger amount of fracture surface oxidation of both the aged and unaged samples tested at lower strain rates was observed than those tested at high strain rates.

Discussion

Tensile test results show the ultimate tensile strength decreases with increasing test temperature. This decrease is apparently due to the annealing and recovery of the dislocation structure at elevated temperature, resulting in a decreased strain hardening capability of the material. These results are consistent with measurements (Fig. 4) that show that the strain-hardening exponent decreases with increasing temperature. Also, the tensile results indicate that the ordered alloys exhibit good high-temperature yield strength. The yield strength, instead of decreasing as in

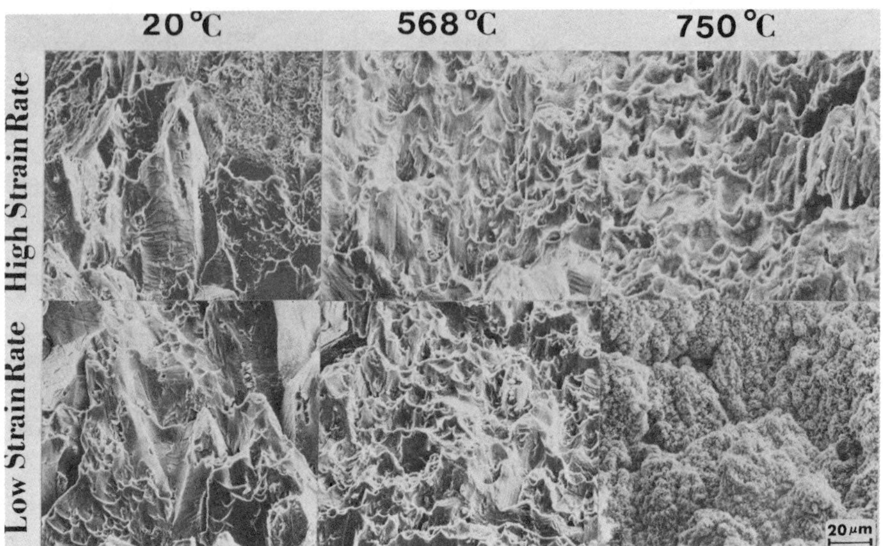

FIG. 9—*Photomicrographs show temperature dependence of fracture mechanism for control specimens. Increasing temperature results in an increase of dimple size. Note severe oxidation of low strain rate specimen tested at 750°C.*

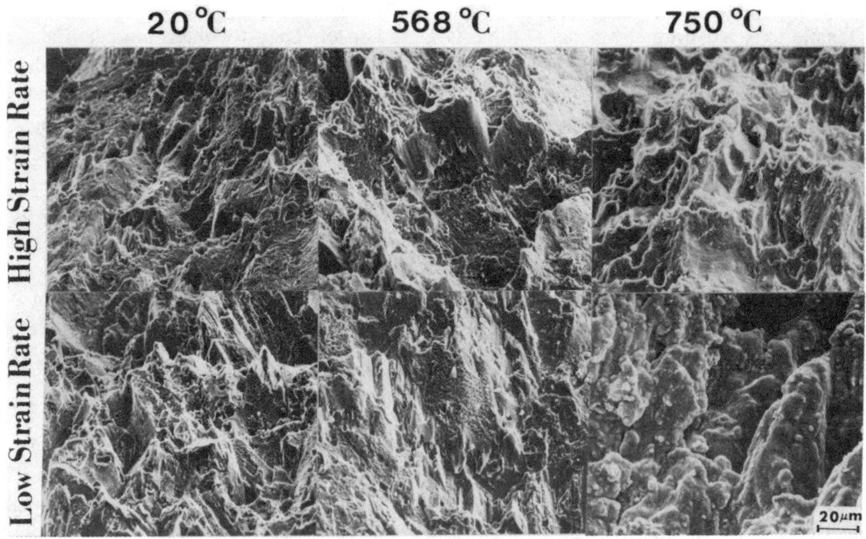

FIG. 10—*Photomicrographs show temperature dependence of fracture mechanism for aged specimens. Increasing temperature results in an increase of dimple size. Note severe oxidation of low strain rate specimen tested at 750°C.*

conventional alloys, increases with temperature (Fig. 2). As a result of this anomalous strength behavior, the iron-base LRO alloys are much stronger than many conventional alloys at elevated temperature. This anomalous temperature dependence of the strength, which has been seen in other ordered alloys, is hypothesized by C. T. Liu [*14*] to be caused by a thermally activated process rather than a change in the degree of order with temperature. The yield strength of the aged sample is independent of test temperature up to 568°C, and the difference in yield strength between high and low strain rate is smaller than that difference in control materials. This suggests that aging has a tendency to reduce the effect of temperature and strain rate on the yield strength.

The elongation versus temperature plot (Fig. 3) shows that the minimum in elongation as a function of test temperature increases as strain rate increases. This is probably caused by shorter test times at elevated temperature, giving less time for oxidation of the surface and ingression of oxygen along grain boundaries to interior surfaces. This hypothesis is consistent with the observation of the microscopic deformation features (Figs. 7 and 8), which show that the intergranular surface microcracks and degree of oxidation increase with increasing strain rate and with increasing temperature. In addition, the decrease in percent elongation with increasing strain rate is due to an increase in intergranular fracture. This difference is greater for control samples. This phenomenon suggests that aging tends to reduce the effect of strain rate on the elongation.

The effect of aging, reducing the effect of strain rate on the yield strength and elongation, is consistent with the results shown in the strain-rate sensitivity versus temperature diagram (Fig. 5). The sensitivity of the unaged group is greater than that of the aged group. The minimum in strain-rate sensitivity is a direct result of dynamic strain aging [*15*].

The microscopic fracture features of the ordered alloys tested at elevated temperature reveal a mixed mode of fracture, intergranular and transgranular fracture. Only the relative amounts of intergranular and transgranular fracture varied as a function of temperature. For temperatures below 568°C, the observations suggest that deformation occurred by dislocation glide. The higher the test temperature, the easier the dislocations overcome the activation barrier to slip. This results in the increase in slip band density with increasing temperature that is demonstrated by the observation of microscopic deformation features (Figs. 7 and 8). In addition, the photomicrographs show that fracture occurs through microvoid nucleation, growth, and coalescence. The vanadium-carbide type particles which exist in grain and grain boundary regions were identified using the energy dispersive analysis of X-ray (EDAX) technique and were found to be the nucleation sites of fracture. At temperatures above 568°C, the diffusion and oxidation mechanisms become significant. The slip band density decreases and dimple size increases as a result of diffusional processes. The migration of oxygen along grain boundaries to interior surfaces resulted in intergranular microcracks observed on the surface. The degree of oxidation of fracture surface was suppressed by increasing the strain rate. Tension tests at elevated temperature suggest that the iron-base LRO alloy have limited resistance to oxidation above 600°C. This limitation will confine the LRO alloys to be used in a vacuum or inert gas environment only. However, the unique strength properties, strain rate insensitivity and stability of the ordered alloys make them an excellent candidate for future development as a high-temperature structural material.

Conclusions

The following conclusions were drawn from the experiment and analysis:

1. Ultimate tensile strength of the $(Fe,Ni)_3V$ LRO alloys decreases as temperature increases.
2. Yield strength of the $(Fe,Ni)_3V$ LRO alloys increases as temperature increases up to T_c then sharp decrease.
3. The $(Fe,Ni)_3V$ LRO alloys exhibit a good ductility in excess of 30% at room and elevated temperature.

4. Long-term aging at 600° C does not cause any significant change in the tensile properties which demonstrates the structure stability of the LRO alloys.

5. The iron-base LRO alloys have limited oxidation resistance at temperatures above 600°C.

References

[1] Stoloff, N. S. and David, R. G., "The Mechanical Properties of Ordered Alloys," *Progress in Materials Science.* Vol. 13, No. 1, 1966, p. 1.

[2] Kear, B. H., Sims, C. F., Stoloff, N. S., and Westbrook, J. H., Eds., Ordered Alloys Structure and Physical Metallurgy, *Proceedings,* Third Bolton Landing Conference, Lake George, N.Y., Claiter Publishing Division, 1970.

[3] Krivoglaz, M. A. and Smirinov, A. A., *The Theory of Order-Disorder in Alloys,* American Elsevier Publishing Company, New York, 1964.

[4] Muto, F. and Takagi,Y., *The Theory of Order-Disorder Transformations in Alloys,* Academic Press, New York, 1956.

[5] Popov, L. E. and Koneva, N. A., in *Order Disorder Transformation in Alloys,* J. Warlimant, Ed., Springer-Verlag, New York, 1974, p. 404.

[6] Marcinkowski, M. J., in *Order Disorder Transformation in Alloys,* J. Warlimant, Ed., Springer-Verlag, New York, 1974, p. 364.

[7] Vidoz, A. E., Lazarenio, D. P., and Chan, R. W., *Acta Metallurgica,* Vol. 11, 1963, p. 17.

[8] Kear, B. H. and Wilsdorf, H., *Transactions of the AIME,* Vol. 224, 1962, p. 382.

[9] Boetlner, R. C., Stoloff, N. S., and Davis, R. G., *Transactions of the AIME,* Vol. 236, 1966, p. 131.

[10] Schulson, E. M., "Order Strengthening as a Method for Reducing Irradiation Creep: An Hypothesis," *Journal of Nuclear Materials,* Vol. 66, 1977, p. 322.

[11] Cottrell, A. H., in *Relation of Properties to Structure, Seminar Series,* American Society of Metals, Cleveland, OH, 1954.

[12] Liu, C. T., "Development of Long-Range-Ordered Alloys," Alloy Development for Irradiation Performance, Proceedings of Program Review Meeting, September 30–October 1, 1980, p. 354.

[13] Lui, C. T. and Inouye, H., "Control of Ordered Structure and Ductility of $(Fe,Co,Ni)_3V$ Alloys," *Metallurgical Transactions* A, Vol. 10, 1979, p. 1515.

[14] Liu, C. T., "Development of Fe-Base Long-Range-Ordered (LRO) Alloys for Fusion Reactor First Wall and Blanket Applications," Proceedings of the Second Topical Meeting on Fusion Reactor Materials, Seattle, WA, August 9–12, 1981, Session 5C.

[15] Lubhan, J. D., "Simultaneous Aging and Deformation in Metals," *Transactions of the AIME,* Vol. 185, 1949, p. 702.

Pressure Vessel Steels

Kenneth R. Lawless,[1] Wayne A. Pavinich,[2] and Arthur L. Lowe, Jr.[3]

Microstructural Characterization of Submerged Arc Weld Metals

REFERENCE: Lawless, K. R., Pavinich, W. A., and Lowe, A. L., Jr., "**Microstructural Characterization of Submerged Arc Weld Metals,**" *Influence of Radiation on Material Properties: 13th International Symposium, ASTM STP 956,* F. A. Garner, C. H. Henager, Jr., and N. Igata, Eds., American Society for Testing and Materials, Philadelphia, 1987, pp. 321–332.

ABSTRACT: This paper presents the results of a detailed microstructural evaluation of Linde 80 submerged arc weld metals typically found in Babcock and Wilcox 177FA Class reactor vessels. Optical metallography was performed on Linde 80 weld metals to characterize the grain structure, and scanning electron microscopy (SEM) was used to characterize the fracture surfaces. These weld metals have a tempered bainitic structure with large amounts of manganese rich inclusions.

Analytical transmission electron microscopy (TEM) was used to characterize the carbides, coarse (0.1- to 2.0-μm) inclusions, and fine (2- to 50-nm) precipitates. Thin foil and extraction techniques were used to characterize particle morphology while convergent beam electron diffraction (CBED) and energy dispersive X-ray (EDS) techniques were carried out to identify the particle chemical composition and crystal structure. These results are compared with published results from simulated reactor vessel welds found in the in-situ annealing program sponsored by the Electric Power Research Institute (EPRI).

KEY WORDS: transmission electron microscopy, optical microscopy, scanning electron microscopy, convergent beam electron diffraction, energy dispersive X-ray techniques, thin foil techniques, extraction techniques, submerged arc weld, reactor vessels, inclusion, copper precipitation, carbides

The microstructure of reactor vessel welds significantly affects the mechanical properties. It is known that the volume fraction of inclusions in a weld will influence the mechanical properties and fracture toughness. Further, the copper content will influence the radiation embrittlement response of the weld metal and the volume fraction of fine copper rich precipitates often found in these materials. It is possible for copper precipitation to occur during reactor vessel thermal processing, irradiation, and possibly during in-situ annealing of the reactor vessel. Therefore, characterization of the microstructure is essential to fully understand differences in mechanical properties from material to material and the change in mechanical properties as a result of exposure to the reactor environment.

The purpose of the work presented in this paper is to begin such a characterization for various unirradiated weld metals which have various copper contents and stress relief times. These results are compared with the results of a simulated reactor vessel weld metal performed as a part of the EPRI thermal annealing program, where microstructure was characterized in the unirradiated, the irradiated, and the irradiated/annealed conditions using energy dispersive X-ray spectroscopy

[1] Professor, Department of Materials Science, University of Virginia, Charlottesville, VA 22901.

[2] Senior research engineer, Babcock & Wilcox Co., Research and Development Division, P.O. Box 11165, Lynchburg, VA 24506-1165.

[3] Advisory engineer, Babcock & Wilcox Co., Nuclear Power Division, P.O. Box 10935, Lynchburg, VA 24506-0935.

(EDS), scanning electron microscopy (SEM), and scanning transmission electron microscopy (STEM) [*1*].

Experimental Procedure

Three Linde 80 submerged arc weld metals from reactor vessel nozzle drop outs, of various copper contents, were selected for study. Table 1 lists the emission spectroscopy analysis for each weld metal. Each material was stress relieved at 607°C (1125°F) for various times and slow cooled at 8.3°C/h (15°F/h) to simulate the typical reactor vessel stress relief cycle. Optical metallography was performed on a Bausch and Lomb RS-1 metallograph. Specimens were mechanically polished and etched with a 2% nital solution to reveal the grain structure. Scanning electron microscopy was performed on representative fracture surfaces using an ETEC Autoscan electron microscope to study the general fracture surface morphology.

The general microstructural features of the specimens were examined with a Philips EM 400T analytical electron microscope operated at 120 kV. Specimens were prepared by sectioning thin slices from the weld areas, mechanically polishing these sections, and cutting 3-mm disks. These disks were electropolished in a 20% solution of sulfuric acid in methanol to provide thin foil specimens suitable for electron microscopy. The thin foil specimens, while providing a good overall image of the microstructure of the weld area, were not suitable for chemical microanalysis, since most of the small precipitates or carbides were completely surrounded by matrix material. In order to obtain good microanalysis of the precipitates, extraction replicas were prepared. The specimens were lightly etched electrolytically in a solution of 10% hydrochloric acid in methanol followed by carbon coating. The carbon film was scribed into ⅛-in. (0.32-cm) squares, then etched further with 2% nital, and the extraction replicas floated free by immersion in distilled water. The replicas were picked up on beryllium grids for examination in the TEM.

Quantitative energy EDS was carried out on the particles in the extraction replicas using the Cliff-Lorimer thin film analysis technique. The elements normally analyzed included aluminum, silicon, sulfur, chromium, manganese, nickel, copper, and molybdenum. It was not possible to determine carbon and oxygen concentrations with our EDS detector. In most cases, the particles examined were sufficiently thin that a correction for X-ray absorption was not made. A severe overlap of the S-K and Mo-L peaks occurs. If there was no visual evidence in the spectrum of a Mo-K peak, analysis was carried out only for sulfur. If the ratio of the L-to-K peak intensities was approximately the theoretical value for molybdenum, analysis was not carried out for sulfur. Otherwise the analysis was made for both sulfur and molybdenum, but the deconvolution was not satisfactory, and results for these elements when both are apparently present are not considered very reliable.

The structure was determined, when the particles were thin enough, by convergent beam electron diffraction (CBED). In most of this work a spot size of 40.0 nm was satisfactory, although smaller spot sizes were used for a few particles.

TABLE 1—*Chemical composition and stress relief times of the weld metals.*

	wt%										
Alloy	C	Mn	P	S	Ni	Cr	Mo	Si	V	Cu	Stress Relief Time, h
WP1	0.12	1.54	0.015	0.016	0.68	0.11	0.42	0.52	0.008	0.37	20 and 50
WP2	0.13	1.47	0.009	0.011	0.62	0.09	0.45	0.61	0.006	0.03	50
WP3	0.13	1.29	0.013	0.014	0.57	0.09	0.39	0.61	0.007	0.32	50

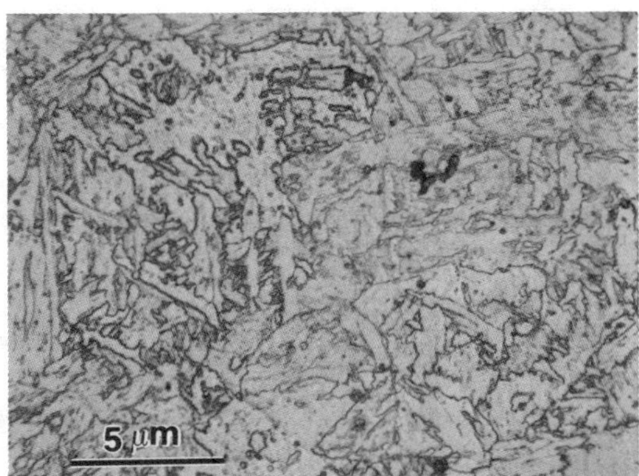

FIG. 1—*Optical micrograph showing typical bainitic structure.*

Results

Grain structure was characterized for each material/stress relief condition. No significant difference in grain size or shape was noted. All the materials exhibited a tempered bainitic structure for all stress relief conditions. Figure 1 is a micrograph typical of all the weld metals.

The SEM investigation of the fracture surfaces revealed a dispersion of large round inclusions with an approximate 2-μm diameter and a spacing of approximately 6 μm. These inclusions are manganese-rich as determined by EDS analysis. Fracture surfaces were essentially equivalent regardless of the material/stress relief combination. Therefore, neither the chemical composition of the material nor the stress relief time affected the morphology of the fracture surfaces. Figure 2 is a scanning electron micrograph showing a typical fracture surface.

The general microstructural features as observed in the thin foils by TEM are shown in Fig. 3 and are characteristic of all the specimens examined. Very large, primarily spherical, precipitates are found randomly distributed throughout each specimen, and large, elongated, or globular precipitates are seen, primarily at grain boundaries, although many are found at intragranular positions as well. Very small precipitates are found within grains but are distributed in a rather irregular fashion. In addition to the characteristic precipitation structure, all of these weld metals

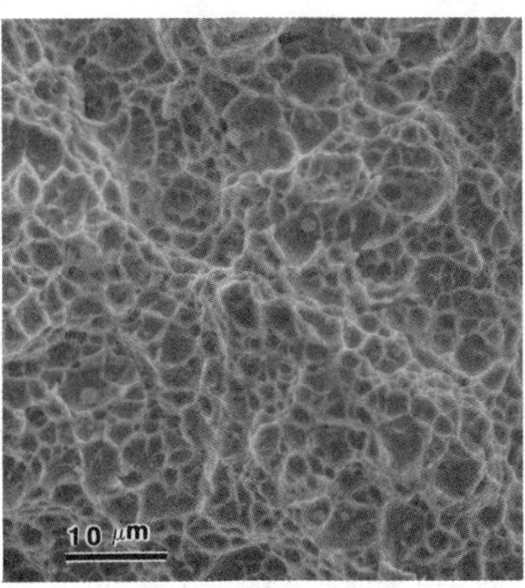

FIG. 2—*Scanning electron micrograph showing typical fracture surface.*

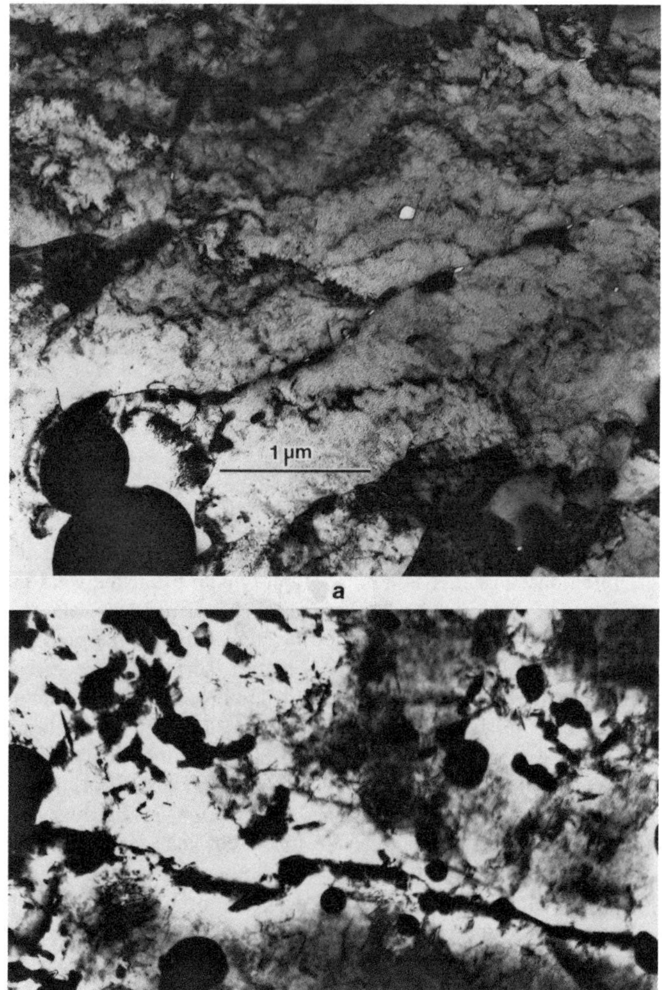

FIG. 3—*TEM of thin foil specimens:* (a) *WP3-031 and* (b) *WP2-027.*

showed a high dislocation density. A few amorphous inclusions were found by EDS analysis to be high in silicon.

In order to analytically determine the structure and chemical composition of the precipitates, extraction replicas were studied, and Figure 4 shows typical relatively low magnification images of four different specimens. Because of the difficulty in reproducing the depth of etching and the extraction of efficiency, changes in precipitate density seen in the micrographs cannot be readily interpreted.

The observed precipitates were identified, when possible, by CBED and EDS, which provided data on the chemical composition of the individual particles. On the basis of the data for several hundred precipitates, four basic types of precipitates were found in all specimens examined. The frequency of distribution for each of the different particle types was not determined in this preliminary study. However, such a characterization is planned in future efforts. Type one precipitates have high manganese and usually high silicon, and occasionally high sulfur. Type two

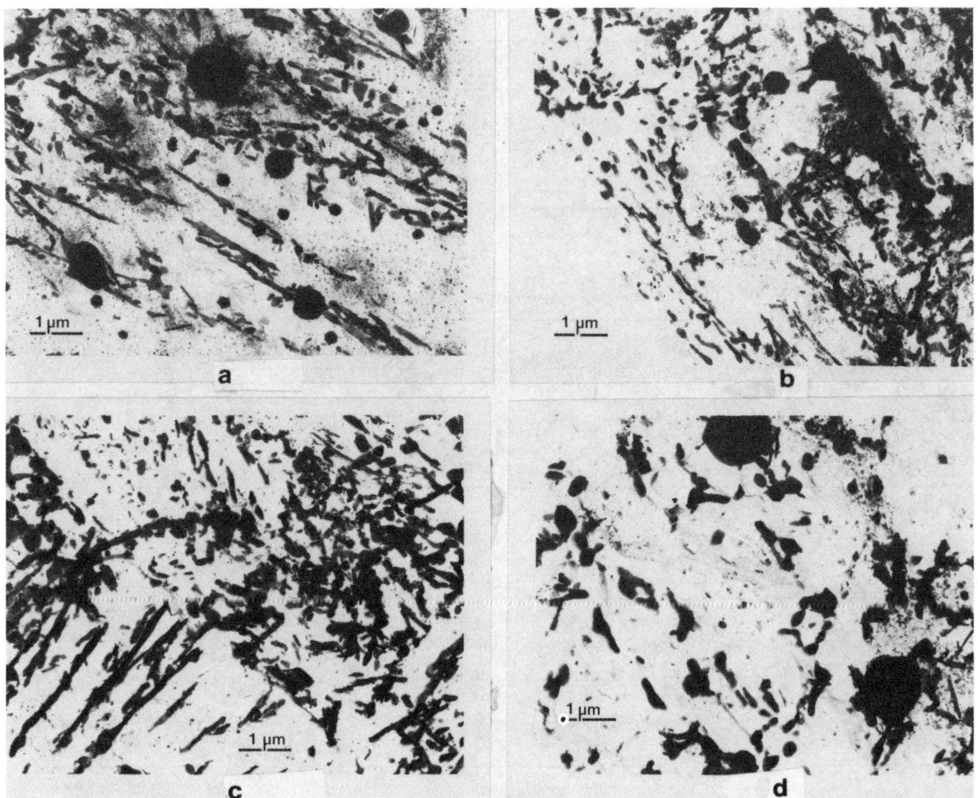

FIG. 4—*Extraction replicas:* (a) *WP1-082,* (b) *WP1-084,* (c) *WP2-027, and* (d) *WP3-031.*

precipitates were high in iron and were M_3C type carbides. Type three precipitates were M_2C molybdenum carbides. Type four precipitates were mostly copper. Occasionally, EDS data were indicative of much more complex structures, or possibly intimate mixtures of several different precipitates, and it was not possible to classify these separately.

Tables 2 through 5 give the average composition of the precipitates analyzed in extraction replicas from four different specimens. Figures 5 through 8 show examples of typical spectra from the four types of precipitates observed.

The large type one precipitates, which are mostly spherical, as seen at A in Fig. 9*a* and in Figs. 4*a, b,* and *d* consist mainly of manganese and silicon, as shown in Table 2, with smaller amounts of almost all of the other elements in the alloy. The particles generally were too thick to identify from diffraction patterns, but it is likely that they are mostly $MnSiO_3$ inclusions. Analysis results indicate that a few of these large particles were manganese sulfides with a high copper content. All of these particles showed the presence of aluminum. The iron content varied considerably between specimens, being greater for the WP1-084 specimens and lowest for the WP1-082 specimens. These large manganese inclusions were not found in the extraction replicas of WP2-027 specimens; the very large particles in these replicas were M_3C carbides. However, the thin foils of WP2-027 showed a small number of these inclusions.

The most common precipitates in all specimens (type two) were large elongated or globular M_3C type orthorhombic carbides with iron and manganese as the major metallic constituents, but

TABLE 2—*Average composition of Mn type precipitates, wt%.*

Specimen	Fe	Mn	Cr	Ni	Si	Cu	Al	S
WP1-082	1.2	57.2	0.2	0.1	31.7	3.7	2.9	3.0
WP1-084	7.1	56.2	0.3	0.2	27.0	4.6	2.0	2.6
WP2-027				no type one particles observed				
WP3-031	3.0	58.8	0.2	0.5	28.4	2.2	5.0	1.9

TABLE 3—*Average composition of M_3C type carbides, wt%*

Specimen	Fe	Mn	Cr	Ni	Si	Cu	Al	Mo
WP1-082	75.9	11.8	1.0	5.0	0	4.2	0	2.0
WP1-084	79.1	13.3	1.0	0.2	0.1	2.7	0	3.5
WP2-027	82.7	7.5	5.1	0.1	0	0.7	0	3.8
WP3-031	79.0	13.6	2.1	0.1	0	1.9	0	3.2

Table 4—*Average composition of M_2C type carbides.*

Specimen	Fe	Mn	Cr	Ni	Si	Cu	Al	Mo
WP1-082	1.6	2.5	1.6	0.1	0.5	1.7	0.2	91.8
WP1-084	5.6	2.9	1.4	0	0.8	4.9	0.4	84.0
WP2-027	2.9	1.0	3.5	0.1	0.2	1.4	0.6	90.3
WP3-031	3.7	1.7	1.5	0	0.3	3.3	0	89.5

TABLE 5—*Average composition of copper precipitates, wt%.*

Specimen	Fe	Mn	Cr	Ni	Si	Cu	Al	S
WP1-082	0.9	0.2	0.2	2.2	0	96.2	0	0.5
WP1-084	3.5	0.3	0.2	0.2	0.2	93.1	0	2.4
WP2-027	3.0	0.1	0.2	0.3	0	95.0	0	1.4
WP3-031	3.3	1.0	0.6	0.3	0.1	91.7	0	3.0

with appreciable amounts of chromium, nickel, and molybdenum as seen in Table 3. These are readily shown in Fig. 4, and are shown at B in Fig. 9*a, b, c,* and *d.* Although most common at grain boundaries, they were also found in the interiors of grains. All of these M_3C precipitates also contained copper, with WP1-082 and WP1-084 showing the largest amounts, and WP2-027 only small amounts as would be expected from the base alloy compositions. It is not known whether the copper is actually incorporated in the carbides, or has perhaps nucleated on the surface of the carbide. The WP2-027 specimen also showed much more chromium, and less manganese than the other specimens. Also, a high nickel content was found in WP1-082 M_3C carbides. A typical CBED pattern from M_3C is shown in Fig. 10*a*.

The type three precipitates were identified as hexagonal M_2C with molybdenum as the major metallic element, but with small amounts of iron, manganese, chromium, and silicon present as shown in Table 4. A considerable amount of copper was present in all of these particles, and

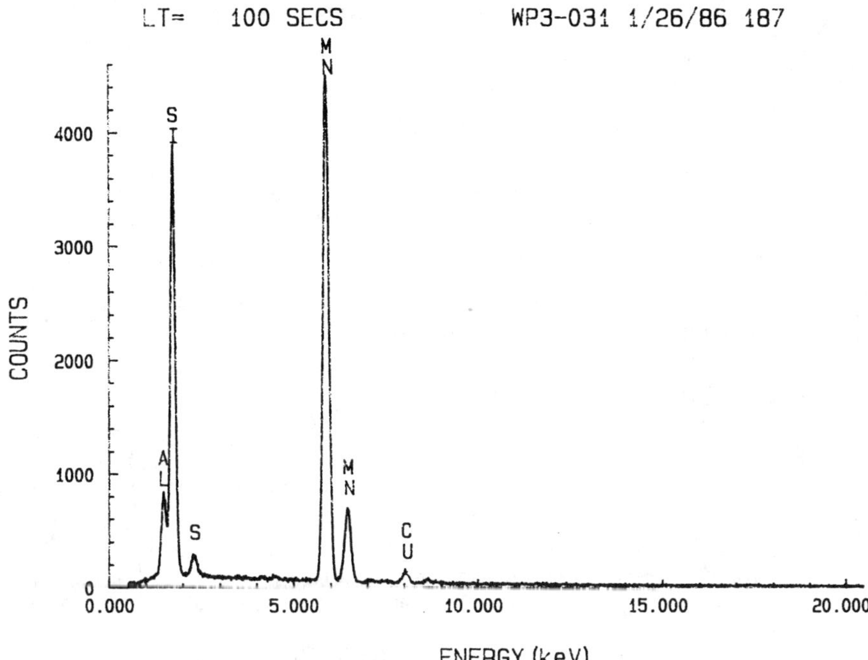

FIG. 5—*Energy dispersive X-ray spectra of the four characteristic precipitates in WP3-031 Mn-Si type.*

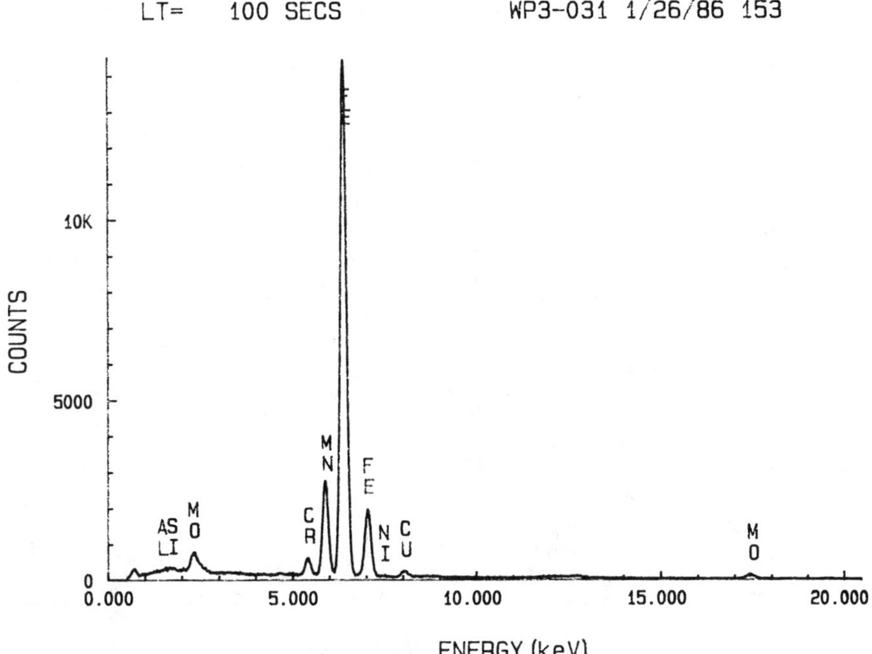

FIG. 6—*Energy dispersive X-ray spectra of the four characteristic precipitates in WP3-031 M_3C type.*

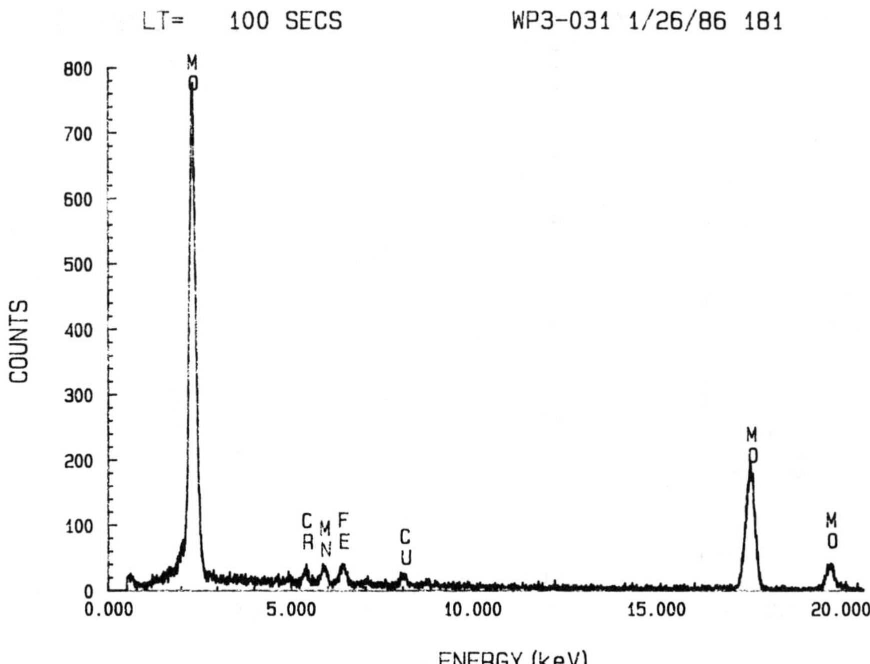

FIG. 7—*Energy dispersive X-ray spectra of the four characteristic precipitates in WP3-031 M_2C type.*

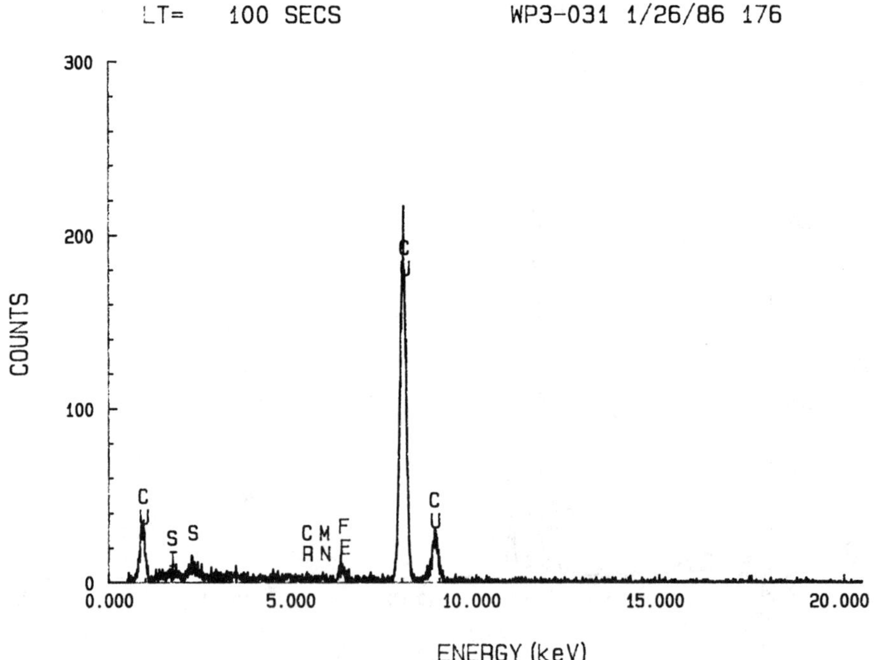

FIG. 8—*Energy dispersive X-ray spectra of the four characteristic precipitates in WP3-031 copper type.*

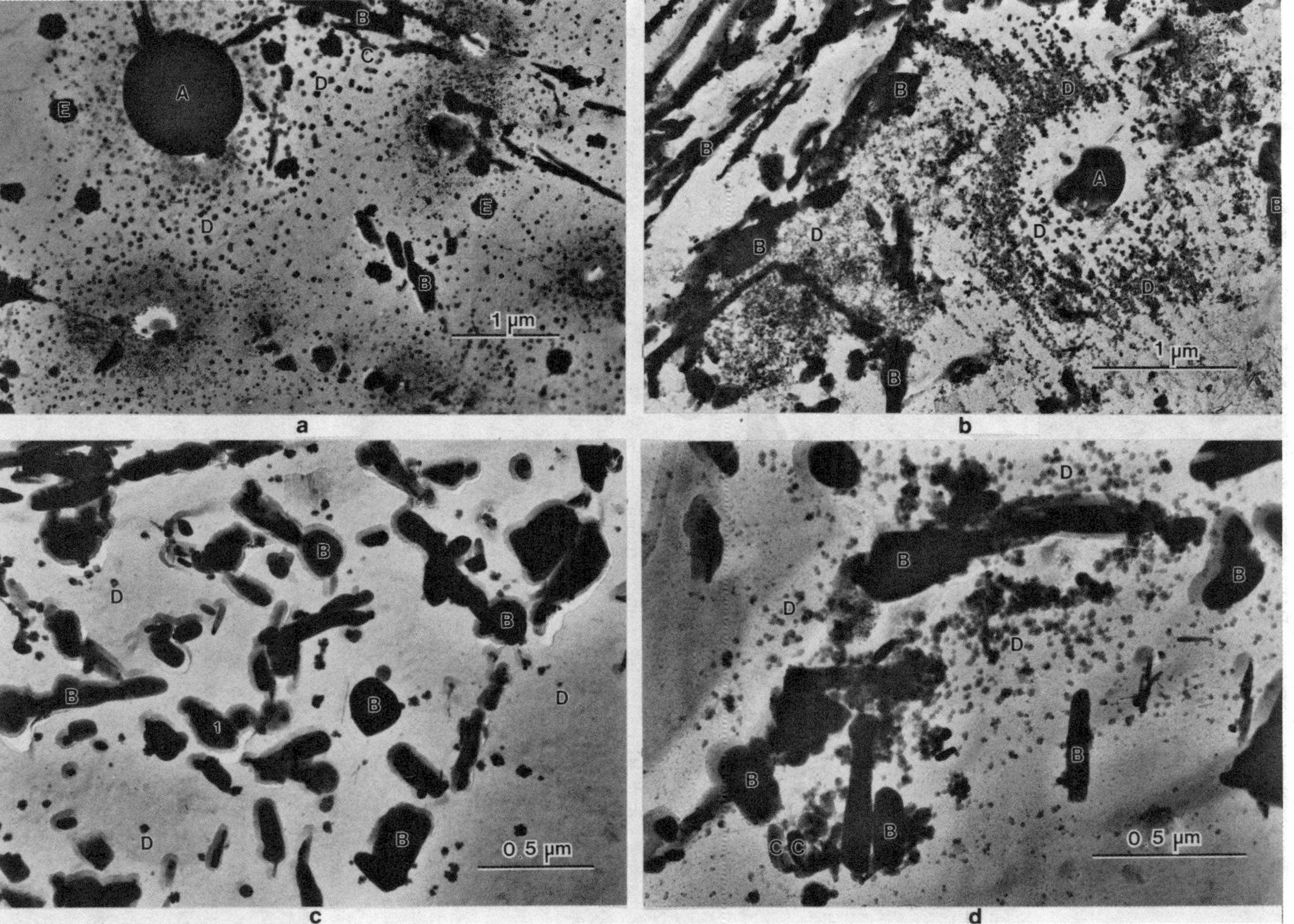

FIG. 9—*Extraction replicas of precipitates showing A—Mn-Si type inclusions, B—M_3C type carbides; C—M_2C type carbides; D and E—copper precipitates:* (a) *WP1-082,* (b) *WP1-084,* (c) *WP2-027, and* (d) *WP3-*

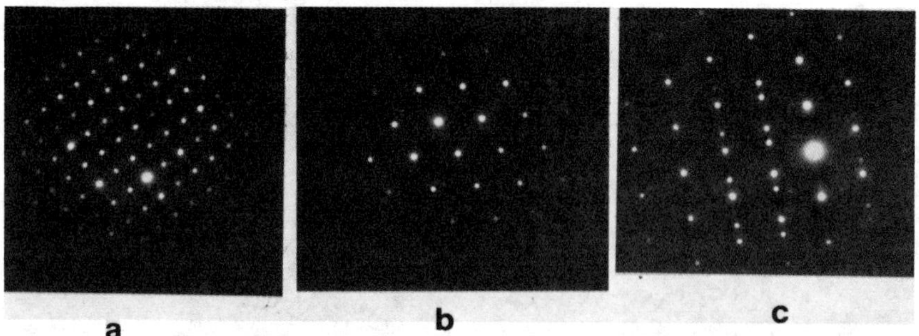

FIG. 10—*Convergent beam electron diffraction of three characteristic precipitates in WP1-084:* (a) *M_3C type,* (b) *Mo_2C, and* (c) *copper twinned.*

some showed small amounts of aluminum and nickel. Most of these precipitates were moderately small, of the order of 100 nm or less in diameter. Some of these are shown at carbon in Fig. 9. WP1-084 M_2C carbides showed a considerably higher iron and copper content than the other specimens, and WP2-027 showed a high chromium content as in the case of the M_3C carbides from this alloy. A typical CBED pattern from the M_2C phase is shown in Fig. 10*b*.

Most of the very small particles, generally 50 nm or less in size, were type four precipitates, which as shown in Table 5, were mostly copper with small amounts of iron and sulfur, and traces of manganese, chromium, nickel, and silicon. One specimen, WP1-082, showed almost ten times as much nickel in the type four precipitates than other specimens, a situation also found for the M_3C carbides in WP1-082. The copper precipitates in WP1-082 appeared mostly as very regular cuboidal particles, whereas those in WP1-084 were much more irregular in shape. Many of these copper particles were internally twinned as shown by the pairs of spots in the CBED pattern in Fig. 10*c*. The density of copper precipitates was highest in WP1-082 and WP1-084, with the next highest density in WP3-031. Only a few scattered copper precipitates were found in WP2-027. The copper appeared to be concentrated around type one or type two precipitates, as seen at D in Fig. 9, and were not distributed homogeneously throughout the matrix material. Copper was also found as the main element present in relatively large particles such as the ones indicated at E in Fig. 9*a*.

Discussion

Comparison of the optical micrographs of the Electric Power Research Institute (EPRI) weld metals and the Babcox & Wilcox (B&W) weld metals shows that they have equivalent bainitic structures. Scanning electron microscopy shows the equivalence of the fracture surface morphology. Analytical TEM reveals a similarity between the EPRI and B&W microstructures. In both weld metals, large manganese-rich spherical inclusions, large elongated M_3C carbides, and relatively small M_2C carbides were identified. However, the similarity of the two weld metals does not continue when the finest precipitates are considered. The small precipitates (20 to 100 nm in diameter) found in the EPRI weld metals are probably Mo_2C whereas those of the B&W weld metals are primarily copper. In fact, copper is present only in the overall X-ray analysis for the EPRI weld metals, which indicates that the copper is primarily in solid solution or in precipitates so small that they cannot be resolved by TEM.

Radiation embrittlement is a complex process of copper precipitation and defect cluster generation [*2,3*]. Copper in solid solution enhances neutron irradiation induced defect cluster generation by either a vacancy trapping mechanism or by reducing the critical void radius. So, as copper precipitates, it is removed from solid solution thus reducing the defect cluster generation effect while enhancing the age hardening of the weld metal. Therefore, it is expected that the EPRI weld metals and B&W weld metals will respond differently to reactor conditions with regards to rate of embrittlement, shifts in transition temperature, drop in upper shelf energy, response to annealing, and rate of re-embrittlement. This suggests that copper in solid solution rather than total copper content should be the key variable in empirical embrittlement correlations.

A major difference in microstructure as a result of different stress relief times was also identified in the present work. Specimen WP1-082, which was subjected to a 50-h stress relief cycle, had M_3C carbides and copper-rich precipitates with nickel contents ten times those of precipitates in Specimen WP1-084, which had a 20-h stress relief cycle. The two specimens were taken from the same weldment and were located adjacent to one another. It would appear that the additional stress relief time has enhanced nickel precipitation at the M_3C carbides and copper precipitates. Nickel has been identified as an element that enhances radiation embrittlement and influences the recovery of the transition temperature after a 399°C (750°F)/168-h heat treatment [*4*]. Therefore, it is likely that the amount of nickel in solid solution will affect the radiation embrittlement response of these weld metals. Thus, the radiation embrittlement response of the weld may be sensitive to stress relief time. Further, it was observed that the copper precipitates of the specimen stress relieved for 50 h were larger, spaced further apart, and more regularly shaped than those of the specimen stress relieved for 20 h. This indicates an overaging effect has taken place. This reaction may introduce an additional sensitivity of neutron irradiation embrittlement to stress relief time. Evaluations as to the role of microstructure in radiation embrittlement are currently being investigated in this on-going program, and future publications will report the results.

Conclusions

A detailed microstructural characterization of production Linde 80 submerged arc weld metals was conducted. Two major conclusions are drawn from this work.

1. Although the general microstructure of the EPRI simulated reactor vessel weld metals are similar to the B&W weld metals, the fine scale microstructures are not equivalent. Large quantities of copper precipitates are evident in the B&W metals while it appears that the copper in the EPRI weld metals is in solid solution. This difference, although subtle, could lead to different responses to neutron irradiation embrittlement and subsequent in-situ annealing.
2. The stress relief time can influence nickel content of the copper rich precipitates and the M_3C carbides. Since nickel is known to enhance neutron irradiation embrittlement and influence the amount of transition recovery after a 399°C (750°F)/168-h heat treatment, stress relief time may significantly affect the response of the weld metal to both the reactor environment and in-situ annealing.

Acknowledgments

This study was sponsored by The Babcock & Wilcox Owners Group as a part of their Reactor Vessel Integrity Program. The authors acknowledge and express appreciation for their support. The authors also acknowledge Bob Ladd of the University of Virginia for thin foil and extraction replica preparation.

References

[*1*] Mager, T. R., *Feasibility of and Methodology for Thermal Annealing an Embrittled Reactor Vessel,* EPRI NP-2712, Vols. 1 and 2, Palo Alto, CA, Nov. 1982.
[*2*] Odette, G. R. and Lombrozo, P., *Physically Based Regression Correlations of Embrittlement Data from Reactor Vessel Surveillance Programs,* EPRI NP-3319, Jan. 1984.

[3] Fisher, S. B., Harbottle, J. E., and Aldrige, N., *Copper Precipitation in Pressure Vessel Steels, CEGB report in 3 Parts,* TPRD/B/0396/N84, TPRD/B/0397/N84, TPRD/B/0398/N84, Central Electricity Generating Board, United Kingdom, 1984.

[4] Hawthorne, J. R., "Significance of Nickel and Copper Content to Radiation Sensitivity and Postirradiation Heat Treatment Recovery of Reactor Vessel Steels," *Effects of Radiation on Materials: Proceedings of the Eleventh International Symposium, STP 782,* American Society for Testing and Materials, Philadelphia, 1982, pp. 375–391.

Allen L. Hiser[1]

Comparison of Irradiation-Induced Transition Temperature Increases from Notch Ductility and Fracture Toughness Tests

REFERENCE: Hiser, A. L., **"Comparison of Irradiation-Induced Transition Temperature Increases from Notch Ductility and Fracture Toughness Tests,"** *Influence of Radiation on Material Properties: 13th International Symposium (Part II), ASTM STP 956,* F. A. Garner, C. H. Henagar, Jr., N. Igata, Eds., American Society for Testing and Materials, Philadelphia, 1987, pp. 333–357.

ABSTRACT: Reactor pressure vessel (RPV) surveillance capsules contain Charpy-V notch ductility C_v specimens, but many do not contain fracture toughness specimens; accordingly, the radiation-induced shift (increase) in the brittle-to-ductile transition region ΔT is based upon the ΔT determined from C_v tests. Since the American Society of Mechanical Engineers (ASME) K_{Ic} and K_{IR} reference fracture toughness curves are shifted by the ΔT from C_v, assurance that this ΔT does not underestimate ΔT associated with the actual irradiated fracture toughness is required to provide confidence that safety margins do not fall below assumed levels.

To assess this behavior, comparisons of ΔTs defined by elastic-plastic fracture toughness and C_v tests have been made using data from RPV base and weld metals in which irradiations were made under test reactor conditions. Using "as-measured" fracture toughness values K_{Jc}, average comparisons between $\Delta T(C_v)$ and $\Delta T(K_{Jc})$ reveal that, on average, $\Delta T(C_v)$ tends to underestimate $\Delta T(K_{Jc})$ for plates by 22°C and to overestimate $\Delta T(K_{Jc})$ for welds by 5°C on average.

When the fracture toughness data are adjusted to account for specimen size using a "β_{Ic} correction," the trends are the same as for K_{Jc}, with the underestimate for plates an average of 10°C and the overestimate for welds 11°C on average.

Comparison of ΔTs at various index levels implies that the C_v curve for irradiated material tends to be shallower than that for unirradiated material. However, the shape of the K_{Jc} curve for irradiated and unirradiated material is the same, but the $K_{\beta c}$ curve for irradiated material is steeper than that for unirradiated material.

KEY WORDS: nuclear pressure vessel steels, transition temperature shift, upper shelf energy, *J* integral, fracture toughness, correlations, Charpy-V notch, embrittlement, cleavage, statistical analysis, upper shelf change, materials irradiation, radiation, irradiation effects, elastic-plastic fracture toughness

As a nuclear power plant operates, the reactor pressure vessel (RPV) is exposed to neutron irradiation caused by the fission process. The effect of this neutron irradiation on the materials constituting the reactor pressure vessel (RPV) (plates welds and forgings) is generally to degrade the material properties. The extent of degradation is highly dependent on the chemical composition, product form, and processing history of the material, as well as the fluence and the irradiation temperature. Generally the degradation is demonstrated through increases in tensile properties (both yield and ultimate), decreases in upper shelf toughness (impact and static), and increases in the brittle-to-ductile transition temperature, that is, the temperature at which the material failure

[1] Mechanical engineer, Materials Engineering Associates, Inc., 9700-B M. L. King, Jr. Highway, Lanham, MD 20706-1837.

mode changes from brittle (cleavage) fracture to a ductile behavior, characterized by tearing of the material. Of these three responses of the material to the irradiation, the increase in the transition temperature is of prime consideration in assessing the potential for failure because of a pressurized thermal shock (PTS) or any other scenario under which relatively low RPV temperatures (much less than the operating temperature of 288°C) may be accompanied by significant pressure in the vessel.

In recognition of such concerns, Appendix G of 10CFR50 (Title 10, Part 50, Code of Federal Regulations) has established a PTS screening criterion of 270°F (132°C) for plates, forgings, and axial weld materials, and 300°F (149°C) for circumferential weld materials. In each case, the criterion gives upper limits on the reference temperature RT_{NDT} as defined in the American Society of Mechanical Engineers (ASME) Boiler and Pressure Vessel Code. RT_{NDT} is also used in an analytical sense to index the K_{IR} (reference stress intensity factor) curve in Appendix G of Section III of the ASME Code. Before operation of the RPV, determination of RT_{NDT} is a relatively simple task, involving tests of drop weight and Charpy-V notch C_v specimens. RT_{NDT} is defined in the ASME code as the greater of T_{NDT} (the nil-ductility transition temperature of dropweight tests) and T_{C_v} − 60°F, where T_{C_v} is either the index at which each C_v test (of three) "shall exhibit at least 35 mil (0.9 mm) lateral expansion and not less than 50 ft · lb (68 J) absorbed energy" or the temperature at which the lower bound C_v curve exhibits 50 ft · lb (68 J) and 35 mil (0.9 mm). To account for irradiation effects, Appendix G of 10CFR50 states that "adjusted reference temperature" means the reference temperature as adjusted for irradiation effects by adding to RT_{NDT} the temperature shift ΔT, measured at the 30-ft · lb (41-J) level, in the average Charpy curve for the irradiated material relative to that for the unirradiated material.

As with the determination of RT_{NDT}, the pre-operation critical stress intensity factor K_{Ic} curve can be determined in a straightforward manner for the RPV materials, assuming that the large test specimens required are available. After irradiation, it is difficult to measure the K_{Ic} curve directly because the specimens that would be required are too large for the space available for surveillance capsules. For calculation of pressure-temperature (P-T) curves, Appendix G of 10 CFR 50 assumes that the K_{Ic} curve of the irradiated material is defined by shifting the K_{IR} curve as indexed by the "adjusted reference temperature" mentioned above. (For accident analyses, the K_{IC} curve in Appendix A of Section XI of the ASME Code is used as the reference curve, which is indexed at the "adjusted reference temperature.") However, it is necessary to establish whether or not the shift in the C_v curve at 41 J is an accurate or conservative estimation of the temperature shift in the K_{Ic} curve in order to ensure that overestimation of the material toughness does not occur. This study combines the available fracture toughness and Charpy-V data (from the same material irradiation conditions) to assess the relationship between temperature differences ΔTs measured by each test type. (Since very large specimens are required to give valid K_{Ic} data, the fracture toughness data used are from an elastic-plastic analysis K_{Jc} and from a "β_{Ic} correction" to the K_{Jc} data $K_{\beta c}$.) In this way, an assessment can be made of the degree of conservatism (or lack thereof) afforded by using ΔT from C_v to estimate ΔT for K.

Materials

The materials used in this study have been characterized over several years, initially at the Naval Research Laboratory (NRL), and more recently at Materials Engineering Associates, Inc. (MEA). Included in the study are an A 508 forging, several heats of A 533 Grade B and A 302 Grade B plate, and submerged-arc welds made with Linde 80, Linde 0091, and Linde 124 fluxes. A listing of the material chemistries and fluences for the subject conditions is given in Table 1. All of the irradiations were conducted in test reactors, under typical accelerated flux conditions of $\sim 8 \times 10^{12}$ n/cm²/s ($E > 1$ MeV) at nominal temperatures of 288°C.

TABLE 1—*Chemistry and fluence values for the subject heats.*

Material	Code	Fluence, 10^{19} n/cm^2, $E > 1$ MeV	C	Mn	P	S	Si	Cr	Ni	Mo	Cu	V	As	Sn	N	B
A 508-2 Forging	BCB[a]	2.8	0.212	0.63	0.006	0.012	0.26	0.38	0.55	0.62	0.036	0.002	0.004	0.011	0.013	0.0002
A 533-B Plate	CAB[a]	1.2, 1.7, 2.1	0.25	1.41	0.008	0.014	0.26	0.11	0.46	0.49	0.12	0.003	0.012	0.008	0.008	0.0007
	CBB[a]	4.4	0.21	1.45	0.006	0.009	0.23	0.05	0.55	0.64	0.013	0.003	0.035	0.015	0.014	0.0004
	3P[b]	2.66, 5.2, 5.2, 2.7, 1.3	0.20	1.26	0.011	0.018	0.25	0.10	0.56	0.45	0.12	[d]	[d]	[d]	[d]	[d]
	02G[c]	1.81	0.23	1.55	0.009	0.014	0.20	0.04	0.67	0.53	0.16	0.003	[d]	[d]	[d]	[d]
	67C	[e]	0.23	1.31	0.025	0.018	0.20	<0.003	0.70	0.51	0.002	[d]	[d]	<0.004	0.009	0.0004
	68A	[e]	0.23	1.31	0.003	0.017	0.22	<0.003	0.70	0.52	0.30	[d]	[d]	0.004	0.010	0.0006
	68C	[e]	0.23	1.31	0.028	0.017	0.22	<0.003	0.70	0.52	0.30	[d]	[d]	0.004	0.010	0.0006
A 302-B Plate	N[a]	2.7	0.24	1.34	0.011	0.023	0.23	0.11	0.18	0.51	0.20	0.001	[d]	0.037	0.008	0.0001
	F23[a,b]	3.7, 2.87, 5.6, 5.6, 2.9, 1.4	0.24	1.34	0.011	0.023	0.23	0.11	0.18	0.51	0.20	0.001	[d]	0.037	[d]	[d]
	6A2	[d]	0.23	1.29	0.002	0.013	0.22	[d]	0.045	0.53	0.28	[d]	[d]	0.004	[d]	[d]
Welds (Linde 80 ϕ)	E19[a]	0.1, 0.8, 2.3	0.12	1.37	0.007	0.013	0.53	0.04	0.59	0.44	0.43	[d]	[d]	0.01	0.011	0.01
	E23[a]	0.7	0.11	1.36	0.008	0.013	0.52	0.04	0.60	0.44	0.24	[d]	[d]	0.01	0.013	0.01
	71W[c]	1.79	0.124	1.58	0.011	0.011	0.54	0.12	0.63	0.45	0.046	0.005	[d]	[d]	[d]	[d]
	W8A[f]	1.5, 2.2, 2.1	0.083	1.33	0.011	0.016	0.77	0.12	0.59	0.47	0.39	0.003	0.001	0.002	[d]	0.000
(Linde 0091 ϕ)	E24[a]	0.7	0.16	1.25	0.005	0.009	0.17	0.04	0.59	0.44	0.37	[d]	[d]	0.01	0.011	0.01
	68W[c]	1.36	0.15	1.38	0.008	0.009	0.16	0.04	0.13	0.60	0.04	0.007	[d]	[d]	[d]	[d]
	69W[c]	1.19	0.14	1.19	0.010	0.009	0.19	0.09	0.10	0.54	0.12	0.005	[d]	[d]	[d]	[d]
	W9A[f]	1.5, 2.2, 2.1	0.19	1.24	0.010	0.008	0.23	0.10	0.70	0.49	0.39	[d]	[d]	0.003	[d]	[d]
(Linde 124 ϕ)	70W[c]	1.95	0.10	1.48	0.011	0.011	0.44	0.13	0.63	0.47	0.056	0.004	[d]	[d]	[d]	[d]
	E4[a]	2.4	0.08	1.38	0.013	0.018	0.49	0.10	0.65	0.44	0.16	0.005	0.008	0.004	[d]	0.001

[a] Reference *1*.
[b] Reference *2*.
[c] Reference *3*.
[d] Not determined.
[e] Not available.
[f] Reference *4*.

The four heats listed in Table 1 for which fluence data are not available (67C, 68A, 68C, and 6A2) have variations in phosphorous, nickel, and copper contents, and will be characterized to give more information about chemistry effects on ΔT comparisons.

Data Analysis Procedures

Notch Ductility C_v

Notch ductility was determined using Charpy-V impact specimens. The Charpy-V energy C_vE data for each heat and condition (irradiated or unirradiated) were fit to a hyperbolic tangent function

$$C_vE = A_0 + A_1 \tanh\left[\frac{T - T_0}{A_2}\right] \tag{1}$$

where A_0, A_1, A_2, and T_0 are determined through a nonlinear regression analysis. Both shelf (upper and lower) transition data are included in the same fit to Eq 1, as opposed to using only transition data for a transition region fit, and so forth. Since few data were determined in a lower shelf region, some of the fits drop to negative C_v levels at low temperatures. A nonnegative lower shelf was not forced, since the purpose for using the chosen equation was to model the data and not to force the data to a model. (The "negative" C_v levels given by some of these fits generally do not have any impact on the results used here.) A nonsloping upper shelf fit was used since justification for a sloping upper shelf was not evident for cases in which sufficient characterization of the upper shelf trends was available.

Once a given set of data had been fit to Eq 1, the upper shelf level and the transition temperatures at various C_v indices were determined from:

$$\text{Upper Shelf Energy (USE)} = A_0 + A_1 \tag{2}$$

$$T_{\text{index}} = T_0 + A_2 \tanh^{-1}\left[\frac{C_vE_{\text{Index}} - A_0}{A_1}\right] \tag{3}$$

where C_vE_{Index} values were selected at 41 J (30 ft · lb per Appendix G of 10CFR50) and 68 J (50 ft · lb).

Fracture Toughness K_{Jc}

The fracture toughness data of interest are only those within the brittle-to-ductile transition region. No data with blunting and stable crack growth of more than 0.15 mm are included. The allowance of such small yet measurable crack growth is consistent with ASTM Test Method for J_{Ic}, a Measure of Fracture Toughness (E 813) and represents an engineering estimate of crack initiation since experimental limitations restrict the ability to locate a true initiation point.

The static fracture toughness data are determined using compact specimens (CS). These specimens were full thickness 25-mm (1 T-) and 12.7-mm (0.5 T-) CS designs, with several differences from the standard ASTM Test Method for Plane-Strain Fracture Toughness of Metallic Materials (E 399) design. The major difference was in enlargement of the notch region to permit placement of razor blades on the "load line" for mounting of a displacement transducer. The enlargement of the notch forced an increased spacing of the pin holes and in some cases a reduction in the pinhole diameter. The pinhole changes were required to maintain sufficient ligaments above and below the holes to prevent bending or bulging of the ligaments.

Because of the small specimen size forced by the constraints of the test reactor irradiation facility, the maximum K_{Ic} levels that could be obtained by ASTM E 399 were below

50 MPa · m$^{1/2}$ (~45 ksi · in$^{1/2}$), much too low to provide a meaningful result. For cases in which only a K_Q value was obtainable, the J integral, an elastic-plastic fracture parameter, was used for data analysis. Specifically, the J value at test termination was taken to be J_{Crit}, since by definition that was the J value at the initiation of "crack growth," albeit fast cleavage fracture. A K_{Jc} value was then calculated from

$$K_{Jc} = \sqrt{EJ_{Crit}/(1 - \nu^2)} \quad (4)$$

where E is Young's modulus at the test temperature, and ν is Poisson's ratio, taken as 0.3 for all of these data. For cases in which a K_{Ic} value could be determined from the test record, the K_{Jc} value from Eq 4 (with $\nu = 0.3$) matched the K_{Ic} value within 5%.

As with the C_v data, the K_{Jc} data were modeled with a mathematical expression, in this case

$$K_{Jc} = B_0 + B_1 e^{(T/B_2)} \quad (5)$$

where B_0, B_1, and B_2 are determined through a nonlinear regression analysis. Since the regression results were not restricted to $B_0 \geqslant 0$, in many cases the resultant curve will give negative K_{Jc} values at low temperatures.

The transition temperatures at various indices were determined from Eq 5 in the following form

$$T_{Index} = B_2 \ln\left[\frac{K_{Index} - B_0}{B_1}\right] \quad (6)$$

with K_{Index} values of 75, 100, and 150 MPa · m$^{1/2}$ used. For comparisons with C_v data at 41 J (30 ft · lb), a K_{Index} value of 100 MPa · m$^{1/2}$ was used. This value is supported by correlations from Rolfe and Novak [5] and Sailors and Corten [6], using 41 J for C_v and 206 800 MPa for Young's modulus.

Fracture Toughness Data with the Irwin β_{Ic} Adjustment ($K_{\beta c}$)

For cases in which a K_{Ic} value could not be obtained from a fracture toughness test, an approximation to K_{Ic} from K_{Jc} was sought. For such cases, Merkle [7] has recommended use of the Irwin β_{Ic} adjustment [8], to give small specimen fracture toughness values that "consistently vary within the same range as the values determined from larger valid LEFM specimens and large structural tests" [7]. Caveats to the use of this adjustment procedure are that cleavage fracture should be the failure mechanism and not ductile tearing, the correct critical conditions are at the onset of unstable cleavage fracture and not the maximum load point, and finally, the testing should be performed in displacement control to accurately locate the cleavage point on the load-displacement record. All of the necessary conditions for using the β_{Ic} adjustments are met by these data. While in some cases cleavage fracture may be preceded by significant stable crack growth (ductile tearing), no such data were used in this study. Only data that do not exceed an engineering estimate of crack initiation in a J-R curve format are used. By definition, K_{Ic} values were not adjusted using this β_{Ic} procedure.

As with the K_{Jc} data, the $K_{\beta c}$ data were modeled with an exponential function

$$K_{\beta c} = C_0 + C_1 e^{(T/C_2)} \quad (7)$$

with C_0, C_1, and C_2 determined through a nonlinear regression analysis. Since C_0 is not limited to nonnegative values, the best-fit curves may yield negative values at low temperatures.

The transition temperature at various indices was determined from Eq 16 as follows

$$T_{Index} = C_2 \ln\left[\frac{K_{Index} - C_0}{C_1}\right] \quad (8)$$

with K_{Index} values of 75 and 100 MPa · m$^{1/2}$. Since many of the $K_{\beta c}$ curves did exceed

75 MPa · m½ but few exceeded 100 MPa · m½, comparisons with C_v = 41 J and K_{Jc} = 100 MPa · m½ were made at $K_{\beta c}$ = 75 MPa · m½.

Test Method Comparisons

The focus in this section is on the change in temperature ΔT at a specific index caused by irradiation, as opposed to the absolute temperature at the index level. Comparisons between ΔTs from notch ductility and fracture toughness are based on product form, with comparisons at a C_v energy level of 41 J and fracture toughness levels from 75 to 150 MPa · m½.

Charpy-V Versus K_{Jc}

Comparison of ΔTs from C_v at 41 J ($\Delta T[C_v$ at 41 J]) and K_{Jc} at 100 MPa · m½ ($\Delta T[K_{Jc}$ at 100 MPa · m½]) reveals an interesting trend. In Fig. 1 for base metals, only 3 of the 18 base metal points represent conservative estimates of $\Delta T(K_{Jc}$ at 100 MPa · m½) by $\Delta T(C_v$ at 41 J). All three of these points are from a single heat of A 533-B plate (the Heavy Section Steel Technology [HSST] 03 plate) tested as a part of the LWR-PV Dosimetry Improvement Program [2]). The largest deviation is for the A 508-2 forging, with $\Delta T(K_{Jc})$ = 102° C and $\Delta T(C_v)$ = 19°C. The base metals (excluding the A 508-2 forging that will be discussed later) have a $\Delta T(C_v)$ that averages 22°C less than $\Delta T(K_{Jc})$.

In contrast to the base metal results, the welds (Fig. 2) illustrate good correspondence (albeit with substantial scatter) between $\Delta T(C_v)$ and $\Delta T(K_{Jc})$, at all ΔT levels. Here, $\Delta T(C_v)$ averages 5°C higher than $\Delta T(K_{Jc})$. While the scatter is large and $\Delta T(C_v)$ tends to exceed $\Delta T(K_{Jc})$, an approximate 1:1 correspondence typifies these data more than the base metal data.

Table 2 gives linear least-squares fits to three different equations relating $\Delta T(C_v)$ to $\Delta T(K_{Jc})$

$$\Delta T(K_{Jc}) = \Delta T(C_v)K_o + K_1 \tag{9}$$

$$\Delta T(K_{Jc}) = \Delta T(C_v) + K_2 \tag{10}$$

$$\Delta T(K_{Jc}) = \Delta T(C_v)K_3 \tag{11}$$

with ±1 σ levels also given. (These analyses were made excluding the A508-2 forging point, which is the only forging data in this data base.) While the latter two equations are simply less

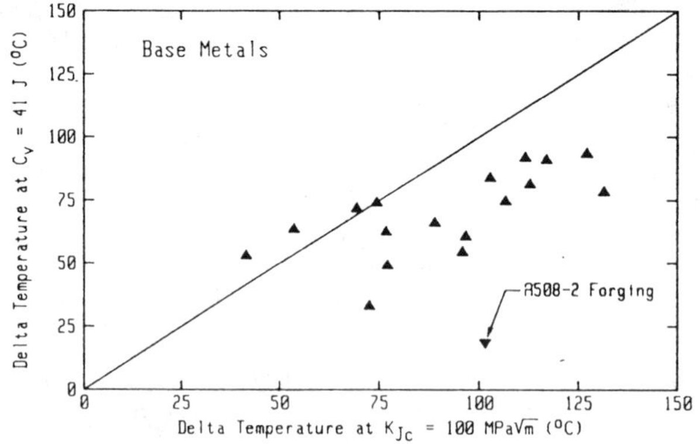

FIG. 1—*Comparison of* ΔT*s from* C_v *at 41 J and* K_{Jc} *at 100 MPa · m½, for base metals. Only three data points exhibit conservative estimates of* $\Delta T(K_{Jc})$ *by* $\Delta T(C_v)$*, with the greatest nonconservative deviation from the A 508-2 forging.*

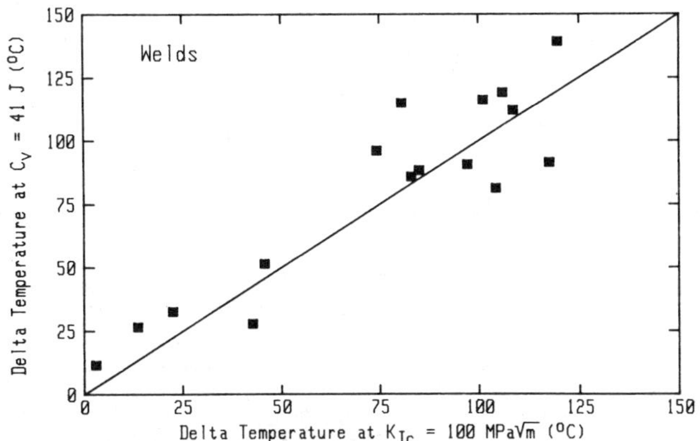

FIG. 2—*Comparison of* ΔT*s from* C_v *at 41 J and* K_{Jc} *at 100 MPa* $\cdot$ *m*$^{1/2}$*, for weld metals. As contrasted with Fig. 1, good agreement exists at all* $\Delta T(K_{Jc})$ *levels, with the tendency towards a conservative estimation of* $\Delta T(K_{Jc})$ *by* $\Delta T(C_v)$.

general subsets of the first, they are less complicated ways for adjusting nonconservative ΔT estimates.

Similar conclusions about $\Delta T(C_v)$ and $\Delta T(K_{Jc})$ can be made at K_{Jc} levels of 75 and 150 MPa $\cdot$ m$^{1/2}$ (Table 2). For example, $\Delta T(C_v)$ for base metal tends to average 22°C less than $\Delta T(K_{Jc})$ while $\Delta T(C_v)$ for weld metal averages 6°C greater than $\Delta T(K_{Jc})$. In fact, the closeness of the K_2 values at each K_{Jc} level implies that on average the K_{Jc} curve for irradiated material is very similar in curvature to that for unirradiated material.

The C_v and K_{Jc} data available for the A 508-2 forging are illustrated in Fig. 3. While variations in the C_v data are such that the unirradiated and irradiated data overlap, the K_{Jc} data exhibit a distinct difference between unirradiated and irradiated specimen behavior. One cautionary note required here is that there are few irradiated K_{Jc} data points (only four), and these few points do exhibit considerable variability. Regardless of the variability consideration, major differences between C_v and K_{Jc} indications of the irradiation degradation are apparent.

Charpy-V Versus $K_{\beta c}$

Comparison of $\Delta T(C_v)$ and that from $K_{\beta c}$, that is, $\Delta T(K_{\beta c})$, reveals different behavior from the prior K_{Jc} comparisons. As illustrated in Fig. 4 for base metals, $\Delta T(K_{\beta c}$ at 75 MPa $\cdot$ m$^{1/2})$ is generally much closer to $\Delta T(C_v$ at 41 J), than was $\Delta T(K_{Jc}$ at 100 MPa $\cdot$ m$^{1/2})$. On average, $\Delta T(C_v)$ is only 10°C less than $\Delta T(K_{\beta c})$ (Table 3), versus 22°C less when compared to $\Delta T(K_{Jc})$. For weld metals (Fig. 5), the overall variation is significantly higher, with $\Delta T(C_v)$ averaging 11°C greater than $\Delta T(K_{\beta c})$. With both plates and welds combined, $\Delta T(C_v)$ is within 1°C of $\Delta T(K_{\beta c})$, on average.

At a $K_{\beta c}$ level of 100 MPa $\cdot$ m$^{1/2}$, the scatter increases significantly with 1 σ limits of $\pm$22°C for base metals and $\pm$47°C for welds (Table 3). In one case, $\Delta T(K_{\beta c}$ at 100 MPa $\cdot$ m$^{1/2})$ is 150°C less than $\Delta T(K_{Jc}$ at 100 MPa $\cdot$ m$^{1/2})$, with a $\Delta T(K_{\beta c})$ of -68°C at this level. This anomaly and in general the poor correspondence is at least partially attributable to an absence of $K_{\beta c}$ data above 80 MPa $\cdot$ m$^{1/2}$. This absence of data in combination with the very large reduction in unirradiated K level caused by the β_{Ic} adjustment (and the much lower yield stress values for unirradiated material versus irradiated material) lead to a question of the reliability of using $\Delta T(K_{\beta c})$ values extrapolated out to the 100 MPa $\cdot$ m$^{1/2}$ level for these data; a recommendation is to not include the latter in consideration of variances between $\Delta T(C_v)$ and $\Delta T(K)$.

TABLE 2—*Statistical comparisons between* $\Delta T(C_v)$ *and* $\Delta T(K_{Jc})$ *(Eqs 9 through 11).*

C_v Level, J	K_{Jc} Level, $MPa \cdot m^{1/2}$	Product Form	$\Delta T(K_{Jc}) = \Delta T(C_v)K_0 + K_1$			$\Delta T(K_{Jc}) = \Delta T(C_v) + K_2$, °C			$\Delta T(K_{Jc}) = \Delta T(C_v)K_3$		
			K_o	K_1, °C	±1 σ, °C	K_2	±1 σ	1-σ Range	K_3	±1 σ	±σ, °C
41	100	base	1.039	19.3	18.6	22.0	18.6	+ 3.4: +40.6	1.303	0.260	19.1
		weld	0.896	3.3	15.6	−5.1	16.1	−21.2: +11.0	0.929	0.171	15.7
		both	0.835	21.2	21.4	8.9	22.0	−13.1: +30.9	1.081	0.280	22.8
41	75	base	1.106	12.6	20.6	19.9	20.7	− 0.8: +40.6	1.278	0.283	20.8
		weld	0.908	1.9	15.3	−5.6	15.7	−21.3: +10.1	0.927	0.167	15.3
		both	0.860	18.0	21.9	7.6	22.3	−14.7: +29.9	1.070	0.281	22.9
41	150	base	1.184	8.9	19.2	21.7	19.5	+ 2.2: +41.2	1.307	0.263	19.3
		weld	0.987	−5.2	29.6	−6.3	29.6	−35.9: +23.3	0.933	0.324	29.6
		both	0.929	13.5	28.2	8.2	28.3	−20.1: +36.5	1.085	0.352	28.7

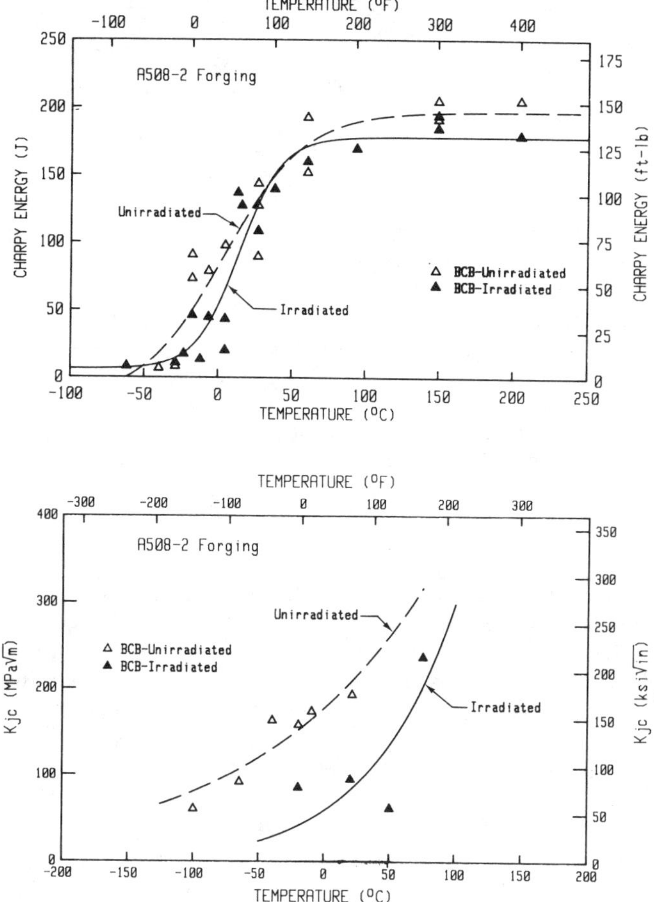

FIG. 3—*Comparison of* C_v *and* K_{Jc} *data for the A508-2 forging. While considerable variability is apparent for each data type, differences in irradiation effect are apparent: the* C_v *data show a small* ΔT *and the* K_{Jc} *data have a much larger* ΔT.

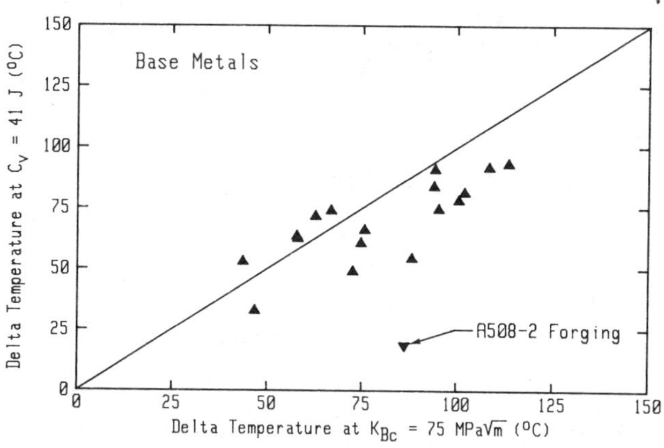

FIG. 4—*Comparison of* ΔT*s from* C_v *at 41 J and* $K_{\beta c}$ *at 75 MPa* $\cdot$ $m^{1/2}$*, for base metals. The situation here is similar to the* K_{Jc} *comparison in Fig. 1, with few* $\Delta T(C_v)$ *values exceeding the* $\Delta T(K_{\beta c})$ *values. Again, the A 508-2 forging point far exceeds any reasonable scatter band.*

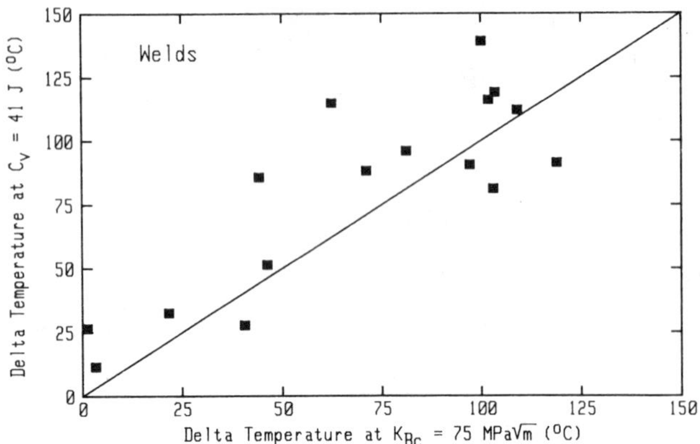

FIG. 5—*Comparison of* ΔTs *from* C_v *at 41 J and* $K_{\beta c}$ *at 75 MPa* · $m^{1/2}$ *for weld metals. As with* K_{Jc} *in Fig. 2, generally a conservative estimate of* $\Delta T(K_{\beta C})$ *is made by* $\Delta T(C_v)$, *but the scatter here is much greater than for the* K_{Jc} *comparison.*

K_{Jc} *Versus* $K_{\beta c}$

As mentioned above, the β_{Ic} adjustment has the effect of reducing high toughness, unirradiated K_{Jc} data much more than high toughness irradiated K_{Jc} data. Therefore, as a general rule, $\Delta T(K_{\beta c})$ will be less than $\Delta T(K_{Jc})$. In 73% of the cases, $\Delta T(K_{\beta c})$ is less than $\Delta T(K_{Jc})$, by as much as 39°C. In cases where $\Delta T(K_{\beta c})$ exceeds $\Delta T(K_{Jc})$, it does so by a maximum of 7°C. This correspondence is apparent in Fig. 6 for base metals and Fig. 7 for weld metals. From Table 4, on average the $\Delta T(K_{\beta c})$ for base metals is 12°C less than $\Delta T(K_{Jc})$, while for weld metals $\Delta T(K_{\beta c})$ is only 6°C less than $\Delta T(K_{Jc})$, on average.

Curse-Fit Versus Eyeball

Most of the C_v (and most of the K_{Jc} and $K_{\beta c}$) data reported by current MEA personnel at MEA and earlier at the NRL have been analyzed using an eyeball or hand-fit method. This method does

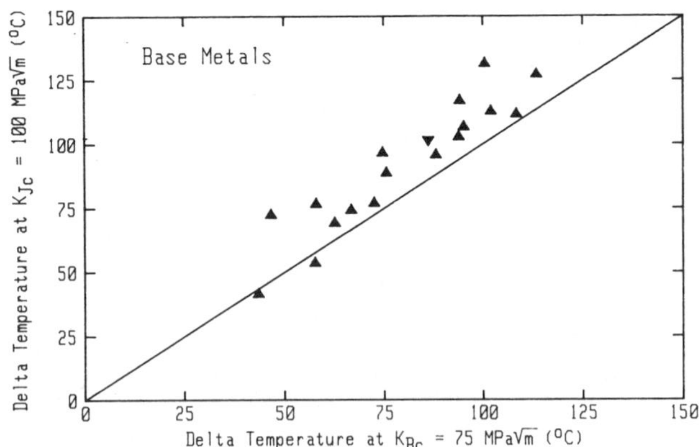

FIG. 6—*Comparison of* ΔTs *from* K_{Jc} *at 100 MPa* · $m^{1/2}$ and $K_{\beta c}$ *at 75 MPa* · $m^{1/2}$, *for base metals. Only in two cases does* $\Delta T(K_{\beta c})$ *exceed* $\Delta T(K_{Jc})$ *at these levels.*

TABLE 3—*Statistical comparisons between* $\Delta T(C_v)$ *and* $\Delta T(K_{\beta c})$.

			$\Delta T(K_{\beta c}) = \Delta T(C_v)K_0 + K_1$			$\Delta T(K_{\beta c}) = \Delta T(C_v) + K_2$, °C			$\Delta T(K_{\beta c}) = \Delta T(C_v)K_3$		
C_v Level, J	$K_{\beta c}$ Level, $MPa \cdot m^{1/2}$	Product Form	K_0	K_1, °C	$\pm 1\sigma$, °C	K_2	$\pm 1\sigma$	1-σ Range	K_3	$\pm 1\sigma$	$\pm 1\sigma$, °C
41	75	base	1.028	8.2	13.3	10.2	13.3	−3.1: +23.5	1.140	0.183	13.5
		weld	0.839	1.7	21.1	−11.3	22.0	−33.3: +10.7	0.856	0.231	21.1
		both	0.806	14.3	20.0	−0.2	20.8	−21.0: +20.6	0.972	0.254	20.7
41	100	base	1.253	−26.6	21.2	−9.0	21.6	−30.6: +12.6	0.890	0.300	22.1
		weld	0.785	−6.2	49.0	−23.5	49.7	−73.2: +26.2	0.722	0.536	49.1
		both	0.821	−2.6	37.6	−16.0	38.0	−54.0: +22.0	0.790	0.462	37.6

TABLE 4—*Statistical comparisons between* $\Delta T(K_{Jc})$ *and* $\Delta T(K_{\beta c})$.

K_{Jc} Level, J	$K_{\beta c}$ Level, MPa · $m^{1/2}$	Product Form	$\Delta T(K_{\beta c}) = \Delta T(K_{Jc})K_0 + K_1$			$\Delta T(K_{\beta c}) = \Delta T(K_{Jc}) + K_2$, °C			$\Delta T(K_{\beta c}) = \Delta T(K_{Jc})K_3$		
			K_o	K_1, °C	±1 σ, °C	K_2	± 1 σ	1-σ Range	K_3	±1 σ	±1 σ, °C
100	75	base	0.793	7.1	8.1	−11.9	9.6	−21.5: −2.3	0.865	0.085	8.3
. . .	. . .	weld	0.972	−4.1	11.5	−6.2	11.6	−17.8: +5.4	0.929	0.135	11.7
. . .	. . .	both	0.898	−0.6	10.3	−9.1	10.8	−19.9: +1.7	0.892	0.114	10.3

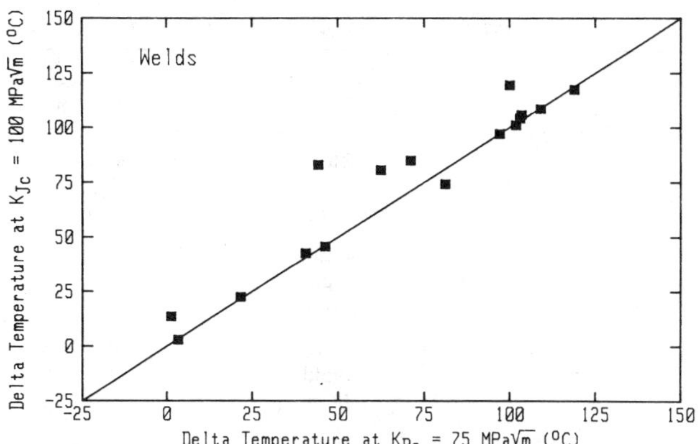

FIG. 7—*Comparison of* ΔT*s from* K_{Jc} *at 100 MPa · m*$^{1/2}$ *and* $K_{\beta c}$ *at 75 MPa · m*$^{1/2}$*, for weld metals. The correspondence here is much improved as compared to the base metal data.*

have advantages, such as the ability of an experienced user to synthesize a curve where insufficient data exist and "weighting" of data based upon historical perspectives. In general, however, the use of standardized computer routines for curve fitting guarantees a repeatable, unbiased fit, which is not user dependent. Since many of these C_v data (and other data not included here) originally were fit using a manual technique and form a unique data base used for assessing composition, fluence, and other effects on radiation sensitivity, comparison of the manual-fitting results with computerized results was considered germane to this study. These comparisons have been made for C_v data in terms of transition temperature (absolute) and ΔT at 41 and 68 J. Additional comparisons were made for K_{Jc} and $K_{\beta c}$ data, with all of the results summarized in Table 5.

As is evident in Fig. 8 for all of the C_v temperatures at 41 J, excellent agreement exists at all levels, from low to high temperatures. In fact, the manual temperature overestimates the curve-fit temperature by less than 2°C on average, with base metals yielding an average overestimate less than 1°C and welds an average overestimate below 3°C (Table 5). The ΔTs (Fig. 9) exhibit similar

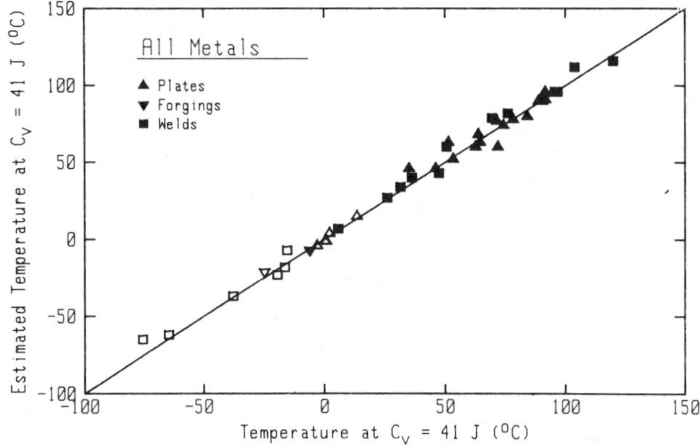

FIG. 8—*Comparison of* C_v *temperatures at 41 J determined from manual fits to the data and computerized fits. The maximum deviation is 12°C for an irradiated A 533-B plate. Filled symbols are irradiated; open symbols are unirradiated. (The computerized-fit data are on the abscissa, with the manual-fit data on the ordinate.)*

TABLE 5—*Comparison of manual-fit and computerized-fit results.*

Test Type Index	Data Type	Product Form	Average Deviation	±1 σ
C_v-41 J	T[a]	base	−0.6°C[b]	±5.0°C
		weld	−2.8	4.9
		both	−1.6	5.0
	ΔT[c]	base	−1.4°C	5.5°C
		weld	−0.1	6.0
		both	−0.9	5.6
C_v-USE	level[d]	base	+0.4 J	2.9 J
		weld	+2.2	3.4
		both	+1.2	3.2
	drop[e]	base	+0.3 J	3.5 J
		weld	+1.4	3.7
		both	+0.8	3.5
K_{Jc}-100 MPa · $m^{1/2}$	T	base	−0.4°C	7.4°C
		weld	+0.7	6.2
		both	+0.1	6.8
	ΔT	base	+2.2°C	8.5°C
		weld	−1.8	6.5
		both	+0.4	7.8
$K_{\beta c}$-75 MPa · $m^{1/2}$	T	base	+0.9°C	3.6°C
		weld	+0.0	5.7
		both	+0.5	4.4
	ΔT	base	+1.7°C	3.7°C
		weld	−4.3	3.9
		both	−0.5	4.7

[a] Transition temperature.
[b] Deviation = (computerized fit) − (manual fit).
[c] Transition temperature increase.
[d] Upper shelf level.
[e] Drop in upper shelf level.

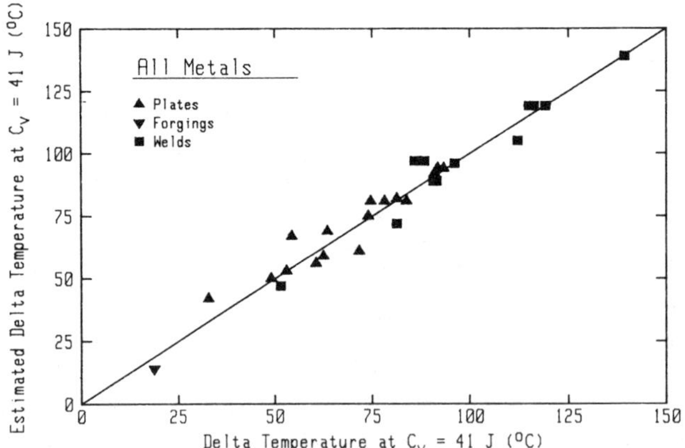

FIG. 9—*Comparison of* ΔT(C_v *at 41 J) determined from manual and computerized fits to the data. In this case, the maximum deviation is 11°C for a Linde 80 weld. (The computerized-fit data are on the abscissa, with the manual-fit data on the ordinate.)*

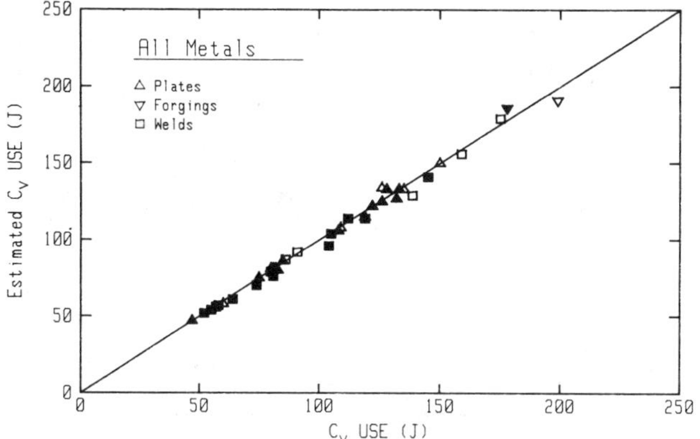

FIG. 10—*Comparison of* C_v *upper shelf level determined from manual and computerized fits to the data. Good agreement is seen at all upper shelf levels. Filled symbols are irradiated; open symbols are unirradiated. (The computerized-fit data are on the abscissa, with the manual-fit data on the ordinate.)*

good agreement, with a 1-σ range of 6°C. Here, the weld ΔTs yield a better average correspondence than the base metal ΔTs.

For C_v upper shelf energy (Fig. 10), excellent agreement again results in comparing manual versus curve-fit levels (1-J overestimate on average), with change in upper shelf level (Fig. 11) yielding a better average correspondence, but slightly greater variability in comparison to Fig. 8. As with earlier data type comparisons, the greatest discrepancy is for the A 508-2 forging, with an underestimate of 16 J for the USE drop. That difference represents an 8-J increase for the unirradiated upper shelf and an 8-J decrease in the irradiated upper shelf, as determined from the curve-fit results in comparison to the manual fit results.

For temperatures at K_{Jc} of 100 MPa · $m^{1/2}$ (Fig. 12), each product form has the manual result within 1°C of the curve-fit result, on average. For the ΔTs, (Fig. 13), the manual result is again within 1°C of the curve-fit result, on average, but the variability is somewhat greater than for the absolute temperature comparisons.

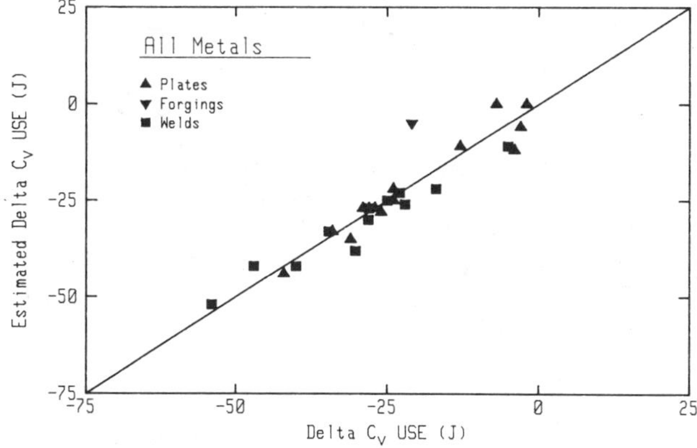

FIG. 11—*Comparison of* C_v *upper shelf drop determined from manual and computerized fits to the data. On average, the manual fit is 1 J less than the computerized fit, with the maximum deviation from the A 508-2 forging, a 16-J discrepancy. (The computerized-fit data are on the abscissa, with the manual-fit data on the ordinate.)*

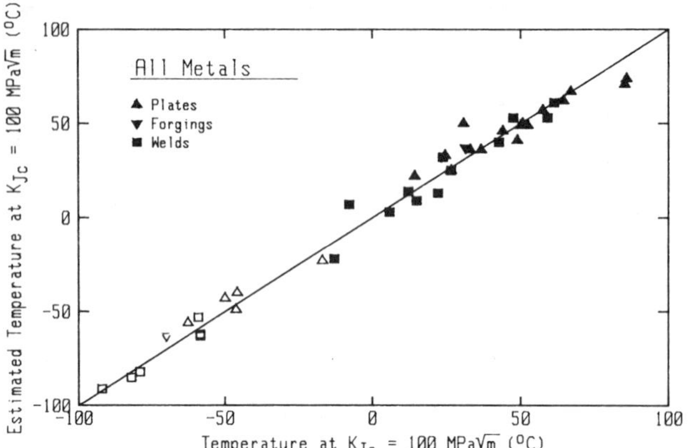

FIG. 12—*Comparison of* K_{Jc} *temperature at 100 MPa* · $m^{1/2}$ *from manual and computerized fits to the data. Good agreement is seen at all temperatures, with a maximum deviation of 20°C for irradiated A 533-B plate. Filled symbols are irradiated; open symbols are unirradiated. (The computerized-fit data are on the abscissa, with the manual-fit data on the ordinate.)*

For the $K_{\beta c}$ data at 75 MPa · $m^{1/2}$, both temperature and ΔT yield manual values within 1°C of the curve-fit values, on average. The standard deviations for the $K_{\beta c}$ comparisons are much less than the K_{Jc} comparisons, attributable to the smaller (and hence more controlled) number of data sets for which $K_{\beta c}$ has been manually curve fit.

Other Comparisons

While considerable interest is forced on the properties of so-called low upper shelf materials with USE levels below 68 J, use of a C_v index at 41-J forces consideration of the C_v curve at levels below the middle of the transition for heats with USE levels above 82 J. For such heats, it is of interest to compare temperature shifts at the 41-J level with shifts at higher energy levels.

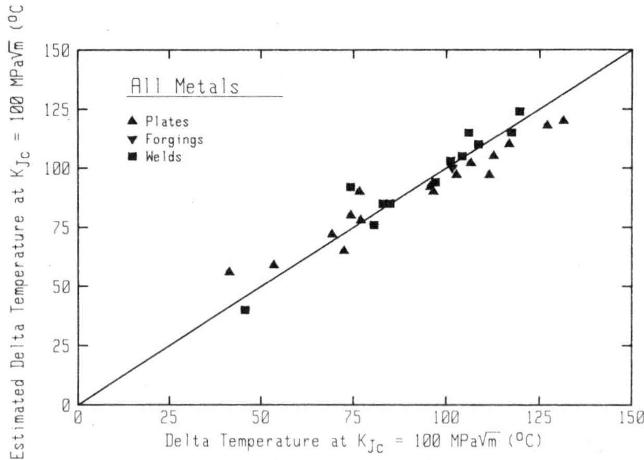

FIG. 13—*Comparison of* ΔT(K_{Jc} *at 100 MPa* · $m^{1/2}$) *determined from manual and computerized fits to the data. Good agreement is observed, on average, with a maximum deviation of 18°C for a Linde 0091 flux weld. (The computerized-fit data are on the abscissa, with the manual-fit data on the ordinate.)*

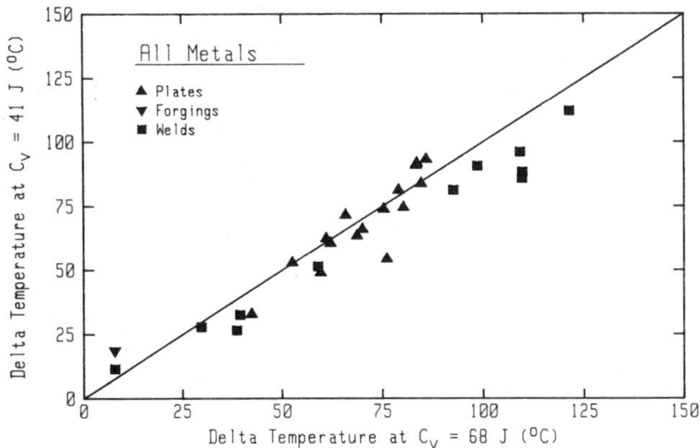

FIG. 14—*Comparison of* $\Delta T(C_v)$ *as determined at 41 and 68 J. On average* $\Delta T(C_v$ *at 68 J) is 5°C higher than* $\Delta T(C_v$ *at 41 J).*

This comparison will give an indication of the relative curve shape between the unirradiated and irradiated conditions, with possibly a better indication of the average ΔT for such higher toughness materials. As an example of such a consideration, $\Delta T(C_v$ at 68 J) is compared with $\Delta T(C_v$ at 41 J). In Fig. 14, on average $\Delta T(C_v$ at 68 J) is 5°C greater than $\Delta T(C_v$ at 41 J), with a 1-σ range of 9°C. This implies that the irradiated C_v curves have a tendency to "lay-over" or flatten-out in comparison to the unirradiated curves.

For the K_{Jc} data, alternative indices of 75 and 150 MPa · $m^{1/2}$ have been considered here. The $\Delta T(K_{Jc}$ at 75 MPa · $m^{1/2}$) is 1°C lower than the $\Delta T(K_{Jc}$ at 100 MPa · $m^{1/2}$), on average, with a 1-σ range of 11°C (Fig. 15). $\Delta T(K_{Jc}$ at 150 MPa · $m^{1/2}$) is on average 1°C lower than the $\Delta T(K_{Jc}$ at 100 MPa · $m^{1/2}$), with a 1-σ range of 18°C (Fig. 16). The variability of the latter comparison here is much greater than that for the former comparison, as the 1-σ ranges indicate. Nevertheless, the lack of variability in average values imply that the K_{Jc} curve does not tend to change shape with radiation.

Unlike the K_{Jc} and C_v data comparisons that show only minor changes in ΔT by changing the index point, the $\Delta T(K_{\beta c}$ at 100 MPa · $m^{1/2}$) is considerably different from that at $\Delta T(K_{\beta c}$ at 75

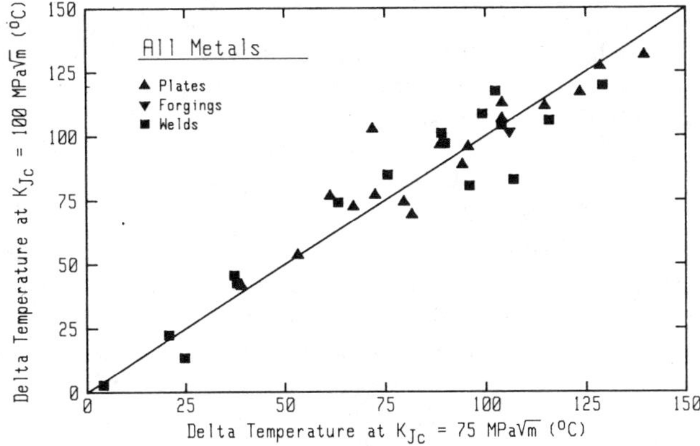

FIG. 15—*Comparison of* $\Delta T(K_{Jc})$ *determined at 75 MPa* · $m^{1/2}$ *and 100 MPa* · $m^{1/2}$. *A high degree of scatter is apparent, but on average these two levels give* ΔT*s about 1°C apart.*

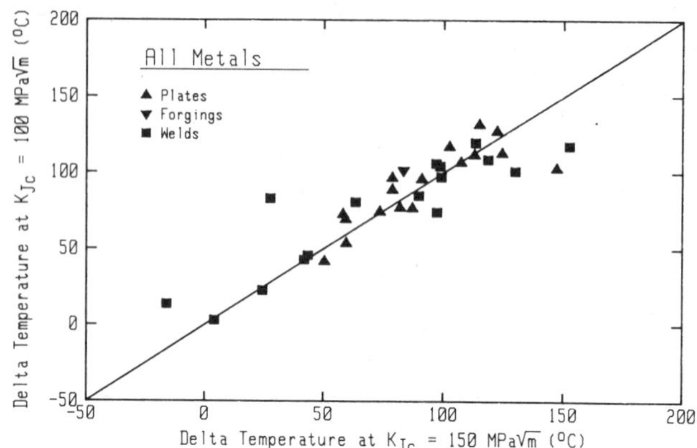

FIG. 16—*Comparison of* $\Delta T(K_{Jc})$ *at 100 and 150 MPa* · *m*$^{1/2}$. *Here, the variability is greater than for the 75 MPa* · *m*$^{1/2}$ *comparison (Fig. 15), but on average the differences are still below 1°C.*

MPa · m$^{1/2}$), on average. Considerable variability is apparent, with an average decrease of 16°C using $\Delta T(K_{\beta c}$ at 100 MPa · m$^{1/2}$), with 1 σ of 27°C. The cause of this extreme variability is the extrapolation of the $K_{\beta c}$ curve fits beyond the available data to 100 MPa · m$^{1/2}$ (little data extend above 80 MPa · m$^{1/2}$).

Comparison of Charpy-V Shifts with Correlations

Many differing and independent efforts have sought to develop procedures for predicting transition temperature shifts induced by 288°C irradiation [*9–19*]. These procedures typically take into account product form (either weld or base metal), atomic or weight percentage of key elements, such as copper, nickel, phosphorus, silicon, molybdenum, manganese and carbon, and most importantly, the neutron fluence for the pertinent material. Some of these procedures are based upon surveillance of experimenal (test reactor) irradiation data only, with no distinction made in some cases. In addition, mean values are sought by some of these procedures, and upper bound values are determined in other cases.

This section compares these data (C_v, K_{Jc}, and $K_{\beta c}$) with predictions of ΔT from the various correlations, to assess the applicability of the correlation to the data (or the data to the correlation). A description of each of the correlations is given in Ref *20*, along with the assumptions and limitations of the equations. Because of flux effects, the correlations that match these data best may not be best suited for power reactor application.

C_v Comparisons

Statistical comparisons of $\Delta T(C_v$ at 41 J) for all of the data with predictions based on the correlations are listed in Table 6. (Any correlation prediction yielding a negative ΔT was set to a zero shift.) Average deviation of the predicted ΔT, and the measured $\Delta T(C_v$ at 41 J) and $\pm$1-σ standard deviations are listed in this table. In addition, the number of data points overpredicted and underpredicted by the correlation are listed. In this table, a positive number indicates an underestimate of the actual $\Delta T(C_v$ at 41 J), with a negative number indicating an overestimate (conservative estimate) of the actual $\Delta T(C_v$ at 41 J).

With each correlation, the ΔT for weld metal is conservatively estimated on average by the correlations, with the closest (average) predictions from the MPC mean curve derived from all

TABLE 6—*Comparison of measured* $\Delta T(C_v$ *at 41 J) with correlation predictions.*

Reference	Base Metals			Weld Metals			All Data		
	Average,[a] °C	1 σ, °C	Over/Under[b]	Average,[a] °C	1 σ, °C	Over/Under[b]	Average,[a] °C	1 σ, °C	Over/Under[b]
Guthrie [9]	+6.3	(12.7)	5/13	−33.9	(23.7)	16/0	−12.6	(27.4)	21/13
Odette/Perrin [10]	+4.7	(10.7)	5/13	−22.5	(20.2)	14/2	−8.1	(20.8)	19/15
MPC [11]									
1[c]	+1.6	(14.7)	5/13	−25.1	(31.8)	12/4	−11.0	(27.5)	17/17
2[d]	+0.8	(17.1)	8/10	−15.3	(25.5)	11/5	−6.8	(22.6)	19/15
3[e]	+3.3	(15.1)	7/11	−28.5	(34.9)	11/5	−11.7	(30.5)	18/16
4[f]	+1.0	(17.3)	8/10	−17.3	(26.3)	11/5	−7.6	(23.5)	19/15
5[g]	−31.1	(17.8)	18/0	−42.5	(26.8)	16/0	−36.4	(22.8)	34/0
6[h]	−27.8	(17.7)	18/0	−42.1	(27.2)	16/0	−34.5	(23.5)	34/0
ASTM E 900 [12]	+0.9	(17.2)	8/10	−17.4	(26.1)	11/5	−7.7	(23.4)	19/15
Varsik [13]	−5.4	(21.1)	11/7	−29.6	(23.0)	14/2	−16.8	(24.9)	25/9
Varsik/Byrne [14]	−5.5	(18.4)	10/8	−33.4	(36.4)	12/4	−18.6	(31.3)	22/12
Rev. Varsik/Byrne [13]	. . .	. . .	. . .	−20.8	(19.7)	13/3	. . .	. . .	. . .
Heller/Lowe [15]	. . .	. . .	. . .	−22.1	(22.9)	6/2	. . .	. . .	. . .
Berggren/Stallman [16]	−40.9	(27.0)	18/0	−40.6	(36.4)	14/2	−40.8	(31.3)	32/2
RG 1.99 Rev. 1 [17]	−51.4	(45.5)	17/1	−46.4	(26.6)	16/0	−49.1	(37.3)	33/1
RG 1.99 Rev. 2 [18][i]	+4.8	(11.9)	5/13	−29.0	(20.5)	16/0	−11.1	(23.6)	21/13
NRC Screening Criteria [19]	+8.3	(12.6)	5/13	−32.4	(38.1)	11/5	−10.8	(34.2)	16/18

[a] (Measured) − (predicted).
[b] Number of sets for which predicted > measured (''over'') and measured > predicted (''under'').
[c] Experimental, welds and plates separately, mean curve.
[d] Experimental, all data, mean curve.
[e] Surveillance and experimental, welds and plates separately, mean curve.
[f] Surveillance and experimental, all data, mean curve.
[g] Experimental, upper bound.
[h] Surveillance and experimental, upper bound.
[i] Without margin.

experimental data. For the base metals, most of the correlations give accurate-to-slightly nonconservative average estimates of $\Delta T(C_v$ at 41 J), with all of the correlations having an average underestimate of less than 10°C. The closest prediction in terms of the average, absolute deviation is the MPC mean curve for all experimental data with the Varsik-Byrne correlation yielding the smallest, average deviation from the conservative correlations. In general, the ΔTs for base metal are well predicted by the correlations.

K_{Jc} *Comparisons*

The $\Delta T(K_{Jc}$ at 100 MPa $\cdot$ m$^{1/2}$ comparisons with the correlation predictions of $\Delta T(C_v$ at 41 J) are listed in Table 7. The major difference between Table 7 and Table 6 is that A508-2 forging data have not been included here because of the large discrepancies between the $\Delta T(K_{Jc})$ and $\Delta T(C_v)$ for this heat noted.

As with $\Delta T(C_v)$, on average the weld metal ΔTs were conservatively estimated by each correlation. The MPC mean curve from all experimental data provided the best match, on average. For base metals, several of the correlations overestimate the $\Delta T(K_{Jc})$ on average, but most of the correlations result in a large underestimate, on average. The closest correlation on average is the MPC upper bound curve based on surveillance and experimental data, with the MPC upper bound curve from experimental data giving the next best match.

$K_{\beta c}$ *Comparisons*

The $\Delta T(K_{\beta c}$ at 75 MPa $\cdot$ m$^{1/2}$) comparisons with the correlation predictions of $\Delta T(C_v$ at 41 J) are given in Table 8. As with the $\Delta T(K_{Jc}$ at 100 MPa $\cdot$ m$^{1/2}$) comparisons, the A508-2 forging data were not used in this table. The results also are comparable to the $\Delta T(K_{Jc})$ comparisons, as base metals are underestimated on average and welds are overestimated on average, by the mean equations. In this case, the Varsik-Byrne correlation gives the most accurate average estimate of the ΔT for base metal, with the MPC mean curve from experimental data yielding the best match with the weld data.

Discussion

One aspect made apparent by the preceding sections is that plates and welds exhibit different trends for relating fracture toughness and notch ductility assessments of irradiation-induced transition temperature shifts (increases). This observation is supported in Tables 2 and 3 (for K_{Jc} and $K_{\beta c}$, respectively), where $\Delta T(C_v)$ for plates underestimates $\Delta T(K_{Jc})$ and $\Delta T(K_{\beta c})$ on average, while $\Delta T(C_v)$ for welds overestimates $\Delta T(K_{Jc})$ and $\Delta T(K_{\beta c})$ on average. (The $\Delta T(K_{\beta c})$ plate data at 100 MPa $\cdot$ m$^{1/2}$ are ignored in this context, since this $K_{\beta c}$ level exceeds the available data in all cases.) Comparison of Figs. 1 and 2 and Figs. 4 and 5 show that $\Delta T(C_v)$ for welds gives a consistent overestimate of $\Delta T(K_{Jc})$ and $\Delta T(K_{\beta c})$ at all ΔT levels. However, $\Delta T(C_v)$ for plates tends to underestimate large $\Delta T(K_{Jc})$ and $\Delta T(K_{\beta c})$ shifts, while giving accurate estimates at a ΔT of 75°C and lower only. This indicates that high base metal embrittlement conditions (as determined from K_{Jc} or $K_{\beta C}$) are not reflected by the base metal notch ductility trends at the same flux and fluence conditions.

Further differences in plate and weld behavior are apparent when comparing correlation (model) predictions of $\Delta T(C_v$ at 41 J) with measured values. For base metals from Table 6, the average correlations give accurate estimates of $\Delta T(C_v$ at 41 J), even though many of the correlations are based upon surveillance conditions while the data in this report are from test reactor irradiations with much higher fluxes. This good agreement may indicate ''flux independence'' for notch ductility determinations from plates.

TABLE 7—*Comparison of measured* $\Delta T(K_{Jc}$ *at 100 MPa* $\cdot m^{1/2})$ *with correlation predictions of* ΔT (C_v *at 41 J*).

Reference	Base Metals			Weld Metals			All Data		
	Average,[a] °C	1 σ, °C	Over/Under[b]	Average,[a] °C	1 σ, °C	Over/Under[b]	Average,[a] °C	1 σ, °C	Over/Under[b]
Guthrie [*9*]	+33.6	(24.9)	2/15	−39.0	(30.4)	15/1	−1.7	(45.8)	17/16
Odette/Perrin [*10*]	+31.8	(21.6)	2/15	−27.6	(24.5)	14/2	+3.0	(37.7)	16/17
MPC [*11*]									
1[c]	+28.6	(24.8)	2/15	−30.2	(36.5)	12/4	+0.1	(42.7)	14/19
2[d]	+27.7	(21.5)	1/16	−20.4	(30.7)	12/4	+4.4	(35.6)	13/20
3[e]	+30.3	(19.7)	0/17	−33.6	(39.5)	12/4	−0.7	(44.5)	12/21
4[f]	+27.9	(21.7)	0/17	−22.4	(31.2)	12/4	+3.5	(36.6)	12/21
5[g]	−6.0	(23.0)	11/6	−47.6	(32.6)	16/0	−26.2	(34.7)	27/6
6[h]	−2.6	(23.0)	10/7	−47.2	(32.7)	16/0	−24.2	(35.7)	26/7
ASTM E 900 [*12*]	+27.9	(21.6)	1/16	−22.5	(31.0)	12/4	+3.4	(36.6)	13/20
Varsik [*13*]	+21.1	(25.3)	3/14	−34.7	(30.1)	15/1	−5.9	(39.4)	18/15
Varsik/Byrne [*14*]	+21.0	(27.2)	4/13	−38.4	(44.3)	13/3	−7.8	(46.9)	17/16
Review of Varsik/Byrne [*13*]	...	...	...	−25.9	(26.6)	13/3	...	...	...
Heller/Lowe [*15*]	...	...	...	−31.8	(31.0)	7/1	...	...	...
Berggren/Stallman [*16*]	−16.5	(30.7)	13/4	−45.7	(43.4)	13/3	−30.7	(39.7)	26/7
RG 1.99 Review 1 [*17*]	−27.6	(45.4)	13/4	−51.5	(31.2)	16/0	−39.2	(40.4)	29/4
RG 1.99 Review 2 [*18*][i]	+31.9	(22.5)	2/15	−34.0	(25.3)	15/1	−0.1	(41.2)	17/16
NRC screening criteria [*19*]	+35.7	(21.3)	1/16	−37.5	(42.1)	12/4	+0.2	(49.4)	13/20

[a] (Measured) − (predicted).
[b] Number of sets for which predicted > measured ("over") and measured > predicted ("under").
[c] Experimental, welds and plates separately, mean curve.
[d] Experimental, all data, mean curve.
[e] Surveillance and experimental, welds and plates separately, mean curve.
[f] Surveillance and experimental, all data, mean curve.
[g] Experimental, upper bound.
[h] Surveillance and experimental, upper bound.
[i] Without margin.

TABLE 8—*Comparison of measured* $\Delta T(K_{\beta c}$ *at 75 MPa* $\cdot$ $m^{1/2})$ *with correlation predictions of* $\Delta T(C_v$ *at 41 J)*.

Reference	Base Metals			Weld Metals			All Data		
	Average,[a] °C	1 σ, °C	Over/Under[b]	Average,[a] °C	1 σ, °C	Over/Under[b]	Average,[a] °C	1 σ, °C	Over/Under[b]
Guthrie [*9*]	+20.8	(21.4)	2/15	−45.2	(30.0)	15/1	−11.2	(42.1)	17/16
Odette/Perrin [*10*]	+19.1	(17.5)	2/15	−33.7	(24.6)	14/2	−6.5	(34.0)	16/17
MPC [*11*]									
1[c]	+15.8	(21.1)	3/14	−36.4	(34.5)	15/1	−9.5	(38.5)	18/15
2[d]	+15.0	(17.2)	1/16	−26.6	(28.6)	12/4	−5.2	(31.3)	13/20
3[e]	+17.6	(15.1)	1/16	−39.8	(37.5)	14/2	−10.2	(40.2)	15/18
4[f]	+15.2	(17.4)	1/16	−28.6	(29.1)	13/3	−6.0	(32.3)	14/19
5[g]	−18.8	(17.5)	16/1	−53.7	(30.6)	16/0	−35.7	(30.1)	32/1
6[h]	−15.4	(17.7)	16/1	−53.3	(30.7)	16/0	−33.8	(31.1)	32/1
ASTM E 900 [*12*]	+15.1	(17.4)	1/16	−28.7	(28.9)	13/3	−6.1	(32.2)	14/19
Varsik [*13*]	+8.4	(21.9)	5/12	−40.8	(29.8)	15/1	−15.5	(35.8)	20/13
Varsik/Byrne [*14*]	+8.3	(23.5)	6/11	−44.6	(45.1)	13/3	−17.4	(44.2)	19/14
Rev. Varsik/Byrne [*13*]	. . .	. . .	. . .	−32.1	(27.3)	13/3	. . .	. . .	. . .
Heller/Lowe [*15*]	. . .	. . .	. . .	−42.9	(32.3)	7/1	. . .	. . .	. . .
Berggren/Stallman [*16*]	−29.2	(26.3)	16/1	−51.9	(41.7)	15/1	−40.2	(35.9	31/2
RG 1.99 Rev. 1 [*17*]	−40.4	(41.6)	16/1	−57.6	(30.1)	16/0	−48.7	(37.0)	32/1
RG 1.99 Rev. 2 [*18*][i]	+19.2	(18.6)	3/14	−40.2	(28.5)	15/1	−9.6	(38.2)	18/15
NRG screening criteria [*19*]	+23.0	(17.1)	0/17	−43.7	(39.9)	14/2	−9.4	(45.1)	14/19

[a] (Measured) − (predicted).
[b] Number of sets for which predicted > measured ("over") and measured > predicted ("under").
[c] Experimental, welds and plates separately, mean curve.
[d] Experimental, all data, mean curve.
[e] Surveillance and experimental, welds and plates separately, mean curve.
[f] Surveillance and experimental, all data, mean curve.
[g] Experimental, upper bound.
[h] Surveillance and experimental, upper bound.
[i] Without margin.

With the large differences observed between $\Delta T(K_{Jc})$ or $\Delta T(K_{\beta c})$ and $\Delta T(C_v)$ from the correlation predictions for plates (Tables 7 and 8), one interpretation is a "real" difference between static K and dynamic C_v assessments of ΔT, or possibly a flux dependence of static fracture toughness. The flux dependence of static fracture toughness would result in greater embrittlement at higher flux for plates, as compared to dynamic C_v assessments of embrittlement.

For weld metals, the average correlations overestimate $\Delta T(C_v)$, $\Delta T(K_{Jc})$, and $\Delta T(K_{\beta c})$ at all measured toughness and energy levels (Tables 6 and 8). This overestimate tends to become greatest at high embrittlement ΔT levels. One interpretation of these findings is a flux dependence, characterized by reduced weld metal embrittlement at high flux levels, from the test reactor data in this report. This postulated flux dependence is reflected in both notch ductility and fracture toughness (K_{JC} and $K_{\beta c}$) assessments of irradiation-induced embrittlement of weld metals. The suggested flux dependence is itself dependent on the assumption that the correlations used here give accurate estimates of weld metal embrittlement trends under power reactor flux conditions.

These trends of possible flux dependence are consistent with observations of Perrin et al. [*10*]. Further data on flux effects will be available in the near future [*21*], at which time a systematic evaluation of fluence and flux effects will be made.

Summary and Conclusions

For RPV safety assessments, knowledge of the fracture toughness K_{Ic} of the RPV materials is required. Since surveillance irradiations of sufficiently large fracture toughness specimens for valid K_{Ic} measurements are not possible, estimates of the irradiated K_{Ic} properties are required. These approximations are made by using pre-irradiation properties RT_{NDT} and the transition temperature shift from Charpy-V specimens [($\Delta T(C_v)$] to index the American Society of Mechanical Engineers (ASME) K_{Ic} and K_{IR} curves. To assess the appropriateness of using $\Delta T(C_v)$ to estimate $\Delta T(K_{Ic})$, comparisons of notch ductility C_v and fracture toughness (K_{Jc} and $K_{\beta c}$) assessments of transition temperature shifts for RPV base metals and welds were made. This study provides the first comprehensive statistical assessment of the suitability of $\Delta T(C_v)$ for estimating $\Delta T(K_{Ic})$. Although these fracture toughness data are not valid according to ASTM E 399, the data collected here K_{Jc} have been adjusted $K_{\beta c}$ to what is thought to be a reasonable approximation to K_{Ic} values. However, this adjustment may over-correct high toughness values, leading to a real K_{Ic} curve, which may lie somewhere between the K_{Jc} and $K_{\beta c}$ curves. Better insight to this will result from the HSST Fifth Irradiation Series [*22*], where large CT specimens (up to 4T-CT) will be tested in an irradiated condition to provide valid K_{Ic} data at high toughness levels.

Also, comparisons between computerized and manual fits to the C_v and K curves were made, with predictions based on various shift correlations compared to measured C_v and K transition curve shifts.

Future work in this study will examine dependence of these observations on chemical composition, with additional data added to the study as more testing is completed.

Initial conclusions from this study are based upon averages of the available data, with variations in behavior quite large in some cases (see the appropriate sections and figures). The initial conclusions are

1. Transition temperature shifts ΔTs measured by fracture toughness methods (K_{Jc} at 100 MPa $\cdot$ m$^{1/2}$) are only slightly greater than ΔTs from notch ductility (C_v at 41 J) tests, on average by 9°C.

2. A product form effect influences the $\Delta T(K_{Jc})$ versus $\Delta T(C_v)$ relationship, whereby $\Delta T(K_{Jc})$ for welds is overestimated by 5°C on average by C_v results and $\Delta T(K_{Jc})$ for plates is underestimated by 22°C on average by C_v results.

3. Adjusting the fracture toughness K_{Jc} data for lack of constraint ($K_{\beta c}$ values) results in an

average match of $\Delta T(K_{\beta c}$ at 75 MPa · $m^{1/2})$ and $\Delta T(C_v$ at 41 J) with 1°C. In this case, weld $\Delta T(C_v)$ values overestimate $\Delta T(K_{\beta c})$ values by an average of 11°C, with plate $\Delta T(C_v)$ values underpredicting $\Delta T(K_{\beta c})$ values by 10°C on average.

4. Comparison of ΔT values at various indices along the C_v, K_{Jc}, and $K_{\beta c}$ curves give some basis for assessing irradiation effect on curve shape. In general, the C_v curves tend to flatten out or lay-over, the K_{Jc} curve has no change in curve shape, and the irradiated $K_{\beta c}$ curve tends to be steeper than the unirradiated curve.

5. Manual and computerized fits to C_v and K data yield temperatures and ΔTs at transition temperature indices within 5°C of one another, on average, for the data base examined. C_v upper shelf energy level and drop values from the manual fits are within 2 J of the computerized fit values, on average.

6. The correlation methods for predicting $\Delta T(C_v)$ tend to overestimate measured $\Delta T(C_v)$ values for weld metals, with base metal $\Delta T(C_v)$ values slightly underestimated by the correlation methods, on average.

7. The $\Delta T(C_v)$ correlation predictions show a large overestimate of $\Delta T(K_{Jc})$ and $\Delta T(K_{\beta c})$ for welds; however, the $\Delta T(C_v)$ correlations underestimate $\Delta T(K_{Jc})$ and $\Delta T(K_{\beta c})$ for plate, on average.

8. While the A 508-2 forging $\Delta T(C_v)$ is predicted reasonably well by the correlations, $\Delta T(K_{Jc})$ and $\Delta T(K_{\beta c})$ for the forging are much larger than the $\Delta T(C_v)$ or any of the correlation predictions. This behavior is atypical of the other heats in the data base, and may indicate a problem limited in magnitude to forgings, or just this single forging. Consideration of this will be made in future work.

9. These results may indicate a flux dependence, whereby weld metals exhibit significantly lower embrittlement with higher flux, while plates exhibit slightly higher embrittlement at higher fluxes. This effect may be magnified for fracture toughness versus notch ductility leading to the product form dependence observed here. This possibly will be studied closer in future work.

The above conclusions must be tempered by the understanding that many of the fracture toughness curves were composed of six or fewer data points. In general, irradiations of fracture toughness specimens result in too few specimens of too small a thickness to provide for complee and unambiguous definition of the fracture toughness behavior with temperature. Several benchmark irradiations of many small thickness specimens of several compositions and product forms would be an excellent supplement to the HSST Fifth Irradiation Series mentioned previously.

References

[1] Hawthorne, J. R., Menke, B. H., Loss, F. J., Watson, H. E., Hiser, A. L., and Gray, R. A., "Evaluation and Prediction of Neutron Embrittlement in Reactor Pressure Vessel Materials," EPRI NP-2782, Electric Power Research Institute, Dec. 1982.

[2] Hawthorne, J. R., Menke, B. H., and Hiser, A. L., "Notch Ductility and Fracture Toughness Degradation of A 302-B and A 533-B Reference Plates from PSF Simulated Surveillance and Through-Wall Irradiation Capsules," USNRC Report NUREG/CR-3295, Vol. 1, MEA-2017, May 1983.

[3] Menke, B. H., McGowan, J. J., Berggren, R. G., Nanstad, R. K., and Miller, K. C., "Effects of Neutron Irradiation on Fracture Toughness of A 533 Grade B Class 1 Plate and Four Submerged-Arc Welds," *Effects of Radiation on Materials: Twelfth International Symposium, STP 870,* American Society for Testing and Materials, Philadelphia, 1984, pp. 1111–1130.

[4] Hawthorne, J. R., "Evaluation of Reembrittlement Rate Following Annealing and Related Investigations of RPV Steels," *Proceedings of the 11th Water Reactor Safety Research Information Meeting,* USNRC Proceedings NUREG/CP-0048, Vol. 4, Jan. 1984, pp. 361–375.

[5] Rolfe, S. T. and Novak, S. R., "Slow-Bend K_{Ic} Testing of Medium-Strength High-Toughness Steels," *Review of Developments in Plane Strain Fracture Toughness Testing, STP 463,* American Society for Testing and Materials, Philadelphia, 1970, pp. 124–159.

[6] Sailors, R. H. and Corten, H. T., "Relationship Between Material Fracture Toughness Using Fracture Mechanics and Transition Temperature Tests," *Fracture Toughness, STP 514,* American Society for Testing and Materials, Philadelphia, 1972, pp. 164–191.

[7] Merkle, J. G., "An Examination of the Size Effects and Data Scatter Observed in Small Specimen Cleavage Fracture Toughness Testing," U.S. NRC Report NUREG/CR-3672, ORNL/TM-9088, Oak Ridge National Laboratory, April 1984.
[8] Irwin, G. R., "Fracture Mode Transition for a Crack Traversing a Plate," *Journal of Basic Engineering,* ASME, Vol. 82, No. 2, 1960, pp. 417–425.
[9] Guthrie, G. L., "LWR Pressure Vessel Surveillance Dosimetry Improvement Program, Quarterly Progress Report for the Period April–June 1982," USNRC Report NUREG/CR-2805, Vol. 2, HEDL-TME-82-19, Hanford Engineering Development Laboratory, Jan. 1983, pp. HEDL-3–4.
[10] Perrin, J. F., Wullaert, R. A., Odette, G. R., and Lombrozo, P. M., "Physically Based Regression Correlations of Embrittlement Data From Reactor Pressure Vessel Surveillance Programs," EPRI NP-3319, Electric Power Research Institute, Jan. 1984.
[11] Metal Properties Council Subcommittee 6 on Nuclear Materials, "Prediction of the Shift in the Brittle-Ductile Transition Temperature of Light-Water Reactor (LWR) Pressure Vessel Materials," *Journal of Testing and Evaluation,* Vol. 11, No. 4, July 1983, pp. 237–260.
[12] ASTM Guide For Predicting Neutron Radiation Damage to Reactor Vessel Materials (E 900), *Annual Book of ASTM Standards,* Vol. 12.02.
[13] Varsik, J. D., Schloss, S. M., and Koziol, J. J., "Evaluation of Irradiation Response of Reactor Pressure Vessel Materials," EPRI NP-2720, Electric Power Research Institute, Nov. 1982.
[14] Varsik, J. D., and Byrne, S. T., "An Empirical Evaluation of the Irradiation Sensitivity of Reactor Pressure Vessel Materials," Effects of Radiation on Structural Materials, *Effects of Radiation on Structural Materials, STP 683,* J. A. Sprague and D. Kramer, Eds., American Society for Testing and Materials, Philadelphia, 1979, pp. 252–266.
[15] Heller, A. S. and Lowe, A. L., Jr., *Correlations for Predicting the Effects of Neutron Radiation on Linde 80 Submerged-Arc Welds,* BAW-1803, Babcock and Wilcox Co., Lynchburg, VA, Jan. 1984.
[16] Berggren, R. G. and Stallmann, F. W., "Statistical Analysis of Pressure Vessel Steel Embrittlement Data," from the ANS Special Session on Correlations and Implications of Neutron Irradiation Embrittlement of Pressure Vessel Steels, Detroit, MI, 12–16 June 1983, *Transactions of the American Nuclear Society,* Vol. 44, 1983, p. 225.
[17] *Effects of Residual Elements on Predicted Radiation Damage to Reactor Vessel Materials,* Regulatory Guide 1.99, Rev. 1, U.S. Nuclear Regulatory Commission, Washington, DC, April 1977.
[18] Draft of Revision 2 to Regulatory Guide 1.99, "Radiation Damage to Reactor Vessel Materials," Working Paper G, Washington, D.C., U.S. Nuclear Regulatory Commission, 14 Aug. 1985.
[19] Dircks, W., NRC Staff Evaluation of Pressurized Thermal Shock, NRC-SECY-82-465, Nov. 1982.
[20] Hiser, A. L., "Correlation of C_v and K_{Ic}/K_{Jc} Transition Temperature Increases Due to Irradiation," U.S. NRC Report NUREG/CR-4395, U.S. Nuclear Regulatory Commission, Washington, D.C., Sept. 1985.
[21] "Structural Integrity of Light Water Reactor Pressure Boundary Components: Four-Year Plan 1984–1988," U.S. NRC Report NUREG/CR-3788, MEA-2047, U.S. Nuclear Regulatory Commission, Washington, D.C., Sept. 1984.
[22] "Heavy-Section Steel Technology Program: Five-Year Plan, FY 1983–1987," U.S. NRC Report NUREG/CR-3595, ORNL/TM-9008, Oak Ridge National Laboratory.

Arthur L. Lowe, Jr.[1] *and Wayne A. Pavinich*[2]

Comparison of the Effect of Power and Test Reactor Irradiations on the Fracture Properties of Submerged-Arc Welds

REFERENCE: Lowe, A. L., Jr. and Pavinich, W. A., "**Comparison of the Effect of Power and Test Reactor Irradiations on the Fracture Properties of Submerged-Arc Welds,**" *Influence of Radiation on Material Properties: 13th International Symposium, ASTM STP 956,* F. A. Garner, C. H. Henager, Jr., and N. Igata, Eds., American Society for Testing and Materials, Philadelphia, 1987, pp. 358–368.

ABSTRACT: The Babcock & Wilcox (B&W) Owners Group Integrated Reactor Vessel Surveillance Program contained two irradiation phases in addition to the plant specific surveillance programs. One phase provided materials for a short-term (test reactor) study of the effects of neutron irradiation on high-copper low-upper-shelf weld metals as a part of the Heavy Section Steel Technology (HSST) Program. This phase was followed with an irradiation study using the long-term (power reactor) radiation source.

These studies were distinguished from other irradiation studies in that production weld metals were used. The chemical composition of the weld metals were similar except for the copper content which ranged from 0.21 to 0.42%. The specimens used for mechanical properties evaluation consisted of tension, Charpy V-notch, and compact fracture specimens.

The power reactor irradiations were conducted in B&W Owner's Group host reactors. The fluence attained in both the test and power reactors was approximately 7×10^{18} neutrons (n)/cm^2. These two neutron sources provide a means for evaluating the influence of neutron environment on material damage.

The Charpy impact toughness is evaluated by the 41 J (30 ft · lb) shift and decrease in upper-shelf energy. The tension data is evaluated using the standard mechanical properties. The fracture toughness data obtained from compact fracture specimens, using *J-R* elastic-plastic fracture test techniques, was evaluated.

These comparisons indicate that test reactor irradiated yield strength data and fracture toughness data is equivalent to that of a power reactor. A comparison of Charpy properties indicates that the data may be sensitive to parameters other than neutron environments.

KEY WORDS: power reactor, test reactor, submerged-arc welds, radiation damage, Charpy impact data, tension test data, compact fracture data, fracture toughness, reactor vessel

The test reactor is used for the study of neutron radiation damage to structural materials, primarily because of the relative ease with which experiments can be performed. The relatively high flux allows irradiations to high fluence levels in short time periods. This permits the study of materials at fluence values that would require years to obtain in a power reactor. The current generation of power reactors were designed and built on the basis of test reactor data. As a safeguard against unexpected material behavior that may result from power reactor irradiation,

[1] Advisory engineer, Babcock & Wilcox Co., Nuclear Power Division, P.O. Box 10935, Lynchburg, VA 24506-0935.

[2] Senior research engineer, Babcock & Wilcox Co., Lynchburg Research Center, P.O. Box 11165, Lynchburg, VA 24506-1165.

each reactor pressure vessel has a materials surveillance program. Periodic withdrawal of capsules containing specimens of irradiation sensitive materials and their subsequent testing and evaluation, provides a system of monitoring material irradiation behavior.

Recent developments in the study of flux and fluence effects on materials have indicated the possibility of a flux affect which could make the use of test reactor materials data unacceptable for the evaluation of materials behavior in a power reactor [*1,2*]. Few opportunities have occurred where the identical materials were irradiated in both test and power reactors to the equivalent fluence and where these effects could be studied.

The B&W Owner's Group integrated reactor vessel surveillance program comprises two irradiation phases in addition to plant specific surveillance programs [*3*]. One phase provided materials and specimens for irradiation and testing by the Heavy Section Steel Technology (HSST) Program sponsored by the Nuclear Regulatory Commission (NRC) at Oak Ridge National Laboratory. The purpose of this study was to determine the response of high-copper, low-upper-shelf weld metals to neutron irradiation. The urgency to obtain data resulted from unexpected Charpy upper-shelf energy reductions as a result of irradiations. The then current prediction techniques indicated some reactor vessels would be in non-compliance with NRC regulations within a short time [*4*]. The other phase provided for a set of irradiation capsules to be placed in two power reactors. These capsules contained specimens made of materials identical to those in the test reactor irradiations. The redundancy of the integrated program was not designed to study the effects of neutron environment on material properties but was to insure that adequate data would be available when needed for the licensing of the participating reactor pressure vessels. Recently, the program was reviewed to determine whether or not sufficient data was available to allow an evaluation of test reactor data versus power reactor data. This paper is a report of the data obtained from the first power reactor capsules compared with corresponding test reactor data.

Materials

The materials used were three production weld metals fabricated with copper-coated Mn-Mo-Mi weld wire and Linde 80 weld flux. These weld metals were taken from nozzle dropouts which were removed from the nozzle shell at the location containing a weld joining two forgings. The chemical composition of the weld metals are presented in Table 1. The weld metals are representative of the high-copper content Linde 80 type weld metals that exhibit low initial Charpy upper-shelf energies.

TABLE 1—*Chemical composition of the Linde 80 submerged-arc weld metals used for comparison of neutron radiation sensitivity.*

Identification		Chemical Composition, wt%								
		C	Mn	P	S	Si	Ni	Cr	Mo	Cu
Weld metal-A	max						0.61			0.25
	avg	0.08	1.45	0.016	0.016	0.51	0.59	0.09	0.38	0.21
	min						0.58			0.18
Weld metal-B	max						0.60			0.31
	avg	0.08	1.45	0.011	0.013	0.49	0.59	0.08	0.38	0.27
	min						0.58			0.22
Weld metal-C	max						0.71			0.39
	avg	0.09	1.55	0.014	0.015	0.55	0.70	0.08	0.41	0.35
	min						0.60			0.31

The specimens of each weld metal for both the test reactor and power reactor irradiations were fabricated from the same weldment. It was often necessary to use one dropout for one phase and a second dropout for the next phase because of the limited weld metal in each dropout. A study of the variation in chemical composition of these weld metals showed that a weld wire and flux combination had a unique value for each element. This characteristic was proven when evaluations were made from different locations of the same weldment [*5*]. Although all the specimens were not machined from the same nozzle dropout, there is no significant difference in chemical composition from one dropout to the other for the same weldment. Therefore, no variation in mechanical properties is expected from dropout to dropout for each weld wire and flux combination.

Specimen Preparation

All specimens used for both the test reactor and the power reactor irradiations were machined by Babcock & Wilcox. Specimens were furnished to the HSST Program as a cooperative effort to obtain test reactor results on these series of materials. The remaining specimens were used in the capsules irradiated in the power reactor.

Irradiation Facilities

The test reactor irradiations were conducted in two different reactors. The University of Buffalo Pool Reactor (Test Reactor 1; TR1) was used for irradiating Charpy specimens to a fluence in the range of approximately 6 to 8 $\times$ 10^{18} neutrons (n)/cm^2 (E>1 MeV) at a nominal fast neutron flux of 8.5 $\times$ 10^{12} n/cm^2/s and irradiation temperature of 288°C (550°F) [*6*]. The Bulk Shielding Reactor (BSR) (Test Reactor 2: TR2) at Oak Ridge National Laboratory was the irradiation site for the Third HSST 4T-CT Irradiation Series. Tension, Charpy impact, and compact fracture toughness specimens were irradiated in this reactor. The target fast-neutron fluence for this series of irradiations was approximatley 1.0 $\times$ 10^{19} n/cm^2 (E>1 MeV) at a nominal radiation temperature of 288°C (550°F) [*7*]. The temperature control of the capsules and fluence accumulation was concentrated on the 4T-CT specimens and, therefore, the smaller specimens received fluences less than the target fluence. The HSST specimens, selected for comparison with those from the power reactor irradiations, received a fluence in the range of 6.0 $\times$ 10^{18} to 1.0 $\times$ 10^{19}. This provides a nominal fluence value of 8.0 $\times$ 10^{18} n/cm^2 which is equivalent to the power reactor nominal fast neutron fluence of 7.0 $\times$ 10^{18} n/cm. The fast flux for these experiments was in the range of 0.72 to 2.12 $\times$ 10^{12} n/cm^2/s with a nominal flux of 1.50 $\times$ 10^{12} n/cm^2/s.

The power reactor irradiations were conducted in Crystal River Unit 3 and Davis Besse Unit 1 as part of the BWOG integrated surveillance program. These reactors have the same neutron environmental characteristics and the data can be compared interchangeably (Power Reactor; PR). These capsules received a nominal fast neutron fluence of 7.0 $\times$ 10^{18} n/cm^2 (E> 1 MeV) and an irradiation temperature of 305°C (580°F). The nominal fast flux for these irradiations was 7.4 $\times$ 10^{10} n/cm^2/s. The test reactor neutron flux was approximately 20$\times$ the flux in the power reactors.

Data Evaluation

To provide uniformity in the data, the test reactor data, with the exception of the tension test data, were re-evaluated using the same techniques used to evaluate the power reactor data. The basic Charpy data were re-evaluated to provide consistency in the interpretation of the data. The fracture toughness data were re-analyzed because of differences that may exist within the analytical techniques used to determine the J_{Ic} and tearing modulus.

Tension Properties

An evaluation of the tensile test results included yield, ultimate, and elongation data. There were no apparent differences between the test reactor and power reactor data. To illustrate this observation the yield strength data for both the unirradiated and irradiated condition are shown in Fig. 1. The scatter of the data is equivalent for both conditions. The slight increase in the strength of the power reactor unirradiated data in the 288 to 315°C (550 to 600°F) temperature range is typical behavior and is a result of the dynamic strain aging which occurs in this temperature range.

Charpy Impact Properties

An evaluation of the Charpy impact test results was performed on the 41 J (30 ft · lb) energy transition data and the upper-shelf energy data since these data are of most importance in the evaluation of reactor pressure vessel integrity. Figure 2 is a presentation of the 41 J (30 ft · lb) energy transition data for each of the three weld metals. The observed shift values are similar for Weld Metals A and B for both power reactor and test reactor irradiations. However, Weld Metal C exhibited greater shifts for power reactor irradiation than for the two test reactor irradiatons, the results of which are nearly identical. The larger shifts for Weld Metal C may be attributed to the high copper and nickel contents. Significantly, the power reactor and test reactor shifts for Weld Metals A and B are similar. There is a significant difference in the initial Charpy energy value from one test data set to the other. This indicates that the relative sensitivity of the materials (that is the amount of the shift) is small with respect to the magnitude of the neutron environment or to the test laboratory environment, provided the same laboratory performs both the unirradiated and irradiated evaluations.

Figure 3 presents the data for the Charpy upper-shelf energy as affected by neutron radiation. For each weld metal, the sensitivity to damage was greatest in the power reactor. The power reactor data compares well with the results from Test Reactor 2. However, the upper-shelf energy

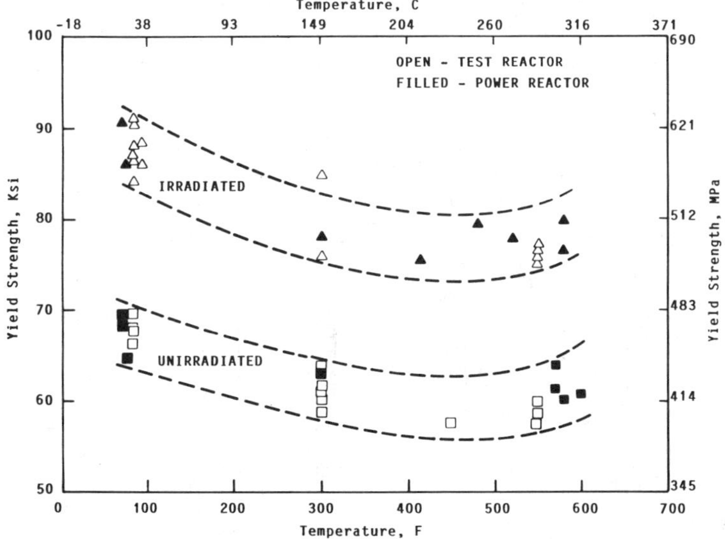

FIG. 1—*Comparison of power reactor and test reactor yield strength data.*

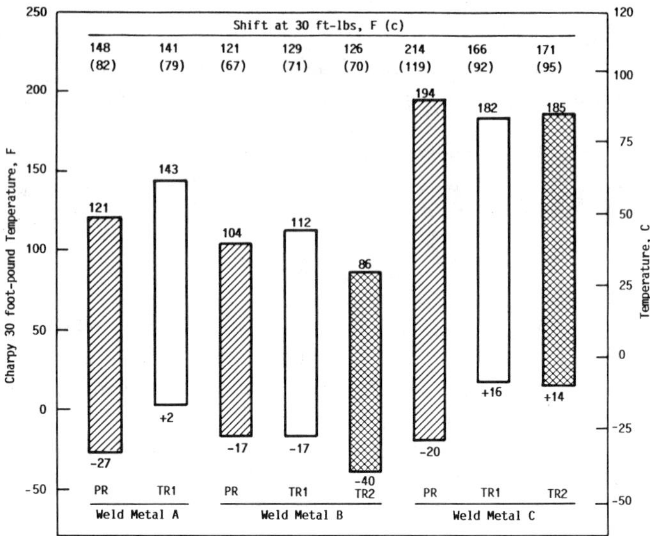

FIG. 2—*Comparison of Charpy transition energy data from power reactor and test reactors.*

decrease for Test Reactor 1 was approximately half of that for either the Test Reactor 2 data or the Power Reactor data. A systematic affect is suspected in the case of the irradiated data from Test Reactor 2 since all unirradiated data are well within expected error.

Three factors can cause the observed upper-shelf energy response for Test Reactor 1. The first is the neutron flux effect, which for Test Reactor 1 the flux is approximately 5.5 times greater than for Test Reactor 2. However, it would be expected that the greater neutron flux would produce greater effects, which is the opposite of what is observed. The other two effects, both of which are not likely because of the reputation of the laboratory performing the tests, are (1) a change in test machine environment and (2) an error in the irradiation temperature. When one considers the options, it is not readily apparent as to the cause of the differences in the data. Also, the transition

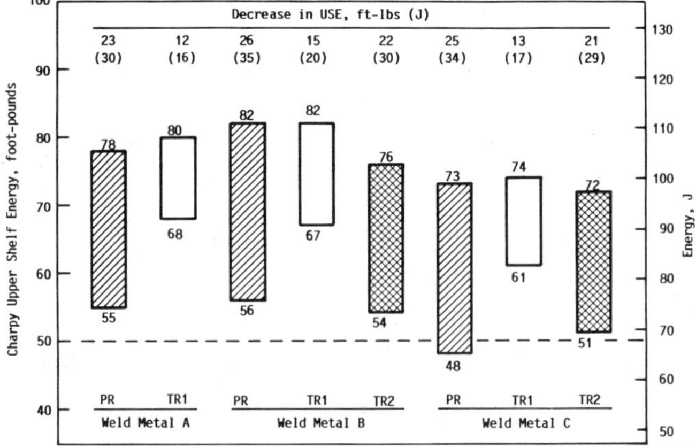

FIG. 3—*Comparison of Charpy upper-shelf energy data from power reactors and test reactors.*

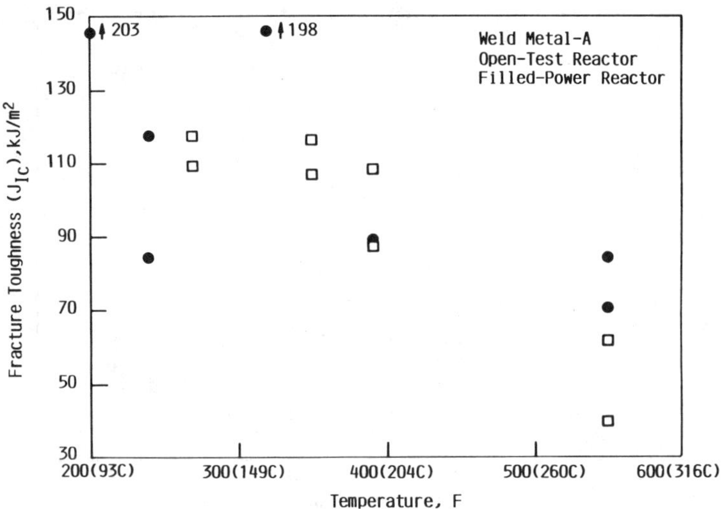

FIG. 4—*Comparison of Weld Metal A—J_{Ic} data.*

energy data does not exhibit similar behavior, a fact which does not support any of the factors described as being possible influences.

Fracture Toughness Properties

The fracture toughness test results were evaluated by comparing the crack initiation values (J_{Ic}) and the tearing modulus (T) for the power reactor and test reactor for each of the three weld metals. The J_{Ic} data for the three weld metals are presented in Figs. 4, 5 and 6. These data exhibit a large amount of scatter with no established trends that can be used to distinguish one data source from the other. The scatter is attributed to the uncertainty in establishing the crack initiation point which is inherent in the analysis procedure as defined in ASTM E813, in conjunction with the

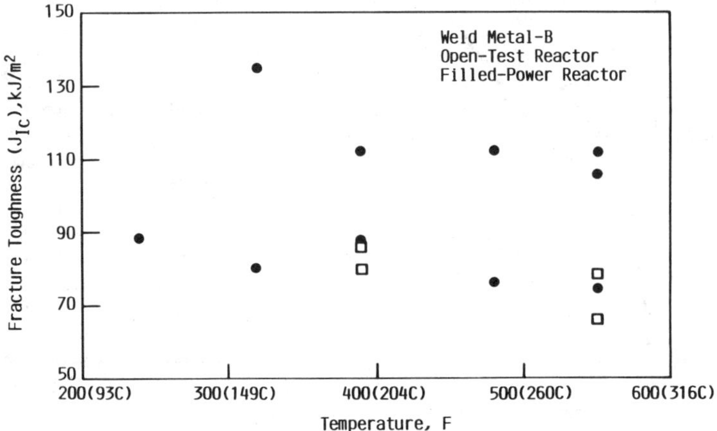

FIG. 5—*Comparison of Weld Metal B—J_{Ic} data.*

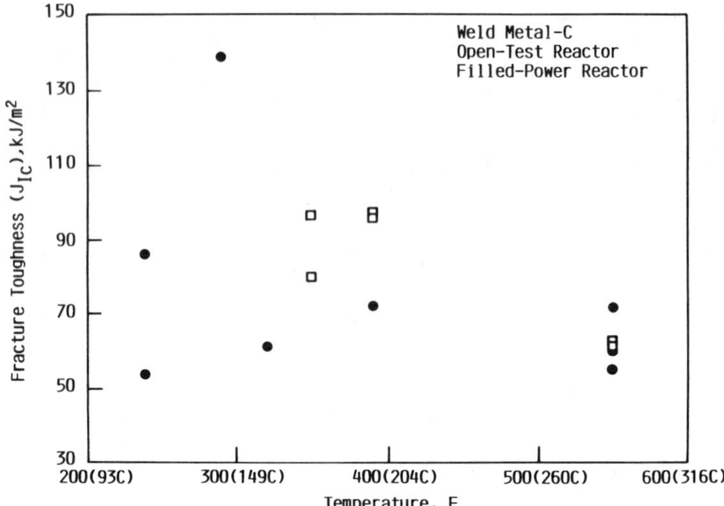

FIG. 6—*Comparison of Weld Metal C—*J_{Ic} *data.*

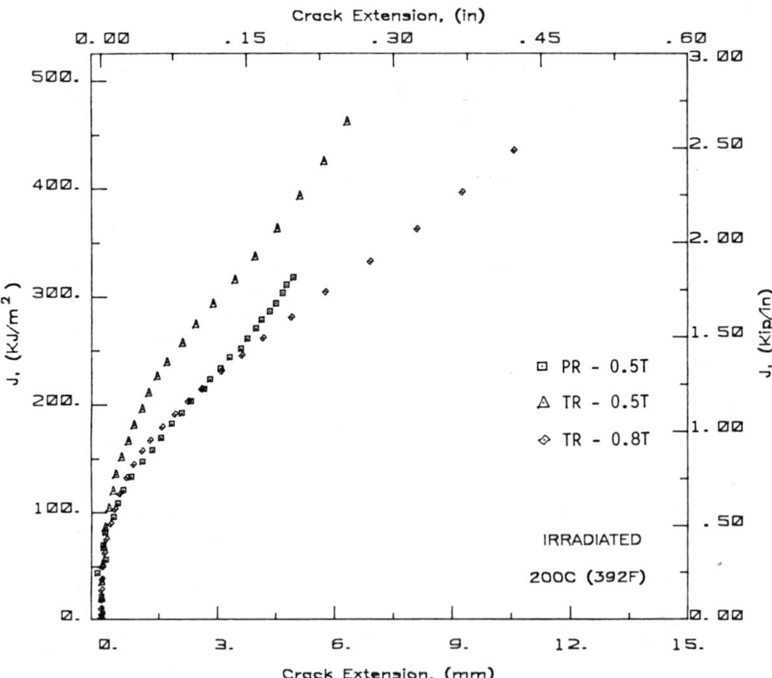

FIG. 7—*Fracture toughness* J-R *curves for Weld Metal A tested at 200°C (392°F).*

inhomogeneity of properties in a multi-pass submerged-arc weld metal. This is evident when the corresponding *J-R* curves were reviewed. The curves for Weld Metal A are shown in Figs. 7 and 8, and for Weld Metal C in Figs. 9 and 10. Although J_{Ic} data exhibits scatter, the corresponding *J-R* curves are equivalent and exhibit similar trends whether irradiated in a test reactor or a power reactor.

The tearing modulus data are shown in Figs. 11, 12, and 13. These data, unlike the J_{Ic} data, establish a trend for both the power reactor and test reactor data. Both the trends and the distribution of the individual data suggest no apparent difference in the two sets of properties. The only significant difference in the data is at 288°C (550°F) where in each case the power reactor data are greater than the comparable data for test reactor irradiations.

Conclusions

This evaluation of the material properties from the three high-copper submerged-arc Linde 80 weld metals showed no significant difference between power reactor and test reactor irradiations with fluence values of 8.0×10^{19} n/cm². If there is an effect on the materials' properties, the testing techniques did not have the sensitivity to define it. In addition, the inhomogeneity of the weld metal may mask any differences which could result from the differences in the two reactor neutron environments. Recent work on the theory of neutron radiation damage to materials indicates that test reactor irradiation damage and power reactor damage may be equivalent at a fluence of approximately 1×10^{18} n/cm² [*8*]. If this theory proves valid then it could explain why there is no apparent difference in the observed effects of neutron radiation on the weld metals. Unfortunately, the program was designed to obtain engineering data and did not contain sufficient sensitivity to detect any small differences which may result from the differences in neutron spectrum characteristics.

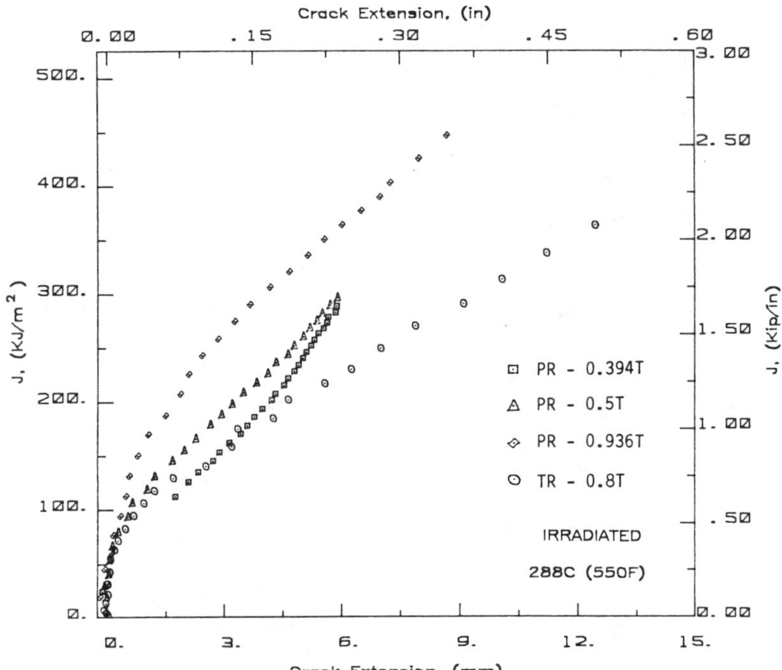

FIG. 8—*Fracture toughness* J-R *curves for Weld Metal A tested at 288°C (550°F).*

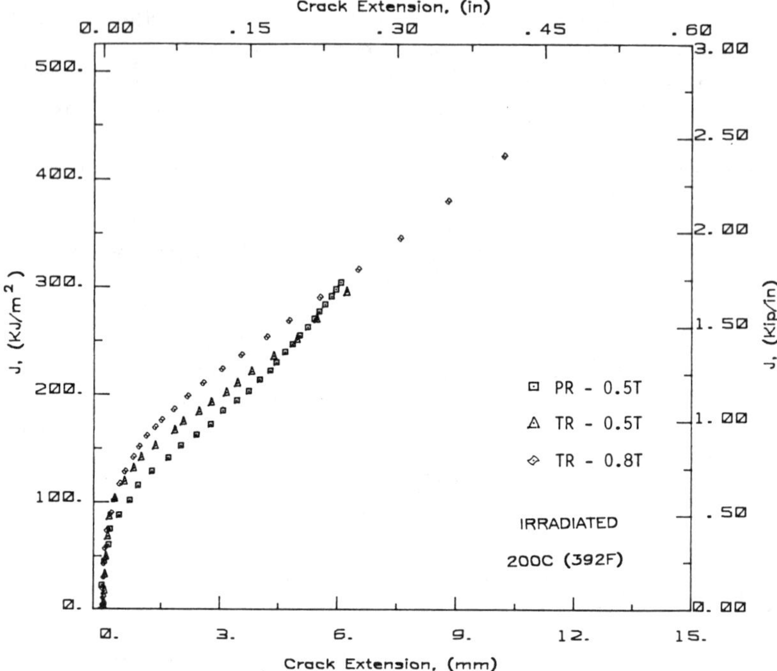

FIG. 9—*Fracture toughness* J-R *curves for Weld Metal C tested at 200°C (392°F).*

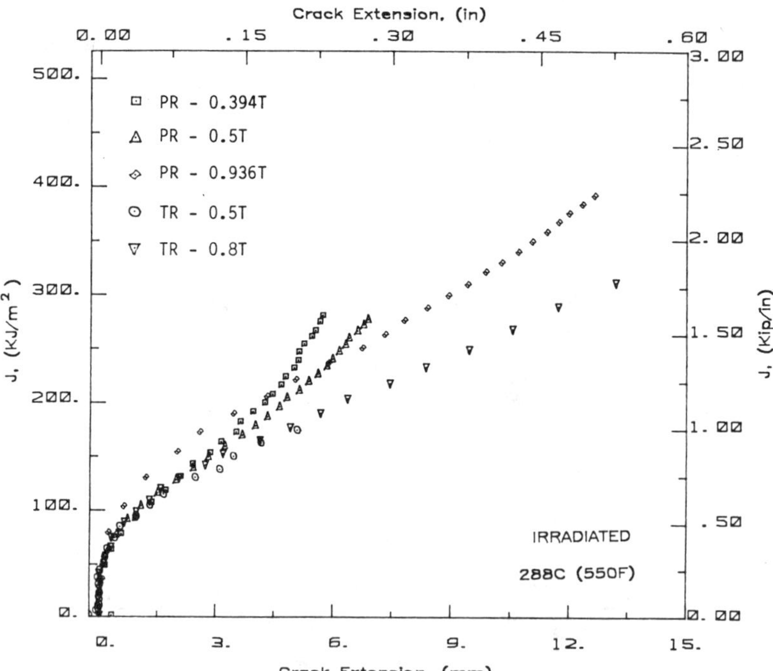

FIG. 10—*Fracture toughness* J-R *curves for Weld Metal C tested at 288°C (550°F).*

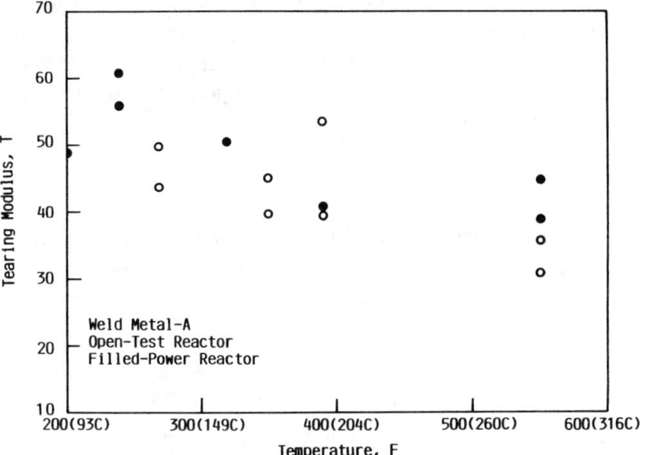

FIG. 11—*Comparison of Weld Metal A tearing modulus data.*

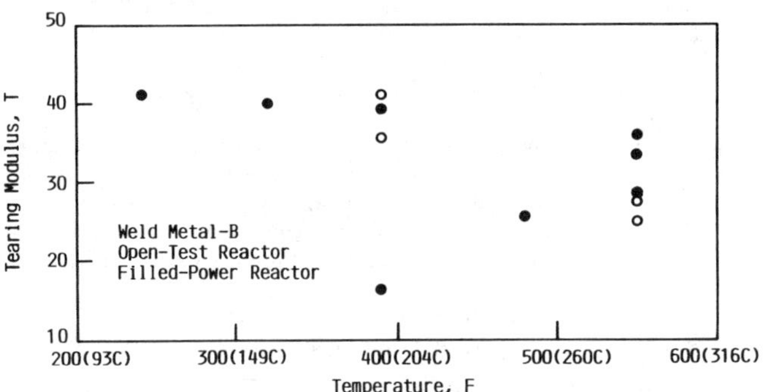

FIG. 12—*Comparison of Weld Metal B tearing modulus data.*

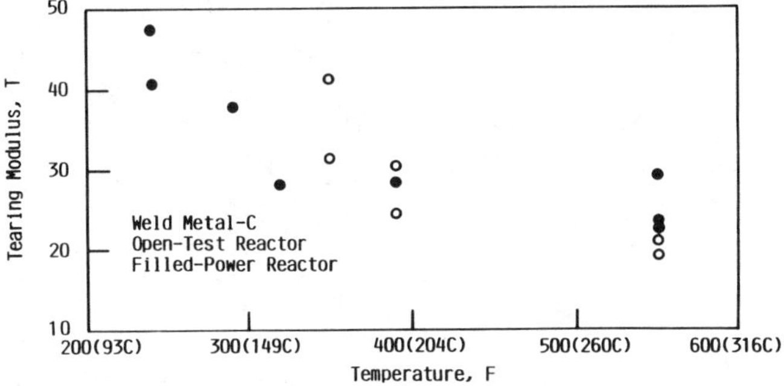

FIG. 13—*Comparison of Weld Metal C tearing modulus data.*

Additional capsules will be removed from the power reactor program with increased fluence values; however, comparable test reactor data may not be available. Any program designed to study the effects of neutron spectrum and flux density characteristics on materials' properties must insure that the sensitivity of the evaluating techniques are adequate to distinguish the differences resulting from the neutron environment.

Acknowledgments

This study was conducted as part of the Babcock & Wilcox Owners Group Reactor Vessel Integrity Program. The authors acknowledge and express appreciation for their support. We also acknowledge the NRC Sponsored Heavy Section Steel Technology Program (Mr. Milton Vagins, Manager) for providing the basic test reactor data.

References

[1] Norris, E. B., "A Service Laboratory Review of the Status and Direction of Reactor Vessel Surveillance," *Radiation Embrittlement and Surveillance of Nuclear Reactor Pressure Vessels: An International Study, STP 819,* American Society for Testing and Materials, Philadelphia, 1983.

[2] Lucas, G. E., et al, "Effects of Composition, Microstructure, and Temperature on Irradiation Hardening of Pressure Vessel Steels," *Proceedings,* 12th ASTM Symposium on Effects of Radiation on Materials, *STP 870,* American Society for Testing and Materials, Philadelphia, 1985.

[3] Lowe, A. L., Jr., Moore, K. E., and Aadland, J. D., "Integrated Reactor Vessel Material Surveillance Program for Babcock and Wilcox—177 FA Plants," *Proceedings,* 12th ASTM Symposium on Effects of Radiation on Materials, *STP 870,* American Society for Testing and Materials, Philadelphia, 1985.

[4] Lowe, A. L., Jr., Zurlippe, C. F., and Palme, H. S., "Surveillance Capsule Results from the Oconee Class Reactors," Joint ASME/CSME Pressure Vessels and Piping Conference, Montreal, Canada, June 25–30, 1978.

[5] Moore, K. E., Heller, A. S., and Lowe, A. L., Jr., "Source and Variability of Copper in Reactor Vessel Weld Metals," *Proceedings,* 12th ASTM Symposium on Effects of Radiation on Materials, *STP 870,* American Society for Testing and Materials, Philadelphia, 1985.

[6] Hawthorne, J. R., "Notch Ductility Degradation of Low Alloy Steels with Low-to-Intermediate Neutron Fluence Exposures," *Proceedings,* Third ASTM-EURATOM Symposium on Reactor Dosimetry, Ispra, Italy, 1–5 October, 1979.

[7] Stallman, F. W. and Kam, F. B., "Neutron Spectral Characterization of the NRC-HSST Experiments," *Proceedings,* Third ASTM-EURATOM Symposium on Reactor Dosimetry, Ispra, Italy, 1–5 October, 1979.

[8] Odette, G. R. and Lucas, G. E., "Radiation Embrittlement of Reactor Pressure Vessel Steels; Mechanisms, Models, and Data Correlations," *Radiation Embrittlement and Surveillance of Nuclear Reactor Pressure Vessels: An International Study, STP 819,* American Society for Testing and Materials, Philadelphia, 1983.

Balakrishna Sastry Viswanathan,[1] Dieter Pachur,[2] and Raju V. Nandedkar[1]

Investigation of Neutron Irradiated Reactor Pressure Vessel Steel by Positron Annihilation and Electron Microscopy

REFERENCE: Viswanathan, B., Pachur, D., and Nandedkar, R. V., "**Investigation of Neutron Irradiated Reactor Pressure Vessel Steel by Positron Annihilation and Electron Microscopy,**" *Influence of Radiation on Material Properties: 13th International Symposium (Part II) ASTM STP 956,* F. A. Garner, C. H. Henager, Jr., and N. Igata, Eds., American Society for Testing and Materials, Philadelphia, 1987, pp. 369–378.

ABSTRACT: Positron lifetime, annihilation line shape, and electron microscopic measurements have been carried out on neutron irradiated pressure vessel steel (HSST Plate 03) meeting ASTM A533-B (Unified Numbering System [UNS] K12539). The positron annihilation parameters for specimens irradiated at 423 K to a dose of 1.5×10^{18} n/cm^2 are found to decrease sharply during the post-irradiation annealing interval, 473 to 623 K. This correlates well with the stage seen in the recovery of Vickers hardness. On the basis of known evidence for carbon migration in α-iron and the interpretation of present results, enhancement and recovery of radiation hardening have been qualitatively understood in terms of formation and dissolution of carbon decorated vacancy complexes.

KEY WORDS: pressure vessel steel, radiation embrittlement, Vickers hardness, positron lifetime, annihilation line shape, dislocation structure, carbon decorated vacancy complex

Radiation induced embrittlement in reactor pressure vessel steels has a considerable impact on the safety of power reactors and is therefore a topic of significant pursuit [*1*]. The degree of embrittlement is related to the alloy composition and the microstructure as well as neutron fluence and irradiation temperature. Recent studies on the embrittlement in pressure vessel steels by irradiation and post-irradiation heat treatment have shown different annealing behavior of various mechanical properties such as Vickers hardness, yield strength, and ultimate tensile strength [*2*]. Existence of defect processes with activation energies ranging from 1.5 to 2.1 eV has been proposed from these studies. Dependence of activation energy on neutron fluence and irradiation temperature have also been observed earlier for certain types of defects [*3,4*]. The relative enhancement of hardness is found to be proportional to the enhancement of nil-ductility temperature (NDT), and it appears that these enhancements are caused by similar irradiation induced defects. However, an unambiguous and a rather unique identification of the underlying defect structure has not been reported so far. The need for correlating microscopic mechanisms, which occur during irradiation and annealing, with mechanical properties of pressure vessel steels cannot be overemphasized. However, questions remain on the issue whether the defect mechanisms depend on the chemical composition of the alloys or more related to the bcc iron structure. Recently, small angle

[1] Scientific officers, respectively, Materials Science Laboratory, Indira Gandhi Centre for Atomic Research, Kalpakkam 603 102, Tamil Nadu, India.

[2] Research physicist, Kernforschungsanlage Jülich, Postfach 1913, D-5160 Jülich, Federal Republic of Germany.

neutron scattering studies on pressure vessel steel specimens irradiated to a neutron fluence of 7×10^{19} n/cm², have assigned the process of recovery to voids with diameter of a few angstroms [5]. Positron annihilation spectroscopy (PAS) is a well established technique [6,7] to provide a more defect-specific information on vacancy type defects of small sizes. In this article, positron annihilation as well as transmission electron microscope (TEM) investigations carried out on neutron irradiated pressure vessel steel are reported. The present study is addressed to the processes operating in the earlier stages of recovery of radiation hardening. Defect structures have been deduced from a qualitative correlation of PAS results and microstructural investigation by TEM.

Principle of Methodology

The physical basis of PAS is briefly outlined as follows. When an energetic positron from a radioactive source such as ^{22}Na, is introduced into a metal, it is thermalized within a few picoseconds (ps). During its lifetime in the metal, the thermalized positron diffuses over several atomic distances before it decays by annihilation. If lattice defects, such as vacancies, vacancy-impurity complexes, or voids are present in the metal, the positron can be trapped in bound states of these defects, and its lifetime is enhanced. The latter is governed by the electron density at the positron site in the metal. Since the average electron density at the trapped positron site is quite defect-specific, lifetime measurement offers a unique method for characterization of the defect structure under study [6]. Other measurements used for the study of defects are those of the line shape of the Doppler broadening spectrum and the two-photon angular correlation profile of the annihilation radiation. In the presence of defects, which act as positron traps, the local electronic structure is altered and a measurably different distribution of electron momenta results [6,7]. The line width of the momentum distribution narrows because of positron trapping at defects, leading to increased peak counts about the centroid of the annihilation line in the Doppler broadening spectrum. Thus, it is possible to do unambiguous characterization of vacancy-type defects by monitoring simultaneously positron lifetime and the annihilation line shape.

Experimental Procedure

Specimens used in the present study were prepared from alloy steel (manganese-nickel-molybdenum) satisfying ASTM specifications for pressure vessel plates: ASTM A533B (Unified Numbering System [UNS] K12539) HSST Plate 03. The material was quenched and tempered for initial hardness. Table 1 shows the chemical composition and material data of the specimens used for investigations. The steel specimens were irradiated at a temperature of 423 K to neutron fluences of 1.5×10^{18} n/cm² and 7×10^{19} n/cm² ($E > 1$ MeV) in the research reactor Forschungsreaktor (FRJ-1) at Jülich. Measurements of irradiation temperature and neutron dose were done in the same manner as described earlier [2]. Positron annihilation measurements and microstructural investigations by TEM were conducted on the specimens both before and after irradiation.

After a waiting period of several days for the radiation induced radioactivity to die out, the specimens for TEM studies were cut in the form of 3-mm-diameter disks using a special punch and die [8]. These disks then electropolished using a jet electropolisher. The electrolytic used was 36% sulfuric acid (H_2SO_4), 54% phosphoric acid (H_3PO_4), and 10% H_2O at 383 K using 105 V. After the perforation, these disks were examined in Philips EM 400 transmission electron microscope using a single tilt holder and operated at 120 kV. Care was taken to minimize the difficulties encountered in the electron microscopy of the present specimens because of their highly magnetic nature.

A positron source of 25 μCi activity of ^{22}NaCl was deposited on a titanium foil of thickness 1.25 μm. The source foil was enclosed between a pair of identical specimens in the usual sample-

TABLE 1—*Chemical composition and material data on the pressure vessel steel used in the present work.*

Parameters	Measurement
Chemical Composition	
carbon	0.22
manganese	1.4
nickel	0.6
chromium	0.12
molybdenum	0.5
copper	0.12
vanadium	0.01
phosphorus	0.012
sulfur	0.008
silicon	0.3
Transition temperature (0.51 MJ/m^2), K	298
Vickers hardness, GN/m^2	2.0
Upper shelf energy, MJ/m^2	1.6
Yield strength, GN/m^2	0.51

NOTE: alloy is manganese-nickel-molybdenum, according to ASTM A533-B, HSST Plate 03.

source sandwich geometry for positron lifetime and line shape measurements. The specimens used for measurements were chemically thinned down to about 100 μm, which was the minimum thickness for the complete absorption of positrons. The positron lifetime spectrometer used was the conventional fast-slow coincidence set up with precision timing systems, and the time resolution of the spectrometer was 220 ps (full width at half maximum [FWHM]) with a high counting efficiency. Details of the set up can be found elsewhere [*9*]. Doppler broadening measurements were performed using a high purity Ge detector, which had an energy resolution of 1.14 keV (FWHM) at 497 keV. All spectra were accumulated to a total of more than 10^6 counts to reduce statistical errors in the derived spectral parameters. Two lineshape parameters, I_v and R parameters, were deduced from the Doppler broadening spectra, and these have the same meaning as described earlier [*7,10*]. The parameter I_v is the integrated counts over a region of ± 1.5 mrad about the centroid of the 511-keV annihilation line, normalized to the total area. While this parameter I_v responds sensitively to regions of low electron momenta probed by the positron, the parameter I_c (corresponding to a region of 5 to 10 mrad) responds to annihilations with high momentum core electrons. Using I_v and I_c, defect-specific R parameter was obtained as

$$R = \frac{|I_v - I_v^{ref}|}{|I_c - I_c^{ref}|} \tag{1}$$

where I_v^{ref} and I_c^{ref} are the respective line shape parameters corresponding to the defect-free reference state of the unirradiated specimen. The parameter R is independent of defect concentration and determines the nature of the dominant positron trap involved [*7,11*]. For the analysis of the lifetime spectra, the source contribution to the annihilation spectrum was determined separately from the measurement of known lifetime of a well annealed metal. After subtracting this source background, the measured lifetime spectra were analyzed to unfold the defect components. The lifetime spectra in the irradiated steel were analyzed with a sum of exponential decays representing the different positron states, using a nonlinear least squares fitting procedure [*10,12*]. The instrumental time resolution, represented by 3 Gaussians, was convoluted with the positron decay functions, and this improved the accuracy of the multicomponent lifetime fit. From the resolved lifetimes and their intensities, a mean lifetime $\overline{\tau} = I_1\tau_1 + I_2\tau_2$ was also computed. Post-irradiation defect recovery was carried out by isochronal annealing performed in vacuum on irradiated specimens in increments of 50 K from 373 to 773 K (annealing time = 1 h). The defect structure corresponding to each temperature was frozen after the annealing step. Positron lifetime and line shape measurements were made at room temperature.

Results and Discussions

Figure 1 shows the difference-profile of the Doppler broadening spectra given by $[N(\theta) - N^{ref}(\theta)]/N^{ref}(0)$ where $N(\theta)$ and $N^{ref}(\theta)$ refer to the normalized momentum profiles measured in the as-irradiated state and unirradiated state of the specimens, respectively, and $N^{ref}(0)$ is the value of $N^{ref}(\theta)$ measured at zero momentum. As seen from Fig. 1 the intensity of the low momentum

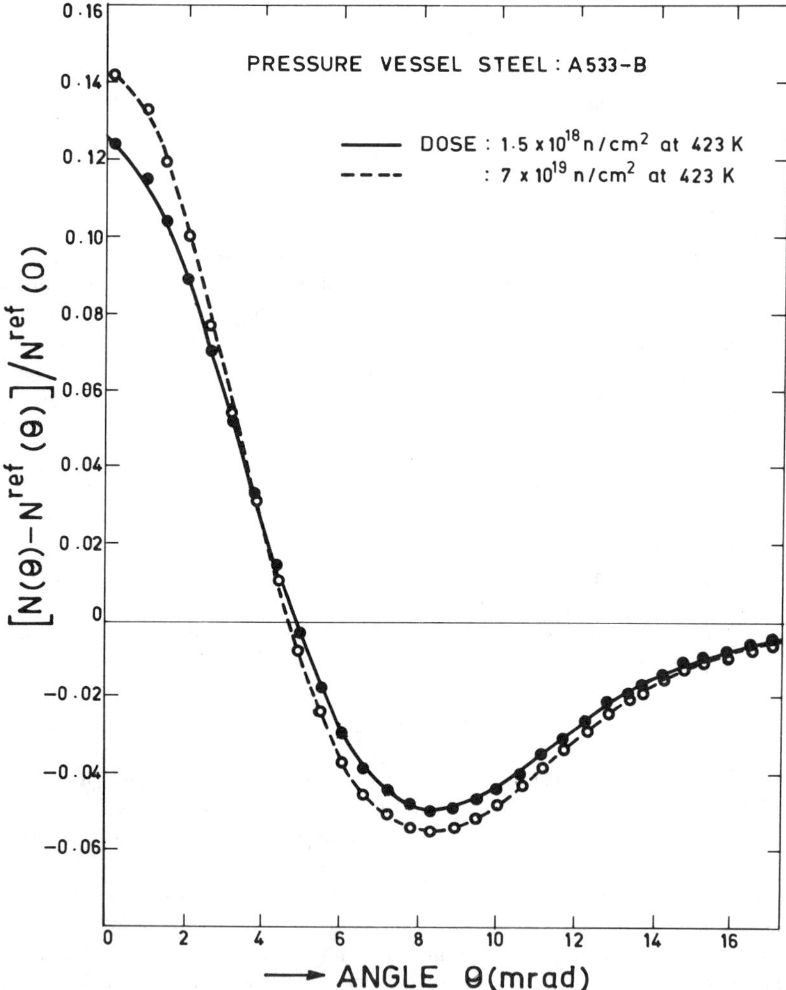

FIG. 1—*Difference profile of the measured Doppler broadening spectra shown for two irradiation doses. See text for details. (In the abscissa of the figure, energy channels have been expressed as momenta in units of mrad.)*

annihilations (0 to 4 mrad) is enhanced at the cost of higher momentum (5 to 15 mrad) core annihilations for the low dose as well as high dose specimens. This is indicative [7] of positron trapping at vacancy-type defects produced by irradiation. Figure 2 shows the variations of the line shape parameter I_v and mean lifetime τ as a function of isochronal annealing temperature in the case of specimens irradiated to a dose of 1.5×10^{18} n/cm². The inset of Fig. 2 shows the variation of Vickers hardness as a function of annealing temperature for comparison. As seen from the figure, I_v remains constant until 473 K, decreases sharply thereafter, and exhibits a plateau-like behavior beyond 573 K. Similar behavior is seen for the mean lifetime τ. The stage seen around 523 K in the recovery of both I_v and τ is in accordance with the recovery behavior of Vickers hardness. During post-irradiation annealing, various defect mechanisms operate in different

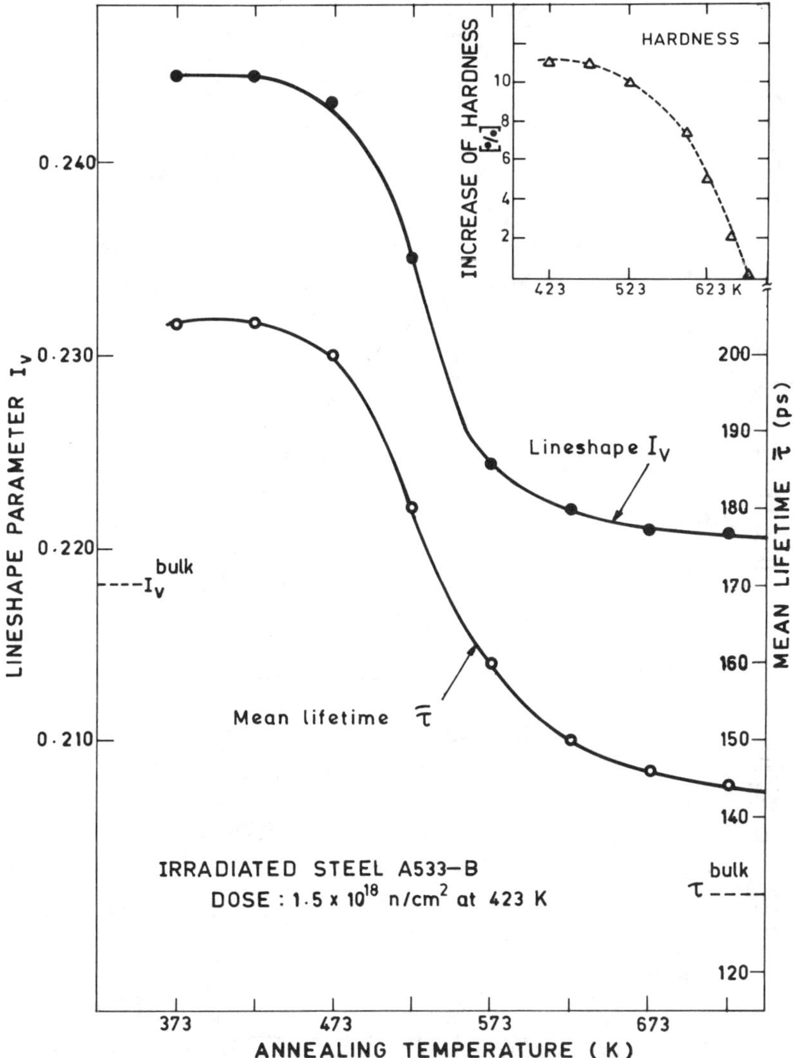

FIG. 2—*Lineshape parameter* I_v *and mean lifetime* $\bar{\tau}$ *versus annealing temperature (annealing time = 1 h). The inset shows the variation of Vickers hardness for comparison.*

temperature ranges [2]. These are characterized by activation energies ranging from 1.5 to 2.1 eV. The agreement seen in Fig. 2 between the annealing results of positron annihilation parameters and that of hardness shows that the lower activation energy processes are detectable by PAS. Even at 723 K, I_v and τ have not recovered to the unirradiated values referred to as I_v^{bulk} and τ^{bulk}, respectively, in Fig. 2. Figure 3 shows the results for the lifetime components τ_1, τ_2, and I_2 as a function of annealing temperature. In the as-irradiated state, the resolved lifetimes are $\tau_1 = 154 \pm 3$ ps with 42% intensity and $\tau_2 = 240 \pm 2$ ps with 58% intensity. The component ($\tau_1 = 154$ ps), which is also reproduced with weaker intensity in the unirradiated specimen, is attributed to positron trapping in dislocations [*10*]. τ_1 stays constant until 473 K and decreases sharply thereafter.

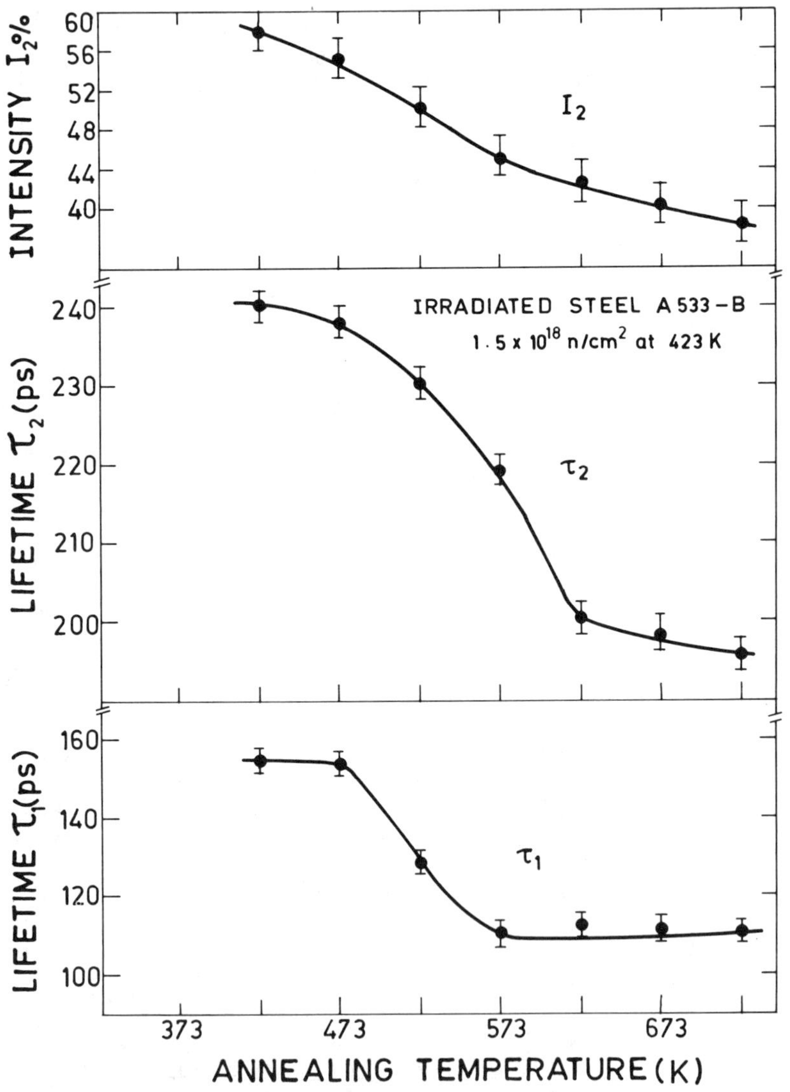

FIG. 3—*Results of an unconstrained two component analysis of positron lifetime spectra.* τ_1, τ_2, *and* I_2 *versus annealing temperature for specimens irradiated to a dose of* 1.5×10^{18} *n/cm*^{2}.

Beyond 573 K, the lifetime (154 ps) corresponding to dislocations vanishes and τ_1 goes over to a value of 110 ps below that of the bulk lifetime (τ^{bulk} = 130 ps). This behavior of τ_1 above 573 K is in accordance with the two-state trapping model [*10*]. The defect complex corresponding to the component τ_2 will be discussed subsequently.

The microstructure of the specimen before irradiation is shown in Fig. 4*a*. The black particles seen in the micrograph are attributed to carbides of the type (Fe_3C) and (Mo_2C). This qualitative identification is based on the morphology as well as the lattice spacing of the second phase. The size of the carbide particles ranged from 0.1 to 0.35 μm. Figures 4*b* and *c* show the typical microstructures of the specimen after irradiation at 423 K to a dose of 1.5 × 10^{18} n/cm^2. The morphology of the carbide after irradiation, seen in Fig. 4*b*, shows a small change. There is a reduction in the average size of the particles by about 10% suggesting a partial dissolution of the second phase because of atomic displacements during irradiation. The dislocation structure formed as a result of irradiation is shown in Fig. 4*c*, and these dislocations are out-of-contrast in Fig. 4*b*. Figure 4*d* shows the microstructure of the irradiated specimen after annealing at 593 K. The observation here is that there is no further change in the average size of the carbide particles. However considerable annealing of the dislocation network is seen at 593 K, as compared to the as-irradiated state. The latter observation by TEM is in agreement with the behavior of positron lifetime τ_1 in Fig. 3.

Next we take up the interpretation of the longer lifetime τ_2 and its intensity I_2. As seen in Fig. 3, τ_2 stays at a value of 240 ps until 473 K, decreases sharply in the interval 523 to 623 K, and tends to reach a value of ~200 ps beyond 623 K. In this interval there is a corresponding reduction of its intensity I_2 from 58 to 40%. It is known from earlier studies [*13,14*] that migration of carbon interstitial occurs above 353 K in α-iron. Based on this known evidence for the carbon migration stage and the present TEM results on the second phase particles, it may be argued that carbon atoms displaced from the carbides during irradiation at 423 K undergo interstitial migration and

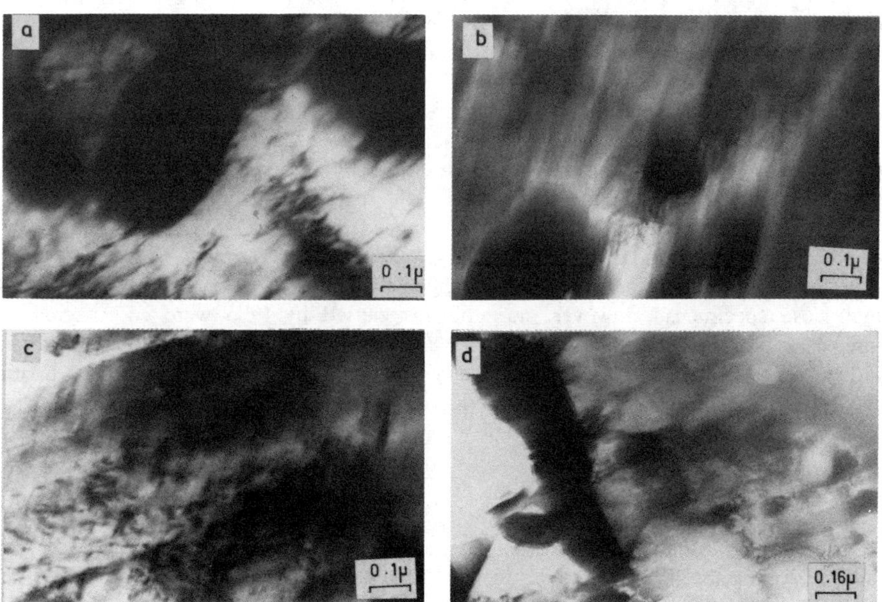

FIG. 4—*Bright field TEM micrographs on the pressure vessel steel:* (a) *unirradiated state,* (b) *and* (c) *irradiated to a dose of 1.5 × 10^{18} n/cm^2, and* (d) *irradiated specimen after annealing at 593 K.*

get trapped to irradiation induced vacancies or clusters of vacancies. The carbon atoms are known to have a strong binding to vacancies and their clusters [*14,15*], and this results in the formation of stable carbon-vacancy agglomerates. The presently observed lifetime, τ_2 = 240 ps until 473 K, can then be understood in terms of positron trapping at these carbon-vacancy clusters. The strong positron trapping observed would suggest that the carbon atom is located off the vacancy center in the cluster.

Calculations reported earlier [*16*] indicate that the carbon atoms are located at a distance 0.73 half-lattice constant from the vacancy center along ⟨100⟩ line, not far from the neighboring octahedral positions. The multiple carbon decoration in the cluster may be easily visualized, as there are six equivalent positions for the carbon around the vacancy. Carbon segregation decreases the free space for positron localization in the trap and increases the effective electron density seen by the trapped positron. This explains qualitatively the presently observed magnitude for the lifetime of the complex (240 ps) being lower than that known for voids (450 ps).

The carbon-vacancy separation [*16*] might imply a typical size of 4 Å for the complex, which is below the present TEM resolution of ~10 Å. This is possibly the reason why this structure was not visible in TEM. As the temperature of anneal is increased, the carbon-vacancy complex dissociates resulting in the shrinkage of the cluster size, and this is in accordance with the observed decrease of τ_2 and I_2. The fact that τ_2 and I_2 decrease rather continuously over the interval 523 to 623 K suggests that the carbon release from the cluster takes place over a broad range of temperature.

The present results on the stability of carbon-vacancy agglomerates are qualitatively similar to those reported in carbon-doped α-iron [*13,14*]. There is however, one notable difference between the present result on the steel alloy and that for α-iron. Whereas complete recovery of irradiation-induced defects is observed below 623 K in α-iron, no such behavior is seen for the steel alloy. Even at 723 K, the lifetime τ_2 shows a value of 195 ps with 38% intensity. Figure 5 shows the variation of the line shape *R*-parameter as a function of annealing temperature. This behavior of the *R*-parameter, similar to that of τ_2, indicates the change of defect structure at 623 K, in support of but independent of the lifetime data. These results possibly suggest that the carbon released from the vacancy complex stabilizes a new structure. It may be seen from Fig. 2 that the recovery of Vickers hardness is not complete even at 623 K. This residual hardening may be caused by a different mechanism [2], which is not yet understood in the present study. Experiments on the effect of irradiation temperature and further elucidation of the role of second phase particles in the alloy are necessary for a complete understanding of radiation hardening in the material.

Before we conclude, a brief reference is made to the measurements on the high dose specimens (7×10^{19} n/cm^2). In these specimens there was a considerable long-lived activity caused by ^{60}Co whose prompt gamma cascades influenced strongly the positron lifetime measurement. Owing to this experimental difficulty, no detailed measurements of the lifetime spectra were carried out on the high dose specimens. However, no problems caused by ^{60}Co were encountered in the measurement of the line shape of the 511-keV annihilation line. The line shape *R*-parameter showed an enhanced value of 0.665, as compared to that in the low dose specimen. This implies that the average size of defects formed in the high dose specimens is larger.

Summary and Conclusions

1. Good correlation has been observed for the post-irradiation recovery behavior of positron lifetime and line shape parameters with that of Vickers hardness in the pressure vessel ASTM A533-B steel.

2. The morphology of the second phase particles in the alloy shows a small change on

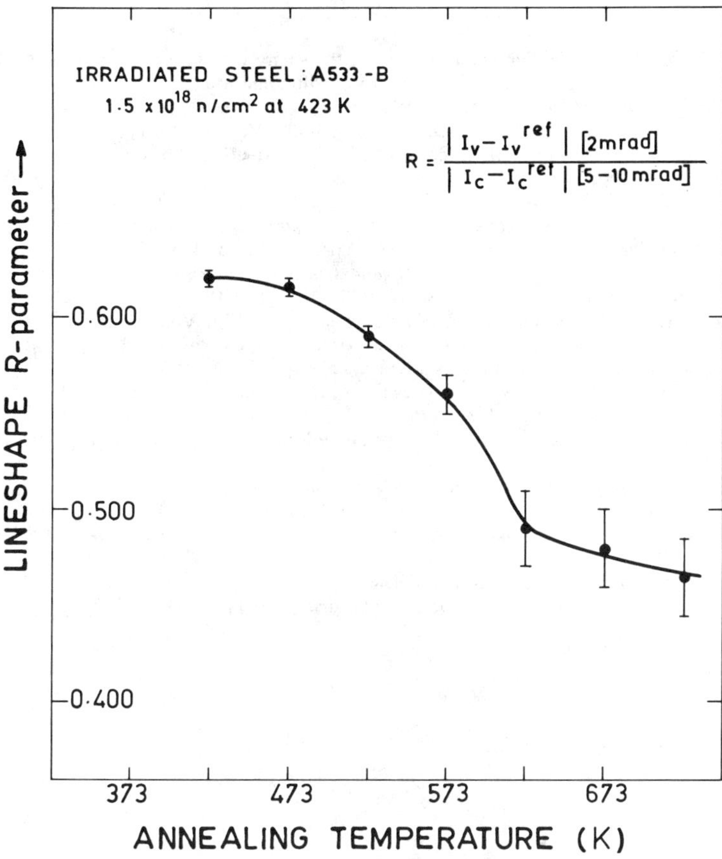

FIG. 5—R-*parameter variation as a function of annealing temperature.*

irradiation, and there is a reduction in the average size of these particles, suggesting partial dissolution caused by irradiation.

3. On the basis of known evidence for carbon migration in α-iron and the present results, enhancement of hardness and its recovery until 623 K have been qualitatively understood in terms of carbon decorated vacancy complexes.

4. A change of defect structure is indicated at 623 K from the behavior of positron lifetime and line shape.

5. Positron annihilation technique appears to be a good complementary tool for assessing the embrittlement potential of pressure vessel steels.

Acknowledgments

Part of the positron annihilation experiments reported in the present study was carried out at the Universitat der Bundeswehr Munchen, Neubiberg, Federal Republic of Germany, during the visit of B. Viswanathan as an Alexander-von-Humboldt fellow. This is gratefully acknowledged, and we also thank Prof. W. Triftshauser for some discussions. Assistance from Mr. K. Varatharajan and Mrs. Padma Gopalan on specimen preparation is acknowledged.

References

[1] *Radiation Embrittlement and Surveillance of Nuclear Reactor Pressure Vessels: An International Study, STP 819,* L. E. Steele, Ed., American Society for Testing and Materials, Philadelphia, 1983.
[2] Pachur, D., *Nuclear Technology,* Vol. 59, 1982, p. 463.
[3] Pachur, D., *American Nuclear Society (ANS) Transactions,* Vol. 38, 1981, p. 306.
[4] Pachur, D., *Effects of Radiation on Materials, STP 725,* American Society for Testing and Materials, Philadelphia, 1981, pp. 5–19.
[5] Schwahn, D., Pachur, D., and Schelten, J., KFA Report, Jül 1543, Oct. 1978.
[6] Siegel, R. W., in *Advanced Techniques for Characterising Microstructures,* F. W. Wiffen and J. A. Spitznagel, Metallurgical Society of AIME, New York, 1982, pp. 413–422.
[7] Triftshauser, W., in *Festkorperprobleme—Advances in Solid State Physics,* H. J. Queisser, Ed., Vol. XV, Pergamon Press/Vieweg, Braunscheveig, 1975, pp. 381–410.
[8] Varatharajan, K. and Sadasivan Nair, R., *Transactions of the IIM,* Vol. 37, 1984, p. 168.
[9] Koegel, G., *Proceedings of the 5th International Conference on Positron Annihilation,* R. R. Hasiguti and K. Fujikawa, Eds., Japan Institute of Metals, 1979, p. 383.
[10] Viswanathan, B., Triftshauser, W., and Koegel, G., *Radiation Effects,* Vol. 78, 1983, p. 231.
[11] Mantl, S. and Triftshauser, W., *Physical Review B,* Vol. 17, 1978, p. 1645.
[12] Kirkegaard, P., Eldrup, M., Mogensen, O. E., and Pedersen, N. J., *Computer Physics Communication,* Vol. 23, 1981, p. 307.
[13] Hautojarvi, P., Johansson, J., Vehanen, A., Yli-Kauppila, J., and Moser, P., *Physical Review Letters,* Vol. 44, 1984, p. 1326.
[14] Hautojarvi, P., Pollanen, L., Vehanen, A., and Yli-Kauppila, J., *Journal Nuclear Materials,* Vol. 114, 1983, p. 250.
[15] Little, E. A., *Journal Nuclear Materials,* Vol. 87, 1979, p. 17.
[16] Johnson, R. A. and Damask, A. C., *Acta Metallurgica,* Vol. 12, 1964, p. 443.

Glenn E. Lucas,[1] *G. Robert Odette,*[1] *Ranen Maiti,*[1] *and J. William Sheckherd*[1]

Tensile Properties of Irradiated Pressure Vessel Steels

REFERENCE: Lucas, G. E., Odette, G. R., Maiti, R., and Sheckherd, J. W., "**Tensile Properties of Irradiated Pressure Vessel Steels,**" *Influence of Radiation on Material Properties: 13th International Symposium (Part II), ASTM STP 956,* F. A. Garner, C. H. Henager, Jr., and N. Igata, Eds., American Society for Testing and Materials, Philadelphia, 1987, pp. 379–394.

ABSTRACT: Yield stress and hardness changes have been demonstrated to be strongly correlated to transition temperature shifts in nuclear pressure vessel steels, and hence such measurements can be a convenient means for evaluating the role of metallurgical and environmental variables in embrittlement. We have irradiated miniature tensile specimens of over 100 alloy (chemistry and thermomechanical treatment) variants at nominal fluxes 0.5 to 5 $\times$ 10^{16} n/m^2 s, temperatures 271 to 325°C, and fluences 0.15 to 3 $\times$ 10^{23} n/m^2. A subset of these specimens has been tested, and the results are reported here. For both commercial and model alloys, good agreement between hardness and yield stress changes were observed. Yield stress changes were strongly influenced by copper and nickel content and by thermomechanical treatment.

KEY WORDS: pressure vessel steels, embrittlement, tensile properties, radiation hardening

For reactor pressure vessel (RPV) steels subject to neutron irradiation, several studies have now demonstrated a strong correlation between changes in yield stress ($\Delta\sigma_y$) or hardness (or both) and shifts in the transition temperature ΔT_{41} defined at the 41 J level in Charpy-V-Notch (CVN) impact tests [*1–5*]. There appear to be two contributions to this relationship [*1*]. The largest comes from a shift in the maximum temperature of elastic fracture T_e, which occurs at about 10 J of absorbed energy. This can be explained in terms of a stress-controlled cleavage fracture mechanism. Cleavage crack propagation occurs when the peak stress ahead of a blunt notch exceeds a critical stress, and near T_e the peak stress scales with the yield stress. Since irradiation does not largely alter the cleavage fracture stress, T_e increases in a way dictated by its initial value and the increase in yield stress, $\Delta\sigma_y$. The other contribution comes from a reduction in upper shelf energy and a consequent decrease in slope of the CVN curve in the transition regime. This shift contribution can be empirically related to the yield stress increase.

Consequently, investigation of yield stress and hardness changes becomes a convenient means of evaluating the role of metallurgical and environmental variables in RPV neutron irradiation embrittlement. To this end we have conducted several irradiation experiments to measure both hardness and yield strength changes in irradiated commercial and model RPV steel [*2,4,6*]. We describe here the most recent irradiation experiments and results obtained to date.

[1] Associate professor, professor, assistant research engineer, and senior development engineer, respectively, Department of Chemical and Nuclear Engineering, University of California, Santa Barbara, CA.

Experimental Procedures

Specimens have been prepared and irradiated in two phases. In the first phase 10 model alloys and 13 commercial alloys were investigated. Miniaturized tensile specimens with the geometry shown in Fig. 1 were machined from 0.63 mm thick sheet for two commercial alloys (designations A and D) and one model alloy (T). Specimens were irradiated in the University of Virginia Reactor at two flux levels (nominally 0.5 and 5 × 10^{16} n/m^2-s), four temperatures (271, 288, 307, and 325°C), and three fluences (nominally 0.25, 0.5 and 1 × 10^{23} n/m^2). Thermal controls were conditioned in He-filled quartz capsules. The experiment is described in more detail elsewhere [*4*]. In addition to miniature tensile specimens, several other specimen types, including microhardness discs, coupons, and miniature CVN specimens, were irradiated. Much of the microhardness data has been reported previously [*4,6*]. We have recently irradiated a full set of microhardness discs with 14 MeV neutrons in the Rotating Target Neutron Source (RTNS-II) at 290°C to an average fluence of 1 × 10^{22} n/m^2. Following irradiation, these discs were subjected to Vickers microhardness testing by procedures described elsewhere [*4*]. We discuss a subset of the Phase I results here for the alloys given in Table 1.

In the second phase, the Phase I alloys were augmented by several sets of additional alloys. These are described in detail elsewhere [*7*]. Only those alloys for which results are reported here will be described below. First, a quantity of alloy V (which already had 10% cold work) was given an additional cold rolling to produce 20% cold-worked material (designation V2). Second, a set of six welds were obtained, and these are described in Table 2. Compositions are those reported by Wang [*8*] for regions of the weld from which specimens were taken. Third, a set of A533B-type steels was supplied by MRK Associates.[2] The compositions, based on wet chemistry analysis, are given in Table 3. Compositions were achieved by appropriate addition of copper, nickel, silicon, phosphorus, and tin to four split melts. The ingots were hot rolled to 25 mm thick at 950°C and water-spray-cooled to room temperature, producing a fully-bainitic, as-received microstructure. Several thermomechanical treatments were then followed for each of the compositions, and these are also described in Table 3. Schedules A and B were established to systematically study the effect of stress relief on a hardening response, and Schedules D and E to examine the effect of cold work. Finally, a set of alloys were supplied by KFA Jülich,[3] in which copper and nickel contents were relatively low and silicon, manganese, phosphorus, chromium, and micro-

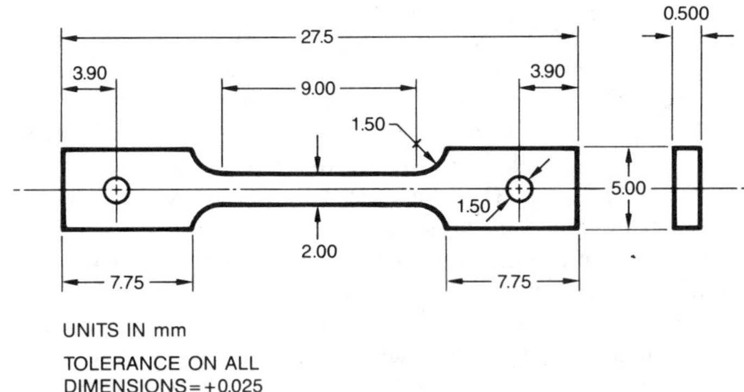

FIG. 1—*Schematic illustration and dimensions of the miniaturized tensile specimen.*

[2] Courtesy M. R. Krishnadev, Laval University.

[3] Courtesy D. Pachur.

TABLE 1—*Description of the Phase I alloys for which results are reported here.*

Alloy Designation	Description	Composition, % by weight										Heat Treatment[b] Schedule
		C	Cu	Ni	Mn	Si	P	S	Cr	Mo	V	
A	A320B plate[a]	0.23	0.2	0.17	1.47	0.26	0.013	0.024	0.05	0.52	0.004	1
B	A533B plate HSST-02	0.23	0.16	0.67	1.55	0.20	0.009	0.014	0.04	0.53	0.003	2
C	EPRI Linde 0080; weld	0.16	0.35	0.6	1.30	0.17	0.005	0.009	0.039	0.44	. . .	3
D	EPRI Linde 0091; weld	0.12	0.4	0.6	1.36	0.51	0.006	0.013	0.044	0.44	. . .	3
H	Model alloy	<0.003	<0.008	<0.03	<0.01	<0.04	<0.009	<0.003	<0.004	. . .	<0.003	4
J	Model alloy	0.1	<0.01	<0.01	<0.03	<0.01	<0.02	<0.02	. . .	. . .	. . .	4
K	Model alloy	0.1	0.15	<0.04	<0.01	<0.07	<0.009	<0.004	<0.004	. . .	<0.004	5
L	Model alloy	0.1	0.28	<0.01	<0.01	<0.01	<0.02	<0.01	<0.01	. . .	<0.01	5
M	Model alloy	0.1	<0.01	0.6	<0.01	<0.05	<0.009	<0.003	<0.004	. . .	<0.003	4
N	Model alloy	0.1	0.16	0.52	<0.008	<0.06	<0.007	<0.004	<0.003	. . .	<0.004	5
P	Model alloy	0.1	0.12	1.15	<0.009	<0.07	<0.007	<0.004	<0.004	. . .	<0.006	5
S	Model alloy	0.1	0.35	0.54	<0.008	<0.07	<0.007	<0.004	<0.003	. . .	<0.007	5
T	Model alloy	0.1	0.27	0.62	1.5	0.64	<0.008	<0.003	<0.008	. . .	<0.006	5
V	Model alloy	0.1	0.15	<0.04	<0.01	<0.07	<0.009	<0.004	<0.004	. . .	<0.004	6

[a] Correlation monitor material.

[b] Heat treatments: (1) Austenitize 899°C, 6 h; water quench; temper 649°C, 6 h; air cool. (2) Austenitize 871°C, 4 h; water quench; temper 663°C, 4 h; furnace cool; stress relief 607°C, 40 h. (3) Submerge-arc weld; post-weld heat treat 620°C; 50 h. (4) Cold roll; recrystallization anneal 857°C, 0.5 h. (5) Schedule 4 + solution anneal 750°C, 1 h; forced air cool. (6) Schedule 4 + 10% cold work.

TABLE 2—*Description of weld metals.*

Alloy Designation	Description	Compositon, % by Weight									Heat Treatment[a] Schedule
		C	Cu	Ni	Mn	Si	P	S	Cr	Mo	
GQ	Quad Cities-1 weld	0.09	0.24	0.54	1.70	0.56	0.014	0.016	0.06	0.43	Y
GC	Quad Cities-2 weld	0.09	0.24	0.56	1.68	0.55	0.014	0.016	0.07	0.40	Y
GD	Dresden-3 weld	0.09	0.28	0.63	1.59	0.51	0.011	0.015	0.08	0.43	Y
GH	Humboldt Bay weld	0.13	0.22	0.07	1.37	0.29	0.014	0.016	0.16	0.46	Z
GM	Milestone Point weld	0.11	0.19	1.02	1.30	0.18	0.017	0.015	0.05	0.51	Z
GT	Tarapur weld	0.12	0.15	0.08	1.61	0.28	0.018	0.014	0.15	0.49	Z

[a] Heat treatment schedules:
Y: Submerged arc weld; post weld heat treat 607°C, 30–50 h; slow cool.
Z: Submerged arc weld; post weld heat treat 621°C, 30–50 h; slow cool.

TABLE 3—*Description of A533B-type alloys for which results are reported here.*

Alloy Designation	Composition, % by Weight											Heat Treatment[a] Schedule
	C	Cu	Ni	Mn	Si	P	S	Sn	Cr	Mo	Al	
LA	0.14	0.40	0.001	1.37	0.22	0.005	0.005	<0.004	0.06	0.54	0.005	A
LB	0.16	0.40	0.175	1.35	0.22	0.005	0.005	<0.004	0.06	0.53	0.006	A
LN	0.16	0.40	0.175	1.35	0.22	0.005	0.005	<0.004	0.06	0.53	0.006	B
LC	0.14	0.41	0.86	1.44	0.23	0.005	0.008	<0.004	0.06	0.55	0.006	A
LO	0.14	0.41	0.86	1.44	0.23	0.005	0.008	<0.004	0.06	0.55	0.006	B
PH	0.14	0.41	0.86	1.44	0.23	0.005	0.008	<0.004	0.06	0.55	0.006	D
PI	0.14	0.41	0.86	1.44	0.23	0.005	0.008	<0.004	0.06	0.55	0.006	E
LE	0.13	0.42	0.85	1.42	0.24	0.025	0.007	0.015	0.06	0.56	0.006	A
LF	0.16	0.42	0.81	1.47	0.73	0.005	0.015	<0.004	0.06	0.56	0.015	A
LD	0.19	0.38	1.25	1.38	0.23	0.005	0.015	<0.004	0.07	0.53	0.01	A
LP	0.19	0.38	1.25	1.38	0.23	0.005	0.015	<0.004	0.07	0.53	0.01	B
LG	0.16	0.005	0.74	1.37	0.22	<0.005	<0.005	<0.004	0.05	0.55	0.01	A
LH	0.16	0.11	0.74	1.39	0.22	<0.005	<0.005	<0.004	0.05	0.55	0.01	A
LI	0.16	0.20	0.74	1.37	0.24	<0.005	<0.005	<0.004	0.05	0.54	0.01	A
LJ	0.14	0.42	0.81	1.34	0.13	<0.005	<0.005	<0.004	0.085	0.56	0.01	A
LK	0.13	0.8[b]	0.81	1.30	0.13	<0.005	<0.005	<0.004	0.085	0.56	0.005	A
LL	0.15	0.21	1.1	1.37	0.23	<0.005	<0.005	<0.004	0.05	0.55	0.01	A

[a] Heat treatment schedules: (A) Austenitize 900°C, 1 h; air cool; temper 664°C, 4 h; air coo_; stress relieve 600°C, 40 h; furnace cool to 300°C; air cool. (B) Austenitize 900°C, 1 h; air cool; temper 664°C, 4 h; air cool. (D) Schedule A + 10% cold work. (E) Schedule A + 20% cold work.

[b] Nominal.

TABLE 4—*Description of alloys supplied by KFA Jülich for which results are reported here.*

Alloy Designation	Composition, % by weight										Heat Treatment[a] Schedule
	C	Cu	Ni	Mn	Si	P	S	Cr	Mo	Al	
FJ	0.20	0.01	0.70	1.31	0.24	0.007	0.007	0.02	0.50	0.02	W
FL	0.20	0.01	0.69	1.30	0.02	0.023	0.007	0.03	0.51	0.03	X

[a] Heat treatment schedules: (W) Austenitize 900°C, ½ h; air cool; temper 650°C, 1.6 h; furnace cool. (X) Austenitize 900°C, ½ h; oil cool; temper 650°C, 1.6 h; furnace cool.

structure were varied systematically. Alloys for which results are reported here and their corresponding compositions and heat treatments are provided in Table 4.

Coupons 37 mm long × 22 mm wide × 0.5 mm thick were fabricated from alloys listed in Tables 1 to 4. From each coupon specimen three miniaturized tensile samples were punched, using a punch-and-die apparatus machined to provide specimens with the geometry given in Fig. 1. The coupon and tensile specimens were then engraved with a specimen and condition code, and chemically milled in phosphoric acid to remove punching burrs and to reduce the dimensions slightly to permit reinsertion of the tensile specimens into the coupons. In addition, a number of other specimen types including one-third sized CVN, notched tensile and small angle neutron-scattering specimens and microhardness coupons were fabricated from selected materials. These are described in detail elsewhere [7].

Specimens were loaded into mild steel capsules which were arranged in a single hot thimble position in the University of Virginia reactor, as shown in Fig. 2. Two capsule geometries were used to obtain a range of fluences. The large capsule was left in the reactor for the entire irradiation (highest fluence), while the small capsules were periodically removed and replaced with fresh ones to achieve intermediate fluence levels. Because of the need to withdraw and replace capsules, the number of thermocouple leads was limited, and hence thermocouples were positioned only at the

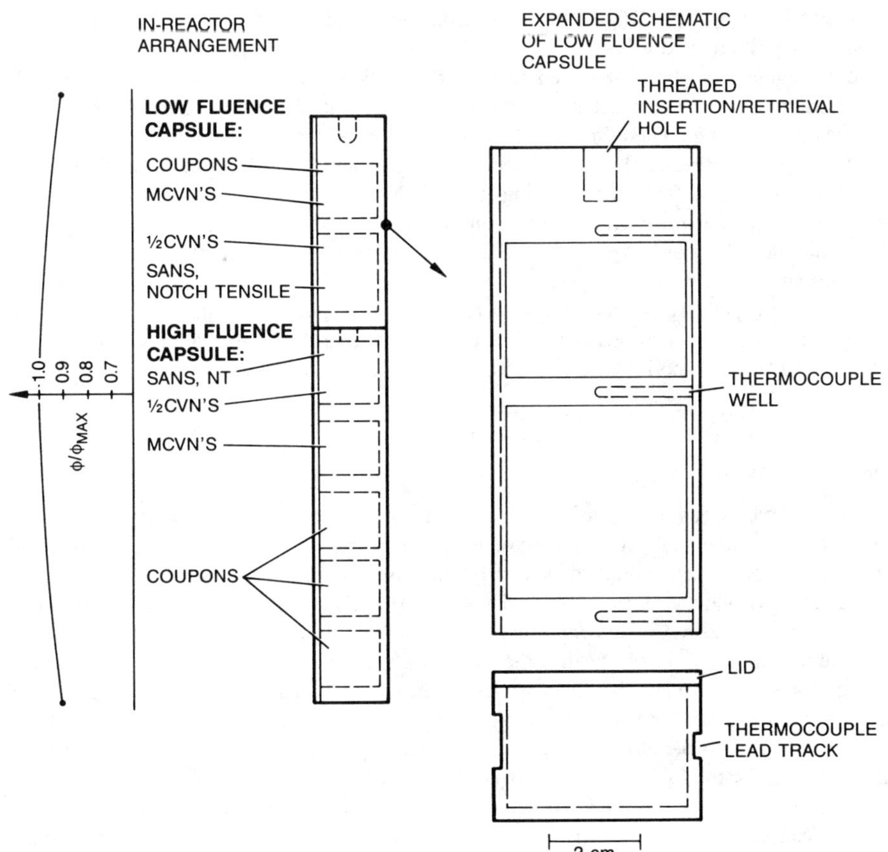

FIG. 2—*Illustration of capsule geometry for high and low fluence irradiations.*

middle of the capsule ribs. Specimen stacks were packed to maintain hard contact among specimens and capsule walls. Temperature drops across the specimen stack were estimated to be as low as 2°C (calculated based on a solid sample) but as high as 10°C (based on data variations from outer to inner specimens). Moreover, there is additional uncertainty between temperatures measured by the thermocouple and adjacent sample temperatures. We are still attempting to refine irradiation temperatures, so the values reported here for Phase II irradiations should be considered preliminary. While flux wires have not yet been evaluated, the nominal fluence for the large capsule is estimated to be 3×10^{23} n/m^2 ($E > 1$ MeV), and estimated fluences for the small capsules were 0.15, 0.45, 0.9, and 1.5×10^{23} n/m^2. These estimates are based on irradiation time and previous flux estimates for the irradiation positions used.

The hot thimbles contained boral shielding to reduce thermal neutron radioactivation, and the capsules were stored for up to a year after irradiation to permit appreciable decay. The capsules were opened in a hot cell at the Westinghouse Research and Development Center and specimens were then shipped to the University of California, Santa Barbara, for testing.

To date only testing of miniaturized tensile specimens has been performed on the Phase II alloys. In addition, tensile tests have been performed on tensile specimens available from the Phase I experiment. Tensile tests were performed in a controlled area on an Instron 1122 load frame at ambient temperature. The cross head speed during testing was 0.0085 mm/s. The specimen and grips were positioned inside a sealed plexiglass box during testing to contain any radioactive debris released by the specimen. The load-displacement data were digitized and recorded on a minicomputer system, and a hard copy record of the analogue signal was also generated. Load-displacement data have been analyzed to date to provide 0.2%-offset yield stress and ultimate tensile strength values. The greatest source of uncertainty in these data arises from uncertainties in dimensional measurements. Specimen dimensions were measured with a dial micrometer to ± 0.013 mm prior to testing. However, because of the chemical milling, gage section dimensions varied by as much as 0.025 mm from end to end. Moreover, operator-to-operator measurement variations were of the order 0.025 mm. Consequently, yield stresses are currently estimated to have uncertainties of the order ± 30 MPa. There are additional specimens for some alloy conditions; hence, subsequent test results may effect a slight revision of the average data given here.

We report here yield stress data obtained for material irradiated nominally at 0.9×10^{23} n/m^2 and 290°C in the Phase II experiments. We also report a subset of the yield stress data obtained to date in the Phase I experiment along with corresponding microhardness data.

Results

Spectral Dependence

A comparison between microhardness changes determined in Phase I alloys irradiated in RTNS-II and in the University of Virginia reactor is shown in Fig. 3. The fission reactor data selected for comparison here were obtained at nearly the same irradiation temperature and at flux and fluence conditions which were nearest to the RTNS-II irradiation conditions. The value of ΔDPH here is the difference in Vickers microhardness determined in the irradiated specimen and in a corresponding thermal control. Within the uncertainty of the microhardness measurements there is a good correspondence between the two sets of ΔDPH data. That is, the hardening in the fusion reactor spectrum scales with that in the high energy neutron spectrum. Moreover, the data suggest that dpa (displacements per atom) is a reasonably good exposure parameter for hardening. Although the fluence level for the fission irradiations is greater by about a factor of 2.5 (for neutron energies greater than 1 MeV) than the RTNS-II irradiation, the degree of hardening is quite similar. Based on a displacement cross section of 2900 barns for iron in RTNS-II [*9*] and 1450 barns in the University of Virginia Reactor [*4*], the damage level is estimated to be 2.9 mdpa on average for the RTNS-II irradiation and 3.7 mdpa for the fission reactor irradiation; that is, the damage level

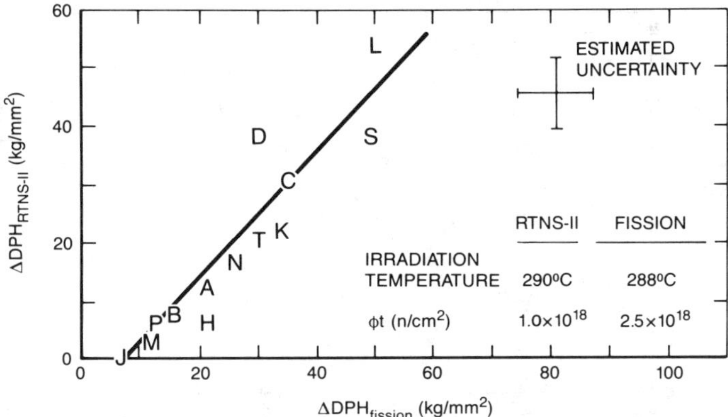

FIG. 3—*Changes in Vickers microhardness (ΔDPH) for materials irradiated in RTNS-II compared to results of irradiations in the University of Virginia reactor.*

in the fission reactor irradiation is only about 25% higher than the RTNS-II irradiation in terms of dpa. This somewhat greater level of displacement damage in the fission reactor irradiations is consistent with the observation that the fission reactor data are systematically higher than the RTNS-II data by about 5 to 10 DPH, based on the fluence dependence of hardening we have observed and reported elsewhere [*6*].

Scaling of hardening with dpa in a pressure vessel steel has also been reported recently by Heinisch et al [*10*]. They irradiated tensile specimens of A302B steel in RTNS-II and a fission reactor (Omega West Reactor—OWR) at 290°C and 90°C. Reasonably good agreement between hardening in RTNS-II and OWR irradiated material at the same level of dpa was observed at 290°C over the range 2×10^{-4} to 2×10^{-2} dpa. However, at 90°C hardening occurred at smaller values of dpa in RTNS-II compared with OWR.

Hardness-Yield Stress Correlation

Figure 4 provides a comparison between observed changes in microhardness and yield stress for Phase I alloys tested to date. Again changes represent differences between measurements on irradiated material and thermal controls. The data for alloys A and D were obtained from specimens irradiated in the Phase I experiment. The irradiated yield data for the model alloys K, L, S, T, and V were obtained from the Phase II experiment for a fluence of $\sim 0.9 \times 10^{23}$ n/m². The corresponding microhardness data are extrapolated to this fluence from actual data obtained at $\sim 0.7 \times 10^{23}$ n/m² in Phase I using a fluence dependence of microhardness change observed in these materials [*6*].

There is a reasonable proportionality between changes in Vickers microhardness and yield stress. The slope of the regression line through the data is approximaely 3.5 MPa/ΔDPH. This is close to the value of 3.6 MPa/ΔDPH reported by Mancuso et al. [*5*] for A533B welds, which we have assumed previously in our modeling work [*4,6*], and it is certainly in agreement with this value within the uncertainty limits of our data. We have previously presented data extracted from the literature which showed a similar dependence between $\Delta\sigma_y$ and ΔDPH for a variety of weld and plate materials [*2*]. If these same data are fit to a regression line which intersects the origin, the slope of the line is also 3.5 MPa/ΔDPH. Moreover, the data of Williams et al. [*3*] also suggest the proportionality constant between microhardness and yield stress changes may be slightly smaller than 3.6. Finally, it should be noted that both the model alloys and engineering alloys

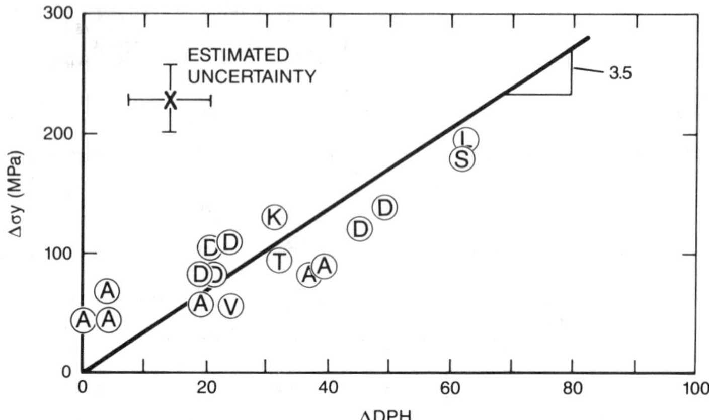

FIG. 4—*Comparison of changes in yield strength and microhardness following irradiation for both model and engineering alloys.*

appear to follow the same trend line; again this has been an assumption made in our previous modeling work, which is in part validated by these data.

This value of 3.5 MPa/ΔDPH is also consistent with several other observations. For instance, the ultimate tensile strength σ_{UTS} generally scales with hardness H as $\sigma_{UTS} = H/3$ [*11*]. Casting this in units of MPa for σ_{UTS} and kg/mm² for DPH, this can be rewritten in terms of *changes* in hardness and strength as $\Delta\sigma_{UTS}$ (MPa) $= 3.26$ ΔDPH (kg/mm²). Based on a limited analysis, we have observed in the irradiated tensile data base that the change in σ_{UTS} is typically about 90% of the change in σ_y. Hence, $\Delta\sigma_y = \Delta\sigma_{UTS}/0.9 = 3.6\ \Delta\text{DPH}$. Moreover, Cahoon et al. [*12*] have suggested that

$$\sigma_y = \frac{H}{3}(0.1)^n \tag{1}$$

where n is the work hardening experiment. Recasting this to calculate $\Delta\sigma_y$ (MPa) from ΔDPH (kg/mm²), one can obtain

$$\Delta\sigma_y = C\ \Delta\text{DPH} = C(\text{DPH}_2 - \text{DPH}_1) \tag{2}$$

where

$$C = \frac{3.26(\text{DPH}_2(0.1)^{n_2} - \text{DPH}_1(0.1)^{n_1})}{\text{DPH}_2 - \text{DPH}_1}$$

Substituting typical measured values of DPH ($\text{DPH}_2 \simeq 240$ and $\text{DPH}_1 \simeq 200$) and $n(n_1 \simeq 0.1$ and $n_2 \simeq 0.075)$ from our irradiated data base, C is determined to have a value of the order 3.5.

The slope of 3.5 MPa/ΔDPH was obtained from the data in Fig. 4 by forcing the fit line through the origin. However, we have observed in general that at low levels of hardening, a change in σ_y can occur without a change in DPH. In fact, the data in Fig. 4 indicate this: a simple linear regression through the data intersects the ordinate at about 30 MPa at a ΔDPH of zero. This behavior may be a consequence of the fact that DPH values correspond to a flow stress at about 8% plastic strain [*11*]. At low levels of hardening, there may be an increase in σ_y without a significant increase in the flow stress at this level of plastic strain (due to dislocations cutting small defects), and hence without an increase in DPH. However, at higher levels of hardening (when defects are larger) the flow stress at 8% strain increases in proportion to the yield stress. Accounting for this difference at low hardening levels might improve the relationship between $\Delta\sigma_y$ and ΔDPH slightly, but given the scatter in the data, a straight-line fit through the origin provides a good approximation.

Dependence of Yield Stress Changes on Composition and Microstructure

In this section we examine yield stress changes evaluated from tensile specimens irradiated at 290°C to ~0.9 × 10^{23} n/m^2 in the Phase II experiment. Thermal controls were not available for all materials. In our previous work on thermal aging [*4,6,13*], however, the model alloys H-V showed some significant (largely transient) softening during aging, but the engineering alloys A-D did not, even at relatively high temperatures and long aging times. Hence, we make the assumption here that aging does not significantly change the yield stress of the more complex alloys (Tables 2 to 4), and for these the change in yield stress, $\Delta\sigma_y$ was determined as the difference in the irradiated and unirradiated values. For the alloys in Table 1, $\Delta\sigma_y$ was determined as the difference in the values obtained on irradiated specimens and material aged at 288°C for 2 × 10^6 s.

Figure 5 shows the variation of $\Delta\sigma_y$ with copper content for model alloys (H, K, L, S, T, V) and for plate material having compositions similar to one another. The data appear to follow two trend lines. Alloys H, K, L, and S are relatively pure, recrystallized ferrite. At zero copper content there appears to be no hardening. Hardening increases with increasing copper, and increases more slowly beyond 0.3 weight percent copper. These results are in good agreement with trends previously observed in microhardness data [*4,6*]. The behavior of the more complex alloys is similar, with several exceptions. The data suggest that there may be some hardening in the absence of copper. Although the level of hardening at low copper is of the order of the uncertainty in the data, a similar response has been observed in microhardness data [*4,6*]. Hardening occurs to a lesser extent at high copper levels in the complex alloys than in the model alloys. This may be due to a combination of microstructure and composition differences. Although we have not yet performed microscopy on all of these aloys, the A533B type alloys generally have a higher dislocation density than the model alloys, and as discussed later higher dislocation densities may delay hardening [*6*]. Moreover, the $\Delta\sigma_y$ for alloy V, which is a model alloy with some cold work, falls close to the line for the complex alloys. Alloy T is a model alloy with a composition similar to A533B (for instance, it contains manganese as well as copper and nickel) and it also exhibits a $\Delta\sigma_y$ similar to the engineering alloys. Finally, it appears that increasing copper content beyond

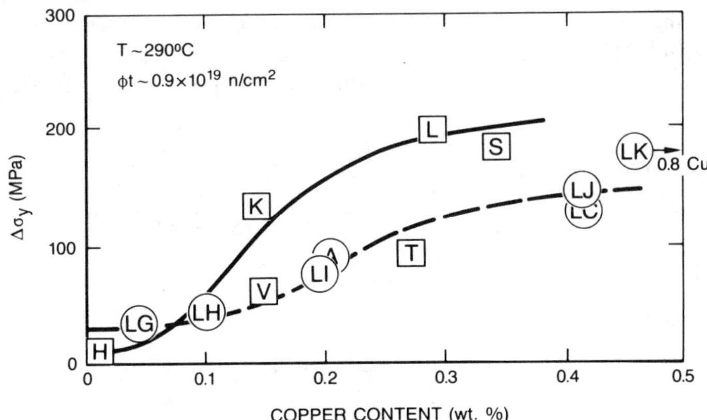

FIG. 5—*Variation of the change in yield strength following irradiation with copper content for simple model alloys and for plate material with similar compositions.*

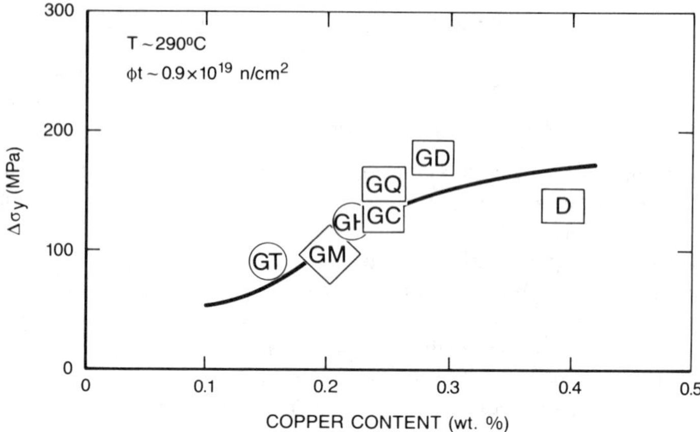

FIG. 6—*Variations of the change in yield strength following irradiation with copper content for weld materials.*

0.4 weight percent (alloy LK) may effect a continued but small increase in $\Delta\sigma_y$. However, it should be noted that at this time, the fraction of the copper in alloy LK that was originally in solution is not yet known.

Yield stress changes in weld material exhibited a similar dependence on copper content. This dependence is shown in Fig. 6. Again microstructural differences among these welds are not well known, so that there is some uncertainty in plotting all these data together. Moreover, alloys GT and GH have lower nickel contents and alloy GM has higher nickel than the other welds. However, as will be discussed next, for the fluence level of these data and for the low copper levels of these welds (GT, GH, GM), the effect of nickel variations may be relatively small. The shape of the curve drawn through these data is based on trends observed and reported elsewhere [*4,6,14*].

The irradiation data shown in Fig. 7 do suggest a systematic increase in $\Delta\sigma_y$ with increasing nickel content. These alloys received identical thermomechanical heat treatments, except for alloy A, which received a similar but not identical heat treatment. This behavior is qualitatively consistent with the variation of chemistry factor observed by Odette et al. [*14*] in their correlation of transition temperature shift to chemistry and fluence for reactor surveillance material. They observed that the chemistry factor increased with increasing nickel, but to a diminishing extent with decreasing copper content. Similar trends in test reactor data have also been reported by Odette and Lucas [*6*]. The data presented here are not inconsistent with a copper dependence of the nickel sensitivity, but such a dependence is not obvious given the uncertainty in the data. Previous work has also

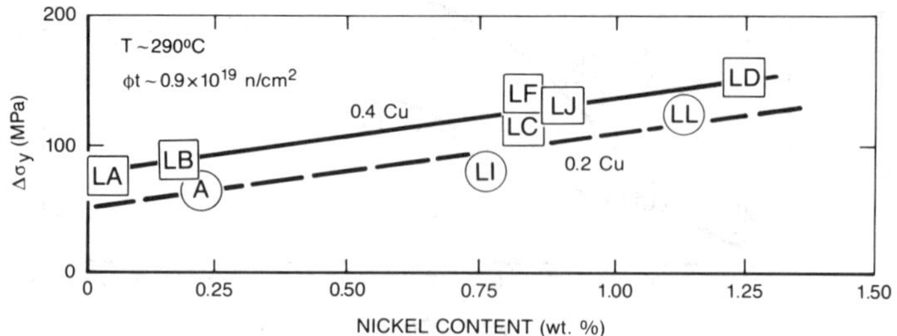

FIG. 7—*Variation of the change in yield strength following irradiation with nickel content for plate materials having similar compositions and thermal histories.*

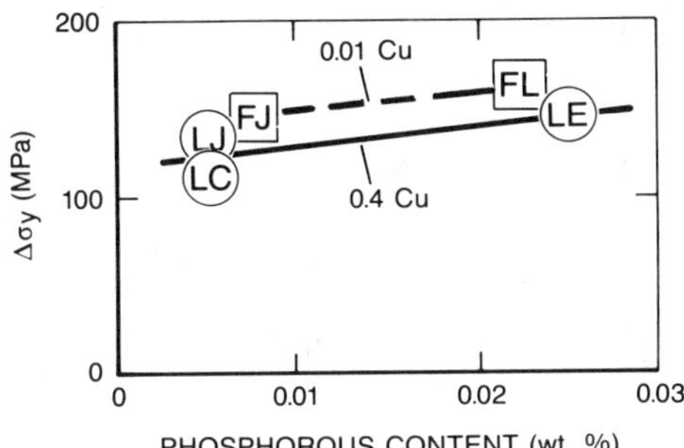

FIG. 8—*Variation of the change in yield strength following irradiation with phosphorus content for two types of plate materials.*

suggested that an independent as well as a synergistic nickel effect increases the transition temperature shift (and thus hardening) [*6*]. However, the independent nickel effect is anticipated to become more significant at higher fluences, and it is expected to be larger for the higher nickel alloys. Again, the data presented here are not inconsistent with an independent nickel effect, but the expected magnitude of the contribution for these data is small relative to the uncertainty of the data. Finally, it should be noted that we do not have data on low copper, high nickel steels.

The yield stress change also appeared to vary slightly with phosphorus. Data which suggest this variation are plotted in Fig. 8. Alloys FJ, FL and alloys LC, LE, LJ have somewhat different fabrication histories and compositions, and hence the absolute values of $\Delta\sigma_y$ between the two alloy sets should not be compared. Moreover, FJ and FL have different cooling conditions after austenitization and different silicon contents, and this may have produced slightly different microstructures, which could account for some of the differences (see below). Nonetheless, within a given alloy set it appears that $\Delta\sigma_y$ increases with increasing phosphorus, although the sensitivity to phosphorus content is relatively small. Other observations suggest that the sensitivity to phosphorus is smaller in alloys with high copper [*6,15*]. However, such a copper dependence of phosphorus sensitivity is not obvious in the data presented in Fig. 8, given the uncertainty in experimental data.

The importance of thermomechanical treatment in the irradiation response is demonstrated in Figs. 9 and 10. Figure 9 illustrates the decrease in $\Delta\sigma_y$ with stress relief annealing for high copper steels containing several levels of nickel. For a given thermal history, $\Delta\sigma_y$ still increases with increasing nickel, but the stress relief treatment substantially reduces the hardening response during irradiation independent of the nickel content. Figure 10 shows the effects of cold working on $\Delta\sigma_y$. The yield strength change for the more complex alloys shows little variation with cold work, while $\Delta\sigma_y$ for the simple model alloy decreases considerably with increasing cold work. The variations of the latter between 10 and 20% cold work (V to V2) is somewhat uncertain, as the thermal control value of σ_y used to calculate $\Delta\sigma_y$ for V2 was estimated from a combination of the unirradiated material data and the known response of V with aging. Nonetheless, it appears that the effect of cold work diminishes after 10%; and the small response of $\Delta\sigma_y$ to cold work for the more complex alloys may be due to their having a relatively high dislocation density prior to cold working.

Discussion

The trends in the yield stress change data reported here are largely consistent with those we have observed previously using microhardness data [*4,6,16*] and with predictions of a physically based model we have developed to describe hardening and embrittlement in neutron-irradiated

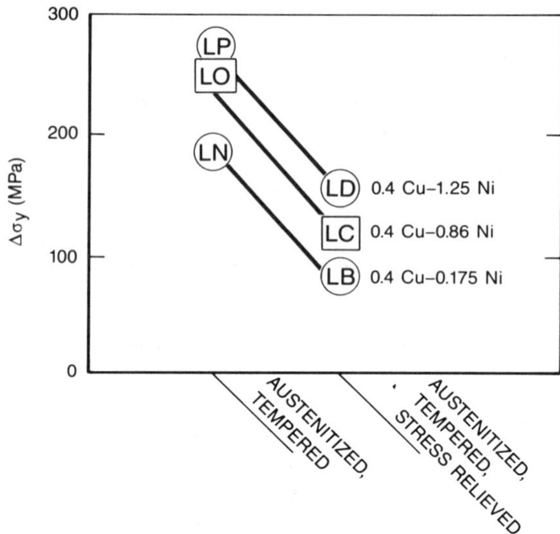

FIG. 9—*Variation of the change in yield strength following irradiation with stress relief for plate materials having different nickel contents.*

pressure vessel steels. The model, which is described in considerable detail elsewhere [*6*], is currently based on the assumption that hardening arises from a radiation-induced defect structure evolution. The predominant defects are postulated to be copper precipitates and radiation-induced extended defects, perhaps microvoids. The precipitation of copper is treated as a radiation-enhanced, diffusion-controlled process. Synergistic nickel effects are incorporated phenomenologically by treating them as if nickel increases the availability of impurity solute (copper and phosphorous) to precipitate during irradiation. Nickel is also believed to enhance the evolution of a stable defect (microvoid) which becomes predominant at high fluences.

The observed increases in $\Delta\sigma_y$ with increasing copper, nickel, and phosphorus content are all consistent with such a model. The decrease in $\Delta\sigma_y$ with the inclusion of a stress relief anneal during processing can also be rationalized in terms of the model, in that some solute copper may be removed by thermally induced copper precipitation during stress relieving. At the stress relieving temperature (625°C), the solubility of copper is only about 0.22 atom percent [*17*] well below the copper content of the test material in Fig. 9. Hence, thermodynamically, copper precipitation is

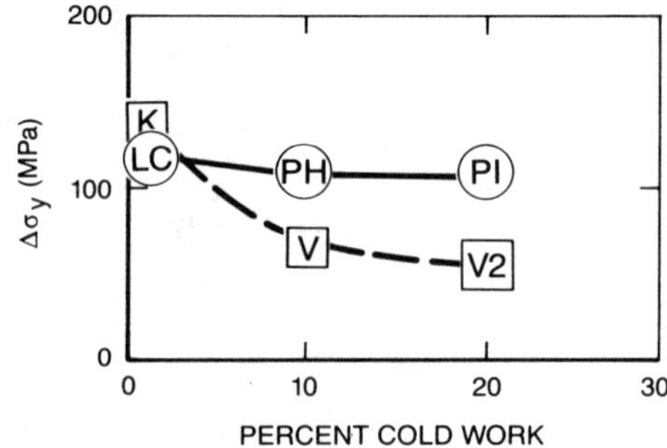

FIG. 10—*Variation of the change in yield strength with cold work for an A533B type alloy and for a simple ferritic model alloy.*

favored. At these stress-relieving conditions and copper levels, the precipitate number density would be expected to be low; and, hence, thermally-induced precipitation hardening would not be large. However, such precipitation would reduce the amount of copper available for subsequent irradiation-induced copper precipitation, which is believed to occur on a much finer scale; hence, the amount of irradiation-induced hardening would be reduced by stress relieving. English [*18*] has reported copper precipitation induced by stress relieving in A533B welds, and he notes that nickel content affects the distribution of copper precipitation during stress relief annealing. The increased hardening with increasing nickel content evident in Fig. 9 suggests that nickel might assist in retaining copper in solution during the pre-irradiation heat treatment. Alternatively, nickel may increase the number density of precipitates during irradiation.

The dependence on cold work can also be rationalized. An increasing dislocation density would increase the rate of point defect loss to sinks during irradiation; in turn, this would reduce the vacancy population and the effective diffusivity of copper, and it would delay hardening by radiation-enhanced copper precipitation. In short, increasing dislocation density would decrease hardening. This appears to be the case for the alloy series K-V-V2 in Fig. 10. With other sinks present, including defect clusters produced by irradiation, the effect of dislocation density would diminish. Moreover, for a large sink density already present, additional cold work may not strongly affect the process. This idea is suggested by the $\Delta\sigma_y$ data for the cold worked, complex alloys LC-PH-PI.

Summary

We have irradiated a large number of alloys of various compositions and thermomechanical treatments in two phases. Irradiation conditions span a range of temperatures (270°C to 325°C), fluxes (0.5 to 5 $\times$ 10^{16} n/m^2-s) and fluences (0.15 to 3 $\times$ 10^{23} n/m^2) in a fission reactor environment. Some limited 14 MeV neutron irradiations have also been performed. To date microhardness and miniature tensile test data have been obtained for most of the irradiation conditions. The observed changes in microhardness appear to correlate well with changes in uniaxial yield stress. The changes in microhardness are similar for fission reactor and 14 MeV neutron irradiations at similar levels of dpa, and, hence, dpa appears to be a reasonable scaling factor for hardening. Similar results have been reported by Heinisch et al. at 290°C [*10*]. Yield stress changes in material irradiated to ~0.9 $\times$ 10^{23} n/m^2 at ~290°C increased with increasing copper, nickel, and phosphorus content, but decreased with increasing cold work or stress relieving during fabrication. Overall, the observed trends are in good agreement with the predictions of physically based models for irradiation hardening and embrittlement.

Acknowledgments

The authors would like to thank P. Hahn, D. Klingensmith, C. Elliott, B. L. Chao, J. Fint, and A. Unruh for their assistance in tensile testing. R. Shogan (Westinghouse) was instrumental in opening our Phase II capsules, and R. Muldar and B. Hostika (University of Virginia) provided assistance in performing the irradiations. The work was supported by the Electric Power Research Institute, Contract No. RP 1021-6.

[*1*] Odette, G. R., Lombrozo, P., Wullaert, R. A., in *Effects of Radiation on Materials: Twelfth International Symposium, ASTM STP 870,* American Society for Testing and Materials, Philadelphia, 1985, p. 840.

[2] Lucas, G. E. and Odette, G. R., in *Proceedings of the International Conference on Nuclear Power Plant Aging, Availability Facts, and Reliability Analysis,* ASM, Metals Park, OH, 1985.

[3] T. Williams, D. Ellis, D. I. Swan, J. McGuire, and S. P. Walley, in *Proceedings of the Second International Symposium on Environmental Degradations of Materials in Nuclear Power Systems,* ANS-TMS-AIME-NACE, Monterey, CA, 1985, p. 393.
[4] Lucas, G. E., Odette, G. R., Lombrozo, P. M., Sheckherd, J. W., in *Effects of Radiation on Materials: Twelfth International Symposium, ASTM STP 870,* American Society for Testing and Materials, Philadelphia, 1985, p. 900.
[5] Mancuso, J., Spitznagel, J., Shogan, R. P., Holland, J., in *Effects of Radiation on Materals: Tenth International Symposium, ASTM STP 725,* American Society for Testing and Materials, Philadelphia, 1981, p. 38.
[6] Odette, G. R. and Lucas, G. E., *Radiation Embrittlement of Nuclear Reactor Pressure Vessel Steels: An International Review (Second Volume), STP 909,* American Society for Testing and Materials, Philadelphia, 1986, pp. 206–241.
[7] Odette, G. R., Lucas, G. E., EPRI Final Report for RP 1021-6, Electric Power Research Institute, Palo Alto, CA, 1987.
[8] Wang, M. T., "Fracture Toughness of Reactor Pressure Vessel Steels," Final Report RP 2180-06, Electric Power Research Institute, 1984.
[9] Greenwood, L., *Journal of Nuclear Materials,* Vol. 103/104, 1981, p. 1433.
[10] Heinisch, H. L., Atkin, S. D., Martinez, C., DAFS Quarterly Report, DOE/ER-0046/25, Department of Energy, Washington, DC, 1986, p. 76.
[11] Tabor, D., *The Hardness of Metals,* Clarendon Press, London, 1951.
[12] Cahoon, J. R., Brighton, W. H., Kutzak, A. R., *Metallurgical Transactions,* Vol. 2, 1971, p. 1979.
[13] Chen, H. R., Masters Thesis, Department of Chemical and Nuclear Engineering, University of California, Santa Barbara, 1985.
[14] Perrin, J. F., Wullaert, R. A., Odette, G. R., Lombrozo, P., EPRI NP-3319, Electric Power Research Institute, Palo Alto, CA, 1982.
[15] Hawthorne, J. R., Water Reactor Information Meeting, Nuclear Regulatory Commission, Washington, DC, October 1983.
[16] Lucas, G. E. and Odette, G. R., *Proceedings of the Second International Symposium on Environmental Degradation of Materials in Nuclear Power Systems,* ANS-TMS-AIME-NACE, Monterey, CA, 1985, p. 345.
[17] Nishigawa, T., Hasebe, M., Ko, M., *Journal of Metallurgy,* Vol. 27, 1979, p. 817.
[18] English, C. A., *Radiation Embrittlement of Nuclear Reactor Pressure Vessel Steels: An International Review (Second Volume), STP 909,* American Society for Testing and Materials, Philadelphia, 1986, pp. 187–205.

Youn H. Jung[1] *and K. Linga Murty*[1]

Effect of Interstitial Impurities on Fracture Characteristics of A533–B Class 1 Pressure Vessel Steel

REFERENCE: Jung, Y. H. and Murty, K. L., "**Effect of Interstitial Impurities on Fracture Characteristics of A533–B Class 1 Pressure Vessel Steel,**" *Influence of Radiation on Material Properties: 13th International Symposium, ASTM STP 956,* F. A. Garner, C. H. Henager, Jr., and N. Igata, Eds., American Society for Testing and Materials, Philadelphia, 1987, pp. 395–407.

ABSTRACT: The effect of interstitial impurities through dynamic strain-aging [DSA] on the mechanical and fracture characteristics of A533–B Class 1 steel was examined using tensile, subsize Charpy three-point bend and single specimen unloading compliance *J*-integral tests. The tensile and fracture properties were studied as a function of the test temperature and applied strain-rate on the as-received pressure vessel [PV] steel material. The effect of neutron radiation exposure on both the tensile and fracture behaviors was investigated. The effect of DSA was noted to result in decreased fracture energies in the upper-shelf region. Neutron irradiation suppressed DSA resulting in improved mechanical and fracture properties of this steel at some temperatures and strain-rates. However, at very high temperatures DSA exhibited embrittlement in addition to that caused by radiation. Temperatures corresponding to fracture energy minima and the applied strain-rates followed an Arrhenius equation with activation energy identifiable with that for diffusion of C and N in steels plus their binding energy with alloying elements such as Mn and V.

KEY WORDS: pressure vessel steels, tensile, fracture, Charpy, fracture toughness, interstitials, radiation, embrittlement, dynamic strain-aging, ductility

Low alloy ferritic steels such as MnMoNi steels are commonly employed in pressure boundary applications and there exists a plethora of experimental results on radiation embrittlement of these steels because of the dire consequences that might result in the event of any structural failures. The decreased upper shelf accompanied by increased ductile-to-brittle-transition-temperature (DBTT) following neutron radiation exposure is known as radiation embrittlement of pressure vessel (PV) steels and these changes are usually monitored from Charpy impact tests on specimens fabricated from base, weld, and heat-affected zone materials. The influence of extrinsic impurities such as copper, nickel, and phosphorus on sensitivity to radiation exposure of nuclear PV steels is well recognized to be included in nuclear regulatory guidelines [*1,2*] and these findings resulted in modern vessel materials and welding methods with reduced amounts of these trace impurities [*3*]. There are very few physically based models elucidating the underlying mechanism of neutron embrittlement of PV steels [*4,5*]. There are still other embrittlement mechanisms known to operate in low alloy ferritic steels, and the influence of interstitial impurities such as C and N on mechanical and fracture properties of unirradiated pressure vessel steels through dynamic strain aging (DSA) has been the subject of recent studies [*6–10*]. It is well known that the simultaneous aging during plastic deformation results in serrated plastic flow at certain temperatures and applied strain rates. The main features in terms of mechanical properties are the significant increase of flow stress and

[1] Research assistant and professor, respectively, North Carolina State University, Raleigh, NC 27695-7909.

TABLE 1—*Chemical composition (in percent by weight) and heat treatment of A533–B steel.*

C	Mn	P	S	Cu	Ni	Mo	Si	V	Al
0.25	1.38	0.01	0.017	0.13	0.61	0.49	0.24	0.04	0.025

Heat Treatment

Austenitized at 1145 K for 1.44×10^4 s,
Tempered at 936 K for 1.44×10^4 s and air cooled to ambient
Stress relieved at 894 K for 7.2×10^4 s,
Furnace cooled to 589 K followed by air cooling to ambient

decreased ductility due to DSA. A conceivable explanation for the DSA has been that the creation of extra dislocations during plastic deformation and the concommitment dragging effect by interstitial impurities on mobile dislocations are responsible for unstable plastic flow in ferritic steels [*11,12*]. When considering nuclear grade pressure vessel steels, concern arose as to whether the brittleness due to interstitial impurities adds to the radiation embrittlement well in the upper-shelf regime. These results would be important in thermal shock, which could occur during a loss-of-coolant accident (LOCA), resulting in a sudden temperature gradient and thereby extensive stresses on the vessel wall. Earlier studies on low carbon mild steel indicate that the concentration of free interstitial impurities is progressively reduced by combining with irradiation induced defects [*13*]. As the concentration of free interstitials decreases, the range of conditions for serrated flow decreases, leading to non-aging steel at high fluences ($\sim 10^{19}$ n/cm^2). Recently, Murty [*9*] pointed out that this suppressive effect of irradiation on DSA at elevated temperatures leads to improved mechanical properties of mild steel following neutron irradiation. This suggests the possible existence of a similar effect in nuclear grade pressure vessel steels. Thus, the main objective of the present study has been to characterize DSA in a typical pressure vessel steel as a function of the test temperature and applied loading rate and to explore the interactive effects between DSA and radiation.

Experimental Procedure

Material, Specimen Preparation, and Irradiation Details

The material used was A533–B Class 1 pressure vessel steel (HSST Plate 04) with the chemical composition shown in Table 1. The plate was austenitized at 1145 K for 1.44×10^4 s, tempered at 936 K for 1.44×10^4 s and air-cooled to ambient temperature, followed by stress relieving at 894 K for 7.2×10^4 s, furnace cooling to 589 K, and air cooling to ambient temperature. The as-received block (330 mm × 406 mm × 64 mm) was taken from the ¼ thickness position of the plate. The reported mechanical properties of the as-received material are given in Table 2.

TABLE 2—*Mechanical properties of A533–B steel.*

Temperature, K	σ_y MPa	σ_{uts}, MPa	ϵ_t, %	C_v, ft · lb
75	455.1	592.6	18.0	49.0
473	428.2	584.6	16.5	. . .
623	389.0	567.0	19.0	. . .

NOTE: σ_y = yield strength, σ_{uts} = tensile strength, ϵ_t = total elongation, and C_v = Charpy energy. 1 ft · lb = 1.355 J.

Additional tests were performed as follows. The tensile and three-point bend specimens were fabricated according to the design shown in Fig. 1, and the orientation of the specimens with respect to the plate is depicted in Fig. 1*a*. The surfaces of the tensile specimens were polished to 6 microns grit after machining followed by stress relieving at 873 K in vacuum for 1.44×10^4 s. Subsize Charpy slow bend and precracked Charpy-size *J*-test specimens were cut out so that the crack plane was parallel to the rolling direction of the plate. A sharpened high speed milling saw was used to produce the standard V-notch root radius of 0.08 mm or less according to ASTM standard E319-78. During the precracking stage, stress intensity in the final 5% of the precrack was estimated to be $K_f/E = 0.043\ \text{mm}^{1/2}$, where K_f is the maximum stress intensity factor for fatigue precracking and E is Young's modulus. Moreover, 15% side-grooving of Charpy V-notch (Fig. 1) profile was imposed on precracked *J*-test specimens to improve the straightness of the crack front.

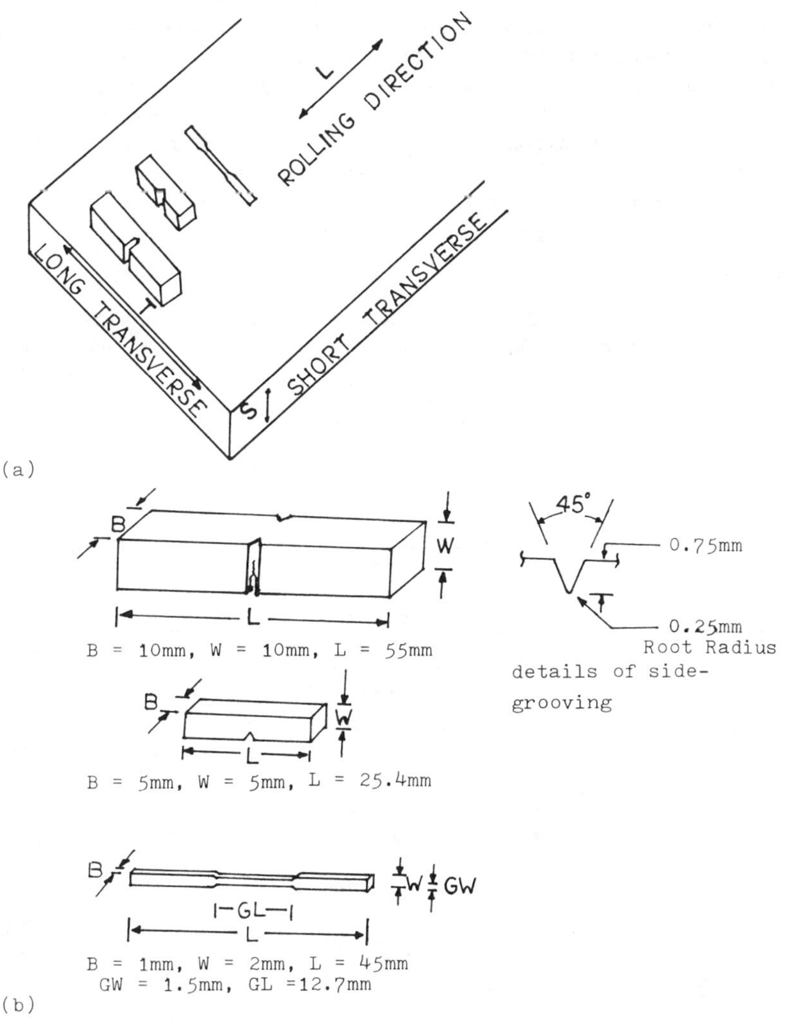

FIG. 1—*Design of tensile and fracture specimens and their orientations with respect to the plate.*

A limited number of tensile, subsize Charpy and Charpy-size *J*-test specimens were irradiated in the PULSTAR experimental reactor at North Carolina State University to a nominal fast neutron (E_n>1 MeV) dose of 5 × 10^{17} n/cm^2. Measured irradiation temperature was 383±5 K. The irradiation dose was estimated from the measured activity of a nickel wire-monitor inserted into a cadmium covered sample holder. Nickel wires were used to follow the nickel-cobalt activation reaction, ^{58}Ni(n,p) ^{58}Co from which the fast flux is evaluated; details may be found in Ref *19*.

Test Setup

All tests were conducted under stroke control mode on a closed loop, INSTRON Model 1350, hydraulic testing machine with a 9 kN load cell. Temperatures were controlled using a three zone control furnace, ATS Model-2961. For single specimen unloading compliance *J*-test, a tension-compression jig with an automatic notch alignment three-point bend fixture was designed. A linear variable differential transformer (LVDT) with ±2.5 mm full range was directly attached to the loading pin to increase the accuracy of unloading compliance measurement and to effectively reduce the machine compliance effect. A schematic of the *J*-test rig is given in Fig. 2. All of the load versus load-line displacement data were monitored using an on-line APPLE-II microcomputer. To increase the resolution of the unloading compliance measurement, a variable amplifier (DAYTRONIC Model-3163) was connected to an LVDT conditioner (DATRONICS Model-3130). The present tension-compression jig significantly reduced the misalignment problems encountered in the usual direct compression type three-point bend tests.

Mechanical Test Details

For each unirradiated/irradiated condition, tensile tests were performed over a sufficient temperature and strain rate range to investigate the strain rate effect. The temperature range covered

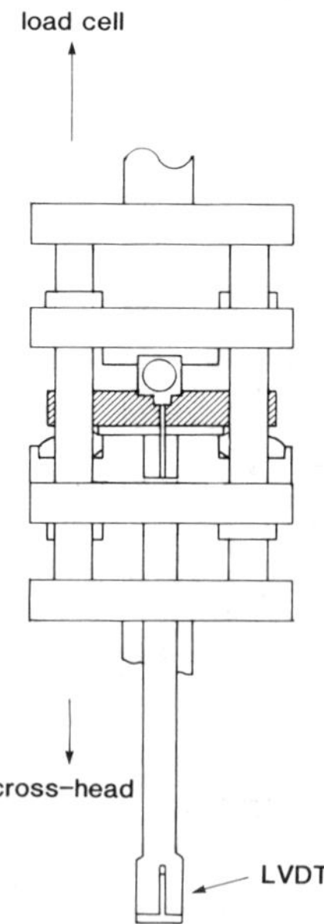

FIG. 2—*Three-point bend test rig.*

was from ambient to 623 K, while nominal strain rates ranged from 5×10^{-5}/s to 1.9×10^{-3}/s.

Three-point slow bend tests using subsize Charpy specimens were conducted over temperatures that ranged from room temperature to ~700 K and over loading rates that ranged from 2×10^{-3} mm/s to 5×10^{-1} mm/s. The compression jig shown in Fig. 2 was used, and load versus displacement data were collected in digital form using the APPLE-II microcomputer for subsequent data analysis using the VAX 11/750.

The single specimen unloading compliance *J*-tests on the unirradiated/irradiated Charpy-size specimens were performed using a specially developed test system and computer program. Prior to each *J*-test, specimen alignment and system reliability were checked by repeated partial unloading compliance measurements applying the load well below the elastic limit. After stabilizing the prescribed test temperature, about 20 unloading steps were used to measure the dynamic crack extension. In addition, heat tinting and the seven point averaging method for actual crack measurement were performed after brittle fracture of the tested specimens at liquid nitrogen temperature.

The value of *J*-integral up to each partial unloading point was calculated using

$$J = \frac{2A}{B_n(W - a_0)} \quad (1)$$

where A is the area under load versus the load-line displacement curve up to each partial unloading step, B_n the net thickness of the side-grooved specimen, W the width, and a_0 the initial crack length. The plane strain elastic compliance function [*14*] was employed to obtain the crack extension from measured unloading compliance.

Extraneous compliance produced from the experimental configuration is inevitable in any mechanical testing. In the present experimental configuration, it was comprised of elastic displacement in the test fixture, supporting roller indentation on the specimen, wrapping effect on the loading pin and friction effects at contact areas. To characterize the extraneous displacement, a series of blunt notch specimens with different notch lengths were tested. The functional form of the extraneous displacement, V_{ex}, was evaluated from [*14*]

$$V_{ex} = \frac{P}{C_1} + \frac{V}{C_2} \quad (2)$$

where P is applied load, V the load-line displacement at each partial unloading step, and C_1, C_2 are the constants which were experimentally determined.

The crack resistance (*J-R*) curve was constructed using the crack extension, Δa, from the corrected compliance and resulting *J*-integral. For the interpretation of the crack resistance curve, the slope of the blunting line was assumed to be proportional to four times the flow stress. The exclusion line was determined from the *J*-controlled crack growth criterion for precracked Charpy specimens [*15*]

$$\Delta a > 0.06(W - a_o) \quad (3)$$

For the employed nominal initial crack length of 5 mm, the exclusion limit was determined as 0.3 mm crack extension. The data points between the blunting and exclusion lines were regarded as effective data points for *J*-controlled crack growth of precracked Charpy-size specimens. The crack initiation fracture toughness J_q was determined as the value of the *J*-integral at the intersection point between the blunting line and the linearly regressed tearing line through the effective data points. A typical *J-R* curve is shown in Fig. 3.

Results and Discussion

A series of engineering stress versus engineering strain curves for unirradiated tensile specimens are included in Fig. 4*a* and *b*. It is clear that the critical temperatures for the appearance and

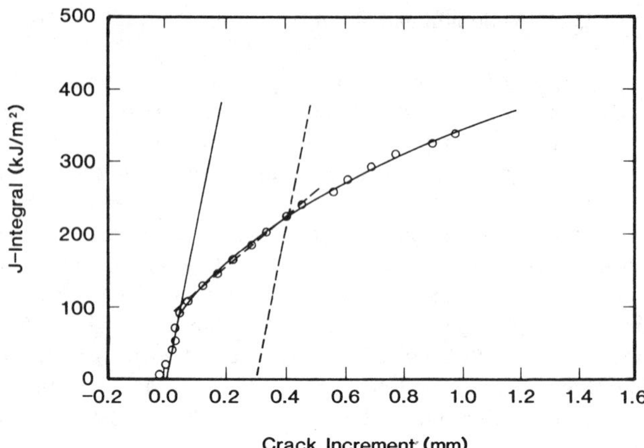

FIG. 3—*Crack resistance curve obtained at 573 K at a loading rate of 0.15 mm/min.*

disappearance of serrated plastic flow are dependent on the applied strain rate. This is a typical indication of dynamic strain aging, which is believed to be triggered by the interaction between solute atoms and moving dislocations [*16*]. These results are similar to those reported by Murty and Hall [*13*] on mild steel, with the exception that the load drops here are quite small. In addition, the magnitudes of Luder's strains are small, and essentially no Luder's band propagation occurred at elevated temperatures. These observations are attributed to the alloying elements in the PV steel

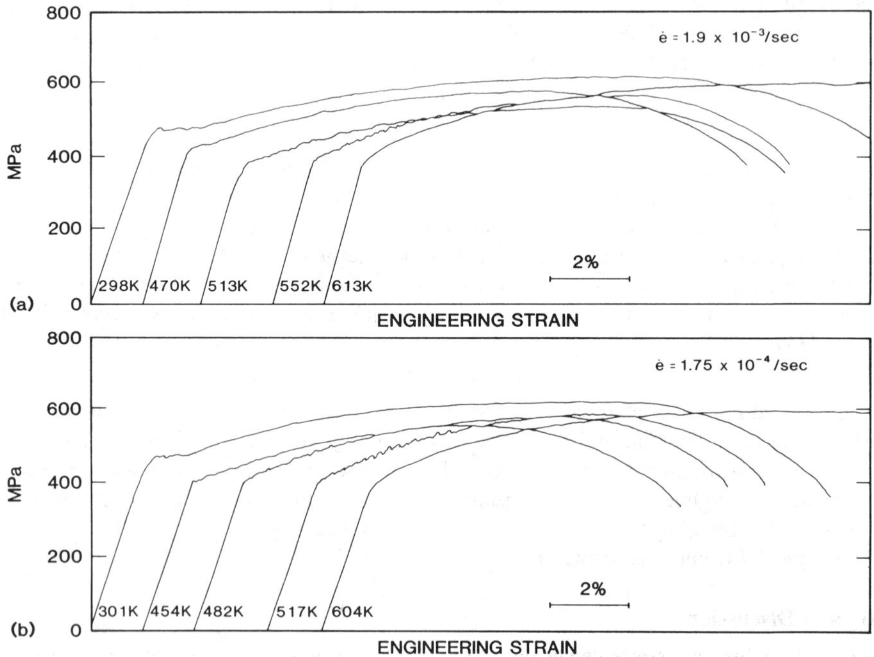

FIG. 4—*Effect of temperature on stress-strain curves at 1.9* × *10^{-3}/s* (a) *and at 1.75* × *10^{-4}/s* (b).

which effectively decrease the dislocation-interstitial interaction [16]. The critical temperatures for the onset and disappearance of serrations followed an Arrhenius relation as shown in Fig. 5

$$\dot{\epsilon} = \epsilon_0 \exp(-Q/RT) \tag{4}$$

where $\dot{\epsilon}$ is the applied strain rate, ϵ_0 the pre-exponent, Q the apparent activation energy, R the universal gas constant and T the temperature in K. Apparent activation energies of 75 kJ/mol for the onset and 149.4 kJ/mol for the disappearance of serrations were obtained. The estimated activation energy for the onset of serrations is, within experimental error, close to that for the diffusion of C and N in α-iron (75 ~ 83 kJ/mol) [12]. The activation energy for the disappearance compares with the earlier results on mild steels by Keh et al. [16] of 126 kJ/mol and Roberts and Owen [17] of 163 kJ/mol. Although Keh et al., associated this activation energy with that for diffusion of interstitial impurities plus the binding energy of interstitial atoms to dislocations (as in the Koster internal friction peak), Roberts and Owen [17] dispute such a contention on the basis of the variation of the activation energy of the Koster peak with carbon concentration. It is clear, however, that both the critical temperatures for the appearance and disappearance of serrated flow depend on the concentration of the interstitial impurities [13].

The results of the three-point slow bend tests on subsize Charpy specimens are shown in Fig. 6 as normalized energy to fracture versus test temperature at two different applied loading rates. The noticeable feature in this figure is that the absorbed energy abruptly dropped at a certain temperature, attained a minimum value and recovered at still higher temperatures. In addition, the energy minima, which are believed to be coincident with the temperature of maximum dynamic strain aging effect, occurred at higher temperatures for higher loading rates as expected. These results clearly demonstrate the undesirable effect of DSA on fracture energy in the upper-shelf region. It should be noted that these effects are not usually discernible in the standard Charpy impact tests where extremely high loading rates are used and thus these drops in the shelf energy are expected to occur at very high temperatures.

As described earlier, single specimen unloading compliance J-integal tests were performed using precracked Charpy size specimens loaded in the three-point bend mode. Figure 7 shows the J-integral versus crack extension Δa at different test temperatures at the nominal test speed of 2.5 $\times$ 10^{-4} mm/s. All the tests were run to a fixed total load-line displacement of about 2 mm.

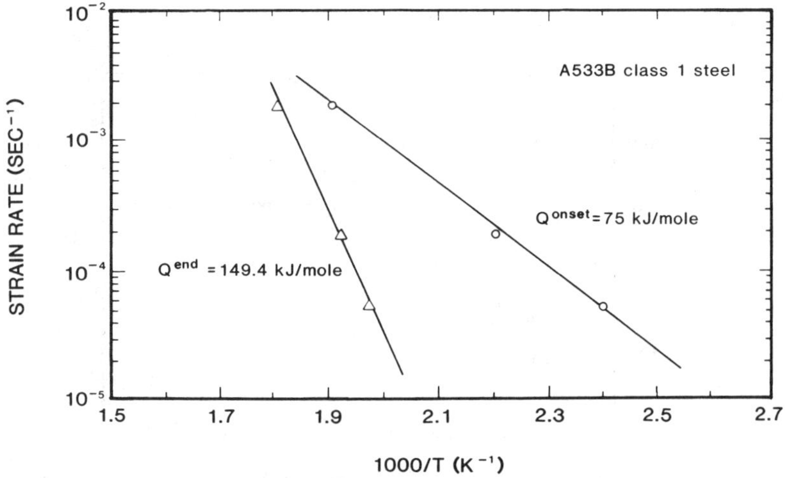

FIG. 5—*Arrhenius plot of strain-rate versus critical temperatures for the appearance and disappearance of serrations.*

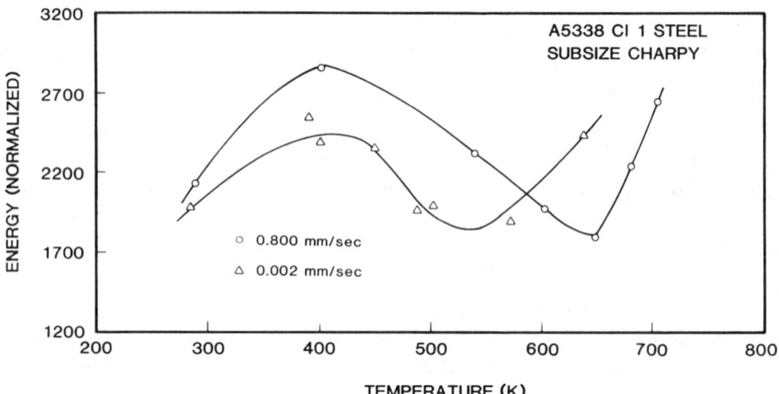

FIG. 6—*Temperature and strain-rate dependences of Charpy fracture energy depicting the effect of dynamic strain aging at elevated temperatures.*

Considering minor effects of the specimens' geometric deviation, the effect of temperature on the resistance curve is apparent. It is clearly noted that as temperature increases from ambient the resistance curve exhibited lower *J*-integral values until a critical temperature is reached (473 K in Fig. 7) and then exhibited higher *J* values with increased temperature. The loading rate effect on crack resistance at a fixed test temperature is depicted in Fig. 8, however, the effect of strain-rate is different from temperatures below the critical temperature at which the *J* values attained a minimum. At lower temperatures, below the *J* minimum, the *J* value increased with increasing strain-rate. Although the *J*-integral cannot have any stringent significance in crack growth after crack initiation on the basis of elastic-plastic fracture mechanics, the above observations constitute sufficient evidence of the effects of loading rate and temperature on the upper-shelf fracture behavior of nuclear grade pressure vessel steels such as A533–B Class 1. The resulting variation of crack initiation fracture toughness (J_q) with temperatue and loading rate is depicted in Fig. 9. It is clear from these results that the temperature at which the J_q minimum occurs increases with applied loading rate as in the tensile (Fig. 5) and subsize Charpy (Fig. 6) test results, implying

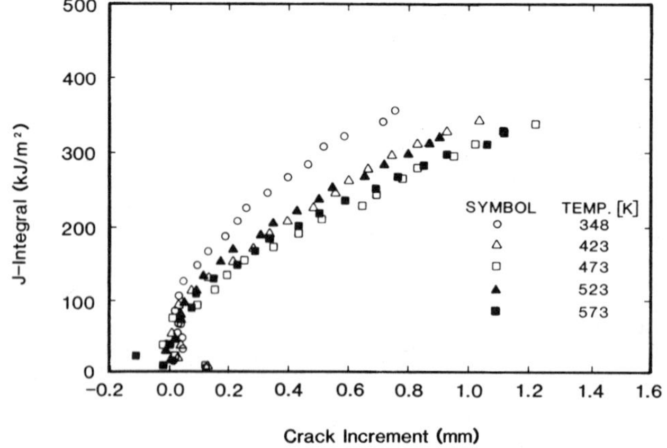

FIG. 7—*Crack resistance curves* (J *versus* Δa) *at various test temperatures.*

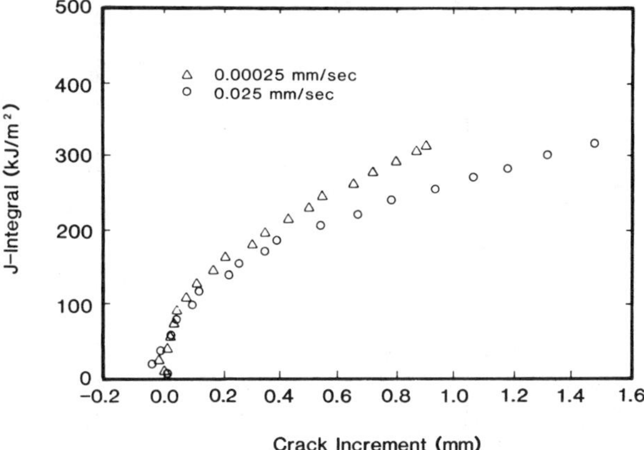

FIG. 8—*Effect of loading rate on crack resistance curve at 573 K.*

that the drop in J_q at elevated temperatures is also due to the interaction of the interstitial impurities with dislocations at the crack front.

Regardless of the differences in the size and notch geometries of the subsize Charpy slow bend and precracked Charpy-size *J*-test specimens, the role of critical strain prevailing in upper-shelf ductile fracture behavior is expected to be the same and thus the kinetics of deformation and fracture would be identical. Evidence for this is obtained through a composite Arrhenius plot of ln(strain rate) versus $1/T$, where the strain-rates were evaluated as loading rates normalized to the length of the bending arm [*18*] and *T* is the critical temperature in K at which toughness minima are noted in the subsize and Charpy size specimens. It is clear from Fig. 10 that the slow bend and J_q results follow the same straight line with a slope corresponding to an activation energy of 129.8 kJ/mol, which is intermediate between the activation energies for the appearance and the disappearance of serrations. This energy value corresponds to the maximum serration height in tensile tests [*16*], and thus the fracture energy minima in three-point bend and *J* tests are regarded as arising from the interaction of the interstitial impurities with dislocations in a well developed plastic zone at the crack front. The higher activation energy of 129.8 kJ/mol is believed to be due

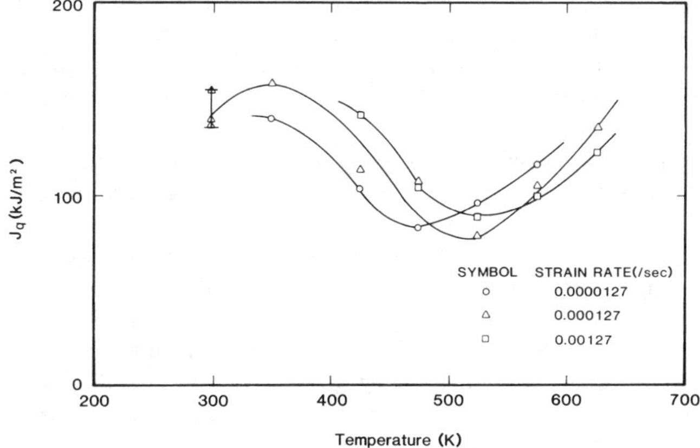

FIG. 9—*Effect of test temperature and strain-rate on crack initiation fracture toughness* (Jq).

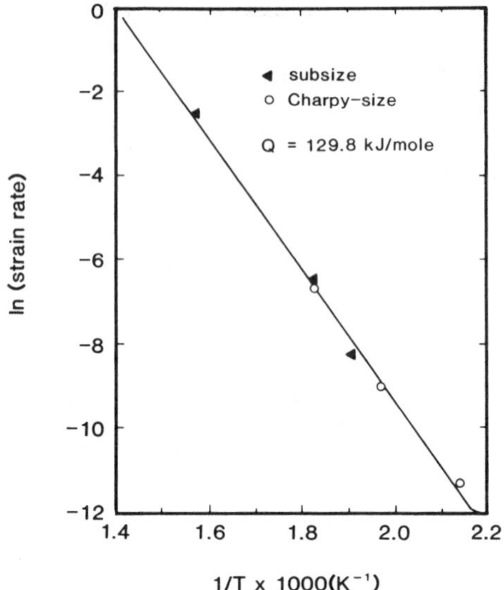

FIG. 10—*Arrhenius plot of strain-rate versus critical temperature for fracture energy minima.*

to the addition of the binding energy of C with Mn and V and to the diffusion energy for C and N in Fe [*16*].

Effect of Neutron Irradiation

As indicated earlier, a limited number of tensile and fracture tests were made on irradiated A533–B steel. The effect of neutron irradiation (5×10^{17} n/cm^2, >1 MeV) on the yield and tensile strengths is depicted in Fig. 11*a,* which indicate the expected radiation hardening at low temperatures. The intervening effect of dynamic strain-aging at elevated temperatures resulted in relative insensitivity of the mechanical strength to radiation exposure. Fig. 11*b* illustrates the effect of irradiation on ductility, and it is interesting to note that the embrittlement due to DSA in the unirradiated material between 400 K and 500 K is eliminated by neutron irradiation, leading to increased ductility at these temperatures following neutron exposure. This synergistic effect of radiation-induced defects and interstitial impurities (DSA) is very similar to that reported earlier by Murty [*18*] in a low carbon mild steel. It should be noted, however, that the interaction of radiation-induced defects and interstitial impurity elements results in a reduced concentration of these interstitials and thus postpones the occurrence of DSA in irradiated materials to higher temperatures [*1*]. This is believed to be the reason for the drop in the ductility of the irradiated material at higher temperatures [≥500 K].

The effect of neutron irradiation on the fracture toughness is depicted in Fig. 12 where J_q is plotted versus test temperature. It is clear that the energy minimum is shifted to a higher temperature following irradiation. In addition, the suppression of DSA following irradiation resulted in slightly increased J values at temperatures below the critical value. At temperatures above the energy minimum, irradiated material exhibited lower toughness values indicative of the radiation embrittlement expected in these steels. Further testing is underway to characterize the effect of strain-rate on the fracture behavior of the irradiated steels.

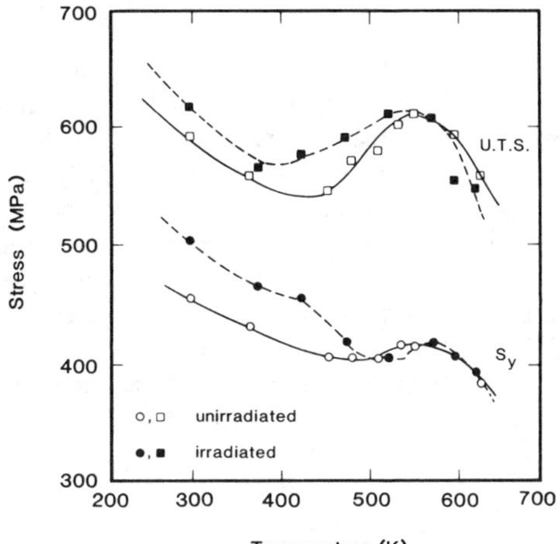

FIG. 11*a*—*Effect of neutron irradiation on temperature dependence of yield and tensile strengths.*

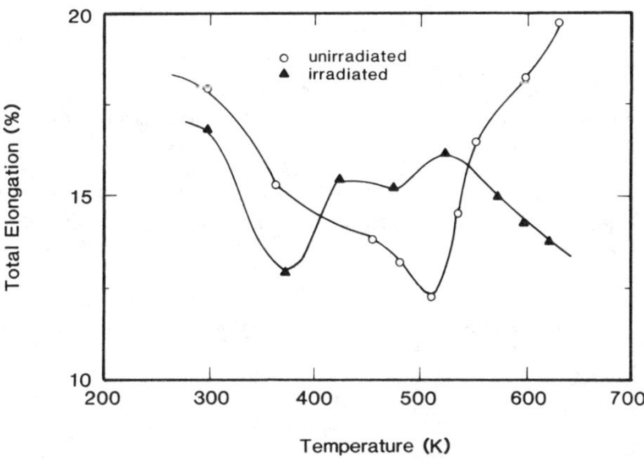

FIG. 11*b*—*Effect of neutron irradiation on temperature dependence of total elongation to fracture.*

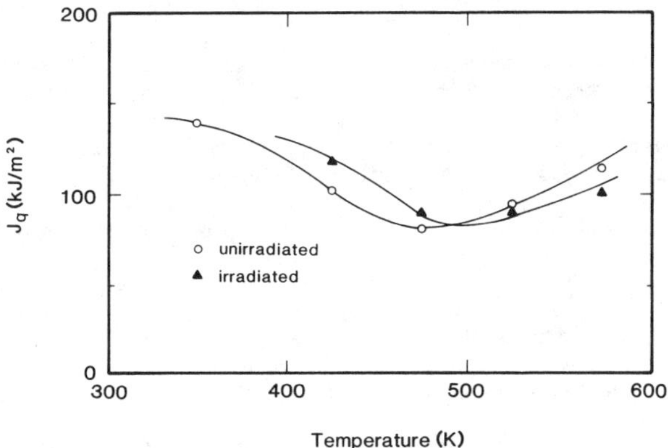

FIG. 12—*Effect of neutron irradiation on temperature dependence of crack initiation fracture toughness.*

Conclusions

Tensile tests on the A533–B Class 1 steel clearly depicted the occurrence of dynamic strain aging, and the activation energies for the appearance and disappearance of serrated flow were consistent with a model of locking of the dislocations by C and N as the primary mechanism. Three-point bend tests on subsize Charpy and *J*-integral tests on Charpy specimens as a function of the test temperature and loading rate clearly exhibited the effect of DSA on the fracture toughness in the upper-shelf regime. Synergistic effects of neutron irradiation and DSA resulted in improved mechanical and fracture characteristics at some temperatures and strain-rates where DSA is suppressed following neutron irradiation. Present results clearly revealed the effect of DSA as a distinct reduction of the upper-shelf energy in addition to radiation embrittlement. These effects, however, are functions of both the test temperature and loading rate.

Acknowledgments

We wish to acknowledge the financial support of Carolina Power and Light Company and Virginia Electric and Power Company. Thanks are due to Mr. Dirk J. Oh for assistance in some of the experimental work and for various discussions.

References

[*1*] Murty, K. L., *Journal of Metals,* October 1985, pp. 34–49.

[*2*] Steele, L. E., "Review and Analysis of Reactor Vessel Surveillance Programs," in *Status of USA Nuclear Reactor Pressure Vessel Surveillance for Radiation Effects, ASTM STP 784,* American Society for Testing and Materials, Philadelphia, 1983, p. 227.

[*3*] Hawthorne, J. R., "Significance of Nickel and Copper Content to Radiation Sensitivity and Post-Irradiation Heat Treatment Recovery of Reactor Vessel Steels," in *Effects of Radiation on Materials: Eleventh Conference, ASTM STP 782,* American Society for Testing and Materials, Philadelphia, 1982, p. 375.

[*4*] Spitznagel, J. A. and Venskytis, F. J., "The Annealing of Copper-Vacancy Aggregates in Neutron-Irradiated Ferritic Pressure Vessel Steels," *Transactions of the American Nuclear Society,* Vol. 12, 1975, p. 161.

[*5*] Odette G. R., *Scripta Metallurgica,* Vol. 17, 1983, p. 1183.

[*6*] Chakravartty, J. K., Wadekar, S. L., Sinha, T. K., and Asundi, M. K., "Dynamic Strain-Aging of A203D Nuclear Structural Steel," *Journal of Nuclear Materials,* Vol. 119, 1983, p. 51.

[*7*] Little, E. A., "Strain-Aging and Neutron Scattering Studies in Irradiated Pressure Vessel Steel," *Twelfth International Symposium on Materials,* American Society for Testing and Materials, Philadelphia, 1984.

[*8*] Ostensson, B., "The Fracture Toughness of Pressure Vessel Steel at Elevated Temperatures," in *Reliability Problems of Reactor Pressure Components,* International Atomic Energy Association, Vol. 1, 1981, p. 303.

[*9*] Murty, K. L., "Interaction of Interstitial Impurities with Radiation-Induced Defects Leading to Improved Elevated Temperature Mechanical Properties of Mild Steel," *Nuclear Technology,* Vol. 67, 1984, p. 124.

[*10*] Rao, S., Paper AE-MA-376, Aktiebolaget Atomenergi, Studswike, Sweden, 1979.

[*11*] Brindley, B. J. and Barnby, J. T., "Dynamic Strain Aging in Mild Steel," *Acta Metallurgica,* Vol. 14, 1966, pp. 1965–80.

[*12*] Blakemore, J. S. and Hall, E. O., "Blue-Brittle Behavior of Mild Steel," *Journal of the Iron and Steel Institute,* Vol. 204, 1966, pp. 817–820.

[*13*] Murty, K. L. and Hall, E. O., "Dynamic Strain-Aging and Neutron Irradiation in Mild Steel," in *Irradiation Effects on the Microstructure and Properties of Metals, ASTM STP 611,* American Society for Testing and Materials, Philadlephia, 1976, pp. 53–71.

[*14*] Neale, B. K., *The Determination of* J_q *the Initiation Elastic-Plastic Toughness Parameter Using Precracked Charpy Specimens* TPRD/B/0012/82, Berkley Nuclear Laboratories, U.K.: Central Electric Generating Board, October, 1979.

[*15*] Hutchinson, J. W., 1979 CSNI Specialists Meeting on Plastic Instability, Washington University, St. Louis, Missouri, Sept. 25–27, 16–28.

[16] Keh, A. S., Nakada, Y., and Leslie, W. C., "Dynamic Strain Aging in Iron and Steels," in *Dislocation Dynamics,* McGraw-Hill, New York, 1968.
[17] Roberts, M. J. and Owen, W. S., *Metallurgica Transactions,* Vol. 1, 1970, p. 3202.
[18] Server, W. L., *Journal of Engineering Materials,* Vol. 100, 1978, p. 183.
[19] Jung, Y. H., "Effect of Dynamic Strain Aging on Fracture Characteristics of A533–B Class 1 Nuclear Pressure Vessel Steel," Doctoral Thesis, North Carolina State University, Raleigh, 1986.

K. S. Sivaramakrishnan,[1] Subrata Chatterjee,[1] S. Anantharaman,[1] U. K. Viswanathan,[1] K. S. Balakrishnan,[1] and Pradip R. Roy[1]

Irradiation Embrittlement of Advanced Pressure Vessel Steels

REFERENCE: Sivaramakrishnam, K. S., Chatterjee, S., Anantharaman, S., Viswanathan, U. K., Balakrishnam, K. S., and Roy, P. R., **"Irradiation Embrittlement of Advanced Pressure Vessel Steels,"** *Influence of Radiation on Material Properties: 13th International Symposium (Part II), ASTM STP 956,* F. A. Garner, C. H. Henager, Jr., and N. Igata, Eds., American Society for Testing and Materials, Philadelphia, 1987, pp. 408–419.

ABSTRACT: Seven commercial steels in the form of plate, forging, or weldments supplied by Japan, France, and the Federal Republic of Germany to the International Atomic Energy Agency (IAEA) Co-ordinated Research Programme on "Analysis of the Behaviour of Advanced Reactor Pressure Vessel Steels under Neutron Irradiation" were irradiated at 290°C to a nominal fast fluence of 1 $\times$ 10^{19} n/cm^2 (>1 MeV) to evaluate the embrittlement sensitivity of these steels under neutron irradiation. These steels were basically of ASTM Types A533 and A508 (Unified Numbering System [UNI] K12521 and K13502) and their weldments, which had extra low copper and phosphorous contents. The copper and phosphorous contents in these individual steels varied from 0.01 to 0.07 wt% and 0.007 to 0.015 wt%, respectively. The nickel contents in these steels varied from 0.56 to 0.93 wt%.

The scope of this study was mainly to evaluate the changes in the notch ductility, tensile properties, and hardness as caused by fast neutron irradiation. This paper reports the results obtained on these advanced pressure vessel steels in the form of base metal, weldments, and weld heat affected zones. The results indicated that these advanced steels were practically insensitive to fast neutron induced embrittlement. The relative embrittlement of these steels were further analyzed to examine the contribution of nickel in presence of copper in causing embrittlement. The impact test results indicated that threshold amounts of 0.04-wt% copper and 0.7-wt% nickel would perhaps be required to cause relatively enhanced embrittlement in these steels, confirming thereby that nickel in presence of copper can cause enhanced fast neutron induced embrittlement.

KEY WORDS: pressure vessel steels, irradiation, irradiation embrittlement, embrittlement sensitivity, notch ductility, tensile properties, hardness, nickel-copper effect, neutron fluence

The International Working Group on Reliability of Reactor Pressure Components (IWG-RRPC) sponsored by the International Atomic Energy Agency (IAEA) launched a Co-ordinated Research Programme to investigate the behavior of advanced pressure vessel steels under neutron irradiation [*1*]. The international program was focused upon the determination of irradiation response of these materials irradiated at 290°C by comparing notch ductility, fracture toughness, tensile properties, and hardness determined both before and after irradiation. In the Indian contribution towards this program, the scope of study was mainly to evaluate the neutron induced change in notch ductility in all these materials and tensile and hardness property changes in some of these materials.

The paper reports the results obtained from these advanced pressure vessel materials in the form of plates, forgings, and weldments and heat affected zones (HAZ).

[1] Head of post irradiation examination section, scientific officer, scientific officer, scientific officer, scientific officer, and associate director of metallurgy group, respectively, Radiometallurgy Division, Bhabha Atomic Research Centre, Trombay, Bombay 400085, India.

TABLE 1—*Nominal chemical composition of the materials (wt%).*

Material	C	Si	Mn	P	S	Ni	Cr	Mo	Cu
JP	0.18	0.22	1.48	0.007	0.007	0.66	0.20	0.57	0.01
FP	0.23	0.22	1.47	0.008	0.003	0.65	0.04	0.52	0.03
JF	0.18	0.26	1.33	0.007	0.005	0.76	0.11	0.50	0.04
FF	0.15	0.25	1.37	0.009	0.008	0.69	0.24	0.47	0.07
JW	0.08	0.36	1.24	0.008	0.007	0.71	0.07	0.48	0.04
JH	0.18	0.22	1.48	0.007	0.007	0.66	0.20	0.57	0.01
GW	0.08	0.16	1.48	0.011	0.009	0.93	0.02	0.62	0.05
FW	0.06	0.54	1.37	0.15	0.011	0.56	0.045	0.52	0.05

Materials

Seven commercial materials were evaluated in this study:

(1) Japanese A533B Class 1 plate (JP),
(2) Japanese A508 Class 3 forging (JF),
(3) Japanese S/A weld (JW),
(4) French A533B Class 1 plate (FP),
(5) French A508 Class 3 forging (FF),
(6) French S/A weld (FW), and
(7) German S/A weld (GW).

The chemical composition and heat treatment details of these materials are given in Tables 1 and 2.

The specimens were prepared according to the recommendation of the standard research program given in IWG-RRPC-78/1 [*1*]. The standard Charpy V-notch impact specimens and subsized round tension specimens were fabricated from quarter thickness locations. The impact specimens had their axis parallel to the surface and perpendicular to the rolling direction with the notch perpendicular to the surface. The tension specimens had their axis parallel to the surface and perpendicular to the rolling direction. The weld deposit and heat affected zone specimens were prepared as per ASTM Practice for Conducting Surveillance Tests for Light Water-Cooled Nuclear Power Reactor Vessels (E 185) requirements.

Irradiation

Irradiations were carried out at 290°C in CIRCUS 40-MW heavy water moderated research reactor to a nominal fluence of 1×10^{19} n/cm^2 (>1 MeV). This corresponded to a fluence of 1.5

TABLE 2—*Heat treatment of the materials.*

Material	Heat Treatment
JP	A: 880°C-8 h, WQ, T: 660°C-6h, a/c, SPWHT: 620°C-26 h, f/c.
FP	A: 900°C-0.7 h, WQ, T: 640°C-10 h a/c, SPWHT: 610°C-24 h, f/c
JF	A: 870/900°C-6.5 h, WQ, T: 635/645°C-7.5 h, a/c, SPWHT
FF	A: 865/88°C-3 h, WQ, T: 630/650°C-5.5 h, a/c, SPWHT: 550°C-35 h/615°C-16 h, f/c.
JW	PHWT: 615 ± 20°C-26 h
JH	. . .
GW	PHWT: 610°C-20 h
FW	PHWT: 617 ± 3°C-8 h

NOTE: A is austenitize; WQ is water quench; T is temper; SPWHT is simulated post weld heat treatment; a/c is air cool; f/c is furnace cool.

$\times 10^{19}$ n/cm^2 (>0.1 MeV). Two impact specimens or three tension specimens were accommodated in one capsule. Eight such capsules were loaded into the reactor as a batch for irradiation. Irradiation capsules occupying the top, middle, and bottom locations contained iron, nickel, and copper flux monitors. The spread in the fluence values along the length of the specimens in the capsules was less than 10%. In order to evaluate the relative neutron embrittlement of these steels, the respective steels were irradiated in single batches at the same irradiation locations in the reactor. The irradiation details are available elsewhere [2]. Table 3 tabulates the number of irradiations and the specimen break down for each irradiation.

Testing of the Materials

Tension tests were carried out on a screw driven Instron universal testing machine modified for in-cell operation. Yield strength, ultimate tensile strength, and elongation were determined on 4 mm diameter and 20 mm gage length specimens.

A pendulum type impact tester of 270*J* total capacity and instrumented impact tester of 360*J* capacity were used to conduct impact tests on standard Charpy-V notch specimens. Both these machines were calibrated as per ASTM Methods for Notched Bar Impact Testing of Metallic Materials (E 23).

Hardness tests were carried out on some arbitrarily chosen broken half of impact specimens using an available microhardness tester with the remotized Leitz MM 5RT metallograph.

Optical metallography was carried out using a remotized Leitz MM 5 RT metallograph. Electron microprobe analyzer was used to check for composition variation in the specimens, if any.

Results

The results of tension tests on unirradiated and irradiated Japanese plate (JP), French plate (FP), and French forging (FF) materials are given in Table 4.

The Charpy V-notch impact test results on unirradiated and irradiated Japanese plate (JP), Japanese forging (JF), Japanese weld (JW), Japanese heat affected zone (JH), French plate (FP), French forging (FF), French weld (FW) and German weld (GW) are shown in Table 5 and Figs. 1 through 8.

Temperature corresponding to 40*J* of absorbed energy was used for the determination of transition temperature. These figures also include the temperature corresponding to 0.9 mm of lateral expansion. Among the materials tested, the unirradiated German weld was found to have the lowest transition temperature, and the unirradiated French plate was found to have highest transition temperature. Irradiated Japanese HAZ was found to have the lowest transition temperature,

TABLE 3—*Irradiation schedule for the program.*

Irradiation Number	Number of Capsules for Irradiation	Material	Number of Specimens	
			Impact	Tension
1	8	JP	14	3
2	8	FP	14	3
3	7	JF	14	. . .
4	8	FF	14	3
5	7	JW	14	. . .
6	7	JH	14	. . .
7	7	FW	14	. . .
8	8	GW	14	. . .

TABLE 4—*Unirradiated and irradiated tensile properties of the materials.*

	Yield Strength, MPa			Ultimate Tensile Strength, MPa			Elongation, %			Reduction in Area, %		
Material	Unirradiated	Irradiated	Increase, %	Unirradiated	Irradiated	Increase, %	Unirradiated	Irradiated	Decrease, %	Unirradiated	Irradiated	Decrease, %
JP	503	556	11	631	700	11	33	29	12	65	62	5
FP	491	579	18	653	711	9	31	26	16	63	58	8
FF	547	633	16	687	729	6	26	24	8	63	54	14

TABLE 5—*Unirradiated and irradiated impact test properties of the materials.*

	Transition Temperature								
	C_{v40J}, °C			$C_{v\,0.9\,mm}$, °C			Upper Shelf Energy J		
Material	Unirradiated	Irradiated	Increase	Unirradiated	Irradiated	Increase	Unirradiated	Irradiated	Decrease
JP	−42	−21	21	−39	−7	32	220	216	4
FP	−19	−2	17	−3	16	19	160	160	0
JF	−29	7	36	−24	24	48	280	216	64
FF	−23	34	57	−6	38	44	184	188	−4
JW	−42	−24	18	−32	−4	28	192	180	12
JH	−52	−36	16	−44	−28	16	220	176	44
FW	−48	−34	14	−40	−18	22	192	172	20
GW	−64	−32	32	−62	−22	40	>240	190	>50

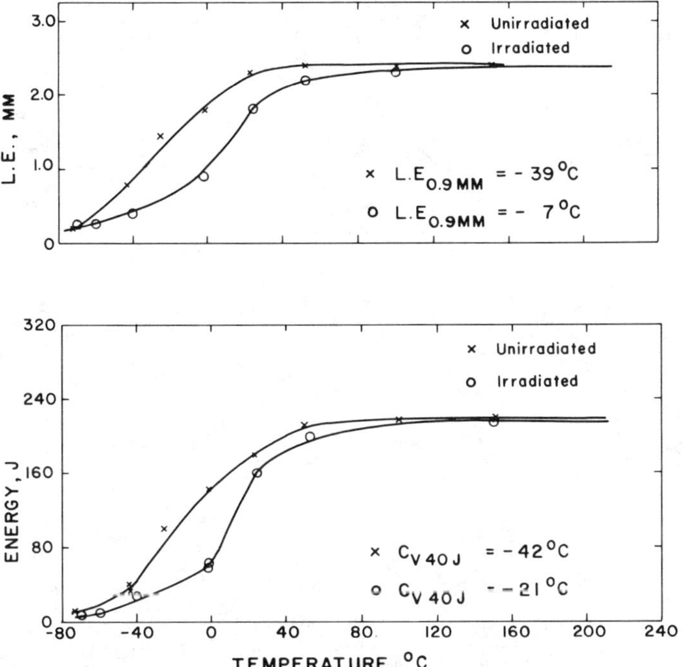

FIG. 1—*Charpy V-notch impact test results on Japanese plate (JP) irradiated at 290°C.*

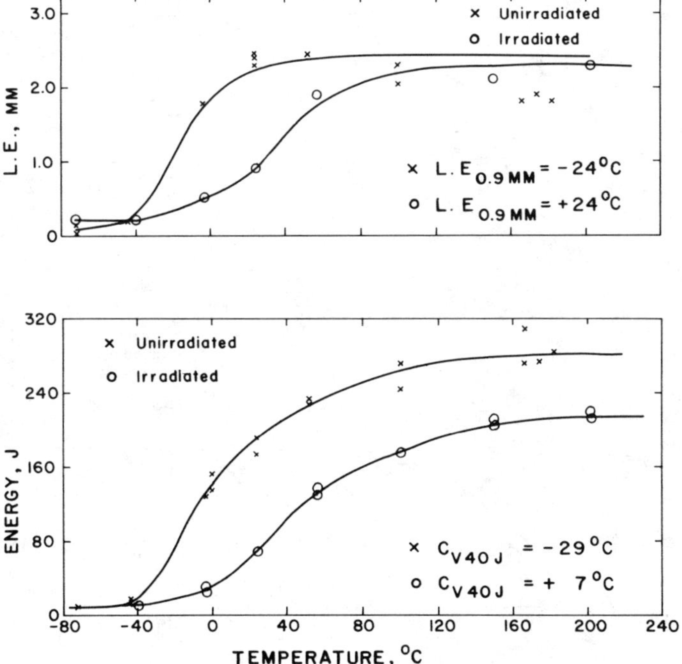

FIG. 2—*Charpy V-notch impact test results on Japanese forging (JF) irradiated at 290°C.*

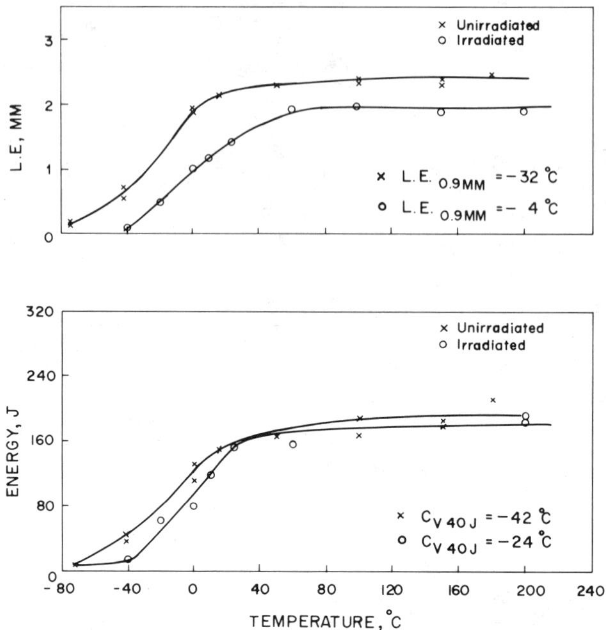

FIG. 3—*Charpy V-notch impact test results on Japanese weld (JW) irradiated at 290°C.*

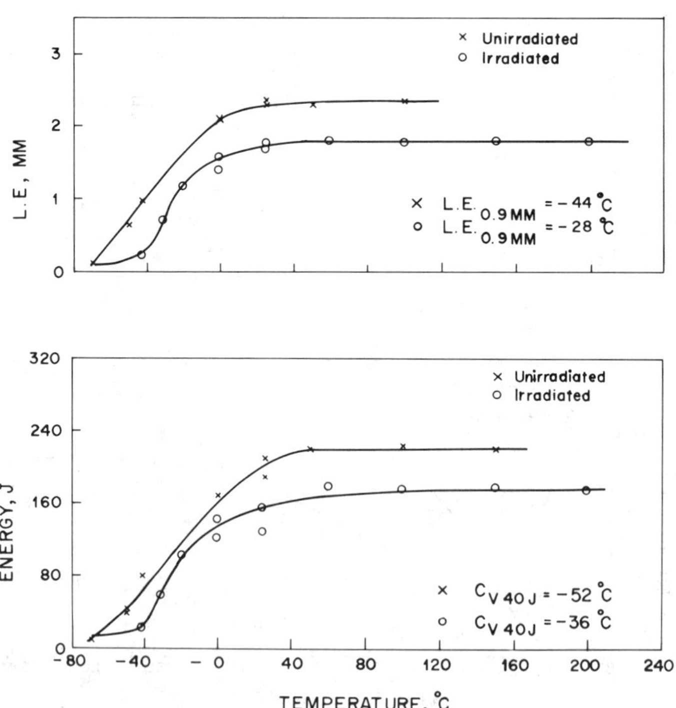

FIG. 4—*Charpy V-notch impact test results on Japanese weld heat affected zone (JH) irradiated at 290°C.*

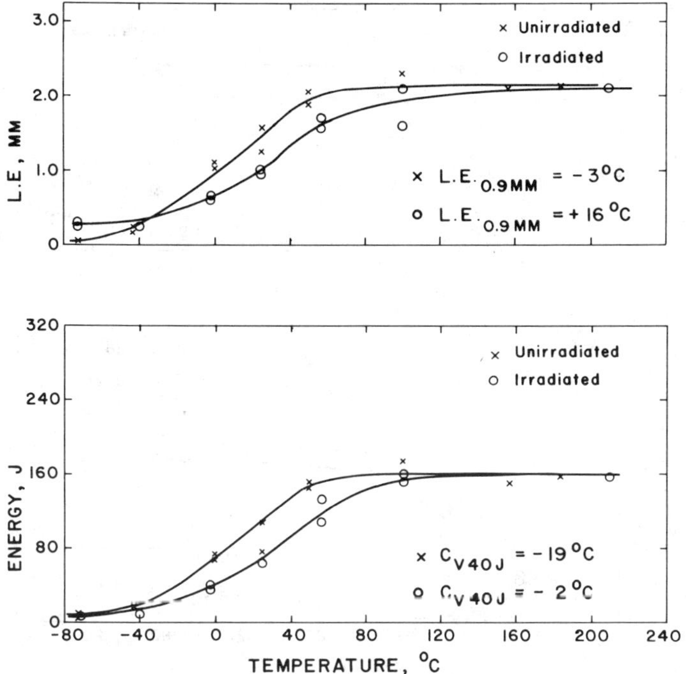

FIG. 5—*Charpy V-notch impact test results on French plate (FP) irradiated at 290°C.*

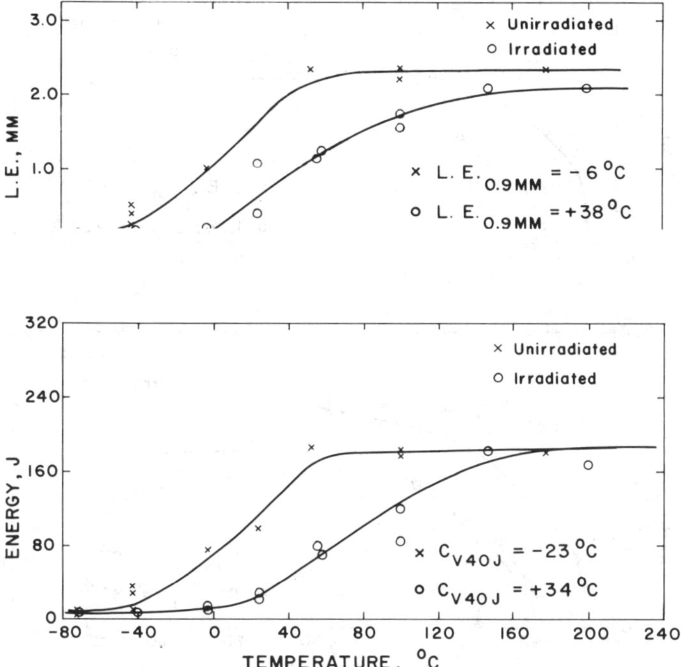

FIG. 6—*Charpy V-notch impact test results on French forging (FF) irradiated at 290°C.*

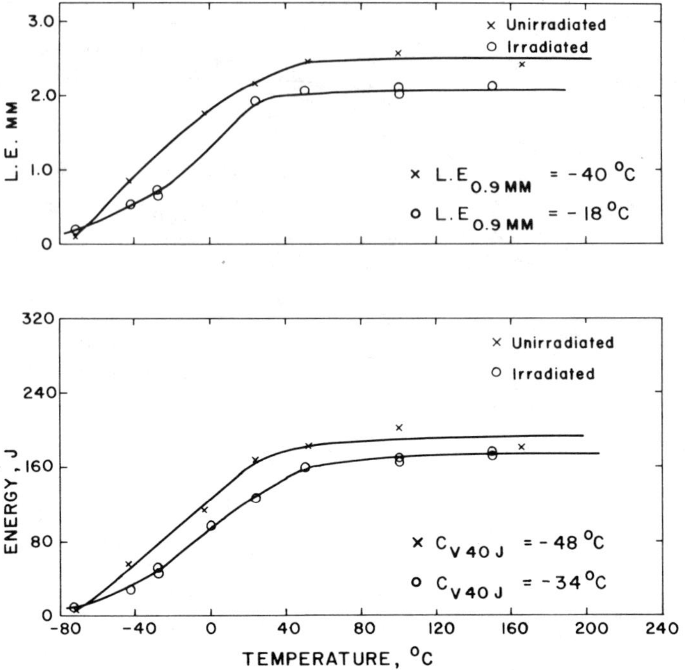

FIG. 7—*Charpy V-notch impact test results on French weld (FW) irradiated at 290°C.*

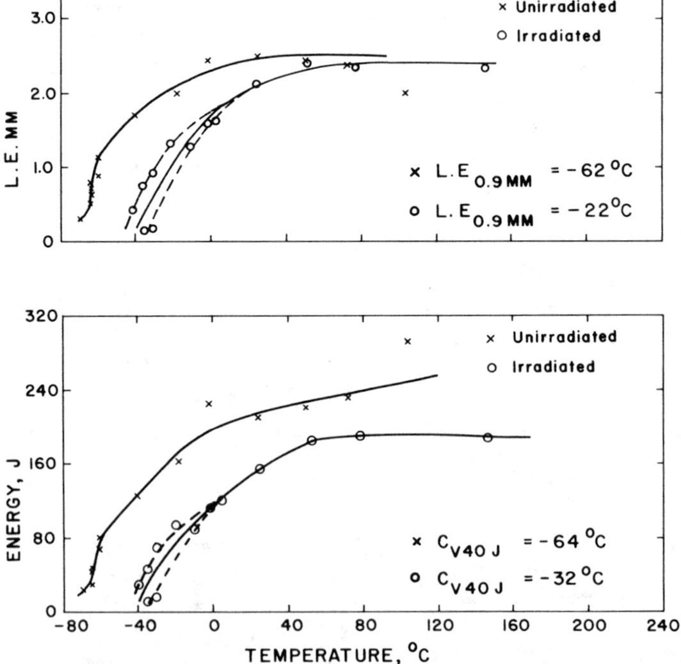

FIG. 8—*Charpy V-notch impact test results on German weld (GW) irradiated at 290°C.*

and irradiated French forging was found to have the highest transition temperature. The radiation induced increase in the transition temperature was lowest for the French weld and highest for the French forging.

Among the unirradiated materials, the German weld was found to have the maximum upper-shelf energy and French forging showed minimum upper shelf energy. The radiation induced reduction in upper shelf energy was maximum for the German weld. These results are schematically presented in Fig. 9.

The results of hardness tests on unirradiated and irradiated French plate, French forging, Japanese plate, and Japanese forgings are given in Table 6. The highest radiation induced increase in hardness was observed for French forging.

Microprobe analysis did not reveal any gross composition variation in specimens compared to the nominal chemical composition.

From the results (Tables 4 through 6), it is very apparent that all the materials showed increases in tensile strength values, decrease in ductility, decrease in percentage reduction in area, increase in hardness, rise in transition temperature, and reduction in upper shelf energy as a result of fast

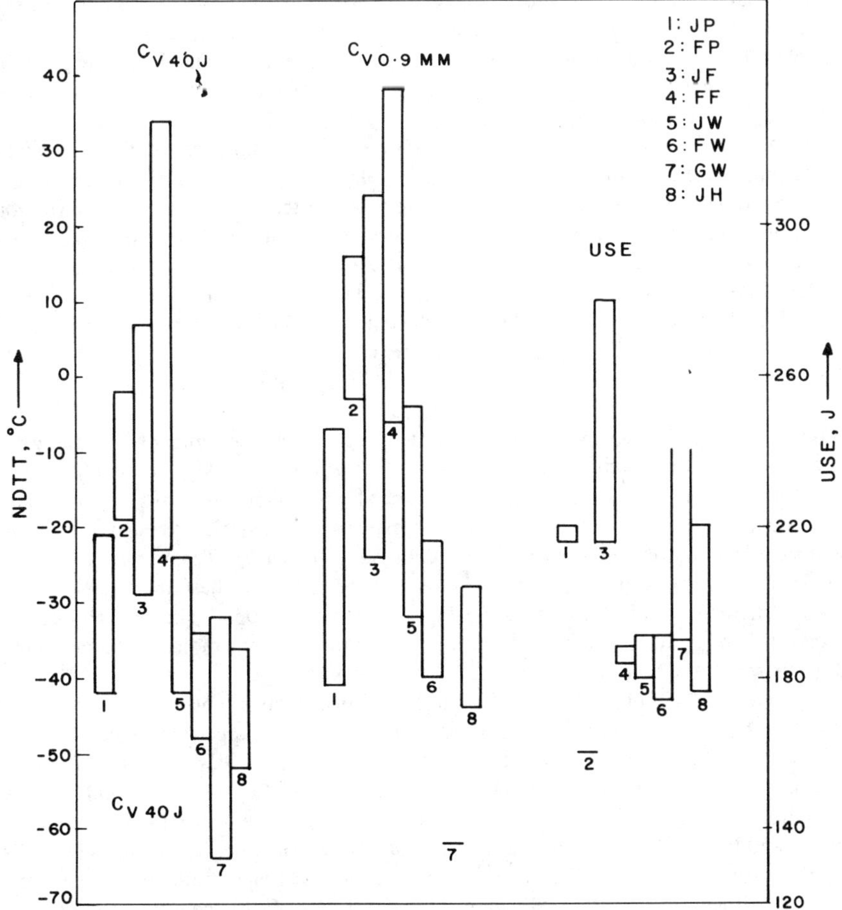

FIG. 9—*Irradiation induced changes in transition temperatures and upper shelf energies of the materials.*

TABLE 6—*Results of average hardness values.*

Material	Hardness		
	Unirradiated VHN	Irradiated VHN	Increase, %
JP	217	236	9
FP	232	251	8
JF	228	245	7
FF	232	264	14

neutron irradiation. All these findings clearly indicated that the materials were embrittled by fast neutron irradiation.

However, the extent of embrittlement was not very significant, the maximum C_{v40J} transition temperature being 34°C for irradiated French forging material. The corresponding shift in transition temperature was 57°C. Except Japanese forging and German weld, the remaining materials, namely, Japanese plate, Japanese weld, Japanese HAZ, French plate, and French weld, showed almost zero radiation sensitivity, the increase in the transition temperatures being typically 20°C or lower. In general, the impact results could be categorized as steels showing increase in C_{v40J} of typically around 17°C and steel showing increase in C_{v40J} of typically around 35°C.

French plate material showed no decrease in upper shelf energy, and Japanese plate, French forging, Japanese weld, and French weld showed small decreases in upper shelf energy because of the fast neutron irradiation.

As is apparent from Table 5, the specific criterion used for defining transition temperature, also affected the individual transition temperature values. In the worst case, a difference of 17°C was observed in the transition temperature value by using 40-*J* criterion against 0.9-mm criterion. The transition temperature as determined by the 40-*J* criterion was always lower than that determined by using the 0.9-mm criterion. In addition, the test results revealed that the determination of the shift in transition temperature as determined from the later criterion in general presented a more pessimistic case. To minimize the uncertainty in the determination of shift in transition temperature arising from the testing of limited number of impact specimens, both the 40 *J* and 0.9-mm transition temperature were considered to evaluate relative radiation induced embrittlement of these steels.

The French forging showed the maximum shift in C_{v40J} transition temperature (Table 5). This material also showed the maximum increase in hardness (Table 6) and yield strength (Table 4). All these findings clearly indicated relatively enhanced embrittlement of this material. Although there was scatter of impact test results for this material, a possible explanation of enhanced embrittlement could be due to higher copper content (0.07 wt%). The impact test results are retabulated for the analysis of the effect of chemical composition of these materials on irradiation induced embrittlement (Table 7). In addition to considering the copper and phosphorus contents, the well established residual elements responsible for neutron induced radiation damage, the nickel contents of these materials were also included in this table.

It was observed that although the French weld had higher copper content (0.05 wt%, but lower nickel content of 0.56 wt%) than that of French plate (0.03 wt% copper but higher nickel content of 0.65 wt%), shift in transition temperature in these two materials was of equivalent magnitude. The German weld having the same amount of copper (0.05 wt%), but high nickel content (0.93 wt%) as compared to the French weld however showed relatively enhanced embrittlement. These observations and the results in Table 7 on all these steels indicated the trend that the simultaneous presence of threshold amounts of copper and nickel were perhaps required to cause enhanced embrittlement in these steels. Steels showing relatively less embrittlement were found to contain copper and nickel amounts lower than 0.04 and 0.7 wt%, respectively. Steels containing higher

TABLE 7—*Irradiation induced changes in transition temperature.*

Material	Chemical Composition			Increase in Transition Temperature	
	P	Cu	Ni	ΔT_{40J}	$\Delta T_{0.9\ mm}$
JF	0.007	0.04	0.76	36	48
FF	0.009	0.07	0.69	57	44
GW	0.015	0.05	0.93	32	40
FP	0.008	0.03	0.65	17	19
FW	0.015	0.05	0.56	14	22
JP	0.007	0.01	0.66	21	32
JW	0.008	0.04	0.71	18	28
JH	0.007	0.01	0.66	16	16

than these threshold contents of copper and nickel had shown relatively enhanced embrittlement. Experimental observations elsewhere somewhat support this [*3,4*]. These observations indicated that nickel contents of 0.7 wt% but not 0.3 wt% could be detrimental to the radiation resistance of ASTM A533 and A508 (United Numbering System [UNS] K12521 and K13502) steels to radiation embrittlement at 288°C, and good radiation resistance could be achieved in high nickel content (0.96 wt% nickel) having low copper content (0.03 wt% copper).

French plate material showed no decrease in upper shelf energy, and Japanese plate, French forging, Japanese weld, and French weld showed negligible decrease in upper shelf energy because of fast neutron irradiation. The maximum decrease in upper shelf energy of more than 100 *J* was shown by the German weld. However, there was no particular relationship between the copper and nickel content and the decrease in the upper shelf energy. Moreover, as all the materials showed quite high upper shelf energy values, use of Charpy size specimens may not be adequate for estimating upper shelf energy as a material property for these materials.

Conclusions

The main goal of this coordinated research program was to demonstrate that careful specifications of pressure vessel steels and weldments using presently available advanced manufacture and welding technology can produce materials that would be practically insensitive to fast neutron induced embrittlement. The results from the Indian participation confirm that view, up to a fast fluence of 1×10^{19} n/cm^2.

The available results indicated that threshold amount of 0.04% copper and 0.7% nickel would perhaps be required to cause relatively enhanced embrittlement in these steels.

References

[*1*] "Co-ordinated Research Programme on Analysis of the Behaviour of Advanced Reactor Pressure Vessel Steels under Neutron Irradiation," IWG-RRPC-78/1, International Atomic Energy Agency, Vienna, Austria, 1977.

[*2*] Sivaramakrishnan, K. S., Chatterjee, S., Anantharaman S., Balakrishnan, K. S., and Viswanathan, U. K., "Irradiation Embrittlement of Reactor Pressure Vessel Steels," BARC-1237, Bhabha Atomic Research Centre, Bombay, India, 1984.

[*3*] Hawthorne, J. R., "Significance of Nickel and Copper Content to Radiation Sensitivity and Post Irradiation Heat Treatment Recovery of Reactor Vessel Steels," *Effects of Radiation on Materials: Proceedings of the Eleventh International Symposium, STP 782,* H. R. Bragger and J. S. Perrin, Eds., American Society for Testing and Materials, Philadelphia, 1982, pp. 375–391.

[*4*] Hawthorne, J. R., Kazoil, J. J., and Byrne, S. T., in *Effects of Radiation on Structural Materials: Proceedings of the Ninth Symposium, STP 683,* American Society for Testing and Materials, Philadelphia, 1979, pp. 235–251.

Christian Brillaud,[1] *F. Hedin,*[2] *and B. Houssin*[3]

A Comparison Between French Surveillance Program Results and Predictions of Irradiation Embrittlement

REFERENCE: Brillaud, C., Hedin, F., and Houssin, B., **"A Comparison Between French Surveillance Program Results and Predictions of Irradiation Embrittlement,"** *Influence of Radiation on Material Properties: 13th International Symposium, ASTM STP 956,* F. A. Garner, C. H. Henager, Jr., and N. Igata, Eds., American Society for Testing and Materials, Philadelphia, 1987, pp. 420–447.

ABSTRACT: At the beginning of 1986, thirty-two 900-MWe and four 1300-MWe nuclear power stations were in service in France. A regulatory surveillance program is in progress concerning the base metal, weld metal, and heat affected zones of every nuclear pressure vessel; a reference metal sample is also present in each reactor. This surveillance program was implemented largely in accordance with U.S. rules. The specifications for the core shell forgings in MnNiMo low alloy steel and the associated welds require low contents in embrittling elements such as copper and phosphorus. The results presented here concern materials containing less than 0.08 wt% Cu and 0.011 wt% P, except for the welds of the first six reactor vessels, which have 0.13 wt% Cu and 0.019 wt% P maximum. The main aim is to obtain the RT_{NDT} shift from Charpy-V tests, in relation to the fluence measured from fissile and activation dosimeters. Tensile, compact, and three-point bend specimens are also included in the capsules.

Results for the first capsules removed after up to four years of operation, corresponding to fluences ($E > 1$ MeV) less than 1.5×10^{19} n · cm^{-2} (2.9×10^{-2} dpa), are presented and evaluated. The mean and upper values of the RT_{NDT} shift are about 20 and 40°C, respectively, and there is no significant difference between parent metal and weld. A new formula to predict embrittlement has been derived from test reactor experiments, because of the strict limits on the residual element contents in French materials when compared with materials for which other formulae have been established. The surveillance program results are discussed and compared with the predictions of these formulae.

KEY WORDS: irradiation embrittlement, surveillance program, pressure vessel steels, predictive formulae

At the beginning of 1986, thirty-two 900-MWe and four 1300-MWe nuclear power stations were in service in France. The two Fessenheim units (900 MWe) have been in service since the end of 1977, and mid 1978, whereas the four Bugey plants (900 MWe) have been in operation since 1979.

Commissioning dates of the first units of the 900-MWe standardized plants (CPY program), which include 28 units, are the following: Tricastin units 1 and 2 (mid 1980), Dampierre units 1 and 3 (end of 1980 and beginning of 1981), and Gravelines units 1 and 3 (mid 1980 and beginning 1981).

[1] Engineer, Electricité de France, SCMI (Service Contrôle de Matériaux Irradies), 37420 Avoine, France.
[2] Electricité de France, SEPTEN (Service Etudes et Projets thermiques et Nucléaires), Lyon, France.
[3] Framatome, Département Matériaux, Paris la Défense, France.

A surveillance program of irradiation embrittlement on base metal, weld, and heat-affected zone (HAZ) for every vessel is being carried out by Electricité De France (EDF). This program is bound by the regulatory rules set out in Article 41 of the French "decree" dated 26 Feb. 1974 concerning the main primary circuit of the nuclear water supply system.

The results given here concern the capsules of the Fessenheim and Bugey units, after 18 months of reactor service, and those of the first 900-MWe program units (after four years of reactor service), corresponding to neutron fluences ($E > 1$ MeV) in the region of 1.5×10^{19} n · cm^{-2} (2.9×10^{-2} dpa).

Main Features of the Surveillance Program

The surveillance program of Fessenheim—Bugey and 900-MWe standardized units—has been defined in order to satisfy the requirements of Article 41, of the French regulations dated 26 Feb. 1974, and as far as possible those of Appendix G and H to the 10 CFR Part 50, and the associated reference documents.

For each unit, the surveillance program includes six capsules containing mechanical testing specimens that are taken out of the following materials (Fig. 1):

- Base metal of irradiated vessel shell(s), taken at ¼ inner thickness.
- Weld metal representative of the weld of the most irradiated core shell, that is, deposited with the same wire/flux combination, and the same welding conditions as the production weld. Specimens are machined at 12.7 mm or over from the weld root.
- Base metal HAZ of selected shell(s), the notch of Charpy-V specimens being at 2.0 mm from the fusion line for the Fessenheim 1 and 2 and Bugey 2 and 3 units, and at 0.8 mm for the next units.
- Reference metal, for comparing the operational characteristics of different units.

These test coupons are submitted to a heat-treatment, fully representative of the components. Two capsules, among these six, are kept aside in case a further annealing-treatment on the vessel might be needed.

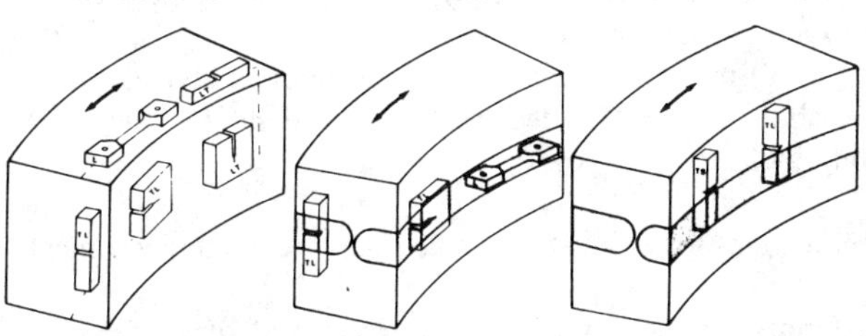

BASE METAL WELD H.A.Z

UNITS	SPECIMEN	BASE METAL	WELD	HAZ	REFERENCE METAL
FESSENHEIM 1 - 2 BUGEY 2 - 3	Charpy V notch	LT	TL	TS	LT
	Tensile	L	L	-	-
	WOL	LT	TL	-	-
BUGEY 4 - 5 900 MW standardized	Charpy V notch	TL	TL	TL	LT
	Tensile	L	L	-	-
	WOL or CT	TL	TL	-	-

FIG. 1—*Taking of surveillance program specimens.*

The withdrawal and eventual exchange schedule, as well as the multi-directional distribution of the capsules in the reactor, are presented in Figs. 2 and 3. Type and number of test specimens are detailed for each capsule in Tables 1, 2, and 3. The irradiation temperature is 286°C.

Dosimetry data for each capsule are necessary for the mechanical testing analysis and are obtained from

(1) two fissile dosimeter blocks: one uranium and one neptunium,
(2) three blocks of five activation dosimeters: copper, nickel, iron, cobalt and cobalt under cadmium, and
(3) two temperature-monitors of 304°C, and another one of 310°C.

Owing to the particularities of the surveillance program for the first units at Fessenheim and Bugey, with respect to those of the 900-MWe program, it can be seen that

1. The exchanges take place during regulatory in-service inspection (first complete in-service inspection after almost 30 months from start-up and every ten years) (Fig. 2).
2. The test specimens have been machined for the base metal in the longitudinal direction with respect to the main forging direction (L-T) for Fessenheim 1 and 2, Bugey 2 and 3, and in the transverse direction (T-L) for the following units (Fig. 1).
3. The capsules are located on the external side of the thermal shield for the Fessenheim and Bugey units, and on the inside of the vessel for the 900-MW plants.

Surveillance Program

The methods used for choosing materials and continuously assessing irradiation embrittlement are described below.

Choice of Materials and Determination of Initial RT_{NDT}

Different materials are as follows:

1. Base metal: low-alloy steel: 16MND 5 type (American Society of Mechanical Engineers [ASME] equivalent to SA 508 c13 specification).
2. Welded joints (SAW) matching the base metal mentioned above.
3. Reference: low alloy Mn-Ni-Mo steel plate (equivalent to ASME SA 533 Grade B c11 specification) 200 mm thick.

The reference specifications for products, and the main specified mechanical properties are given in Table 4.

The choice of materials for the surveillance program and the definition of initial RT_{NDT} are carried out by the fabricator in agreement with EDF.

The choice criteria have changed over time as follows:

1. Fessenheim 1 and 2 and Bugey 2 and 3: no criteria for choosing between base metal of the two irradiated core shells (C1 and C2),
2. Bugey 4, 5, and 900-MWe standardized plants: materials that have to be considered are those in the core areas for which the RT_{NDT} shift is above 28°C (50°F) in agreement with Appendix G to 10 CFR 50: nozzle shell (part) course named B, weld joint named B/C1, core shell named C1, weld joint C1/C2, in part lower shell C2. Base metal and weld joint selected are those that

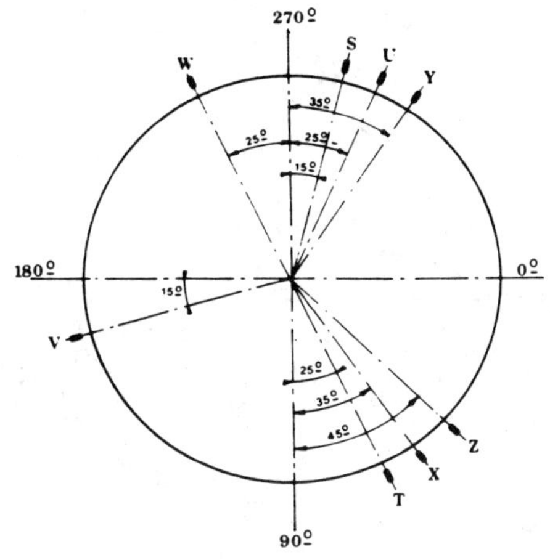

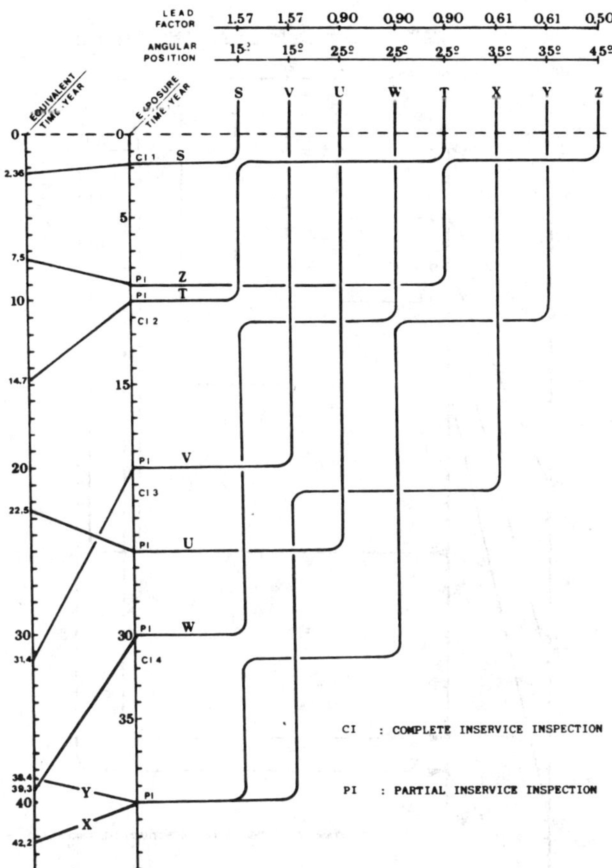

FIG. 2—*Capsules withdrawal and exchange program for the Fessenheim 1 and 2 and Bugey 2, 3, 4, and 5 units.*

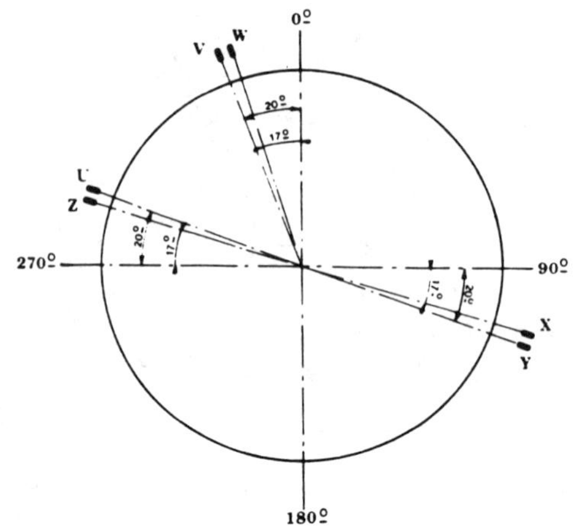

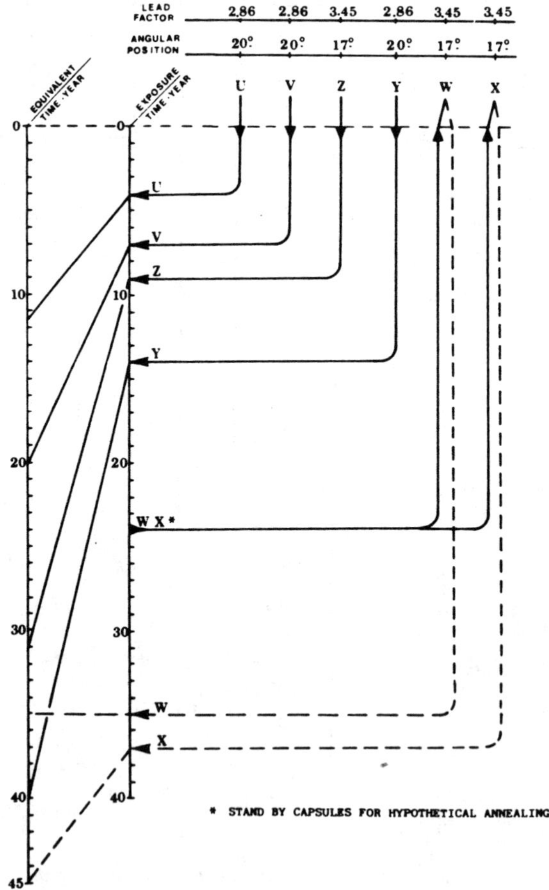

FIG. 3—*Capsules withdrawal program for 900-MW standardized plants.*

TABLE 1—*Share out of test specimen in capsules of surveillance program: Fessenheim 1-2 and Bugey 2-3.*

Capsules	Specimen	Base Metal C1	Base Metal C2	Weld C1/C2	HAZ	Reference Metal
S	Charpy-V notch	0	10	10	12	0
	tensile	0	3	0	0	0
	WOL	0	4	0	0	0
	unmachined sample	0	1	0	0	0
T	Charpy-V notch	0	10	10	12	0
	tensile	0	0	3	0	0
	WOL	0	0	4	0	0
	unmachined sample	1	0	0	0	0
U	Charpy-V notch	11	11	0	0	10
	tensile	0	3	0	0	0
	WOL	0	4	0	0	0
	unmachined sample	0	1	0	0	0
V	Charpy-V notch	10	0	10	12	0
	tensile	0	0	3	0	0
	WOL	0	0	4	0	0
	unmachined sample	1	0	0	0	0
W	Charpy-V notch	10	0	10	12	0
	tensile	3	0	0	0	0
	WOL	4	0	0	0	0
	unmachined sample	1	0	0	0	0
X	Charpy-V notch	11	11	0	0	10
	tensile	3	0	0	0	0
	WOL	4	0	0	0	0
	unmachined sample	0	1	0	0	0
Y	Charpy-V notch	11	11	0	0	10
	tensile	0	3	0	0	0
	WOL	0	4	0	0	0
	unmachine sample	1	0	0	0	0
Z	Charpy-V notch	11	11	0	0	10
	tensile	3	0	0	0	0
	WOL	4	0	0	0	0
	unmachined sample	1	0	0	0	0

have the higher end of life (40 years) RT_{NDT}, considering the latter's initial value, copper content of metal in question (ASME curve, Section XI, Appendix A, 1974, see Fig. 4) and the fluence.

The criteria for the initial RT_{NDT} were based on the ASME Code, Section III: NDTT is determined by drop weight tests, in accordance with ASTM Method for Conducting Drop-Weight Test to Determine Nil-Ductility Transition Temperature of Ferritic Steels (E 208) using the two pass procedure for the crack starter deposition.

Three Charpy-V impact toughness specimens are tested at one given temperature, starting at NDTT + 33°C. If, at this temperature, all three specimens give a Charpy-V energy, and a lateral expansion higher or equal to 68.8 J (8.6 daJ/cm²), and 0.9 mm, respectively, the RT_{NDT} is taken equal to NDTT.

Otherwise, the test temperature is increased by 5°C increments, until the requirements are fulfilled. The RT_{NDT} is then taken equal to TT, the temperature at which the requirements are reached, minus 33°C.

TABLE 2—*Share out of test specimen in capsules of surveillance program: Bugey 4-5.*

Capsules	Specimen	Base Metal C1 or C2	Weld C1/C2	HAZ	Reference Metal
	Charpy-V notch	12	12	12	12
S	tensile	3	0	0	0
	WOL	4	0	0	0
	Charpy-V notch	12	12	12	12
V	tensile	0	3	0	0
	WOL	0	4	0	0
	Charpy-V notch	24	12	12	0
U	tensile	3	0	0	0
	WOL	4	0	0	0
	Charpy-V notch	12	12	12	12
W	tensile	0	3	0	0
	WOL	0	4	0	0
	Charpy-V notch	12	12	12	12
T	tensile	0	3	0	0
	WOL	0	4	0	0
	Charpy-V notch	24	12	12	0
X	tensile	0	3	0	0
	WOL	0	4	0	0
	Charpy-V notch	24	12	12	0
Y	tensile	3	0	0	0
	WOL	4	0	0	0
	Charpy-V notch	24	12	12	0
Z	tensile	3	0	0	0
	WOL	4	0	0	0

The estimated end of life RT_{NDT} on the inside wall, for a design life-time of 40 years (or 32 years for equivalent nominal power), takes into account copper content and fluences for Fessenheim, Bugey units, and 900-MWe standardized plants equals to 4.8×10^{19} n · cm^{-2} ($E > 1$ MeV),5 × 10^{19} n · cm^{-2} ($E > 1$ MeV), and 5.1×10^{19} n · cm^{-2} ($E > 1$ MeV), respectively.

Monitoring of Irradiation Effects

EDF has the regulatory mission to carry out the survey of the shifts in the characteristics of vessel materials induced by irradiation damage. In this way, pre- and post-irradiation experiments are done in Service Contrôle de Matériaux Irradies (SCMI) laboratories of the *Groupe des Laboratoires*.

As far as the pre-irradiation experiments are concerned, some test specimens of chosen materials, which are taken, yield the following information:

- Charpy-V notch transition curve, using the European tup and according to AFNOR NF A 03 161 standard,
- lateral expansion curve,
- brittle fracture appearance curve, and
- tensile properties at room and 300°C temperatures.

TABLE 3—*Share out of test specimen in capsules of surveillance program: 900 MW standardized plants.*

Capsules	Specimen	Base Metal C1 or C2	Weld C1/C2	HAZ	Reference Metal
U	Charpy-V notch	15	15	15	15
	tensile	5	4	0	0
	CT	6	6	0	0
	3-point bend	1	0	0	0
V	Charpy-V notch	15	15	15	15
	tensile	5	4	0	0
	CT	6	6	0	0
	3-point bend	1	0	0	0
W	Charpy-V notch	30	15	15	0
	tensile	5	4	0	0
	CT	6	6	0	0
	3-point bend	1	0	0	0
X	Charpy-V notch	30	15	15	0
	tensile	5	4	0	0
	CT	6	6	0	0
	3-point bend	1	0	0	0
Y	Charpy-V notch	15	15	15	15
	tensile	5	4	0	0
	CT	6	6	0	0
	3-point bend	1	0	0	0
Z	Charpy-V notch	30	15	15	0
	tensile	5	4	0	0
	CT	6	6	0	0
	3-point bend	1	0	0	0

After drawing and dismantling the capsules, the following data are determined:

- Charpy-V notch transition curve,
- lateral expansion and brittle fracture curves, and
- tensile properties at 300°C.

At the present time, specimens for fracture mechanics tests (WOL, compact, and three-point bending) are saved, awaiting confirmation. Then specimen sizes are appropriate for testing procedures.

Dosimeter analysis is entrusted to the *Centre d'Etudes Nucléaires de Grenoble* (CENG) by EDF; the CENG ascertains thermal and fast fluxes, fluence from flux values, and the unit operational graph [*1*].

A statistical method is used for plotting the Charpy-V curves: average curve (modified hyperbolic tangent: $th\ (x) + g\ (x)$) and upper and lower ($+$ or $-$ $2s$) curves may be be plotted. Scatter for base metal and HAZ at 56 J (7-daJ/cm^2) level is about ±20°C. The scatter is less important for welds (±15°C).

From the results, the shifts are also given as follows:

- the transition temperature TK 7 for an impact energy of 56 J (7 daJ/cm^2) and
- the temperature corresponding to a lateral expansion of 0.9 mm.

TABLE 4—*Chemical and mechanical specifications for base metal and weld of vessels: Fessenheim-Bugey 900-MW program.*

	P, wt%		Cu, wt%		Ni, wt%		S, wt%	
Base Metal	Heat	Product	Heat	Product	Heat	Product	Heat	Product
Fessenheim-Bugey units	0.020 max		0.15 max		0.40/0.80	0.37/0.83	0.020 max	
900-MW standardized plants	0.012 max	0.015 max	0.10 max		0.40/0.80	0.37/0.83	0.015 max	0.020 max

Weld	P, wt%	Cu, wt%	Ni, wt%	S, wt%
Fessenheim-Bugey 2-3-4	0.025 max	[a]	[a]	0.025 max
Bugey 5	0.025 max	[a]	0.45/0.75	0.025 max
900-MW standardized plants	0.015 max	0.10 max	1.2 max	0.025 max

	20°C R 0.002 min MPa	343°C R 0.002 min MPa	Rm, MPa	A, %	Z, %	KCV mean −20°C, J	KCV mean 0°C, J[c]
Base Metal and Weld	345	285	551/667	18	38	32	56

[a] Check value but no requirement.
[b] Mean value on three test specimens; minimum individual value: 22.4 J (2.8 daJ/cm^2).
[c] Mean value on three test specimens; minimum individual value: 40 J (5.0 daJ/cm^2).

The shift of *TT* is the highest of the two previous values; this level is considered equal to RT_{NDT}. Further information are also obtained as follows:

- the FATT shift, the upper shelf change and
- the change in yield strength at 300°C.

Lastly, predicted values must be shown to be conservative estimates of measured data.

Summary of the Results

Initial State

Table 4 shows the specifications concerning chemical composition (phosphorus, copper, nickel, and sulfur) and mechanical properties for base metal and welds for the 34 vessels of the 900-MWe units.

The amounts of embrittling elements, such as phosphorus, copper, and nickel, for base metal and welds of the 34 vessels of the 900-MW units and reference metal are shown in Fig. 5. Mean and associated scatter are shown in Tables 5 and 6.

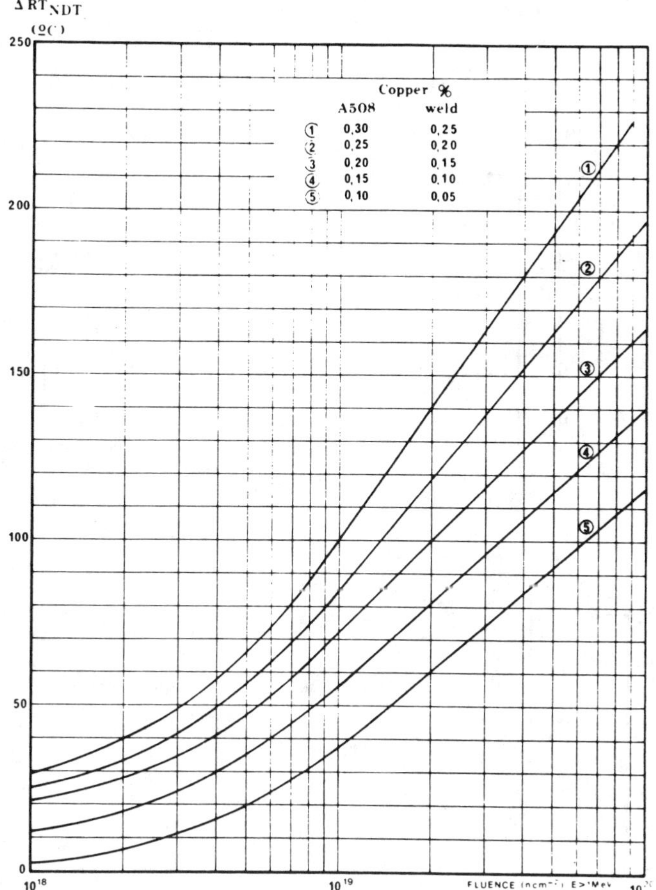

FIG. 4—RT_{NDT} *shift for choice of materials in the 900-MV units surveillance program ASME curve, Section XI, Appendix A, 1974.*

Welds of the first six Fessenheim and Bugey units have maximum copper content of 0.13% and phosphorus contents of 0.019%, whereas 900-MW standard welds have lower maximum copper and phosphorus contents, 0.060 and 0.014%, respectively. The copper, phosphorus, and nickel contents of the reference metal are 0.030, 0.006, and 0.66%, respectively.

Pre-irradiation RT_{NDT}

The pre-irradiation RT_{NDT} (°C) distributions for base metal, weld, and reference metal are shown in Table 7.

For the units concerned by the results of the surveillance program presented here, Table 8 gives phosphorus, copper, nickel, and sulfur contents of the pre-irradiation test specimens, and Table 9 shows the different levels for TK 7, initial RT_{NDT}, upper shelf of the Charpy-V notch transition curve, and 300°C yield strength with offset 0.2% ($Re^{300}_{0.002}$).

It may be noted that before irradiation, transition temperature TK 7 is rather low: from −57 to −16°C for base metals and from −57°C to −25°C for welds.

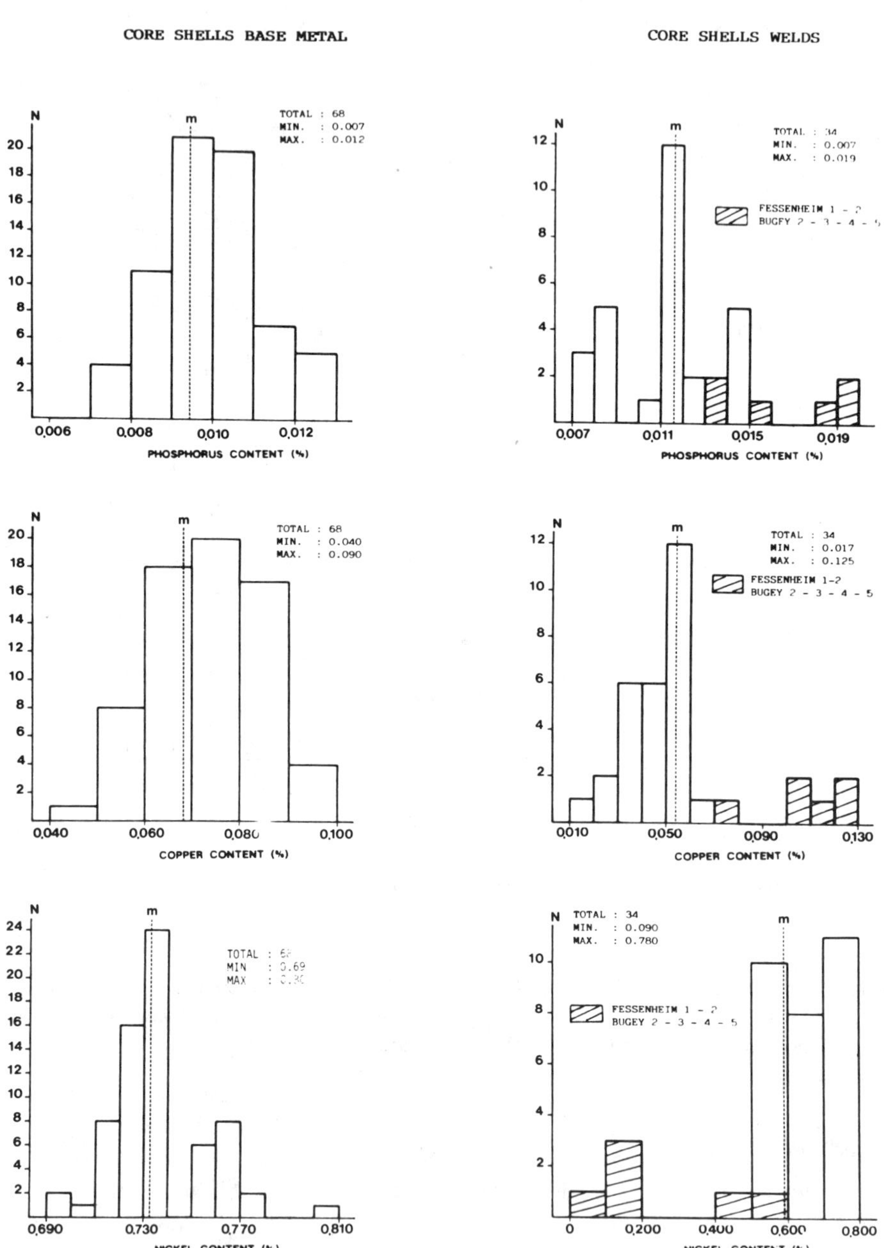

FIG. 5—*Histograms of copper, phosphorus, nickel contents (wt%) of base metal and weld of 900-MW vessels.*

TABLE 5—*Base metal core shells (product analysis), 34 vessels of 900-MW units.*

Material	Mean Value	Standard Deviation
P, wt %	0.010	0.002
Cu, wt %	0.070	0.014
Ni, wt %	0.71	0.03

TABLE 6—*Weld metal: submerged arc welding.*

	Fessenheim-Bugey Units (6)		900-MW Standardized Units (28)	
Weld Metal	Mean Value	Standard Deviation	Mean Value	Standard Deviation
P, wt %	0.016	0.003	0.011	0.002
Cu, wt %	0.106	0.018	0.043	0.012
Ni, wt %	0.25	0.18	0.66	0.07

TABLE 7—*Pre-irradiation distributions.*

Material	Mean	Standard Deviation
Base metal	−24	8.3
Weld	−45	6.3
Reference metal	−20	. . .

TABLE 8—*Copper, phosphorus, nickel, and sulfur contents (wt%) of pre-irradiation test specimens.*

Units	Cu, %	P, %	Ni, %	S, %
		BASE METAL		
Fessenheim 1	0.080	0.009	0.84	0.014
Fessenheim 2	0.070	0.008	0.69	0.007
Bugey 2	0.080	0.009	0.65	0.009
Bugey 3	0.050	0.008	0.72	0.009
Bugey 5	0.050	0.006	0.71	0.005
Tricastin 1	0.060	0.007	0.69	0.005
Tricastin 2	0.066	0.008	0.72	0.006
Dampierre 1	0.064	0.006	0.69	0.006
Dampierre 2	0.070	0.008	0.74	0.008
Gravelines 1	0.057	0.008	0.69	0.005
Gravelines 3	0.040	0.007	0.69	0.007
		WELD		
Fessenheim 1	0.100	0.019	0.09	0.016
Fessenheim 2	0.090	0.017	0.10	0.015
Bugey 2	0.120	0.012	0.07	0.020
Bugey 3	0.100	0.016	0.49	0.012
Bugey 4	0.130	0.017	0.08	0.013
Bugey 5	0.100	0.015	0.51	0.011
Tricastin 1	0.030	0.009	0.66	0.010
Tricastin 2	0.031	0.008	0.66	0.009
Dampierre 1	0.033	0.009	0.64	0.011
Dampierre 2	0.030	0.009	0.66	0.009
Gravelines 1	0.033	0.008	0.64	0.009
Gravelines 3	0.040	0.014	0.55	0.013
		REFERENCE METAL		
All units	0.030	0.006	0.66	. . .

TABLE 9—*Initial mechanical characteristics for base metal, weld, and reference metal: TK7, initial* RT_{NDT} *upper shelf and* $R_{0.002}^{300°C}$.

Base Metal	TK7, °C	Initial RT_{NDT}, °C	USE, J	$R_{0.002}^{300}$, MPa
Fessenheim 1 shell C2[a]	−51	−32	168	430
Fessenheim 2 shell C2[a]	−34	−32	196	431
Bugey 2 shell C2[a]	−42	−37	196	417
Bugey 3 shell C2[a]	−35	−22	204	437
Bugey 4[b]	−39	−27	172	475
Bugey 5[b]	−45	−12	184	461
Tricastin 1[b]	−57	−37	196	400
Tricastin 2[b]	−54	−17	208	444
Dampierre 1[b]	−51	−42	200	417
Dampierre 2[b]	−16	−22	140	422
Gravelines 1[b]	−55	−32	192	416
Gravelines 3[b]	−20	−17	152	445
REFERENCE METAL: TK7 = −46°C; USE = 184 J				
WELD				
Fessenheim 1	−41	−20[c]	128	402
Fessenheim 2[d]	−43	−20	144	400
Bugey 2[d]	−23	−20	144	374
Bugey 3	−25	−20	136	500
Bugey 4	−47	−20	168	447
Bugey 5	−17	−20	140	489
Tricastin 1[d]	−41	−47	155	530
Tricastin 2[d]	−57	<−52	156	485
Dampierre 1[d]	−47	<−52	144	473
Dampierre 2[d]	−33	<−52	144	504
Gravelines 1[d]	−45	−42	140	463
Gravelines 3	−42	−32	148	448

[a] Chargy V notch, longitudinal.
[b] Charpy-V notch, traverse.
[c] RT_{NDT} value estimated from Charpy-V notch specimens results.
[d] Same wire, flux combination.

TABLE 10—*Surveillance program results on base metal (mean values).*

Units	Fluence (E > 1 MeV) ncm^{-2} × 10^{19}	Fluence (E > 1 MeV) dpa × 10^{-2}	ΔTK7, °C	ΔL.Ex., °C	ΔFATT, °C	$\Delta R_{0.002}^{300}$, MPa	USE Variation, J
Fessenheim 1 shell C2	0.30	0.42	17	16	0	24	...
Fessenheim 2 shell C2	0.33	0.47	0	9	0	16	...
Bugey 2 shell C2	0.30	0.40	23	30	5	28	...
Bugey 3 shell C2	0.45	0.68	12	12	0	17	...
Bugey 4	0.30	0.43	34[a]	33[a]	28[a]	0	...
Bugey 5	0.30	0.45	14	11	25	12	+24
Tricastin 1	1.53	2.94	28	39	36	39	−32
Tricastin 2	1.52	2.95	21	24	24	0	−12
Dampierre 1	1.33	2.57	3	7	4	25	+16
Dampierre 2	1.36	2.62	8	5	10	65	...
Gravelines 1	1.31	2.52	14	12	11	56	...
Gravelines 3	1.40	2.69	1	3	9	25	...

[a] Over estimated value due to a sampling anomaly.

TABLE 11—*Surveillance program results on weld (mean values).*

Units	Fluence (E > 1 MeV)		ΔTK 7, °C	ΔL.Ex., °C	ΔFATT, °C	$\Delta R^{300}_{0.002}$, MPa	USE Variation, J
	$ncm^{-2} \times 10^{19}$	dpa $\times 10^{-2}$					
Fessenheim 1	0.30	0.42	12	12	0	. . .	+32
Fessenheim 2	0.33	0.47	12	21	7	. . .	−16
Bugey 2	0.30	0.40	8	7	0	. . .	+8
Bugey 3	0.45	0.68	27	26	22	. . .	−16
Bugey 4	0.30	0.43	13	15	11	. . .	. . .
Bugey 5	0.30	0.45	9	11	23	. . .	+20
Tricastin 1	1.53	2.94	26	25	32	14	−27
Tricastin 2	1.52	2.95	24	24	25	21	−24
Dampierre 1	1.33	2.57	29	25	22	46	−16
Dampierre 2	1.36	2.62	11	12	12	26	−12
Gravelines 1	1.31	2.52	10	10	17	24	+8
Gravelines 3	1.40	2.69	33	38	21	37	−16

Charpy-V upper shelf energy (USE) for base metal and welds is rather high although sometimes slightly lower for welds (USE mean = 146 J) than for base metal (USE mean = 184 J).

Post-Irradiation Results

Tables 10, 11, 12 and 13 show the results now available concerning the capsules of the surveillance program regarding the twelve units concerned: base metal, weld, HAZ, and reference metal—TK 7, lateral expansion (L.Ex.), FATT, USE, and $R^{300}_{0.002}$.

Typical Charpy-V energy, brittle fracture, and lateral expansion curves before and after irradiation (Tricastin 2) are shown in Fig. 6 (base metal and weld).

Inadequacy of Currently Proposed Calculative Procedures for the Prediction of Irradiation Embrittlement

Some welded joints of the oldest reactor vessels operating in France (Fessenheim and Bugey plants) contain quite high phosphorus contents (0.019%) when compared with modern material

TABLE 12—*Surveillance program results on HAZ (mean values).*

Units	Fluence (E > 1 MeV)		ΔTK7, °C	ΔL.Ex., °C	ΔFATT, °C	USE Variation, J
	$ncm^{-2} \times 10^{19}$	dpa $\times 10^{-2}$				
Fessenheim 1 shell C2	0.30	0.42	19	12	0	−8
Fessenheim 2 shell C2	0.33	0.47	16	30	3	−8
Bugey 2 shell C2	0.30	0.40	17	16	5	. . .
Bugey 3 shell C2	0.45	0.68	16	11	3	. . .
Bugey 4	0.30	0.43	22	21	23	. . .
Bugey 5	0.30	0.45	0	0	9	+28
Tricastin 1	1.53	2.94	0	0	15	−8
Tricastin 2	1.52	2.95	28	27	34	−32
Dampierre 1	1.33	2.57	27	29	16	−12
Dampierre 2	1.36	2.62	19	20	24	−12
Gravelines 1	1.31	2.52	21	14	0	. . .
Gravelines 3	1.40	2.69	44	34	23	. . .

TABLE 13—*Surveillance program results on reference metal (mean values).*

Units	Fluence (E > 1 MeV) $ncm^{-2} \times 10^{19}$	Fluence (E > 1 MeV) $dpa \times 10^{-2}$	ΔTK7, °C	ΔL.Ex., °C	ΔFATT, °C	USE Variation, J
Bugey 4	0.30	0.43	8	15	9	...
Bugey 5	0.30	0.45	20	21	7	+12
Tricastin 1	1.53	2.94	13	13	16	−16
Tricastin 2	1.52	2.95	20	21	16	+10
Dampierre 1	1.33	2.57	23	22	19	−8
Dampierre 2	1.36	2.62	4	4	1	+20
Gravelines 1	1.31	2.52	19	21	15	−12
Gravelines 3	1.40	2.69	16	13	18	+8

specifications. Consequently, it is important to know the precise influence attributed to the phosphorus element in the prediction of irradiation embrittlement, the more so, since new formulae, for instance RG 1.99, Revision 2 [2], do not take it into account.

The data used for the last predictive procedure [3,4] are characterized for the welds by a small number of data having copper contents less than 0.16%, and in these cases the nickel contents are less than 0.2%. Thus, they do not cover the chemical compositions of the welded joints of the French reactor vessels, having copper contents ranging from 0.03 to 0.13%.

Finally, the quantity called "Margin" that is to be added to obtain conservative upper bound values of adjusted reference temperature in RG 1.99, Revision 2 is 31°C for welds and is significantly higher than for the base metal (19°C). This fact does not seem to be corroborated by the present results of the surveillance program or test reactor experiments and leads to overpredict the neutron induced shifts for weldments of 900-MW standardized plants. This discrepancy was also pointed out by Davies recently [5]. On the other hand, the old R.G. 1.99, Revision 1 [6] is no more appropriate for predicting the shifts in transition temperatures of the welds of the oldest vessels since the phosphorus influence seems to be overestimated, and the lower sensitivity to irradiation embrittlement caused by low or moderate nickel contents of some welds of the Fessenheim and Bugey plants cannot be estimated from this guide. The RG 1.99 formulae were developed from SA 533 Grade B C11 steel plates and associated welds; the chemical compositions are equivalent to those of forgings in 16 MND 5 (SA 508 c13) forging but may exhibit some small differences related to steel making processes or segregations and, concerning welds, to the welding process or filler metals and fluxes used. All these reasons were strong motivations for trying to develop specific relationships for products representative of French reactor vessels. Therefore, a data bank of irradiation embrittlement results was constituted with results obtained both from test reactor (TR) experiments and surveillance programs (SP) of power reactors.

Description of the Data Base

Data collecting was oriented toward actual forgings of industrial heats in manganese-nickel-molybdenum low alloy seel (French 16 MND 5, standard ASME: SA 508 C13 specification) and associated welds. The range in chemical variations for the 25 forging data and 23 weld data are given in Ref 7. The total number of collected data is 78.

The distribution fluence data are presented in Fig. 7. The fluences of the surveillance program results are concentrated in the low fluence region of the data bank caused by the youth of the French nuclear plants. The distributions in copper, nickel, and phosphorus of materials irradiated in test reactor experiments are presented in the histogram of Fig. 7, with the histograms of the

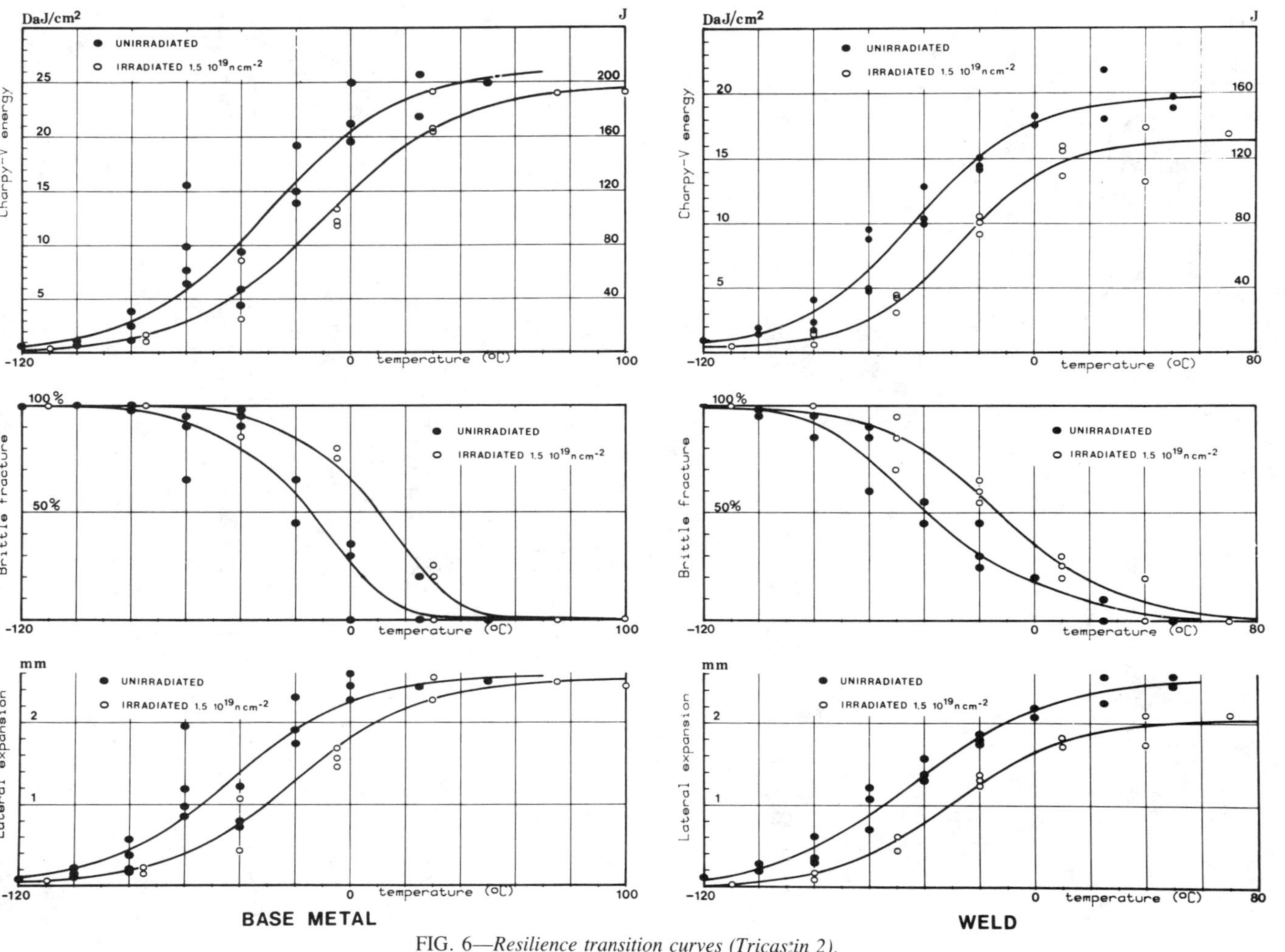

FIG. 6—*Resilience transition curves (Tricastin 2).*

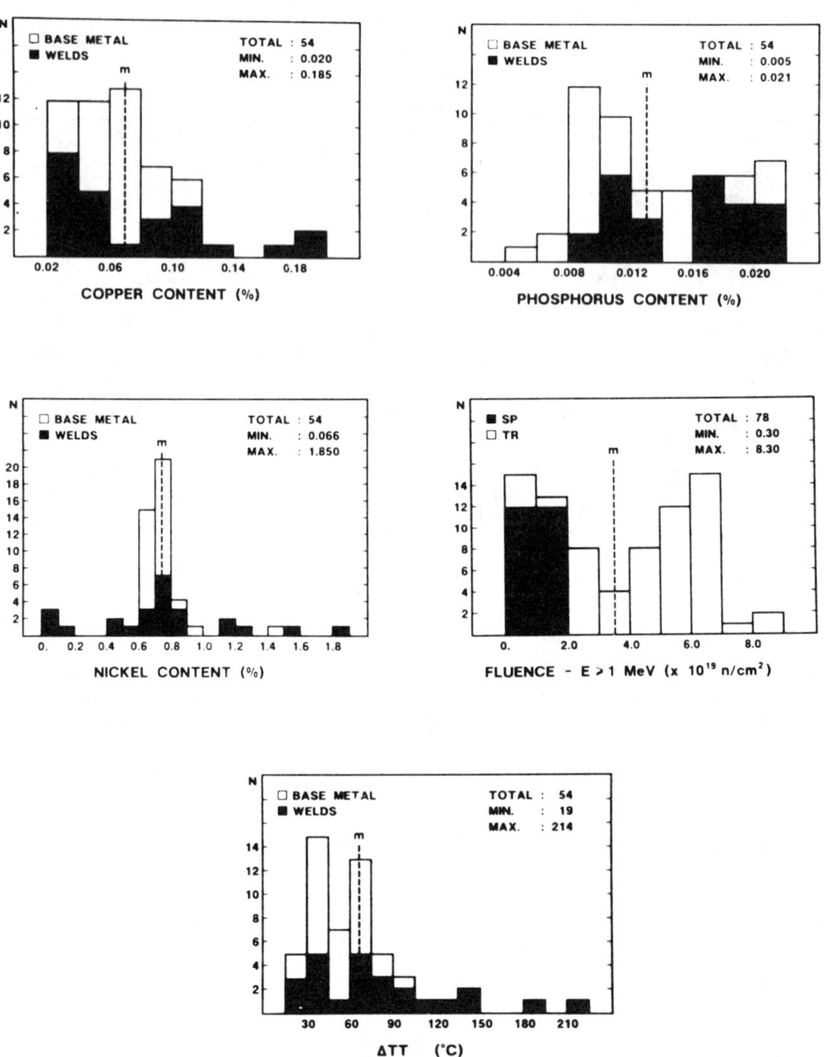

FIG. 7—*Histograms of copper, phosphorus, nickel contents and* ΔTT *and fluence of the data bank.*

ΔTT shifts in transition temperatures. The chemical compositions of the surveillance program results are given in Table 8, and the shifts ΔTT in Tables 10 through 13. Interesting observations can be made about the important number of data (welds) having copper contents less than 0.12% and the wide range in nickel contents provided by the weld data in contrast with the narrow variation in nickel content of the base metal around the mean value of 0.7%. More uniform distributions are observed for the copper and phosphorus elements.

Data Analysis

First, it was assumed that the data could be represented by a function characterized by the product of a chemistry factor (*CF*) and a fluence factor. The fluence factor was postulated to be a power law of the fluence ϕ and that the shift ΔTT in Charpy-V notch transition temperature could be represented by the following relationship

$$\Delta TT(°C): CF \times \phi^n \tag{1}$$

Two studies by multiple linear regression analysis techniques were undertaken to determine the chemistry factor expression and the best adjustment of the n exponent that fit the data well. The first and oldest approach was performed on a mixed data bank containing the 54 test reactor data and the first 12 surveillance program results obtained at the 3×10^{18} n · cm^{-2}, $E > 1$ MeV (4.20×10^{-3} dpa) fluence level (Tables 10 and 11). The second approach was made more recently [7] using the 54 test reactor data only.

FIM (*Fragilisation par Irradiation Moyenne*) and FIS (*Fragilisation par Irradiation Supérieure*) relationships (TR + SP) were as follows:

From the analysis, it was postulated that the structure of the chemistry factor was similar to those proposed in a previous work [8]. The result of this study based on 30 data sets (14 welds and 16 base metals in 16MND 5) of materials irradiated at fluences close to 5×10^{19} n · cm^{-2}($E > 1$ MeV) was a CF of the following type:

$$CF = A + B \times (\text{Cu} - \text{Cu}^*) + C \times \text{Ni}^2 \times \text{Cu} + D(\text{P} - \text{P}^*) \tag{2}$$

Where A, B, C, and D are constants of fit, Cu* and P* are thresholds under which the contributions of the respective elements are neglected, such that

$$(\text{Cu} - \text{Cu}^*) = 0 \text{ if Cu} < \text{Cu}^*$$
$$(\text{P} - \text{P}^*) = 0 \text{ if P} < \text{P}^*$$

The analysis was conducted on the 66 data previously mentioned and used in the first approach.

The ranges of variation in the different element contents (wt%) were 0.005 to 0.021% for the phosphorus, 0.020 to 0.185% for copper, and 0.08 to 1.85% for nickel.

The statistical analyses were performed with different variables to be predicted: one using Expression 1 for ΔTT and the other using $\Delta TT/\phi^n$ as variable. In the first case, the estimated standard deviation on ΔTT is constant while in the second case it becomes a function of the fluence. The results of the analysis indicate the first expression to be more realistic especially for the high fluence data because of too much overpredictive results. The result of the fitting procedure minimizing the estimated standard deviation, leads to the final regression equation, called FIM, for predicting the mean shift ΔTT in transition temperature

$$\text{FIM (°C)} = (17.3) + 1537(\text{P} - 0.008) + 238\,(\text{Cu} - 0.08) + 191\,\text{Ni}^2\text{Cu})\left(\frac{\phi}{10^{19}}\right)^{0.35} \tag{3}$$

where Cu, P, Ni are in wt%, with

$$(\text{P} - 0.008) = 0 \text{ if P} < 0.008\%$$
$$(\text{Cu} - 0.08) = 0 \text{ if Cu} < 0.08\%$$

and

$$\phi: \text{n} \cdot \text{cm}^{-2}\ (E > 1 \text{ MeV})\ (\text{unit of } 10^{19})$$

The multiple correlation coefficient of the regression is high (0.99), and the estimated standard deviation is 9.8°C.

The adjustment of the regression coefficients shows a clear maximum in variance for the copper value of 0.08% while the variance evolution versus phosphorus content curve is flat with no significant change between 0.008 and 0.012%. The value of 0.008% was kept as the phosphorus threshold in the FIM formula to pursue the analogy with the formula given in RG 1.99.1.

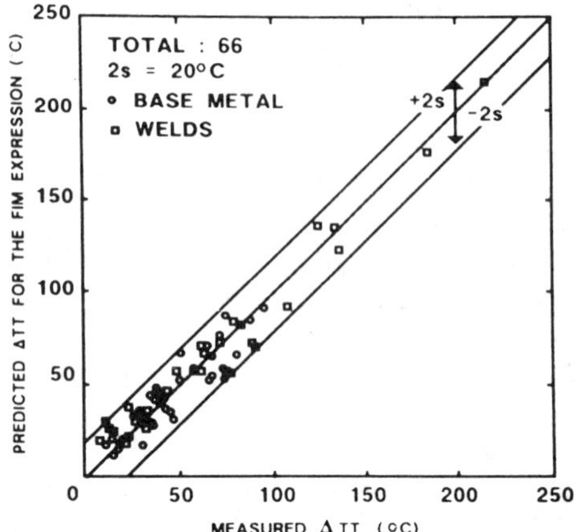

FIG. 8—*Comparison of the* ΔTT *predicted by FIM formula to the measured values.*

A comparison between the ΔTT values predicted by FIM and the experimentally measured shifts is presented in Fig. 8. Examination of this figure does not show any particular grouping of the weld and base metal data. Separate statistical analyses of the weld and base metal populations do not lead to an improvement of the quality of the prediction confirming the visual observation that a unique formula can be used for welds and base meals.

From the analysis of differences between calculated and measured values, an upper bound, FIS, was developed for the population examined. The FIS equation is

$$\text{FIS}\ (^\circ\text{C}) = 8 + (24 + 1537(\text{P} - 0.008) + 238\,(\text{Cu} - 0.08) + 191\ \text{Ni}^2\text{Cu})\left(\frac{\phi}{10^{19}}\right)^{0.35} \quad (4)$$

where

Cu, P, Ni: wt%

with

$$(\text{P} - 0.008) = 0 \text{ if P} < 0.008\%$$
$$(\text{Cu} - 0.08) = 0 \text{ if Cu} < 0.08\%$$

and

$$\phi\text{: n} \cdot \text{cm}^{-2}\ (E > 1\ \text{MeV})\ (\text{unit of } 10^{19})$$

The phosphorus influence in FIM and FIS is about half the effect taken into account by the RG 1.99 Revision 1.

The C.F.3 Relationship (TR) was as follows:

Because of a possible source of bias in statistical analyses that could be introduced by the concentration of SP results in the low fluence region (Fig. 7) new analyses were conducted on TR results only.

After a graphical analysis of the data, multiple linear and nonlinear regression analyses were tested to fit the data with different chemistry factor expressions. Figure 9 shows the comparison between predicted ΔTT by the $CF3$ formula to the measured values.

This work, reported in Ref *7,* allows the following conclusions to be drawn:

- Copper, nickel, and phosphorus were found to be the only elements inducing a detrimental effect on the resistance to neutron irradiation embrittlement.

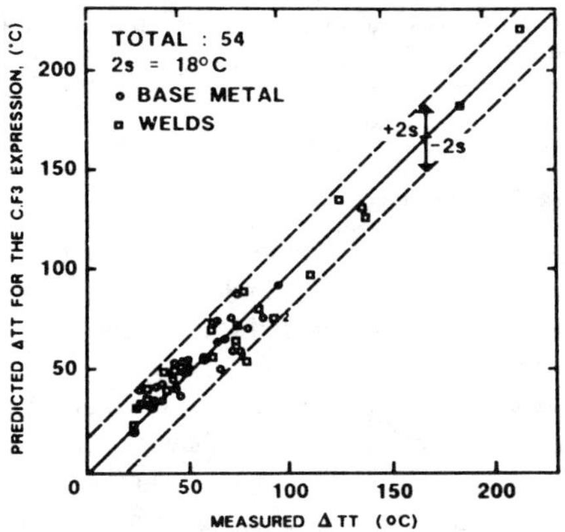

FIG. 9—*Comparison of the* ΔTT *predicted by the CF3 formula to the measured values.*

• An expression was found to predict the magnitude of embrittlement induced by accelerated neutron irradiation in test reactor, in the range of chemical composition and fluence defined by the field in copper-nickel indicated on the Fig. 10, a phosphorus content less than 0.021% and fluences less than 8×10^{19} n · cm^{-2} ($E > 1$ MeV). The expression is defined by the regression equation (Fig. 8)

$$\Delta TT\ (°C) = CF_3 \times \phi^{0.40}$$

with $s = 9.1$°C and $r = 0.973$.

The definition of the CF_3 chemistry factor is given by the following relationship

$$CF_3 = 21.9 + 203\ (Cu - 0.07) + 376\ Cu^{1.5} \times Ni^{2.5} + 1700\ (P - 0.012)$$

where

Cu, Ni, P: wt%

with

$$(Cu - 0.07) = 0 \text{ if } Cu < 0.07\%$$
$$(P - 0.012) = 0 \text{ if } P < 0.012\%$$

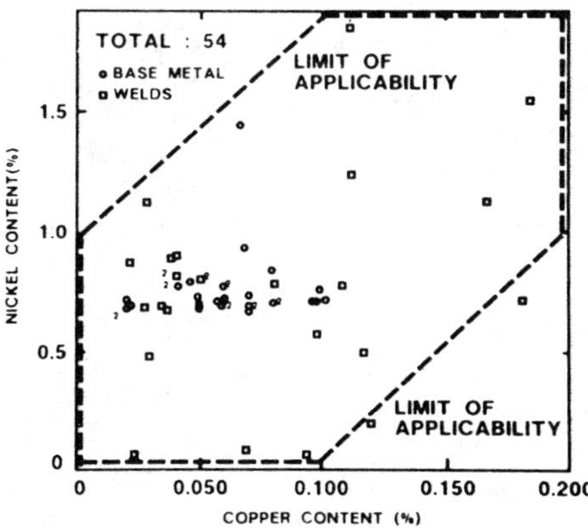

FIG. 10—*Range of applicability of the CF3 formula in the copper nickel diagram.*

The test reactor data analysis leads to a slightly higher fluence exponent than the one obtained on the mixed data bank. This is probably a consequence of the higher neutron fluxes in the test reactors.

Analysis of the Surveillance Program Results

First, it can be noted that for a given irradiation time in the reactor, measured fluence levels for various capsules are similar; this explains a similarity of behavior from a neutronic operating point of view. For an 18 months irradiation-period (Fessenheim, Bugey), fluences ($E > 1$ MeV) are rather low: 3×10^{18} n · cm^{-2} (4.2×10^{-3} dpa).

From a general point of view, mean RT_{NDT} shifts are of about 20°C; because of the scatter, an upper value for RT_{NDT} shift is of about 40°C, for a fluence ($E > 1$ MeV) near 1.5×10^{19} n · cm^{-2} (2.9×10^{-2} dpa) and a lead factor (flux specimen/flux wall) of 2.86 (Fig. 11). There is no significant difference on RT_{NDT} shift between base metal and weld or between base metal and HAZ.

There is good relationship between ΔTK7 and ΔL.Ex., particularly for weld caused by a rather low scatter, probably related to microstructural effect.

It seems that fluence ($E > 1$ MeV) levels of about 1.5×10^{19} n · cm^{-2} (2.9×10^{-2} dpa) are too low for showing a significant decrease in the upper shelf energy level.

As far as the reference metal is concerned, mean value and scatter of RT_{NDT} are close to those measured for the base metal; measured scatter levels (ΔRT_{NDT}) for fluences ($E > 1$ MeV) of about 1.5×10^{19} n · cm^{-2} (2.9×10^{-2} dpa) range from 4 to 23°C.

A slight increase of about 6% in the yield strength is seen; this applies to the base metal, as well as the weld, although the scatter for the welds is half that of the base metal. As things stand today, it does not seem possible to establish a correlation between this variation in the yield strength and the embrittlement of the material. Table 14 shows the ΔRT_{NDT} levels where the calculations are based on various predictive formulae: RG 1.99.1, RG 1.99.2 (project) FIM, FIS, CF_3, Guthrie, and Guthrie $2s$.

Figures 12 and 13 show the differences between the predicted ΔTT and their measured values.

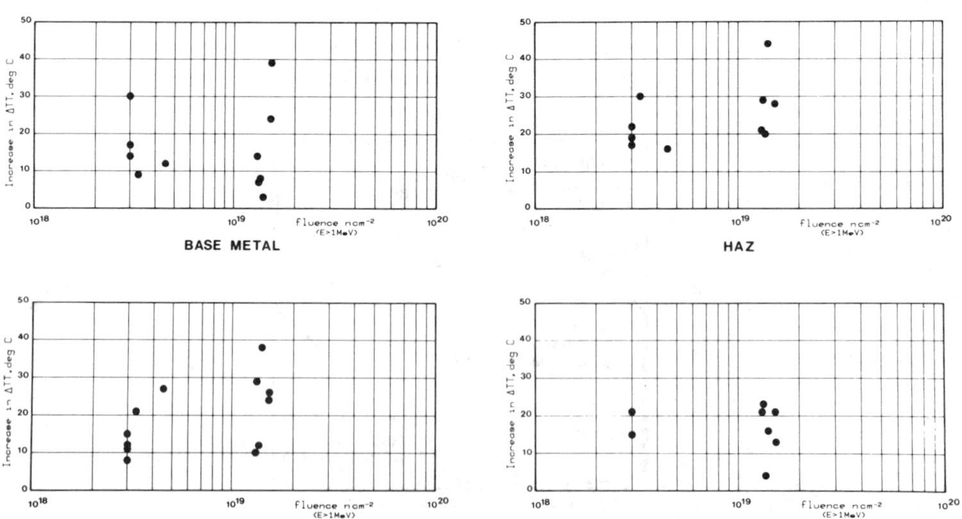

FIG. 11—*Surveillance program results: increase in* ΔTT *versus fluence.*

Base Metal

The results yield the following comments for the base metal:

1. The RG 1.99.1 fits the results, with differences between measured and estimated values ranging from -20 to $+20$°C.
2. The RG 1.99.2 with the added quantity, "Margin," gives a satisfactory estimate of the results, near the FIS formulation for fluences ($E > 1$ MeV) around 1.5×10^{19} n·cm^{-2} (2.9×10^{-2} dpa),
3. The FIS and Guthrie 2*s* formulae give a fair margin as compared with experimental results, with a stronger conservation tendency for Guthrie 2*s* (from 8 to 45°C) than for FIS (from -1 to 36°C).

Weld

For welds

1. The RG 1.99.1 gives better estimates than for the base metal (the nickel content is lower in the weld metal, but the copper and phosphorus contents are higher). For example, in Bugey 2 where nickel levels are 0.65% in the base metal and 0.07% in the weld metal, the differences between calculated then measure RT_{NDT} are -17 to $+22$°C.
2. The RG 1.99.2 (project) gives the most conservative way of looking at the results: the difference between calculated and measured values stretches from 17 to 60°C, even for fluences ($E > 1$ MeV) of 3×10^{18} n · cm^{-2} (4.2×10^{-3} dpa).

This formulation overestimates the ΔRT_{NDT}, notably for the welded joints of the surveillance program. This seems due to the nature of the data banks from which it was established, containing very little data with copper levels less than 0.16%, while most of the welds of the surveillance programs had copper levels around 0.030%.

3. The FIS formulation gives a good fit of the data at 3×10^{18} n · cm^{-2} (4.2×10^{-3} dpa) and at 1.5×10^{19} n · cm^{-2} (2.9×10^{-2} dpa); the differences between calculated and measured values ranging from 10 to 30°C. The Guthrie 2*s* formula gives more conservative results at 3×10^{18} n · cm^{-2} (4.2×10^{-3} dpa) (difference between 20 and 40°C) than at 1.5×10^{19} n · cm^{-2} (2.9×10^{-2} dpa) (varying from -1 to 25°C).

As far as the influence of phosphorus is concerned, comparing predicted and measured ΔRT_{NDT} for base metal and weld in for Gravelines 3 (having same copper content, a similar nickel content and twice as much phosphorus in the weld) it is noted that

1. Guthrie 2*s* gives similar values (about 37°C), which means a much lower conservative rating in the case of a high phosphorus level.
2. The RG 1.99.1 and FIS that take phosphorus into account yield better estimates of the test results.
3. The RG 1.99.2 gives the most important difference between calculated and measured values (26°C).

For the weld metal, the weight given to copper in the Guthrie 2*s* formula gives a much more conservative approach for the Fessenheim and Bugey plants than for the CPY units. This remark is valid for the base metal, since chemical analyses are similar for the weld metals of Bugey 2 and the latter CPY units.

TABLE 14—RT_{NDT} *shifts predicted from RG 1.99.1, RG 1.99.2 (project), FIM, FIS, CF3, Guthrie, Guthrie* 2s *compared to the measured values on base metal, weld, and reference metal of surveillance program.*

	Irradiation			ΔRT_{NDT} Shift, °C							
		Fluence ($E > 1$ MeV)									
Unit	Period, Year	$ncm^{-2} \times 10^{19}$	dpa $\times 10^{-2}$	Actual ΔTT	RG 1.99.1	RG 1.99.2 (project)[a]	FIM[a]	FIS	CF3	Guthrie	Guthrie 2s
Fessenheim 1											
base metal C2	1.5	0.30	0.42	17	13	38	19	32	36	20	47
weld				12	35	40	25	38	43	16	43
Fessenheim 2											
base metal C2	1.5	0.33	0.47	9	13	34	16	29	34	16	43
weld				21	30	38	23	35	40	14	41
Bugey 2											
base metal C2	1.5	0.30	0.40	30	13	38	16	29	34	18	45
weld				8	30	44	22	34	38	20	47
Bugey 3											
base metal	2.5	0.45	0.68	12	15	26	17	30	35	12	39
weld				27	37	78	29	42	45	24	51
Bugey 4											
weld	1.5	0.30	0.43	15	41	48	28	41	44	22	49
reference metal				15	12	14	13	25	32	4	31
Bugey 5											
base metal	1.5	0.30	0.45	14	12	22	14	27	33	10	37
weld				11	29	72	25	37	40	22	49
reference metal				21	12	14	13	25	32	4	31

Tricastin 1											
base metal	4	1.53	2.94	39	27	42	26	42	46	20	47
weld				26	31	50	25	40	45	7	34
reference metal				13	27	24	23	39	45	7	34
Tricastin 2											
base metal	4	1.52	2.95	24	27	44	27	43	47	23	50
weld				24	27	52	23	39	45	7	34
reference metal				21	27	24	23	39	45	7	34
Dampierre 1											
base metal	4	1.33	2.57	7	25	43	25	41	45	21	48
weld				29	28	54	24	39	43	8	35
reference metal				23	25	24	22	37	43	6	33
Dampierre 2											
base metal	4	1.36	2.62	8	26	45	27	43	46	25	52
weld				12	29	50	24	39	43	6	33
reference metal				4	26	24	22	37	43	6	33
Gravelines 1											
base metal	4	1.31	2.52	14	25	40	25	40	45	18	45
weld				10	25	54	22	37	43	8	35
reference metal				21	25	24	22	37	43	6	33
Gravelines 3											
base metal	4	1.4	2.69	3	26	32	23	39	44	11	38
weld				38	46	64	32	48	48	10	37
reference metal				16	26	24	22	38	44	7	34

[a] With added the ''margin'' quantity.

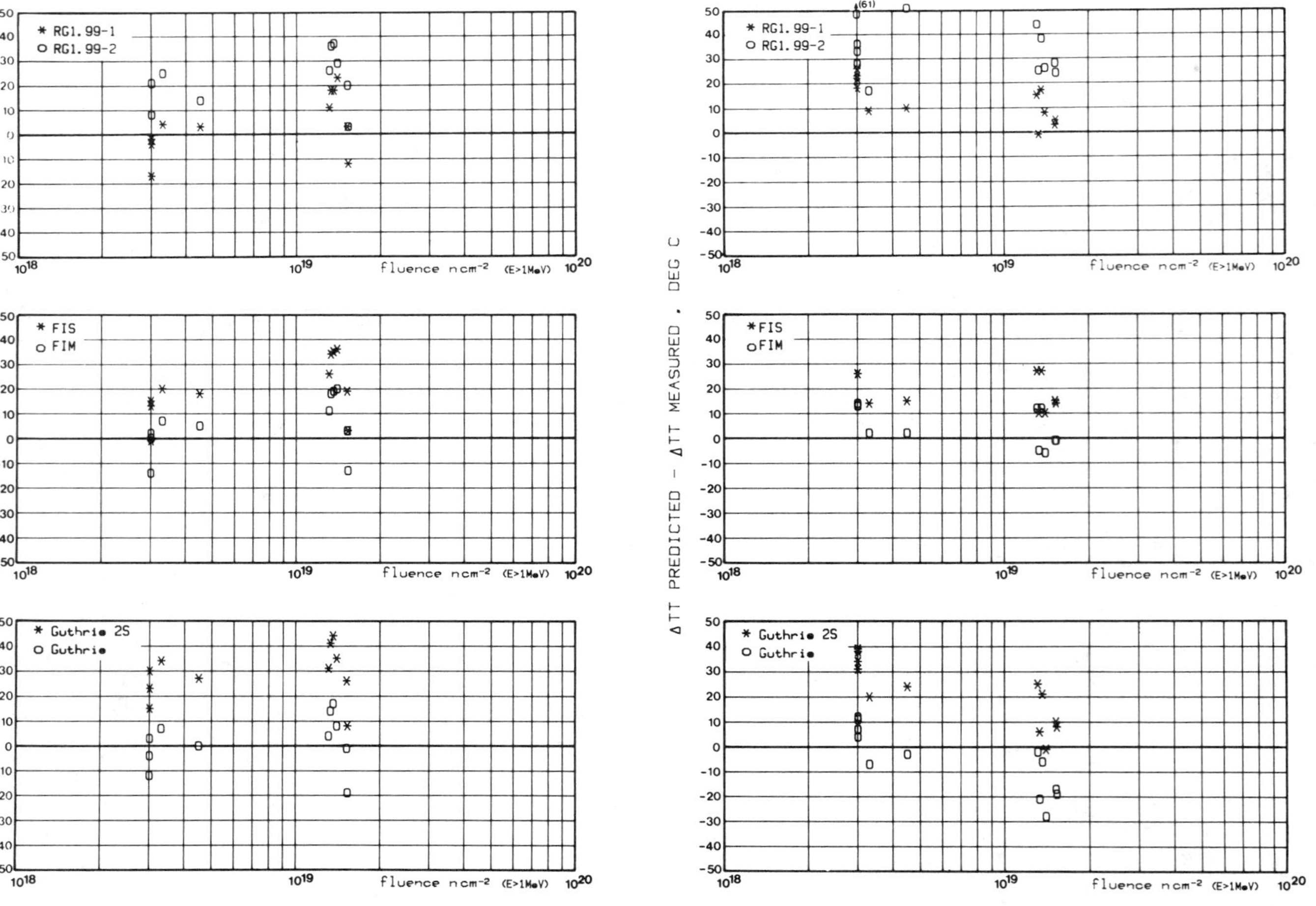

FIG. 12—*Surveillance program results: difference between predicted and measured* ΔTT *for base metal and weld.*

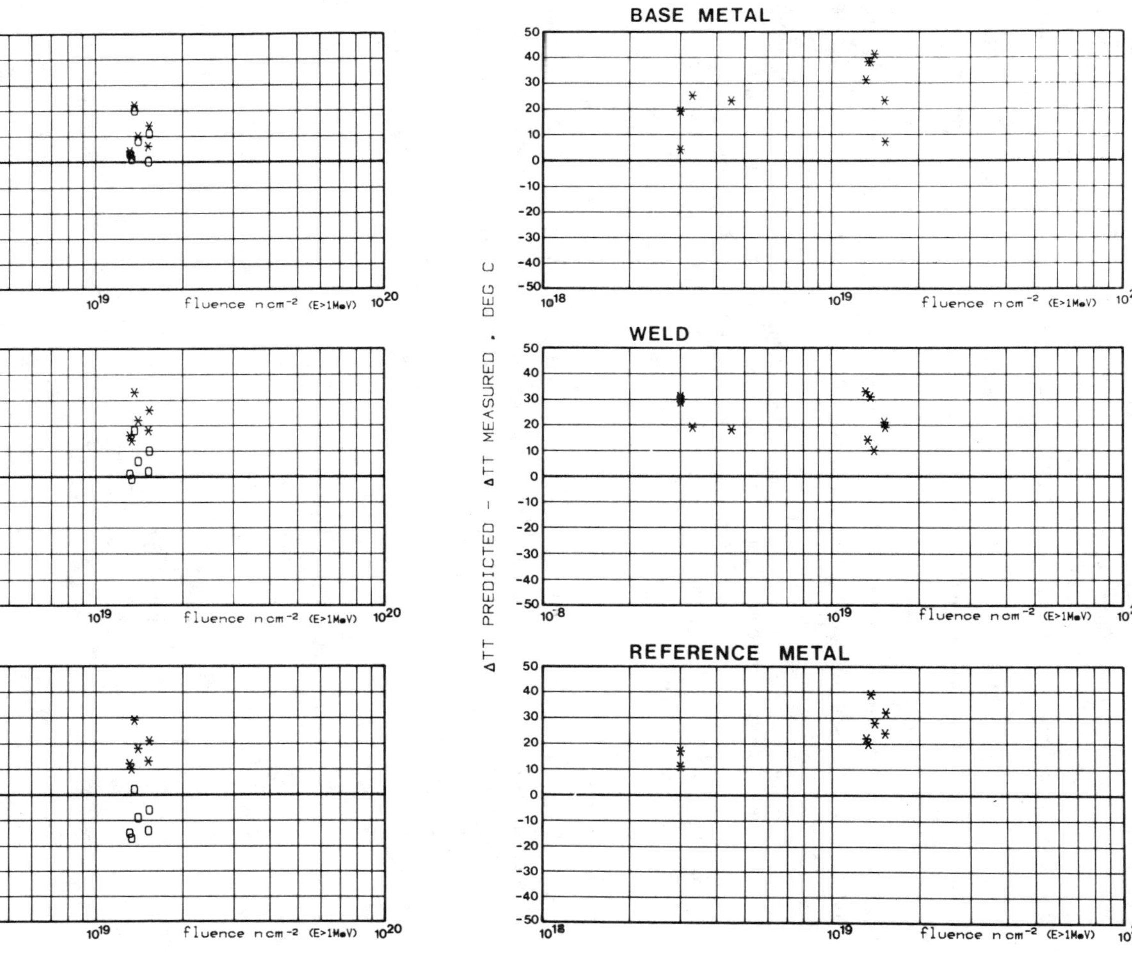

FIG. 13—*Surveillance program results: difference between predicted and measured* ΔTT *for reference metal and for the CF3 formula.*

The effect of the fluence ($E > 1$ MeV) on embrittlement, is shown on Fig. 11:

1. At 3×10^{18} n · cm^{-2} (4.2×10^{-3} dpa), the weak effects of fluence and the intrinsic scatter of the measured results make it difficult to reach a definite conclusion; the lower scatter observed for the weld metal as compared with to the base metal, illustrates this remark (see Tables 10 and 11).

2. At 1.5×10^{19} n · cm^{-2} (2.9×10^{-2} dpa), the average of the test results shows a slight embrittlement, nevertheless, this average embrittlement is of the order of the scatter level for the base metal (20°C).

In the light of these observations, it is premature to make out any precise influence of the fluence on the embrittlement for the materials investigated at fluence levels ($E > 1$ MeV) around 1.5×10^{19} n · cm^{-2} (2.9×10^{-2} dpa).

The next results expected from the first CPY units will be from capsules withdrawn during 1987 after seven years in the reactor (equivalent 20 years of vessel life) corresponding to a fluence around 2.3×10^{19} n · cm^{-2} ($E > 1$ MeV) and in the 1989/90 years for fluence of about 3.8×10^{19} n · cm^{-2} ($E > 1$ MeV). These results should allow one to confirm the validity of the selected models, as far as fluence influence is concerned (Fig. 14).

Conclusion

The results available at the moment, concerning the surveillance program of the irradiation effects on the core shells and welds of 900-MW reactor vessels, show a moderate embrittlement; the average and maximum values of the RT_{NDT} shift are 20 and 40°C, respectively.

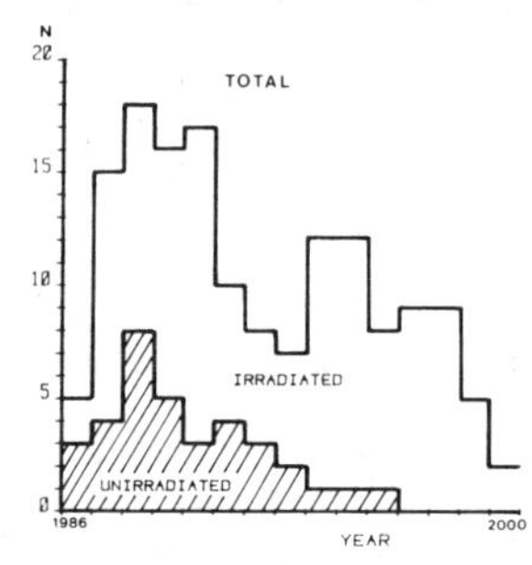

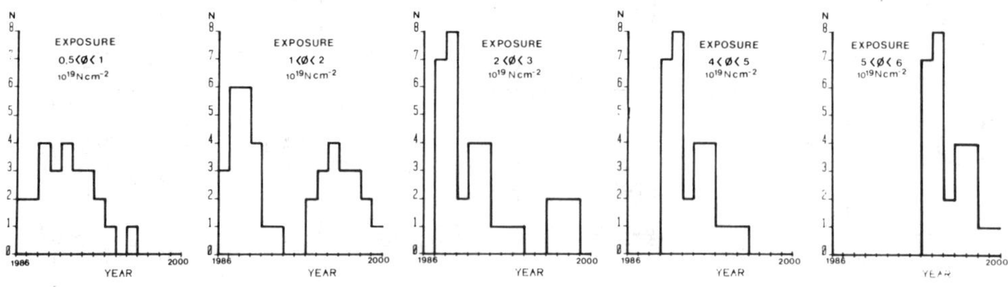

FIG. 14—*Capsules schedule versus fluence for the withdrawals to come.*

These results can be explained by

1. The rather low levels of residual embrittling elements (%Cu maximum = 0.07, %P maximum = 0.014) except for the welds of the six first vessels (%Cu maximum = 0.13, %P maximum = 0.019).
2. Fluence levels ($E > 1$ MeV) of about 1.5×10^{19} n · cm^{-2} (2.9×10^{-2} dpa); resulting in no significant difference between base metal and weld for RT_{NDT} measured shift.

The scatter of the results around the mean embrittlement level, mainly after an 18-month irradiation period, makes it more difficult. The comparison to the different predictions of

(1) either mainly, irradiation results in test reactors are not numerous for the low fluences, or
(2) materials not typical of French practice (%Cu maximum: 0.13 for instance).

Therefore, further work was deemed necessary by both the utility and the fabricator to take these parameters into account.

Future surveillance program results in the next few years will provide more data that should allow a more precise assessment of the influence of flux and fluence on irradiation embrittlement. Results for 20 years (2.3×10^{19} n · cm^{-2} [$E > 1$ MeV]) and 30 years (3.8×10^{19} n · cm^{-2} [$E > 1$ Mev]) irradiation will be available in 1987 and 1989, respectively.

Finally, the maximum predicted levels for end-of-life RT_{NDT} (40 years), based on the FIS formula, are 54°C (base metal) for the 28 vessels of the 900-MW standardized units and 84°C (weld) for Fessenheim and Bugey units.

Acknowledgment

The authors wish to thank F. Faure for reading the manuscript and making useful comments.

References

[*1*] Bevilacqua, A., Bournay, P., Lloret, R., Poitou, M., and Servajean, J. B., "Dosimetry Associated with the Surveillance Program of EDF Reactors Pressure Vessel," *4th ASTM-EURATOM International Symposium,* National Bureau of Standards, Washington, DC, 22–26 March 1982.

[*2*] "Proposed Revision 2 to Regulatory Guide 1.99: Radiation Damage to Reactor Vessel Materials," Draft, U.S. Nuclear Regulatory Commission, Washington DC, Feb. 1986.

[*3*] Guthrie, G. L., "Charpy Trend Curves Based on 177 PWR Data Points," *LWR Pressure Vessel Surveillance Dosimetry Improvement Program,* NUREG/CR-3391, Vol. 2, prepared by Hanford Engineering Development Laboratory, HEDL-TME 83-22, April 1984.

[*4*] Perrin, J. F., Wullaert, R. A., Odette, G. R., Lombrozo, P. M., and Marston, T. U., "Physically Based Regression Correlations of Embrittlement Data from Reactor Pressure Vessel Surveillance Programs," Electric Power Research Institute, NP-3319, 1984.

[*5*] Davies, L. M. and Ingham, T., "Overview of Studies in the UK on Neutron Irradiation Embrittlement of Pressure Vessels Steels," IAEA Specialists' Meeting on Radiation Embrittlement and Surveillance of Reactor Pressure Vessel Steels, Vienna, Austria, 8–10 Oct. 1984, to be published.

[*6*] "Effects of Residual Elements on Predicted Radiation Damage to Reactor Vessel Materials," Regulatory-Guide 1.99, Rev. 1, U.S. Nuclear Regulatory Commission, Washington DC, April 1977.

[*7*] Houssin, B., Brasseur, D., and Meyzaud, Y., "Prediction of Irradiation Embrittlement of PWR Vessel Materials: Effect of Cu, P, Ni and Fluence," *ASM International Conference on Fatigue, Corrosion Cracking, Fracture Mechanics, Failure Analysis,* 2–6 Dec. 1985, Salt Lake City, UT, to be published.

[*8*] Guionnet, C., Houssin, B., Brasseur, D., Lefort, A., Gros, D., and Perdreau, R., *Effects of Radiation on Materials: Eleventh International Symposium, STP 782,* H. R. Brager and J. B. Perrin, Eds., American Society for Testing and Materials, Philadelphia, 1982, pp. 392–411.

Wayne A. Pavinich[1] *and Arthur L. Lowe, Jr.*[2]

The Effect of Thermal Annealing on the Fracture Properties of a Submerged-Arc Weld Metal

REFERENCE: Pavinich, W. A. and Lowe, A. L., Jr., "**The Effect of Thermal Annealing on the Fracture Properties of a Submerged-Arc Weld Metal,**" *Influence of Radiation on Material Properties: 13th International Symposium (Part II), ASTM STP 956,* F. A. Garner, C. H. Henager, Jr., and N. Igata, Eds., American Society for Testing and Materials, Philadelphia, 1987, pp. 448–460.

ABSTRACT: Charpy impact and compact fracture specimens were tested to determine the response of actual reactor vessel weld metals to the in-situ annealing cycle (454°C for 168 h) suggested by the Electric Power Research Institute (EPRI). This annealing cycle was developed to recover fracture properties lost as a result of radiation damage. The Linde 80 submerged-arc weld metals from the Babcock and Wilcox (B&W) Owners Group Integrated Reactor Vessel Material Surveillance Program were irradiated in power reactors. Resultant data are compared with the data from the EPRI sponsored in-situ annealing program. Comparison of the power reactor and test reactor data show similar responses to annealing. One-hundred percent recovery of the upper shelf and transition temperature was achieved after annealing for both reactor types. The EPRI *J-R* data showed only partial recovery with annealing. However, the B&W submerged-arc weld metals exhibited four responses to annealing: 1) complete recovery, 2) partial recovery, 3) negligible effect of either irradiation or annealing on the *J-R* fracture toughness, and 4) an increase in *J-R* fracture toughness to greater than the unirradiated value.

These varied responses indicate that the mechanism(s) of annealing recovery may be more complex than heretofore believed. A qualitative rationale, based on a defect cluster mechanism and a copper precipitation mechanism, is employed to explain the various annealing responses.

KEY WORDS: in-situ annealing, submerged-arc weld, radiation damage, reactor vessel, *J-R* curve, Charpy impact testing

Mechanical properties of reactor vessel steels are altered as a result of defects produced by neutron irradiation. Specifically, the Charpy transition temperature, yield and tensile strengths increase while fracture toughness, uniform and total elongations decrease with increasing neutron fluence. The reduction in fracture toughness and mechanical properties causes the calculation of more restrictive temperature-pressure operating limitations of the reactor and, therefore, the operating flexibility. This characteristic is expected to produce unacceptable operating limitations for reactor vessels fabricated with the high-copper Linde 80 submerged-arc welds contained in earlier reactors. The low initial toughness of these weld metals, caused by a large volume fraction of inclusions, is exacerbated by the high copper content which accelerates radiation embrittlement. Consequently, the reactor vessel operating limitations are reduced to unacceptable levels. In-situ annealing of the reactor vessel has been proposed as a method to recover fracture toughness and

[1] Senior Research Engineer, Babcock & Wilcox Co., Research and Development Division, P.O. Box 11165, Lynchburg, VA 24506-1165.

[2] Advisory Engineer, Babcock & Wilcox Co., Nuclear Power Division, P.O. Box 10935, Lynchburg, Va 24506-0935.

mechanical properties of the affected reactor vessel materials by the reduction or the elimination of the neutron irradiation produced defects [*1*].

It has been hypothesized that reactor vessel radiation embrittlement is actually comprised of two mechanisms [*2,3*]. The first involves a copper precipitation mechanism which is dominant at high temperature and low neutron flux. The second is a radiation damage defect cluster mechanism which is dominant at lower temperatures and high neutron flux. Evidence of copper precipitation has been reported elsewhere. Grant et al demonstrated with field ion microscopy (FIM) and an atom probe that copper atoms cluster as a result of irradiation [*4*]. Central Electricity Generating Board (CEGB) studies have shown that the neutron irradiation induced shift of the transition temperature was larger for data obtained in relatively low flux, low fluence Magnox reactors than similar data obtained in high flux test reactors [*5*]. These data coincide with the small angle neutron scattering (SANS) experiments which demonstrated that Magnox reactor irradiated materials contained a greater amount of copper precipitation than the same material irradiated in a test reactor. In this case, age hardening is the primary mechanism of embrittlement, where an increase in the irradiation temperature accelerates the embrittlement process resulting in a reduced time to peak hardness. Norris [*6*] and Williams et al [*7*] observed that increases in irradiation temperature caused decreases in the change in transition temperature for power reactor data and test reactor data, respectively. In these studies, the radiation damage defect cluster mechanism is apparently dominant. Increases in irradiation temperature cause defect annihilation thus slowing the embrittlement process. As pointed out by Williams et al [*7*], the embrittlement process is quite complex and can lead experimenters to different conclusions as to the role of neutron flux, neutron fluence, and chemical composition on this process for different irradiation environments. These observations raise a question as to the validity of using simulated power reactor vessel materials irradiated in test reactors to develop data for making evaluations of power reactor vessel materials response to the process of in-situ annealing.

Research funded by EPRI [*1*] has shown that a 100% recovery of the Charpy upper-shelf energy can be achieved if the irradiated steel is annealed at 454°C (850°F) for 168 h. This conclusion is based primarily on Charpy impact data obtained from simulated reactor vessel weld metals irradiated in a high flux test reactor facility. Fracture toughness data obtained in the same program produced results wherein the tearing modulus was fully recovered and the J_{Ic} values only partially recovered as a result of the in-situ annealing cycle.

The goal of the work described in this paper is to address this validity question by comparing the EPRI findings with fracture toughness and Charpy impact data derived from several production Linde 80 weld metals which were irradiated in power reactors. The program as originally defined was to provide confirmatory data in support of test reactor annealing results. However, the results obtained in this work indicate that the process of in-situ annealing of submerged-arc weld metals is more complex than was previously believed. Further work will be required to answer the questions arising from this program.

Experimental Procedure

Several different Linde 80 weld metals from Babcock & Wilcox (B&W) Integrated Reactor Vessel Material Surveillance Program irradiation capsules were annealed at 454°C (850°F) for 168 h to duplicate the EPRI recommended in-situ annealing cycle. Table 1 itemizes the chemical composition of each weld metal. The irradiation history of each material is presented in Table 2.

Charpy specimens were tested in accordance with ASTM E23. Transition curves for five of the weld metals were plotted for the unirradiated and the irradiated conditions. One to three Charpy specimens per weld metal were annealed and tested. The data was compared to the irradiated and unirradiated data for each of the five weld metals. All fracture toughness tests were conducted in

TABLE 1—*Chemical composition of the submerged-arc weld metals (% by weight).*

	Weld Metal						
Component	A	B	C	D	E	F	G
C	0.090	0.075	0.090	0.0907	0.0786	0.0763	0.080
Mn	1.619	1.675	1.494	1.6267	1.472	1.452	1.448
P	0.0135	0.015	0.0161	0.0184	0.0157	0.0158	0.0113
S	0.0151	0.0128	0.0158	0.0093	0.0154	0.0157	0.0129
Si	0.04575	0.0490	0.5138	0.540	0.5429	0.5127	0.4876
Cr	0.1034	0.130	0.060	0.1067	0.0655	0.0861	0.080
Ni	0.6638	0.625	0.5863	0.5933	0.5881	0.5933	0.5893
Mo	0.400	0.390	0.3925	0.3987	0.4029	0.378	0.3831
Cu	0.3275	0.275	0.275	0.4187	0.3176	0.2143	0.270
V	0.0055	0.006	0.0058	0.0096	0.006	0.0059	0.0081
Al	0.0054	0.005	0.005	0.0185	0.0051	0.0111	0.0082
Ti	0.0015	0.002	0.002	0.002	0.0019	0.0019	0.0020
Co	0.019	0.178	0.0201	0.0189	0.0205	0.0142	0.0172
Cb	0.0046	0.0055	0.0041	0.0053	0.0045	0.0043	0.0049
Ta	0.0076	0.0118	0.0076	0.0073	0.0067	0.0100	0.0079
Sn	0.0088	0.0060	0.0096	0.0050	0.0081	0.0031	0.0029
B	0.0010	0.0010	0.0010	0.0010	0.0010	0.0010	0.0010
Pb	0.0021	0.0025	0.0020	0.0025	0.0028	0.0018	0.0017
As	0.0144	0.0070	0.0089	0.0093	0.0097	0.0076	0.0072
Zr	0.0030	0.0033	0.0040	0.0037	0.0037	0.0034	0.0035
W	0.0161	0.0185	0.0205	0.0165	0.0211	0.0145	0.0160

accordance with ASTM E813. *J-R* curves of the unirradiated, irradiated, and irradiated/annealed weld metal were plotted for comparison.

Results

All the Charpy impact data exhibited similar characteristics as shown in Figs. 1 through 5. In each case a 100% recovery of the upper shelf and transition temperature was achieved as a result of annealing the material at 454°C (850°F) for 168 h. These observations are consistent with data developed in the EPRI program where a complete recovery of the Charpy transition and upper-shelf properties was achieved after annealing [*1*]. Generally, the fracture toughness data did not exhibit the same response to annealing as the Charpy data. The behavior of the fracture toughness data can be described in four categories.

TABLE 2—*Irradiation history of the submerged-arc weld metals.*

Weld Metal	Fast Neutron Fluence, n/cm^2, $E > 1.0$ MeV	Fast Neutron Flux, n/cm^2/s, $E > 1.0$ MeV
A	7.0×10^{18}	7.4×10^{10}
B	12.9×10^{18}	1.2×10^{11}
C	4.0 and 6.6×10^{18}	1.0×10^{11}
D	7.0×10^{18}	7.4×10^{10}
E	7.0×10^{18}	7.4×10^{10}
F	7.0×10^{18}	7.4×10^{10}
G	7.0×10^{18}	7.4×10^{10}

NOTE—Irradiation Temperature = 304°C (580°F)

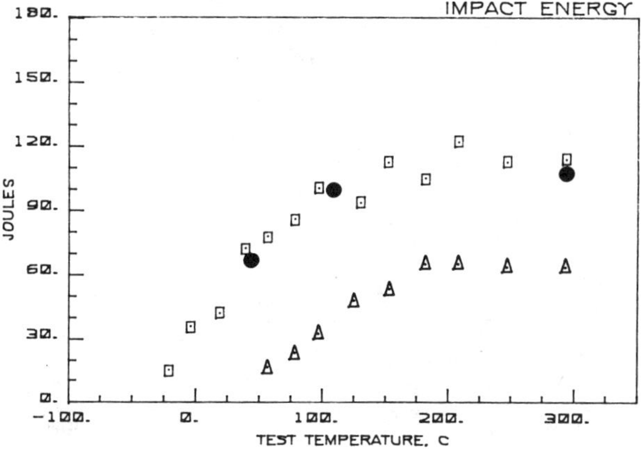

FIG. 1—*Charpy transition curves—Weld Metal A.*

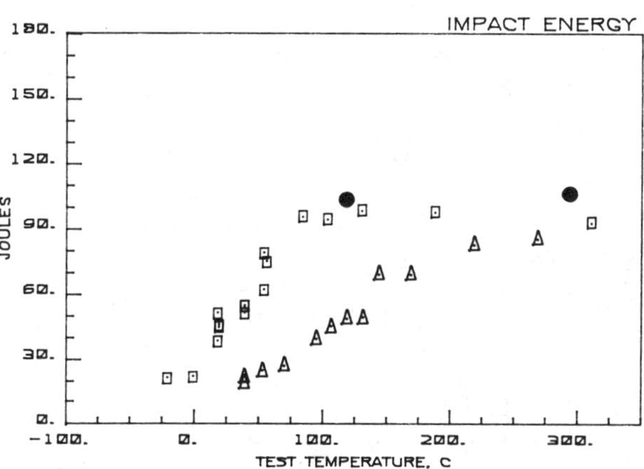

FIG. 2—*Charpy transition curves—Weld Metal B.*

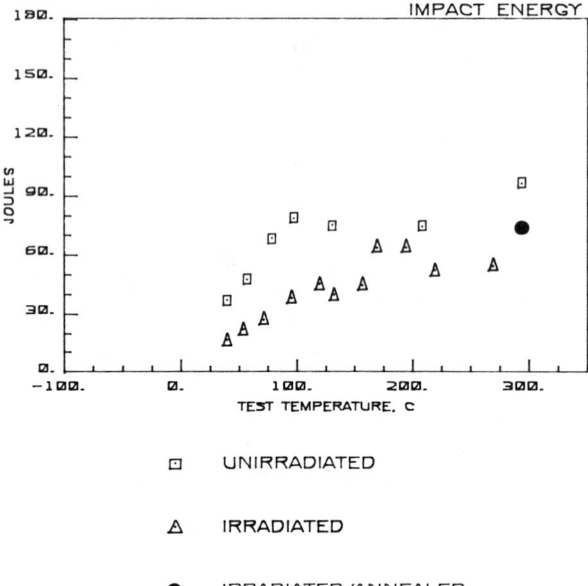

FIG. 3—*Charpy transition curves—Weld Metal D.*

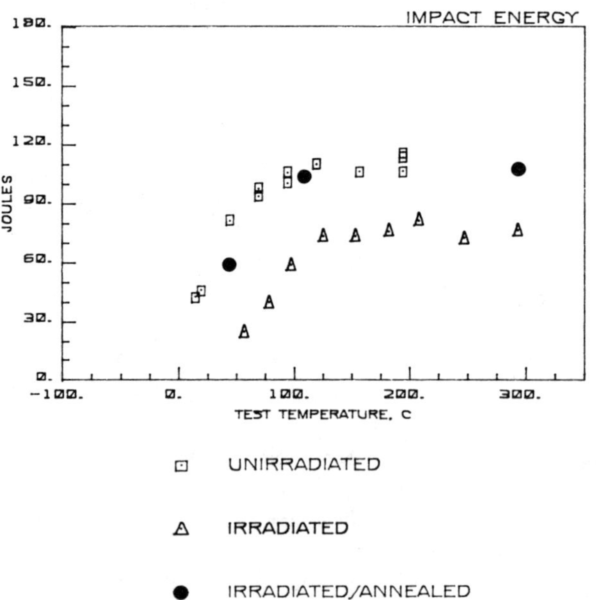

FIG. 4—*Charpy transition curves—Weld Metal F.*

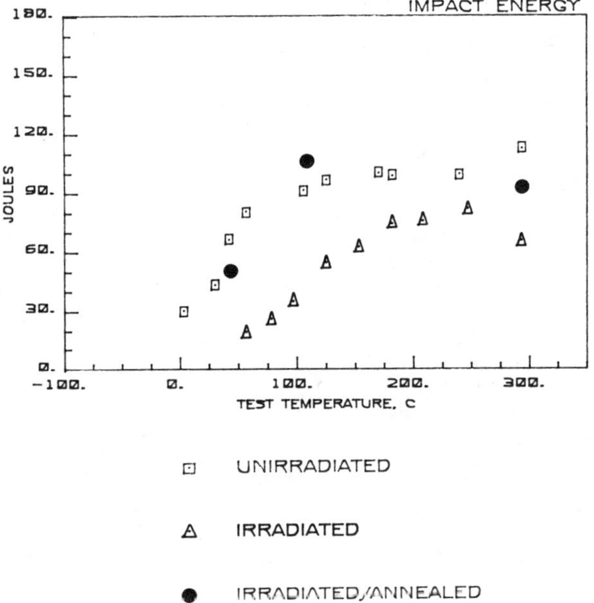

FIG. 5—*Charpy transition curves—Weld Metal G.*

1. Complete recovery of the *J-R* curve after annealing. Figures 6 and 7 show a 30% reduction in the *J-R* curves after irradiation and *J-R* curve which are equivalent to the unirradiated curves as a result of annealing.

2. Partial recovery of the *J-R* curve after annealing. Figures 8 and 9 show reductions of the *J-R* curves as a result of neutron irradiation and increases in the *J-R* curves to approximately 75% of the unirradiated curves after annealing.

3. Negligible effect of irradiation or annealing on the *J-R* curve. Figure 10 shows a weld metal whose fracture properties exhibit little change as a result of irradiation or the subsequent annealing process.

4. Increases in *J-R* curves above unirradiated values as a result of annealing. Figures 11 and 12 show that irradiation causes a decrease in the *J-R* curve on the order of 30 to 50%. However, both the irradiated/annealed and unirradiated/annealed *J-R* curves increased markedly above unirradiated values.

These responses of the *J-R* curves to annealing imply that the annealing process causes some other microstructural changes in addition to irradiation induced defect annihilation. Metallography and scanning electron microscopy (SEM) were performed in an attempt to identify microstructural changes that would increase fracture resistance with the annealing cycle. No definable changes were observed using these techniques. Fine scale microstructural changes, such as a copper precipitation (2 to 20 nm diameter), may be causing the observed behavior. However, more sophisticated techniques such as scanning transmission electron microscopy (STEM), SANS, or FIM will be required to identify them.

The behavior of the *J-R* data for the B&W production Linde 80 submerged-arc weld metals is more varied than the EPRI program data from a simulated reactor vessel weld metal which exhibited only a partial recovery of fracture toughness properties with annealing.

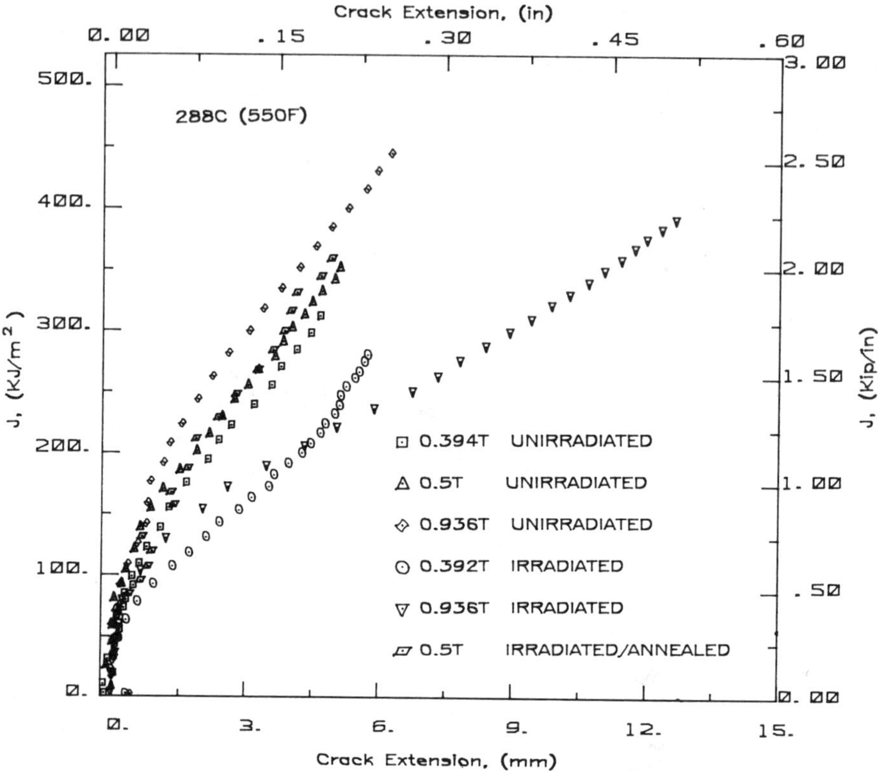

FIG. 6—*Complete recovery of* J-R *curves after annealing—Weld Metal A.*

Discussion

It is evident from the data presented that the response of Charpy energies to annealing is not indicative of the response of the fracture toughness for this series of weld metals. The large difference in the strain rates of the two different test methods is probably a contributing factor.

The *J-R* data show a varied response to annealing. This implies that the mechanism(s) of annealing recovery may be more complex than heretofore believed. Radiation embrittlement is a complex process probably involving a defect cluster mechanism and a fine scale copper precipitation mechanism as previously mentioned. Therefore, the embrittlement recovered by annealing is related to the amount of defect recombination and the amount of enhanced aging or overaging caused by the annealing cycle. According to the work of other researchers [*3*], the precipitation mechanism operates over extended periods of time. Accordingly, the variation of the response of fracture toughness parameters to annealing can be qualitatively explained. In Case 1, where 100% recovery of properties is achieved, precipitation of a copper rich phase is probably complete or nearly complete prior to irradiation due to heat treatments such as post weld stress relieving. Annealing does not cause significant additional precipitation or overaging, while defect clusters are totally annihilated thus causing the full recovery but no increase in the *J-R* curve magnitudes over the unirradiated material.

The partial recovery of Case 2 can be attributed to incomplete copper precipitation prior to irradiation, followed by additional precipitation combined with defect cluster production during irradiation, and finally followed by some or no precipitation and defect annihilation during

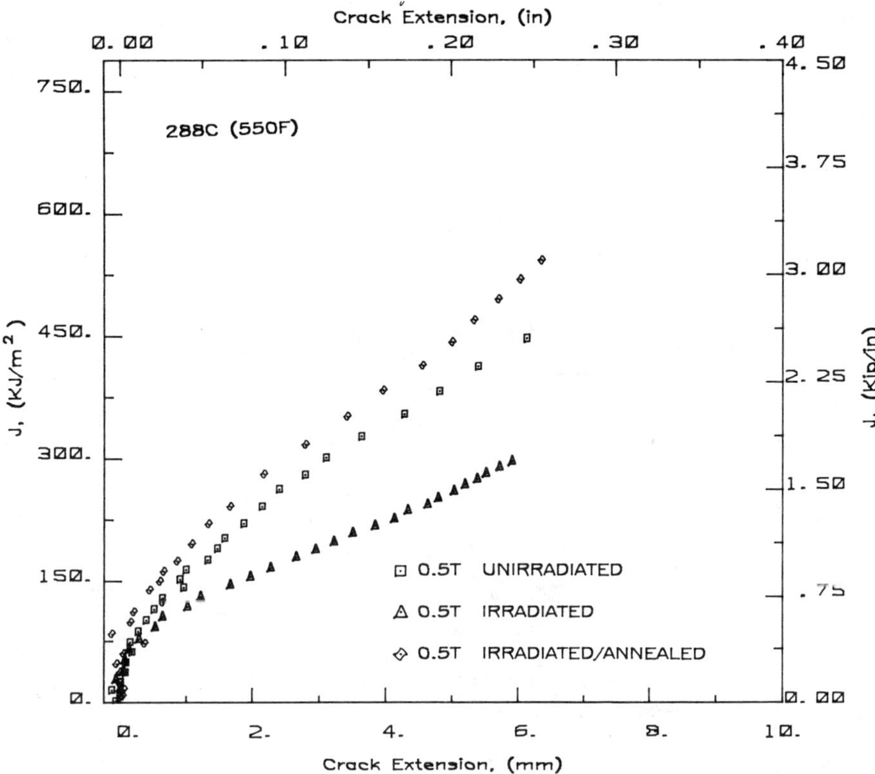

FIG. 7—*Complete recovery of* J-R *curves after annealing—Weld Metal F.*

annealing. In this case, copper precipitation during irradiation causes embrittlement which is not recovered during annealing.

The materials of Case 3 are apparently insensitive to either fast neutron induced radiation embrittlement or embrittlement due to copper precipitation. Perhaps copper precipitation was complete prior to irradiation as a result of thermal processing and stress relief cycles experienced during fabrication. In Case 4, significant overaging due to annealing could have caused the observed increases in the *J-R* curves over the unirradiated *J-R* curves, for both the irradiated/annealed and unirradiated/annealed conditions.

Although these explanations may be subject to debate, any evaluation of radiation embrittlement and annealing response must consider copper precipitation and when it occurs. The amount of copper in solid solution or in precipitate form is affected by prior thermal processing history or heat treatment, neutron irradition enhanced precipitation, and aging or over-aging during the annealing cycle. Therefore, it is essential to quantify the amount of precipitation in the unirradiated, irradiated, and irradiated/annealed conditions to fully understand the material response to in-situ annealing.

Acknowledgment

This study was sponsored by The Babcock & Wilcox Owners Group as a part of their Reactor Vessel Integrity Program. The authors acknowledge and express appreciation for their support.

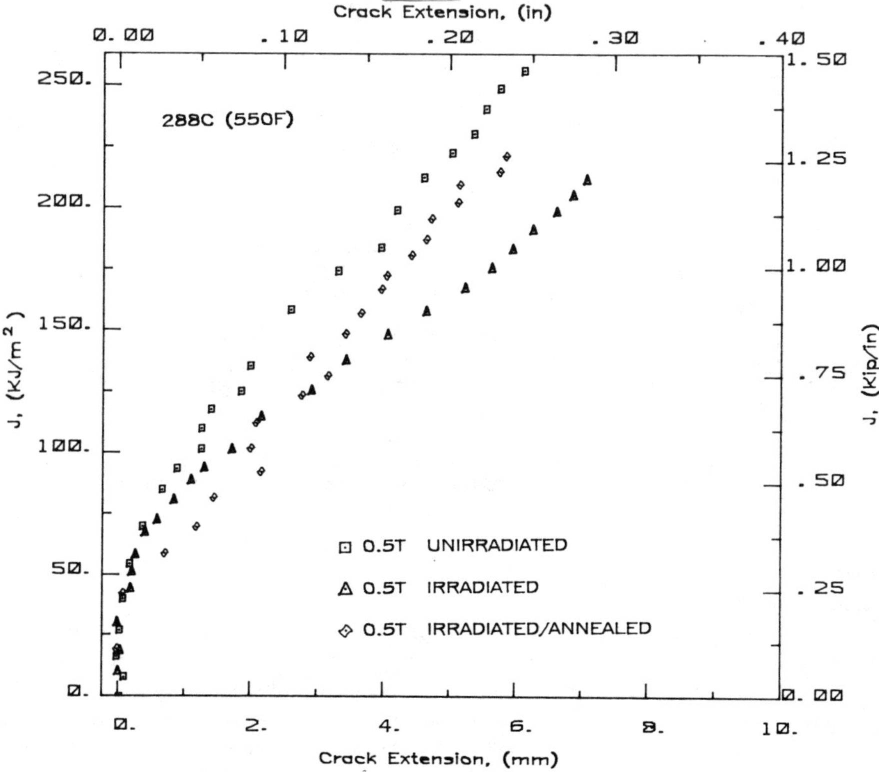

FIG. 8—*Partial recovery of* J-R *curves after annealing—Weld Metal D.*

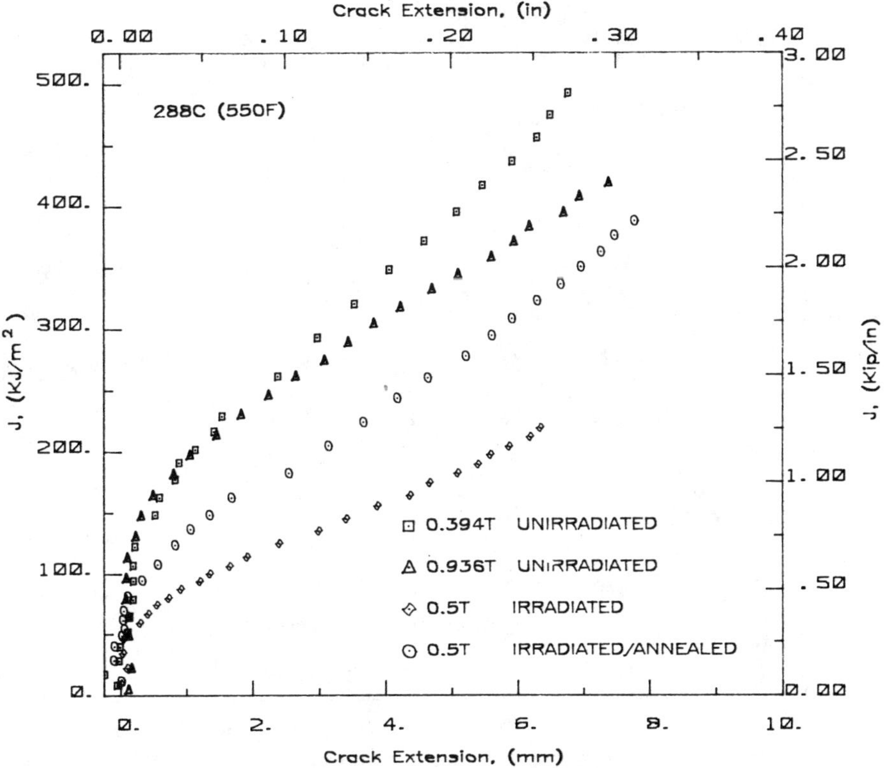

FIG. 9—*Partial recovery of* J-R *curves after annealing—Weld Metal E.*

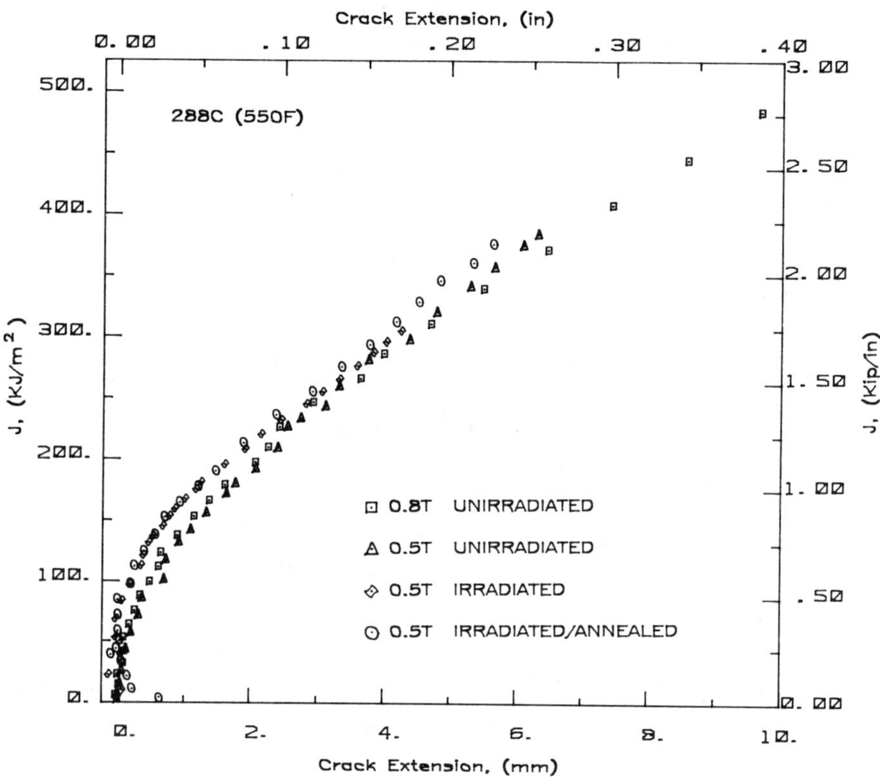

FIG. 10—*Negligible effect of irradiation or annealing on the* J-R *curves—Weld Metal G.*

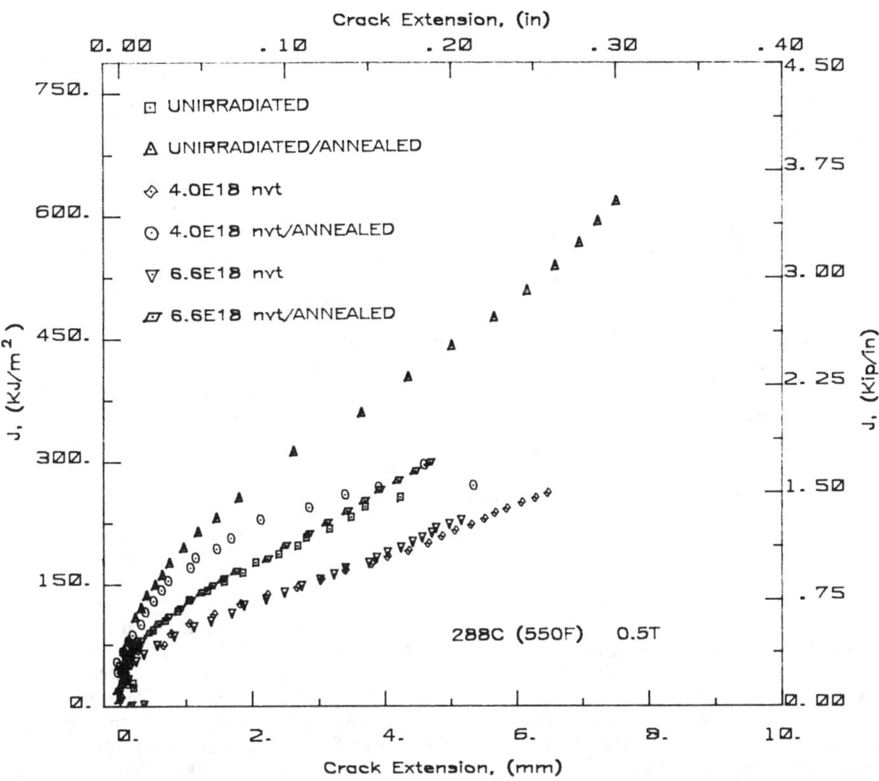

FIG. 11—*Increases in* J-R *curves above unirradiated values as a result of annealing—Weld Metal C—288°C.*

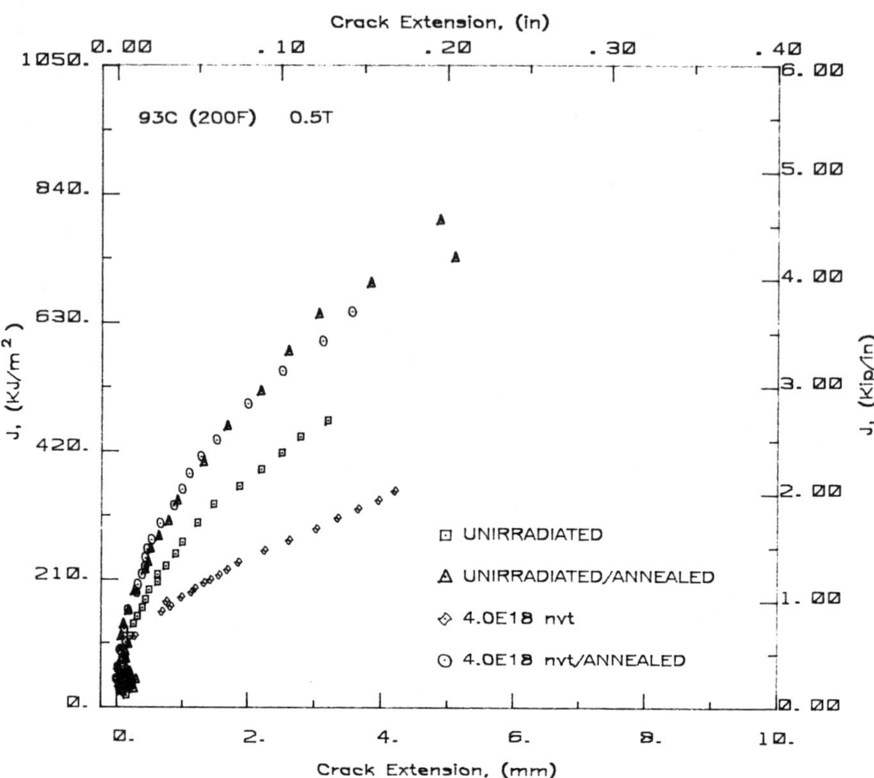

FIG. 12—*Increases in* J-R *curves above unirradiated values as a result of annealing—Weld Metal C—93°C.*

References

[*1*] Mager, T. R., "Feasibility of and Methodology for Thermal Annealing an Embrittled Reactor Vessel," EPRI NP-2712, Vol. 1 and 2, Electric Power Research Institute, Nov. 1982.

[*2*] Odette, G. R. et al, "Physically Based Regression Correlations of Embrittlement Data from Reactor Vessel Surveillance Programs," EPRI NP-3319, Electric Power Research Institute, Jan. 1984.

[*3*] Fisher, S. B. et al, "Copper Precipitation in Pressure Vessel Steels," CEGB Report in three parts, TPRD/B/0396/N84, TPRD/B/0397/N84, TPRD/B/0398/N84, Central Electricity Generating Board, 1984.

[*4*] Grant, S. P., Earp, S. L., Brenner, S. S., and Burke, M. G., "Phenomenological Modeling of Radiation Embrittlement in Light Water Reactor Vessels with Atom Probe and Statistical Analysis," *Proceedings,* Second International Symposium on Environmental Degradation of Materials in Nuclear Power Systems—Water Reactors, Sept. 1985.

[*5*] Priest, R. H., Charnock, W., and Neale, B. K., "The Fracture Toughness of Accelerated Irradiated Submerged-Arc Weld Metal, in *Effects of Radiation on Materials: Twelfth Conference, STP 870,* American Society for Testing and Materials, Philadelphia, 1984.

[*6*] Norris, E. G., "A Service Laboratory's View of the Status and Direction of Reactor Vessel Surveillance," in *Radiation Embrittlement and Surveillance of Nuclear Reactor Pressure Vessels: An International Study, STP 819,* American Society for Testing and Materials, Philadelphia, 1983.

[*7*] Williams, et al, "The Influence of Copper, Nickel, and Irradiation Temperature on the Irradiation Shift of Low Alloy Steels," *Proceedings,* Second International Symposium on Environmental Degradation of Materials in Nuclear Power Systems—Water Reactors, Sept. 1985.

J. Russell Hawthorne[1]

Steel Impurity Element Effects on Postirradiation Properties Recovery by Annealing

REFERENCE: Hawthorne, J. R., "**Steel Impurity Element Effects on Postirradiation Properties Recovery by Annealing,**" *Influence of Radiation on Material Properties: 13th International Symposium (Part II), ASTM STP 956,* F. A. Garner, C. H. Henager, Jr., and N. Igata, Eds., American Society for Testing and Materials, Philadelphia, 1987, pp. 461–479.

ABSTRACT: The influence of copper content and phosphorus content on notch ductility recovery by 399°C postirradiation heat treatment was explored for A 533-B and A 302-B pressure vessel steels. Charpy-*V* (C_v) specimens for the investigation were obtained from ten plates produced from four (4-way split) laboratory melts. The plates were 15.2 mm thick but were heat treated to reproduce the microstructure of 150-mm and thicker A 302-B plates at the quarter-thickness location.

The C_v specimens were irradiated in two assemblies at 288°C (550°F) to a fluence of ~2.5 × 10^{19} n/cm^2 in a light-water-cooled and moderated test reactor. Notch ductility properties in the as-irradiated and 399°C, 168 h postirradiation-annealed conditions were determined. In addition, separate sets of specimens were thermally conditioned at 288°C and at 288°C followed by 399°C to benchmark the effects of temperature in the absence of irradiation.

The results indicate that copper has a significant influence on the magnitude of residual embrittlement after annealing. In contrast, phosphorus contents in the range of 0.002 to 0.025% were found not to have an effect on residual embrittlement either in high or low copper steels. Essentially full recovery in 41-J transition temperature was observed for high phosphorus, low copper content steels. Effects of nickel alloying on recovery behavior were also investigated through data comparisons for A 302-B versus A 533-B plates.

KEY WORDS: radiation embrittlement, notch ductility, pressure vessel steels, embrittlement relief, annealing, A 302-B steel, steel impurities, Charpy-*V* properties, nuclear radiation, postirradiation heat treatment, A 533-B steel

Studies of heat treatment for the mitigation of radiation-induced embrittlement are finding increasing evidence of an effect of material composition on the amount of notch ductility recovery obtained [*1,2*]. Indications of a composition effect on recovery can be found in current data banks on 399°C annealing and in data banks on 454°C annealing. The large scatter observed in percentage recovery with 399°C annealing is a particularly clear indication of a dependency of some type. In view of these indications, it has been concluded that composition must be treated as a potentially critical variable influencing both recovery rate and the degree of residual embrittlement after the anneal.

Metallurgical influences on the annealing process have not been investigated in detail previously [*3*], but obviously, their identification forms a prerequisite to the weighing of commercial annealing applications. This point is reinforced by the fact that reactor surveillance programs are limited in

[1] Research metallurgist, Materials Engineering Associates, Inc., 9700-B Martin Luther King, Jr. Highway, Lanham, MD 20706-1837.

material are included in a surveillance program. As a further limitation, surveillance capsules normally contain 15 or fewer Charpy-*V* (C_v) specimens and, at most, three tension test specimens of any one material. Thus, excess specimens are not available in the quantities required to probe material annealing behavior even for those few materials chosen for surveillance. Laboratory tests thus must be relied upon for the bulk of information on metallurgical variable effects (such as composition effects) on annealing behavior. This is also true for studies of reirradiation behavior after the anneal.

The magnitude and direction of individual effects on annealing recovery with a given heat treatment are expected to vary among elements. The reason is simply that more than one radiation effects mechanism is suspected of acting in the embrittlement process at elevated temperature. Two impurity elements of primary concern are copper and phosphorus; these elements have been identified as key factors governing the irradiation response of steels and weld metals at typical reactor vessel service temperatures (~288°C) [*4*]. Nickel alloying is an additional concern. Experiments have demonstrated that nickel reinforces the detrimental effect of copper on radiation resistance [*1*].

The present investigation was undertaken for the Nuclear Regulatory Commission (NRC) to explore the role of the three elements in apparent annealing behavior and annealing effectiveness. The investigation employed laboratory melts having closely controlled composition variations about the base compositions for A 302-B steel (0.4% nickel, maximum) and A 533-B steel (0.4 to 0.7% nickel). A total of ten composition variations were selected for the exploratory effort. An annealing temperature of 399°C was chosen for the study because a lower temperature heat treatment appears academic to most vessel applications. On the other hand, a higher temperature heat treatment conceivably could produce too great a recovery for meaningful material comparisons. Also, annealing at temperature higher than 399°C (for example, 454°C) has produced anomalous results in some investigations, which still bear explanation. A heat treatment duration of 168 h was employed here and is a reasonable annealing period for most plant applications. In fact this was the target heat treatment period for the Army SM-1A and Belgian BR-3 reactor vessels.

Materials

The test materials (12.7- or 15-mm-thick plates) are identified by composition and heat treatment in Table 1. Prior investigations of their relative radiation embrittlement sensitivities have been reported [*1,5*]. Plate microstructures were similar in character, consisting of tempered upper bainite with traces of free ferrite [*1,5*]. Preirradiation tensile strengths were also very similar (Table 2); accordingly, material comparisons for judgments of annealing response were much simplified.

Material Irradiation

The materials were irradiated in the form of Charpy-*V* (C_v) specimens in two irradiation assemblies, designated CEA-1 and CEA-2. The exposure temperature was controlled at a nominal 288°C; the target neutron fluence was $\sim 2.5 \times 10^{19}$ n/cm^2 ($E > 1$ MeV). The neutron flux was on the order of 8.5×10^{12} n/cm$^2 \cdot$ s^{-1} ($E > 1$ MeV). Fluence values given in this report are average values based on a calculated neutron energy spectrum and analyses of iron and nickel dosimeters included in the specimen arrays. Individual specimen fluences were all within 6% of the reported (average) fluence for the irradiation assembly involved. For the irradiation facility used the dpa equivalent value can be obtained by dpa $= 1.6 \times 10^{-22}$ fluence ($E > 1$ MeV). Also fluence ($E > 0.1$ MeV) is 2.91 times the fluence ($E > 1$ MeV). Estimated uncertainties for the fluence ($E > 1$ MeV) values based on the counting technique and assumed cross section components for the iron and nickel dosimeters are $\pm 8\%$ and $\pm 7\%$, respectively, at the 1σ confidence level. Irradiation temperatures were monitored throughout the exposure period with thermocouples. Irradiation

TABLE 1—*Chemical compositions of plates from laboratory (4-way split) melts of steel.*

Plate/ Cast	Source Melt	Chemical Composition, wt%											
		C	Mn	P	S	Si	Cu	Ni	Cr	Mo	Sn	N	B
67B[a]	1	0.23	1.31	0.015	0.018	0.20	0.002	0.70	<0.003	0.51	<0.004	0.009	0.0004
67C[a]	1	0.23	1.31	0.025	0.018	0.20	0.002	0.70	<0.003	0.51	<0.004	0.009	0.0004
68A[a]	2	0.23	1.31	0.003	0.017	0.22	0.30	0.70	<0.003	0.52	0.004	0.010	0.0006
68B[a]	2	0.23	1.31	0.016	0.017	0.22	0.30	0.70	<0.003	0.52	0.004	0.010	0.0006
68C[a]	2	0.23	1.31	0.028	0.017	0.22	0.30	0.70	<0.003	0.52	0.004	0.010	0.0006
5C[b]	3	0.24	1.30	0.002	0.014	0.21	0.16	0.27	0.003	0.53	0.004	...	...
5D[b]	3	0.24	1.30	0.002	0.014	0.21	0.16	0.68	0.003	0.53	0.004	...	...
6A[b]	4	0.23	1.29	0.002	0.013	0.22	0.28	0.35	...	0.53	0.004	...	...
6B[b]	4	0.23	1.29	0.002	0.013	0.22	0.28	0.27	...	0.53	0.004	...	...
6C[b]	4	0.23	1.29	0.002	0.013	0.22	0.28	0.59	...	0.53	0.004	...	...

[a] Heat treatment: 899°C for 1 h, AC; 649°C for 1 h, WQ; and 677°C for 5 h, WQ.
[b] Heat treatment: 899°C for 1 h, AC; and 649°C for 1 h, WQ.

TABLE 2—*Tensile properties at ambient temperature (preirradiation condition, 6.4-mm gage diameter).*

Plate Code	Yield Strength 0.2% Offset		Tensile Strength		Elongation in 25.5 mm, %	Reduction in Area, %
	MPa	ksi	MPa	ksi		
67B[a]	465	67.37	624	90.43	...	...
67C[a]	481[b]	69.77	624	90.43	...	...
68A[a]	460[b]	66.77	621	90.02	25.9	64.6
68B[a]	449[b]	65.16	637	92.43	...	...
68C[a]	477[b]	69.17	646	93.63	...	...
5C	536[c]	77.8	667	96.8	22.2	63.2
5D	550[c]	79.8	690	100.0	22.9	66.2
6A	550[c]	79.7	685	99.4	22.3	65.1
6B	561[c]	81.3	697[d]	101.1[d]	22.4	63.2
6C	587[c]	85.1	719[d]	104.3[d]	22.4	61.7

[a] Specimen blanks removed from austenitized plate and then tempered.
[b] Single determinations.
[c] Average of duplicate tests.
[d] Exceeds A 533-B specification range of 550 (minimum) to 690 (maximum) MPa.

temperatures were within 8°C of the controlled temperature. Exposure periods were on the order of 800 h.

Specimens of five materials were contained in each assembly and were commingled to preclude exposure differences. Temperature and fluence gradients within the assembly active zones were sufficiently small to be ignored in the analysis of data. In support of this judgment, data scatter was low.

Postirradiation annealing was accomplished in a recirculating air furnace where specimen temperatures were held to within ±1°C of the specified temperature. Furnace cooling (slow) from the heat treatment temperature was employed, rather than air cooling, to better simulate an actual vessel anneal. Cooling to ambient required several hours.

Unirradiated specimens representing thermal controls were "aged" at 288°C for a period comparable to the in-reactor exposure time period, that is, the time at full reactor power of 2 MW. For five of the materials, additional specimen sets were given a duplex aging treatment of 288°C for 800 h followed by 399°C for 168 h to simulate the irradiation and postirradiation annealing thermal cycles. Thermally aged specimens (and unirradiated condition reference specimens) were tested on the same impact machine used for the irradiated specimens.

Postirradiation Testing

Radiation sensitivity and annealing recovery relative to transition behavior was judged from C_v 41-J transition temperature changes. Only limited tests of the irradiated condition were performed to maximize the number of specimens available for annealing. In general, 8 to 10 specimens were used to establish as-irradiated condition properties. To guide the selection of postirradiation test temperatures, prior irradiation results cited in Refs *1* and *5* were utilized. This approach proved quite successful as will be evident below.

Results

The experimental results are summarized in Tables 3 and 4 and are illustrated in the data figures. Additional details of this study are given in Ref *6*.

TABLE 3—*Notch ductility determinations with irradiation assemblies VRS-1 and CEA-1.*

Plate Code[a]	P, %	Cu, %	Plate Sections	Experiment	Fluence[b], × 10^{19}	Charpy-V 41-J Temperature, °C (°F)						
						Initial	Irradiated	ΔT	Annealed[c]	ΔT_{rec}	ΔT_{res}	rec, %
67B	0.015	0.002	1 to 3	VRS-1 (UBR-39)	2.50	−7 (20)	38 (100)	44 (80)	... ...	... ...	... ...	... ...
			5 to 6	CEA-1 (UBR-59)	2.55	7 (45)	68 (155)	61 (110)	7 (45)	61 (110)	0 (0)	100.0
67C	0.025	0.002	1 to 3	VRS-1 (UBR-39)	2.50	−23 (−10)	49 (120)	72 (130)	... ...	... ...	... ...	... ...
			8 to 13	CEA-1 (UBR-59)	2.55	−7 (20)	74 (165)	81 (145)	−7 (20)	74 (165)	0 (0)	100.0
68A	0.003	0.30	1 to 3	VRS-1 (UBR-39)	2.50	−18 (0)	126 (260)	144 (260)	... ...	... ...	... ..	... ..
			8 to 13	CEA-1 (UBR-59)	2.55	−18 (0)	143 (290)	161 (290)	49 (120)	94 (170)	67 (120)	58.6
68B	0.016	0.30	1 to 3	VRS-1 (UBR-39)	2.50	−18 (0)	143 (290)	161 (290)	... ...	... ...	... ...	... ...
			5 to 6	CEA-1 (UBR-59)	2.55	−18 (0)	160 (320)	178 (320)	49 (120)	111 (200)	67 (120)	62.5
68C	0.028	0.30	1 to 3	VRS-1 (UBR-39)	2.50	−7 (20)	154 (310)	161 (290)	... ...	... ...	... ...	... ...
			8 to 13	CEA-1 (UBR-59)	2.55	−7 (20)	174 (345)	181 (325)	57 (135)	117 (210)	64 (115)	64.6

[a] A 533-B Steel (0.70% nickel).
[b] n/cm^2, $E > 1$ MeV (calculated spectrum) at 288°C.
[c] 399°C (750°F), 168 h anneal.
[d] Annealed USE > initial USE.
[e] Data scatter, value approximate.
[f] Percent recovery (rec) based on Sections 1–3 USE.

TABLE 4—*Notch ductility determinations for A 302-B and A 533-B plates in irradiation assemblies CNI-1, -2 and CEA-2 (Part 2).*

Plate Code[a]	Ni, %	Cu, %	Plate Sections	Experiment	Fluence[b], × 10^{19}	Charpy-V Upper Shelf Energy, J (ft · lb)						
						Initial	Irradiated	ΔUSE	Annealed[c]	ΔUSE_{rec}	ΔUSE_{res}	rec, %
5C	0.27	0.16	1 to 8	CNI-1 (UBR-33)	2.40	140 (103)	140 (103)	~0 (~0)	140 (103)	[d] . . .	[d] . . .	[d] . . .
			9 to 11	CEA-2 (UBR-63)	2.78	140 (103)	113 (83)	27 (20)	~134 (~99)	~21 (~16)	~6 (~4)	80.0
5D	0.68	0.16	1 to 8	CNI-1 (UBR-33)	2.40	144 (106)	140 (103)	~4 (~3)	144 (106)	~4 (~3)	0 0	100
			9 to 11	CEA-2 (UBR-63)	2.78	144 (106)	111 (82)	33 (24)	151 (111)	40 (29)	0 0	100
6A	0.05	0.28	1 to 8	CNI-2 (UBR-34)	2.60	144 (106)	114 (109)	30 (22)	136 (100)	22 (16)	8 (6)	72.7
			9 to 11	CEA-2 (UBR-63)	2.78	144 (106)	109 (80)	35 (26)	136 (100)	27 (20)	8 (6)	76.9
6B	0.27	0.28	1 to 8	CNI-2 (UBR-34)	2.60	144 (106)	114 (84)	30 (22)	136 (100)	22 (16)	8 (6)	72.7
			9 to 11	CEA-2 (UBR-63)	2.78	144 (106)	102 (75)	42 (31)	132 (97)	30 (22)	12 (9)	62.9[e]
6C	0.69	0.28	1 to 8	CNI-2 (UBR-34)	2.60	152 (112)	104 (77)	48 (35)	144 (106)	40 (29)	8 (6)	82.9
			9 to 11	CEA-2 (UBR-63)	2.78	137 (101)	102 (75)	35 (26)	124 (91)	22 (16)	13 (10)	61.5

[a] 0.002% P.
[b] n/cm^2, $E > 1$ MeV (calculated spectrum) at 288°C.
[c] 399°C, (750°F) 168-h anneal.
[d] Not applicable.
[e] Approximate caused by initial USE uncertainty.

Unirradiated Condition Tests

The plate sections used for the original properties determinations for the materials were not immediately adjacent to those sections used for the present investigation. Accordingly as a precaution, check tests were performed to verify unirradiated condition properties. In some cases, significant properties differences versus the original set of data were revealed (see data for plate Codes 67B and 67C as examples). An across-plate properties gradient is the obvious cause; test procedures were not responsible since the same equipment was used throughout for these codes. The variations demonstrate one type of "error," which can contribute to data bank scatter but which has an easy cure.

Thermal Aged Condition Tests

The unirradiated specimens aged at 288°C for 800 h indicated a trend toward slight improvement of notch ductility properties in those cases where changes were found. Upper shelf improvement was the most apparent change. In contrast, specimens given the duplex 288°C, 800 h + 399°C, 168 h aging treatment showed tendencies toward a slight degradation of notch ductility properties, that is, the data points fell to the right and somewhat below the data for the 288°C aging treatment. In the transition region, the duplex aging data remained within the data scatter band for the unirradiated (reference) condition. For the upper shelf regime however, the values for duplex aging were lower than those for the unirradiated condition, especially in the case of plate Codes 6A, 6B, and 6C. In this report, results for the irradiated and postirradiation annealed conditions are referenced to the preirradiation condition data (as opposed to thermal aging data) for consistency.

Experiment CEA-1

The materials in this experiment represented five variations about the base composition for A533-B steel. The variations permitted a critical assessment of the effect of phosphorus (3 levels) and the effect of copper (2 levels) on recovery.

Figures 1 to 5 illustrate the experimental results from this investigation. The lower portion of each graph shows the data from experiment CEA-1. The upper portion of each graph contains reference data from the earlier radiation sensitivity test series [5]. The upper graph also contains limited data produced from samples thermally aged at 288°C for 800 h. For this series of steels, the thermal treatment alone did not produce a marked change in properties.

Figure 6 compares observations for the as-irradiated condition against prior determinations for the materials for a slightly lower fluence exposure.[2] Transition temperature elevations tend to be somewhat higher than those observed earlier. More importantly, the relative transition temperature increases among the five materials show a good correspondence. In turn, the new data fully support the earlier Materials Engineering Associates, Inc., (MEA) determination that the contribution of phosphorus to radiation sensitivity development is inversely dependent on the amount of copper present. That is, the phosphorus influence is most pronounced when the copper content is low [5].

Figure 7 provides a summary of the annealing results. The data reveal a clear distinction between copper and phosphorus relative to their influence over annealing response. For the materials having a high copper content (~0.3% copper), the residual embrittlement was the same in all cases. Transition temperature recoveries were on the order of 60%, and upper shelf recoveries were on the order of 64 to 84%. Notice that recovery in the high copper plates was independent of the phosphorus content (0.003 versus 0.016 versus 0.028% phosphorus). When copper content was low, essentially full recovery was observed. Jointly, the results provide the conclusion that residual

[2] For this report, average fluences were computed to the second decimal place for comparison purposes. However, fluence differences so described are not believed critical to metal behavior.

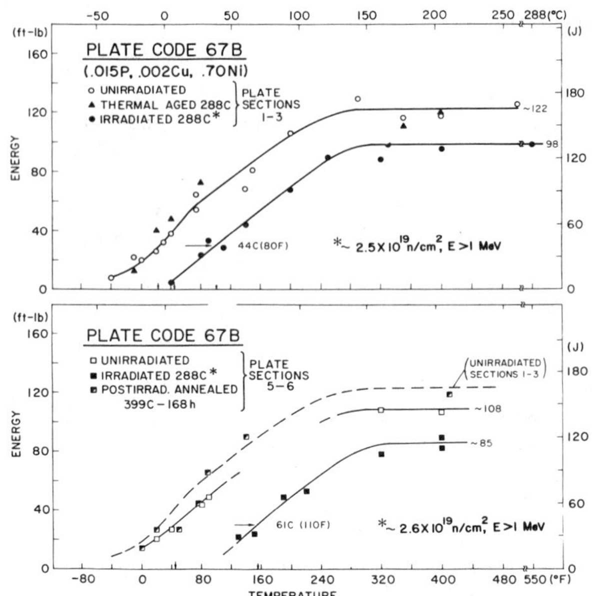

FIG. 1—*Charpy-V notch ductility of plate code 67B before and after 288°C irradiation and after 399°C postirradiation annealing for 168 h. (In this figure and in Figs. 2 to 5, the upper graph shows results from MEA irradiation test VRS-1 conducted earlier, and the lower graph shows results from irradiation assembly CEA-1. The neutron fluences were 2.50 and 2.55 × 10^{19} n/cm^2, respectively. Unirradiated condition check test results and thermal aging results are also given in Figs. 1 to 5.)*

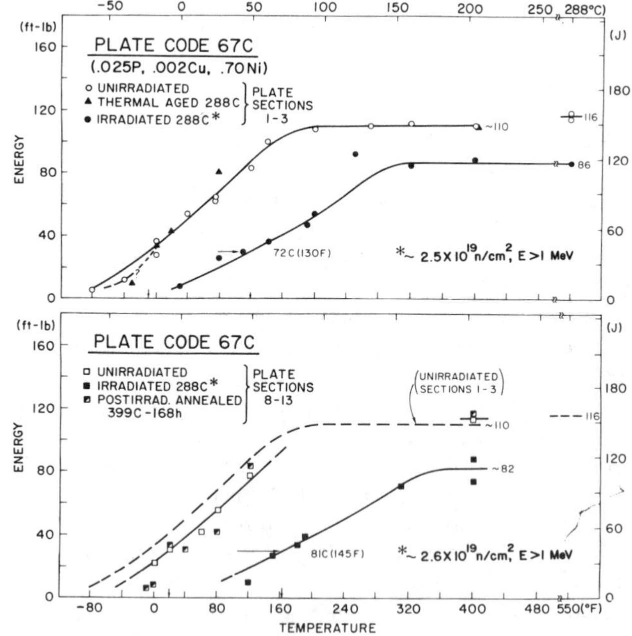

FIG. 2—*Charpy-V notch ductility of plate Code 67C before and after 288°C irradiation and after 399°C postirradiation annealing for 168 h.*

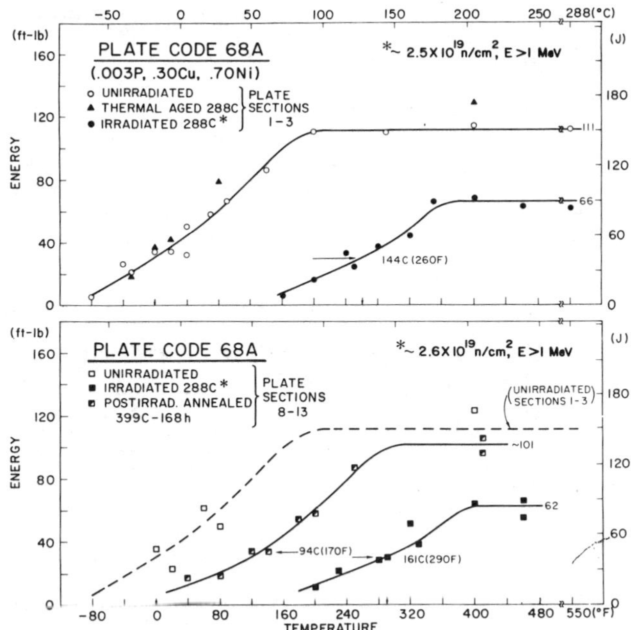

FIG. 3—*Charpy-V notch ductility of plate code 68A before and after 288°C irradiation and after 399°C postirradiation annealing for 168 h.*

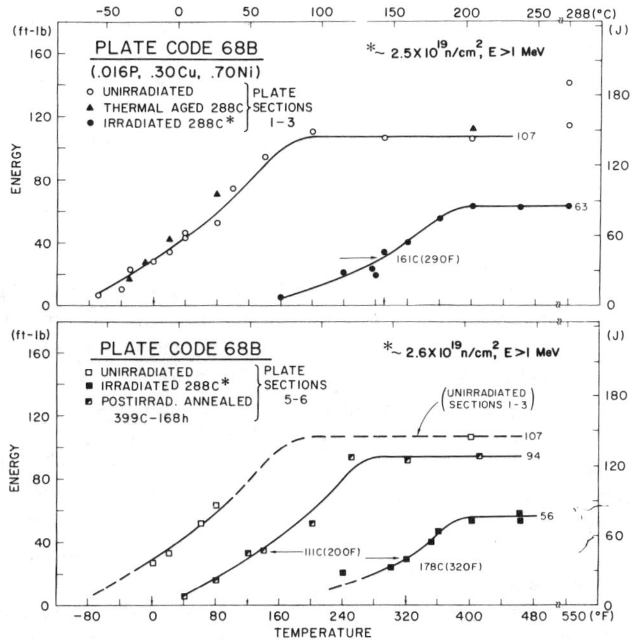

FIG. 4—*Charpy-V notch ductility of plate code 68B before and after 288°C irradiation and after 399°C postirradiation annealing for 168 h.*

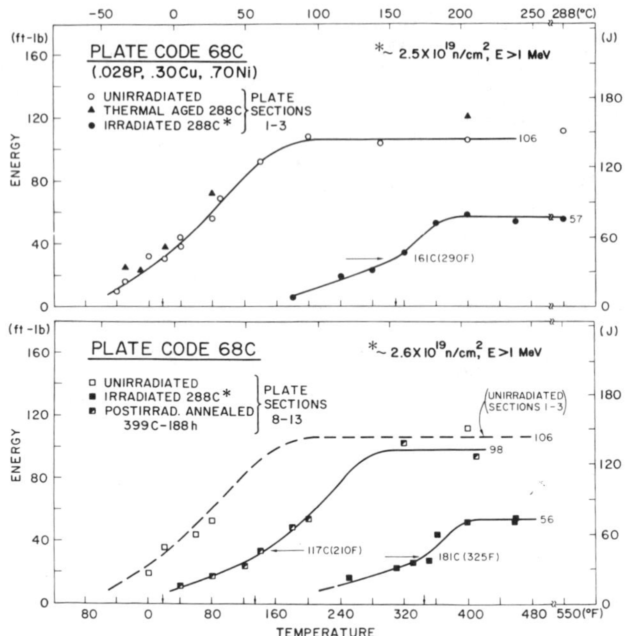

FIG. 5—*Charpy-V notch ductility of plate code 68C before and after 288°C irradition and after 399°C postirradiation annealing for 168 h.*

embrittlement is a function of copper content but not phosphorus content, at least for fluences on the order of 2.5×10^{19} n/cm^2.

Of further interest, the transition temperature recovery for plate Codes 67C (high phosphorus, low copper) was of the same magnitude as that for the plate Code 68A (high copper, low phosphorus). However, in terms of percentage, the latter showed only 58.6% recovery.

Experiment CEA-2

Materials included in this experiment were chosen to evaluate the effects of variable copper content and variable nickel content for the case of a low phosphorus content. The target radiation exposure of this assembly matched that of CEA-1, to facilitate cross comparisons of the data.

Figures 8 to 12 present the results. The upper portion of each graph shows the results of the earlier irradiation test of the material. Check test data for the unirradiated condition are shown also. The middle illustration presents the irradiation and postirradiation anneal data from CEA-2. Data for the 399°C, 168 h postirradiation-annealed condition developed with the earlier irradiation experiment are also indicated here. The lower portion of the figure compares the thermal aging data to the unirradiated condition trend line reproduced from the upper illustration.

Concerning the check test data, results for the plate Code 5D and the plate Code 6C fall within but toward the right-hand side of the data scatter band (see dashed trend line). In contrast, the results for the aged condition do superimpose nicely on the original trend curve for each of the materials. The inferred bias of the check test data therefore was ignored in developing the analyses.

The radiation-induced elevations in 41-J transition temperature observed with the CEA-2 assembly and the earlier irradiation test are illustrated in Fig. 13. Referring to the results for plate Codes 5C and 5D, the earlier and recent findings in general appear comparable. (For plate Code

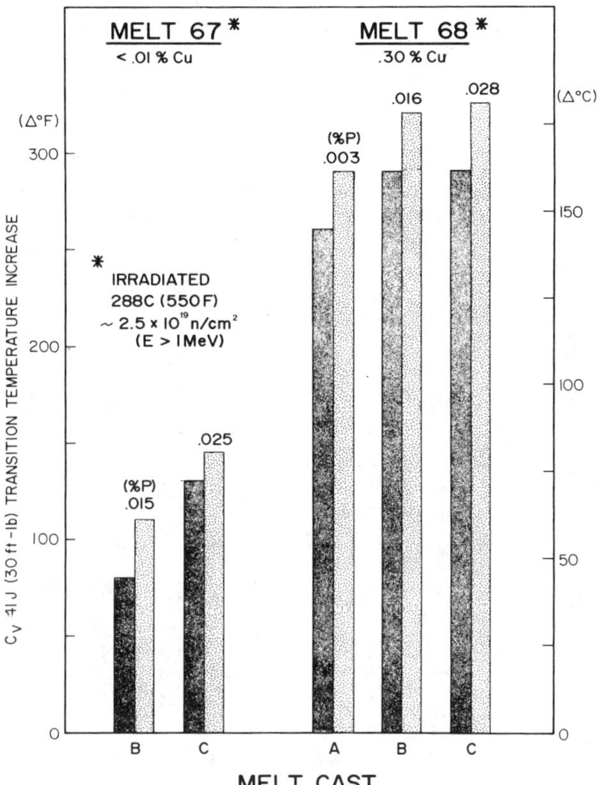

FIG. 6—*Comparison of* C_v *41-J transition temperature elevations of A 533-B plates with 288°C irradiation. Left-hand bars indicate observations for an earlier test series. The right-hand bars are results from the current study.*

5D, the difference is 14°C.) The data figures for these plates, however, show a significant difference in the shape of the transition curves for the two experimental tests. An explanation for the apparent anomaly cannot be offered; biased data scatter may be responsible. Good agreement between postirradiation trend curves was observed with the remaining three materials.

As expected, the high copper, high nickel plate Code 6C exhibited the greatest radiation-induced elevation in 41-J temperature. However, the magnitude of the elevation is less than expected on the basis of the performance of plate Code 68A having the same nominal composition (Table 1). Whether or not the difference in plate heat treatments is responsible is an open question.

Reinforcing the earlier assessment of nickel content effects in the presence of a high copper content, the radiation embrittlement sensitivities of plate Codes 6A and 6B appear equal. Thus, nickel in amounts of 0.27% or less is not a factor in radiation sensitivity. On the other hand, the latest data for plate Codes 5C and 5D indicate that the nickel reinforcement of the copper effect is much less in 0.16% copper material than in 0.28% copper material.

For all five materials, a good reproducibility of data and data trends is described by the two sets of annealing data. The upper shelf data for plate Code 6C could be taken as an exception; however, the unirradiated condition data for the two test areas over the plate show a similar separation of upper shelf values.

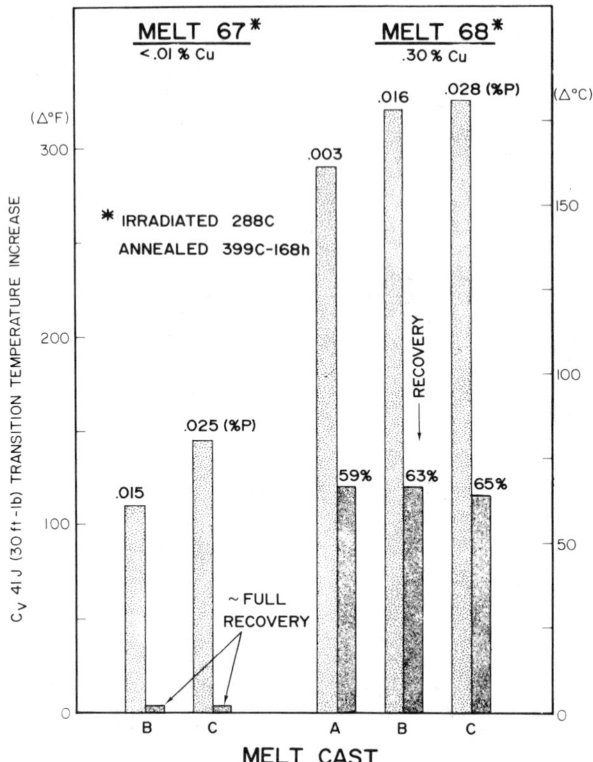

FIG. 7—*Charpy-V 41-J transition temperature changes for A 533-B plates with 288°C irradiation (left-hand bars) and with 399°C postirradiation annealing (right-hand bars).*

Figure 14 summarizes the determinations on transition temperature recovery by annealing. Incomplete recovery was found for the 0.16% copper materials. Even greater residual embrittlement and a lower percentage recovery was found for the 0.28% copper plates. This is consistent with the trend observations of experiment CEA-1 above. Nickel content does not appear to be a factor in the recovery of the 0.16% copper materials; however, with the 0.28% copper plate 6C, the residual embrittlement after annealing is clearly greater than that of plate Code 6B. This can only be ascribed to its higher nickel content (0.68% versus 0.27% nickel).

If one takes into account the effect of the duplex heat treatment on the unirradiated material, the postirradiation anneal produced essentially full upper shelf recovery for all five composition variations. On the other hand, incomplete recovery would be the conclusion if the unirradiated condition is used as the base line for judging recovery. Obviously, more needs to be known about the metallugical effects to reactor pressure vessel materials by duplex heat treatments in the absence of irradiation.

Summary

The investigations have revealed that copper, but not phosphorus, has a critical influence on notch ductility recovery by 399°C, 168 h postirradiation annealing. The results indicate also that a high (0.68%) nickel content can be detrimental to the recovery of high copper content steels in

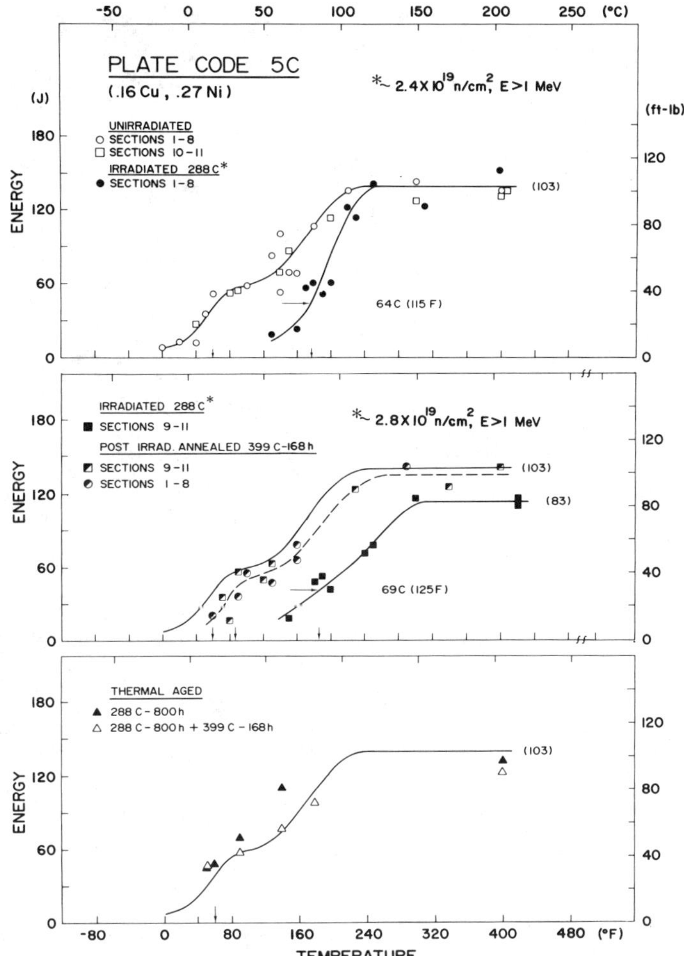

FIG. 8—*Charpy-V notch ductility of plate Code 5C before and after 288°C irradiation and after 399°C postirradiation annealing for 168 h. Data for two thermal aging conditions are also shown. (In this figure and in Figs. 9 to 12, the upper illustration shows the results from an earlier irradiation test and the middle illustration shows results from experiment CEA-2; the neutron fluences were as indicated).*

terms of residual embrittlement and percentage recovery. Nickel content levels usually found in A 302-B steels, however, do not appear critical to transition temperature recovery.

A potential for full recovery in high phosphorus, low copper A 533-B steel is shown. In contrast, recovery on the order of 60% in transition and 75% in upper shelf energy appears typical for high copper steel.

Long-term thermal aging at 288°C was not observed to degrade the notch ductility of any of the materials investigated. On the other hand, a duplex aging of 288°C for 800 h followed by a 168-h period at 399°C caused a noticeable change in notch ductility for some of the materials investigated, especially a reduced upper shelf energy level. The reduction in upper shelf energy by thermal aging of unirradiated material could account for a significant portion of the "incomplete" recovery in upper shelf energy seen in postirradiation annealing.

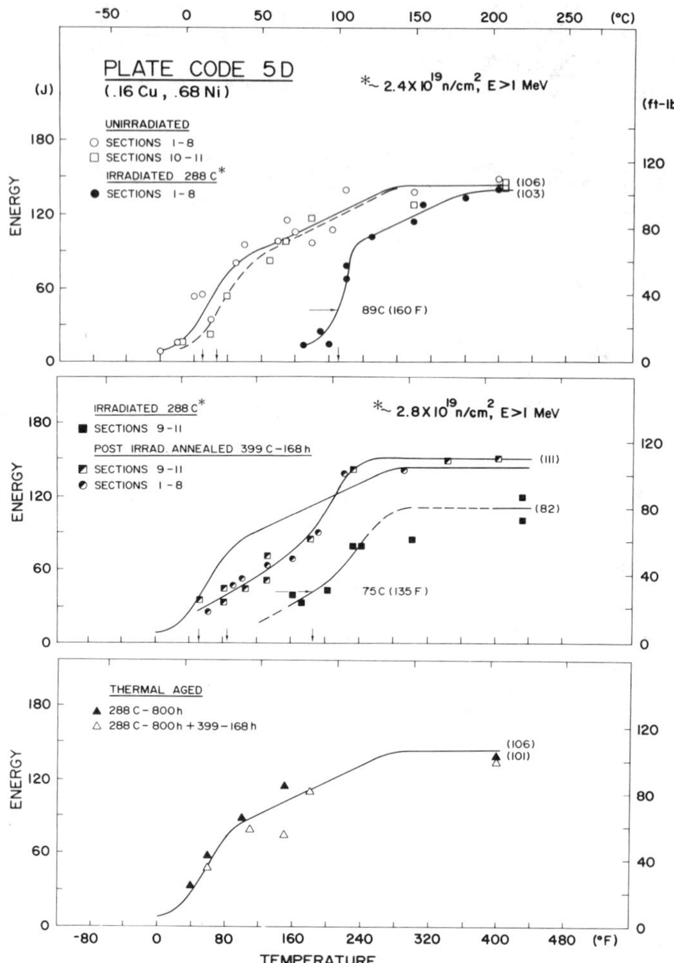

FIG. 9—*Charpy-V notch ductility of plate Code 5D before and after 288°C irradiation and after 399°C postirradiation annealing for 168 h. Data for two thermal aging conditions are also shown.*

Results for the as-irradiated condition of the test materials match well the prior results for the materials that first revealed the inverse dependence on copper content of the phosphorus effect on radiation resistance. That is, even though phosphorus does not appear to have a role in recovery, this impurity element does have a strong detrimental effect on elevated temperature radiation resistance in low copper content steels [5]. Likewise, the new data reinforce prior determinations of an adverse influence of a high nickel content on the radiation resistance of high copper steels.

Acknowledgments

This series of investigations was sponsored by the Nuclear Regulatory Commission, Materials Engineering Branch (M. Vagins, project manager).

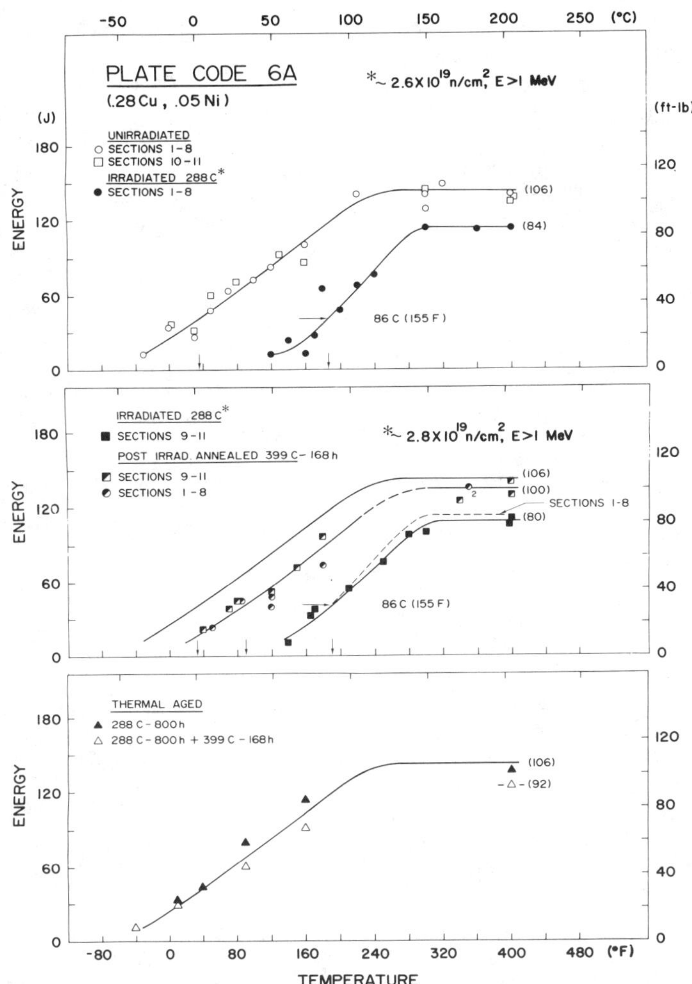

FIG. 10—*Charpy-V notch ductility of plate Code 6A before and after 288°C irradiation and after 399°C postirradiation annealing for 168 h. Data for two thermal aging conditions are also shown.*

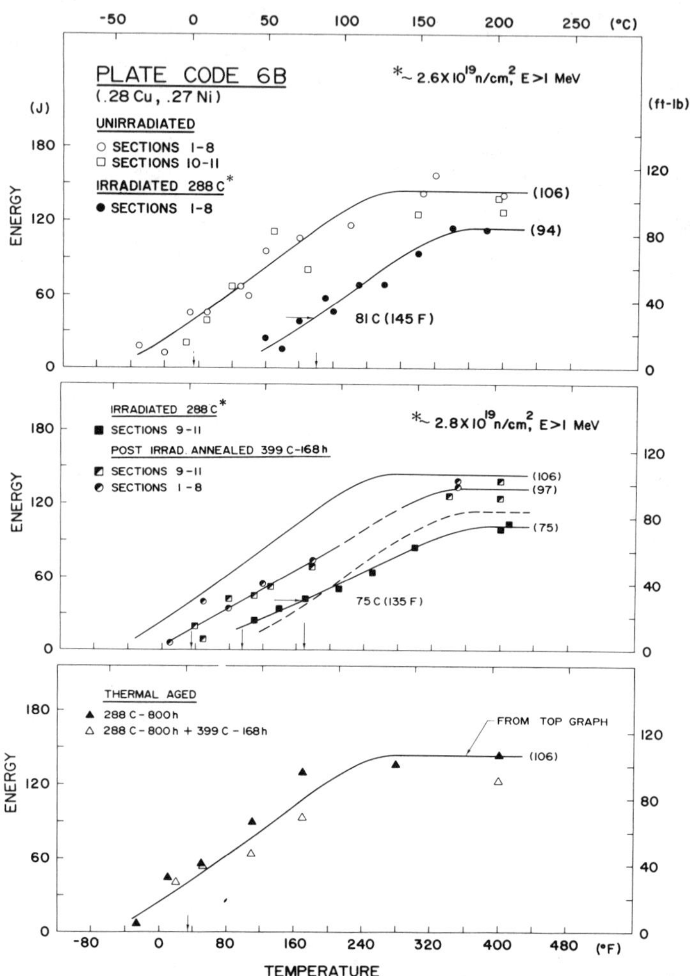

FIG. 11—*Charpy-V notch ductility of plate Code 6B before and after 288°C irradiation and after 399°C postirradiation annealing for 168 h. Data for two thermal aging conditions are also shown.*

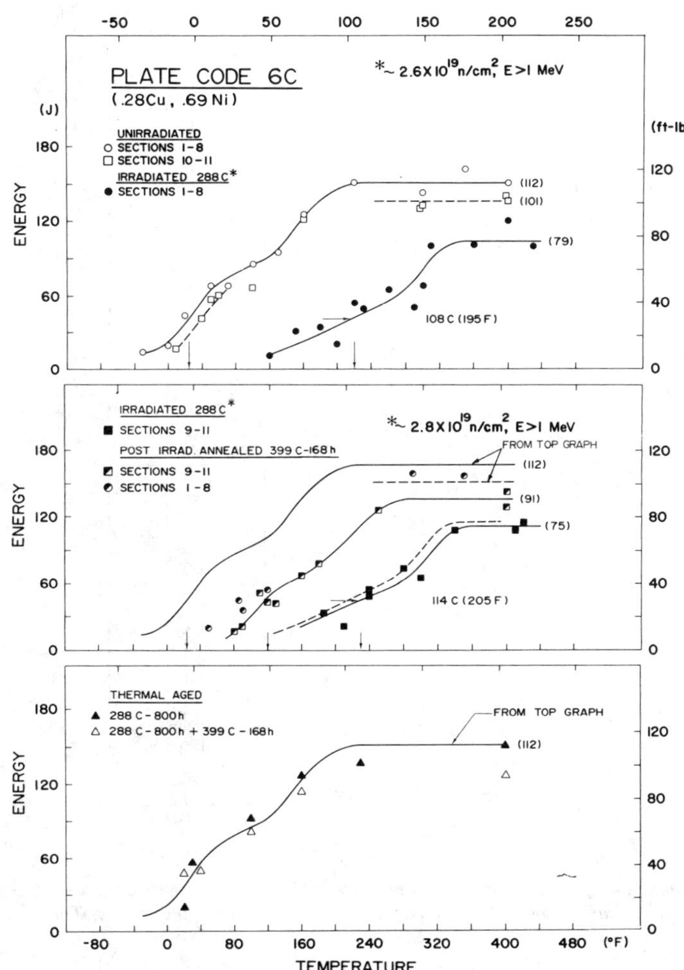

FIG. 12—*Charpy-V notch ductility of plate Code 6C before and after 288°C irradiation and after 399°C postirradiation annealing for 168 h. Data for two thermal aging conditions are also shown.*

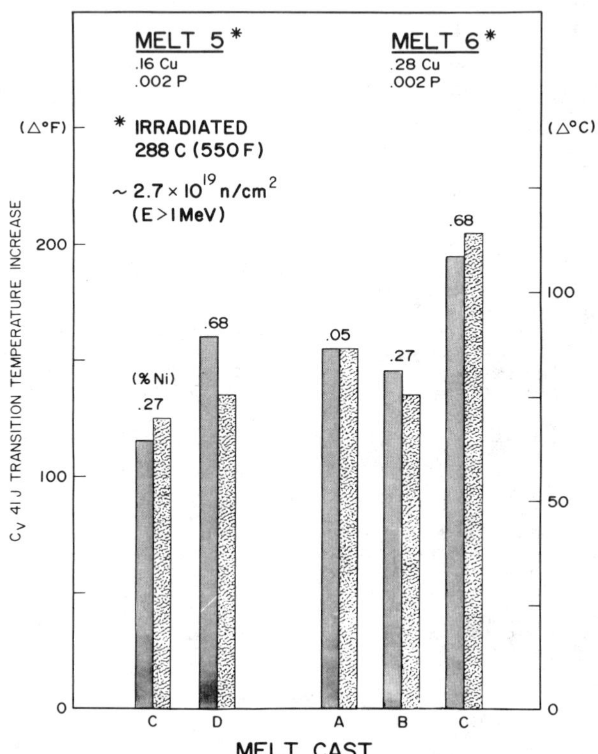

FIG. 13—*Comparison of* C_v *41-J transition temperature elevations of A 302-B and A 533-B plates with 288°C irradiation. Left-hand bars indicate observations for an earlier test series. The right-hand bars are results from the current study.*

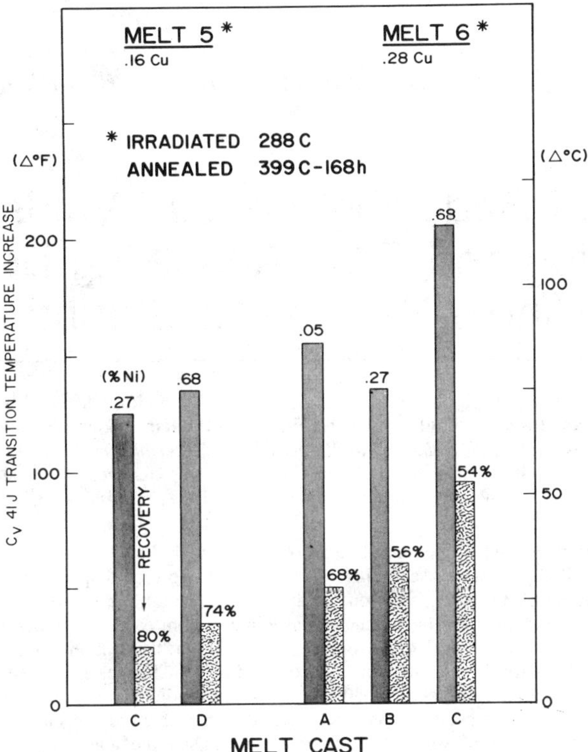

FIG. 14—*Charpy-V 41-J transition temperature changes for A 302-B and A 533-B plates with 288°C irradiation (left-hand bars) and with 399°C postirradiation annealing (right-hand bars).*

References

[*1*] Hawthorne, J. R., *Effects of Radiation on Materials: Eleventh Conference, STP 782,* H. Brager and J. Perrin, Eds., American Society for Testing and Materials, Philadelphia, 1982, pp. 375–391.

[*2*] Hawthorne, J. R., "Exploratory Assessment of Postirradiation Heat Treatment Variables in Notch Ductility Recovery of A 533-B Steel," USNRC Report NUREG/CR-3229, Washington, DC, April 1983.

[*3*] Hawthorne, J. R., "Survey of Postirradiation Heat Treatment as a Means to Mitigate Radiation Embrittlement of Reactor Vessel Steels," USNRC Report NUREG/CR-0486, Washington, DC, Feb. 1979.

[*4*] Potapovs, U. and Hawthorne, J. R., *Nuclear Applications,* Vol. 6, No. 1, Jan. 1969, pp. 22–46.

[*5*] Hawthorne, J. R., "Exploratory Studies of Element Interactions and Composition Dependencies in Radiation Sensitivity Development," USNRC Report NUREG/CR-4437, Washington, DC, Oct. 1985.

[*6*] Hawthorne, J. R., "Influence of Steel Composition on Embrittlement Relief by Postirradiation Heat Treatment," MEA-2168, Materials Engineering Associates, Lanham, MD, in preparation.

Isang E. Ukpong,[1] *Aaron D. Krawitz,*[2] *David F. R. Mildner,*[3] *and H. P. Leighly, Jr.*[4]

A Study of Radiation Induced Voids and Precipitates and Their Annealing Behavior by Small-Angle Neutron Scattering

REFERENCE: Ukpong, I. E., Krawitz, A. D., Mildner, D. F. R., and Leighly, H. P., Jr., "**A Study of Radiation Induced Voids and Precipitates and Their Annealing Behavior by Small-Angle Neutron Scattering,**" *Influence of Radiation on Material Properties: 13th International Symposium (Part II), ASTM STP 956,* F. A. Garner, C. H. Henager, Jr., and N. Igata, Eds., American Society for Testing and Materials, Philadelphia, 1987, pp. 480–493.

ABSTRACT: We have studied the annealing response of an irradiated pressure vessel steel, ASTM A302B, using small-angle neutron scattering. We have used two sets of specimens in this study, one irradiated to a fluence of 3 to 4 $\times$ 10^{19} n/cm^2 at 288°C, and the other a parallel set of the same steel in the unirradiated condition. We have annealed specimens from each set isothermally at 300, 350, 400, and 450°C for increasing periods of time up to five h, and have observed the variations in the SANS spectra in both irradiated and unirradiated specimens. Changes in the small-angle neutron scattering (SANS) spectra commence at 300°C for the unirradiated material and at 400°C in the irradiated material. The observed response suggests that the changes are caused basically by the precipitation of copper out of solid solution in α-iron during the annealing of unirradiated and irradiated specimens; in the case of the irradiated specimens, the release of damage-induced defects enhances the precipitation of the copper.

KEY WORDS: nuclear pressure vessel steels, precipitation, small-angle neutron scattering (SANS), annealing damage, radiation damage, precipitation

High energy neutrons cause atoms to be displaced from their lattice sites to become self interstitials leaving behind an excess of vacancies. The vacancies and self interstitials may recombine so that the lattice reverts to its original state, or the defects may be attracted to sinks, such as grain boundaries. Alternatively, these defects may form clusters such as interstitial or vacancy loops, or the vacancies may create three-dimensional clusters, that is, voids. Accompanying this defect generation is a loss in ductility together with an increase in the mechanical properties [*1,2*] (strength and hardness). Swelling may also result from void formation [*3*]. In steel this loss of ductility can be described quantitatively by an increase in the ductile-to-brittle transition temperature as determined by the Charpy impact test. In addition, radiation damage reduces the high-temperature shelf energy of the Charpy impact test curve [*3*]. These changes in mechanical

[1] Graduate research assistant in mechanical and aerospace engineering, University of Missouri-Columbia, 25 Atekong Dr., P.O. Box 1194 Calabar, C. R. State, Nigeria.

[2] Professor of mechanical and aerospace engineering, University of Missouri-Columbia, Columbia, MO 65211.

[3] Senior research scientist at University of Missouri Research Reactor and associate professor of physics, University of Missouri-Columbia, Columbia, MO 65211.

[4] Professor of metallurgical engineering, University of Missouri-Rolla, Rolla, MO 65401.

behavior are of particular interest in the case of steel used in nuclear reactor pressure vessels because of possible embrittlement-induced failure during operation.

There are two factors that affect the properties of the pressure vessel steels: the microstructure [*4,5*], which can be altered by radiation damage, and the composition. A typical pressure vessel steel contains about 0.25 wt% carbon, which produces a mixed ferrite-pearlite microstructure on slow cooling from the austenite range. Among the metallic elements present in steel, copper is the most deleterious, probably because of its very limited solid solubility which results in the formation of precipitates. Long-term retention of the pearlite at a temperature below the eutectoid temperature leads to a spheroidite structure, spherical cementite in a ferrite matrix. In addition to carbon these steels contain a number of other alloying elements all of which are in solid solution in the ferrite. Odette [*6*] has proposed that copper or copper-microvoid precipitates form, having a diameter of about 0.3 nm, which cannot be observed by conventional transmission electron microscopy (TEM). These precipitates cause embrittlement of the steel. Experimentally the copper content in pressure vessel steel has been found responsible for much of the radiation induced shift in the transition temperature [*7–16*]; the higher copper content steels show a substantial increase in yield strength compared to those with an increasing phosphorus content. Smidt and Sprague [*16*] have studied several pressure vessel steels with controlled amounts of copper and phosphorus using compression tests and TEM. They found that radiation strengthening is greatly enhanced by the presence of copper. They also found, using TEM, that copper causes the formation of many smaller defect aggregates during irradiation. These defects grow to a visible size at low fluences, suggesting that small clusters of copper atoms serve as nucleation sites for the defect aggregates. Ohmae and Zhiebold [*17*] have shown that when (Mn,Fe)S inclusions are present, more copper is found on the upper shelf fracture surfaces than on the lower shelf fracture surfaces. Almost all the copper segregates to the sulfides, affecting the upper shelf energy but not the transition temperature. Hawthorne [*18*] studied the effect of nickel on the properties of copper bearing steels and found that nickel in excess of 0.3% increases the transition temperature and also increases the amount of annealing required to recover the initial transition temperature. The addition of copper similarly increases the transition temperature. However, nickel increases the upper shelf energy for both irradiated and annealed specimens.

Small Angle Neutron Scattering

Small-angle neutron scattering (SANS) has become an important technique in the study of metallurgical materials [*19*]. Structural features in the spatial range of 10 to 1000 Å may be observed by the analysis of the scattering pattern of neutrons in the wave-vector range below 1 $Å^{-1}$ produced by fluctuations in the nuclear scattering length density within the specimen. Kostorz [*19*] and Gerold and Kostorz [*20*] have reviewed the results of this technique and its applications to materials science, which include the study of phase separation in binary and ternary systems, density and concentration fluctuations in single-phase systems, and studies of various structural defects such as voids and dislocations. The scattered intensity is recorded as a function of the scattering angle θ around the incident beam but is better represented in terms of the scattering vector Q, given by $Q = (4\pi/\lambda) \sin \theta/2$ where λ is the wavelength of the radiation. For small angles, as used in this analysis, Q may be given by $2\pi\theta/\lambda$, so that figures showing scattered intensity $I(Q)$ as a function of Q represent scattering patterns as a function of scattering angle θ after radial averaging over all azimuthal angles.

A small-angle neutron scattering pattern is essentially a Fourier transform of the spatial extent of the inhomogeneities within the specimen, so that differences in the spectra reflect average differences between the specimens on a microscopic scale. For a system of identical particles, the probability of scattering into a unit solid angle depends on the number of such particles, the square of the particle volume, the square of the difference in scattering density of the particles from the

TABLE 1—*Composition of ASTM Steel A302B (See Footnote 4).*

Component	Percent
C	0.24
Mn	1.34
P	0.011
S	0.023
Mo	0.51
Si	0.23
Cu	0.2
Ni	0.18
Al (soluble)	0.04
Al (insoluble)	0.002
Cr	0.11
Sn	0.037
Bi	0.0001
N	0.008
Ti	0.015
V	0.001

surrounding matrix, and the structure factor of the particle. This must be integrated over the number distribution of such particles to give an ensemble average, so that it is difficult to obtain precise microscopic information from the scattering technique. In general, as the scattering particles grow in volume, the SANS spectrum shifts to lower values of scattering vector. However other techniques, such as TEM, show properties of individual particles as opposed to bulk properties as given by SANS, and these results may be used to interpret the general features of the scattering system more precisely.

Experimental Materials

For this investigation we have used a standard pressure vessel containment steel, ASTM A302B, produced by the U.S. Steel Corporation and provided by the Westinghouse Electric Corporation. The composition of this particular steel is given in Table 1. The original plate was 230 × 180 × 15 cm (92 × 72 × 6 in.) and has undergone a 1:1 transverse to longitudinal rolling to avoid directionality. After rolling, the plate was heat treated using the procedure shown in Table 2. The purpose of this heat treatment is to produce a homogeneous steel before the water quench, which itself produces a fine pearlite-ferrite structure. The subsequent heating at 649°C for 6 h allows the breakdown of the pearlite into spheroidite. Specimens have dimensions of about 10 by 10 by 1

TABLE 2—*Processing of ASTM A302B steel.*

Plate at 594°C
Put in furnace
Heated at 35°C/h to 900°C
Held at 900°C for 4 h
Water quenched to 150°C
Charged into furnace at 370 to 400°C
Heated to 649°C
Held at 649° for 6 h
Air cooled to ambient temperature

mm were cut from both irradiated and unirradiated steel. The irradiation[5] was performed at about 288°C to a fluence of 3 to 4 $\times$ 10^{19} n/cm^2 ($E > 1$ MeV) at the University of Buffalo Nuclear Reactor. The impact property changes resulting from irradiation are given in Table 3 [*21*]. Tensile property changes are given in Table 4 [*21*].

Experimental Procedure

We have used the small-angle neutron scattering spectrometer (SANS) [*22*] at the University of Missouri Research Reactor (MURR) for these various measurements. We have performed preliminary experiments in order to assess reproducibility and reliability of the experimental measurements and to determine the experimental conditions necessary to assure accurate data.

1. The first measurements on an irradiated specimen showed a very high scattered intensity (after correction for background scattering introduced by the instrument and surroundings). We have decided to make SANS measurements on each specimen for a minimum of 12 h to obtain adequate statistical data to observe small changes for an intercomparison between specimens annealed at different temperatures.
2. There is little multiple neutron scattering for a 1-mm-thick specimen, but when two or more specimens are stacked together, the scattered intensity and the transmission values obtained suggest that there is significant multiple scattering to distort the measurements; hence we have used 1-mm specimens for all subsequent experiments.
3. We found no information on the effect of surface preparation on the resulting SANS data, so we made measurements on parallel specimens with as-received surfaces and with metallographically polished surfaces. We find no statistically significant difference in the resulting SANS spectra.
4. We considered the possibility of effects caused by surface oxidation. Specimens of SAE 1018 steel were annealed with and without oxidation protection for 6 h at 500°C, longer than any proposed annealing cycle and at a temperature higher than the proposed maximum annealing temperature. We observed no differences in the spectra for the two specimens, indicating that oxidation is not a problem. Nevertheless, during the subsequent experimental annealing, we placed the specimens in stainless steel heat-treatment envelopes.
5. We have obtained a SANS spectrum for each of the as-received specimens, both unirradiated and irradiated. From a pool of seven unirradiated and seven irradiated specimens, we chose the four closest spectra of each for use in this study.

[5] Lott, R., private communication.

TABLE 3—*Pre- and post-irradiation Charpy-V properties of A302B steel* [21].

Experiment/ Fluence, $\times 10^{19}$ n/cm^2 ($E > 1$ MeV)	Charpy Transition Temperature			Charpy Upper Shelf			
	41 J (30 ft · lb) Index, °C	68 J (50 ft · lb) Index, °C	0.9 mm (35 mil) Index, °C	Energy J	Energy ft · lb	Lateral Expansion mm	Lateral Expansion mils
Unirradiated (TL)	15	. . .	24	58	43	1.07	42
Irradiated 2.7	96	. . .	. . .	47	35	0.81	32
Irradiated change	81	. . .	. . .	11	8	0.26	10
Unirradiated (LT)	−7	18	2	110	81	1.63	64
Irradiated 3.6	90	101	96	81	60	1.22	48
Irradiated change	97	83	94	29	21	0.41	16

TABLE 4—*Tensile properties of A302B steel* [21].

Condition	Test, °C	Yield Strength, MPa	Tensile Strength, MPa	Reduction of Area, %	Elongation, % in 25.4 mm
Unirradiated	22	465	625	49.2	25.2
Irradiated (288°C)	24	578	721	41.4	17.3

We examined 1-mm-thick specimens of irradiated and unirradiated ASTM A302B steel using the MURR-SANS spectrometer, and recorded spectra for specimens in the unannealed condition and after each of several isothermal annealing steps. The temperatures selected for the isothermal annealing are 300, 350, 400 and 450°C, which are above the temperature of the specimens during neutron irradiation (~288°C). The time intervals vary from 0.25 to 5 h and are chosen to reveal those structural changes caused by annealing which occur in the specimens.

We have made a SANS measurement for each specimen for a minimum of 12 h in order to obtain adequate statistics. Though there is evidence of a small amount of magnetic scattering, none of the data shown is for a specimen with an applied magnetic field. (The saturation magnetization of pure iron is 7.5 kg, so that a magnetic field of at least this strength must be used to observe the effects of magnetic small-angle neutron scattering.) Thus all spectra shown here are symmetric around the beam center line, and may be radially averaged. Each data set is corrected for transmission, and the beam-open background is subtracted, so that all spectra can be compared directly. A full description of the procedure used for correction and averaging of the data is given in Ref *22*.

Experimental Results and Discussion

Figures 1 through 4 show the small-angle neutron scattering patterns for the unirradiated specimens after annealing at 300, 350, 400, and 450°C, respectively. The transmission-corrected

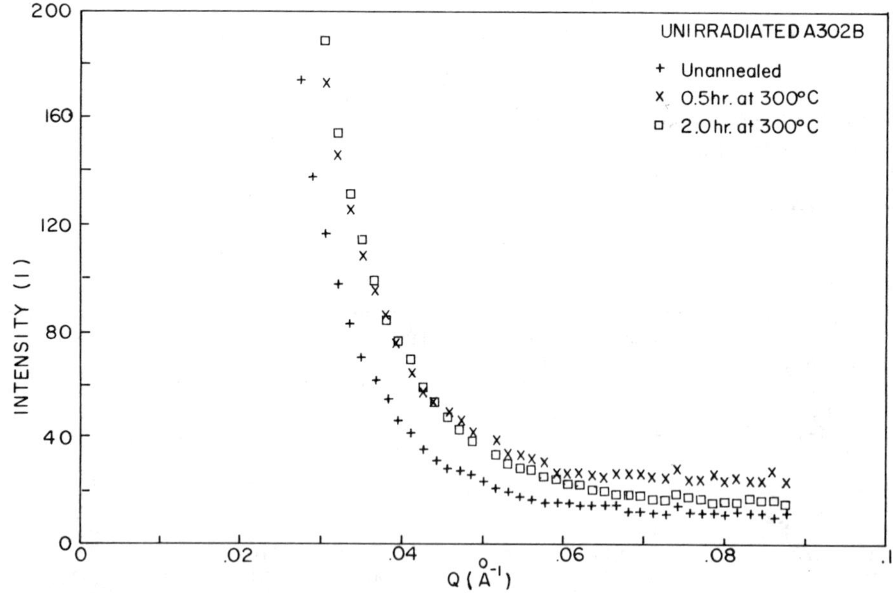

FIG. 1—*Annealing response of unirradiated A302B at 300°C.*

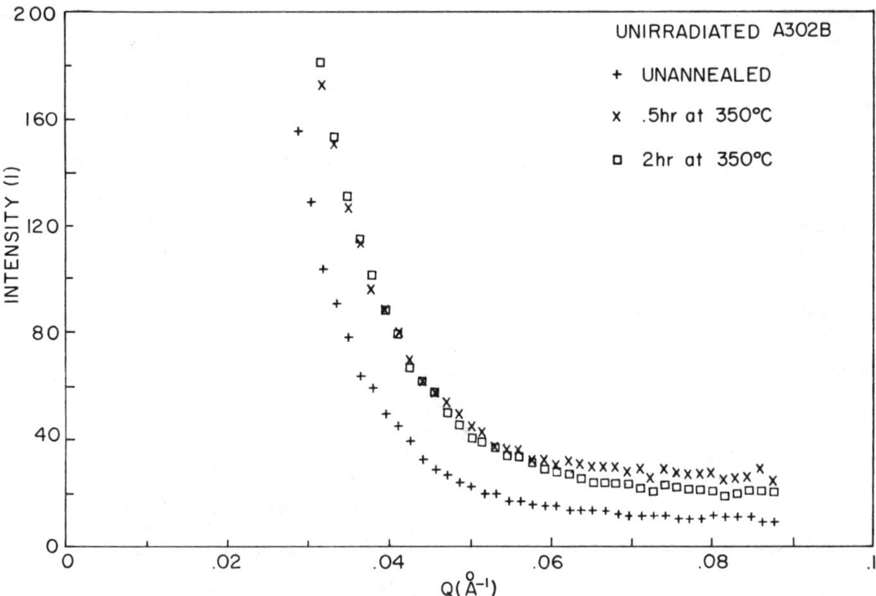

FIG. 2—*Annealing response of unirradiated A302B at 350°C.*

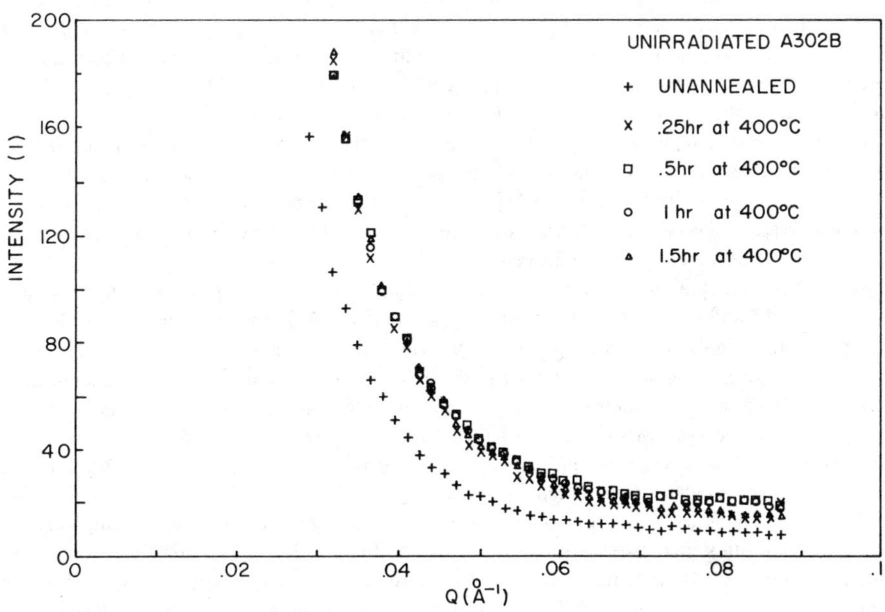

FIG. 3—*Annealing response of unirradiated A302B at 400°C.*

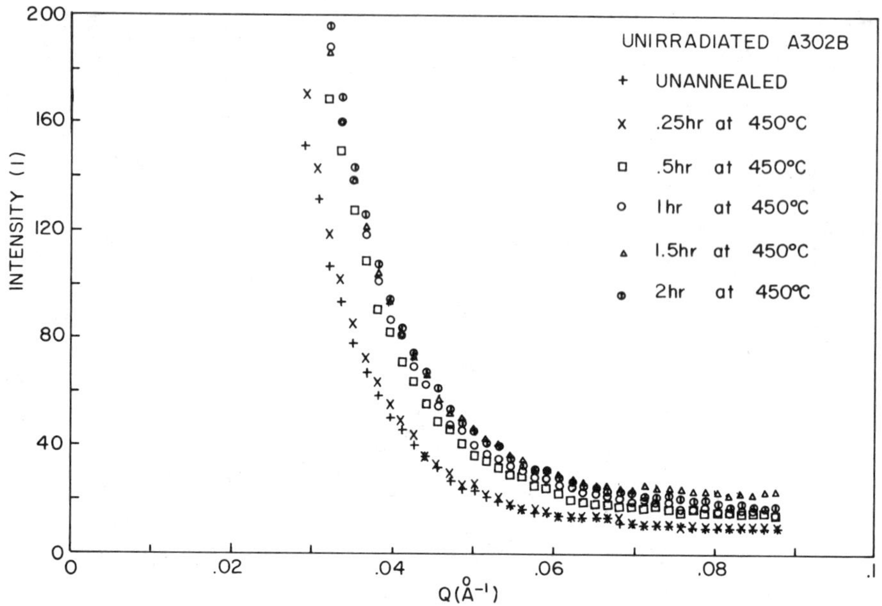

FIG. 4—*Annealing response of unirradiated A302B at 450°C.*

spectra for all of the unannealed specimens coincide as expected. Subsequent annealing causes the SANS spectra to increase and to shift to larger values of *Q;* this indicates nucleation of inhomogeneities throughout the material. At longer annealing times, this enhanced intensity tends to shift to lower values of scattering vector, corresponding to growth of the inhomogeneities. Comparison of Figs. 1 and 2 shows that this effect is greater for annealing at 350°C than at 300°C, which is consistent with diffusion-controlled growth, as expected. The spectra for specimens annealed at 350, 400, and 450°C coincide approximately, particularly at lower values of scattering vectors, indicating that short-term structural effects occur very rapidly.

Copper is the only substitutional solute present that exceeds the solubility limit in α-iron. The heat treatment at 649°C for 6 h causes all the copper to go into solid solution (at 649°C, the solubility of copper in α-iron is 0.28% [*23*]). There is no information on the cooling of this steel to room temperature, however, Figures 1 through 4 suggest that the cooling rate was fast enough to prevent equilibrium from being achieved.

We believe that copper precipitates from solid solution in the subsequent annealing treatments used in this study and causes the observed changes in the SANS specta. Lucas and Odette [*24*] have proposed that, in radiation damaged steel, the surfaces of vacancy clusters formed in cascades have a strong binding interaction for impurities, such as copper, hence producing precipitates. In an effort to substantiate this interpretation we have analyzed the trends in the data by focussing on two regions in the spectra (at values of Q around 0.04 $Å^{-1}$ and $Q \approx 0.08$ $Å^{-1}$) and making intercomparisons, with a view to understanding the gross features in the metallurgical system. Figure 5 shows the values of the intensity around $Q = 0.04$ $Å^{-1}$ and $Q = 0.08$ $Å^{-1}$ for the unirradiated specimens as a function of annealing time. At $Q = 0.04$ $Å^{-1}$, the intensity for all annealing temperatures increases to a maximum value after 0.5 h and essentially remains constant thereafter. At $Q = 0.08$ $Å^{-1}$, the intensity also increases to a maximum in about 0.5 h for the specimens annealed at 300 and 400°C, but decreases for longer annealing times. For the 450°C anneal, the intensity increases gradually to a maximum value after 2.0 h. This is an indication

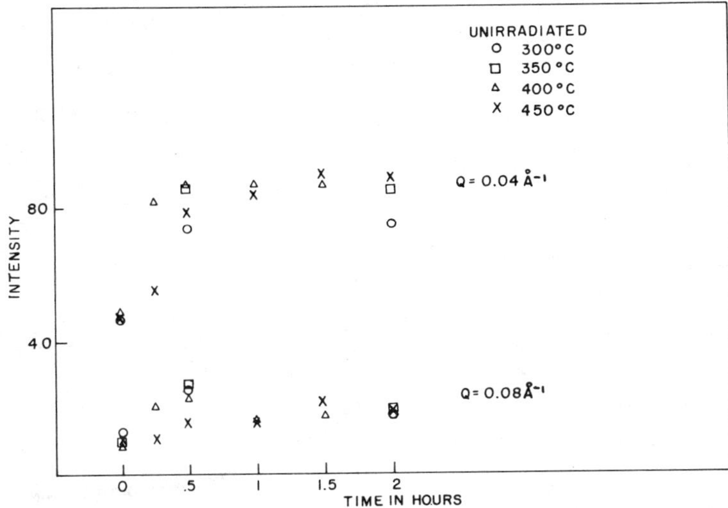

FIG. 5—*Intensity versus annealing time for unirradiated ASTM A302B at* Q = *0.04Å*$^{-1}$ *and* Q = *0.08Å*$^{-1}$.

that the precipitation of copper occurs very rapidly at the lower annealing temperatures because of the large supersaturation and adequate diffusivity of copper. At 450°C, the degree of supersaturation, and thus the driving force, is less so that the increase in intensity with annealing time is slower [*23*].

Figures 6 through 9 show the SANS spectra for the irradiated specimens annealed at 300, 350, 400, and 450°C, respectively. The spectra for the specimens annealed at 300 and at 350°C do not

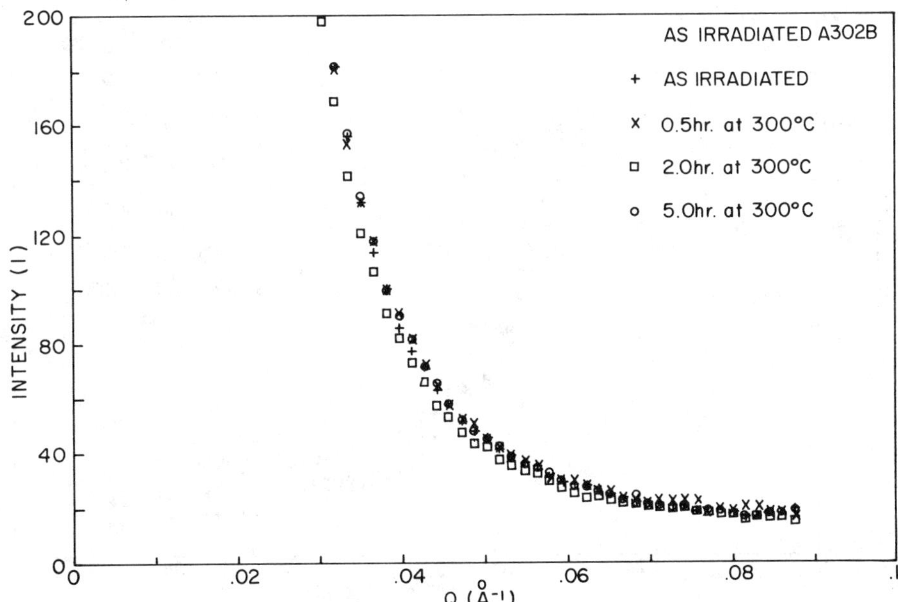

FIG. 6—*Annealing response of irradiated A302B at 300°C.*

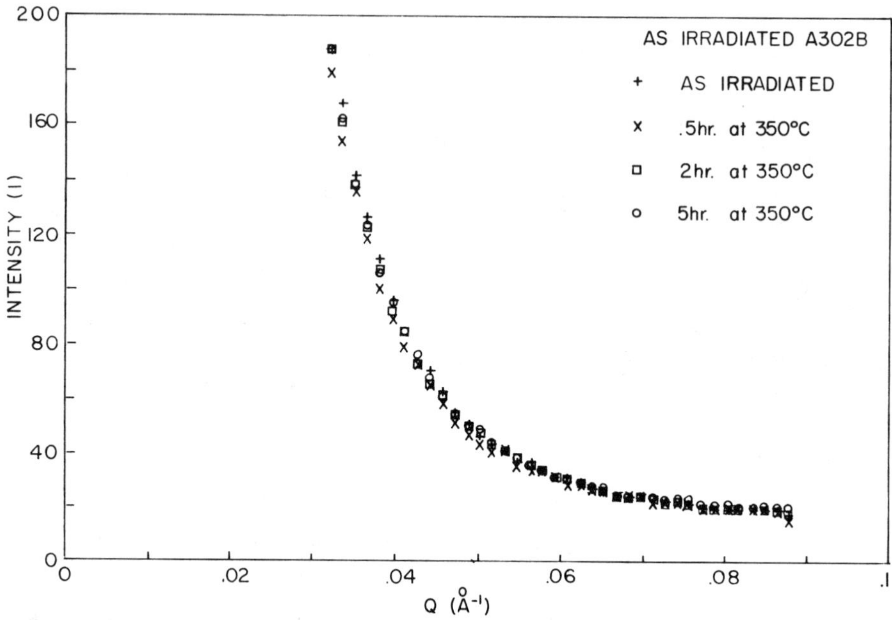

FIG. 7—*Annealing response of irradiated A302B at 350°C.*

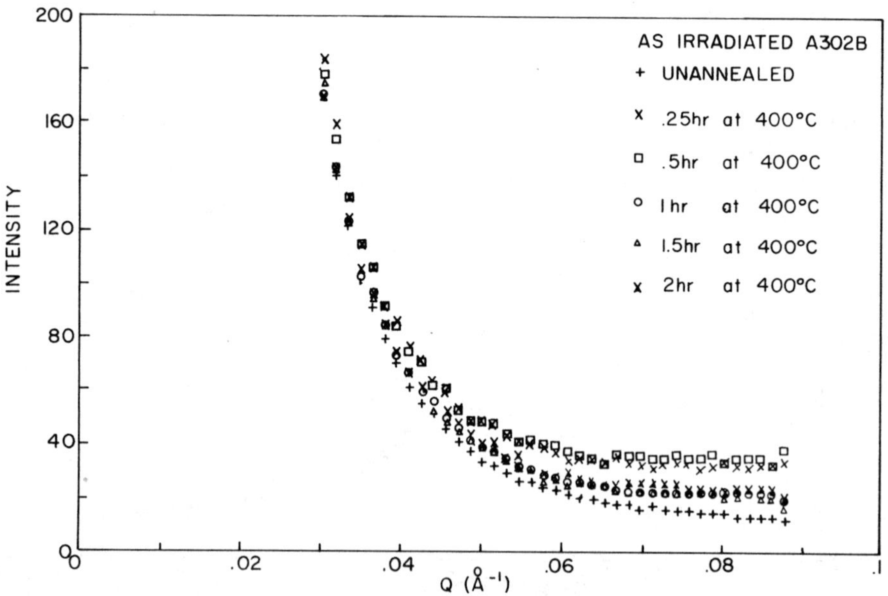

FIG. 8—*Annealing response of irradiated A302B at 400°C.*

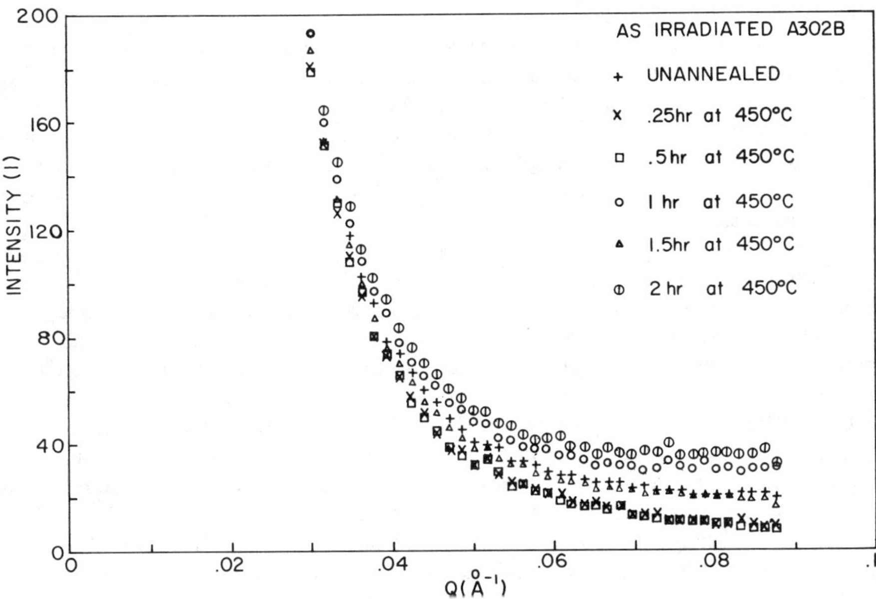

FIG. 9—*Annealing response of irradiated A302B at 450°C.*

change with annealing time, while those obtained from the 400 to 450°C anneals change substantially and in a complex way. For the specimens annealed at 300° and 350°C, the major metallurgical changes have occurred already because of the combined effects of radiation damage and long duration at the irradiation temperature. In contrast to the unirradiated specimens, the spectra for the unannealed, irradiated specimens essentially superimpose with the spectra obtained after annealing; this indicates that the major changes in the structure have already occurred because of radiation damage.

Figure 10 shows the intensities for values of $Q = 0.04$ Å^{-1} and $Q = 0.08$ Å^{-1} for the irradiated

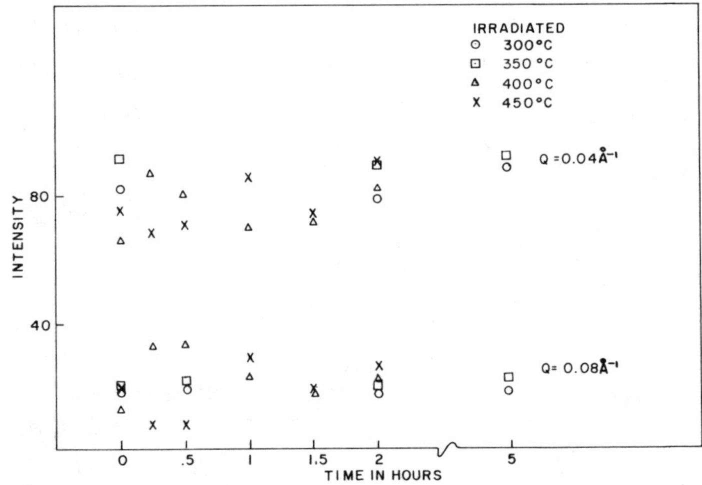

FIG. 10—*Intensity versus annealing time for irradiated ASTM A302B at* Q = *0.04Å*$^{-1}$ *and* Q = *0.08Å*$^{-1}$.

specimens as a function of annealing time. For annealing at 300 and 350°C, there are essentially no changes in the specimens throughout the annealing as stated above; this parallels the data obtained from the unirradiated specimens exclusive of the initial rise. This is because the annealing temperature is only slightly greater than the irradiation temperature. For the specimen annealed at 400°C there is an initial increase in intensity followed by a decrease and finally another increase. This pattern is followed for both values of Q. We find the same changes for the specimen annealed at 450°C as those described for the specimen annealed 400°C, except that these changes are compressed into a shorter time span.

Comparison of the effect of annealing at 300 and 350°C for the irradiated and unirradiated specimens reveals that irradiation causes the spectra to increase at lower values of Q, while at higher annealing temperatures the reverse is true; the spectra of the irradiated specimens shift to lower values of Q. There is also a coincidence in intensity for the unirradiated and irradiated spectra at higher values of Q. The changes in the spectra for the irradiated specimens annealed at 400 and 450°C are much greater than for the unirradiated specimens at the same annealing temperatures. This can be seen by comparing Figs. 5 and 10. At $Q = 0.04 \text{Å}^{-1}$, the intensity for the unirradiated specimens rises rather quickly to a fairly constant value while, for the irradiated specimens, there are changes in the intensity with time, which suggest that annealing is altering the structure. Similar changes occur for these higher temperature anneals when $Q = 0.08 \text{Å}^{-1}$.

In an effort to distinguish between metallurgical and irradiation effects for the specimens annealed at 400 and 450°C, we have subtracted the difference between the spectrum for the unannealed/unirradiated specimen and the spectrum after annealing from the irradiated/annealed spectrum. Figures 11 and 12 show the resulting difference spectra. We also show for comparison the spectrum for the unirradiated/unannealed specimen on each figure, as well as the spectrum for the irradiated/unannealed specimen. There is a strong similarity between SANS curves for specimens annealed at these two temperatures. For longer times, the specimen annealed at 450°C shows a greater reduction in the intensity particularly at the larger values of Q, indicating a greater change in the microstructure of the steel. Annealing at the higher temperatures allows the defects

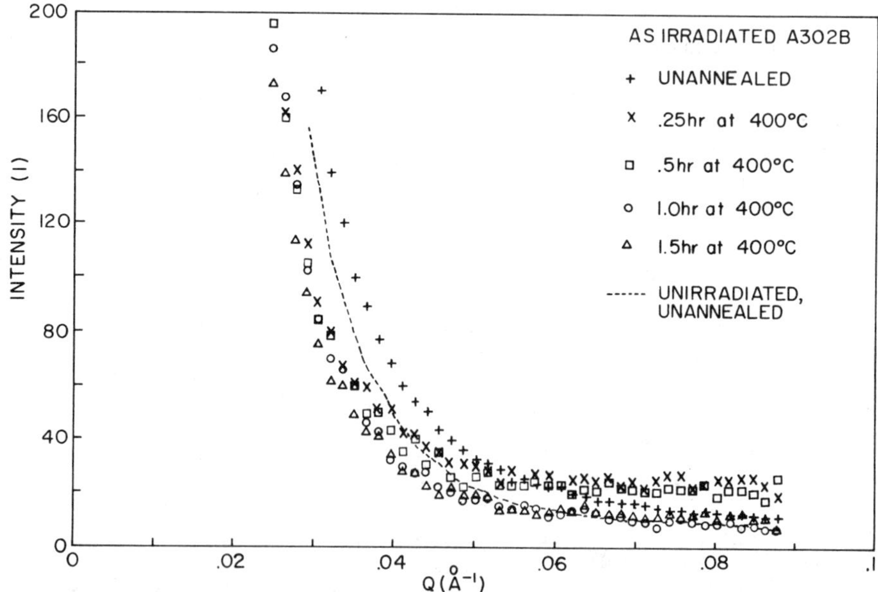

FIG. 11—*Results of anneals at 400°C after correction for metallurgical effects.*

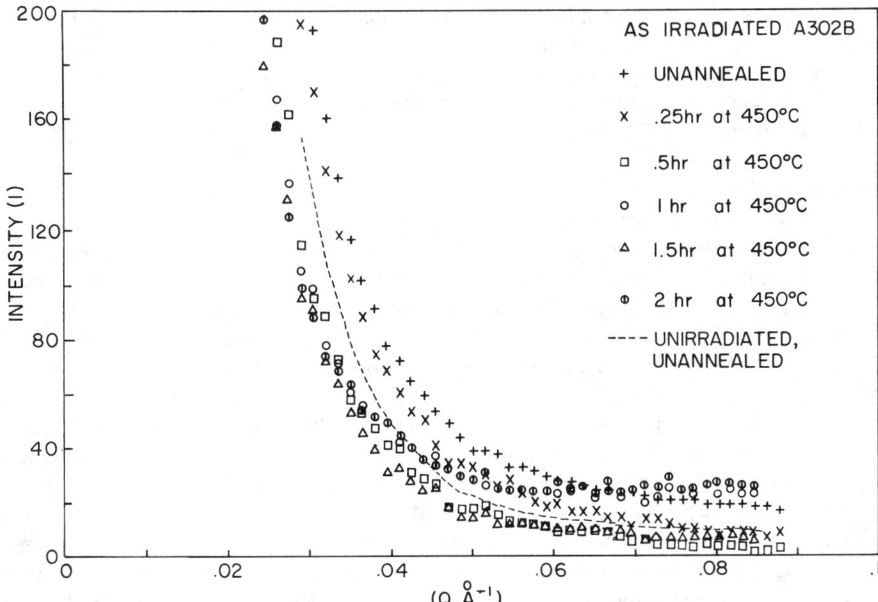

FIG. 12—*Results of anneals at 450°C after correction for metallurgical effects.*

and defect clusters introduced by radiation damage to become mobile at these temperatures, thus accelerating the rate of annealing.

Figure 13 gives the intensity at $Q = 0.04\text{Å}^{-1}$ and $Q = 0.08\text{Å}^{-1}$ after the metallurgical effects have been removed and shows the changes that are primarily caused by radiation damage annealing. At $Q = 0.04\text{Å}^{-1}$, the intensity decreases rapidly to a level below that of the unirradiated/

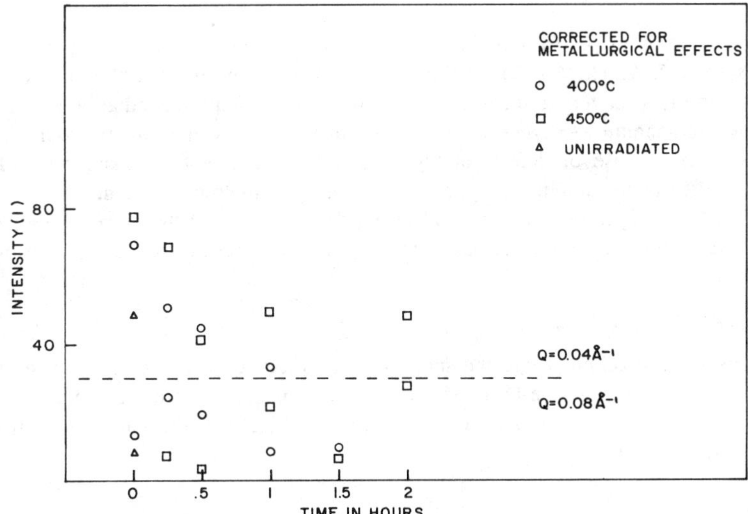

FIG. 13—*Intensity versus annealing time for ASTM A302B corrected for metallurgical effects when* Q = *0.04Å*$^{-1}$ *and* Q = *0.08Å*$^{-1}$.

unannealed specimen, indicating a rapid loss in the radiation caused defects. Because the intensity at this value of Q has declined to such a low level, it is proposed that the copper atoms have been able to cluster to a much larger size because of diffusion enhanced by radiation defects [24–28]. Some of these effects caused by copper precipitation are masked by the extra defects introduced by radiation damage. Annealing has removed these defects, revealing the effects caused by copper precipitation.

A second possibility for explaining the observed results is the continuation of the spheroidization process, which was started during the heat treatment performed to establish the microstructure, as given in Table 2. However, because the annealing temperatures used in this experiment are lower than the heat treatment temperature, changes in the SANS spectra for the unirradiated specimens cannot be attributed to additional spheroidization. Even with the increased defect concentration caused by radiation damage, it is unlikely that this is a source of the changes in the SANS spectra.

Schwahn et al. [29] performed a similar study on an ASTM A533B pressure vessel steel, which contains 0.12% copper and 0.22% carbon and a second steel, which has additions of vanadium (0.15%) and nickel (1.5%) but with a carbon content of 0.12%. They used 1-cm-thick specimens of these steels, which have been irradiated to fluence of 5×10^{19} n/cm^2 at 150°C. These authors have performed SANS measurements on specimens isochronally annealed for 4 h at 300, 350, 400, and 450°C. They showed that the scattered intensity for the ASTM A533B steel after irradiation tends to increase. The initial annealing at 300°C shows a marked increase in intensity while subsequent annealing leads to decreases; and after annealing at 400°C the spectrum coincides with the unirradiated spectrum. It appears that annealing at 300°C produces an effect similar to our observations in the annealing of the unirradiated specimen (see Fig. 1). The steel containing vanadium and nickel shows less change in intensity because of irradiation and annealing. There are small changes in intensity caused by irradiation, but subsequent isochronal annealing does not cause any further changes. A difficulty with this study is that their specimen thickness (1 cm) gives neutron transmissions of only 24%, indicating a large amount of multiple scattering which tends to smear out any observed features. The specimen thickness (1 mm) employed in the present study gave a transmission of 75% with relatively little multiple scattering.

Conclusions

The precipitation of copper from the α-iron solid solution upon the annealing of the unirradiated pressure vessel steel, ASTM A302B, in the temperature range of 300 to 450°C causes changes in the SANS spectra in a scattering vector range below 0.1A$^{0-1}$. Neutron irradiation of the same steel at an elevated temperature (~288°C) leads to similar effects, which are resistant to annealing below 400°C. The release of defects at higher annealing temperatures enhances diffusion and substantially increases the amount of copper precipitating from solid solution. This causes greater changes in the SANS spectra to occur than during the annealing of the unirradiated steel. These copper precipitates tend to be trapped at dislocation sites and cause the embrittlement of the steel.

Acknowledgments

The authors wish to acknowledge the support of the Electric Power Research Institute for this research, to express the gratitude to T. Marsden for his encouragement, and to acknowledge the assistance of R. Lott and S. Woods of Westinghouse Electric Corporation for providing the specimens used in this work.

References

[1] Smidt, F. A., Jr. and Bement, A. L., *Dislocation Dynamics,* A. R. Rosenfeld et al., Eds., McGraw Hill, New York, 1968, p. 409.

[2] Little, E. A. and Harris, D. R., *Metals Science,* Vol. 4, 1970, p. 188.
[3] Wiffen, F. W., *Radiation-Induced Voids in Metals,* J. W. Corbett and L. C. Ianiello, Eds., U.S. AEC, 1972, p. 386.
[4] Carpenter, G. F., Knopf, N. R., and Byron, E. S., *Nuclear Science and Engineering,* Vol. 19, 1964, p. 18.
[5] Parish, M. L. and Seman, D. J., "Fracture Toughness of Non-irradiated and Irradiated Reactor Vessel Steels," WAPD-TM-565, Westinghouse Atomic Power Dept., Pittsburgh, PA, March 1968.
[6] Odette, G. R., *Scripta Metallurgica,* Vol. 17, No. 10, Oct. 1983, p. 1183.
[7] Bush, S. H., *Journal of Testing Evaluation.* Vol. 2, No. 6, Nov. 1974, p. 435–462.
[8] Hawthorne, J. R. and Potapovs, U., *Irradiation Effects in Structural Alloys for Thermal and Fast Reactors, STP 457,* American Society for Testing and Materials, Philadelphia, 1969, p. 113.
[9] Potapovs, U. and Hawthorne, J. R., *Nuclear Applications,* Vol. 6, No. 1, Jan. 1969, p. 27.
[10] Hawthorne, J. R., *Irradiation Effects on Structural Alloys for Nuclear Reactor Applications, STP 484,* American Society for Testing and Materials, Philadelphia, 1971, p. 96.
[11] Palme, H. S., "Radiation Embrittlement Sensitivity of Reactor Pressure Vessel Steels," BAW-10056 Topical Report, Babcock and Wilcox, Lynchburg, VA, March 1973.
[12] Hawthorne, J. R., "Radiation Embrittlement Resistance of Advanced Ni-Cr-Mo Steel Plates, Forgings and Weldments," NRL Report 7573, Naval Research Laboratory, Washington, DC, 13 Sept. 1973.
[13] Hawthorne, J. R. and Fortner, E., *Journal of Engineering Industry, Series B,* Transactions *ASME,* Vol. 94, 1973, p. 807.
[14] Grant, S. P. and Fortner, E., *Metals Engineering Quarterly,* Vol. 12, No. 3, Aug. 1972, p. 17.
[15] Hawthorne, J. R., "Contributions of Selected Residual Elements to the Radiation Sensitivity of Steel Forgings," NRL Report 7526, NRL, Washington DC, 18 Jan. 1973.
[16] Smidt, F. A., Jr. and Sprague, J. A., *Effects of Radiation on Substructure and Mechanical Properties of Metals and Alloys, STP 529,* American Society for Testing and Materials, Philadelphia, 1973, pp. 78–91.
[17] Ohmae, K. and Zhiebold, T. O., *Journal of Nuclear Materials,* Vol. 43, 1972, p. 254.
[18] Hawthorne, J. R., *Effects of Radiation on Materials: Eleventh International Symposium, STP 782,* H. R. Brager and J. S. Perrin, Eds., American Society for Testing and Materials, Philadelphia, 1982, pp. 375–391.
[19] G. Kostorz, *Treatise on Materials Science and Technology,* Vol. 15, Academic Press Inc. (London) Ltd., 1979, p. 227.
[20] Gerold, V. K., Kostorz, G., *Journal of Applied Crystallography,* Vol. 11, 1978, p. 376.
[21] Hawthorne, J. R., "Evaluation and Prediction of Neutron Embrittlement in Reactor Pressure Vessel Materials," EPRI-NP-2782, 1983.
[22] Mildner, D. F. R., Berliner, R., Pringle, O. A., and King, J. S., *Journal of Applied Crystallography,* Vol. 14, 1981, p. 370.
[23] Speich, G. R., Gula, J. A., and Fisher, R. M., *The Electron Microprobe,* T. D. McKinley, K. F. J. Heinrich, and D. B. Willy, Eds., John Wiley and Co., New York, 1966, p. 525.
[24] Lucas, G. E. and Odette, G. R., *Proceeding of the Second International Symposium on Environmental Degradation of Materials in Nuclear Power Systems Water Reactors,* American Nuclear Society, 1986, p. 121, p. 345.
[25] Lucas, G. E., Odette, G. R., Lombrozo, P. M., and Sheckford, J. W., *Effects of Radiation on Materials: Twelfth International Symposium, STP 870,* American Society for Testing Materials, Philadelphia, 1985, pp. 900–930.
[26] Odette, G. R. and Lucas, G. E., *Radiation Embrittlement of Nuclear Reactor Pressure Vessel Steels: An International Review, STP 909,* American Society for Testing and Materials, Philadelphia, 1986, pp. 206–241.
[27] Frisius, F., Kampman, R., Beaven, P. A., and Wagner, R., *Proceedings of the Conference on Dimensional Stability and Mechanical Behaviour of Irradiated Metals and Alloys/Vol I,* British Nuclear Energy Society, 1983, p. 171.
[28] Fisher, S. B., Harbotte, J. E., and Aldridge, N. B., *Proceedings of the Conference on Dimensional Stability and Mechanical Behaviour of Irradiated Metals and Alloys/Vol I,* British Nuclear Energy Society, 1983, p. 87.
[29] Schwahn, D., Packur, D., and Schelten, J., *Neutron Scattering on Neutron Irradiated Steel,* Kernforschungsanlage, Jülich, Oct. 1978, p. 1543.

Subrata Chatterjee,[1] S. Anantharaman,[1] K. S. Sivaramakrishnan,[1] and Pradip R. Roy[1]

Irradiation Embrittlement Evaluation Without Lateral Expansion Measurements

REFERENCE: Chatterjee, S., Anantharaman, S., Sivaramakrishnan, K. S., and Roy, P. R., "**Irradiation Embrittlement Evaluation Without Lateral Expansion Measurements,**" *Influence of Radiation Material Properties: 13th International Symposium (Part II), ASTM STP 956,* F. A. Garner, C. H. Henager, Jr., and N. Igata, Eds., American Society for Testing and Materials, Philadelphia, 1987, pp. 494–504.

ABSTRACT: It is progressively felt that the lateral expansion criterion should be included for the determination of the transition temperature for irradiation embrittlement studies as this criterion compensates for the effects of irradiation induced increases in the strength. Lateral expansion measurements on irradiated specimens require specially designed jigs and time consuming remote measurements to be made inside a shielded cell. Results generated on eighth A533 and A508 category of pressure vessel steels and weldments containing copper contents up to 0.125 wt%, phosphorous contents up to 0.015 wt%, and subjected to fast neutron fluences (>1 MeV) of the order of 1 × 10^{19} n/cm^2 at 290°C were analyzed to find out whether fixed energy criterion alone could be used instead of lateral expansion criterion. The adjusted reference temperature RT_{NDT} of these irradiated steels and weldments, as obtained from the unirradiated reference temperature RT_{NDT} according to the American Society of Mechanical Engineers (ASME) Code, Section III, and the temperature shift at 41*J* absorbed energy (ΔT_{41J}), were compared with the reference temperature RT_{NDT} of the irradiated steels and weldments as obtained directly as per ASME Code, Section III. It was observed that in practically all these materials, the latter values were conservative. It was found that for these steels and weldments, if 68*J* absorbed energy criterion is used, no lateral expansion measurements need to be carried out.

KEY WORDS: irradiation embrittlement, lateral expansion, absorbed energy, Charpy V-notch requirements, RT_{NDT}, base metal, weld, weld heat affected zone, transition temperature, transition temperature shift, fluence

For the evaluation of irradiation induced embrittlement in ferritic steels, the Charpy V-notch requirements need to include lateral expansion rather than absorbed energy alone, because the former provides protection from the variation in yield strength produced by irradiation. As such the requirement of 1 mm of lateral expansion, or alternatively 0.9 mm of lateral expansion and 68*J* of absorbed energy at RT_{NDT} + 33°C throughout the life of the component, is required to provide assurance of adequate fracture toughness at upper shelf temperature [*1,2*]. It is not surprising, therefore, that criterion, such as temperatures corresponding to 41 and 68*J* of absorbed energy, and 0.9 mm of lateral expansion, have been widely used in the study of irradiation embrittlement of ferritic steels [*3*] and in the surveillance of reactor pressure vessels using ferritic steels as per ASTM Practice for Conducting Surveillance Tests for Light Water-Cooled Nuclear Power Reactor Vessels (E 185).

[1] Scientific officer, scientific officer, head of postirradiation examination section, and associate director of metallurgy group, respectively, Radiometallurgy Division, Bhabha Atomic Research Center, Trombay, Bombay 400085, India.

As a part of Tarapur Atomic Power Station (TAPS) pressure vessel surveillance program, absorbed energy versus temperature and lateral expansion versus temperature curves were generated on specimens fabricated from base, weld, and heat affected zones with A302B (nickel modified) as the base material. In addition, absorbed energy versus temperature and lateral expansion versus temperature curves were also generated on specimens from A533 B plate, A508 forging, and their weldments of controlled copper and phosphorus, as part of the Indian contribution to the International Atomic Energy Agency (IAEA) Co-ordinated Research Programme on the Analysis of Behaviour of Advanced Reactor Pressure Vessel Steels under Neutron Irradiation.

The test results were analyzed and the utility of absorbed energy, and lateral expansion criteria for establishing transition temperatures for the study of irradiation embrittlement were critically examined.

This paper presents the results thus obtained.

Materials

The following materials were included in this study:

(1) A302B (nickel modified) plate designated as B,
(2) A302B (nickel modified) weld designated as W,
(3) A302B (nickel modified) heat affected zone (HAZ) designated as H,
(4) Japanese A533-B Cl-1 plate designated as JP,
(5) Japanese A508-Cl-3 forging designated as JF,
(6) French A533-B Cl-1 plate designated as FP,
(7) French A508-Cl-3 forging designated as FF, and
(8) French S/A weld designated as FW.

The chemical compositions of these materials are given in Table 1.

All the plate and forging materials were quenched and tempered, and were subjected to simulated post-weld heat treatment. The welds were subjected to a post-weld heat treatment. The Charpy V-notch specimens were prepared from quarter thickness locations.

Irradiation

A302B (nickel modified) surveillance specimens were irradiated at the wall and shroud specimen surveillance locations of the TAPS reactor at 290°C. Since this reactor is a boiling water reactor, no separate temperature monitors were kept. The flux and fluence values at the wall and shroud locations were 7×10^{10} n/cm^2/s and 2.2×10^{18} n/cm^2, and 4.8×10^{11} n/cm^2/s and 1.5×10^{19} n/cm^2, respectively.

All the remaining specimens were irradiated to a fluence of 1×10^{19} n/cm^2 at 290 ± 10°C in the CIRUS reactor. Low-temperature eutectic alloys were used for monitoring the irradiation temperature. The irradiation details are available elsewhere [*4,5*].

Table 2 gives the compilation of the number of Charpy V-notch specimens tested in unirradiated and irradiated conditions.

Approach

The absorbed energy and the corresponding lateral expansion values undergo a similar change over the same general temperature range. Therefore a linear regression fit was applied to the corresponding data points of absorbed energy and lateral expansion measurements for all the individual unirradiated and irradiated materials. The data points for this analysis on these unirradiated

TABLE 1—*Chemical composition of base metals and weldments, wt%.*

Material	C	Si	Mn	P	S	Ni	Cr	Mo	V	Cu	Co
JP	0.18	0.22	1.48	0.007	0.007	0.66	0.20	0.57	0.01	0.01	. . .
JF	0.18	0.26	1.33	0.007	0.005	0.76	0.11	0.50	0.01	0.04	0.008
FP	0.23	0.22	1.47	0.008	0.003	0.65	0.04	0.52	. . .	0.03	. . .
FF	0.15	0.25	1.37	0.009	0.008	0.69	0.24	0.47	0.01	0.07	0.02
FW	0.056	0.54	1.37	0.015	0.011	0.56	0.045	0.52	0.01	0.05	0.023
B	0.2	0.18	1.10	0.008	. . .	0.44	. . .	0.29	. . .	0.125	. . .
W	0.2	0.15	1.16	0.007	. . .	0.38	. . .	0.49	. . .	0.093	. . .
H	0.2	0.32	1.12	0.008	. . .	0.49	. . .	0.42	. . .	0.120	. . .

TABLE 2—*Number of specimens used.*

Material	Charpy-V Specimens	Fluence, 1 MeV, n/cm[2][a]
B	7	0
	32	2.2×10^{18}
	7	1.5×10^{19}
W	6	0
	10	2.2×10^{18}
	7	1.5×10^{19}
H	8	0
	12	2.2×10^{18}
	8	1.5×10^{19}
JP	10	0
	9	1×10^{19}
JF	16	0
	12	1×10^{19}
FP	14	0
	13	1×10^{19}
FF	11	0
	14	1×10^{19}
FW	7	0
	11	1×10^{19}

[a] Fluence variations among specimens ~8%.

and irradiated materials have been taken from the respective absorbed energy versus temperature and lateral expansion versus temperature measurements. The test procedures followed were in accordance with ASTM Methods for Notched Bar Impact Testing of Metallic Materials (E 23) and Methods and Definitions for Mechanical Testing of Steel Products (A 370). The lateral expansions were measured with a remotized micrometer with special end measuring faces of 10 mm diameter to meet ASTM A 370 requirements. The test temperatures included the transition and upper shelf temperature regions. The individual regression equations so obtained were used to obtain the corresponding lateral expansion values at 41J of absorbed energy and at 68J of absorbed energy. Conversely, these regression equations were also used to obtain the absorbed energy values corresponding to 0.9 mm of lateral expansion.

The transition temperatures obtained at 41J of absorbed energy (T_{41J}), at 68J of absorbed energy (T_{68J}) and at 0.9 mm of lateral expansion ($T_{0.9mm}$) of these materials for the unirradiated and irradiated conditions were compared by applying linear regression fit between the possible combinations of $T_{0.9mm}$ and T_{68J}, $T_{0.9mm}$, and T_{41J}, and T_{68J} and T_{41J}. The degree of scatter between the corresponding radiation induced shift in the transition temperatures was also obtained.

Finally, the adjusted reference temperature RT_{NDT} of these irradiated materials, as obtained from the unirradiated reference temperature RT_{NDT} according to ASME Code, Section III, and the temperature shift at 41J (ΔT_{41J}), were compared with the reference temperature RT_{NDT} of the irradiated materials as obtained directly as per ASME Code, Section III. All the RT_{NDT} values were obtained from the absorbed energy versus temperature curves.

Results

The regression constants for all the materials and the corresponding correlation coefficients are given in Table 3. It can be seen that in most of the cases, the correlation coefficient r^2 is very

TABLE 3—*Coefficients of regression equation* C_v = M (LE) + N, *including* r^2, *for the base metals and weldments.*

Material	Number of Specimens Used	M	N	r^2
JP	10	100.47	−32.86	0.97
JF	16	94.50	14.01	0.70
FP	14	76.09	−5.41	0.99
FF	11	79.82	−4.59	1.00
FW	7	77.73	−6.57	0.99
JP(1)	9	100.44	−18.99	0.99
JF(1)	12	94.70	−18.92	0.95
FP(1)	13	85.89	−14.87	0.97
FF(1)	14	76.11	−5.48	0.96
FW(1)	11	90.07	−21.80	0.98
B	7	83.18	−16.68	0.97
B(2)	11	64.93	3.36	0.99
B(2)	11	61.38	6.12	0.99
B(2)	10	57.56	6.22	0.96
B(3)	7	67.94	2.19	1.00
W	6	64.37	−6.99	1.00
W(2)	10	59.95	0.42	0.98
W(3)	7	52.19	11.14	0.97
H	8	87.75	−10.94	0.95
H(2)	12	67.00	3.28	0.96
H(3)	8	84.01	−7.11	0.96

NOTE: Material without parentheses are unirradiated. (1) 1×10^{19} n/cm², >1 MeV. (2) 2.2×10^{18} n/cm², >1 MeV. (3) 1.5×10^{19} n/cm², >1 MeV.

close to unity. The data points for this analysis have been taken from the respective Charpy energy versus temperature and lateral expansion versus temperature measurements [*4,5*].

The corresponding lateral expansion values at absorbed energy of 41*J* (LE_{41J}) and absorbed energy of 68*J* (LE_{68J}), and the absorbed energy at 0.9 mm lateral expansion ($C_{v0.9\ mm}$) as obtained from the respective linear regression equations are given in Table 4. It can be seen that for practically all the cases, the lateral expansion values at absorbed energy of 68*J* are higher than 0.9 mm, and the lateral expansion values at absorbed energy of 41*J* are typically around 0.6 mm. Table 4 also includes the room temperature yield strength as determined for these unirradiated and irradiated materials.

Figures 1, 2, and 3 give the relationship between $T_{0.9mm}$ and T_{68J}, $T_{0.9mm}$ and T_{41J}, and T_{68J} and T_{41J} for these unirradiated and irradiated materials. Figure 1 reveals that $T_{0.9mm}$ is 5°C below the equality line. Figure 2 reveals that $T_{0.9mm}$ is 15°C above the equality line. Figure 3 reveals that T_{68J} is 20°C above the equality line. These figures illustrate the variation in transition temperatures in these materials as obtained from different commonly used criteria.

Figures 4, 5, and 6 give the relationship between the shift in the transition temperatures (ΔT) as affected by irradiation. Figure 4 reveals that except for one data point, $\Delta T_{0.9mm}$ values lie in a scatter band of 9°C above and 5°C below the equality line. Figure 5 reveals that except for two data points, $\Delta T_{0.9mm}$ values lie in a scatter band 19°C above the equality line. Figure 6 reveals that except for one data point, ΔT_{68J} values lie in a scatter band 19°C above and 5°C below the equality line. In these figures the few distinct data points falling outside the bounds of the scatter bands belong to FF(1) and W(3) materials (Table 5).

TABLE 4—*Intercomparison of Charpy energy versus temperature and lateral expansion versus temperature curves.*

Material	Yield Strength, MPa	$(C_v)_{0.9mm}$, J	$(LE)_{41J}$, mm	$(LE)_{68J}$, mm
JP	478	57.6	0.74	1.00
JF	450	99.0	0.29	0.57
FP	480	63.1	0.61	0.97
FF	547	67.3	0.57	0.91
FW	508	. . .	0.61	0.96
JP(1)	556	71.4	0.77	1.12
JF(1)	. . .	. . .	0.63	0.92
FP(1)	579	62.4	0.65	0.97
FF(1)	633	63.0	0.61	0.97
FW(1)	. . .	. . .	0.70	1.00
B	419	58.2	0.69	1.02
B(2)	476	61.8	0.58	1.00
B(2)	514	61.4	0.59	1.01
B(2)	493	58.0	0.60	1.07
B(3)	500	63.3	0.57	0.97
W	462	50.9	0.75	1.17
W(2)	524	54.4	0.61	1.06
W(3)	557	58.1	0.57	1.09
H	405	68.0	0.59	0.90
H(2)	464	63.4	0.56	0.97
H(3)	526	68.5	0.57	0.89

NOTE: Materials without parentheses are unirradiated. (1) 1×10^{19} n/cm², >1 MeV. (2) 2.2×10^{18} n/cm², >1 MeV. (3) 1.5×10^{19} n/cm², >1 MeV.

Discussion

The shift in transition temperatures, ΔT_{41J}, are only 40°C for the surveillance specimens and only 30°C for the other plate and weld specimens. Such low radiation sensitivities could be attributed to the extra low copper content (typically 0.04 wt%) for the plate and weld specimens [*6*]. For the surveillance specimens the obtained low radiation sensitivity can be attributed to the low copper content (typically 0.1 wt%) and the recovery that possibly could take place during prolonged irradiation inside the reactor [*7*].

From Fig. 1, it is clear that for all these materials, in their unirradiated and irradiated conditions, T_{68J} values are always higher than $T_{0.9mm}$ values, and these values are within 5°C. Table 4 reveals that the lateral expansion values corresponding to 68*J* absorbed energy are almost always more than 0.9 mm. Thus, it appears that T_{68J} could be used for the conservative estimation of transition temperature of these materials in their unirradiated and irradiated conditions, in absence of lateral expansion measurements.

Table 5 reveals that the shift in transition temperatures in these materials as determined by the different criteria are not very different, and particulary $\Delta T_{0.9mm}$ and ΔT_{68J} values are in fairly good agreement. Moreover, the relative amount of scatter between transition temperature and shift in transition temperature (Figs. 1 and 4, Figs. 2 and 5, and Figs. 3 and 6) are very much of the same order, the scatter in values being 5 and 14°C, 15 and 15°C, and 20 and 24°C, respectively. Therefore, the adjusted reference temperature RT_{NDT} of these irradiated materials derived from the unirradiated reference temperature RT_{NDT} as per ASME Code Section III and the ΔT_{41J} are compared with the reference temperature RT_{NDT} of the irradiated materials obtained directly as per ASME

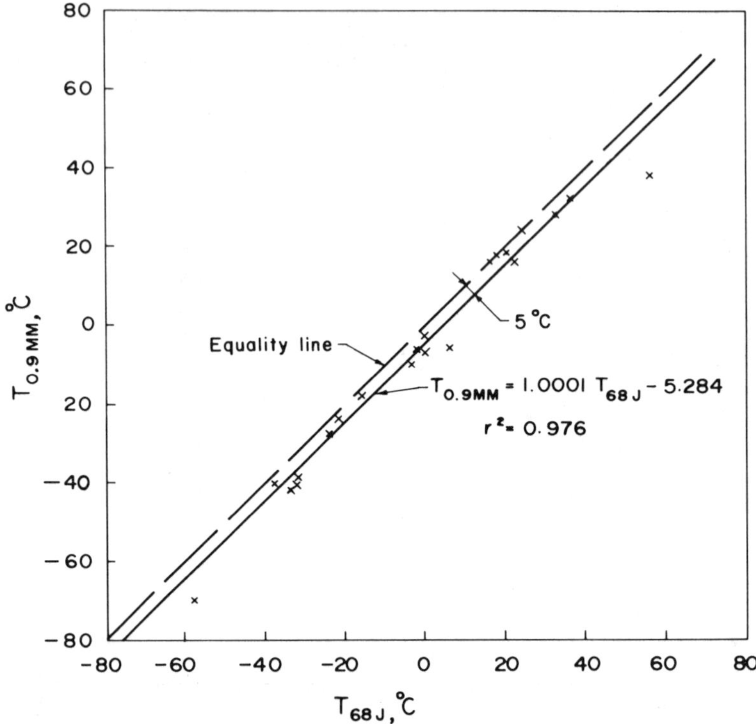

FIG. 1—*Correlation between lateral expansion 0.9 mm and Charpy 68J transition temperatures.*

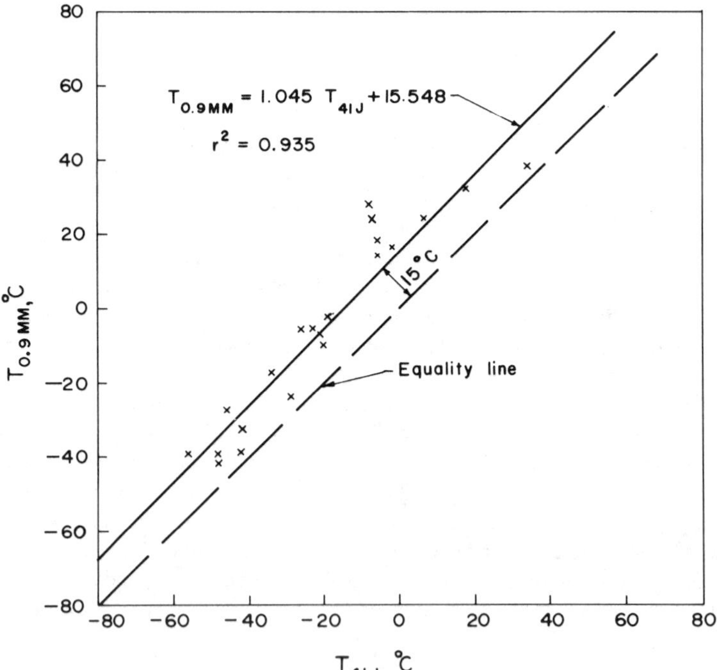

FIG. 2—*Correlation between lateral expansion 0.9 mm and Charpy 41J transition temperatures.*

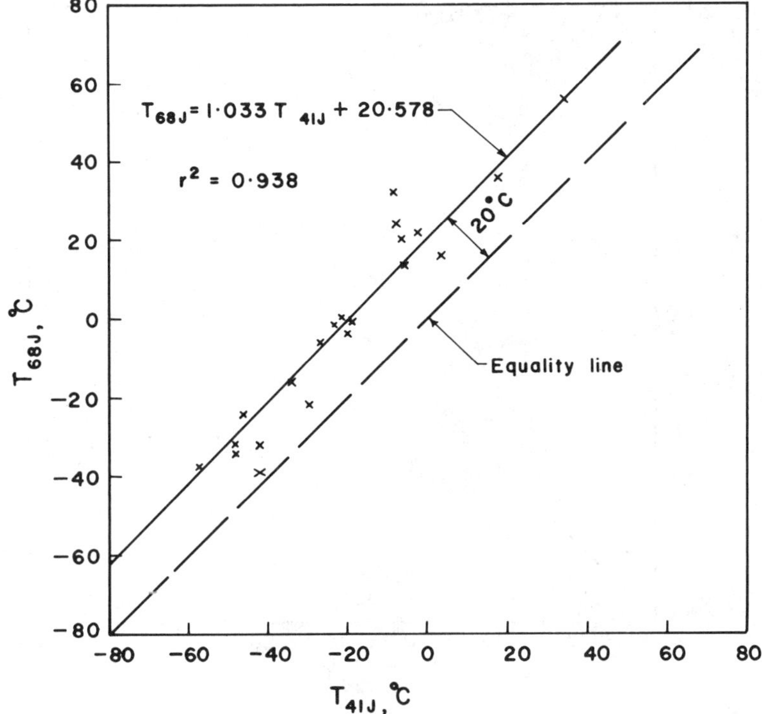

FIG. 3—*Correlation between Charpy 68J and Charpy 41J transition temperatures.*

TABLE 5—*Radiation induced shift in transition temperature as determined by different criteria.*

Material	ΔT_{41J}, °C	ΔT_{68J}, °C	$\Delta T_{0.9mm}$, °C
JP(1)	21	32	32
JF(1)	36	46	48
FP(1)	17	22	19
FF(1)	57	58	44
FW(1)	14	16	22
B(2)	14	18	24
B(3)	38	40	42
W(2)	22	40	36
W(3)	40	66	70
H(2)	26	20	30
H(3)	36	34	42

NOTE: (1) 1 × 10^{19} n/cm², >1 MeV. (2) 2.2 × 10^{18} n/cm², >1 MeV. (3) 1.5 × 10^{19} n/cm², >1 MeV.

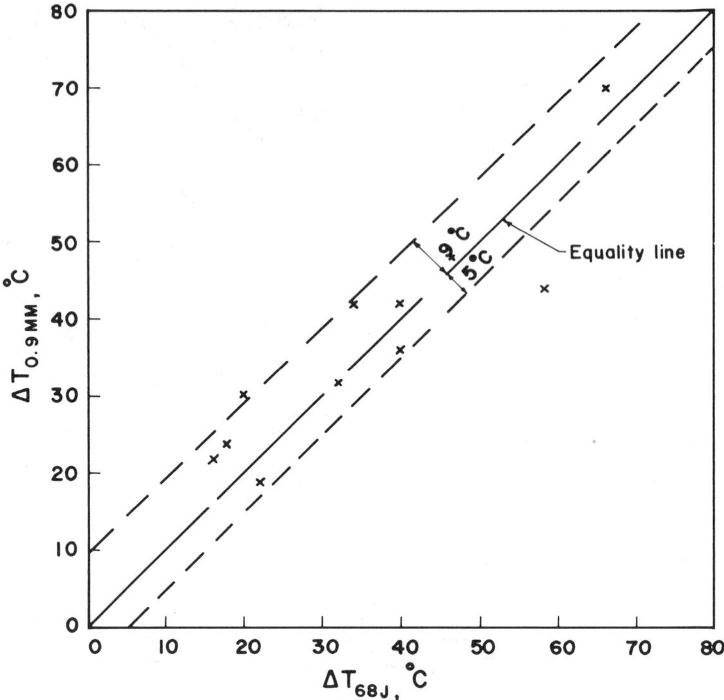

FIG. 4—*Correlation between lateral expansion 0.9 mm and Charpy 68*J *transition temperature shifts.*

Code Section III. Table 6 reveals that in all except for two materials, the latter results are conservative. Keeping in view of the fact that the recent work on A508 Cl 2, A533B Cl 1, and 508 Cl 3 equivalent steels have exhibited that the procedure of determining the adjusted reference temperature on the basis of $C_{v_{41J}}$, following ASTM E 185 procedures is itself conservative [*7*], the results imply that in all likelihood, estimation of RT_{NDT} of irradiated materials from the Charpy energy versus the temperature curve should be adequate, and use of either ASTM E 185 or ASME Code, Section III, would be adequate.

On the basis of these observations and in view of the fact that lateral expansion measurements of irradiated materials involve specially designed jigs and time consuming remote measurements inside a shielded cell, it may be worthwhile to recommend that T_{68J} could be used and that no lateral expansion measurements need to be carried out on these materials, if used up to a fluence (>1 MeV) of 1×10^{19} n/cm^2.

Conclusion

For low radiation sensitive A533B category of plate and A508 category of forging materials and weldments, 68*J* absorbed energy criterion alone could probably be used instead of 68*J* absorbed energy and 0.9 mm lateral expansion dual criteria for the determination of unirradiated and irradiated transition temperatures and the irradiation induced shifts in transition temperatures, up to a fluence (>1 MeV) of the order of 1×10^{19} n/cm^2, which is the typical life time fluence of many pressure vessel steels.

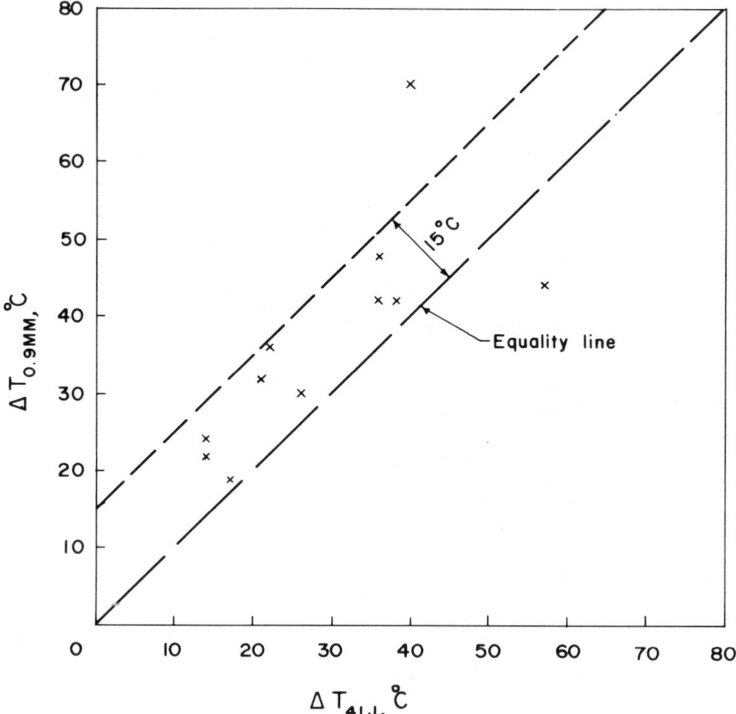

FIG. 5—*Correlation between lateral expansion 0.9 mm and Charpy 41*J *transition temperature shifts.*

TABLE 6—*Adjusted reference temperature and reference temperature of the irradiated base metals and weldments.*

Material	RT_{NDT} (Unirradiated)	ΔT_{41J}, °C	$RT_{NDT(adjusted)}$ (Irradiated) °C	RT_{NDT} (Irradiated) °C	[a]Difference
JP	−65	21(1)	−44(1)	−33(1)	+11
JF	−55	36(1)	−19(1)	−9(1)	+10
FP	−33	17(1)	−16(1)	−11(1)	+5
FF	−35	57(1)	22(1)	23(1)	+1
FW	−65	14(1)	−51(1)	−49(1)	+2
B	−37	14(2)	−23(2)	−19(2)	+4
B	−37	38(2)	1(2)	3(2)	+2
W	−67	22(2)	−45(2)	−27(2)	+18
W	−67	40(3)	−27(3)	−1(3)	+26
H	−91	26(2)	−65(2)	−71(2)	−6
H	−91	36(3)	−55(3)	−57(3)	−2

NOTE: (1) 1 × 10^{19} n/cm², >1 MeV. (2) 2.2 × 10^{18} n/cm², >1 MeV. (3) 1.5 × 10^{10} n/cm², >1 MeV.
[a]Difference = $RT_{NDT(irradiated)} - RT_{NDT(adjusted)(irradiated)}$.

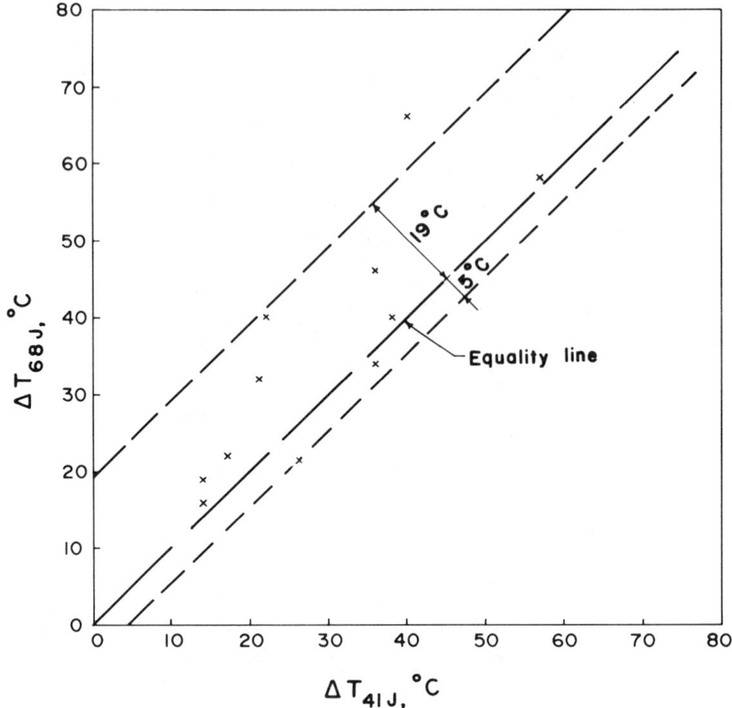

FIG. 6—*Correlation between Charpy 68*J *and Charpy 41*J *transition temperature shifts.*

References

[*1*] "PVRC Recommendations on Toughness Requirements for Ferritic Materials," *Welding Research Council Bulletin No. 175,* PVRC Ad HOC Task Group on Toughness Requirements, Aug. 1972.

[*2*] ASME Boiler and Pressure Vessel Code Section III, Sub-Section NB, American Society of Mechanical Engineers, New York, 1983.

[*3*] Davies, L. M., Ingham, T., and Squires. R. L., "Evaluation of Advanced Reactor Pressure Vessel Steels under Neutron Irradiation," *Effects of Radiation on Materials: Proceedings of the Eleventh International Symposium, STP 782,* H. R. Brager and J. S. Perrin, Eds., American Society for Testing and Materials, Philadelphia, 1982, pp. 433–463.

[*4*] Sivaramakrishnan, K. S., Chatterjee, S., Anantharaman, S., Balakrishnan, K. S., Viswanathan, U. K., and Roy, P. R., "Boiling Water Reactor (BWR) Pressure Vessel Surveillance Program," *Radiation Embrittlement of Nuclear Reactor Pressure Vessel Steels: An International Review (Second Volume), STP 909,* L. E. Steel, Ed., American Society for Testing and Materials, Philadelphia, 1986, pp. 106–117.

[*5*] Sivaramakrishnan, K. S., Chatterjee, S., Anantharaman, S., Balakrishnan, K. S., and Viswanathan, U. K., "Irradiation Embrittlement of Reactor Pressure Vessel Steels," presented to IAEA Program on Analysis Behaviour of Advanced Pressure Vessel Steels under Neutron Irradiation, Vienna, Oct. 1984.

[*6*] Steele, L. E., Davies, L. M., Ingham, T., Brumovsky, M., "Results of the International Atomic Energy Agency (IAEA) Coordinated Research Program on Irradiation Effect on Advanced Pressure Vessel Steels," *Effects of Radiation on Materials, Twelfth International Symposium, STP 870,* F. A. Garner and J. S. Perrin, Eds., American Society for Testing and Materials, Philadelphia, 1985, pp. 863–899.

[*7*] Föhl, J., Leitz, Ch., and Anders, D., "Irradiation Experiments in the Testing Nuclear Power Plant VAK," *Effects of Radiation on Materials: Eleventh Conference, STP 782,* H. R. Bragger and J. S. Perrin, Eds., American Society for Testing and Materials, Philadelphia, 1982, pp. 520–549.

William N. McElroy,[1] Raymond Gold,[1] Robert L. Simons,[1] and J. H. Roberts[2]

Trend Curve Data Development and Testing

REFERENCE: McElroy, W. N., Gold, R., Simons, R. L., and Roberts, J. H., "**Trend Curve Data Development and Testing,**" *Influence of Radiation on Material Properties: 13th International Symposium, ASTM STP 956,* F. A. Garner, C. H. Henager, Jr., and N. Igata, Eds., American Society for Testing and Materials, Philadelphia, 1987, pp. 505–534.

ABSTRACT: Existing trend curves do not account for previous and more recently observed test and power reactor flux-level, thermal neutron and γ-ray field-induced effects. Any agreement between measured data and trend curve predictions that does not adequately represent the important neutron environmental and temperature effects as well as the microstructural damage processes, therefore, could be fortuitous. Empirically derived end-of-life (EOL) and life-extension-range (LER) trend curves are presented and discussed in this paper for high temperature [~288°C (550°F)] irradiation of two weld, two plate, and two forging pressure vessel (PV) steels and low-temperature [~60°C (140°F)] irradiation of one support structure-type steel.

Preliminary results of a comprehensive study of the effects of environmental variables (neutron spectrum, exposure, exposure rate, and the thermal and γ-ray fluxes associated with surveillance capsules and PV through-wall gradients) were used to develop these trend curves.

Pressurized water reactor (PWR) and boiling water reactor (BWR) plant-specific results together with those of the Poolside Facility (PSF) of the Oak Ridge Research Reactor (ORR) at the Oak Ridge National Laboratory (ORNL) and other research reactor experiments support the existence of a significant material-dependent flux-level effect for PV and support structure steels; that is, a steel may show a decrease, an increase, or no change in the measured Charpy shift with changes in flux-level. Further, the actual behavior of a material can change significantly as a function of neutron exposure; also, thermal neutron and γ-ray effects can contribute to observed changes in property, especially near steel-water interface positions with high thermal-to-fast-neutron ratios.

KEY WORDS: pressure and support structure steels, embrittlement, transition temperature, trend curves, radiation damage, neutron spectrum, neutron exposure, exposure rate, flux level, thermal neutron, and γ-ray effects

Existing trend curves [*1–3*] do not account for previous [*4–6*] and more recently observed [*7–12*] test and power reactor flux-level, thermal neutron and γ-ray field-induced effects. Any agreement between measured data and trend curve predictions that does not adequately represent the important neutron environmental and temperature effects as well as the microstructural damage processes [*13*], therefore, could be fortuitous.

Simons [*8*] and McElroy et al [*9,10*] have furthered the work of Alberman [*5,6*], Norris [*4*], Lucas et al [*13*] and others by studying the effects of the environmental variables of neutron spectrum, exposure, exposure rate, and the thermal and γ-ray fluxes associated with surveillance capsules and PV through-wall gradients. The results of these studies, together with those of Hawthorne [*12*] support the existence of a significant material-dependent flux-level effect for pressure vessel (PV) and support structure steels; that is, a steel may show a decrease, an increase, or no change in the measured Charpy shift with changes in flux level. Further, the actual behavior

[1] Westinghouse Hanford Company, Hanford Engineering Development Laboratory, Richland, WA 99352.
[2] Metrology Control Corporation (MC²), Richland, WA 99352.

of a material can change significantly as a function of the neutron exposure; also, thermal neutron and γ-ray effects can contribute to observed changes in property, especially near steel-water interface positions with high thermal-to-fast-neutron ratios.

In this paper, preliminary results of the study of the effects of environmental variables (neutron spectrum, exposure, exposure rate, and thermal and γ-ray fluxes) on the embrittlement of two weld, two plate, and two forging PV steels irradiated at ~288°C (550°F) and one support structure-type steel irradiated at or below 60°C (140°F) are presented and discussed. Empirically derived through-wall Charpy shift gradient curves for the six PV steel materials and end-of-life (EOL) and life-extension-range (LER) trend curve model equations for the six PV steel materials and the one support structure-type steel are under development and study. The applications and implications on the use of the measured in-wall Charpy shift attenuation curves as well as the derived trend curve model equations are presented and discussed. Additional information on the subject matter of this paper is provided elsewhere [*14*].

Trend Curve Modeling and Environmental Variable Effects

Norris and Alberman Studies

Norris [*4*] developed and tested two damage models (one with and one without a rate-dependent saturation factor) by using data from a single heat of A302B (ASTM reference plate) steel irradiated in seven power reactor surveillance capsules. His rate-dependent model is based on an equation form suggested by G. L. Guthrie in an early draft of the ASTM E706 (IE) Standard on ''Damage Correlation for Reactor Vessel Surveillance'' [*15,16*].

Using his A302B steel rate-dependent model equation, Norris found that very large predicted changes were possible in the Charpy shift (up to a factor of ~3.25 decrease) for a change in flux level of 2×10^{10} neutrons (n)/cm$^2 \cdot$ s [typical for a pressurized water reactor (PWR) vessel inside diameter [ID] flux] to 1×10^9 n/cm$^2 \cdot$ s [typical for a boiling water reactor (BWR) vessel ID flux]. The change was a factor of ~1.34 for a factor of two decrease in flux in going from the ID to the ¼-T location for a typical PWR. These results were obtained for a neutron exposure of 1×10^{19} n/cm^2, $E > 1$ MeV. At an exposure of 1×10^{18} n/cm^2, $E > 1$ MeV, the factors reduced to ~2.17 and ~1.13, respectively.

Alberman's [*5,6*] findings for specimens irradiated at ~100°C for a nonboron-containing A533B steel strongly support the use of the EURATOM displacements per atom (dpa) in iron as one of, if not, the best exposure parameters for the correlation of PV and support structure steel irradiation effects data. The ASTM E693 dpa standard cross section has been used to calculate all dpa exposure parameter values for the damage analysis studies of this paper. The ASTM E693 dpa cross section provides calculated dpa values that are, essentially, the same as those obtained with the EURATOM dpa cross section [*17*]. Thus, the use of dpa as the exposure parameter allows for the correlation of changes in measured property changes that are associated with neutron spectral differences.

In a similar manner, it is essential to use the dpa/s displacement rate to account for any changes in mechanical property that are associated with flux-level. That is, any systematic bias that is associated with neutron spectral changes from one irradiation position to another must be accounted for to separate out other variable effects related to neutron exposure, exposure rate, and thermal neutron and γ-ray field-induced effects.

Hanford Engineering Development Laboratory (HEDL) Studies

Thermal Neutron and Gamma-Ray Field Environmental Variable Effects—McElroy et al. [*10*] indicates that there is a significant improvement (reduction) in the standard deviation of the fit for weld Charpy shift trend curves that include the effect of low-energy thermal neutrons. That is, for

weld Charpy shift trend curves that include the effect of low-energy thermal neutrons. That is, for a 30-point pressurized water reactor (PWR) surveillance capsule weld data set, improvements of the amount observed could occur at a frequency of no more than ~4% by chance. This conclusion is based on the work of Guthrie et al [*18*]. Gold and McElroy [*11*] have stated that

> This conclusion regarding a thermal neutron effect is not a unique interpretation of the data. Indeed, a collection of systematic effects caused by flux level, helium production, and gamma-ray heating cannot be ruled out. The intensity of the gamma-ray field found in PV environments is highly correlated with thermal neutron intensity. Consequently, the thermal neutron effect recently reported [*10*] may actually arise from a combination of effects, including annealing from gamma-ray heating. In this event, one must recognize that gamma-ray heating at surveillance capsule locations is considerably higher than that which is attained within a pressure vessel. Therefore, the annealing rate from gamma heating at the surveillance capsule location would be considerably higher than the annealing rate from gamma heating within the pressure vessel. Hence, gamma-ray heating could be another factor responsible for introducing a systematic bias in trend curve analyses that use surveillance capsule data bases. In this case, the effect of gamma-ray heating would be nonconservative [*11*].

Gold et al [*19*] have used absolute electron spectral measurements obtained with the Janus probe γ-ray spectrometer to quantitatively assess the displacement rate produced by the γ-ray field in light water reactor (LWR)-PV environments. The γ-ray displacement results are presented for the ¼-T, ½-T, and ¾-T locations of the Pool Critical Assembly (PCA) simulated PV mockup. Compared with neutron-induced displacement rates, the calculated γ-ray field-induced displacement rates are negligible at all locations. The ratio of γ-ray field-induced to neutron-induced displacement rates never exceeds roughly 5×10^{-3}.

While this effect is generally small, however, for Alberman's [*6*] study of the effect of thermal neutrons in very high thermal/fast (T/F) ratios of up to ~2000 or more, the γ-ray field contribution to the total displacement rate might be considerably larger than that resulting from neutrons. It is important, therefore, that the γ-ray field-induced displacement rate be determined and added to the low-energy neutron-induced iron (n,γ) recoil and high-energy neutron-induced displacement rates. This should be done for any damage correlation studies that make use of Alberman's A537 linear steel data base.

PSF Experiment Environmental Variable Effects and Data Base

The results of a comprehensive study of the comparison of predicted (calculated) to measured (experimental) (C/E) Charpy shift ratios using seven plate and weld trend curve model equations and different exposure parameters have been reported by McElroy et al [*9*] for high-temperature (288°C) irradiation of two weld, two plate, and two forging PV materials irradiated in the simulated surveillance capsule (SSC) and the simulated pressure vessel capsules (SPVC) of the Oak Ridge Research Reactor-Poolside Facility (ORR-PSF) at Oak Ridge National Laboratory (ORNL).

These ORR, SSC, and SPVC irradiations have been designated as the PSF Experiment [*7,10,16–23*]. Together with other test reactor data, Simons [*8*] has used the PSF physics-dosimetry-metallurgy data base (consensus evaluation of Ref *16*) to develop and test physically based trend curve model equations for each of the six PSF weld, plate, and forging materials. For one of the materials (A302B, ASTM reference plate), Simons has used both low- (<116°C) and high- (~288°C) temperature data to help calibrate his model equations.

For the present study, an evaluated set of reference nil ductility transition temperature ($\Delta RT\text{-}NDT$) values was selected based on a study of the combined Charpy (C_V), yield stress (YS), and ultimate stress (US) results, as reported by Hawthorne et al [*21,22*]. That is, on the basis of a comparison of the six weld, plate, and forging material C_V, YS, and US results, the $\Delta RT\text{-}NDT$ Charpy shift could be taken to be approximately proportional to ΔYS and ΔUS for the six alloy

TABLE 1—*Evaluated Charpy shift metallurgy results for an irradiation temperature of 288°C (550°F) for the PSF experiment.*

Position and Chemistry	Measured Δ*RT-NDT* Values (°F)[a]					
	A533B Weld, Code R	A533B Weld, Code EC	A533B Plate, Code 3PU	A302B Plate, Code F23	A508 Forging, Code K	A508 Forging, Code MO
SSC-1	409 (2)[c] [227][b]	186 (3) [103]	114 (6) [63.3]	145 (3) [80.6]	102 (10) [56.7]	28.9 (18) [16.1]
SSC-2	496 (4) [276]	212 (1) [118]	147 (5) [81.7]	165 (5) [91.7]	164 (3) [91.1]	55.9 (19) [31.1]
O-T	515 (0) [286]	205 (0) [114]	135 (0) [75.0]	146 (0) [81.1]	130 (0) [72.2]	45.0 (0) [25.0]
¼-T	459 (1) [255]	184 (6) [102]	113 (7) [62.8]	121 (4) [67.2]	102 (18) [56.7]	29.8 (15) [16.6]
½-T	424 (1) [236]	176 (7) [97.8]	85.7 (11) [47.6]	101 (8) [56.1]	99.5 (3) [55.3]	20.4 (17) [11.3]
Cu % by weight	0.23	.24	0.12	0.20	0.12	0.05
Ni % by weight	1.58	0.64	0.56	0.18	0.96	0.75
Mn % by weight	1.54	1.57	1.26	1.34	0.72	1.43
Mo % by weight	0.34	0.48	0.45	0.51	0.63	0.53
Cr % by weight	0.12	0.02	0.10	0.11	0.45	TBD
P % by weight	0.009	0.007	0.011	0.011	0.009	0.008
Si % by weight	0.45	0.52	0.25	0.23	0.16	0.28
S % by weight	0.008	0.011	0.018	0.023	0.004	0.008
C % by weight	0.05	0.11	0.20	0.24	0.18	0.20
B % by weight	TBD	TBD	TBD	TBD	TBD	TBD
N % by weight	TBD	TBD	TBD	TBD	TBD	TBD
Ni/Cu Ratio	6.87	2.67	4.67	0.90	8.0	15.0

[a] These are measured, but evaluated Δ*RT-NDT* values based on a study of the combined Charpy (C_v), yield stress (*YS*) and ultimate stress (*US*) results as reported by Hawthorne et al [*21,22*]. It was concluded, on the basis of the comparison of the Code R weld material C_v, *YS*, and *US* results, that is could be assumed that the Δ*RT-NDT* α *YS* and Δ*RT-NDT* α *US* for the six alloy materials used in the PSF Experiment. Consequently, measured changes in C_v, *YS*, and *US* were established relative to the PSF-OT position for each of the six materials. These relative ratios were then averaged for the C_v, *YS*, and *US* measurements to produce a reevaluated, but self-consistent set of Δ*RT-NDT* values with assigned uncertainties.

[b] The square bracketed values are the Δ*RT-NDT* values in °C.

[c] Relative error estimates (to the O-T surface position) are given as (1σ) percent uncertainties and are the values in parentheses.

materials used in the PSF Experiment.[3] Consequently, measured changes in C_V, *YS*, and *US* were established relative to the PSF-OT (surface) position for each of the six materials. These relative ratios were then averaged for the C_V, *YS*, and *US* measurements to produce a re-evaluated, but self-consistent, set of Δ*RT-NDT* values with assigned uncertainties. At least, to a first approximation, and for a single material, this removed chemistry and microstructure as controlling variables for the subsequent study of the effects of the environmental variables of neutron spectrum, exposure, exposure rate, and thermal neutrons.

This evaluated Charpy shift data base for two weld, two plate, and two forging materials is tabulated in Table 1 for the PSF SSC-1, SSC-2, O-T, ¼-T, and ½-T positions. The chemistry for

[3] Lucas et al have found that the Charpy shift is approximately proportional to the change in yield strength for both weld and plate materials irradiated in power reactors [*13*]; that is, under certain conditions.

the six materials is also given. The corresponding evaluated physics-dosimetry results for the five positions and six materials are tabulated in Table 2. Evaluated exposure parameter values for thermal [T($E < 0.414$ eV) at 288°C] and fast [F($E > 1$ MeV)] flux and fluence, T/F ratio, dpa, dpa/s, and irradiation time are also given.

Associated with the PSF Experiment high-temperature SSC and SPVC irradiations, was the irradiation of a low temperature ~38°C (100°F) simulated void box capsule (SVBC) placed at the back surface of the PSF void box [*24*]. This SVBC contained a series of support structure-type steel materials, including two A537 steels with significantly different chemistry (0.26% Cu–0.21% Ni and 0.11% Cu–0.20% Ni) [*24*]. The one 0.26% Cu–0.21% Ni A537 steel has the same Cu and Ni concentrations as the A537 liner steel studied by Alberman and discussed elsewhere [*6*]. In a subsequent section, Alberman's data are used together with the SVBC results to develop a low-temperature EOL and LER trend curve model equation for the A 537 steel.

Measured Charpy Shift In-Wall Radial Gradient Curves for Six PV Steel Materials

In Ref *1,* Randall discusses the basis for his Revision (Rev. 2 of Regulatory (Reg.) Guide 1.99. As stated, the Guide is being updated to reflect recent studies of the physical basis for neuron radiation damage and efforts to correlate damage to chemical composition and fluence. Revision 2 contains several significant changes: welds and base metal are treated separately; nickel content is added as a variable and phosphorus is removed; and the exponent in the fluence factor is reduced, especially at high fluences. Randall's trend curve formulas for plate and weld materials are based on an evaluation and synthesis of the trend curve formulas developed by Guthrie [*3*] and Odette [*2*].

In Rev. 2, the equation used for PV wall fluence attenuation by Randall is

$$\text{Fluence}(x) \;\alpha\; e^{ax} \tag{1}$$

With a value of $a = 0.24$, as selected by Randall, Eq 1 can be re-written

$$\text{Fluence}(x) = \text{Fluence}(\text{Surface}) \cdot e^{-0.24x}. \tag{2}$$

Randall represented the Charpy shift attenuation by the relationship

$$\text{Charpy shift}(x) \;\alpha\; e^{ax \cdot N} \tag{3}$$

With a value of $N = 0.28$, as selected by Randall, Eq 3 can be re-written

$$\text{Charpy shift}(x) = \text{Charpy Shift}(\text{Surface}) \cdot e^{-0.067x} \tag{4}$$

where Eq 4 assumes a simple property change power law function with an exponent N of 0.28, and x is the depth in the wall in inches, measured from the inside surface. Equation 2 is based on transport calculations by Guthrie et al [*25–27*] for the dpa attenuation through an 8.0-inch (203.20 mm) vessel wall.

The PSF Experiment-measured Charpy shift radial gradients are plotted in Figs. 1 and 2 for two welds, two plates, and two forging PV steel materials irradiated for ~2 years in the SPVC at an irradiation temperature of ~288°C (550°F) in the PSF of the ORR at ORNL. The metallurgical specimens at the O-T, ¼-T, and ½-T positions received neutron fluence ($E > 1$ MeV) exposures of $\sim 4 \times 10^{19}$, $\sim 2 \times 10^{19}$, and $\sim 1 \times 10^{19}$ n/cm^2, respectively.

These measured in-wall PV attenuation curves are compared with Randall's Eq 4 predictions, the small-dashed curves, in Figs. 1 and 2. It is apparent that the Eq 4 predictive formula is adequate for obtaining rough approximations, but it cannot be used for reliable plant-specific predictions. That is, any agreement between the measured data and the predictions that do not adequately represent the important neutron environmental and microstructural damage processes could be fortuitous.

TABLE 2—*Evaluated physics-dosimetry results for the PSF experiment.*

Position and Material	Evaluated Exposure Parameter Values[a]							
	Thermal (T) Flux	Thermal (T) Fluence	Fast (T) Flux	Fast (T) Fluence	T/F	dpa	dpa/s	Irradiation Time
A533B Weld, Code R								
SSC-1	44.8	1.72	65.1	2.52	0.69	3.70	96.3	3.842 + 06
SSC-2	41.1	3.46	66.7	5.31	0.62	7.82	92.9	8.420 + 06
O-T	32.6	16.6	7.41	3.85	4.40	5.85	11.5	5.097 + 07
¼-T	2.26	1.15	4.22	2.19	0.53	3.70	7.26	5.097 + 07
½-T	0.341	0.174	2.12	1.10	0.16	2.20	4.32	5.097 + 07
A533B Weld, Code EC								
SSC-1	48.9	1.88	45.2	1.75	1.08	2.74	71.3	3.842 + 06
SSC-2	45.0	3.79	46.4	3.69	0.97	5.78	68.6	8.420 + 06
O-T	7.73	3.94	5.72	2.97	1.35	4.80	9.42	5.097 + 07
¼-T	0.747	0.381	3.12	1.62	0.24	2.95	5.79	5.097 + 07
½-T	0.226	0.115	1.54	0.800	0.15	1.73	3.39	5.097 + 07
A533B Plate, Code 3PU								
SSC-1	43.2	1.66	64.3	2.49	0.67	3.65	95.0	3.842 + 06
SSC-2	39.9	3.36	65.9	5.24	0.61	7.70	91.4	8.420 + 06
O-T	31.2	15.9	7.08	3.68	4.41	5.56	10.9	5.097 + 07
¼-T	2.14	1.09	3.95	2.05	0.54	3.43	6.73	5.097 + 07
½-T	0.334	0.170	1.94	1.01	0.17	1.99	3.90	5.097 + 07
A302B Plate, Code F23								
SSC-1	46.9	1.80	70.3	2.72	0.67	4.00	104	3.842 + 06
SSC-2	43.1	3.63	72.0	5.73	0.60	8.44	100	8.420 + 06
O-T	33.9	17.3	7.76	4.03	4.37	6.15	12.1	5.097 + 07
¼-T	2.32	1.18	4.35	2.26	0.53	3.83	7.51	5.097 + 07
½-T	0.341	0.174	2.16	1.12	0.16	2.24	4.39	5.097 + 07
A508 Forging, Code K								
SSC-1	47.6	1.83	44.7	1.73	1.07	2.70	70.2	3.842 + 06
SSC-2	43.7	3.68	45.9	3.65	0.95	5.69	67.6	8.420 + 06
O-T	7.42	3.78	5.47	2.84	1.36	4.56	8.95	5.097 + 07
¼-T	0.708	0.361	2.93	1.52	0.24	2.73	5.36	5.097 + 07
½-T	0.220	0.112	1.40	0.729	0.16	1.57	3.08	5.097 + 07
A508 Forging, Code MO								
SSC-1	51.3	1.97	48.4	1.89	1.06	2.94	76.5	3.842 + 06
SSC-2	47.1	3.97	50.0	3.98	0.94	6.21	73.8	8.420 + 06
O-T	8.04	4.1	5.99	3.11	1.34	5.04	9.89	5.097 + 07
¼-T	0.771	0.393	3.21	1.67	0.24	3.05	5.98	5.097 + 07
½-T	0.226	0.115	1.58	0.820	0.14	1.77	3.47	5.097 + 07

[a] The thermal ($E < 0.414$ eV) and fast ($E > 1$ MeV) flux and fluence values are given in units of 10^{11} n/cm$^2 \cdot$ s and 10^{19} n/cm^2, respectively. Dpa and dpa/s values are given as displaced atoms per target atom of iron based on the ASTM E693 dpa standard. The dpa and dpa/s are given in units of 10^{-2} and 10^{-10}, respectively. The irradiation time is given in seconds, where 3.842 + 06 = 3.842×10^6. The absolute (1σ) error estimates for most of the individual exposure parameter values are in the 5 to 15% range and are not listed separately. The error estimates for the thermal values would be somewhat higher, in the range of 10 to 25% (1σ). The relative (to the O-T position) error estimates are much less and all should be in the 3 to 10% range, depending on the exposure parameter.

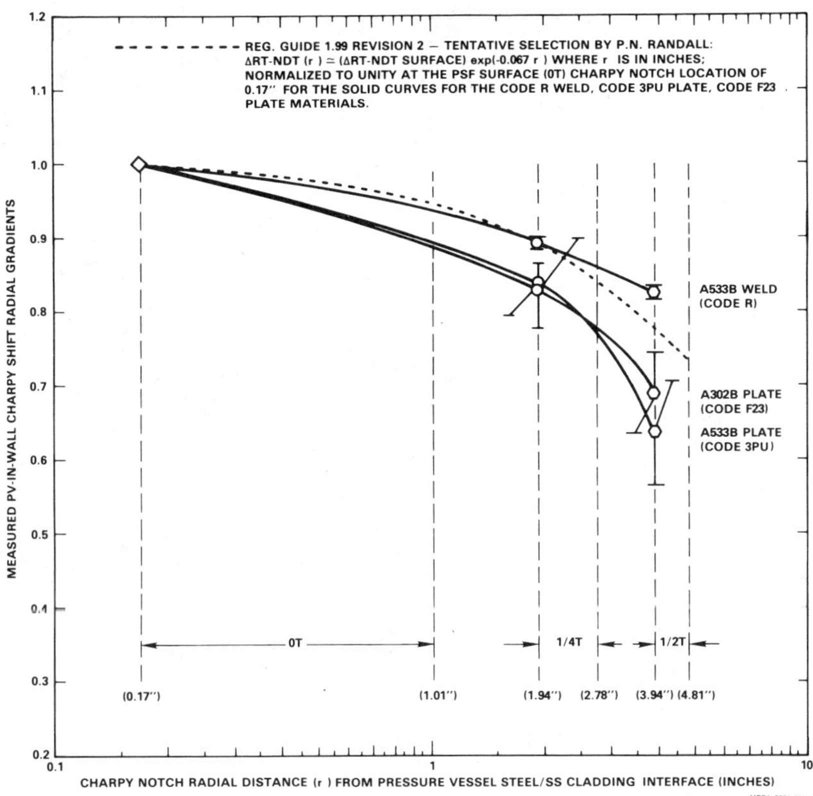

FIG. 1—*PSF experiment-measured PV-in-wall* ΔRT-NDT *shift radial gradients for one weld and two plate materials irradiated at 288°C (550°F).*

Trend Curve Development and Testing

Trend Curve Equation Modeling and Variables—The PSF Metallurgical Blind Test comparisons [*16,28*] show that the various PV steel fracture toughness and embrittlement prediction formulas are adequate to obtain rough property change approximations, but that none capture the complex and not-well-understood correlation between radiation embrittlement and fluence, fluence rate, neutron spectrum, chemistry, heat treatment, and other factors. For support structure steels, the PSF Experiment has provided low-flux and low-fluence bounding physics-dosimetry-metallurgy data for future Charpy trend curve data development and testing studies [*24*]. This and other test reactor irradiation studies have demonstrated beyond any doubt that test reactor data are essential and both supplement and complement PWR and BWR power reactor physics-dosimetry-metallurgy plant-specific and generic data; that is, test reactor information will continue to be needed to understand what the controlling variables and mechanisms are and to help in correcting power reactor results for different variable effects and systematic biases [*12,13,17,29–46*].

Simons [*8*], McElroy et al [*9*], and Gold and McElroy [*11*] have concluded, that for a decrease in flux level, a steel material can show three different characteristics: 1) a decrease in shift, 2) an increase in shift, or 3) no change in shift at all. Further, the actual behavior of a material can change significantly as a function of the neutron exposure; also, thermal neutron and γ-ray effects

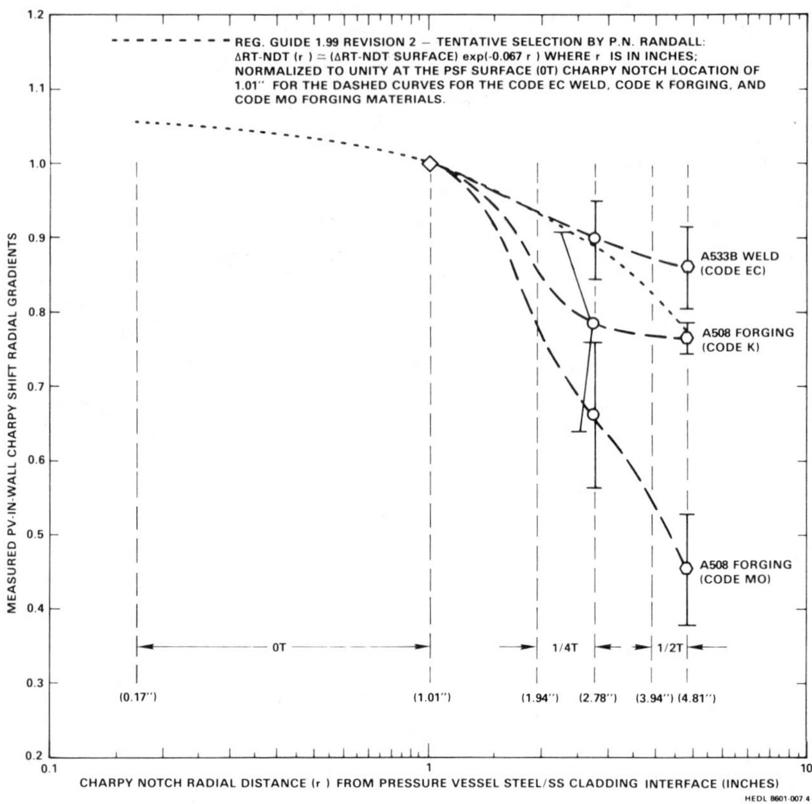

FIG. 2—*PSF experiment-measured PV-in-wall* ΔRT-NDT *shift radial gradients for one weld and two forging materials irradiated at 288°C (550°F).*

can contribute to observed changes in property, especially near steel-water interface positions with high thermal-to-fast-neutron ratios.

Simons' physically based model [*8*] permits the development and testing of material-dependent trend curve models that can have one or more of the three identified characteristics. A separate paper on this work was presented by Simons at this symposium.

Lucas, Odette, Lombrozo, and Sheckherd [*13*] have developed a physically based model in which the embrittlement is due primarily to an irradiation-enhanced copper precipitation with a secondary contribution from radiation-induced defect cluster (microvoid) formation. Their plotted curves for the copper precipitation component of the model suggest factors of about 1.2 to 3.2 increases in shift for an order of magnitude decrease in flux level (from 5×10^{12} to 5×10^{11} n/cm$^2 \cdot$ s) at a fluence ($E > 1$ MeV) of 10^{19} n/cm^2. The factor of 1.2 increase corresponds to higher Cu (0.4% by weight) while the 3.2 value corresponds to lower Cu (0.05% by weight) steel. The precipitation model also predicts little or no flux effect for fluences ($E > 1$ MeV) above $\sim 10^{19}$ n/cm^2 for flux levels at or below ~ 1 to 5×10^{11} n/cm$^2 \cdot$ s, depending on the Cu content. For the microvoid component, their model predicts a factor of ~ 3 decrease in shift for a factor of 10 decrease in flux level. They state

> That at low fluxes, microvid are produced at a low rate and anneal out quickly and hence there is little contribution from this source; but at high fluxes the microvoid population can become significant.

Given the proper sets of parameters, it would appear that their combined copper precipitation and microvoid model would show that a material could exhibit an increase, a decrease, or no change in shift with a decrease in flux level. A result not stated by these authors, but one that is consistent with the findings of the present study.

In earlier work [*2*], and based on the study of power reactor surveillance data, Odette et al have stated

> Within the limited range of the data ($\phi \sim 3 \times 10^{11}$ to 3×10^{12} n/cm$^2 \cdot$ s) and $\phi t \lesssim 5 \times 10^{19}$ n/cm^2 ($t \lesssim 2 \times 10^8$s), it appears appropriate to neglect a flux variable.

This conclusion is inconsistent with the present study results (Table 1) that makes use of the PSF Experiment data base with $1.4 \times 10^{11} \leq \phi \leq 7.2 \times 10^{12}$ n/cm$^2 \cdot$ s and $0.73 \times 10^{19} \leq \phi t \leq 5.7 \times 10^{19}$ n/cm^2 ($4 \times 10^6 \leq t \leq 5 \times 10^7$ s).

Odette's et al conclusion to neglect the exposure rate variable was based on the examination of a very large number of variations of their model equations using nonlinear regression procedures. Since they used a PWR and BWR data base with a large number of materials with different chemistry and microstructure, it is not at all surprising that they were unable to observe or quantify a material dependent exposure rate effect. On the other hand Norris [*4*], who concentrated on the analyses of data sets from single heats of materials (irradiated in surveillance capsules), was able to observe and partially quantify the effect for selected PV steels. Because of his pioneering work in recognizing the significance of this effect, and since it had eluded definitive detection by others, it would be most appropriate to name this the "Norris Effect."[4]

Derivation and Characteristics of Material-Dependent EOL and LER Trend Curves

Results of the application of an empirical approach for establishing EOL and LER material-dependent trend curves that permit the separation of the effects of the environmental variables of neutron exposure, exposure rate, and thermal neutron-induced helium production will now be considered. The material-dependent trend curve model equation selected for this work is

$$\Delta RT\text{-}NDT = X1 + X2 \ln \dot{He} + X3 \ln \dot{D} + X4 \ln D \quad (5)$$

where $\Delta RT\text{-}NDT$ is the Charpy shift (°F), D is the iron dpa[5] neutron exposure (dpa/0.016), $\dot{D}$ is the neutron exposure rate (dpa/s), $\dot{He}$ is the helium atom production rate, and *X1, X2, X3,* and *X4* are a set of empirically derived constants.

The constants of Eq 5 are material dependent and will depend on the chemistry and microstructure of each heat of steel material. Thus, Eq 5's applicability is limited to single heats of steel materials. At least to a first approximation, this removes the variable effect of chemistry and microstructure and allows for the separate study of the effects of environmental variables for each material. The applicability of Eq 5 is further limited to the EOL and LER of $\sim 1 < D < 6$, which corresponds to a neutron fluence Φ ($E > 1$ MeV), of $\sim 1 \times 10^{19} < \Phi < \sim 6 \times 10^{19}$ n/cm^2.

In Eq 5, *X1* and *X4* represent the intercept and slope of each material-dependent trend curve equation at some fixed reference values of $\dot{He}$ and $\dot{D}$, which are used to determine the values of constants *X2* and *X3*. The effect of $\dot{He}$ and $\dot{D}$ is, therefore, to simply move the $X1 + X4 \ln D$ trend curve either up or down, depending on the chemistry and microstructural properties of the given heat of steel.

If *X3* is found to be zero, then the steel has the characteristic of not showing a change in property with changes in flux level (represented here by changes in dpa/s). If *X3* is negative, then the steel has the characteristic of showing a decrease in shift with a decrease in flux level. On the

[4] Norris died on Feb. 3, 1986, without having received recognition for his work from his contemporaries.

[5] Including that caused by iron recoil atoms from the Fe (n,γ) reaction and from any γ-ray field-induced displacements in iron.

other hand, if *X3* is positive, the material has the characteristic of showing an increase in shift with a decrease in flux level. Depending on the selected reference value of $\dot{D}$, however, the actual sign of the constant *X3* could be reversed from that stated above.

If the constant *X2* is zero, then the steel has the characteristic of not showing a change in property with a change in helium production rate. Based on Alberman's work [5], it is assumed: 1) that boron is the primary contributor to the production of helium; 2) in the absence of boron, thermal neutrons have little or no effect on the measured property changes; and 3) the effect of helium is similar to that of Cu, Ni, or any other chemistry term. That is, an increase in the helium production rate will move the Eq 5 trend curve up, just the same as with an increase in the concentration of Cu, Ni, and certain other chemistry terms.[6]

Finally, for this study it is further assumed that any displaced iron atoms resulting from direct γ-ray field-induced interactions can simply be added to the neutron-induced displacements; that is, as an additive dpa value to the exposure term *D*. This also applies for the dpa/s exposure rate, $\dot{D}$.

PSF Experiment SPVC Results

Results of the application of Eq 5 for two welds, two plates, and two forgings irradiated in the PSF Experiment are presented in Fig. 3 through 8. The heavy stripped curved in each figure

[6] It is important to note that the effect of helium appears to saturate at values of boron content ≳ 5 ppm [5].

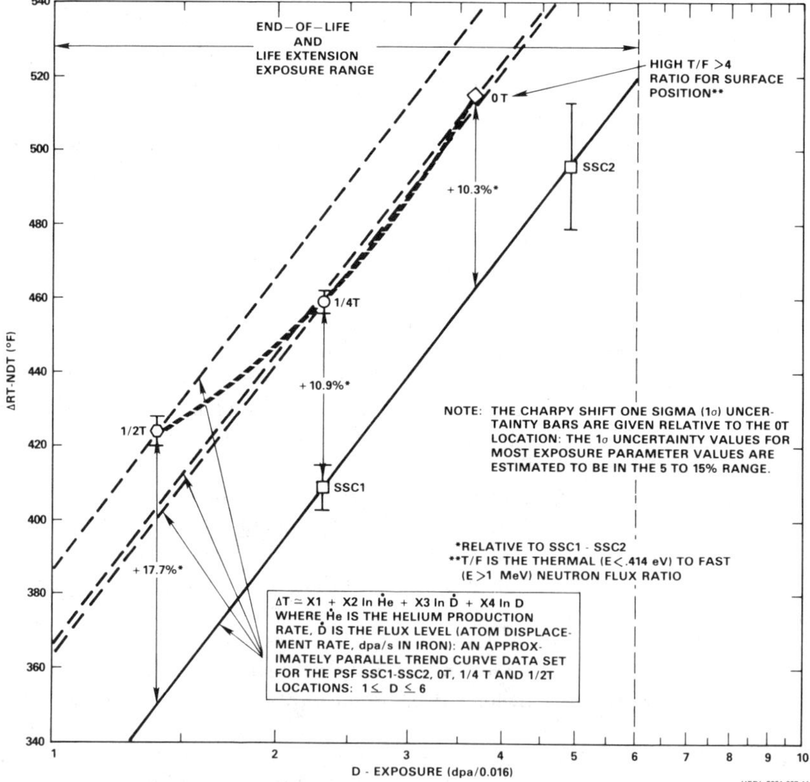

FIG. 3—*PSF experiment-measured Charpy shift versus neutron exposure for A533B Code R (UK ref.) weld material-environmental variable effects.*

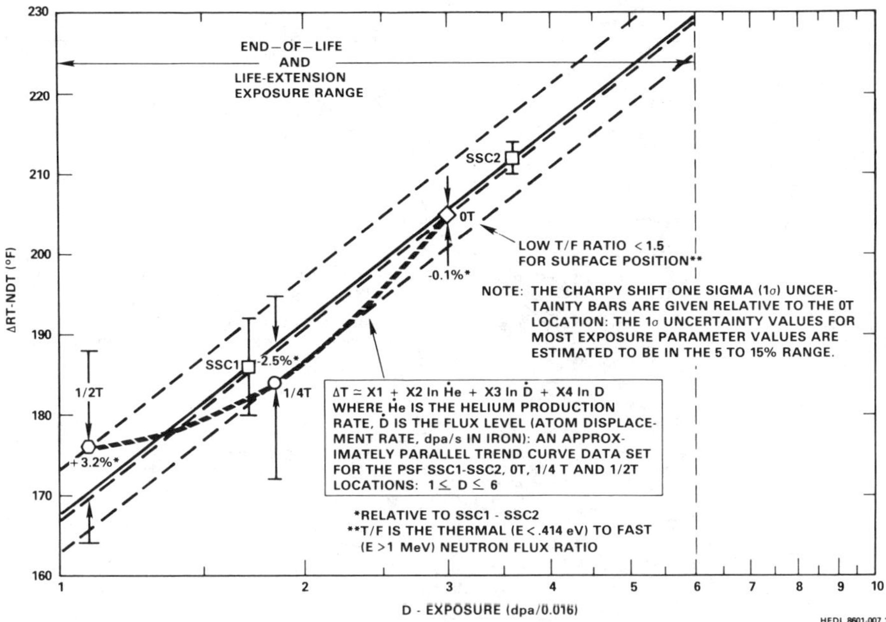

FIG. 4—*PSF experiment-measured Charpy shift versus neutron exposure for A533B Code EC (EPRI ref.) weld material-environmental variable effects.*

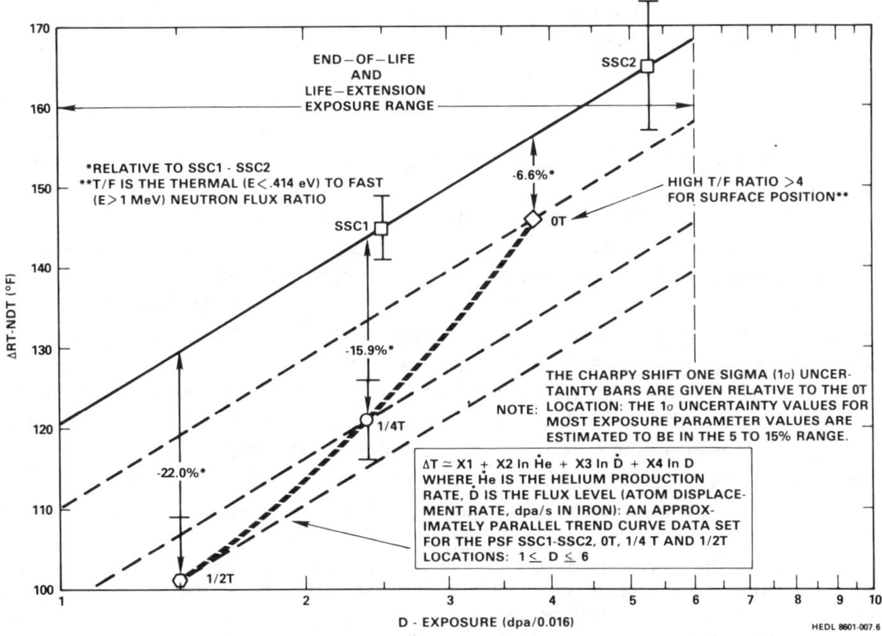

FIG. 5—*PSF experiment-measured Charpy shift versus neutron exposure for A302B Code F23 (ASTM ref.) plate material-environmental variable effects.*

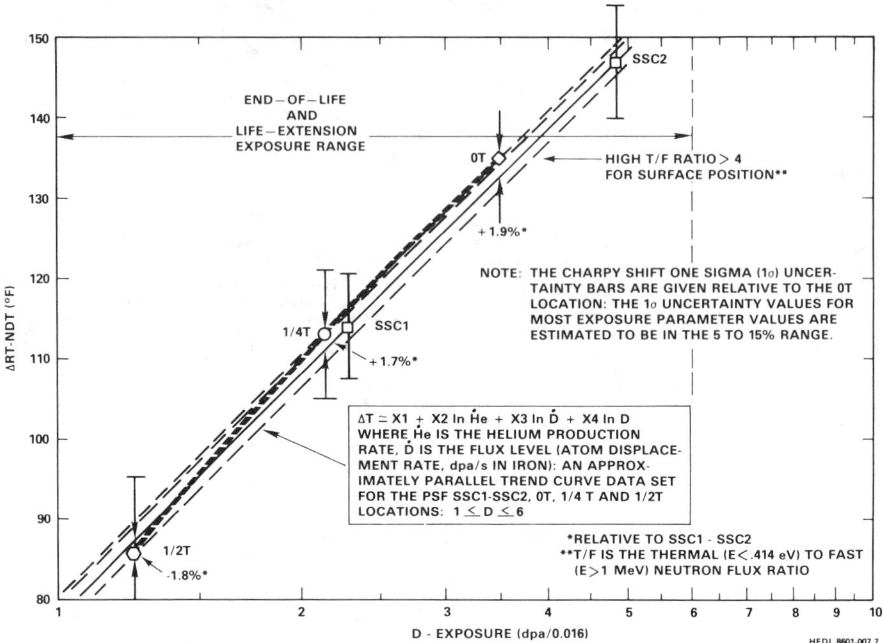

FIG. 6—*PSF experiment-measured Charpy shift versus neutron exposure for A533B Code 3PU (HSST-03 ref.) plate material-environmental variable effects.*

represents the trend curve that would usually be drawn by most experimenters through the three SPVC in-vessel-wall data points at the ½-T, ¼-T, and O-T positions; that is, when no account is taken for differences in flux levels and helium production rates and the constants *X3* and *X2* in Eq 5 are set equal to zero. Further, observed differences from some type of smoothed curve fit for the Charpy shift between the three ½-T, ¼-T, and O-T and the two simulated surveillance capsule (SSC-1 and SSC-2) positions would be ignored and, most often, would be attributed to experimental scatter and uncertainties.

A more rational interpretation, however, of the six sets of PSF Experimental Charpy shift data is that the differences between the SSC and SPVC are real and can be attributed, mainly, to a flux-level effect with a smaller contribution caused by the production rate of helium.

Interpreted this way, it is easily seen that the UK A533B reference weld (Code R) material has the characteristic of a significant increase in the Charpy shift with a decrease in flux level for all EOL and LER exposures. Further, a careful analysis of the Fig. 1 and corresponding Tables 1 and 2 data base suggests that thermal neutron interactions with boron to produce helium in the Code R material may contribute up to ~2% to the measured Charpy shift. This further suggests that the boron content of this steel is very low (<<5 ppm).[7] The coefficient *X2* in Eq 5 cannot be determined without knowing the exact boron content of this steel. Plans have been made to determine the boron content of this as well as the other five steels identified in Figs. 4 through 8.

[7] Only three of the PSF Experiment PV steels were exposed in the O-T front position with its relatively high (T/F > 4) thermal-to-fast-neutron ratio: the Code R weld and the Codes 3PU and F23 plates. The other three materials were exposed in the O-T back position with its much lower (T/F ~1.4) ratio, see Table 2. Using the PSF Experiment data base, therefore, it was not possible to infer anything about the effect of boron-induced helium production in these latter three materials.

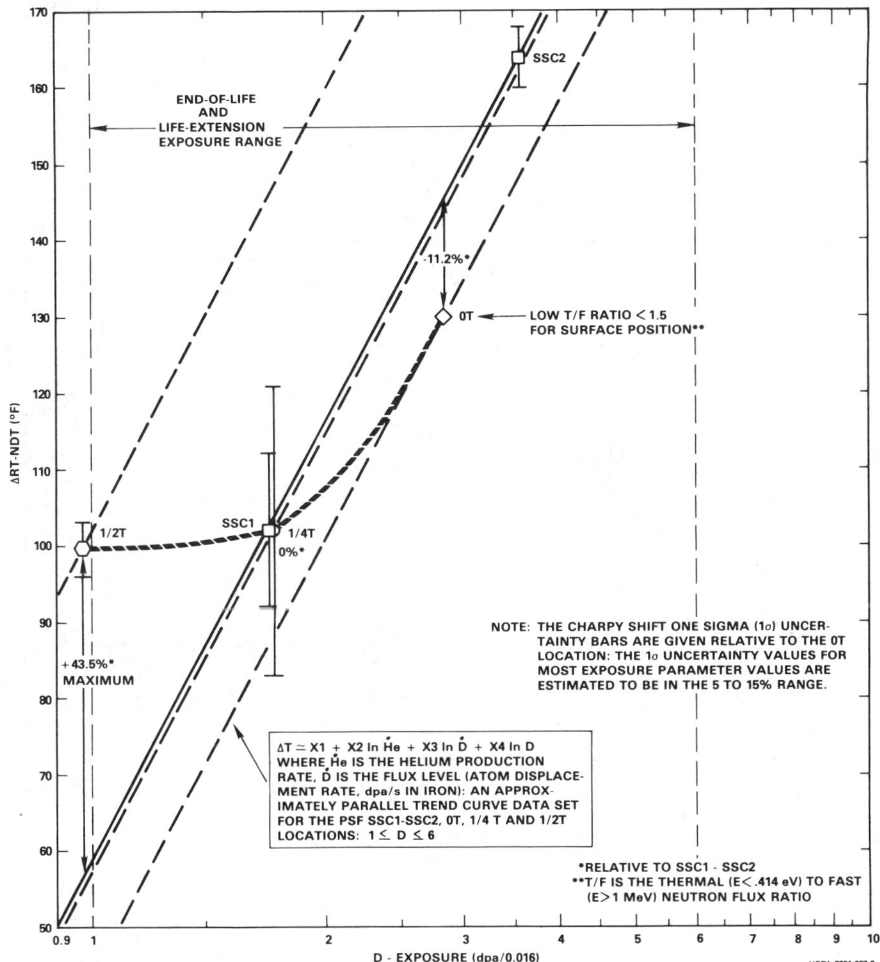

FIG. 7—*PSF experiment-measured Charpy shift versus neutron exposure for A508 Code K (KFA ref.) forging material-environmental variable effects.*

If position-to-position differences in the effect of helium are assumed to be small for the Code R steel and *X2* is set equal to zero, values for the other coefficients *X1*, *X3*, and *X4* can be determined. This has been done for an exposure value D corresponding to the SSC-1 and SPVC ¼-T exposures, which were essentially identical.

Further, the dpa per unit fluence ($E > 1$ MeV) spectral index values are essentially the same for the ¼-T and SSC positions. Thus, even if dpa were not the appropriate neutron spectrum data correlation exposure parameter, the effect of spectral differences between the ¼-T and SSC positions would be negligible for the PSF Experiment. This is also the case for the other five irradiated weld, plate, and forging materials.

The derived values of *X1*, *X3*, and *X4* for the Code R material are tabulated in Table 3. Using the Table 3 values of *X1*, *X3*, and *X4* in Eq 5, the predicted-to-measured C/E ratio values of the Charpy shift for the Code R material for the SSC-1, SSC-2, O-T, ¼-T, and ½-T positions are 1.00, 1.00, 0.98, 1.00, and 0.96, respectively.

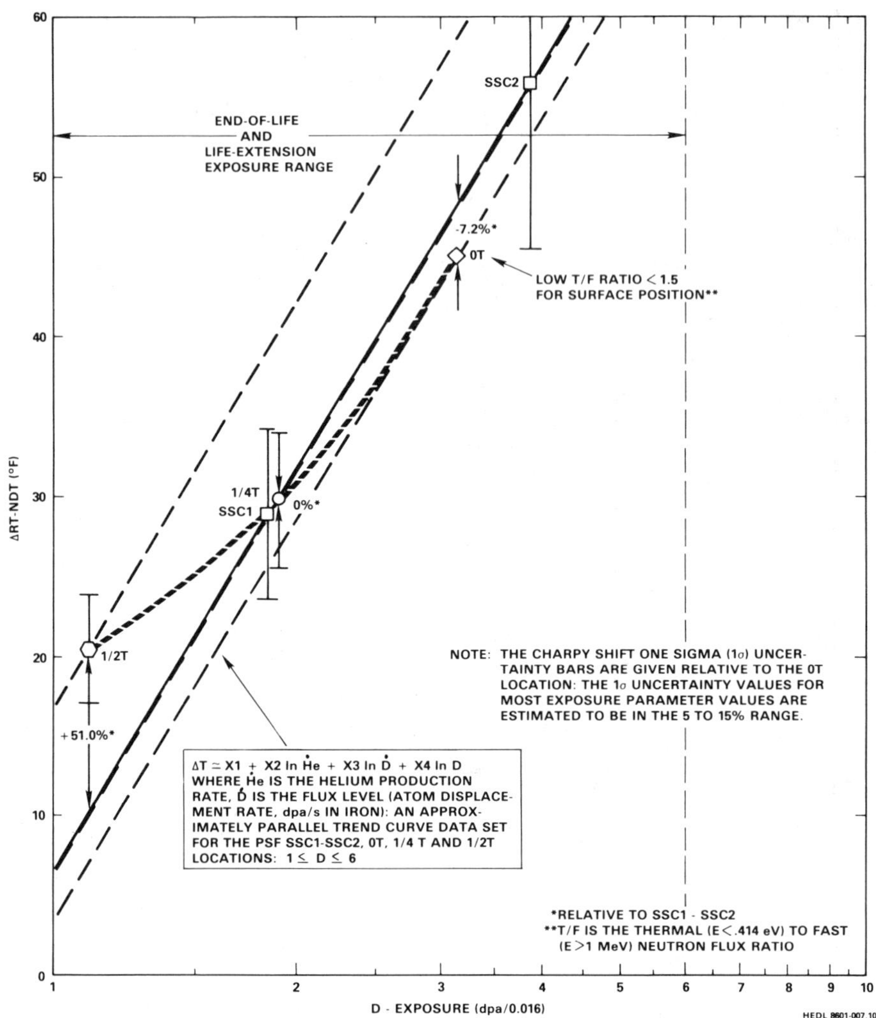

FIG. 8—*PSF experiment-measured Charpy shift versus neutron exposure for A508 Code MO (Mol ref.) forging material-environmental effects.*

This same procedure has been used to derive values of the constants for the five other materials. The values for exposure values *D* corresponding to the ¼-T as well as other selected data points, depending on the material, Figs. 4 through 8, are given in Table 3.

The values of *X3* for the ¼-T data points have been set equal to zero for the Code EC, Code 3PU, Code K, and Code MO materials because of the small differences of −2.5%, +1.7%, 0%, 0%, respectively, between the ¼-T and SSC curves of Figs. 4, 6, 7, and 8. This indicates the lack of an observable or significant flux level effect for the A533B steel Code EC Weld and Code 3 PU plate materials in the EOL and LER of $1 < D < 6$ for the flux levels encountered in the PSF Experiment. A comparison of the PSF Experiment Code 3PU plate (heavy section steel technology: HSST, Plate 03) results with those reported by Pachur [*36–38*] for the FRJ2 experiments, however, suggests that the Code 3PU material does exhibit a significant, exposure-

TABLE 3—*Empirically derived material-dependent trend curve model equations for an irradiation temperature of 288°C (550°F).*

British A533B Ref. Weld (Code R Material, Fig. 3)

$$\Delta\text{RT-NDT (°F)} = 399.8 + X2 \ln \dot{He} - 19.34 \ln \dot{D} + 116.30 \ln D \quad (6)$$

where setting $X2 = 0$ introduces an approximate -2% bias at the O-T position in the calculated shift values, and the Eq 6 use is limited to the exposure range: $1 \leq D \leq 6$.

EPRI A533B Ref. Weld (Code EC Material, Fig. 4)

$$\Delta\text{ RT-NDT (°F)} = 167.2 + 34.81 \ln D \quad (7)$$

where because of the small differences between measured data points, $X2$ and $X3$ have been arbitrarily set equal to zero, and the Eq 7 use is limited to the exposure range: $1 \leq D \leq 6$.

ASTM A302B Ref. Plate (Code F23 Material, Fig. 5)

$$\Delta\text{RT-NDT (°F)} = 80.16 + X2 \ln \dot{He} + 8.675 \ln \dot{D} + 26.78 \ln D \quad (8)$$

where setting $X2 = 0$ introduces an approximate -6% bias at the O-T position in the calculated shift values, and the Eq 8 use is limited to the exposure range: $1 \leq D \leq 6$.

HSST A533B Ref. Plate 03 (Code 3PU Material, Fig. 6)

$$\Delta\text{RT-NDT (°F)} = 77.56 + 44.19 \ln D \quad (9)$$

where because of the small differences between measured data points, $X2$ and $X3$ have been arbitrarily set equal to zero, and the Eq 9 use is limited to the exposure range; $1 \leq D \leq 6$.

German A508 Ref. Forging (Code K Material, Fig. 7)

$$\Delta\text{RT-NDT (°F)} = 58.40 + X3 \ln \dot{D} + 83.21 \ln \text{D} \quad (10)$$

where $X2$ has been arbitrarily set equal to zero and $X3$ appears to be a multi-valued function of exposure D:

$$\Delta\text{RT-NDT (°F)} = 58.40 + 83.21 \ln D \quad (10a)$$

(For: $D = 1.68$; ¼-T position)

$$\Delta\text{RT-NDT (°F)} = 26.26 + 7.594 \ln \dot{D} + 83.21 \ln D \quad (10b)$$

(For: $2.85 \leq D \leq 6$; O-T position)

$$\Delta\text{RT-NDT (°F)} = 161.51 - 13.73 \ln \dot{D} + 83.21 \ln D \quad (10c)$$

(For: $D = 0.981$; ½T -position)

Belgium A508 Ref. Forging (Code MO Material, Fig. 8)

$$\Delta\text{RT-NDT (°F)} = 6.912 + X3 \ln \dot{D} + 36.12 \ln \dot{D} \quad (11)$$

where $X2$ has been arbitrarily set equal to zero and $X3$ appears to be a multi-valued function of exposure, D:

$$\Delta\text{RT-NDT (°F)} = 6.912 + 36.12 \ln D \quad (11a)$$

(For: $D = 1.91$; ¼-T position)

$$\Delta\text{RT-NDT (°F)} = -0.23 + 1.652 \ln \dot{D} + 36.12 \ln D \quad (11b)$$

(For: $3.15 \leq D \leq 6$; O-T position)

$$\Delta\text{RT-NDT (°F)} = 20.75 - 3.203 \ln \dot{D} + 36.12 \ln D \quad (11c)$$

(For: D = 1.11; ½-T position)

dependent, flux-level effect at high neutron exposure rates, with the characteristic of an increase in shift with a decrease in flux level. Further, Pachur looked for a flux-level effect for the HSST plate 03 at 150°C, and at two very different flux levels of $\sim 6 \times 10^{11}$ and $\sim 3 \times 10^{13}$, and found no effect. It would appear, therefore, that he selected a set of environmental conditions for which this material shows no observable or significant flux-level effect.

Norris' study (discussed in a previous section) for the HSST A533B (02) plate suggests that this material exhibits a flux-level effect at low neutron exposures ($0.1 < D < 1$) for flux level ($E > 1$ MeV) in the range of $\sim 10^9$ to 2×10^{10} n/cm$^2 \cdot$ s; which is about one to two orders of magnitude lower than that encounered in the PSF Experiment and about three orders of magnitude less than in the FRJ2 experiment. Here, Norris found that the A533B (reference plate 02) exhibited the characteristic of a decrease in shift with a decrease in flux level. It is noted that the Point Beach 2 A533B 02 plate (correlation monitor material) data of Fig. 9 appears to exhibit this same characteristic.

For the A508 steel Code K and Code MO forging materials, either material appears to show a flux-level effect at their ¼-T dpa exposure values. At the O-T exposure values, however, both materials appear to show the same characteristic of a small and significant decrease in shift with a decrease in flux level, Table 3, Eqs 10*b* and 11*b*. Then at the ½-T dpa exposure values, both materials show the reverse characteristic of an increase in shift with a decrease in flux level; but with a very dramatic increase in shift (in the 40% to 50% range), see Table 3, Eqs 10*c* and 11*c*. This reversal of characteristics, with an apparent pivoting about the ¼-T dpa exposure value, indicates that three forms of Eq 5 must be used to represent the more complex behavior of the Codes K and MO materials; Eqs 10, 10*b,* and 10*c* for the German steel and Eqs 11, 11*b,* and 11*c* for the Belgium steel.

The last high-temperature-irradiated material to be considered is the A302 ASTM reference plate, Code F23 material, Fig. 5. This material shows the favorable characteristic of a very significant decrease in the Charpy shift with a decrease in flux level. An analysis of the Fig. 5 and corresponding Tables 1 and 2 data base suggest that thermal neutron interactions with boron to produce helium in the Code F23 material may contribute up to ~6% to the measured Charpy shift. As with the Code R material, this suggests that the boron content of this steel is very low (<<5 ppm) and the coefficient *X2* in Eq 5 cannot be determined without knowing the boron content of this steel. If position-to-position differences in the effect of helium are assumed to be

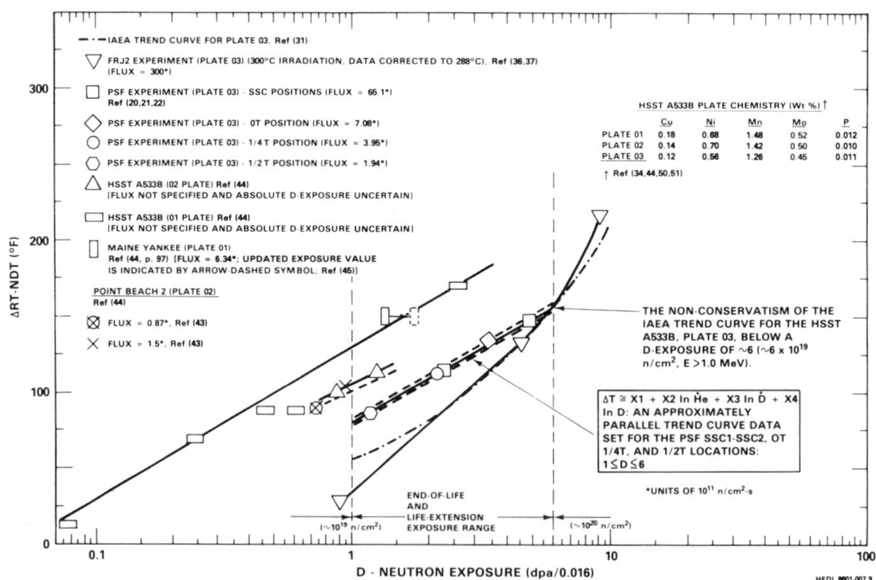

FIG. 9—*Charpy (41 J) transition temperature shift neutron exposure and flux-level dependency for HSST A533B reference plate (01, 02, and 03) materials irradiated at 288°C (550°F).*

small and $X2$ is set equal to zero, values for the other coefficients $X1$, $X3$, and $X4$ can be obtained. This has been done for an exposure value D corresponding to the ¼-T exposure.

The derived values of $X1$, $X3$, and $X4$ for the Code F23 material are given in Table 3. Using these values in Eq 5, the C/E ratio values of the Charpy shift for the Code F23 material for the SSC-1, SSC-2, O-T, ¼-T, and ½-T positions are 1.00, 1.00, 0.94, 1.00, 1.01, respectively. These Eq 5 results are very consistent with the results previously reported by Norris [*4*] and Simons [*8*] for the A302 ASTM reference (ref.) plate. They are not, however, consistent with the previously reported results of Hawthorne [*39*]. See the Fig. 10 dotted Naval Research Laboratory (NRL) trend for 288°C irradiations for the ASTM A302B ref. plate material. (See next section.)

ASTM A302B Ref. Plate—Data Consistency

Acceptance or rejection of the applicability of Simon's [*8*] physically based equations and the more empirically derived Table 3 equations will require further study and additional experimental results. Figure 10 presents the results of a preliminary study of the consistency of available low- and high-temperature irradiated ASTM A302B ref. plate Charpy shift data.

The data base used is limited to data on ASTM A302B ref. steel plate irradiated at temperatures <250°C (482°F) and at ~288°C (550°F). The low-temperature irradiations of the A302B steel were accomplished in research reactors, including an SSC and three simulated pressure vessel mockups IRL-3, -4, and -5, at the Industrial Research Laboratory (IRL) [*40–43*]. For the ASTM A302B plate material, the low-temperature (<116°C) IRL and high-temperature (~288°C) PSF Experiments provided a unique opportunity to study the consistency of through-wall embrittlement data.

The shaded area in Fig. 10 represents the NRL trend band for <250°C research reactor irradiations [*39*]. The dotted curve represents the NRL trend for 288°C irradiations, and this curve

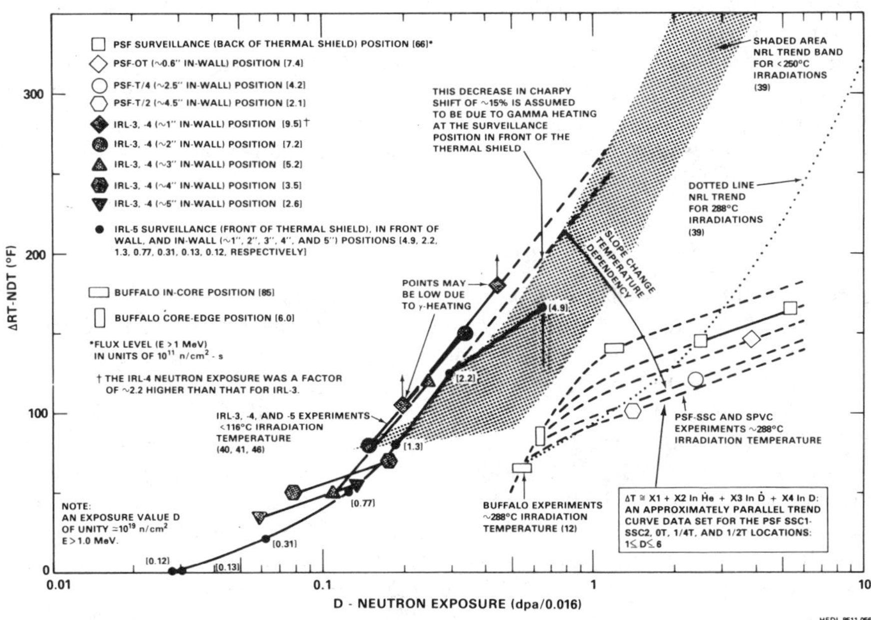

FIG. 10—*Charpy (41 J) transition temperature shift neutron exposure and flux-level dependency for ASTM A302B reference plate materials irradiated at 288°C (550°F) and less than 250°C (482°F).*

is based on both research and power reactor data available up to ~1974. The most recent PSF Experiment, Fig. 5, and Buffalo Experiment results of Hawthorne [*12*] are shown by the lower-right set of dashed curves. These curves are represented, approximately, by Eq 8 of Table 3 for the dpa exposure range of $1 \leq D \leq 6$.

A most striking result here is that, with the selection of appropriate parameters and constants, the physically based model equation of Simon's [*8*] will correctly predict the dramatic convergence of the set of parallel lines of Eq 8 at the value of $D \sim 0.6$ ($\sim 6 \times 10^{18}$ n/cm^2, $E > 1$ MeV). That is, Simon's model does support the absence of a significant flux-level effect for the A302B plate steel at a fast fluence of $\sim 6 \times 10^{18}$ n/cm^2.

The disagreement between the dotted NRL trend for 288°C irradiations and the PSF Experiment Eq 8 results is thought to be associated with the use by Hawthorne of the Yankee Rowe power reactor surveillance capsule data, which was irradiated at a temperature much lower than 288°C, perhaps near or just below 250°C. This is the most likely explanation, since the slope of the NRL dotted curve is almost the same as the "shaded area NRL trend band for <250°C irradiation [*39*]." The "slope change temperature dependency" between the low-temperature (<116°C) IRL and high-temperature (~288°C) PSF Experiments is represented by the arc-angle difference between these two sets of A302B experimental data, see Fig. 10.

The IRL-3, -4, and -5 measured Charpy shift versus neutron dpa exposure data points are represented by the partially shaded and solid symbols plotted just to the left and in the NRL trend band for <250°C irradiations. The estimated neutron flux level for each data point is given in the legend or just to the right of each symbol. Flux levels are also given for each PSF and Buffalo experiment data point. As discussed by Simons, these IRL and other selected low-temperature reseach reactor ASTM A302B ref. steel data are consistent with those obtained at higher temperature (~288°C) in the PSF and Buffalo experiments. That is, they exhibit the same characteristic of a decrease in Charpy shift with a decrease in flux level. Further, it appears, that γ-ray heating at the accelerated SSC position in front of the thermal shield has caused the IRL-5 point [4.9] to deviate below the expected approximately parallel set of dashed-lines behavior (Fig. 10) attributed to constant temperature point-to-point variations in flux level.

The findings of this and Simon's study strongly support conclusions of the previously reported work of Norris [*4*]. Further, all of the available research and power reactor derived ASTM A302B ref. plate data appear to be self-consistent and can be correlated when the separate effects of the environmental variables of neutron spectrum, exposure rate, and thermal neutron and γ-ray field-induced effects are considered.

HSST A533A Ref. Plate—Data Consistency

Figure 9 presents the results of a preliminary study of the consistency of available high-temperature (~288°C) irradiated HSST A533A Plate 03 and some selected Plate 02 and 01 data. When available, flux values are given for each data point. Chemistry concentrations for Cu, Ni, Mn, Mo, and P for the three plates are also given.

The PSF Experiment, Fig. 6, results are shown together with the FRJ2 Experiment results reported by Pachur [*36,37*]. The flux level for the FRJ2 irradiation position is ~5 to 150× as high as those for the PSF ½-T to SSC positions. The combined PSF and FRJ2 results suggest that the plate 03 material does exhibit a significant, exposure-dependent, flux-level effect at high exposure rates (neutron flux $> 65 \times 10^{11}$ n/cm$^2 \cdot$ s, $E > 1$ MeV). Further the material appears to have the characteristic of showing a significant increase (by a factor of two to three) in Charpy shift for a drop in flux level from $\sim 3 \times 10^{13}$ to $\sim 6.5 \times 10^{12}$ at a dpa exposure of $D = 1$ ($\sim 10^{19}$ n/cm^2, $E > 1$ MeV). At an exposure of $D = 6$ ($6 \times \sim 10^{19}$ n/cm^2, $E > 1$ MeV), the effect is no longer apparent, and it appears to reverse itself at still higher exposures.

The International Atomic Energy Agency (IAEA) coordinated research programme's measured mean trend curve for the HSST 03 Plate material is represented by the dash-dot curve of Fig. 11 [*31*]. There appears to be a definite non-conservatism associated with using the IAEA mean trend curve for low flux-level power reactor applications for the $D \leq 6$ exposure range. That is, the IAEA testing of the 03 Plate material at very high flux levels (in the low 10^{13} n/cm$^2 \cdot$ s range) may have introduced a significant bias in the reported results because of a previously unobserved flux-level effect that is very neutron exposure dependent.

The HSST Plates 01 and 02 measured Charpy shift curves appear to have about the same slope as the PSF Experiment 03 plate. Williams et al have studied the influence of Cu, Ni, and temperature on the irradiation shift of low alloy steels [*32*]. Of particular interest to them was the effect of nickel in possibly controlling and increasing the dose exponent for lower Cu steels. Further, they state that "high levels of nickel also appear to reduce the tendency for the dose exponent to reduce as dose increases."

With essentially the same Ni, Mn, Mo, and P chemistry, and the listed variations in Cu content between 0.12 to 0.18% by weight (Fig. 9), these three high Ni-low-Cu HSST plate data sets appear to be self-consistent and support the use of Eq 5 for flux levels below $\sim 65 \times 10^{12}$, but with different values of the constant *X1*, and perhaps *X3*, for each of the three materials. As previously discussed, Eq 9 of Table 3 represents the behavior of the HSST plate 03 material irradiated at ~288°C in the PSF Experiment.

The Fig. 10 PSF results for the 03 Plate are compared in Fig. 9 with previously reported Maine Yankee (MY) A533B weld and plate Charpy shift data. These data are based on results obtained from two MY accelerated (AC1 and AC2) and one wall surveillance capsules [*47–49*]. The NRL research reactor data points are those reported in the references. Randall's Reg. Guide 1.99, Rev. 2, fluence dependency [*1*] is represented by the dotted curve. This curve is normalized

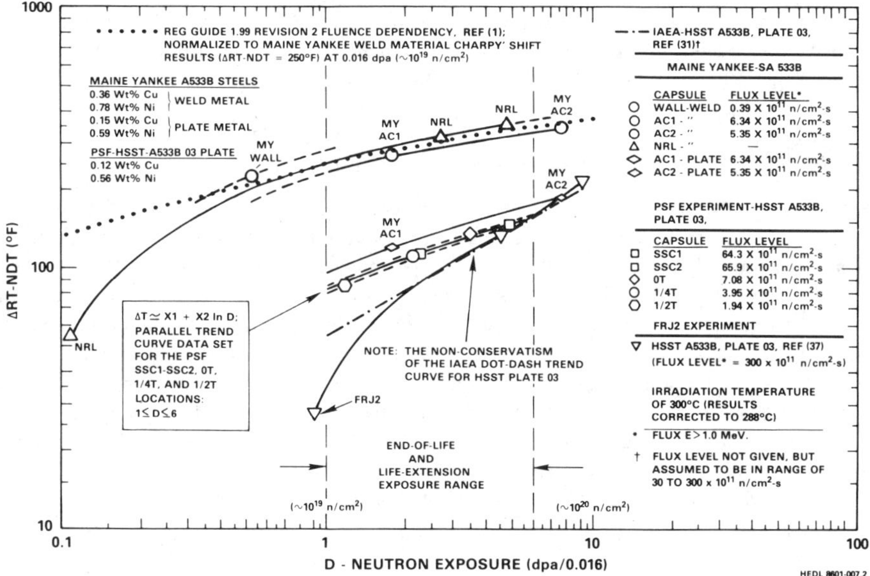

FIG. 11—*Charpy (41 J) transition temperature shift neutron exposure and flux-level dependency for HSST A533 (Plate 03) and Maine Yankee (MY) A533 weld and plate materials irradiated at 288°C (550°F).*

to the MY weld material Charpy shift results ($\Delta RT\text{-}NDT$ = 250°F) at 0.016 dpa (~10^{19} n/cm^2, $E > 1$ MeV).

Relative to the two MY accelerated capsule and NRL data points, the MY wall capsule weld data point is higher than what might be expected on the bases of Randall's Reg. Guide, Rev. 2, the IAEA mean, and the PSF Experiment curves, Eq 9 of Table 3. It is possible, therefore, that this difference indicates that the MY A533B weld material may have the characteristic of an increase in Charpy shift with a decrease in flux level; that is, for a factor of ~15 decrease in flux level. On the other hand, this apparent unfavorable behavior could actually be reversed from that shown in Fig. 11 because of higher temperatures in the accelerated position than the wall capsule position caused by higher gamma-heating rates in the accelerated positions.

It would be of interest to determine the flux level of the NRL set of data points shown in Fig. 9 for the A533B weld material to see if it were close to that of the MY AC1 and AC2 capsules. If it were, then the combined MY and NRL data set for the A533B weld material would appear to show about the same (but shifted up and to the left) overall behavior as the HSST A533B Plate 03.

As a final comment, the consistency shown in Figs. 10 and 11 for the HSST 03, 02, and 01 plates and the MY A533B plate material is very apparent. These four plates have similar chemistry, with the MY Cu concentration being 0.15% as compared with values of 0.12%, 0.14%, and 0.18% for the HSST 03, 02, and 01 plates, respectively. The two MY A533B plate data points are shown as MY AC1 and AC2 diamond symbols in the figure. Whether or not this material has a flux-level dependency cannot be determined by the available data. It would appear, however, to have characteristics similar to that of the HSST A533B Plate 03, Eq 9 of Table 3, for the neutron dpa exposure range of $1 \leq D \leq 6$ and flux levels at or below ~6×10^{11} n/cm$^2 \cdot$ s, $E > 1$ MeV.

Just as with the discussion of the consistency of the ASTM A302B ref. plate data, the available results for the HSST A533B ref. plates (01, 02, and 03), appear to be very self-consistent and can be correlated when the separate effects of the environmental variable of neutron spectrum, exposure, exposure rate, and thermal neutron and γ-ray field-induced effects are considered. In order to complete this plate-to-plate correlation, however, the Cu, Ni, Mn, Mo, P, S, C, N, B, and so forth, dependency of the constants *X1* through *X4*, Eq 5, would have to be studied and established. This is beyond the scope of the present study.

A537 Steel PSF-SVBC and EL.3 Heavy Water Reactor Test Results

Figure 12 presents the results of a preliminary study of the consistency of available low temperature (≤60°C) irradiated A537 liner and support structure-type steel. Alberman's [*6*] EL.3 heavy water reactor and Perrin's [*24*] PSF-SVBC experiment-measured Charpy shift versus neutron dpa exposure values for irradiations at ~60°C and ~38°C, respectively, are shown. Alberman's Cd-covered and bare steel irradiations in the EL.3 reactor are identified by the circle and square and diamond and triangle symbols, respectively. The dpa/0.016 exposure values D and dpa/s ($\dot{D}$) values, as well as the thermal-to-fast ratio (T/F), are of interest for this discussion. Alberman used two irradiations, one consisting of 12 bare and the other of 13 Cd-shielded C_V specimens. The irradiations provided EOL fluences (in the range of ~10^{17} to ~ 2×10^{18} n/cm^2) expected to be received by the A537 steel liner material.

Alberman determined the dpa/s ($\dot{D}$) exposure rate for these irradiations by the relationship,

$$D = [(835\ \Phi^{f}_{\text{fer}} + 13\ \Phi^{th}) \times 10^{-24}/t] \tag{12}$$

where Φ^{f}_{fer} is the iron-equivalent fission fluence, Φ^{th} is the thermal neutron fluence and t is the time in seconds. Equation 12 may be rewritten as

$$D = D \cdot t = 835 \times 10^{-24}\,(\Phi^{f}_{\text{fer}} + 0.0156\,\Phi^{th}) \tag{13}$$

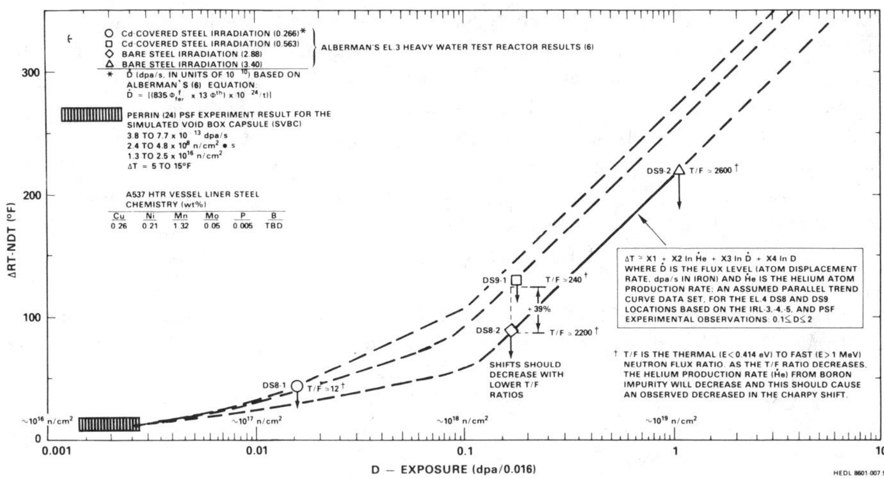

FIG. 12—*EL.3 heavy water reactor experiment-measured Charpy shift versus neutron exposure for A537 HTR vessel liner steel material irradiated at* <*60°C-environmental variable effects.*

to represent the dpa exposure term for the correlation of his measured Charpy shift data set. He elected, however, not to use Eq 13, but to empirically determine the coefficient of the thermal neutron term in Eq 13. The result for the Charpy shift was

$$\Delta TT\ (°\text{C}) = 145 \cdot \Phi_{\text{eq}}^{1/2} \tag{14a}$$

where

$$\Phi_{\text{eq}} = (\Phi^f_{\text{fer}} + 0.0045\ \Phi^{th})\ \text{n/cm}^2 \tag{14b}$$

and Φ_{eq} is an equivalent fluence and 0.0045 represents the relative contribution of thermal neutrons to the damage exposure. Alberman found that use of Eq 14 for the four irradiations (the four plotted open symbol data points in Fig. 12) gave predicted values of the measured shifts that were valid to better than 4%. He concluded

> The influence of thermal neutrons is slightly less marked than predicted by the theoretical model Eq 12, but remains on the same order of magnitude, that is, around 0.5%. The analysis of damage on the basis of Eq 14 thus shows, for irradiation of a liner steel in a spectrum characterized by $\Phi^{th}/\Phi^f_{\text{fer}} \simeq 1000$, more than 70% of the embrittlement is caused by the thermal neutrons, thus demonstrating the significance of the experiment.

Based on the analyses of the PSF, IRL, Buffalo, FRJ2, and other experiments discussed in this paper, there is another interpretation of Alberman's data, with dramatically different conclusions.

First, with reference to the PSF Experiment A533B (Code R) results, Fig. 3, and the A302B (Code F23) results, Fig. 5, it is not possible to correlate both sets of data with an exposure parameter term such as Eq 14*b*. That is, the Fig. 3 Code R material O-T data point (with its higher T/F ratio of 4.4) could be correlated with the SSC-2 point (with its smaller T/F ratio of 0.62) by shifting the O-T point to the right, the procedure used by Alberman. This will not work, however, for the Fig. 5 A302B material with similar O-T and SSC-2 T/F ratios of 4.41 and 0.61, respectively, see Table 2. If this were attempted, there would be an increase rather than a decrease in the separation between these two data points; that is, a worsening of the correlation.

Further, Alberman found that in the absence of boron, an A533B steel would not show any readily observable effect of a thermal neutron irradiation [*5*]. Consequently, the combined results of Alberman's EL.3 and the PSF Experiments suggest that the major effect of thermal neutrons may not be associated with just displacement damage, but also with the production of helium. Interpreted in this manner, it appears that Alberman should have used his Eq 13 for the neutron exposure term parameter for his A537 data correlation studies. If he had done this, he would have found that he needed to establish a correlation that would account for changes in both helium production rate and flux level; that is, a formulation such as Eq 5.

Using Alberman's data, Fig. 12, and Eq 5 with the same analysis approach used for the six PSF Experiment steel materials, an EOL and LER equation,

$$\Delta\text{RT-NDT}(°\text{F}) = 215.73 + X2 \ln \dot{He} + X3 \ln \dot{D} + 71.02 \ln D \quad (15)$$

has been established for the HTR A537 liner steel. The derived values of dpa exposure (D) and exposure rate ($\dot{D}$) are based on the use of Eq 13 and Eq 12, respectively. The values of thermal-to-fast neutron ratio (T/F) were calculated using the relationship $\Phi > 1\ \text{MeV}/\Phi^{f}_{\text{fer}} \simeq 0.90$ and the individually reported values of Φ^{f}_{fer} and Φ^{th}, found in Ref *6*. For this study, $T = \Phi^{th}$ and $F = \Phi > 1$ MeV. Equation 15 is applicable for the dpa exposure range of $\sim 0.1 \leq D \leq 2$ (or $\sim 10^{18}$ to 2×10^{19} n/cm^2, $E > 1$ MeV).

By using Perrin's single PSF-SVBC data point for the A537 (Code 7D) steel with the same values of Cu and Ni of 0.26% and 0.21%, respectively, the Fig. 12 higher exposure value data points of Alberman's have been extrapolated down to dpa exposures (D) in the ~0.0015 to 0.003 range. These extrapolated and converging curves are represented by the three dashed curves in Fig. 12. Their shapes below $D \simeq 0.1$, are based on IRL-3, -4, and -5 experimental observations for the low-temperature irradiation of the ASTM A302B ref. plate material, Fig. 10.

Above $D \simeq 0.1$, these three dashed curves form a set of three parallel trend curves with a slope $X4 = 71.02$ determined by a straightline fit between the DS8-2 and DS9-2 data points. This single straightline should provide a reasonably good representation of the behavior of the A537 steel since the T/F ratios for the two points are about the same (2200 and 2600), as are the dpa/s values (2.88 and 3.40).

Without knowing the boron content of the A537 steel, it is not possible to determine the constant $X2$ in Eq 15. Further, it is not possible to separate out the effects of helium and flux level with the available experimental data. All that can be done here, is to place a lower bound on the effect of flux level. To accomplish this it is noted in Fig. 12 that for the DS8-1 and DS8-2 data points, as well as the other points, that as the T/F ratio decreases, the helium production rate ($\dot{He}$) from the boron impurity will decrease. This should cause an observed decrease in the Charpy shift. Consequently, the observed +39% difference between the DS9-1 and DS8-2 dashed curves should represent a lower bound on the actual flux-level effect for the A537 liner steel. On this basis, the value of the constant $X3$ in Eq 15 can be determined by assuming that $X2 = 0$ or that the contribution of the term $X2 \ln \dot{He}$ is much smaller than $X3 \ln \dot{D}$. The result of this exercise is

$$\Delta RT\text{-}NDT\ (°\text{F}) = 239.90 - 22.85 \ln \dot{D} + 71.02 \ln D \quad (16)$$

where the use of Eq 16 is limited to the exposure range of $0.1 \leq D \leq 2$.

A complicating factor of this study of the behavior of the A537 steel in high T/F ratios is that no account has been taken of the possible contribution of γ-ray field-induced displacements in iron. This was discussed in the introductory part of the HEDL Studies Section. Additional experimental results will be required to resolve this issue. For the present, it will be assumed that any γ-ray field-induced displacements in iron in high T/F ratio irradiation positions (DS9-1, DS9-2, and DS8-2 with T/F values of ~240, ~2600, and ~2200, respectively) would provide simple additive increments to the total dpa exposure and exposure rate values. The net effect of this would be to shift each of the DS9-1, DS9-2, and DS8-2 data points to the right, with the DS8-2 and DS9-2 points showing the largest shifts. Also, the difference in the $\dot{D}$ exposure rates between the DS8-1, DS9-1, DS8-2, and DS9-2 positions would all be enhanced. Certainly, the exact effect of γ-ray field-induced displacements in iron is not presently known, so little more can be said on this subject.

The interpretation offered here of Alberman's thermal and fast neutron irradiation of the A537 liner steel, when coupled with the results of the PSF Experiment, appears to be self-consistent; it provides an alternate method of correlation that allows for the consideration of the separate effects of the environmental variables of neutron spectrum, exposure, exposure rate, and thermal and γ-ray field-induced effects.

If this interpretation were used then the conclusion of Alberman's study "that more than 70% of the embrittlement is caused by the thermal neutrons" would have to be altered. That is, a partition of the percentage of the measured Charpy shift from thermal neutrons (primarily from helium production in the boron in the A537 steel) and that from neutron and γ-ray field dpa exposure rate changes would be required.

Application and Implications of the Use of EOL and LER Trend Curves

The aforementioned HEDL studies strongly support the existence of a significant material-dependent flux-level effect for PV and support structure steels. That is, a steel may show a decrease, an increase, or no change in the measured Charpy shift with changes in the dpa exposure rate (flux level). Further, the actual behavior of a material can change significantly as a function of the neutron dpa exposure; also, thermal neutron and γ-ray effects can contribute to observed changes in property, especially near steel-water interface positions with high-thermal-to-fast-neutron ratios.

As stated by Norris [*4*], "the importance of this mechanism, if it truly exists, cannot be overemphasized, as illustrated by the examples" given in his paper. In a similar fashion, examples are given in Table 4 of the application of selected Table 3 model equations. The Table 4 results are plotted in Fig. 13 for the British A533B ref. Weld (Code R), the ASTM A302B ref. Plate (Code F23), the German A508 Forging (Code K), the Belgium A508 Forging (Code MO), and the A537 HTR liner steel. For each material, relative values of the Charpy shift are given; that is, the data have been normalized to values of unity at the maxium predicted Charpy shift values of Table 4.

With reference to Fig. 13 and Table 3, Eqs 10*a*, 10*b*, and 10*c* and Eqs 11*a*, 11*b*, and 11*c*, both low-Cu, high-Ni A508 forging materials appear to show a reversal of behavior for a dpa exposure in the range of D equals 1.68 to 1.91 (~0.027 to 0.030 dpa), see Figs. 7 and 8. As stated before, both materials appear to have the unfavorable EOL characteristics of showing an increase in shift with a decrease in flux for values of *D* less than ~1.8 and the favorable EOL characteristic of a decrease in shift with a decrease in flux for values of *D* above ~1.8. At *D* = ~1.8, neither material appears to show an observable flux-level effect. The A537 high-temperature reactor (HTR) liner steel, Eq 16 and Fig. 12, appears to have the unfavorable EOL characteristic of showing an increase in shift with a decrease in flux level for values of $D \geq 0.1$ (or 0.0016 dpa). Finally, the British A533B ref. Weld, Eq 6 and Fig. 3, shows the behavior of an increase in shift with a decrease in flux level while the ASTM A302B ref. Plate, Eq 8 and Fig. 5, shows the opposite effect.

Equations 6, 8, 10*a*, 10*b*, 10*c*, and 11*a*, 11*b*, 11*c* of Table 3 as well as Eq 16 do not represent a unique or the only interpretation of the embrittlement characteristics of these five steel materials. Certainly, there are others. For instance, Simons [*8*] has developed model equations for the British A533B and the ASTM A302B steels as well as for the Electric Power Research Institute (EPRI) A533B ref. Weld, Eq 7 of Table 3. As previously stated, he used a different evaluated metallurgical data base (consensus evaluation of Ref *16*) than that given in Table 1 of this paper. Although his model equations were based on physical principles, he still had to select values for a number of critical parameters and constants. Further, he was not able to quantify, with the available

TABLE 4—*Comparison of Charpy shift predictions for selected model equations and different low alloy steels.*[a]

Material	Eq[a]	Charpy Shift (°C)											
		3×10^{13}[b] (430)[k]	6×10^{12}[b] (86)	6×10^{11}[b] (8.6)	6×10^{10}[b] (0.86)	2×10^{10}[c] (0.29)	1×10^{10}[d] (0.18)	5×10^{9}[e] (0.11)	1.3×10^{9}[f] (0.006)	1×10^{9}[g] (0.014)	5×10^{8}[h] (0.0088)	2.5×10^{8}[i] (0.0053)	1.3×10^{8}[j] (0.0032)
British A533B Weld*	(6)	157	174	199	224	236	241	246	251	268	273	278	283
Code R		(238)[l]	(314)	(358)	(403)	(424)	(433)	(442)	(452)	(482)	(491)	(501)	(511)
ASTM A302 Plate*	(8)	74	66	55	44	38	36	34	32	24	22	19	17
Code F23		(133)	(119)	(99)	(79)	(69)	(64)	(61)	(57)	(43)	(39)	(35)	(30)
German A508 Forging*	(10*b*)[m]	89	82	72	62	58	56	54	52	45	43	41	39
Code K		(159)	(147)	(130)	(112)	(104)	(100)	(97)	(93)	(81)	(77)	(74)	(70)
	(10*c*)	18	31	48	66	74	78	82	85	97	101	105	109
		(33)	(55)	(87)	(119)	(134)	(140)	(147)	(154)	(175)	(181)	(188)	(195)
Belgium A508 Forging*	(11*b*)[n]	29	27	25	23	22	21	21	20	19	19	18	18
Code MO		(51)	(49)	(45)	(41)	(39)	(38)	(38)	(37)	(34)	(33)	(32)	(32)
	(11*c*)	0.7	3.6	7.7	12	14	15	15.5	16.4	19	20	21	22
		(1.3)	(6.5)	(14)	(21)	(25)	(26)	(28)	(29)	(34)	(36)	(38)	(39)
A537 HTR**	(16)[o]	0	0	15	44	58	64	70	77	97	103	109	115
Linear Steel		(0)	(0)	(27)	(80)	(105)	(116)	(127)	(139)	(174)	(185)	(196)	(208)

* Irradiated at ~288°C (550°F).
** Irradiated at ~60°C (140°F).
[a] From Table 3 and for an EOL fluence ($E > 1$ MeV) of 1×10^{19} n/cm^2 (dpa $\simeq 0.016$; $D = 1$), unless indicated otherwise.
[b] Selection of flux levels (n/cm$^2 \cdot$ s) encountered in test reactor irradiations; surveillance capsules range from about 8×10^{11} down to about 1×10^{9}.
[c] Typical PWR vessel ID flux level (n/cm$^2 \cdot$ s).
[d] Typical PWR ¼-T flux level.
[e] Typical PWR ½-T flux level.
[f] Typical PWR ¾-T flux level.
[g] Typical BWR vessel ID flux level.
[h] Typical BWR ¼-T flux level.
[i] Typical BWR ½-T flux level.
[j] Typical BWR ¾-T flux level.
[k] The parenthesized values are the exposure rate $\dot{D}$ in units of 10^{-10} dpa/s = 1.
[l] The parenthesized values are the Charpy shift in°F.
[m] For an LER fluence of 2.85×10^{19} n/cm^2 (dpa $\simeq 0.046$; $D = 2.85$).
[n] For an LER fluence of 3.15×10^{19} n/cm^2 (dpa $\simeq 0.050$; $D = 3.15$).
[o] For an EOL fluence of 0.1×10^{19} n/cm^2 (dpa $\simeq 0.0016$; $D = 0.1$).

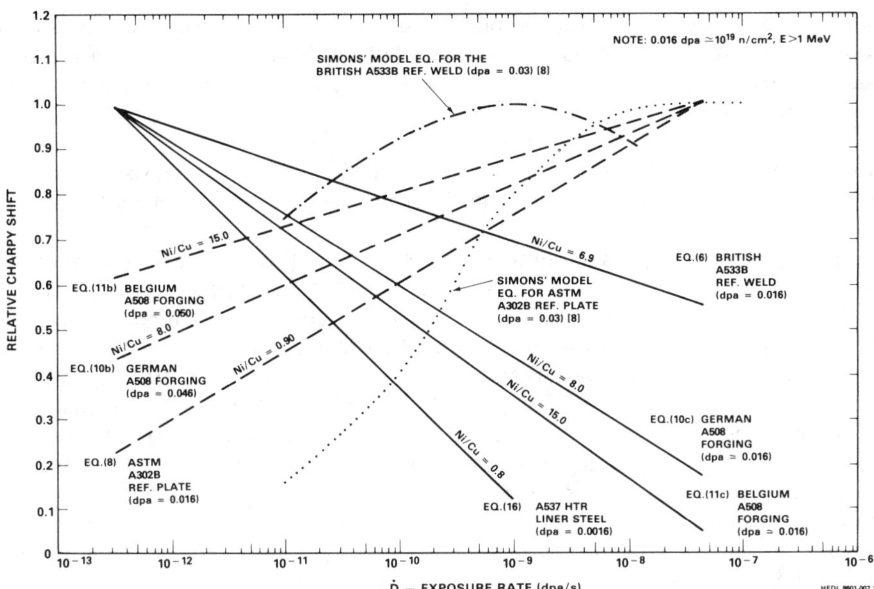

FIG. 13—*Comparison of relative Charpy shift predictions for selected model equations and different low alloy steels—data normalized to unity at the maximum predicted Charpy shift values of Table 4. [For an irradiation temperature of ~288°C (550°F) for all materials except the A537 HTR liner steel that was irradiated at ≤60°C (140°F).]*

experimental data, the exact flux-level behavior of the HSST A533B 03 Plate or the two A508 Forging materials. He concluded, therefore, that at a dpa exposure of ~0.03, none of these three steels showed a significant flux level effect. His findings are consistent with those of this study.

Simons model equation results for the British A533B ref. weld and the ASTM A302B ref. plate are plotted in Fig. 13 and are shown as dot-dash and dotted curves, respectively. His results have been normalized to unity at his maximum predicted Charpy shift value for each material. His A302B results show a drop-off similar to the Eq 8 straight line (dashed) curve, but with a significant model dependent oscillation above and below a pivot point near an exposure rate of $\sim 6 \times 10^{-10}$ dpa/s. On the other hand, and for comparison with Eq 6 results of this study, his A533B results show a similar increase in shift with a decrease in exposure rate from 10^{-8} down to 10^{-9} dpa/s, after which, his model equation suggests a reversal of behavior with a decrease in shift for a decrease in exposure rates for the 10^{-9} to 10^{-11} dpa/s range.

It is very apparent that neither Simons' two model equations, nor Eqs 6 and 8 can be used to precisely (or truly) represent the embrittlement behavior of these different PV steels over a six to seven decade range of exposure rates of 10^{-7} down to 10^{-13} dpa/s. These model equations simply provide an alternative and new interpretation of some very important, but very limited test reactor data. More experimental data must be obtained to resolve the important issues raised by this, Guthrie's [*7*], Gold and McElroy's [*11*], McElroy et al [*9*], Simons' [*8*], and Norris's [*4*] studies. In this regard, it would be of considerable interest to evaluate the applicability and consistency of the present HEDL results with those of the Italian and French trend curve data development and testings studies of Maricchiolo et al [*52*] and Brilland et al [*53*], respectively. Such an effort is beyond the scope of the present study.

The quantification of the actual effects of the different environmental variables for a given steel material is and will continue to be a very challenging task. No claim is made here or in Simons'

[*8*] paper for the accuracy or general applicability of the model equations that have been developed and tested for the seven steel materials. In a similar fashion, Norris [*4*] made no such claim for his model equation. It is worthwhile, however, to briefly discuss some of the possible implications of the use of these equations:

- First, the results presented in Table 4 and Fig. 13 indicated that for the exposure rate range from $\sim 10^{-13}$ to $\sim 10^{-7}$ dpa/s, some steels might show factors of change of up to 5 to 10 in measured Charpy shifts.
- Secondly, the exposure rates most often encountered in test reactor irradiations would be in the $\sim 4 \times 10^{-8}$ to $\sim 10^{-10}$ dpa/s range. The range for PWR surveillance capsule irradiations would be from $\sim 1 \times 10^{-9}$ down to $\sim 5 \times 10^{-11}$ dpa/s and for BWRs from $\sim 5 \times 10^{-10}$ down to $\sim 1 \times 10^{-12}$ dpa/s. These ranges should be compared with the exposure rate range in the actual PV of commercial power plants, which, from Table 4 varies from $\sim 3 \times 10^{-11}$ down to 6×10^{-13} for PWR and from $\sim 1.4 \times 10^{-12}$ down to 3.2×10^{-13} dpa/s for BWR. Consequently, the conventional meaning of a lead factor for a surveillance capsule derived plant specific trend curve could lose all of its significance, depending on the PV steel and its exposure rate embrittlement characteristics; that is, the lead factor could increase, decrease, or stay the same.
- Thirdly, the in-wall surface to ¾-T change in embrittlement caused by exposure rate changes might not exceed about 10 to 30%; and this can be either an increase or decrease depending on the steel material, see Table 4.
- Finally, as shown in Fig. 13, there appears to be a definite correlation of the relative slopes of Eqs 6, 8, 10*b*, 10*c*, 11*b*, and 11*c* with the Ni/Cu ratio. For steels with the characteristic of an increase in shift with a decrease in exposure rate, the relative slope increases with higher Ni/Cu ratios. For materials with the opposite behavior, however, the relative slope increases as the Ni/Cu ratio becomes smaller. It is also interesting to note that the two materials that do not show a significant exposure rate effect have Ni/Cu ratios of 2.67 and 4.67, that fall in between the ratios of the other four materials. The Ni/Cu ratio, therefore, might prove to be an approximate indicator of the exposure rate dependency of selected steels for a specified set of conditions. These observations appear to be consistent with the recently reported results of the study by Williams et al [*32*] on the influence of copper, nickel, and irradiation temperature on the irradiation shift of low alloy steels.

Summary

Existing trend curves do not account for previous and more recently observed test and power reactor flux-level, thermal neutron, and γ-ray field-induced effects. Any agreement between measured data and trend curve predictions that does not adequately represent the important neutron environmental and temperature effects as well as the microstructural damage processes, therefore, could be fortuitous.

Sorting out the effects of different variables on the metallurgical properties of different steels is not easily done because first, they are frequently manifested simultaneously and second, it is not known which of the above variables is the controlling factor in a given experiment or situation.

Simons, McElroy, Gold, and Roberts have furthered the work of Alberman, Norris, Lucas et al, and others by studying the effects of the environmental variables of neutron spectrum, exposure, exposure rate, and the thermal and γ-ray fluxes associated with surveillance capsules and PV through-wall gradients. The results of these studies, together with those of Hawthorne support the existence of a significant material-dependent flux-level effect for PV and support structure steels; that is, a steel may show a decrease, an increase, or no change in the measured Charpy shift with changes in flux level. Further, the actual behavior of a material can change significantly as a function of the neutron exposure; also, thermal neutron and γ-ray effects can contribute to observed changes in property, especially near steel-water interface positions with high thermal-to-fast-neutron ratios.

In this paper, preliminary results of the continuing study of the effects of environmental variables (neutron spectrum, exposure, exposure rate, and thermal and γ-ray fluxes) on the embrittlement of two weld, two plate, and two forging pressure vessel steels irradiated at ~288°C (550°F) and one support structure-type steel irradiated at or below 60°C (140°F) are presented and discussed. Empirically derived through-wall Charpy shift gradient curves for the six PV steel materials and EOL and LER trend curve model equations for the six PV steel materials and one support structure-type steel are under development and study. The application and implications on the use of the measured in-wall Charpy shift attenuation curves as well as the derived trend curve model equations are presented and discussed.

As stated by Gold and McElroy in another paper at this symposium, the existence of a flux-level effect has important implications for the U.S. commercial nuclear power industry, since accelerated locations have almost invariably been used in PV surveillance programs. These accelerated PV surveillance capsules have provided lead factors that have been applied to obtain projections of PV embrittlement. In fact, accelerated PV capsules comprise the largest existing data base for trend curve analyses. Consequently, it is clear that a flux-level effect would imply that some correction would be necessary in the application and interpretation of lead factors. Otherwise, the application of lead factors could not always ensure a conservative extrapolation. At the same time, it is apparent that any reduction in embrittlement afforded from low leakage cores, which are now being adopted in some U.S. power plants, must be quantified in terms of a flux-level effect, lest the predicted gain be under- or over-estimated.

Acknowledgments

This work has been accomplished under the auspices of a Department of Energy (DOE) technology transfer Agreement No. 85-WHSC 9010 between the Metrology Control Corporation (MC^2) and Westinghouse Hanford Company (WHC) and under the Nuclear Regulatory Commission's (NRC) Light Water Reactor Pressure Vessel Surveillance Dosimetry Improvement Program (LWR-PV-SDIP).

References

[*1*] Randall, P. N., ''Basis for Revision 2 of U.S. NRC Regulatory Guide 1.99,'' *Proceedings,* 10th MPA Seminar, Stuttgart, FRG, Oct. 10, 1984.

[*2*] Perrin, J. S., Wullaert, R. A., Odette, G. R., and Lombroso, M. P., ''Physically Based Regression Correlations of Embrittlement Data from Reactor PV Surveillance Programs,'' EPRI NP-3319, Electric Power Research Institute, Palo Alto, CA, Jan. 1984.

[*3*] Guthrie, G. L., ''Charpy Trend Curves Based on 177 PWR Data Points,'' LWR Pressure Vessel Dosimetry Improvement Program Quarterly Progress Report, April 1983–June 1983, NUREG/CR-3391, Vol. 2, HEDL-TME 83-22, Nuclear Regulatory Commission, Washington, DC, April 1984.

[*4*] Norris, E. B., ''A Service Laboratory's View of the Status and Direction of Reactor Vessel Surveillance,'' in *Radiation Embrittlement and Surveillance of Nuclear Reactor Pressure Vessels: An International Study, ASTM STP 819,* L. E. Steele, Ed., American Society for Testing and Materials, Philadelphia, 1983, pp. 194–204.

[*5*] Alberman, A. A. et al, ''Damage Function for the Mechanical Properties of Steels,'' *Nuclear Technology,* Vol. 36, p. 336, 1977.

[*6*] Alberman, A. A. et al, ''Influence des Neutrons Thermiques sur la Fragilisation de l'Acier de Peau d'Etancheite des Reacteurs a Haute Temperature,'' *Proceedings,* 4th ASTM-EURATOM Symposium on Reactor Dosimetry, Gaithersburg, MD, March 22–26, 1982, NUREG/CP-0029, Nuclear Regulatory Commission, Washington, DC, Vol. 2, July 1982, p. 839; and *Nulear Technology,* Vol. 66, Sept. 1984, p. 639.

[*7*] Guthrie, G. L., ''HEDL Analysis of the Poolside Facility Experiment,'' LWR Pressure Vessel Surveillance Dosimetry Improvement Program Semiannual Progress Report, April 1984–Sept. 1984, NUREG/CR-3746, Vol. 2, HEDL-TME 84-21, Nuclear Regulatory Commission, Washington, DC, March 1985.

[8] Simons, R. L., "Damage Rate and Spectrum Effects in Ferritic Steel Δ*NDTT* Data," LWR Pressure Vessel Surveillance Dosimetry Improvement Program Progress Report, Oct. 1984–Sept. 1985, NUREG/CR-4307, Vol. 1, HEDL-TME 85-14, Nuclear Regulatory Commission, Washington, DC, Nov. 1985.

[9] McElroy, W. N., Gold, R., Lippincott, E. P., Simons, R. L., and Anderson, S. L., "Trend Curve Data Development and Testing," LWR Pressure Vessel Surveillance Dosimetry Improvement Program Progress Report, Oct. 1984–Sept. 1985, NUREG/CR-4307, Vol. 1, HEDL-TME 85-14, Nuclear Regulatory Commission, Washington, DC, pp. HEDL-58–HEDL-74, Nov. 1985.

[10] McElroy, W. N. et al, "Trend Curve Exposure Parameter Data Development and Testing," HEDL-SA-3126 and *Proceedings,* 5th ASTM-EURATOM Symposium on Reactor Dosimetry, Geesthacht, FRG, Sept. 24–28, 1984.

[11] Gold, R. and McElroy, W. N., "Current Limitations of Trend Curve Analysis for the Prediction of Reactor Pressure Vessel Embrittlement," LWR Pressure Vessel Surveillance Dosimetry Improvement Program Quarterly Progress Report, Oct. 1984–Sept. 1985, NUREG/CR 4307, Vol. 1, HEDL-TME 85-14, Nuclear Regulatory Commission, Washington, DC, Nov. 1985.

[12] Hawthorne, J. R., "Radiation Sensitivity and Annealing Parameter Studies," Materials Engineering Research/Pressure Vessel Research, Section 20, *Proceedings,* NRC 13th WRSR Information Meeting, NBS, Gaithersburg, MD, Oct. 22–25, 1985.

[13] Lucas, G. E., Odette, G. R., Lombrozo, P. M., and Sheckherd, J. W., "Effects of Composition, Microstructure, and Temperature on Irradiation Hardening of Pressure Vessel Steels," in *Effects of Radiation on Materials: Twelfth International Symposium, ASTM STP 870,* F. A. Gardner and J. S. Perrin, Eds., American Society for Testing and Materials, Philadelphia, 1985, pp. 900–930.

[14] McElroy, W. N., Gold, R., Simons, R. L., and Roberts, J. H., "Trend Curve Data Development and Testing," Metrology Control Corporation, MCR-TR-86-10, HEDL-7591, Aug. 1986.

[15] ASTM E706, "Master Matrix for LWR PV Surveillance Standards," *Annual Book of ASTM Standards,* American Society for Testing and Materials, Philadelphia, PA, Sec. 12, Vol. 12.02, current edition.

[16] McElroy, W. N., Eds., "LWR Pressure Vessel Surveillance Dosimetry Improvement Program: PSF Experiments Summary and Blind Test Results," NUREG/CR-3320, Vol. 1, HEDL-TME 86-8, and Appendix A.1, Nuclear Regulatory Commission, Washington, DC, July 1986, p. A. 1–4.

[17] ASTM E693, "Standard Practice for Characterizing Neutron Exposures in Ferritic Steels in Terms of Displacements per Atom (dpa)," *1979 Annual Book of ASTM Standards,* American Society for Testing and Materials, Philadelhia, PA, current edition.

[18] Guthrie, G. L., McElroy, W. N., and Simons, R. L., "Effect of Thermal Neutrons in Irradiation Embrittlement of PWR Pressure Vessel Plates and Welds," LWR Pressure Vessel Surveillance Dosimetry Improvement Program Quarterly Progress Report, April 1983–June 1983, NUREG/CR-3391, Vol. 2, HEDL-TME 83-22, Nuclear Regulatory Commission, Washington, DC, April 1984.

[19] Gold, R., Roberts, J. H., and Doran, D. G., "Determination of Gamma-Ray Displacement Rates," LWR Pressure Vessel Surveillance Dosimetry Improvement Program Progress Report, Oct. 1984–Sept. 1985, NUREG/CR-4307, Vol. 1, HEDL-TME 85-14, Nuclear Regulatory Commission, Washington, DC, pp. HEDL-22–HEDL-38, Nov. 1985.

[20] McElroy, W. N. et al, "LWR Pressure Vessel Surveillance Dosimetry Improvement Program—1984 Annual Report, Oct. 1, 1983–Sept. 30, 1984," NUREG/CR-3746, Vol. 3, HEDL-TME 84-31, Nuclear Regulatory Commission, Washington, DC, March 1985.

[21] Hawthorne, J. R., Menke, B. H., and Hiser, A. L., "LWR Pressure Vessel Surveillance Dosimetry Improvement Program: Notch Ductility and Fracture Toughness Degradation of A302-B and A533-B Reference Plates from PSF Simulated Surveillance and Through-Wall Irradiation Capsules," NUREG/CR-3295, MEA-2017, Vol. 1, Nuclear Regulatory Commission, Washington, DC, April 1984.

[22] Hawthorne, J. R., and Menke, B. H., "LWR Pressure Vessel Surveillance Dosimetry Improvement Program: Postirradiation Notch Ductility and Tensile Strength Determinations for PSF Simulated Surveillance and Through-Wall Specimen Capsules," NUREG/CR-3295, MEA-2017, Vol. 2, Nuclear Regulatory Commission, Washington, DC, April 1984.

[23] Stallmann, F. W., "Statistical Evaluation of the Metallurgical Test Data in the ORR-PSF-PVS Irradiation Experiment," NUREG/CR-3815, ORNL/TM-9207, Nuclear Regulatory Commission, Washington, DC, Aug. 1984.

[24] Perrin, J. S., "Simulated Void Box Capsule (SVBC) Charpy Impact Test Results," EPRI NP-4630, NUREG/CR-3320, Vol. 5, Electric Power Research Institute, Palo Alto, CA, Aug. 1986.

[25] Guthrie, G. L., McElroy, W. N., and Anderson, S. L., "A Preliminary Study of the Use of Fuel Management Techniques for Slowing Pressure Vessel Embrittlement," *Proceedings,* 4th ASTM-EURATOM Symposium on Reactor Dosimetry, Gaithersburg, MD, March 22–26, 1982, NUREG/CP-0029, Nuclear Regulatory Commission, Washington, DC, Vol. 1, July 1982, pp. 111–120.

[26] Guthrie, G. L., McElroy, W. N., and Anderson, S. L., "Investigations of Effects of Reactor Core Loadings on PV Neutron Exposure," LWR Pressure Vessel Surveillance Dosimetry Improvement

Program: Quarterly Progress Report, Oct.–Dec. 1981, NUREG/CR-2345, Vol. 4, HEDL-TME 81-36, Nuclear Regulatory Commission, Washington, DC, Section E and Appendix, Oct. 1982.

[27] McElroy, W. N. et al, "Verification of Effects of Fuel Management Schemes on the Condition of Pressure Vessels and their Support Structures," HEDL-SA-2791 and *Proceedings,* 10th Water Reactor Safety Research Information Meeting, Gaithersburg, MD, Oct. 12–15, 1982, NUREG/CP-0041, Vol. 4, pp. 184–227, Jan. 1983.

[28] Kam, F. B. K., Stallman, F. W., Guthrie, G., and McElroy, W. N., "LWR Surveillance Dosimetry Improvement Program: PSF Metallurgical Blind Test Results," *Transactions,* 12th Water Reactor Safety Research Information Meeting, Gaithersburg, MD, NUREG/CP-0057, Nuclear Regulatory Commission, Washington, DC, Oct. 1984.

[29] Alberman, A. A. et al, "DOMPAC Dosimetry Experiment Neutron Simulation of the Pressure Vessel of a Pressurized-Water Reactor, Characterization of Irradiation Damage," CEA-R-5217, Centre d'Etudes Nucleaires de Saclay, France, May 1983.

[30] Davies, L. M. et al, "Analysis of the Behavior of Advanced Reactor Pressure Vessel Steels Under Neutron Irradiation—The UK Programme," Report from the UK for the IAEA Coordinated Research Programme on the Analysis of the Behavior of Advanced Reactor Pressure Vessel Steels Under Neutron Irradiations, Unnumbered UKAEA-Harwell Report, Harwell, UK, April 1983.

[31] Davis, L. M. and Squires, R. L., "Comparison of Mechanical Test Results from the IAEA-Coordinated Research Programme and Surveillance Dosimetry Improvement Programme," *Proceedings,* 2nd International Conference on Environmental Degradation of Nuclear Reactor Materials, Monterey, CA, Sept. 12–14, 1985.

[32] Williams, T. J. et al, "The Influence of Copper, Nickel, and Irradiation Temperature on the Irradiation Shift of Low Alloy Steels," *Proceedings,* 2nd International Conference on Environmental Degradation of Nuclear Reactor Materials, Monterey, CA, Sept. 12–14 1985.

[33] Stahlkopf, K. E. and Marston, T. U., "A Comprehensive Approach to Radiation Embrittlement Analysis," *Proceedings,* IAEA Specialists' Meeting on Irradiation Embrittlement, Thermal Annealing and Surveillance of Reactor Pressure Vessels, Vienna, Austria, Feb. 26–March 1, 1979, IWG-RRPC-79/2, International Atomic Energy Agency, Vienna, Austria, Dec. 1979.

[34] Steele, L. E., Ed., in *Radiation Embrittlement and Surveillance of Nuclear Reactor Pressure Vessels: An International Study, ASTM STP 819,* American Society for Testing and Materials, Philadelphia, PA, Nov. 1983.

[35] Steele, L. E., Davis, L. M., Ingham, T., and Brumovsky, M., "Results of IAEA-Coordinated Research Programs on Irradiation Effects on Advanced Pressure Vessel Steels," *Proceedings,* 12th Conference on Effects of Radiation on Materials, *ASTM STP 870,* Williamsburg, VA, June 1984.

[36] Pachur, D. and Sievers, G., "Development Programs on Irradiation Embrittlement of Low Alloy Pressure Vessel Steels in the Federal Republic of Germany," in *Properties of Reactor Structural Alloys After Neutron or Particle Irradiation, ASTM STP 570,* American Society for Testing and Materials, Philadelphia, PA, pp. 555–564, 1975.

[37] Pachur, D., "Apparent Embrittlement Saturation and Radiation Mechanisms of Reactor Pressure Vessel Steels," in *Effects of Radiation on Materials: Tenth Conference, ASTM STP 725,* American Society for Testing and Materials, Philadelphia, PA, pp. 5–19, 1981.

[38] Pachur, D., "Mechanical Properties of Neutron-Irradiated Reactor Pressure Vessel Steel Dependent on Radiation Mechanisms," from the ANS Special Session on Correlations and Implications of Neutron Irradiation Embrittlement of Pressure Vessel Steels," Detroit, MI, June 12–16, 1983, *Transactions of the American Nuclear Society,* Vol. 44, p. 229, 1983.

[39] Hawthorne, J. R., in *Radiation Effects Information Generated on the ASTM Reference Correlation Monitor Steels, ASTM DS54,* American Society for Testing and Materials, Philadelphia, PA, 1974.

[40] Serpan, C. Z., Jr., and Hawthorne, J. R., "Through-Thickness Notch Ductility and Tension Properties as a Function of Neutron Exposure to a Simulated Pressure Vessel Wall of A302-B Steel," *Nuclear Engineering and Design,* Vol. 6, pp. 78–88, North-Holland Publishing Company, Amsterdam, The Netherlands, 1967.

[41] Serpan, C. Z., Jr., "Reliability of Fluence-Embrittlement Projections for Pressure Vessel Surveillance Analysis," *Nuclear Technology,* Vol. 12, pp. 108–118, Sept. 1971.

[42] Simons, R. L., "Re-evaluation of Ferritic Steel ΔDBTT Data Used in Damage Function Analysis," LWR Pressure Vessel Surveillance Dosimetry Improvement Program: Quarterly Progress Report, Oct.–Dec. 1978, NUREG/CR-0720, HEDL-TME 79-18, Nuclear Regulatory Commission, Washington, DC, pp. HEDL-15 to HEDL-47, July 1980.

[43] Simons, R. L., "Re-evaluation of Dosimetry for Test Reactor Irradiations Used to Obtain ΔNDTT Data for Pressure Vessel Steels," LWR Pressure Vessel Surveillance Dosimetry Improvement Program: Semi-Annual Progress Report, Oct.–Dec. 1982, NUREG/CR-2805, Vol. 4, HEDL-TME 82-21, Nuclear Regulatory Commission, Washington, DC, pp. HEDL-14–HEDL-22, July 1982.

[44] Steele, L. E., Ed., in *Status of USA Nuclear Reactor Pressure Vessel Surveillance for Radiation Effects, ASTM STP 784,* American Society for Testing and Materials, Philadelphia, PA, Jan. 1983.

[45] McElroy, W. N., Ed., ''LWR Pressure Vessel Surveillance Dosimetry Improvement Program: LWR Power Reactor Surveillance Physics-Dosimetry Data Base Compendium,'' NUREG/CR-3319, HEDL-TME 85-3, Hanford Engineering Development Laboratory, Richland, WA, Aug. 1985.

[46] Serpan, C. Z., Jr., and McElroy, W. N., ''Damage Function Analysis of Neutron Energy and Spectrum Effects Upon the Radiation Embrittlement of Steels,'' NRL 6925, Naval Research Laboratory, Washington, DC, July 1969.

[47] Wullaert, R. A. and Shuckherd, J. W., ''Evaluation of the First Maine Yankee Accelerated Surveillance Capsule,'' CR75-317, Effects Technology, Inc., Goleta Heights, CA, Aug. 15, 1975.

[48] Yanichko, S. E. et al, ''Analysis of the Maine Yankee Reactor Vessel Second Accelerated Surveillance Capsule,'' WCAP-9875, EPRI Project RP1021-3, Westinghouse Electric Corp., Pittsburgh, PA, March 1981.

[49] Perrin, J. S. et al, ''Maine Yankee Nuclear Plant Reactor Pressure Vessel Surveillance Program: Capsule 263,'' BCL-585-21, Battelle Memorial Institute, Columbus, OH, Dec. 21, 1980.

[50] Loss, F. J., Ed., ''Structural Integrity of Water Reactor Pressure Boundary Components, Progress Report Ending 29 Feb. 1976,'' NRL Report 8006, Naval Research Laboratory, Washington, DC, Aug. 26, 1976.

[51] Spelzman, W. J., Berggren, R. G., and Jones, P. N., Jr., ''ORNL Characterization of the HSST Program Plates 01, 02, and 03,'' NUREG/CR-4092, ORNL/TM-9491, Nuclear Regulatory Commission, Washington, DC, April 1985.

[52] Maricchiolo, C., Milella, P. P., and Pini, A., ''Prediction of Reference Transition Temperature Increase Due to Neutron Irradiation Exposure,'' *Proceedings,* IAEA Specialist's Meeting on Radiation Embrittlement and Surveillance of Reactor Pressure Vessel Steels, Vienna, Austria, Oct. 8–10, 1984.

[53] Brilland, C., Hedin, F., and Houssin, B., ''A Comparison Between French Surveillance Program Results and Predictions of Irradiation Embrittlement,'' *Proceedings,* 13th ASTM International Symposium on the Effects of Radiation on Materials, Seattle, WA, June 23–25, 1986.

Robert L. Simons[1]

Damage Rate and Spectrum Effects in Ferritic Steel ΔNDTT Data

REFERENCE: Simons, R. L., "**Damage Rate and Spectrum Effects in Ferritic Steel ΔNDTT Data,**" *Influence of Radiation on Material Properties: 13th International Symposium (Part II), ASTM STP 956,* F. A. Garner, C. H. Henager, Jr., and N. Igata, Eds., American Society for Testing and Materials, Philadelphia, 1987, pp. 535–551.

ABSTRACT: A model was developed for irradiation-induced hardening in pressure vessel steels by cascade-induced nucleation and free defect diffusion of copper to coherent copper clusters. The model was used as a functional framework for fitting data on macroscopic property change (ΔNDTT) as a function of damage dose and dose rate. Several defect cross sections having different neutron energy dependences were tried. The cross section for Frenkel pair production, based on resistivity measurements at 4 K, was found to give a comparable or slightly better fit to the data than dpa. Damage rate is an important parameter to consider in correlation of ΔNDTT data, and damage rate can affect the nucleation as well as the growth of the copper precipitation.

KEY WORDS: defect cross sections, displaced atoms, damage rate, precipitation hardening, nucleation, growth

To ensure the service integrity of light-water reactor (LWR) containment vessels, researchers have developed trend curves for macroscopic property change (ΔNDTT) as a function of neutron fluence (>1 MeV) or displacements per atom (dpa) to predict the condition of the reactor vessel wall [*1,2*]. Equations are generally developed for a narrow irradiation temperature range and are based primarily on data from existing LWR surveillance programs.

Norris [*3*] studied the implication of damage rate effects in A302B and A533B steels. Although he concluded damage rate effects could prove to have significant impact on the correlation and projection of LWR surveillance data to the end-of-life reactor pressure vessel conditions, he thought that, at that time, data uncertainties were too high to statistically support the proposition that a damage rate effect existed. More recent model development by Lucas et al. [*4*] includes damage rate effects in the nucleation and growth of copper precipitates and microvoids. Steele et al. [*5*] also reviewed recent work sponsored by the International Atomic Energy Agency (IAEA) to identify spectrum effects and damage rate effects. They found that experimental data are consistent with simple annihilation models, but there was insufficient flux and spectrum variation to verify the model predictions. McElroy et al.[2] recently reviewed experimental efforts over the last 15 years to identify flux effects and found that the effects were often glossed over or misinterpreted.

The objective of this work is to devise a physically based model that incorporates spectrum and rate-dependent terms in order to take advantage of selected sets of data to infer the dependencies of ΔNDTT on neutron spectrum and damage rate. Three areas of trend curve model development will be considered: (1) hardening models that relate the irradiation-induced microstructure to

[1] Senior scientist, Hanford Engineering Development Laboratory, P.O. Box 1970, Richland, WA 99352.
[2] McElroy, W. N. et al, in this publication, pp. 505–534.

ΔNDTT, (2) spectrum variation in other parameters besides $\phi t > 1$ MeV and dpa, and (3) damage rate effects. The rate effects arise from the evolution of obstacle size and density in the hardening model. The analysis will be limited to ASTM A302B reference plate steel irradiated at both "low" (<513 K [240°C]) and "high" (561 K [288°C]) temperatures and five other plate, forging, and weld materials irradiated at 561 K. The high-temperature data were obtained in the poolside facility (PSF) irradiation of a simulated surveillance capsule (SSC) and simulated pressure vessel (SPV). The low-temperature irradiations of the A302B steel were done in research reactors, including SSC and SPV mockups at the Industrial Research Laboratory (IRL) [*6,7*]. For the ASTM A302B reference plate material, this provides a unique opportunity to study through-wall embrittlement data in two temperature regimes.

Hardening Model

The macroscopic property change (ΔNDTT) is related to the existing microstructure. Odette [*8*] demonstrated empirically that ΔNDTT at 40.7-J (30-ft · lb) absorbed energy is directly proportional to the change in 0.2% yield strength $\Delta\sigma$. The proportionality constant is between 0.5 and 0.7 when ΔNDTT is in °C, and yield strength is expressed as MPa. He proposed that the dominant microstructure hardening mechanism is coherent copper clusters that nucleate and grow early in the irradiation. Both copper and nickel are found to be important elements in correlating ΔNDTT data; hence the obstacles may contain nickel as well. This was substantiated by field ion microscopy studies by Grant et al. [*9*] in which they observed clusters containing copper, nickel, manganese, and iron. Helium may also be important for stabilizing vacancy clusters. If coherent precipitates are causing the material to harden, then the hardening model of Russell and Brown [*10*] is appropriate. In their work, the change in yield stress was correlated with an Orowan-type equation modified to account for differences in the elastic modulus of the solution matrix and the precipitate. The form of the equation is

$$\Delta\text{NDTT} \propto \Delta\sigma = 0.8\sqrt{3}\,\frac{\mu\mathbf{b}}{L}\left[1 - \left(\frac{E_{Cu}}{E_{Fe}}\right)^2\right]^{\frac{1}{2}} \tag{1}$$

where the $\sqrt{3}$ factor converts from shear to tensile stress by the Von Mises Criterion, μ is the shear modulus of iron (temperature dependent), **b** is the Burgers vector (**b** = 0.248 nm), L is the mean distance between obstacles in the slip plane, and the last term accounts for the difference between the energy of the dislocation in the precipitate and the energy of the dislocation in the iron matrix in terms of the elastic moduli of the two materials.

The value of L is geometrically derived from the microstructure and is equal to $1/\sqrt{Nd}$, where N is the volume density of obstacles and d is their diameter. Expressions for both N and d can be derived from simple rate theory models. To arrive at a closed form solution, it must be assumed that they are independent; that is, nucleation is completed before significant growth of the obstacles occurs. The data used in this analysis support this assumption.

The ratio E_{Cu}/E_{Fe} proposed by Russell and Brown is precipitate-size dependent. The maximum value for the ratio is equal to the ratio of the Cu-to-Fe elastic moduli (0.6). The ratio should also depend on the density of copper atoms and vacancies in the cluster. Russell and Brown rationalized the use of a logarithmic function of obstacle diameter d to describe the ratio E_{Cu}/E_{Fe}. Their function reduces to the form

$$E_{Cu}/E_{Fe} = 1 - A\ \ln\,(d/\ell\mathbf{b}) \tag{2}$$

where A and ℓ are adjustable parameters (ℓ controls the threshold size for obstacle hardening) and **b** is the Burgers vector. Russell and Brown found A = 0.133 and ℓ = 5, when the precipitate size is expressed in terms of the diameter.

In this work, Eq 1 was modified by an empirical factor [arc $\sin(E_{Cu}/E_{Fe})$]. This multiplying factor causes the hardening to increase more rapidly near the threshold size.

Obstacle Density

Field ion microscopy observations have shown that copper atoms may be associated with the vacancy clusters [*11*]. Thus, it is inferred that vacancy clusters produced in the displacement cascade are stabilized by copper atoms (or possibly helium atoms) and then continue to grow by attracting copper atoms. The net production rate of vacancy clusters $\dot{N}$ depends on the production rate in the cascade less those annihilated by cascade overlap and those lost by temperature-dependent processes. The rate is described by

$$\dot{N} = \alpha H - \beta G N - \nu N \tag{3}$$

where α and β are constants, H is the cluster production rate, G is the cascade overlap rate (assumed to be proportional to the total defect production rate), and ν is the thermal annihilation rate. The thermal annihilation can take place by two means: (1) thermal emission of vacancies given by

$$\nu_e = 4\pi r_o D_v C^0 \tag{4}$$

and (2) absorption by diffusing iron interstitials given by

$$\nu_a = 4\pi r_o D_i C_i \tag{5}$$

where r_o is the cluster radius, D is the diffusion coefficient for either vacancies v or interstitials i, C^0 is the equilibrium vacancy concentration, and C_i is the mobile interstitial concentration. Solving Eq 3 gives the exposure dependence of the obstacle density

$$N = \frac{(\alpha/\beta)\,(H/G)}{1 + \nu/\beta G}\left\{1 - \exp^{-}\left[\left(1 + \frac{\nu}{\beta G}\right)\beta G t\right]\right\} \tag{6}$$

The product of the diffusion constant and concentration $D_v C_v$ for a vacancy cluster can take on two extremes. The first condition is for a well-annealed material with low defect sink density such that recombination of defects is dominant and

$$D_i C_i \propto D_v C_v \propto \sqrt{D_v F} \tag{7}$$

where F is the free defect production rate. Equation 7 assumes that the defect density from thermal emission from sinks is much less than the density produced by displacement of atoms.

The second condition occurs when there is a high density of sinks for point defects, as produced in the preirradiation thermomechanical conditioning of the steel or as a result of irradiation. In this case, the product DC is

$$D_i C_i \propto D_v C_v \propto D_v F \tag{8}$$

Consequently, the obstacle site density can have three rate dependences through the term $\nu/\beta G$ in Eq 6 for the conditions discussed above. For the purpose of simplifying the discussion, it is assumed that F and G are proportional to one another. When emission of point defects dominates (Eq 4), the damage rate term shows a $1/F$ rate dependence. For absorption of point defects and a recombination dominant microstructure (Eqs 5 and 7), the denominator shows a $1/\sqrt{F}$ rate dependence. Finally with absorption of point defects and a point defect sink dominant microstructure (Eqs 5 and 8), there is no rate dependence. Actually the exponent for rate dependence may vary continuously between the extremes because the mobility of the defects changes with temperature.

Obstacle Size

For spherical obstacles, the growth rate is given by

$$\frac{dd}{dt} = \frac{4\Omega}{d}\left(D_vC_vC_c - \frac{\pi d^3}{6}N\right) \tag{9}$$

where the product D_vC_v is the irradiation-enhanced diffusion constant for the defect migrating by a vacancy diffusion mechanism, C_c is the initial concentration of chemical elements in the lattice that contributes to the growth of the precipitate, Ω is the atomic volume, d is the diameter of the obstacle as a function of time, and N is the concentration of clusters. The latter term in the parentheses accounts for the depletion of chemical species contributing to the growth of the cluster. Equation 9 can be integrated to a closed form solution as done by Odette [*8*]. However, the resultant equation cannot be solved explicitly for the cluster diameter. In an interim analysis, it will be assumed that depletion of the chemical species is not important.[3] In addition, it is assumed that the product DC is approximately independent of time. In actual fact, DC decreases slowly with time. With these approximations in mind, the integration of Eq 9 gives

$$d = \sqrt{d_o^2 + 8\Omega D_vC_vC_vt} \tag{10}$$

Substituting either Eqs 7 or 8 into Eq 10 for D_vC_v gives the size dependence of the obstacle for two different microstructural conditions. For recombination-dominant conditions, the diameter has a $1/\sqrt[4]{F}$ damage rate dependence

$$d = \sqrt{d_0^2 + 8\Omega C_c\left(\sqrt{\frac{D_v}{F}}\right)Ft} \tag{11}$$

However, for a sink-dominant microstructure, there should be no significant damage rate dependence, that is

$$d = \sqrt{d_o^2 + 8\Omega C_cD_vFt} \tag{12}$$

Combining the above equations, the general form for precipitate hardening used in this work is as follows:

$$\Delta\text{NDTT} = 0.48\sqrt{3}\,\mu\mathbf{b}\sqrt{\frac{C_0(H/G)}{1 + \frac{C_1}{G^n}}\left\{1 - \exp\left[-C_2\left(1 + \frac{C_1}{G^n}\right)Gt\right]\right\}}\sqrt{d_o^2 + \frac{C_3}{F^m}Ft}$$
$$\times\,[1 - (E_{Cu}/E_{Fe})^2]^{\frac{1}{2}} \arcsin(E_{Cu}/E_{Fe}) \tag{13}$$

where from Eq 2, $E_{Cu}/E_{Fe} = 1 - A\ln(d/C_4b)$ and where the constants C_0, C_1, C_2, C_3, and C_4 are determined by fitting the equation to the data. The concentration of chemical species C_c has been incorporated into the constant C_3. Thus Eq 13 is applicable only to a single material. The values of the constants n and m investigated included 0, 0.5, and 1.0. The appropriate values depend, in principle, on the prevailing microstructure and temperature. The ratio E_{Cu}/E_{Fe} accounts for differences in energy of the dislocation in the precipitate and the iron matrix (Eq 2). In Eq 13 the parameters H, G, and F are all spectrum dependent. As stated before, H is assumed to equal the vacancy cluster production rate, G is associated with annihilation of clusters by cascade overlap, and F is associated with cluster growth by absorbing free defects.

[3] Results of the analysis justify this assumption. The calculated size and density of the obstacles corresponds to only a few percent of the available copper.

Data Tabulation

There were three basic data sets on ASTM A302B reference plate F analyzed. The first two sets included research reactor data previously used to develop a damage function for low temperature [<513 K (240°C)] irradiations [*6,12*]. The second set is the physics-dosimetry-metallurgy data base for the low-temperature [<389 K (116°C)] IRL-SSC-SPV tests. The third set is from the PSF-SSC-SPV experiment run at 561 K (288°C) [*13*]. The first two sets have the largest spectral variation, including light-water reactor, heavy-water reactor, and graphite-moderated reactor spectra. The light-water reactor data, therefore, includes results from three SSC-SPV experiments (Table 1). The displacement rate varies by a factor of 300 over these data sets.

The neutron spectra used for the low-temperature data set was taken from Serpan and Menkes' compilation of neutron spectra [*14*] used in the damage function analysis by McElroy et al. [*12*] and a reevaluated analysis of the spectra in the SSC-SPV experiment in the Industrial Research Reactor test (IRL-5) [*7*].

Heinisch and Mann [*15*] calculated neutron cross sections for copper based on computer simulations of displacement cascades. They include production of Frenkel pairs (fppa), interstitial clusters (ic), their size (that is, average number of interstitials per cluster), mobile interstitials (mi), vacancy clusters (vc), their size, mobile vacancies (mv), and lobe (or subcascade) production. ENDF/B-V nuclear cross sections were used in their calculations. These calculations were repeated for this work using iron neutron cross-sectional data from ENDF/B-V and the primary recoil energy-dependent defect functions for copper. The defect functions were normalized to give measured values for total Frenkel pairs determined from resistivity measurements in iron at liquid helium temperature.

The calculated exposure parameters are tabulated in Table 1. Also included are fluence $E > 1$ MeV, dpa, irradiation time, irradiation temperature, and ΔNDTT for A302B steel. The PSF parameters are for the spectral set location at the center of the capsule. The actual PSF SSC-SPV data and dpa doses are shown in Table 2. These values are the consensus evaluation values from the PSF blind test results [*13*].

Data Analysis and Results

The low-temperature ΔNDTT data were used to determine which set of defect production cross sections best fits the defect production parameters H, G, and F. The constant C_4 is an obstacle hardening parameter, so it may also be determined from these data. In all analyses, the constant C_2 was driven to a large value, which indicated that the site density was saturated; consequently, C_2 is not important in the correlation. This leaves three constants plus selection of the parameters n and m that control the damage rate effect. After several trial fits to the low-temperature ΔNDTT data on A302B, the best results were obtained with $n = 1$ and $m = 0$. This implies that the thermal annihilation of obstacle sites was controlled by emission of defects and the growth of the obstacles occurred in a sink dominant microstructure.

Table 3 shows the various combinations of H, G, and F that were tried and the respective standard deviation of the fits. The lowest standard deviation is obtained with the Frenkel pair per atom (fppa) function. The function is significantly better than all other combinations tried except the dpa function. The fppa standard deviation was only 8% smaller than the dpa variance, which is not statistically significant. The fppa function differs from the dpa function by having an enhanced low-energy component. This arises because the decreased density of defects in a low-energy cascade results in less recombination. The fppa function is consistent with the damage function unfolded by McElroy et al. [*12*] using the same data set. Their damage function also showed an enhanced low-energy damage component. The interstitial clusters (ic) and mobile interstitials (mi) show fairly low variances.

TABLE 1—*Integrated damage exposure per atom for ΔNDTT of A302B Reference Plate F determined from irradiation experiments.*

Spectrum	$\phi t > 1$, 10^{19} n/cm²	dpa	Frenkel Pairs	Interstitial Clusters	Interstitial Size (I)	Interstitial Mobile	Vacancy Clusters	Vacancy Size (V)	Vacancy Mobile	Lobes	Time, s	T_{IRR}, °C	ΔNDTT (A302B), °C
CVTR 10-L	0.691	1.87E−02	1.07E−02	5.10E−04	5.1	5.65E−03	4.52E−04	5.5	4.77E−03	4.22E−05	2.11E+07	<240	133
LITR C-53	1.44	2.17E−02	9.91E−03	9.17E−04	5.3	2.72E−03	6.96E−04	6.1	1.60E−03	8.14E−05	4.85E+06	<116	119
LITR C-53	2.01	3.02E−02	1.38E−02	1.27E−03	5.3	3.78E−03	9.67E−04	6.1	2.22E−03	1.13E−04	6.74E+06	<116	167
LITR C-49	1.05	1.53E−02	6.89E−03	6.82E−04	5.3	1.49E−03	5.12E−04	7.0	6.10E−04	5.82E−05	4.76E+06	<116	114
LITR C-28	1.37	1.86E−02	7.93E−03	8.52E−04	5.3	1.53E−03	6.30E−04	7.4	5.47E−04	7.77E−05	1.91E+06	<116	122
LITR C-18	1.94	2.72E−02	1.16E−02	1.24E−03	5.3	2.28E−03	9.22E−04	7.4	8.04E−04	1.19E−04	3.91E+06	<116	142
LITR C-55	2.56	3.53E−02	1.51E−02	1.61E−03	5.3	3.02E−03	1.20E−03	7.4	1.09E−03	1.43E−04	6.47E+06	<93	161
BGR W-44	0.816	1.50E−02	7.13E−03	5.95E−04	5.2	1.40E−03	4.65E−04	6.2	1.40E−03	4.65E−05	2.50E+07	<138	114
IRL3 4⅝ in.	0.232	3.15E−03	1.34E−03	1.46E−04	5.4	2.46E−04	1.07E−04	7.5	8.27E−05	1.37E−05	2.44E+06	<116	58
IRL3 5⅝ in.	0.175	2.36E−03	1.00E−03	1.09E−04	5.3	1.85E−04	8.07E−05	7.7	5.94E−05	9.99E−06	2.44E+06	<116	44
IRL3 6⅝ in.	0.126	1.74E−03	7.38E−04	8.00E−05	5.3	1.39E−04	5.92E−05	7.7	4.49E−05	7.11E−06	2.44E+06	<116	28
IRL 3 7⅝ in.	0.0861	1.25E−03	5.34E−04	5.72E−05	5.3	1.04E−04	4.25E−05	7.7	3.42E−05	4.90E−06	2.44E+06	<116	28
IRL3 8⅝ in.	0.0632	9.39E−04	4.04E−04	4.27E−05	5.3	8.11E−05	3.18E−05	7.6	2.74E−05	3.55E−06	244E+06	<116	19
LITR C-43	3.21	4.43E−02	1.89E−02	2.03E−03	5.3	3.64E−03	1.51E−03	7.5	1.26E−03	1.85E−04	4.36E+06	<116	172
HWCTR gray rod	0.616	1.21E−02	6.11E−03	4.36E−04	5.3	2.38E−03	3.46E−04	5.7	1.83E−03	3.83E−05	1.72E+06	<240	106
IRL5-1	0.748	1.04E−02	4.47E−03	4.73E−04	5.4	9.02E−04	3.51E−04	7.0	3.63E−04	4.35E−05	1.54E+07	<116	92
IRL5-2	0.331	4.65E−03	1.98E−03	2.14E−04	5.3	3.71E−04	1.58E−04	7.6	1.23E−04	1.95E−05	1.54E+07	<116	69
IRL5-3	0.199	2.99E−03	1.28E−03	1.37E−04	5.3	2.43E−04	1.02E−04	7.8	7.66E−05	1.19E−05	1.54E+07	<116	44
IRL5-4	0.119	2.00E−03	8.73E−04	8.86E−05	5.3	1.94E−04	7.90E−05	7.6	6.25E−05	8.98E−06	1.54E+07	<116	28
IRL5-5	0.0478	9.87E−04	4.31E−04	4.37E−05	5.2	9.77E−05	3.31E−05	7.4	3.45E−05	3.32E−06	1.54E+07	<116	11
IRL5-6	0.0200	4.94E−04	2.19E−04	2.15E−05	5.2	5.23E−05	1.64E−05	7.4	1.92E−05	1.49E−06	1.54E+07	<116	0
IRL5-7	0.0178	4.45E−04	1.97E−04	1.94E−05	5.2	4.71E−05	1.48E−05	7.4	1.73E−05	1.34E−06	1.54E+07	<116	0
PSF SSC-1	2.64	4.09E−02	1.75E−02	1.87E−03	5.3	3.46E−03	1.39E−03	7.7	1.12E−03	1.54E−04	3.84E+06	288	[a]
PSF SSC-2	5.65	8.82E−02	3.79E−02	4.01E−03	5.3	7.56E−03	2.99E−03	7.7	2.48E−03	3.30E−04	8.42E+06	288	[a]
PSF O-T	4.25	6.80E−02	2.94E−02	3.08E−03	5.3	6.09E−03	2.30E−03	7.5	2.17E−03	2.53E−04	5.10E+07	288	[a]
PSF Q-T	2.28	4.16E−02	1.80E−02	1.88E−03	5.3	4.74E−03	1.41E−03	7.7	1.23E−03	1.44E−04	5.10E+07	288	[a]
PSF H-T	1.09	2.39E−02	1.04E−02	1.07E−03	5.3	2.25E−03	8.06E−04	7.7	7.42E−04	7.52E−05	5.10E+07	288	[a]

[a] See Table 2.

TABLE 2—*Integrated damage exposures for ΔNDTT data from the PSF experiment.*

Material	Location	ΔNDTT[a], °C	dpa	$\phi t > 1$ MeV, 10^{19} n/cm²	Time, s
A302B(F23)	SSC-1	81	0.0400	2.72	3.842E+06
	SSC-2	93	0.0844	5.73	8.420E+06
	O-T	78	0.0615	4.03	5.097E+07
	Q-T	61	0.0383	2.26	5.097E+07
	H-T	51	0.0224	1.12	5.097E+07
A533B(3P)	SSC-1	68	0.0365	2.49	3.842E+06
	SSC-2	82	0.0770	5.24	8.420E+06
	O-T	73	0.0556	3.68	5.097E+07
	Q-T	69	0.0343	2.05	5.097E+07
	H-T	53	0.0199	1.01	5.097E+07
K Forging	SSC-1	58	0.0270	1.73	3.842E+06
	SSC-2	101	0.0569	3.65	8.420E+06
	O-T	76	0.0456	2.84	5.097E+07
	Q-T	74	0.0273	1.52	5.097E+07
	H-T	60	0.0157	0.73	5.097E+07
A508B(MO) forging	SSC-1	17	0.0294	1.89	3.842E+06
	SSC-2	39	0.0621	3.98	8.420E+06
	O-T	27	0.0504	3.11	5.097E+07
	Q-T	22	0.0305	1.67	5.097E+07
	H-T	17	0.0177	0.82	5.097E+07
Weld EC	SSC-1	110	0.0274	1.75	3.842E+06
	SSC-2	121	0.0578	3.69	8.420E+06
	O-T	117	0.0480	2.97	5.097E+07
	Q-T	95	0.0295	1.62	5.097E+07
	H-T	89	0.0173	0.80	5.097E+07
Weld R	SSC-1	226	0.0370	2.52	3.842E+06
	SSC-2	297	0.0782	5.31	8.420E+06
	O-T	290	0.0585	3.85	5.097E+07
	Q-T	261	0.0370	2.19	5.097E+07
	H-T	240	0.0220	1.10	5.097E+07

[a] Consensus evaluation (CE) values from Ref *14*.

The same set of data was fit to an empirical function of the form

$$\Delta \text{NDTT} = A(X - X_0)^B \tag{14}$$

where X is an exposure function (for example, dpa, fppa, and so forth) and A, B, and X_0 are fitted parameters. The best fit was obtained with the dpa and fppa functions. The remaining damage functions, including $\phi t > 1$ MeV, showed only slightly poorer fits to the data. However, the fppa

TABLE 3—*Standard deviation of data fit to Eq 13 and various combinations of spectral parameters.*

H	*G*	*F*	Standard Deviation, σ
dpa	dpa	dpa	0.00919 (9.5)[a]
fppa	fppa	fppa	0.00781 (8.8)
Vc	dpa	mv	0.0339 (18.4)
Ic	dpa	mi	0.0142 (11.9)
lobes	lobes	mv	0.0383 (19.6)

[a] Values in parentheses are percent standard deviation.

TABLE 4—*Standard deviation of data fit to Eq 14 and various spectral parameters.*

X	Standard Deviation, σ
dpa	0.0178 (13.3)[a]
fppa	0.0186 (13.6)
ic	0.0220 (14.8)
Lobes	0.0214 (14.6)
mi	0.0367 (19.2)
mv	0.0584 (24.2)
vc	0.0236 (15.4)
$\phi t > 1$	0.0224 (15.5)

[a] Values in parentheses are percent standard deviation.

and dpa functions in Table 3 show more than a factor of 1.5 improvement over the results with the empirical function shown in Table 4.

Figure 1 shows the measured and calculated ΔNDTT for the damage-rate-dependent Eq 13 using fppa, and Fig. 2 shows measured and calculated ΔNDTT for Eq 14. The absolute values of the error in the calculated ΔNDTT do not differ greatly between the two plots, but a substantial number of the calculated values in Fig. 1 are shifted onto the exact correlation line when damage rate is considered.

The obstacle threshold diameter determined from the low temperature A302B daa was 0.58 nm. (This is about one half the value deduced by Russell and Brown [*10*], who were dealing with larger concentrations of copper and incoherent fcc ε-copper precipitates. In contrast, the irradiation-produced obstacles are probably copper/vacancy clusters on the order of 0.6 to 2.0 nm in diameter.) With the constant C_4 determined from the low temperature data used in the analysis of high-

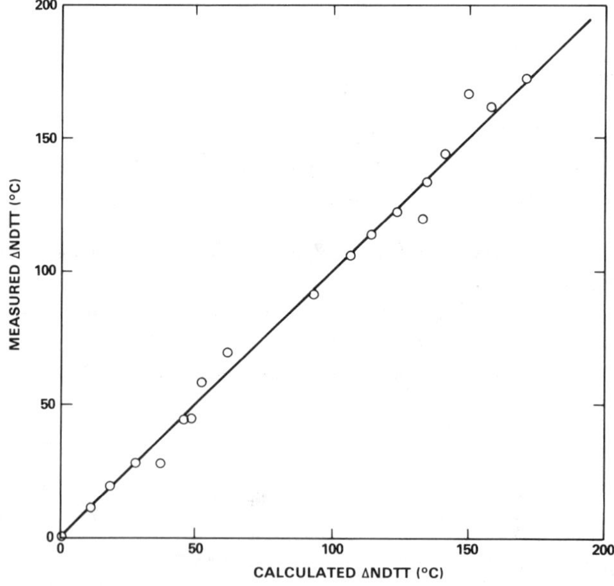

FIG. 1—*Comparison of measured and calculated ΔNDTT data using Eq 13 and fppa.*

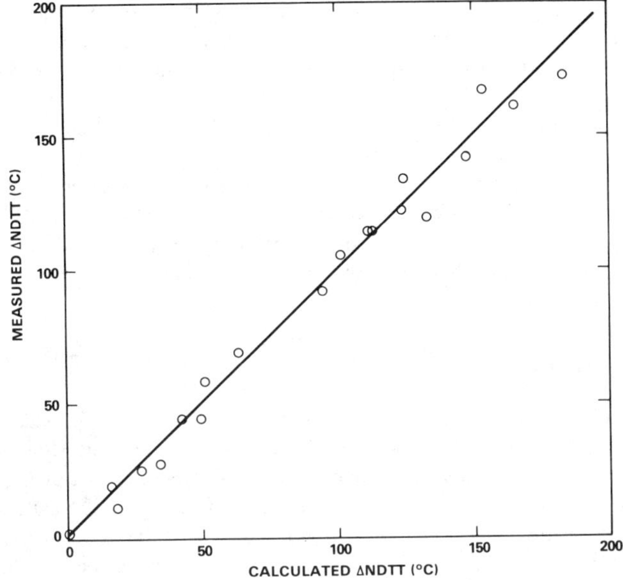

FIG. 2—*Comparison of measured and calculated ΔNDTT data using Eq 14 and fppa.*

temperature PSF data, three constants remain. Lack of data made it necessary to assume that the damage mechanism was the same at both temperatures. Since the PSF-SSC-SPV experiment included only five damage exposures, this leaves at most two degrees of freedom with which to choose n and m for Eq 13.

The standard deviation for each equation fit to the high-temperature data is tabulated in Table 5. Two empirical power law functions were used in the analysis in addition to Eq 13. These were

$$\Delta \text{NDTT} = AX^{B} \tag{15}$$

and

$$\Delta \text{NDTT} = AX^{B + C\ln X} \tag{16}$$

TABLE 5—*Percent standard deviation of data from PSF-SPV fit to various equations.*

	Standard Deviation, %				
			Eq 13		
Material[a]	Eq 15[b]	Eq 16	$n=m=0$	$n=m=½$	$m=0, n=1$
A302B Plate	12.5	14.2	15.0	6.2	3.8[c]
A533B Plate	4.8	4.0[c]	4.5	5.5	9.8
K Forging	13.4[c]	13.9	17.8	14.7	19.5
MO Forging	16.3	13.4[c]	19.7	18.4	26.3
EC Weld	6.8	8.2	8.6	5.4[c]	11.7
R Weld	7.3	8.1	9.8	4.9[c]	23.3

[a] Reference *17*.
[b] Equation 15 has three degrees of freedom, Eqs 13 and 16 have two degrees of freedom.
[c] Lowest percent standard deviation for the material.

where X is the damage exposure and A, B, and C are fitted parameters. Equation 16 is the form used successfully by Guthrie to correlate plate and weld data from pressure vessel surveillance data [*1*]. The remaining equations correspond to various combinations of n and m in Eq 13.

The best correlation for A302B is the same form found for the low temperature data, with a standard deviation only 30 to 50% of those achieved with the empirical equations. For the A533B plate and the two forging materials, the empirical equations give the lowest standard deviations. These standard deviations, however, are not too different from those achieved with Eq 13 with values of $n = m = 0$ for the Heavy Section Steel Technology (HSST) A533B plate; $n = m = ½$ for the K forging; and $n = m = 0$ or ½ for the MO forging. Both weld data sets showed a 25 to 50% reduction in the standard deviation by including a damage rate effect with m and n equal to one half.

Implications of the Model

A data correlation better than that achieved using the ASTM E693 dpa standard cross section (ASTM Master Mix for Light-Water Reactor Pressure Vessel Surveillance Standards [E 706] was obtained for the low-temperature ΔNDTT data on ASTM A302B reference plate steel. The Frenkel pair production cross section, which gave the improved correlation of ΔNDTT with exposure, is based on resistivity measurements at 4 K [*17*]. The assumption is that the resistivity is proportional to the total defects produced. The resulting defect production cross section shows an enhanced efficiency, relative to the dpa cross section, for retaining damage produced by low energy (<10 keV) primary knock-on atom (PKA) recoil events. The impact on damage production is that softer spectra, such as in D_2O moderated reactors, will have a higher proportion of defect survival than harder neutron spectra.

For harder spectra, such as light-water reactors or 14-MeV neutrons, the Frenkel pair damage efficiency is approximately one half that of the calculated standard dpa in bcc iron; that is, the two defect cross sections are directly proportional for higher energy PKA recoil events (>10 keV). This is illustrated in Fig. 3, which shows data on change in yield strength in A302B steel after irradiation in PSF and by 14-MeV neutrons (unpublished data from H. L. Heinisch) at an irradiation temperature of 563 K (290°C). The 14-MeV neutron irradiations were performed at an intermediate damage rate between the SCC and SPV and fall between the extrapolation of the two PSF data sets. This occurs in spite of the fact that the average dpa cross section for 14-MeV neutrons is about an order of magnitude larger than those for the PSF spectra.

In the present analysis of PSF data, it was assumed that dpa was an adequate spectrum correlation parameter and that subsequent data scatter could be explained by introducing rate effects. The adequacy of dpa is supported by the fact that the ratio dpa/fppa is constant for all PSF spectra and only fppa gave a better correlation of ΔNDTT than dpa at low irradiation temperatures.

The analysis of the PSF data did not result in a single equation that consistently gave a superior fit to the data. The A302B and weld steel gave the best fit using the damage-rate-dependent equation. However, the best fit for A302B steel was with $m = 0$ and $n = 1$, while welds showed the best fit with $m = n = ½$. If only Eq 13 is considered, the best overall fit occurred with $m = n = ½$ for all but the A302B and A533B steels. (According to the model, $m = n = ½$ implies a recombination-dominant microstructure, as in an annealed material.) No attempt has been made to determine the chemical compositional dependence of the fitted constants.

The A302B data supported the same rate function at low and high temperatures. Figure 4 shows the relative rate dependence for both temperatures at 0.03 dpa. The symbols span the damage rate range for the data used. In comparison, the range of damage rates for LWR surveillance capsules is shown by the vertical dashed lines of Fig. 4. Since the damage rate for the vessel walls is lower and the net damage implied is lower, surveillance capsule data should provide a conservative estimate of the condition of a pressure vessel wall having properties similar to the ASTM reference

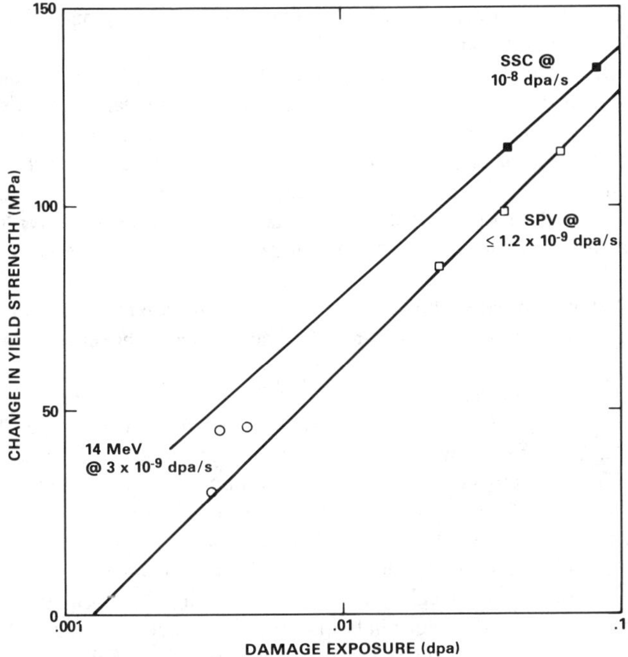

FIG. 3—*Irradiation-induced change yield strength in A302B steel from irradiation in PSF and 14-MeV neutrons.*

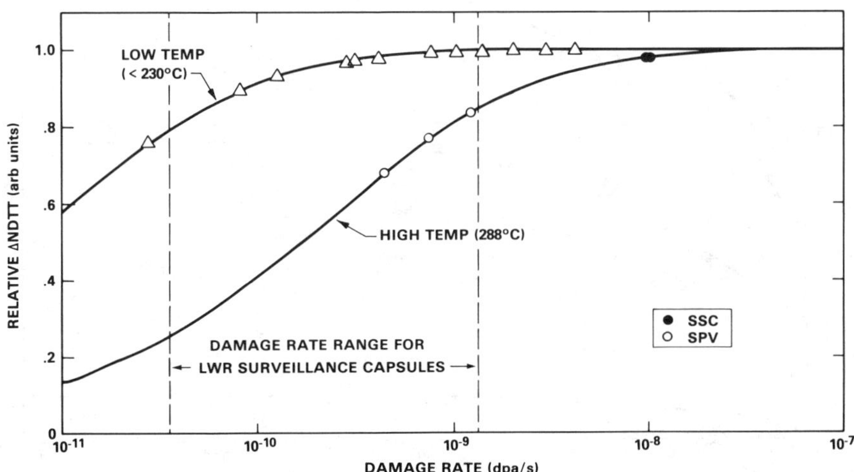

FIG. 4—*Calculated damage rate sensitivity of A302B steel for low- and high-temperature irradiations at 0.03 dpa. The symbols are calculated values for the damage rates used in this analysis.*

A302B steel plate at a neutron exposure of 0.03 dpa. However, this conclusion is based on only a narrow range of damage rates and chemical compositions.

Figure 5 shows the calculated rate dependence of the low irradiation temperature A302B data. Most of the data fall in the range 10^{-10} to 10^{-8} dpa/s. The constant exposure curves show an independence of damage rate above approximately 10^{-9} dpa/s. Figure 6 shows the same data and model plotted as a function of dpa with constant rate lines included. The data show a small rate dependence affecting ΔNDTT in the data range measured; however, extending the model to lower rates (10^{-11} dpa/s) reveals a high sensitivity. While this extension to lower rates may not be conservative, the actual high damage rate ΔNDTT measurements are conservative relative to lower rates.

Figures 7 and 8 show similar parametric curves for the high-temperature damage rate dependence of ΔNDTT in A302B steel irradiated in PSF. The rate independent level is shifted to $\sim 10^{-8}$ dpa/s.

Both weld data sets support the same damage rate dependence in the PSF irradiation. Figure 9 shows the damage rate dependence of the R and EC weld data at 0.03 dpa. The deduced rate dependences agree within 3 to 6% over the damage rate range shown. If this functional dependence is significant, the LWR surveillance data could show a minimum-to-maximum spread in ΔNDTT of 20% for the lower copper ($\leq$0.24 wt%) steels studied herein.

The most significant result is that highly accelerated testing may produce nonconservative data. That is, there can be smaller property changes at high damage rates than at low rates. This is illustrated in Fig. 10, which shows calculated ΔNDTT for the *R* weld material versus dpa. The SSC data show lower property change trends than the lower rate SPV data. Note that the high- and low-damage rates (10^{-11} and 10^{-7}) cross over at low dpa. This type of behavior could contribute to the apparent scatter in a larger data base, particularly if the cross-over point is chemistry sensitive.

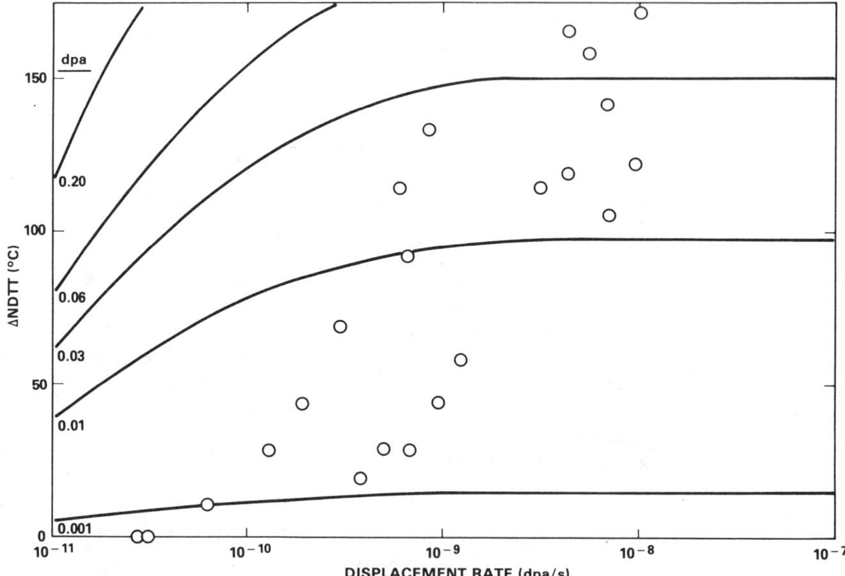

FIG. 5—*Calculated damage rate dependence of ΔNDTT in A302B steel for low-temperature irradiations. The curves are for constant dpa doses. The plotted points show the range of the data field used in the analysis.*

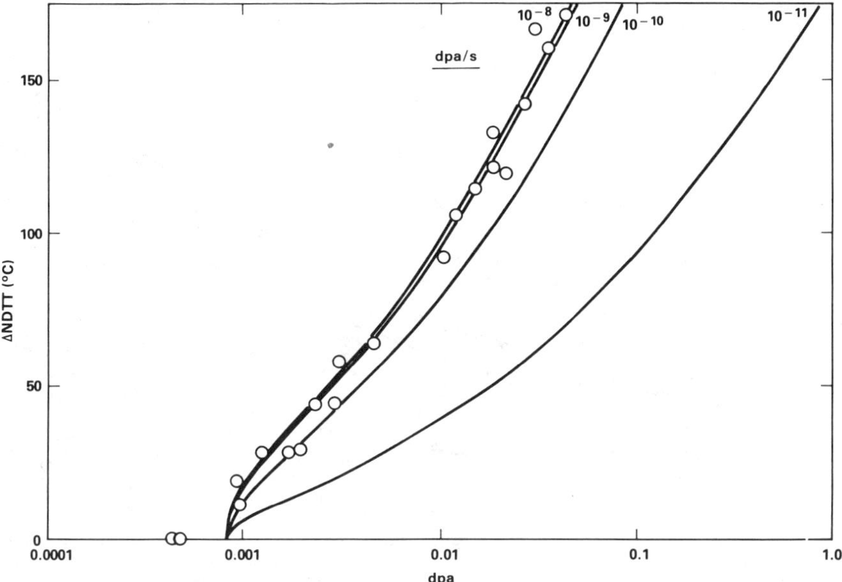

FIG. 6—*Calculated dpa dose dependence of ΔNDTT in A302B steel for low temperature irradiations. The curves are for constant dpa dose rates. The plotted points show the data field used in the analysis.*

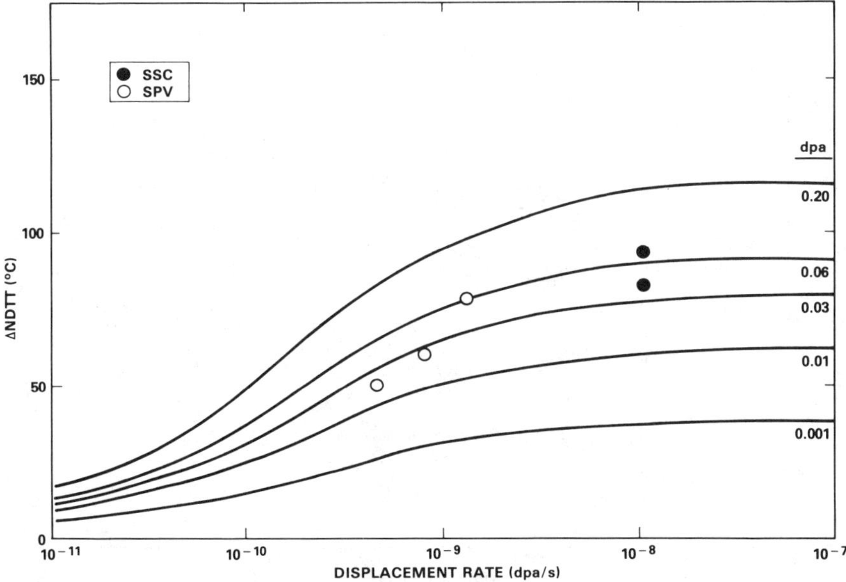

FIG. 7—*Calculated damage rate dependence of ΔNDTT in A302B steel for high-temperature irradiations. The curves are for constant dpa doses. The plotted points show the data field used in the analysis.*

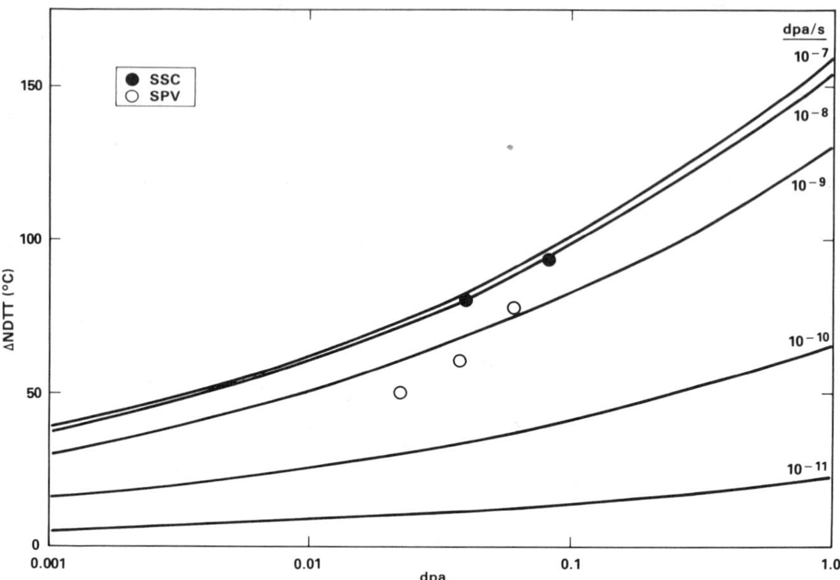

FIG. 8—*Calculated dpa dose dependence of ΔNDTT in A302B steel for high-temperature irradiations. The curves are for constant dpa dose rates. The plotted points show the data field used in the analysis.*

In Fig. 11 the ΔNDTT data are plotted as a function of damage rate with parametric curves for different dpa values. Note that the quarter thickness location and SSC-1 have approximately the same exposure but a factor of ten difference in damage rate, and they both fall on the same exposure curve.

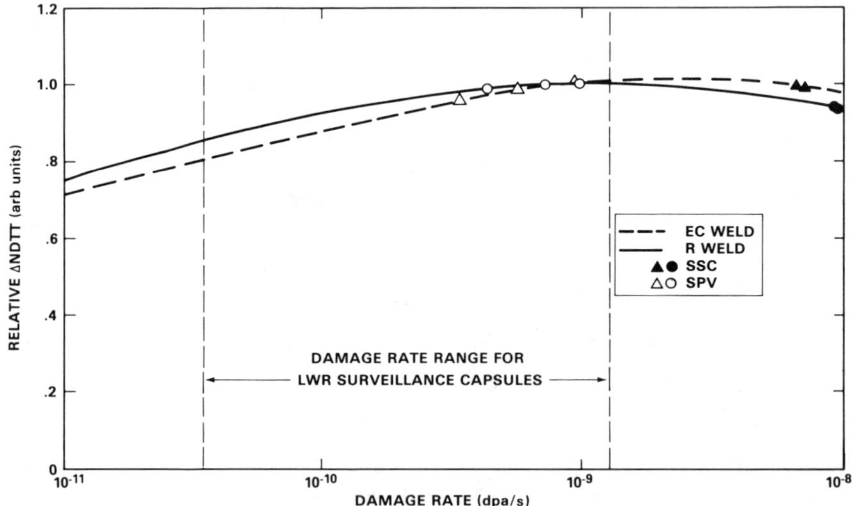

FIG. 9—*Calculated damage rate sensitivity of EC and R welds irradiated in PSF at 0.03 dpa. The symbols are calculated values for the damage rates used in this analysis.*

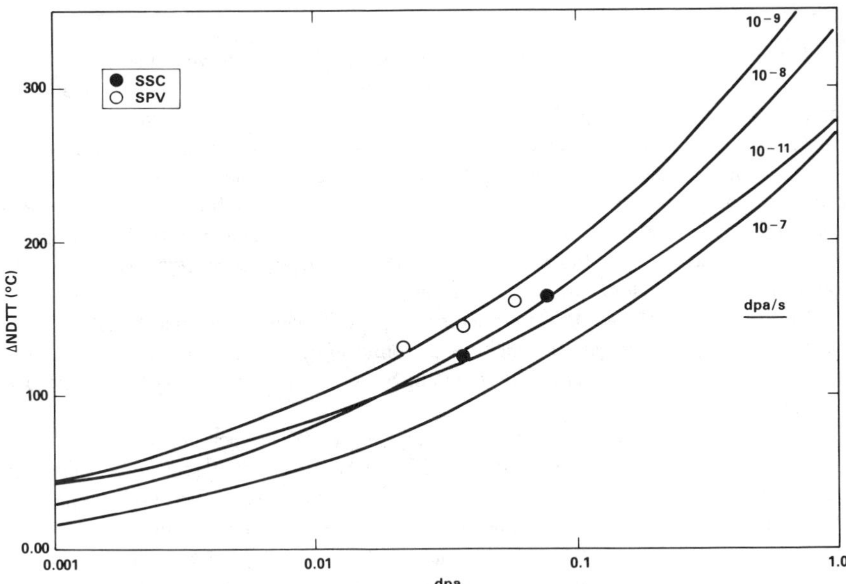

FIG. 10 *Calculated dpa dependence for R weld steel for the high-temperature irradiations. The curves are for constant dpa dose rates. The plotted points show the data field used in the analysis.*

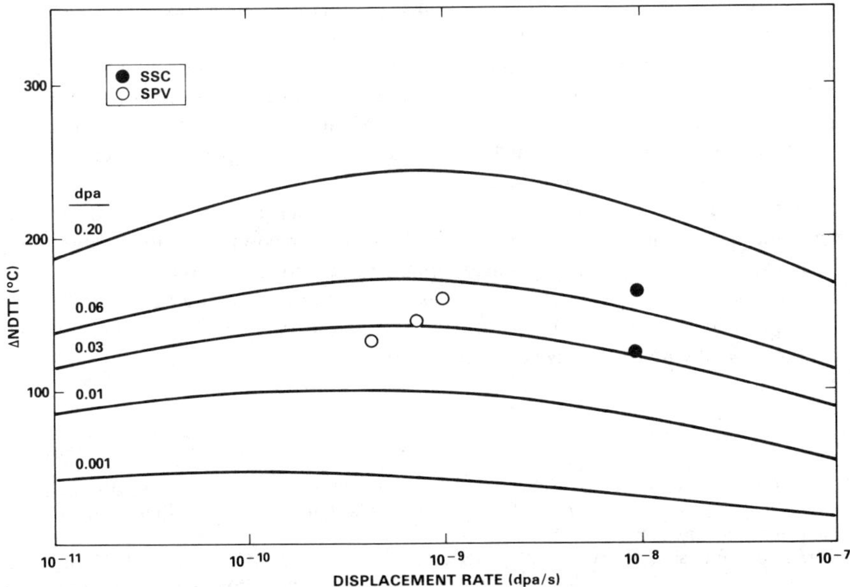

FIG. 11—*Calculated damage rate dependence of ΔNDTT in the R weld steel for high-temperature irradiations. The curves are for constant dpa doses. The plotted points show the data field used in the analysis.*

Conclusions

A correlation equation for irradiation-induced hardening in pressure vessel steels was developed to specifically address damage rate and neutron spectrum effects. The best correlation of low-temperature ΔNDTT data on ASTM A302B Reference plate steel, which had a relatively wide spectrum variation, was obtained with a defect cross section for Frenkel pair production (fppa). This cross section is characterized by an enhanced defect production by low-energy neutrons relative to high-energy neutrons. The damage rate dependence observed in the data implies that the primary effect is on nucleation of obstacle sites and is associated with thermal emission of point defects from the clusters. Introducing a damage rate dependence gives over a factor of two reduction in the variance compared with an empirical power law.

At high temperatures only the PSF data were analyzed. It was found that the A302B data were best fit with the same damage rate dependence as found in the low-temperature data. The damage rate parameters deduced for the remaining data suggests a recombination-dominant microstructure existed during irradiation.

In the case of the R and EC welds (with 0.23 and 0.24 wt% copper, respectively), the damage rate sensitivity indicates that accelerated surveillance (or low-flux test reactor) data might be expected to give a conservative end-of-life material condition. This conservatism may not hold for high damage rate test reactor spectra or at all dpa exposures. The remaining materials (A533B and the two forgings) did not support, within data scatter, a strong correlation with damage rate.

In general, the existence of damage rate effects will depend on the condition of the material. Damage rate effects may become negligible for some materials after extended irradiation induces a high density of point defect sinks.

There are three models involved in this analysis: hardening, nucleation of sites, and growth of obstacles. An inadequacy in one of the models can cause a reduction in the quality of the correlation or may be compensated for by another model.

For example, the hardening model used in this work is a modified version of Russell and Brown's model. The modification steepens the sensitivity to the obstacle size threshold; the threshold was deduced from the low-temperature data. The hardening model, if inaccurate, may be distorting the rate theory models. This needs to be addressed in further work.

Multiple hardening mechanisms have not been addressed in this work. Smidt and Sprague [*18*] observed loops, voids, and blackspots in addition to the preirradiation-induced dislocation structure after irradiating A302B and binary alloys to 0.8 dpa. From their work, it appears that significant hardening from loop formation and growth will occur only at dpa exposures higher than of current interest to LWR pressure vessel surveillance programs.

The principal difficulty encountered in this analysis is a lack of experimental data. The three-parameter damage rate equation leaves only two degrees of freedom. From that, one must select a rate dependence on the obstacle site density m and a rate dependence on growth n for Eq 13. A proper analysis must consider a wider range of spectra, damage rates, and fluence than are offered in the PSF experiment. Furthermore, the effect of chemistry (including helium) variation on neutron exposure and damage rate needs to be explored.

References

[*1*] Guthrie, G. L., "Charpy Trend Curves Based on 177 PWR Data Points," *LWR-PV-SDIP Quarterly Progress Report, April 1983–June 1983,* NUREG/CR-3391, Vol. 2, HEDL-TME 83-22, NRC, Washington, DC, pp. HEDL-3–HEDL-15, March 1984.

[*2*] Perrin, J. S., Wullaert, R. A., Odette, G. R., and Lombroso, M. P., *Physically Based Regression Correlations of Embrittlement Data from Reactor PV Surveillance Programs,* EPRI NP-3319, Electric Power Research Institute, Palo Alto, CA, Jan. 1984.

[3] Norris, E. B., "A Service Laboratory's View of the Status and Direction of Reactor Vessel Surveillance," *Radiation Embrittlement and Surveillance of Nuclear Reactor Pressure Vessels, STP 819*. L. E. Steel, Ed., American Society for Testing and Materials, Philadelphia, 1983, pp. 194–204.

[4] Lucas, G. E., Odette, G. R., Lombrozo, P. M., and Sheckherd, J. W., "Effects of Composition, Microstructures and Temperatures on Irradiation Hardening of Pressure Vessel Steels," *Effects of Radiation on Materials: 12th International Symposium, STP 870,* American Society for Testing and Materials, Philadelphia, 1985, pp. 900–930.

[5] Steele, L. E., Davies, L. M., Ingham, T., and Brumovsky, M., "Results of International Atomic Energy Agency (IAEA) Coordinated Research Programs on Irradiation Effects on Advanced Pressure Vessel Steels," *Effects of Radiation on Materials: 12th International Symposium, STP 870,* American Society for Testing and Materials, Philadelphia, pp. 863–899

[6] Serpan, Jr., C. Z., "Reliability of Fluence-Embrittlement Projections for Pressure Vessel Surveillance Analysis," *Nuclear Technology,* Vol. 12, Sept. 1971, pp. 108–118.

[7] Simons, R. L., "Re-evaluation of Dosimetry for Test Reactor Irradiation Used to Obtain ΔNDTT Data for Pressure Vessel Steels," *LWR-PV-SDIP Semi-Annual Progress Report, October–December 1982,* NUREG/CR-2805, Vol. 4, HEDL-TME 82-21, NRC, Washington, DC, July 1982, pp. HEDL-14–HEDL-22.

[8] Odette, G. R., "On the Dominant Mechanism of Irradiation Embrittlement of Reactor Pressure Vessel Steels," *Scripta Metallurgica,* Vol. 17, 1983, pp. 1183–1188.

[9] Grant, S. P., Earp, S. L., Brenner, S. S., and Burke, M. G., "Phenomenological Modeling of Radiation Embrittlement in Light Water Reactor Vessels with Atom Probe and Statistical Analysis," *Proceedings of the 2nd International Conference on Environmental Degradation of Nuclear Reactor Materials-Water Reactors,* ISBN 0-89448-124-Y, ANS, La Grange Park, IL, 1986, pp. 385–392.

[10] Russell, K. C. and Brown, L. M., "A Dispersion Strengthening Model Based on Differing Elastic Module Applied to the Iron-Copper System," *Acta Metallurgica,* Vol. 20, July 1972.

[11] Brenner, S. S., Wagner, R., and Spitznagel, J. A., "Field-Ion Microscope Detection of Ultra-Fine Defects in Neutron-Irradiated Fe-0.34% Cu Alloy," *Metallurgical Transactions,* Vol. 9A, 1978, p. 1971.

[12] McElroy, W. N., Dahl, R. E., Jr., and Serpan, C. Z., Jr., "Damage Functions and Data Correlations," *Nuclear Applications Technology,* Vol. 7, No. 6, Dec. 1969, pp. 561–571.

[13] McElroy, W. N., Ed., *LWR-PV-SDIP: PSF Experiments Summary and Blind Test Results,* Vol. 1, NUREG/CR-3320, HEDL-TME 85-4, NRC, Washington, DC, 1985.

[14] Serpan, C. Z., Jr., and Menke, B. H., *Nuclear Reactor Neutron Energy Spectra, ASTM DS 52,* American Society for Testing and Materials, Philadelphia, 1974.

[15] Heinisch, H. L. and Mann, F. M., "Neutron Cross Sections for Defect Production by High-Energy Development Cascades in Copper," *Journal of Nuclear Materials,* Vols. 122 and 123, 1984, pp. 1023–1027.

[16] Hawthorne, J. R. and Menke, B. H., "LWR-PV-SDIP: Postirradiation Notch Ductility and Tensile Strength Determinations for PSF Simulated Surveillance and Through-Wall Specimen Capsules," NUREG/CR-3295, MEA-2017, Vol. 2, NRC, Washington, DC, April 1984.

[17] Simons, R. L., "Correlation of Irradiation Effects Data Using Primary Recoil Spectra," *DAFS Quarterly Progress Report, July–September 1980,* DOE/ER-0046/3, U. S. Department of Energy, Washington, DC, Nov. 1980, pp. 41–75.

[18] Smidt, F. A. and Sprague, J. A., "Property Changes Resulting from Impurity-Defect Interactions in Iron and Pressure Vessel Steel Alloys," *Effects of Radiation on the Substructure and Mechanical Properties of Metals and Alloys, STP 529,* American Society for Testing and Materials, Philadelphia, 1973, pp. 78–91.

Raymond Gold[1] and William N. McElroy[1]

Current Limitations of Trend Curve Analysis for the Prediction of Reactor PV Embrittlement

REFERENCE: Gold, R. and McElroy, W. N., **"Current Limitations of Trend Curve Analysis for the Prediction of Reactor PV Embrittlement,"** *Influence of Radiation on Material Properties: 13th International Symposium (Part II), ASTM STP 956,* F. A. Garner, C. H. Henager, Jr., and N. Igata, Eds., American Society for Testing and Materials, Philadelphia, 1987, pp. 552–568.

ABSTRACT: In operating light water reactor (LWR) commercial power plants, neutron radiation induces embrittlement of the pressure vessel (PV) and its support structures. As a consequence, LWR-PV integrity is a primary safety consideration. LWR-PV integrity is a significant economic consideration, since the PV and its support structures are nonreplaceable power plant components and embrittlement of these components can, therefore, limit the effective operating lifetime of the plant.

To define the effects of neutron radiation damage on LWR pressure-temperature operating limits and to assess fracture toughness of power reactor PV, trend curves for the prediction of PV embrittlement have been developed. These trend curves are very general PV embrittlement curves that are used to evaluate current PV status as well as to predict the future state of the PV. In such trend curves, the two main measures of radiation damage are the adjusted reference nil-ductility temperatue $ART_{NDT}(RT_{NDT}$ initial $+ \Delta RT_{NDT})$ and the decrease in upper-shelf energy level determined from Charpy V notch impact tests. Current measures of neutron exposure most commonly used in trend curve analyses are fluence >1 MeV and displacements per atom (dpa) in iron.

Since trend curves play such a crucial role in the assessment of PV embrittlement of operating commercial LWR power plants, a critical appraisal of trend curve analysis is essential. To this end, current limitations in trend curve analysis for the prediction of reactor PV embrittlement are examined. It is concluded that a number of systematic effects can arise because environmental differences exist between test reactors, surveillance capsule locations, and the actual irradiation conditions that accrue within the PV of an operating LWR commercial power plant. An irradiation test program is advanced to investigate these systematic effects and to produce the requisite data needed to correct for such systematic biases in trend curve analysis.

KEY WORDS: radiation damage, neutron irradiation, dosimetry, pressure vessel surveillance, Charpy shift, trend curve analysis

Neutron radiation induces embrittlement of the pressure vessel (PV) and its support structures in operating light water reactor (LWR) commercial power plants. Since the PV and its support structures are nonreplaceable power plant components, embrittlement of these components can limit the effective operating lifetime of the plant. In recognition of this safety issue, the U.S. Nuclear Regulatory Commission (NRC) established the LWR-PV Surveillance Dosimetry Improvement Program (SDIP) in 1977 for improving, maintaining, and standardizing neutron dosimetry, damage correlation, and the associated reactor analysis procedures used for predicting the integrated effect of neutron exposure to LWR-PVs and their support structures. A vigorous research effort

[1] Fellow scientist and program manager, respectively, Westinghouse Hanford Co., Hanford Engineering Development Laboratory, Richland, WA 99352.

attacking the same measurement and analysis problems has gone forward worldwide, and strong cooperative links have been established between the NRC-supported activities at Hanford Engineering Development Laboratory (HEDL), Oak Ridge National Laboratory (ORNL), Materials Engineering Associates (MEA), and National Bureau of Standards (NBS) and those supported by Centre d' Etude de l' Energie Nucleaire/Studiencentrum voor Kernenergie (CEN/SCK) (Mol, Belgium), Electric Power Research Institute (EPRI) (Palo Alto, CA), Kernforschungsanlage (KFA) (Jülich, Germany), and several U.K. laboratories. The *major benefit* of this program has been and continues to be a significant improvement in the accuracy of the assessment of the current metallurgical condition and the remaining safe operating lifetime of LWR-PVs [*1*].

Neutron-induced PV embrittlement has been recognized as a serious consideration for many years, as attested to by surveillance dosimetry programs instituted over the years in U.S. LWR commercial power plants [*2*]. While considerable investigation and study have already been conducted over the years on neutron-induced embrittlement of PV steels, the details and subtleties of this problem apparently still continue to unfold. The complexity of this phenomenon cannot be overemphasized. To illustrate this complexity, many scientific disciplines are required to attack this problem. These efforts can be broadly classified into three main disciplines, namely

- Neutron metrology or dosimetry;
- Reactor physics; and
- Material science or metallurgy.

To further illustrate the profound nature of this problem, many factors have been identified as basic contributors to radiation-induced PV embrittlement. Some of these factors are summarized in Table 1. It should be stressed that each of these factors can comprise many variables. For example, Factor 1 of Table 1 concerning composition and microstructure possesses, perhaps, the most variables. Moreover, Table 1 does not purport to be an exhaustive list of contributory factors, since, for example, factors related to the actual physical or metallurgical tests of steel property changes have not been included here.

Owing to the complexity of this embrittlement phenomenon, experimental and calculational strategies have been developed in the LWR-PV-SDIP, which are in turn being made available for use by the U.S. nuclear power industry as ASTM Standards. In fact, a primary objective of the multi-laboratory LWR-PV-SDIP is to prepare an updated and improved set of dosimetry, damage correlation, and associated reactor analysis ASTM Standards for LWR-PV irradiation surveillance programs.

While a detailed review of all of these efforts would carry us too far afield, some insight into the full extent of these activities can be gained by examining the ASTM Master Matrix for these standards [*3*], which is shown in Fig. 1. Federal Regulation 10CFR50 [*4*] already calls for

TABLE 1—*Physics, dosimetry, and metallurgy factors contributing to pressure vessel embrittlement.*

1. Steel chemical composition and microstructure
2. Steel irradiation temperature
3. Power plant configurations and dimensions—core edge to surveillance to vessel wall to support structure positions
4. Core power distribution
5. Reactor operating history
6. Reactor physics computations
7. Selection of neutron exposure units
8. Dosimetry measurements
9. Neutron spectral effects
10. Neutron dose rate effects

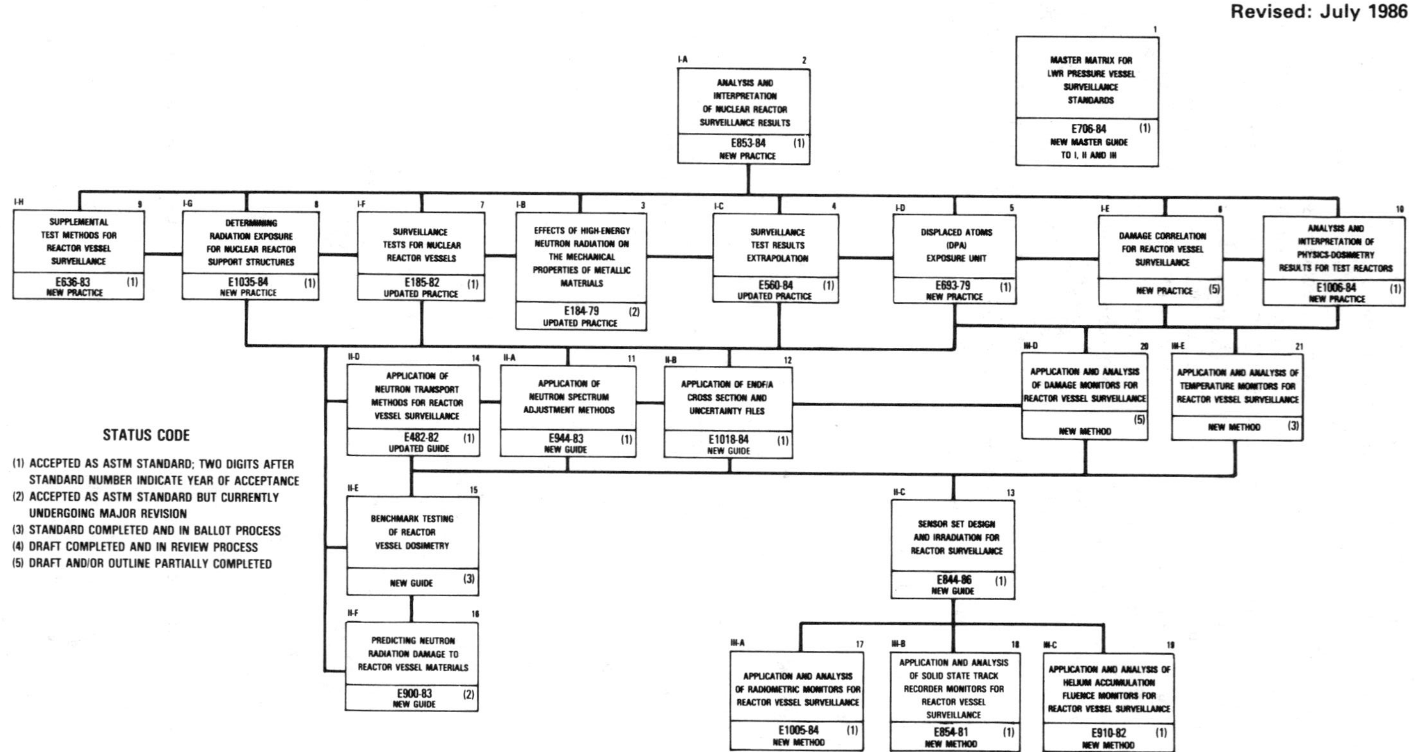

FIG. 1—*ASTM standards for surveillance of LWR nuclear reactor pressure vessels and support structures.*

adherence to several ASTM Standards in LWR-PV irradiation surveillance. Revised and new standards in preparation under this matrix will be carefully structured to be up-to-date, flexible, and, above all, consistent so that they can provide guidance to the U.S. nuclear power industry in meeting regulatory requirements.

Beyond these needs will be the consideration of what additional criteria will be required for design changes, licensing, regulation, surveillance, and research for the safe operation of plants that are operated beyond their present design life; *i.e.*, the definition of the requirements for new and expanded physics-dosimetry-metallurgy information that will be needed to support emerging and new plant life-extension programs (in the range up to, say 50 years or more). One perspective on these activities is forecasted in Fig. 2.

In order to define the effects of neutron radiation damage on LWR pressure-temperature operating limits and to assess fracture toughness of power reactor PV, trend curves for the prediction of PV embrittlement must be used. Appendices A, G, and H of Federal Regulation 10CFR50 [*4*] and the *U.S. Nuclear Regulatory Commission Regulatory Guide 1.99,* [*5*], which provide the appropriate procedures to be followed, necessitate plant-specific assessment and projection to end-of-life (EOL) of radiation-induced PV embrittlement. In the absence of verified plant-specific trend curves, very general PV embrittlement curves have been developed and used to make the required projections. In such trend curves, the two main measures of radiation damage are the adjusted reference nil-ductility temperature ART_{NDT}(RT_{NDT} initial + ΔRT_{NDT}) and the decrease in upper-shelf energy level determined from Chary V notch impact tests. Current measures of neutron exposure most commonly used in trend curve analyses are fluence >1 MeV and displacements per atom (dpa). The applicability and conservatism of general trend curve predictions are checked and verified by plant-specific surveillance program data during the operating service life of a given pressure vessel.

The importance of determining and specifying the accuracy of these predictions and projections has increased significantly as a result of new NRC regulations regarding required protection against pressurized thermal shock (PTS) events in pressurized water reactors (PWRs) [*6*]. The screening criterion proposed by NRC is a "reference temperature" of 270°F for plate materials and axial welds and 300°F for circumferential welds. Below these temperatures, the risk from PTS events

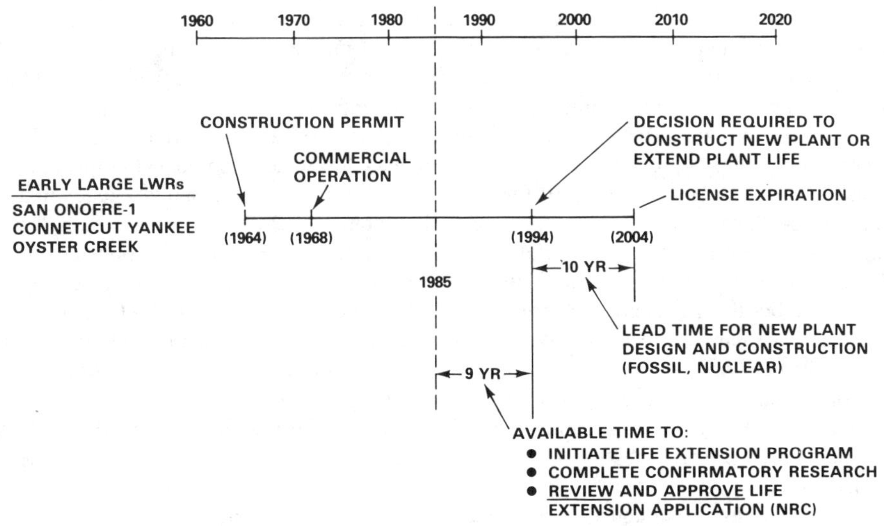

FIG. 2—*A scenario for the early initiation of LWR extension research.*

would be considered acceptable. The risk above that level also might prove to be acceptable, but a demonstration would require plant-specific evaluations and, possibly, modifications to existing equipment, systems, and procedures.

From this discussion, it is apparent that trend curves play a central role in the assessment of PV embrittlement of operating LWR power plants. Consequently, it is imperative that the limitations of trend curve analyses be clearly delineated. To this end, limitations in the development as well as the application of trend curves are considered in the next section. In this light, the current status of trend curve development is then examined, especially from the viewpoint of any deficiencies that may exist for predictions in actual LWR operating power plants. An initial attempt to develop an irradiation test matrix that overcomes some identified deficiencies is described in the final section.

Limitations of Current Trend Curve Analyses

Mathematical Formulation

Difficulties that arise in the generation of trend curves surfaced at a special session on PTS and reactor materials which was held at the 1984 annual meeting of the American Nuclear Society (ANS) [*7*]. One team of experts reported that a definitive correlation existed between copper concentration and ΔRT_{NDT}. In support of their contention, they introduced a physical model in which copper precipitates acted to stabilize damage sites. Still another team of researchers found no statistically significant evidence to support any correlation between copper content and ΔRT_{NDT} in a large weld group under study. Further discussions centered on the effects of nickel, with some groups reporting a correlation of nickel content with ΔRT_{NDT} and other groups finding no basis for such a correlation. Still other groups maintained the existence of a cross correlation between copper content, nickel content, and ΔRT_{NDT}.

These differences of view imply the existence of systematic effects that either are not recognized or are not fully appreciated. The origin of such difficulties can range from the trivial to the profound. For example, it could be as simple as one team working with base metal as opposed to another team working with weldments. Or it could be more subtle; both teams might use the same material, but the history of the material used by each team could be different (*e.g.*, one team might have used more annealed material than the other team). Even more subtle systematic effects may be responsible, such as a flux-level effect or a saturation phenomenon; these possibilities are examined in greater detail in separate presentations at the present symposium [*8,9*].

Although macroscopic effects are emphasized here, one should not overlook the subtlety that can exist at the microscopic level. Recent work conducted to understand annealing phenomena in solid-state track recorders (SSTR) [*10*] furnishes a good example of the subtlety that can arise at the microscopic level. Through these very general theoretical considerations, it became apparent that tracks produced in dielectric materials by charged particles depend not only on annealing after track formation, but the very temperature that exists in the medium during the actual track formation process. At about the same time, experimental results were first observed in SSTR that indicated the existence of such an effect [*11*]. While these first efforts were not clearly understood, further work has now confirmed the dependence of track characteristics on the temperature of the medium during the track formation process [*12*]. This phenomenon is now called the "registration temperature effect." As a consequence of these SSTR efforts, one can conclude that this microscopic effect actually exists in all materials undergoing radiation damage, not just dielectric materials.

Hence, from a more profound viewpoint that recognizes the full complexity of radiation damage function analysis [*13*], the differences of view that arose at the special ANS session are not surprising. In view of this complexity, one must accept at the outset that trend curve analyses are only approximate descriptions of very complex processes. As a consequence, considerable care must be exercised in the development and application of trend curves.

A significant example of this viewpoint lies in an issue that has just started to emerge in trend curve analysis. It concerns the assumption of separability between the chemistry and the exposure dose dependence of ΔRT_{NDT}. Indeed, in the generation of trend curves, it has almost universally been assumed that

$$\Delta RT_{NDT} = F_1(C) \cdot F_2(D) \tag{1}$$

where F_1 is a function of the important chemistry variables (C), and F_2 is a function of the neutron exposure dose variables, D. While this assumption has been adopted, no doubt, because of the convenience and simplicity it introduces in least squares statistical analyses, to our knowledge separability of these two classes of variables has never been rigorously proved. On the contrary, many instances have arisen that indicate that this assumption may not be valid. Recent analyses of the poolside facility (PSF) experiment also tend to illustrate this point [*14,15*].

Further insight into the physical plausibility of this assumption can be gained from a heuristic extension of Odette's treatment of microvoid density [*16*]. In this treatment, the microvoid density N_{mv} is give by

$$\dot{N}_{mv} = G_{mv} - N_{mv}/\tau_{mv} \tag{2}$$

where the production term of microvoids G_{mv} is given by

$$G_{mv} = \phi\sigma_{mv} \tag{3}$$

with ϕ the scalar neutron flux and σ_{mv} the microvoid productio cross section. The term N_{mv}/τ_{mv} in Eq 2 represents the thermal annealing rate, where τ_{mv} is the microvoid thermal annealing time.

This equation does not account for the possibility that microvoids could be stabilized by chemical variables such as copper, nickel, or helium content or some combination of these in such a way as to prevent or deter annealing. Such a speculation can be investigated by introducing a stabilization term into Eq 2 of the form $+N_{mv}/\tau_s$, where τ_s is the stabilization time. Consequently, a more general description of the microvoid density could be written as

$$\dot{N}_{mv} = G_{mv} - N_{mv}/\tau_{mv} + N_{mv}/\tau_s \tag{4a}$$

or

$$N_{mv} = G_{mv} - \left(\frac{1}{\tau_{mv}} - \frac{1}{\tau_s}\right) N_{mv} \tag{4b}$$

Here the stabilization time τ_s would obviously depend on the chemical composition and microstructure of the given steel, so that τ_s must generally be assumed to depend on all chemical variables.

Equation 4b can be written in the form

$$\dot{N}_{mv} = G_{mv} - (1 - \alpha)\, N_{mv}/\tau_{mv} \tag{5a}$$

where

$$\alpha = \tau_{mv}/\tau_s \tag{5b}$$

Implicit in this description is that $\tau_{mv} \leq \tau_s$; otherwise, another net production term would be added in Eq 4. Consequently, the parameter α satisfies the condition

$$0 \leq \alpha \leq 1 \tag{6}$$

The solution of Eq 5a is given by

$$N_{mv} = \frac{\phi\sigma_{mv}\tau_{mv}}{(1 - \alpha)}\left\{1 - \exp\left[-(1 - \alpha)\,\phi t/\phi\tau_{mv}\right]\right\} \tag{7}$$

Equation 7 provides some very simple physical implications. Since the parameter $\alpha = \tau_{mv}/\tau_s$ generally depends on chemistry variables, this time-dependent representation of the microvoid density obviously does not satisfy any separability criterion. From Eq 7, one finds a saturation value of the microvoid density, N^s_{mv}, which is given by

$$\lim_{t \to \infty} N_{mv} = N^s_{mv} = \frac{\phi\sigma_{mv}\tau_{mv}}{(1 - \alpha)} \tag{8}$$

Here the saturated microvoid density, N^s_{mv}, depends not only on flux (ϕ), but chemistry variables as well. In fact, since one would expect τ_s to decrease, or α to increase, with increasing content of trace constituents such as copper, nickel, or helium, then the saturation value would also increase with increasing content of these trace constituents. The attainment of saturation occurs at a fluence value that also depends on chemical variables through both α and τ_{mv}. In so far as the α-dependence is concerned, increasing trace constituents would shift the onset of saturation to higher fluences.

Equation 7 also implies the existence of a flux-level effect. This effect can be illustrated in terms of the neutron exposure dose, D, which can be defined as

$$D = \int_0^\infty \int_0^t \phi(E_n, t')\,dE_n dt' \tag{9}$$

where the neutron flux depends generally on both neutron energy, E_n, and time, t'. Here t is the time duration of the irradiation. For steady state irradiations of duration t, Eq 9 reduces to

$$D = \phi \cdot t \tag{10}$$

Consequently for steady-state irradiations, Eq 7 becomes

$$N_{mv} = \frac{\phi\sigma_{mv} \cdot \tau_{mv}}{(1 - \alpha)}\left\{1 - \exp\left[-(1 - \alpha)D/\phi\tau_{mv}\right]\right\} \tag{11}$$

Thus, even for the conceptually simple case of steady state irradiations, as described by Eq 11, one finds that N_{mv} depends on both D and ϕ. Moreover, since

$$\partial N_{mv}/\partial\phi = (N_{mv} - \sigma_{mv} \cdot D \cdot \exp\left[-(1 - \alpha)D/\phi\tau_{mv}\right])/\phi \tag{12}$$

one finds that for the same irradiation dose, D, the microvoid density can increase with increasing ϕ, decrease with increasing ϕ, or remain constant with increasing ϕ. Hence this heuristic formulation not only implies the non-separability of chemical and environmental variables, but also suggests the existence of a dose rate effect.

On the basis of even this oversimplified description, it is not surprising to learn that flux-level effects have been inferred with a power reactor data base [*17*], as well as in a specific test reactor experiment, namely in the PSF metallurgical test [*18*]. While a number of materials were irradiated in the PSF experiment, the most readily observable flux-level effects were discerned for the ASTM A302B Reference plate and the Code R A533B Weld Material [*8,9*]. One of the first indications of a flux-level effect observed at HEDL were with the British Code R Reference weld material [*19*], a highly radiation-sensitive standard material that provided Charpy shift measurements of a few percent accuracy. On the other hand, measurements attained with the other four materials were of considerably less accuracy and a flux-level effect was, therefore, difficult to resolve for these materials. This experience underscores the need for higher quality data bases in trend curve analyses.

Variable Effects, Extrapolation, and Lead Factors

The existence of a flux-level effect has important implications for the U.S. commercial nuclear power industry, since accelerated locations have almost invariably been used in PV surveillance programs. These accelerated PV surveillance capsules have provided lead factors that have been applied to obtain projections of PV embrittlement. In fact, accelerated PV capsules comprise the largest existing data base for trend curve analyses. Consequently, it is clear that a flux-level effect would imply that some correction would be necessary in the application and interpretation of lead factors. Otherwise, the application of lead factors could not always ensure a conservative extrapolation. At the same time, it is apparent that any reduction in embrittlement afforded from low leakage cores which are now being adopted in some U.S. power plants, must be quantified in terms of a flux-level effect, lest the predicted gain be underestimated or overestimated.

The flux-level effect discussed here illustrates a general limitation of trend curve analysis that arises through the inadequacy of the data base. Data bases used for trend curve analyses have various origins. Surveillance capsule measurements comprise the largest available data pool and, therefore, have been used most extensively. However, none of these data bases represents the specific conditions of radiation exposure that exist within an actual pressure vessel. As a consequence, trend curves developed by least-squares analyses of these data bases can deviate systematically from the radiation damage that actually accrues in a pressure vessel. This systematic deviation stems from the lack of the data base to truly represent the irradition conditions that actually arise in the pressure vessel of operating power plants.

The flux-level effect discussed above is just one of a number of systematic effects that can arise because of inadequacy in the data base. Indeed, the neutron spectral dependence of PV embrittlement has been recognized for some time [*3*]. In recognition of this fact, current trend curve analyses employ, for the neutron exposure dose, either the fast neutron fluence (usually above 0.1 MeV or 1.0 MeV) or dpa [*20*]. For low-temperature (<230°C) irradiation of the ASTM A302B Reference plate, Simons [*9*] has shown that Frenkel pairs per atom (fppa) may be a better spectrum damage index than dpa for the existing ASTM A302B research reactor-derived physics-dosimetry-metallurgy data base. At higher temperature (~288°C), however, dpa and fppa appear to be equally good indices. Further, recent analyses reveal that even a correlation with thermal neutron intensity may exist [*15*].

This recent conclusion regarding a thermal neutron effect is not a unique interpretation of the data. Indeed, a collection of systematic effects caused by flux level, helium production, and gamma-ray heating cannot be ruled out. The intensity of the gamma-ray field found in PV environments is highly correlated with thermal neutron intensity. Consequently, the thermal neutron effect recently reported [*15*] may actually arise from a combination of effects, including annealing from gamma-ray energy deposition. In this event, one must recognize that gamma-ray-induced energy deposition at surveillance capsule locations is considerably higher than that which is attained within a pressure vessel. Therefore, the annealing rate from gamma-ray-induced energy deposition at the surveillance capsule location could be considerably higher than the annealing rate from gamma-ray-induced energy deposition within the pressure vessel. Hence, gamma-ray-induced energy deposition could be another factor responsible for introducing a systematic bias in trend curve analyses that use surveillance capsule data bases.

While the systematic effects derived from this model are nonconservative, it must be stressed that other systematic effects can and do exist. Hence, one should not conclude that all systematic effects need be nonconservative. It would be naive to reach such a conclusion based solely on an analysis of the heuristic model considered here. In particular, in separate presentations at this symposium [*8,9*], it is shown that the flux-level effect can range from conservative to nonconservative depending on the material under consideration. In fact, the more detailed descriptions developed by McElroy et al [*8*] and Simons [*9*] allows a microvoid density that can be (1) lower, (2) higher, or (3) even show no change at higher flux levels, depending on the material properties of the steel under consideration.

From these considerations, it is clear that the present day understanding of the phenomenological processes underlying radiation-induced embrittlement of pressure vessels must be improved. It is also equally clear that use of this improved knowledge in trend curve analyses would be pointless unless differences that exist in environmental conditions between the pressure vessel and the data base are explicitly taken into account. Incorporation of such improvement should provide, in principle, a more rigorous basis for trend curve analyses. Using such advanced trend curve analyses together with plant-specific data, bounds for pressure vessel neutron exposure can be realistically set that provide a proper margin of safety without excessive conservatism, which would otherwise penalize the U.S. commercial nuclear power industry.

Current Status of Trend Curve Analysis

As a part of the LWR-PV-SDIP [*21,22*], statistically based data correlation studies have been made by HEDL and other program participants using existing PWR and boiling water reactor (BWR) physics-dosimetry-metallurgical data in anticipation of the analysis of new fracture toughness and embrittlement data from the Bulk-Shielding Reactor–Heavy Section Steel Technology (BSR-HSST), State University of New York-Nuclear Science and Technology Facility (SUNY-NSTF), Oak Ridge (Research) Reactor Poolside Facility (ORR-PSF), and other experiments.

At the last ASTM-EURATOM international symposium on reactor dosimetry, Simons presented results of evaluation and reevaluation of exposure units and values for 47 PWR and BWR surveillance capsule reports for Westinghouse (W), Babcock and Wilcox (B&W), Combustion Engineering (CE), and General Electric Power plants [*23*]. Using a consistent set of auxiliary data and dosimetry-adjusted reactor physics results, the revised fluence values for $E > 1$ MeV averaged 25% higher than the originally reported values. The fluence values (new/old) ranged from a low of 0.80 to a high of 2.38. These HEDL-derived FERRET-SAND II exposure parameter values have subsequently been used by HEDL in PWR and BWR trend curve analyses.

Randall has discussed the basis for Revision 2 of the NRC's *Regulatory Guide 1.99* [*24*]. As stated, the *Regulatory Guide* is being updated to reflect recent studies of the physical basis for neutron radiation damage and efforts to correlate damage to chemical composition and fluence. Revision 2 contains several significant changes. Welds and base metal are treated separately. Nickel content is added as a variable and phosphorus is removed. The exponent in the fluence factor is reduced, especially at high fluences; and guidance is given for calculating attenuation of damage through the vessel wall.

The effects of changes in different variables and use of different exposure parameter models for predicting the Charpy shift for the 30-point-PSF-weld, plate, and forging data base and a 30-point-PWR-weld data base have been discussed in considerable detail [*15,25*]. The main comments and conclusions of Guthrie's study [*25*], which is based on the use of PSF and test reactor data, are as follows:

1. In surveying the previously existing data available for the alloys in the PSF experiment, it has become apparent that the fluence exponent is dependent on irradiation temperature and flux level. For the A302B alloy, the PWR surveillance data fell consistently below the higher flux level low-intensity test reactor (LITR) data and showed a lower value for the fluence exponent. The overall scatter of the existing data is such that it is not clear that Charpy tests or *K* tests can be used to uncover fine details in radiation damage mechanisms.

2. Because of the possible rate effect (which was predicted by G. R. Odette in his PSF blind test submission), the PWR surveillance trend-curve laws cannot be expected to work as well in the PSF as might be expected from their stated standard deviations.

3. In applying existing Charpy shift laws to the PSF C_v data, we find that the largest observed shift occurred for the Rolls Royce A533B weld (Code R), which had a high nickel content (1.58%), which is well outside the range of the data base used to develop the HEDL PWR Charpy shift equations [*26*].

4. There appears to be a rate effect in the PSF Charpy and compression data. The fluence exponent appears to increase with increased flux and appears to decrease with increased copper.

5. The similarity of the spectra at the separate irradiation positions severely limits the possible comments about damage functions.

6. No extra thermal neutron effect, beyond that already represented in the ASTM dpa cross section, was identifiable in the PSF data.

The main comments and conclusions of the study by McElroy et al [*15*] (which is based on the use of PSF, PWR, and BWR data) are:

1. There is a significant improvement (reduction) in the standard deviation of the fit for weld Charpy shift trend curves that include the effect of low-energy thermal neutrons. For the 30-point-weld data set, the degree of improvement observed could occur at a frequency of no more than approximately 4% by chance.

2. A knowledge of the actual boron content of PV steels and the use of a trend curve that makes use of an exposure parameter dose term, which includes the total production of dpa and helium in iron, could make significant improvements in lowering the standard deviation of the fit for the existing PWR surveillance-capsule, metallurgical weld data base.

3. Based on a trend-curve model that includes the possible effect of thermal neutrons for both PWR and BWR power plants, up to about 80% of the SS clad/PV steel wall interface and surveillance-capsule specimen dose term values could be attributed to helium production in PV steels, depending on the particular surveillance capsule design, Charpy specimen placement, steel boron content, and power plant operating conditions.

4. Existing PWR and BWR surveillance-capsule-derived embrittlement trend curves (based on the use of only fast fluence ($E > 1$ MeV) or dpa for the exposure term) cannot be expected to give reliable predictions of the combined fast and thermal neutron contributions to the Charpy shift at the SS clad/PV steel wall interface, ¼-*T*, ½-*T*, ¾-*T*, or 1-*T* locations. (It is noted that the PSF experiment provides physics-dosimetry-metallurgy data for predicting the Charpy shift in PV steels at deep in-wall locations, such as the ¼-*T*, ½-*T*, and ¾-*T* positions, where the thermal-to-fast neutron fluence (*T*/*F*) ratios are in the very low range of ~0.14 to ~0.53. However, even for these very low ratios, helium from both boron and steel high energy (n,α) reactions may still contribute 5% to 30% to the exposure parameter dose term value.)

5. None of the Charpy shift trend-curve equations studied, Table 1 of Ref *15*, except perhaps the one based on the use of an exposure parameter of fluence $E > 0.1$ MeV, appear to properly bound all the six PV-steel-observed PSF damage gradient curves. Based on the French simulated PV-wall DOMPAC Experiment [*22,27*], Alberman concluded that for low-temperature (<100°C) irradiations, fast fluence ($E > 1$ MeV) is too "optimistic" and is not, therefore, a conservative neutron exposure parameter and that, at low temperature, 95% of the measured damage (based on tungsten and graphite DM results) comes from neutrons with energy $E > 0.1$ MeV. This led him to conclude that the exposure parameter, fluence ($E > 0.1$ MeV), is perhaps "pessimistic," but has the advantage of being the lower threshold of all (displacement) damage models and thus it takes into account all neutrons that create (displacement) damage.

6. The plant specific weld data sets used in the PWR and BWR data base studies, except for one, do not support a saturation effect at high fluences above $\sim 1 \times 10^{19}$ n/cm^2 ($E > 1$ MeV). Consequently, the existing *Regulatory Guide 1.99* [*5*] upper-bound (truncated) trend-curve model

shape (or plant-specific curves) may have to be used for high fluence embrittlement predictions for PV steel welds, and perhaps forging and plates.

7. Any significant thermal neutron contribution to PV steel embrittlement is, most probably, a result of (n,α) reactions in boron-10 rather than by neutron-induced Fe(n,γ) recoil reactions.

8. It appears that the current ASTM E693 [*20*] dpa cross section should not be used to correlate highly thermalized light-water-moderated or heavy-water-moderated power or test reactor irradiation effects data because it significantly overestimates the low-energy thermal neutron dpa contribution.

9. The PV-wall SS clad/PV steel interface surface thermal-to-fast neutron fluence (T/F) ratio for PWR and BWR power plants is expected to be in the range of 2 to 6 on the basis of surveillance capsule measurements, Westinghouse transport calculations, General Electric measurements, and PSF experiment physics-dosimetry results.

10. Individual Charpy specimens (with natural boron content ranging from ~0.4 up to perhaps 5 wt ppm) in PWR and BWR surveillance capsules will be subject to ℄ neutron exposures with T/F ratios in the range of ~0.5 to 5, depending on the surveillance capsule design, its placement, and the reactor operating conditions. The T/F variation for individual Charpy specimens, therefore, could be an important parameter for the correlation of a set of Charpy specimen results and derived ΔRT_{NDT} values.

11. From this study, that of Grant and Earp [*28*], and others as discussed at the ASTM-EURATOM reactor dosimetry symposium in Geesthacht [*15*], a final conclusion is the following: the PSF experiment and PWR and BWR surveillance program results clearly show that comparison of the effects of radiation damage on yield strength, hardness, RT_{NDT} and upper shelf energy (USE) will be needed to aid in improving and refining our knowledge of trend curves and PV wall damage gradients. Implicit in this work is the establishment of separate trend curves for welds, forgings, and plates which will give an increased understanding and accuracy in projections of the present and future metallurgical condition of PV steels.

Test Matrix Formulation

While it is our intent to develop a preliminary test matrix that addresses neutron-induced embrittlement of LWR-PV, an often overlooked aspect of such efforts is the quality of the measurements. Charpy data are often beset with large fluctuations of statistical or otherwise unknown origin that undermine not only the data base, but any analyses based thereon. Although development of a new set of Charpy data would certainly add to the data base, the quality of such data is deemed more important at this time. Consequently, our recommended first priority is for high-quality data. Next in priority would be the type and quantity of measurements. Our priorities are based on the view that the underlying phenomenological processes are more readily resolved and better understood in terms of the quality of the data base rather than the size of that data base. The aforementioned references to the observation of a flux level effect for the PSF experiment provide rather convincing support for this viewpoint. Indeed, we cannot overemphasize the need for high-quality data at the present time. Considering this aspect of the problem, it is essential that high-quality Charpy, tensile, hardness, transmission electron microscopy (TEM), small-angle neutron scattering (SANS), and field ion microscopy (FIM) experimental results be obtained and reported, as well as those related to the physics-dosimetry measurements and data analysis.

In view of the many damage-effect variables that exist in neutron-induced embrittlement of LWR-PV steels, selection of the most relevant variables is an extremely difficult process. Nevertheless, such a selection process is mandatory. In fact, since the range of the selected variables actually define the domain of the test matrix, it is clear that the size of the test matrix will grow rapidly as the number of selected damage-effect variables is increased. Because of the expensive nature of irradiation tests of this type, one must clearly limit the number of variables to keep overall funding requirements at realistic levels.

In order to start the selection process, Table 2 displays our choice of the most relevant damage-effect variables. Here we have partitioned variables into three main classes, namely material properties, environmental irradiation conditions, and material effects. Even if considerations are restricted to those variables cited in Table 2, a test matrix comprising all these variables would still be too large to implement.

In order to stay within budgetary constraints and still generate data that bear upon the PV embrittlement process, one can restrict consideration to submatrices of the larger overall test matrix. In this event, those variables that are not treated within a given submatrix must be held constant. Clearly, for those variables which are held constant, values must be prescribed that are representative of the range of values that actually exist in the pressure vessels of operating LWR power plants. Otherwise, the data generated would not be applicable for trend curve analyses of operating LWR power plants.

From this point of view, U.S. irradiation test programs already exist [*22,28–33*] that address distinct submatrices of this overall test matrix. The submatrices addressed in the ongoing irradiation programs deal chiefly with material effects and material properties, *i.e.,* Columns 1 and 3 of Table 2. These already existing efforts focus primarily on the following phenomena: compositional effects; dependence on impurities or alloying elements or both; annealing recovery; irradiation-anneal-reirradiation characteristics; and dose rate (*i.e.,* flux-level) effects.

As a consequence of these already existing efforts, we have chosen to restrict our considerations here to a submatrix example involving environmental irradiation variables only (*i.e.,* Column 2 of Table 2). Rather than focusing on material properties and effects, this submatrix will concentrate on the investigations of systematic biases that can arise because of differences in environmental conditions between current data bases and the actual conditions that exist in LWR pressure vessels. For this submatrix to be of realistic proportions, one can consider no more than five or so environmental variables. Hence, we have limited our considerations to the five environmental irradiation variables shown in Table 3.

In general, the purpose of such an environmental irradiation submatrix is to define the overall dependence of the Charpy shift on all relevant environmental variables. From this submatrix viewpoint, the functional form of the Charpy shift trend curve can be written as

$$\Delta RT_{\mathrm{NDT}} = F(x_i, \ldots x_m; a_i, \ldots a_m), \tag{13}$$

where $\{x_i\}$ are the relevant environmental variables and $\{a_i\}$ are a set of parameters. Here the set

TABLE 2—*Selected damage effect variables.*

Material Properties	Environmental Irradiation Conditions	Material Effects
Type of steel	fast neutron fluence	mechanical treatment heat treatment
Impurities: Cu, Ni, Mn, Mo, Cr, P, Si, S, C, B, N	neutron flux level thermal[a] neutron fluence	annealing pre-irradiation
Microstructure	thermal-to-fast neutron ratio Gamma-ray fluence Gamma-ray flux level Gamma-ray heating temperature (including any gradients) irradiation time	during-irradiation postirradiation (time and temperature dependence) recovery

[a] Implicit here is the contribution from epithermal as well as thermal neutrons.

TABLE 3—*Environmental irradiation submatrix.*

Material Properties[a]	Environmental Irradiation Variables[a]
Weld (2)	flux level (3)
Plate (1)	dose term (3)
Forging[b] (1)	thermal-to-fast neutron fluence (3)
	temperature (3)
	gamma heating[c] (3)

[a] The number in parentheses following the variable is the recommended number of different values to be used for this given variable in the test submatrix.

[b] A possible future option, which is not included in the present submatrix.

[c] This variable is identified here because of the possibility of systematic effects associated with gamma-ray induced temperature gradients.

of parameters $\{a_i\}$ represents all remaining variables that are not treated within the environmental irradiation submatrix, such as those enumerated in Columns 1 and 3 of Table 2.

In order to examine the detailed dependence of ΔRT_{NDT} on environmental variables, materials of high radiation sensitivity must be chosen for the environmental irradiation test submatrix. The R material of the PSF test is an excellent example of such a material. With materials of this type, an absolute or relative accuracy, or both, of Charpy shift measurements as good as a few percent can be attained. To quantify the behavior of systematic environmental effects of the order of 10% to 20%, such an improved accuracy level is mandatory. While Table 3 indicates that only one plate and two weld materials are to be included in the environmental irradiation submatrix, it is essential that the radiation sensitivity of these materials be high enough to furnish Charpy shift and other property measurements of required accuracy levels. A forging is also identified in Table 3 as a future option, but it is not included in the present discussion.

As already noted in the general submatrix approach, the important parameters that lie outside the submatrix must be assigned constant values that are representative of LWR-PV commercial power plant irradiations. For our environmental variable submatrix, the remaining parameters have already been identified in Columns 1 and 3 of Table 2. Examples of representative values for these remaining parameters are given in Table 4. These values are based on the preliminary

TABLE 4—*Representative values of parameters for the environmental submatrix.*

Parameter	Steel[a]	Composition (% by weight)										
Material Type		Cu	Ni	Mn	Mo	Cr	P	Si	S	C	B	N
Weld	A533	0.36	1.0	1.5	0.5	0.1	0.01	0.3	0.01	0.2	[b]	[b]
Weld	A533	0.25	1.0	1.5	0.5	0.1	0.01	0.3	0.01	0.2	[b]	[b]
Plate	A533	0.25	1.0	1.5	0.5	0.1	0.01	0.3	0.01	0.2	[b]	[b]
Forging[c]	A533	0.12	1.0	1.5	0.5	0.1	0.01	0.3	0.01	0.2	[b]	[b]
Microstructure				selected to be representative of a typical PV steel								
Heat treatment				selected to be representative of a typical PV steel								
Annealing				selected to be representative of a typical PV steel								

[a] Steels have been chosen that should possess large Charpy shifts, such as those attained with the Rolls-Royce Code R weld of the PSF experiment.

[b] To be determined, but representative of actual PV steels. Also, other minor alloying constituents, such as V, Al, *etc.*, should be maintained as constant as possible and be representative of actual steels.

[c] A possible future option, which is not included in the present submatrix.

analysis of PSF experimental results, the power reactor data base, and recent irradiation test results obtained in the UK [34].

In spite of the restrictions that have been adopted, the effort to implement this submatrix can still be quite formidable. Let us assume that only three metallurgical tests are employed, namely Charpy, yield strength, and hardness. Since two welds and one plate material have been selected for this submatrix, one would need nine test specimens for each of the environmental irradiation conditions specified. In order to simplify these estimates, each Charpy measurement will be taken to represent one test specimen even though a set of, say, 8 to 12 C_v samples is required to conduct the measurement. Using three values for each of the five variables recommended in Table 3 requires a total of 3^5, or 243, different irradiations. Consequently, one would require a total of roughly 2200 irradiation test specimens.

Actually, more than 2200 specimens would be required because of the nature of the thermal neutron tests. Recent trend curve analyses have demonstrated that improvements can be effected if the dose term is generalized to a linear combination of different spectral fluence components (*i.e.*, thermal, intermediate, fast, dpa, *etc.*). As an example, the dose D can be expressed in the form

$$D = aT + bI + cF \tag{14}$$

where T, I, and F are the thermal, intermediate, and fast neutron fluence with a, b, and c appropriate constants.

A simple test of the validity of this representation involves the commutativity of the different components of the dose. Let us first consider a simplified example that is obtained by limiting the partition of the irradiation dose into just two groups, namely thermal and fast neutrons. For this test, let S represent the available set of metallurgical specimens, which is divided into four subsets: S_1, S_2, S_3, S_4. The size of these subsets is chosen to provide an adequate measurement base. The target dose is fixed in the environmental submatrix given in Table 3 at one of the three designated values. These values should be chosen based on relevant results from the surveillance capsule data base or the PSF experiment or both. All four subsets should be irradiated to this target dose, using a different sequence of thermal and fast neutron irradiations. The dose commutativity submatrix for these sequential irradiation tests is shown in Table 5.

This dose commutativity submatrix will examine not only the commutativity assumption, but equally important issues, such as whether or not the thermal neutron component of the dose enters as a defect stabilization mechanism or can help produce trend curve saturation. In addition, the validity of the coefficients used in the linear combination to represent the dose are also tested. In this manner, one can directly investigate the (relative) contribution of these different neutron energy groups to the embrittlement process.

Inclusion of this dose commutativity submatrix increases the total number of irradiation test specimens by 729 from roughly 2200 up to 3000. Moreover, the simplest kind of dose commutativity submatrix has been considered here since it corresponds to the partitioning of the dose into just

TABLE 5—*Dose commutativity submatrix.*[a]

Subset	Thermal Neutron Irradiation	Fast Neutron Irradiation
S_1	First	Second
S_2	Second	First
S_3	First	None
S_4	None	First

[a] All subsets are subjected to the same target dose.

two energy groups. Extension of this commutativity submatrix by partitioning into more than two groups, as would arise by inclusion of such components as intermediate neutron energy fluence I or dpa, would produce a substantial expansion of this commutativity submatrix. Other directions could also be considered, such as the inclusion of different materials in order to examine dependence on impurities and alloying elements. In this case, information could be obtained on the chemistry dependence of the coefficients that arise in the partitioning process (*i.e.*, *a*, *b*, *c*, etc.). However, this would also produce a submatrix of significantly expanded proportions.

Our current view is not to pursue either of these latter two options at this time, since some work is already in progress and the analysis of existing data remains to be completed [*22,30–37*]. The submatrix, without these options, represents an irradiation test program of environmental factors that is more manageable in scope and one that can more readily be implemented. The need or justification or both to pursue these as well as other options will become more apparent when the results of this more modest irradiation program are already in hand and can then be compared with results from already existing programs. Of higher priority at this time is the need to more fully understand environmental effects from (1) flux-level; (2) dose; (3) thermal neutrons; and (4) gamma-ray heating and temperature.

Each of these effects can produce a systematic deviation in trend curve analysis. Hence, the immediate goal of the irradiation test program defined by this submatrix is to generate the information needed to correct trend curve analyses for systematic biases introduced by these environmental effects.

This completes our initial attempt to formulate a test submatrix that treats environmental variables. Implementation of such an irradiation test program would provide the basis for understanding systematic effects from environmental radiation conditions. This knowledge would permit, in time, a realistic evaluation of the limitations of current data bases. As a consequence, significant systematic effects could be included in trend curve analyses that would lead to more accurate assessment and prediction of PV embrittlement in operating U.S. LWR commercial power plants.

Acknowledgments

This work was performed under the auspices of the Materials Engineering Branch, Division of Engineering Technology, Office of Nuclear Regulatory Research of the U.S. Nuclear Regulatory Commission.

References

[*1*] McElroy, W. N. et al, "LWR-PV-SDIP Semiannual Progress Report, April–September 1984," NUREG/CR-3746, Vol. 2, HEDL-TME 84-21, NRC, Washington, DC, March 1985.

[*2*] Steele, L. E., Ed., *Status of USA Nuclear Reactor Pressure Vessel Surveillance for Radiation Effects, ASTM STP 784,* American Society for Testing and Materials, Philadelphia, PA, January 1983.

[*3*] "Master Matrix for Light-Water Reactor Pressure Vessel Surveillance Standards," (E706-84), *Annual Book of ASTM Standards,* Volume 12.02.

[*4*] "Domestic Licensing of Production and Utilization Facilities," "General Design Criteria for Nuclear Power Plants" (Appendix A), "Fracture Toughness Requirements" (Appendix G), "Reactor Vessel Material Surveillance Program Requirements" (Appendix H), *Code of Federal Regulations* (10CFR50), U.S. Government Printing Office, Washington, DC.

[*5*] Regulatory Guide 1.99, "Effects of Residual Elements on Predicted Radiation Damage to Reactor Vessel Materials," Rev. 1, Nuclear Regulatory Commission, Washington, DC, April 1977.

[*6*] Dircks, W. J., *Pressurized Thermal Shock (PTS), and Enclosure A,* "NRC Staff Evaluation of PTS," SECY-82-465, Nuclear Regulatory Commission, Washington, DC, November 1982.

[*7*] Mager, T. R., "Pressured Thermal Shock and Reactor Materials," *Transactions of the American Nuclear Society,* Vol. 46, pp. 376–381, 1984.

[8] McElroy, W. N., Gold, R., Lippincott, E. P., Simons, R. L., Anderson, S. L., and Manahan, M. P., "Trend Curve Data Development and Testing," Effects of Radiation on Materials, 13th International ASTM Symposium, Seattle, WA, 23–25 June 1986.

[9] Simons, R. L., "Damage Rate and Spectrum Effects in Ferritic Steel ΔNDTT Data," Effects of Radiation on Materials, 13th International ASTM Symposium, Seattle, WA, 23–25 June 1986.

[10] Gold, R., Roberts, J. H., and Ruddy, F. H., "Annealing Phenomena in Solid-State Track Recorders," *Nuclear Tracks,* Vol. 5, 1981, pp. 253–264.

[11] O'Sullivan, D., Thompson, A., Adams, J. A., and Beahm, L. P., "New Results on the Investigation of the Variation of Nuclear Track Detector Response with Temperature," *12th International Conference on Solid State Nuclear Track Detectors, Acapulco, Mexico, 4–10 September 1983,* Pergamon Press, Oxford, UK, 1984.

[12] Thompson, A. and O'Sullivan, D., "The Ionization Dependence of the Registration Temperature Effect in Solid-State Nuclear Track Detectors," 13th International Conference on Solid State Nuclear Track Detectors, Rome, 23–27 September 1985.

[13] Gold, R., Lippincott, E. P., McElroy, W. N., and Simons, R. L., "Radiation Damage Function Analysis," *Effects of Radiation on Structural Materials—Proceedings of the 9th International Symposium, ASTM STP 683,* J. A. Sprague and D. Dramer, Eds., American Society for Testing and Materials, Philadelphia, PA, 1979, pp. 380–401.

[14] Guthrie, G. L., "HEDL Analysis of the Poolside Facility Experiment," LWR-PV-SDIP: Semiannual Progress Report, April 1984–September 1984, NUREG/CR-3746, Vol. 2, HEDL-TME 84-21, Nuclear Regulatory Commission, Washington, DC, March 1985, pp. HEDL-28–HEDL-52.

[15] McElroy, W. N. et al, "Trend Curve Exposure Parameter Data Development and Testing," HEDL-SA-3126 and Proc. of the 5th ASTM-EURATOM Symposium on Reactor Dosimetry, Geesthacht, FRG, September 24–28, 1984.

[16] Perrin, J. S., Wullaert, R. A., Odette, G. R., and Lombroso, M. P., "Physically Based Regression Correlations of Embrittlement Data from Reactor PV Surveillance Programs," EPRI NP-3319, Electric Power Research Institute, Palo Alto, CA, January 1984.

[17] Norris, E. B., "A Service Laboratory's View of the Status and Direction Reactor Vessel Surveillance," *Radiation Embrittlement and Surveillance of Nuclear Reactor Pressure Vessels: An International Study, ASTM STP 819,* L. E. Steele, Ed., American Society for Testing and Materials, Philadelphia, PA, 1983, pp. 194–204.

[18] Guthrie, G. L., Lippincott, E. P., and McGarry, E. D., "LWR-PV-SDIP: PSF Blind Test Workshop Minutes," HEDL-7467, Hanford Engineering Development Laboratory, Richland, WA, April 9–10, 1984.

[19] Davis, L. M. and Squires, R. L., "Comparison of Mechanical Test Results from the IAEA-Coordinated Research Programme and Surveillance Dosimetry Improvement Programme," Proceedings of the 2nd International Conference on Environmental Degradation of Nuclear Reactor Materials, Monterey, CA, September 12–14, 1985.

[20] "Standard Practice for Characterizing Neutron Exposures in Ferritic Steels in Terms of Displacements per Atom (dpa)," (E693), *Annual Book of ASTM Standards, 1979,* American Society for Testing and Materials, Philadelphia, PA.

[21] McElroy, W. N. et al, "LWR-PV-SDIP 1983 Annual Report," NUREG/CR-3391, Vol. 3, HEDL-TME 83-23, Nuclear Regulatory Commission, Washington, DC, January 1984.

[22] McElroy, W. N. et al, "LWR-PV-SDIP 1984 Annual Report, October 1, 1983–September 30, 1984," NUREG/CR-3746, Vol. 3, HEDL-TME 84-31, Nuclear Regulatory Commission, Washington, DC, March 1985.

[23] Simons, R. L., Kellogg, L. S., Lippincott, E. P., and McElroy, W. N., "Re-evaluation of the Physics-Dosimetry from PWR and BWR Pressure Vessel Surveillance Programs," HEDL-SA-3210 and *Proceedings of the 5th ASTM-EURATOM Symposium on Reactor Dosimetry,* Geesthacht, FRG, September 24–28, 1984.

[24] Randall, P. N., "Basis for Revision 2 of US NRC Regulatory Guide 1.99," *Proceedings of the 10th MPA Seminar,* Stuttgart, FRG, October 10, 1984.

[25] Guthrie, G. L., "HEDL Analysis of the PSF Experiment," Attachment to *Minutes of the 14th LWR-PV-SDIP Meeting,* HEDL-7511, Hanford Engineering Development Laboratory, Richland, WA, pp. A7-1, October 1–5, 1984.

[26] Guthrie, G. L., "Charpy Trend Curves Based on 177 PWR Data Points," LWR-PV-SDIP: Quarterly Progress Report, April 1983–June 1983, NUREG/CR-3391, Vol. 2, HEDL-TME 83-22, Nuclear Regulatory Commission, Washington, DC, April 1984, pp. HEDL-3–HEDL-15.

[27] Alberman, A. A., et al, "DOMPAC Dosimetry Experiment Neutron Simulation of the Pressure Vessel of a Pressurized-Water Reactor, Characterization of Irradiation Damage," CEA-R-5217, Centre d'Etudes Nucleaires de Saclay, France, May 1983.

[28] Grant, S. P. and Earp, S. L., "Methods for Extending Life of a PWR Reactor Vessel After Long-Term Exposure to Fast Neutron Radiation, Effects of Radiation on Materials: *12th International Symposium, STP 870,* American Society for Testing and Materials, Philadelphia, 1985, pp. 1027–1045.

[29] Kam, F. B. K., "Characterization of the Fourth HSST Series of Neutron Spectral Metallurgical Irradiation Capsules," presented to the 4th ASTM-EURATOM Symposium on Reactor Dosimetry, Gaithersburg, MD, March 22–26, 1982. (Preprints available.)

[30] Kam, F. B. K. et al, "Neutron Exposure Parameters for the Fourth HSST Series of Metallurgical Irradiation Capsules," *Proceedings of the 4th ASTM-EURATOM Symposium on Reactor Dosimetry,* Gaithersburg, MD, March 22–26, 1982, NUREG/CP-0029, Nuclear Regulatory Commission, Washington, DC, Vol. 2, July 1982, pp. 1023–1033.

[31] Materials Engineering Associates, "Structural Integrity of Light Water Reactor Pressure Boundary Integrity: 4-Year Plan, 1984–88," NUREG/CR-3788, MEA-2047, Nuclear Regulatory Commission, Washington, DC, September 1984.

[32] Odette, G. R., Lombrozo, P. M., and Wullaert, R. A., "Relationship Between Irradiation-Induced Changes in Strength and Embrittlement of Light Water Reactor Pressure Vessel Steels," *Effects of Radiation on Materials: 12th International Symposium, STP 870,* American Society for Testing and Materials, Philadelphia, 1985, pp. 840–862.

[33] Lucas, G. E., Odette, G. R., Lombrozo, P. M., and Sheckherd, J. W., "Effects of Composition, Microstructure, and Temperature on Irradiation Hardening of Pressure Vessel Steels," *Effects of Radiation on Materials: 12th International Symposium, ASTM STP 870,* F. A. Garner and J. S. Perrin, Eds., American Society for Testing and Materials, Philadelphia, 1985, pp. 900–930.

[34] Fisher, S. B., Harbottle, J. E., and Aldridge, N. B., *Microstructural Changes Related to Irradiation Hardening in Pressure Vessel Steels, Dimensional Stability, and Mechanical Behavior of Irradiated Metals and Alloys,* British Nuclear Energy Society, London, UK, 1984.

[35] Stallmann, F. W., Baldwin, C. A. and Kam, F. B. K., "Neutron Spectral Characterization of the 4th Nuclear Regulatory Commission Heavy Section Steel Technology 1T-CT Irradiation Experiment: Dosimetry and Uncertainty Analysis," NUREG/CR-3333, ORNL/TM-8789, Nuclear Regulatory Commission, Washington, DC, July 1983.

[36] Pachur, D., "Mechanical Properties of Neutron-Irradiated Reactor Pressure Vessel Steel Dependent on Radiation Mechanisms," from the ANS Special Session on Correlations and Implications of Neutron Irradiation Embrittlement of Pressure Vessel Steels, Detroit, MI, June 12–16, 1983, *Transactions of the American Nuclear Society,* Vol. 44, 1983, p. 229.

[37] Williams, T. J., Thomas, A. F., Berrisford, R. A., Austin, M., Squires, R. L., and Venable, J. H., *Influence of Neutron Exposure, Chemical Composition and Metallurgical Condition on the Irradiation Shift of Reactor Pressure Vessel Steels, ASTM STP 782,* American Society for Testing and Materials, Philadelphia, PA, 1982, p. 343.

John J. McGowan[1] *and Randy K. Nanstad*[2]

A Statistical Analysis of Fracture Toughness of Irradiated Low-Alloy Steel Plate and Welds

REFERENCE: McGowan, J. J. and Nanstad, R. K., "**A Statistical Analysis of Fracture Toughness of Irradiated Low-Alloy Steel Plate and Welds,**" *Influence of Radiation on Material Properties: 13th International Symposium (Part II), ASTM STP 956,* F. A. Garner, C. H. Henager, Jr., and N. Igata, Eds., American Society for Testing and Materials, Philadelphia, 1987, pp. 569–589.

ABSTRACT: Studies of the effects of neutron irradiation on fracture toughness properties of steels have generally included a minimum number of tests for each material condition. The present study attempts to apply statistical analyses with multiple testing at selected temperatures to assess the accuracy and reliability of the results. Fracture toughness test specimens were irradiated in the Bulk Shielding Reactor at Oak Ridge National Laboratory at 288°C to target neutron fluences of 2×10^{23} n/m^2 (>1 MeV). The materials were ASTM A533 Grade B Class 1 plate (HSST Plate 02) and four submerged-arc welds representing current nuclear pressure vessel fabrication practice. Both unirradiated and irradiated specimens were tested by two separate laboratories, and multiple tests were conducted at selected temperatures. Statistical analyses permitted determination of material and test variability and an interlaboratory comparison. Behavior in both the transition and upper-shelf regions was studied.

KEY WORDS: irradiation, statistical analyses, fracture, fluences, welds, plates

Numerous studies of the effects of impurities and fast neutron irradiation on fracture toughness of nuclear reactor pressure vessel materials (plates, forgings, and welds) have been reported in the literature [*1–10*]. Most such studies have included a minimum number of tests for each material or combination of irradiation parameters or both. Statistical analysis of neutron irradiation effects has been possible only by using a large body of data representing a variety of materials and neutron exposure conditions. However, such analytical methods do not address the question of accuracy of each datum in the large set, or the accuracy and reliability of results from a single small data set such as might be obtained in a nuclear reactor pressure vessel surveillance study. The present studies address this question by multiple testing at two laboratories of typical nuclear pressure vessel materials (both irradiated and unirradiated) and statistical analyses of the test results. Multiple tests were conducted at each of several test temperatures for each material, standard deviations and coefficients of variation were determined, and results from the two laboratories were compared. This irradiation program was conducted on four submerged-arc welds and one plate of A533 Grade B Class 1 pressure vessel steel. The welds were made by commercial vendors using current welding practices and contained relatively low copper levels. The target fast neutron fluence was 2×10^{23} n/m^2 (>1 MeV), and the target irradiation temperature was 288°C. Charpy-*V*-notch impact (CVN), tensile, and 25.4-mm-thick compact (ITCS) fracture toughness specimens were

[1] Formerly of Oak Ridge National Laboratory, now associate professor with Virginia Military Institute, Department of Engineering, Lexington, VA.

[2] Group leader, Metals and Ceramics Division, Oak Ridge National Laboratory, Oak Ridge, TN 37831.

used in this study. Previous papers [*11,12*] reported some of the fracture toughness results from this study but did not include statistical analyses.

Materials

The plate material was from a 305-mm-thick plate of ASTM A533 Grade B Class 1 manganese-molybdenum-nickel steel produced by Lukens Steel Company for the Heavy-Section Steel Technology (HSST) Program [*13*]. This plate material was designated HSST Plate 02, and portions of it have been used in many investigations. All test specimens were prepared in the transverse (T-L) orientation.

All four submerged-arc weldments were made in ASTM A533 Grade B Class 1 plate. Two of the submerged-arc weldments were supplied by the Electric Power Research Institute (EPRI) and were fabricated by Combustion Engineering, Inc. HSST weld 68W was produced as a 178-mm-thick submerged-arc weldment using Linde 0091 flux and 4.8-mm-diameter MIL-B-4 (low-copper and -phosphorus) wire and was designated "CGS" in EPRI studies [*14*]. HSST weld 69W was produced as a 300-mm-thick submerged-arc weldment using Linde 0091 flux and 4.8-mm-diameter MIL-B-4 wire and was designated "CHS" in EPRI studies [*14*]. Both welds were postweld heat treated 25 h at 621°C. The other two welds were supplied by Babcock and Wilcox Company [*15*]. Both welds were produced as 174-mm-thick submerged-arc weldments using Mn-Mo-Ni (SFA-5.23EF2N) filler wire. HSST weld 70W was fabricated using Linde 0124 flux, and HSST weld 71W was fabricated using Linde 0080 flux. Both welds were postweld heat treated 48 h at 607°C.

The four weldments are considered "current practice" and are of relatively low-copper content. Chemical compositions of the five materials are presented in Table 1.

Material Irradiation

All irradiations were conducted at two faces of the Bulk Shielding Reactor (BSR) at Oak Ridge National Laboratory (ORNL), a 2-MW pool-type reactor. Thermal shields of 42.5-mm-thick stainless steel were used between the reactor core and the specimen capsules to reduce gamma heating. Each capsule contained 60 ITCS fracture toughness specimens, 80 to 90 Charpy-*V*-notch impact test specimens, and 10 to 20 tension test specimens in an arrangement shown in Fig. 1.

Specimen temperatures during irradiation were controlled by a combination of electrical heaters and controlled sweep gas composition (helium and nitrogen). Specimen temperatures during irradiation were controlled at 288 $\pm$ 5°C. Irradiation times for the capsules ranged from 4300 to 5400 h.

Complete spatial maps of damage exposure parameter values were determined for each capsule. The dosimetry in the capsules consisted of multiple foil sets and gradient wires. The dosimetry and spectral adjustment procedures are presented elsewhere [*16,17*]. Uncertainties for the damage exposure parameters were less than 8% (standard deviation).

Test and Analysis Procedures

All weld test specimens were prepared with the loading direction perpendicular to the weld centerline and crack propagation direction parallel to the surface. All fracture toughness specimens tested in this study were 25.4-mm-thick compact specimens (ITCS) modified for load-line displacement. Those specimens tested in the ductile-shelf region were side grooved 10% of the specimen thickness on each side, while those tested in the transition region were not side grooved.

The testing of the fracture toughness specimens was apportioned between Materials Engineering Associates (MEA), Lanham, MD, and ORNL, Oak Ridge, TN, such that both laboratories tested roughly equivalent numbers of specimens. Both laboratories used the single specimen compliance

TABLE 1—*Chemical compositions of plate and submerged-arc welds.*

Material	Composition, wt%									
	C	Mn	P	S	Si	Cr	Ni	Mo	Cu	V
Plate 02, 0.14% Cu 0.67% Ni	0.23	1.55	0.009	0.014	0.20	0.04	0.67	0.53	0.14	0.003
Weld 68W, 0.04% Cu, 0.13% Ni	0.15	1.38	0.008	0.009	0.16	0.04	0.13	0.60	0.04	0.007
Weld 69W, 0.12% Cu, 0.10% Ni	0.14	1.19	0.010	0.009	0.19	0.09	0.10	0.54	0.12	0.005
Weld 70W, 0.056% Cu, 0.63% Ni	0.10	1.48	0.011	0.011	0.44	0.13	0.63	0.47	0.056	0.004
Weld 71W, 0.046% Cu, 0.63% Ni	0.124	1.58	0.011	0.011	0.54	0.12	0.63	0.45	0.046	0.005

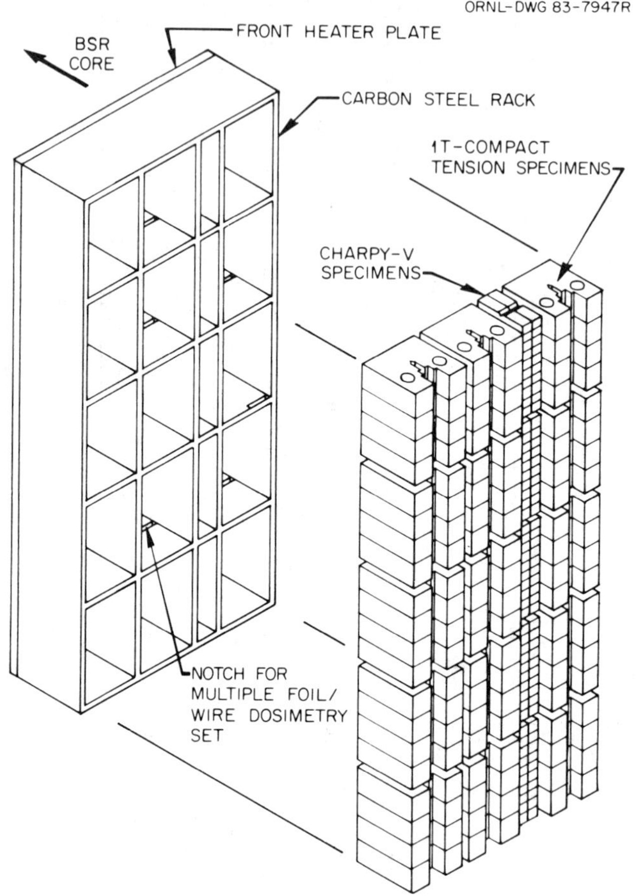

FIG. 1—*Fourth HSST irradiation series capsule specimen rack assembly.*

(SSC) technique for determining the *J*-integral resistance curve. The SSC technique provides a method to determine the amount of specimen crack extension by means of small unloadings (<15% for maximum load) conducted at regular intervals throughout the test.

Values of the *J* integral were calculated using the modified version of the *J* integral, known as J_M, as proposed by Ernst [*18*]. Deformation theory *J* (J_D) is the formulation of the *J* integral specified for use in the ASTM Test for J_{Ic}, a Measure of Fracture Toughness, (E 813) and in the tentative ASTM *J*-Integral Resistance (*J-R*) Curve Test Procedure [*19*]. Modified *J* (J_M) was used in this program because Ernst has shown it to be more specimen-size independent when the crack extension exceeds the *J*-controlled crack growth regime. A typical *J-R* curve produced with the SSC technique is illustrated in Fig. 2. The *J-R* curve format of Fig. 2 is in accordance with that of ASTM E 813, that is, J_{Ic}, the elastic-plastic initiation toughness, is defined by the intersection of the blunting line with the linear regression fit (data between the 0.15- and the 1.5-mm exclusion lines). Using the slope of the linear regression *dJ*/*da* the tearing modulus can be determined

$$T = (E/\sigma_f^2)*dJ/da \tag{1}$$

where σ_f = (yield strength + ultimate strength)/2, and E = Young's modulus. To convert the

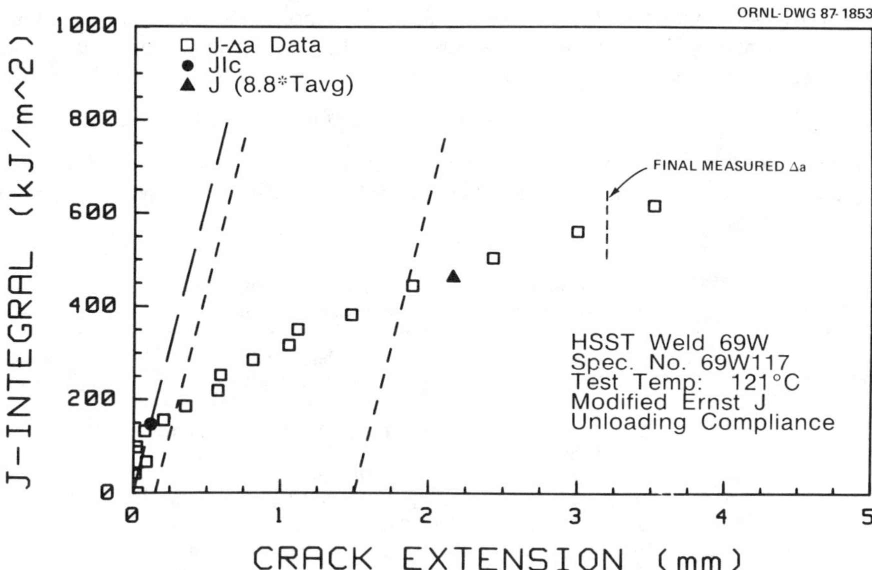

FIG. 2—*ASTM E 813 procedure for determination of* J_{Ic} *using the SSC technique.*

elastic-plastic J_{Ic} initiation toughness to a quasi-elastic K_{Jc} value, the following relationship is used

$$K_{Jc} = (E^*J_{Jc})^{0.5} \tag{2}$$

where E(GPa) = 207.2 − 0.057*T(°C). For specimens that experience cleavage fracture, the value of J_M at the onset of cleavage is used to calculate K_{Jc}.

ASTM E 813 gives an equation for converting J_{Ic} to K_{Jc}, which is similar to Eq 2, but includes a plane strain term. In the ductile-shelf region and the upper-transition region, the behavior is nearly plane stress because the specimens experienced near-net-section yielding; therefore this kind of formulation is appropriate. In the lower transition region, however, the specimen through-thickness constraint is greater and more nearly plane strain. In this case it is unclear whether the plane strain term should be added to Eq 2. Most of the data produced in this program were more nearly plane stress and, for consistency of presentation, all K_{Jc} values were calculated using Eq 2. For low-alloy steels, inclusion of the plane strain term gives K_{Jc} values about 5% greater.

At toughness levels too great to allow measurement of a valid K_{Ic} with a small specimen, K_{Jc} values can be determined as stated above; however, these data tend to be of greater magnitude than that which would be obtained with a larger specimen capable of measuring the linear elastic toughness level by ASTM Test Method for Plane-Strain Fracture Toughness of Metallic Materials (E 399) criteria. A similar observation was made by Irwin [*20*] in that the plane stress fracture toughness (K_c) overestimates K_{Ic}. Irwin developed an empirical relationship from which K_{Ic} could be estimated once K_c was known. Recently, Merkle [*21*] has shown that reasonable estimates of K_{Ic} can be determined by using K_{Jc} as a substitute for K_c in Irwin's relationship. In this study, K_{Jc} values in the transition region were corrected to a parameter called $K_{\beta c}$ using the Irwin-Merkle β_{Ic} correction to provide another means of determining the irradiation-induced transition-temperature shift.

Two distinct types of curve fit were used in this study. Linear regressions using standard least-square techniques were used to determine the K_{Jc} versus temperature behavior and the tearing modulus versus temperature behavior in the ductile-shelf regime. Exponential fits using nonlinear least-square techniques [22] were used to determine the K_{Jc} versus temperature behavior and the $K_{\beta c}$ versus temperature behavior in the transition region. The form of the exponential fit was

$$K_{Jc} = A + B^*[\exp(C^*T)] \tag{3}$$

with the values of the constants A, B, and C chosen to optimize the fit. Statistical analyses consisted of determination of mean values, standard deviations, and coefficients of variation using standard techniques.

TABLE 2—*Comparison of 1TCS test results for unirradiated pressure vessel plate and submerged-arc welds.*

Test Temperature, °C	Mean K_{Jc}, MPa · $m^{1/2}$			Standard Deviation, MPa · $m^{1/2}$			Number of Tests	
	ORNL	MEA	Combined	ORNL	MEA	Combined	ORNL	MEA
			A533 GRADE B CLASS 1 (PLATE 02)					
−95	49.5	49.4	49.5	10.9	8.0	7.9	2	3
−40	71.3	90.4	80.9	2.9	1.4	10.6	3	3
−5	124.2	130.5	129.0	. . .	6.6	6.2	1	3
23	241.8	230.4	236.1	36.9	24.0	28.6	3	3
50	180.9	203.4	194.4	35.5	8.2	22.4	2	3
121	169.9	179.3	175.5	26.0	9.6	15.6	2	3
204	180.2	168.5	173.2	18.8	2.8	11.6	2	3
288	162.2	153.6	157.0	4.8	2.4	5.6	2	3
			SUBMERGED-ARC WELD (68W)					
−80	90.8	150.8	110.8	45.5	. . .	47.3	2	1
−30	212.3	292.7	252.5	. . .	. . .	56.9	1	1
121	201.8	176.8	189.3	. . .	. . .	17.7	1	1
288	198.6	158.4	178.5	. . .	. . .	28.4	1	1
			SUBMERGED-ARC WELD (69W)					
−100	41.7	53.4	47.5	1.3	9.5	8.8	2	2
−60	61.7	67.2	63.9	6.3	18.5	10.7	3	2
−5	95.7	114.2	106.8	0.3	21.5	18.3	2	3
22	184.1	158.0	171.1	38.9	29.8	34.1	3	3
50	174.9	178.0	177.0	. . .	8.5	6.3	1	2
121	176.3	170.9	173.6	7.1	1.3	5.2	2	2
203	143.0	173.0	163.0	. . .	24.1	24.3	1	2
288	155.1	137.9	146.5	12.1	2.3	12.2	2	2
			SUBMERGED-ARC WELD (70W)					
−135	41.6	. . .	. . .	8.1	. . .	. . .	2	. . .
−115	. . .	56.0	. . .	. . .	16.8	. . .	. . .	2
−65	109.7	103.7	106.7	1.4	28.9	17.1	2	2
50	148.2	191.8	170.0	. . .	. . .	30.8	1	1
121	167.2	164.7	166.3	6.7	. . .	5.0	2	1
204	155.5	146.2	152.4	4.6	. . .	6.3	2	1
288	148.0	135.0	143.7	0.3	. . .	7.5	2	1
			SUBMERGED-ARC WELD (71W)					
−93	47.8	60.6	56.3	. . .	1.1	7.4	1	2
−50	85.8	95.5	90.7	. . .	. . .	6.9	1	1
−10	157.8	. . .	. . .	66.8	. . .	. . .	2	. . .
50	142.3	158.2	150.3	. . .	. . .	11.2	1	1
121	161.9	158.0	159.3	. . .	9.3	7.0	1	2
204	154.5	129.6	137.9	. . .	4.0	14.7	1	2
288	128.8	121.0	123.6	. . .	7.8	7.1	1	2

Results and Discussion

The fracture toughness results K_{Jc} in the unirradiated and irradiated conditions are summarized in Tables 2 and 3, respectively. These same results are shown for Plate 02, welds 68W, 69W, 70W, and 71W in Figs. 3 through 7, respectively. The means for the two laboratories are generally within one standard deviation of each other, indicating good reproducibility between the laboratories. The standard deviations are larger in the upper-transition region than in the lower-transition or ductile-shelf regions. This is expected because of the random nature of the brittle-cleavage phenomenon. A series of ITCS fracture toughness tests (26 tests run at a single temperature)

TABLE 3—*Comparison of 1TCS test results for irradiated pressure vessel plate and submerged-arc welds.*

Test Temperature, °C	Mean K_{Jc}, MPa · $m^{1/2}$			Standard Deviation, MPa · $m^{1/2}$			Number of Tests	
	ORNL	MEA	Combined	ORNL	MEA	Combined	ORNL	MEA
			A533 Grade B Class 1 (Plate 02)					
−25	46.7	44.04	45.3	11.5	3.20	7.7	3	2
50	85.9	91.9	88.9	29.6	21.2	24.5	5	5
80	157.7	140.4	149.1	40.8	31.5	35.5	5	5
121	170.3	180.2	175.3	1.6	5.9	6.7	3	3
204	155.5	167.7	162.8	4.3	5.9	8.2	2	3
288	164.6	131.2	147.9	13.3	6.7	20.6	3	3
			Submerged-Arc Weld (68W)					
130	45.0	45.5	45.3	10.7	7.6	7.6	2	3
−75	126.8	109.1	116.2	10.2	35.1	27.1	2	3
−50	171.7	145.1	151.8	. . .	9.9	15.5	1	3
25	. . .	199.9	178.5	. . .	12.0	. . .	. . .	2
121	174.0	207.0	190.5	16.8	2.1	21.4	2	2
204	159.5	196.4	177.9	6.7	16.1	23.6	2	2
288	162.0	154.9	159.6	21.1	. . .	15.5	2	1
			Submerged-Arc Weld (69W)					
−35	60.6	65.4	61.8	8.1	. . .	7.0	3	1
10	73.8	112.6	89.3	3.7	1.3	21.4	3	2
50	156.8	161.7	159.2	54.4	40.6	39.3	2	2
121	119.2	159.3	139.3	20.1	14.1	27.1	2	2
204	123.6	138.2	130.9	15.3	20.5	17.0	2	2
288	114.8	109.4	112.1	24.3	3.9	14.6	2	2
			Submerged-Arc Weld (70W)					
−125	37.4	38.2	37.8	0.7	1.2	0.9	2	2
−50	86.0	101.2	93.6	12.4	1.6	11.4	2	2
−25	111.9	123.3	118.8	68.6	7.3	35.2	2	3
31	183.0	158.2	170.6	. . .	. . .	17.5	1	1
121	148.9	184.1	166.5	14.1	14.1	23.3	2	2
204	158.5	155.7	157.1	3.4	0.0	2.5	2	2
288	138.8	140.7	139.7	3.9	1.1	2.6	2	2
			Submerged-Arc Weld (71W)					
−70	64.9	68.6	65.9	2.9	. . .	3.0	4	1
−30	87.0	73.2	80.1	16.5	0.6	12.4	2	2
0	142.7	136.6	139.6	26.0	2.1	15.4	2	2
121	136.9	149.3	143.1	15.1	3.6	11.5	2	2
204	145.6	134.4	140.0	5.4	3.5	7.5	2	2
288	135.7	103.7	125.0	0.6	. . .	18.5	2	1

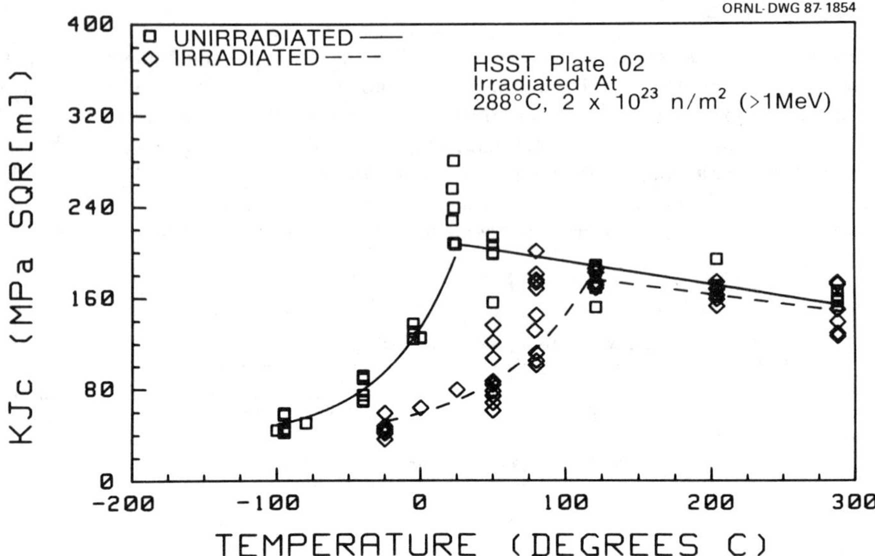

FIG. 3—*Variation of* K_{Jc} *with temperature and irradiation for Plate 02.*

performed by the authors [23] on A533 Grade B Class 1 steel plate (HSST Plate 13A) indicated that the standard deviation of K_{Jc} in the transition region could be approximated by half the mean K_{Jc} minus the lower-shelf value. This indicates that the variation increases as the toughness increases. In light of these results at only one temperature, the variations between the laboratories in Tables 2 and 3 seem reasonable, even after irradiation. On the ductile shelf the fracture process was dominated by ductile processes and, therefore, less variation is expected.

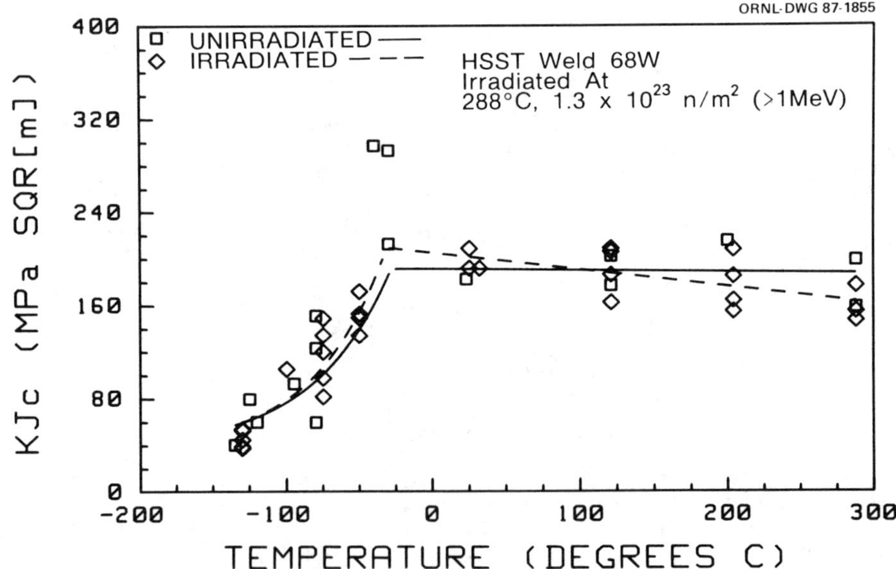

FIG. 4—*Variation of* K_{Jc} *with temperature and irradiation for Weld 68W.*

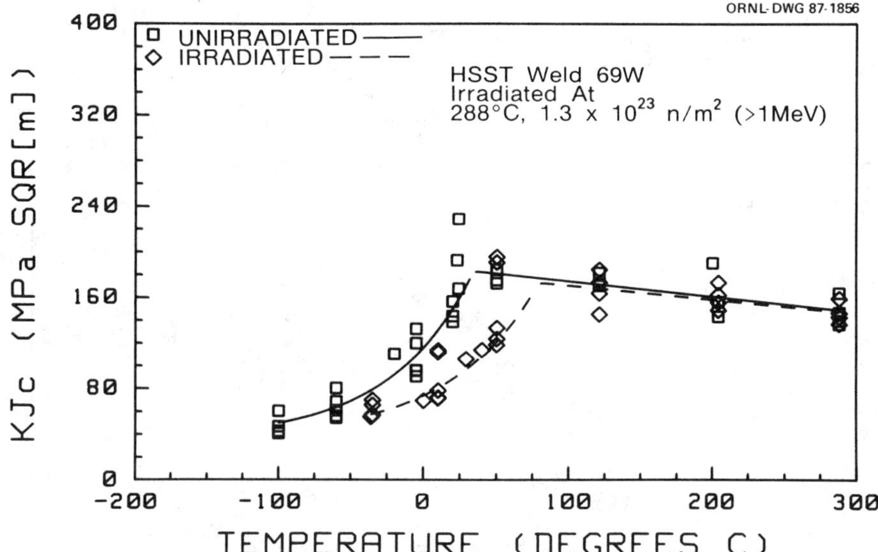

FIG. 5—*Variation of* K_{Jc} *with temperature and irradiation for Weld 69W.*

Table 4 provides a comparison of the coefficients of variation observed for unirradiated CVN and K_{Jc} results. The values shown in the table for each material in the transition region are averages of all transition temperature results. Similarly, averages are reflected for the ductile shelf. Except for the case of Weld 68W, for which the number of fracture toughness specimens was very limited, the variations of CVN and K_{Jc} results in the transition region are comparable. On the ductile shelf the variations in K_{Jc} are somewhat greater than those for CVN. The table amplifies, however, the

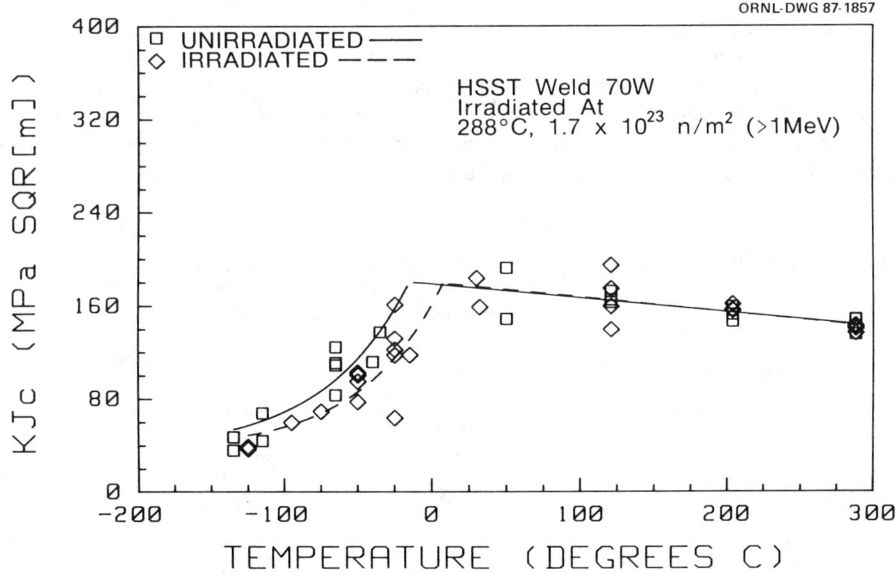

FIG. 6—*Variation of* K_{Jc} *with temperature and irradiation for Weld 70W.*

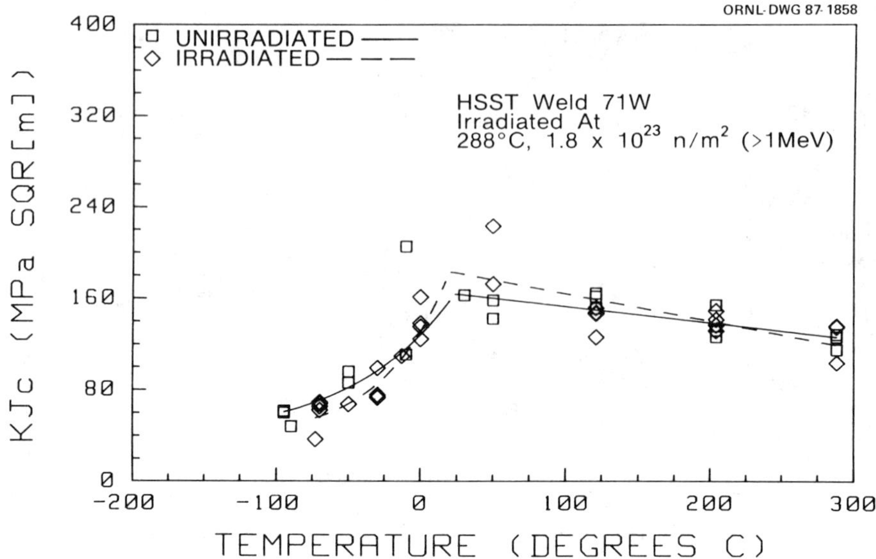

FIG. 7—*Variation of* K_{JC} *with temperature and irradiation for Weld 71W.*

previous observation that variations are much greater in the transition region than on the ductile shelf.

For the irradiated fracture toughness results, Table 5 compares unirradiated and irradiated K_{Jc} results for HSST Plate 02 in the transition region and ductile shelf. The variations at the lower temperatures, in the transition region, are much greater for the irradiated tests. On the ductile shelf, there appears to be slightly more variation for the irradiated tests.

Some of the fracture toughness variations in K_{Jc} results for irradiated materials are due to variations in fluence throughout the capsule. An attempt to account for this variation was made by fitting the transition temperature shift versus fluence with a power law function. This function was then used to correct each specimen's test temperature (in the transition region) to represent the mean fluence for each material. The only results affected by this procedure were those for Plate 02, as shown in Fig. 8. The other materials had such low irradiation damage that fluence variations had little effect on the fracture toughness.

TABLE 4—*Coefficients of variation for unirradiated Charpy impact and fracture toughness tests.*

	Coefficients of Variation, %			
	Charpy Impact		Fracture Toughness	
Material	Transition Region	Ductile Shelf	Transition Region	Ductile Shelf
Plate 02	16	6	11	9
68W	24	10	43	16
69W	22	5	18	8
70W	20	6	22	8
71W	17	7	21	7

TABLE 5—*Statistical variations in fracture toughness results for HSST Plate 02.*

Test Temperature, °C	Number of Tests	Mean K_{Jc}, MPA · $m^{1/2}$	Standard Deviation, MPa · $m^{1/2}$	Coefficient of Variation, %
		HSST PLATE 02, UNIRRADIATED		
−95	5	50	7.9	16
−40	6	81	10.6[a]	13
23	6	236	28.6	12
121	5	175	15.6	9
288	5	157	5.6[b]	4
		HSST PLATE 02, IRRADIATED		
−25	5	45	7.7	17
50	10	89	24.5	28
80	10	149	35.5[b]	24
121	6	175	6.7	4
288	6	148	20.6[a]	14

[a] Difference between ORNL and MEA mean K_{Jc} values greater than two standard deviations.
[b] Difference between ORNL and MEA mean K_{Jc} values greater than one standard deviation.

The fracture toughness in terms of the thickness-corrected parameter $K_{\beta c}$ is shown in Figs. 9 through 13 for Plate 02, Welds 68W, 69W, 70W, and 71W, respectively. (Note that in Fig. 9 the fluence-corrected temperature is used and that the fracture toughness scale was changed for Figs. 9 through 13,)

A comparison of the irradiation effects on the fracture toughness parameters, previously measured tensile parameters [*24*] and previously measured Charpy-*V*-notch [*25*] parameters is shown in Table 6. The transition temperature shift is determined by fitting an exponential curve fit through

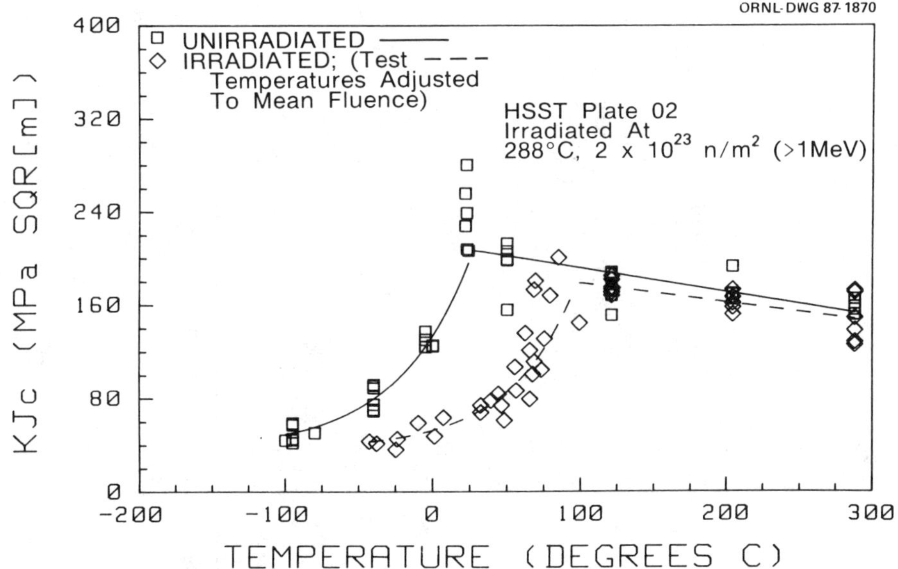

FIG. 8—*Variation of* K_{Jc} *with temperature and irradiation for Plate 02; temperature of irradiated specimens corrected for fluence variations to the mean fluence of* 2×10^{23} *n/m²*.

TABLE 6—*Summary of changes in material properties caused by irradiation.*

Material	Mean Fluence $n/m^2 \times 10^{23}$ (>1 MeV)	Flow Strength Increase[a], %	Transition Temperature Increase, °C			Ductile-Shelf Change, %		Tearing Modulus Change,[a] %
			CVN at 41 J	K_{Jc} at 125 MPa · $m^{1/2}$	$K_{\beta c}$ at 90 MPa · $m^{1/2}$	CVN	K_{Jc}[a]	
Plate 02, 0.14% Cu, 0.67% Ni	2.0	24	68	82	66	−16	−5	−49
Weld 68W, 0.04% Cu, 0.13% Ni	1.4	10	6	−6	−2	9	−7	−11
Weld 69W, 0.12% Cu, 0.10% Ni	1.2	12	34	47	39	−3	−2	−29
Weld 70W, 0.056% Cu, 0.63% Ni	1.6	10	31	25	21	−10	0	−32
Weld 71W, 0.046% Cu, 0.63% Ni	1.6	10	26	−1	. . .	8	1	−28

[a] Average change from 22 to 288°C.

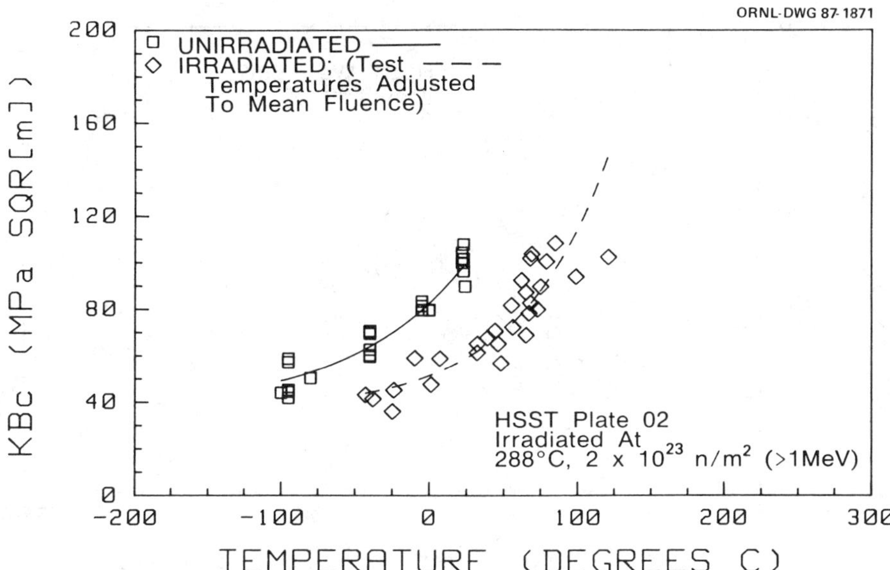

FIG. 9—*Variation of* $K_{\beta c}$ *with temperature and irradiation for Plate 02; temperature of irradiated specimens adjusted for fluence variations to the mean fluence of* 2×10^{23} *n/m*^{2}.

the K_{Jc} or $K_{\beta c}$ data versus temperature, and then determining the difference in temperature given by the two different fits at the desired K value. A value of 125 MPa · m$^{1/2}$ is used for K_{Jc} and a value of 90 MPa · m$^{1/2}$ is used for $K_{\beta c}$. Comparison of the CVN shift with the K_{Jc} and the $K_{\beta c}$ shifts give similar results since the unirradiated and irradiated CVN, K_{Jc}, and $K_{\beta c}$ transition curves are essentially parallel. Modified test temperatures for Plate 02, calculated as described above, were used to determine the irradiated curve.

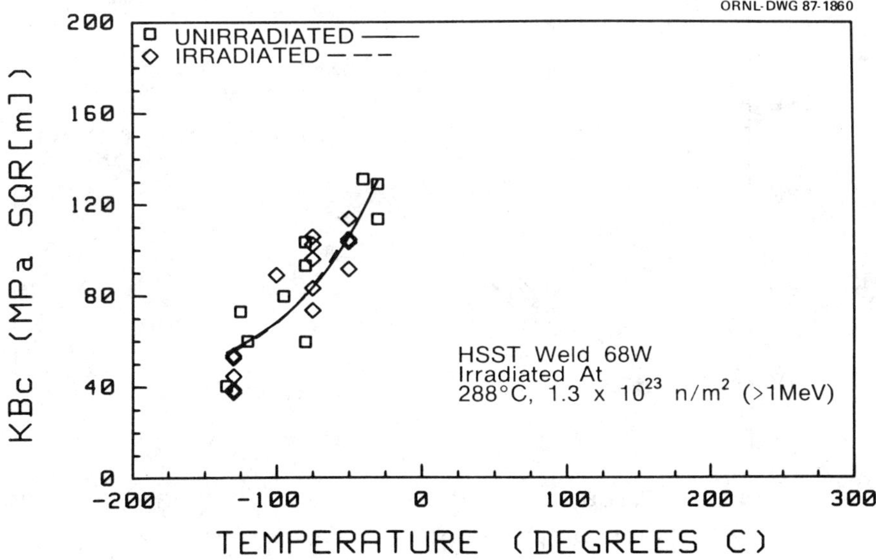

FIG. 10—*Variation of* $K_{\beta c}$ *with temperature and irradiation for Weld 68W.*

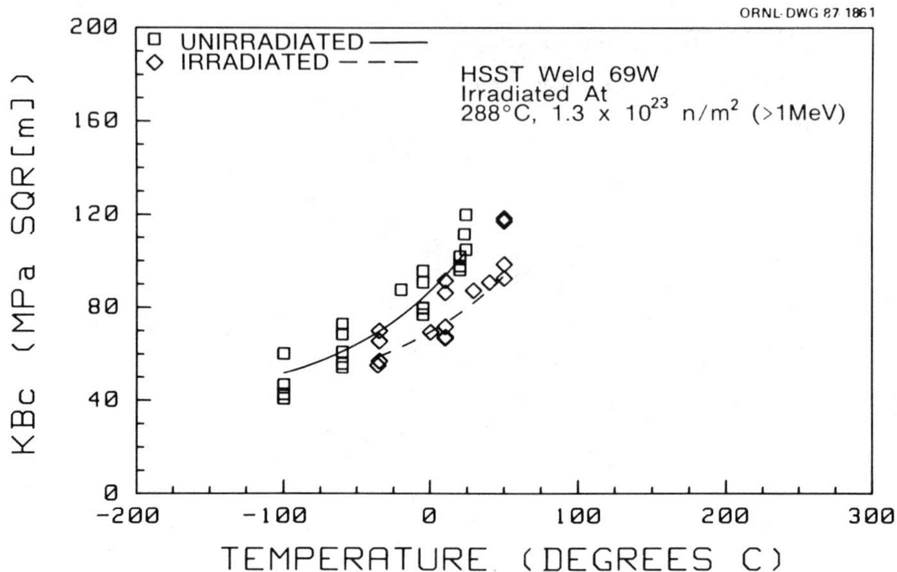

FIG. 11—*Variation of* $K_{\beta c}$ *with temperature and irradiation for Weld 69W.*

Table 7 shows a comparison of the statistical confidence associated with the mean transition temperature increases of all materials for both CVN and K_{Jc} test results. The 95% confidence intervals for fracture toughness are generally greater than those for the Charpy impact energy.

It is clear that the transition temperature shifts measured by both K_{Jc} and CVN are reduced by decreasing both the copper and the nickel contents. Reducing either one separately does not produce as substantial a change in the shift as reducing both. Only the welds can be compared

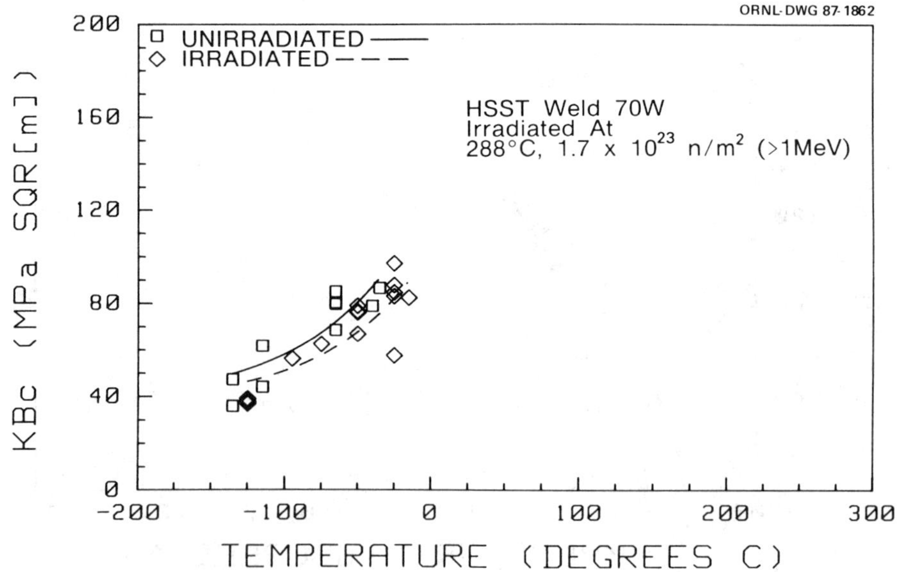

FIG. 12—*Variation of* $K_{\beta c}$ *with temperature and irradiation for Weld 70W.*

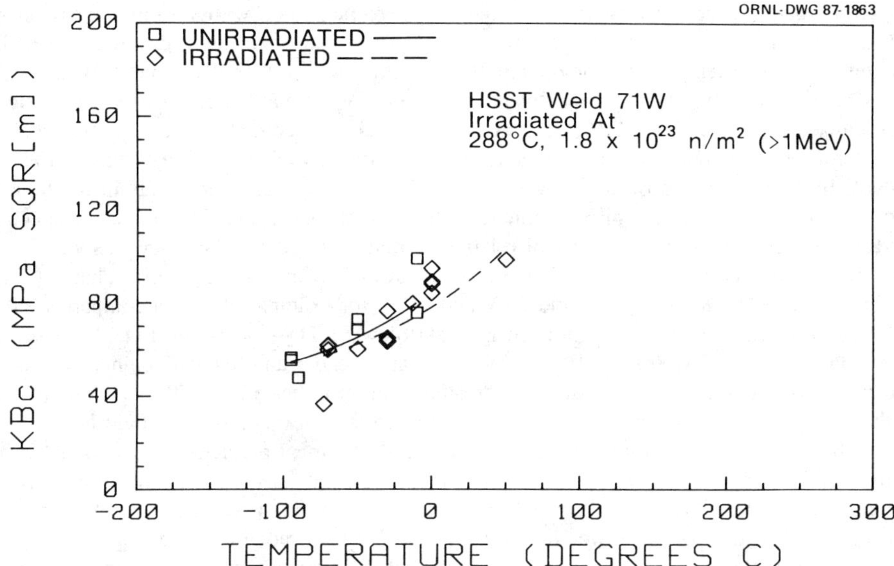

FIG. 13—*Variation of* $K_{\beta c}$ *with temperature and irradiation for Weld 71W.*

directly. The highest shift is evidenced by the results from Weld 69W with the highest copper and lowest nickel, indicating that copper alone has a substantial effect. However, increased nickel with lowered copper levels (Welds 70W and 71W) also produced a substantial but lesser shift. The lowest shift was produced by Weld 68W, where both the copper and nickel levels were low. Although the magnitude of the transition temperature shifts (K_{Jc}, $K_{\beta c}$, or CVN) varies somewhat, all three support these observations.

There are a number of factors that could obscure the conclusions regarding effects of composition. One factor is the fluence difference between the various welds. To evaluate the sensitivity of results to neutron fluence, the methods of Refs *2* and *7* were used to calculate predicted shifts for Weld 69W assuming the higher fluence experienced by welds 70W and 71W. Those methods predict a shift of only 3 to 4°C greater than that at the actual fluence. Thus, the difference in fluence for the welds of this study do not substantially obscure the observations. A second factor

TABLE 7—*Comparison of 95% confidence intervals for transition temperature shifts of Charpy impact and fracture toughness tests.*

	Transition Temperature Increase, °C			
	CVN		K_{Jc}	
Material	At 41 J	95% Confidence Interval	at 125 MPa · $m^{1/2}$	95% Confidence Interval
Plate 02	68	±4	81	±8
Weld 68W	10	±13	−6	±10
Weld 69W	27	±11	47	±10
Weld 70W	23	±8	25	±18
Weld 71W	23	±10	−1	±25

is material condition before irradiation. The welds were fabricated with various fluxes and weld wires and given different postweld heat treatments. That resulted in quite different unirradiated strengths. The room temperature yield strengths of welds 68W, 69W, 70W, and 71W were about 542, 638, 478, and 469 MPA, respectively. The ratios of ultimate strength to yield strength at room temperature were about 1.16, 1.13, 1.24, and 1.28, respectively. Additionally, other than the differences in copper and nickel contents, the silicon contents of 70W and 71W were substantially higher than those for 68W and 69W. Fourth, the scatter of results in the transition region for both CVN and fracture toughness tests is quite large, and the number of specimens tested at any given temperature is small relative to that required for a comprehensive statistical evaluation. Thus, curve fits can be significantly skewed with resulting large changes in the calculated shifts. The K_{Jc} shift for Weld 70W and 71W, for example, does not compare well with the CVN shift. The curve for unirradiated K_{Jc} data of weld 71W, shown in Fig. 7, has a much lower slope than the curve fits for the other welds. Large scatter and few data points have skewed the curve toward that for the irradiated data, resulting in no indicated shift at 125 MPa · $m^{1/2}$. It is felt that differences of 15°C in K_{Jc} curve shift determinations in this program could be expected.

Table 8 shows a comparison of several radiation embrittlement predictive calculations with the results of this study. The predictive methods were developed only with Charpy impact results. Furthermore, the trend curve formulations are largely based on surveillance results (except for those of Ref *7*) and represent lower fluxes and much longer irradiation times than those for the present experimental study. The values shown for Ref *2* do not include the specified "margin," and those for Ref *7* do not include the bounds specified for conservatism. It should be noted that the Regulatory Guides [*1,2*] are generally conservative. Also, relative to the predictions of the Regulatory Guide; the draft regulatory guide [*2*] (without "margin") predicts lower shifts for welds 68W and 69W but predicts higher shifts for 70W and 71W. The draft guide incorporates nickel content while Revision 1 of the guide does not. Thus, the draft guide predicts higher shifts for the two welds with higher nickel contents. It is interesting to note the wide disparity in shift predictions for the four welds. Using the experimental trend curves of Ref *17*, shown parenthetically in Table 8, the predictions increased for welds and decreased for Plate 02.

In each of Figs. 3 through 8, the ductile-shelf behavior of K_{Jc} shows a moderate decrease with temperature, for both the unirradiated and irradiated conditions. However there is no significant difference between the unirradiated or irradiated values on the ductile shelf. As shown in Table 6, the reduction in the ductile-shelf K_{Jc} is very limited for all the materials studied, including Plate 02. This correlates well with the results from the Charpy-*V*-notch impact tests. It appears for this range of copper and nickel compositions, there is little effect of irradiation on ductile-shelf behavior.

TABLE 8—*Comparison of several radiation embrittlement prediction expressions with results of present study.*

		Charpy Impact Transition Temperature Increase, °C							
			Reference						
Material	Fluence, 10^{23} n/m² (>1 MeV)	Observed ΔTT	[*1*]	[*2*][a]	[*3*]	[*4*]	[*5*]	[*6*]	[*7*]
Plate HSST-02	1.8	68	78	65	72	59	66	65	72 (55)
Weld HSST-68W	1.4	10	26	22	−44	20	17	20	−5 (13)
Weld HSST-69W	1.2	27	55	36	23	35	29	37	34 (41)
Weld HSST-70W	1.6	23	39	52	31	46	76	82	3 (21)
Weld HSST-71W	1.6	23	39	43	26	38	74	82	−2 (17)

[a] Without "margin."

[b] Primary values shown are based on surveillance trend curves while the values shown in parentheses are based on experimental trend curves, in both cases for plate and welds as applicable.

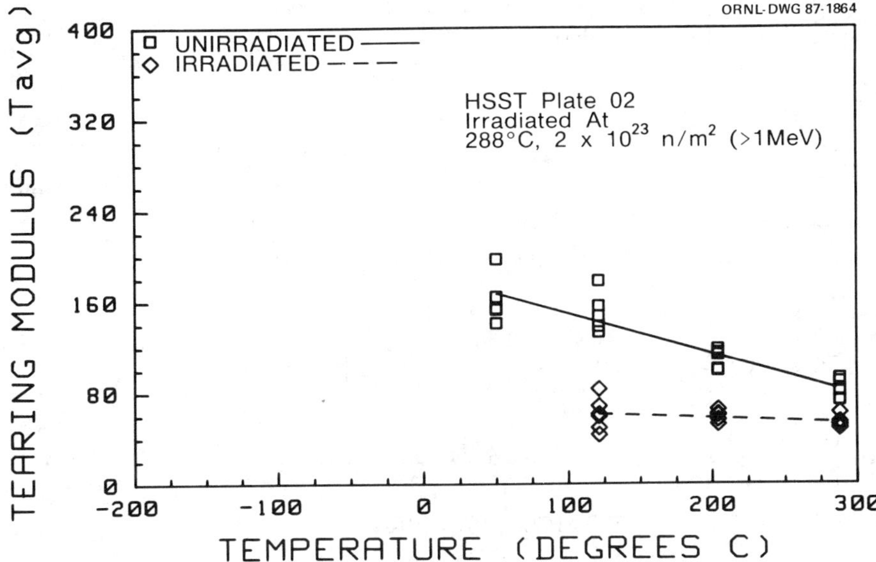

FIG. 14—*Variation of tearing modulus with temperature and irradiation for Plate 02.*

The effects of irradiation on tearing modulus are shown in Figs. 14 through 18 for Plate 02 and Welds 68W, 69W, 70W, and 71W, respectively. Note that in each of these figures, the tearing modulus decreases in a roughly linear fashion with temperature, whether the material is irradiated or not. Primarily, the *J*-integral results have indicated that reduced levels of copper and nickel will reduce the degradation of the ductile-shelf tearing modulus; although there may still be some decrease in the tearing modulus caused by the increase in the yield stress (see Eq 1). As noted in Table 6, significant decreases (>25%) in the tearing modulus were observed in all materials other

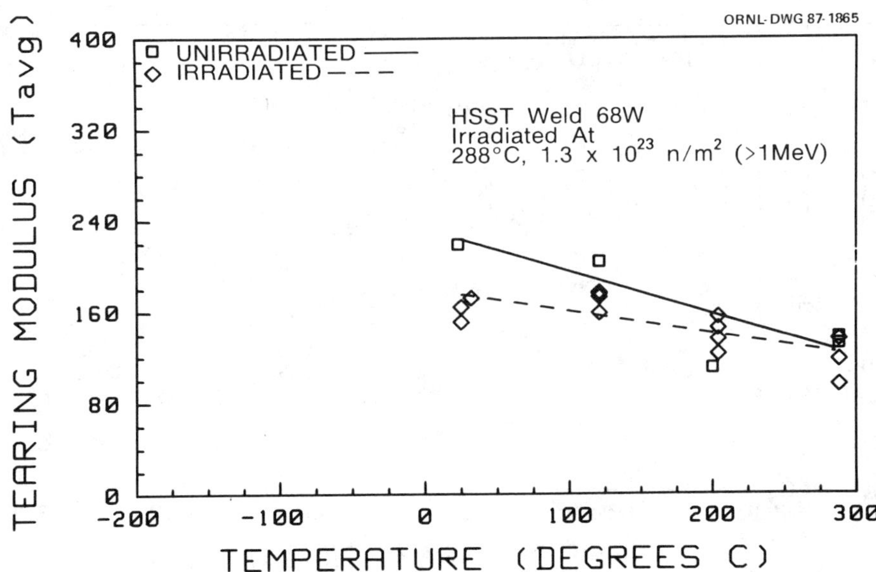

FIG. 15—*Variation of tearing modulus with temperature and irradiation for Weld 68W.*

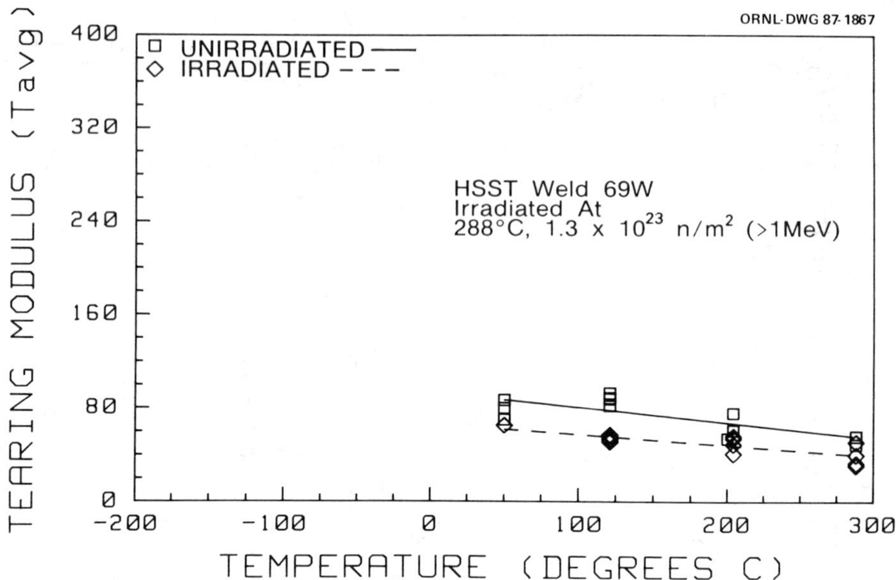

FIG. 16—*Variation of tearing modulus with temperature and irradiation for Weld 69W.*

than Weld 68W. Only when both copper and nickel were reduced was the tearing modulus reduction reduced.

Conclusions

1. There was good agreement between the fracture toughness results from the two laboratories, ORNL and MEA.
2. Ductile-shelf elastic-plastic initiation fracture toughness is essentially insensitive to radiation

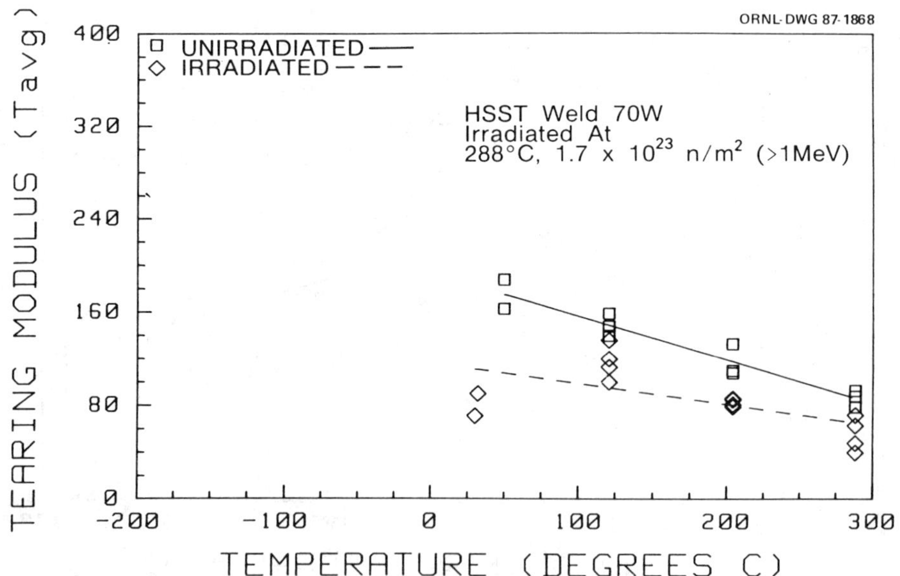

FIG. 17—*Variation of tearing modulus with temperature and irradiation for Weld 70W.*

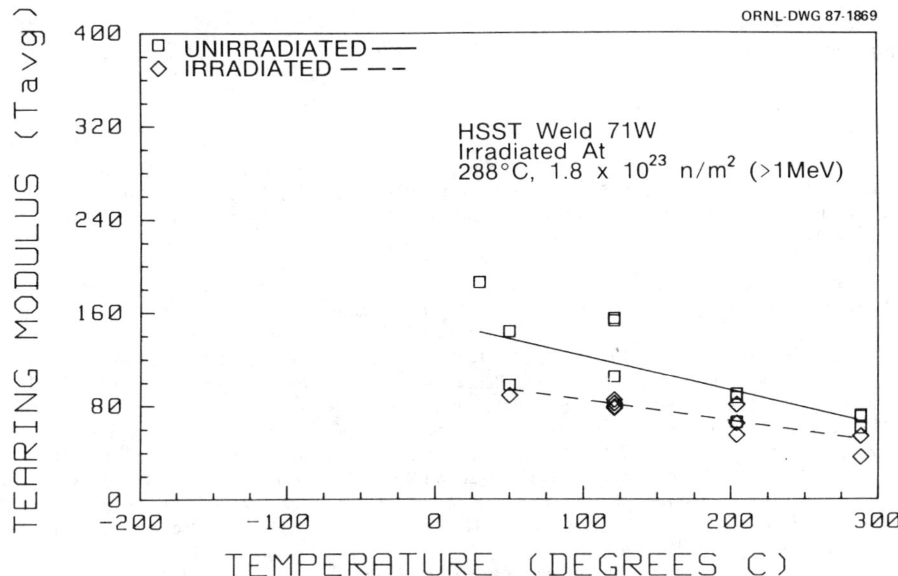

FIG. 18—*Variation of tearing modulus with temperature and irradiation for Weld 71W.*

damage and transition temperature shifts are relatively small for the current-practice welds characterized in this study.

3. Transition temperature shifts to higher temperature with irradiation are directly related to the copper and nickel contents of the material, with copper producing the greatest effect.

4. Ductile-shelf initiation fracture toughness and tearing modulus decrease linearly with temperature.

5. Irradiation degradation of the tearing modulus is related to the copper and nickel contents of the material.

6. Transition temperature changes are similar, whether estimated from Charpy-*V*-notch data indexed to the 41-J level, K_{Jc} data indexed to the 125-MPa · m$^{1/2}$ level, or $K_{\beta c}$ data indexed to the 90-MPa · m$^{1/2}$ level.

7. The scatter of fracture toughness K_{Jc}, with small specimens in the transition region is similar to that observed with Charpy impact tests.

8. A small data set can lead to large uncertainties regarding irradiation-induced changes in transition-temperature fracture toughness.

Acknowledgments

We wish to thank M. Vagins of the U.S. Nuclear Regulatory Commission for financially supporting this study. We appreciate the assistance of the Electric Power Research Institute and Babcock and Wilcox Company in providing the weldments. We wish to thank F. B. K. Kam and his coworkers for the dosimetry analyses; J. W. Woods and D. Heatherly for construction and operation of the irradiation capsules; technicians T. N. Jones, T. D. Owings, and R. L. Swain for conducting the tests at ORNL; and L. Lamont, B. H. Menke, and K. C. Miller for conducting the tests at MEA. We also thank J. L. Bishop for preparation of the manuscript.

This research was sponsored by the Office of Nuclear Regulatory Research, Division of Engineering Safety, U.S. Nuclear Regulatory Commission, under Interagency Agreement DOE

0551-0551-A2 with the U.S. Department of Energy under Contract DE-AC05840R21400 with Martin Marietta Energy Systems, Inc.

References

[1] "Effects of Residual Elements on Predicted Radiation Damage to Reactor Vessel Materials," Regulatory Guide 1.99 (Revision 1), U.S. Nuclear Regulatory Commission, Washington, DC, April 1977.

[2] Randall, P. N., "Effects of Residual Elements on Predicted Radiation Damage to Reactor Vessel Materials," Regulatory Guide 1.99 (Revision 2 Draft), U.S. Nuclear Regulatory Commission, Washington, DC, Jan. 1984.

[3] Varsik, J. D. and Byrne, S. T., "An Empirical Evaluation of the Irradiation Sensitivity of Reactor Pressure Vessel Materials," in *Effects of Irradiation on Structural Materials, STP 683,* American Society for Testing and Materials, Philadelphia, 1979, pp. 252–266.

[4] Perrin, J. F., Wullaert, R. A., Odette, G. R., and Lombrozo, P. M., *Physically Based Regression Correlations of Embrittlement Data from Reactor Pressure Vessel Surveillance Programs,* EPRI NP-3319, Electric Power Research Institute, Palo Alto, CA, Jan. 1984.

[5] Varsik, J. D., Schloss, S. M., and Koziol, J. J., *Evaluation of Irradiation Response of Reactor Pressure Vessel Materials,* EPRI NP-2720, Electric Power Research Institute, Palo Alto, CA, Nov. 1982.

[6] Guthrie, G. L., "Charpy Trend Curves Based on 1717 PWR Data Points," *LWR Pressure Vessel Surveillance Dosimetry Improvement Program, Quarterly Progress Report for the Period April–June 1983,* NUREG/CR-3391, Vol. 2, HEDL-TME-83-22, Hanford Engineering Development Laboratory, Richland, WA, pp. HEDL-3-4.

[7] Metal Properties Council Subcommittee 6 on Nuclear Materials, "Prediction of the Shift in the Brittle-Ductile Transition Temperature of Light-Water Reactor (LWR) Pressure Vessel Materials," *Journal of Testing and Evaluation,* Vol. 11, No. 4, July 1983, pp. 237–260.

[8] Potapovs, U. and Hawthorne, J. R., "The Effect of Residual Elements on the Response of Selected Pressure Vessel Steels and Weldments to Irradiation at 550°F," *Nuclear Applications,* Vol. 6, No. 1, 1969, pp. 27–46.

[9] Guionnet, C., Houssin, B., Brasseur, D., Lefort, A., Gros, D., and Perdreau, R., "Radiation Embrittlement of PWR Reactor Vessel Weld Metals: Nickel and Copper Synergism Effects," *Effects of Radiation on Materials: Proceedings of the Eleventh International Symposium, STP 782,* American Society for Testing and Materials, Philadelphia, 1982, pp. 392–411.

[10] Leitz, C., Gerscha, A., Hofmann, G., and Strobel, H. J., "Comparative Irradiation Study of Reactor Pressure Vessel Steel Weld Metals," *Effects of Radiation on Materials: Proceedings of the Eleventh International Symposium, STP 782,* American Society for Testing and Materials, Philadelphia, 1982, pp. 412–432.

[11] Menke, B. H., McGowan, J. M., Berggren, R. G., Nanstad, R. K., and Miller, K. C., "Effects of Neutron Irradiation on Fracture Toughness of A533-B Class 1 Plate and Four Submerged-Arc Welds," *Effects of Radiation on Materials: Twelfth International Symposium, STP 870,* F. A. Garner and J. S. Perrin, Eds., American Society for Testing and Materials, Philadelphia, 1985, pp. 1111–1130.

[12] Berggren, R. G., McGowan, J. J., Menke, B. H., and Nanstad, R. K., "Irradiation Effects on Fracture Toughness of Four Nuclear Reactor Pressure Vessel Submerged-Arc Welds," *Nuclear Engineering and Design,* Vol. 89, 1985, pp. 181–198.

[13] Childress, C. E., *Fabrication History of the First Two 12-inch Thick ASTM A533 Grade B, Class 1 Steel Plates of the Heavy Section Steel Technology Program, Documentary Report 1,* ORNL-4313, Oak Ridge National Laboratory, Oak Ridge, TN, Feb. 1969.

[14] Marston, T. V., Borden, M. P., Fox, J. H., and Reardon, L. D., *Fracture Toughness of Ferritic Materials in Light Water Nuclear Reactor Vessels,* MML-75-152, Combustion Engineering, Inc., Chattanooga, TN, Oct. 1975.

[15] Lowe, A. L., Jr. and Quresi, J. I., *Fabrication of Weldments Using Linde 80 and Linde 124 Weld Fluxes for HSST-4 Irradiation Program,* BAW-1537, Babcock and Wilcox Company, Lynchburg, VA, June 1981.

[16] Baldwin, C. A., *Neutron Spectral Characterization Calculations for the Fourth Nuclear Regulatory Commission Heavy Section Steel Technology 1T-CT Irradiation Experiments,* NUREG/CR-3311, ORNL/TM-8782, Oak Ridge National Laboratory, Oak Ridge, TN, June 1983.

[17] Stallman, F. W., Baldwin, C. A., and Kam, F. B. K., *Neutron Spectral Characterization Calculations for the Fourth Nuclear Regulatory Commission Heavy Section Steel Technology 1T-CT Irradition Experiments: Dosimetry and Uncertainty Analysis,* NUREG/CR-3333, ORNL/TM-8789, Oak Ridge National Laboratory, Oak Ridge, TN, July 1983.

[18] Ernst, H. A., "Material Resistance and Instability Beyond J-Controlled Crack Growth," *Elastic-Plastic*

Fracture, Inelastic Analysis, STP 803, American Society for Testing and Materials, Philadelphia, 1983, pp. 191–213.
[*19*] Albrecht, P., Andrews, W. R., Gudas, J. P., Joyce, J. A., Loss, F. L., McCabe, D. E., Schmidt, D. W., and Van Der Sluys, W. A., "Tentative Test Procedure for Determining the Plain Strain J_I-R Curve," *Journal of Testing and Evaluation,* Vol. 10, No. 6, Nov. 1982, pp. 245–251.
[*20*] Irwin, G. R., "Fracture Mode Transition for a Crack Traversing a Plate," *Journal of Basic Engineering,* Vol. 82, No. 2, 1960, pp. 417–25.
[*21*] Merkle, J. G., *An Examination of the Size Effects and Data Scatter Observed in Small-Specimen Cleavage Fracture Toughness Testing,* NUREG/CR-3672, ORNL/TM-9088, Oak Ridge National Laboratory, Oak Ridge, TN, April 1984.
[*22*] Marquardt, D., "An Algorithm for Least Squares Estimation of Nonlinear Parameters," *Journal of the Society for Industrial and Applied Mathematics,* Vol. 11, No. 2, 1963.
[*23*] McGowan, J. J. and Nanstad, R. K., research as yet unpublished.
[*24*] McGowan, J. J., *Tensile Properties of Irradiated Nuclear Grade Pressure Vessel Plate and Welds for the Fourth HSST Irradiation Series,* NUREG/CR-3978, ORNL/TM-9516, Oak Ridge National Laboratory, Oak Ridge, TN, Jan. 1985.
[*25*] Berggren, R. G., Hawthorne, J. R., and Nanstad, R. K., "Analysis of Charpy V-Notch Impact Toughness of Irradiated A533-B Class 1 plate and Four Submerged-Arc Welds," *Effects of Radiation on Materials: Twelfth International Symposium, STP 870,* F. A. Garner and J. S. Perrin, Eds., American Society for Testing and Materials, Philadelphia, 1985, pp. 1094–1110.

Radiation Damage in Nonmetals

Hisayuki Matsui,[1] *Mikio Horiki,*[1] *Masayoshi Tamaki,*[1] *and Tomoo Kirihara*[1]

Fission Fragment Damage in Ceramic Nuclear Fuels

REFERENCE: Matsui, H., Horiki, M., Tamaki, M., and Kirihara, T., "**Fission Fragment Damage in Ceramic Nuclear Fuels,**" *Influence of Radiation on Material Properties: 13th International Symposium (Part II), ASTM STP 956,* F. A. Garner, C. H. Henager, Jr., and N. Igata, Eds., American Society for Testing and Materials, Philadelphia, 1987, pp. 593–601.

ABSTRACT: Post-irradiation examinations of several ceramic nuclear fuels are summarized, emphasizing the identification and characterization of fission induced defects. The lattice parameter and electrical resistivity were significantly altered with fission damage in all fuels. A remarkable reduction, however, was found in the irradiation induced changes of both physical properties in uranium dioxide (UO_2), uranium nitrite (UN), and uranium sesquicarbide (U_2C_3) at high fission doses, probably because of a preferential annihilation of extended defects in the damage process. An effect of the extended defects in irradiated single crystalline UC was a greater increased resistivity at low temperature, indicating a contribution of interstitial-type defects. The irradiation induced physical properties, on the other hand, were annealed out, within a few recovery steps in all the fuels, exhibiting some dependence on the extent of the fission damage. It was of great interest that a drastic change was observed in the magnetic property changes in some magnetic substances, such as uranium monosulfide (US) and UN. All the magnetic parameters changed with fission damage and recovered to the original values in successive annealing processes.

KEY WORDS: fission damage, ceramic nuclear fuels, lattice parameters, electrical resistivity, magnetic properties, recovery annealing

Knowledge of the effects of fission (fragment) damage on the physical properties in nuclear fuels is essential to understand the irradiation behavior and hence is useful to predict the irradiation performance of the practical fuels in nuclear reactors. However, little work has been performed on the fundamental processes of fission damage, even for the well-known oxide fuel (uranium dioxide, UO_2) used in the light water reactor. In the potential use of carbide-type fuels for fast breeder reactor, more studies must be done to obtain a fundamental understanding of the irradiation effects, in order to establish the performance behavior of the nuclear fuels.

In our laboratory, post-irradiation examinations have been performed on several ceramic nuclear fuels, such as uranium carbides (uranium monocarbide UC, uranium dioxide UC_2, and uranium sesquicarbide [U_2C_3]), nitrides (UN and β-U_2N_3), and others (US and uranium monophosphide [UP]), as well as UO_2 for a practical interest. The changes in the lattice parameter and electrical resistivity at room temperature have been extensively examined after reactor (thermal neutron) irradiations. Recently, we have focused mainly on the fission damage in single crystalline uranium monocarbide (UC).

Though several pioneering works have been previously reported by other investigators for UC [*1–6*], UN [*5,6*] and UO_2 [*7,8*], those results have not been in good agreement with each other. In the present report, we summarize our results of the post-irradiation examinations on the above ceramic nuclear fuels and discuss the irradiation induced defects produced in a relatively earlier

[1] Department of Nuclear Engineering, Nagoya University, Furo-cho, Chikusa-ku, Nagoya 464, Japan.

stage of the fission damage. Recent studies on single crystalline UC and on the magnetic properties of ferromagnetic uranium monosulfide (US) are also presented to interpret the mechanisms of the fission damage.

Procedure

Specimens

Carbides and nitrides of uranium in the present study were prepared by arc-melting and plasma-jet melting, while others were sintered specimens obtained from Japan Atomic Energy Research Institute (JAERI). We obtained single crystalline UC by a high-temperature annealing of the plasma-fused polycrystal. Chemical analysis was mainly carried out for light elements, such as carbon, nitrogen, and oxygen (C, N, and O), because they significantly affected the physical properties of the uranium compounds investigated here.

The size of the specimens was small, frequently with dimensions of 1 by 1 by 5 mm in a rectangular form, being convenient for measurements of the principal physical properties. These were irradiated in the Japan Research Reactors (JRR) and in some cases in the Japan Material Test Reactor (JMTR) and the Kyoto University Reactor (KUR). Both JRR and KUR had slightly different neutron spectra, but the thermal neutron flux and the irradiation temperature were nearly identical. On the other hand, the neutron flux and the temperature were high, 10^{18} n/m^{2}s and about 400°C, respectively, in JMTR.

Measurements

An improved double crystal X-ray diffractometer (using germanium bent-crystals) was used to measure the lattice parameter. Measurement of the electrical resistivity was done by the conventional four probe method, both in a cryostat (for low temperature) and in a quartz tube furnace (for high temperature). The magnetic properties were measured using a torsion balance in a temperature range between 4.2 or 77 and 273 K. In the latter two cases, we applied a microcomputer controlled system in the measurements. In that manner, enough data points were obtained to yield precise results by least square fitting.

Results and Discussions

Lattice Parameter

The change in lattice expansion caused by fission damage is shown in Fig. 1 for polycrystalline UC (arc-melt, $UC_{0.98}$), UN (plasma-fused, $UN_{0.99}$), US (sintered, $US_{1.002}$), UO_2 (sintered, $UO_{2.004}$), and U_2C_3 (arc-melted). In all cases, the lattice parameter initially increased with fission dose, fissions/m^3 (or f/m^3, a common unit for all nuclear fuels), attained a maximum and then decreased with further irradiations. A similar trend was observed in the nuclear fuels with the sodium chloride (NaCl)-type crystal structure, UC, UN, and US. However, a remarkable reduction of the lattice expansion at high dose was found in UN in the present study, similar to that previously reported for sintered specimens [5]. A major part of the reduction was caused by a preferential annihilation of extended defects with each other. In the irradiated UN, however, survival of excess vacancies after saturation of the extended defects might also contribute to the shrinkage of the lattice spacing at high fission doses.

The effective displacement volume of a single fission event could be evaluated in each compound from the dose dependence of the lattice parameter. A nearly constant value (4×10^{-23} m^3) was obtained for the NaCl-type compounds, while for UO_2 (calcium fluoride [CaF_2])-type) a large volume (6×10^{-22} m^3 at the first saturation step) was derived. The same trend as UO_2 has been

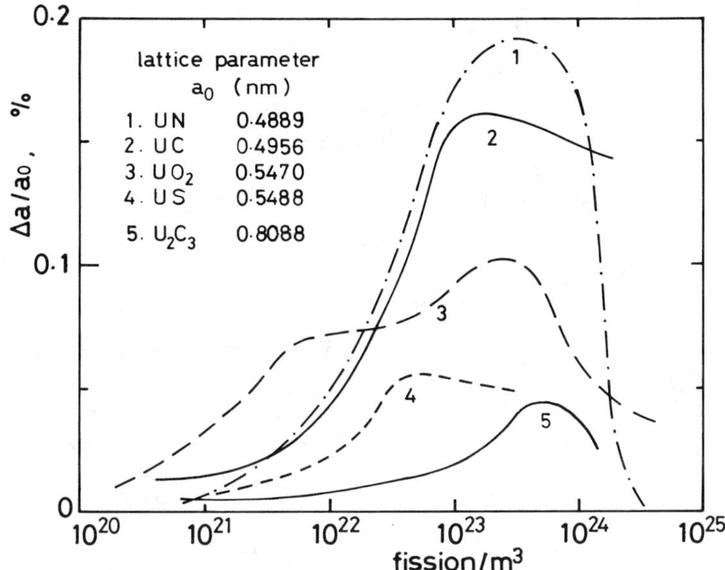

FIG. 1—*Change of lattice parameter in ceramic nuclear fuels (polycrystal with cubic crystal structure) with fission dose.*

reported for other oxide fuels, such as plutonium oxide (PuO_2) [*9*] and thorium oxide (ThO_2) [*10*]. Uranium sesquicarbide (U_2C_3, a complicated bcc structure with a large lattice constant, a_0 = 0.8088 nm) did not show saturation until a high fission dose, hence the effective volume could be small, though the irradiations were done at a high temperature (400°C in JMTR). In low-temperature irradiations of this material, there was no saturation until 5×10^{23} f/m^3. The lattice parameters, a_0 and c_0 of the bct structure, of α-UC_2 showed a different type of change, which was however consistent with an increase of the electrical resistivity [*11*]. In the initial stage of fission damage, the *c*-axis shrank, while no change was found in the *a*-axis, and a remarkable increase occurred at about 10^{22} f/m^3 in both axes.

In the case of UC, the saturation value of the lattice expansion seemed to be different with the nonstoichiometry; the higher the carbon content, the lower the saturated lattice expansion (see also Ref *1*). Temperature during irradiation also played a role in the reduction of the lattice expansion, because the increase was small in the JMTR irradiations compared with in JRRs. As a matter of fact, there was a missing recovery step below 400°C in the annealing process of such specimens.

The change of lattice parameter in single crystal UC (plasma-fused, $UC_{1.00}$) was similar to that in the polycrystals. Using a single crystalline UC specimen irradiated to 1.9×10^{22} f/m^3, we examined the recovery behavior of the fission-induced lattice expansion. As shown in Fig. 2, there were clearly three steps of the recovery. These steps were revealed in all the fuels investigated here, though the temperatures and the extents of the recovery steps were not identical in each fuel. When the fission dose was high in a particular compound, the recovery shifted to lower temperatures. Considering the results of a quenching study for UC (both for single and polycrystals [*12*]), the low (200°C) and high (750°C) temperature step should be correlated to the migration of carbon and uranium vacancies, respectively. The middle step, at about 500°C in Fig. 2, was only observed as a result of fission damage. Therefore, it was concluded that some sort of an extended defect was introduced by the fission damage. A comparatively low temperature (500°C) and low activation energy (1.0 ± 0.1 eV) at the middle step indicated that rather simple type extended defects (probably the interstitial clusters) might be attributed to that recovery step.

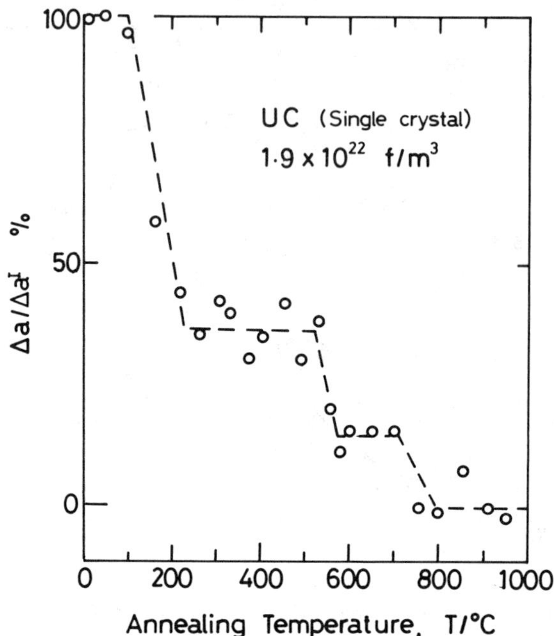

FIG. 2—*Annealing of fission-induced lattice expansion in single crystalline UC irradiated to 1.9 × 10²² f/m³.*

Electrical Resistivity

In the change of the electrical resistivity by fission damage, a quite similar trend to that found in the lattice parameter was observed in all compounds except the semiconductive UO_2. An example is shown in Fig. 3 for single crystalline UC. The measuremens were done at room temperature (285 K). The resistivity began to increase steeply at 1×10^{22} f/m³, where the extended defects had been obsevable as black dots in a transmission electron microscopy study [*13*]. It was noted here that the incremental increase of the resistivity at saturation was much larger in single crystals ($\Delta\rho/\rho_0 = 170\%$) than in polycrystals (75%) [*14*]. It was also noted that the high-temperature irradiations (JMTR) reduced the incremental increase (40%) [*15*] at saturation. Again in this case, a mutual annihilation of the irradiation induced extended defects reduced the resistivity increase at high fission doses. Comparing with the large in-growth resistivity of UC, on the other hand, other compounds indicated much smaller incremental increases at saturation ($\Delta\rho/\rho_0$ was about 10 and 5% in UN [*16*] and US [*17*], respectively). A remarkable reduction of the increased resistivity was again revealed in UN and U_2C_3 at high fission doses, as expected from the lattice shrinkage mentioned in the preceding section. In addition, a value of the effective displacement volume, nearly identical to that derived from the lattice expansion results, was derived for the fuels with the NaCl-type structure from that dose dependence of the electrical resistivity. Only α-UC_2 showed a different irradiation behavior from the others: the first saturation was at around 10^{22} f/m³ followed by a further rapid increase of the resistivity up to 10^{24} f/m³, as also expected from an anomalous lattice expansion [*11*].

Dependence of the low-temperature electrical resistivity is shown in Fig. 4 for the case of single crystal UC irradiated to 9.2×10^{20} f/m³ at room temperature. A special feature was found in the difference of the resistivity between pre- and post-irradiation. The difference was larger at low temperature (below 130 K) than at room temperature. A single crystal UC specimen with a high

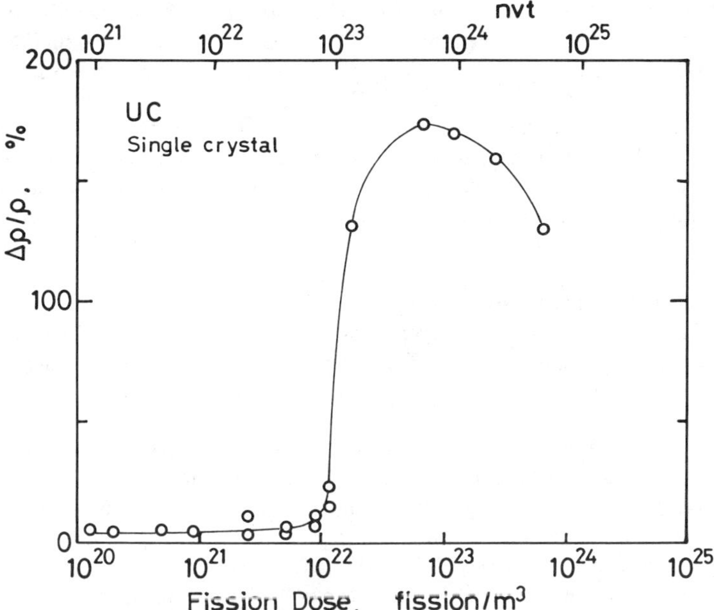

FIG. 3—*In-growth electrical resistivity caused by fission damage in single crystalline UC.*

fission dose (2×10^{22} f/m^3) also previously indicated a more pronounced in-growth resistivity at a lower temperature [*18*]. If only vacancies were introduced, say by quenching, the resistivity difference would be inverse: smaller at lower temperature than at room temperature. The induced change in the low-temperature electrical resistivity by fission damage was highly pronounced in magnetic substances, such as US and UN, which will be treated in the next section.

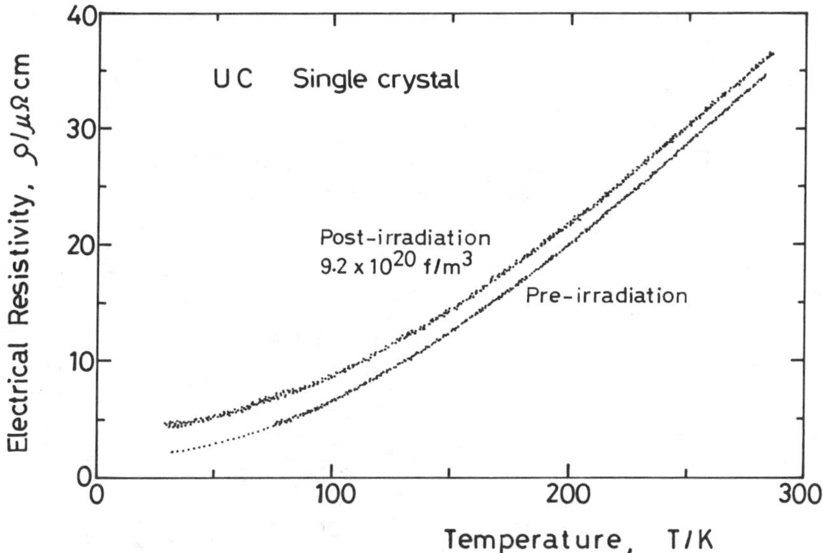

FIG. 4—*Temperature dependence of electrical resistivity below room temperature of UC (single crystal) with and without fission damage (9.2×10^{20} f/m^3).*

An interesting feature was found in the annealing behavior of the increased resistivity of the irradiated UC single crysals. As shown in Fig. 5, the recovery behavior was similar to the lattice parameter recovery (as shown in Fig. 2), when the resistivity measurements were done at room temperature. There existed three steps of recovery below 1000°C. The low (250°C) and high (800°C) temperature steps were due to the movement of carbon and uranium vacancies, respectively, that has been confirmed by quenching studies for both single and polycrystalline UC [*12*]. An extra step at 500°C was only and always observed in the neutron irradiated UC. On the other hand, in the measurements at lower temperature (for instance, at liquid nitrogen temperature), only two steps were detected, at 500 and 800°C. The low-temperature step at 250°C was very ambiguous or missing.

These phenomena have been also found in quenched UC. In quenched UC, two recovery steps appeared at 200 and 700°C in the room temperature measurements (see Ref *10* for the polycrystalline UC). Whereas, in the measurements at liquid nitrogen temperature, only a single step was observed at 700°C. Again the low-temperature recovery step at 200°C was missing in the case of the quenched UC.

When vacancies were introduced (by quenching) before neutron irradiation, the extent of the middle step in the irradiated UC became smaller with increasing quenching temperature, and again the low-temperature step was missing. The extended defects produced by fission damage must annihilate with the preexisting quenched-in vacancies. Therefore, the type of the extended defects should be interstitial in nature. In conclusion, the defects that contributed to the middle step in the recovery of the neutron irradiated UC were insterstitial clusters that yielded the extra resistivity at lower temperature and the elongation of the crystal lattice as well.

Magnetic Property

The effect of fission damage on the magnetic properties is also interesting in magnetic substances, such as uranium monosulfide (US, ferromagnet) and uranium mononitride (UN, antiferromagnet). All the magnetic parameters were drastically altered: a reduction of the magnetic transition point, either the Curie point T_C or the Néel point T_N, and the change in the magnetization and electrical resistivity caused an introduction of magnetic (short range) disorder. As an example, we show the

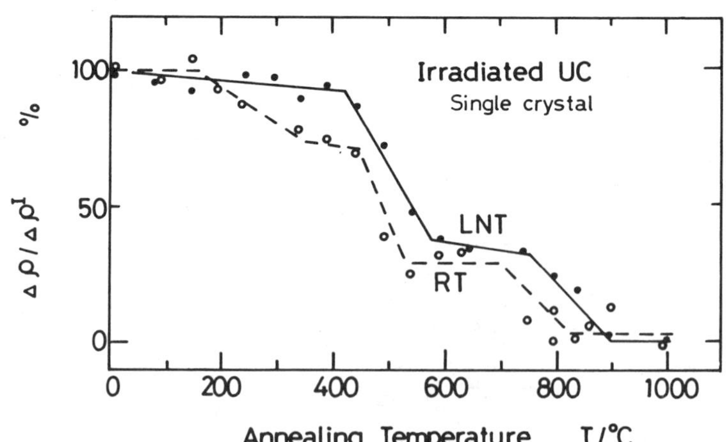

FIG. 5—*Annealing of fission-induced resistivity in UC (single crystal with 5.3 × 10^{21} f/m³) measured at 77 K (LNT) and 285 K (RT).*

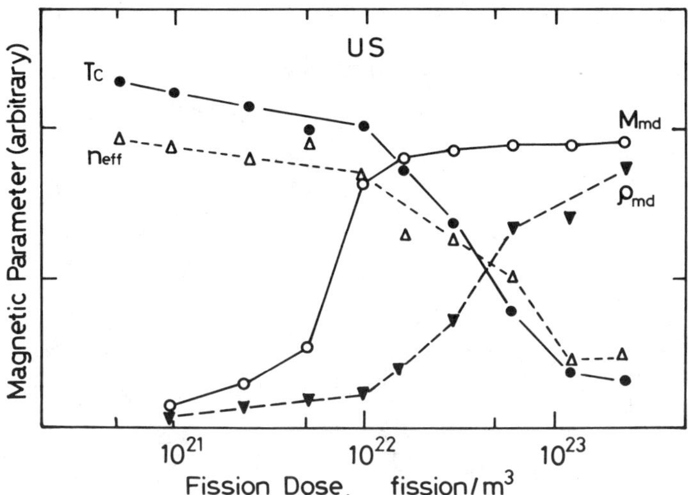

FIG. 6—*Change of magnetic parameters* (T_C, n_{eff}, M_{md}, and ρ_{md}) *in ferromagnetic US with fission dose.*

results on ferromagnetic US (T_C = 180 K) in Fig. 6 (M_{md} and ρ_{md} are the induced disordered magnetization and resistivity, respectively, and n_{eff} is the effective Bohr magneton numbers in the paramagnetic state). All parameters changed with fission dose, similar to that found in other physical properties.

Such behavior was also previously revealed in β-U_2N_3 (T_C = 188 K), UP (T_N = 123 K), and U_2C_3 (T_N = 57 K) [*19*]. Antiferromagnetic UN (T_N = 51.6 K) showed a more pronounced reduction of the transition point (37.4 K at 1.8 × 10^{24} f/m^3) and, in addition, a partial recovery in T_N occurred at a high fission dose (48.6 K at 3.6 × 10^{24} f/m^3) [*20*], again consistent with the dose dependences of other properties. Therefore, the origin of the fission induced change in the magnetic properties should be the same as mentioned previously. In fact, the shift of the magnetic transition points (either T_C and T_N) correspond very well to an elongation of the lattice parameter. The same behavior had been found in the UNC compounds, in which T_N shifted to lower temperatures in accordance with an expanding lattice parameter by an addition of the paramagnetic UC to the antiferromagnetic UN [*21*]. On the other hand, nonmagnetic materials, such as UC and α-UC_2, revealed an increase in the magnetic susceptibility, probably because of an accumulation of the fission products as magnetic impurities.

The fission induced magnetic changes were annealed out in successive pulse annealings at 50°C intervals up to 1200°C, in the same manner as in the electrical resistivity. In Fig. 7, we show the temperature dependence of the disordered resistivity of US irradiated to 2.4 × 10^{23} f/m^3 after some annealing steps. The resistivity varied strongly with annealing temperature in the magnetic state. In accordance with the recovery annealing of the disordered resistivity, T_C ($d\rho/dT$ is at a maximum) shifted to the original temperature. It is noticed here that the electrical resistivity obeyed a quadratic temperature dependence ($\rho = aT^2 + \rho_I^0$) below the Curie point in each case. In addition to a small annealing step at about 200°C, there was a large and wide recovery step between 400 and 800°C, which might include two annealing processes.

The disordered magnetization and all other magnetic parameters well recovered at the same temperature as above. But the temperature dependence of in-recovered magnetization was somewhat strange in US with high fission damage. Figure 8 illustrates the change of the induced magnetization with annealing. The measurements were performed at 0.4 *T* (tesla) that was applied after the specimen was cooled down to 4.2 K (zero-field cooling). The level of the magnetization became

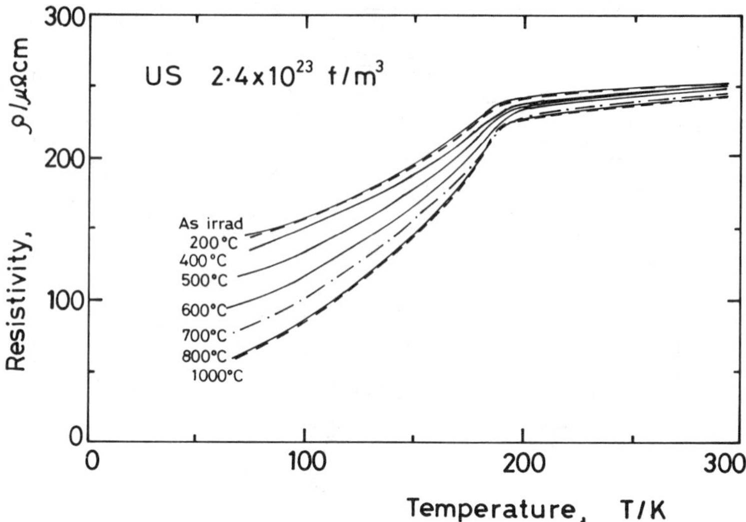

FIG. 7—*Annealing process of low-temperature electrical resistivity of US irradiated to 2.4 × 10^{23} f/m³.*

high with annealing and at 1000°C all the magnetic parameters returned back to the original values. But there were two peaks in the magnetization in low-temperature (400 and 600°C) annealed US. On the other hand, only a single peak was detected in US irradiated to lower fission doses, as was seen in the 800°C-annealed specimen in that figure. It is known that the lattice defects have a pinning effect on the magnetic domain walls in ferromagnetic substances, resulting in a reduced magnetization. Therefore, two types of defect could contribute to the reduction of the magnetization in US with high fission damage, where the lattice expansion and the increased electrical resistivity saturated. Here again the induced vacancies and the extended defects (the interstitial clusters)

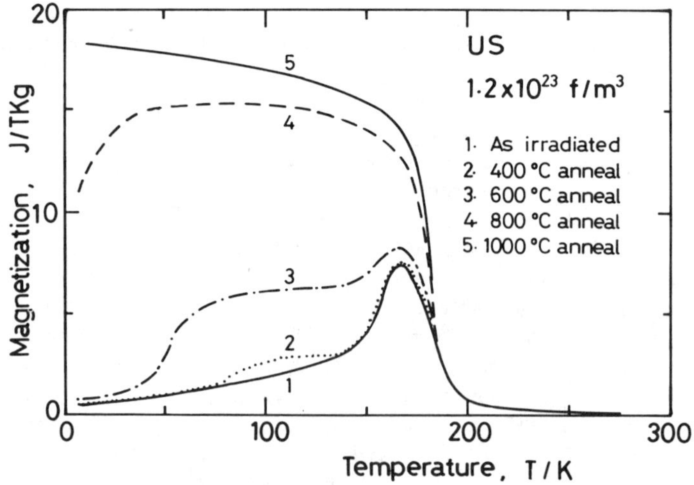

FIG. 8—*Low-temperature behavior of fission-induced (disordered) magnetization in US with high fission damage (1.2 × 10^{23} f/m³) at a variety of annealing temperatures (pulse annealing).*

could be assumed. Both could pin the magnetic domain walls, but the pinning effect might be different, resulting in the separate peaks in the magnetization curves shown in Fig. 8.

Acknowledgment

We are indebted to the colleagues in JAERI (especially to Dr. S. Nasu and H. Watanabe) and Dr. T. Tamai (KUR) for the reactor irradiations of the specimens. Financial support by the Japan Ministry of Education, Science and Cultures is also acknowledged.

References

[*1*] Childs, B. G., Ogilvie, A., Ruckman, J, C., and Whitton, J. L., *Radiation Damage in Reactor Materials,* International Atomic Energy Agency, Vienna, Austria, 1963, pp. 241–274.
[*2*] Childs, B. G., Ruckman, J. C., and Buxton, K., *Carbides in Nuclear Energy,* Vol. 2, L. E. Russell, Ed., Macmillan & Co. Ltd., London, 1964, pp. 849–863.
[*3*] Griffiths, L. B., *Transactions of the British Ceramic Society,* Vol. 62, No. 3, March 1963, pp. 189–195.
[*4*] Bloch, J. and Mustelier, J. P., *Journal of Nuclear Materials,* Vol. 17, No. 4, Dec. 1965, pp. 350–355.
[*5*] Dienst, E., "Radiation, Detection and Annealing of Crystal Lattice Point Defects in Neutron Irradiation of Ceramic Materials," Canadian Report AECL-4375, Whiteshell Nuclear Research Establishment, Pinawa, Manitoba, Canada, Feb. 1973.
[*6*] Rogers, M. D. and Adam, J., "Radiation Damage and Its Thermal Recovery in Uranium Monocarbide and Uranium Mononitride," AERE-R 4046, Atomic Energy Research Establishment, Harwell, June 1962.
[*7*] Wait, E., *Radiation Effects in Ceramic Fuels,* Vol. 2, B. T. Bradbury and B. R. T. Frost, Eds., Gordon and Breech, London, 1967, pp. 180.
[*8*] Bloch, J., *Journal of Nuclear Materials,* Vol. 3, No. 2, March 1961, pp. 237–238.
[*9*] Kapshukov, I. I., *Pu-1975 and Other Actinides,* H. Blank and R. Lindner, Eds., North-Holland, Amsterdam, The Netherlands, 1976, pp. 861–869.
[*10*] Akabori, H. and Shiba, K., *Journal of Nuclear Materials,* Vol. 101, No. 1/2, Oct. 1981, pp. 184–191.
[*11*] Matsui, H., Takada, T., Kirihara, T., and Nasu, S., *Journal of Nuclear Materials,* Vol. 56, No. 3, June 1975, pp. 275–278.
[*12*] Matsui, H. and Matzke, Hj., *Journal of Nuclear Materials,* Vol. 89, No. 1, March 1980, pp. 41–52.
[*13*] Eyre, B. L. and Sole, M. J., *Journal of Nuclear Materials,* Vol. 18, No. 3, March 1966, p. 314–322.
[*14*] Matsui, H., *Journal of Nuclear Materials,* Vol. 56, No. 2, May 1975, pp. 161–168.
[*15*] Matsui, H., Horiki, M., and Kirihara, T., *Journal of Nuclear Science and Technology,* Vol. 18, No. 12, Dec. 1981, pp. 922–929.
[*16*] Tamaki, M., Matsui, H., Ohnuki, A., Matsumoto, G., and Kirihara, T., *Radiation Effects,* Vol. 91, No. 1/2, Dec. 1985, pp. 61–69.
[*17*] Matsui, H., Nakashima, S., Katori, K., Tamaki, M., and Kirihara, T., *Journal of Nuclear Materials,* Vol. 110, No. 2/3, Oct. 1982, pp. 208–214.
[*18*] Matsui, H., Horiki, M., Ohya, N., Kato, T., and Osada, M., *Journal of Nuclear Materials,* Vol. 115, No. 1, March 1983, pp. 128–130.
[*19*] Matsui, H., *Journal of Nuclear Science and Technology,* Vol. 18, No. 11, Nov. 1981, pp. 895–897.
[*20*] Tamaki, M., Ohnuki, A., Matsui, H., Matsumoto, G., and Kirihara, T., *Physica B + C,* Vol. 102B + C, No. 1/3, Dec. 1980, pp. 258–261.
[*21*] Tamaki, M., Matsui, H., Matsumoto, G., and Kirihara, T., *Physica Status Solidi (a),* Vol. 79, No. 1, Sept. 1983, pp. K93–98.

Masayoshi Tamaki,[1] Hisayuki Matsui,[1] and Tomoo Kirihara[1]

Fission Fragment Induced Hardening in Uranium Mononitride

REFERENCE: Tamaki, M., Matsui, H., and Kirihara, T., "**Fission Fragment Induced Hardening in Uranium Mononitride,**" *Influence of Radiation on Material Properties: 13th International Symposium (Part II), ASTM STP 956,* F. A. Garner, C. H. Henager, Jr., and N. Igata, Eds., American Society for Testing and Materials, Philadelphia, 1987, pp. 602–608.

ABSTRACT: Fission fragment-induced hardening of uranium mononitride (UN) was examined by the Knoop indentation technique. Measurements were done at a variety of indentation loads (25 to 500 g) and for specimens with various fission doses (2.7×10^{21} to 5.3 to 10^{24} f/m^3). The hardness showed a gradual increase with fission dose until 10^{23} f/m^3. After that, however, for irradiations in JRRs (a lower thermal neutron flux and low irradiation temperature), the hardness attained a maximum at around 5×10^{23} f/m^3 followed by a successive decrease. By contrast, in the Japan Material Test Reactor (JMTR), in which the thermal neutron flux and irradiation temperature were 7×10^{17} n/m^{2}s and 400°C, respectively, the hardness increased monotonically up to 5.3×10^{24} f/m^3. In the thermal annealing process, on the other hand, the hardness exhibited two recovery steps at 650 and 950°C, to which two types of extended defects caused by fission fragment damage could be attributed. It is concluded here that dislocation loops and vacancy clusters contributed to the fission fragment-induced hardening of UN.

KEY WORDS: irradiation hardening, uranium mononitride, neutron irradiation, fission fragment damage, Knoop hardness, thermal annealing, interstitial cluster, vacancy cluster, defect identification, post-irradiation examination

It has been considered that fission-induced defects act as pinning points of dislocations. This is one of the origins for irradiation hardening, as discussed in terms of the depleted zone by Seeger [*1*]. The hardness is a useful measure to evaluate dimensional stability of nuclear fuel during reactor irradiation, being connected to other mechanical properties, such as plasticity and elasticity. Irradiation hardening is also related to the pellet-cladding mechanical interaction (PCMI). However, no result of the irradiation hardening for fissile materials have been reported so far, except for uranium dioxide (UO_2) by Bates [*2*] and Kim [*3*] and Kim and Kirihara [*4*]. In the present investigation, the hardening of uranium mononitride (UN, a candidate fuel component for the fast breeder reactor) with fission fragment damage was examined by the Knoop indentation method.

Experiments

A fused UN button was prepared by melting high-purity uranium metal in a nitrogen plasma-jet furnace under 0.5 MPa of nitrogen gas. Specimens for hardness testing were cut in dimensions 2 by 2 by 10 mm^3 from the button by a diamond-wheel slicing machine. After polishing the specimens to a suitable finish, they were annealed at 1300°C for 6 h in vacuum less than 1×10^{-3} Pa. From chemical analysis, the oxygen content was evaluated to be <1000 ppm. These

[1] Research associate, research associate, and Emeritus professor, respectively, Department of Nuclear Engineering, Faculty of Engineering, Nagoya University, Nagoya, 464 Japan.

specimens were irradiated in the Japan Research Reactor (JRR) 3, 4, and the Japan Material Test Reactor (JMTR) at the Japan Atomic Energy Research Institute (JAERI), as listed in Table 1. The irradiation temperature was estimated to be $<100°C$ in JRRs and 400°C in JMTR. After decay of the radioactivity for a few weeks to months, the hardness testing was performed in a hot cell because of high radioactivity. After removing the surface layer of the specimens by the mechanical polishing, the measurements were done at room temperature by the hardness tester (modified AKASHI MH-I) using a remote controlled TV monitor for observing the dimensions of the Knoop indentation. The hardness number was calculated by the following equation

$$\text{Knoop hardness number } (H_K) = 14.229\ P_K/d_K^2 \tag{1}$$

where P_K = indentation load (kg), and d_K = long diagonal length of indentation (μm). A mean value of over ten measurements was taken as the hardness number.

In order to investigate the annealing behavior of the fission-induced hardness, pulse anneals were performed on UN irradiated to 1.6×10^{23} f/m^3. The annealing was done by 100°C steps up to 1500°C, holding for 1 h at each step. After each annealing step, the hardness testing was carried out at room temperature with 140-g indentation load.

Results

Dependence on Indentation Load

Figure 1 shows the load dependence of the Knoop hardness for UN with various fission doses in JMTR. The measured values were somewhat scattered because of a technical difficulty in handling the highly radioactive specimens. However, it appears that the load dependences can be represented by two linear parts having a knee at a load of around 30 g. It was not easy to determine which value was intrinsic in the present study. This difficulty is often encountered in ceramic materials and, in fact, was observed in the case of UO_2 [*3,4*]. Some literature data for unirradiated UN are also included in Fig. 1. The value reported by Tripler et al. (Knoop) [*5*], Speider and Keller (Knoop) [*6*], and Sari and Matzke (Vickers) [*7*] are somewhat lower than the present work. This is probably because of the different characteristics of UN because all the specimens in the literature were sintered.

The Knoop hardness revealed a large increase after irradiation. Figure 2 represents the Knoop hardness increase (ΔH_K: difference of Knoop hardness number between pre- and post-irradiation) as a function of indentation load that were evaluated from Fig. 1. A curious drop was observed at indentation loads between 50 and 60 g. Above that range, the hardness increase showed nearly

TABLE 1—*Irradiation condition of uranium mononitride specimens.*

Specimen Name	Fission Dose, f/m^3	Irradiation Position	Neutron Flux, n/m^2 s
R4-BS-27	2.7×10^{21}	JRR-4 T pipe	8×10^{17}
R4-CS-14	1.5×10^{22}	JRR-4 T pipe	8×10^{17}
R3-ES-9	3.5×10^{22}	JRR-3 VG-7-1	2.7×10^{15}
R3-HS-25	1.6×10^{23}	JRR-3 VG-7-2	1.2×10^{16}
R3-JS-15	3.7×10^{23}	JRR-3 VR-1-1	2.0×10^{16}
R3-MS-21	3.6×10^{24}	JRR-3 VR-1-4	1.8×10^{17}
MTR-1S-2	4.6×10^{21}	JMTR HR-1	7×10^{17}
MTR-2S-2	1.1×10^{23}	JMTR HR-1	7×10^{17}
MTR-3S-2	1.3×10^{24}	JMTR HR-1	7×10^{17}
MTR-3S-2	5.3×10^{24}	JMTR HR-1	7×10^{17}

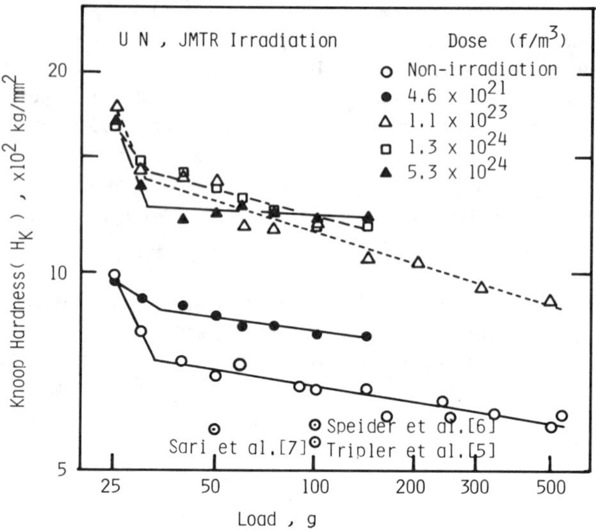

FIG. 1—*Load dependence of Knoop hardness of UN irradiated in JMTR at various fission doses.*

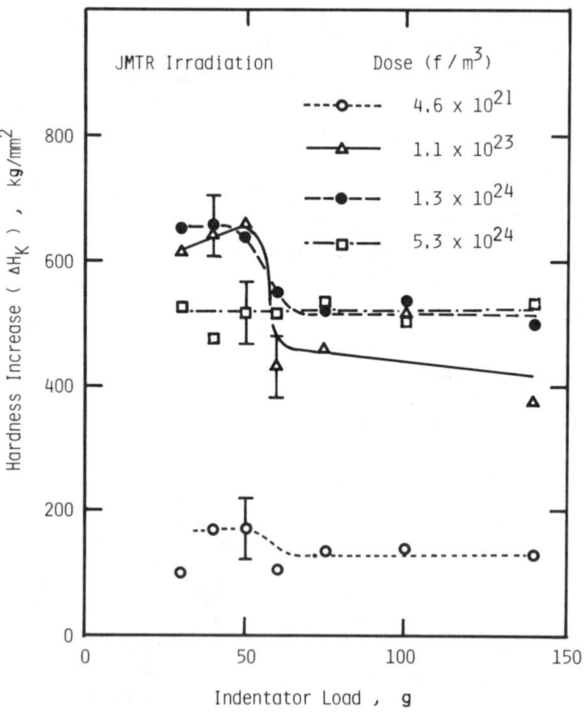

FIG. 2—*Load dependence of Knoop hardness increase that was calculated from Fig. 1.*

TABLE 2—*Knoop hardness number of neutron-irradiated UN with indentation load of 140 g.*

Fission Dose, f/m^3	Knoop Hardness Number, kg/mm^2, ±40	Remarks (Reactor)
2.7×10^{21}	747	JRR-4
1.5×10^{22}	797	JRR-4
3.5×10^{22}	1058	JRR-3
1.6×10^{23}	1118	JRR-3
3.7×10^{23}	1144	JRR-3
3.6×10^{24}	1006	JRR-3
4.6×10^{21}	805	JMTR
1.1×10^{23}	1055	JMTR
1.3×10^{24}	1173	JMTR
5.3×10^{24}	1208	JMTR
Before irradiation	677	

a constant value at each irradiation dose. Hence, in the present study, the indentation load of 140 g was chosen for Knoop hardness number comparisons since the indentation markers were clear and the measured values showed less scatter.

Dependence on Fission Dose

The Knoop hardness number H_K, at 140-g load, of neutron-irradiated UN with various fission doses are listed in Table 2 for the cases of both JRRs and JMTR irradiation. The variations of fission-induced hardening ΔH_K are illustrated in Fig. 3, together with the results of the fission-induced lattice strain η, which was evaluated from the X-ray diffraction profile [*8,9*]. The induced hardening in both types of reactor irradiations showed a gradual increase with fission dose up to 10^{23} f/m^3. In the JRRs irradiations (<100°C), however, hardening showed a maximum at around 5×10^{23} f/m^3 and a subsequent 30% decrease at 3.6×10^{24} f/m^3. By contrast a monotonic increase was found up to 5.3×10^{24} f/m^3 in the case of the JMTR irradiation (400°C). It should be noticed here that the lattice strains increased steeply at high fission doses.

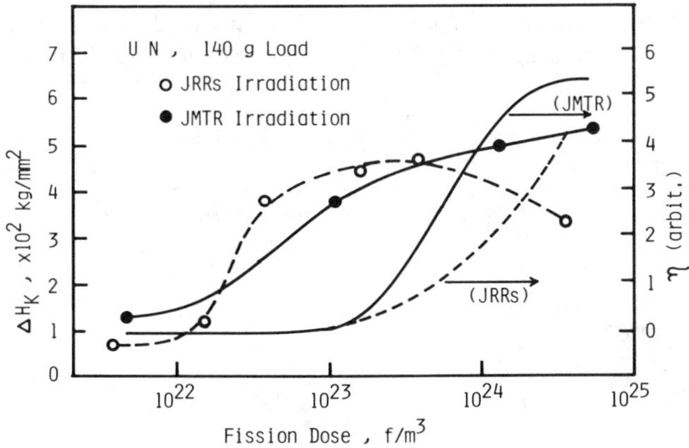

FIG. 3—*Change of Knoop hardness in UN irradiated in JRR-3 and JMTR for various fission doses.*

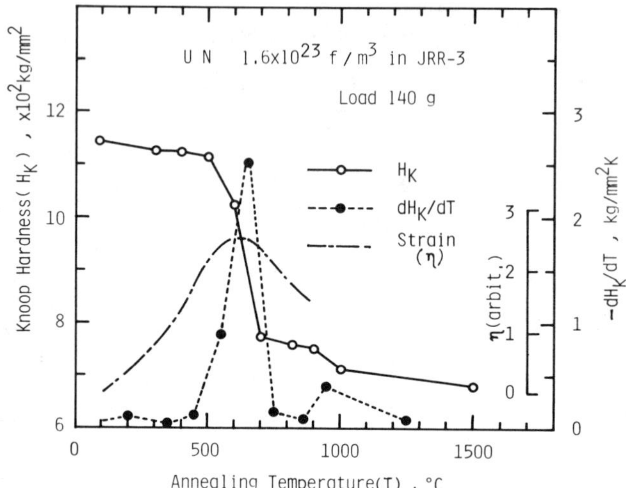

FIG. 4—*Recovery upon annealing of hardness of UN irradiated to 1.6 × 10^{23} f/m³ in JRR-3 (*$-dH_K/dt$ *is temperature derivative of* H_K *and* η *is lattice strain* [8,9]).

Recovery of Fission-Induced Hardening

In Fig. 4, we represent the recovery of the hardness by step annealings of UN irradiated to 1.6 × 10^{23} f/m³ in JRR-3, including the change of the lattice strain η. As seen in the temperature derivatives of the Knoop hardness (dH_K/dT), two recovery steps were clearly observed at 650 and 950°C. Table 3 compares the recovery temperatures for hardness with those obtained from measurements of the electrical resistivity and lattice parameter of UN with similar fission damage [*8,9*]. No distinct recovery in the hardness occurred in the temperature range below 500°C, while three recovery steps were observed below 500°C in both the lattice parameter and electrical resistivity.

Discussion

Change of Defect-Induced Hardness

From post-irradiation examination of the electrical resistivity, lattice parameter and magnetic properties of UN with fission fragment damage [*8,9*], an identification of fission-induced defects

TABLE 3—*Recovery temperature in increased Knoop hardness of irradiated UN at exposure to 1.6 × 10^{23} f/m³ compared with the case of electrical resistivity and lattice parameter* [8,9]

Property	Recovery Temperature, °C, ±10				
	Steps				
	I	II	III	IV	V
Knoop hardness	. . .	. . .	. . .	650	950
Electrical resistivity[a]	135	280	525	700	. . .
Lattice parameter[a]	100	225	425	650	800

[a] Irradiated to 1.4 × 10^{23} f/m³.

TABLE 4—*Characterization of fission fragment-induced defects in uranium mononitride* [9].

Step	Recovery Temperature, °C	Activation Energy eV	Recovery Mechanism[a]
1	100 to 135	0.2	I-V recombination (or I clustering) by I migration
2	225 to 280	0.3	I clustering (or I-V recombination) by I migration
3	425 to 525	0.9	V clustering by V migration
4	650 to 700	2.1	V-I_C annihilation by V migration
5	800 to 950	. . .	V_C dissociation and V-I_C annihilation by V migration

[a] is Interstitial, V is vacancy, I_C is interstitial cluster and V_C is vacancy cluster.

and mechanisms of defect recovery have been concluded as summarized in Table 4. It can be considered that the annealing step at 650°C is attributed to an annihilation of interstitial clusters with migrating vacancies, while the annealing step at 950° is attributed to an annihilation of vacancy clusters (or voids) by dissociation of the vacancy clusters and migration of vacancies to sinks.

The lack of observed recovery below 500°C indicates that the hardness is not very sensitive to simple defects as compared with the case of the lattice expansion and an electrical resistivity increase [*8,9*]. This is consistent with the results of a transmission electron microscope (TEM) investigation on neutron irradiated uranium monocarbide (UC) by Eyre and Sole [*10*] that both the size and density of the clusters showed no remarkable change by annealing up to 700°C.

Dose Dependence

Figure 5 illustrates schematically the change of the hardness with fission dose in comparison with those of the electrical resistivity, lattice parameter and lattice strain [*8,9*], and magnetic properties [*11*] in the case of JRRs irradiations. The hardness of UN increased with fission dose, and a small reduction was observed at above 10^{24} f/m^3, while the lattice parameter and electrical resistivity showed a steep reduction to the original value at that dose. The H_K and mag curves attained a maximum change at the same fission dose. The lattice strain, on the other hand, showed its steep increase only above approximately 10^{24} f/m^3. This fact may indicate that clustering of simple defects produced at low fission doses occurs at that fission dose. Therefore, it is concluded that the hardness as well as magnetic properties are sensitive to the defect clusters (interstitial loop and voids), but not the lattice parameter and electrical resistivity. A similar trend was observed between lattice parameter and hardness in fission-damaged UO_2 [*3,4*]. The difference in the hardness change between JRRs and JMTR irradiations might arise from an interaction of the simple interstitial type defects (mobile in the temperature range from 100 to 400°C) with other immobile defects (vacancies, clusters, and voids).

Conclusions

Fission fragment-induced hardening was examined in uranium mononitride with various fission doses in a range from 2.7×10^{21} to 5.3×10^{24} f/m^3. The hardness increased gradually with fission dose until 10^{23} f/m^3, and attained a maximum at around 5×10^{23} f/m^3 followed by a successive decrease in JRRs irradiation, whereas in JMTR the hardness increased monotonically up to 5.3×10^{24} f/m^3. In the annealing process, the hardness exhibited two recovery steps at 650 and 950°C. It is concluded that interstitial clusters (dislocation loops) and vacancy clusters (voids) contributed to the fission fragment-induced hardening of UN.

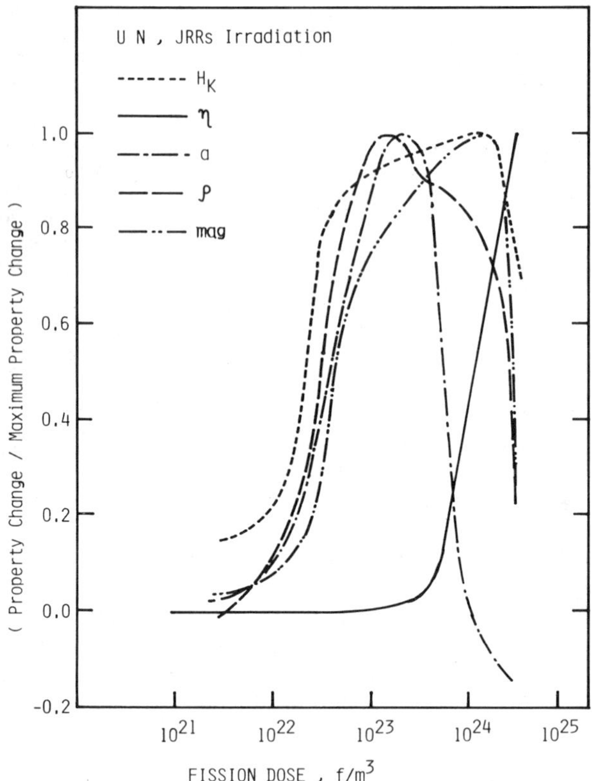

FIG. 5—*Schematic dose dependences of reduced changes (change/maximum change) in properties of UN with neutron irradiation* [8,9,11] H_K *is Knoop hardness,* a *is lattice parameter,* ρ *is electrical resistivity, mag is magnetic properties, and* η *is lattice strain).*

References

[*1*] Seeger, A. K., *Second United Nation International Conference on Peaceful Uses of Atomic Energy,* 1958, a/CONF/p/998.

[*2*] Bates, J. L., HW-77799, 1963.

[*3*] Kim, S. S., "Micro-Hardness and Indentation Creep of Irradiated UO_2," Doctor Thesis of Nagoya University, Nagoya, Japan, 1983.

[*4*] Kim S. S. and Kirihara, T., *Journal Nuclear Science and Technology,* Vol. 20, No. 12, Dec. 1983, pp. 1023–1031.

[*5*] Tripler, A. B., Snyder, M. J., and Duckworth, W. H., "Further Studies of Sintered Refractory Uranium Compounds," BMI-1313, Battelle Memorial Institute, Columbus, OH, 1959.

[*6*] Speider, E. O. and Keller, D. J., "Fabrication and Properties of Uranium Mononitride," BMI-1633, Battelle Memorial Institute, Columbus, OH, 1963.

[*7*] Sari, C. and Matzke, Hj., work at Institute for Transuranium Elements, Karlsruhe, West Germany, private communication from Hj. Matzke.

[*8*] Tamaki, M., Matsumoto, S., Ishimaru, K., Matsumoto, G., and Kirihara, T., *Journal of Nuclear Materials,* Vol. 108/109, 1982, pp. 671–677.

[*9*] Tamaki, M., "Fission Fragment Damage in Uranium Mononitride," Doctor Thesis of Nagoya University, Nagoya, Japan, 1983.

[*10*] Eyre, B. L. and Sole, M. J., *Journal of Nuclear Materials,* Vol. 18, 1966, pp. 314–322.

[*11*] Tamaki, M., Matsui, H., Ohnuki, A., Matsumoto, G., and Kirihara, T., *Radiation Effects,* Vol. 91, 1985, pp. 61–69.

Suzanna G. Burnay[1] and James W. Hitchon[1]

The Assessment of Seal Performance and Lifetime Prediction in a Nuclear Environment

REFERENCE: Burnay, S. G. and Hitchon, J. W., "**The Assessment of Seal Performance and Lifetime Prediction in a Nuclear Environment,**" *Influence of Radiation on Material Properties: 13th International Symposium (Part II), ASTM STP 956,* F. A. Garner, C. H. Henager, Jr., and N. Igata, Eds., American Society for Testing and Materials, Philadelphia, 1987, pp. 609–614.

ABSTRACT: A method of seal lifetime prediction has been developed using time-temperature superposition to detemine the functional relationships between compression set and the accelerating variables, that is, temperature, dose, and dose rate. Using this technique, seal lifetimes under service conditions can be predicted even when the elastomer exhibits dose rate effects or synergism between radiation and temperature.

KEY WORDS: dose rate effects, elastomers, modelling, radiation effects, synergism, temperature effects

In many nuclear applications, elastomeric materials are used as seals under conditions in which they are likely to be exposed to relatively low dose rates for long periods of time. The seals may also be exposed to elevated temperatures and possibly chemical contamination during their service life. In some cases, replacement of such seals would entail a major shutdown of plant, therefore it is necessary to know the safe working life of a seal so that unnecessary shutdowns are avoided. In assessing the long-term behavior of these materials, accelerated aging tests need to be used to enable estimates of seal lifetime under the anticipated service conditions. It is now well established that many polymers can exhibit dose rate effects or synergism between radiation and temperature or chemical contamination [*1*]. It is therefore necessary for any prediction of seal lifetimes based on accelerated testing to take into account such factors.

Elastomers rely for their sealing properties on the pressure exerted by the compressed seal on the faces of the seal seating. The decay of this sealing force with time is a well known phenomenon that can be accelerated by elevated temperatures, ionizing radiation or some chemical contaminants. The relaxation of stress in the seal can be temporary, because of a physical rearrangement of the molecular structure of the elastomer, or it can be permanent, if crosslinking occurs. As the sealing force decreases, at some point leakage past the seal will increase to the stage at which the seal is deemed to have failed. For routine seal assessment, direct measurement of leakage across a seal is rarely used as it is a time-consuming procedure. More indirect methods are generally used, for example, measurements of compression set and retained sealing force. The simplest of these techniques experimentally is the measurement of compression set; previous work comparing seal leakage, retained sealing force and set measurements have shown that the use of the compression set test for routine assessment of seal performance is fully justified [*2*].

This paper describes the predictive model that has been developed for lifetime prediction in elastomer seals. The model was originally developed with two elastomers, a polyurethane and a

[1] Materials Development Division, AERE Harwell, Didcot, Oxon OX11 ORA, United Kingdom.

fluoropolymer, for which a large amount of compression set data were available. The model has recently been used successfully in low dose rate experiments on the fluoropolymer in a temperature regime well outside the range originally investigated.

Lifetime Prediction

The model is based on the use of the superposition technique. In thermal aging of polymeric materials, the principle of time-temperature superposition is widely used in accelerated tests [*3*]. The present model is an extension of this technique to include radiation dose rate as one of the superposition parameters.

In accelerated testing, the most common accelerating technique is the constant overstress technique, that is, the stress variable is raised to a constant value and the damage parameter is monitored as a function of time [*4*]. In radiation aging, the stress variables of interest are the temperature and dose rate; for seals, the compression set is used as the damage parameter. If the aging processes are accelerated by the same factor by an increase in the stress variable, then curves of the damage parameter versus log (time) will be parallel and can be superimposed by a horizontal shift on the log (time) axis (Fig. 1). The multiplying factor used to produce this shift is referred to as the shift factor for the particular stress variable. Superposition of the curves will only occur if the degradation processes underlying the measured damage parameter have all been equally accelerated. In elastomeric materials, these processes may be complex but are frequently dominated by one main mechanism, particularly in the case of thermal degradation. Once superposition has been obtained, the functional relationships between the shift factors and the stress variables can be determined.

When temperature is the stress variable, the thermal shift factors a_T are often found to approximate to an Arrhenius relationship

$$a_T \propto \exp(-E_a/RT) \tag{1}$$

where E_a is the activation energy, R is the gas constant, and T is the absolute temperature. If the data are referenced to a temperature T_{ref} such that $a_T = 1$ at $T = T_{\text{ref}}$, then the shift factor is given by

$$a_T = \exp \frac{E_a}{R} \left(\frac{1}{T_{\text{ref}}} - \frac{1}{T} \right) \tag{2}$$

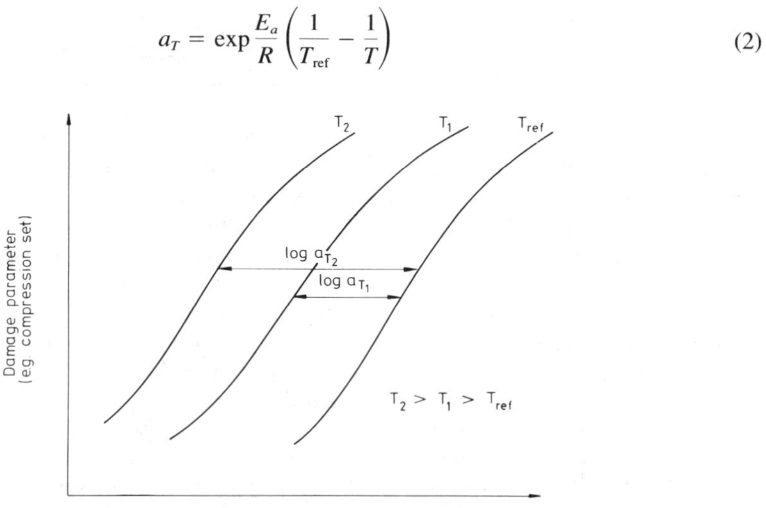

FIG. 1—*Definition of thermal shift factors* a_T *used in superposition model.*

This type of behavior is also observed in elastomers under radiation aging conditions where the additional stress variable of dose rate is also present. In this case, the activation energy E_a is often observed to be dependent on dose rate (Fig. 2), decreasing with increasing dose rate, as has been observed in the polyurethane and fluoropolymer previously studied [2]. In these elastomers, the empirical relationship between E_a and the dose rate D has been found to take the form

$$E_a = C_1 - C_2 \log_{10} D \tag{3}$$

where C_1 and C_2 are empirical constants that must be determined experimentally.

At a constant dose rate, all of the experimental data obtained at different temperatures can be superimposed using the shift factors a_T to form a single master curve referenced to T_{ref}. If such master curves are obtained at different dose rates, all referenced to the same temperature, then superposition can also be used on the master curves to define a dose rate shift factor a_R, referenced to the unirradiated master curve. These shift factors are also found to have a simple relationship

$$a_R \simeq C_3 D^x \tag{4}$$

where x is an empirical constant, usually in the range of 0.5 to 1.0 [*5–8*]. If $x = 1$, then no dose rate effects will be observed. However in the polyurethane and fluoropolymer, x was found to be 0.75 and 0.9 respectively, showing that dose rate effects exist in these materials [*2*]. Since by definition $a_R = 1$ for unirradiated material, the shift factors are more accurately described as

$$a_R = 1 + C_3 D^x \tag{5}$$

Provided that the functional relationships between a_T, a_R, E_a, and D are known, predictions can be made of the behavior of an elastomer under conditions of temperature and dose rate appropriate to the service environment. This would normally entail extrapolation towards lower temperatures and dose rates than have been examined in the accelerated tests. The value of the damage parameter

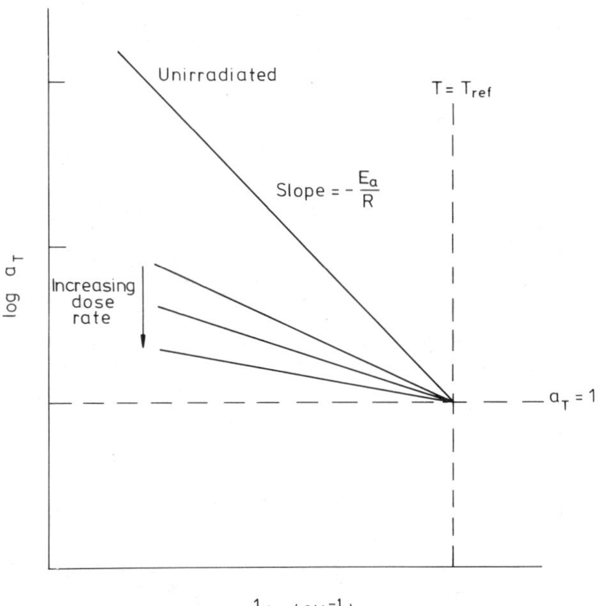

FIG. 2—*Temperature dependence of thermal shift factors in the superposition model (schematic).*

at which a seal is deemed to have reached its useful life will depend on the application, for example, for static O-ring seals, a conservative failure criterion might be the attainment of 80% compression set. The service lifetime t_f will then be defined as

$$t_f = \frac{t_m}{a_T a_R} \tag{6}$$

where t_m is the time to failure on the master curve at $T = T_{ref}$, and $D = 0$, with a_T and a_R as defined in Eqs 2 and 5. Plots of t_f as a function of dose rate and temperature will then take the form shown in Fig. 3.

The time-temperature-dose rate superposition model works very successfully with the polyurethane material previously studied, which has a relatively simple type of degradation behavior. Under both thermal and radiation aging, oxidative degradation is the dominant process, but under irradiation, the oxidation reactions are limited by the rate at which oxygen can diffuse into the sample [2]. With this type of mechanism, the empirical constants C_1 and C_2 in Eq 3 are likely to be dependent on the sample geometry as well as the elastomer composition. An Arrenhius plot of the shift factors a_T (Fig. 4) shows that straight line plots are obtained at each dose rate; it should be noted that the point $a_T = 1$, $T = T_{ref}$ is common to all four sets of data.

The fluoropolymer studied exhibits very clearly one aspect of the model where problems can arise. If a change in degradation mechanism occurs in the temperature regime examined, then it is frequently found that either the damage versus log (time) curves cannot be superimposed, or, if superposition is possible over a limited range, then the shift factors a_T do not show a simple temperature dependence. Figure 5 illustrates this for the fluoropolymer where, for temperatures >250°C, marked deviations from a straight line Arrhenius plot are observed under irradiation. Therefore for this particular material, the model can only be used at temperatures <250°C; at higher temperatures the model is no longer appropriate. However, at lower temperatures the model has proved to be very successful. Earlier experiments carried out on the fluoropolymer at 150 to

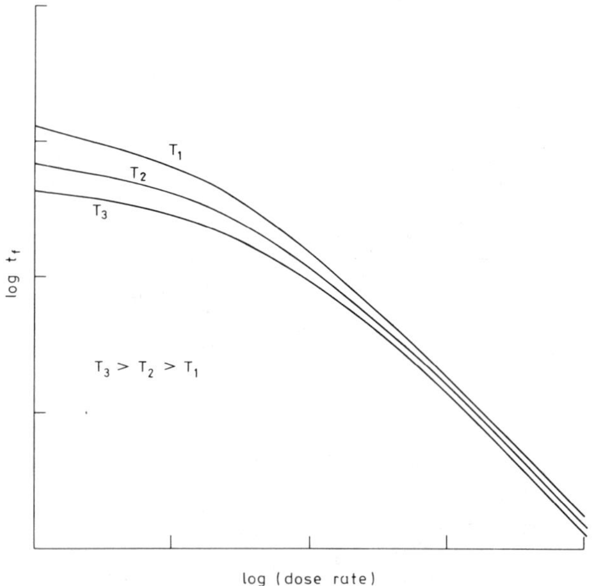

FIG. 3—*Dose rate dependence of seal lifetime using the superposition model (schematic).*

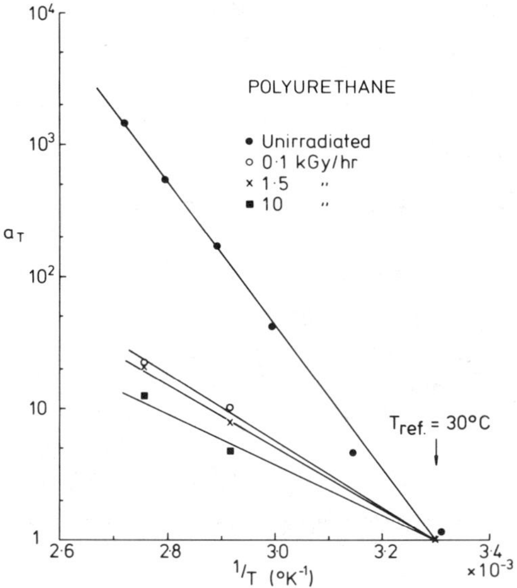

FIG. 4—*Temperature dependence of thermal shift factors* a_T *for polyurethane.*

300°C and at dose rates of 0.1 to 10 kGy/h[2] were used to determine values for the empirical constants C_1, C_2, C_3, and x. The behavior at the much lower temperature and dose rate of 80°C and 13 Gy/h predicted by the model has been found to fit experimental data very well (Fig. 6).

Currently the model is being applied to other elastomeric materials to check whether the behavior observed in the polyurethane and fluoropolymer studied so far is more generally applicable.

Conclusions

A semi-empirical model of radiation aging in elastomeric materials has been developed. The model does not depend on detailed knowledge of the mechanisms underlying the degradation

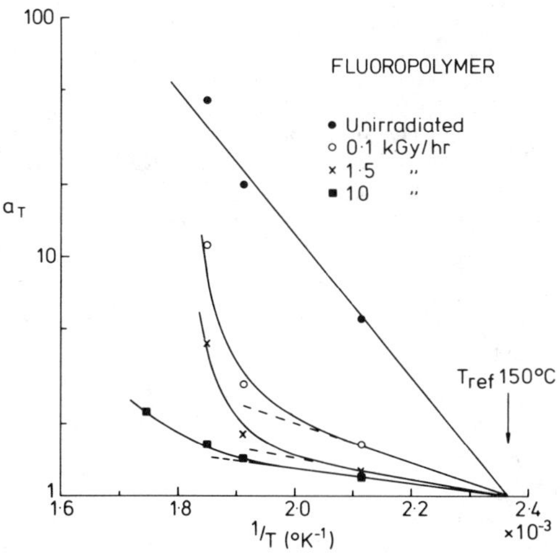

FIG. 5—*Temperature dependence of thermal shift factors* a_T *for fluoropolymer.*

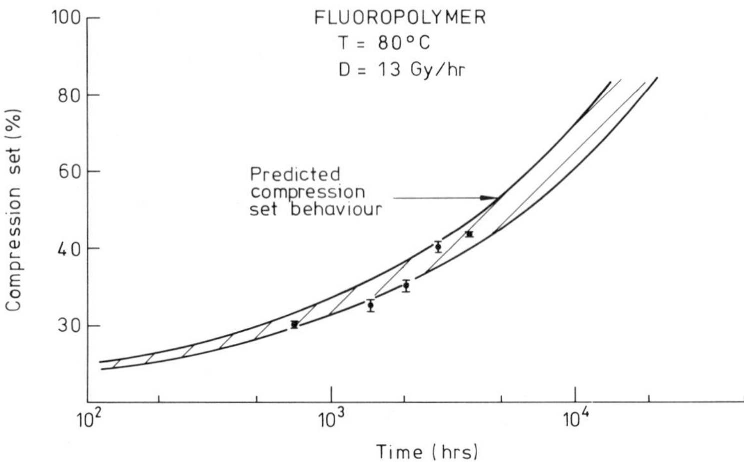

FIG. 6—*Comparison between experimental data and predictions for the model for low dose rate experiments in fluoropolymer.*

process but uses a time-temperature-dose rate superposition process to determine experimentally the functional relationships between the shift factors a_T and a_R, the activation energy, dose rate, and temperature. The model can take into account the presence of dose rate effects and synergism between radiation and temperature. The model has been successfully applied to two elastomers with markedly different degradation behavior, and further work is in progress on other elastomers to examine whether the model is more generally applicable. Extrapolation from high dose rate, high temperature accelerated tests to low dose rate, low temperature conditions using this technique has been demonstrated successfully for the fluoropolymer.

References

[1] Clough, R. L., Gillen, K. T., Campan, J.-L., Gaussens, G., Schönbacher, H., Seguchi, T., Wilski, H., and Machi, S., *Nuclear Safety,* Vol. 25, 1984, p. 238.
[2] Burnay, S. G. and Hitchon, J. W., *Journal of Nuclear Materials,* Vol. 13 , 1985, p. 197.
[3] Ward, I. M., *Mechanical Properties of Solid Polymers,* Wiley-Interscience, New York, 1971, Chapter 7.
[4] Gillen, K. T. and Mead, K. E., SAND 79-1561, 1979.
[5] Gillen, K. T., SAND 79-0930C, 1979.
[6] Machi, S., *Radiation Physics and Chemistry,* Vol. 18, 1981, p. 125.
[7] Arakawa, K., Seguchi, T., Watanabe, Y., Hayakawa, N., Kuriyama, I., and Machi, S., *Journal of Polymer Science: Polymer Chemistry,* Vol. 19, 1981, p. 2123.
[8] Ito, M., *Radiation Physics and Chemistry,* Vol. 18, 1981, p. 653.

D. Clive Phillips,[1] *Diana I. Johnson,*[1] *Suzanna G. Burnay,*[1] *and James W. Hitchon*[1]

The Effects of Radiation on Organic Matrix Waste Forms

REFERENCE: Phillips, D. C., Johnson, D. I., Burnay, S. G., and Hitchon, J. W., **"The Effects of Radiation on Organic Matrix Waste Forms,"** *Influence of Radiation on Material Properties: 13th International Symposium, ASTM STP 956,* F. A. Garner, C. H. Henager, Jr., and N. Igata, Eds., American Society for Testing and Materials, Philadelphia, 1987, pp. 615–635.

ABSTRACT: This report summarizes the results of a program carried out between July 1981 and March 1985 on behalf of the Commission of European Communities (CEC) and the Department of Environment (DoE). The effects of gamma and alpha irradiation have been studied on a range of bitumen and polymer matrix waste forms.

Under irradiation, bitumens swell because of the nucleation and growth of bubbles of radiolytic gas. Swelling depends on dose rate and specimen size, as well as total dose, but accelerated tests can be carried out by irradiating specimens of varied size under the same conditions of dose rate × specimen depth squared. Swelling commences at around 50 kGy and saturates at 300 kGy. The magnitude of swelling depends on the type of bitumen. For bitumenisates used in Europe the magnitude is ≤10%. The effects of alpha radiation are more severe than equivalent doses of gamma radiation but are unlikely to have a significant effect on physical properties in practice. Irradiation to 10 MGy and swelling have no significant effect on leach rates.

The behavior of polymer waste forms under gamma irradiation is more complicated. The most pronounced effects occur with waste forms containing wet wastes. Some of these can shrink by substantial amounts on irradiation caused primarily by water loss. Others exhibit small amounts of swelling or shrinkage and the exact behavior appears to depend on a combination of bubble growth, polymer cross linking, and water desorption. Data are presented on the effects of radiation on leaching and on radiolytic gas evolution.

KEY WORDS: bitumenisates, thermosetting resins, radiation effects, alpha irradiation, swelling, gas evolution, leach testing

This paper summarizes the results of a program carried out between July 1981 and March 1985 on behalf of the Commission of the European Communities (CEC) and the Department of the Environment (DoE). Detailed accounts of aspects of this work will appear as further reports [*1–4*], but this report provides an overview and will help place those individual reports in context.

The research was part of a wider European program aimed at the characterization of low and medium level radioactive waste forms [*5,6*]. At the start of the program a working group of experts identified ten reference waste forms of importance in the European context. The overall objective of the program carried out at Harwell was the assessment of the effects of radiation. All ten reference waste forms (RWFs) were investigated together with an additional waste form (RWF17) of interest to the United Kingdom. These are listed in Table 1 together with approximate activity levels. The highest levels imply approximate integrated doses of the order of 7 MGy at 100 years and ~20 MGy at 1000 years, with correspondingly lower doses for the less active materials (1 Gy = 100 rad). The total dose and dose rates that will be received by different types of waste

[1] Materials Development Division, AERE Harwell, Didcot, Oxon OX11 ORA, United Kingdom.

TABLE 1—*Waste forms studied in radiation characterization program.*

Specimen	Waste/Matrix	Typical Activity, Ci m^{-3} β^{γ}	α
RWF1	BWR evaporator concentrates/cement or pozzolana cement (1/1 and 1/2, respectively)	0.55	. . .
RWF2	PWR evaporator concentrates/cement or pozzolana cement (2/1 and 2/2, respectively)	0.5 to 50	$<10^{-3}$
RWF3	PWR evaporator concentrates/polyester or epoxide resins	0.5 to 50	$<10^{-3}$
RWF4	organic ion exchanger/polystyrene	$<10^{3}$	$<5 \times 10^{-6}$
RWF5	organic ion exchanger/polyester or epoxide resins	5 to 500	$<10^{-6}$
RWF6	Magnox fuel pond water sludge/cement	35 to 70	1 to 2
RWF7	reprocessing concentrates/bitumen	60	2
RWF8	reprocessing concentrates/cement	10 to 40	~1
RWF9	reprocessing sludges/bitumen	70 to 850	0.3 to 0.8
RWF10	incinerator slags		
WF17	mixed ion exchangers/polyester or vinyl ester resins		

forms vary widely depending on the radionuclide inventory, but it cannot be assumed that a low activity waste form will receive a low dose, unless it can be guaranteed that it will be shielded from neighboring, higher activity, waste forms during storage. Consequently irradiation experiments were carried out to high doses on all the materials studied.

The aims of the program were both to generate data of immediate use in engineering assessments of safety and to establish an adequate scientific understanding for the confident prediction of long-term behavior from accelerated laboratory tests. Earlier results from the program have been presented in several conference papers [*7–10*] and reports [*5,6,11,12*].

The waste forms had cement, slag, polymer, and bitumen matrices. In general the effects of radiation are more severe in organic materials than in inorganic materials, and for that reason more emphasis was placed in the overall Harwell program on the organic matrix waste forms. This report describes the results of the work on organic matrix waste forms, and the work on cement and slag waste forms is described elsewhere [*13*].

Scope of Program

Several different types of experiment were planned at the outset and then progressed during the program. Much of the work was carried out on inactive simulates because of the ease of handling, but where appropriate specimens were spiked with low levels of radionuclides for leaching experiments. Also, specimens were produced containing large quantities of the alpha emitter ^{241}Am for accelerated alpha damage experiments. Some limited experiments were also carried out on fully active, real waste forms, to verify the results obtained from simulates. In addition to experiments on waste forms, some experiments were carried out on the matrix materials alone. The basic experiments and the reasons for carrying them out are described below, and Table 2 summarizes the experiments carried out on particular waste forms. Detailed specifications of the waste forms are given in Refs *5* and *6*.

Dimensional Stability and Integrity

It was known from earlier work at Harwell that gamma irradiated bitumenisates can swell. This is due to the nucleation and growth of bubbles of radiolytic gas and depends on dose rate and

TABLE 2—*Irradiation experiments carried out on organic waste forms during the program:* X *denotes experiment carried out and* O *denotes material supplied but testing not carried out.*

Experiment	Waste Form Supplier	Waste Form						
		RWF3	RWF4	RWF5	RWF5(H)	RWF7	RWF9	WF17
Dimensional stability, mechanical integrity, and weight change	UKAEA, Harwell		X		X	X		X
	CEGB							
	CEA	X		X			X	
Gas evolution	UKAEA, Harwell	X		X		X		
	CEA							
Leaching	UKAEA, Harwell	X		X		X		X
	CEGB							
	CEA	O		O				
Alpha damage	UKAEA, Harwell					X		
Fully active	CEA	O		O			X	

specimen size, as well as total dose, the swelling increasing with increasing specimen size and increasing dose rate. Earlier work on bitumenisates under development in the United Kingdom has shown that the swelling of large specimens under low dose rate conditions could be predicted from accelerated tests employing high dose rates on small specimens, through two models, the IL^2 and bubble growth models [*11,12*]. The IL^2 model predicts that a family of right cylindrical specimens in containers of different depth L, irradiated at different dose rates I, will swell similar amounts provided IL^2 is the same for all; the bubble growth model gives a more complete physical description of the processes involved and defines conditions under which swelling will occur. It was not known how polymer matrix waste forms would behave when irradiated as bulk specimens as opposed to the small test specimens, which have been studied in most polymer irradiation research. It is important to know this both to ensure that the container is not at risk and also to be able to produce, in accelerated tests, materials that have the correct radiation-aged structures for other measurements, such as radionuclide release during leaching. Accordingly inactive specimens of the polymer and bitumen waste forms were produced in containers of a range of sizes; gamma irradiated in a ^{60}Co cell, at ambient temperature, over a range of dose rates but under the same IL^2 conditions. Dimensional changes were measured by a water displacement technique, in order both to determine the magnitude of any swelling or shrinkage and to ascertain whether the IL^2 model is generally applicable. The weight changes of these specimens were also measured, and mechanical test specimens were cut out of the irradiated materials for measurements such as flexural strength and strain to failure. Similar measurements were also carried out on identical control samples which were stored in the radiation cells, but shielded from the radiation.

Gas Evolution

The quantities and compositions of gases produced during gamma irradiation were determined by mass spectrometry. Specimens of crushed waste forms were gamma-irradiated in a ^{60}Co cell in sealed stainless steel specimen holders at a dose rate of 4 to 5 kGy h^{-1}, in both oxygen and argon atmospheres. This provides information of use in the design of vented containers and store ventilation, and in the interpretation of radiation degradation processes.

Leach Testing

Leach tests were carried out on gamma-irradiated and unirradiated control specimens to determine the effects of radiation on radionuclide release. The waste forms were manufactured using simulated waste that was spiked with low levels of ^{137}Cs. Leaching was carried out at ambient temperature on 2-cm-right cylinders containing approximately 35 μCi g^{-1} of ^{137}Cs using the ISO test [*9*]. Specimens were irradiated in a spent fuel storage pond at 10 to 15 kGy h^{-1} to various dose levels; specimen temperature during irradiation reached approximately 30°C at 10 kGy h^{-1} and approximately 50°C at 15 kGy h^{-1}.

Alpha Irradiation

Only the bitumen waste forms contain significant quantities of alpha emitters; therefore accelerated alpha-irradiation experiments were carried out only on those waste forms. Simulates of RWF7 were manufactured containing ^{241}Am at activity levels of approximately 1 Ci L^{-1}. This results in an alpha dose rate of ~0.1 kGy h^{-1}, which is three orders of magnitude higher than the α dose rate in the real waste form. Some specimens were also produced containing in addition ^{137}Cs. Measurements were made of swelling, leaching, changes in bitumen mechanical properties, and of microstructure. This work is fully reported elsewhere [*1*].

Fully Active Studies

Much of the program was carried out on inactive or spiked simulates because of the greater safety and ease with which these materials can be studied. Most of the simulates were produced by the waste-producing organizations to ensure accurate simulates. To check the relevance of the experiments on simulates, some experiments have also been carried out on real waste forms. Specimens of thermosetting resin and bitumenous real waste forms were supplied by the Commissariat a l'Energie Atomique, France (CEA). Only the bitumenous waste forms from the CEA were received in time for study. These have been subjected to gamma irradiation and have undergone leach testing, and have been shown to behave similarly to the inactive simulates. The thermosetting resin waste forms have been stored, and it is hoped to study these in the next CEC program.

Results of Bitumen Waste Forms

Dimension and Weight Changes

Gamma irradiation experiments were carried out on inactive, simulated bitumenisates, RWF7 and RWF9, and on their matrix bitumens, using specimens cast into open-topped steel containers of a range of sizes from 4 cm to 17.5 cm in diameter, at dose rates such that IL^2 was the same for all, the dose rates ranging from 0.25 to 5.5 kGy h^{-1}. The irradiations were carried out in a ^{60}Co cell at approximately 22°C, at an IL^2 of 80 kGy cm^2 h^{-1}, which corresponds to the conditions experienced by a 200-L drum of waste containing approximately 1 Ci L^{-1} activity. The RWF7 simulate was manufactured at Harwell using a published flow sheet for the process. Details of this and of the background to the experiments have been published in Refs *11* and *12*. The RWF9 simulate was provided by the CEA. On irradiation the weight changes that occurred were negligible, but significant swelling occurred. Figure 1 shows the results for RWF7 and its matrix bitumen R85/40, and Figs. 2 and 3 show the results for RWF9 and its matrix R90/40.

Figure 1 shows that the behavior of RWF7 is similar to that of other bitumenisates studied in the United Kingdom and that the behavior of R85/40 is identical to that observed in an earlier United Kingdom program [*11*]. On irradiation, swelling commences at about 50 kGy and saturates

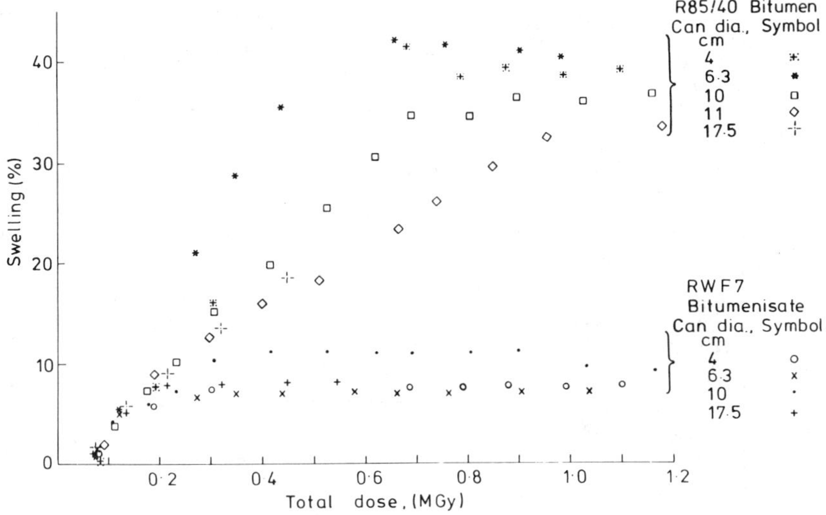

FIG. 1—*γ-irradiation induced swelling of unfilled bitumen R85/40 and bitumenisate RWF7.*

at about 300 kGy. The data for the different container sizes and dose rates agree reasonably well, confirming that the IL^2 model is applicable. The unfilled bitumen swells by 40 $^v/_o$ (volume %), but the swelling of the bitumenisate is much less at around 10 $^v/_o$.

The data for bitumen R90/40 are more variable but also fit the overall pattern of behavior observed for other bitumens. The saturation swelling, at a mean value of approximately 10 $^v/_o$, is much less than the 30 to 40 $^v/_o$ of R85/40, which is surprising as their physical characteristics are

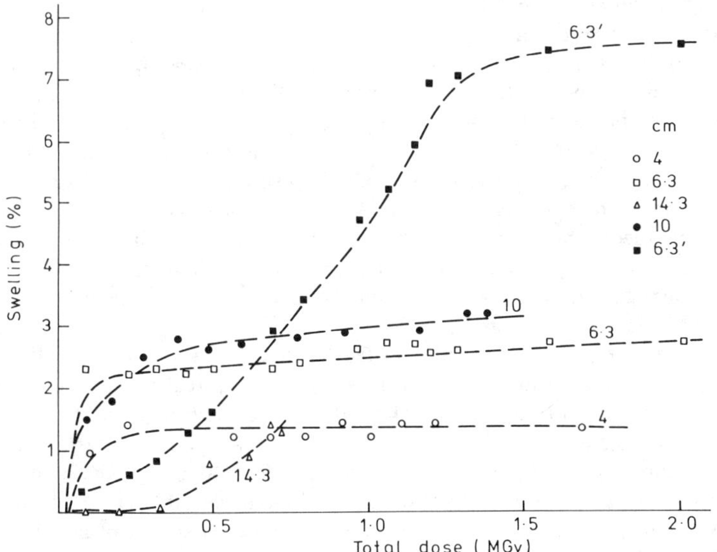

FIG. 2—*Swelling of bitumenisate RWF9 as a function of specimen size and γ-irradiation dose.*

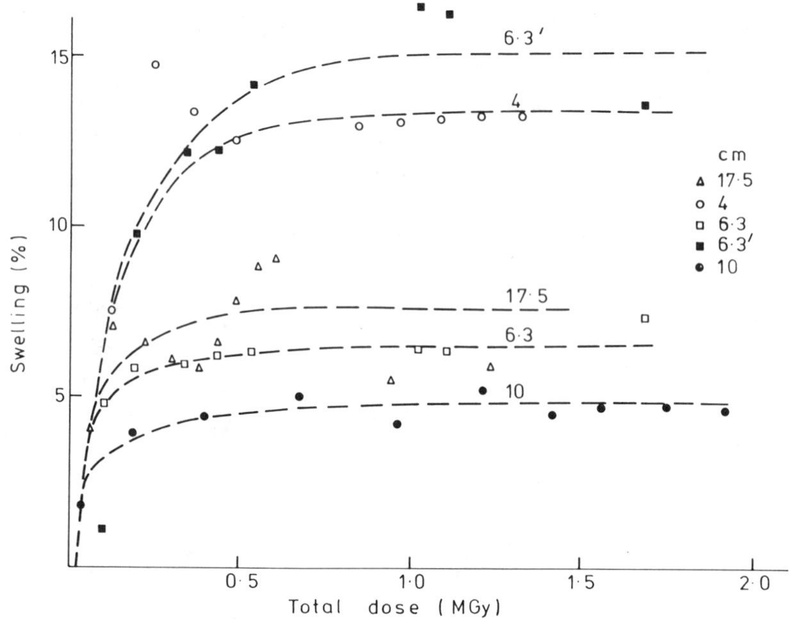

FIG. 3—*Swelling of R90/40 bitumen as a function of specimen size and γ-irradiation dose.*

similar. The behavior of RWF9 differs from that of all the other bitumenisates that have been studied [*11*] in that two types of swelling behavior are observed, the 6.3′- and 14.3-cm cans behaving differently from the others. The reasons for this anomaly are not known but may be due to differences in processing. The data shown in Figs. 2 and 3 were obtained from specimens that were not all in an identical condition to the material supplied by the CEA. In order to extend the range of can sizes that could be irradiated, or to make swelling measurements easier, some of the French material was remelted at Harwell and cast into different cans. For R90/40 (Fig. 3), only the can labeled 6.3′ cm was in the as-received condition; all other specimens had been remelted. For RWF9 in Fig. 2 only the 14.3-cm specimen was in the as-received condition. The 6.3′ cm specimen had been remelted, to provide a smoother top surface, but left in its original container. The other three specimens had been melted and poured into other containers.

There are strong theoretical grounds for believing that the initial, unirradiated microstructure of bitumenisates and particularly the distribution of trapped gas bubbles can affect subsequent swelling [*11,12*]. The differences in behavior may be due to this effect since the remelting and pouring may have affected the microstructures.

Because of the demonstrated applicability of the IL^2 model, the saturation swelling observed in these experiments is expected to be that which would occur in real waste forms during storage. This is 10 $^v/_o$ or less and will not present a hazard to the container provided this volume is left free on filling. For completeness, Table 3 shows saturation swelling data for these and other bitumens and bitumenisates [*9*]. The blown bitumens, R85/40 and R90/40, are clearly more suitable than some other bitumens, such as the distilled bitumens, M25 and M35, which might be more attractive from processing considerations. Although a good understanding now exists of bitumenisate swelling, the anomalous behavior of RWF9 suggests that further investigation would be prudent.

TABLE 3—*Bitumen saturation swelling data, %.*

Bitumen Type	Waste Content, Weight %		
	0	24	40 to 51[a]
M25	. . .	>80	20 to 40
M35	40 to 70	. . .	12
R85/40	30 to 40	. . .	8 to 11
R90/40	5 to 15	. . .	1 to 3 or 7 to 8

[a] Maximum waste loading achievable.

Gas Evolution During Gamma Irradiation

Measurements of the volume of gas evolved and its composition as a function of gamma dose in oxygen and in argon have been made from crushed specimens of RWF7. During gamma-irradiation in oxygen, gas evolution from RWF7 consists of 95% hydrogen (H_2) with some carbon monoxide (CO), carbon dioxide (CO_2), and low molecular weight hydrocarbons at total doses ~0.5 MGy. At higher doses ~3 MGy, the relative proportions change drastically, becoming 50% H_2, 5% CO, and 45% CO_2. Under an argon atmosphere, a higher proportion of hydrocarbons are evolved under irradiation, ~4.5% at 0.5 MGy increasing to 7% at 2.5 MGy The rates of gas evolution for the main constituents are summarized in Table 4 for both oxygen and argon irradiations. CO and CO_2 evolution rates increase with gamma dose under oxygen but decrease in argon. Hydrogen evolution rates are independent of dose in oxygen but increase slightly in argon.

The Effects of Gamma Irradiation on Leaching

Figure 4 shows typical data for the effect of gamma radiation on the leaching of RWF7. Specimens were manufactued at Harwell containing approximatley 40 μCi g^{-1} of ^{137}Cs. The specimens were cast in slabs and then cut out of the slabs as 2-cm right cylinders. Leaching was measured using the International Standards Organization (ISO) leach test at 22°C. Irradiation was carried out at 10 kGy h^{-1} at ~28°C. Irradiation to 1 MGy appears to reduce the rate of leaching with an increase at 10 MGy. The results of these and other leaching experiments have been analyzed in detail [*1*]. The data appear to fit a diffusion model with surface contamination and time dependent sorption terms, that is

$$\frac{\Sigma a_n}{A_o}\frac{V}{S} = 2\sqrt{\frac{Dt}{(K_d\rho + 1)\pi}} + \Sigma_o$$

where Σa_n is the cumulative activity leached after a time t; A_o is the initial activity present in the

TABLE 4—*Gas evolution rates (mL g^{-1} kGy^{-1}) from γ-irradiated RWF7 bitumenisates.*

Gas Evolved	Oxygen Atmosphere		Argon Atmosphere	
	At 0.6 MGy	At 3 MGy	At 0.5 MGy	At 2.6 MGy
H_2	3.9×10^{-4}	3.9×10^{-4}	1.5×10^{-4}	2.2×10^{-4}
CO	1.3×10^{-5}	3.5×10^{-5}	2.1×10^{-5}	1.8×10^{-5}
CO_2	2.6×10^{-5}	3.0×10^{-4}	1.1×10^{-5}	6.7×10^{-6}
CH_4	$<1 \times 10^{-6}$	1×10^{-6}	5.2×10^{-6}	7.5×10^{-6}
C_nH_{2n+2} ($n \geq 2$)	$<1 \times 10^{-6}$	5×10^{-6}	1×10^{-6}	2.2×10^{-6}

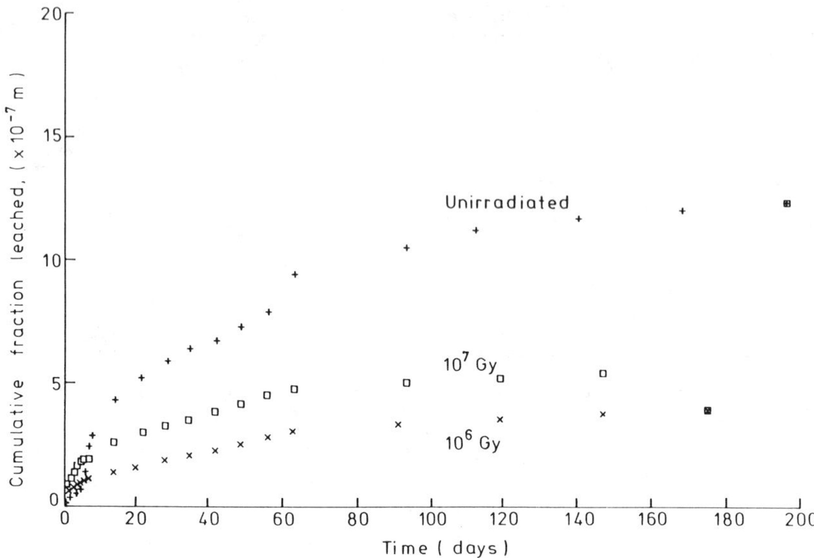

FIG. 4—*The effect of γ-irradiation on the cumulative fraction of* ^{137}*Cs leached from RWF7.*

specimen; V/S is the ratio of specimen volume to geometric surface area; D is the diffusion coefficient; K_d is the geometric distribution coefficient for ^{137}Cs (and is a function of the leachant renewal interval); ρ is the specimen density and Σ_o is a surface contamination factor. Figure 5 shows the improvement over a simple diffusion model. Analyzing the data in this way shows that the diffusion coefficient varies with radiation dose. The variation of K_d with leachant renewal interval was obtained from separate sorption experiments and was also found to vary with radiation dose. Typical values of diffusion and distribution coefficients are given in Table 5. Observations

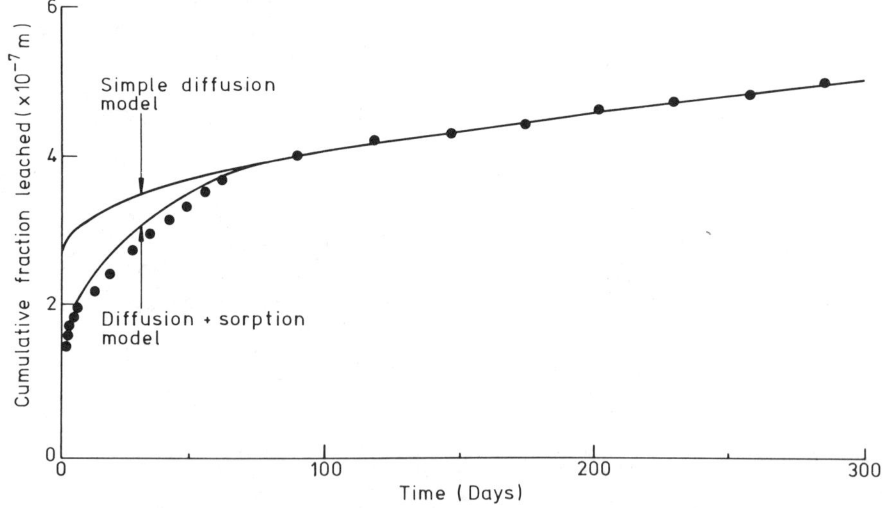

FIG. 5—*Comparison between theoretical models and experimental data for the leaching of* ^{137}*Cs from bitumenisate RWF7.*

Table 5—*Diffusion and distribution coefficients for ^{137}Cs in RWF7.*

γ-Dose (Gy)	Diffusion Coefficient D, m^2 day^{-1}	Distribution Coefficient K_d, m^3 kg^{-1}		
		Δt = 1 day	Δt = 7 days	Δt = 28 days
0	2.5×10^{-13} 1.5×10^{-13}	2.7×10^{-3}	0.0145	0.09
10^6	3.0×10^{-14} 4.0×10^{-14}	4.2×10^{-3}	0.019	0.10
10^7	2.0×10^{-13} 1.3×10^{-13}	7×10^{-3}	0.036	0.40

of the microstructure of leached specimens indicate that water diffuses into the bitumenisate, dissolving the contained salts and causing swelling, which then disrupts the surface of the bitumen.

Alpha Irradiation

Accelerated alpha irradiation has been carried out on RWF7 simulate by incorporating ^{241}Am at a level equivalent to 1 Ci L^{-1}; this provides a dose rate of 0.1 kGy h^{-1}, which is a factor approximately 1000 greater than in the real waste form. A range of measurements were carried out and where appropriate comparative gamma irradiations were carried out at the same dose rate as the alpha irradiations so that the relative effects of α and γ could be ascertained.

The ^{241}Am was added to the simulated, high water content waste prepared according to a published flow sheet. The bitumenisate was then prepared by slowly adding the concentrate, which is a thin slurry containing 23% by weight of solids, to the bitumen in a heated Z-blade mixer at 120°C. After mixing, the final water content was <0.5% and the solids content 30% by weight. Several batches were prepared including one containing ^{241}Am at 0.80 mCi g^{-1} for swelling experiments and microstructural studies, and one containing ^{241}Am at 0.77 mCi g^{-1} and ^{137}Cs at 19.2 μCi g^{-1} for leaching experiments. Experiments included measurements of swelling; determination of distribution of ^{241}Am using α-autoradiography; studies of microstructure through scanning electron microscopy; measurements of changes in mechanical properties by a needle penetration technique; leaching of ^{241}Am and ^{137}Cs, and sorption of ^{137}Cs.

In order to keep activity levels acceptably low in the glove box in which experiments were carried out, the α-swelling experiments were carried out on much smaller specimens than the gamma radiation experiments described earlier. The alpha loaded bitumenisates were cast into open-topped, aluminum cans of 2 cm diameter, and Fig. 6 shows typical swelling data for specimens containing bitumenisates of different depths. At the dose rate of 90 Gy h^{-1} of these experiments gamma radiation does not cause swelling in specimens of these sizes, however alpha irradiation can cause significant swelling, the amount depending on the depth of the bitumenisate in the cans. A direct measurement has not been made of the rate of gas evolution during alpha irradiation, but in order to fit the theoretical bubble growth model [*11*] to the experimental data, it is necessary to infer that the gas production rate during alpha irradiation is 4.8×10^{-3} mL g^{-1} kGy^{-1}, which is a factor of approximately ten times higher than during gamma irradiation. There is some supporting evidence in the literature that gas evolution rates from bitumenisates are much higher during α irradiation than during γ irradiation [*14*]. A comparison between the saturation swelling values observed in these experiments, and the predictions of the bubble growth model using this gas production rate [*1*], is shown in Fig. 7. The agreement is good and the deviation at

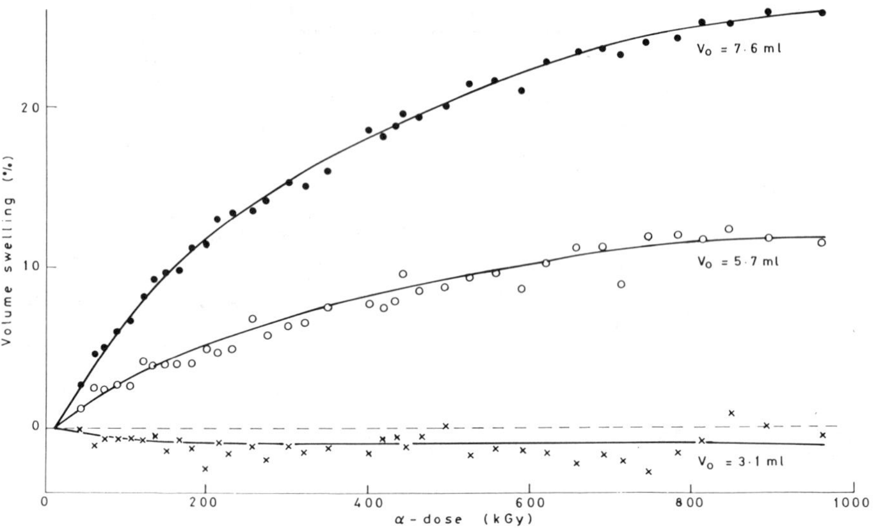

FIG. 6—*Swelling of RWF7 bitumenisate during α-irradiation at a dose rate of 90 Gy/h.*

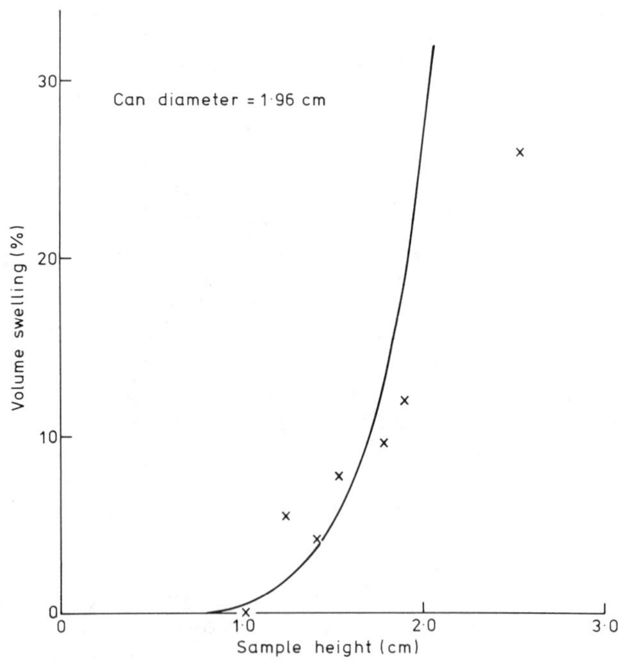

FIG. 7—*Predicted and measured saturation swelling during α-irradiation of RWF7 for different can depths (gas production rate = 4.8 × 10^{-3} mL g^{-1} kGy^{-1}).*

high swelling is attributed to the observed occurrence of large cracks that allow gas release. The results of these experiments have been used with the bubble growth model to assess the behavior of 200- and 500-L drums of real waste, and it has been shown that the alpha emitters will not cause any swelling in practice.

Penetration measurements of the hardness of the bitumenisates have shown that for equivalent doses, alpha irradiation increases hardness substantially more than gamma irradiation.

Freestanding specimens of RWF7 containing ^{241}Am do not swell during self-irradiation because of the large surface area available for loss of evolved gas. For this reason, leaching experiments were carried out on specimens that were first cast into containers (as in the α-swelling experiments), then self-irradiated so that they would swell, and then removed from the containers for leach testing; and also on specimens that had been irradiated as free standing cylinders and had thus not swollen. Figure 8 shows typical data for leaching of ^{241}Am obtained from specimens α-irradiated to 560 kGy; the specimens irradiated in containers and decanned before leaching swelled approximately 5% during irradiation, while the cut cylinders had not swollen during irradiation. During leaching, as the bitumenisates absorbed water they began to swell, as has been observed by other workers. The leaching of unirradiated material could be fitted by a (time)$^{1/2}$ model up to around 200 days of water immersion, at which time deviation from the model occurred attributable to the occurrence of substantial swelling resulting from the absorbed water. A similar effect occurred with the irradiated material but the time over which a (time)$^{1/2}$ model could be fitted was much smaller, as shown in Fig. 8, and deviation occurred at around 100 days. Scanning electron microscopy observations of the microstructure of these specimens have revealed that for the irradiated specimens the absorption of water leads to cracking of the surface at relatively short times, because of its lower ductility as a result of α-irradiation. It is this fissuring of the surface that leads to deviation from the (time)$^{1/2}$ model.

Detailed analyses in terms of the diffusion model described earlier have been made of the leaching data from these experiments and data for diffusion are shown in Table 6. The principal conclusions are that the diffusion coefficients obtained for ^{241}Am in these experiments are similar to those observed in full-scale (200-L) leach tests of RWF7 simulate at Saclay [*15*] and in small-scale fully active specimens of RWF7 at Mol [*16*], and that there is a decrease in diffusion coefficient on α-irradiation, similar to that observed on gamma irradiation. A problem of

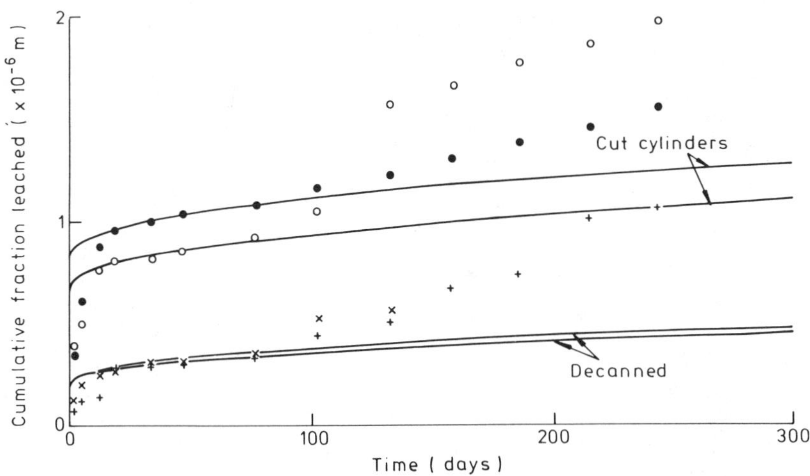

FIG. 8—*Leaching of ^{241}Am from specimens of RWF7 self-α-irradiated to a total dose of 560 kGy.*

TABLE 6—*Leaching data for ^{241}Am and ^{137}Cs from α-irradiated RWF7.*

α-Dose, Gy	Irradiation Swelling, %	Specimen Type	*D*, m^2 day^{-1}	
			^{241}Am	^{137}Cs
0	0	decanned	6×10^{-15}	$<5 \times 10^{-15}$
	0	cut	6×10^{-15}	
5.6×10^5	2.7	decanned	4×10^{-16}	$<5 \times 10^{-15}$
	4.8	decanned	3×10^{-16}	
5.6×10^5	0	cut	5.5×10^{-16}	$<5 \times 10^{-15}$
	0	cut	5.0×10^{-16}	

interpretation that remains to be resolved has emerged from a study of the concentration profile of ^{241}Am in the leach specimens, as measured by α autoradiography. This indicates that the concentration profile can be modelled by a diffusion coefficient for ^{241}Am of $\sim 5 \times 10^{-9}$ m^2 day^{-1}, which is similar to the value for water diffusion, but very much higher than the diffusion coefficient of $\sim 6 \times 10^{-15}$ m^2 day^{-1} obtained from leachate analysis. However, since the ^{241}Am concentrations observed in the leachates are of the same order of magnitude as the solubility limits, it is likely that the data obtained from standard leach tests do not give a true picture of the behavior of ^{241}Am during leaching.

An important general observation from this work is that where the properties of a bitumenisate are affected by gamma radiation, a larger effect is observed at an equivalent dose of α irradiation. Thus care must be taken when using γ irradiation to simulate α irradiation.

Fully Active Studies

Fully active specimens of RWF9 were supplied by the CEA. The six specimens had each been cast into 25-mm-diameter aluminum cans at Cadarache to produce right cylinders. At Harwell

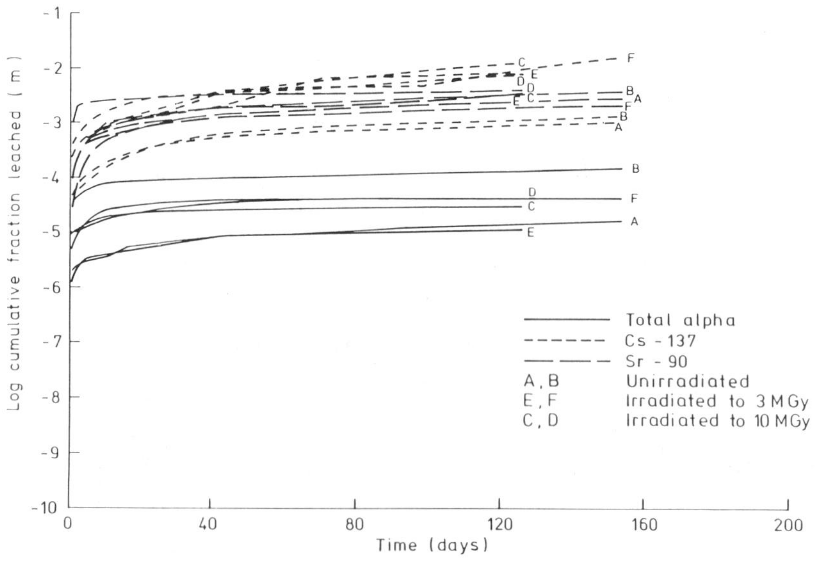

FIG. 9—*Effect of γ-irradiation on leaching of fully active specimens of RWF9. Dose rate = 4 kGy h^{-1}.*

they were divided into three pairs of which one pair was irradiated to 3 MGy (E,F), one pair was irradiated to 10 MGy (C,D), and the third pair (A,B) were kept as unirradiated controls. Irradiation was carried out in the spent fuel ponds at approximately 4 kGy h^{-1}, and the specimens attained a temperature of around 30°C. Under these conditions the irradiated specimens swelled an observable amount. Before leaching, the cans were dissolved in a sodium hydroxide solution, and the specimens were washed in deionized water. Leaching was carried out at 25°C for a total of 168 days. Each leachate was acidified with nitric acid in order to minimize precipitation of radionuclides on the wall of the vessel. Aliquots were taken for γ spectroscopy for fission and activation products, β-counting following radiochemical separation for ^{90}Sr, and α counting. Information about the initial specific activities of the specimens supplied by the CEA indicates that ^{90}Sr (3.3 μCi/g), ^{137}Cs (79.6 μCi/g), and ^{106}Ru (52.8 μCi/g) were the main $\beta\gamma$-emitting radionuclides present and that the total α-activity was 16.2 μCi/g.

Apart from two specimens for which ^{106}Ru was identified after the first day's leaching, only ^{90}Sr, ^{137}Cs, and α-emitters were detected in the leachates. Figure 9 shows the cumulative leach data. Irradiation had no significant, degrading, effect on leach rates, a result in agreement with the data on irradiated simulated RWF7. This result has increased significance in as much as the irradiated specimens had swollen, and these results also show that radiation swelling does not affect leach rates significantly.

Results of Polymer Waste Forms

Dimension and Weight Changes

The polymer matrix waste forms that have been studied can be divided into two types: those which contain substantial quantities of water; these are RWF4, RWF5, and RWR17, which contain wet ion exchangers; and those which contain nominally dry wastes, of which RWF3 was the only example, containing PWR evaporator concentrate. Several different types of matrix were involved including epoxide, polyester, and vinyl ester resins and polystyrene. The behavior of these waste forms under irradiation was more complicated than that of the bitumenisates partly because of the presence of water, which is physically encapsulated as droplets in the matrices and not chemically bound. A considerable amount of data was generated in the program in order to clarify the patterns of behavior. Only the principal observations are described here and are illustrated with typical data.

As with the bitumenisates, specimens in open-topped containers of a range of sizes were irradiated in a ^{60}Co cell over a range of dose rates, but under the same IL^2 conditions, and weight and volume changes were recorded. The specimens included simulated waste forms, the matrix materials without encapsulated waste and some matrices containing encapsulated water without associated waste. In addition to the irradiated specimens, control specimens were stored under similar unirradiated conditions.

All the specimens that contained water lost weight, whether irradiated or unirradiated; this was due to water loss, and none of the dry materials showed a significant weight change. The rate of water loss depended on the matrix material, ranging from ~1% over 100 days for epoxide to ~10% over 100 days for polyester. For some materials the rate of water loss was greater when irradiated while for others it was greater when unirradiated, typical data for a vinyl ester resin waste form are shown in Fig. 10. The rate of water loss in these experiments depends not only on the water diffusion characteristics of the matrices, and possibly on the reaction of radiolytic products, but also on the shrinkage of the waste form away from the walls of the container during manufacture, leading to increased surface area for water loss. Water diffusion data have been generated for the different matrix materials, through measurement of weight increase on water immersion and of weight loss from pre-soaked specimens stored in air. Table 7 shows typical

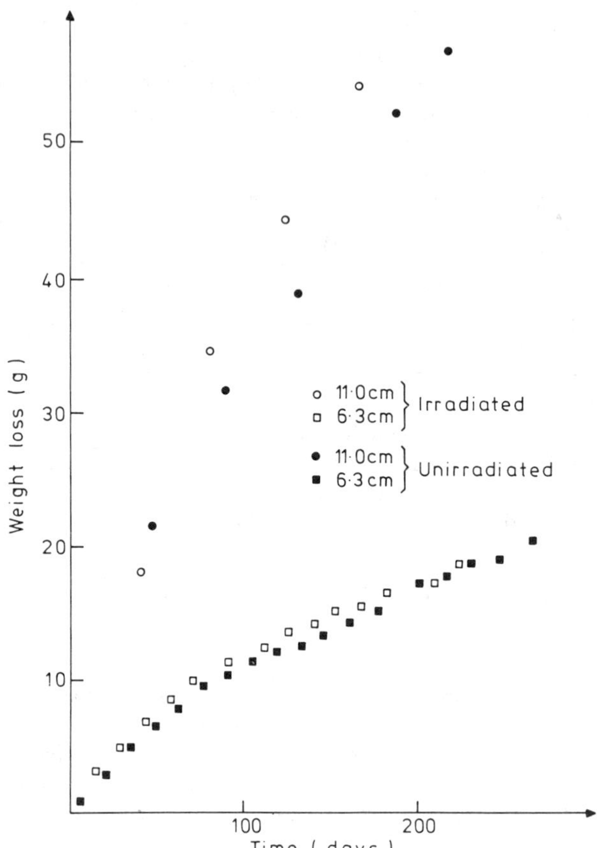

FIG. 10—*Weight change as a function of specimen size and γ-irradiation dose for RWF17 (vinyl ester resin).*

data, which will be of value in modelling the water-loss processes. The results show that water uptake and desorption are dependent on the amount of water in the specimen, but that irradiation to 10 MGy has little effect on water permeation rates.

All the specimens changed volume when irradiated. The dry epoxide resins swelled slightly while the dry polyester and vinyl ester resins and polystyrene sometimes swelled and sometimes shrank slightly. These changes were small amounting to at most $\pm 2\%$ at a total dose of 10 MGy. There was no clear pattern of behavior relating different sizes and dose rates of any given material, but since the effects were small, experimental error could have masked any trends. The epoxide resins containing water or wet waste also swelled slightly, up to approximately 2% at 10 MGy.

The most significant effects were displayed by the polyester and vinyl ester resins, and polystyrene, containing water or wet wastes. The majority of these shrank, some quite considerably to a maximum of around 8% at 10 MGy. Typical data, for the vinyl ester resin of RWF17, are shown in Fig. 11. The shrinkage data for some of these, irradiated under similar IL^2 conditions, lie close together as shown in Fig. 11 for the vinyl ester waste form. This might mean that they fit, by coincidence, some sort of IL^2 model, but there is no obvious physical argument for this and it is more probable that they show similar effects at similar doses independent of dose rate and size, over the ranges studied in these experiments. The large shrinkage of these specimens is

Table 7—*Results of water uptake experiments.*

Waste Form	Resin	γ Dose, MGy	Initial Water Content, %	Mean Solubility, g cm^{-3}	Mean Diffusion Coefficient, cm^2 day^{-1}
RWF 5(H)[a]	Polyester A	10	7.7	0.215	2.44×10^{-2}
		10	dried to constant wt	0.275	1.52×10^{-2}
	Polyester B	0	14.95	0.124	14.70×10^{-2}
		0	dried to constant wt	0.263	2.25×10^{-2}
RWF 17	Polyester	0	21.5[c]	0.073	62.00×10^{-2}
			dried to constant wt	0.193	3.13×10^{-2}
		10	23.63[c]	0.050	73.00×10^{-2}
			dried to constant wt	0.221	2.87×10^{-2}
	Vinyl Ester	0	22.93[c]	0.076	48.00×10^{-2}
			dried to constant wt	0.282	2.09×10^{-2}
		10	24.17[c]	0.028	215.00×10^{-2}
			dried to constant wt	0.272	2.50×10^{-2}
RWF 4	Polystyrene	0	18.92[c]	0.173	8.24×10^{-2}
		10	dried to constant wt	0.379	1.46×10^{-2}
			20.55[c]	0.029	59.75×10^{-2}
			dried to constant wt	0.120	4.46×10^{-2}
RWF 4(H)[b]	Polystyrene	0	17.73[c]	0.155	10.63×10^{-2}
		10	dried to constant wt	0.397	1.66×10^{-2}
			25.17[c]	0.010	76.90×10^{-2}
			dried to constant wt	0.111	4.09×10^{-2}

[a] These specimens had been made and irradiated then stored for a long period before being used in water uptake experiments, this accounts for the low residual water content of these specimens.

[b] These specimens use polystyrene as the matrix resin but contain the same waste simulate as the RWF 5(H) specimens.

[c] The water contents calculated from the initial weight and the dried weights of these specimens are higher than would be expected from observed weight losses during irradiation and storage. It is possible that some unreacted monomer was also present in the specimens and that this was volatilized during drying.

associated with considerable water loss and in most cases there is a good correlation between the volume of water lost and the volume shrinkage. These specimens also shrank when unirradiated, the volume changes being about a half of those under irradiation. It was observed that these specimens had also shrunk significantly during manufacture, while curing, so as to leave a gap between the solid waste form and the wall of the container. It is believed that this permitted much easier water loss than from the specimens which did not display curing shrinkage, such as the water-filled epoxide, and some of the other water-filled thermosets (Fig. 12).

The behaviour of RWF3 was unusual. During irradiation some specimens of the polyester version of this waste form exuded waste at low doses ~50 kGy (Fig. 13) and lost weight and shrank whether irradiated or unirradiated; the epoxide version of RWF3 was unaffected by radiation. The dimension and weight changes were unexpected because, according to its specification, RWF3 contains dry PWR evaporator concentrate with little water. Further investigation revealed that the specimens contained large quantities of water that may have been included by mistake in manufacturing the simulate. It is anticipated that the exudation of waste and volume shrinkage would not occur in correctly manufactured material, and the results indicate the importance of good quality control.

A quantitative theoretical modelling of the processes that occur during the irradiation of polymer

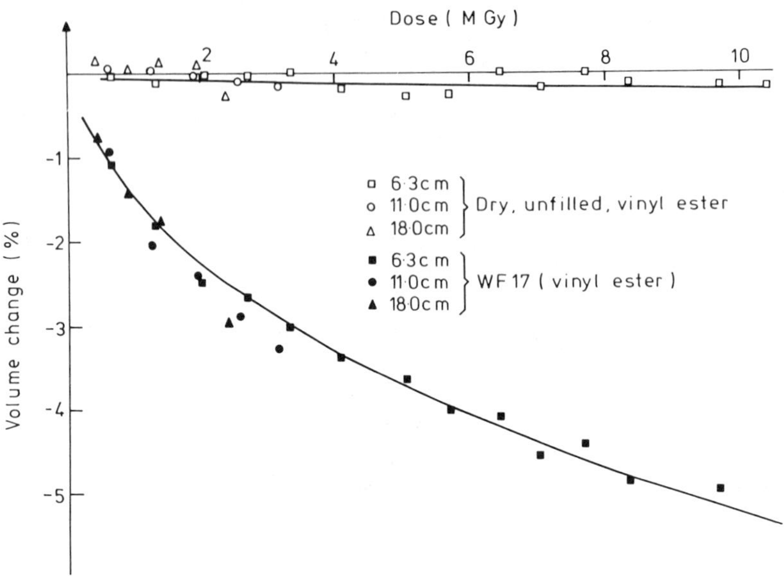

FIG. 11—*Volume change as a function of specimen size and γ-irradiation dose for vinyl ester resin and RWF17 (vinyl ester).*

matrix waste forms containing wet wastes has not yet been developed. It is believed that at least three processes are involved. The main process is water loss which causes shrinkage; this effect is greatest when the waste form has shrunk during curing producing a gap between the waste form and the container wall, resulting in a much larger surface area for water loss. A small amount of shrinkage can also occur in some systems because of further cross-linking of the polymer; this effect is observed when, for example, dry polyester specimens are irradiated. A small amount of

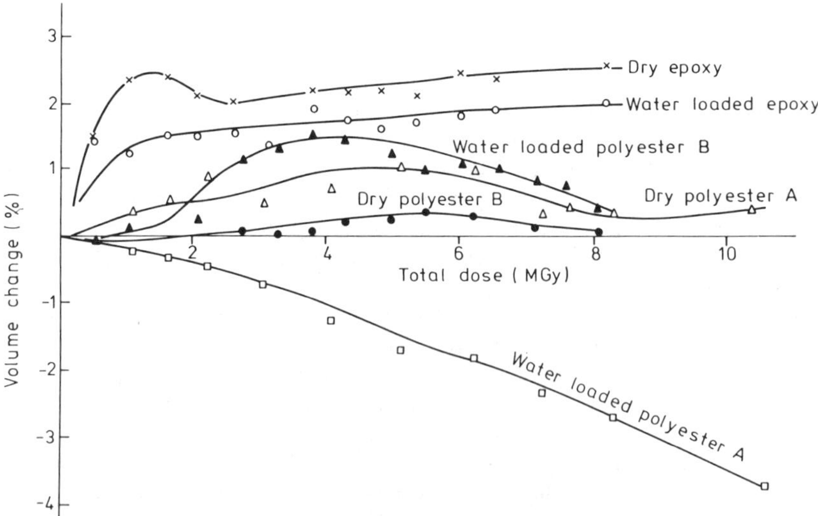

FIG. 12—*Volume change for 10-cm cans of thermosetting resin γ-irradiated at 800 Gy h*$^{-1}$.

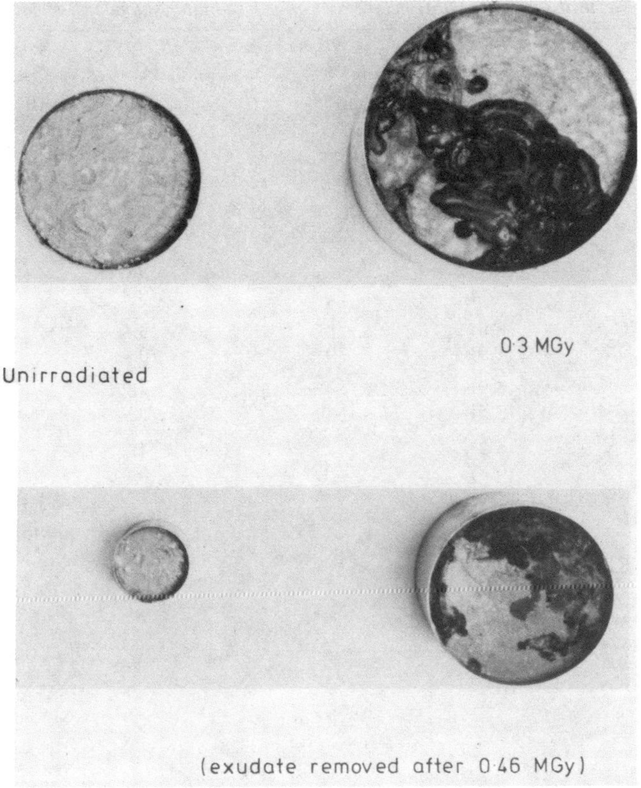

FIG. 13—*Exudation of waste during the γ-irradiation of RWF3 (polyester).*

swelling can occur due, perhaps, to bubble growth; this may account for the swelling of dry epoxide resin under irradiation, and Murphy has recently developed an extension of the bubble growth model, which fits reasonably the epoxide swelling data generated in this program [*17*].

The volume changes observed in these experiments do not pose a hazard to the container, but further theoretical and experimental work is necessary before it will be possible to predict with confidence the behavior of full-scale waste forms under low dose rates from small-scale accelerated tests. In one specimen, which displayed large radiation shrinkage, a large, annular crack was observed. This would provide an increased surface area for leaching. Some consideration will need to be given to the implication of cure shrinkage and water loss. It is not clear what would happen in a full-scale, vented drum; however, one possibility is that the waste form would separate from the wall of the container to leave an annular gap, which would slowly fill with water desorbing from the waste form. The credibility of this happening and its implications need to be assessed.

Mechanical Integrity

The flexural strength, modulus, and strain to failure of matrices and simulated waste forms have been measured before and after irradiation. Test specimens were cut out of the large blocks of irradiated material, rather than by irradiating small test specimens, so that the irradiated material

TABLE 8—*Flexural strength (MPa) before and after irradiation, standard deviations ( ).*

Waste Form	Radiation Dose	Epoxide Resin	Vinyl Ester Resin	Polyester Resin[a] (I)	Polyester Resin[a] (II)	Polyester Resin[a] (III)	Polystryene Resin
No Waste	0	74.9 (4.9)	105.1 (9.3)	62.6 (5.3)	47.5 (8.4)	...	25.2 (2.6)
	10 MGy	67.5 (6.2)	77.5 (11.2)	51.4 (4.5)	22.4 (3.6)	...	27.4 (6.5)
RWF3	0	36.1 (2.5)	...	...	3.6 (0.4)	...	...
	5 MGy	38.0 (1.9)	...	...	3.4 (0.4)	...	...
RWF5	0	15.9 (2.5)	...	...	4.2 (0.9)	...	...
	5 MGy	14.2 (0.8)	...	...	4.7 (0.6)	...	...
RWF5(H)	0	...	...	...	...	8.2 (0.6)	...
	10 MGy	...	...	...	...	5.4 (0.5)	...
RWF4	0	...	...	...	...	...	1.93 (0.4)
	10 MGy	...	...	...	...	...	3.9 (0.9)
RWF17	0	...	9.3 (0.7)	6.3 (0.5)	...	...	...
	10 MGy	...	6.2 (0.7)	5.1 (0.5)	...	...	...

[a] (I) Polyester resin used for RWF17. (II) Polyester resin used for RWF3 and RWF5. (III) Polyester resin used for RWF5(H.)

would have the correct, radiation-aged microstructures. Table 8 shows some flexural strength data. The strengths of waste forms containing wet wastes were substantially lower than those of the matrices apart from the epoxide version of RWF3. The strengths after irradiation are not greatly affected and remain comparable with those of cement matrix waste forms.

Gas Evolution

Measurements of the gases generated during the gamma radiolysis of the polyester and epoxide matrices have been made. In the polyester matrix material, gas evolution is mainly CO and CO_2, with some H_2 and low hydrocarbons. In an oxygen atmosphere, gas composition is 46% CO and 49% CO_2 at low doses ($\sim 10^5$ Gy) changing to 35% CO and 74% CO_2 at high doses ($\sim$4 MGy). In an argon atmosphere, the proportions are 25% CO and 59% CO_2 at 10^5 Gy, changing to 37% CO and 55% CO_2 at 2 MGy. The gas evolution rates are summarized in Table 9. CO_2 evolution rates are independent of dose in both oxygen and argon whereas CO evolution increases with dose in argon but decreases in oxygen. The magnitude of the evolution rates for CO and CO_2 in an inert atmosphere indicate that these gases arise primarily from radiation degradation of the polymer chain rather than by oxidation.

TABLE 9—*Gas evolution rates (mL g^{-1} kGy^{-1}) from γ-irradiated polyester resin (matrix material of RWF 3 and 5).*

	Oxygen Atmosphere		Argon Atmosphere	
Gas Evolved	At 0.1 MGy	At 3 MGy	At 0.1 MGy	At 2 MGy
H_2	6.6×10^{-5}	4.0×10^{-5}	3.0×10^{-5}	3.0×10^{-5}
CO	7.7×10^{-4}	3.1×10^{-4}	1.6×10^{-4}	2.9×10^{-4}
CO_2	8.8×10^{-4}	8.8×10^{-4}	2.9×10^{-4}	2.9×10^{-4}
CH_4	$<10^{-6}$	2.9×10^{-5}	8.6×10^{-7}	6.6×10^{-6}
Hydrocarbons ($C > 2$)	1.3×10^{-5}	4.6×10^{-5}	4.0×10^{-5}	1.1×10^{-5}

In the epoxide resin, overall gas evolution rates are lower than in the polyester, and a higher proportion of H_2 is observed, typically ~14% with 41% CO and 45% CO_2. In an inert atmosphere, CO and CO_2 production is drastically reduced, H_2 becoming the dominant gas evolved.

Leaching

Spiked epoxide and polyester simulates of RWF3 and RWF5 were manufactured at Harwell for γ-irradiation followed by leaching. These specimens in the form of 2 cm diameter cylinders, were manufactured by casting into polytetrafluoroethylene (PTFE) molds for easy removal. They were spiked with ^{137}Cs and were gamma-irradiated to 10^5, 10^6, and 10^7 Gy at approximately 10 kGy h^{-1} in the Harwell spent fuel storage ponds. Similar ^{137}Cs spiked specimens of WF17 simulates encapsulated in polyester and vinyl ester, supplied by the CEGB, were also irradiated to the same total doses. These specimens had all surfaces exposed. After irradiation, all of the specimens underwent leach testing using the ISO static leach test.

Figures 14 and 15 show some of the results of these experiments. All the waste forms containing ion exchange resins, that is, wet wastes, behaved as in Fig. 14, which shows the results

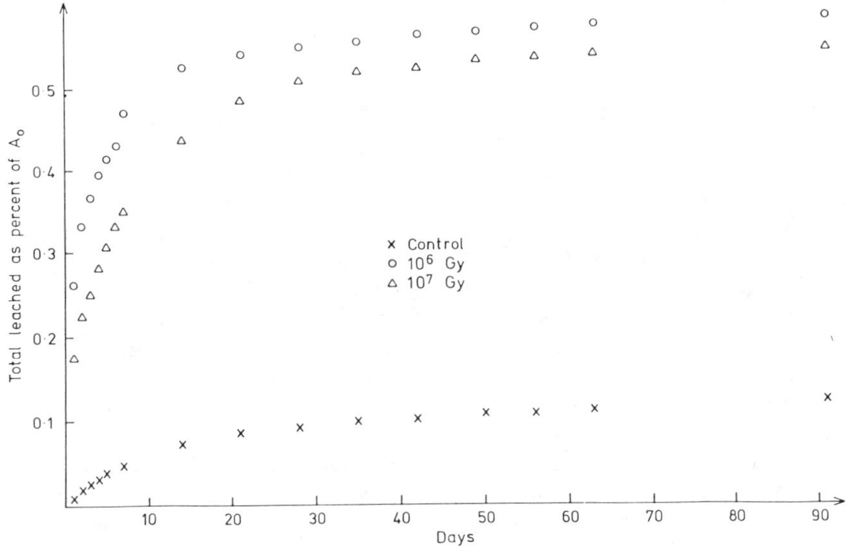

FIG. 14—*Effect of γ-irradiation on the leaching of ^{137}Cs from RWF17 (ion exchange resins in vinyl ester resin).*

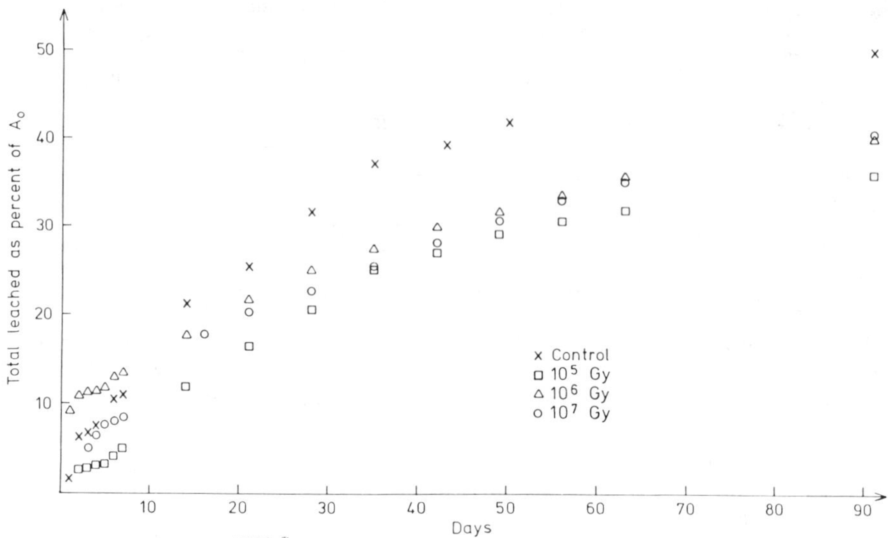

FIG. 15—*Effect of γ-irradiation on the leaching of* [137]*Cs from RWF3 (evaporator concentrates in polyester resin).*

for the vinyl ester version of WF17. Irradiation appears to result in an increase in the amount leached at short times, but the rate of leaching at long times (>25 days) is approximately the same for unirradiated and irradiated specimens. This effect is believed to be due to water loss during irradiation. The water evaporating from the specimen surface leaves a ^{137}Cs-rich surface layer, which is readily leached in the first few days of leach testing. Figure 15 shows the behavior of RWF3, which in this case contained dry waste. Irradiation may have reduced the rate of leaching from this waste form, but the effect is not large.

Conclusions

1. Bitumenisates swell during irradiation, swelling commencing at around 50 kGy, and saturating at around 300 KGy. Use of a blown grade of bitumen and a high waste loading should guarantee that irradiation swelling does not pose a hazard to the integrity of the container providing that up to 10% of the available volume of the container remains unfilled.
2. Comparisons between α and γ irradiation of RWF7 bitumenisate have shown that where properties change on γ irradiation, a greater effect is observed for an equivalent α-dose.
3. The radiation damage produced by the alpha active component in bitumenized wastes should not in itself cause significant radiation swelling of the full-scale waste form.
4. Irradiation to 10 MGy and irradiation swelling of 4 to 7 v/o does not significantly influence the leach rates of bitumenisates for ^{137}Cs, ^{90}Sr, and α emitters.
5. The behavior of thermoset resin based waste forms is more complex than that of bitumen based waste forms, but it is unlikely that any substantial swelling would occur.
6. The behavior of thermoset resin based waste forms during irradiation depends upon the degree of cure shrinkage for the particular specimen under testing. Specimens where cure shrinkage is high tend to shrink during irradiation whereas those whose cure shrinkage is low may swell slightly.
7. Gamma irradiation may increase the initial leach rates for thermoset waste forms containing ion exchange resins, but the effects on long-term behavior are small.

8. Gas evolution rates are typically $<10^{-2}$ mL g^{-1} kGy^{-1} and decrease in the order polyester > epoxide > RWF7 bitumenisate.

Acknowledgments

We acknowledge the experimental assistance of our colleagues Mr. J. R. Dyson, Mr. G. Worrall, and Mr. M. Brownsword, and the role played by Mr. R. A. J. Sambell in coordinating with the CEC. This work has been carried out under funding from the Commission of the European Communities through Contracts 243-81-13-WAS-UK and 299-83-13-WAS-UK and the U.K. Department of the Environment. The results will be used in the formulation of U.K. Government policy, but at this stage they do not necessarily represent U.K. Government policy. This report will form part of a larger, joint final report on the characterization program funded by the CEC, which will be published in the EUR series.

References

[*1*] Burnay, S. G. and Dyson, J. R., "Comparative Evaluation of α and γ Radiation Effects in a Bitumenisate," AERE-R11857, Oxon, England, 1986.

[*2*] Johnson, D. I., Burnay, S. G., and Phillips, D. C., "The Effects of Gamma Radiation on Polymer Matrix Waste Forms," AERE-R12081, Oxon, England, 1986.

[*3*] Burnay, S. G., Johnson, D. I., Brownsword, M., and Phillips, D. C., "The Effects of Gamma Irradiation on Leaching of ^{137}Cs Organic Waste Forms," AERE-R12192, Oxon, England, 1986.

[*4*] Johnson, D. I., Hitchon, J. W., and Phillips, D. C., "Further Observations of Radiation Effects in Bitumenisates," AERE-R12292, Oxon, England, 1986.

[*5*] Sambell, R. A. J., Ed., "Characterisation of Low and Medium Level Radioactive Waste Forms," EUR 8663 EN, 1983.

[*6*] Vejmelka, P. and Sambell, R. A. J., Eds. and Compilers, "Characterisation of Low and Medium Level Radioactive Waste Forms—Joint Annual Progress Report 1982," Report EUR 9523 EN, 1984.

[*7*] Phillips, D. C., Johnson, D. I., Burnay, S. G., and Hitchon, J. W., "The Effects of Radiation on Organic Matrix Waste Forms," *Testing, Evaluation and Shallow Land Burial of Low and Medium Radioactive Waste Forms,* W. Krischer and R. Simon, Eds., EUR-8979, Commission of European Communities, 1984.

[*8*] McHugh, G., Spindler, W. E., and Mattingley, N. J., "Radiation Effects in Hydraulic Cement Matrix Waste Forms," *Testing, Evaluation and Shallow Land Burial of Low and Medium Radioactive Waste Forms,* W. Krischer and R. Simon, Eds., EUR-8979, Commission of European Communities, 1984.

[*9*] Phillips, D. C., Johnson, D. I., Burnay, S. G., and Hitchon, J. W., "The Effects of Radiation on Intermediate Level Organic Matrix Waste Forms," *Radioactive Waste Management,* British Nuclear Energy Society, London, 1985.

[*10*] Phillips, D. C., de Angelis, G., and Koester, R., "Radiation, Thermal and Mechanical Effects on Low and Medium Active Conditioned Waste," *Second European Conference on Radioactive Waste Management and Disposal,* Luxembourg, April 1985, to be published by the Commission of European Communities.

[*11*] Phillips, D. C., Hitchon, J. W., Johnson, D. I., and Matthews, J. R., "The Radiation Swelling of Bitumens and Bitumenised Wastes," *Journal of Nuclear Materials,* Vol. 125, 1984, pp. 202–218.

[*12*] Phillips, D. C., Hitchon, J. W., Johnson, D. I., and Matthews, J. R., "Irradiation Swelling of Bitumenised Radioactive Wastes," AERE-R10925, Oxon, England, August 1983.

[*13*] McHugh, G., Sambell, R. A. J., Spindler, W. E., and Mattingley, N. J., "The Effects of Radiation on Cement Waste Forms," AERE R12298, Oxon, England, 1986.

[*14*] Koziewicz, S. T., *Nuclear Chemical Waste Manual,* Vol. 1, 1980, p. 139.

[*15*] Nominé, J. C. and Vejmelka, P., *Testing, Evaluation and Shallow Land Burial of Low and Medium Radioactive Waste Forms,* W. Krischer and R. Simon, Eds., EUR 8979, 1984, p. 7.

[*16*] R. de Batist, Timmermans, W., Reyneders, R., and Hild, W., *Testing, Evaluation and Shallow Land Burial of Low and Medium Radioactive Waste Forms,* W. Krischer and R. Simon, Eds., EUR-8979, Commission of European Communities, 1984, p. 43.

[*17*] Murphy, S. M., "A Theoretical Model of the Swelling of Gamma-Irradiated Epoxy Resin," AERE-R11616, April 1985.

James O. Henrie[1] *and Dann J. Flesher*[1]

Hydrogen Control in the Handling, Shipping, and Storage of Wet Radioactive Waste

REFERENCE: Henrie, J. O. and Flesher, D. J., "**Hydrogen Control in the Handling, Shipping, and Storage of Wet Radioactive Waste,**" *Influence of Radiation on Material Properties: 13th International Symposium (Part II), ASTM STP 956,* F. A. Garner, C. H. Henager, Jr., and N. Igata, Eds., American Society for Testing and Materials, Philadelphia, 1987, pp. 636–646.

ABSTRACT: Under contract with the Department of Energy, Rockwell Hanford Operations (Rockwell) conducted a series of studies related to Three Mile Island (TMI) post loss of coolant accident (LOCA) waste handling safety. The key factor in these studies was the control of radiolytic gases, specifically hydrogen and oxygen.

Success was achieved in the prediction of gas generation rates and in the development and application of gas control technologies. That success led to the publication of a "lessons learned" document, providing (1) a step-by-step guide in evaluating potential radiolysis hazards in waste storage and shipment activities and (2) a set of corrective options for preventing or controlling the hazards. The evaluation steps developed in the document are outlined in a functional logic diagram. Several procedures are described which provide safe and effective controls for hydrogen and oxygen levels in containers scheduled for shipment or interim storage. A method is provided to pre-evaluate each control option and select the most cost effective procedure for each case.

KEY WORDS: radiolysis, hydrogen, oxygen, safety, catalyst, recombiner, storage, shipping, waste, wet wastes, gas control, flammable, pressure

The production of flammable gases by the radiolysis of water and organic materials is proportional to the amount of ionizing radiation absorbed by the water or the organic material. In closed systems containing organic or wet radioactive materials, the buildup of flammable mixtures of gases can be expected and has been observed. Since the presence of flammable gases poses a number of potential hazards, careful evaluation and control of these materials is essential in ensuring an acceptable degree of safety.

The importance of flammable gas control was highlighted by the massive stabilization and cleanup operations subsequent to the Three Mile Island Unit 2 (TMI-2) loss of coolant accident (LOCA). The canisters planned for the transport of the TMI-2 core debris will be subject to radiolytic gas generation. The calculated (probable-maximum) hydrogen-oxygen gas generation rate for these containers is 0.11 L/h for a canister loaded with fine, wet core debris [*1*].

To ensure that flammable gas mixtures will not exist in these canisters, Rockwell Hanford Operations (Rockwell) conducted a demonstration test program in support of the U.S. Department of Energy (DOE), EG&G Idaho, Inc., and GPU Nuclear Corp., to determine which catalysts and bed configurations would be acceptable and near optimum under these wet conditions. These studies demonstrated the simplicity and effectiveness of catalytic recombination of hydrogen and oxygen, even under very wet conditions. The stoichiometric reaction of these gases in the container

[1] Principal engineer and senior engineer respectively, Rockwell Hanford Operations, P.O. Box 800, 2E/2700E/A206, Richland, WA 99352. Rockwell Hanford Operations is operated by Rockwell International, Inc., for the U.S. Department of Energy under Contract De-Ac06-77RL01030.

atmosphere maintains the minimum constituent (either hydrogen or oxygen) below 1%, and ensures that flammable mixtures (>4.1% OH_2 and 5.0% O_2) will not be attained. Other control procedures were also evident from the test results.

The U.S. Nuclear Regulatory Commission (NRC) Office of Inspection and Enforcement (I&E) has established requirements for the shipment of waste materials subject to hydrogen gas generation [2]. This I&E notice establishes a safe-shipping time period that is twice the expected shipping and handling period (from canister purging and closing to completion of shipment). During this period, the concentration of hydrogen gas must be limited to 5% by volume (or 0.063 g-mol/ft^3 when at pressures below one atmosphere), or the concentration of oxygen gas must be limited to 5% by volume.

The document, RHO-WM-EV-9P, Rev. 1 [3], from which this paper is abstracted, was prepared to convey the pertinent information concerning the handling and shipping of wet radioactive wastes that resulted from the TMI-2 experience. On a broader base, it provides engineering tools, procedures, and precautions that are intended to ensure the safe handling, shipping and storage of wet radioactive wastes for a variety of conditions and applications. A stepwise procedure is presented that permits the individual investigator to evaluate the potential for flammable gas generation, and to minimize potential hazards, with the intent of meeting the referenced NRC requirements.

Evaluating Specific Problems

Gas Sources and Their Evaluation

There are two major sources of flammable gas generation: (1) reaction between metals and water, which takes oxygen from the water to oxidize the metal and releases the remaining hydrogen as gas; and (2) long-term exposure of water or organic materials to ionizing radiation (radiolysis). Metal-water reactions occurring at near-ambient conditions are generally slow and self-limiting for most structural metals and become significant only at extreme temperatures.

The most important source of hydrogen-oxygen generation for typical wet radioactive waste storage and shipping is radiolysis. When water or organic materials are brought into prolonged contact or proximity to sources of radiation, radiolysis will occur.

The various radiolysis processes are schematically shown in Fig. 1. A technical review of each specific case should indicate which of these reaction steps predominate. Figure 2 is a functional logic diagram that describes the necessary steps developed in this study for conducting orderly, detailed evaluations of the radiolysis hazard and for selecting corrective actions.

Engineering Methods and Procedures

Methods for assessing hazardous situations and specific procedures for corrective actions are required at facilities with hazardous radioactive waste containers. These methods and corrective actions should be easy to understand and prioritized to ensure timely implementation.

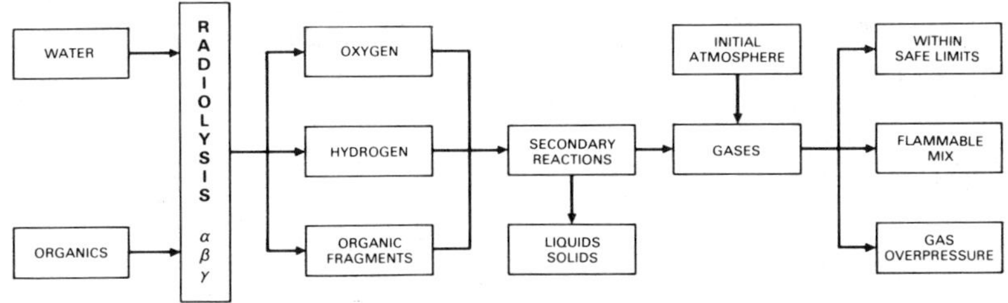

FIG. 1—*Schematic of radiolysis effects for a container of radioactive hydrogenous waste.*

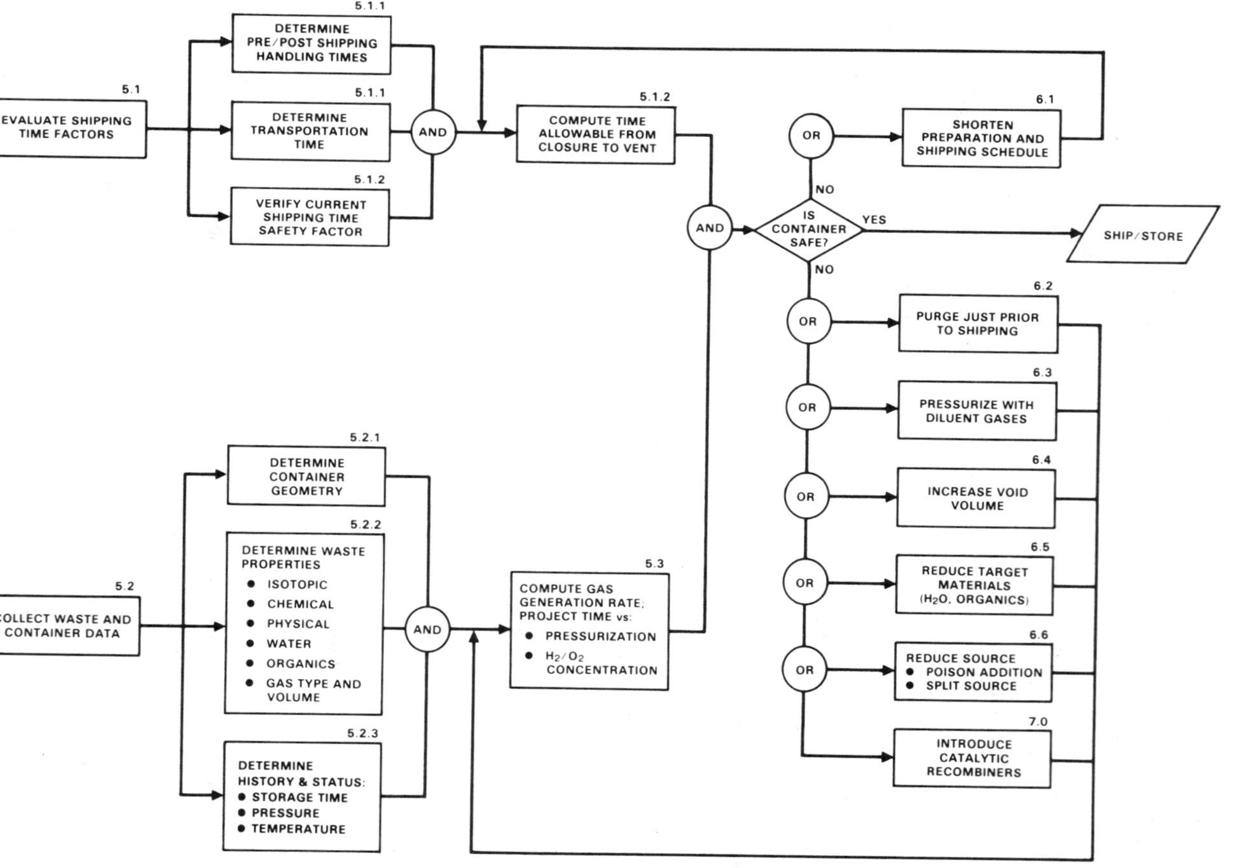

FIG. 2—*Safety logic diagram for shipping and storing containers of radioactive waste (the numbers above blocks refer to sections of the report RHO-VM-EV-9P).*

Screening procedures used to identify potential problem areas should include methods of predicting (computing) gas (H_2, O_2) generation rates and methods of confirming those gas generation rates.

Corrective procedures for existing containers should be evaluated and should include as a minimum: venting of the gases, diluting with inert gases, and using hydrogen-oxygen recombiners.

Preventative actions for future container designs and waste handling operations should include reducing the expected or existing gas generation rate, providing administrative controls, and adding hydrogen-oxygen recombiners to new vessels before loading with wet wastes.

Radiolytic Gas Generation

The basic equation for computing radiolytic gas generation is as follows:

$$\text{gas generated} = E \times F \times G \times C$$

where

E = the total ionizing radiation energy within the container,
F = the fraction of the total ionizing radiation that is absorbed by the target (hydrogen-containing material),
G = the number of molecules of gas generated per unit (100 eV) of ionizing radiation absorbed by the target material, and
C = a conversion factor based on the units of measurement.

In addition, other parameters may be required or will be of value in determining the safety of each container. Methods of determining each of these values are presented in Table 1. Primarily these will be those of the ideal gas law $PV = nRt$, the volume and mass of waste present, and the internal volume of the container.

TABLE 1—*Methods table.*[a]

Factor	Units	Methods
Gas generated in time t	Litres (L)	1. physical measurement and analysis 2. compute from E, F, and G data
E, total ionizing energy in time t	watt-hour, joules (J), greys (Gy)	1. compute from known isotopic content 2. compute from measured source terms 3. estimate from comparable cases 4. estimate by calorimetry means
F, fraction of energy absorbed by target material	no units	1. estimate based on comparable cases 2. estimate based on radiation types, energies, waste particle size, quantities of target water or organic materials exposed, quantities of nonhydrogen-generating materials (absorbers), and waste container size and shape
G, gas formed per unit of energy absorbed	molecules/100 eV	1. use previously verified (tested) values 2. use values from comparable cases, with corrections for back reactions based on chemical makeup
C, conversion factor	based on units of E	1 J = 1 W 1 Gy = 100 rads = 1 J/kg 100 eV = 1 L = 2.688×10^{22} mol

[a] See Ref *3* for details on each of the above methods.

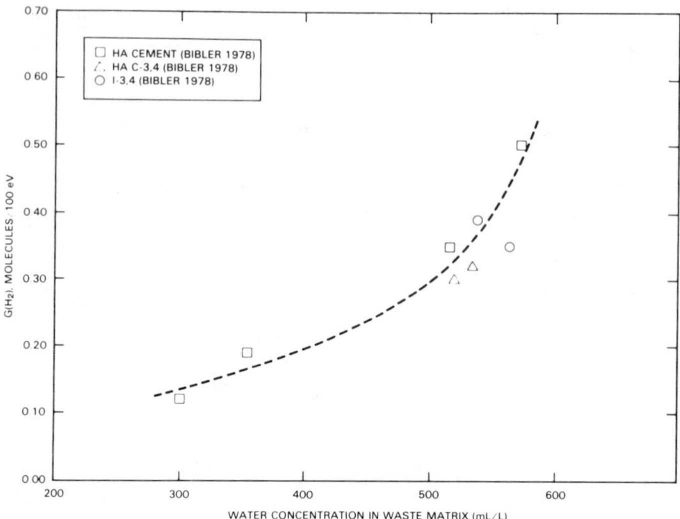

FIG. 3—*Hydrogen gas generation rate versus water concentration in grouts (Alpha radiation).*

Empirical Gas Generation Data

The data from the literature on hydrogen and oxygen generation have been grouped and plotted. The data were found to fall in families characterized by their chemical (inorganic or organic) and physical attributes. The $G(H_2)$ values for each family of tests, such as those for nitrate and nitrite salts, were found to cluster based on easily measured parameters such as water content or density. Thus, these parameters can provide good initial estimates or ranges for $G(H_2)$ values. Figures 3 through 6, selected from the full report, indicate how first estimates of $G(H_2)$ can be made based on simple physical parameters [*3*].

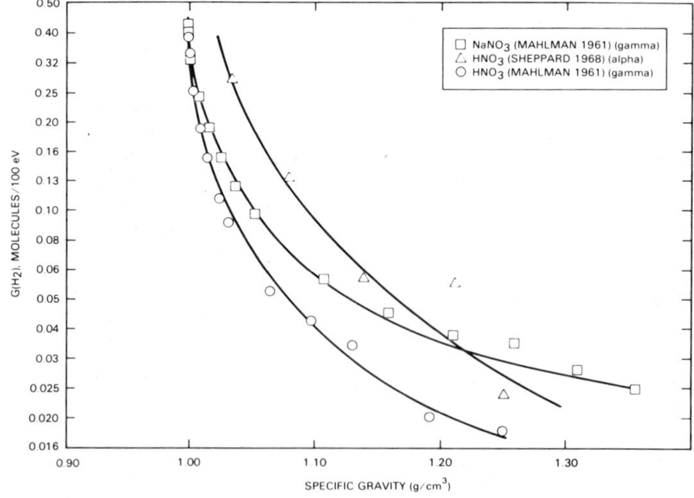

FIG. 4—*Hydrogen gas generation rate versus specific gravity for NO_x solutions (Alpha-Gamma radiation).*

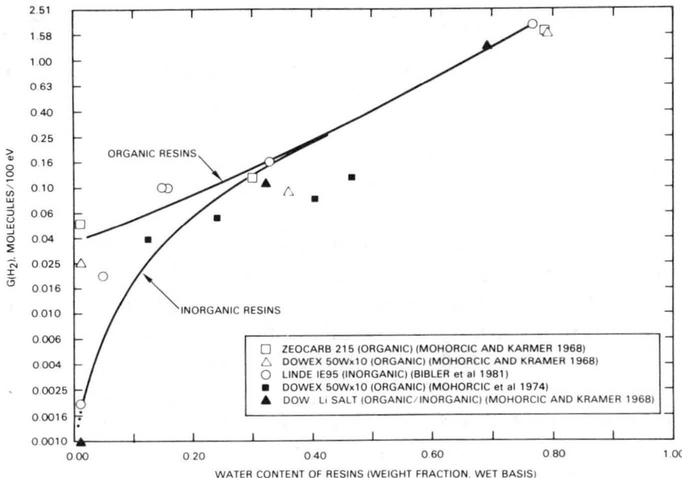

FIG. 5—*Hydrogen gas generation rate versus specific gravity of resins (Gamma radiation).*

Applying the Technology

Reference *3* provides details for the individual steps in using Table 1 and the functional logic diagram (Fig. 2). The stepwise procedures help the engineer select the best option for gas control in a specific situation. Details for gathering valid data for the gas generation calculations are also provided.

Corrective Procedures

The evaluation process identifies those containers that can be safely shipped and those containers that will require some other action before shipment. Corrective procedures are presented in the

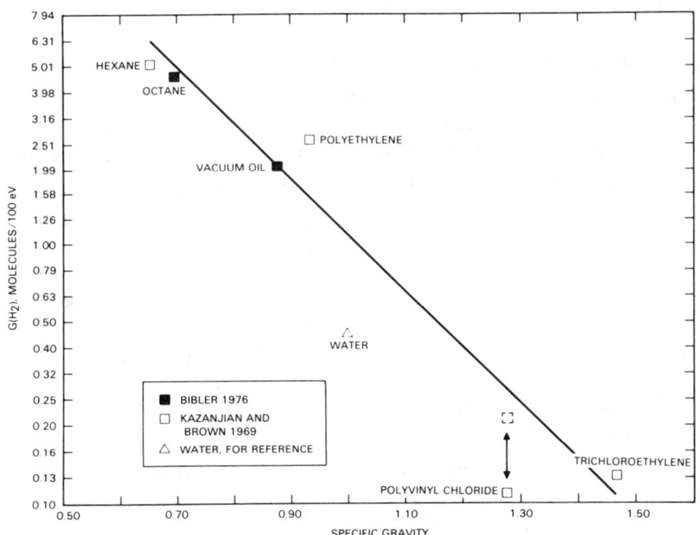

FIG. 6—*Hydrogen gas generation rate versus specific gravity of organics (Beta-Gamma radiation).*

order of increasing difficulty or cost. Treated iteratively, they should provide the optimum route to safe shipping or storage conditions.

The following are brief descriptions of the options in blocks 6.1 through 6.6 of the logic diagram (Fig. 2):

6.1 Where time to flammability or time to overpressurization exceeds allowed shipping time by a small margin, it may be possible to accelerate handling activities to accommodate a safe shipping schedule to meet the NRC guidelines of Ref *2*.

6.2 The canister preparation time is effectively reduced if a vent-purge step (using an inert gas, if appropriate) is performed as close to the actual shipping point as practical.

6.3 The flammable limits for hydrogen and oxygen are given in volume or molar percent. By increasing the pressure of inert gases within the caniser from one atmosphere to two atmospheres, the time to flammability is effectively doubled. The limitation on this procedure is canister or cask overpressure.

6.4 The time to flammability and time to overpressurization are both dependent on the void volume of the container. Two approaches to increasing the void volume are removal of liquids and venting.

(a) Water or other liquid inside the container may be removed by draining or vacuum pumping to provide more vapor space. Removal of this liquor will have the added beneficial effect of reducing the radiolysis target and the possible added effect of reduction of the radioactive source (see 6.5 and 6.6).

(b) The cask design may permit the direct venting of the atmosphere of the radioactive material container to the atmosphere of the cask (the inner containment vessel of a cask usually must provide double containment to use this option). Assuming reasonable diffusion rates through the vent, this effectively increases the void volume.

6.5 The amount of water or organic material available as a target or producer of hydrogen-oxygen gases can be reduced. This reduces the F value.

6.6 The reduction of radioactive source terms is possible by either removal of some of the source or by poisoning of the source. Source removal parallels that of the target removal and is most easily accomplished if the source is dissolved in the liquid portion of the waste.

By using chemical "poisons," the value of $G(H_2)$, the net hydrogen generation rate may be reduced. Generation rate is reduced by the presence of several soluble species (nitrates, for example). The addition of other salts (nitrites, and so forth) can increase back reactions, or otherwise act as radiolysis inhibitors. Such agents can be added as poisons to reduce radiolysis where initially no salts or only weak solutions exist in the liquid portion of the waste (Fig. 4).

Catalytic Recombiners

The catalytic recombiner option (Fig. 2, Block 6.7) provides the benefits of passive, maintenance-free hydrogen-oxygen control. Catalysts of the type used in TMI-2 studies provide continuous control of the concentration of those gases or the minimum constituent gas (usually oxygen in initially inerted systems) to levels well below flammable concentrations.

Catalytic recombiners remove hydrogen-oxygen gases in the stoichiometric 2:1 ratio. Where nonstoichiometric net gas generation occurs, such as in cases where the oxygen is being scavenged by waste components, the hydrogen can build up in excess. If the oxygen is sufficiently limited, there is little hazard from flammability; however, the excess hydrogen can cause overpressurization of the container and provides a potential for ignition upon venting to the atmosphere or upon in-leakage to the container.

Catalysts

Recent progress has been made in the development of passive catalytic recombination systems. In the passive system, the hydrogen-oxygen gases diffuse to the catalyst and are recombined to form water vapor, which then diffuses to and condenses on the colder surfaces in the system. Pilot plant work has identified an effective catalyst blend for use under very wet conditions [*1*]. Mixtures of Engelhard Deoxo D (Nuclear-Grade A 16430) and silica base (No. 85-42) or silicone-coated Atomic Energy of Canada Limited (AECL) catalysts were found to be very effective recombiners and strongly synergistic under very wet (but not submerged) conditions. The Engelhard catalyst, particularly under reasonably dry conditions, provides highly efficient recombination of hydrogen-oxygen gas while the AECL catalyst provides rapid recovery of the entire catalyst bed from very wet conditions.

The catalyst currently recommended for hydrogen control in the handling, shipping, and storage of wet radioactive waste is a 50-50 blend of the two following materials:

- Engelhard Deoxo Type 18467 Catalyst,[2] a paladium-on-alumina catalyst, previously designated as Engelhard Deoxo D, Nuclear-Grade A16430.
- AECL No. 85-42 Silica Base Catalyst,[3] a hydrophobic, platinum-on-silica catalyst.

Catalyst Bed Development

A highly effective catalyst bed design utilizes thin (approximatly 1-cm thick), disk-shaped beds of catalyst. Beds at each end of the cylindrical canister hold about 100 g of catalyst. This design provides a safety margin of several times the minimum quantity of catalyst required for gas control in the TMI-2 core debris canister.

An 8-mesh (8 wires/in.), stainless steel retaining screen utilizing 0.035-in. (0.089-cm) diameter wire is recommended to cover the bed. This size screen provides adequate containment of the catalyst pellets (⅛-in. [0.318-cm] minimum size), heavy wire construction for maximum bed protection, and a large, effective open area for gas diffusion.

Catalyst Bed Location Features

Catalyst beds were built symmetrically into both ends of new TMI-2 core debris canisters to ensure gas-exposed (nonsubmerged) catalysts in any canister position. This design [*1*] thus prevents radiolytic gas buildup even under postulated accident scenarios, where the cask is put in an upside down position.

Other designs, where most of the fluids are removed, may require only a single bed, placed such that submersion is not possible.

Catalyst Bed Retrofits

The engineer assigned to retrofit a catalyst system to a specific container can readily evaluate the specific features of the existing container and review the minimum requirements for the catalyst bed(s). The best approach will become readily apparent, and the design will usually be a matter of selection between options.

[2] Manufactured by Engelhard Corporation, 2655 U.S. Route #22, Union, NJ 97083.

[3] Manufactured by Atomic Energy of Canada Limited, Commercial Operations, Chalk River Nuclear Laboratories, Chalk River, Ontario, Canada, K0J1J0.

Catalyst Bed Design Criteria

Each catalytic unit located in a canister of wet radioactive material for flammable gas control should meet the following criteria:

- One catalyst bed must be exposed to the gas/vapor space at all times.
- Very wet systems should contain catalyst beds composed of 50% Engelhard Deoxo D and 50% silica base AECL catalysts or a test-proven equivalent. (This catalyst mix is wet resistant and recovers rapidly after being submerged in water.)
- The amount of exposed catalyst required is proportional to the gas generation rate. The recommended ratio of the bulk volume of the mixed catalyst to the gas generation rate, in mL of catalyst per mL of H_2 + O_2 gas produced per hour, is 1.0. Thus, a generation rate of 50 mL/h requires a 50-mL bed of the mixed catalyst. Where multiple beds are needed to preclude submersion, each bed should contain this volume of the mixed catalyst.
- The recommended ratio of bed volume in mL to exposed (screened) area in cm^2 is 1.0. This results in a 1-cm-thick bed of the mixed catalyst when screened on one side.

The catalytic recombiners designed for the TMI-2 core debris canisters are consistent with these criteria. These mixed-bed recombiners are projected to maintain the oxygen concentration below 0.5%, or the hydrogen concentration below 1.0%, even under the very wet conditions in a maximum-loaded TMI-2 canister. Tests conducted at temperatures below freezing showed that the catalyst beds, as designed, would remove hydrogen and oyxgen gases at the design rate for at least a few weeks under these temperature extremes.

Safety

The hazards of handling hydrogen gas have been evaluated over many decades, and the literature provides tables for specific mixtures with air, oxygen, and so forth [*5,6*]. Hydrogen gas has the following characteristics:

- Extreme flammability in concentrations of 4.1 to 74.2 vol% in air.
- Requires an oxidizing agent in minimum quantity (usually oxygen, minimum 5.0%) to be flammable.
- Easily ignited and burns with a hot, nearly invisible flame.

These characteristics lead to a single conclusion: avoid any conditions that can produce combustion within a container. This precaution also extends to gases leaking or being vented from a container, and to potential accumulation in buildings or other enclosures in which canisters of waste may be stored.

Hydrogen Reactions with Various Oxidants

The Bureau of Mines provides considerable information on this subject [*7*], including the following limits of flammability of hydrogen in various oxidants at 25°C and atmospheric pressure shown in Table 2.

There are also flammability limits for the oxidants. For instance, the lower flammability limit of oxygen in hydrogen is 5%. Oxygen is more likely than hydrogen to be consumed (scavenged) by chemical reactions with other materials in a waste canister. Also, the oxygen gas generation rate is only half that of hydrogen in the water radiolysis process. These factors often make the control of oxygen more advantageous than the control of hydrogen. Under this control condition,

TABLE 2—*Flammability limits of hydrogen in various oxidants at 25°C and atmospheric pressure.*

Oxidant	Lower Limit	Upper Limit
Oxygen	4.0	95
Air	4.0	75
Chlorine	4.1	89
N_2O	3.0	84
NO	6.6	66

NOTE: Detonations can occur at higher concentrations (>17% hydrogen in air).

the initial atmosphere must be inerted and air in-leakage excluded, usually by maintaining the canister internal pressure above atmosphereic pressure. In addition, assurance must be obtained that no oxygen producing reactions will occcur, such as those which would convert nitrates to nitrites, or that no other oxidizers such as nitric oxide (NO) or nitrogen dioxide (N_2O) will be produced.

Catalysts

Catalysts introduced suddenly into an atmosphere of a flammable hydrogen-oxygen mixture can become an ignition source. Therefore, in retrofit situations, containers should be vented and purged before the initial introduction of a catalyst bed to prevent the possibility of ignition. Where gas concentrations (H_2 and O_2) are under the flammability limits, the recombination reaction proceeds slowly and rapid heating or burning does not occur.

Overpressure Versus Flammability

As either of the key problems of flammability or overpressure is addressed, the investigator should ensure that conditions are not introduced that produce a potential hazard from the other. Each corrective action has the potential for affecting both, and complete evaluations and projections should be made before taking the corrective action.

Also, shipping and long-term storage provide significant differences in basic parameters, especially the time factor. Evaluations should be made for both overpressurization and flammability under all projected canister handling and storage conditions.

Summary

This paper provides a background source document for engineers who face radioactive waste transportation or storage situations. Conditions encountered may vary from those as isolated and rigorous as the TMI-2 core debris removal to as frequent and modest as the disposal of small quantities of radioactive hospital laboratory waste. The same set of basic rules and computations can be applied to provide assurance against hydrogen buildup in disposal containers.

References

[*1*] Henrie, J. O. and Appel, J. N., *Evaluation of Special Safety Issues Associated with Handling the Three Mile Island Unit 2 Core Debris,* RHO-WM-EV-7, Rockwell Hanford Operations, Richland, WA, 1985; also GEND-051, EG&G Idaho, Inc., Idaho Falls, ID.

[2] Jordan, E. L., *Clarification of Conditions for Waste Shipments Subject to Hydrogen Gas Generation,* IN 84-72, U.S. Nuclear Regulatory Commission, Washington, DC, 1984.
[3] Henrie, J. O., Flesher, D. J., Quinn, G. J., and Greenberg, J., *Hydrogen Control in the Handling, Shipping, and Storage of Wet Radioactive Waste,* RHO-WM-EV-9P, Rev. 1, Rockwell Hanford Operations, Richland, WA, 1986; also GEND-052, EG&G Idaho, Inc., Idaho Falls, ID.
[4] ASTM, *Annual Book of ASTM Standards,* Related Standards: E 459-72, E 511-73, E 422-83, E 457-72, E 639-78, C 976-82, and C 518-76, American Society for Testing and Materials, Philadelphia, PA, 1985.
[5] Perry, J. H., Ed., *Chemical Engineer's Handbook,* 3rd ed., McGraw-Hill Book Company, New York, 1950.
[6] Weast, R. C., Ed., *CRC Handbook of Chemistry and Physics,* 48th ed., The Chemical Rubber Co., Cleveland, OH, 1967–1968.
[7] Zabetakis, M. G., *Flammability Characteristics of Combustible Gases and Vapors,* Bulletin 627, Bureau of Mines, U.S. Department of the Interior, Washington, DC, 1965.

Bibliography

Bibler, N. E., *Radiolytic Gas Production During Long-Term Storage of Nuclear Wastes,* DP-MS-76-51, E. I. du Pont de Nemours & Co., Savannah River Laboratory, Aiken, SC, 1976.
Bibler, N. E., *Gas Production from Alpha Radiolysis of Concrete Containing TRU Incinerator Ash,* Progress Report 2, DPST-78-150-2, E. I. du Pont de Nemours & Co., Savannah River Laboratory, Aiken, SC, 1978.
Bibler, N. E., Wallace, R. M., and Ebra, M. A., *Effects of High Radiation Doses on Linde Ion SIV IE-95,* DP-MS-81-18, E. I. du Pont de Nemours & Co., Savannah River Laboratory, Aiken, SC, 1981.
Flaherty, J. E., Fujita, A., Deltete, C. P., and Quinn, G. J., *A Calculational Technique to Predict Gas Generation in Sealed Radioactive Waste Containers,* GEND 041, EG&G Idaho, Inc., Idaho Falls, ID, 1986.
Friedman, H. A., Dole, L. R., Gilliam, T. M., and Rogers, G. C., *Radiolytic Gas Generation Rates from Hanford RHO-CAW Sludge and Double-Shell Slurry Immobilization in Grout,* ORNL/TM-9412, Oak Ridge National Laboratory, Oak Ridge, TN, 1985.
Gaumann, T., and Hoigné, J., Eds., *Aspects of Hydrocarbon Radiolysis,* Academic Press, London, England, 1968.
Henrie, J. O. and Postma, A. K., *Analysis of Three Mile Island Unit 2 Hydrogen Burn,* RHO-RE-SA-8, Rockwell Hanford Operations, Richland, WA, 1983; also GEND-023, Vol. IV, EG&G Idaho, Inc., Idaho Falls, ID.

Syotaro Hayashi,[1] Yukio Wada,[1] Yoshiki Chigusa,[2] Kunio Fujiwara,[2] Yasuji Hattori,[2] and Yasuo Matsuda[2]

Photobleaching Effects on Radiation-Induced Loss for Silica Glass Image Fiber

REFERENCE: Hayashi, S., Wada, Y., Chigusa, Y., Fujiwara, K., Hattori, Y., and Matsuda, Y., **"Photobleaching Effects on Radiation-Induced Loss for Silica Glass Image Fiber,"** *Influence of Radiation on Material Properties: 13th International Symposium (Part II), ASTM STP 956,* F. A. Garner, C. H. Henager, Jr., and N. Igata, Eds., American Society for Testing and Materials, Philadelphia, 1987, pp. 647–653.

ABSTRACT: The photobleaching effect (annealing effect on light absorption by light energy) on silica glass image fiber was investigated. We evaluated photobleaching characteristics using He-Cd laser, a xenon lamp, and a D_2 lamp as photobleaching light sources. Fibers were irradiated at two levels of dosage, 300 rad/h and 10^4 rad/h. The thermal effect on photobleaching was also evaluated by comparing the effect at 28°C and that at 70°C. The investigations demonstrated the superiority of the xenon lamp and thermal dependency of photobleaching. Furthermore through xenon lamp bleaching, we have succeeded in elongating the usable life span of image fiber in a 300-rad/h gamma-ray environment to more than five times that for nonbleached fiber.

KEY WORDS: optical fibers, losses, radiation

Several analytical measuring systems have been developed for remote maintenance in testing facilities for recycling fast reactor fuel. These systems employ silica glass image fiber for signal transmission and as remote sensors. When silica glass image fiber is used in a radiated environment, the light transmission of the image fiber becomes degraded. However, it has been widely expected that by launching strong light into image fiber, that attenuation increase can be recovered, thereby elongating the usable life span of fiber.

This annealing effect on the light absorption by light energy is called the photobleaching effect. We investigated the influence of light sources, light power, irradiation dosage, and temperature upon photobleaching. In this paper, these photobleaching characteristics are detailed.

Experiment

Test Fibers

Image fibers that consist of pure silica cores and fluorine containing silica cladding were used as test fibers. Fiber parameters are shown in Table 1.

[1] General manager and senior engineer, respectively, Power Reactor and Nuclear Fuel Development Corporation, 4-33 Tokai-Mura, Naka-Gun, Ibaraki-Ken, Japan.

[2] Engineer, engineer, chief research associate, senior engineer, and engineer, respectively, Sumitomo Electric Industries, Ltd., 1, Taya-cho, Totsika-Ku, Yokohama, 244 Japan.

TABLE 1—*Parameters of test fibers.*

Fiber	Fiber Diameter, mm	Number of Picture Elements	Core Material	Cladding Material
Fiber 1	0.2	140	pure silica	fluorine containing silica
Fiber 2	1.0	4000	pure silica	fluorine containing silica

Conditions for Gamma-Ray Irradiation and Photobleaching

$^{60}C_0$ was employed as the gamma-ray source. In order to evaluate light source, light power, temperature, and dosage dependency for the photobleaching effect, He-Cd laser, xenon lamp, and D_2 lamp were employed as photobleaching light sources. Table 2 shows the experimental conditions.

The Measuring System

Figure 1 shows the method of photobleaching and the measuring system during irradiation. The irradiated fiber length is 10 m for Fiber 1 and 8.4 for Fiber 2. For photobleaching, the light was continuously launched into the fiber. Measuring the attenuation increase, the fiber was then removed from the light source and one end of the fiber was coupled to the light source of a spectrometer

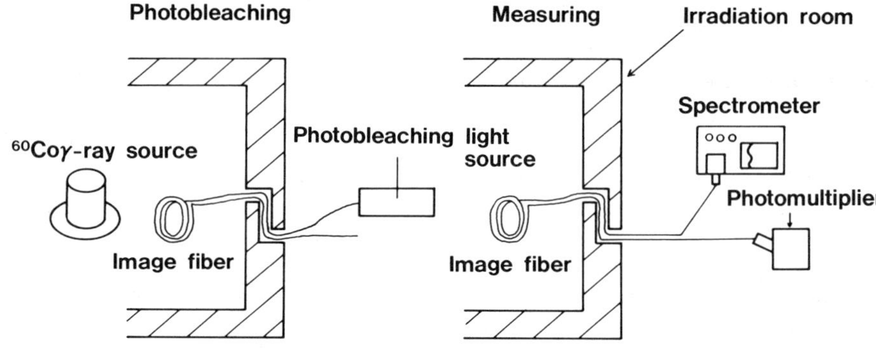

FIG. 1—*Photobleaching and measuring systems.*

TABLE 2—*Experimental conditions.*

Sample	Light Source	Initial Light Power, mW/mm²	Dosage, rad/h	Temperature, °C
Fiber 1	He-Cd laser	164 72	300, 10^4	28
	xenon lamp	0.80 0.39	300, 10^4	28, 70 28
	D_2 lamp	under 0.001	300	
	dark	. . .	300, 10^4	28
Fiber 2	xenon lamp	0.77	300	
	dark	. . .	300	

TABLE 3—*Experimental results for Fiber 1.*

Light source			Xenon Lamp	He-Cd Laser	D_2 Lamp	Dark
Output spectrum, μm (through fiber)			0.3 ~ 1.1	peak 0.442 0.36 ~ 0.65	0.2 ~ 0.7	. . .
Initial output light power (for 10 m length fiber), mW/mm²			0.8	164	under 0.001	. . .
Attentuation increase, dB/10 m	At 0.32 μ (10^4 rad/h × 2 h irradiation)		8.9	9.7	. . .	16
	at 0.30 μm (300 rad/h × 1 h irradiation)	Temperature 28°C	1.8	2.3	2.6	2.9
		70°C	0.6	. . .	. . .	. . .

and the other end was coupled to a photomultiplier. After measuring, the fiber was removed from the measuring equipment and then returned to the photobleaching light source.

Results

Table 3 shows the experimental results for Fiber 1. Figures 2 and 3 show the light power dependency of the spectral attenuation in the ultraviolet (UV) region for photobleaching. In addition Figures 4 and 5 show the characteristics of the attenuation increase in the UV region during irradiation for Fiber 2.

Discussion

The output wavelengths of the employed light sources are shown in Table 3. The radiation-enhanced absorption is considered to be decreased by launching the light around the absorption

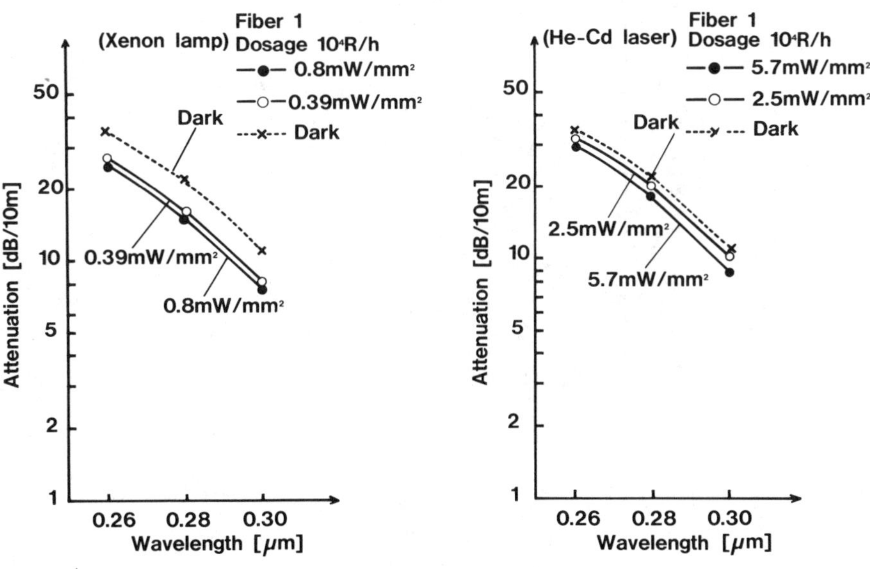

FIG. 2—*Light power dependency of the spectral attenuation (10^4 rad/h × 3 h irradiation).*

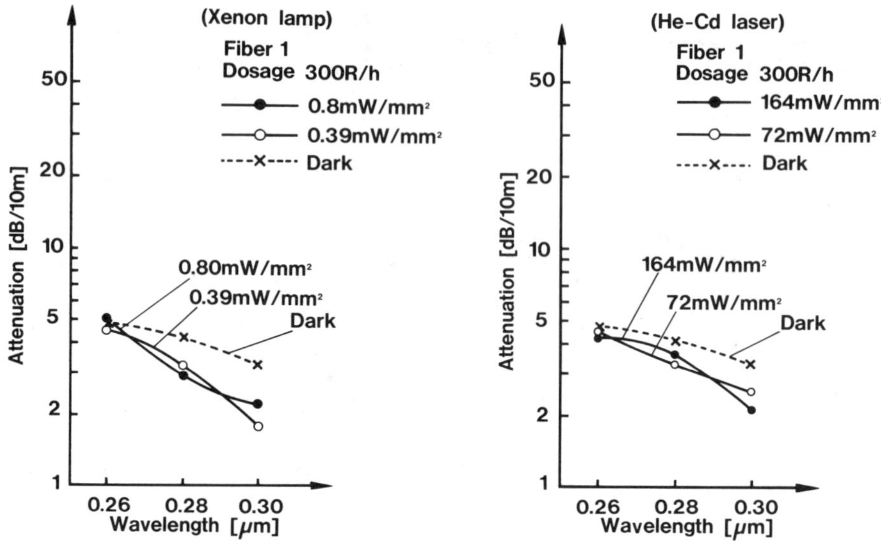

FIG. 3—*Light power dependency of the spectral attenuation (300 rad/h × 3 h irradiation).*

peak. It is hypothesized that this is due to the following mechanism. In Fig. 6, electrons, which the absorption is caused by, are at the trapped energy state. And electrons at the trapped energy level are excited to the excitation state by the light energy and then are excited to the conduction band by the thermal energy. These excited electrons move in the glass and recombine with holes [*1*].

In the wavelength regions of Figs. 2 through 4, no marked absorption peaks appear, and only a flare of the absorption band that is in the shorter wavelength region than 0.26 μm appears. From

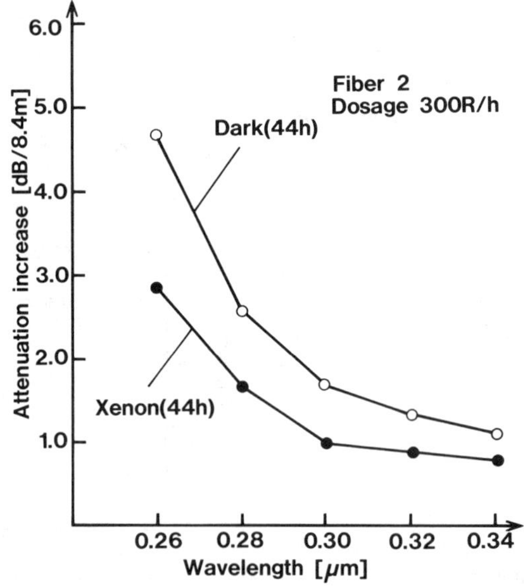

FIG. 4—*Wavelength dependency of attenuation increase for Fiber 2.*

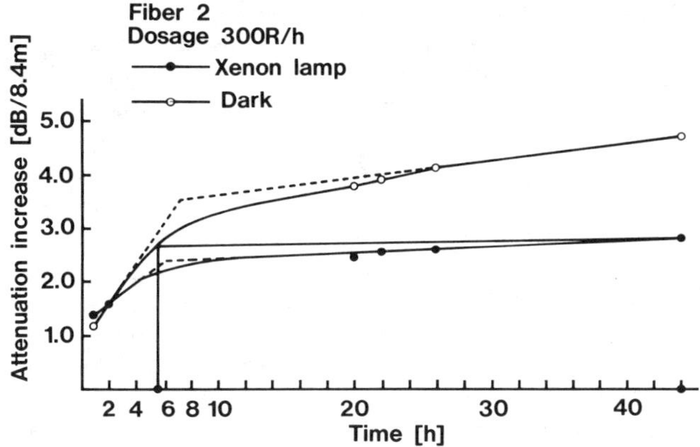

FIG. 5—*Growth of the radiation-induced attenuation at 0.25 μm for Fiber 2.*

this fact, it is clear that the absorption peak exists in the wavelength region shorter than 0.26 μm, and it is assumed that this absorption flare is due to E_1' center [2]. Therefore, the D_2 lamp must have the most effective wavelength region of these three light sources. However, it was found that the xenon lamp was most effective for the photobleaching in spite of the gamma-ray irradiation dosage.

Xenon lamp's wavelength region is between those of the other two sources, and this fact indicates the necessity of considering the light power dependency for photobleaching. Comparing the output light power of the xenon lamp and He-Cd laser in Table 3, He-Cd laser power is 205 times greater than that of the xenon lamp, but the photobleaching effect is greater with xenon

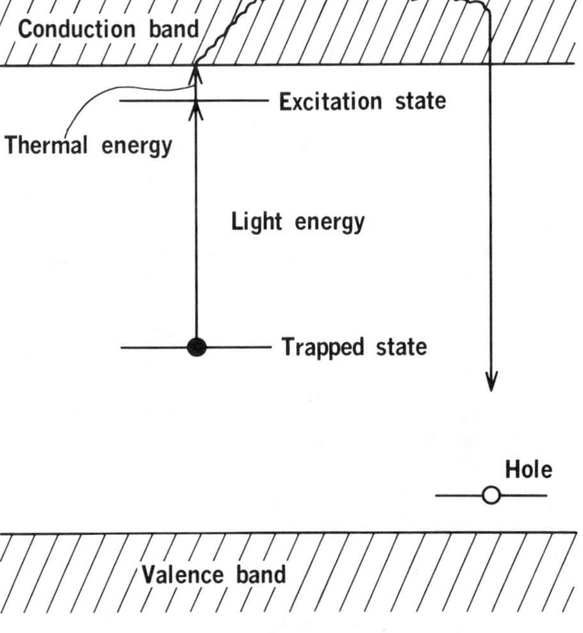

FIG. 6—*The model for photobleaching.*

lamp than with the He-Cd laser. Furthermore, Figs. 2 and 3 show little power dependency between medium and high light power for 10^4 rad/h and for 300-rad/h irradiation. Therefore, it was concluded that the light power (0.8 mW/mm^2) of $\lambda = 0.3 \sim 1.1$ μm is sufficient for photobleaching in the case of 10^4-rad/h irradiation, and in the case of 300-rad/h irradiation. Figure 7 shows the output light power spectra of xenon lamp and He-Cd laser through Fiber 1. Comparing the light power spectra of these light sources, the He-Cd laser does not have an output spectrum in the $\lambda = 0.3 \sim 0.36$ μm region. Therefore, it is assumed that the xenon lamp is more effective than the He-Cd laser for photobleaching because of the existence of the spectra in the $\lambda = 0.3 \sim 0.36$ μm region whose total power is about 10-μW/mm^2 (original output power for a 10-m-length fiber). Moreover it is also supposed that the reason why the D_2 lamp whose light power is only 0.11% of the xenon lamp is the least effective of these three light sources may be its lack of light power. The output power at the peak wavelength of the He-Cd laser is 167 times greater than that in the flare wavelength region, and furthermore the output light power in $\lambda = 0.36 \sim 0.4$ μm of He-Cd laser is greater than that of xenon lamp.

From these facts, it is hypothesized that almost none of the light power at the peak wavelength of He-Cd laser is useful because that power is of saturated level for color centers, which can be bleached by the peak light power of He-Cd laser. Rather, the flare spectra $\lambda = 0.36 \sim 0.4$ μm may be more effective for photobleaching.

The results of a photobleaching (with xenon lamp) experiment, which was done in nearly practical conditions for spectral analysis in a radiation environment, are shown in Figs. 4 and 5. The luminescent spectrum of boron, which is one of the analyzed chemical elements, is near 0.25 μm. And for the spectral analysis with image fiber, it is desired to elongate the analyzable life span for boron. Comparing degradation at 0.26 μm in Fig. 5, it is obvious that the analyzable life span of the spectral analysis with image fiber is elongated more than five times by photobleaching with the xenon lamp.

A comparison of photobleaching effect at 28 and 70°C with the xenon lamp is shown in Table 3.

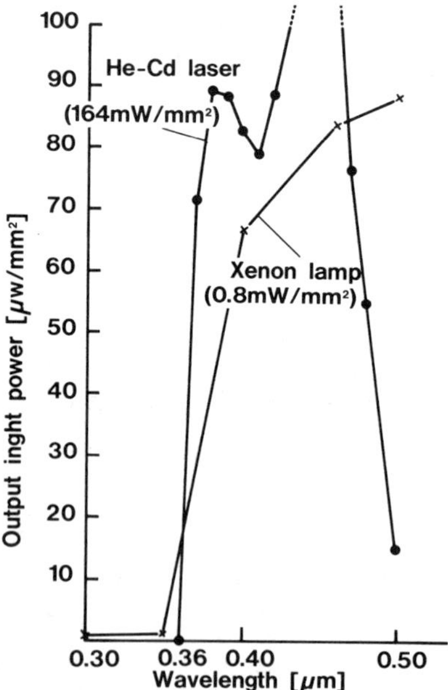

FIG. 7—*Original output spectra through Fiber 1.*

Superiority of photobleaching at high temperature was found. It is proposed that the heating helps the transition to the conduction band of electrons that were excited to the excitation level by the light energy.

Conclusions

1. It was found that the xenon lamp, He-Cd laser, and D_2 lamp have the photobleaching effect on the degradation of light transmission in the UV region.
2. The xenon lamp is the most effective of these three light sources. It is supposed that this superiority is due to the existence of $0.3 \sim 0.36$ μm spectra whose total power is about 10-μW/mm^2 (original output power for 10-m-length fiber).
3. By investigating the light power dependency of the photobleaching effect, it was found that 0.8 mW/mm^2 (original output power for 10 m length fiber) for $0.3 \sim 1.1$-μm region is sufficient for 10^4 rad/h, and for 300-rad/h irradiation.
4. We have succeeded in elongating the analyzable life span of image fiber to more than five times in a 300-rad/h irradiation environment by photobleaching with the xenon lamp.
5. It was found that the photobleaching effect depends upon the environmental temperature.

Acknowledgment

Acknowledgments are due to Mr. Okamoto and other staff members in the Radiation Center of Osaka Prefecture for their helpful suggestions in carrying out the irradiation tests.

References

[*1*] *Glass Technology Handbook,* Asakura Shoten, p. 141.
[2] Nagasawa, K., Hoshi, H., Oki, Y., and Yahagi, K., *JIEE, Digest,* Vol. 1, No. 26, EIM-85-26, p. 8.

Francis Lefevre,[1] Gilles Thevenot,[1] Bernard Rasneur,[2] and Fernande Botter[2]

Irradiation Devices for Reactor Materials: Results Obtained from Irradiated Lithium Aluminate at the OSIRIS Reactor

REFERENCE: Lefevre, F., Thevenot, G., Rasneur, B., and Botter, F., "**Irradiation Devices for Reactor Materials: Results Obtained from Irradiated Lithium Aluminate at the OSIRIS Reactor,**" *Influence of Radiation on Material Properties: 13th International Symposium, ASTM STP 956,* F. A. Garner, C. H. Henager, Jr., and N. Igata, Eds., American Society for Testing and Materials, Philadelphia, 1987, pp. 654–668.

ABSTRACT: The study of radiation behavior is of prime importance in the investigation of fusion reactor materials. With this purpose, the "Services des Piles de Saclay" (SPS) has been using the experimental OSIRIS pool reactor, having a nominal power of 70 MW, to develop irradiation devices. The devices enable (1) the study of the crack growth on a sample under alternative stress in a neutron flux, (2) the study of stress relaxation on materials under irradiation, (3) the study of creep and growth of cylindrical test pieces in a neutron flux, (4) the irradiation of large amounts of protection ceramics of fusion reactors so as to characterize their mechanical properties, and (5) the irradiation of lithium containing ceramics, which are possible materials for the breeding blanket of a fusion reactor. The last device was used in the ALICE 1 experiment, carried out in collaboration with the "Division d'Etudes de Séparation Isotopique et de Chimie Physique" (DESICP), for the irradiation of 130 specimens of γ $LiAlO_2$, of natural lithium 6 abundance and of various textures. The unclad samples were irradiated at 400 and 600°C, in a maximum fast flux of 2×10^{14} $n \cdot cm^{-2} \cdot s^{-1}$ ($E >$ 1 MeV) and maximum perturbed thermal flux 6×10^{13} $n \cdot cm^{-2} \cdot s^{-1}$ for 26 days (from 16 Oct. to 13 Nov. 1984). The amount of tritium generated was 330 Ci. The ^6Li burnup could amount to 15 to 20%. The first post irradiation examinations showed general good behavior of the samples. There were no changes in the dimensions of the samples, nor significant thermal conductivity variations. Retained tritium and helium amounts varied depending on the ceramic texture and irradiation temperature. Small variations in the sound velocity and cracks were observed only in two 600°C levels, corresponding to containers that received mechanical shocks during the dismantling of the experimental device.

KEY WORDS: irradiation devices, ceramics radiation behavior, controlled fusion reactor, crack propagation

Studies on controlled fusion reactors of the Tokamak type require the examination of the radiation effects on the behavior of various potential materials. In the first part of this paper, the devices adapted to these materials studies and used in the OSIRIS reactor are presented. In the second part, an irradiation experiment on ceramics that are candidates for breeding material is described, and the first results are given.

[1] IRDI/DERPE/SPS, Centre d'études nucléaires de Saclay, 91191 Gif sur Yvette, Cedex, France.
[2] IRDI/DESICP/SPCM/SPCS, Centre d'études nucléaires de Saclay, 91191 Gif sur Yvette, Cedex, France.

The OSIRIS Reactor and the COLIBRI Device

The OSIRIS Reactor

The open core OSIRIS pool reactor generates, for a nominal power of 70 MW, high neutron fluxes:

- 4.5×10^{18} n/m^2 s in fast flux and
- 3.5×10^{18} n/m^2 s in thermal flux

with a maximum gamma heating level of 8 W/g. The fuel, manufactured in France, is a "Caramel" type made of uranium dioxide (UO_2) plates. Uranium enrichment is low (6 to 8%). The plates are clad with zircaloy.

The present configuration of the reactor core allows five experimental in-core positions, several positions in the beryllium reflector and many peripherical positions in the pool (Fig. 1). The neutronic characteristics of these various positions give a large range of fluxes.

The irradiations, most frequently, concern the studies on structure materials and various nuclear fuels, activation analysis, production of radioisotopes, silicon doping, and so forth. But taking into account the capabilities and characteristics of the OSIRIS reactor, it is easy to comply with irradiation requests for the research program on controlled fusion reactors. In this program, the irradiations essentially concern structural materials, protection ceramics, and breeding materials and require a high flux, obtained in the core of the OSIRIS reactor. In order to address these

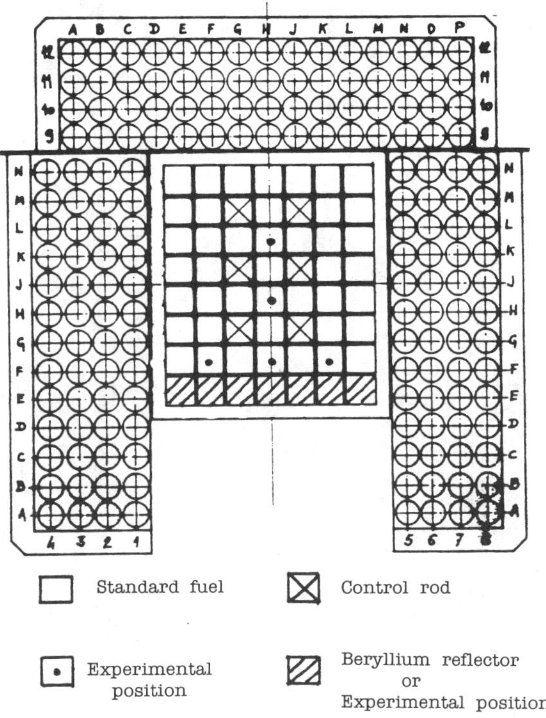

FIG. 1—*Irradiation positions in the OSIRIS reactor.*

problems, we have been making use of the COLIBRI device, which is the basis of particular devices developed for the fusion program.

The COLIBRI Device

More than 100 COLIBRI rigs have been irradiated, for times exceeding 12 000 h in some cases. In its in-flux part, the COLIBRI device (Fig. 2) consists of two stainless steel containers placed one inside the other. The gap between these two volumes corresponds to a gas jet whose thermal conductivity can be varied during irradiation by adjusting the proportions of a helium and nitrogen mixture [*1*].

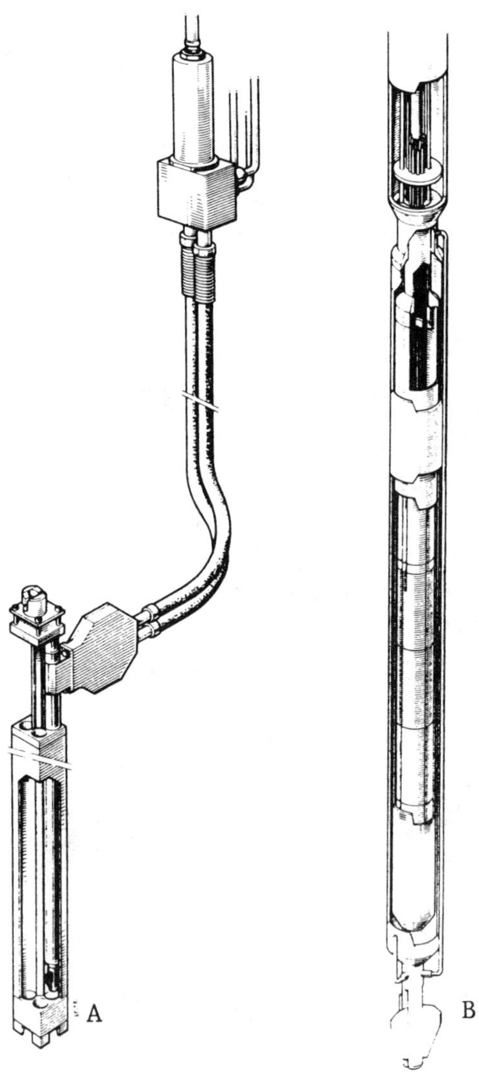

FIG. 2—*COLIBRI device:* (a) *in-pile installation and* (b) *useful in-flux part.*

Irradiations can be done from 250 to 1200°C, and a single device is able to work at two temperature levels. Moreover, when it is necessary in order to improve the operating or to maintain a very stable temperature, electrical heaters can be set on the sample-holder in the inside of the device, or placed on the stainless steel of the COLIBRI inner pipe. The in-core positions allow placement of one to four devices, with a useful diameter of 67 to 25 mm.

Devices Developed for the Fusion Program

Study of the Fatigue Crack Growth

A COLIBRI type with electrical heaters, allows the study of the fatigue crack growth on a CT18 (compact tension) specimen, whose thickness is only 5 mm in order to reduce the transverse thermal gradient.

The sample-holder, reloadable in hot cells, holds two specimens; one is put in alternative tensile stress, and the other is unstressed and used as a control sample. These two samples are set in the same flux and temperature conditions (250 to 450°C). The temperature is controlled to ±5°C by electrical heaters and the homogeneity is obtained by immersion in a sodium-potassium alloy (NaK).

As shown on the Fig. 3, on the stressed specimen, the alternative tensile stress is obtained by a 10^4 N jack with a frequency of one cycle per minute. The continuous measurement of the notch opening (compliance method) is carried out at the lip level by a probe system connected to a linear variable differential transformer (LVDT) sensor.

This measurement method is complemented by a pneumatic system based on the loss of leaktightness of pressurized holes bored in the perpendicular plane of the crack. This system gives a direct measurement of the length of the crack, though it is discontinuous.

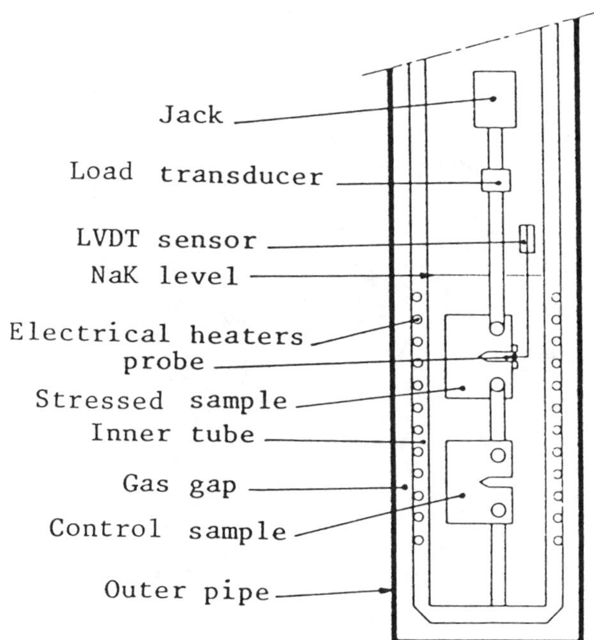

FIG. 3—*Study of the crack growth diagram of the in-flux part of the device.*

Study of Relaxation

A COLIBRI device, equipped with electrical heaters, and a sample-holder permit the study of stress relaxation of materials. The samples are thin plates; the stress is obtained by bending the plate and maintaining it in position during the irradiation with the system shown on Fig. 4 (three points bending). A measurement of the residual deformation in a hot laboratory gives the relaxation under irradiation.

The sample-holder, with a useful diameter of 60 mm, can hold 20 samples on five different levels. Each level has at least two flux integrators and four thermocouples.

To keep the temperature of samples at 350 ± 5°C, the whole sample-holder is immersed in NaK, and the temperature regulation is ensured by the COLIBRI electrical heaters whose position and power are adapted to this type of loading. Also, this device can be reloaded in hot cells with new samples.

Study of Creep Under Traction

Following the COLIBRI device design with electrical heaters, this rig permits the study under irradiation of axial creep of materials subjected to a constant traction load. It is a hot cell unloading and reloading device that allows sample metrology studies in the hot laboratory between the irradiation cycles. The sample-holder contains two trains of three tubular test-pieces with a maximum diameter of 12 mm subjected to two different loads up to 1000 N. As shown on the Fig. 5, the test pieces are put under stress by pressurized bellows; an automatic system regulates the pressure to ensure a constant stress. The homogenization of the test-piece temperature is ensured by the NaK in which they are immersed; the temperature is roughly reached with the gas gap and adjusted at ±5°C with the electrical heaters.

A new sample-holder allowing a larger load on one train of three test pieces is now being studied.

Irradiation of Insulating Ceramics

A COLIBRI device of useful diameter of 30 mm allows the neutron irradiation of insulating ceramics with the aim of checking the evolution of their mechanical and dielectrical properties. This rig, partially shown in Fig. 6, permits irradiation of a large number of samples (200) of

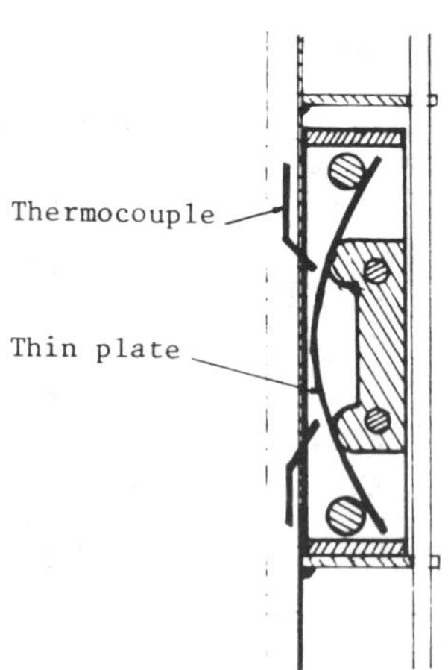

FIG. 4—*Study of the stress relaxation sample in the bending system.*

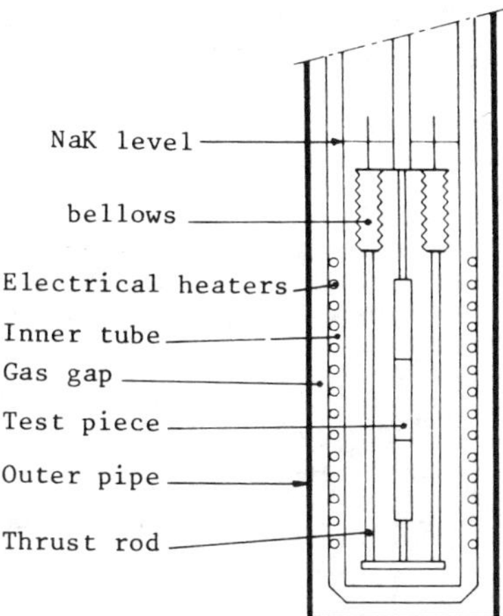

FIG. 5—*Study of creep under traction. Diagram of the in-flux part of the device.*

various geometries (parallelepiped, cylinder, disk) and of various natures in a high purity helium environment. The temperature, regulated by the gas gap, is 500 ± 10°C. The sample-holder is made of graphite barrels with a double wall. Thus, the samples are protected from any contact with the metallic parts of the device. In addition, thermocouples and flux integrators are placed in the graphite crossbars. The samples, made of high purity and low activation material, can be easily carried and handled in a glove-box after irradiation. This device can be used again after having received a new sample-holder.

Irradiation of Lithium Containing Ceramics

Many studies are also made on breeding materials, especially lithium containing ceramics. The study of their behavior after irradiation can be done in a COLIBRI rig of a diameter smaller than

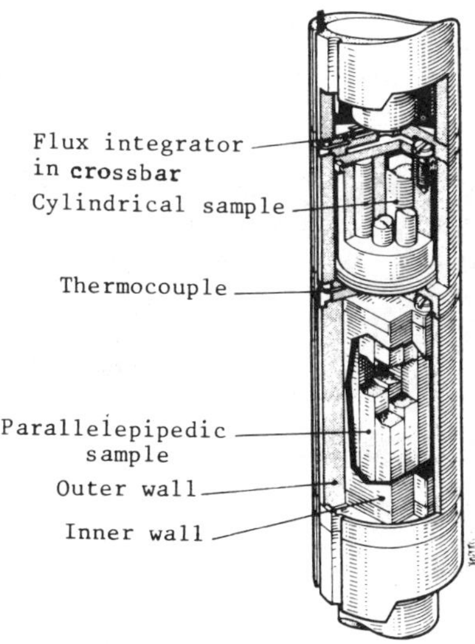

FIG. 6—*Insulating ceramics irradiation detail of the sample-holder.*

TABLE 1—*Characteristics of the various textural groups.*

	Textural Groups					
Parameters	6	4	1	3	5	2
Ultimate compressive strength[a] (CS), MPa						
20°C	56	67	186	217	287	489
at the irradiation temperature range (ITR), °C	26	30 to 40	85 to 112	103 to 127	136 to 168	232 to 286
Grain diameter ϕ, μm	13	3	0.35	0.47	0.40	0.68
Porosity	0.23	0.285	0.29	0.26	0.24	0.16
Pore radius, μm	1.55	0.4	0.048	0.055	0.45	0.043
Specific surface area, ($m^2 \cdot g^{-1}$)	0.2	0.7	5.7	5.3	5.8	2.7
Young's modulus *Y*, GPa						
20°C	64	46	45	54	61	92
ITR, °C[b]	60	43	42	50	57	87

[a] The compressive strength is related to the grain diameter and the porosity by $CS = \frac{2e^{-10\epsilon}}{\sqrt{\phi}}$.

[b] The Young modulus is related to the porosity by $Y = Yo(1 - \epsilon)\left(\frac{(1 - \epsilon)}{0.7}\right)^2$.

or equal to 25 mm, specially designed for tritium confinement. Particularly, the entire in-pile part has a double controlled containment, and the circuit to be tritiated is completely welded and has no organic joint. To avoid any tritium accumulation in the device, which would enhance its diffusion to the outside, all the circuits are periodically swept with helium. The monitoring of the experiment is done in a special glove-box, with powerful ventilation.

The samples, clad or unclad, are cylinders or disks of various dimensions, placed in barrels (Fig. 7). The temperature from 400°C up to 800°C ± 10°C is regulated by a gas gap. The sample environment is high purity helium. This device can accommodate a tritium production of 500 Ci.

In particular, with this rig, the ALICE 01 experiment on lithium aluminate ($LiAlO_2$) has been performed.

ALICE 01 Experiment

Specimen Characteristics

The specimens have been machined out of compacts of porous γ $LiAlO_2$ corresponding to six different groups of very homogeneous textures, listed in Table 1. The specimen dimensions have been chosen to meet the requirements of the post irradiation tests, given in the Fig. 8 and Table 2.

TABLE 2—*Post irradiation examinations (PIE) planned for ALICE 01.*

The weight } The metrology }	the porosity }	Young's modulus
The ultrasound velocity	}	
Mechanical tests		
Thermal tests		
Scanning electronic microscopy (SEM)		
X-rays		
Residual T_2 and ^4He extraction		
Burnup measurement by $^6Li/^7Li$ mass spectrometry determination		

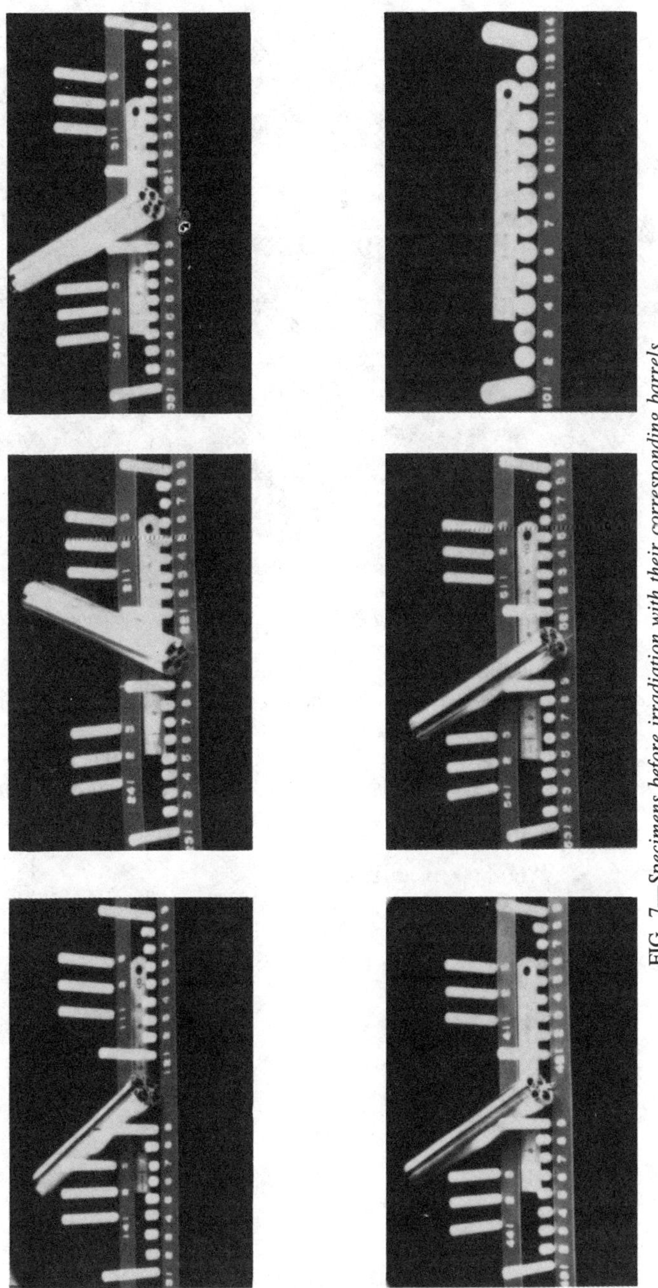

FIG. 7—*Specimens before irradiation with their corresponding barrels.*

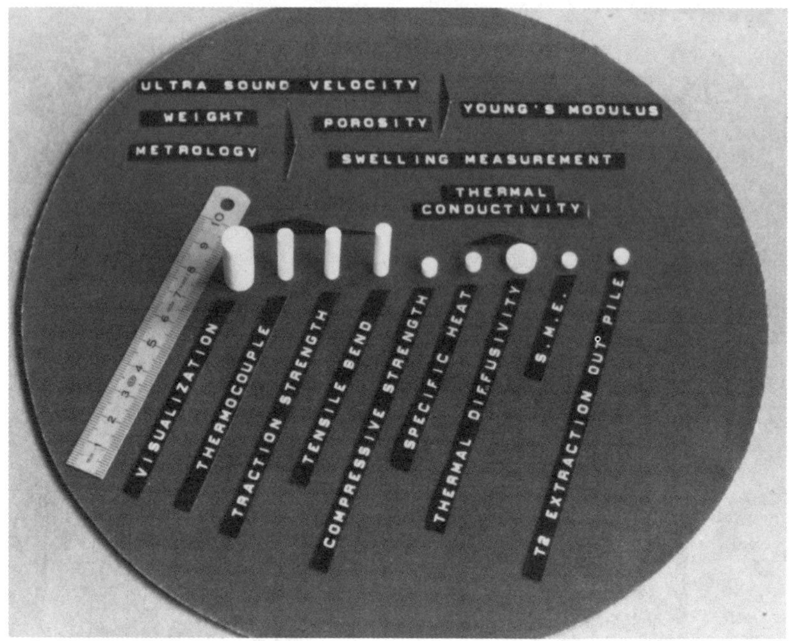

FIG. 8—*Porous $LiAlO_2$ samples of different geometries for post irradiation examinations.*

Mechanical characteristics of the different groups are shown on Figs. 9 and 10. Young's modulus is given as a function of porosity, and ultimate compressive strength is represented in the porosity-grain diameter diagram [2].

Nonclad specimens (132) were inserted in six levels of the rig. Figure 7 shows the specimens and their respective barrels. Five barrels (Fig. 11) have four housings for specimen insertion.

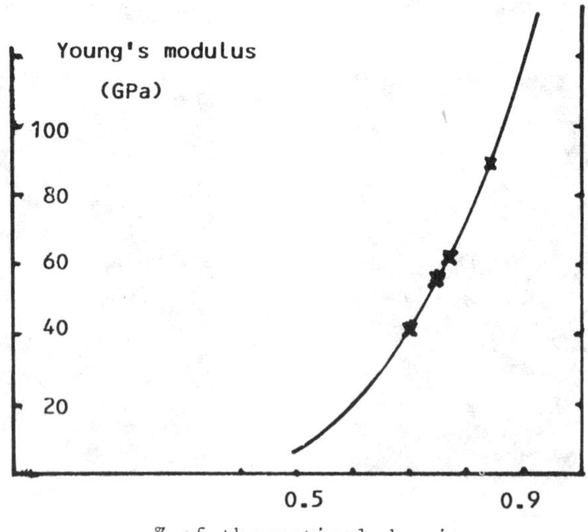

FIG. 9—*Mechanical characteristics (Young's modulus).*

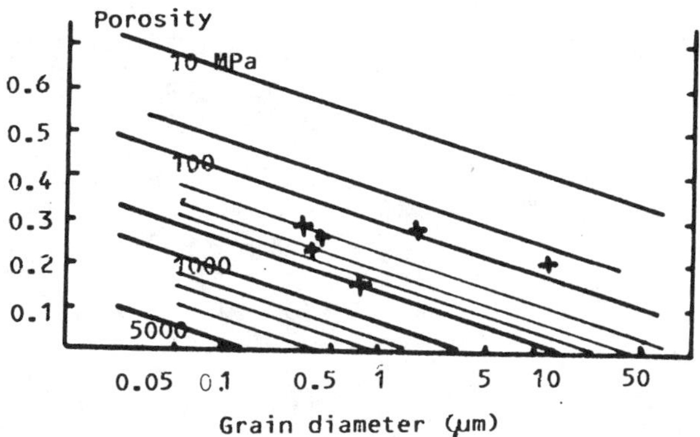

FIG. 10—*Mechanical characteristics (ultimate compressive strength).*

Dimensions are 5.25 mm in diameter and 91 mm in height; one barrel, not shown on the photograph, has a simple housing 10.25 mm in diameter, 91 mm in height. Barrels 5.25 mm in diameter have four stainless steel centering wires 0.05 mm in diameter for centering the 4.99 mm diameter specimens. Because of the sensitivity to air moisture of lithium aluminate [*3*], all specimens have been annealed 1 h at 800°C before being dry weighed; the sound velocity and the dimensions of the specimens have been measured and the samples were stored in tight flasks before insertion in the barrels. The installation of the barrels in the rig took 32 h. The whole rig was then degassed 42 h at room temperature and 40 h at 200°C.

Characteristics of the Irradiation

Figure 12 shows the neutron energy spectrum corresponding to the ALICE 01 experiment. The flux is perturbed by the neutron absorption of the ceramic.

Table 3 gives the fluence values along the rig. At the maximum flux level, the thermal power released by the neutronic reaction on ^6Li was about 80 W/cm^3, the γ power was 8 W/g in $LiAlO_2$,

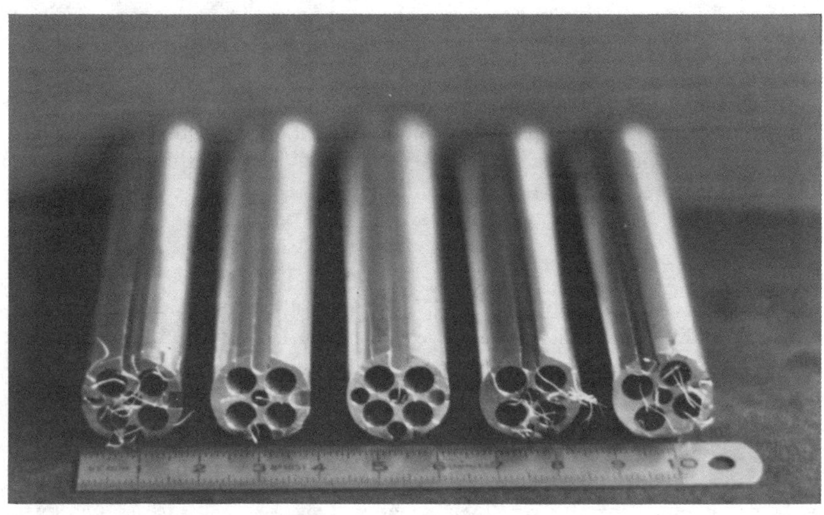

FIG. 11—*Five of the six barrels used in ALICE 01 experiment.*

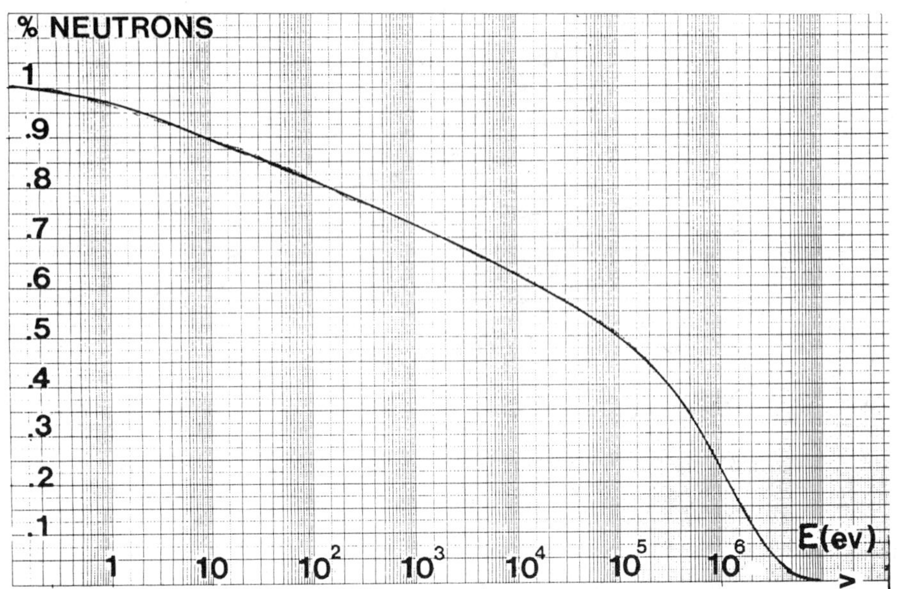

FIG. 12—*Disturbed neutron energy spectrum for the ALICE 01 experiment.*

therefore the total power dissipated in the specimens of about 96 W/cm^3. The mean temperature chosen for the samples was about 400°C in the upper part and 600°C in the lower part of the rig. Their exact values are given on Table 3. Except during two reactor shutdowns, the temperatures measured at each level remained very stable. It may be concluded that thermal conductivity did not change much. This has to be confirmed by thermal post irradiation tests.

Two reactor shutdowns occurred, 15 min and 33 h long; and temperature drops as high as 400°C/min affected the 600°C samples. Irradiation started 16 Oct. 1984, and ended 13 Nov. 1984, corresponding to 25.7 full power days.

Table 3 gives the amount of tritium generated calculated for each barrel, corresponding to about 330 Ci generated from the total 82 g of natural lithium γ $LiAlO_2$. The burnup was about 16% of the ^{6}Li.

During the irradiation, the gas atmosphere around the samples was changed two or three times a day by helium purging. The experimental conditions for tritium extraction were optimized.

More than half of the generated tritium was removed, as checked in the stack and confirmed by first post irradiation measurements of residual tritium.

Specimens after Irradiation

The COLIBRI device was dismantled in the ''Laboratoire d'Etudes des Combustibles Irradiés'' (LECI) at Saclay, and the first post irradiation tests were carried out also in the LECI (weighing at 0.13 relative hygrometry, metrology, sound velocity, residual T_2, and ^{4}He extractions).

During dismantling, the lower levels were submitted to mechanical shocks, especially Levels 5 and 6. Conjugated thermal and mechanical shocks seem to be responsible for the breakage of several samples in Levels 5 and 6. Figure 13(a) shows a bent and broken sample of Level 5 and Group 1.

But, the other samples generally kept their integrity; Figs. 13b and c show two examples of samples from Levels 2 and 4, respectively, of textural Groups 4 and 3.

TABLE 3—*ALICE 1 irradiation conditions (133 unclad γ $LiAlO_2$ samples).*

Level Number	Arrangement	Samples: Weight, g	Samples: Textural Group[a]	Temperatures, °C: Core	Temperatures, °C: Surface	Temperatures, °C: Drop Consecutive to Reactor Shutdowns	Flux, $n \cdot s^{-1} \cdot m^{-2}$	Fluence, $n \cdot m^{-2}$: Fast, $\times 10^{24}$	Fluence, $n \cdot m^{-2}$: Thermal, $\times 10^{23}$	Generated Tritium, Ci or ($\times 3.7 \times 10^{10}$ Bq)	Li Burnup in % of $^6Li + ^7Li$
1	4 housing / 5 mmφ (levels 1–5)	14	2-3-5-6	410	340	200°C per min (levels 1–3)		1.78 2.19 2.55	4.58 6.06 7.44	32	0.5
2	*Per level*	13	1-4	390	320		fast 2.1×10^{18} at the maximum level (levels 2–4)	2.74 3.22 3.63	10.17 12.00 13.54	57	0.96
3	10 cylinders 30 mm high 14 pellets 5 and 2.5 mm high	14	2-3-5	370	250		perturbed thermal 6×10^{17} at the maximum level	3.70 4.10 4.44	12.56 13.76 14.80	71	1.11
4		14	2-3-5	535	415	400°C per min (levels 4–6)		4.59 4.70 4.66	12.81 13.24 13.37	71	1.11
5		13	1-4-5	610	520			4.64 4.30 3.84	14.70 13.90 12.09	67	1.13
6	1 housing / 10 mmφ 2 cylinders 11 pellets	13.5	1-2-3 4-5-6	. . .	580			3.67 3.10 2.47	10.51 8.74 6.75	41	0.67

[a] See Table 1 for a description of textural groupings.

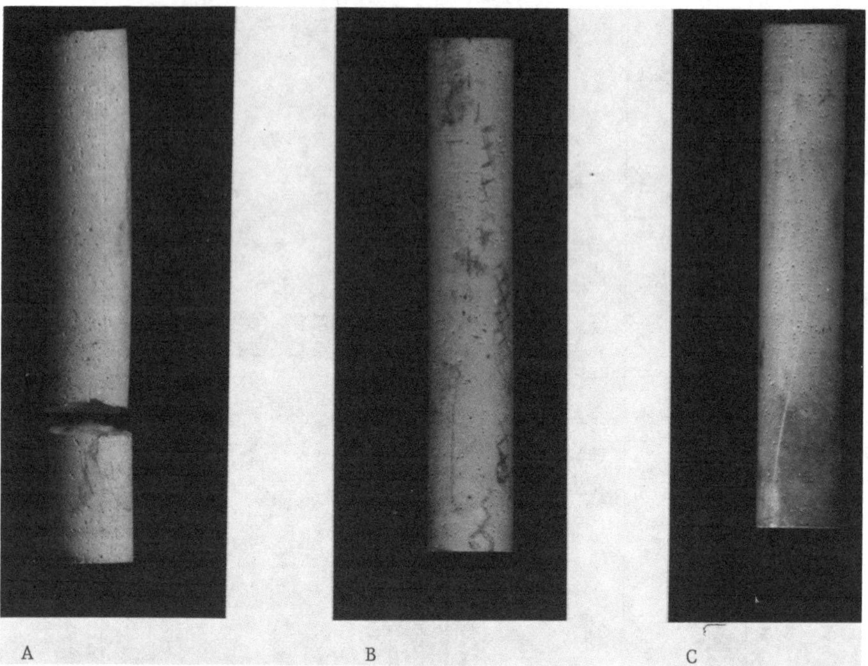

FIG. 13—*Samples after irradiation:* (a) *Level 5, Group 1;* (b) *Level 2, Group 4; and* (c) *Level 4, Group 3.*

In particular, samples of low porosity in Level 4 which withstood the maximum flux at the higher temperature, kept their integrity and showed no length variations, although they presented a little decrease of their Young's modulus or a loss of weight.

Transit times of 6 μs through a 30-mm-long samples, increased by 0.1 to 0.2 μs at 400°C and by 0.15 (Group 2) and ≃ 0.3 μs (Groups 3 and 5) at 600°C. Weight losses, negligible for all specimen groups at 400°C, would correspond to more than the loss of (T_2O + 2 He) at Level 4, at 600°C.

Measurements of retained tritium and helium were performed after irradiation.[3] The tritium was extracted in a pure argon sweep gas; the tritiated water was collected in water bubbles, the exiting gas was oxidized and then collected in water bubblers. Solutions were analyzed by liquid scintillation techniques. The helium was removed under vacuum, separated from residual water, followed by a mass spectrometric helium analysis. In this procedure, kinetic information was obtained at various temperatures. Retained helium was little influenced by the irradiation temperature but, retained tritium was strongly influenced. The ratio of T/He varied from 2 to 155 (Table 4). Besides, the percentage of tritiated water was found more important at the low irradiation temperature. Another observation is related to the influence of the texture on the tritium release: at the higher irradiation temperature the amount of retained tritium was larger for small grains, but for the lower irradiation temperature, retained tritium was found higher for the large grain samples. The effect of fluence and of the porosity of the material was slightly sensitive.

There was no difference in X-ray examinations of two textures of ceramics before and after irradiation: small grain size of textural Group 1 irradiated at Level 2 and larger grain size of textural Group 4 irradiated at Level 5.

[3] Bonnal, J. P. private communication.

TABLE 4—*Residual gases.*

			Residual Gases					
				TRITIUM Retained			Helium Retained	
Level Number	Sample Arrangement	Mean Temperature, °C	Textural Group Tested[a]	Total T, %	As Water, %	As Gas, %	He, %	T/He Ratio
1		400	3	17	97.5	2.5	4	4.2
2		400	4	62	99.6	0.4	0.4	155
			1	44	98	2.0	4.4	10
3		400	5	70	99.9	0.1	3	23
4		600	5	8	98	2	1.4	5.7
5		600	5	2.8	65.5	35.3	1.2	2.3
			4	0.6	82.5	17.5	0.3	2
6		600	no test	no test	no test	no test	no test	no test

[a] See Table 1 for description of textural groupings.

Samples of group 5 γ $LiAlO_2$, exhibited both the best kinetic of tritium release in "in-pile" runs [*4*], and a good irradiation behavior, so that this type of aluminate could represent a valuable candidate for a fusion reactor blanket material.

Conclusions

Within the framework of the European Communities Fusion Technology Program, other more classical irradiation experiments are also carried out in the OSIRIS reactor. At the present time, the utilization of the described devices is underway in the OSIRIS reactor as well as device adaptations allowing better accommodation with the requests of the fusion program.

Concerning the ALICE 01 experiment, other post irradiation examinations must be completed. However, this irradiation experiment has already proven the good behavior of high density γ $LiAlO_2$. ALICE 02 experiment is being prepared, with the best textural γ $LiAlO_2$ specimens, including both natural and enriched lithium samples, and is operated at about 600 and 700°C.

Acknowledgments

The ALICE 01 experiment was performed within the framework of the European Communities Fusion Technology Program.

References

[*1*] Bauchede, J., "COLIBRI Device With Gas Gap Thermal Regulation," Euratom Workshop, CEA—Report P, No. 152, Oct. 1967.

[*2*] Rasneur, B., "Tritium Breeding Porous Ceramic: γ $LiAlO_2$," *Fusion Technology 1984,* Vol. 2, Pergamon Press, New York, pp. 1017–1022.

[*3*] Rasneur, B., "Tritium breeding material: γ $LiAlO_2$," *Fusion Technology,* July 1985, p. 1909.

[*4*] Roth, E., Abassin, J. J., Botter, F., Briec, M., Chenebault, P., Masson, M., Rasneur, B., and Roux, N., "Irradiation of Lithium Aluminate and Tritium Extraction," *Journal of Nuclear Materials,* Vol. 133–134, 1985, pp. 238–241.

Katsunori Abe,[1] Clinton M. Logan,[2] Keiji Saneyoshi,[3] and Frank W. Clinard, Jr.[4]

Irradiation Effects in KAPTON Polyimide Film from 14-MeV Neutrons and Cobalt-60 Gamma Rays

REFERENCE: Abe, K., Logan, C. M., Saneyoshi, K., and Clinard, F. W., Jr., **"Irradiation Effects in KAPTON Polyimide Film from 14-MeV Neutrons and Cobalt-60 Gamma Rays,"** *Influence of Radiation on Material Properties: 13th International Symposium (Part II), ASTM STP 956,* F. A. Garner, C. H. Henager, Jr., and N. Igata, Eds., American Society for Testing and Materials, Philadelphia, 1987, pp. 669–681.

ABSTRACT: Polyimide films are insulating plastic materials that are useful over a wide temperature range and the most radiation resistant of the polymers. In this paper, mechanical property changes of the polyimide film, KAPTON, caused separately by 14-MeV neutrons and by cobalt-60 gamma rays are examined in detail. Miniature tension specimens of 3-mil-thick film were irradiated at Rotating Target Neutron Source-II and at the Cobalt-60 Pool of Lawrence Livermore National Laboratory at room temperature to a maximum neutron fluence of 1.6×10^{22} n/m^2 and a maximum gamma dose of 8.6×10^7 Gy, respectively. Neutron fluences greater than 1×10^{21} n/m^2 and gamma dose greater than 2.2×10^7 Gy caused a reduction in total elongation of the film. At our highest tested neutron fluence of 1.6×10^{22} n/m^2, the elongation dropped to less than one-fifth and the fracture stress dropped to one-third of the unirradiated value. It was found that 14-MeV neutrons were about eight times more effective than cobalt-60 gamma rays in producing mechanical property changes when compared on the basis of absorbed dose. Color changes induced by irradiation were also examined.

KEY WORDS: polyimide, irradiation effects, neutron irradiation, gamma ray irradiation, mechanical properties

Among polymers, polyimide films are insulating plastic materials that are useful over a wide temperature range and resistant to thermal and radiation effects. Polypyromellitimide (PPMI) is one of the polyimides and its chemical formula is $(C_{22}H_{10}N_2O_5)_n$. KAPTON is a commercial PPMI manufactured by Du Pont Company and is known to be resistant to radiation levels greater than 10^7 Gy (10^9 rad) (Fig. 1) [*1,2*]. These films serve as electrical insulation in a wide variety of irradiation environments. In the fusion power program, polyimides may be used as an insulating material in superconducting magnets, as reactor designs become more compact and doses become higher [*3*].

Mechanical property changes in polyimides induced by radiation exposure, such as gamma rays, electrons, and fission neutrons, have been studied [*4*]. Fission neutron irradiation effects inevitably incude the influence of gamma rays. Compressive strength of organic composites including

[1] Professor, The Research Institute for Iron, Steel and Other Metals, Tohoku University, Sendai 980, Japan.
[2] Group leader, Lawrence Livermore National Laboratory, P.O. Box 808/L-122, Livermore, CA 94550.
[3] Research scientist, Tokyo Institute of Technology, Yokohama 29, Japan.
[4] Senior scientist, Los Alamos National Laboratory, Los Alamos, NM 87545.

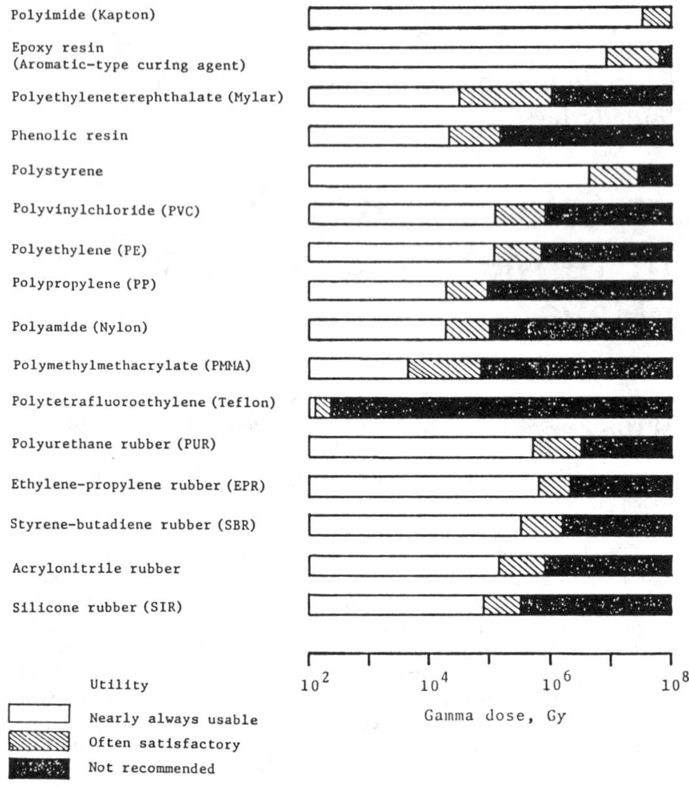

FIG. 1—*Radiation resistance of various polymers* [1,2].

polyimides was recently studied after 14-MeV neutron irradiation at RTNS-II at room temperature [*5*]. Flexural strength of polyimide-based fiber-reinforced material was studied after fast neutron irradiation at Intense Pulsed Neutron Source (IPNS) at cryogenic temperatures [*6*]. These data are important from a practical point of view in superconducting magnet design in the fusion program. There are no data on effects caused by neutrons in the absence of a gamma field in unreinforced polyimide.

In this report, tensile properties of KAPTON film are presented in detail after irradiation at room temperature with 14-MeV neutrons and cobalt-60 gamma rays. The purpose of this study was to distinguish the radiation damage caused by fusion neutrons and gamma rays in the polyimide itself. It is important to establish the differences in order to account for spectral effects [*6*]. Neutron and gamma rays caused color changes together with mechanical property changes in the film. The relationship of mechanical property changes to optical property changes is also studied.

A miniature specimen tension test technique originally prepared for fusion-fission correlation studies in metals and alloys was used [*7*]. This technique was very useful because we could utilize the steep flux gradient at RTNS II to simultaneously obtain specimens over a wide fluence range and because tensile properties are more sensitive to irradiation change than any other stress mode such as compression and bending in polymers with high plasticity.

Experimental Procedure

Specimens for tensile testing and optical absorption were made from KAPTON-H film of 3-mil nominal thickness. The measured film thickness was 73 μm. Tension specimens were 4 by 10

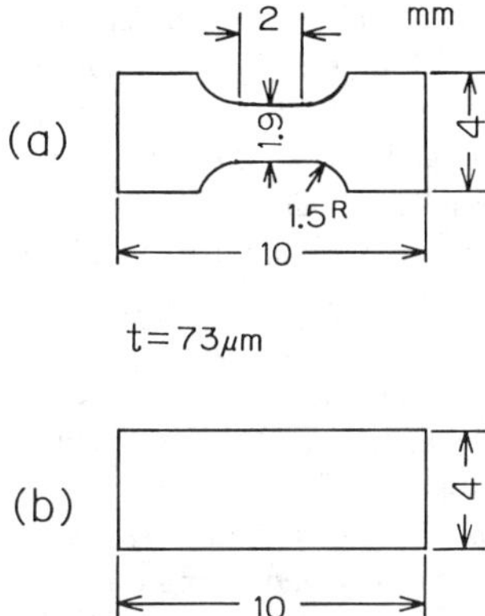

FIG. 2—*Specimen for the* (a) *tensile and* (b) *optical test.*

mm with reduced gage section of 2 mm length and 1.9 mm width as shown in Fig. 2. Tensile direction was taken parallel to the stripe pattern on the surface of the as-received film.

The 14-MeV neutron irradiation was done at Lawrence Livermore National laboratory's RTNS-II facility. Several bundles of specimens were mounted with iron dosimetry foils together with specimens of metals and alloys for separate experiments, in a vacuum tight vessel made of aluminum alloy (6061-T6) as shown in Fig. 3*a*. The thickness of the front window of the vessel was only 0.18 mm in order to locate the specimens in the high flux region close to the target.

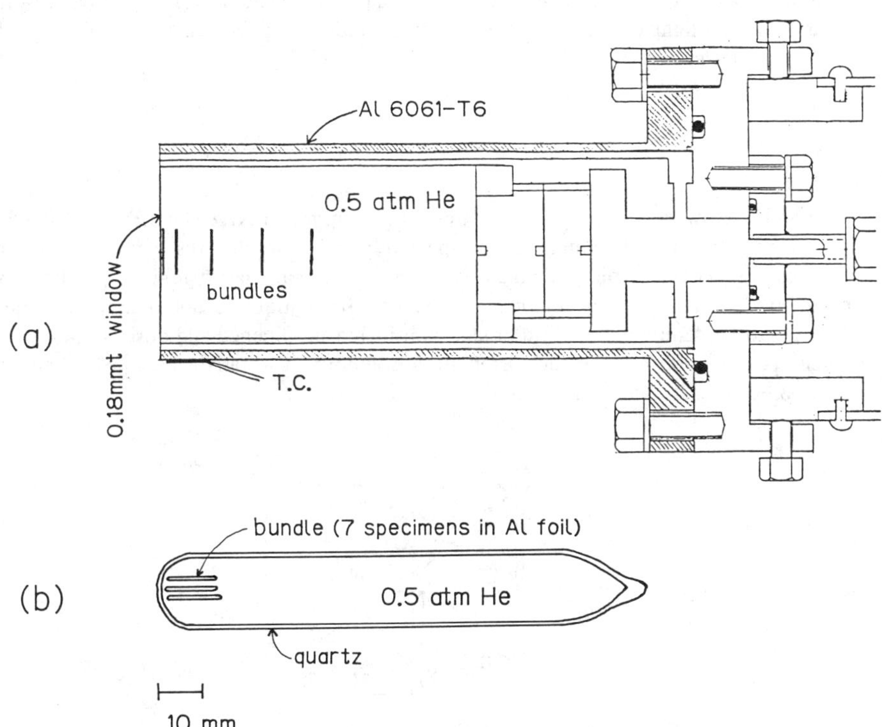

FIG. 3—*Irradiation vessel used at* (a) *RTNS-II and* (b) *Cobalt-60 Pool.*

Each bundle of five tension specimens and two optical specimens was wrapped in aluminum foil. The vessel was filled with 0.5-atm helium gas after vacuum pumping. The irradiation temperature was monitored with the thermocouple attached to the vessel and maintained at 293 to 298 K. The irradiation period was 9.0×10^5 s, and various fluences were obtained depending on the distance of the mounting places in the vessel from the rotating target: 8.9×10^{18}, 1.2×10^{20}, 6.5×10^{20}, 1.1×10^{21}, 2.7×10^{21}, 8.1×10^{21}, and 1.6×10^{22} n/m^2. The two lower fluences were obtained through irradiation in air outside the vessel. The influence of recoil atoms from aluminum wrapping foils on the irradiation effects was found to be negligible since the specimen adjacent to the foil showed the same property as the specimen not adjacent to it.

The gamma ray irradiation at room temperature was done at the Cobalt-60 Pool at Lawrence Livermore National Laboratory. Bundles of specimens were sealed with 0.5-atm-helium gas in a quartz capsule as shown in Fig. 3*b*. The average dose rate was 600 Gy/min. The dose levels were 6.9×10^6, 2.2×10^7, and 8.6×10^7 Gy depending on the cumulative irradiation period of 6.9×10^5, 2.2×10^6, and 8.6×10^6 s, respectively.

Helium gas was chosen as the irradiation atmosphere in both experiments. Previous work has shown that the damage behavior of polymers depends on the atmosphere during irradiation [*8*]. These experiments were designed to study the influence of different radiation without influence from the surrounding atmosphere.

Post-irradiation tension tests were performed with an Instron-type machine. Shimazu Autograph DSC 500, at room temperature with a cross-head speed of 0.4 mm/min, corresponding to the strain rate of 3.33×10^{-3}/s. Miniature specimen grips that were used for testing are shown in Fig. 4. The cross-sectional change wih tensile deformation was estimated by measuring directly the change of width and length of markings on the specimen surface during deformation. The rigidity of the tensile machine, including loading cell, pull rod, and grips, was determined by stretching a hard metal. This rigidity was used to evaluate the elastic modulus. Optical absorption of visible and infrared light was measured for neutron-irradiated specimens using a CRAY Model 14 spectrophotometer at wavelength from 400 to 1000 nm.

Results

Stress-Strain Behavior

Figures 5 and 6 show typical engineering stress-strain curves of KAPTON film irradiated with 14-MeV neutrons and cobalt-60 gamma rays, respectively. For unirradiated specimens, engineering stress-strain curves showed typical characteristic regions: linear elastic region, parabolic work hardening region, linear hardening region, and fracture. Irradiation did not change these general features. It caused hardening and embrittlement with increasing fluence and dose. Reproducibility among three specimens tested for each irradiation condition was satisfactory. Absolute scatter among data points at each fluence decreased with increasing fluence.

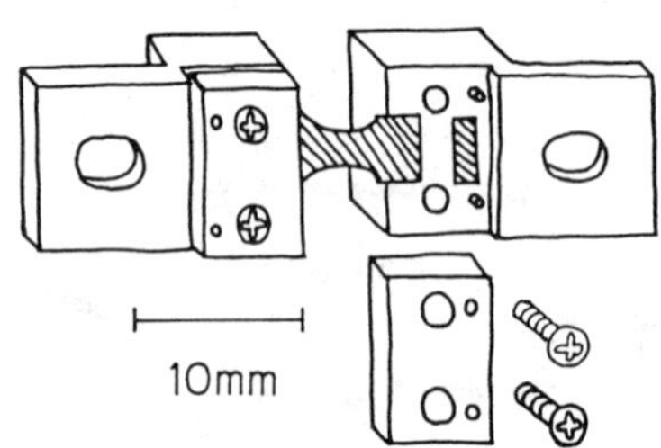

FIG. 4—*Miniature specimen grips for tensile testing.*

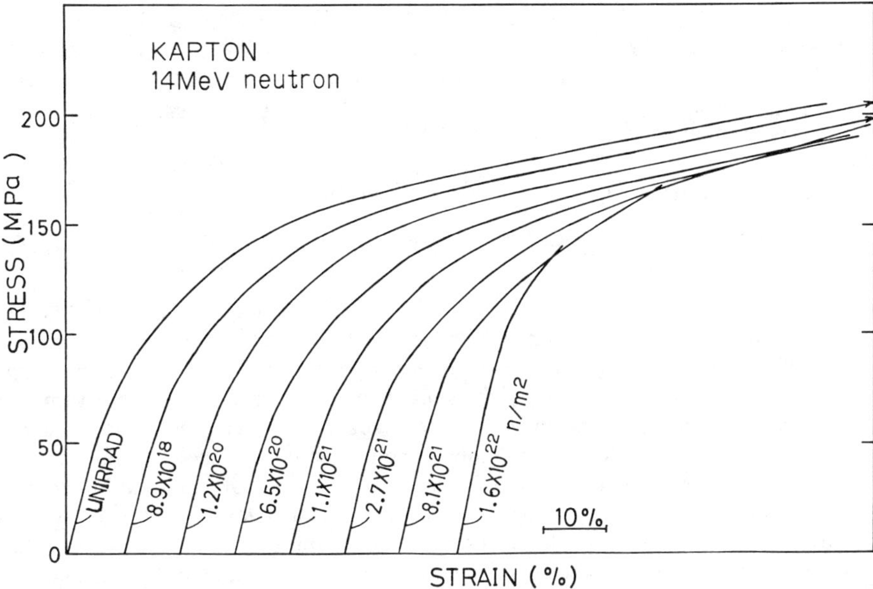

FIG. 5—*Engineering stress-strain curves at room temperature of KAPTON irradiated with 14-MeV neutrons.*

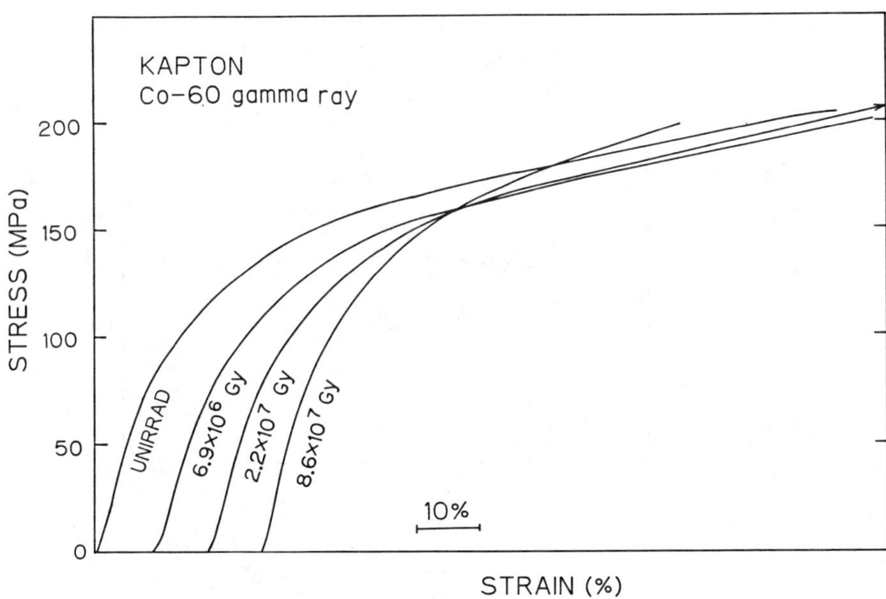

FIG. 6—*Engineering stress-strain curves at room temperature of KAPTON irradiated with Cobalt-60 gamma rays.*

An example of dimensional change of unirradiated KAPTON under room temperature tensile deformation is shown in Fig. 7. The length l and the cross-sectional area S under load were determined and compared with their original values lo and So. The condition that the volume is constant during plastic deformation, that is, $l \times s = lo \times So$, holds in the case of crystals such as metals. This condition does not hold for KAPTON as shown in Fig. 7. The true stress should be estimated from load and cross-sectional area at any strain. The fracture stress, that is, the true stress at which the specimen broke, was determined using the data of Fig. 7.

Tensile Properties

The change of elastic modulus, flow stress, and work hardening rate with 14-MeV neutron fluence is shown in Fig. 8. The elastic modulus and the stress to produce 1% plastic strain increased slightly at fluences above 10^{21} n/m^2. The work hardening rate was nearly constant. The work hardening rate was calculated from the stress increment between 1 and 5% strain, and is expressed in terms of the stress increment per 100% strain. In the case of gamma irradiation, a slight increase in elastic modulus occurred only at the highest dose as shown in Fig. 9.

Figures 10 and 11 show the change in total elongation (that is, elastic plus plastic) and fracture stress with 14-MeV neutrons and gamma rays. The average total elongation was about 120% for unirradiated and low irradiation conditions ($<6.5 \times 10^{20}$ n/m^2 or $<2.2 \times 10^7$ Gy). At neutron fluences greater than 10^{21} n/m^2, the elongation decreased with fluence, ultimately to less than 20% at 1.6×10^{22} n/m^2. These data suggest that the acceptable neutron fluence ranges from 10^{21} to 10^{22} n/m^2 depending on whether the design requires large plasticity or can accommodate only moderate strain before fracture. Fracture stress is strongly influenced by neutron irradiation,

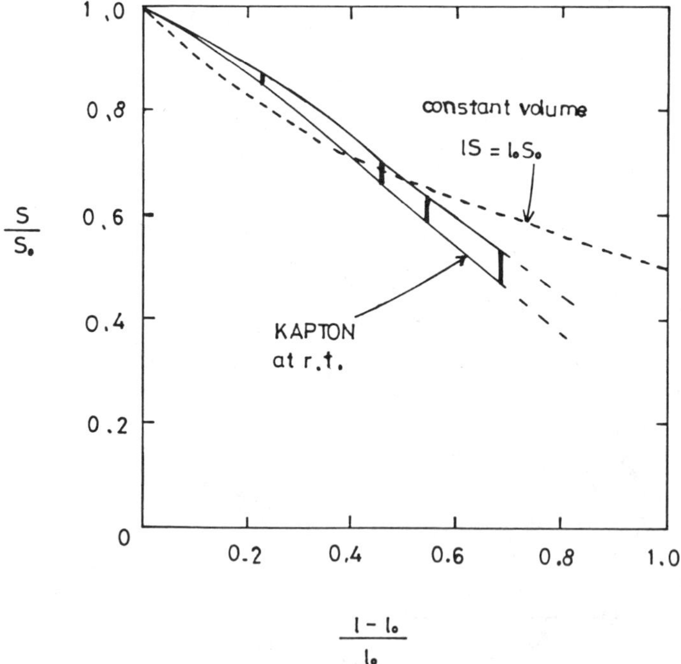

FIG. 7—*The relationship between the changes in length and cross-sectional area under room temperature deformation of KAPTON.*

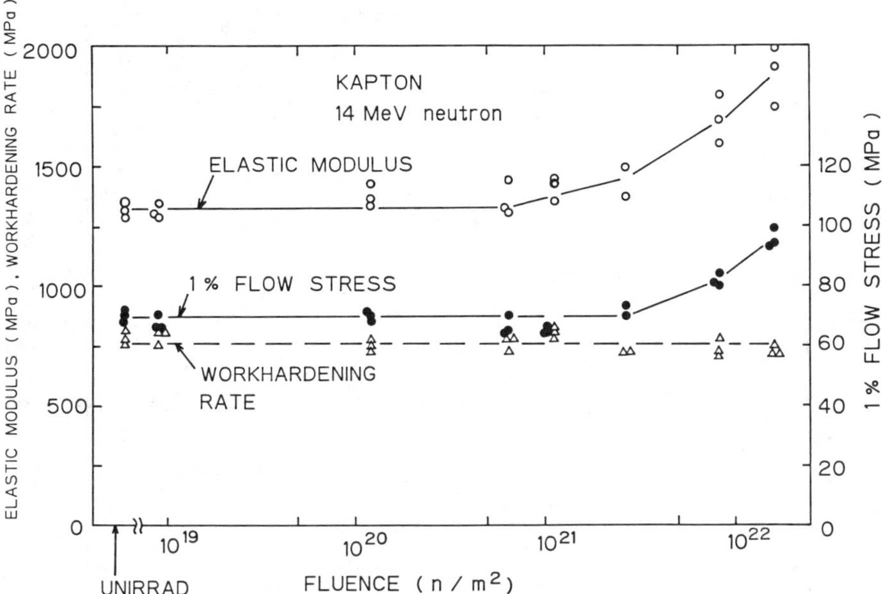

FIG. 8—*The change in elastic modulus, flow stress at 1% plastic strain, and work hardening rate with 14-MeV neutron irradiation in KAPTON.*

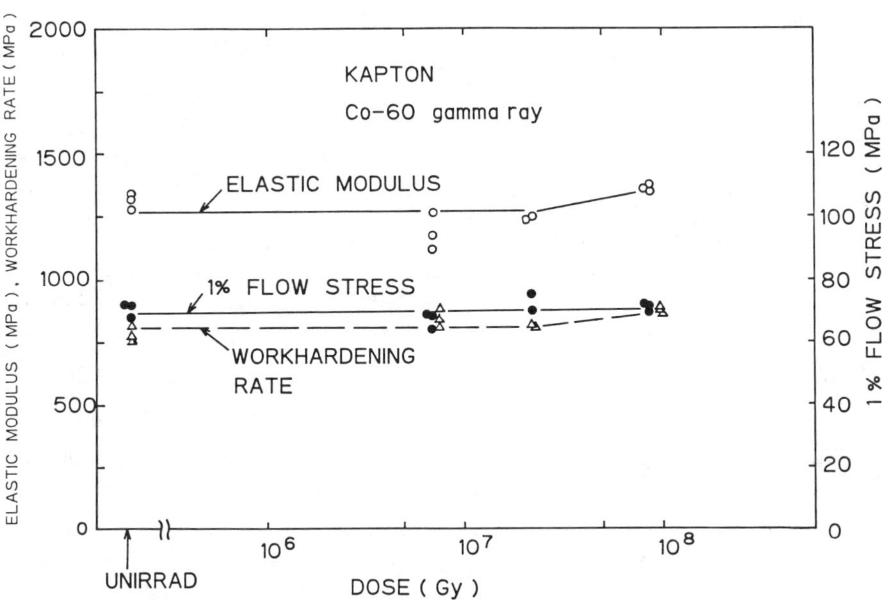

FIG. 9—*The change in elastic modulus, flow stress at 1% plastic strain, and work hardening rate with Cobalt-60 gamma ray irradiation in KAPTON.*

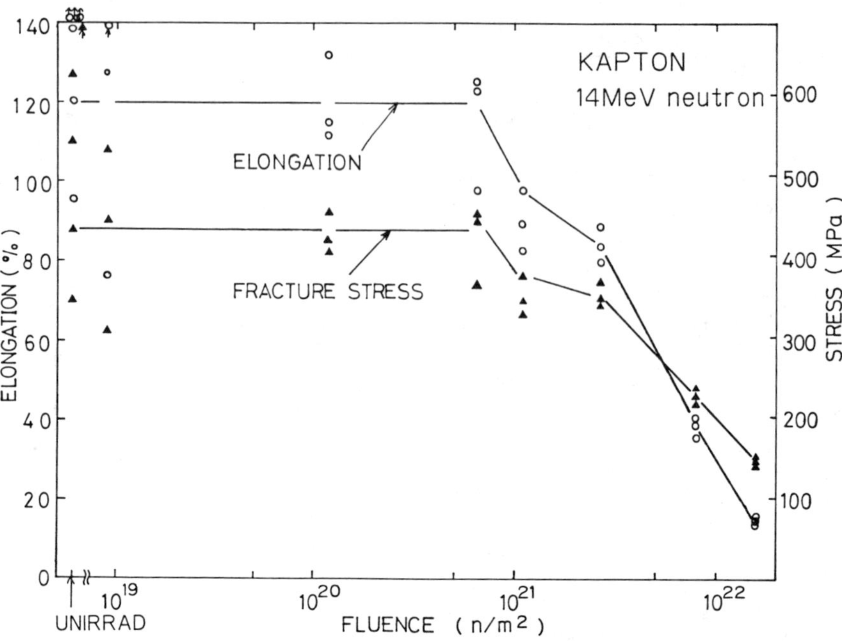

FIG. 10—*The change in total elongation and fracture stress with 14-MeV neutron irradiation in KAPTON.*

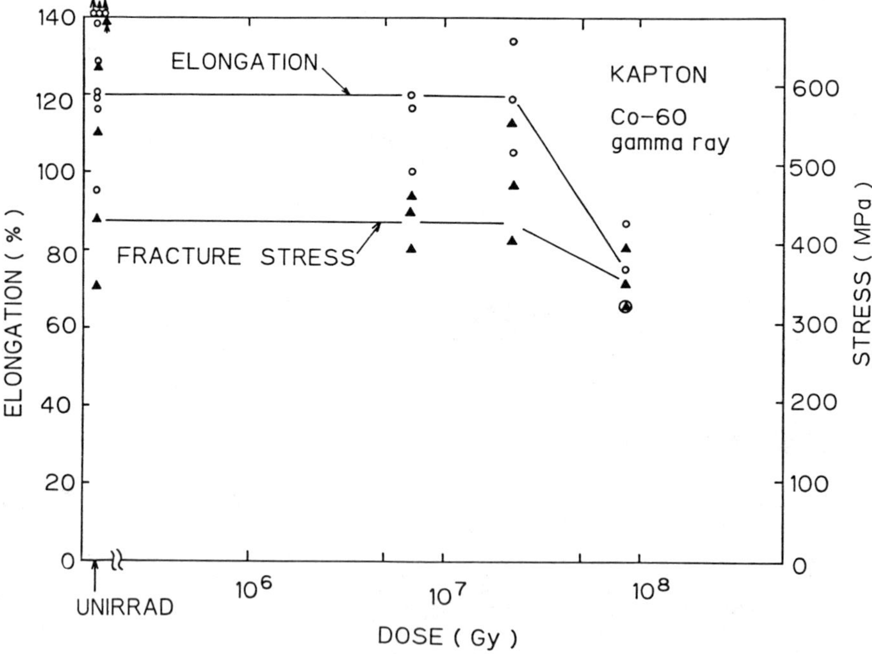

FIG. 11—*The change in total elongation and fracture stress with Cobalt-60 gamma ray irradiation in KAPTON.*

dropping to one-third of the unirradiated value at the highest tested neutron fluence. In the case of gamma rays, a loss of ductility was observed only at the highest dose levels. In both irradiation conditions, irradiation embrittlement is well correlated with the reduction in fracture stress.

Optical Properties

The material darkened in color and the transparency decreased monotonically with increasing 14-MeV neutron fluence. Figure 12 shows the change in the optical absorption spectrum with neutron fluence. The effect of irradiation depended on wavelength. Figure 13 shows the relation of absorbance at 600 nm (orange color), and at 800 nm (infrared region) with neutron fluence. In the case of 14-MeV neutron irradiation, the fluence dependence of 600-nm absorbance corresponds well with the mechanical property change as seen in Figs. 10 and 13.

In the case of gamma irradiation, optical absorption measurements are pending. However, it is possible to report visual color changes for comparison with mechanical property changes. This subject is discussed later.

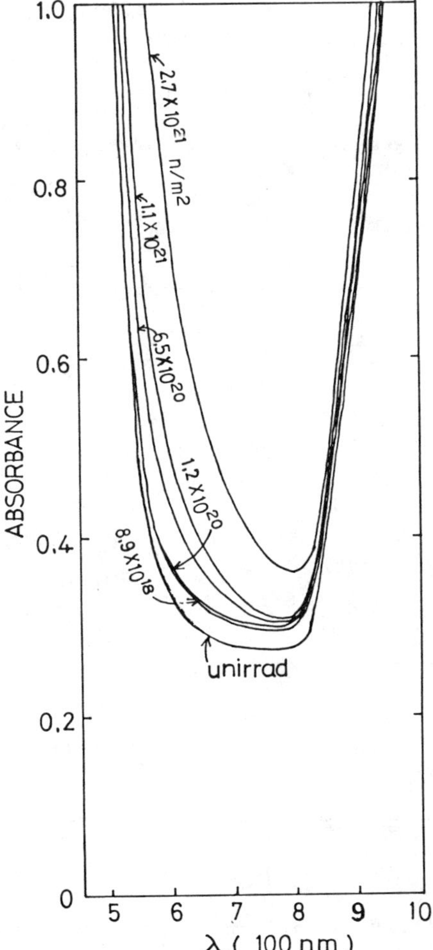

FIG. 12—*The change in the optical absorption spectrum of KAPTON with 14-MeV neutron irradiation.*

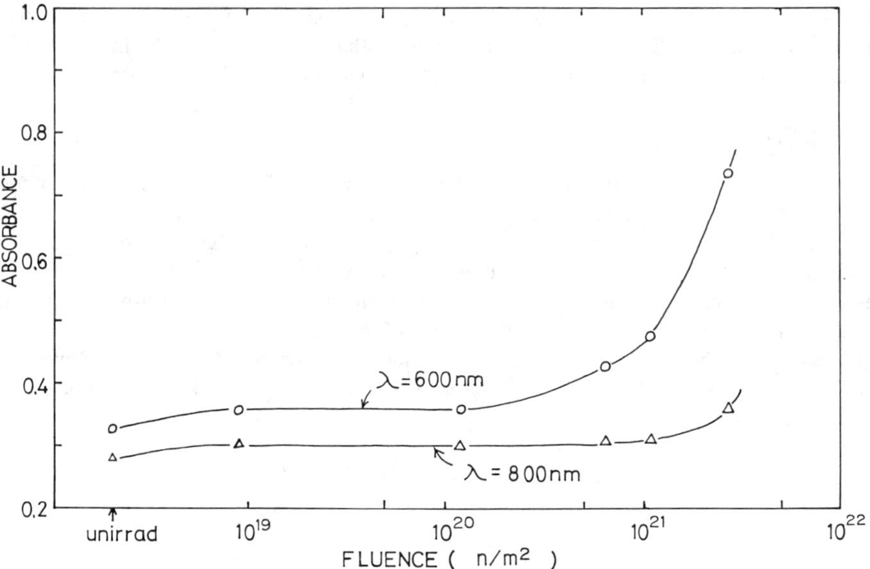

FIG. 13—*The relationship between absorbance and 14-MeV neutron fluence in KAPTON.*

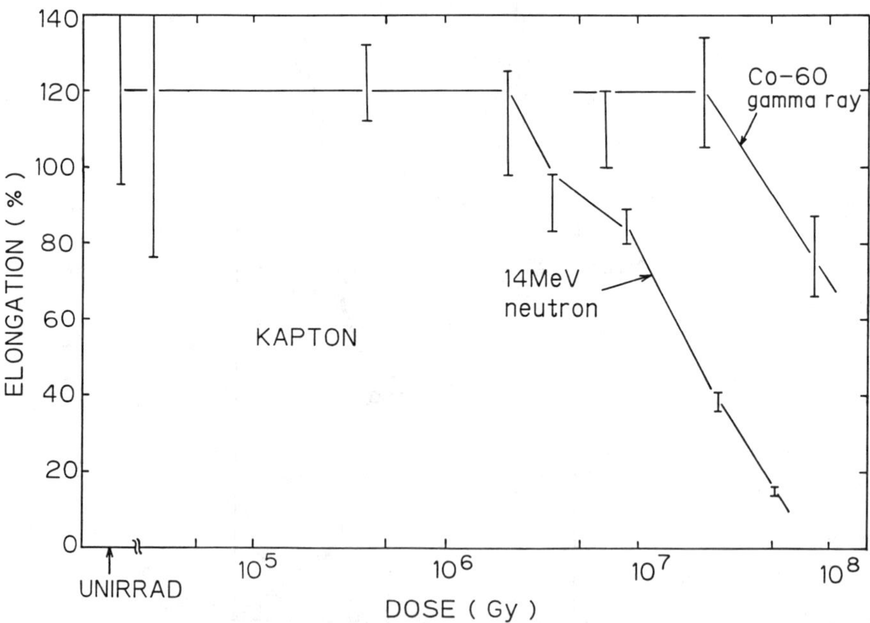

FIG. 14—*The comparison of irradiation embrittlement caused by 14-MeV neutrons and Cobalt-60 gamma rays on the basis of absorbed dose.*

Discussion

14-MeV Neutron/Gamma Ray Correlation

In order to correlate the 14-MeV neutron data with Cobalt-60 gamma ray data, the following elemental conversion factors for 14-MeV neutron fluence to absorbed dose was used: C = 2.03×10^{-15}, H = 4.74×10^{-14}, N = 3.13×10^{-15}, and O = 1.74×10^{-15} Gy/(n/m^2). The contribution of the individual elements leads to a KAPTON conversion factor of 3.25×10^{-15} Gy/(n/m^2) for 14-MeV neutrons.

By using this conversion factor, the changes in total elongation caused by 14-MeV neutrons and by gamma rays can be expressed on the basis of absorbed dose as shown in Fig. 14. The average elongation after 8.6×10^7 Gy gamma ray irradiation is 76%. This elongation value is expected after 14-MeV neutron irradiation of 1.1×10^7 Gy from Fig. 14. That is, 14-MeV neutrons give rise to irradiation embrittlement about eight times more effectively than gamma rays. Since the gamma ray dose levels for the specimens irradiated at RTNS-II are less than 10^4 Gy, the property change by 14-MeV neutrons in Fig. 14 is due purely to neutrons. Figure 14 shows a very important result because these data were obtained utilizing the same testing technique, and the irradiation conditions were pure 14-MeV neutrons and pure gamma rays.

From Fig. 14, the practical conversion factor to give an elongation value of 76% is estimated to be about 2.6×10^{-14} Gy/(n/m^2). This kind of practical conversion factor is, of course, dependent on the properties considered. In fact, the practical conversion factor to give color change is thought to be somewhat larger than that for elongation change, as discussed later. The present result shows that gamma irradiation data such as given in Ref *9* cannot be simply used to predict neutron irradiation behavior without careful assessment of specific properties and neutron spectra.

In the case of inorganic materials and metals, irradiation effects depend on the details of energy transfer because of the role of atom displacement. On the other hand, irradiation effects in organic materials are generally considered not to depend on the details of energy transfer, but to depend mainly on absorbed dose. However, the present results show clearly that irradiation effects in KAPTON depend on the details of energy transfer and not simply on absorbed dose. It was previously reported that degradation efficiency of polystyrene for fast neutrons was larger than that for gamma rays [*10*]. Recently, irradiation work on composite materials including polyimides has been published [*11–16*]. Further irradiation experiments to distinguish energy spectral effects on each component will be useful to understand these data.

Mechanical Property-Optical Property Correlation

Since optical absorption measurements have not yet been done on gamma irradiated specimens, color change determined visually is compared qualitatively. The degree of color change arising from 8.6×10^7 Gy gamma irradiation is between that caused by 1.2×10^{20} n/m^2 and 6.5×10^{20} n/m^2 neutron irradiation. If we take the mean value of 4×10^{20} n/m^2 and the conversion factor of 3.25×10^{-15} Gy/(n/m^2), then we have a neutron dose of 1.3×10^6 Gy to produce the same color change as a dose of 8.6×10^7 Gy of gamma rays. Therefore, 14-MeV neutrons cause color change about 70 times more effectively than gamma rays. In the case of mechanical property degradation, 14-MeV neutrons are about eight times more effective than gamma rays. Therefore, the mechanism producing optical property change is different from that producing mechanical property change. In other words, the good correspondence between mechanical property change and orange color absorption in specimens exposed to 14-MeV neutrons is probably unique to that radiation environment.

Irradiation Embrittlement in Polyimide

Polyimide incorporates multiple bonds along the backbone of the polymer chain, which results in high resistance to thermal and radiation degradation because several covalent bonds must be broken to break the chain. Stretching and uncoiling of the chains is responsible for the volume change during the deformation. First, the main reason for embrittlement in KAPTON is the reduction of the fracture stress as seen in Fig. 10 and 11. Second, 14-MeV neutrons are about eight times more effective than cobalt-60 gamma rays in producing mechanical property change when compared on the basis of absorbed dose. These two facts suggest that irradition embrittlement in a polyimide KAPTON is related to the breaking of the main chain of the polymer. It is probable that atomic displacement by high energy neutrons can enhance the breaking. Scanning electron microscope (SEM) observations like those in Ref *16* and absorption experiments over a wide range of wave lengths are needed to study the mechanisms. Tension tests on specimens irradiated with 14-MeV neutrons at cryogenic temperatures in progress will also be helpful.

Conclusions

1. Significant decrease in elongation begins to occur at a 14-MeV neutron fluence of 10^{21} n/m^2, dropping to less than one-eighth of the unirradiated value at 1.6×10^{22} n/m^2.
2. Fracture stress is strongly influenced by irradiation, dropping to one-third of the unirradiated value at 1.6×10^{22} n/m^2.
3. Comparing the separate data with neutrons and gamma rays, it was found that on the basis of absorbed dose 14-MeV neutrons are about eight times more effective then cobalt-60 gamma rays in producing mechanical property changes.
4. Color change corresponded well with the mechanical property change in the case of 14-MeV neutrons, but did not correspond in the case of gamma irradiation.

Acknowledgments

Authors K. A. and K. S. would like to express their appreciation to Professors K. Kawamura and K. Sumita for promoting the utilization of RTNS-II. One of the authors (K. A.) would like to express his appreciation to Professor T. Uchida for encouraging the work, and to Monbusho and Yoshida Foundation for Science and Technology for financial support during his stay at LLNL. Gamma irradiations were arranged through the courtesy of C. Graham and D. Trombino.

This work was performed under auspices of the U.S. Department of Energy and Japan (Monbusho) by Lawrence Livermore National Laboratory under Contract W-7405-ENG-48.

References

[*1*] Schonbacher, H. and Stolarz-Lzycka, A., CERN 70-04, 1979, pp. 79–98.
[*2*] Beynel, P., Maier, P., and Schonbacher, H., CERN 82-10, 1982.
[*3*] Scott, J. L., Clinard, F. W., Jr., and Wiffen, F. W., *Journal of Nuclear Materials,* Vols. 133 and 134, 1985, pp. 156–163.
[*4*] The Radiation Effects Information Center Report, REIC 21.
[*5*] Ida, T. and Sumita, K., 1982 Annual Research Report of Japanese Contributions for Japan-US Collaboration on RTNS-II Utilization, pp. 13–20.
[*6*] Tucker, D. S., Clinard, F. W., Jr., Hurley, G. F., and Fowler, J. D., *Journal of Nuclear Materials,* Vols. 133 and 134, 1985, pp. 805–809.
[*7*] Abe, K., 1984 Annual Research Report of Japanese Contribution for Japan-US Collaboration on RTNS-II Utilization, pp. 76–89.
[*8*] Yamaoka, Y. and Tagawa, S. T., *Journal of Atomic Energy Society of Japan,* Vol. 26, 1984, pp. 739–747.

[*9*] Coltman, R. R., Jr. and Klabunde, C. E., *Journal of Nuclear Materials,* Vols. 103 and 104, 1981, pp. 717–722.

[*10*] Egusa, S., Ishigure, K., and Tabata, Y., *Macromolecules,* Vol. 13, 1981, p. 171.

[*11*] Coltman, R. R., Jr., *Journal of Nuclear Materials,* Vols. 108 and 109, 1982, pp. 559–571.

[*12*] Egusa, S., Kirk, M. A., Birtcher, R. C., Hagiwara, M., and Kawanishi, S., *Journal of Nuclear Materials,* Vol. 119, 1983, pp. 146–153.

[*13*] Egusa, S., Kirk, M. A., and Birtcher, R. C., *Journal of Nuclear Materials,* Vol. 126, 1984, pp. 152–159.

[*14*] Egusa, S., Kirk, M. A., Birtcher, R. C., Hagiwara, M., and Kawanishi, S., *Nuclear Instruments and Methods in Physics Research B1,* 1984, pp. 610–616.

[*15*] Hurley, G. F. and Coltman, R. R., Jr., *Journal of Nuclear Material,* Vols. 122 and 123, 1984, pp. 1327–1337.

[*16*] Hagiwara, M., Udagawa, A., Kawanishi, S., Egusa, S., and Takeda, N., *Journal of Nuclear Materials,* Vols. 133 and 134, 1985, pp. 810–814.

Rajendran S. Michael,[1] *Steve Frank,*[1] *Dusan Stulik,*[1] *and J. Thomas Dickinson*[2]

Changes in Surface Morphology and Microcrack Initiation in Polymers Under Simultaneous Exposure to Stress and Fast Atom Bombardment

REFERENCE: Michael, R. S., Frank, S., Stulik, D., and Dickinson, J. T., "**Changes in Surface Morphology and Microcrack Initiation in Polymers Under Simultaneous Exposure to Stress and Fast Atom Bombardment,**" *Influence of Radiation on Material Properties: 13th International Symposium (Part II), ASTM STP 956,* F. A. Garner, C. H. Henager, Jr., and N. Igata, Eds., American Society for Testing and Materials, Philadelphia, 1987, pp. 682–687.

ABSTRACT: We present studies of the changes in surface morphology due to simultaneous exposure of polymers to stress and fast atom bombardment. The polymers examined were Teflon,[3] Kapton,[3] Nylon,[3] and Kevlar-49.[3] The incident particles were 6 keV xenon atoms. We show that in the presence of mechanical stress these polymers show topographical changes at particle doses considerably lower than similar changes produced on unstressed material. Applied stress also promotes the formation of surface microcracks which could greatly reduce the mechanical strength of the material.

KEY WORDS: polymers, fast atom bombardment, stress, surface morphology

Bombardment of material surfaces with energetic particles (electrons, ions, or neutral atoms) leads to emission of secondary particles (electrons, ions, atoms, and molecules) through various physical processes such as electron stimulated desorption and physical/chemical sputtering. The latter was first observed in 1852 when Grove published the first experimental evidence of sputter deposition of materials due to ion bombardment of a target [*1*]. Vast amounts of experimental data on sputtering of metals, alloys, and simple inorganic materials have been collected and the observed effects have attracted considerable theoretical effort to describe the process quantitatively [*2*]. In single element substrates and non-reactive incident particles, the dissipation of collision energy can only alter the physical structure of the material [*3*]. In the case of alloys and inorganic compounds, the observed changes can be considerably more complex, including preferential sputtering [*4*] and electronic excitations which can lead to loss of anions in ionic materials [*5*]. In the case of molecular solids, considerable changes in the chemistry of the bombarded surfaces can occur. Also, the chemical bonds and composition of the sample play major roles in the process of secondary particle emission and the resulting chemical changes due to bombardment [*6*].

Changes in surface morphology are also observed with increasing total particle bombarding dose on polymer surfaces. Our knowledge of the mechanisms of these changes is severely limited, including details of the physical and chemical structure of the solid *before* exposure to bombardment.

[1] Department of Chemistry, Washington State University, Pullman, WA 99164.
[2] Department of Physics, Washington State University, Pullman, WA 99164.
[3] Registered Trade-names of E. I. Dupont de Nemours and Co.

This limited description of the polymer surface is in part due to the tendency for such surfaces to be so readily damaged by radiation. Nevertheless, we feel that examining the overall changes in the material from exposure to various types of bombardment is of importance, particularly due to the technological interest in applications of polymers in increasingly severe environments.

Our previous studies [*7,8*] of the changes in the surface morphology in Teflon (polytetrafluoroethyene) exposed to fast atom bombardment indicated that such changes differ substantially from what occurs on metal substrates [*9*]. The changes observed in Teflon start with tiny surface fissures which grow into small surface cracks, forming a dense crack network, which eventually forms filamentary structure and cone-like formations. It was also shown [*8*] that bombardment of Teflon in the presence of mechanical stress greatly increased the rate at which the surface morphology changed, including more rapid crack formation, when compared to the unstressed material.

In this paper, we examine these effects in Teflon as well as Kapton, Nylon, and Kevlar-49.

Procedure

The materials were all obtained from commercial sources. The Teflon and Kapton were in the form of thin sheets, 3 mm and 0.5 mm in thickness and Nylon and Kevlar-49 fibers were 120 μm and 10 μm in diameter, respectively. Fast Xe atoms with 6 keV kinetic energy were produced using a saddle-field type Fast Atom Bombardment Gun (Ion Tech. Ltd.) and allowed to impinge on both stressed and unstressed polymer samples. The angle of incidence in most cases was 600° with respect to the surface normal. During bombardment the pressure in the target chamber was 1.3×10^{-6} torr. The samples were bombarded with various doses of Xe atoms ranging from 1×10^{16} to 7×10^{16} atoms/cm^2. Stressed samples were mounted next to unstressed samples of the same material. Stress was achieved by means of bending the thin sheets of the polymer in tight radii. In the case of the Nylon and Kevlar-49 fibers, these were bent over a stainless steel wire (1.2 mm in diameter). All specimens were partially covered with aluminum foil to block the incident atom beam so that the exposed regions could be readily compared with the original surfaces. After exposure to the fast atom beam the samples were coated with 300 Å of gold and examined in an ETEC U-1 Scanning Electron Microscope (SEM).

Results and Discussion

In Fig. 1*a* we show an SEM photograph of the sample holder with stressed and unstressed Teflon specimens. The bending deformation produces a surface in tension, although a certain amount of relaxation of the stress which was exposed to the fast atom beam. At low magnification, differences in texture on the surface of these samples between exposed and unexposed regions could already be seen. Closer examination of the bombarded area on the stressed specimen shows several rows of crack-like features in the surface with the major cracks oriented approximately normal to the direction of tension. The density of microcracking was observed to be much higher in regions where the beam was normal to the surface as opposed to glancing incidence.

No such structure can be seen in the area of the sample that was covered with aluminum foil which shows that the cracks are not simply due to mechanical deformation alone. A higher magnification view of the region outlined in the center of Fig. 1*a* is shown in Fig. 1*b*. Comparison of this structure with previous results obtained on unstressed Teflon samples [*8,10*] shows that such large cracks in the bombarded surface are atypical. Thus, we can conclude that the combination of stress *and* bombardment is responsible for the formation of these larger flaws.

The walls of the cracks have a columnar structure and in some parts of the crack, filamentary formation can be seen. The major cracks are very deep which suggests that stress resulted in opening and propagation of the crack from initial flaws created by bombardment.

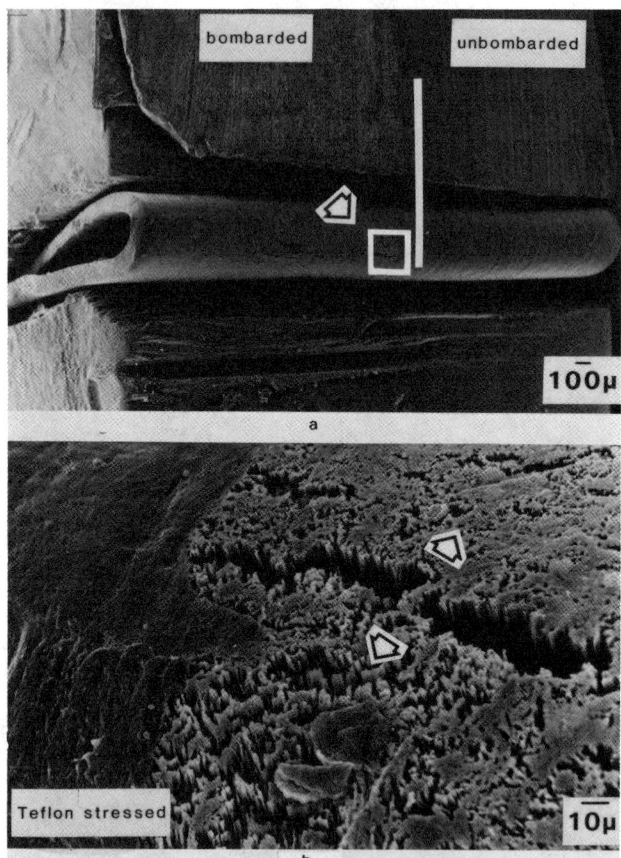

FIG. 1—*Crack development in stressed Teflon®:* (a) *Sample holder with stressed and unstressed Teflon samples.* (b) *Detail of surface morphology development on stressed sample (bombarding dose: 7.2 × 10^{16} Xe atoms/cm²).*

Figure 2 shows a comparison of the surface morphology developed on stressed and unstressed Kapton exposed to fast Xe atom bombardment. Kapton is known to be a heat and radiation resistant polymer material [*11*]. In comparison with Teflon, the development of visible surface effects on the Kapton for both the stressed and unstressed specimens required a substantially higher bombardment dose ($>10^{16}$ atoms/cm²). Likewise, in Kapton, the stressed sample developed damage (shown in an advanced stage in Fig. 2*b*) at considerably lower doses than in the case of unstressed material. At the stress levels and fast atom beam doses used in these studies, we did not observe crack initiation.

Nylon fibers under stress and bombardment exhibit surface crack initiation, as shown in Fig. 3*a*. The part of the bombarded surface marked by the white square is shown at higher magnification in Fig. 3*b*. If we compare the structures shown here with the developed morphology in Teflon (Fig. 1*b*) we see that they have similar features, namely columnar structures with a tendency to separate into free-standing filaments, and similarly shaped microcracks. The doses required to achieve comparable damage are considerably larger in the case of Nylon compared with Teflon. If the dose used to create the damage in Nylon shown in Fig. 3 were applied to Teflon, it would

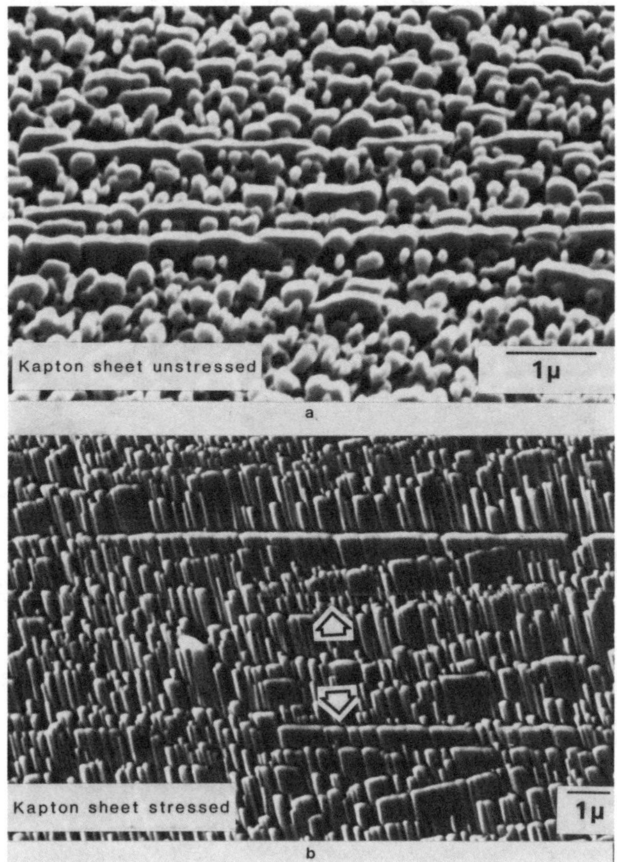

FIG. 2—*Surface morphology development on Kapton:* (a) *Unstressed sample (bombarding dose: 2.8 × 10^{17} Xe atoms/cm²).* (b) *Stressed sample (bombarding dose: 2.0 × 10^{17} Xe atoms/cm²)—arrows indicate the direction of applied stress.*

create an advanced degree of cone formation on the Teflon surface [*10*]. Estimates of the sputtering yields of Nylon versus Teflon were substantially lower for Nylon. This suggests that the ability of the incident particle to break bonds and move atoms and molecules both *from* the surface and *on* the surface is a critical part of the damage mechanism.

Figure 4*a* shows the surface morphology development on an unstressed Kevlar-49 fiber from fast Xe atom bombardment. Following a dose of 2.8 × 10^{16} atoms/cm², the relatively smooth fiber surface is transformed into a dramatic array of ridges which are oriented perpendicular to the direction of the incident particle beam which was also along the axis of the fiber. A stressed Kevlar-49 fiber (Fig. 4*b*) exposed to the same dose shows a more advanced degree of development, where in some areas of the bombarded region there is a transition from ridges to columnar structures. The major axes of the columns are oriented parallel to the fiber axis and therefore parallel to the direction of incidence of Xe atoms. We are currently determining if it is the fiber orientation or the direction of incidence that produces the orientation of these patterns. Kevlar is a highly structured material with strong orientation of crystalline regions [*12*], which may be the dominant cause of the structures we are seeing here.

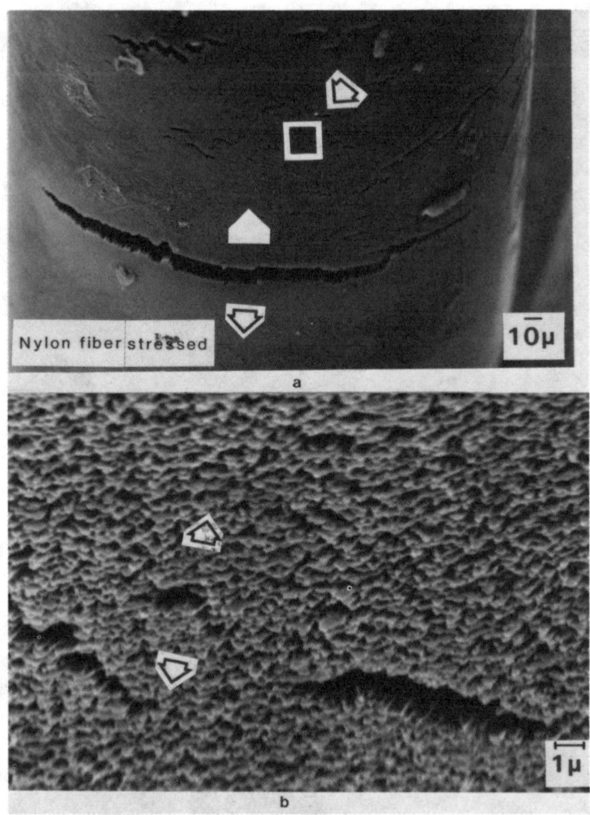

FIG. 3—*Crack formation in Nylon fiber (bombarding dose: 2.8 × 10^{17} Xe atoms/cm²):* (a) *Low magnification. Stress is perpendicular to major crack.* (b) *Detail of surface morphology development and crack initiation.*

Conclusions

We believe that the mechanisms for the creation of the wide variety of surface changes shown here involve physical sputtering (removal of molecular fragments) *and* movement of molecules on the surface. The details of these processes are obviously yet to be revealed. However, the substantial increase in the rate of formation of surface damage when the material is under stress as well as the appearance of surface microcracks indicates that the physical state of the near surface layer can play an important role. In other experiments [*13–15*], we are in fact studying the effects of focused radiation into stressed crack tips in polymers where the radiation induced bond breaking leads directly to crack propagation. Our immediate goal is to develop a more detailed quantitative and theoretical description of these effects of morphological changes, microcrack formation, and bombardment induced crack propagation in polymers.

References

[*1*] Gove, W. R., *Transactions of the Royal Society of London,* Vol. 142, 1985, p. 87.
[*2*] Behrisch, R., Ed., *Sputtering by Particle Bombardment,* Vol. I–III, Springer Verlag, Berlin, 1980–85.
[*3*] Baranova, E. C., Gusev, V. M., Martynenko, Yu. V., Starinin, C. V., and Haibullin, I. B., *Radiation Effects,* Vol. 18, 1973, p. 21.

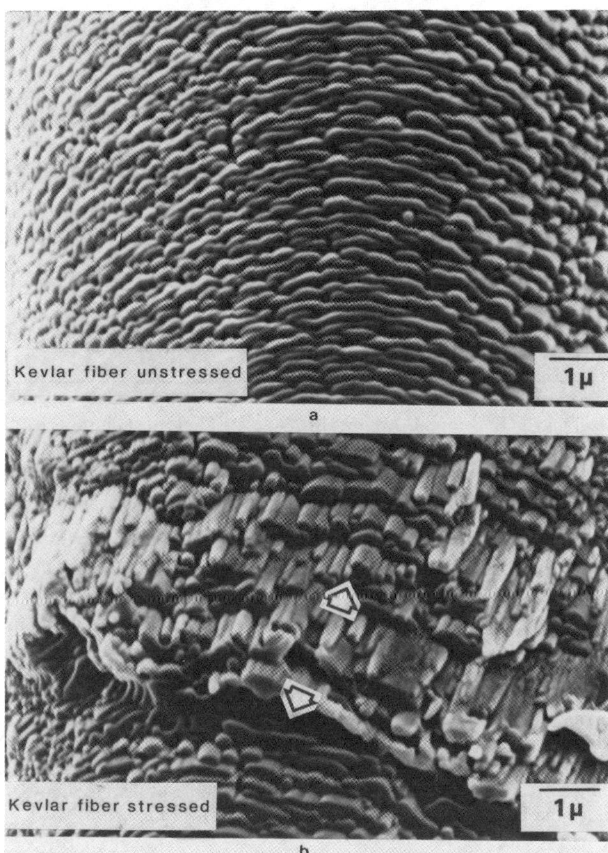

FIG. 4—*Surface morphology development on Kevlar fiber (bombarding dose 2.8* $\times$ *10*^{16} *Xe atoms/cm*^{2}*);* (a) *Unstressed fiber.* (b) *Stressed fiber (arrows indicate the direction of applied stress).*

[4] Harper, J. M. E. and Gambino, R. J., *Journal of Vacuum Science and Technology,* Vol. 16, 1979, p. 1901.

[5] Brenig, W. and Menzel, D., Eds., *Desorption Induced by Electronic Transitions—DIET II,* Springer Verlag, Berlin, 1985.

[6] Orth, R. G., Jonkman, H. T., Stulik, D., and Michl, J., *Proceedings of the 30th Annual Conference on Mass Spectrometry and Allied Topics,* June 1982, Honolulu, American Mass Spectrometry Society, p. 212.

[7] Michael, R. and Stulik, D., *Radiation Effects Letters,* Vol. 87, 1985, p. 9.

[8] Michael, R. and Stulik, D., *Nuclear Instruments and Methods in Physics Research,* Vol. B14, 1986, p. 278.

[9] Auciello, O. and Kelly, R., Eds., *Ion Bombardment Modification of Surfaces,* Elsevier, Amsterdam, The Netherlands, 1984.

[10] Michael, R. and Stulik, D., *Journal of Vacuum Science and Technology,* Vol. A4, 1986, p. 1861.

[11] Sweeting, O. J., Eds., *The Science and Technology of Polymer Films,* Vol. II, Wiley, New York, 1971, Chapter 16.

[12] Morgan, R. J., Pruneda, C. O., and Steele, W. J., *Journal of Polymer Science: Polymer Physics Edition,* Vol. 21, 1983, p. 1757.

[13] Dickinson, J. T., Klakken, M. L., Miles, M. H., and Jensen, L. C., *Journal of Polymer Science: Polymer Physics Edition,* Vol. 23, 1985, p. 2273.

[14] Dickinson, J. T., Jensen, L. C., and Klakken, M. L., *Journal of Vacuum Science and Technology,* to be published.

Dorinda A. Sparacio[1] *and Mary A. Amini*[2]

Effect of Gamma Radiation on the Permeability of Opthalmic Preservatives Through Fluorine Surface-Treated Low-Density Polyethylene Bottles

REFERENCE: Sparacio, D. A. and Amini, M. A., "**Effect of Gamma Radiation on the Permeability of Opthalmic Preservatives Through Fluorine Surface-Treated Low-Density Polyethylene Bottles,**" *Influence of Radiation on Material Properties: 13th International Symposium (Part II), ASTM STP 956,* F. A. Garner, C. H. Henager, and N. Igata, Eds., American Society for Testing and Materials, Philadelphia, 1987, pp. 688–697.

ABSTRACT: Efforts to improve the barrier properties of polyethylene have led to the development of fluorine surface treatments in which the surface of polyethylene is changed from nonpolar to polar by replacing some of the hydrogen atoms in the polyethylene chains with fluorine atoms. The purpose of this study was to determine the effect of gamma radiation on the permeability of benzyl alcohol, phenylethyl alcohol, and 3-phenyl-1-propanol through fluorine treated and untreated 15-cm^3 Boston round bottles, which were injection blow molded from low-density polyethylene (LDPE). Using a weight loss technique, it was concluded that the fluorine treatment alone improved the barrier properties, especially at elevated temperatures. Irradiation of the bottles at the 5.0- and 7.5-Mrad level also improved their barrier properties; irradiation at the 2.5-Mrad level improved their barrier properties except for untreated bottles that were stored at 50°C. Results are analyzed based on the molecular structure of the LDPE with and without fluorine surface treatment.

KEY WORDS: low-density polyethylene, permeability, opthalmic solution preservatives, fluorine surface treatment, gamma radiation, Cobalt-60 decay, sterilization, benzyl alcohol, phenylethyl alcohol, 3-phenyl-1-propanol

Introduction

Permeability

The use of plastics, especially polyethylene, in packaging for pharmaceuticals is continually increasing. An important factor that must be considered in the selection of a packaging material is its permeability to components of the product and the environment. Polyethylene has a relatively low permeability for water vapor and polar organic vapors, but a relatively high permeability for oxygen and nonpolar organic vapors. Many studies, only the most comprehensive of which have been cited, have been done to determine the transmission rates of organic compounds through polyethylene films and containers [*1–5*]. The general conclusion of these studies is that the permeability of organic compounds through polyethylene decreases as the size and polarity of the penetrant increase, with alcohol having about the lowest permeabilities. In the last study cited, it

[1] Research scientist, Colgate-Palmolive Company, Research and Development Division, 909 River Road, Piscataway, NJ 08854.

[2] Associate director, Center for Packaging Engineering, Rutgers—The State University of New Jersey, Piscataway, NJ 08854.

was found that benzyl alcohol and benzoic acid had the lowest permeability coefficients of the six structurally related aromatic compounds studied. This was attributed to the intermolecular hydrogen bonds that produced dimers and trimers whose permeabilities were restricted because of their size. Other studies have been performed to determine the solubility of organic compounds in polyethylene and the degree of polymer/product interactions in pharmaceutical containers. Although related to permeability, no effort has been made in this paper to analyze the results of those studies. However, it is important to note that many organic molecules have been shown to swell polyethylene, increasing the permeability of both the swelling agent and other molecules that are present in the product or the environment.

Preservatives are often added to opthalmic solutions. Many of these preservatives are organic compounds that can permeate through the walls of a plastic container. It is, therefore, important to measure their transmission rates through a candidate container in order to determine its shelf life. Three compounds that were used as opthalmic preservatives in a study by Russell and Stock [*6*] were benzyl alcohol ($C_6H_5CH_2OH$), phenylethyl alcohol ($C_6H_5CH_2CH_2OH$), and 3-phenyl-1-propanol ($C_6H_5CH_2CH_2CH_2OH$). This is a homologous series of organic compounds that differ only by the number of methyl groups between the phenyl and hydroxyl groups. The percentage of nonpolar groups differs, as does their size. Benzyl alcohol is the smallest and least nonpolar; 3-phenyl-1-propanol is the largest and most nonpolar. The first objective of the current study was to evaluate the effects of size and polarity on the permeation rates of these three compounds through polyethylene containers.

Fluorine Surface Treatment

Efforts to improve the barrier properties of polyethylene to organic compounds have led to the development of fluorine surface treatments in which the surface of polyethylene is changed from nonpolar to polar by replacing some of the hydrogen atoms in the polyethylene chain with fluorine atoms. Pinsky et al. [*7*] tested the permeability of nine organic chemicals and six commercial products through fluorine treated containers. They found a decrease in the permeability of nonpolar penetrants but no improvement for polar penetrants. There was no change in physical properties except for an improvement in ink receptivity.

Two methods of fluorine treatment of polyethylene containers are in commercial use today. The process developed by Air Products and Chemicals, Inc. uses fluorine in the blow gas during the blow molding process to treat the inner surface of the containers. The process developed by the Linde Division of the Union Carbide Corporation post-treats the molded containers in a mixture of fluorine and nitrogen, treating both the inner and outer surfaces. The use of the Linde process has been approved for food contact applications.

The reaction that takes place on the surface of the polyethylene involves the replacement of one hydrogen atom on every other carbon atom in the chain's backbone with a fluorine atom, creating a polar, cross-linked surface on the nonpolar polyethylene. This surface chemically resembles polyvinyl fluoride. The reaction usually occurs to a depth of 2×10^{-8} to 4×10^{-8} m (200 to 400 Å). The second objective of the current study was to determine the effect of fluorine treatment of polyethylene containers on the permeation rates of polar compounds that had different degrees of polarity.

Sterilization

It has been shown [*8–11*] that the transmission rates of certain volatile components of opthalmic solutions autoclaved in polyethylene containers that were marketed as being autoclavable were high when compared to the same solutions autoclaved in glass containers. The high transmission rates were a combined result of the choice of packaging material and sterilization method. The morphology of the polyethylene containers were modified during exposure to high temperatures experienced during autoclaving. One technique that will decrease permeation rates during autoclaving (processing in an atmosphere similar to the concentration of the volatile components of the product) is not readily adaptable to production practices. Because of these unacceptable permeation losses during autoclaving, most opthalmic solutions are currently aseptically filled into ethylene oxide sterilized polyethylene containers.

The pharmaceutical and medical device industry has been replacing autoclaving (the heat can cause deformation, degradation, and morphological changes) and ethylene oxide sterilization (ethylene oxide is toxic, explosive, and readily absorbed by the polymer) with the use of high energy electrons (accelerated) and gamma rays produced from Cobalt-60 decay for sterilization of products and packages. Radiation doses are usually in the range of 1 to 5 megarads (Mrad). Cobalt-60 produces gamma rays of 1.14 MeV, and accelerated electrons produce energy between 0.25 and 10 MeV. This sterilization process allows the packaged product to be sterilized in one step after filling (post-sterilization). However, all plastics are affected to some extent by ionizing radiation. Alterations in the molecular structure of the polymer can cause physical and chemical changes, sometimes increasing or decreasing permeability coefficients. The size of the dose, the environment of the sample during irradiation, and the nature of the plastic will affect the outcome. The two major mechanisms operative are chain scission, a random rupturing of bonds that reduces the molecular weight of the polymer, and cross-linking, which forms primary bonds between polymer chains to produce a three-dimensional network structure. Usually, both mechanisms take place, but one will predominate. In general, cross-linking will predominate in polymers with carbon-carbon chain backbones if one or more carbons have hydrogens attached to them. When tetra-substituted carbons are involved, scission is predominant. It is also generally assumed that repeated sterilization is cumulative and additive. Since sterilization may have an effect on the packaging material, it is possible that the unsterilized package may adequately protect a product, but the sterilized package may not provide the required barrier properties. Thus, knowledge of the effects of radiation on materials is critical in the selection of packages and sterilization methods for products that must be sterile when used by the consumer.

Low- and high-density polyethylene are relatively resistant to single dose radiation sterilization and can withstand doses up to 100 Mrad without substantial changes in their mechanical properties [*12*]. The predominant mechanism is cross-linking. Gases liberated during irradiation are composed of 98% hydrogen. Sobolev et al. [*13*] found that doses of 10 Mrad did not affect the permeability of polyethylene films to nitrogen, oxygen, carbon dioxide, or methyl bromide. A 50% drop in permeability occurred after a dose of 100 Mrad. The results were linked to a reduction in the diffusion coefficient when the polymer was highly cross-linked during irradiation. Bent [*14*] studied the permeation of twelve organic solvents through 4-oz. (118-cm^3) polyethylene bottles with and without irradiation (26-Mrad dose). The testing was performed at 21, 38, 54, and 74°C. His results showed that irradiation caused a small increase in the permeability of penetrants with high to medium polarity.

Polyvinyl fluoride (PVF) is less sensitive to radiation than is polytetrafluoroethylene (PTFE), which undergoes chain scission, liberating fluorine atoms that break more carbon-carbon bonds, causing degradation at a low dose. PVF tends to cross-link on exposure and is capable of multiple sterilization. Since the surface of the fluorine treated containers in this study was similar to polyvinyl fluoride, and the remainder of the container was polyethylene, it would be assumed that cross-linking should be the predominant mechanism occurring during the irradiation of the containers. This should lead to a decrease in permeation rates, if enough cross links are formed. The third objective of the current study was to determine the effect of gamma radiation on the permeability of untreated and fluorine surface treated polyethylene bottles.

Experimental Procedures

Materials

The 15-cm[3] Boston round bottles were blow molded by the Dougherty Brothers Company, Buena, NJ, from Alathon® low-density polyethylene resin (Alathon® is a registered trademark of E. I. du Pont de Nemours and Company, Inc.). Polyethylene plugs and polystyrene screw caps were also supplied by the Dougherty Brothers Company. Half of the bottles were fluorine treated by the Linde Division of the Union Carbide Corporation at their pilot plant in Keasby, NJ, using their SMP-fluorination process at standard treating conditions. Equal numbers of empty treated and untreated bottles were separated into three groups and sent to the Ethicon Corporation for Cobalt-60 irradiation sterilization. Each group was exposed to 1, 2, or 3 passes of 2.5-Mrad radiation. The plugs and caps were neither fluorine treated nor irradiated.

The penetrants used in this study were benzyl alcohol (BA), phenylethyl alcohol (PE), and 3 phenyl-1-propanol (PP). They were supplied by the Eastman Kodak Company.

Specimen Preparation

All bottles were filled with 15 mL of the penetrant to be studied. The plug was inserted, and a hot soldering iron was used to close the orifice at the end of the plug and to fuse the interface between the plug and the bottle neck finish. The cap was screwed on and hand tightened. Devcon 5 minute epoxy was used to seal the edge of the cap on the neck finish of the bottle. This procedure was used to minimize leakage through the closure, since permeation through the bottle walls was the only mechanism of interest. At least four specimens were tested for each combination of surface treatment/sterilization treatment/permanent/storage condition.

Weight Loss Experiments

The specimens were stored either in the laboratory at approximately 22°C and approximately 40% relative humidity (RH) or in a convection oven at 50°C and approximately 10% RH. Since the concentration of the permeant chemicals was negligible in the environment, the partial pressure differential driving force caused the chemicals to permeate out of the containers, resulting in weight loss. Each bottle was weighed periodically on a Sartorius analytical balance to the nearest 0.1 mg. Weighings were discontinued when a steady state transmission (weight loss) rate could be determined from the linear portion of a plot of container weight versus time.

Bottles of each type were also filled with dessicant (indicating Drierite, calcium sulfate obtained from the Fischer Scientific Company), closed, sealed, and tested by the same weight change method. In this case, the steady state transmission (weight gain) rate of water vapor from the storage environment into the containers was measured to serve as a control.

Calculations

The calculations were based on the ASTM Test for Permeability of Thermoplastic Containers to Packaged Reagents or Proprietary Products (D 2684). Container weight in grams was plotted versus time in days. Linear regression techniques were used to determine the slope of the line that best fit the points for each bottle tested. The slope gave the transmission rate of the permeant through the container in milligrams per day. The average transmission rate caused by water entry into the corresponding control condition was subtracted from the transmission rate of each bottle containing penetrant to give the corrected transmission rate for that specimen.

Results and Discussion

Initial studies showed weight gains, instead of weight losses, in some bottles containing the organic compounds. This phenomenon has been observed in other studies [*1,7*]. It was attributed to entry of water molecules into the containers, since the concentration of water vapor inside the containers was lower than that in their environment. The alcohols being studied easily form hydrogen bonds with water molecules, allowing water to dissolve into the alcohol. In essence, these alcohols acted as desiccants. Because of this phenomenon, experiments were set up to measure the water vapor transmission rates (WVTR) of the containers. The results of those tests are shown in Fig. 1. The WVTR's of the irradiated containers were significantly lower than those for the containers that were not irradiated; however, there was no effect on the WVTR when the containers were fluorine treated. This would be expected, since polyethylene is nonpolar and, therefore, a good water vapor barrier. Adding a polar surface should not provide a better barrier to entry of polar water molecules. Also, the WVTRs were unchanged by multiple exposures to gamma radiation for the range of radiation levels studied; once the initial cross-linking had occurred, any additional cross-linking that may have occurred after repeated exposure to radiation had no effect on the diffusion of small water vapor molecules. The WVTRs at 50°C were slightly less than or equal to those at 22°C. The relative humidity of the environment of the packages inside the oven (10% RH, 50°C) was significantly lower than that in the laboratory (40% RH, 22°C). The expected increase in transmission rate caused by the increased storage temperature was almost exactly balanced by the decrease caused by the lower driving force for water vapor transmission inside the oven.

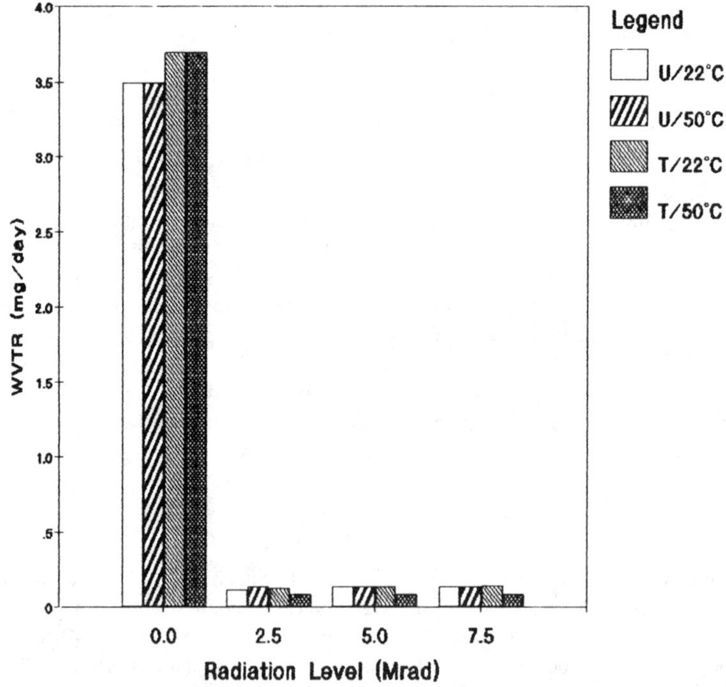

FIG. 1—*Water vapor transmission rate into the container.*

The results of measuring the transmission rates of BA, PE, and PP through the containers are shown in Figs. 2 to 4. These charts give the average transmission rates (BATR, PETR, and PPTR) of the containers after correcting for water vapor entry into the containers. In order to clarify the data for the samples with very low transmission rates, these charts were replotted with expanded ordinates in Figs. 5 to 7. When the average water vapor transmission rate into the containers was subtracted from the transmission rate of the organic compound out of a container for a given surface treatment/sterilization treatment/storage temperature combination, weight gains were still found in a few containers (shown as negative transmission rates out of the containers in the charts). It was assumed that the organic chemicals plasticized the polyethylene containers, allowing a higher WVTR into the plasticized containers filled with organic chemicals than into the unplasticized containers filled with desiccant. Even with the very small weight gains in some containers, the trends in the permeability rates of the organic compounds are obvious from this study. All of the observed effects were magnified by storage at 50°C.

If the progression from BA to PE to PP is studied, the size of the molecule increases while the polarity decreases. At 50°C the trend in transmission rates was obviously BATR > PETR > PPTR for the untreated containers, except for the containers irradiated at 2.5 Mrad. The same trend was found, but to a lesser extent, in 9 of the other 15 combinations of surface treatment, sterilization treatment, and storage temperature. This leads to the conclusion that, in general, size was more important than polarity for the systems studied; the transmission rates decreased as the size of the penetrant was increased. However, except at 50°C for the unirradiated, untreated

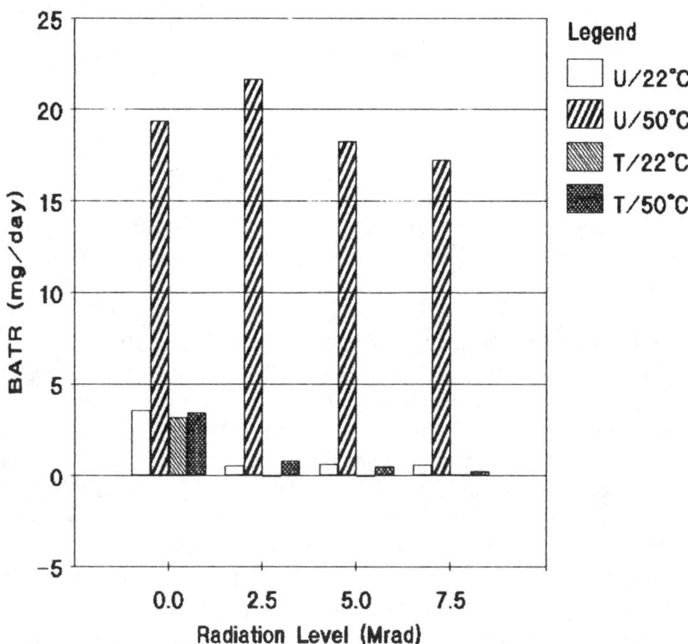

FIG. 2—*Benzyl alcohol transmission rate out of the container.*

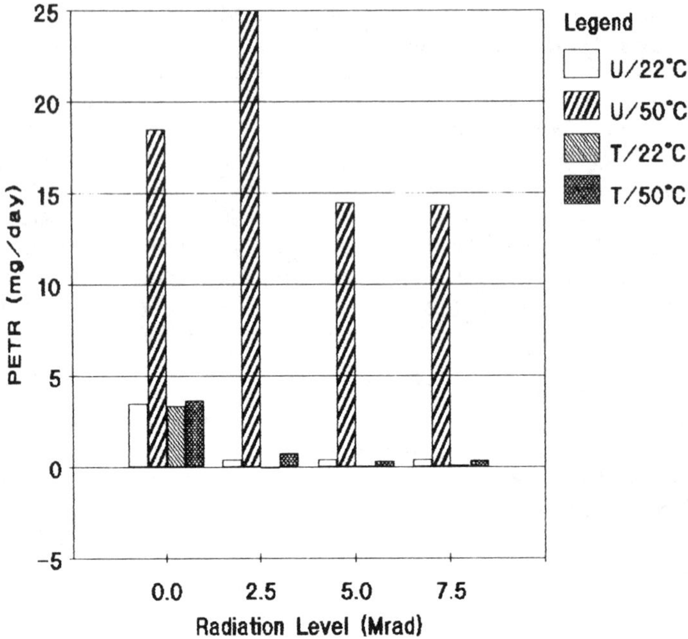

FIG. 3—*Phenylethyl alcohol transmission rate out of the container.*

containers, the results were closer to BATR ≈ PETR ≈ PPTR. The two effects (size and polarity) were actually approximately equal. The expected decrease in permeation rate caused by increased penetrant size was almost exactly balanced by the expected increase in permeation rate caused by decreased penetrant polarity.

At 50°C the fluorine treatment alone reduced permeabilities significantly (by factors of approximately 6 for BA, 5 for PE, and 4 for PP). Besides being polar, the fluorinated surfaces were also cross-linked, inhibiting the permeability of all relatively large organic molecules to approximately equal extents. Swelling, which is accompanied by increased permeation rates, was reduced by having the cross-linked fluorinated layer on the inside of the containers. The solvents could not dissolve into the container in large enough quantities to significantly swell the material. At 22°C the fluorine treatment reduced the permeability of BA by 11%, PE by 4%, and PP by 6%. The effect of swelling was negligible at room temperature, and adding a polar surface would not be expected to decrease permeation rates of polar compounds. Actually, fluorination reduced all transmission rates to approximately the same value, that is all samples had transmission rates between 3.1 and 3.6 mg/day for both 22 and 50°C.

The permeation rates of the irradiated bottles (treated and untreated) stored at 22°C (−0.10 to +0.58 mg/day) were significantly lower than those of the unirradiated bottles. The fluorinated surface was not adversely affected by irradiation. Just as for water vapor transmission, these data indicate that cross-linking was the predominant mechanism that occurred during irradiation, which, in turn, reduced the permeability of the containers to organic compounds. At 50°C the transmission

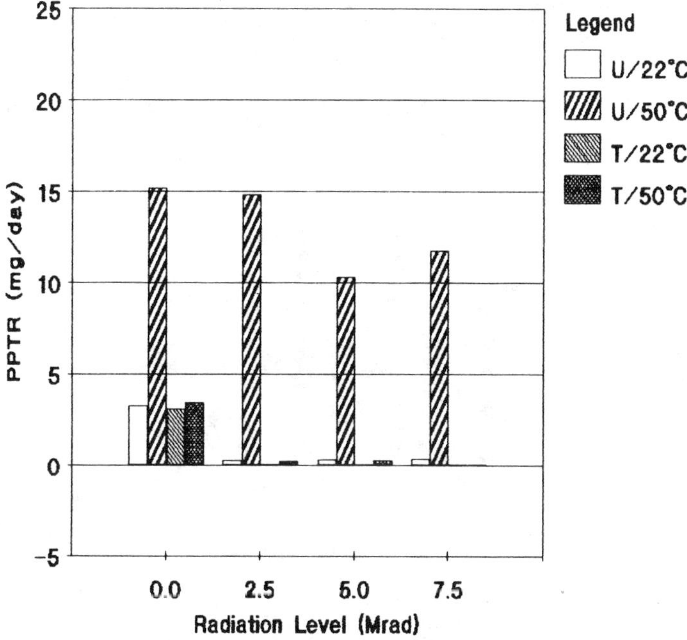

FIG. 4—*3-phenyl-1-propanol transmission rate out of the container.*

rates of the untreated specimens, which were irradiated at 2.5 Mrad, were equal to or higher than those of the unirradiated specimens for the untreated containers.

Even if a small amount of cross-linking had been initiated (10 Mrad only causes formation of approximately one cross-link in every 600 monomer units of polyethylene [*13*], it was not enough to prevent swelling. Also, although the predominant mechanism was cross-linking, some chain scission also occurred (Charlesby and Pinner [*15*] obtained a ratio of chain scission to cross-linking of 0.3 for polyethylene). The effects of this chain scission, slight decreases in density and percent crystallinity in addition to a decrease in molecular weight, were emphasized at elevated temperatures. Bent [*14*] explained this phenomenon by diffusion induced melting (disorder) at elevated temperatures, where organic molecules cause swelling and breaking down of crystallites.

Except for PP, further irradiation (total of 5.0 Mrad or 7.5 Mrad) of the fluorine treated containers decreased the transmission rates of the containers stored at 50°C by 50 to 75% when compared to either the 0.0-Mrad or the 2.5-Mrad samples. Additional cross-links were formed with each 2.5 Mrad dose. For PP, the most nonpolar of the penetrants studied, swelling of the nonpolar polyethylene occurred even after 5.0 Mrad. It was not until exposure to 7.5 Mrad that sufficient cross-links were formed to minimize swelling. Because of the very low transmission rates of both the treated and untreated specimens at room temperature, this effect was not noticable; further irradiation either resulted in no change in or a slight increase in transmission rate. At 50°C, further irradiation decreased the transmission rates by 20 to 40%; enough cross-links were formed that swelling was limited.

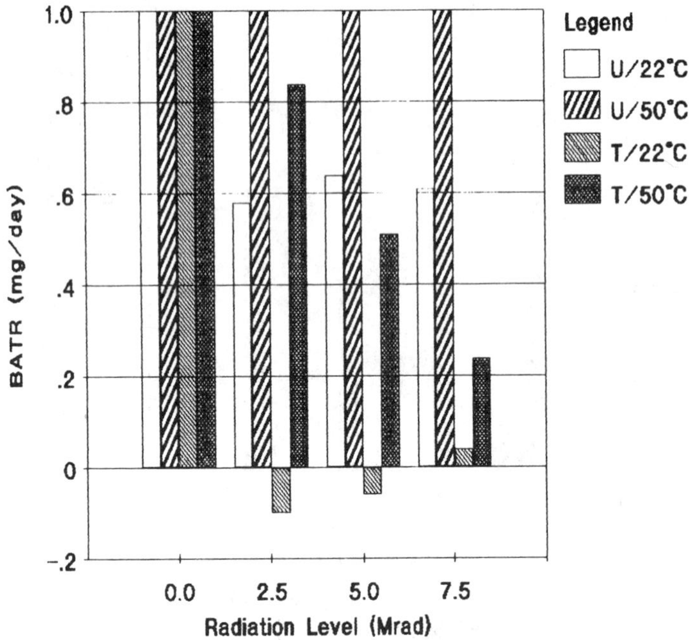

FIG. 5—*Benzyl alcohol transmission rate out of container.*

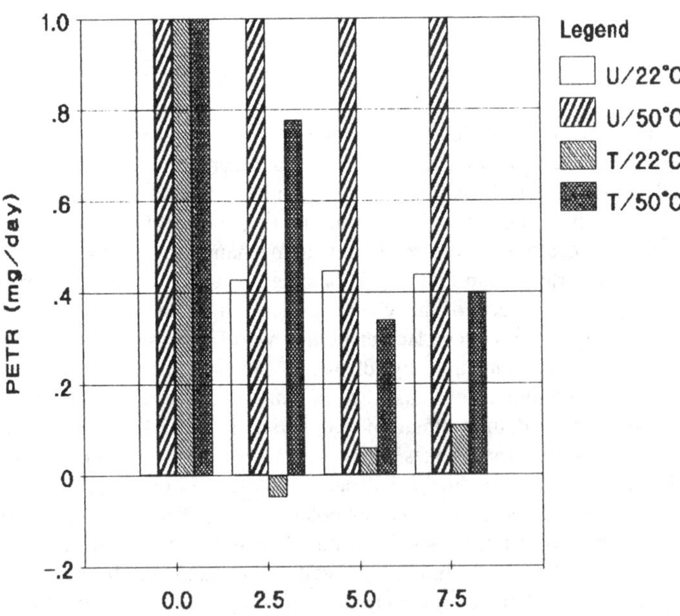

FIG. 6—*Phenylethyl alcohol transmission rate out of the container.*

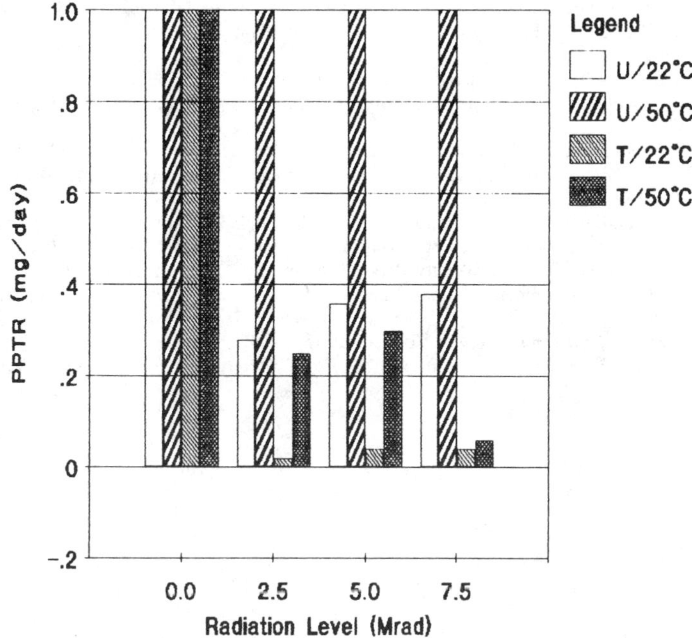

FIG. 7—*3-phenyl-1-propanol transmission rate out of the container.*

Conclusions

In general, penetrant size was more important to transmission rate than penetrant polarity. The transmission rates decreased as the size of the penetrant was increased. This phenomenon was most obvious for the untreated containers that were stored at 50°C. Fluorine surface treatment of low-density polyethylene improved its barrier properties to benzyl alcohol, phenyl ethyl alcohol, and 3 phenyl-1-propanol, especially at 50°C. Subsequent irradiation of the fluorine treated containers at levels of 2.5, 5.0, or 7.5 Mrad did not adversely affect the barrier properties and, in fact, improved the barrier properties in most cases by combining a cross-linked, polar surface with additional radiation initiated cross-linking. Irradiation of the untreated containers also improved their barrier properties, except when they were stored at 50°C.

Acknowledgments

This paper is based on the M. S. thesis of the same title by Dorinda A. Sparacio, Mechanics and Materials Science Department, Rutgers, The State University of New Jersey, New Brunswick, NJ, May 1984.

The authors wish to thank the Dougherty Brothers Company for providing sample containers, the Linde Division of the Union Carbide Corporation for fluorine surface treating the containers, and the Ethicon Corporation for irradiating the containers.

References

[*1*] Pinsky, J., Nielsen, A., and Parliman, J., *Modern Packaging,* Vol. 28, No. 2, Oct. 1954, p. 145.

[*2*] Pinsky, J., *Modern Plastics,* Vol. 34, No. 4, April 1957, p. 145.

[*3*] Salame, M. and Pinsky, J., *Modern Packaging,* Vol. 36, No. 3, Nov. 1962, p. 153.

[*4*] Gonzales, M., Nematollahi, J., Guess, W., and Autian, J., *Journal of Pharmacological Science,* Vol. 56, 1967, p. 1288.

[5] Rogers, C. E., Stannett, B., and Szwarc, M., *Tappi Monograph 23,* 1962, p. 78.
[6] Russell, J. and Stock, B. H., *Australasian Journal of Pharmacy,* Supplement 40, Vol. 47, 1966, p. 537.
[7] Pinsky, J., Adonkis, A., and Nielsen, A., *Modern Packaging,* Vol. 33, No. 6, Feb. 1960, p. 130.
[8] Pollack, A. E., Roberts, M. S., and Schumann, F., *American Journal of Hospital Pharmacy,* Vol. 27, 1970, p. 638.
[9] Goss, J., Gregerson, P., and Pollack, A. E., *American Journal of Pharmacy,* Vol. 25, 1968, p. 346.
[10] Russell, J. and Stock, B. H., *Australasian Journal of Pharmacy,* Supplement 40, Vol. 47, 1966, p. 537.
[11] Busse, M. and Hughes, D., *Pharmacological Journal,* Vol. 338, Sept. 1968.
[12] Plester, D., ''The Effects of Radiation Sterilization on Plastics,'' in *Industrial Sterilization,* G. B. Phillips and W. S. Miller, Eds., Duke University Press, Durham, NC, 1973, p. 141.
[13] Sobolev, I., Meyer, J., Stannett, V., and Szwarc, M., *Journal of Polymer Science,* Vol. 17, 1955, p. 417.
[14] Bent, H., *Journal of Polymer Science,* Vol. 24, 1957, p. 387.
[15] Charlesby, A. and Pinner, S., *Proceedings of the Royal Society (London),* Series A, Vol. 249, 1959, p. 367.

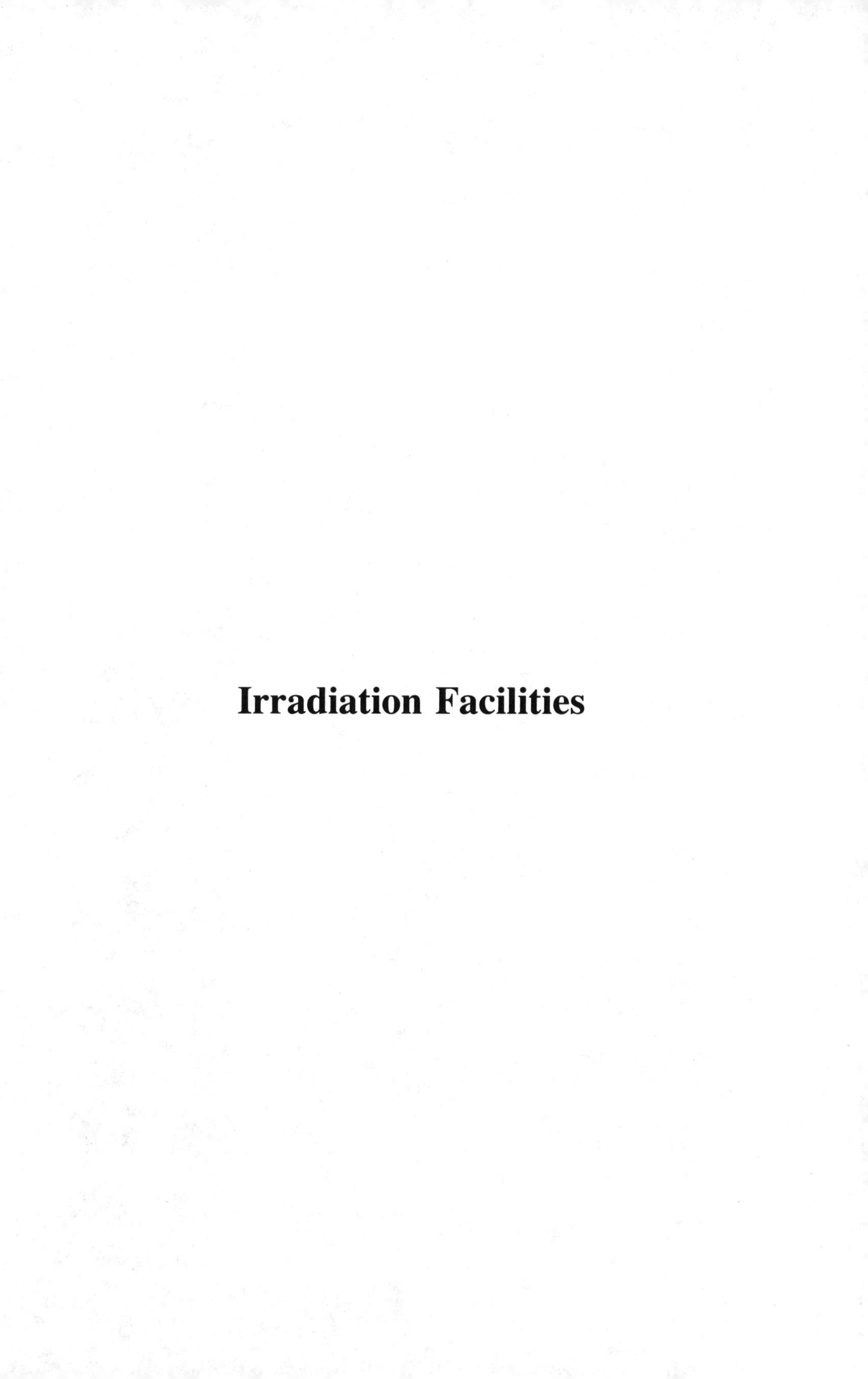

Irradiation Facilities

H. R. Kerchner,[1] R. R. Coltman, Jr.,[2] C. E. Klabunde,[1] and F. W. Young, Jr.[1]

The Low-Temperature Neutron Irradiation Facility at ORNL

REFERENCE: Kerchner H. R., Coltman, R. R., Jr., Klabunde, C. E., and Young, F. W., Jr., **"The Low-Temperature Neutron Irradiation Facility at ORNL,"** *Influence of Radiation on Material Properties: 13th International Symposium (Part II), ASTM STP 956,* F. A. Garner, C. H. Henager, Jr., and N. Igata, Eds., American Society for Testing and Materials, Philadelphia, 1987, pp. 701–704.

ABSTRACT: The Low-Temperature Neutron Irradiation Facility (LTNIF) is now operating at Oak Ridge National Laboratory. The facility provides high radiation intensities and special environmental and testing conditions for qualified experiments at no cost to users. A general description and major specifications of the LTNIF are presented along with the results of performance tests. In addition, the hardware and other considerations required to perform experiments in the LTNIF are described.

KEY WORDS: radiation effects, neutron irradiation, nuclear reactors, cryogenics, user facility

In May 1983 the Division of Materials Sciences, Office of Basic Energy Sciences of the Department of Energy authorized the establishment of the Low-Temperature Neutron Irradiation Facility (LTNIF) at the Bulk Shielding Reactor (BSR) in the Oak Ridge National Laboratory (ORNL). Work on the design and construction of the facility proceeded from that time until it was completed in February 1986. (Work on auxiliary equipment continues.) This facility is available for qualified experiments at no cost to users.

The LTNIF provides a combination of high radiation intensities and special environmental and testing conditions that have not previously been attainable. Irradiations are possible at temperatures between 3.2 K and 800 K. The facility has been optimized initially for a large flux of fast neutrons (2×10^{17} neutrons n/m^{2}s, $E > 0.1$ MeV) at moderate gamma-ray intensity. Radiation modifiers will be constructed as needed to provide modified fast neutron spectra, thermal neutrons, or gamma rays. This facility is expected to be used for a wide variety of experiments on radiation effects in materials. The only other currently operating facilities for reactor irradiations at low temperature are at the Kyoto University Reactor (KUR) and at the Munich Research Reactor (FRM). The KUR facility provides 3×10^{14} n/m^{2}s at temperatures near 10 K.[3] The FRM facility characteristics are similar to what is described here except for smaller dimensions of its irradiation and test chambers.[4]

Bulk Shielding Reactor

The core of the BSR is a rectangular array of square fuel elements resting upon a grid plate, which is supported by a bridge and carriage assembly. This assembly allows the core to be

[1] Research staff members, Solid State Division, Oak Ridge National Laboratory, Oak Ridge, TN 37831.

[2] Retired research staff member, Solid State Division, Oak Ridge National Laboratory, Oak Ridge, TN 37831.

[3] T. Shibata, I. Iwata, H. Yoshida, M. Nakagawa, and M. Okada, "Some Aspects of Low Temperature Irradiation Facility of Kyoto University Reactor," Annual Reports of Research Reactor Institute Kyoto University, Vol. 2, 1969, pp. 89–92.

[4] N. Riehl, W. Schilling, and H. Meissner, *Research Reactor Journal,* Vol. 3, No. 1, Oct. 1962, pp. 9–13.

positioned at almost any location in a 6 × 12 m open pool of demineralized water. The exposed core is fully accessible on three faces. The reactor can be operated for indefinite periods at any power up to 2 MW. Primary control of the BSR is available for the operation of the LTNIF.

Two features of the BSR will aid LTNIF experimenters. First, the core fuel loading provides sufficient excess reactivity to permit restarting after any operating history—xenon poisoning is not an operational limitation. This is an asset for experiments that require frequent on-off reactor operations. Second, the open pool provides convenient locations for ambient-temperature and cryogenic storage and test facilities for radioactive specimens. These facilities are located close to the irradiation cryostat, and they make use of pool water for biological shielding. A layout of the LTNIF is shown in Fig. 1.

Irradiation Cryostat

The irradiation cryostat sketched in Fig. 2 is capable of cooling specimens to temperatures as low as 3.2 K during irradiation and testing. At 5 K, 20 W of nuclear heating (generated in about 50 g of metal at full reactor power) can be removed from experiment assemblies during irradiations. The irradiation chamber is a 37-mm-diameter tube inside the lower, rectangular parallelepiped section of the cryostat, which nests inside a U-shaped gamma-ray shield in place of a missing fuel element. Just above the irradiation chamber is a 197-mm-diameter test chamber for *in situ* testing

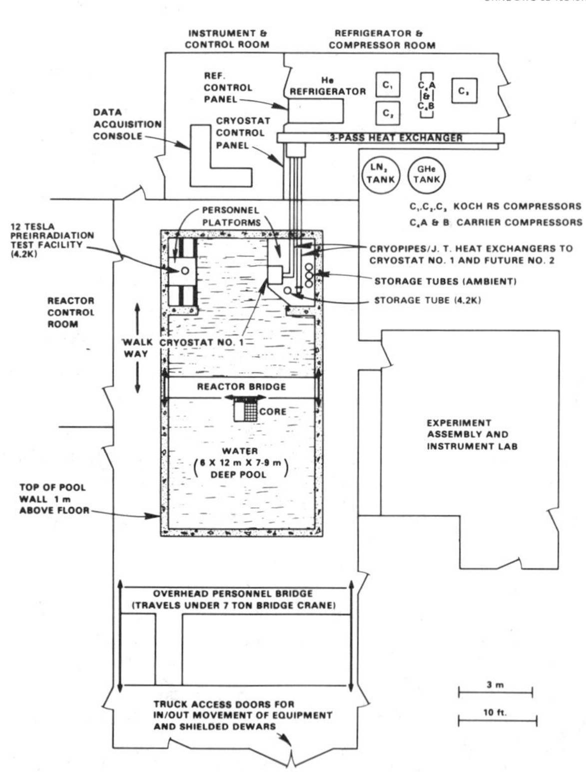

FIG. 1—*Layout plan for the LTNIF at ORNL's Bulk Shielding Reactor.*

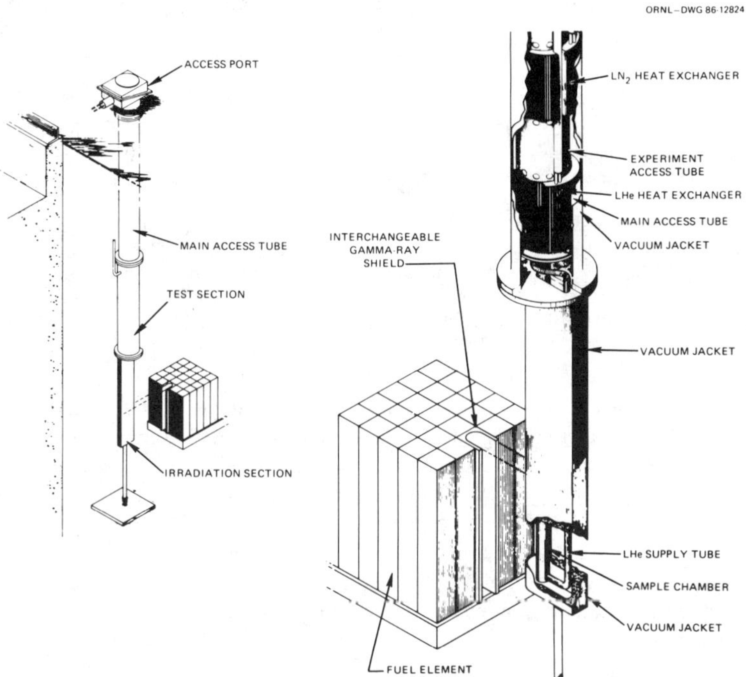

FIG. 2—*Irradiation cryostat.*

of irradiated samples. Liquid helium and liquid nitrogen are delivered by the refrigeration system to heat exchangers surrounding the test chamber. Depending upon the desired irradiation temperature, gaseous or liquid (condensed on the helium heat exchanger) helium cools the sample by circulating through the irradiation chamber in a natural convection loop. Because the test chamber is outside the irradiation zone, specimens can be irradiated to high fluences and then tested in the absence of nuclear heating, using unirradiated test devices that may be sensitive to radiation (for example, a superconducting magnet).

The configuration described above optimizes the fast neutron flux at moderate gamma and thermal-neutron flux. Preliminary measurements of the radiation characteristics are listed in Table 1. It is expected that irradiations with highly thermalized neutrons and other modified spectra will be of interest to experimenters. A useful aspect of the open-pool construction of the BSR is the ability to move the core and insert radiation modifiers easily into and out of position between the fixed cryostat and a core face.

Other Features of the LTNIF

Although the design and construction of experiments are primarily the responsibility of the user, certain auxiliary equipment is on hand and available for experimenters' use. A computer-based, on-line data acquisition system is provided with hard-wired electrical terminals available at the cryostat top. Data can be tabulated, plotted, or transmitted to a distant computer over telephone lines for later analysis. Provisions are available for transfer of irradiated specimens at 4.2 K into the user's test device or into vessels for shipment to other laboratories. Temporary storage at 4.2 K or 308 K of radioactive samples or experiment assembles is available in the BSR pool. Short-

TABLE 1—*Preliminary radiation characteristics at the full BSR power of 2 MW.*

Radiation	Intensity
Fast neutrons (E > 0.15 MeV)	1.4×10^{17} n/m^{2}s[a]
Thermal neutrons	2.8×10^{17} n/m^{2}s[a]
Gamma rays (in Al)	0.33 W/g[b]

[a] Measured in the cryostat at ambient temperature and analyses by L. R. Greenwood (Footnote 5).

[b] The nuclear heating of an aluminum specimen was measured, and the heating by fast neutrons and by the decay of aluminum nuclei activated by the neutron flux was subtracted.

term storage is often essential to allow decay of short lived radioactive isotopes that may be present. A 12 T superconducting magnet is on hand and will soon be mounted for use either in a poolside experiment dewar or in the test section of the irradiation cryostat. An ambient temperature irradiation facility (ATNIF) is available for irradiation at temperatures from ambient up to a maximum of about 800 K. A 200 kV electron microscope is being procured, and a liquid helium stage and transfer device to move specimens from the irradiation cryostat to the microscope without warmup is under design. After completion of this microscope facility it will be possible to obtain direct microscopic structural information about radiation-induced defects and their thermal annealing as it occurs in the microscope.

Requirements for Experiments

The difficulty of carrying out experiments will vary greatly depending upon their complexity. The first experiments should be as simple as reasonable; for example, cryogenic irradiation with the specimens being tested at room temperature in the user's laboratory. The LTNIF staff will suspend the user's specimens in an aluminum wire basket for the irradiation, remove them and store them for the user in such experiments. When *in situ* testing is desired, the user will need to construct a test assembly that can be suspended in the test chamber and a coaxial nesting specimen holder that can be lowered to the irradiation zone for irradiation and raised to the test chamber for testing. Consultation with the LTNIF staff will be required for proper design of experiment and test assemblies. The LTNIF staff will also assist users in planning for low temperature transfer to cryostats in other major facilities (for example, for characterization of irradiated samples by neutron diffraction).

All experiment proposals must be reviewed both for scientific merit (by a user's committee) and for reactor safety (by ORNL staff members). A proposal must include a brief description of the basis for the experiment and certain crucial technical details. Proposal forms are available on request. Prospective users should contact one of the authors.

Acknowledgments

This work is sponsored by the Division of Materials Sciences, U.S. Department of Energy under Contract DE-AC05-840R21400 with Martin Marietta Energy Systems, Inc.

[5] L. R. Greenwood, private communication.

Virendra P. Gupta,[1] James S. Herring,[1] Robert E. Korenke,[1] and Yale D. Harker[1]

Irradiation Facilities at the Idaho National Engineering Laboratory

REFERENCE: Gupta, V. P., Herring, J. S., Korenke, R. E., and Harker, Y. D., "**Irradiation Facilities at the Idaho National Engineering Laboratory,**" *Influence of Radiation on Material Properties: 13th International Symposium (Part II), ASTM STP 956,* F. A. Garner, C. H. Henager, Jr., and N. Igata, Eds., American Society for Testing and Materials, Philadelphia, 1987, pp. 705–717.

ABSTRACT: Although there is a growing need for neutron and gamma irradiations by governmental and industrial organizations in the United States and in other countries the number of facilities providing such irradiations are limited. At the Idaho National Engineering Laboratory, there are several unique irradiation facilities producing high neutron and gamma radiation environments. These facilities could be readily used for nuclear research, materials testing, radiation hardening studies on electronic components/circuitry and sensors, and production of neutron transmutation doped (NTD) silicon and special radioisotopes. In addition, a neutron radiography unit, suitable for examining irradiated materials and assemblies, is also available. This paper provides a description of the irradiation facilities and the neutron radiography unit as well as examples of their unique applications.

KEY WORDS: Idaho National Engineering Laboratory, reactors, intense gamma radiation fields, neutron radiography, NTD silicon, radiation hardening, material testing, isotope production

There is a growing need for high flux neutron and gamma irradiations by governmental and industrial organizations in the United States and in other countries. However, the number of facilities providing such irradiations are limited. The Idaho National Engineering Laboratory (INEL), located on the Snake River Plain in southeastern Idaho (Fig. 1), has been the site for about 50 testing and developmental reactors since 1949. There are several unique irradiation facilities at the INEL producing intense neutron and gamma radiation fields. These facilities include the Advanced Test Reactor (ATR), the ATR Critical Facility (ATRC), the Advanced Reactivity Measurement Facility(ARMF), the Coupled Fast Reactivity Measurement Facility (CFRMF), the Power Burst Facility (PBF), and the ATR Gamma Radiation Facilities. The CFRMF is also utilized as a source for a neutron radiograph system.

EG&G Idaho, the prime operating contractor at the INEL, is responsible to the U.S. Department of Energy (DOE) for operation of these facilities. These intense radiation field facilities could be readily used for research (for example, nuclear research, material testing, and radiation hardening studies), and for production of neutron transmutation-doped silicon and special radioisotopes. This paper presents a brief overview of these facilities, their capabilities, and unique applications.

[1] Scientific specialist, senior engineering specialist, senior program specialist, and scientific specialist, respectively, Idaho National Engineering Laboratory, EG&G Idaho, Inc., P.O. Box 1625, Idaho Falls, ID 83415.

FIG. 1—*Location of the INEL in relation to surrounding areas.*

Advanced Test Reactor Facilities

ATR Facility

The highest powered INEL reactor, the ATR, has been used for materials development and high-flux nuclear research since 1967. The ATR is a pressurized light-water reactor, uses highly enriched U-235 fuel, and is rated at 250 MW_{th}. Depending on experimental requirements, the normal operating power of the ATR is 100 to 170 MW_{th}. This reactor was primarily designed for use in developing advanced nuclear fuel systems and materials for power reactor programs. The ATR is fueled by 40 plate-type curved fuel elements. These 40 elements are arranged in a serpentine annulus between and around the nine flux-trap positions. Figure 2 shows a horizontal cross-sectional view of the ATR core and location of irradiation facilities.

The normal operating schedule of the ATR in support of reactor materials development consists of 15 to 30 days of operation, followed by 2 to 8 days of shutdown for replacement of experiments and refueling. When in operation, the reactor is at constant power 24 h/day. During the past six yeas, the ATR has averaged 75% on-stream time.

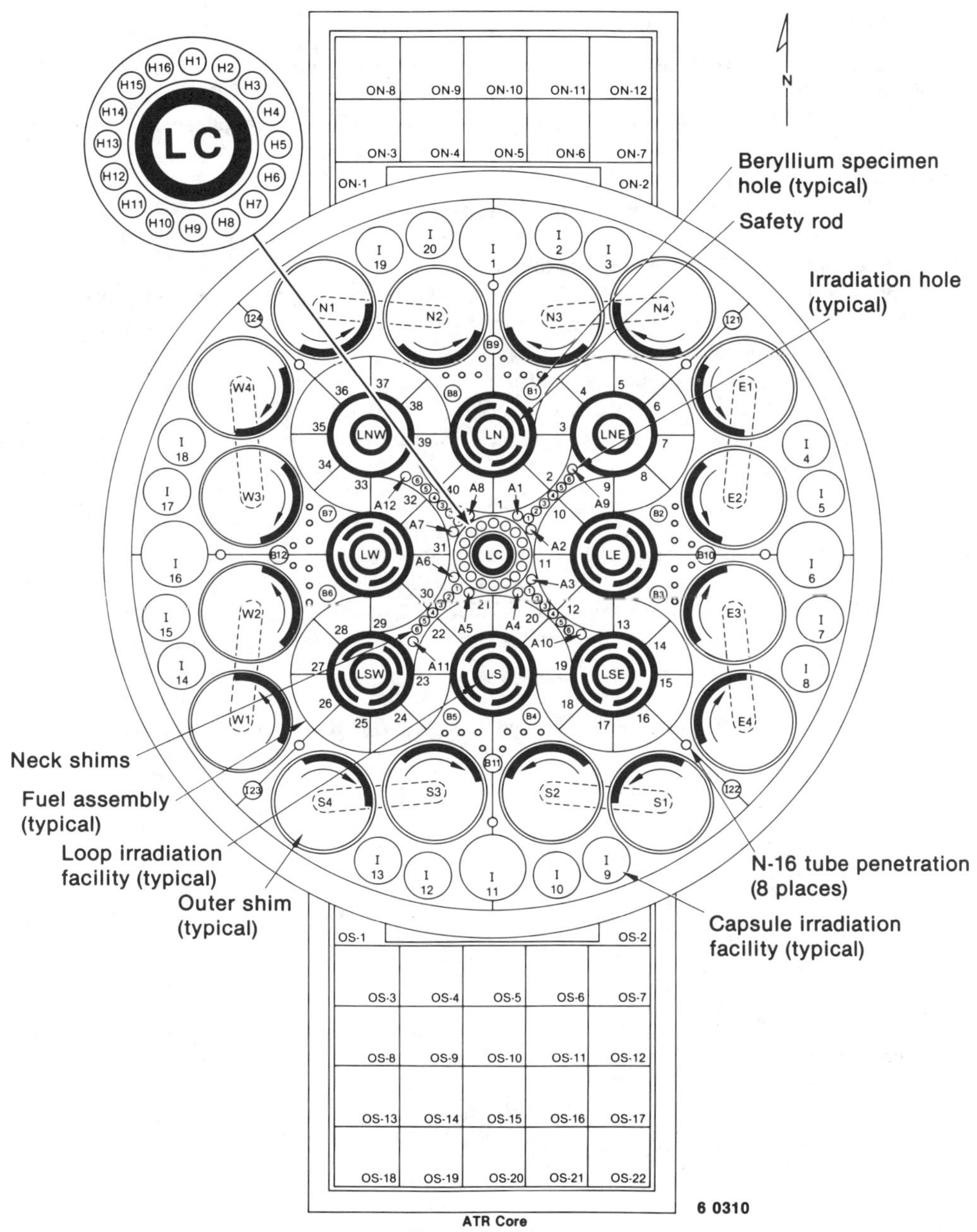

FIG. 2—*ATR horizontal cross-sectional view.*

ATR Applications

The experiment facilities consist of nine flux-trap positions (designated L in Fig. 2) designed to provide high neutron flux over the entire core length. Dozens of holes of various diameters are available in and near the ATR core for capsules. Many holes are contained in two-capsule irradiation tanks that are attached to the outside of the core-reflector tank that is, the OS—and ON—locations (Fig. 2).

Each of the nine flux-trap positions is occupied by an in-pile pressure tube assembly. Each in-pile tube is part of a pressurized, water-cooled test loop; a closed piping system provides bottom reentrant coolant flow through a pressure tube located within the reactor vessel. All nine in-pile loops are fully utilized, being dedicated to specific DOE research and development experiments. But the capsule irradiation positions in the reactor core and reflector are not fully occupied and are available for additional experiments.

Table 1 specifies diameters, peak neutron flux, and gamma heating values at various capsule positions shown in Fig. 2. In the capsule irradiation tanks, filler blocks with holes of various diameters are interchangeable in most of the locations. Large holes are available in the beryllium reflector. Small holes such as the "A" and "B" holes are used mainly for long-term irradiations of material specimens.

Capsules may be of any length up to 1.21 m. Capsules with instrument leads are typically about 0.6 or 0.9 m. Uninstrumented capsules are typically about 0.15 m and are stacked in aluminum tubes. If required, the temperature of the capsules is controlled with a flowing mixture of helium and argon. The unique capabilities (high-flux and large capsule holes) of the ATR can be readily adapted via hardware to produce various special radioisotopes and provide for material testing and nuclear research.

TABLE 1—*Approximate peak flux values for various ATR capsule positions at maximum allowable power, 250 MW.*

Position	Diameter, mm	Thermal Flux, n/cm² · s at 2200 m/s	Fast Flux, n/cm² · s $E_n > 1$ MeV	Gamma Heating, W/g · °C
B Holes				
B-1 thru B-8	22.22	5.75×10^{14}	1.85×10^{14}	14.5
H Holes				
H-1 thru H-16	15.87	4.28×10^{14}	3.83×10^{14}	19.0
A Holes				
A-1 thru A-8	15.87	4.28×10^{14}	3.83×10^{14}	19.0
A-9 thru A-12	15.87	4.64×10^{14}	5.25×10^{14}	20.0
I Holes				
large (I-1, I-6, I-11, I-16)	127.00	3.89×10^{13}	3.05×10^{12}	1.5
small (I-2, -3, -4, -5, -7, -8, -9, -10, -12, -13, -14, -15, -17, -18, -19, -20)	82.55	7.71×10^{13}	3.01×10^{12}	1.5
Outer Holes				
ON-4	[a]	9.85×10^{12}	2.82×10^{11}	0.34
ON-5	[a]	8.74×10^{12}	2.50×10^{11}	0.40
ON-9	[a]	3.83×10^{12}	0.88×10^{11}	0.16
OS-5	[a]	7.86×10^{12}	2.33×10^{11}	0.32
OS-7	[a]	7.30×10^{12}	2.52×10^{11}	0.26
OS-10	[a]	2.98×10^{12}	0.77×10^{11}	0.12
OS-15	[a]	1.24×10^{12}	0.27×10^{11}	0.045
OS-20	[a]	0.56×10^{12}	0.08×10^{11}	0.024

[a] Depending on reconfiguration of irradiation tank.

ATRC Facility and Applications

The Advanced Test Reactor Critical Facility (ATRC) is a low-power, full-size nuclear duplicate of the ATR, designed to test prototypical experiments before irradiation of the actual experiments in the ATR. The ATRC provides valuable reactor physics data, including (1) control element worths and calibrations, (2) excess reactivities, and thus charge lifetimes, (3) thermal and fast neutron flux distribution, (4) gamma heat generation rates, (5) fuel loading requirements, (6) effects of the insertion and removal of experiments and experiment void reactivities, and (7) temperature and void reactivity coefficients.

The ATRC is a swimming pool type reactor (Fig. 3) located in an extension of the ATR canal. A typical operating power level for the ATRC is 300 W. The ATRC core is identical to the core

FIG. 3—*ATRC facility with control bridge removed and working platform in place.*

of the ATR except that irradiation holes B-9 through 12 and I-21 through 24 do not exist in the ATRC. The outer irradiation tank facilities are also not present on the ATRC. Before each ATR cycle, the experiments proposed for ATR are measured in the ATRC or the reactivity effects are calculated based on prior benchmark measurements taken in the ATRC.

Of the various irradiation locations available in the ATRC, the nine flux-trap positions are specifically suitable for doping of silicon. Since ATRC is a DOE funded facility, the intent is not to compete with existing commercial and university producers of NTD silicon in the United States, but rather to exploit the ability of the ATRC to accommodate larger diameter ingots that are not available elsewhere.

ARMF and CFRMF Facilities

General Description

The Advanced Reactivity Measurement Facility (ARMF) and the Coupled Fast Reactivity Measurement Facility (CFRMF) are two very similar small reactors located at the Test Reactor Area at the INEL. They are both swimming-pool type reactors having light-water moderated core fueled by fully enriched U-235. The major test region in the ARMF is a 0.15 by 0.15-m central water hole, and the major test region in the CFRMF is a 0.15 by 0.15-m filter assembly designed to produce a fast neutron spectrum with essentially no thermal neutrons. These reactors were originally designed for precise reactivity worth mesurements on "small specimens," but their capabilities have been expanded beyond the original design concept. The reactor power is precisely maintained at a constant and preset level by means of a DC servo controlled regulating rod system. Both reactors can be operated at power levels up to 100 kW, and the CFRMF can also be used as a source for a neutron radiography facility. The ARMF can perform reactivity measurements on specimens, such as fuel pins and shim rods, where the length of such assemblies is greater than the length of the ARMF core. Neutron flux levels up to 8×10^{11} n/cm$^2 \cdot$ s in a fast reactor type spectrum as well as in a thermal spectrum can be achieved in these reactors. Maximum gamma dose rate inside the fast neutron region of the CFRMF is about 6 Mrad/h (60 kGy/h).

ARMF Applications

The ARMF is used for the nondestructive analysis of reactor fuel pins or plates, control rods, and smaller specimens of composite materials. This reactor is ideally suited for this type of work because its core and spectrum can readily be tailored to fit the needs of the measurements. Examples of applications are

(1) determining the distribution and total content of boron in irradiated commercial power reactor shim rods,

(2) determining the distribution and total fuel content in irradiated fast reactor fuel pins,

(3) measuring the effective neutron cross sections of composite materials such as geological core drill specimens and irradiated test specimens, and

(4) irradiation of electronic components and circuits in the central 15 by 15-cm position (Fig. 4).

CFRMF Applications

The CFRMF has been modified to include a coupled fast flux region as shown in the cutaway pictorial diagram (Fig. 5). The fast zone is constructed of materials that keep the zone water tight and "filter" or "tailor" the neutron energy spectrum. The neutron spectrum of the fast zone has

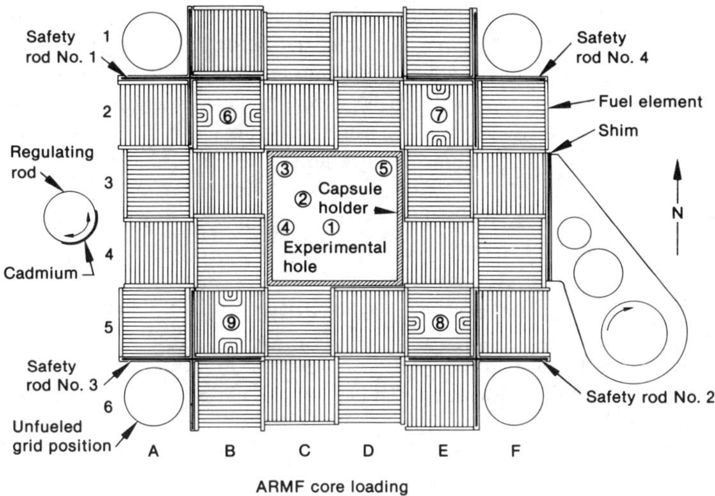

FIG. 4—*ARMF core cross-sectional diagram.*

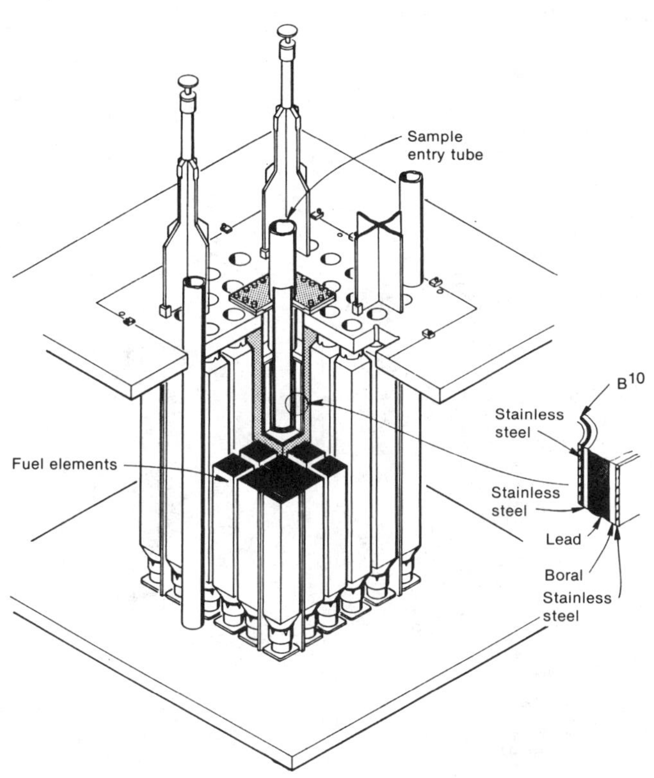

FIG. 5—*Cutaway pictorial diagram showing general assembly of the CFRMF.*

a mean energy of 0.7 MeV. This facility has found much application to fast neutron physics for activation measurement of materials in the fast breeder reactor program. The CFRMF can accept test assemblies and electronic components up to 38 mm in diameter in its central entry tube. There are two positions in the driver core that can be used for thermal neutron activations. One is a Flexo-Rabbit Transfer System used for short-term irradiations, which provides the capability for fast chemical separations or counting immediately after irradiation. The neutron flux at this position is approximately 5×10^{11} n/cm^2 · s. The other is a dry tube that extends into one of the corner positions of the core. The flux at this position is approximately 2×10^{11} n/cm^2 · s.

Examples of applications are

(1) irradiation of electronic components and electrical insulator materials to determine fast neutron damage,

(2) measurement of fast neutron cross sections, and

(3) neutron activation analysis for trace elements and fissile materials in both fast and thermal neutron fields.

Neutron Radiography Facility

The CFRMF is also utilized as a source for neutron radiography (NR), as shown in Fig. 6. In NR, neutron absorption is dependent upon nuclear cross sections giving materials of near equal density great differences in neutron absorption. Thus NR is complimentary to X-ray radiography (XR) in some cases, and superior in others. Inspection of radioactive objects, such as an irradiated fuel rod as illustrated in Fig. 6, is not possible with XR caused by the radiation emitted by the

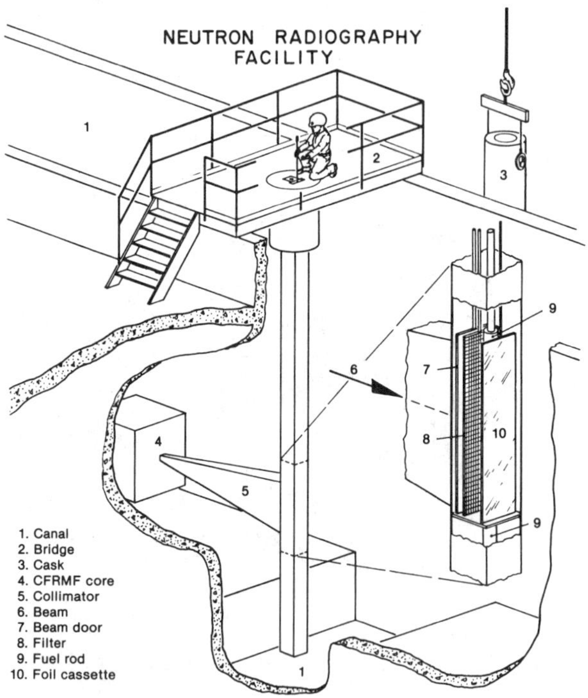

FIG. 6—*Neutron Radiography Facility.*

object. With NR, however, this radiation presents no problem because the imaging process used in this NR facility is sensitive to neutrons only.

The NR facility is a dry tube facility. The dry tube extends to the bottom of the canal and into a hole extending 2.45 m below the canal floor. A tapered collimator passes neutrons from the CFRMF core to the specimens to be radiographed. Maximum specimen length that can be accommodated is 4.6 m. Specimens longer than 0.91 cm require multiple radiographs to cover the entire length. The facility has the capability of rotating the objects so that they can be viewed from different angles. The resolution has been measured to be less than 0.1 mm. A typical radiograph is shown in Fig. 7.

Power Burst Facility

General Description

The PBF was designed to provide experimental data to aid in defining the behavior of nuclear fuels in off-normal operating conditions. The PBF core (Fig. 8) is a right-circular annular cylinder, 1.32 m in diameter and 0.91 m high, that encloses a centrally located vertical test space 0.21 m in diameter. The PBF fuel consists of stainless steel rods containing ceramic fuel pellets and is cooled by a low-pressure primary coolant system. The PBF reactor can be operated in three modes: (1) a steady state mode with power levels up to 28 MW, (2) a natural power burst mode that yields reactor pulses as short as 5.6 ms (FWHM) and peak powers as large as 270 GW, and (3) a shaped burst mode, for example, a square wave. These modes result in energy densities in test

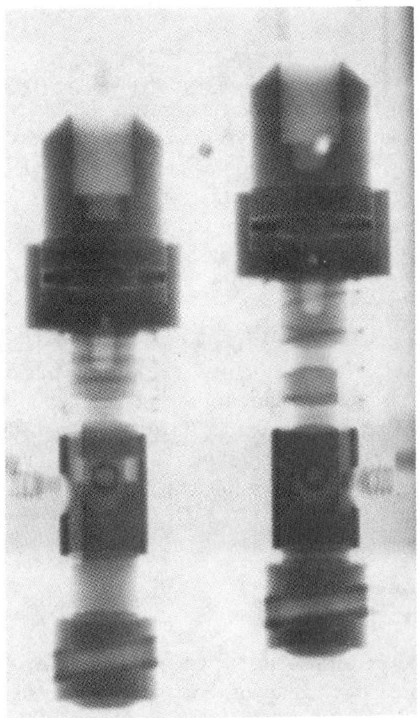

FIG. 7—*Neutron radiograph of specimen holder.*

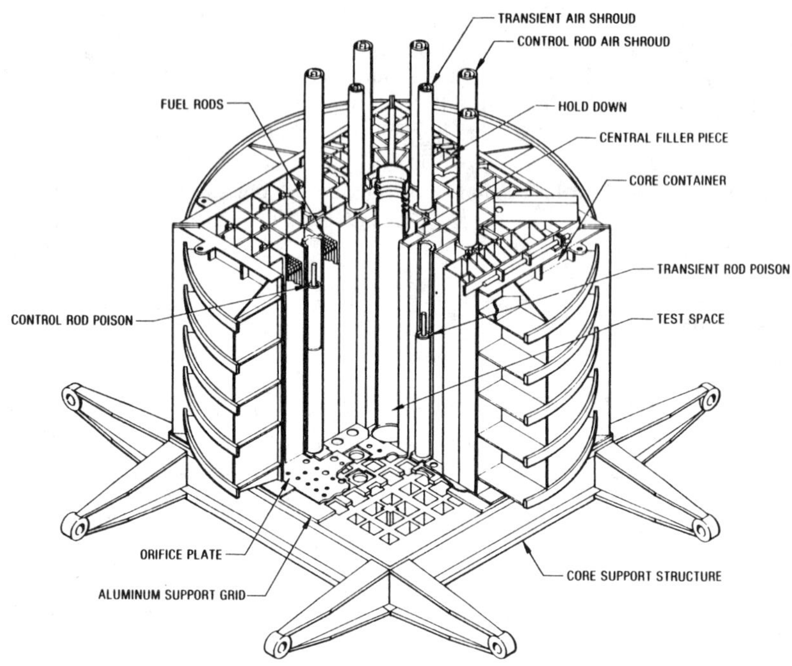

FIG. 8—*Cutaway view of PBF core.*

fuel rod clusters for many types of postulated reactor accidents. The peak neutron flux of about 7×10^{17} n/cm$^2 \cdot$ s can be realized in the central flux trap region.

PBF utilizes a computer-controlled data acquisition and reduction system (the PBF/DARS). The design of the PBF/DARS allows (1) production of most of the hardcopy plots needed for on-line, quick-look reports during a test and (2) reduction of most of the detailed test data within three days following a test. Further processing of the data is performed on a large Cyber 176 system at the INEL Computer Science Center in Idaho Falls.

PBF Applications

Although PBF was used exclusively for thermal fuel experiments over the past ten years, it has been made available for nuclear effects testing following completion of the thermal fuel tests. The facility can produce a high neutron and gamma fluxes similar to those generated by a nuclear explosion or by long-term high radiation environments. This capability can be applied to radiation hardness testing of electronic and optical equipment for defense and aerospace applications. In addition to the in-pile central test space (~17 L), larger equipment can be tested at a lower flux in the space above the in-pile tube.

Gamma Facilities and Their Applications

Gamma irradiations can be performed in the Gamma Facilities, located in the fuel handling canal of the Advanced Test Reactor. Photos of these facilities are seen in Figs. 9 and 10. Intense gamma radiation is supplied to these facilities by spent (that is, irradiated) fuel elements shortly after their discharge from the ATR. The Large Gamma Facility can accommodate test objects up

FIG. 9—*Large gamma facility test canister.*

to 0.45 m in diameter and 1.2 m high. Depending on the density and self-shielding characteristics of the test object, gamma dose rates up to 1 Mrad/h (10 kGy/h) can be achieved. The test object can be supplied with cooling water, circulating gas, electrical power and instrumentation during the irradiation.

The test article is loaded into the Large Gamma Facility (Fig. 9) by hoisting the test canister out of the canal and removing the bolted top flange. Following insertion of the test article, the canister is lowered back into the fuel element holding fixture, which is located on the canal floor. When the test canister is in place, the required number of spent fuel elements are loaded into the holding fixture.

FIG. 10—*Small gamma facility.*

One of the small gamma irradiation facilities is shown in Fig. 10. The small gamma facilities consist of 125-mm-diameter stainless steel tubes that extend from the surface of the canal water into the ATR spent fuel storage rack. Depending on the arrangement of fuel elements in the positions adjacent to the irradiation tube, gamma dose rates of up to 3 Mrad/h (30 kGy/h) can be achieved. As with the Large Gamma Facility, the test article can be supplied with cooling water, circulating gas, electrical power, and instrumentation leads during irradiation. The smaller gamma facilities offer the additional flexibility of not requiring fuel movement when the test article is inserted or removed from the tube. This allows for more rapid turnaround between irradiations. If relatively short instrumentation or power leads are required, the power supplies or monitoring equipment or both can be located in the upper, relatively well-shielded, portion of the irradiation tube, provided all equipment is less than 110 mm in diameter.

When spent fuel is used as the radiation source, the gamma spectrum is both the large and small gamma facilities is characteristic of fission product decay. If a more monoenergetic source is required for the experiment, cobalt-60 pins can be placed in the fuel element positions. Cobalt-60 is being produced in quantity in the Advanced Test Reactor. As was the case with the ARMF and CFRMF, the Gamma Facilities are simple and inexpensive to use.

Summary

Various reactor facilities at the INEL have unique design features so that very reproducible power levels can be established on a run-to-run basis with minimum perturbation to the neutron energy spectrum and flux distribution. The availability of a range of neutron flux levels (10^8 to 10^{17}) at these facilities and a neutron radiography system makes them very useful for a broad range of experimental investigations important to fast and thermal reactor physics, material testing, radiation hardness studies, and dosimetry. Also, availability of intense gamma fields (up to 3 Mrad/h) over large volumes provides very suitable facilities for irradiating large electronic circuits and optical sensors.

An overview of the INEL facilities available for neutron or gamma irradiations and their capabilities has been presented in this paper. The access and use of these facilities can be arranged through the Idaho Operations Office, U.S. Department of Energy, Idaho Falls, ID 83402. We are open to discussions of possible collaboration suitable to the needs of potential user programs.

Acknowledgment

This work was performed under the auspices of the U.S. Department of Energy under DOE Contract DE-AC07-76-ID01570.

Walter F. Sommer,[1] Wolfgang Lohmann,[2] Karl Graf,[2] Ivan K. Taylor,[1] and Raymond M. Chavez[1]

Operating Experience at the Los Alamos Spallation Radiation Effects Facility at LAMPF

REFERENCE: Sommer, W. F., Lohmann, W., Graf, K., Taylor, I. K., and Chavez, R. M., **"Operating Experience at the Los Alamos Spallation Radiation Effects Facility at LAMPF,"** *Influence of Radiation on Material Properties: 13th International Symposium (Part II), ASTM STP 956,* F. A. Garner, C. H. Henager, Jr., and N. Igata, Eds., American Society for Testing and Materials, Philadelphia, 1987, pp. 718–729.

ABSTRACT: Operation of a new irradiation facility at the beam stop (Target Station A-6) at the Clinton P. Anderson Los Alamos Meson Physics Facility (LAMPF) began in May 1985. The facility is now fully operational. A closed-loop water system, a closed-loop helium system, remote handling procedures for activated materials, experiment change-out procedures, and experiment control equipment are all in place.

Experience dictated a change in irradiation capsules from a system that used metal seals to a system that is completely welded. The seals were found to be unreliable and difficult to replace by remote means.

Several materials have been irradiated in the direct 760-MeV proton beam. Material property changes as well as helium production in a variety of materials are being investigated.

A special sample holder that can accommodate transmission electron microscopy (TEM) specimens, isolate them from the cooling medium, maintain fixed temperature, and retain essentially a stress-free state has been developed. Remote handling retrieval of these specimens is being developed.

Activation foil measurements are being made to determine the secondary particle flux and spectrum (charged particles and neutrons) that result from interaction of the direct proton beam with targets at the beam stop. Twelve independent irradiation ports, each with an irradiation volume of 0.12 by 0.25 by 0.50 m, are available for exposing material to this particle flux, primarily neutrons. Two irradiations were completed in this area during the period May through December 1985. The neutron spectrum here resembles a fission spectrum with the addition of neutrons in the MeV energy range (high-energy tail). The maximum neutron flux is about 6×10^{17} n/m^2 s at one of the 12 ports and drops by a factor of about 10 as a minimum at another of the 12 ports.

Three independent ports for proton irradiations are in place. Each has an irradiation volume of about 150 cm^3. The proton beam has a Gaussian intensity profile; the maximum proton flux at the center of the beam is 1.2×10^{14} protons/cm$^2 \cdot$ s. The beam spot has a diameter of approximately 5 cm at 2σ. Four irradiations were completed in this area during the period May through December 1985.

KEY WORDS: protons, spallation neutrons, irradiation facility, irradiation techniques, Los Alamos Meson Physics Facility (LAMPF)

A new facility, in which materials may be exposed to a proton beam of energy ~760 MeV and a neutron flux that results from the interaction of the proton beam with bulk targets and the beam

[1] Associate group leader, senior technician, and senior designer, respectively, Los Alamos National Laboratory, Los Alamos, NM 87545.

[2] Research scientist and engineer, respectively, KFA Jülich GmbH, P.O. Box 1913, D5170, Jülich, West Germany.

stop at the Los Alamos Meson Physics Facility (LAMPF), began operation in May 1985. Several experiments, under controlled temperature conditions, were completed during three LAMPF run cycles between May 1985 and Jan. 1986. An experiment capsule, a closed-loop helium temperature-control system, a closed-loop water temperature control system, and remote-handling procedures for radioactive material have been developed and used successfully. It was also shown that a complete experiment changeover could be accomplished in less than one day. This is of importance because during a nine-week run cycle, the LAMPF accelerator is only idle for one day each two weeks for routine maintenance and some experiments require sample changes in a time frame shorter than nine weeks. Further refinement of the facility will take place during a shut-down period in spring 1986. In the text below, we attempt to summarize our recent experience in using the facility and also to describe the equipment that we feel has proven to be reliable.

Facility Description

The Los Alamos Spallation Radiation Effects Facility (LASREF) is located at the beam-stop area (Target Station A-6) at LAMPF. The design of this facility is modular in that each component and experiment is attached to its own biological shield plug (insert). This can be seen by noting a schematic (Fig. 1) and a photograph (Fig. 2). The biological shield is necessary for containment of the radiation produced along the beamline, at the isotope production bulk targets, and at the beam stop. Services, such as cooling water and instrumentation cabling, are routed through stepped slots in the biological shield to the top of the facility. Instrumentation cabling is further routed

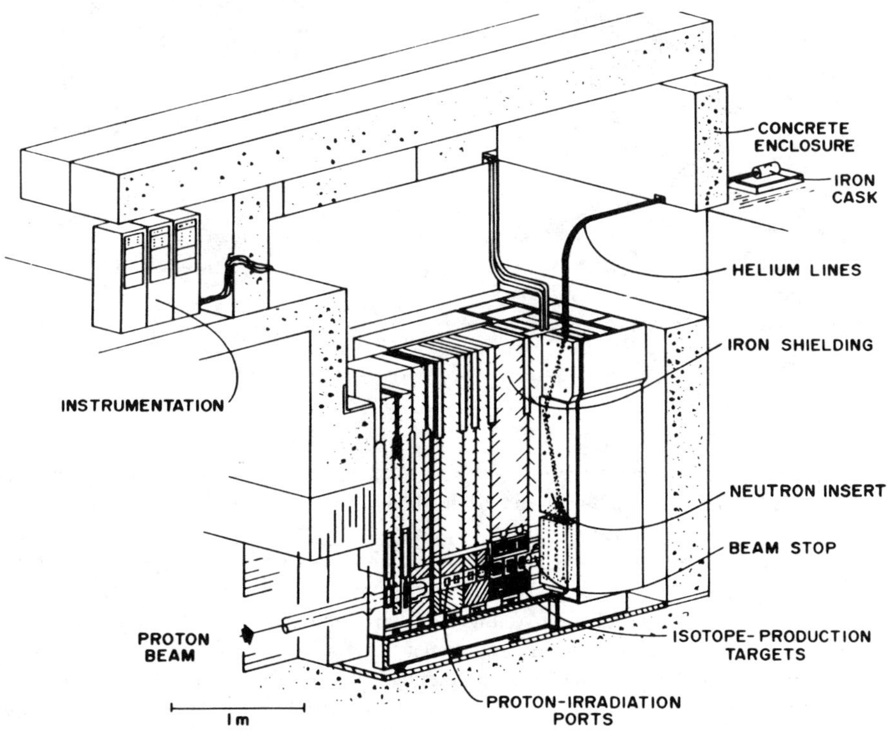

FIG. 1—*Schematic of the Los Alamos Spallation Radiation Effects Facility (LASREF) at LAMPF.*

FIG. 2—*A proton-irradiation insert and capsule.*

through a maze in a secondary containment concrete hut to an area that is accessible when beam is on, where instrumentation cabinets can be located. This can be seen in schematic form in Fig. 1.

Three independent insert locations (proton irradiation inserts) are available for placing capsules in the direct proton beam. Their relationship to other components at Target Station A-6 is shown in a side view in Fig. 3. The irradiation volume for each is about 150 cm^3. It is possible to modify the insert to accommodate more sensitive equipment in a position removed from the beamline where both shielding and distance decrease the particle radiation flux.

Twelve independent insert locations (neutron irradiation inserts) are available for placing capsules in the neutron flux. Figure 4 shows their relationship to other components at Target Station A-6 in a top view and Fig. 5 shows, schematically, their relationship to the neutron-producing targets.

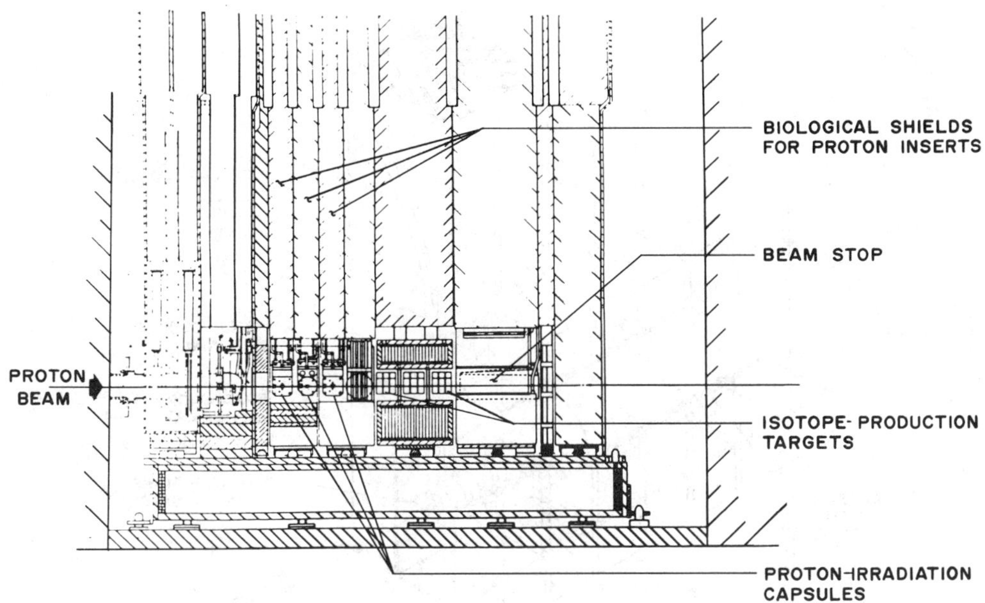

FIG. 3—*Proton-irradiation insert locations relative to other components in Target Station A-6.*

(Protons, in comparable number, are also produced by spallation reactions in the targets but ~90% are absorbed in the target material [*1,2*]. This is also the case for other secondary charged particles produced in the targets.) Each insert location has a 12- by 25- by 50-cm volume available for irradiation capsules whose geometry and complexity is left to the discretion of the experimenter.

Although LAMPF accelerates protons to a maximum energy of 800 MeV, interactions with two graphite targets upstream from the facility lowers the energy to about 760 MeV. At a nominal

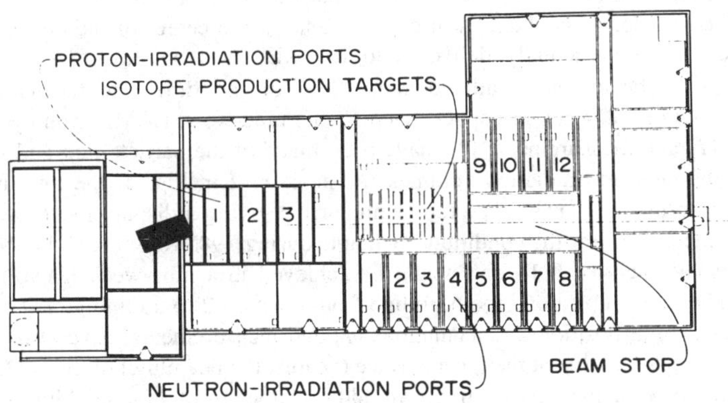

FIG. 4—*Neutron-irradiation insert locations relative to other components in Target Station A-6. This is a top view.*

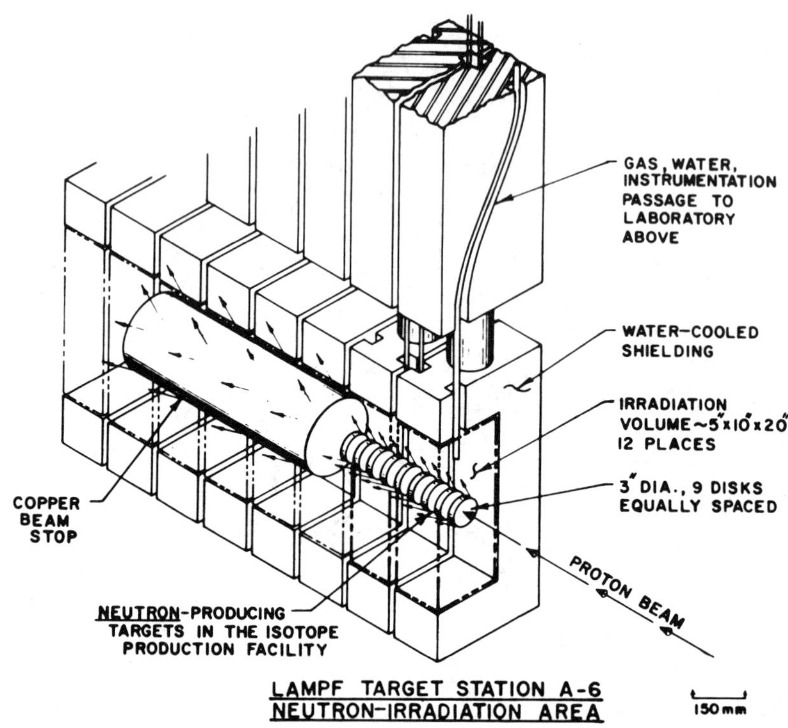

FIG. 5—*Schematic of the relation of neutron-irradiation inserts, 1 through 8, to the neutron-producing targets in Target Station A-6.*

total beam current of 1 mA and a Gaussian-like beam spot size of diameter 5 cm to 2σ (σ is one standard deviation), it has been calculated [*3*] that a displacement rate of 8.4×10^{-7} displacements per atom per second (dpa/s) is obtained in copper metal at the center of the beam. During 1985, the average beam current actually delivered to LASREF was on the order of 0.8 mA and, considering that the beam was available about 80% of the time, the resultant total atomic displacements would be 2.9 dpa in copper metal in one nine-week LAMPF run cycle.

Calculations [*1*] and measurements [*2*][3] have been made of the particle flux and spectrum (Fig. 6) incident on the neutron-irradiation volumes. At position 2 in Fig. 4, the total neutron flux is about 5.5×10^{17} n/m^2 s at 1 mA of proton beam current giving a displacement rate of 4×10^{-8} dpa/s in copper metal [*3*]. For the conditions available during 1985 (0.8 mA and 80% availability), a total atom displacement of 0.14 dpa would be achieved in a nine-week LAMPF run cycle in copper metal. The flux drops along the beamline from position 2 to a minimum at positions 8 and 12, ~10x lower than at position 2. Calculations [*1*] and measurements[3] have also shown that the secondary charged-particle flux at the inner surface (nearest the beamline) of the neutron-irradiation volumes is about 10% of the neutron flux. An additional 1 cm of iron shielding further reduces the charged-particle flux to about 1% of the neutron flux.

A real-time signal of the average beam current being delivered to LASREF is continually available. The signal is compatible with standard data-acquisition equipment.

[3] Davidson, D. R. et al., in this publication, pp. 730–740.

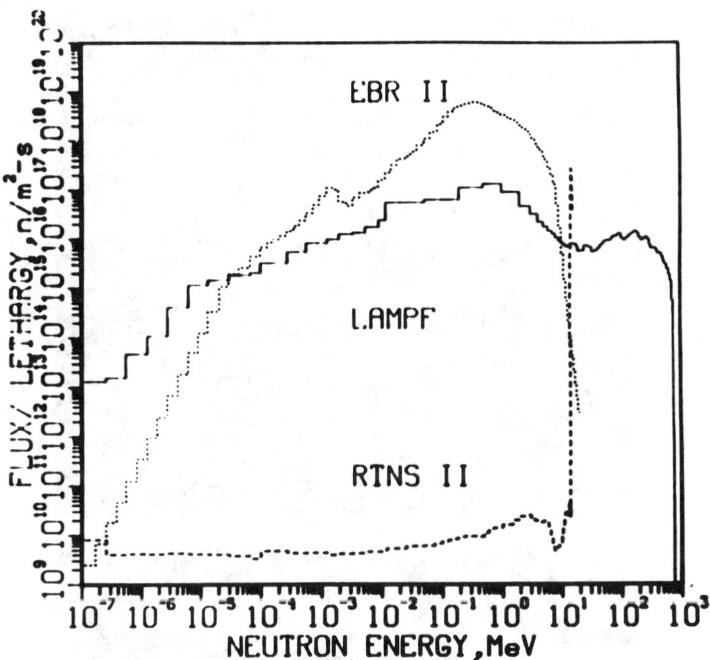

FIG. 6—*Neutron flux and spectrum at Target Station A-6 at neutron-irradiation insert location 2.*

Hardware Description and Performance

Capsules that can survive the radiation environment, provide an atmosphere for controlled conditions of temperature and stress, and allow remote handling and removal of specimens without disturbing the material has been developed and used successfully during 1985. In addition, a ''rabbit'' system [*4*] that is capable of placing several small capsules (10.3 mm diameter by 21.6 mm long) loaded with foils for activation analyses to determine neutron and charged-particle flux and spectrum was developed and used successfully.

Figure 7 is a pictorial view of a proton irradiation capsule, and the cross-sectional drawing (Fig. 8) shows some of the important details. Note that samples are attached to a finger-like sample holder (materials are selected for appropriate thermal expansion compensation, or one end is left free for unperturbed thermal expansion). This sample holder is attached to a plug with a specially designed weld prep, which in turn is inserted into a tube that directs coolant (in this case, water from a closed-loop water system) over the samples. The proton beam is directed onto the tubes, centered between the plenums, and centered vertically on the group of five tubes. Since the proton beam has a Gaussian intensity profile, there is a gradient in accumulated dose for samples placed in tubes vertically away from the center. This allows some fluence-dependent observations from a single irradiation. The intensity gradient also is present, of course, in the horizontal direction so care must be taken in properly dimensioning the sample gage length so that this gradient does not cause significant ambiguity in the results. There are a total of 13 tubes, in which we have attached 8 samples each; thus a single capsule yields 104 irradiated samples.

Removal of the samples is facilitated by the special weld prep on the plug. By remote means, a special slow-speed grinder removes the weld metal and the entire sample assembly can then be removed from the capsule. This procedure eliminates the excessive vibration and possible mechanical

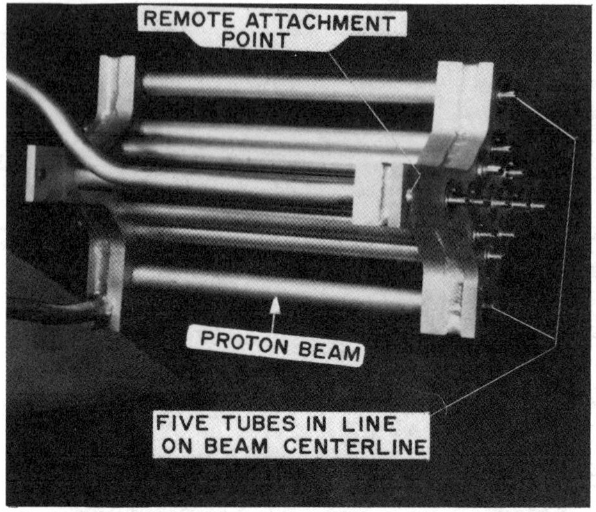

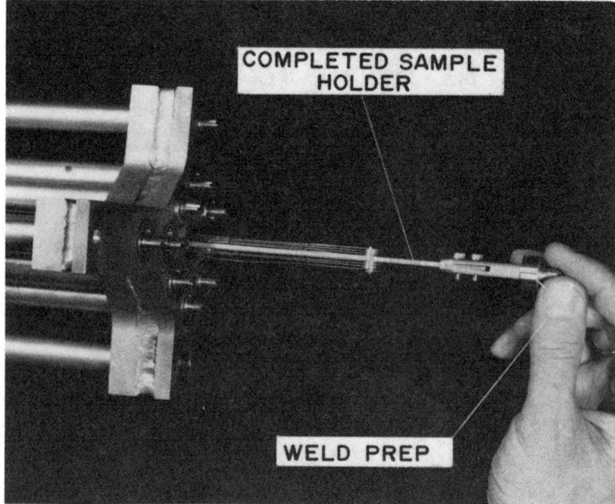

FIG. 7—*Proton-irradiation capsule.*

damage that we encountered in an earlier design that used a bolted flange; removal of the bolts was found to be difficult and metal seals were found to be unreliable.

Another capsule has been developed for use with the closed-loop helium system. Since more than 200 W/cm^3 (in iron) are deposited by beam heating, we found that helium gas alone was not sufficient to maintain a suitable temperature gradient across the samples. The major contribution to the heat load came from the tubes that direct the flow rather than the samples. We therefore designed a double-tube capsule (Fig. 9) in which water flows in the annulus between the tubes and extracts the energy deposited in the tubes and helium, preheated to the temperature needed for the experiment, flows over the samples and extracts the energy deposited in the samples. Installation and remote handling of the specimens is accomplished in a manner identical to that described above.

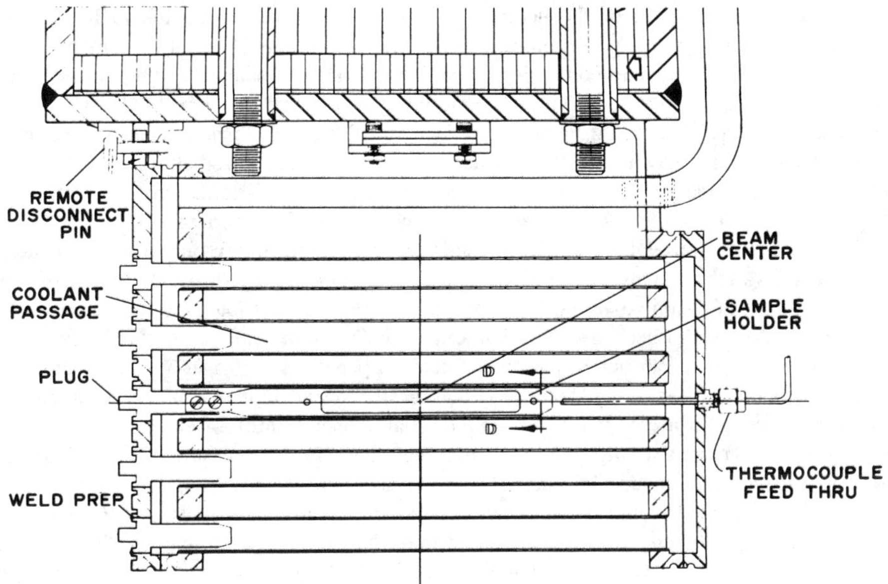

FIG. 8—*Cross section of a proton-irradiation capsule.*

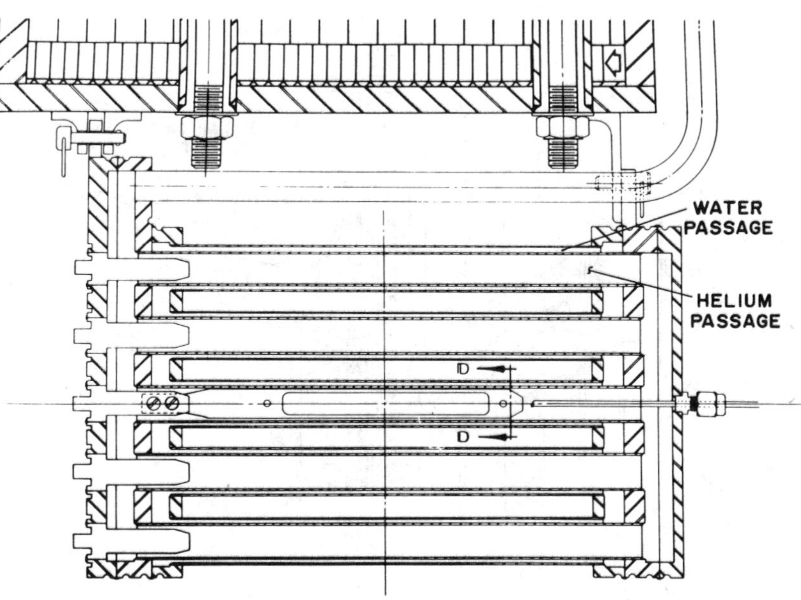

FIG. 9—*Cross section of a proton-irradiation capsule for use with helium or other gas.*

In order to provide a map of the neutron and charged-particle flux in the facility (complex because of the distributed source introduced by the various targets [Fig. 5]), a "rabbit" system capable of placing small capsules containing activation foils (Fig. 10) above and below the beamline and at four locations radially out from the beamline was built and developed [*4*]. The "rabbits" are small cylinders, 10.3 mm in diameter by 21.6 mm long, which house the activation foils. They are pneumatically driven into the irradiation space and, after irradiation, into shield casks outside the concrete enclosure by pressurized helium gas. Several irradiations, in neutron insert Positions 2 and 6, were completed and are now being analyzed[3] as a comparison to earlier work [*1,2*]. The insert that houses the "rabbit" system can be placed in any one of the 12 irradiation locations. It is available for use by the radiation damage community; a nuclear chemistry group studying activation of cosmic material has used the system to learn about material interaction with high-energy (> 100 MeV) neutrons of which ample numbers are available in the facility.

A special capsule that can determine mechanical properties, such as flow stress, fatigue, and creep while the material is being irradiated, was designed and built by the KFA-Jülich group. Central to the successful use of this device is the reliable operation of a stepping motor that loads the samples, a linear variable differential transformer (LVDT) that measures length changes and a load cell that measures load. The internal mechanism of the device is shown in Fig. 11. As a reliability test, we placed identical components on a neutron irradiation insert in Position 7. The motor was used to raise and lower a deadweight onto the load cell while the LVDT recorded the cyclic motion. We found that the devices would properly function in this environment for adequate times. The environment approximates the radiation levels we expect at the proton irradiation insert locations in the areas, outside the beamline, where the devices are planned for use. We found that it is necessary to use mineral-insulated cabling on all inserts used in the facility.

Handling of Radioactive Inserts: Experiment and Sample Changes

It was demonstrated, using existing LAMPF remote handling equipment and other LAMPF systems, that an insert and, hence, an experiment change could be accomplished in a 6-h period.

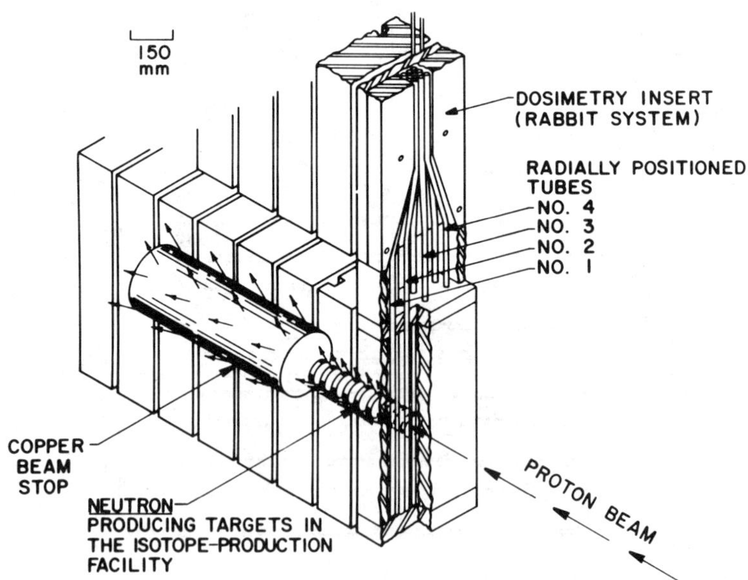

FIG. 10—*"Rabbit" system in place in the neutron-irradiation area.*

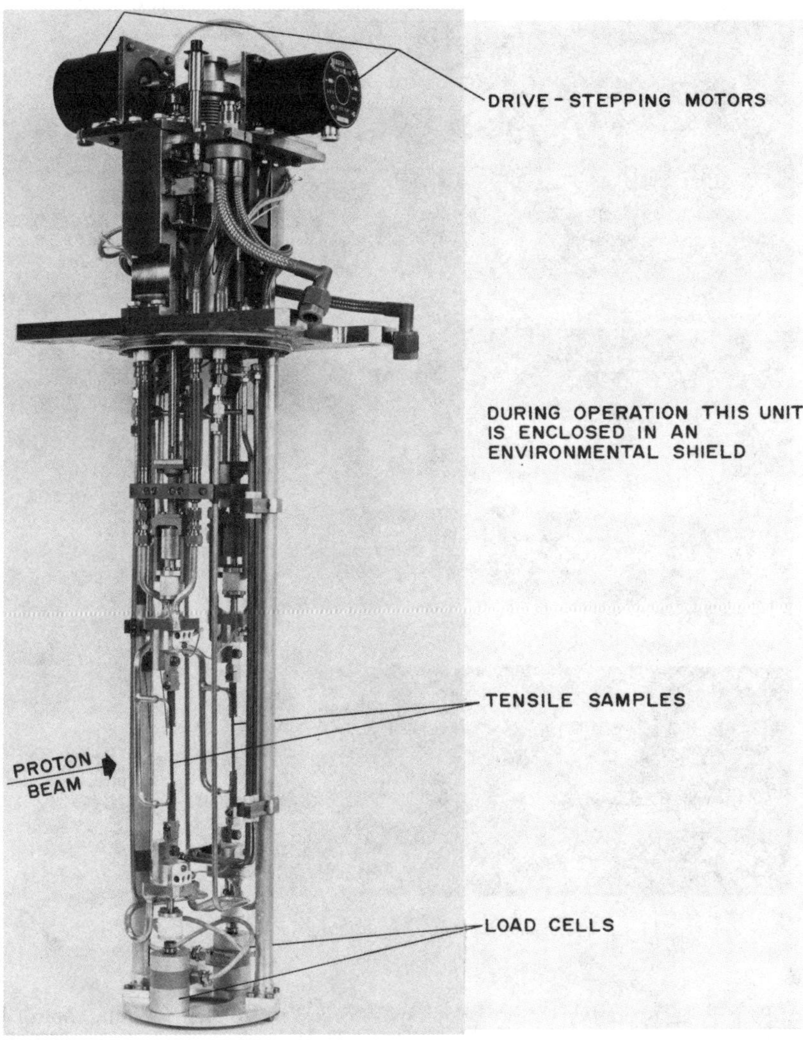

FIG. 11—*Internal mechanism of the in-situ mechanical testing device.*

To facilitate this work, the new experiment is mounted on a new, unirradiated insert and instrumentation, and services are completed and terminated at the top of the insert with the entire unit supported on a special rack in a vertical attitude. Remote handling procedures for sample handling and preparation, for example, of transmission electron microscopy (TEM) samples, are now being developed. With regard to TEM, we developed a special holder that can accommodate 3-mm disks, place them in a tube in a proton-irradiation capsule, isolate them from the cooling medium, and maintain temperature control. This is advantageous when only microstructural evolution study by TEM is desired since it simplifies sample preparation of radioactive material. This device is shown in Fig. 12.

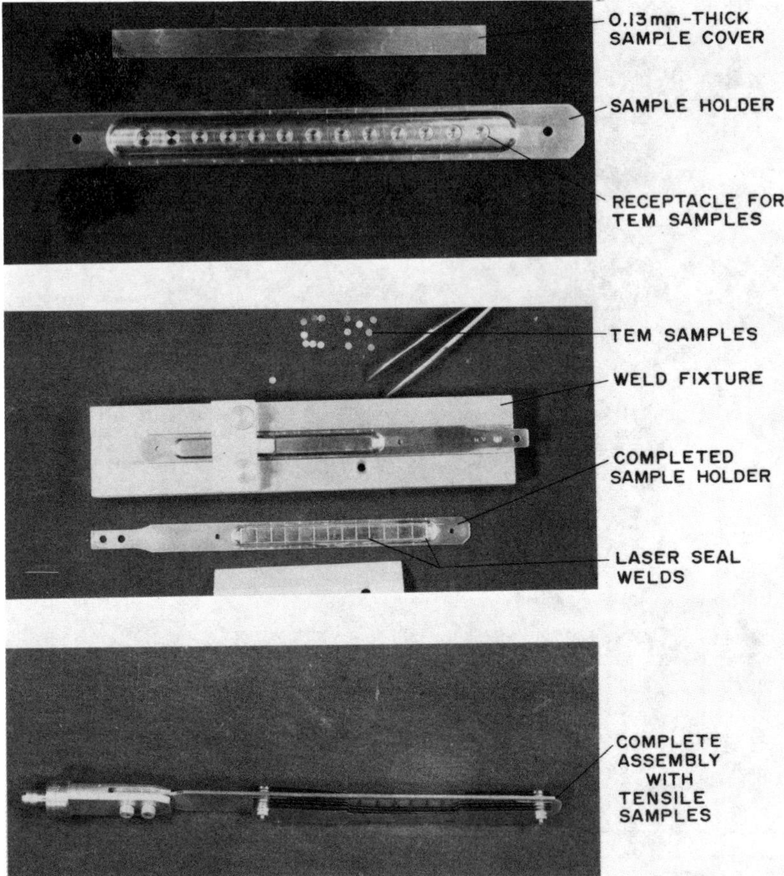

FIG. 12—*Special fixture for irradiating TEM specimens.*

Summary

The LASREF is an operating facility. Equipment necessary for temperature control has been developed. Remote handling capability has been demonstrated. Irradiation capsules have proven reliability. Inquiries relating to the use of the facility can be directed to the authors.

Acknowledgments

The authors thank A. Horsewell for helpful critique and C. Mueller, B. Ortiz, F. M. Gehrmann, S. Lopez, T. Montoya, F. Roybal, A. Maestas, L. Martinez, and the LAMPF Remote Handling and Targeting Section, whose expert technical assistance led to the successful commissioning of our facility.

References

[*1*] Davidson, D. R., Little, R. C., Sommer, W. F., Bradbury, J. N., and Prael, R. E., "Characterization of the Radiation Environment at a New Proposed Irradiation Facility at LAMPF," *Journal of Nuclear Materials,* Vols. 122 & 123, 1984, pp. 989–994. M. S. Wechsler and W. F. Sommer, "The Radiation Damage Facility at the LAMPF A-6 Target Station," J. Nucl. Mater. 122– 123, 1078–1084 (1984).

[2] Davidson, D. R., Greenwood, L. R., Reedy, R. C., and Sommer, W. F., "Measured Radiation Environment at the Clinton P. Anderson Meson Physics Facility (LAMPF) Irradiation Facility," *Effects of Radiation on Materials: Twelfth International Symposium, STP 870,* F. A. Garner and J. S. Perrin, Eds., American Society for Testing and Materials, Philadelphia, 1985, pp. 1199–1208.

[3] Wechsler, M. S., Davidson, D. R., Greenwood, L. R., and Sommer, W. F., "Calculation of Displacement and Helium Production at the Clinton P. Anderson Meson Physics Facility (LAMPF) Irradiation Facility," *Effects of Radiation on Materials: Twelfth International Symposium, STP 870,* F. A. Garner and J. S. Perrin, Eds., American Society for Testing and Materials, Philadelphia, 1985, pp. 1189–1198.

[4] Davidson, D. R., Sommer, W. F., Taylor, I. K., Brown, R. D., and Martinez, L., " 'Rabbit' System for Activation Foil Irradiations at the Los Alamos Spallation Radiation Effects Facility at LAMPF," LA-UR-85-3630 Revision, abstract accepted for the Second International Conference on Fusion Reactor Materials, Chicago, IL, 13–17 April 1986.

Dorothy R. Davidson,[1] Robert C. Reedy,[1] Lawrence R. Greenwood,[2] Walter F. Sommer,[1] and Monroe S. Wechsler[3]

Additional Measurements of the Radiation Environment at the Los Alamos Spallation Radiation Effects Facility at LAMPF

REFERENCE: Davidson, D. R., Reedy, R. C., Greenwood, L. R., Sommer, W. F., and Wechsler, M. S., "**Additional Measurements of the Radiation Environment at the Los Alamos Spallation Radiation Effects Facility at LAMPF,**" *Influence of Radiation on Material Properties: 13th International Symposium, ASTM STP 956,* F. A. Garner, C. H. Henager, Jr., and N. Igata, Eds., American Society for Testing and Materials, Philadelphia, 1987, pp. 730–740.

ABSTRACT: Foil activation dosimetry experiments were conducted in a "rabbit" system at the completed Los Alamos Spallation Radiation Effects Facility (LASREF). The "rabbit" system contains four tubes spaced radially outward 0.12, 0.18, 0.27, and 0.38 m off beam centerline. Foils were irradiated for 3 to 62 h to measure the neutron flux and energy spectrum radially from beam centerline, along the beamline, and the effect of the isotope production (IP) target loadings on the neutron flux in the neutron irradiation locations. Irradiations showed a decrease in the radial flux by a factor of 6 in 0.15 m of iron outside the IP targets. An enhancement was seen in the 24-keV energy region outside 0.15 m. There was little difference in the shape of the spectra outside the IP targets and the beam stop with the exception of the high energy tail (energies above 20 MeV). The decrease in the high energy tail outside the beam stop is due to the degradation of the energy of the proton beam in the IP targets. Irradiations outside the beam stop with zero and eight IP targets gave the same spectral shape with the exception of the high energy tail. The magnitude of the integral flux decreased by a factor of two when eight IP targets were present. Irradiations with five "rabbits" stacked on top of each other showed no difference in the integral flux below, on and above beam centerline.

KEY WORDS: neutron, proton, integral flux, differential flux, foil, activation, dosimetry, spectrum, cross section, spallation

The Los Alamos Spallation Radiation Effects Facility (LASREF) began operation in May 1985, at the Los Alamos Meson Physics Facility (LAMPF). The facility is located at the beam stop area (Target Station A-6) at LAMPF. Monte Carlo calculations [*1*] were done previously to predict the neutron flux and spectrum at the facility using a geometry that approximated the final facility. Measurements of the neutron flux and energy spectrum, using activation foil techniques, were then made at the existing beam stop area [*2*]. For these measurements we were limited to only one location. Also, the geometry was somewhat different from what LASREF will eventually have. Thus, at the onset of the operation of LASREF, a special insert ("rabbit" system) was built and placed in service so that multiple activation foil measurements in several locations could be made. This work was necessary to facilitate further design of experiments at LASREF and gave a map of the neutron flux and spectrum at the facility.

[1] Nuclear engineer, nuclear chemist, and associate group leader, respectively, Los Alamos National Laboratory, Los Alamos, NM 87545.

[2] Nuclear physicist, Argonne National Laboratory, Argonne, IL 60439.

[3] Professor of nuclear engineering, Iowa State University, Ames, IA 50010.

The expected radiation damage parameters at LASREF have been reported [*3,4*]. We are also reporting our experience in operating the facility [*5*], as well as giving a description of radiation-hardened equipment that we have developed.

"Rabbit" System Description and Measurement

A schematic, in isometric form, of LASREF is shown in Fig. 1. Note the relation of the "rabbit" insert to the neutron-producing isotope production (IP) targets and the copper beam stop. Note also the location, outside the concrete "hut" surrounding the facility, of the iron casks used for retrieval of the activation foils. Figure 2 shows, in schematic form, a more detailed view of the relation of the "rabbit" system to the IP targets and the LAMPF beam stop. The "rabbit" insert contains four tubes spaced radially outward 0.12, 0.18, 0.27, and 0.38 m off beam centerline.

Aluminum capsules ("rabbits"), 10.16 mm in diameter and 19.05 mm long, are pneumatically driven by helium gas to the irradiation location. They are placed into the system and later retrieved at the cask position, outside the hut, as shown in Fig. 1; this allows experiments to be irradiated and terminated without interruption of LAMPF operations. Each of the four tubes can accept five capsules; when properly placed the capsules can record the neutron flux and energy spectrum above, on, and below beam centerline. Each capsule can contain numerous foils. As in previous work [*2*], the high purity foils were chosen to obtain a clear picture of the neutron flux and spectrum. Capsules contained iron, cobalt, nickel, titanium, vanadium, scandium, aluminum, zinc, gold, and copper foils for the neutron measurements. Secondary protons are produced by spallation reactions as the primary beam interactions with the targets. Measurements of this proton flux were made using iron, titanium, and copper foils. Foils were punched 5.6 mm in diameter and 0.013

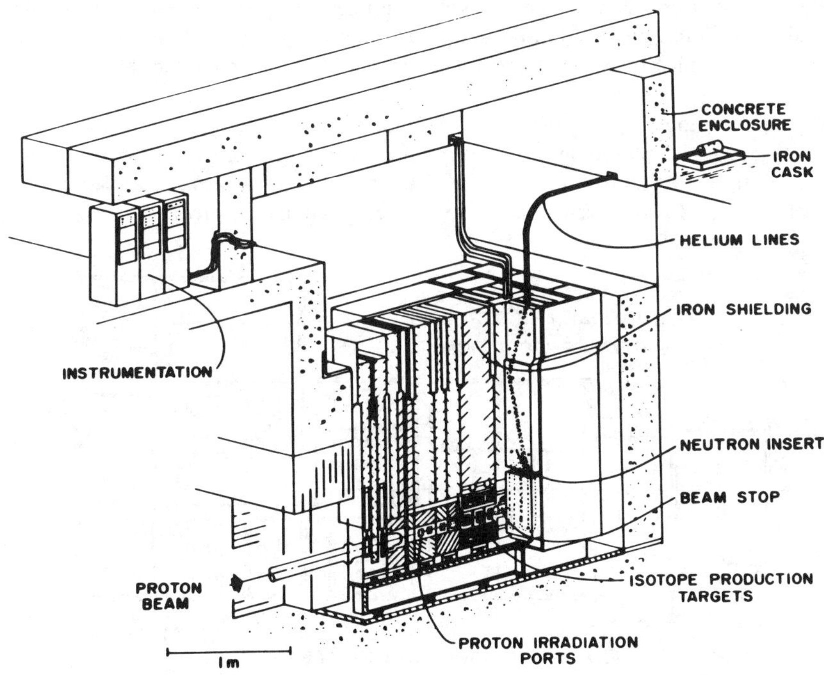

FIG. 1—*Schematic of LASREF showing "rabbit" system beside the beam stop and the iron cask outside the concrete hut for retrieval of "rabbits."*

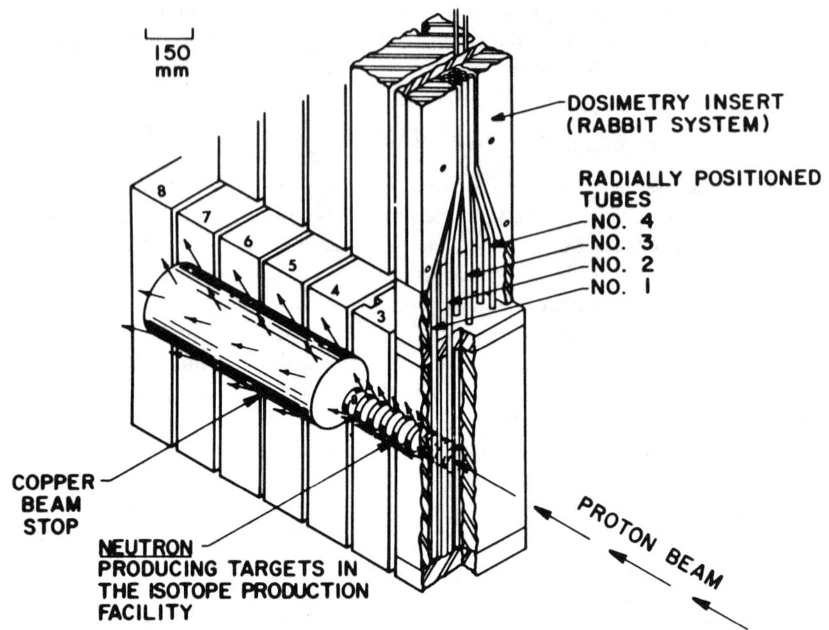

FIG. 2—*"Rabbit" system in Neutron Irradiation Location 2 showing inlet and outlet helium lines.*

to 0.127 mm thick, with a center hole. They were strung in a specific order on an aluminum wire to assure identification. Foils of iron, cobalt, gold, and scandium were wrapped in 0.03-mm thick gadolinium foil and placed in the bottom of the "rabbit" to determine the thermal-to-epithermal flux.

Figure 3 shows a top view of LASREF and denotes the numbering system and location of the 12 inserts that may be placed in the facility and receive neutron irradiation. For the present experiment, Positions 2 and 6 were used; the "rabbit" system was designed for use in any of 12 locations. Locations 2 and 6 were chosen because calculations [*1*] predicted the highest neutron

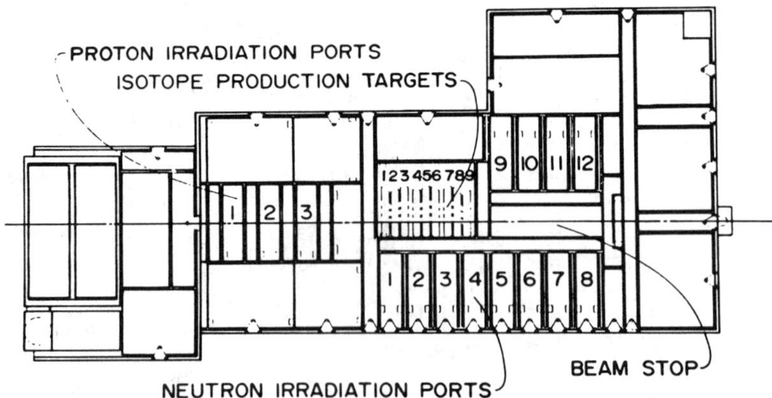

FIG. 3—*Top view of LASREF indicating 12 neutron and 3 proton irradiation locations and their locations relative to the IP targets and beam stop.*

flux at these locations for the target geometry described below. A total of 13 irradiations were made at LASREF during which foils were irradiated for 3 to 62 h; the results from only four of these irradiations (936-2, 936-5, 936-6, 936-7) are discussed in this report. Table 1 describes the IP target loading and exposure achieved in μA-h for the four irradiations. ''Rabbits'' were stacked on top of two solid aluminum spacers so that ''rabbits'' were at beam centerline unless otherwise stated.

Activity Measurement and Spectral Analysis

Foils were mounted on aluminum counting plates and counted with a calibrated high-purity germanium crystal (HPGe). The gamma-ray spectra were analyzed using the SPECANL computer code [*6*]. Recorded beam current histories were used to correct the activity for decay during irradiation. The corrected activities along with a calculated spectrum, activation cross sections, and variance-covariance information were input into the STAY'SL computer code [*7*]. STAY'SL uses a least-squares technique to calculate the most probable neutron energy spectrum. Strong correlations were assumed between nearby groups in the flux and cross sections. Cross sections were taken from several sources. Below 20 MeV, cross sections were from ENDF/V-B [*8*]. Between 20 and 44 MeV, cross sections were based on experimental and calculated data tested in a Be(*d,n*) neutron field [*9*]. Proton spallation cross sections were used to extend the activation cross section file to 800 MeV [*10–12*].

TABLE 1—*Irradiations in ''rabbit'' system at LASREF.*

1. Irradiation 936-2 (''Rabbit'' system in Neutron Irradiation Location 2.
   Proton dose on beamline targets: 9.14 × 10^4 μA-h.)
   IP targets in place:
   2—CsCl
   3—MnCl
   4—CsCl
   5—Mo
   7—RbBr

2. Irradiation 936-5 (''Rabbit'' system in Neutron Irradiation Location 2.
   Proton dose on beamline targets: 1.87 × 10^4 μA-h.)
   IP targets in place:
   2—ZnO
   3—CsCl
   4—CsCl
   5—MnCl
   6—Mo
   8—In

3. Irradiation 936-6 (''Rabbit'' system in Neutron Irradiation Location 6.
   Proton dose on beamline targets: 1.08 × 10^4 μA-h.)
   IP targets in place: None

4. Irradiation 936-7 (''Rabbit'' system in Neutron Irradiation Location 6.
   Proton dose on beamline targets: 2.5 × 10^4 μA-h.)
   IP Targets in place:
   2—ZnO
   3—RbBr
   4—CsCl
   5—MnCl
   6—Mo
   7—Al
   8—In
   9—RbBr

The reactions analyzed and their corrected activities are listed in Table 2 for Irradiation 936-5, Tube 1 next to the IP targets. The energy limits indicate a substantial contribution from high energy neutrons (energy above 44 MeV). This contribution is particularly evident for spallation reactions such as Al(n,x)^{7}Be, Cu(n,x)^{46}Sc, Cu(n,x)^{51}Cr, and Cu(n,x)^{52}Mn. Other reactions, such as ^{58}Ni(n,p)^{58}Co, ^{59}Co(n,p)^{59}Fe, and ^{59}Co($n,2n$)^{58}Co, have no significant response above 36 MeV. The listed differences show reasonable agreement between most of the measured and calculated (STAY'SL) activities. Reactions ^{58}Ni(n,x)^{5}Ni, Cu(n,x)^{60}Co, and Cu(n,x)^{59}Fe have large, positive errors. Their response is due primarily to high-energy neutrons where cross-section uncertainties are large and cross sections are based on proton, not neutron, measurements. The error for these two reactions was positive and large for all irradiations. The measured activity for ^{27}Al(n,x)^{22}Na was low by 64.7% from the calculated activity. The measured activity was consistently low for all irradiations. This response is not well understood.

Irradiation 936-2 was done next to the IP targets with two targets in the beam to measure differences in the neutron flux above and below beam centerline. Five "rabbits" each with iron, nickel, and cobalt foils were stacked on top of each other in Tube 1; "Rabbit" 1 was one on the bottom and "Rabbit" 3 at beam centerline. The corrected activities listed in Table 3 show no difference in flux with location other than statistical fluctuations in the data.

Irradiations 936-5, 936-6, and 936-7 were made to measure the neutron flux and energy spectrum along the beamline, radially from the beam centerline, and to determine the effect of the IP target loading on the spallation neutron flux. The differential neutron flux for Irradiation 936-5 next to the IP targets is shown in Fig. 4. Above 0.01 MeV, the spectra are similar for Tubes 1, 2, and 3. Tube 4 shows a decrease in the high-energy flux due to spallation and an enhancement at approximatley 50 keV. This enhancement is due to the 24-keV transmission window in the iron cross section. Neutrons born in this region and those scattered into it will enhance the flux. The peak energy is greater than 24 keV because of the limited reactions (Co(n,γ)^{60}Co and Fe(n,γ)^{59}Fe) that are responsive to neutrons below 0.1 MeV. The integral fluxes for Tubes 1, 2, 3, and 4 are 2.8×10^{17}, 1.3×10^{17}, 5.0×10^{16}, and 1.5×10^{17}, n/m^{2}-s-mA, respectively. Figure 5 shows

TABLE 2—*Saturated activities for an irradiation next to the IP targets with five (Irradiation 936-5) targets in the beam. The value* M *is the measured activity and* A *the STAY'SL adjusted activity.*

Reaction	Activity, atom/atom-s	Difference, % (M-A/M) × 100	90% Energy Limits, MeV
^{59}Co(n,γ)^{60}Co[a]	1.01×10^{-11}	7.4	3.0×10^{-5} to 1.0×10^{-4}
^{58}Fe (n,γ)^{59}Fe	4.13×10^{-13}	5.5	1.0×10^{-4} to 2.0×10^{-1}
^{58}Fe(n,γ)^{59}Fe[a]	2.77×10^{-13}	−35.7	1.0×10^{-4} to 2.0×10^{-1}
Fe(n,x)^{54}Mn	1.74×10^{-12}	−4.4	3 to 400
Fe(n,x)^{51}Cr	9.12×10^{-13}	−3.4	32 to 440
^{58}Ni(n,p)^{58}Co	4.19×10^{-13}	−23.7	2 to 20
^{58}Ni(n,x)^{57}Ni	2.71×10^{-14}	70.0	140 to 200
^{60}Ni(n,p)^{60}Co	4.58×10^{-14}	−13.4	6 to 180
^{59}Co(n,p)^{59}Fe	1.94×10^{-14}	−0.6	3 to 24
^{59}Co($n,2n$)^{58}Co	2.18×10^{-13}	6.1	12 to 36
^{59}Co($n,3n$)^{57}Co	8.28×10^{-14}	29.0	20 to 60
Al(n,x)^{22}Na	1.41×10^{-14}	−64.7	28 to 400
Al(n,x)^{7}Be	1.36×10^{-15}	0.4	48 to 560
Cu(n,x)^{46}Sc	8.46×10^{-16}	7.9	180 to 560
Cu(n,x)^{51}Cr	1.01×10^{-14}	18.1	120 to 540
Cu(n,x)^{52}Mn	3.54×10^{-15}	−4.0	110 to 520
Cu(n,x)^{60} Co	2.71×10^{-14}	66.7	36 to 460
Cu(n,x)^{59}Fe	2.43×10^{-15}	65.2	72 to 500

[a] Gadolinium-wrapped foils.

TABLE 3—*Corrected activities for irradiation next to the IP targets with two targets (Irradiation 936-2) in the beam. "Rabbits" are stacked on top of each other in tube one. "Rabbit" three is at beam centerline.*

Reaction	"Rabbit" 1	"Rabbit" 2	"Rabbit" 3	"Rabbit" 4	"Rabbit" 5
$Fe(n,x)^{54}Mn$	1.5×10^{-12}	1.5×10^{-12}	1.6×10^{-12}	1.4×10^{-12}	1.4×10^{-12}
$Fe(n,x)^{51}Cr$	7.6×10^{-13}	7.7×10^{-13}	7.6×10^{-13}	7.3×10^{-13}	7.2×10^{-13}
$^{58}Fe(n,\gamma)^{59}Fe$	3.2×10^{-13}	3.3×10^{-13}	3.2×10^{-13}	3.1×10^{-13}	3.3×10^{-13}
$^{59}Co(n,p)^{59}Fe$	9.2×10^{-15}	1.2×10^{-14}	1.1×10^{-14}	1.1×10^{-14}	9.6×10^{-15}
$^{59}Co(n,\gamma)^{60}Co$	7.2×10^{-12}	9.0×10^{-12}	9.0×10^{-12}	9.0×10^{-12}	8.1×10^{-12}
$^{59}Co(n,2n)^{58}Co$	1.1×10^{-13}	1.4×10^{-13}	1.5×10^{-13}	1.5×10^{-13}	1.2×10^{-13}
$^{59}Co(n,3n)^{57}Co$	4.3×10^{-14}	5.4×10^{-14}	5.4×10^{-14}	5.4×10^{-14}	4.8×10^{-14}
$^{58}Ni(n,p)^{58}Co$	5.0×10^{-13}	4.0×10^{-13}	4.1×10^{-13}	3.8×10^{-13}	3.8×10^{-13}
$^{60}Ni(n,p)^{60}Co$	6.9×10^{-14}	5.1×10^{-14}	6.0×10^{-14}	5.2×10^{-14}	4.9×10^{-14}

the normalized activity for Irradiation 936-5 radially from beam centerline. The activities are normalized to the corresponding activity in Tube 1. The reactions $^{58}Fe(n,\gamma)^{59}Fe$, $^{58}Ni(n,p)^{58}Co$, $Fe(n,x)^{51}Cr$, and $Cu(n,x)^{52}Mn$ were chosen because of the neutron energy limits of the reaction. These energy groups are 1×10^{-4} to 2×10^{-1}, 2 to 20, 36 to 420, and 110 to 520 MeV, respectively, as shown in the last column of Table 2. Reaction $^{65}Cu(p,n)^{65}Zn$ was added to compare the decrease in the activation for a proton-induced reaction (secondary protons) to the neutron-induced reactions. The activity for $^{58}Fe(n,\gamma)^{59}Fe$ decreases radially at the slowest rate for the reactions chosen. $Fe(n,x)^{51}Cr$ and $^{58}Ni(n,p)^{58}Co$ decrease at an intermediate rate to low and high energy reactions. The high-energy neutron and proton activities decrease at the same rate between

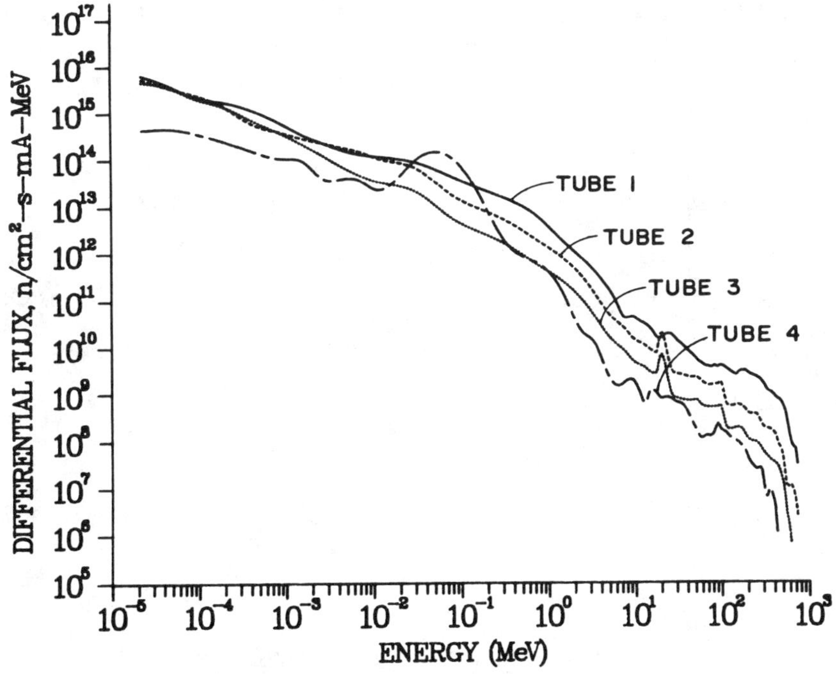

FIG. 4—*Differential flux for the dosimetry insert next to the IP targets (Irradiation 936-5, Tubes 1 to 4) that are in the beam.*

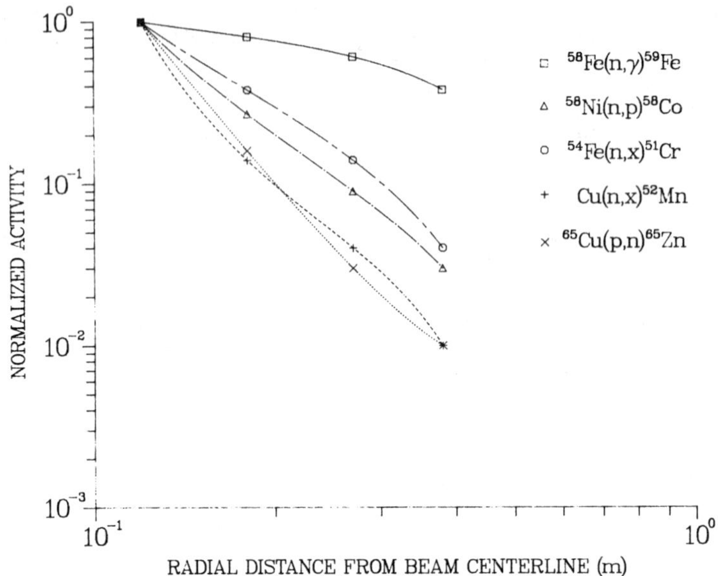

FIG. 5—*Normalized activity versus radial distance from beam centerline for dosimetry insert next to the IP targets (Irradiation 936-5 that are in the beam.*

tubes one and four. Figure 6 shows the activity in atoms/atom-s for the same reactions. Note that the proton-induced activity decreases by two orders of magnitude between Tubes 1 and 4.

Tubes 1 and 2 show a similar spectral shape for irradiations next to the beam stop with zero (Irradiation 936-6) and eight (Irradiation 936-7) IP targets in the beam (Figs. 7 and 8, respectively).

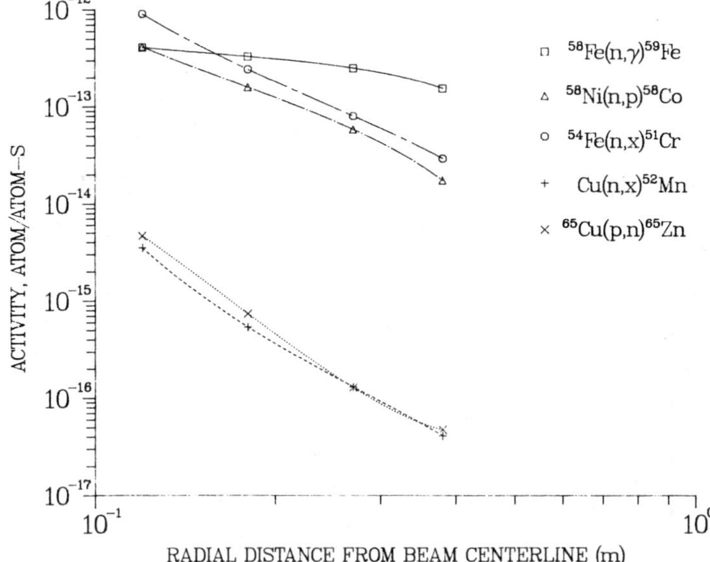

FIG. 6—*Activity versus radial distance from beam centerline for dosimetry insert next to the IP targets (Irradiation 936-5) that are in the beam.*

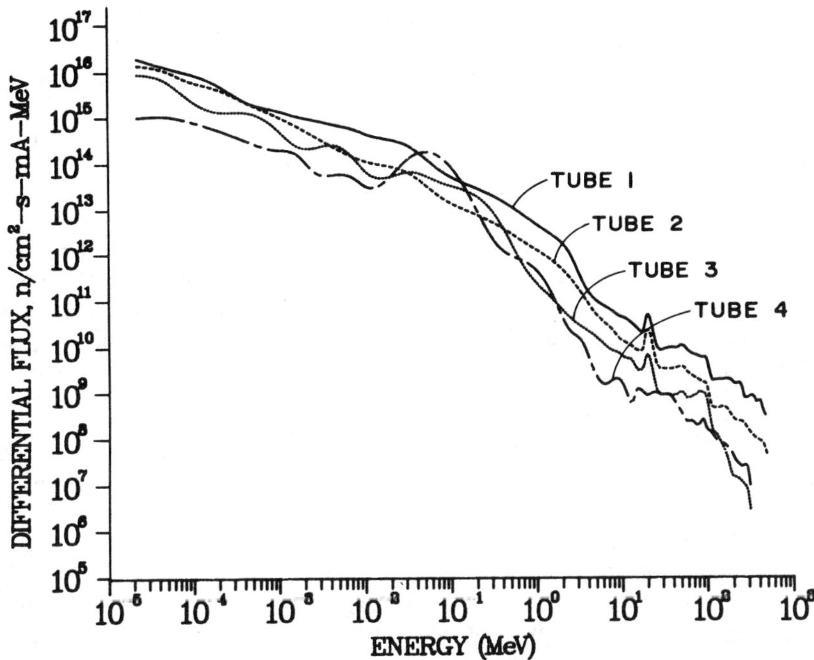

FIG. 7—*Differential flux for the dosimetry insert next to beam stop with no (Irradiation 936-6) IP targets in the beam.*

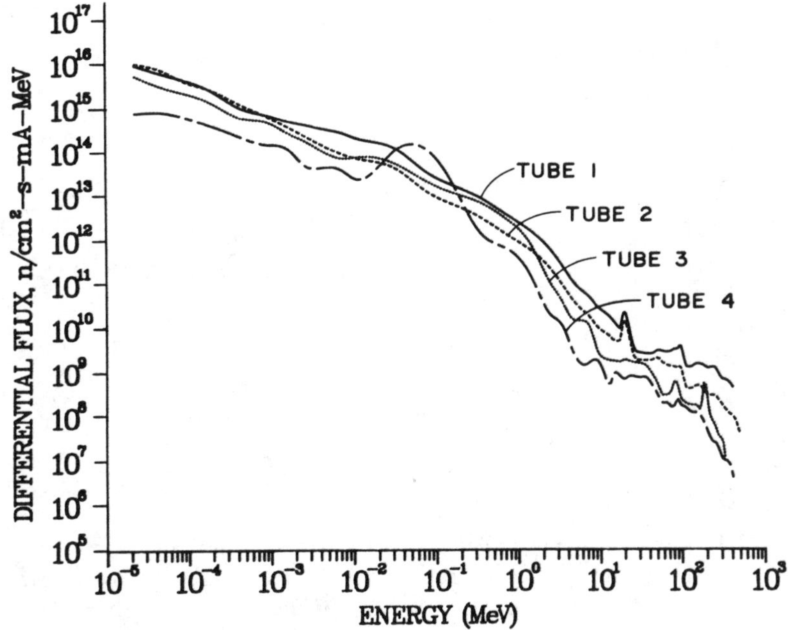

FIG. 8—*Differential flux for the dosimetry insert next to the beam stop with eight (Irradiation 936-7) IP targets in the beam.*

Tube 3 shows a slightly enhancement above 0.01 MeV for both irradiations. This enhancement is increased for Tube 4. Again the peak is shifted to higher energies because of the limited reactions analyzed in this energy region. The scatter in the spectra above 40 MeV is due to uncertainties in the cross sections and input spectra. The integral fluxes for Irradiation 936-6 are 4.6×10^{17}, 1.6×10^{17}, 1.4×10^{17}, and 1.8×10^{17} n/m^2-s-mA for Tubes 1, 2, 3 and 4, respectively. The integral fluxes for Irradiation 936-7 are 2.2×10^{17}, 9.9×10^{16}, 1.2×10^{17}, and 1.5×10^{17} n/m^2-s-mA for Tubes 1, 2, 3, and 4, respectively.

The fluxes for Irradiations 936-5 and 936-7, Tube 1, are shown in Fig. 9 to compare the energy spectra along the beamline. The spectra are similar below 20 MeV. The integral fluxes are 2.8×10^{17} and 2.2×10^{17} n/m^2-s-mA for Irradiations 936-5 and 936-7, respectively. The difference in the spectra outside the IP targets and the beam stop lies in the high-energy tail (energies above 20 MeV). In this energy region, the flux decreases by a factor of two for Irradiation 936-7 compared to Irradiation 936-5. Fewer high-enegy neutrons for Irradiation 936-7 result from the degradation of the energy of the proton beam in the IP targets prior to interaction with the beam stop. Table 4 lists the group energy fluxes for Irradiations 936-5 through 936-7.

The effect of IP target loading is shown in Fig. 10 for Irradiations 936-6 and 936-7, Tube 1. The shapes of the spectra are the same within statistial variations. This indicates that the shape is primarily determined by the target interacting with the photon beam (*i.e.*, the copper beam stop). The magnitude of the integral flux changes from $4.6 \times 10^{17} \pm 8\%$ to $2.2 \times 10^{17} \pm 10\%$ n/m^2-s-mA in going from no IP targets (Irradiation 936-6) to eight (Irradiation 936-7). The high energy tail (neutron energy above 20 MeV) is 40% lower for Irradiation 936-7 than 936-6. A comparison of the group energy fluxes is listed in Table 4.

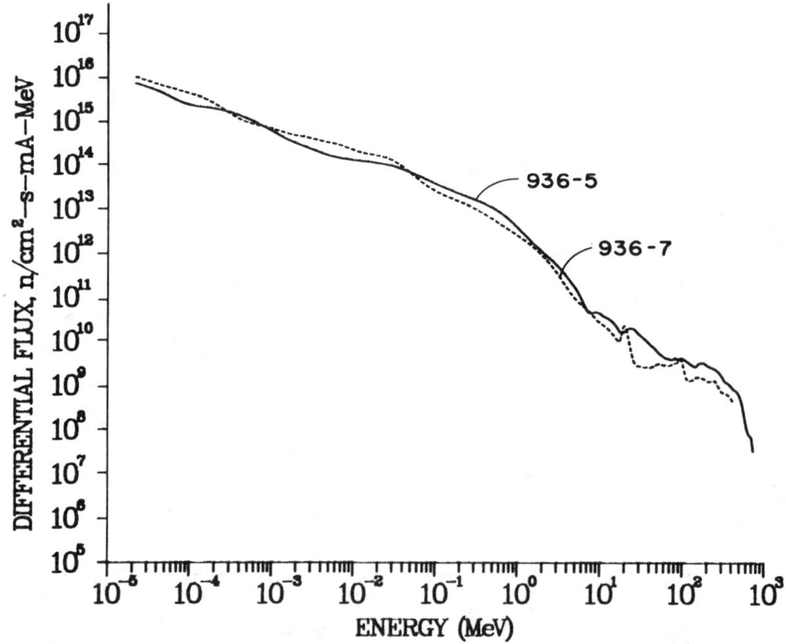

FIG. 9—*Comparison of the differential flux in Tube 1 for irradiations next to the IP targets (Irradiation 936-5) and next to the beam stop (Irradiation 936-7).*

TABLE 4—*Group fluxes for irradiations next to the IP targets (Irradiation 936-5), next to beam stop with no IP targets (Irradiation 936-6) in the beam, and next to the beam stop with eight IP targets (Irradiation 936-7) in the beam.*

Energy, MeV	936-5 n/m²-s-mA	936-6 n/m²-s-m/A	936-7 n/m²-s-mA
<0.1	9.7×10^{16} (±12)	2.7×10^{17} (±13)	1.2×10^{17} (±15)
0.1–1.0	1.1×10^{17} (±8)	1.3×10^{17} (±8)	6.5×10^{16} (±8)
1–10	3.7×10^{16} (±10)	4.8×10^{16} (±10)	2.8×10^{16} (±12)
10–20	2.6×10^{16} (±17)	3.4×10^{15} (±15)	1.7×10^{15} (±10)
20–40	3.0×10^{15} (±17)	3.0×10^{15} (±19)	1.1×10^{15} (±14)
40–100	2.9×10^{15} (±18)	4.8×10^{15} (±20)	2.1×10^{15} (±17)
100–200	3.3×10^{15} (±28)	2.1×10^{15} (±20)	1.5×10^{15} (±26)
>200	4.8×10^{15} (±20)	1.3×10^{15} (±19)	2.0×10^{15} (±16)
Total	2.8×10^{17} (±8)	4.6×10^{17} (±8)	2.2×10^{17} (±7)

Conclusions

"Rabbits" with high-purity activation foils were successfully irradiated to measure the neutron flux and energy spectrum as a function of radial distance from the beam centerline and above and below beam centerline. The neutron flux in Neutron Irradiation Location 2 decreased by a factor of 6 in iron moving radially from beam centerline from Tube 1 to Tube 3. Between 0.27 and 0.38 m, Irradiation 936-5 showed an increase in the integral neutron flux due to the enhancement in the 24-keV region. Irradiations outside the beam stop with zero (Irradiation 936-6) and eight (Irradiation 936-7) IP targets in the beam showed little effect of the IP target loading on the spectral shape, but they did show a decrease by a factor of 2 in the magnitude of the flux. This

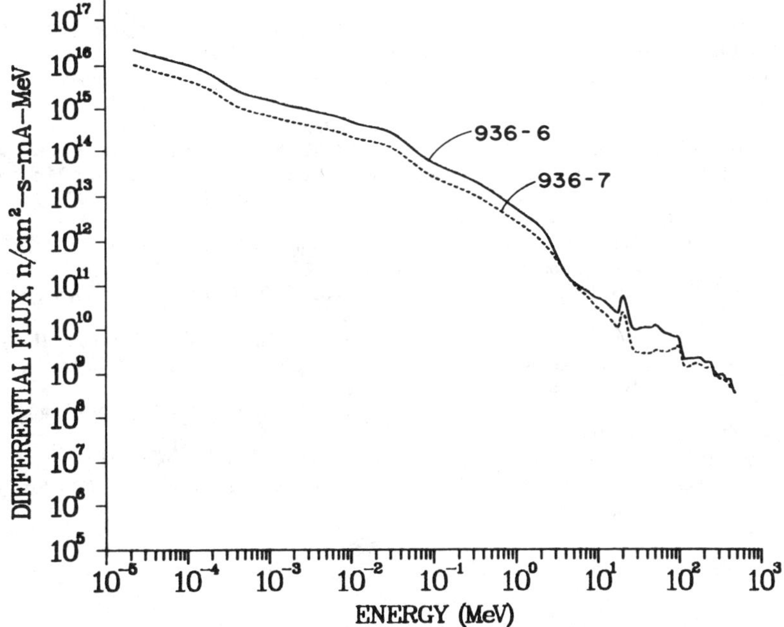

FIG. 10—*Comparison of the differential flux in Tube 1 next to the beam stop with no IP targets (Irradiation 936-6) and eight IP targets (Irradiation 936-7) in the beam.*

decrease was particularly evident in the high energy tail (energy above 20 MeV) for the irradiation with eight IP targets. For similar IP target loadings, the integral flux outside the IP targets was similar to that outside the beam stop, yet the flux above 20 MeV was a factor of 2 greater outside the IP targets. This is due to degradation of the energy of the proton beam in the IP targets prior to interaction in the copper beam stop. Irradiations placing five "rabbits" on top of one another in Tube 1 showed no difference in the activities below, on, and above beam centerline. As a result of the irradiations, a modified dosimetry package has been chosen based on known cross sections and reactions. The foil package will include iron, nickel, cobalt, and copper foils and will be irradiated with all experiments at LASREF.

Acknowledgments

We thank R. Chavez and T. Montoya for assistance with drawings and photographs.

References

[*1*] Davidson, D. R., Sommer, W. F., Bradbury, J. N., Prael, R. E., and Little, R. C., LA-UR-83-33, Los Alamos National Laboratory, Los Alamos, NM, 1983.

[*2*] Davidson, D. R., Greenwood, L. R., Reedy, R. C., and Sommer, W. F., "Measured Radiation Environment at the Clinton P. Anderson Los Alamos Meson Physics Facility (LAMPF) Irradiation Facility," *Effects of Radiation on Materials: Twelfth International Symposium, ASTM STP 870,* F. A. Garner and J. S. Perrin, Eds., American Society for Testing and Materials, Philadelphia, 1985, pp. 1199–1208.

[*3*] Wechsler, M. S. and Sommer, W. F., *Journal of Nuclear Materials,* Vol. 123, 1984, p. 1078.

[*4*] Wechsler, M. S., Davidson, D. R., Greenwood, L. R., and Sommer, W. F., "Calculation of Displacement and Helium Production at the Clinton P. Anderson Los Alamos Meson Physics Facility (LAMPF) Irradiation Facility," *Effects of Radiation on Materials: Twelfth International Symposium, STP 870,* F. A. Garner and J. S. Perrin, Eds., American Society for Testing and Materials, Philadelphia, 1985, pp. 1189–1198.

[*5*] Sommer, W. F., Lohmann, W., Graft, K., Taylor, I. K., and Chavez, R. M., "Operating Experience at the Los Alamos Spallation Radiation Effects Facility." *Influence of Radiation on Materials: Thirteenth International Symposium STP 956,* American Society for Testing and Materials, Philadelphia, 1987, pp. 718–729.

[*6*] Gunnink, R. and Niday, J. B., "Computerized Quantitative Analysis by Gamma-Ray Spectrometry. Vol. 1. Description of the GAMANAL Program," UCRL-51061, Lawrence Livermore Laboratory, Livermore, CA 94550. (Modified at Los Alamos National Laboratory, Los Alamos, NM 87545.)

[*7*] Perey, F. G., ORNL/TM-6062, Oak Ridge National Laboratory Oak Ridge, TN, 1977. (The code has been modified by L. R. Greenwood, Argonne National Laboratory, Argonne, IL 60439.

[*8*] Kinsey, R., "ENDF-201 ENDF/B Summary Documentation," BNL-NCS-17541 ENDF 201, 3rd ed., ENDF/V-B, Brookhaven National Laboratory, July 1979.

[*9*] Greenwood, L R., Heinrich, R. R., Saltmarsh, M. J., and Fulmer, C. B., *Nuclear Science Engineering,* Vol. 72, 1979, p. 175.

[*10*] Greenwood, L. R. and Smither, R. K., "Measurement of Spallation Cross Sections at IPNS, Damage Analysis and Fundamental Studies," in Quarterly Progress Report, DOE/ER-0046/14, Department of Energy, Washington, D. C., August 1983, pp. 19–24.

[*11*] Tobailem, J., "Cross Sections of the Nuclear Reactions Induced by Protons, Deuterons, and Alpha Particles," CEA-N1466, 1971.

[*12*] Michel, R. and Stuck, R., "On the Production of Cosmogenic Nuclides in Meteorites by Primary Galactic Particles: Cross Sections and Calculations," *Journal of Geophysical Research,* Vol. 89, Supplement, 1984, pp. B673–684.

Dosimetry of Radiation Environments

Lawrence R. Greenwood[1]

Recent Research in Neutron Dosimetry and Damage Analysis for Materials Irradiations

REFERENCE: Greenwood, L. R., "**Recent Research in Neutron Dosimetry and Damage Analysis for Materials Irradiations,**" *Influence of Radiation on Material Properties: 13th International Symposium (Part II), ASTM STP 956,* F. A. Garner, C. H. Henager, Jr., and N. Igata, Eds., American Society for Testing and Materials, Philadelphia, 1987, pp. 743–749.

ABSTRACT: Neutron fluence measurements and damage calculations are being routinely performed for fusion materials irradiations in fission reactors, at 14 MeV sources, and at high-energy accelerator neutron sources. Neutron activation cross sections are being developed for dosimetry at fission reactors, fusion reactors, and high-energy accelerator neutron sources. Integral and differential measurements are being combined to determine new cross sections and to adjust discrepant data. These data are needed to adequately define the neutron exposure and radiation damage in materials. The production of long-lived radioisotopes is being measured at 14 MeV for fusion reactors and reactions are being developed for plasma diagnostics. At high energies, spallation cross sections are being measured for dosimetry and damage analysis at accelerator neutron sources. Helium production cross sections are being tested and a new thermal helium effect has been discovered in copper similar to the well-known effect in nickel. Damage calculations are routinely provided by the SPECTER computer code and new calculations are being performed to extend our damage cross sections to higher energies for accelerator neutron sources and to include alloys and compounds.

KEY WORDS: neutron dosimetry, damage calculations, fusion materials, irradiation, helium production

Fusion materials irradiations require accurate measurements of the neutron exposure and calculations of the induced damage in the materials of interest. The neutron flux and energy spectrum are measured by neutron activation analysis in various materials folowed by spectral adjustment of calculated neutron spectra. Atomic displacements and gas production are then calculated using our SPECTER computer code [*1*]. The ability of these techniques depends on our knowledge of nuclear cross sections and their uncertainties. Consequently, much of our recent research is devoted to the measurement of neutron cross sections and the development of computer codes and procedures. These measurements span a large range of neutron energies since fusion materials studies are currently in progress in both mixed-spectrum and fast fission reactors, 14 MeV accelerator neutron sources, and higher-energy accelerator neutron sources. Data and procedures are also being developed for use at future fusion reactors.

The present paper summarizes some of our current cross-section measurements and improvements in radiation damage calculations. Specifically, neutron activation cross-section measurements are presented for 22 reactions and production cross-section measurements are presented for the long-lived isotopes ^{94}Nb and ^{91g}Nb from Mo and ^{94}Mo in the energy range from 14.5 to 14.9 MeV. Data is also presented for an enhanced thermal-neutron helium production in copper. Cross sections have been deduced for all of the important stages of this reaction from ^{63}Cu to ^{65}Zn which produces

[1] Physicist, Chemical Technology Division, Argonne National Laboratory, 9700 South Cass Ave., Argonne, IL 60439.

the extra helium. Results are presented for a new computer code, SPECOMP, which calculates displacement damage in alloys, insulators, and other compounds.

Cross-Section Measurements Near 14 MeV

Neutron activation cross section measurements near 14 MeV have been completed for 22 separate reactions using the Rotating Target Neutron Source II (RTNS II) at Lawrence Livermore National Laboratory. Samples were placed at five different angles between 0 and 60° at five different distances between 5 and 380 cm. The resultant cross sections are listed in Table 1. All of the measurements have a relative accuracy of 1.5%, except as stated. All of the measurements are correlated since the materials were irradiated simultaneously; however, the gamma measurements are relatively uncorrelated since all samples were measured several times on separately calibrated detectors and the results averaged. The neutron spectra were calculated at each angle, as shown in Fig. 1. Previous data [*2–4*] were then averaged over the calculated spectra prior to the comparisons shown in Table 1. Corrections were made for absorption, finite solid angles, and room-return effects. The absolute values are normalized to the ^{93}Nb($n,2n$) reaction which is known to about ±4% with a flat energy dependence near 14 MeV [*5*] . Nuclear decay data was taken from Ref *6*.

Clearly it is not possible in the confines of the present paper to discuss each of the 22 reactions in detail or to compare each one with the numerous previous measurements. Consequently, we have only attempted to compare our results with the best-available evaluations of these reactions. Wherever possible we have compared our results to the ENDF/B-V Dosimetry Data File [*2*]. However, some of our reactions are not contained in this file, and we have thus used either ENDF/B-IV or other recent references [*3,7*]. As can be seen, several of the reactions have sizable differences with the ENDF/B-V Dosimetry File. The largest differences are for Ti(n,x)^{47}Sc (30%), ^{60}Ni(n,p)^{60}Co (21%), and ^{58}Ni(n,p)^{58}Co (13%). It is also important to note that several of the reactions which were thought to be well-known in ENDF/B-V indeed show good agreement, such as ^{27}Al(n,α)^{24}Na, ^{54}Fe(n,p)^{54}Mn, and ^{197}Au($n,2n$)^{196}Au. In the case of ^{54}Fe (n,α)^{51}Cr our values are in good agreement with the suggested curve in BNL 325 [*4*] and with a recent evaluation by Evain, et al [*7*] of 87.9 mb at 14.7 MeV. For the ^{59}Co(n,p)^{59}Fe reaction, our values near 47 mb are substantially lower than the suggested curve in BNL 325 [*4*] or the recent value of 56.6 mb from Ref *6* at 14.7 MeV. However, previous measurements for this reaction show a large scatter [*4,7*]. Measurements for ($n,2n$) reactions on Y, Zr, Ag, and Tm are in good agreement with recent measurements [*3*]. The measurements of the ^{54}Fe(n,p) reaction are particularly useful since they are being used to monitor long irradiations at RTNS II thanks to the relatively long half-life of ^{54}Mn (312 days). The data can also be used for dosimetry at future fusion reactors.

Fusion Reactor Applications

One of the concerns regarding fusion reactor materials is the activation of very long-lived isotopes, especially regarding maintenance and disposal. In addition to the above applications, some of these reactions can also be used for neutron dosimetry and fusion plasma diagnostics. Unfortunately, many of the production cross sections near 14 MeV are very poorly known. Consequently, we have undertaken to measure some of these reactions using high-fluence neutron irradiations at RTNS II. The reactions ^{27}Al($n,2n$)^{26}Al(7.2 × 10^{5}y) and ^{54}Fe($n,2n$)^{53}Fe to ^{53}Mn (3.7 × 10^{6}y) have been reported previously [*8,9*]. New measurements have been completed for the production of ^{94}Nb (2.03 × 10^4 y) from ^{94}Mo and ^{Nat}Mo [*10*]. The results are shown in Table 2. The measurements were made at the RTNS II for an exposure of 81 days over a period of 7 months. This long exposure then permitted us to directly measure the gamma emissions from ^{94}Nb using Ge gamma spectroscopy. Measurements are also reported in Table 2 for the reactions ^{92}Mo

TABLE 1—*Cross section measurements (mb) near 14 MeV.*[a]

Reaction MeV	0° 14.90	15° 14.85	30° 14.80	45° 14.65	60° 14.50
$^{27}Al(n,\alpha)^{24}Na$	108	107	111	113	117
ENDF	113	114	115	117	119
$^{45}Sc(n,2n)^{44}Sc$	128	...	...	...	117
ENDF[b]	124	...	...	...	112
$^{45}Sc(n,\alpha)^{42}K$[c]	54	...	...	...	58
$Ti(n,x)^{46}Sc$	24.4	24.7	24.5	24.4	24.6
ENDF	26.6	26.5	26.4	26.2	25.8
$Ti(n,x)^{47}Sc$	23.5	23.5	22.3	21.2	19.4
ENDF	18.2	17.9	17.5	16.7	15.8
$^{48}Ti(n,p)^{48}Sc$	66.3	67.5	67.6	69.4	66.0
ENDF	61.9	62.1	62.2	62.6	63.0
$^{55}Mn(n,2n)^{54}Mn$	840	827	807	825	791
ENDF	786	784	779	770	759
$^{54}Fe(n,p)^{54}Mn$	296	298	303	311	331
ENDF	289	292	298	309	324
$^{54}Fe(n,\alpha)^{51}Cr$	92.7	93.3	92.6	92.5	91.2
BNL-325[b]	95.8	95.7	95.3	94.8	94.1
$^{59}Co(n,p)^{59}Fe$	45.9	45.9	46.1	47.5	48.9
BNL-325	62.2	62.5	62.8	63.3	63.8
$^{59}Co(n,2n)^{58}Co$	803	796	800	789	771
ENDF	819	818	814	806	794
$^{58}Ni(n,p)^{58}Co$	295	295	303	312	329
ENDF	335	339	346	360	378
$^{58}Ni(n,2n)^{57}Ni$	40.9	41.2	40.0	38.2	33.1
ENDF	36.7	36.2	35.4	33.7	31.5
$^{60}Ni(n,p)^{60}Co$	132	132	135	138	142
ENDF	108	109	110	113	116
$^{63}Cu(n,\alpha)^{60}Co$[d]	40.1	38.8	38.4	40.4	41.2
ENDF	38.4	38.7	39.1	40.0	40.9
$^{89}Y(n,2n)^{88}Y$	991	...	...	...	929
LASL[b]	942	...	...	...	887
$^{89}Y(n,\alpha)^{86}Rb$[e]	5.8	...	...	...	5.5
$Zr(n,x)^{89}Zr$	846	850	824	796	737
LASL	799	791	776	745	703
$^{96}Zr(n,2n)^{95}Zr$	1593	1599	1581	1580	1608
$^{107}Ag(n,2n)^{106}Ag$	567	...	551	...	547
LASL	612	...	605	...	581
$^{169}Tm(n,2n)^{168}Tm$	1946	...	...	...	1939
LASL	2034	...	...	...	2022
$^{197}Au(n,2n)^{196}Au$	2174	2172	2171	2154	2151
ENDF	2109	2113	2118	2130	2142

[a] Values normalized to ^{93}Nb $(n,2n)$ at 463 mb (±4%).
[b] ENDF/B-V = Ref *2;* LASL = Ref *3;* BNL-325 = Ref *4.*
[c] ±10%.
[d] ±3%.
[e] ±13%.

$(n,2n)^{91}Mo$ to ^{91m}Nb, $^{98}Mo(n,\alpha)^{95}Zr$, and $^{95}Mo(n,p)^{95}Nb$. The reaction to ^{91m}Nb is also of interest since this isotope decays to the ground state ^{91g}Nb which has a half-life of about 700 years. Our data provides a good estimate of the production cross section to this neglected isotope, as shown in Table 2. These new measurements allow us to calculate the production of ^{94}Nb and ^{91g}Nb in fusion reactor materials. A first-wall (STARFIRE) irradiation of Mo to 20 MW-y/m² will produce about 70 μCi/g of ^{94}Nb and 11 mCi/g of ^{91g}Nb [*10*]. Measurements on other long-lived isotopes

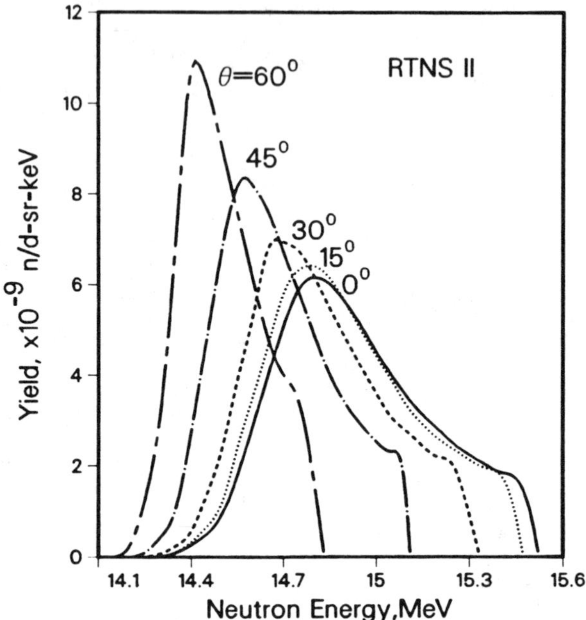

FIG. 1—*Calculated neutron energy spectra are shown for various angles with respect to the incident 360 keV deuteron beam on a TiT_2 target.*

including ^{55}Fe (2.7 y), ^{63}Ni (100 y), ^{59}Ni (7.6×10^4 y), and ^{93}Mo (3500 y) are now in progress using a variety of techniques including radiochemistry, liquid scintillation counting, and accelerator mass spectrometry.

New Thermal Helium Effect In Copper

A new thermal helium production mechanism has been recently discovered in copper [*11*] similar to the well-known helium effect in nickel. In a three-stage process ^{63}Cu captures a neutron to form ^{64}Cu, which then partly decays to ^{64}Zn. The ^{64}Zn then captures a neutron to form ^{65}Zn,

TABLE 2—*Measured cross sections (mb) for Mo.*

Reaction	14.55	14.60	14.78	14.80	±%[a]
$^{94}Mo(n,p)^{94}Nb$	57.2	. . .	53.1	. . .	10
$^{Nat}Mo(n,x)^{94}Nb$ [b]	. . .	7.9	. . .	7.8	11
$^{95}Mo(n,x)^{94}Nb$ [c]	. . .	16.3	. . .	18.3	15
$^{95}Mo(n,p)^{95}Nb$	40.4	. . .	37.1	. . .	6
$^{92}Mo(n,x)^{91m}Nb$ [d]	157.	153.	145.	145	7
$^{98}Mo(n,\alpha)^{95}Zr$	6.56	6.56	6.24	6.21	6
$^{92}Mo(n,x)^{91g}Nb$ [e]	≈300	. . .	. . .	. . .	. . .
$^{Nat}Mo(n,x)^{91g}Nb$ [e]	≈45	. . .	. . .	. . .	. . .

[a] Major sources of uncertainty includes neutron fluence (5%), ^{94}Nb half-life (8%), efficiency (1.5%), statistics (1%), and deconvolution of ^{95}Mo (12%) for ^{94}Nb, (2%) for ^{95}Nb.

[b] Sum of reactions from $^{94,95,96}Mo$.

[c] Sum of $(n,d + np + pn)$ reactions.

[d] Sum of $(n,2n + d + np + pn)$ reactions.

[e] Values include contribution from ^{91m}Nb.

TABLE 3—*Thermal cross sections(b) for He production in Cu.*

Reaction	Data[a]	ENDF/B-V
$^{64}Cu(n,\gamma)^{65}Cu$	270 ± 170	<6000
$^{65}Zn(n,abs)$[b]	66 ± 8	. . .
$^{65}Zn(n,\alpha)^{62}Ni$	4.7 ± 0.5	250 ± 150

[a] Data measured in HFIR with 7% epithermal flux.
[b] Total absorption includes (n,γ), (n,p), and (n,α).

which we have shown to have a large thermal cross section for the production of helium. A combination of radiometric and mass spectrometric data from Argonne and helium measurements from Rockwell International has allowed us to uniquely determine all key cross sections involved in this reaction process. These results are pesented in Table 3. The helium data and calculations are shown in Fig. 2.

This reaction is of interest to fusion materials irradiations involving either copper or zinc since the process will produce more helium than anticipated. Similar to the thermal effect in nickel, this effect on copper can be used to enhance the helium-to-displacement damage ratio, thereby improving the simulation of fusion reactor damage for irradiations in mixed-spectrum (part thermal, part fast) fission reactors such as the High Flux Isotopes Reactor (HFIR) at Oak Ridge National Laboratory where many fusion materials irradiations are now in progress.

The effect in copper is not as pronounced as in nickel since three steps are required rather than two. Nevertheless, this process will equal the fast helium production in only a month in HFIR and will produce ten times more helium after a year. A crude equation describing this effect in HFIR is, as follows:

$$\text{Enhanced } ^4\text{He (appm)} = 0.669 \times \phi^{2.58} \tag{1}$$

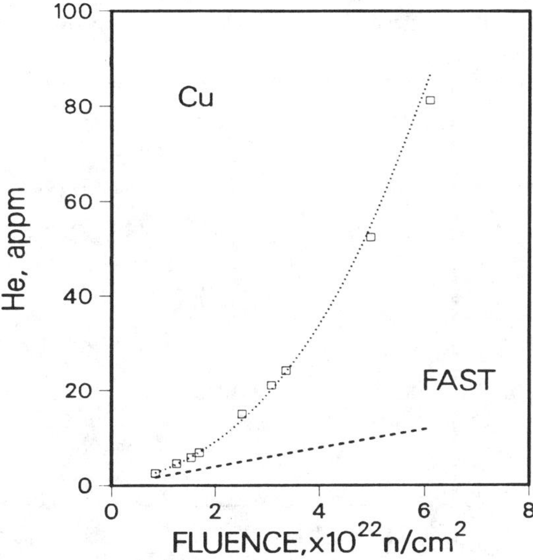

FIG. 2—*Measured and predicted helium production in copper irradiated in HFIR as a function of the thermal neutron fluence. The dashed line shows the contribution from fast neutrons (the fast neutron fluence* >*0.1 MeV is about 0.8 times the thermal fluence shown). The dotted line through the data is based on calculations using the data in Table 3.*

where ϕ is the thermal neutron fluence in units of 10^{22} n/cm^2. It must also be noted that our data show that the fast neutron helium production is 31% higher than in ENDF/B-V [*2*]. The thermal reaction on Zn will also produce some extra displacement damage in the ratio of 1 displacement per atom (dpa) per 492 appm extra helium; however, this is insignificant relative to the fast neutron damage, unlike the substantial damage effect in nickel [*12,13*]. Further measurements are in progress and we hope to determine the cross sections more accurately and to establish a more general equation for estimating the effect.

New SPECOMP Computer Code for Damage in Compounds

The SPECTER [*1*] computer code is routinely used to calculate displacement damage and gas production in 38 elements of interest to the fusion materials program. However, the code does not calculate damage in alloys, compounds, or insulator materials since one has to consider all possible interactions of recoil atoms with atoms in the matrix [*14*]. A new addition to the SPECTER computer code package has recently been developed in collaboration with M. Lazo (University of New Mexico, Albuquerque) which performs damage calculations for compound materials. The code SPECOMP is based on the SPECTER primary recoil libraries. Given the composition of any mixture of elements in SPECTER, SPECOMP integrates over the primary recoil distributions times the secondary recoil functions using a Lindhard model for each combination of recoil atom and matrix atom according to the prescription of Parkin, et al [*14*]. Sample calculations for $LiAlO_2$ are shown in Fig. 3. Displacement cross sections are also shown for the elemental mixture of Li, Al, and O. As can be seen, the difference between simply adding the elemental damage and performing exact calculations for the compound are about 30–40% for lithium aluminate. However, for compounds such as Al_2O_3 and SiO_2 the differences are quite small. This behaviour agrees with predictions that the compound effect is small where the masses involved are similar [*14*]. Calculations for other alloys, compounds, and insulator materials are in progress. We plan to publish a more complete description of the code with damage functions for various materials and to make the code available for general use. One of the advantages of this approach with SPECTER and SPECOMP is that it is not necessary to recalculate the recoil atom energy distributions.

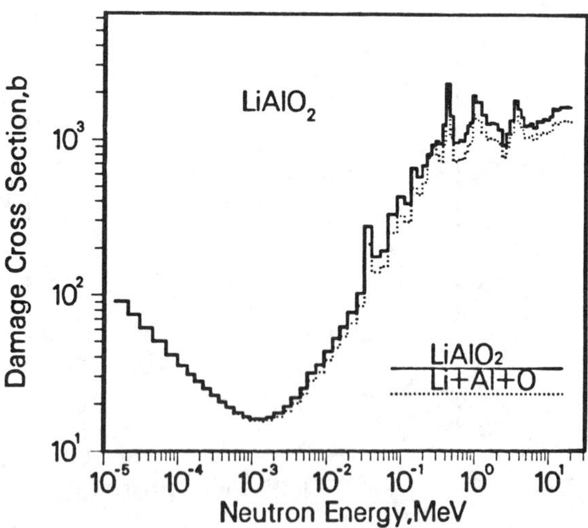

FIG. 3—*Calculated displacement damage cross section for $LiAlO_2$. The exact calculations are compared to the elemental sum of Li, Al, and O cross sections. As can be seen, the differences are typically 30–40% for lithium aluminate.*

Consequently, the codes are relatively small and can be run cheaply and quickly on small computers and they require no access to the ENDF data files.

Acknowledgments

The author would like to acknowledge the assistance of M. Guinan (Lawrence Livermore National Laboratory) and D. Kneff (Rockwell International) in the 14 MeV cross-section measurements at RTNS II, D. Doran and H. Heinisch (Hanford Engineering Development Laboratory) in the Mo measurements, and M. Lazo (University of New Mexico, Albuquerque) in the development of the SPECOMP computer code. This work was supported by the Office of Fusion Energy, U.S. Department of Energy.

References

[*1*] Greenwood, L. R. and Smither, R. K., "SPECTER: Neutron Damage Calculations for Materials Irradiations," ANL/FPP/TM-197, January 1985.

[*2*] "Evaluated Neutron Data File, Part B, Version V," Dosimetry Data File, Brookhaven National Laboratory, 1979.

[*3*] Bayhurst, B. P., Gilmore, J. S., Prestwood, R. J., Wilhelmy, J. B., Jarmie, N., Erkkila, B. H., and Hardekopf, R. A., *Physical Review*, Vol. C12, (1975), pp. 451–467.

[*4*] "Neutron Cross Sections," Vol. 2, BNL-325, 1976.

[*5*] Nethaway, D. R., *Journal of Inorganic Nuclear Chemistry*, Vol. 40, 1978, p. 1285.

[*6*] Kocher, D. C., "Radioactive Decay Data Tables," DOE/TIC-11026, 1981.

[*7*] Evain, B. P., Smith, D. L., and Lucchese, P., ANL/NDM-89, April 1985.

[*8*] Smither R. K. and Greenwood, L. R., *Journal of Nuclear Materials*, Vol. 122, 1984, pp. 1071–1077.

[*9*] Smither R. K. and Greenwod, L. R., "Damage Analysis and Fundamental Studies," Quarterly Progress Report, DOE/ER-0046/17, May 1984.

[*10*] Greenwood, L. R., Doran, D. G., and Heinisch, H. L., "Production of ^{91}Nb, ^{94}Nb, and ^{95}Nb from Mo by 14.5–14.8 MeV Neutrons," in *Physical Review*, Vol. 35, 1987, pp. 76–80.

[*11*] Kneff, D. W., Greenwood, L. R., Oliver, B. M., Skowronski, R. P., and Callis, E. L., Helium Production in Copper by a Thermal Three-Stage Reaction," *Radiation Effects*, Vols. 92–96, 1986, pp. 553–556.

[*12*] Greenwood, L. R., *Journal of Nuclear Materials*, Vol. 115, 1983, pp. 137–142.

[*13*] Greenwood, L. R., Kneff, D. W., Skowronski, R. P., and Mann, F. M., *Journal of Nuclear Materials*, Vol. 122, 1984, pp. 1002–1010.

[*14*] Parkin, D. M. and Coulter, C. A., *Journal of Nuclear Materials*, Vol. 86, 1979, pp. 611–615.

Willem L. Zijp,[1] *Éva M. Zsolnay,*[2] *and Henk J. Nolthenius*[1]

Improvement of Accuracy Assessment in Radiation Damage Predictions

REFERENCE: Zijp, W. L., Zsolnay, É. M., and Nolthenius, H. J., "**Improvement of Accuracy Assessment in Radiation Damage Predictions,**" *Influence of Radiation on Material Properties: 13th International Symposium (Part II), ASTM STP 956,* F. A. Garner, C. H. Henager, Jr., and N. Igata, Eds., American Society for Testing and Materials, Philadelphia, 1987, pp. 750–760.

ABSTRACT: This paper gives the first results of the interlaboratory exercise, REAL-84. The exercise has as its aim the improvement of the assessment of the accuracy in predictions of radiation damage in steel, in a calculation exercise in which the best available input data and the best available calculation methods are used. In this report the results of 39 solutions from 10 participants, which were received before Feb. 1986, are discussed.

KEY WORDS: steels, physical radiation effects, forecasting, damaging neutron fluence, neutron spectra, cross sections, data covariances, group constants, spectra unfolding

The REAL-84 exercise is a follow-up of the REAL-80 exercise [*1*] and is organized by the Nuclear Data Section of the International Atomic Energy Agency (IAEA).

The main aim of the exercise is to improve the assessment of accuracies in radiation damage predictions by various laboratories using good quality input data and proper calculation methods. The emphasis lies on radiation damage to reactor pressure vessels and related nuclear technology. Therefore, the neutron energy range of interest is below 20 MeV. The long-term aim of REAL-84 is to strive towards establishment of standardized metrology procedures and recommended nuclear data for use in spectrum adjustments and damage parameter calculations. The short-term aim is the improvement of the available data particularly in spectra and cross-sectional covariance information. In addition, the exercise will allow assessment and validation of the accuracy of the methods and computer codes used. The joint effort of the participants of the exercise will contribute in solving some basic mathematical and physical problems that occur in neutron spectrum adjustment procedures for radiation damage purposes.

The exercise has been described in detail in information sheets [*2,3*] that were distributed to candidate participants. A magnetic tape with input data and utility programs was distributed to interested experts at the end of April 1985.

The first responses of the exercise were studied in Nov. 1985. In Feb. 1986, 39 solution spectra have been received from 10 participants (Table 1).

The exercise comprised calculations for seven input spectrum data sets: that is, three data sets for thermal reactors, two data sets for fusion spectra, one data set for a fission neutron spectrum for ^{235}U, and one data set for a fast reactor spectrum (Table 2).

The detection reactions applied for the adjustment of input spectra are shown in Table 3.

[1] Chief of Radiation Metrology Group and professional engineer, respectively, Netherlands Energy Research Foundation ECN, Westerduinweg 3, 1755 LE Petten, The Netherlands.

[2] Associate professor, Nuclear Reactor of the Technical University of Budapest, Hungary.

TABLE 1—*Participation to the REAL-84 exercise.*

	Code	Institute
Set 1	NRMI	Nuclear Radiation Measurements and Instrumentation, Department of Nuclear Engineering, Faculty of Engineering, University of Tokyo, Japan.
Set 2	ORNL	Engineering Physics and Mathematics Division, Oak Ridge National Laboratory, Oak Ridge, TN
Set 3	ANL	Chemical Technology Division, Argonne National Laboratory, Argonne, IL
Set 4	IJS	Institut "Jozef Stefan," Ljubljana, Yugoslavia
Set 5	KFKI	Hungarian Academy of Sciences, Central Research Institute for Physics, Budapest, Hungary
Set 6	ECN	Netherlands Energy Research Foundation ECN, Physics Department, Petten, The Netherlands
Set 7	PTB	Physikalisch-Technische Bundesanstalt, Braunschweig, Federal Republic of Germany
Set 8 Set 9	BME	Nuclear Reactor of the Technical University, Budapest, Hungary
Set 10	IPM	Institute of Preventive Medicine, Division for Dosimetry, Bratislava, Czechoslovakia
Set 11	ORNL	Operations Division, Oak Ridge National Laboratory, Oak Ridge, TN

This report gives information on the results and a discussion of a few related topics. In a number of cases the solutions that were received have led to new questions on data and methods. In letters to the participants additional information was requested. Many details need to be studied before the final report on this exercise can be issued.

Aspects of the Exercise and Quality of the Input Data

The participants were asked to use adjustment codes that explicitly can treat covariance matrices, in order to

TABLE 2—*Spectra studied.*

Spectrum Code	Spectrum Description	N_r [a]	V_r [b]	C_r [c]	N_{sg} [d]	N_{cg} [e]
ANO	pressure vessel cavity of the Arkansas Power and Light reactor (Arkansas Nuclear One-1)	6	+	+	55	16
PS1	Oak Ridge Research Reactor Poolside Facility in the metallurgical irradiation experiment; position simulated surveillance capsule	10	+	−	37	37
PS2	Oak Ridge Research Reactor Poolside Facility in the metallurgical irradiation experiment; ¼T position in the simulated pressure vessel capsule	6	+	−	37	37
RTN	fusion simulation spectrum measured at the RTNS-II, a 14-MeV neutron source at Lawrence Livermore Laboratory; a pretty fair simulation of a fusion first wall spectrum	12	+	+	60	60[f]
TAN	accelerator spectrum Be(*d,n*) with deuteron energies of 16 MeV	18	+	+	39	39[f]
U35	a fission neutron spectrum for ^{235}U	22	+	+	24	24
CFR	neutron spectrum in the center of the coupled fast reactivity measurement facility (CFRMF)	23	+	−	26	26

[a] Number of reaction rates.
[b] Availability of reaction rate variance information.
[c] Availability of reaction rate covariance information.
[d] Number of spectrum groups.
[e] Number of spectrum covariance groups.
[f] Expert's estimates, not based on calculation.

Table 3—*List of input reaction rates.*

Detection Reaction	Neutron Spectrum Case						
	ANO	PS1	PS2	TAN	RTN	U35	CFR
$^{6}Li(n,\alpha)^{3}H$							+
$^{10}B(n,\alpha)^{9}Li$							+
$^{27}Al(n,p)^{27}Mg$						+	+
$^{27}Al(n,\alpha)^{24}Na$				+	+	+	+
$^{45}Sc(n,\gamma)^{46}Sc$		+	+	+	+		+
$^{46}Ti(n,p)^{46}Sc$	+	+	+	+	+	+	+
$^{47}Ti\ (n,p)^{47}Sc$				+	+	+	+
$^{48}Ti(n,p)^{48}Sc$				+	+	+	+
$^{54}Fe(n,p)^{54}Mn$	+	+	+	+	+	+	+
$^{55}Mn(n,2n)^{54}Mn$						+	
$^{56}Fe(n,p)^{56}Mn$				+		+	
$^{58}Fe(n,\gamma)^{59}Fe$		+	+				+
$^{58}Ni(n,p)^{58}Co$	+	+	+	+	+	+	+
$^{58}Ni(n,2n)^{57}Ni$				+	+	+	
$^{59}Co(n,\gamma)^{60}Co$		+	+	+	+		+
$^{59}Co(n,\alpha)^{56}Mn$				+		+	
$^{59}Co(n,2n)^{58}Co$				+	+		
$^{60}Ni(n,p)^{60}Co$				+	+		
$^{63}Cu(n,\gamma)^{64}Cu$						+	+
$^{63}Cu(n,\alpha)^{60}Co$	+	+				+	
$^{115}In(n,\gamma)^{116}In^{m}$						+	+
$^{115}In(n,n')^{115}In^{m}$				+		+	+
$^{127}I(n,2n)^{126}I$						+	
$^{197}Au(n,\gamma)^{198}Au$				+	+	+	+
$^{232}Th(n,\gamma)^{233}Th$							+
$^{232}Th(n,f)FP$						+	+
$^{235}U(n,f)FP$		+		+		+	+
$^{237}Np(n,f)FP$	+	+				+	+
$^{238}U(n,\gamma)^{239}U$				+			+
$^{238}U(n,f)FP$	+	+		+		+	+
$^{239}Pu(n,f)FP$						+	+

(1) perform good neutron spectrum adjustments,
(2) evaluate displacement rates and gas production rates for steel,
(3) provide uncertainty and correlation data and
(4) specify the procedures followed.

Since it was required to treat the given covariance information, the number of participants was limited.

In the preparation of the input set for the magnetic tape of the exercise it was tried to obtain data of good quality for the various spectra.

Emphasis was given to the use of realistic input data. Therefore, several discussions with the experts in this field took place, and their special contributions were added to the available literature data in the definition of the contents of the input data sets.

Because of the tight time schedule the enquiries and tests for the input set had to be finished rather early, with the consequence that a few shortcomings remained in the input data.

On the other hand, there were several objective difficulties, for example, lack of good experimental and physics based data and procedures, which lead to consequences to be taken into account during the evaluation of the results.

Input Neutron Spectra

The input spectra that were applied in the exercise are considered to be of good quality, but the spectrum information in the low energy range is in most cases highly uncertain (Fig. 1). This high uncertainty is sometimes caused by a too broad group structure in the important energy range (for example, PS2, TAN, RTN, and U35), and sometimes it is due to spectrum data below 0.5 to 1 eV (for example, ANO, PS1, and CFR) not being available. Similar problems, for example, broad group structure, can also be encountered in the high energy region of some spectra. This situation may lead to difficulties in the adjustment and in the calculation of the spectrum characteristics such as fluence rate values and dpa rates.

Cross-Sectional Data

The reaction cross-sectional data for the exercise were derived from the best available up-to-date compilations, such as ENDF-B/V (version 2) and IRDF85. Nevertheless, during the preparation of the input sets a number of reactions could not be incorporated because the cross-sectional data were not readily available. This lack of reaction cross-sectional data together with their uncertainties might be the reason for the restricted number of detection reactions that would be used in several cases in our input sets (for example, $^{45}Sc(n,2n)^{44}Sc$, $^{52}Cr(n,p)^{52}V$, $^{54}Fe(n,\alpha)^{51}Cr$, $^{64}Zn(n,\gamma)^{65}Zn$, $^{59}Co(n,p)^{59}Fe$, $^{89}Y(n,2n)^{88}Y$, $^{93}Nb(n,\gamma)^{94}Nb$, $^{93}Nb(n,2n)^{92}Nb^m$, $^{109}Ag(n,\gamma)^{110}Ag^m$, $^{169}Tm(n,2n)^{168}Tm$, $^{197}Au(n,2n)^{196}Au$, and $^{238}U(n,2n)^{237}U$.

Especially for the fusion spectra experimentally determined, reaction rates had to be deleted because of lack of these data.

Too large cross-sectional uncertainties, or even discrepant cross-sectional values, were found by several participants for a number of reactions, for example, $^{47}Ti(n,p)^{47}Sc$, $^{58}Fe(n,\gamma)^{59}Fe$, $^{58}Ni(n,2n)^{57}Ni$, and $^{127}I(n,2n)^{126}I$; furthermore the $^{115}In(n,\gamma)^{116}In^m$ cross section had to be calculated from the total cross-sectional data given for $^{115}In(n,\gamma)^{116}In$ in the ENDF/B-V Version 2 that was used.

The Damage Cross Sections

In the input data set damage cross-sectional values for three different materials (iron, chromium, and nickel) were supplied. The displacement cross-sectional values for steel alloyed by these materials were calculated in order to determine the displacement rate for the specified type of steel (71% iron, 18% chromium, and 11% nickel). One can argue about the question whether the linear combination of the components' displacement cross sections will constitute the total displacement cross section for the given type of steel. A special evaluation for this type of damage cross section seems to be necessary.

Quality of Input Reaction Data

The input data sets for the thermal reactor neutron spectra have in all three cases a limited number of reaction rates. In the ANO case the thermal neutron response is even missing. The data sets show that on one hand experimental reaction rate data are limited, while the other hand a significant effort has been made to obtain realistic covariance information for the input data and to use an advanced adjustment method. From this point of view an intensified effort for improvement of the metrology situation seems to be justified. Also the improvement of the uncertainty data of the measured reaction or fission rates needs to be given in a number of cases more attention.

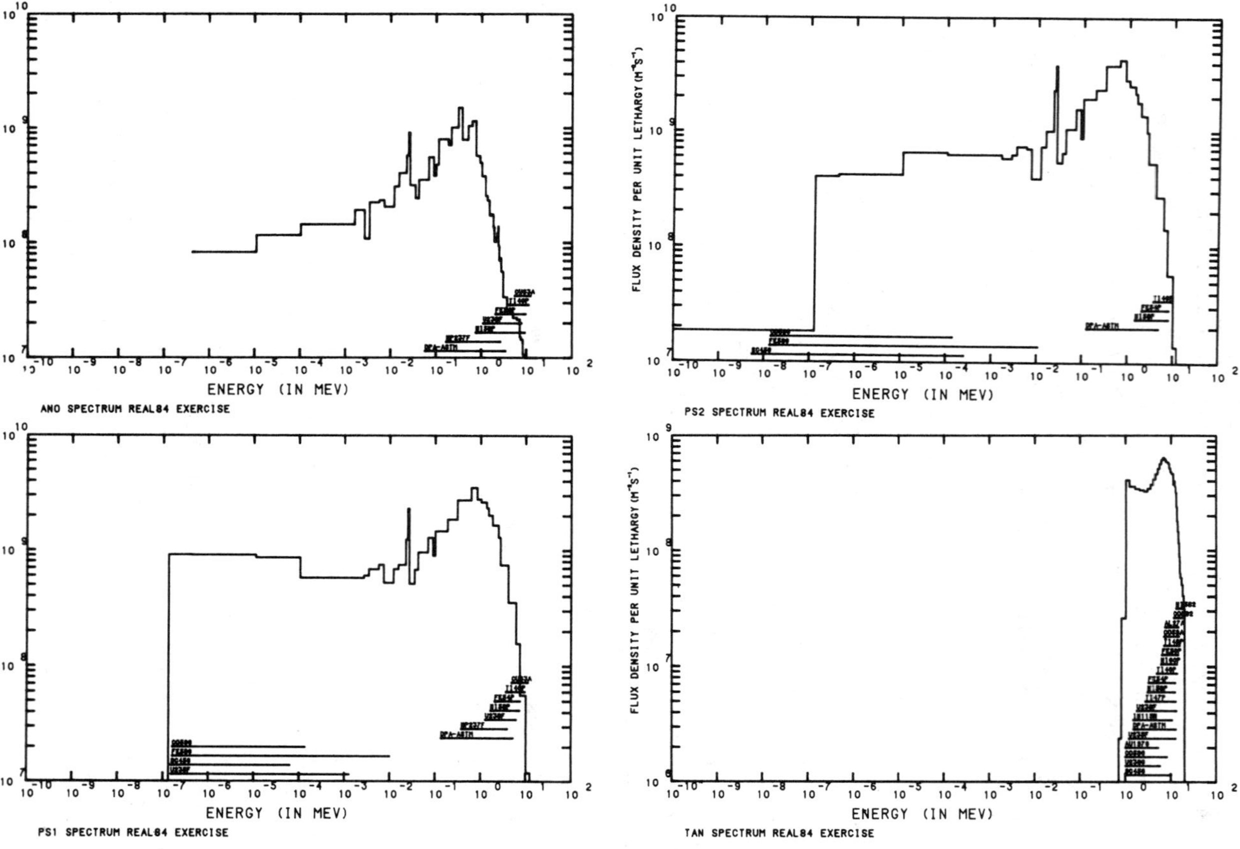

FLUX DENSITY PER UNIT LETHARGY (M-2S-1)
ENERGY (IN MEV)
ANO SPECTRUM REAL84 EXERCISE
PS2 SPECTRUM REAL84 EXERCISE
PS1 SPECTRUM REAL84 EXERCISE
TAN SPECTRUM REAL84 EXERCISE

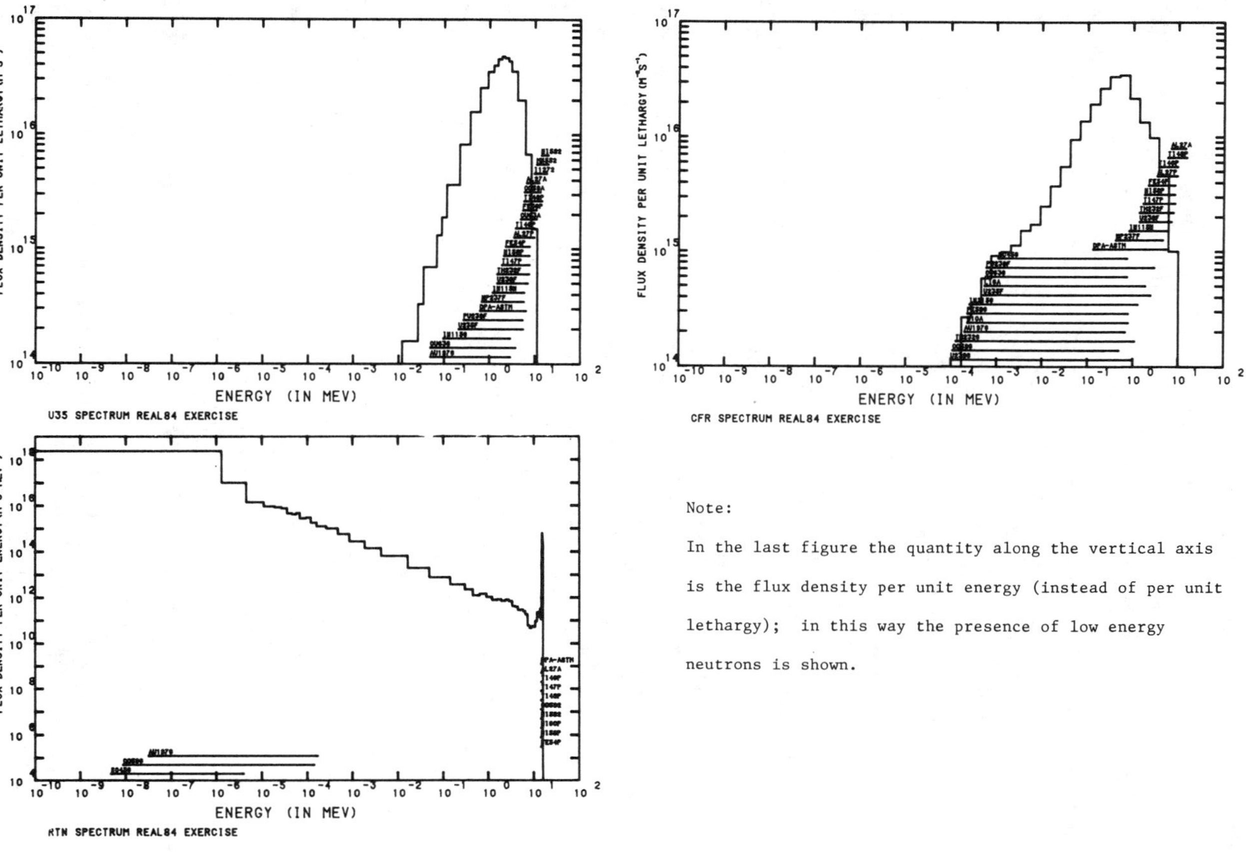

FIG. 1—*Neutron spectrum shapes invesigated in the REAL-84 exercise.*

Role of Group Structure of Input Data

The calculated neutron spectra were available in various group structures (from 16 to 55 groups). In most cases these group structures were not optimum for direct use in adjustment calculations. The suitable choice of the group structure can especially be important in cases when a weighting spectrum is derived for cross-sectional conversion purposes. The existing original group structure may lead to significant differences between measured and calculated reaction rates.

Computer programs for the conversion of covariance information from one to another group structure have not generally been available until now; therefore, interpolation problems may arise if the boundaries of the two group structures do not coincide.

Cross-Sectional Covariances

Cross-sectional covariance information is now available for the metrology reactions in the IRDF85 and ENDF/B-V libraries. For several cases the variance values are so large that the reaction has a small statistical weight in the asjudtment procedure.

Experimental uncertainty information for the damage producing reactions is not available. Uncertainty values have been arbitrarily chosen for the gas production and displacement cross sections so that they have only meaning for comparison purposes. The uncertainty values chosen are 10% for iron, 12% for nickel, and 18% for chromium. It has been assumed that the correlation between the group damage cross sections can be described by means of a special Gaussian function with a specified width parameter.

Uncertainties of the Input Data

For a number of groups the input neutron spectra have standard deviations larger than 50%; also the cross-sectional data show large uncertainties for certain groups. The estimates of the reaction rate uncertainties of the experimental data sometimes seem to be large too. These (too) large uncertainties may lead to

(1) unpredictable performance of the adjustment procedures (for example, negative values for fluence rates or the application of a strong asymmetric probability distribution in the adjustment);
(2) an unlikely value for the least squares sum; and
(3) reduced influence of the adjustment.

This effect should be considered especially for the cross-sectional uncertainties.

Results

Adjustment Codes and Utility Programs Applied

The participants applied various adjustment codes and procedures to calculate the neutron spectrum and characteristic integral (damage) data, together with their uncertainties. Some characteristic output data are shown in Table 4. Adjustment codes NEUPAC-JLOG, LEPRICON, various versions of STAY'SL and the code SANDBP were used. In most cases no detailed information on the differences between the STAY'SL versions is available.

The cross sections and their uncertainty information were converted by the various participants from the ENDF/B-V format to the group structure of interest with aid of the computer programs, NJOY, PUFF-2, or UNC33. One particpant systematically used the standard deviations of the ENDF/B-V file in combination with Gaussian shaped correlation contributions.

For the conversion of covariance matrices (required by the change in group structure) no special programs were listed by the participants.

TABLE 4—*Spread in results for important output parameters. The symbol* n *denotes the number of solutions for each case.*

Spectrum	ANO, %	PS1, %	PS2, %	TAN, %	RTN, %	U35, %	CFR, %
ϕ (>0.1 MeV)	3.3 (n = 10)[a]	5.0 (n = 4)	15.4 (n = 5)	8.5 (n = 4)	2.1 (n = 5)[a]	1.9 (n = 6)	4.0 (n = 3)
R_{dpa} (iron)	3.9 (n = 10)	3.2 (n = 3)	13.1 (n = 5)	[b]	[b]	2.4 (n = 6)	2.3 (n = 3)
R_{He} (steel)	3.6 (n = 7)	1.4 (n = 2)	7.6 (n = 3)	[b]	[b]	2.5 (n = 5)	4.5 (n = 2)
R_H (steel)	2.8 (n = 7)	0.3 (n = 2)	9.6 (n = 3)	[b]	[b]	2.6 (n = 5)	4.1 (n = 2)

[a] One outlier not taken into account.
[b] Only one solution received.

TABLE 5—*Range of reported variation coefficients in important output parameters, %*

Variation Coefficient in	Spectrum						
	ANO	PS1	PS2	TAN	RTN	U35	CFR
				Number of Responses			
	10	3	5	4	6	4	3
ϕ (>0.1 MeV)	9.3 to 12.1	5.9 to 6.3	5.5 to 7.6	2.3 to 2.4	3.2 to 4.5	0.1 to 1.4	0.5 to 0.6
R_{dpa} (iron)	7.7 to 13.3	8.9 to 11.5	5.5 to 12.0	. . .	10.5 to 10.6	7.2 to 10.1	7.2 to 10.8
R_{He} (steel)	5.7 to 20.8	9.4 to 13.1	5.5 to 12.6	. . .	7.9[a]	8.3 to 10.6	8.6 to 10.5
R_H (steel)	5.7 to 19.8	8.5 to 11.6	5.3 to 11.7	. . .	7.8[a]	7.6 to 10.3	7.9 to 10.2

[a] Only one value available.

Uncertainties

For a number of parameters the uncertainty values of integral (damage) parameters determined by the different participants show a large spread (Table 5). For the most important parameters differences by a factor of 2 do sometimes occur between the uncertainties from the different laboratories.

Inconsistencies

For several input data sets that were applied in this exercise inconsistencies were observed. Some participants modified the input data in such a way that a consistent input data set was obtained. For the modification no unique procedure was used.

The following methods were encountered:

- deletion of a detection reaction from the input data;
- an increase of the uncertainty of the experimental reaction rate; and
- modification of the input spectrum.

Not in all cases did these methods give the required results. For one case (spectrum PS2) most participants were not convinced of the quality of the input data, and no justified correction methods were found. These types of inconsistencies lead to remarkable "remedy" procedures during adjustments, which may have real impact on the solution. For this reason attention should be given to procedures for tracing inconsistencies in the input data, to methods for identifying their sources, and to ways for the elimination of significant inconsistencies.

Summary of Damage Predictions

The result of the REAL-84 calculational procedure is a set of parameters together with their uncertainties that describe the characteristics of the irradiation of the materials of interest (iron and steel). The parameters currently used for this purpose are neutron spectrum characteristics and typical parameters that are considered to have a more direct relation to the damage processes. These parameters comprise the displacement rate and the production rate of helium and hydrogen atoms. Of course, for the actual irradiation experiments an integration of the parameters over the time is of interest. Good quality gas production cross sections were available in the exercise and also the displacement cross sections are considered to be of good quality. However, uncertainty data for these cross sections are absent.

In the case of steel we may expect extra uncertainties because of the method applied for derivation of the damage cross section for steel from the data for the pure metal components.

It should be realized that the calculated uncertainties of the damage parameters in this exercise are based partly on experimental uncertainty values for spectrum reaction rates and activation fission cross sections and partly on postulated uncertainties of gas and damage cross sections. Therefore it may be somewhat difficult to draw firm conclusions, which are generally valid for all spectrum situations. However, the authors believe that the general uncertainty pattern of the calculations gives a qualitatively realistic picture of the state of the art of damage predictions for irradiated steels.

Preliminary Conclusions

1. The set of best available input data showed some shortcomings:

- For some reactions the evaluated cross-sectional files (such as ENDF/B-V (Version 2) and IRDF) contain discrepancies in the cross-sectional values;

- for some spectra the input set of experimental reaction rates is too small; and
- the displacement and gas production cross-section data are not accompanied by covariance information.

2. Reaction rate measurements do not offer any difficulties in practice. However, it is recommended to use extended detector sets and to pay more attention to the influence of foil covers, especially when the influence of thermal and intermediate neutrons is of interest.

3. Good methods are available for the adjustment of input spectra with adequate treatment of covariance information. These methods require more effort, even with a large computer available, and a high level of competence.

4. The various adjustment programs applying the generalized least squares method (or an equivalent method) and the subsequent calculation of integral damage parameters (like displacement rate and gas production rate) show a few percent spread in the results (but 13% for the PS2 spectrum).

5. The final uncertainties in the integral damage parameters comprise a nonnegligible contribution because of postulated uncertainties in the damage cross sections (both for displacement production and for gas production by (n,α) and (n,p) reactions).

The hypothetical uncertainties in these damage cross sections proposed for this exercise are 10% for iron, 12% for nickel, and 18% for chromium. Also an idealized type of correlation between the group values for the damage cross sections was assumed. The uncertainties in the final integral damage parameters are more determined by the uncertainty level of the damage cross sections than was chosen by the uncertainty level of the adjusted spectrum.

6. If hypothetical displacement cross sections without uncertainties would have been chosen, then the final values for the uncertainties in the displacements would have been reduced in most cases by at least a factor of 2.

7. Neutron metrologists are able to predict important damage parameters (production of displacements and gas atoms) with a good reproducibility. This reproducibility is advised by differences in algorithms of the spectrum adjustment codes and by differences in the input data. The spread in the calculated damage parameters is below 4% (except the difficult case of PS2).

8. Important is the finding that the participants' values for the standard deviations in the integral damage parameters for iron show a spread of less than 14%. For the other damage parameter the spread is about a factor of 2.

Acknowledgment

The work of É. M. Zsolnay was partly sponsored by the International Atomic Energy Agency (IAEA) Project 3639/RB.

References

[1] Zijp, W. L., Zsolnay, É. M., Nolthenius, H. J., Szondi, E. J., Verhaag, G. C. H. M., Cullen, D. E., and Ertek, C., "Final report on the REAL-80 exercise," Report ECN-128, ECN, Petten, The Netherlands, Feb. 1983.

[2] Zijp, W. L., Zsolnay, É. M., and Cullen, D. E., "Information Sheet for the REAL-84 Exercise," Report INDC(NDS)-166, International Atomic Energy Agency, Vienna, Austria, 1 March 1985.

[3] Szondi, E. J. and Nolthenius, H. J., "Additional Data and Information for the REAL-84 Exercise," Report INDC(NDS)-167, International Atomic Energy Agency, Vienna, Austria, 11 March 1985.

Hiroshi Sekimoto[1]

Comparison of Calculated Integral Values Using Measured and Calculated Neutron Spectra for Fusion Neutronics Analyses

REFERENCE: Sekimoto, H., "**Comparison of Calculated Integral Values Using Measured and Calculated Neutron Spectra for Fusion Neutronics Analyses,**" *Effects of Radiation on Material Properties: 13th International Symposium (Part II), ASTM STP 956,* F. A. Garner, C. H. Henager, Jr., and N. Igata, Eds., American Society for Testing and Materials, Philadelphia, 1987, pp. 761–768.

ABSTRACT: The kerma heat production density, tritium production density, and dose in a lithium-fluoride pile with a deuterium-tritium neutron source were calculated with a data processing code, UFO, from the pulse height distribution of a miniature NE213 neutron spectrometer, and compared with the values calculated with a Monte Carlo code, MORSE-CV. Both of the UFO and MORSE-CV values agreed within the statistical error (less than 6%) of the MORSE-CV calculations, except for the outer-most point in the pile. The MORSE-CV values were slightly smaller than the UFO values for almost all cases, and this tendency increased with increasing distance from the neutron source.

KEY WORDS: nuclear fusion, integral values, kerma heat production, tritium production, dose, neutron spectrum, deuterium-tritium neutron source, fusion blanket, lithium-fluoride, neutron spectrometer, data processing code, unfolding, Monte Carlo code, group cross section set, covariance

Designing a fusion reactor blanket requires accurate evaluation of the distributions of heat generation, tritium production, and radiation damage and induced activity of the material in it. In order to estimate the accuracy of the calculation of these quantities, several integral experiments in the past have been performed. However they met difficulties in measuring these quantities precisely, and in identifying the cause of the differences between the measured and calculated values. In the present study, these quantities were calculated using the pulse height distribution of a miniature NE213 spectrometer, and compared with the corresponding values calculated with a Monte Carlo code. The accuracy of a neutron transport calculation using a specified data base can be evaluated for this type of neutronics problem. The kerma heat generation rate, tritium production rate from lithium-7, and dose rate contributed by the neutrons above 1.0 MeV were chosen as the calculated quantities.

Analytical Method

The quantities I are integrated values of the neutron spectra, and their expectation values were given by the following equations

$$I^m = \sum_h W_h \Phi_h$$

for measured values, and

[1] Associate professor, Research Laboratory for Nuclear Reactors, Tokyo Institute of Technology, O-okayama, Meguro-ku, Tokyo 152, Japan.

$$I^c = \sum_g w_g \phi_g \Delta E_g$$

for calculated values

where

W_h = weighting function at hth energy point to estimate I from the measured values,
w_g = weighting function for the gth energy group to estimate I from the calculated spectrum,
ϕ_g = average neutron fluence for gth energy group, and
ΔE_g = energy width of the gth energy group, and
Φ_h = the value that minimizes

$$\epsilon^2 = \sum_i \frac{1}{s_i^2}\left(c_i - \sum_h R_{ih}\Phi_h\right)^2 + \frac{\tau^2}{h_{\max}}\sum_h \left(\frac{\Phi_h}{q_h}\right)^2$$

satisfying the FERDOR principle [*1*] where

$q_h = \min_i \frac{c_i + s_i}{R_{ih}}$,
c_i = counts per ith channel,
s_i = standard deviation of c_i,
R_{ih} = spectrometer response for ith channel of a neutron with hth energy, and
τ^2 = unfolding oscillation damping factor.

The error for each integral value was determined as follows: For the measured values, the variances were given by

$$\sigma^{m2} = \sum_{h,h'} W_h C^m_{hh'} W_{h'}$$

where $C^m_{hh'}$ is the covariance of Φ_h and was evaluated with UFO code, a modified version of FORIST [*2*]. The variances for the calculated values were given by

$$\sigma^{c2} = \sum_{g,g'} w_g \Delta E_g C^c_{gg'} w_{g'} \Delta E_{g'}$$

where $C^c_{gg'}$ is the covariance of Φ_g and was calculated with MORSE-CV Monte Carlo code [*3*].

Experimental Procedure

An experiment for a lithium-fluoride cuboidal pile with a deuterium-tritium neutron source was analyzed. The pile was a cubical with a 50 cm side length, and constructed with 10- by 10- by 2.5-cm ceramic blocks. The lithium fluoride block having a density of 1.93 ~ 1.96 g/cm^3 was obtained with the fabrication process presented in Ref *4*. The isotopic concentration of ^{6}Li in the lithium was 7.539 ± 0.017% measured with a Varian-MAT CH-5 mass spectrometer, and the total amount of the chemical impurities (sodium fluoride [NaF], potassium fluoride [KF], and so forth) was less than 2%. The target position of the neutron source was set at a distance of 4.0 cm from the front surface of the pile. The direction of the accelerated beam was perpendicular to the line from the target to the center of the front surface. The angular distribution of the neutron number density of the source was isotropic, and the energy-angle relation was calculated using the kinematics of the $T(d,n)$ helium reaction with a 110-keV acceleration voltage. The measurement was performed with a miniature NE213 spectrometer at several positions along the centerline in the pile. More detailed descriptions of the pile, spectrometer, electronics, and experimental setup are presented in Ref *4–7*, respectively.

Monte Carlo Calculation

For cross-sectional input, the GICXFNS group cross-sectional set [*8*] was used, which was processed from the ENDF/B-IV library at the Japan Atomic Energy Research Institute. The calculation was performed for 61 energy groups (16.399 to 0.964 MeV).

The modified point detector estimator of Carter and Cashwell [*9*] was employed to estimate the spectrum. This estimator considers an imaginary sphere around the estimation point and assumes uniform collisions within the sphere. The radius of the sphere was chosen to be 10 cm for the present calculation. The effects of the radius on the neutron spectrum were studied in Ref *4*, and appeared to be small.

The total number of random-walk histories were 1.6×10^4, 4.9×10^4, 4.9×10^4, and 1.96×10^5 for d = 16.4, 26.4, 31.4, and 46.4 cm, respectively, where d is the distance of the measuring position from the front surface of the pile.

Weighting Functions

The macroscopic kerma factors for lithium fluoride were calculated using the GICXFNS group cross sectional set [*8*]. The group cross sections for $^7Li(n,n'\alpha)T$ reaction were calculated using the JENDL-2 library [*10*]. For the dose per unit neutron fluence for different neutron energy, the values accepted in the Japanese regulation were employed. The values of these weighting functions are shown in Fig. 1. Since the cross section of $^7Li(n,n'\alpha)T$ is microscopic, the density of ^7Li, 4.11×10^{-22} $(n \cdot cm^2)^{-1}$, must be multiplied to evaluate tritium production density. The weighting functions for the tritium production density and dose are almost constant above ~6 and ~2 MeV, respectively, and decrease rapidly below these energies. The weighting function for the heat production increases gradually with increasing neutron energies.

Results and Discussion

The neutron energy spectra measured and calculated at d = 16.4 and 31.4 cm are shown in Fig. 2, where d is the distance from the front surface of the pile. These spectra were all normalized

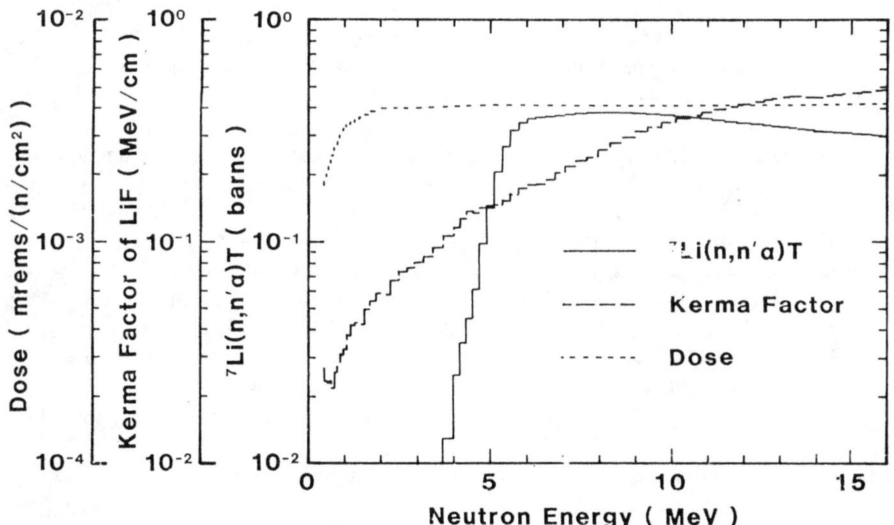

FIG. 1—*Weighting functions of integral values.*

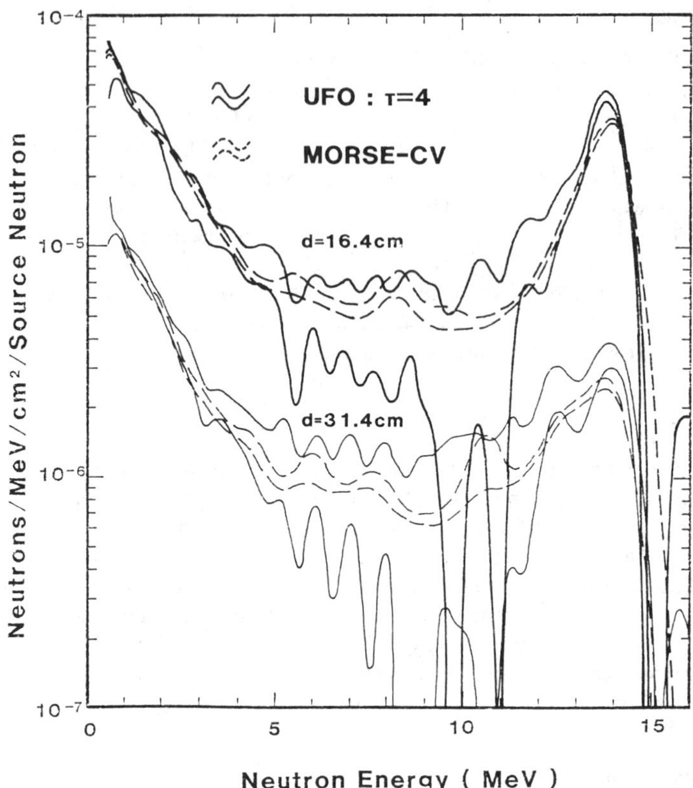

FIG. 2—*Measured and calculated neutron spectra per unit source neutron in a lithium-fluoride pile.*

to one source neutron and smoothed with the spectrometer resolution function. The oscillation damping factor τ for the UFO calculation was chosen to be the FERDOR value, 4.0 [*1*]. The two curves at each measurement position indicate the uncertainty. The principal component of this uncertainty is the statistical error. The response function error causes the oscillation of the spectrum. The detailed discussions on the spectra are given in Ref *4*.

The expectation values and relative standard deviations of the kerma heat production density, tritium production density, and dose are presented in Tables 1 through 3. The UFO values were calculated for $\tau = 0$ and 4.

The relative standard deviation for $\tau = 0$ was extraordinarily large, being more than 30% for some cases. It exhibited random change with different cases. On the other hand, $\tau = 4$ took smaller values, about 1% for the kerma heat production, less than 1% for the tritium production and about 6% for the dose. The smooth curve fitting of the UFO values and the good agreement with the MORSE-CV values suggest that $\tau = 4$ may present better results. Hence, we will only consider $\tau = 4$ for the UFO calculations in the following discussions.

The expectation values obtained with UFO for $\tau = 4$ and with MORSE-CV are shown in Figs. 3 through 5.

All three integral values show similar change as a function of distance *d*. The changes of the kerma heat production density and tritium production density were almost identical. Compared to these two curves, the dose decreased more slowly with the increasing distance. This can be attributed to the spectrum softening along increasing distance.

TABLE 1—*Kerma heat production density in a lithium-fluoride pile (upper figures are absolute [MeV/cm³/source neutron], lower figures are fractional standard deviation).*

	UFO			UFO-MORSE
d, cm	$\tau = 4$	$\tau = 0$	MORSE	UFO
16.4	4.227×10^{-5} 0.009	4.413×10^{-5} 0.305	4.059×10^{-5} 0.035	0.040
26.4	1.069×10^{-5} 0.012	1.077×10^{-5} 0.067	1.026×10^{-5} 0.042	0.040
31.4	5.691×10^{-6} 0.013	5.569×10^{-6} 0.310	5.367×10^{-6} 0.058	0.057
46.4	8.869×10^{-7} 0.015	8.827×10^{-7} 0.097	7.308×10^{-7} 0.036	0.176

TABLE 2—*Tritium production density from lithium-7 in a lithium-fluoride pile (upper figures are absolute [MeV/cm³/source neutron]), lower figures are fractional standard deviation).*

	UFO			UFO-MORSE
d, cm	$\tau = 4$	$\tau = 0$	MORSE	UFO
16.4	1.337×10^{-6} 0.005	1.328×10^{-6} 0.057	1.283×10^{-6} 0.039	0.040
26.4	3.400×10^{-7} 0.008	3.424×10^{-7} 0.058	3.339×10^{-7} 0.046	0.018
31.4	1.826×10^{-7} 0.008	1.847×10^{-7} 0.135	1.719×10^{-7} 0.069	0.059
46.4	2.906×10^{-8} 0.008	2.913×10^{-8} 0.016	2.390×10^{-8} 0.040	0.178

TABLE 3—*Dose in a lithium-fluoride pile (upper figures are absolute [mrem/source neutron]), lower figures are fractional standard deviation).*

	UFO			UFO-MORSE
d, cm	$\tau = 4$	$\tau = 0$	MORSE	UFO
16.4	6.953×10^{-7} 0.053	6.909×10^{-7} 0.307	7.068×10^{-7} 0.030	−0.017
26.4	2.035×10^{-7} 0.059	2.087×10^{-7} 0.270	1.985×10^{-7} 0.032	0.033
31.4	1.120×10^{-7} 0.060	1.073×10^{-7} 0.317	1.116×10^{-7} 0.045	0.003
46.4	1.895×10^{-8} 0.067	1.865×10^{-8} 0.328	1.673×10^{-8} 0.038	0.117

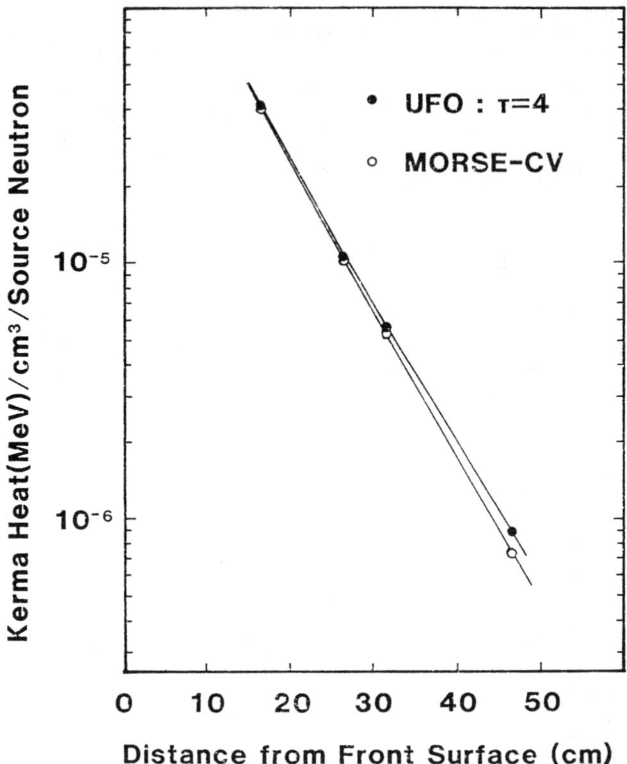

FIG. 3—*Kerma heat production density in a lithium-fluoride pile.*

The UFO and MORSE-CV values showed good agreement for d = 16.4, 26.4, and 31.4 cm. The agreement for the dose value was especially good, and the deviations were within the statistical errors expected by the UFO and MORSE-CV calculations. The deviations for the kerma heat production density and tritium production density were also within the statistical errors of the MORSE-CV values.

The MORSE-CV values were slightly smaller than the UFO values in almost all cases. This tendency increased with increasing distance. The UFO value at d = 46.4 cm might be overestimated because of room-returned neutrons, since the distance from the back surface was only 3.5 cm. The Monte Carlo code could not estimate precisely the contribution of the room-returned neutrons because of the poor statistics.

The error of the UFO spectrum was larger than the MORSE-CV spectrum as shown in Fig. 2. However, the errors of the kerma heat production and tritium production for the UFO calculation were smaller than for the MORSE-CV calculation. This may be attributed to the difference of the energy correlation between these spectra. The correlation for the UFO spectrum was more negative than for the MORSE-CV spectrum. The MORSE-CV values were more scattered than the UFO values, reflecting the larger errors.

Conclusion

The multipurpose data analysis code, UFO, for NE213 neutron spectrometer was developed. The pulse height data of the miniature spectrometer measured in the lithium-fluoride pile were

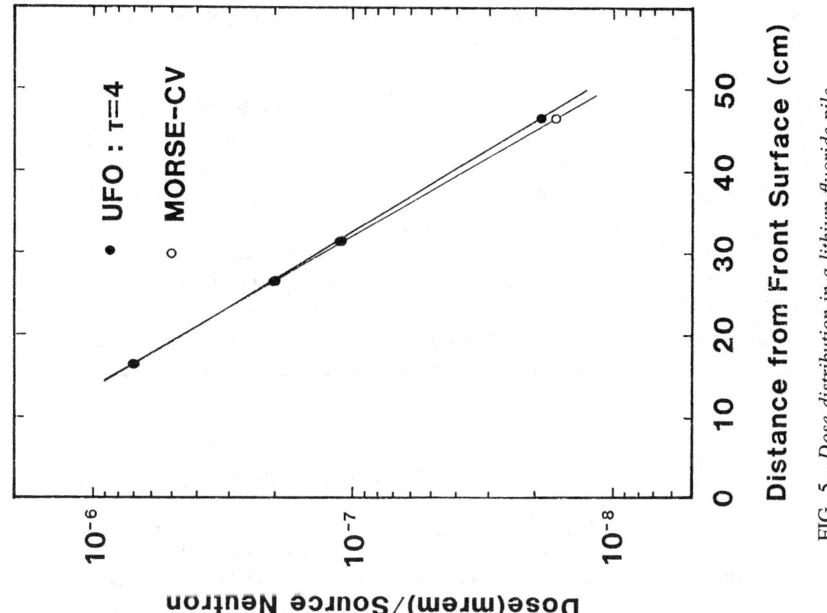

FIG. 5—*Dose distribution in a lithium-fluoride pile.*

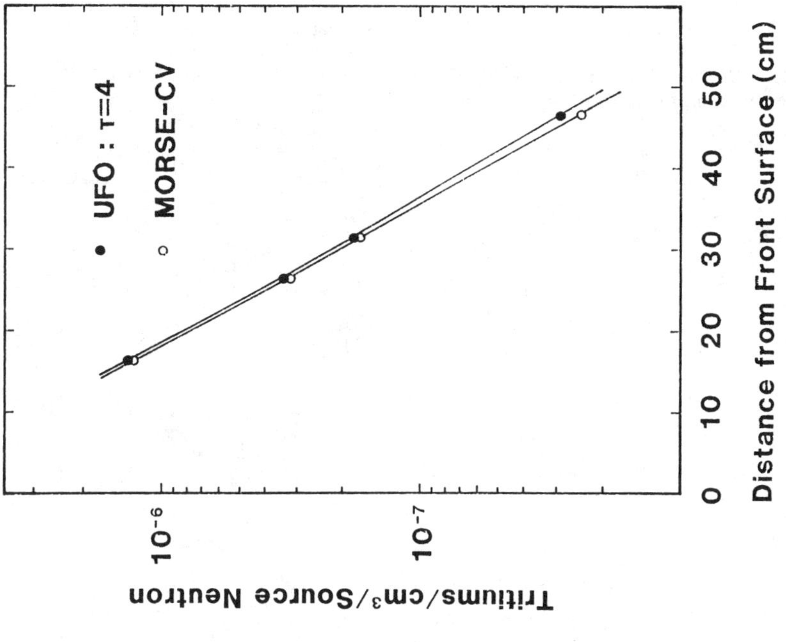

FIG. 4—*Tritium production density from lithium-7 in a lithium-fluoride pile.*

analyzed with this code to estimate the kerma heat production density, tritium production density, and dose, which were compared with the values calculated with the MORSE-CV Monte Carlo code.

The MORSE-CV values agreed with the UFO values within the MORSE-CV errors (less than 6%) except d = 46.4 cm, which was ony 3.6 cm from the back surface, even though there were large differences between the measured and calculated spectra for certain energy ranges. It may be attributed to the fact that the response function shape of the spectrometer at a certain pulse height is similar to the weighting function. The MORSE-CV values were, however, slightly smaller than the UFO values for almost all cases, and this tendency increased with increasing distance.

The UFO calculation was performed for $\tau = 0$ (without any constraints on the spectrum) in addition to $\tau = 4$ (FEROR value). The error for this calculation became extraordinarily large.

The spectral error for UFO was larger than MORSE-CV, but the errors for the kerma heat production and tritium production calculated with UFO were smaller than MORSE-CV.

References

[1] Kendrick, H. and Sperling, S. M., ''An Introduction to the Principles and Use of the FERDOR Unfolding Code,'' GA-9882, Gulf Radiation Technology, San Diego, CA, 1970.

[2] Johnson, R. H., *Nuclear Science and Engineering,* Vol. 73, 1980, pp. 93–97.

[3] Sekimoto, H., *Nuclear Science and Engineering,* Vol. 94, 1986, pp. 277–281.

[4] Lee, D., Sekimoto, H., and Yamamuro, N., *Journal of Nuclear Science and Technology,* Vol. 22, 1985, pp. 28–37.

[5] Sekimoto, H., Ohtsuka, M., and Yamamuro, N., *Nuclear Instruments and Methods,* Vol. 189, 1981, pp. 469–476.

[6] Sekimoto, H., Lee, D., Hojo, K., Hojo, T., Oishi, K., Noura, T., Ohtsuka, M., and Yamamuro, N., *Journal of Nuclear Materials,* Vol. 133 and 134, 1985, pp. 882–886.

[7] Sekimoto, H., Hojo, K., Hojo, T., and Oishi, K., *Nuclear Instruments and Methods in Physics Research,* Vol. A234, 1985, pp. 148–151.

[8] Seki, Y., Kawasaki, H., Maekawa, H., Oyama, Y., Ikeda, Y. and Nakamura, T., ''Calculation of Absolute Fission-Rate Distributions Measured in Graphite-Reflected Lithium Oxide Blanket Assembly,'' JAERI-M 83-061, Japan Atomic Energy Research Institute, Tokai, Ibaraki, Japan, 1983.

[9] Carter, L. L. and Cashwell, E. D., ''Particle Transport Simulation with the Monte Carlo Method,'' TID-26607, U.S. Energy Research and Development Administration Technical Information Center, 1975.

[10] Shibata, K., ''Neutron Nuclear Data of ^{7}Li Adopted in JENDL-2,'' JAERI-M 84-163, Japan Atomic Energy Research Institute, Tokai, Ibaraki, 1984.

Stephen B. Brumbach,[1] *Peter J. Collins,*[1] *and Brian M. Oliver*[2]

Measurements of $^{10}B(n,He)$ Reaction Rates in a Mockup Control Rod in ZPPR

REFERENCE: Brumbach, S. B., Collins, P. J., and Oliver, B. M., "**Measurements of ^{10}B (*n*,He) Reaction Rates in a Mockup Control Rod in ZPPR,**" *Influence of Radiation on Material Properties: 13th International Symposium, ASTM STP 956,* F. A. Garner, C. H. Henager, Jr., and N. Igata, Eds., American Society for Testing and Materials, Philadelphia, 1987, pp. 769–780.

ABSTRACT: Results are reported for $^{10}B(n,He)$ reaction rate measurements inside B_4C pellets contained in a mockup control rod in the Zero Power Plutonium Reactor (ZPPR). The helium accumulation fluence monitor (HAFM) technique was used, with absolute amounts of helium determined by isotope-dilution mass spectrometry. Rates of ^{235}U fission were measured in foils irradiated between B_4C pellets. Fission rate measurements used gamma ray counting techniques. Both helium production and fission rates were calculated using three-dimensional nodal transport methods. Measured helium production rates decreased about 18% within the first 63 mm of the pins, and decreased 10 to 12% going from outer to inner pins in this small, 16-pin control assembly. The calculated and measured reaction rate profiles agreed within the 1 to 2% measurement uncertainties. Helium production rates were underestimated by calculations by about 8%.

KEY WORDS: boron carbide, helium production, boron capture, control rod damage, fast reactors, reaction rate measurements

Knowledge of the $^{10}B(n,He)$ reaction rate in a control rod in a fast neutron spectrum is important in control rod design studies for fast reactors. Production of helium (and also lithium) in the neutron absorption reactions in ^{10}B results in control rod swelling, buildup of gas pressure, and a reduction in thermal conductivity which can limit the lifetime of the control rod. Methods for calculating the $^{10}B(n,He)$ rates in control rods have been discussed by Rowlands, et al [*1*] and by McFarlane and Collins [*2*]. The calculations used to predict boron capture rates generate neutron fluxes in different regions of the control rod based on the composition and geometry of the rod and surrounding core regions. Calculations are particularly challenging near a control rod surface because of rapid changes in both neutron spectrum and flux. Because the calculations are difficult, a technique for measuring boron capture rates is of interest to provide a test of the accuracy of the calculations. For the test to be useful, it is important that the measurements be performed in a configuration which reproduces the neutronic characteristics of a control rod assembly in a reactor core.

Previous measurements of reaction rates in control rods have been sparse. One technique utilized foils of metallic uranium (normally ^{235}U) placed between pellets of absorber material in a mockup control rod irradiated in a critical assembly. Measured ^{235}U fission rates can be used to estimate ^{10}B capture rates using calculated values for the reaction rate ratio. Foil irradiation results have been reported by Broomfield, et al [*3*].

This paper reports the first direct measurement of ^{10}B capture rate in a mockup control rod in a critical assembly. The experiment used the helium accumulation fluence monitor (HAFM) technique.

[1] Physicist and physicist, respectively, Argonne National Laboratory, Idaho Falls, ID 83403.
[2] Staff scientist, Rockwell International, Canoga Park, CA 91303.

The HAFM technique has been used by Farrar, Oliver and co-workers [*4–6*] to measure the $^{10}B(n,He)$ reaction rates in a variety of fast neutron spectra. In the HAFM technique, ^{10}B in a small, sealed, stainless steel capsule is irradiated producing helium. Following irradiation, the capsule and contents are vaporized in a vacuum system and the amount of helium released is measured by isotope-dilution mass spectrometry. The applicability of the HAFM technique in the Zero Power Plutonium Reactor (ZPPR) was demonstrated previously in the fuel regions of the ZPPR-13C assembly [*7*]. Following this successful demonstration, an irradiation of ^{10}B-containing HAFMs was conducted in a mockup pin-type control rod in the ZPPR-12 Metal Blanket (MB) assembly. This ZPPR-12MB irradiation was intended as a demonstration of the HAFM method in a B_4C control assembly, with more detailed measurements in a larger, more prototypic control assembly, planned for the future.

Foils of ^{235}U were also placed between pellets of the control rod in pins symmetrically equivalent to pins containing HAFMs, and following irradiation their fission rates were measured. The purpose of the foil measurements was to compare the foil and HAFM techniques for their ability to verify calculated ^{10}B capture rates. Lithium-fluoride-containing HAFMs, enriched in the isotope ^{6}Li, have been used along with ^{10}B HAFMs in previous experiments [*4–7*] and served here as a check on the boron measurements.

Experiment Description

The ZPPR-12MB Assembly

The ZPPR critical assembly is a split-table device and each half contains an array of square cross section tubes. Each tube can accommodate a drawer which contains plates of fissile, fertile, coolant, and structural material enabling construction of a full-size replica of a fast reactor. The ZPPR-12MB assembly was rather small with a core volume of 352 liters. The composition was typical of a liquid-metal reactor (LMR) with mixed plutonium-uranium oxide fuel. Because of its small size, the ZPPR-12MB core had a relatively high fissile enrichment and consequently had a neutron spectrum somewhat harder than a typical LMFBR. The two halves of the assembly were reflected about the interface. The loading pattern is shown in Fig. 1. Each square in Fig. 1 represents one matrix tube, about 55 mm square. There were 125 fuel drawers in each assembly half and the total core height was 914 mm. Sodium coolant and fuel were sealed in stainless steel cans. The fuel was in the form of a Pu-U-Mo alloy. Outside the core were radial and axial blankets of depleted uranium metal and reflectors of stainless steel.

Mockup Control Rod

For the $^{10}B(n,He)$ reaction rate measurements, the central fuel drawer in one assembly half was removed and replaced with a mockup control rod assembly, representing a half-inserted control rod. The control rod consisted of a stack of three stainless steel calandria, each 304.8 mm long and 51.7 mm square containing 16 tubes in a square array. A calandria (containing solid steel rods) is shown in Fig. 2. In the ZPPR-12MB experiments, each tube contained a stack of B_4C pellets with the boron enriched to 92% ^{10}B. Each pellet was 25.4 mm long and 9.52 mm in diameter. The calandria closest to the axial center of the assembly did not contain sodium and did not have the normal stainless-steel jacket. The two calandria away from the axial center were filled with sodium and had stainless-steel jackets. The drawer opposite the calandria was a standard fuel drawer rather than a sodium-filled channel that one might expect opposite a half-inserted control rod. This control rod mockup does not reproduce in detail the geometry expected in a power-reactor control assembly, but it does provide some of the neutronic environments encountered in conventional control assemblies and does provide an appropriate configuration for testing the ability of the HAFM technique to provide $^{10}B(n,He)$ reaction rate values.

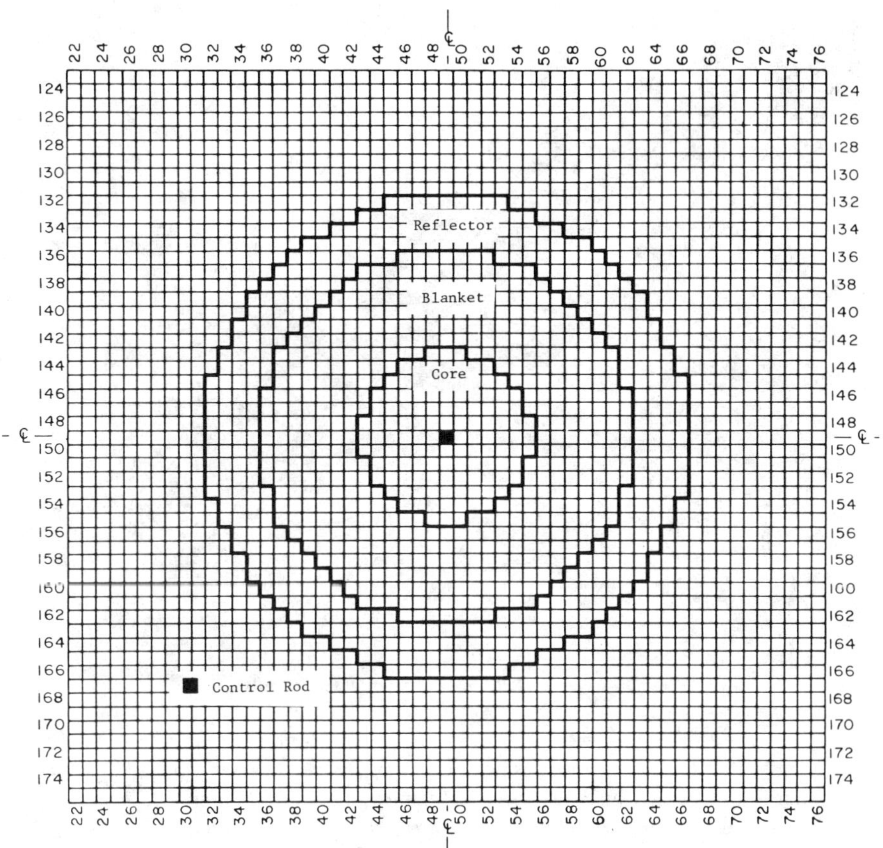

FIG. 1—*Loading pattern of the ZPPR-12MB assembly.*

To compensate for the reactivity lost when removing fuel and adding boron, blanket drawers at the core/blanket boundary were converted to fuel drawers.

HAFM Capsules and Foils

The HAFM capsules were manufactured at Rockwell from thin-walled (0.076 mm), 2.36 mm outside diameter (OD), Type 304 stainless-steel tubing. A Type 303 stainless-steel plug, 1.45 mm long, was electron beam welded on one end. After filling with measured masses of either ^{10}B or ^{6}LiF, a second end plug was welded in place, sealing the capsule under vacuum. Capsule lengths were about 12.4 mm. Sample masses were about 37 mg for the ^{10}B and about 43 mg for the ^{6}LiF. The ^{10}B enrichment was 93.10% and the ^{6}Li enrichment was 99.10%.

To accommodate the HAFMs, the B_4C pellets were bored with 2.54 mm diameter holes. The holes were oriented either axially along the centerline of the pellet (12.7 mm deep) or radially through the 9.5 mm diameter of the pellet, 6.4 mm from one face of the pellet. The holes were made using a spark erosion technique.

Because the inside diameter of a calandria tube is 9.87 mm and the HAFM is about 12.4 mm long, a radially oriented HAFM required holes to be cut in the wall of the calandria tube. To

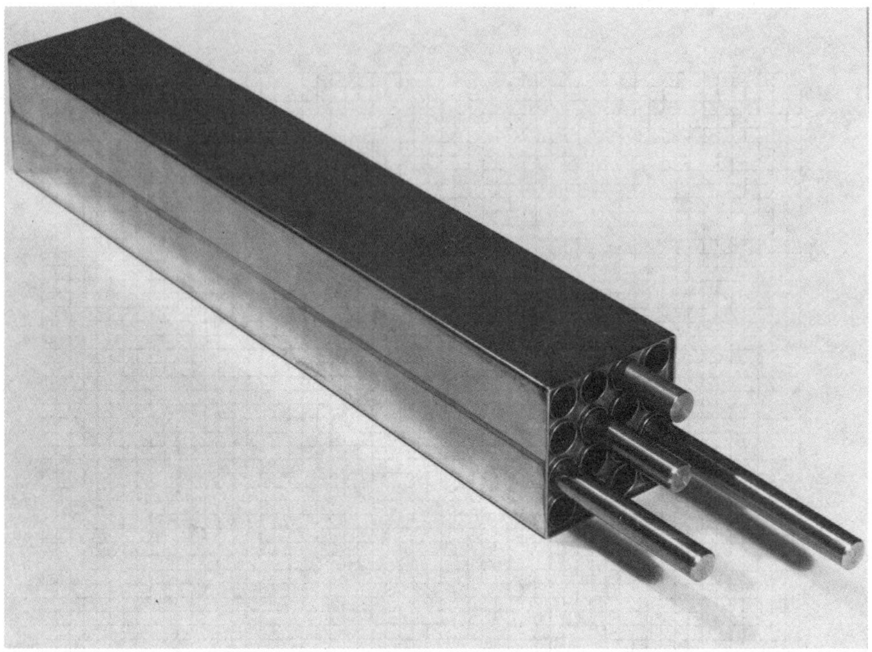

FIG. 2—*Calandria with stainless steel pins.*

insert the HAFM through the calandria tube and pellet, it was necessary to use a calandria which did not contain sodium and which had its outer stainless steel jacket removed. Also, only exterior tubes on the outer calandria edge were practical for placing radially oriented HAFMs.

A cross-sectional view of a pin calandria showing the pin numbering scheme is given in Fig. 3. There are three different pin environments; corner pins (1, 4, 13, and 16), edge pins (2, 3, 8, 12, 15, 14, 9, and 5) and interior pins (6, 7, 10, and 11). Axially oriented HAFMs were irradiated in the first three pellets in each of the three environments in pins 5, 10, and 13. Radially oriented HAFMs were placed in pin 9 which had an environment equivalent to pin 5.

Empty HAFMs, serving as blanks, were irradiated near the tip of the rod in pins 1 and 15. Lithium fluoride-containing HAFMs were irradiated in the first two pellets in pin 8 and in the first three pellets in pin 4.

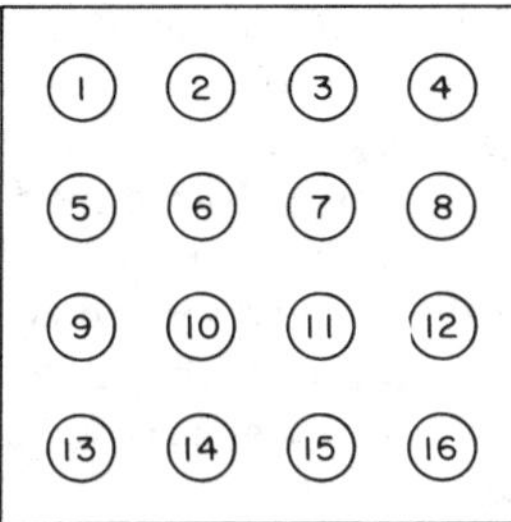

FIG. 3—*Pin numbering scheme for calandria.*

The uranium foils contained about 93% ^{235}U, were 0.13 mm thick and about 8.8 mm in diameter. The ^{235}U mass was typically 115 mg. The foils were covered with 0.0025 mm-thick aluminum discs to prevent fission product loss.

Foils could only be irradiated between pellets or at the very front of the calandria. Foils were placed in pins 11, 12, and 16 which had environments equivalent to the pins containing HAFMs and with axial locations as close as possible to the axial locations of the HAFMs.

Larger foils of ^{235}U, 12.5 mm in diameter and mounted in stainless steel holders, were irradiated in fuel drawers along the x and y axes of the assembly to provide flux shape informtion and to provide a normalization for rates measured in the control rod. Typical masses for these foils were 230 mg.

The HAFMs and foils were irradiated for about 2000 Wh.

Helium Analysis and Results

Following irradiation, the 19 boron, lithium, and empty HAFMs were returned to Rockwell and were analyzed for their helium content. Analysis was by isotope-dilution mass spectrometry [*4*]. Each analysis involved vaporizing the complete capsule and its contents in a resistance-heated graphite crucible in one of the mass spectrometer system's high-temperature vacuum furnaces. The absolute amount of ^{4}He releases was then determined from a measurement of the $^{4}He/^{3}He$ ratio in which the amount of ^{3}He "spike," about 2×10^{13} atoms, was accurately known.

The ^{3}He spikes used for the ratio measurements were obtained by expanding and partitioning a known quantity of ^{3}He gas through a succession of calibrated volumes [*8*]. The mass spectrometer was repeatedly calibrated for mass sensitivity during each series of measurements by analyzing known mixtures of ^{3}He and ^{4}He. Small amounts of ^{3}He ($\sim 5 \times 10^{9}$ atoms) in the ^{6}LiF HAFMs, resulting from the decay of generated tritium, were accounted for in the analysis.

Table 1 gives the measured helium production in the irradiated ^{10}B and ^{6}LiF HAFMs from

TABLE 1—*Measured helium production rates and comparisons with calculations in ZPPR-12MB.*

Pin Number	Pin Type	Isotope	Distance from Rod Tip, mm	Measured Reaction Rate[a]	*C/E*
13	corner	^{10}B	0 to 12.7	5.834	0.939
13	corner	^{10}B	25.4 to 38.1	5.219	0.921
13	corner	^{10}B	63.5 to 76.2	4.816	0.938
9[b]	edge	^{10}B	6.4	5.455	0.951
9[b]	edge	^{10}B	31.8	4.771	0.961
9[b]	edge	^{10}B	69.8	4.481	0.960
5	edge	^{10}B	0 to 12.7	5.497	0.944
5	edge	^{10}B	25.4 to 38.1	4.887	0.938
5	edge	^{10}B	63.5 to 76.2	4.590	0.938
10	interior	^{10}B	0 to 12.7	5.257	0.938
10	interior	^{10}B	25.4 to 38.1	4.598	0.954
10	interior	^{10}B	63.5 to 76.2	4.211	0.977
4	corner	^{6}Li	0 to 12.7	3.380	0.951
4	corner	^{6}Li	25.4 to 38.1	3.081	0.952
4	corner	^{6}Li	63.5 to 76.2	2.949	0.937
8	edge	^{6}Li	0 to 12.7	3.059	1.015
8	edge	^{6}Li	25.4 to 38.1	2.897	0.982

[a] Units of 10^{-17} reactions per atom per second at a reactor power of 1 W.

[b] Radially oriented HAFM. All other HAFMs were oriented axially.

ZPPR-12MB. The total number of helium atoms produced was about 1×10^{12} for the ^{10}B and 0.3×10^{12} for the lithium. The atom fractions were about 0.4×10^{-9} for ^{10}B and 0.3×10^{-9} for ^{6}Li. The helium production rates are in reactions per atom per second normalized to a reactor power of 1 W as measured by an ex-core power monitor. Measured helium in the two empty HAFMs was zero within measurement uncertainties.

The helium measurements were corrected for background helium released during the vaporization process by the graphite crucibles and vacuum furnace. This background correction averaged about 6×10^{9} atoms. The ^{6}LiF helium reaction rates were also corrected for helium generation from ^{19}F(*n*,He) reactions. This correction was calculated to be about 0.2% based on numerous earlier HAFM measurements.

Uncertainties in the helium measurements presented in Table 1 are about 0.8% for the ^{10}B HAFMs and about 1.8% for the ^{6}LiF HAFMs. These uncertainties come largely from two sources: (1) random and systematic uncertainties in the mass spectrometer analysis system, and (2) random uncertainty in the background helium originating from the vacuum furnaces and graphite crucibles used to vaporize each HAFM capsule. Extensive analysis of the mass spectrometer system [*8*] has indicated a systematic uncertainty of about 0.3% and a random uncertainty of about 0.5%. An estimate of the random uncertainty associated with the helium release from the vacuum furnace and graphite crucibles was obtained from the helium data from numerous empty crucibles analyzed along with the HAFMs during the present series of measurements. Analysis of these empty-crucible data indicated an uncertainty of about 5×10^{9} atoms. Combining the various uncertainties in quadrature gives the estimated total uncertainty of 0.8% for the ^{10}B HAFMs and 1.8% for the ^{6}LiF HAFMs.

The results in Table 1 show, as expected, a decrease in the ^{10}B and ^{6}Li reaction rates with increasing axial depth into the control rod pins. This decrease is about 18% in all pins for the 63.4 mm change in depth (distance between HAFM centers). Also, at all axial depths, rates are highest in the corner pin, lowest in the interior pin and intermediate in the edge pins. The variation in boron capture within a plane of the control rod is 10% to 12% at all three axial depths. Relative reaction rates, based on unity for the highest value (pin 13, pellet 1 for ^{10}B, pin 4, pellet 1 for ^{6}Li) are shown in Fig. 4. The ^{10}B and ^{6}Li rates are normalized separately.

It should be noted that the measured (*n*,He) reaction rates are all spatial averages over the effective length of the HAFM. The ^{10}B-filled length of the HAFMs is about 9.5 mm, so the ^{10}B just spans the pellet diameter in the radially oriented HAFMs. Because the measured values are spatial averages, the 9.5 mm sample length limits the spatial resolution of the present measurements.

Foil Counting and Results

The absolute number of fissions which occurred in the ^{235}U foils was determined by counting gamma rays from fission products in a calibrated Ge(Li) counting system. The counting system and the calibration procedures are described elsewhere [*9,10,11*]. Typical uncertainties in relative fission rate values are 0.6%, while uncertainties in absolute values are about 1.5%. Measured ^{235}U fission rates are given in Table 2 for the foils irradiated in the mockup control rod. The units are the number of fissions per atom of ^{235}U/s normalized to a reactor power of 1 W.

As was the case for the HAFMs, the ^{235}U(*n,f*) rates decrease with increasing depth into the control rod and rates are highest in the corner pin, lowest in the interior pin, and intermediate in the edge pin. There appears to be a larger variation in the fission rate in the interior pin than in the corner pin with the edge pin again intermediate. Results for the foils at the outer surfaces of the pins are difficult to interpret because they do not sample the flux or spectrum within the pin.

Calculated ^{235}U (n,f), ^{10}B(n,He) and ^{6}Li (n,He) Reaction Rates

Reaction rate values were calculated for ^{235}U(*n,f*), ^{10}B(*n*,He) and ^{6}Li (*n*,He) at the foil and HAFM locations in ZPPR-12MB. The calculations used the Argonne National Laboratory system

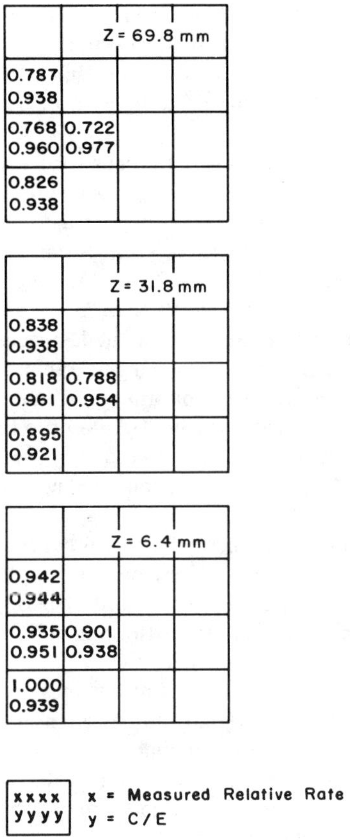

FIG. 4—*Relative boron capture rates and* C/E *values.*

TABLE 2—*Measured fission rates and comparisons with calculations in the ZPPR-12MB control rod.*

Pin Number	Pin Type	Distance from Rod Tip, mm	Measured Fission Rate[a]	*C/E*
16	corner	0	5.632	0.977
16	corner	25.4	4.809	1.014
16	corner	50.8	4.610	1.009
16	corner	76.2	4.483	0.999
12	edge	0	5.517	0.959
12	edge	25.4	4.638	1.018
12	edge	50.8	4.413	1.020
12	edge	76.2	4.253	1.019
11	interior	0	5.393	0.946
11	interior	25.4	4.360	1.049
11	interior	50.8	4.147	1.052
11	interior	76.2	4.032	1.041

[a] Units of 10^{-17} fissions per atom per second at a reactor power of 1 W.

of reactor codes. Cross sections were obtained from the evaluated nuclear data files ENDF/B-IV general-purpose file. Calculated helium generation in the ^{10}B and ^{6}Li was limited to the dominant nonthreshold reactions $^{10}B(n,\alpha)^{7}Li$ and $^{6}Li(n,\alpha)^{3}H$. Helium generation from threshold reations in either ^{10}B or ^{6}Li, for example, $^{10}B(n,2\alpha)^{3}H$ or $^{6}Li(n,n'\alpha)^{2}H$, was calculated to be small (<0.2%), and was neglected.

Homogenized cross sections in 28 energy groups for the various ZPPR drawers were produced by the codes MC^{2}-2 and SDX [*12*]. The atom densities for each drawer type were obtained from the known average masses of the materials contained in those drawers and from the known drawer volume. The reaction rates were calculated in three-dimensional (x,y,z) geometry with the nodal transport version of the code DIF3D [*13*]. Onc-fourth xy plane symmetry was assumed with full z-dimensional modelling to accommodate the half-inserted control rod. There was one node per drawer in the xy plane. In the z dimension, the node spacing was 100 mm for the first 300 mm in each half, and then 150 mm for the remaining 150 mm of core, and 450 mm of axial blanket. The control rod cross sections assumed uniform mix of all isotopes in the node.

Calculated reaction rate values at a specific HAFM or foil location were obtained by polynomial interpolation, consistent with the nodal solution. Values were determined assuming all material is at the center of the HAFM. Calculated reaction rates were not integrated over the dimensions of the HAFMs or foils.

All calculated reaction rates were normalized to provide an average value of unity for the ratio of the calculated (C) to measured (E) ^{235}U fission rates for the 16 measurements of $^{235}U(n,f)$ in the fuel zone. This normalization procedure was used because relative reaction rates (profiles) and reaction rate ratios are normally of interest. Also, it is difficult to measure absolute power in ZPPR to better than about 5%.

The measured $^{235}U(n,f)$ values and the corresponding C/E values for the measurements in the fuel region are given in Table 3. The standard deviation in C/E for the 16 fission rates is 0.73%. This is only a little larger than the uncertainty in the measured values (0.6%). The C/E values increase monotonically with increasing radius with values about 1% above average at the core edge and about 1% below average near the core center. This variation is similar to that found in the core without any control rod.

Comparison of Measured and Calculated Reaction Rates

The values for the C/E ratio for $^{10}B(n,He)$ and $^{6}Li(n,He)$ measured in the mockup control rod are presented in Table 1. The C/E values for boron are also shown in Fig. 4. For the axially

TABLE 3—*Measured* $^{235}U(n,f)$ *rates and comparisons with calculations in the fuel region of ZPPR-12MB.*

Matrix Location	E[a]	C/E	Matrix Location	E[a]	C/E
	Positive x axis			Negative x axis	
149–50	5.637	0.986	149–48	5.604	0.992
149–51	6.098	0.998	149–47	6.092	0.999
149–52	5.923	0.998	149–46	5.878	1.006
149–53	5.438	1.000	149–45	5.401	1.007
149–54	4.735	1.003	149–44	4.711	1.008
149–55	3.853	1.011	148–43	3.793	1.007
	Positive y axis			Negative y axis	
148–49	5.804	0.998	150–49	5.757	0.996
147–49	6.095	0.997	151–49	6.047	1.005

[a] Measured value in units of 10^{-17} fissions per atom per second at a reactor power of approximately 1 W.

oriented HAFMs, there is no clear tend in *C/E* value with axial displacement from the assembly center, although the interior pin results suggest a possible decrease in *C/E* with increasing axial depth. There is no clear trend in *C/E* value going from the corner to edge to interior pin environments. For the equivalent environments in pins 5 and 9, *C/E* values for the axially oriented HAFMs are on average about 1.8% higher than for the radially oriented HAFMs.

The relative boron capture rates within the various pins, based on unity for the HAFM closest to the rod tip, are shown in Table 4 for both the measurement and the calculation. The axial profiles of the capture rates are predicted within 2% except for the deepest HAFM in pin 10 where there is a 4% difference.

For ^{6}Li, the *C/E* values are consistent within pins 4 and 8 but are about 5% different between pins. The *C/E* difference between pins 4 and 8 is not understood.

The *C/E* values for $^{235}U(n,f)$ rates in the mockup control rod are presented in Table 2. It is clear that there is a problem with the foils at the pin surface compared to the foils in the interior of the pin. As already mentioned, the surface foil environment is difficult to model in a calculation, in part because of the streaming path caused by the small gap at the assembly interface. If the foils at the pin surfaces are ignored, the *C/E* values within each pin are the same within uncertainties. For the interior pin (Number 11), the foil *C/E* values are 2% to 3% higher than for the other two pins. The relative fission rates within the various pins, based on unity for the foils between the first and second pellets (25.4 mm from the rod tip) are shown in Table 5. The calculation predicts an axial decrease of about 8% going from 25.4 mm to 76.2 mm from the rod tip, which is in good agreement with the measured profiles.

Discussion

The HAFM measurements in ZPPR-12MB demonstrated the usefulness of the technique for obtaining direct boron capture measurements within B_4C pellets of a control rod. While the ZPPR-12MB control rod mockup was not a realistic model of a power reactor control assembly, the neutronic environments expected in a control assembly were represented. A representative neutronic environment is important in obtaining a realistic test of the calculations because the calculations generate the neutron spectra and neutron fluxes in the control rod based on the geometry and composition of the entire reactor. The measured boron capture rates showed the expected trends of decreasing rate with increasing axial depth in the rod and decreasing rates going from exterior to interior pins.

There was a difference of about 2% in capture rates in equivalent pins using axially oriented and radially oriented HAFMs. The HAFM midpoints were equally displaced from the rod tip in both orientations. Given the estimated measurement uncertainty of about 1% for relative capture rates, there appears to be little difference between measurements which integrate reaction rates radially and axially, at least for the geometry and spectrum of ZPPR-12MB.

The boron capture rates calculated using three-dimensional nodal transport methods gave *C/E* ratios which were consistent within 2% at all locations in the control rod, with the exception of the deepest HAFM location (pin 10), indicating that there was good agreement between calculation and measurement for the relative reaction rates (profiles) within the rod.

Fission rates in ^{235}U were measured and calculated for foils placed between pellets of the control rod and at the rod surface. Results for foils at the rod surface were difficult to interpret. Measurements in the rod interior showed trends similar to those observed for boron capture. The *C/E* ratio for fission was consistent within each pin but was 2% to 3% higher in the interior pin than in the other two pins. The consistent *C/E* within the pins means that the calculated axial reaction rate profiles agree well with the measurements. While there are not enough data from ZPPR-12MB to be conclusive, it appears that ^{235}U measurements are potentially useful for testing the calculated ^{10}B reaction rate profiles in a control rod.

In addition to comparisons of measured and calculated reaction rate profiles, two other *C/E*

TABLE 4—*Measured and calculated relative boron capture rates within pins in ZPPR-12MB.*

Distance from Rod Tip, mm	Corner Pin, 13		Edge Pin, 9[a]		Edge Pin, 5		Interior Pin, 10	
	Measured	Calculated	Measured	Calculated	Measured	Calculated	Measured	Calculated
0–12.7	1.000	1.000	1.000	1.000	1.000	1.000	1.000	1.000
25.4–38.1	0.895	0.878	0.875	0.884	0.889	0.884	0.875	0.889
63.5–76.2	0.826	0.825	0.821	0.830	0.835	0.830	0.801	0.834

[a] Radially oriented HAFM. All other HAFMs axially oriented.

TABLE 5—*Measured and calculated relative fission rates within pins in ZPPR-12MB.*

Distance from Rod Tip, mm	Corner Pin, 16		Edge Pin, 12		Interior Pin, 11	
	Measured	Calculated	Measured	Calculated	Measured	Calculated
25.4	1.000	1.000	1.000	1.000	1.000	1.000
50.8	0.959	0.954	0.952	0.954	0.951	0.953
76.2	0.932	0.918	0.917	0.918	0.925	0.918

comparisons are of inerest. One is the boron capture in the rod relative to the fission in the core regions. This is of interest because it allows estimates to be made of control rod reaction rates relative to the power in the fuel regions. The second is the *C/E* for the reaction rate ratio of boron capture to uranium fission within the control rod. This second comparison is of interest in comparing the HAFM and foil techniques for their ability to verify calculated reaction rates in the control rod. In ZPPR-12MB, the average *C/E* value for boron capture in the rod to fission in the core was 0.945 ± 0.012 while the average *C/E* for capture in the rod to fission in the rod (omitting fission values at the rod surface) was 0.922 ± 0.022.

Deviations from unity of the *C/E* values have contributions from errors introduced by the calculation methods, as well as from errors in cross sections. These two error sources cannot be easily separated. However, the *C/E* ratios for boron capture to uranium fission obtained in this work are consistent with earlier work by Oliver and Farrar [*5,6*] in well characterized benchmark fast neutron fields. In these earlier measurements, the deviation from unity in *C/E* was attributed to inaccuracies in the ENDF cross sections for boron capture above about 0.1 MeV.[3] Previous measurements [*7*] using the HAFM technique in the core region of the ZPPR-13C gave a *C/E* for the boron capture to uranium fission ratio of 0.930 ± 0.012.

All calculations produced for this report used ENDF/B version IV cross sections. The effect of using version V cross sections was investigated in an *rz*-geometry model of the reactor and control rod. With version V cross sections, the average *C/E* value for ^{235}U fission in the control rod increased from 1.025 to 1.035 and the average *C/E* for ^{10}B capture decreased from 0.945 to 0.928. As before, these version V *C/E* values are based on a *C/E* of unity for ^{235}U fission in the core region.

Conclusions

The measurements in ZPPR-12MB indicate that the HAFM technique can provide useful, direct measurements of the $^{10}B(n,He)$ reaction rates within a control rod. Boron capture rates were underestimated by calculations by about 8% in the spectrum of the control rod. This underestimation using ENDF-B/IV cross sections is consistent with results from other fast neutron spectra. A future experiment in a larger, more prototypic control rod assembly is planned.

Acknowledgments

Helpful contributions to this work were made by W. R. Marley of Rockwell International who provided technical support, by D. W. Maddison and J. M. Gasidlo of Argonne National Laboratory who performed the uranium fission rate measurements, by G. L. Grasseschi of Argonne National Laboratory who assisted in the calculations, and by H. Farrar IV of Rockwell International who aided in the initial planning of the experiments.

This work was supported by the U.S. Department of Energy, Nuclear Energy Programs under contracts W-31-109-ENG-38 and DE-AT03-83SF11902.

[3] Schenter, R. E. et al., in this publication, pp. 781–787.

References

[*1*] Rowlands, J. L., et al, "The Development and Validation of Control-Rod Calculation Methods," International Symposium on Fast Reactor Physics, Aix-en-Provence, France, Sept. 24–28, 1979, IAEA-SM-244/36, International Atomic Energy Agency, 1980, p. 83.

[*2*] McFarlane, H. F. and Collins, P. J., "Control Rods in LMFBRs: A Physics Assessment," ANL-82-13, Argonne National Laboratory, 1982.

[*3*] Broomfield, A. M., et al, "The Mozart Control Rod Experiments and Their Interpretation," *Proceedings,* International Symposium on Physics of Fast Reactors, Tokyo, Oct. 16–19, 1973, Vol. I, p. 312.

[*4*] Farrar, H., IV, et al, "Helium Production Cross Section of Boron for Fast-Reactor Neutron Spectra," *Nuclear Technology,* Vol. 25, 1975, p. 305.

[*5*] Oliver, B. M., et al, "Spectrum-Integrated Helium Generation Cross Sections for ^{6}Li and ^{10}B in the ΣΣ and Fission Cavity Standard Neutron Fields," *Proceedings,* 4th ASTM-EURATOM Symposium of Reactor Dosimetry, NUREG/CP-0029, Vol. 2, 1982, pp. 889–901.

[*6*] Oliver, B. M., et al, "Spectrum-Integrated Helium Generation Cross Sections for ^{6}Li and ^{10}B in the Intermediate-Energy Standard Neutron Field," *Proceedings,* 5th ASTM-EURATOM Symposium on Reactor Dosimetry, D. Reidel Publishing Co., Dodrecht, Holland, Vol. 2, 1985, pp. 877–885.

[*7*] Oliver, B. M., et al, "Measurements of the ^{10}B Reaction Rate in ZPPR," Presented to the International Conference on Nuclear Data for Basic and Applied Sciences, May 13–17, 1985, Sante Fe, New Mexico.

[*8*] Oliver, B. M., et al "Helium Concentration in the Earth's Lower Atmosphere," *Geochimica et. Cosmochimica Acta,* 48, 1984, p. 1759.

[*9*] Maddison, D. W., "Computer-Controlled Four-Detector Gamma-Ray Counting System," in *Computers in Activation Analysis and Gamma-Ray Spectroscopy,* DOE Symposium, Series 49, 1979 p. 708.

[*10*] Maddison, D. W., "Development of a Program for Reduction of Large Numbers of Gamma-Ray Spectra," in *Computers in Activation Analysis and Gamma-Ray Spectroscopy,* DOE Symposium, Series 49, 1979, p. 230.

[*11*] Brumbach, S. B. and Maddison, D. W., "Reaction Rate Calibration Techniques at ZPPR for ^{239}Pu Fission, ^{235}U Fission, ^{238}U Fission, and ^{238}U Capture," ANL-82-38, Argonne National Laboratory, 1982.

[*12*] McKnight, R. D., et al, "Validation Studies of ENDF/MC^{2}-2/SDX Cell Homogenization Path," *Proceedings,* Topical Meeting on Advances in Reactor Physics and Core Thermal Hydraulics, NUREG/CR-0034, Vol. 1, 1982, p. 406.

[*13*] Lawrence, R. D., "Three-Dimensional Nodal Diffusion and Transport Methods for the Analysis of Fast-Reactor Critical Experiments," *Proceedings,* Topical Meeting on Reactor Physics and Shielding, Sept. 17–19, 1984, Chicago, Illinois, Vol. II, p. 814.

Robert E. Schenter,[1] B. M. Oliver,[2] and Harry Farrar, IV[2]

Spectrum Integrated (*n*,He) Cross Section Comparison and Least Squares Analysis for ^{6}Li and ^{10}B in Benchmark Fields

REFERENCE: Schenter, R. E., Oliver, B. M., and Farrar, H., IV, **"Spectrum Integrated (*n*,He) Cross Section Comparison and Least Squares Analysis for ^{6}Li and ^{10}B in Benchmark Fields,"** *Influence of Radiation on Material Properties: 13th International Symposium (Part II), ASTM STP 956*, F. A. Garner, C. H. Henager, Jr., and N. Igata, Eds., American Society for Testing and Materials, Philadelphia, 1987, pp. 781–787.

ABSTRACT: Spectrum integrated cross sections for ^{6}Li and ^{10}B from five benchmark fast reactor neutron fields are compared with calculated values obtained using the ENDF/B-V Cross Section Files. The benchmark fields include the Coupled Fast Reactivity Measurements Facility (CFRMF) at the Idaho National Engineering Laboratory, the 10% Enriched U-235 Critical Assembly (BIG-10) at Los Alamos National Laboratory, the Sigma Sigma and Fission Cavity fields of the BR-1 reactor at CEN/SCK, and the Intermediate-Energy Standard Neutron Field (ISNF) at the National Bureau of Standards.

Results from least square analyses using the FERRET computer code to obtain adjusted cross section values and their uncertainties are presented. Input to these calculations include the above five benchmark data sets. These analyses indicate a need for revision in the ENDF/B-V files for the ^{10}B cross section for energies above 50 keV.

KEY WORDS: ^{10}B, ^{6}Li, integral cross sections, benchmark fields, neutrons, adjusted cross sections

Direct measurements of the ^{6}Li(*n*,He) and ^{10}B(*n*,He) reaction rates have been reported earlier for five benchmark fast reactor fields [*1–3*]. These measurements were conducted in the Coupled Fast Reactivity Measurements Facility (CFRMF) at the Idaho National Engineering Laboratory (INEL), the 10% enriched U-235 critical assembly (BIG-10) at the Los Alamos National Laboratory (LANL), the Sigma Sigma and Fission Cavity fields of the BR1 reactor at the CEN/SCK laboratories in Mol, Belgium, and the Intermediate-Energy Standard Neutron Field (ISNF) at the U S. National Bureau of Standards (NBS).

Results from these measurements form an important data source that can be used to integrally test past and future evaluated cross-section libraries such as ENDF/B-V, JEF-1, JEF-2, ENDF/B-VI, JENDL-2, and JENDL-3. In this paper, results of such tests are given for the ENDF/B-V files. In addition, the above described measurement values are used as input to least squares analyses using the FERRET data analysis code [*4*]. The output from the FERRET code is an adjusted microscopic cross section with uncertainties that can be compared directly with the energy dependent curves of the above indicated libraries.

Experimental Reaction Rate Measurements

A brief description of the five facilities and experimental details are given below:

[1] Fellow scientist, Westinghouse Hanford Co., P.O. Box 1970, Richland, WA 99352.

[2] Scientist and manager, respectively, Rockwell International, 6633 Canoga Ave., Canoga Park, CA 91303.

CFRMF

The CFRMF at INEL is a zoned-cooled critical assembly with a fast-neutron zone in the center of an enriched ^{235}U, water-moderated thermal zone. Three irradiations were conducted in the CFRMF [*5*]. The first irradiation lasted 16.0 h at a power level of 9.75 kW, with the second and third lasting 22.72 and 29.0 h, respectively, at a power level of 6 kW.

BIG-10

The BIG-10 facility at LANL is a fast critical assembly consisting of a cylindrical central region of 10% enriched ^{235}U surrounded by a second region of alternating annular metal plates of 93% enriched and natural uranium [*6*]. Two irradiations were conducted in BIG-10, lasting for 14.67 and 18.33 h, respectively, both at a power level of 1.3 kW.

Sigma Sigma and Fission Cavity

The Sigma Sigma and Fission Cavity are both located in thermal columns of the BR1 reactor at CEN/SCK. Sigma Sigma is a thermal-fast coupled spherical assembly located within a 50-cm diameter spherical cavity in the horizontal graphite thermal column of the BR1 [*7*]. The spherical source shell consists of a natural boron carbide shell, 1-cm inside diameter (ID), surrounded by an outer shell of natural uranium. A single 103-h irradiation was conducted in Sigma Sigma.

The Fission Cavity facility is contained in a separate 1-m spherical cavity situated in a vertical graphite column of the BR1. The irradiation volume consists of a cylindrical region in the center of the cavity, which is surrounded by a thin enriched ^{235}U foil driver. A single irradiation was carried out in the fission cavity concurrent with the Sigma Sigma irradiation.

ISNF

The ISNF facility at NBS is similar in concept to the Sigma Sigma facility at CEN/SCK. The facility operates in a 30-cm diameter spherical cavity in the graphite thermal column of the research reactor at NBS [*8*]. The working volume of the facility is a spherical 5-cm diameter region surrounded by a spherical shell of enriched ^{10}B. Eight enriched ^{235}U driver disks are located symmetrically outside the ^{10}B shell. A single irradition was conducted in ISNF lasting for 45.8 h.

For all the measurements the same experimental technique for measuring the helium content was employed. This technique involved irradiating both bare and encapsulated specimens of natural boron, enriched ^{10}B, and enriched ^{6}LiF, and subsequently measuring the irradiation-generated helium in the specimens by means of a high sensitivity gas mass spectrometer system. Helium analysis was performed at Rockwell International by vaporizing the complete sample (including capsule). Details of the mass spectrometer, including the helium "spiking" systems used for the isotope dilution, have been reported elsewhere [*9*].

A summary of the measured ^{6}Li and ^{10}B reaction rates from the five benchmark irradiations is given in Table 1. The measured reaction rate values for the Sigma Sigma, Fission Cavity, and ISNF irradiations are the same as those reported earlier [*2,3*]. The values for BIG-10 and CFRMF, however, are changed slightly from their reported values [*1*] as a result of small changes in several corrections that had been originally applied to the data, and a small change in the absolute calibration of the helium mass spectrometer system.

Uncertainties in the measured reaction rates, in percent, are given in parentheses, and comprise both random and systematic sources. In general, the largest source of uncertainty was from corrections applied to the measured helium data to account for neutron perturbations such as flux gradients and neutron self-shielding. Neutron self-shielding was the largest for the ^{10}B specimens,

TABLE 1—*Measured ^{6}Li (n,He) and ^{10}B(n,He) reaction rates in benchmark fields.*

Benchmark Facility	Measured Neutron Flux, 10^9 n/cm^{2}s		Measured Reaction Rate, 10^{-15} s^{-1}			
			^{6}Li		^{10}B	
CFRMF	78.4	(2.6)[a]	74.8	(1.1)	147.0	(2.4)
BIG-10	81.4	(2.5)	77.8	(1.3)	111.0	(1.4)
Sigma Sigma	2.51	(3.0)	2.33	(1.3)	4.25	(2.3)
Fission Cavity	1.04	(3.0)	0.475	(2.7)	0.563	(2.9)
ISNF	0.762	(2.6)	0.633	(1.5)	1.40	(2.0)

[a] Values in parentheses are uncertainties in percent.

ranging from ~1% in the Fission Cavity, to ~9% in Sigma Sigma. Uncertainy in the self-shielding corrections was assumed to be 20% of the correction value. Uncertainty in the helium measurements themselves was generally less than 1%, as were all other sources of uncertainty. Helium loss from the specimens (stainless steel capsules or bare crystal) has been investigated in the much more severe environment of EBR-II and found to be entirely negligible. Consequently, for the results reported here it can also be disregarded.

The measured neutron flux values for the five irradiations are also given in Table 1. The values in parentheses are the estimated flux uncertainties in percent. For all the facilities, the flux values were obtained independently from fission chamber measurements conduced under identical experimental conditions. A review of the fission chamber flux values, and their uncertainties has recently been conducted by Grundl [*10*]. This review involved relating the fission chamber data to the ^{252}Cf source at NBS by means of a flux transfer using the ^{239}Pu fission rate. As a result of this review, the flux values for the BIG-10 and CFRMF irradiations in Table 1 are changed slightly (−2.3% and +0.7%, respectively) from the values reported earlier [*1*]. The flux values for Sigma Sigma, the Fission Cavity, and ISNF remain unchanged.

C/E Comparisons

Table 2 presents the results of "C/E" comparisons for the ^{6}Li(n,He) and ^{10}B(n,He) reaction rates for the five benchmark fields. Calculations of the reaction rates were made using the ENDF/B-V data files for the microscopic cross section data and multigroup neutron spectrum values obtained from Grundl [*10*]. In addition "C/E" ratios for ^{235}U(n,f) are given in Table 2 to indicate the source of the deviations from 1.00 for the ^{6}Li and ^{10}B cases since the cross section for ^{235}U(n,f) is accurately known on ENDF/B-V.

The results given in Table 2 show a consistent and relatively large discrepancy between the experimental results for ^{6}Li and ^{10}B and the calculations using ENDF/B-V data. The calculations

TABLE 2—*Comparison of C/E values for ^{6}Li and ^{10}B in five benchmark fields.*

Reaction	C/E Ratios				
	CFRMF	BIG-10	Sigma Sigma	Fission Cavity	ISNF
^{6}Li(n,He)	0.95	0.90	0.93	1.00	0.95
^{10}B(n,He)	0.85	0.85	0.88	0.91	0.91
^{235}U(n,f)	1.00	1.00	. . .	. . .	1.00

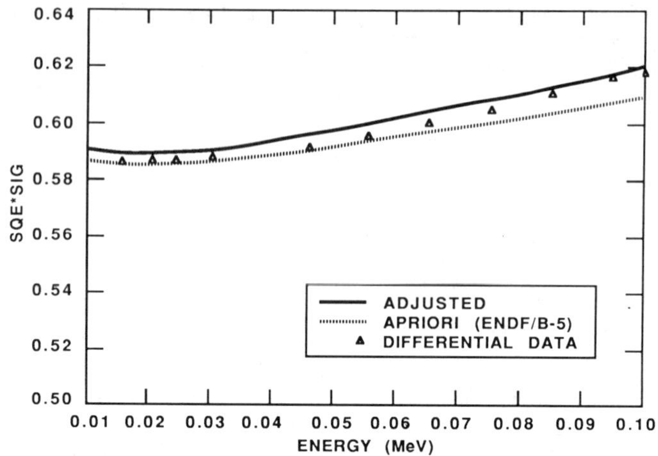

FIG. 1—*Adjusted a priori and differential B10* (n,*He*) *cross sections for 0.01 to 0.10 MeV.*

underpredict the measurements by as much as 10% for ^{6}Li (BIG-10) and in a range of 9% to 15% for ^{10}B.

Least Squares Analyses

Starting with "a priori" input curves from ENDF/B-V adjusted pointwise curves for the ^{10}B(*n*,He) and ^{6}Li(*n*,He) cross sections were obtained using the FERRET data analysis code [*4*]. All five experimental integral reaction rate data values and their uncertainties were included as input in a single computer run. In addition covariance matrices (uncertainty data) for the a priori curve and the multigroup spectrum data were input to the calculations. The covariance matrix for the a priori case (ENDF/B-V) was obtained by reviewing recent results of differential data analyses reported by Carlson et al. [*11*] and comparing with ENDF/B-V. The covariance matrices for the integral spectra were obtained from Grundl [*10*].

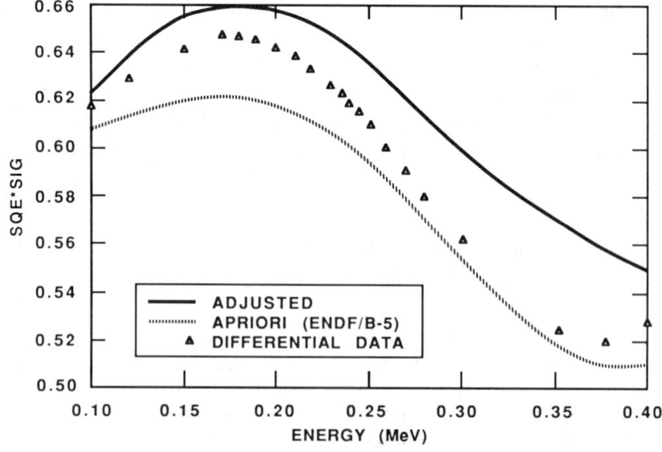

FIG. 2—*Adjusted a priori and differential B10* (n,*He*) *cross sections for 0.10 to 0.40 MeV.*

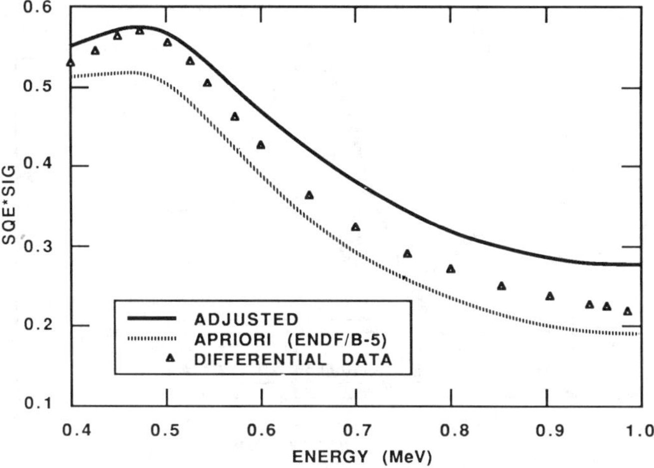

FIG. 3—*Adjusted a priori and differential B10* (n,*He*) *cross sections for 0.4 to 1.0 MeV.*

Results of these calculations are given in the solid curves ("ADJUSTED") shown in Figs. 1 through 6 for the ^{10}B(n,He) and ^{6}Li(n,He) cases. The energy range 0.01 to 1.0 MeV is presented since for energies less than 0.01 MeV both cross sections are very well known (uncertainty less than 0.5%), and also most of the integral response is below 1.0 MeV. The dashed curves labeled "APRIORI" are the ENDF/B-V input data, and the triangle data points ("DIFFERENTIAL DATA") were obtained from Carlson et al. [*11*]. They represent a least squares analysis of previous experimental differential data. Uncertainty information was not given by Carlson et al. [*11*]. However, since these reactions are standards for ENDF/B-V up to 100 KeV and these are "preliminary ENDF/B-VI" results, it is expected they have very small uncertainties. For the ^{10}B case both the adjusted curves from integral (FERRET calculations) and differential data analyses show significant increases from ENDF/B-V for energies above about 0.05 MeV. For ^{6}Li the integral result (FERRET calculation) shows an increase and the differential data show almost no change from ENDF/B-V.

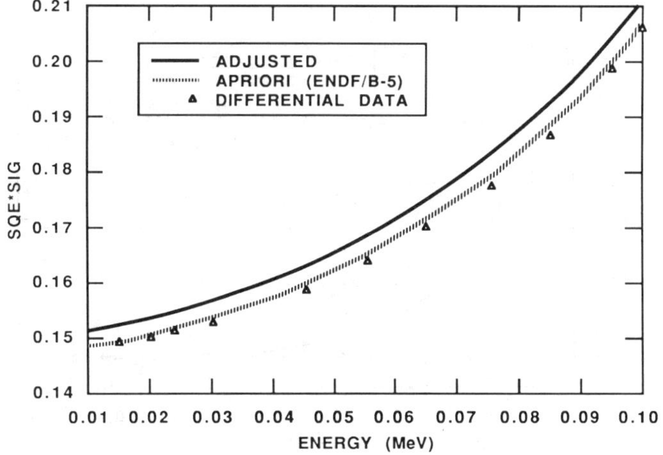

FIG. 4—*Adjusted a priori and differential Li6* (n,t) *cross sections for 0.01 to 0.10 MeV.*

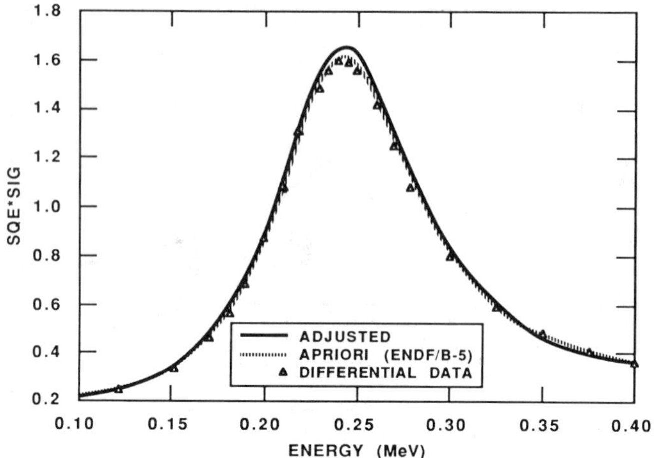

FIG. 5—*Adjusted a priori and differential Li6* (n,t) *cross sections for 0.10 to 0.40 MeV.*

Conclusions

Use of the experimental integral data result in both a ''C/E'' summary and a least squares adjustment solution that indicates a need to increase the $^{10}B(n,He)$ cross sections relative to ENDF/B-V. Most of the increase for ^{10}B occurs above 50 keV, and this is confirmed by comparisons to the differential data results of Carlson et al. [*11*]. For ^{6}Li the integral results also imply an increase in the cross sections above 10 KeV with a smaller quantitative adjustment. However, the differential data results do not support the change in the ^{6}Li values.

References

[*1*] Farrar, H. IV, Oliver, B. M., and Lippincott, E. P., ''Helium Generation Reaction Rates for ^{6}Li and ^{10}B in Benchmark Facilities,'' *Proceedings of the 3rd ASTM-EURATOM Symposium on Reactor Dosimetry*, EUR 6813 EN-FR, Vol. 1, 1980, p. 552.

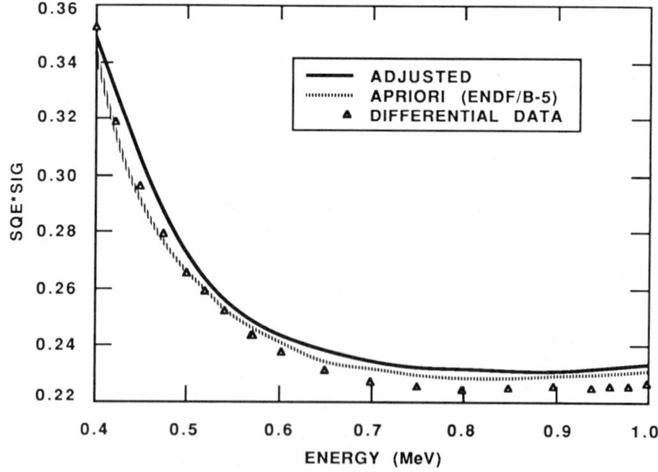

FIG. 6—*Adjusted a priori and differential Li6* (n,t) *cross sections for 0.4 to 1.0 MeV.*

[2] Oliver, B. M., Farrar, H., IV, Lippincott, E. P., and Fabry, A., "Spectrum-Integrated Helium Generation Cross Sections for ^{6}Li and ^{10}B in the Sigma Sigma and Fission Cavity Standard Neutron Fields," *Proceedings of the 4th ASTM-EURATOM Symposium on Reactor Dosimetry,* NUREG/CP-0029, Vol. 2, 1982, p. 889.

[3] Oliver, B. M., Farrar, H., IV, Gilliam, D. M, and Lippincott, E. P., "Spectrum-Integrated Helium Generation Cross Sections for ^{6}Li and ^{10}B in the Intermediate-Energy Standard Neutron Field," *Proceedings of the 5th ASTM-EURATOM Symposium on Reactor Dosimetry,* Vol. 2, D. Reidel Publishing Co., Dordrecht, Holland, 1985, p. 877.

[4] Schmittroth, F., "FERRET Data Analysis Code," HEDL-TME 79-40, Hanford Engineering Development Laboratory, Richland, WA, Sept. 1979.

[5] Rogers, J. W., Millsap, D. A., and Harker, Y. D., "CFRMF Neutron Field Spectral Characterization," *Nuclear Technology,* Vol. 25, 1975, p. 330.

[6] Dowdy, E. J., Lozito, E. J., and Plassman, E. A., "The Central Neutron Spectrum of the Fast Critical Assembly BIG-TEN," *Nuclear Technology,* Vol. 25, 1975, p. 381.

[7] Fabry, A., DeLeeuw, G., and DeLeeuw, S., "The Secondary Intermediate-Energy Standard Neutron Field," *Nuclear Technology,* Vol. 25, 1975, p. 349.

[8] Eisenhauer, C. M. and Grundl, J. A., "Neutron Transport Calculations for the Intermediate-Energy Standard Neutron Field (ISNF) at the National Bureau of Standards," *Proceedings of the International Symposium on Neutron Standards and Application,* NBS Special Publication 493, U.S. Department of Commerce, Washington, DC, 1977.

[9] Oliver, B. M., Bradley, J. G., and Farrar, H., IV, "Helium Concentration in the Earth's Lower Atmosphere," *Geochim et Cosmochim, Acta,* Vol. 48, 1984, p. 1759.

[10] Grundl, J. A., "Examination of ^{10}B(n,He) and ^{6}Li(n,He) Cross Section Measurements in Reactor Physics Benchmarks, (11)," *Proceedings of the International Conference on Nuclear Data for Basic and Applied Sciences,* Santa Fe, NM, 13–17 May 1985, in press.

[11] Carlson, A. D., Poenitz, W. P., Hale, G. M., and Peele, R. W., "The Neutron Cross Section Standards Evaluations for ENDF/B-VI," *Proceedings of the International Conference on Nuclear Data for Basic and Applied Sciences,* Santa Fe, NM, 13–17 May 1985, in press.

Author Index

Subject Index

D

H

I

K

L

M

R

S

T

U

V

W

X–Z